Fuels and Petrochemicals Division 2013

Core Programming Topic at the 2013 AIChE Spring Meeting &
9th Global Congress on Process Safety

San Antonio, Texas, USA
28 April – 2 May 2013

Volume 1 of 2

ISBN: 978-1-62748-464-0

Printed from e-media with permission by:

Curran Associates, Inc.
57 Morehouse Lane
Red Hook, NY 12571

Printed by Curran Associates, Inc. (2013)

For permission requests, please contact AIChE
at the address below.

AIChE
3 Park Avenue
New York, NY 10016-5991

Phone: (203) 702-7660
Fax: (203) 775-5177

www.aiche.org

Additional copies of this publication are available from:

Curran Associates, Inc.
57 Morehouse Lane
Red Hook, NY 12571 USA
Phone: 845-758-0400
Fax: 845-758-2634
Email: curran@proceedings.com
Web: www.proceedings.com

TABLE OF CONTENTS

VOLUME 1

VOLUME 2

() Rebirth of Manufacturing in the U.S.: What are the Implications for Chemical Engineers?

Dan Romasko, Tesoro Corporation, San Antonio, TX

Technology advances have increased production of oil and natural gas from reserves that were previously unprofitable. This rapid increase in production is restructuring the energy industry and providing more cost advantaged feedstocks and fuel to the nation's manufacturing base, particularly the petrochemical industry. This improved cost structure leads to increased competitiveness in the global markets and supports growth in the domestic industry. Along with this growth comes the need for skilled engineers at a time when the demographic changes point toward a declining supply of engineering talent.

A UNIFIED PID METHODOLOGY TO MEET PLANT OBJECTIVES

Gregory K. McMillan, CDI Process & Industrial, Austin, TX and Héctor H. Torres, Eastman, Tlaxcala, México

Introduction

Plant objectives for basic and advanced regulatory control can be quite diverse. PID control plays an important role in maximizing safety, environmental and equipment protection, process efficiency and process capacity. There can be many different sources of process variability that control loops need to deal with, such as raw material and recycle composition, weather, utility temperature and pressure, operators, interactions, optimization, on-off actions, startups, transitions, shutdowns, measurement and process noise, and limit cycles. Process nonlinearities, process non-self-regulation, dead time dominance, slow measurements and valves, analyzer cycles times, deadband, and threshold sensitivity or resolution limits can make the task much more difficult. The techniques for applying PID control to achieve plant objectives are largely individualistic than systematic. Each application tends to be treated as unique requiring special expertise and tuning. The result has been over 100 tuning rules to meet different perceived and actual needs of the process. Consultants are often required and PID features are not effectively utilized. In some plants PID loops have been pushed into model predictive control (MPC) because PID function application practices are largely heuristic and undocumented.

A unified approach to tuning has been found that enables a common and simplified method for setting PID tuning parameters. Key features can be used to eliminate the need for retuning to deal with different dynamics and objectives. This paper shows a methodology that integrates a unified tuning approach and key features that minimizes implementation and maintenance efforts. A step by step method will be used that will address the myriad of dynamics and objectives in PID applications.

Types of Process Responses

There are 4 major types of process responses. Early tuning rules were developed with one type of process response in mind. Tuning rules developed in the 1990s have expanded the scope of processes addressed but few effectively deal with all types of responses. The tuning requirements are so different for each type of response that users can readily end up confused. The first step to resolving this deficiency is recognition of the distinctive characteristics of each type of process response.

Response Methodology and Terminology

The effect of automation system dynamics will be included and distinguished in the process response. A first order plus dead time (FOPDT) approximation is used that consists of a dead time, a time constant, and a gain. For a near-integrating process the gain is divided by the time constant to given an integrating process gain used for integrating processes. The identification or approximation of a secondary time constant is added to improve the setting of the rate time for the derivative mode.

In the literature these parameters to define the dynamic response are referred to as process dead time, process time constants, and process gain. Since in industrial applications these parameters are determined by the automation system besides process dynamics, more general terms will be used.

The total loop dead time is the sum of all the pure time delays plus the fraction of the small time constants that is equivalent dead time. All time constants smaller than the largest (primary time constant) and next largest if a secondary time constant is identified create an effective dead time. The delays and time constants as shown in Figure 1 in the final control element (e.g. control valve or variable speed drive), process, measurement, and controller must be included.

In the literature primary and secondary time constants are considered to be in the process as shown in Figure 1. In industrial applications these time constant can be in the automation system. In this paper the largest time constant in Figure 1 is termed the open loop time constant.

The open loop gain is the product of the final control element, process, and measurement gains shown in Figure 1. For self-regulating processes, the open loop gain is dimensionless (%/%). For integrating processes, the resulting gain called an integrating process gain has units of 1/sec (%/sec/%). Processes with a large process time constant are termed near-integrating. The near-integrating process gain is attained by dividing a self-regulating process gain by the primary process time constant.

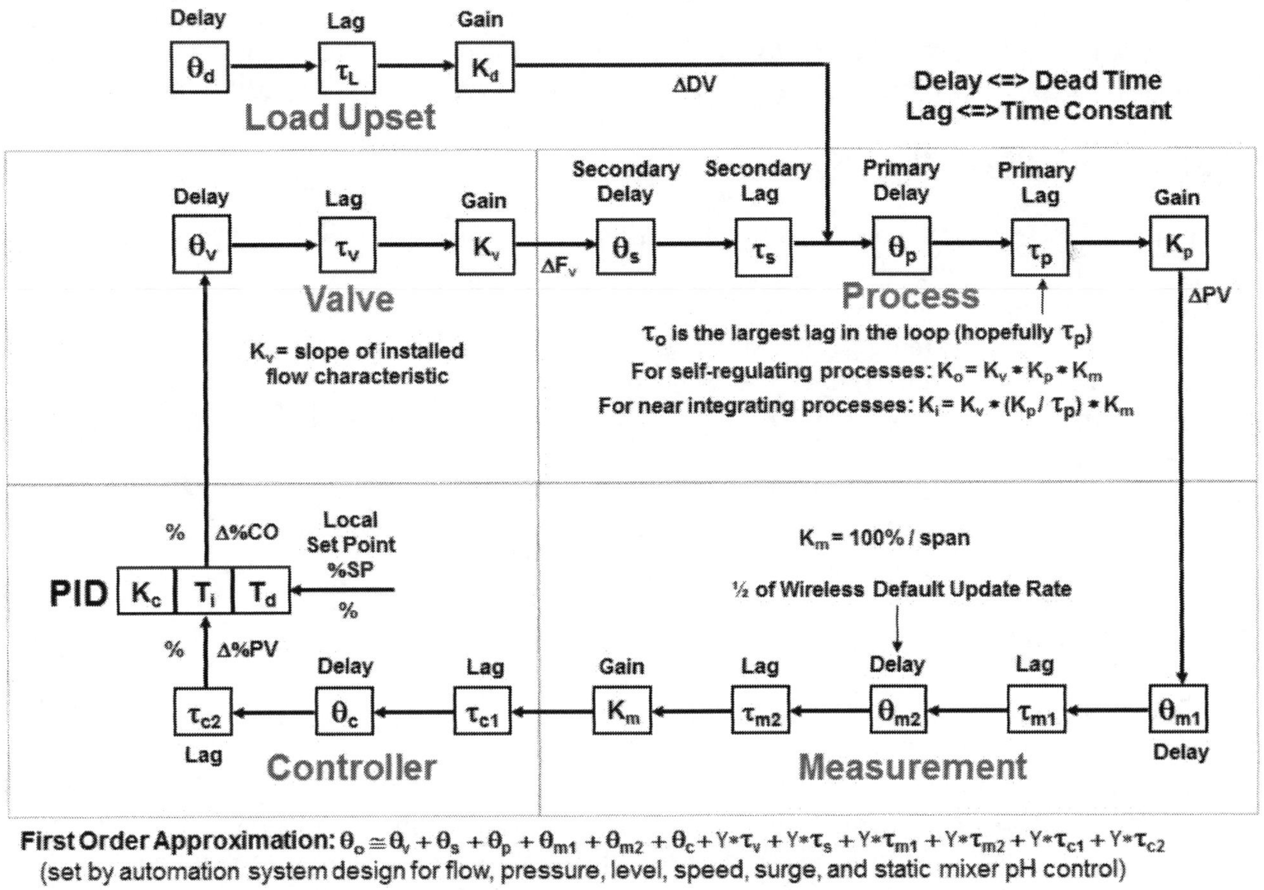

Figure 1 – Block diagram reveals the myriad sources of dynamics in the FOPDT response

Dead time Dominant Self-Regulating Responses

A self-regulating process will reach a new steady state if a manual change in the PID output is made provided there are no disturbances. The steady state is the result of internal self-regulation from a negative feedback process time constant. The actual time to steady state observed will depend upon the open loop time constant (largest time constant in the loop), which may be in the automation system. If the total loop dead time exceeds the open loop time constant, the process is dead time dominant.

The process dynamic response of conveyors, sheets and spin lines is essentially a pure dead time from the transportation delay. However, if die bolt heaters are used instead of hydraulic actuators to control sheet thickness, the thermal lag introduced can be large enough that the process is no longer dead time dominant.

Blenders, desuperheaters, extruders, static mixers, and plug flow reactors have a negligible process time constant due to the lack of back mixing. However, the temperature response is often not dead time dominant due to the thermal lags in the equipment heat transfer surface and thermowell.

The addition of a large wireless default update rate in a flow loop can make the flow loop dead time dominant. A large wireless default update rate in a static mixed pH loop or the use of an at-line or off-line analyzer with a large cycle time can make these composition loops dead time dominant. For these cases where the dead time dominance is due to a discontinuous update of an automation system component, we will see that a PID feature can be enabled to prevent instability and simplify tuning.

Lag Dominant Self-Regulating Response

If the self-regulating process response has an open loop time constant larger than the total loop dead time, the process is termed lag dominant. Continuous operations with mixing such as columns, crystallizers, evaporators, neutralizers, and reactors have a lag dominant response. Liquid flow has a self-regulating response but the process time constant is less than the total loop dead time. A signal filter used to reduce noise can make the process lag dominant. A large automation system lag can make a response lag dominant that would have been dead time dominant based on the process time constant.

If the open loop time constant is much larger than the loop dead time, the process response seen by the PID is a ramp in the control region of 4 dead times. A PID tuned for disturbance rejection will have done nearly all of its corrections in 4 dead times. So far as what the PID sees, the process response is integrating. Such processes are termed “near-integrating” and can use integrating process tuning rules. Temperature and composition control loops on continuous processes are near-integrating.

Integrating Response

Processes that have no negative feedback process time constant and hence no steady state are termed integrating processes. The process response integrates any unbalance in the process. Since the unbalance is nearly never exactly zero, the process will always ramp when the PID is in manual. For most integrating processes the ramp rate is extremely slow (e.g. 0.000001 %/sec). The extreme exception is gas pressure where the ramp rate can be a million to a billion times faster. An electrical furnace had an integrating process gain of 1 %/sec/%. A waste heat incinerator gas pressure had an integrating process gain of 1000 %/sec/% (too fast for any digital controller).

The most common example of an integrating process is level. Gas pressure has an integrating response when changes in operating pressure are a small fraction of the pressure drop across inlet or outlet valves and consequently do not appreciably affect the flow into or out of the system. Batch operations cause an integrating response because there is no discharge flow until the operation is complete. Batch temperature is integrating unless the heat loss changes appreciably with temperature. Batch composition and batch pH has an integrating response unless the reactant or reagent is rapidly consumed in a reaction.

Runaway Response

Processes with a positive feedback time constant can accelerate. These processes are termed open loop unstable. If the PID is put in manual, the process can accelerate and run off-scale. Exothermic reactions where the reaction rate greatly increases with temperature can have runaway response if the heat release exceeds the cooling capability. A large secondary time constant (e.g. heat transfer surface or thermowell lags) or dead time approaching the positive feedback time constant can cause the process to be unstable regardless of PID tuning.

Process Objectives

The main task of the PID is to determine the transfer of variability from a controlled variable such as temperature to the manipulated variable such as coolant flow. The process input most often manipulated is a process or utility flow. For maximum disturbance rejection, the transfer of variability is maximized. This is generally the case for vessel and column temperature, composition, gas pressure, and pH control. An insufficiently aggressive transfer of variability can cause product degradation or an emission, uncontrollable reactions, a relief valve or rupture disc to blow, and a safety instrumented system (SIS) to activate. Safety, equipment and environmental protection, on-stream time, and yield depend upon disturbance rejection.

To meet these objectives the integrated and peak error must be minimized. The ultimate limit to how small these errors are depend upon the total loop dead time (θ_o) and the open loop time constant (τ_o). Figure 2 shows that the ultimate limit to the peak error is proportional to the ratio of dead time to the 63% response time which is the dead time plus the open loop time constant. The peak error is the altitude of the right triangle whose base is the dead time since no correction can be enacted and seen till one dead time. A mirror image of this right triangle placed back to back with the first triangle shows the total recovery from the disturbance. The ultimate limit to the integrated error is the area of the two triangles which ends up being the dead time to 63% response time multiplied by the dead time.

In some applications the transfer must be minimized because changes in the manipulated flow affect other loops and operations. The most common application is surge tank level control where the level PID manipulates the discharge flow to the degree only needed to keep the tank level within low and high limits. The objective is to minimize the disruption of feed changes to downstream operations. This objective is also effective for a distillate receiver level PID manipulating distillate flow. This goal is often termed "maximization of the absorption of variability".

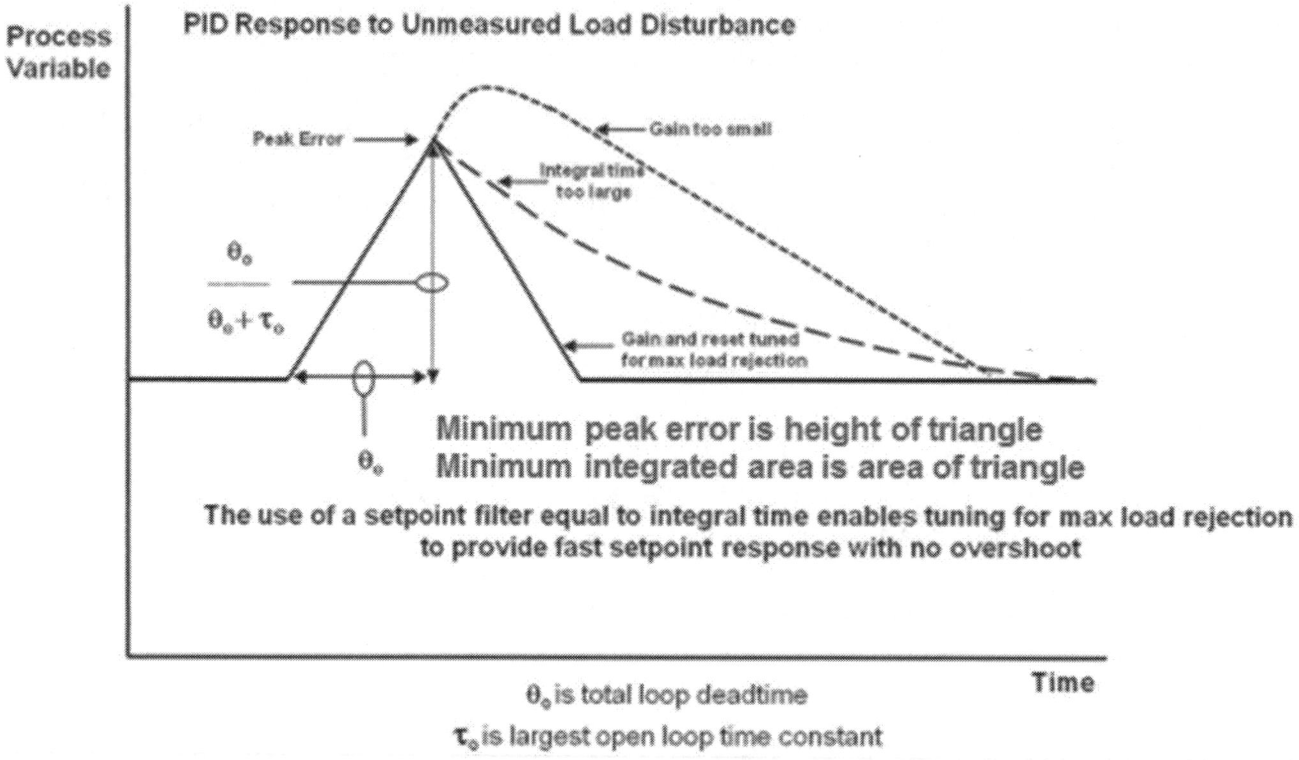

Figure 2 – The ultimate limit to the peak and integrated errors depends upon the total loop dead time and open loop time constant

Aggressive control action can also upset other loops. For interactions that cannot be reduced sufficiently by decoupling techniques, the transfer of variability for the least important loops is minimized to avoid upsetting the more important loops.

When loops are manipulating flows affecting a unit operation, the transfer of variability to these flows is coordinated so the stoichiometric ratio is not momentarily upset from one flow being faster or slower than another flow. In this case the transfer of variability in all loops is regulated to match the slowest loop. This coordination of manipulated flows is particularly important for operations without back mixing to smooth out transients. Coordination is consequently critical for blending and plug flow liquid and gas reactions (e.g. catalyst bed gas reactors).

PID loops must not increase variability. A PID loop can oscillate from overly aggressive action and interactions and in integrating and runaway processes from inadequate immediate action due to a gain setting that is too low. A PID can create variability when there are automation system difficulties.

Automation System Difficulties

The response of the automation system can introduce excessive discontinuities, lag, noise, and dead time. Here we look at the components of the system for the sources of these difficulties. There is only room in this paper to note some of the more significant deficiencies.

Final Control Element

A control valve can have significant stiction from excessive friction often described as a resolution limit. Actuators and positioners can also have a threshold sensitivity limit. One or more integrators in the process or PID in a loop with a resolution or threshold sensitivity limit will cause a limit cycle that cannot be eliminated by tuning. A limit cycle is a sustained equal amplitude oscillation.

A control valve or damper can have significant backlash from play in linkages and connections seen as a dead band upon reversal of direction of the signal. Two or more integrators in the loop will cause a limit cycle than cannot be eliminated by tuning. Thus, an integrating process and a PID controller with integral action can cause a limit cycle. A self-regulating process with a cascade loop where the primary and secondary PID each have integral action will exhibit a limit cycle.

A variable speed drive (VSD) with a resolution limit due to insufficient number of bits in the analog to digital signal input card will result in a limit cycle for one or more integrators. If a dead band is introduced in the VSD setup, limit cycle will develop for two or more integrators.

A loop with a slow final control element that cannot keep up with the rates of change of the controller output will burst into oscillations. The oscillations may not show up during loop tests where typically small changes are made. The burst of oscillations may only occur for large fast disturbances and large setpoint changes causing the rate of change of the PID output to be faster than the valve slewing rate or VSD rate limited response. Large valves and dampers and VSD with marginally sized motors can be too slow.

Measurement

The sensor design, location, and operating conditions can result in excessive measurement noise. Differential head meters and vortex meters are especially sensitive to velocity profile and hence the upstream and downstream piping configuration. Most flow meters get noisy as you approach the low flow limit. pH electrodes are particularly vulnerable to noise from electromagnetic interference and concentration fluctuations.

The measurement can also introduce an excessive lag into the control loop. Fouled electrodes and thermowells can have an excessive lag from a decrease in the mass transfer and heat transfer coefficients, respectively. Pressure and differential pressure transmitters on compressor control can become too slow by simply increasing the transmitter damping setting.

If the measurement lag becomes by far the largest time constant in the loop an insidious situation develops where the trend charts may look smoother because the measurement is showing an extremely attenuated (filtered) view of the actual process oscillations. The controller gain may even be able to be increased further misleading the user into thinking the measurement lag is beneficial. The key is the loop period should show an increase due to the increase in loop dead time.

Wireless measurements can introduce excessive dead time in fast loops by a slow default update rate. At-line analyzers and especially off-line analyzers can cause excessive dead time in almost any loop from the sample transport and processing, analysis cycle time, and multiplex time.

Controller

The controller can introduce an excessive dead time from too large of a module execution time or signal filter time. Even without considering the effect of the increase in dead time on the ultimate limit to errors and tuning, a signal filter and module execution time larger than 10% of the reset time will increase the integrated errors by more than 10% for an unmeasured disturbance.

Too slow of a secondary controller tuning can cause a cascade loop to burst into oscillations when the primary controller output is changing faster the secondary loop can respond. Insufficient gain action can cause slow rolling oscillations in an integrating process and instability in a runaway process.

Unified Tuning

The key breakthrough in thinking is to use the lambda tuning rules for integrating processes for lag dominant self-regulating and runaway processes. The lambda tuning rules automatically prevent the product of the PID gain and reset time from being less than the twice the inverse of the integrating process gain that would result in slow rolling oscillations. The main decision is whether to maximize the transfer of variability or maximize the absorption of the variability. Key features are used rather than tuning to address the automation system difficulties and to meet different process objectives.

Maximization of the Transfer of Variability

The maximum dead time for all possible operating conditions is used as the arrest time, which is the time to start to turn back an excursion from an unmeasured disturbance. The arrest time is lambda. This minimum lambda is the dead time as shown in Figure 2. The minimum lambda plus the maximum integrating process gain (fastest ramp rate) is used in the tuning equations. The lambda tuning rule then gives the same gain and reset time as the Ziegler Nichols Reaction Curve method. For temperature, composition, and pH loops, the rate time should be set equal to the secondary time constant. If this time constant cannot be identified, ½ of the total loop dead time can be used as an approximation. This gives a rate time that is about 1/6 of the reset time. Note that for the ISA Standard Form PID, the rate time must always be less than the reset time to avoid instability (see Appendix A).

If a lower controller gain is used. The equivalent lambda needs to back calculated and used in the calculation of the reset time to prevent slow rolling oscillations. For runaway processes, the PID gain must always be greater than the inverse of the process gain to prevent a runaway.

Maximization of the Absorption of Variability

When the absorption of variability must be maximized, lambda is chosen to be as large as possible. The maximum arrest time lambda depends upon the integrating process gain, the allowable change in the process variable and the available change in the manipulated flow. The following equations show how to calculate lambda for a generic application and then a surge tank level loop.

The maximum lambda (λ) is the maximum allowable % excursion ($\Delta\%PV_{\max}$) divided by the maximum possible PV ramp rate. The maximum possible ramp rate is the PV rate of change per percent output change multiplied by the maximum available percent output change ($\Delta\%CO_{\max}$).

$$\lambda = \frac{\left|\Delta\%PV_{\max}\right|}{\left|\frac{\frac{\Delta\%PV}{\Delta t}}{\Delta\%CO}\right| * \left|\Delta\%CO_{\max}\right|} \tag{1}$$

Realizing that the integrating process gain is the PV ramp rate per percent output change:

$$\lambda = \frac{1}{K_i} * \frac{\left|\Delta\%PV_{\max}\right|}{\left|\Delta\%CO_{\max}\right|} \tag{2}$$

An equivalent setpoint rate limit on the controller output (e.g. flow controller setpoint):

$$\left|\Delta\%CO / \Delta t\right|_{\max} = \frac{\left|\Delta\%CO_{\max}\right|}{\lambda} = K_i * \frac{\left|\Delta\%CO_{\max}\right|^2}{\left|\Delta\%PV_{\max}\right|} \tag{3}$$

For a PV limit ($\%PV_{Limit}$) and corresponding CO limit ($\%CO_{Limit}$) we have:

$$\lambda = \frac{1}{K_i} * \left|\frac{\%PV_{Limit} - \%SP}{\%CO_{Limit} - \%CO}\right| \tag{4}$$

The above calculation would be done for high and low operating limits and various setpoints. The smallest of the arrest times would be used in tuning.

We can obtain the more detailed requirements for surge level tank level control by computing the integrating process gain for level. The integrating process gain is the product of the valve, process, and measurement gains:

$$K_i = K_v * K_p * K_m \tag{5}$$

The valve gain or variable speed drive gain for a linear installed characteristic or flow loop is:

$$K_v = \frac{\Delta F_v}{\Delta\%CO} = \frac{\Delta F_{\max}}{100\%} \tag{6}$$

The level process gain for mass flow is (omit density term for volumetric flow):

$$K_p = \frac{1}{A * \rho} \tag{7}$$

The level measurement gain is:

$$K_m = \frac{\Delta\%PV}{\Delta L} = \frac{100\%}{\Delta L_{\max}} \qquad (8)$$

Substituting in the valve, process, and measurement gains, the integrating process gain is:

$$K_i = K_v * K_p * K_m = \frac{\Delta F_{\max}}{100\%} * \frac{1}{A*\rho} * \frac{100\%}{\Delta L_{\max}} = \frac{\Delta F_{\max}}{\Delta L_{\max}} * \frac{1}{A*\rho} \qquad (9)$$

The consequential arrest time for a level loop is:

$$\lambda = \frac{A*\rho*\Delta L_{\max}}{\Delta F_{\max}} * \frac{\Delta\%PV_{\max}}{\Delta\%CO_{\max}} \qquad (10)$$

An equivalent setpoint rate limit on the controller output (e.g. flow controller setpoint):

$$\left|\Delta\%CO/\Delta t\right|_{\max} = \frac{\left|\Delta\%CO_{\max}\right|}{\lambda} = \frac{\Delta F_{\max}}{A*\rho*\Delta L_{\max}} * \frac{\left|\Delta\%CO_{\max}\right|^2}{\left|\Delta\%PV_{\max}\right|} \qquad (11)$$

Where:

$A =$ cross sectional area (m^2)
$\Delta F_{\max} =$ maximum change in valve or variable speed drive flow (e.g. flow span) (kg/sec)
$\Delta L_{\max} =$ maximum change in level (e.g. level span) (m)
$K_i =$ integrating process gain (1/sec)
$K_m =$ measurement gain (%/m for level)
$K_p =$ process gain (m/kg for level)
$K_v =$ valve or variable speed drive gain (kg/sec/%)
$\lambda =$ arrest time (sec)
$\rho =$ fluid density (kg/m^3)
$\%CO =$ operating point controller output (%)
$\%CO_{Limit} =$ controller output limit (%)
$\Delta\%CO_{\max} =$ maximum available change in controller output (%)
$\%PV_{Limit} =$ process variable limit (%)
$\Delta\%PV_{\max} =$ maximum allowable change in process variable (%)
$\%SP =$ operating point setpoint (%)

Key PID Features

There are seven key related PID control features; structure, external-reset feedback, feedforward, setpoint filter and lead-lag, setpoint rate limits, output tracking, and an enhancement for wireless. These key features synergistically provide capabilities that have not been fully realized. See Appendix A for conversion of tuning settings for different PID Forms and tuning setting units.

PID Structure

Most modern DCS offer a choice of structures shown in Table 1.

Table 1 - Major PID Structure Choices

(1) PID action on error ($\beta = 1$ and $\gamma = 1$)
(2) PI action on error, D action on PV ($\beta = 1$ and $\gamma = 0$)
(3) I action on error, PD action on PV ($\beta = 0$ and $\gamma = 0$)
(4) PD action on error, no I action ($\beta = 1$ and $\gamma = 1$)
(5) P action on error, D action on PV, no I action ($\beta = 1$ and $\gamma = 0$)
(6) ID action on error, no P action ($\gamma = 1$)
(7) I action on error, D action on PV, no P action ($\gamma = 0$)
(8) Two degrees of freedom controller (β and γ adjustable 0 to 1)

The structure choices use β and γ set point weighting factors. A controller that has both factors adjustable is called a "two degree freedom controller." Other structures have the β and γ factors set equal to 0 or 1. The user can also omit a mode entirely to get P-only, I-only, ID, and PD control with various assigned factors. PI control is achieved by simply setting the derivative (rate) time to zero. In general, the user must not set the controller gain equal to zero in an attempt to get I-only or ID control or set the integral (reset) time to zero in an attempt to get P-only or PD control. Note that the use of P-only or PD control requires additional choices of how to set the bias and its ramp time.

In choosing the correct structure it is important to know the relative contribution of the proportional, integral, and derivative modes. Figure 3 shows the contribution of each PID mode for a setpoint change. There is no feedback from the process in this figure (e.g. block valve closed), so we can more clearly see the contribution of each mode.

Structure 1 (PID on error) provides the fastest approach to a new setpoint by virtue of a step and a kick in the controller output from the setpoint change as seen in the second response in Figure 3. For small setpoint changes and low controller gains, the step and kick can help get through significant valve backlash and stick-slip to get the valve moving. The kick appears to be a spike on trend charts with a large time spans. The abrupt change in output is often seen as disruptive by operators when they make setpoint changes. If the burst of flow through the control valve does not affect other users of the process or utility fluid, the step and kick is more of a psychological than a process concern.

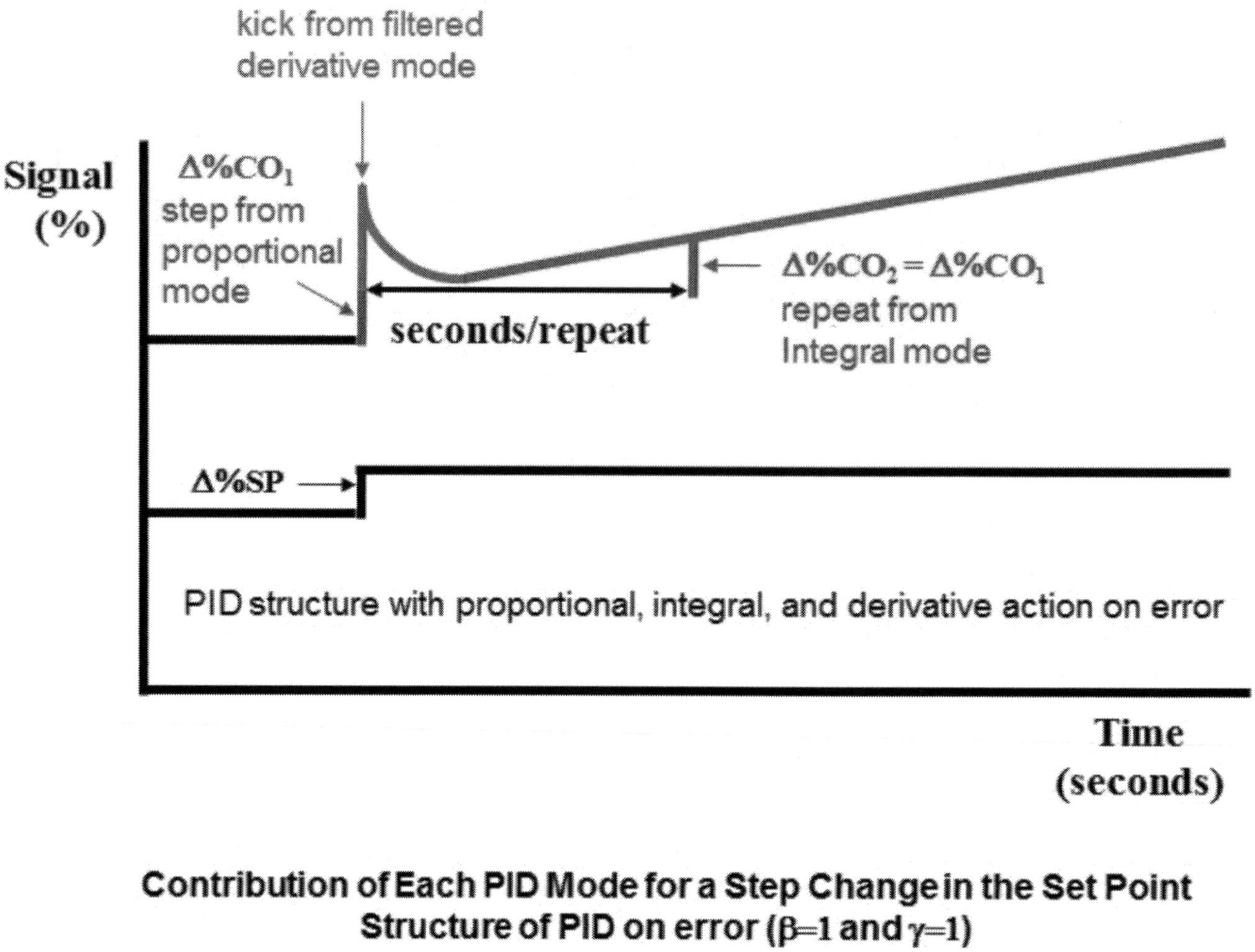

Figure 3 – Contribution of each PID mode for a step change in setpoint for a PID on error structure

Structure 2 (PI on error, D on PV) is the structure most often used. Structure 2 eliminates the kick from derivative action for a setpoint change by setting gamma to zero (γ=0). The increase in rise time going from structure 1 to 2 is negligible for the more important loops, such as column and vessel temperature where derivative action is used. The step in the output from the proportional mode on a setpoint change is large for a high controller gain. Increases in process gain or dead time will increase the overshoot unless the controller gain is decreased accordingly. If the elimination of setpoint overshoot is much more important than rise time (e.g. bioreactor temperature and pH shifts), then structure 3 may be best.

Structure 3 (I on error, PD on PV) eliminates overshoot but with quite a sacrifice in speed of approach to the setpoint. For near and true integrating and runaway processes, structure 3 may have a rise time an order of magnitude larger than the rise time for structure 1 for the same tuning settings and process dynamics. A setpoint filter set equal to the reset time provides the same response.

Structure 4 (PD on error, no I) is used on processes adversely affected by integral action. The temperature control of severely exothermic polymerization reactors use structure 4 because integrating action in the controller increases the risk of a runaway. If integral action is used, the reset time should be increased by a factor of 10 for these positive feedback processes to be safe. Users may not be aware of this requirement leading to overshoot that can trigger a runaway. The bias for structure 4 is set equal to the normal PD controller output when the PV is at setpoint.

Structure 4 is also used on batch processes that respond in only one direction. For example, in bringing a batch pH up to a setpoint by the addition of a base where the base is not consumed in a reaction, the batch will only respond in the direction of increasing pH. The pH will overshoot setpoint if integral action is used. If split ranging is added with an acid reagent, there will be some wasted reagent due to cross neutralization of reagents and limit cycling across the split range point from stiction that is greatest near the closed position. For structure 4 and a single reagent, the bias is set for zero reagent addition when the PV is at setpoint.

Structure 5 (P on error, D on PV, no I) is used for many of the same reasons as structure 4.

Structure 6 (ID on error, no P) is used for valve position controllers (VPC) to eliminate the interaction with process controller whose valve position is being optimized. The VPC could be optimizing the coarse adjustment from a large control valve in parallel with a small control valve manipulated as fine adjustment by the process controller. In Figure 4 the process controller FC215 adjusts the small valve input to the process. The output to the small valve is also used as a PV of the VPC ZC215. The small valve provides precision and the large valve gives rangeability.

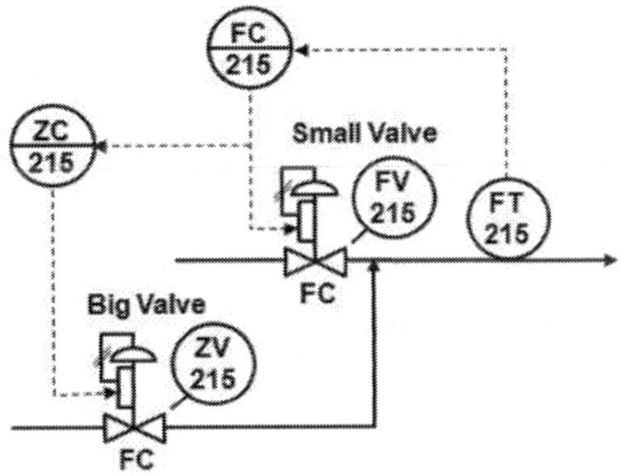

Figure 4 - Valve position control increases precision and rangeability

The VPC could be optimizing utility supply pressure or temperature to minimize energy use or optimizing feed rate to maximize production rate. Normally, the rate time is zero. The tuning of this integral-only controller is problematic. Tuning rules often cited suggest the integral time should be larger than 10 times the product of the gain and reset time of the process controller and 10 times the residence time of the process to eliminate interaction. However, the VPC response for this tuning is too slow to prevent the process controller from getting into trouble for large disturbances. Feedforward action can be added to help. In addition external-reset feedback, setpoint directional rate limits, and an enhanced PID developed for wireless can be used as a flexible and easy to tune solution to deal with unknowns. This solution is described in the Setpoint Rate Limits section.

Structure 7 (I on error, D on PV, no P) is used for the same reasons as structure 6. As with structure 2, the spike from rate action for setpoint changes is eliminated.

Structure 8 (two degrees of freedom via adjustable β and γ) is used to provide a balance between a fast rise time and overshoot. The same effect can be accomplished by the use of a setpoint lead-lag as discussed in the setpoint filter section. This structure can be used to get the structures with proportional or derivative action on PV by zeroing the respective β and γ set point factors.

External-Reset Feedback

PID controllers with the positive feedback implementation of integral action as shown in Figure 5 enable the use of an external reset feedback. When the switch in Figure 5 is in the position to get an external reset feedback to the integral mode from the process variable (PV) of a setpoint being manipulated whether an Analog Output block or a secondary PID, the PID output is prevented from changing faster than the PV of the control valve, VSD, or secondary loop can respond. There is an increasingly larger scope of potential benefits discovered as described in subsequent features and the summary of the unified methodology. The only requirement is that the external reset feedback be properly connected to a fast update of the PV for the valve, VSD, or loop being manipulated. Updates of actual valve position, VSD speed, or secondary loop PV should be at least twice as fast as the the execution time of the controller with external reset feedback.

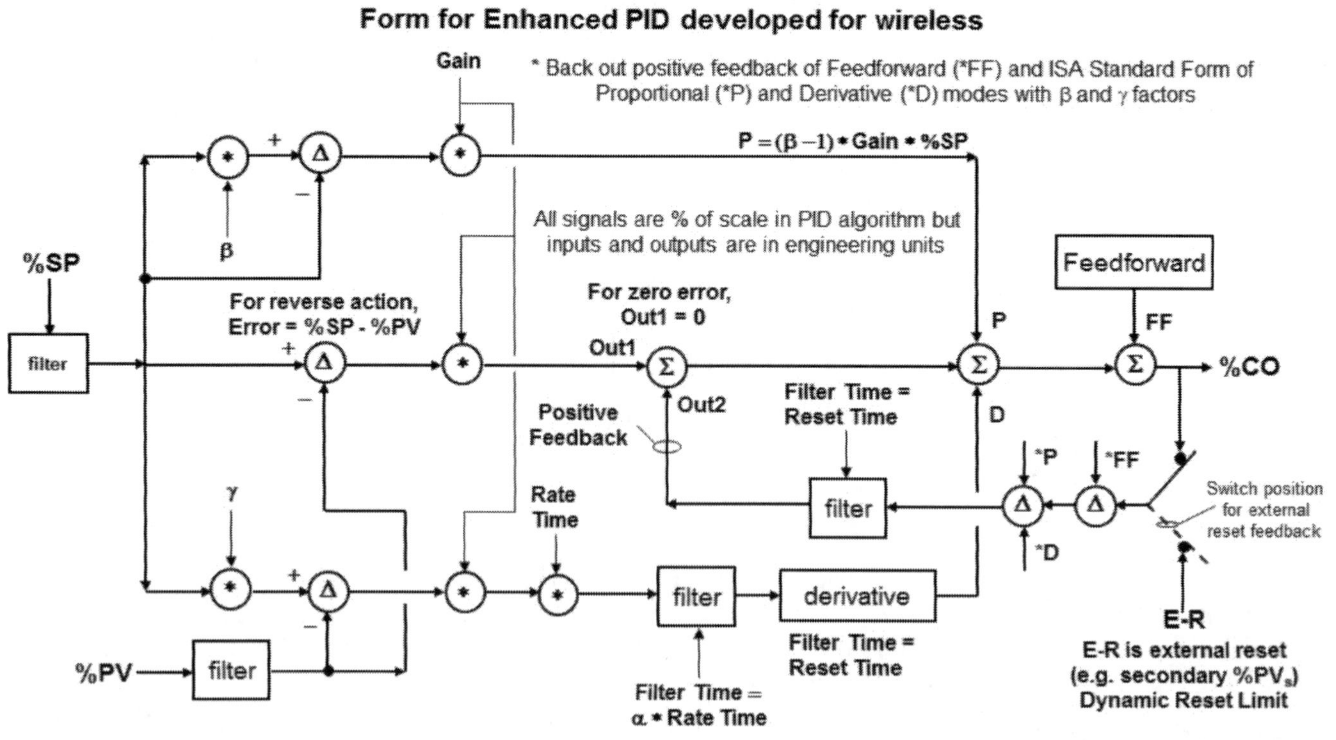

Figure 5 – ISA Standard Form PID with External Reset Feedback

Feedforward

Feedforward control is particularly effective when a disturbance is large, fast, and precisely measured. While theoretically a feedforward multiplier is more effective for changes in the slope of a plot of the manipulated flow versus the measured disturbance, for many practical reasons such as scaling and nonlinearity problems, a feedforward summer is used for feedforward control especially on vessels and columns. The most common feedforward signal is flow because the changes in feed flows are faster than changes in composition and temperature and the flow is the most common process input. A flow feedforward summer is implemented via an external ratio and bias station so the operator can set the desired ratio and see the actual ratio. On startup, the operator may run on ratio control with the process controller in manual until the equipment reaches operating conditions. The initial flow ratios can be set based on the process flow diagram (PFD). The PFD can be put live online on the operator graphics to show the actual ratios used. The technique also enables a plant wide feedforward system to rapidly increase and decrease production rates.

Setpoint Filter and Lead-Lag

A setpoint filter can provide the desired closed loop time constant important for the coordination of loops. The use of secondary PID flow setpoint enables stoichiometric ratios to be enforced without a transient unbalance for production rate changes and setpoint changes and disturbances to the primary PID. External reset feedback must be turned on in the primary PID to prevent it from trying to change the secondary PID set point faster than filter will allow the secondary loop to respond. Turning external reset on eliminates the need to retune the primary PID when the secondary PID filter is added or tuned. If coordination of loops is not necessary, a setpoint filter should not be added to a secondary loop because it will slow down the rejection of primary loop disturbances.

The rise time (time to reach setpoint) and overshoot for a setpoint change can be optimized by the use of a setpoint lead-lag. In this case the setpoint lag time is set equal to the reset time and the setpoint lead time is set equal to about 20% of the lag time. If preventing overshoot is more important than reducing rise time for a setpoint change, the lead time can be decreased.

Setpoint Rate Limit

The setpoint rate limit option in analog output (AO) and PID function blocks in a modern DCS can provide move suppression, a tuning parameter found to be very useful in model predictive control. Since there is often an up and down rate limit, directional move suppression can be used provided the ability needed to enable a slow approach to an optimum and a fast getaway to prevent an abnormal condition. This technique is used to provide a fast opening and slow closing anti-surge control valve. There are many other applications as we will see in the unified methodology. If external reset feedback is turned on in the PID manipulating the setpoint, the PID does not have to be retuned when the rate limits are added or tuned in the manipulated AO or PID. These rate limits can be written to online.

Output Tracking

Disturbances can be too fast and the consequences too severe to rely just on PID control. Here output tracking is used to get the loop out of trouble for equipment and environmental protection. The most common example is compressor surge prevention, where the PID output will track a surge valve position large enough to prevent surge when a crossing of the surge curve is detected or a precipitous drop in compressor suction flow occurs indicating the start of surge. The PID stays in output tracking holding the protective output for several seconds to ensure the system has stabilized.

The fastest possible setpoint response is obtained from a full throttle approach (bang-bang control method) where the PID is put in output tracking at the output limit to drive the PV to the setpoint. The PID output stays at the limit until the PV one dead time into the future is predicted to reach setpoint. A dead time (DT) block with its parameter set equal to the loop dead time is used to create an old PV that is subtracted from the new PV to create a delta PV. The delta PV added to the new PV that is the DT block input gives the PV one dead time into the future. When the future PV is projected to reach setpoint, the PID tracks a final resting value (FRV) captured form previous batches or startups and holds this FRV for one dead time before releasing the PID back to feedback control.

Enhanced PID for Wireless

The enhanced PID option developed for wireless enables the following features:

1. Positive feedback implementation of reset with external-reset feedback
2. Immediate response to a setpoint change or feedforward signal or mode change
3. Suspension of integral action until there is a change in PV
4. Integral action is the exponential response of the positive feedback filter to the change in controller output for the time interval since last update
5. Derivative action is the PV or error change depending upon PID structure divided by the time interval since the last update (instead of PID execution time) multiplied by the gain and rate time

The enhanced PID will go into a holding mode for a loss of communication, stuck control valve, or failure of an electrode, analyzer, or sample system providing a smooth continuation upon restoration of communication and analyzer results. For offline analyzers, late lab results do not cause a problem. The enhanced PID eliminates the need for retuning as discontinuous delays from wireless devices and analyzers are increased. For interacting loops and valve position control and final control elements with resolution limits and deadband, a threshold sensitivity setting enables the enhanced PID to eliminate reactions to insignificant changes that could cause fighting between loops or perpetuate a limit cycle. The threshold sensitivity is set just larger than the PV noise amplitude after filtering.

A Unified Methodology

The following is a generalized step by step procedure to address a wide spectrum of automation system difficulties and application objectives. Note that PID features vary considerably from one DCS manufacturer to another. Any setting of options or tuning must be based on information from the supplier and should be thoroughly tested before and carefully observed after commissioning.

1. Set the output limits to keep the manipulated setpoints in the desired operating range. For variable speed drives set the process PID low output limit so the speed cannot cause the discharge head to approach the static head in order to prevent excessive sensitivity to pressure and to prevent reverse flow. In general, set the anti-reset windup limit to match the output limit. If the output scale is engineering units, the output limits and anti-reset windup must be based on the

output scale range and units.

2. Choose the best structure for your application. Generally the best choice is structure 2 with PI on error and D on PV. For a single direction response (e.g. batch heating or neutralization), use structure 4 or structure 5 so that there is no integral action. For a highly exothermic reaction, you might want structure 5 to help prevent a runaway from integral action.

3. Set the signal filter noise just large enough to keep the controller output fluctuations from exceeding the resolution limit or deadband of the final control element.

4. For near-integrating, true integrating, and runway processes use the lambda integrating process tuning rules. To maximize the transfer of variability from the process variable to the manipulated variable, set the lambda (arrest time) equal to the maximum dead time and use the largest integrating process gain for all possible operating conditions in the tuning. To maximize the absorption of variability (e.g. surge tank level) use the minimum arrest time computed from Equations 1 through 10 for all possible operating conditions. If you decrease the PID gain, proportionally increase the PID reset time to prevent slow rolling oscillations. For runaway processes, the PID gain must not be less than the inverse of the open loop gain. When changing from the Series Form to the ISA Standard Form in newer DCS, convert the tuning settings based on the differences in Form. In the ISA Standard Form the rate time should not be greater than ¼ the reset time. Be sure to realize and convert any difference in Form and tuning setting units between different PID vintages and suppliers (see Appendix A). You must take into account the units and inverse relationship between gain and proportional band and reset time in seconds and minutes per repeat. Not addressing these differences can result in settings off by orders of magnitude.

5. For self-regulating processes with the open loop time constant less than 4 times the dead time, use the lambda self-regulating tuning rules. To maximize the transfer of variability from the process variable to the manipulated variable set the lambda (closed loop time constant) equal to the maximum dead time and use the largest process gain and smallest time constant for all possible operating conditions in the tuning. Schedule tuning settings preferably with an adaptive tuner with the specified lambda to deal with changes in the dynamics.

6. Turn on external reset feedback. Make sure the external reset feedback signal is correctly propagated back to the PID (e.g. BKCAL signal) especially if there are split range, signal characterizer, or signal selector blocks on the PID output.

7. For final control elements that are slow or that have deadband or resolution limit, use a fast readback of the valve position or variable frequency drive speed as the external reset feedback to prevent a burst of oscillations from the PID output changing faster than the final control element can respond.

8. For final control elements that create limit cycles from resolution limits and deadband, use a fast readback of the valve position or variable frequency drive speed to stop the limit cycles.

9. For cascade control, use the PV of the secondary loop as the external reset feedback to prevent a burst of oscillations from violation of the cascade rule where the secondary loop must be significantly faster than the primary loop.

10. For setpoint filters of secondary loops for coordination of flow loops, use the PV of the secondary loop as the external reset feedback to prevent the need to retune the PID.

11. For setpoint rate limits use the PV of the analog output block or secondary loop as the external reset feedback to prevent the need to retune the PID. Add setpoint rate limits to minimize the interaction between loops and in valve position control and to provide directional move suppression to enable a fast getaway for abnormal conditions and a slow approach to optimum. For valve position control, use an enhanced PID developed for wireless with a threshold sensitivity limit to ignore insignificant changes in the valve position to be optimized.

12. Add output tracking for equipment protection and a full throttle (bang-bang control) strategy for the fastest possible time to reach setpoint on startup and for batch operations.

13. Add output tracking logic to momentarily track an output that insures equipment and environmental protection. For compressor surge protection track a sufficiently large opening of the surge valves. To prevent a RCRA pH violation, track a rapidly incrementing reagent valve position to prevent an effluent excursion below 2 pH or above 12 pH.

14. Add feedforward control for large and fast measured disturbances. For flow feedforward, use a ratio and bias station so the operator can enter a desired flow ratio and see the actual flow ratio. Setup the PID to provide a bias correction to the manipulated flow. Add dynamic compensation (dead time and lead-lag blocks) to the feedforward so the manipulated flow arrives at the same point in the process at the same time as the measured disturbance.

15. For wireless devices or analyzers that introduce a large discontinuous PV update delay use an enhanced PID to eliminate the need to retune the controller to prevent oscillations. If the delay is much larger than the 63% process response time, the PID gain can be set as large as the inverse of the open loop gain for self-regulating processes.

16. For valve position control (VPC), make sure the key PID features in Table 3 are used. For examples of VPC for process optimization, see Table 4. The use of an enhanced PID enables the use of integral action in the VPC and simplifies the tuning of the VPC. While model predictive control can provide a more sophisticated optimization, the VPC can be implemented by a simple configuration change that involves setting up a PID as the VPC. The minimal implementation cost and time is attractive for small optimizations.

Table 3 – Key Feature Function and Advantages for Valve Position Control

Feature	Function	Advantage 1	Advantage 2
Direction Velocity Limits	Limit VPC Action Speed Based on Direction	Prevent Running Out of Valve	Minimize Disruption to Process
Dynamic Reset Limit	Limit VPC Action Speed to Process Response	Direction Velocity Limits	Prevent Burst of Oscillations
Adaptive Tuning	Automatically Identify and Schedule Tuning	Eliminate Manual Tuning	Compensation of Nonlinearity
Feedforward	Preemptively Set VPC Out for Upset	Prevent Running Out of Valve	Minimize Disruption
Enhanced PID (PIDPlus)	Suspend Integral Action until PV Update	Eliminate Limit Cycles from Stiction & Backlash	Minimize Oscillations from Interaction & PV Update Delay

Table 4 – Examples of Valve Position Control for Process Optimization

Optimization	VPC PID PV	VPC PID SP	VPC PID Out
Minimize Prime Mover Energy	Reactor Feed Flow PID Out	Max Throttle Position	Compressor or Pump Pressure SP
Minimize Boiler Fuel Cost	Steam Flow PID Out	Max Throttle Position	Boiler Pressure SP
Minimize Boiler Fuel Cost	Equipment Temperature PID Out	Max Throttle Position	Boiler Pressure SP
Minimize Chiller or CTW Energy	Equipment Temperature PID Out	Max Throttle Position	Chiller or CTW Temperature SP
Minimize Purchased Reagent or Fuel Cost	Purchased Reagent or Fuel Flow PID Out	Min Throttle Position	Waste Reagent Or Fuel Flow SP
Minimize Total Reagent Use	Final Neutralization Stage pH PID Out	Min Throttle Position	First Neutralization Stage pH PID SP
Maximize Reactor Production Rate	Reactor or Condenser Temperature PID Out	Max Throttle Position	Feed Flow or Reaction Temperature SP
Maximize Reactor Production Rate	Reactor Vent Pressure PID Out	Max Throttle Position	Feed Flow or Reaction Temperature SP
Maximize Column Production Rate	Reboiler or Condenser Flow PID Out	Max Throttle Position	Feed Flow or Column Pressure SP
Maximize Ratio or Feedforward Accuracy	Process Feedback Correction PID Out	50% (Zero Correction)	Flow Ratio or Feedforward Gain

Not mentioned is the benefit of using cascade control. Fast secondary loops can correct for many disturbances before they affect the primary loop. Secondary loops also reduce nonlinearities seen by the primary loop. A secondary flow loop isolates the nonlinearity of the installed characteristic of the control valve and enables flow feedforward. A secondary jacket or coil temperature loop isolates a crystallizer, evaporator, or reactor temperature loop from cooling or heating system nonlinearities.

Conclusion

The use of key PID features, a unified PID tuning method, and an adaptive tuner enables an effective application of PID control for wide spectrum of process responses. Drastically different sources of variability, automation system difficulties and changing process objectives do not require PID retuning. PID features can be used to tailor the PID response to different application requirements on a much more understandable basis by the user. The role of the PID can be expanded to include process optimization besides regulation.

Resources

1. McMillan, G. "PID Tuning Rules", *InTech*, Jan.-Feb. 2012.
2. McMillan, G. *Good Tuning: A Pocket Guide, 3rd Edition* ISA, 2012.
3. McMillan, G. "Full Throttle Batch and Startup Response" *Control*, May 2006.
4. Shinskey, G. "The Power of External-Reset Feedback" *Control*, May 2006.
5. McMillan, G. "What Is the Key PID Feature for Basic and Advanced Control?" Control Talk blog, *Control*, Apr. 26, 2012.
6. McMillan, G. and Torres, H. "Effective Use of Key PID Features" Paper presented at ISA Automation Week, Orlando, 2012.
7. McMillan, G. "Don't Overlook PID in APC" *Control*, Nov. 2011.
8. Yu, C. and Luyben, W. "Analysis of Valve-Position Control for a Dual Input Process" *Ind. Eng. Chem. Fund.*, pp. 344-350, Vol. 25, No. 3, 1986.
9. Harriott, P. *Process Control* McGraw-Hill, 1964.
10. Bohl, A. and McAvoy, T. "Linear Feedback vs. Time Optimal Control. II. The Regulator Problems", *Ind. Eng. Chem.*, Process Des. Dev., Vol. 15, No. 1, 1976.
11. McMillan, G. *Essentials of Modern Measurements and Final Elements in the Process Industries* ISA, 2010.
12. McMillan, G. "Advances in Flow and Level Measurements Enhance Process Knowledge, Control" *InTech*, Jan.-Feb. 2010.
13. McMillan, G. "Feedforward Control Enables Flexible, Sustainable Manufacturing" *InTech*, Mar.-Apr. 2011.
14. McMillan, G. "Key Design Components of Final Control Elements" *InTech*, Mar.-Apr. 2010.
15. McMillan, G. "What's Your Flow Control Valve Telling You?" *Control Design*, May 2004.
16. McMillan, G. "Improve Control Loop Performance" *Chemical Processing*, Oct. 2007.
17. Shinskey, G. "Flow and Pressure Control Using Variable-Speed Motors" *Proceedings of* Control *Conference*, pp. 161-167, Chicago, 1980.
18. Al-Khalifah, M. and McMillan, G. "Control Valves versus Variable Speed Drive for Flow Control" Paper presented at ISA Automation Week, Orlando, 2012.
19. Smith, C. "Watch Out with Variable Speed Pumping" *Chemical Processing*, May 2008.
20. McMillan, G. "Unlocking the Secret Profiles of Batch Reactors" *Control*, Jul. 2008.
21. McMillan, G. "Is Wireless Process Control Ready for Prime Time" *Control*, May 2009.
22. McMillan, G. "Wireless - Overcoming Challenges of PID Control, Analyzer Applications" *InTech*, Jul.-Aug. 2010.

Appendix A – PID Controller Forms

Conversion of Settings to Correct Units and Form

The methodology in this paper is based on the ISA Standard Form with specific tuning setting units where the controller gain is dimensionless, the reset time is seconds (seconds per repeat), and the rate time is seconds. PID tuning settings must first be checked for units and if necessary undergo a units conversion before being used.

(1) Convert proportional band (%) to controller gain (%/%) (dimensionless)

Gain = 100 % / Proportional Band

(2) Convert reset setting in repeats per minute to reset time in seconds per repeat

Seconds per repeat = 60 / repeats per minute

(3) Convert rate time in minutes to rate time in seconds

Seconds = 60 * minutes

(4) After the tuning setting units are verified, convert the Series or Parallel Form tuning settings to the ISA Standard Form tuning settings per equations below. The primed tuning settings are for the Series Form, the double primed tuning settings are for the Parallel Form, and the unprimed tuning settings are for the ISA Standard Form.

To convert from Series to ISA Standard Form controller gain:

$$K_c = \frac{T_i' + T_d'}{T_i'} * K_c' \qquad \text{(A.1)}$$

To convert from Series to ISA Standard Form reset (integral) time:

$$T_i = \frac{T_i' + T_d'}{T_i'} * T_i' = T_i' + T_d' \qquad \text{(A.2)}$$

To convert from Series to ISA Standard Form rate time:

$$T_d = \frac{T_i'}{T_i' + T_d'} * T_d' \qquad \text{(A.3)}$$

Note that if the rate time is zero, the ISA Standard and Series Form settings are identical. When using the ISA Standard Form, if the rate time is greater than ¼ the reset time the response can become oscillatory. If the rate time exceeds the reset time, the response can become unstable from a reversal of action form these modes. The Series Form inherently prevents this instability by increasing the effective reset time as the rate time is increased.

We can convert from ISA Standard Form to the Series Form using the following equations if the reset time is equal to or greater than 4 times the rate time ($T_i \geq 4 * T_d$).

$$K_c' = \frac{K_c}{2} * \left[1 + \left(1 - 4 * \frac{T_d}{T_i} \right)^{0.5} \right] \quad \text{(A.4)}$$

$$T_i' = \frac{T_i}{2} * \left[1 + \left(1 - 4 * \frac{T_d}{T_i} \right)^{0.5} \right] \quad \text{(A.5)}$$

$$T_d' = \frac{T_i}{2} * \left[1 - \left(1 - 4 * \frac{T_d}{T_i} \right)^{0.5} \right] \quad \text{(A.6)}$$

The Parallel form is used in some older DCS and PLC to isolate the proportional mode tuning setting from the other the modes. The gain setting does not affect the contribution from the integral and derivative modes and is sometimes called non-interacting. We can convert from the Parallel to the ISA Standard Form using the following equations.

$$K_c = K_c'' \quad \text{(A.7)}$$

$$T_i = K_c'' * T_i'' \quad \text{(A.8)}$$

$$T_d = \frac{T_d''}{K_c''} \quad \text{(A.9)}$$

We can convert from the ISA Standard Form to the Parallel Form by the following equations:

$$K_c'' = K_c \quad \text{(A.10)}$$

$$T_i'' = \frac{T_i}{K_c} \quad \text{(A.11)}$$

$$T_d'' = K_c * T_d \quad \text{(A.12)}$$

Where:

K_c = controller gain for ISA Standard Form (%/%) (dimensionless)
$K_c^{'}$ = controller gain for Series Form (%/%) (dimensionless)
$K_c^{''}$ = controller gain for Parallel Form (%/%) (dimensionless)
T_i = integral time (reset time) for ISA Standard Form (seconds)
$T_i^{'}$ = integral time (reset time) for Series Form (seconds)
$T_i^{''}$ = integral time (reset time) for Parallel Form (seconds)
T_d = derivative time (rate time) for ISA Standard Form (seconds)
$T_d^{'}$ = derivative time (rate time) for Series Form (seconds)
$T_d^{''}$ = derivative time (rate time) for Parallel Form (seconds)

Positive Feedback Implementation of Integral Mode

The positive feedback implementation of the integral mode effectively yields the equations above as seen in literature for a PI controller using Laplace transforms. Instead of an integrator, a filter whose input is the controller output and whose output is added to the contribution from the proportional mode in the positive feedback implementation. The filter time is the integral time setting. When external reset feedback is enabled, the input to the filter is switched from the controller output to the external reset feedback signal.

$$O(s) = K_c * E(s) + \frac{1}{1+T_i * s} * O(s) \quad \text{(A.13)}$$

$$O(s) * \left(1 - \frac{1}{1+T_i * s}\right) = K_c * E(s) \quad \text{(A.14)}$$

$$O(s) * \left(\frac{T_i * s}{1+T_i * s}\right) = K_c * E(s) \quad \text{(A.15)}$$

$$\frac{O(s)}{E(s)} = K_c * \left(\frac{1+T_i * s}{T_i * s}\right) = K_c + \frac{K_c}{T_i * s} \quad \text{(A.16)}$$

Where:

$E(s)$ = Laplace transform of controller error (%)
$O(s)$ = Laplace transform of controller output (%)

Fundamentals of Model Based Loop Tuning

James Beall

Principal Control Consultant

Emerson Process Management

13AIChE Spring Meeting, San Antonio, TX

EMERSON. Process Management

James Beall

- Principal Process Control Consultant, Emerson Process Management
- 31 Years experience in process control
  - 12 Years Emerson Process Management
  - 19 Years at Eastman Chemical Company
- Chairman ISA 75.25 Committee - Control Valve Performance Testing
- James.Beall@emerson.com 903-235-7935

13AIChE Spring Meeting, San Antonio, TX

EMERSON. Process Management

"PID = Proportional, Integral, Derivative"

- **P = proportional** - output contribution is "proportional" to difference between Set Point and PV (error). Therefore, output contribution is "gain" times error. Will have "offset" between Set Point and PV.
- **Integral -** output contribution is "integration" of the difference between Set Point and PV (error). Therefore, always tries to make PV = Set Point (eliminates offset).
- **Derivative -** output contribution is based on the derivative (rate of change) of the difference between Set Point and PV. Actually usually based on PV, not error.

13AIChE Spring Meeting, San Antonio, TX

EMERSON. Process Management

Caution!!!

- Controllers have different units for their tuning constants. For this presentation:
  - Proportional action will be expressed in **gain**
  - Integral action will be expressed as integral time, **Tr,** in **seconds/repeat**
  - Derivative in units of **seconds.**
- Controllers have different forms (Parallel, Series, Standard) of the PID algorithm. This can significantly affect the actual tuning values to achieve the desired response. The tuning rules in this presentation are for Standard and Series.

13AIChE Spring Meeting, San Antonio, TX

EMERSON. Process Management

Tuning Methods

- First tuning method due to Ziegler & Nichols (1942)
  - Ultimate Gain/Period
  - Reaction Curve
- "little black books"
- Default tuning (gain=1.0, Reset=1 min)
- Many people still do not use any method preferring to "tune-by-feel"
  - Classical control skills now rare
- Most older tuning methods try for "as fast as possible"
- Net result is each loop tuned independently
  - process dynamics not coordinated

13 AIChE Spring Meeting, San Antonio, TX

EMERSON. Process Management

Tuning Procedure

- Determine basic process type
  - Self-regulating
  - Integrating
- Determine process dynamics
- Choose desired closed loop response time, "Lambda"
- Calculate tuning constants

13 AIChE Spring Meeting, San Antonio, TX

EMERSON. Process Management

Lambda Tuning - Self Regulating Process Open Loop Response

Controller Output in manual

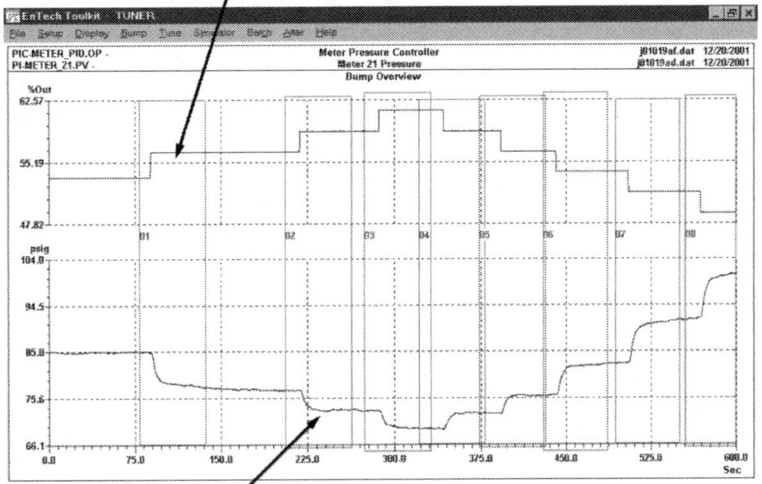

Process Variable

13 AIChE
Spring Meeting, San Antonio, TX

EMERSON
Process Management

Characterize the Process Response

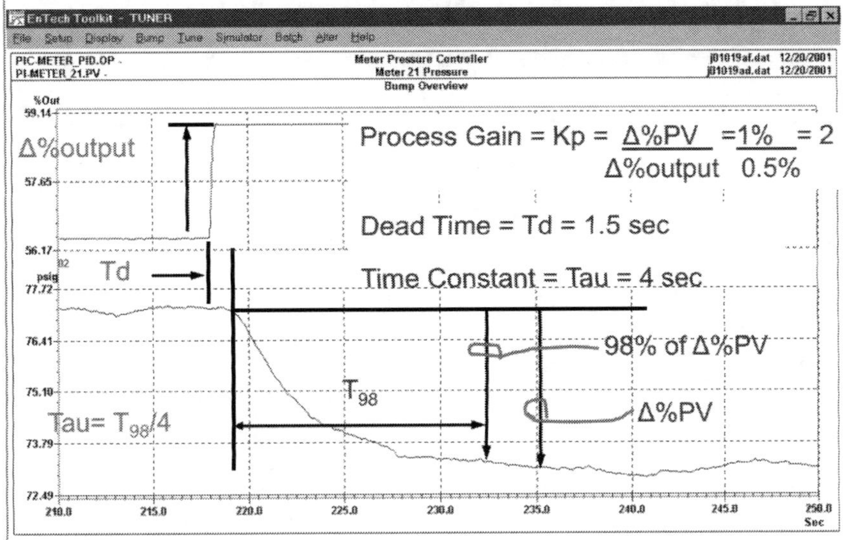

13 AIChE
Spring Meeting, San Antonio, TX

EMERSON
Process Management

Lambda Tuning Rules - Self Regulating Process

- Choose Lambda (λ), the "closed loop time constant"
  - A recommended starting point to ensure robustness is 3 * (larger of Td or Tau). This results in stable tuning if Deadtime and Process Gain double.
  - The closed loop "Time to Steady State" (T_{98}) (for a set point change) is approximately 4*Lambda (assuming "P" is on error)
- Tr = Reset Time = Tau (units are time/repeat)
- Kc = Controller Gain = $\frac{Tr}{Kp\,(\lambda + Td)}$
- NOTE – Tr stays the same, only Kc changes with Lambda

13 AIChE Spring Meeting, San Antonio, TX

EMERSON. Process Management

Lambda Tuning Example

- Choose Lambda = 3 (4 seconds) = 12 seconds
- Integral Time Tr = Tau = 4 seconds/repeat
- Kc = Tr / Kp (λ + Td)

  = 4 seconds /(2*(12 seconds + 1.5 seconds)) = 0.148

Lambda	Gain	Integral Time
12 sec	0.15	4 sec/rep
8 sec	0.21	4 sec/rep
4 sec	0.36	4 sec/rep
1.5 sec	0.67	4 sec/rep

Faster (Lambda ↓) — Higher Gain (Gain ↓)

13 AIChE Spring Meeting, San Antonio, TX

EMERSON. Process Management

Response – Lambda = 12 sec , 8 sec

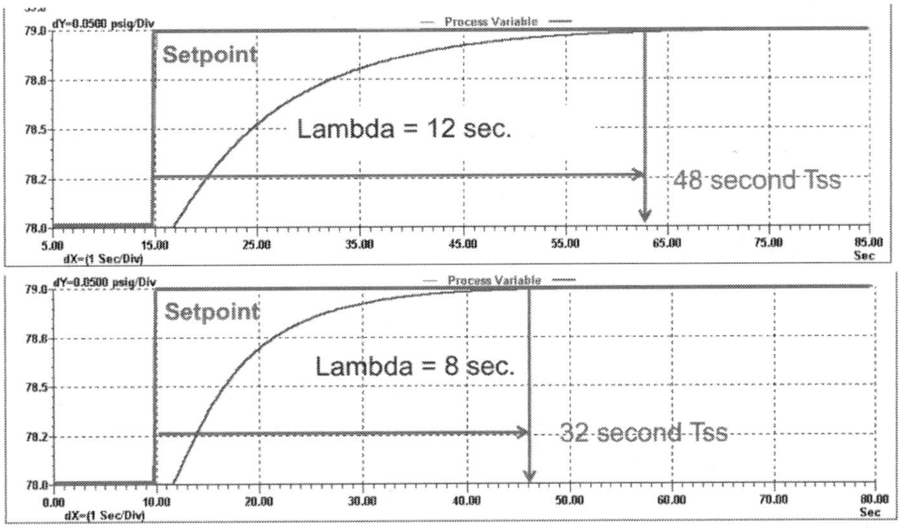

13AIChE
Spring Meeting, San Antonio, TX

EMERSON.
Process Management

Response – Lambda = 4 sec , 1.5 sec

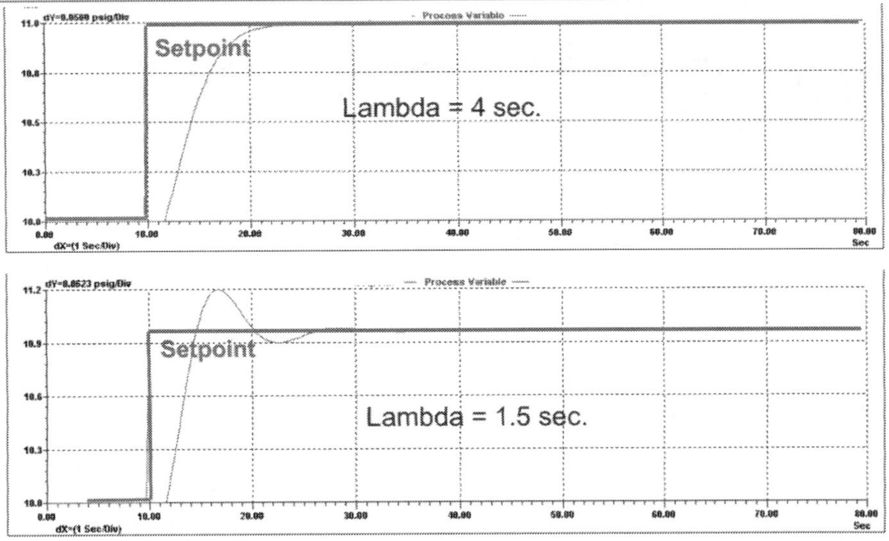

13AIChE
Spring Meeting, San Antonio, TX

EMERSON.
Process Management

Lambda Tuning - Is it slow? Compared to what?

Typical "Tune by Feel"

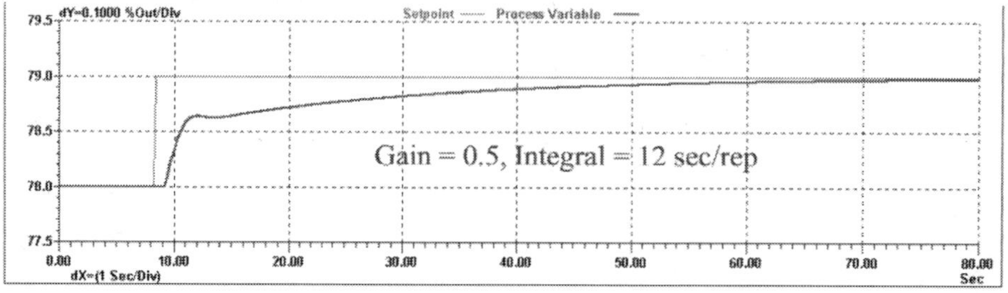

Lambda = 6.6 seconds (normal robustness)

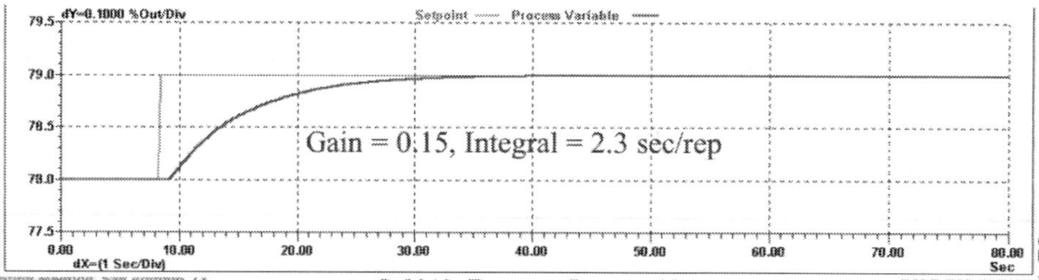

Integrating Process – Identify Open Loop Response

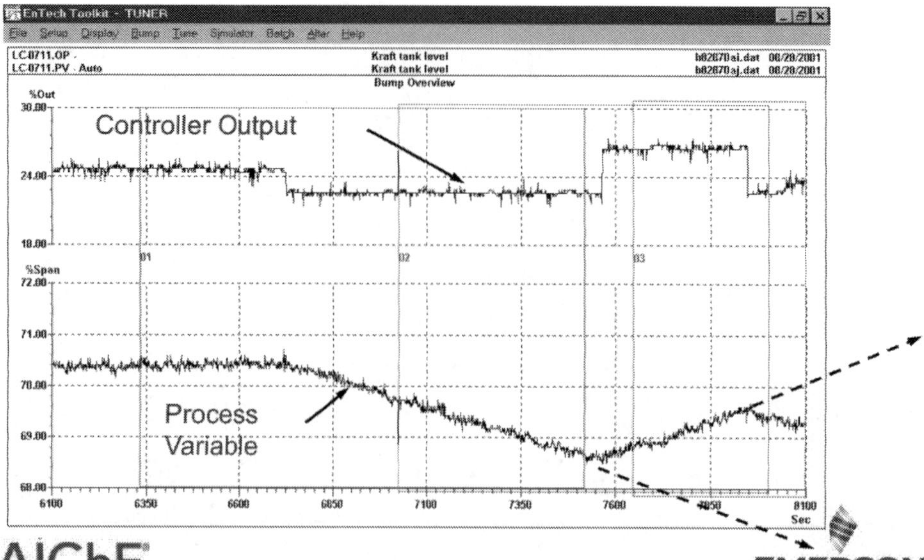

Lambda Tuning for Integrating Processes – Setpoint (Closed Loop) Response

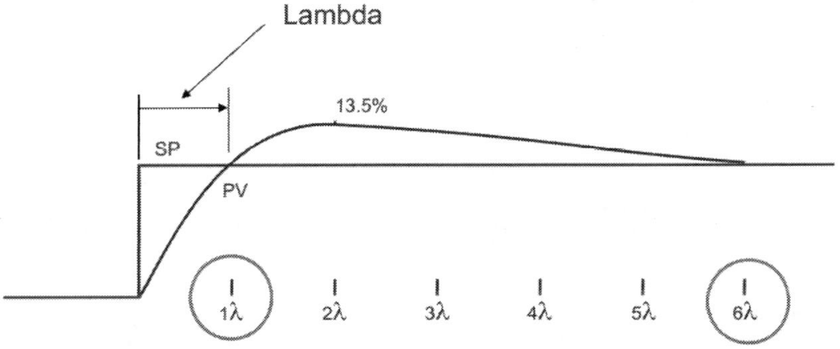

13AIChE Spring Meeting, San Antonio, TX

EMERSON. Process Management

Lambda Tuning for Integrating Processes - Load Disturbance Response

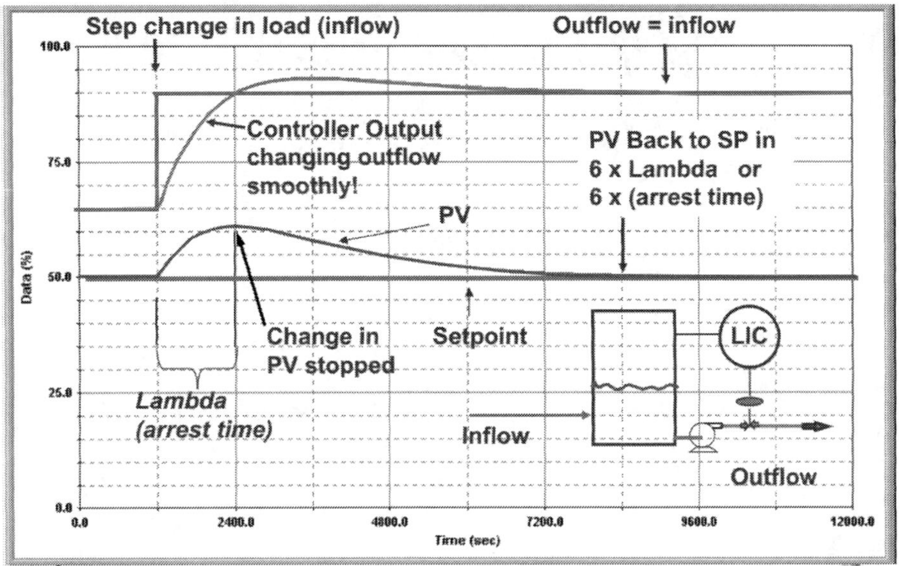

13AIChE Spring Meeting, San Antonio, TX

EMERSON. Process Management

Characterize the Process Response

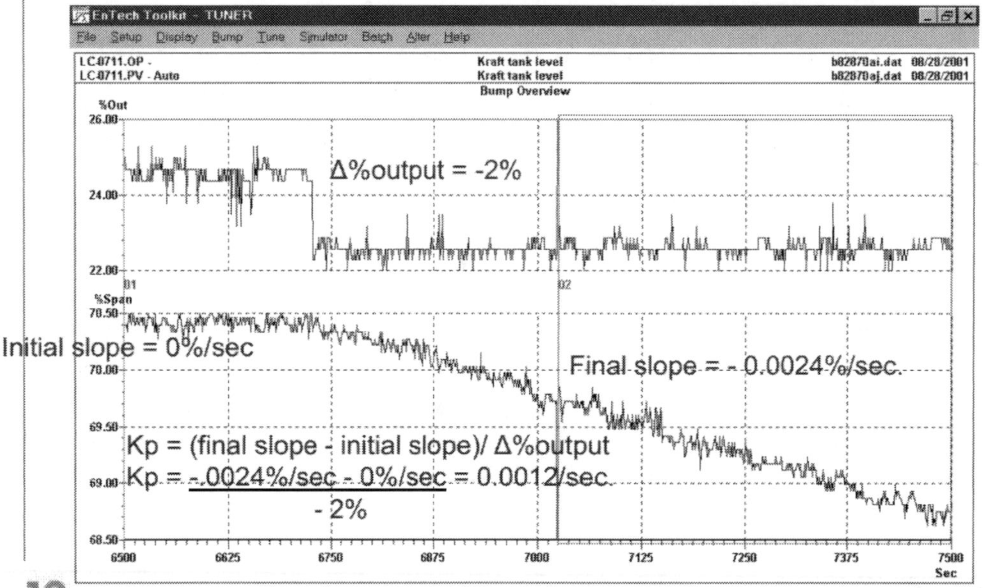

Lambda Tuning - Integrating Process

- THIS RULE DOES NOT INCLUDE DEAD TIME OR A LAG. Contact the author for applications with dead time or a significant lag.
- Choose Lambda (λ) (more on this later!)
- $Tr = (2^* \lambda)$
- $Kc = \frac{Tr}{Kp(\lambda)^2}$
- NOTE – Gain AND Reset Time change with Lambda

Lambda Tuning Rules - Integrating Process

- Example – Choose Lambda = 10 minutes = 600 sec
- Tr = (2* λ) = (2*600 sec) = 1200 sec/rep
- $Kc = \dfrac{Tr}{Kp(\lambda)^2}$ = 1200 sec/((.0012/sec)*(600sec)$^{2)}$
- Kc = 2.78

Lambda	Gain	Integral Time
600 sec	2.78	1200 sec/rep
300 sec	5.56	600 sec/rep

Faster Higher Gain Faster (shorter) Reset Time

13 AIChE Spring Meeting, San Antonio, TX

EMERSON. Process Management

Lambda = 10 min. for Integrating Process

Kc = 2.78, Tr = 1200 sec./rep

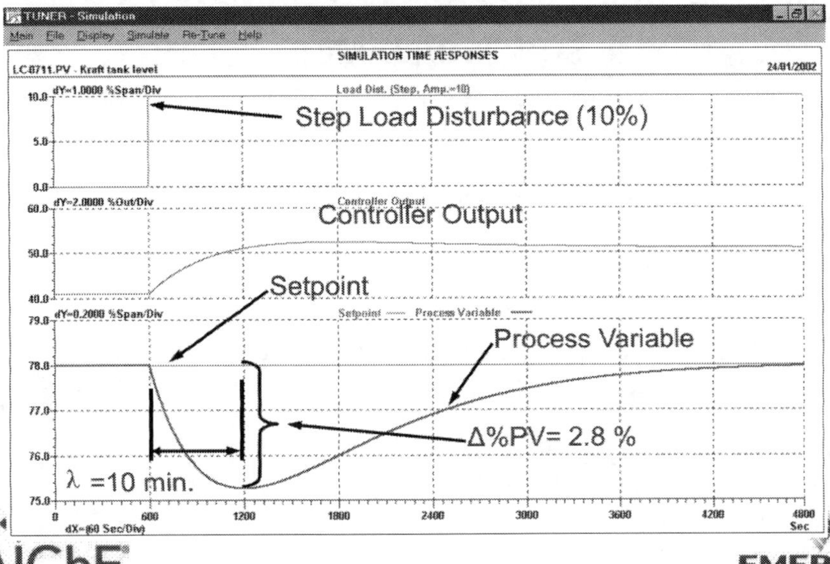

13 AIChE Spring Meeting, San Antonio, TX

EMERSON. Process Management

Lambda = 5 min. for Integrating Process

Kc = 5.56, Tr = 600 sec./rep

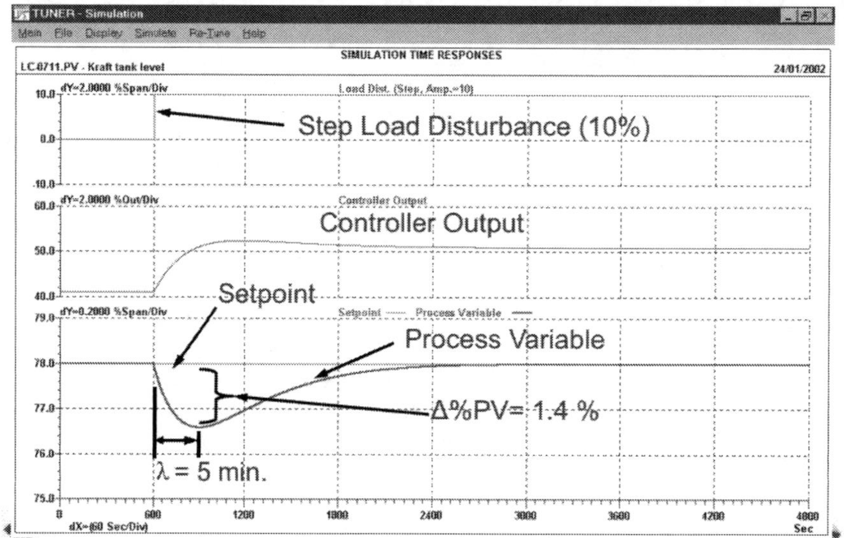

Attenuate Variability with Control/Equipment

- "Capacity" in the process can be used to attenuate or absorb variability
- Primary source of process capacity is level control
- To utilize level control as a capacity tune the controller only "fast" enough to hold the PV within the allowable level deviation (ALD) for a maximum load change

Utilize Level Control as Variability Sinks

- Choose the arrest time (Lambda) "slow" enough to provide a variability sink yet maintain level within the allowable variation
- Lambda α $\frac{\textbf{ALV}}{\textbf{Kp * MLD}}$
  - ALV = Allowable Level Variation
  - Kp = Integrating process gain
  - MLD = Maximum Load Disturbance (in % of controller output)

13 AIChE Spring Meeting, San Antonio, TX

EMERSON. Process Management

Example of Lambda Calculation

- Previous Example Kp= 0.0012/sec
- Let ALV = +- 15%
- Let MLD = 25%
- Lambda = 1333 seconds = 22.2 minutes
- **Tr = 2666 seconds; Kc =1.25**

13 AIChE Spring Meeting, San Antonio, TX

EMERSON. Process Management

Example of Lambda Calculation

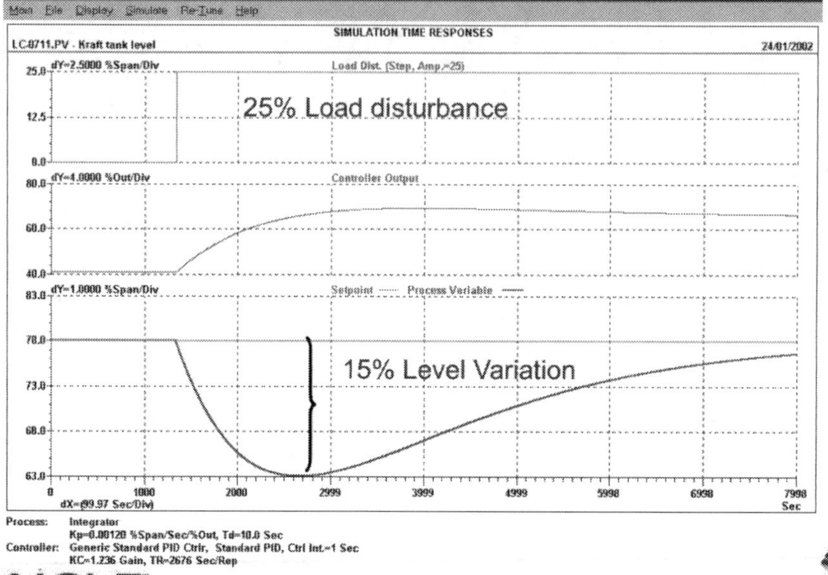

13 AIChE Spring Meeting, San Antonio, TX

EMERSON. Process Management

Results – Integrating Process Tuning

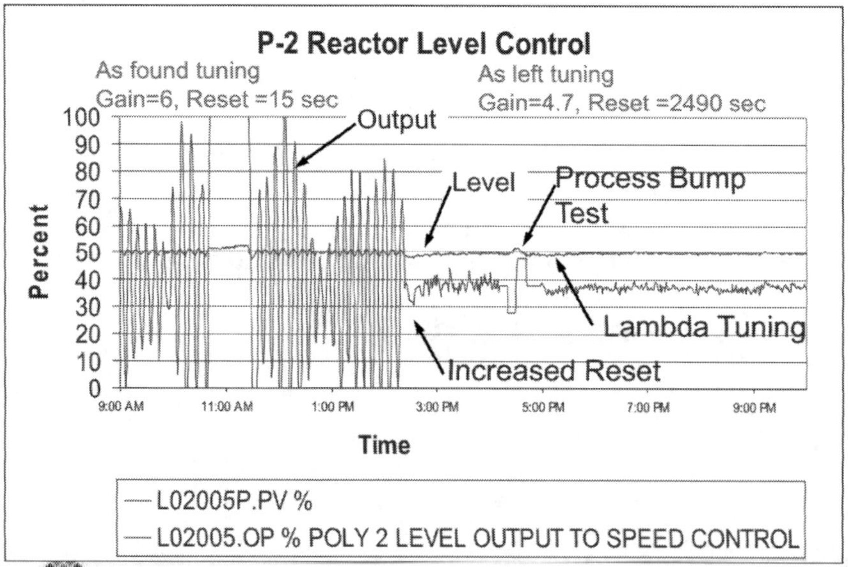

13 AIChE Spring Meeting, San Antonio, TX

EMERSON. Process Management

*Process Dynamics – *Model Based Tuning May Include Derivative*

- Pure Gain
- 1st Order
- *2nd Order, Over Damped
- *2nd Order, Under Damped
- 2nd Order, Lead
- 2nd Order, Lead with Overshoot
- 2nd Order, Non-Minimum Phase
- Integrator
- *Integrator, 1st Order Lag
- Integrator, 1st Order Lead
- Integrator, Non-Minimum Phase

13 AIChE Spring Meeting, San Antonio, TX

EMERSON. Process Management

Summary Tuning Procedure

- Determine basic process type
  - Self-regulating
  - Integrating
- Determine process dynamics
- Choose desired closed loop response time, "Lambda"
- Calculate tuning constants
- NOTE – For better load regulation, self-regulation loops with long lags can be tuned as in integrator

13 AIChE Spring Meeting, San Antonio, TX

EMERSON. Process Management

Questions?

James.Beall@Emerson.com

903-235-7935

13 AIChE
Spring Meeting, San Antonio, TX

EMERSON.
Process Management

DETECTING LOSS OF FLAME IN OIL REFINERY FIRED HEATERS USING ADVANCED PRESSURE DIAGNOSTICS

Presented at AIChE Spring Meeting, San Antonio, TX, April 28-May 2, 2013

John P. Miller – Emerson Process Management – Chanhassen, MN, USA
Randy S. Stier – Valero Energy Corporation – San Antonio, TX, USA

Biography

John Miller is a principal engineer of pressure technology for the Rosemount business unit of Emerson Process Management. John works in developing and promoting applications for Rosemount Advanced Diagnostics pressure transmitters and in the research and development of new products. John has a BS in Mechanical Engineering and an MS in Electrical Engineering both from the University of Minnesota. He is inventor on 24 patents and other pending applications.

Randy Stier is a fired heater engineer with a strong track record of improving oil refinery profits through reliability, availability and energy related projects. A chemical engineer by training, Randy earned a BSChE degree from The University of Michigan and an MSChE degree from the Illinois Institute of Technology. Randy is currently Valero's Director of Refinery Heater Best Practices and Chairman of the API Subcommittee on Heat Transfer Equipment.

Abstract

Maintaining stable combustion is critical to the safe operation of fired heaters. When the ratio of air flow to fuel flow to a burner transitions from a point inside the burner's operating envelope to a point outside the burner's operating envelope, the combustion process becomes unstable and then stops. If this loss of flame is not recognized and acted upon properly, an explosion may occur.

One possible way to detect loss of flame may be to monitor pulsations in the difference between the flue gas pressure inside a fired heater and atmospheric pressure at a given elevation (commonly referred to as "draft"). During stable combustion, the draft will pulsate slightly (see Figure 1). This is commonly referred to as combustion rumble. As a burner's air/fuel ratio transitions from inside to outside the burner's operating envelope, the draft pulsations may exhibit a sharp increase in amplitude and frequency, caused by instability in the burner flames. Flame instability is often a precursor to flame out.

Draft transmitters with Advanced Diagnostics Statistical Process Monitoring (SPM) capability monitor the pulsations in the draft measurement and detect the increase in amplitude or frequency associated with flame instability. SPM data may be integrated with heater operations. An alarm limit on the SPM standard deviation may be configured to alert a heater operator to potential combustion problems or flame out.

Valero has worked with Rosemount to evaluate the use of SPM technology for detection of flame instability or flame out. Recent testing was performed on five different burners at three different burner vendors. The technology was also evaluated on a heater at the Valero Texas City Refinery.

Experience gained from these tests may be used to determine where to set the alarm limit for detecting flame instability. There is a correlation between the volumetric firing rate of a burner/heater system and the draft standard deviation. This correlation can be exploited for selecting an appropriate alarm limit in an operating fired heater. Burner vendor tests indicate that a draft standard deviation alarm limit of 30 thousandths inches of water column (m-inWC) may be appropriate.

Additional work is required to validate these observations but initial results indicate the use of advanced pressure diagnostics may be an effective way to detect loss of flame.

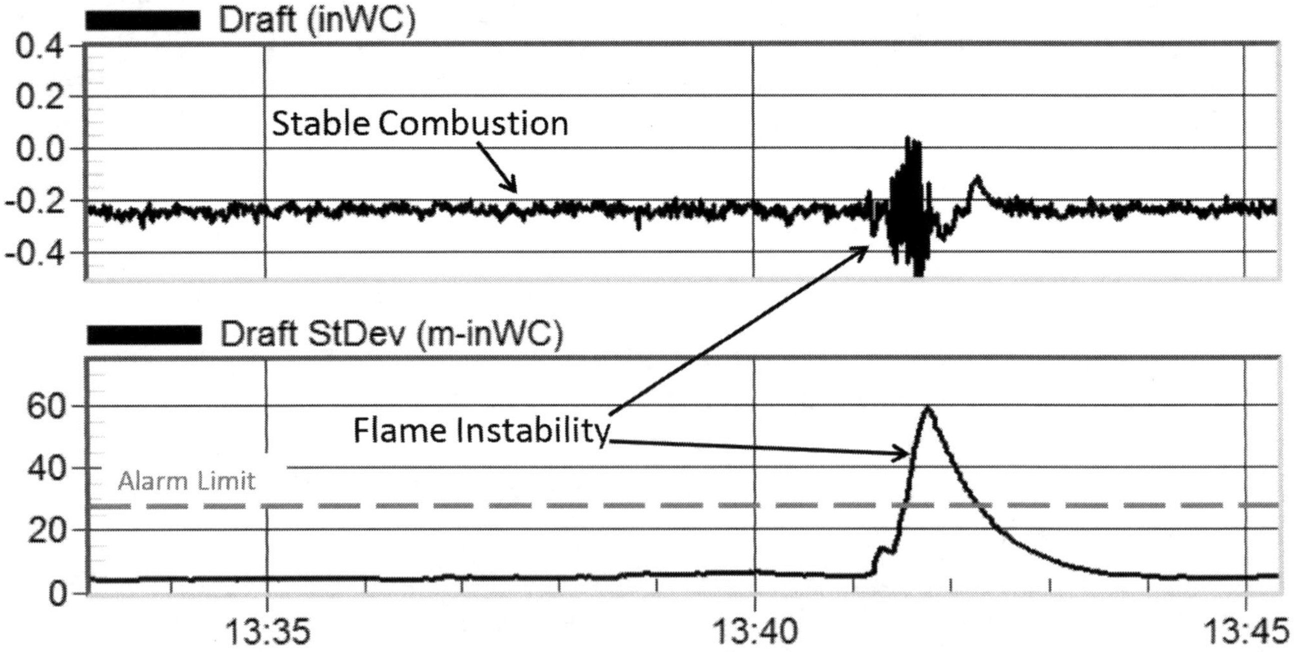

Figure 1 – Detection of Flame Instability with Advanced Pressure Diagnostics

Fired Heaters and Proper Combustion

Fired heaters are a class of indirect-fired equipment used to heat process fluids in oil refineries and other process industries. Many refineries burn fuel gas in their fired heaters in order to get the energy needed to heat the process fluids. These process fluids flow through tubes mounted throughout the fired heaters. Energy is transferred to the process fluids flowing through the heater tubes by conduction, convection, and radiant heat transfer. Process fluids can enter and exit the tubes as a gas, a liquid, or both. Fired heaters account for 2/3 to 3/4 of the energy consumed in a refinery.

Fired heater burners are designed to operate with more combustion air than needed in order to react all of the fuel gas. They are designed this way because burners are not perfect mixing devices. If a burner is only fed the amount of air theoretically required, not all of the fuel gas will find the available oxygen. The fuel gas without oxygen will not burn and combustion will be incomplete.

Maintaining stable combustion is critical to the safe operation of fired heaters. When the ratio of air flow to fuel flow to a burner transitions from a point inside the burner's operating envelope to a point outside the burner's operating envelope, the combustion process will become unstable and then stop. The transition of a burner's air/fuel ratio from inside to outside the burner's operating envelope may be caused by a number of different initiating events including a significant change in fuel gas composition, a sudden decrease in air flow, a sudden increase in fuel flow or a shift in the boundaries of the burner operating envelope.

Incomplete combustion (also known as substoichiometric combustion), caused by too little combustion air, is undesirable in a fired heater for the following reasons:

- A low oxygen-to-fuel gas ratio causes long flames which may touch the tubes. The term for this is flame impingement. If there is flame impingement, it may cause the tubes and other heater parts, such as the tube supports and refractory, to overheat.
- Energy is wasted when unburned fuel gas leaves the heater via the stack.
- Unburned fuel gas has the potential to ignite elsewhere in the fired heater if a source of oxygen is found. This is called "afterburning" and can result in damage to components in the fired heater.
- A low oxygen-to-fuel gas ratio may cause flame instability and potentially flame out.

Refinery fired heaters often use waste gas (hydrocarbon gas produced as a bi-products of other refining processes) as the fuel gas source. Burning waste gas is advantageous because it costs only a fraction per BTU as other fuel sources. The disadvantage of waste gas is that there can be significant variability in the composition, and therefore, a large variability in BTU per standard cubic foot. If the fuel gas BTU content suddenly increases, the heater may go into a state where there is not enough available oxygen for all the fuel. Figure 2 shows an example sequence of events that may result.

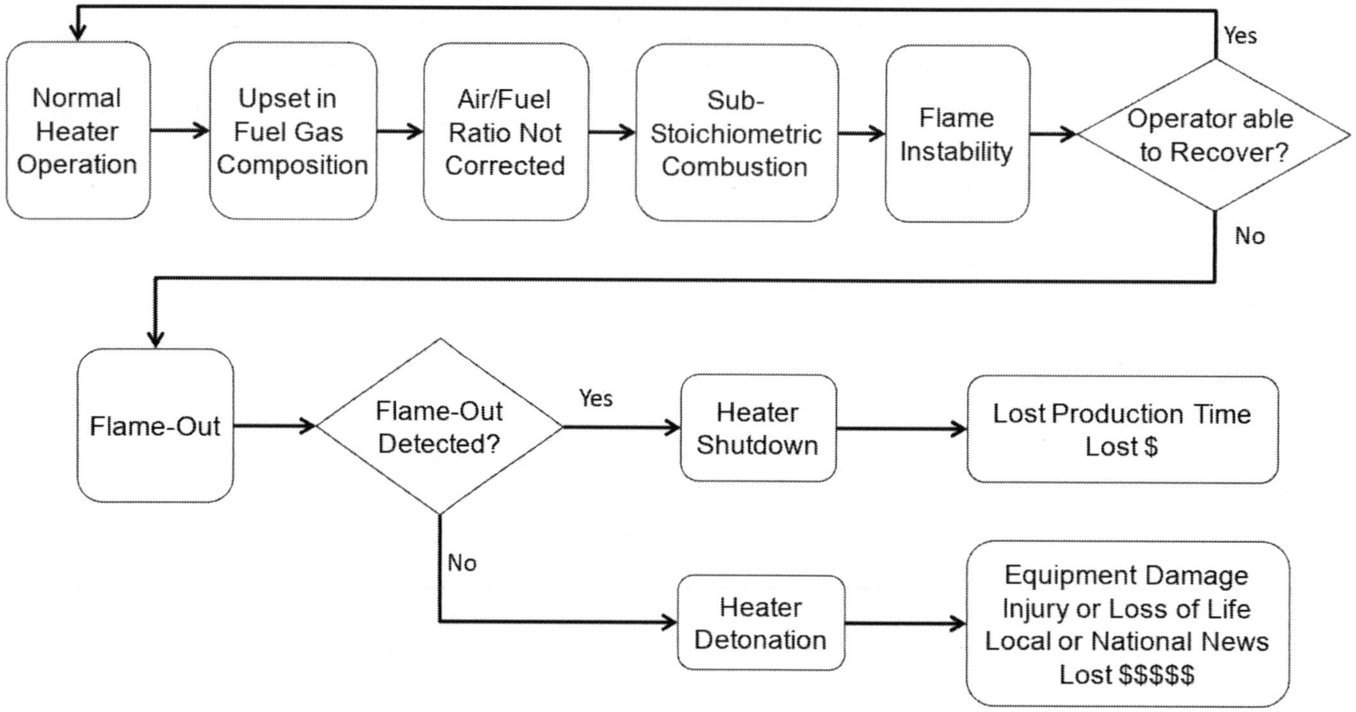

Figure 2 – Example Sequence of Events before and after a Flame out

If a heater experiences an upset in the fuel gas composition (e.g. a surge in the BTU content of the fuel gas or waste gas) and if the air/fuel ratio controller is not able to keep up, the burners may begin operating in substoichiometric, or incomplete, combustion. Depending upon the type of burner, the flames may start going unstable (Flame Instability). If the air/fuel ratio is not corrected (either manually by an operator, by via automatic O_2 control), the burner's flame may be extinguished shortly thereafter. After a flame out, a heater must be shut down to purge out the unburned fuel gas, resulting in lost production time for that operating unit. If a flame out is not detected, it is an especially dangerous situation because unburned fuel and air continues to be fed into the firebox. The slightest spark or ignition source could light this air/fuel mixture, causing a detonation of the heater, resulting in major equipment damage, and possibly, injury or loss of life.

Some refinery fired heaters with are equipped with flame scanners that can detect a loss of flame. However, flame scanners become problematic in multi-burner heaters because a flame scanner cannot detect if just a single burner is extinguished. Thus, most multi-burner heaters do not use flame scanners. For these heaters, current industry practice is to train operators to manually recognize loss of flame via operational variables, such as a sudden drop in firebox temperature and a sudden increase in O_2 or combustibles in the flue gas.

Draft Measurement and Flame Instability

The industry best practice calls for fired heaters to have a draft measurement. Draft is defined as the pressure difference between the ambient atmosphere outside of the heater and the inside of the heater firebox where combustion occurs ($Draft = P_{atmosphere} - P_{firebox}$). Figure 3 illustrates draft measurement on a fired heater. The red line shows the absolute pressure inside the heater at various elevations ($P_{firebox}$), while the blue line shows the corresponding ambient pressure ($P_{atmosphere}$).

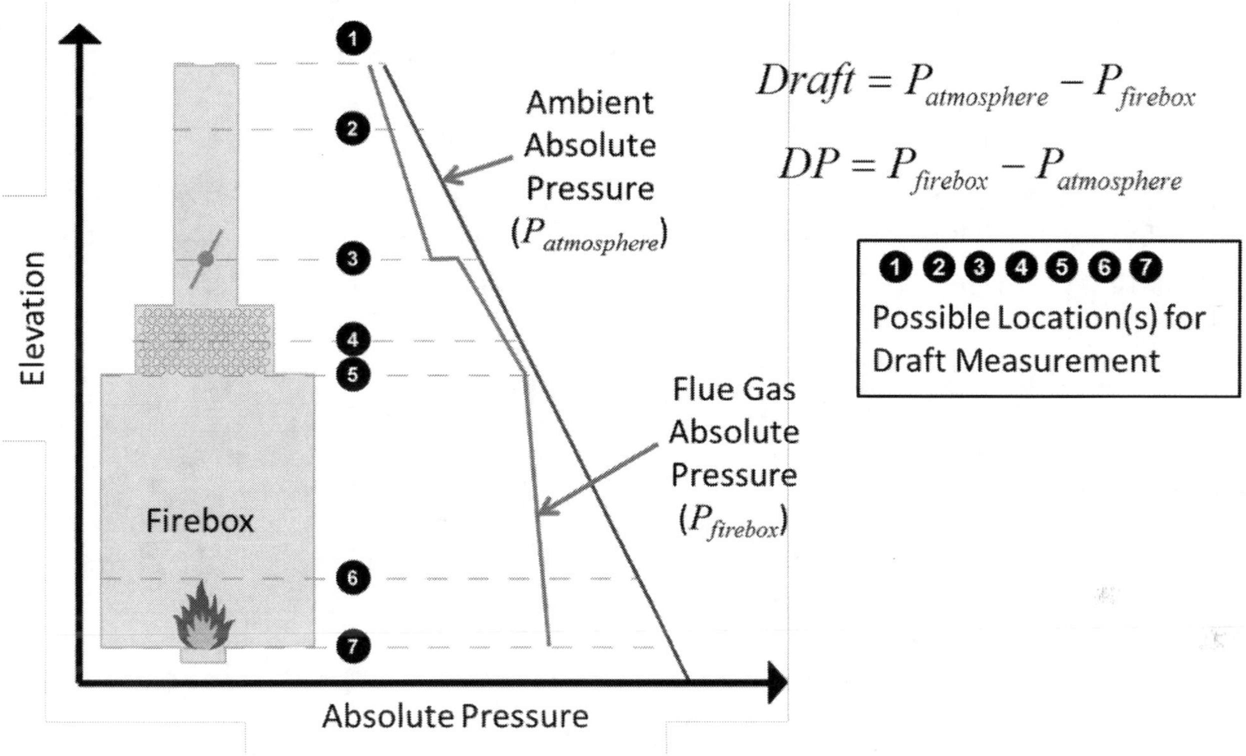

Figure 3 – Draft Measurement on a Fired Heater

Draft can be measured at any one of multiple elevations on the heater including at the stack (1 and 2), near the stack damper (3), at the convection section (4), at the bridgewall (5), near the burner (6), and at the floor of the heater (7). Typically, draft is measured at only one of these elevations. During normal operation, the firebox pressure is less than atmospheric pressure. If the firebox pressure goes higher, it is an indicator that there may be problems. Depending upon the elevation, heater design, and heater firing rate, the draft can be anywhere from 0.1 inWC (inches of water column) to 1.0 inWC. Within the industry, sometimes the draft measurement is expressed as a differential pressure ($DP = P_{firebox} - P_{atmosphere}$). Under this convention, the draft value is negative during normal combustion, and goes positive only if there are combustion problems (e.g. flame instability). Throughout the paper, we utilize this convention, showing a negative draft value in all of the trends.

As described previously, substoichiometric combustion may lead to instability in the burner flames. If someone is standing near the heater, flame instability can be heard audibly, as a "huffing" or "panting" in the heater. It is known in the industry that flame instability causes pulsations (or oscillations) in the firebox draft measurement. Sometimes flame instability is confirmed by an operator visually observing a draft gauge swinging between the positive and negative. Thermo acoustic coupling [1, 2, 3, and 4] is one mechanism that may contribute to flame instability and draft oscillations. Industry has investigated utilizing patterns in the draft oscillations for detecting flame instability [5].

Figure 4 shows an example of draft oscillations that occur under flame instability. Under normal heater combustion, the draft measurement is negative and relatively steady. However, when the flames go unstable, the draft starts oscillating, and may periodically go into positive. In the example below, during Healthy Combustion, the draft is approximately -0.2 inWC. But during Flame Instability the draft starts oscillating at a frequency of about 3 Hz, with a peak-to-peak amplitude of about 0.6 inWC. The draft even periodically goes as high as +0.2 inWC.

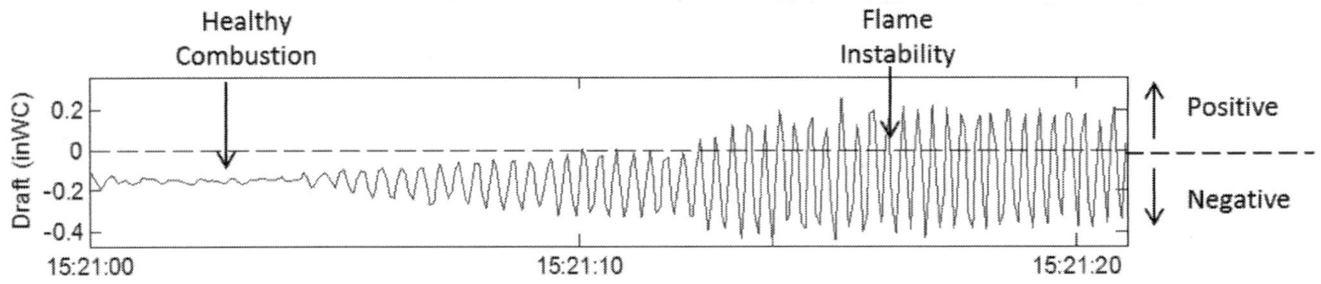

Figure 4 – Example Draft Measurement Signal during Flame Instability

However, the draft oscillations associated with flame instability are not normally seen in the draft measurement because of typical damping, filtering, or sampling, which may be implemented in the draft transmitter and/or in the Distributed Control System (DCS). Figure 5 shows the effect of transmitter damping (PV damping) on the flame instability draft oscillations. Many pressure transmitters are pre-configured with a PV damping of approximately 0.4 seconds. As shown in the middle trend, this damping flattens out the draft oscillations significantly. Sometimes a draft measurement is configured with a much higher damping (e.g. 5 seconds as shown in bottom trend) to smooth out the impact of combustion rumble on the signal, and make the draft steady when a heater is operating normally. A damping this high makes the flame instability draft pulsations impossible to see. With either the 0.4 second or 5 second PV damping, neither an operator nor an automatic control system can see the pulsations and positive differential pressure associated with flame instability.

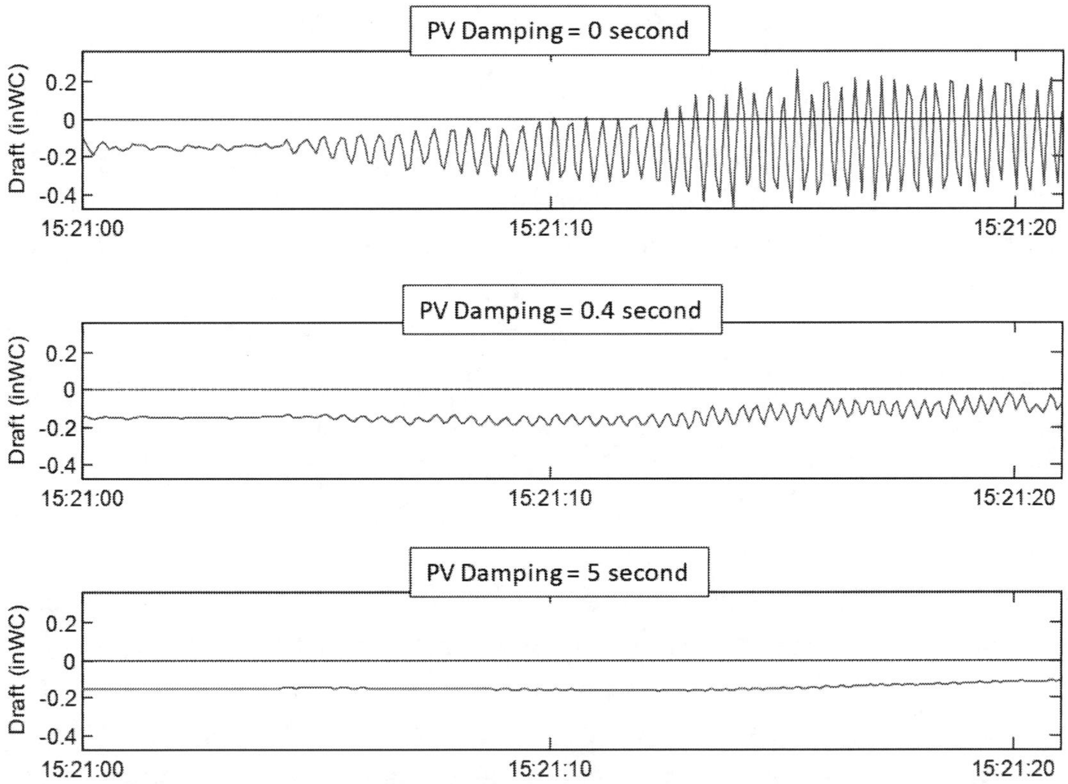

Figure 5 – Effect of Draft Transmitter Damping on Flame Instability Oscillations

The other limitation to observing the draft pulsations associated with flame instability is the damping and slow sampling rate of a typical DCS. A DCS may contain additional damping or filtering to further smooth out the draft measurement. A DCS also typically has a sampling rate too slow to see these draft oscillations. For example, a DCS may sample the pressure only once per second. Figure 6 illustrates a 1-second sampling of the draft measurement during flame instability. Because this is much slower than the 3 Hz oscillations of the draft, none of the positive pressure peaks are detected.

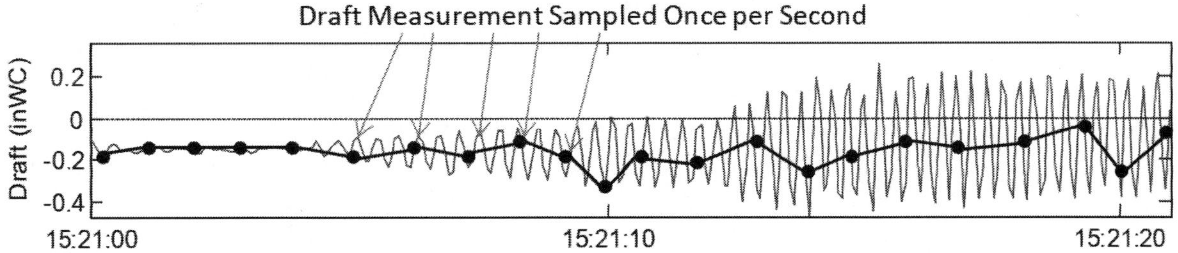

Figure 6 – Slower Sampling in the Distributed Control System (DCS)

Thus, because of limitations of transmitter PV damping, and DCS sampling rate, the draft oscillations associated with flame instability are normally not seen by a heater operator.

Advanced Pressure Diagnostics

Over the last twenty years, smart field devices have become increasingly more prevalent in the process industry. Smart field devices provide not just the basic process measurement (e.g. pressure, differential pressure, draft, temperature, or flow) but they provide a wide range of additional information, such as second and third process variables, device identification and health, and device diagnostics information. One example of such smart field devices is the Rosemount 3051S Advanced Diagnostics Pressure Transmitter with Statistical Process Monitoring (SPM) capability. SPM is available on most types of pressure measurements within the 3051S family, including absolute and gauge pressure, differential pressure (DP), DP flow, DP level, and draft. As used herein, the term "pressure transmitter" encompasses all of these pressure measurement types including fired heater draft measurement.

SPM technology allows a plant engineer or operator to see the variation (or noise) in a process that is normally filtered out when the variable is measured and displayed in the DCS. Because the pressure transmitter is closer to the process than the DCS, it can sample the pressure at a much faster rate, and turn any process variation into valuable information. The Rosemount 3051S pressure transmitter samples the pressure at 22 Hz, much faster than the typical DCS sampling rate of once every 1-2 seconds. The diagnostics transmitter calculates mean and standard deviation of the pressure and sends this information to the host system via digital communication protocol, such as HART or Foundation Fieldbus.

Figure 7 illustrates the architecture and implementation of a pressure transmitter with Advanced Diagnostics SPM capability.

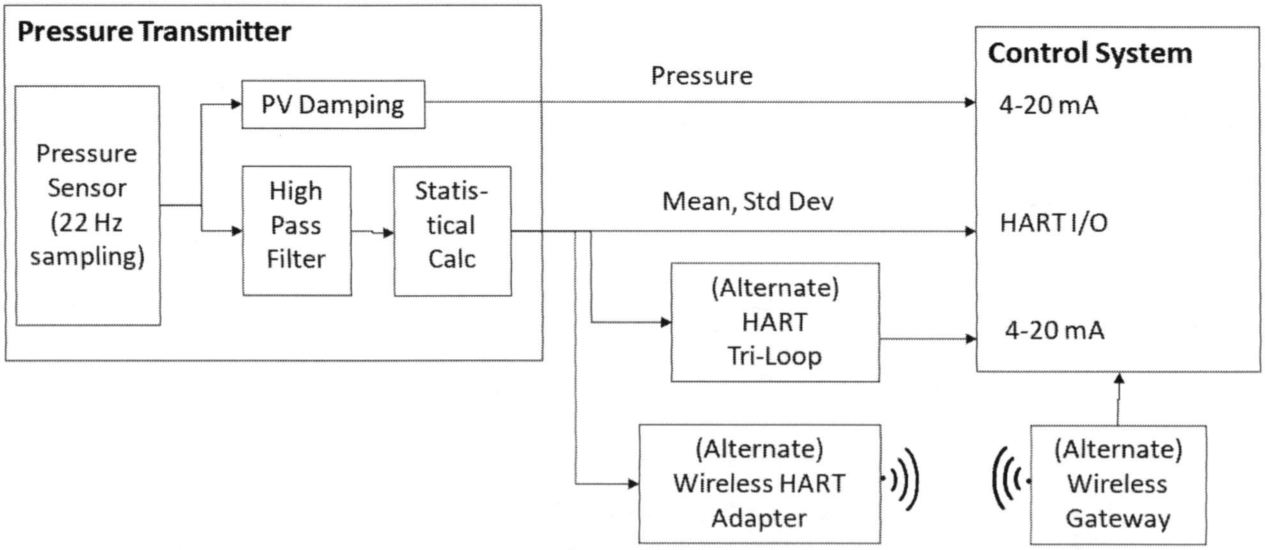

Figure 7 – Architecture and Implementation of Advanced Pressure Diagnostics

The pressure sensor measures the pressure at 22 Hz, applies the normal PV Damping, and then sends the measured pressure to the Control System via 4-20 mA. On a parallel path, a High Pass Filter and Statistical Calculations module takes the un-damped pressure and calculates mean and standard deviation. The statistical values are made available to the control system as secondary variables via HART I/O.

If a control system does not have HART I/O ability, the statistical values can still be accessed via a HART Tri-Loop (e.g. Rosemount 333) which converts the HART data to a 4-20 mA. Alternatively, a wireless HART adapter (e.g. 775 Smart Wireless THUM adapter) can be used to access the advanced diagnostics information, and transmit the information wirelessly to a gateway (e.g. 1420 Smart Wireless Gateway), which then integrates the diagnostics information with the Control System.

Figure 8 further illustrates Statistical Process Monitoring technology. On the left, is the process variable as measured by the pressure transmitter, with different amount of process variation (Normal, Noisy, and Quiet). On the right are the variables that the advanced diagnostics pressure transmitter sends to the control system.

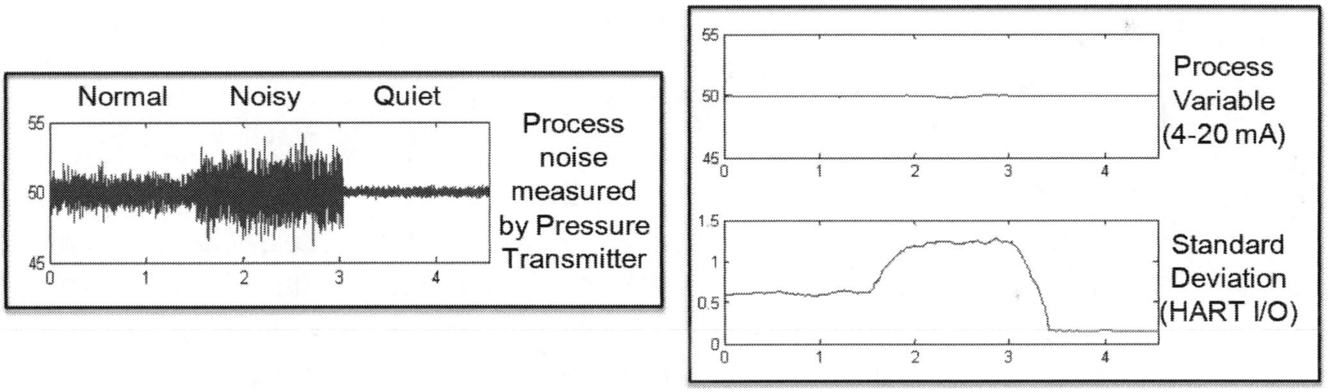

Figure 8 – Statistical Process Monitoring in a Pressure Transmitter

Note that the process variable (4-20 mA) is relatively flat, because the PV damping has removed all of the process variation. However, the standard deviation is a measure of how much process variation is in the original signal. The standard deviation is higher when the process is noisier, and it is smaller when the process is quieter.

As shown in Figure 7, a high-pass filter is used prior to the Statistical Calculations module. The purpose of the high-pass filter is to ensure that the SPM analyzes only the process variation at higher frequencies (approximately 1 Hz and higher). When there is just a slow change in the pressure measurement (such as the heater draft), this should not be interpreted as an increase in the process variation. Figure 9 illustrates the effect of using the high-pass filter. The top trend (a) represents the actual process behavior, including both a nominal (average) pressure, and random process noise. The middle trend (b) represents the pressure reading that is transmitted to the DCS via the 4-20 mA. The PV damping has smoothed out all the noise, and preserved the nominal pressure value. The bottom trend (c) represents the high-frequency noise that is analyzed by the SPM. The high-pass filter allows the SPM to analyze the pressure variation that is removed by the PV damping.

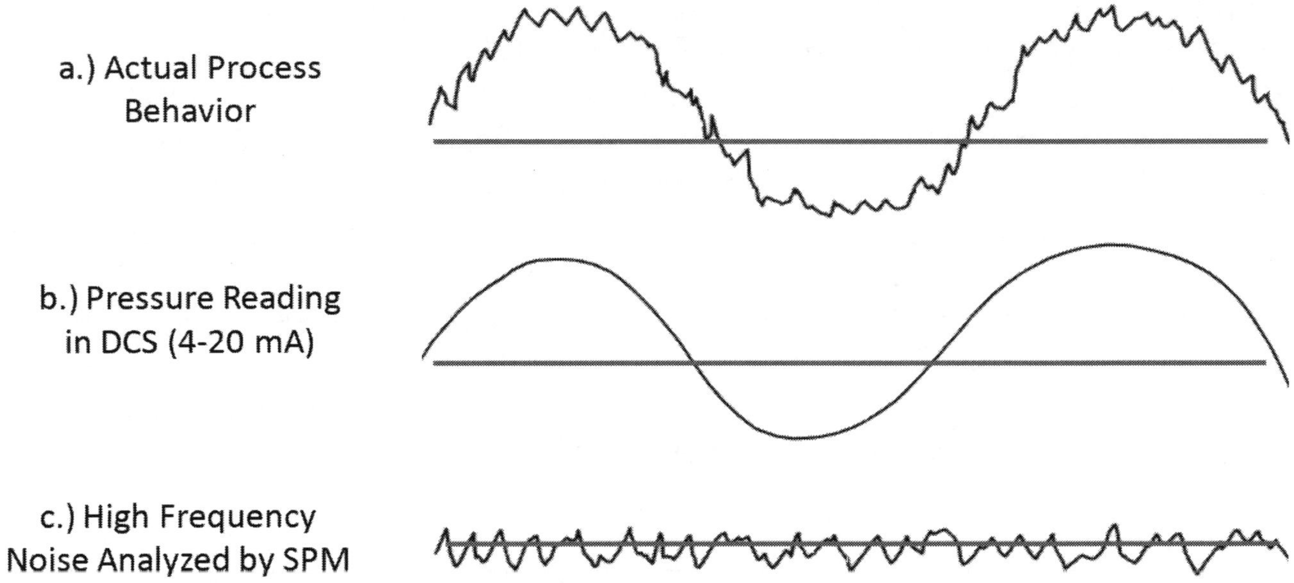

Figure 9 – Effect of High-Pass Filter on the SPM Algorithm

When the advanced diagnostics standard deviation is viewed in the control system, it is possible to detect abnormal conditions that would otherwise be unobservable using only the standard 4-20 mA pressure signal. Many different abnormal conditions have been shown to be detectable using advanced pressure diagnostics SPM technology, including plugged impulse lines [6, 7], entrained gas in liquid flows [8], increased liquid loading in natural gas wells [9], and distillation column flooding [10, 11].

Flame Instability Detection with SPM

As published previously [12, 13, 14, 15, 16], SPM technology can also be utilized to detect the draft oscillations associated with flame instability. Referring back to Figure 1, the increase in draft oscillations is detected as a change in the SPM standard deviation. This change in standard deviation occurs regardless of the transmitter PV damping (Figure 10). A heater operator, control system, or burner management system may utilize the SPM information as an indicator of instability in the burner flames either to make adjustments to prevent a flame out, or to recognize that the heater has flamed out and initiate a shut-down.

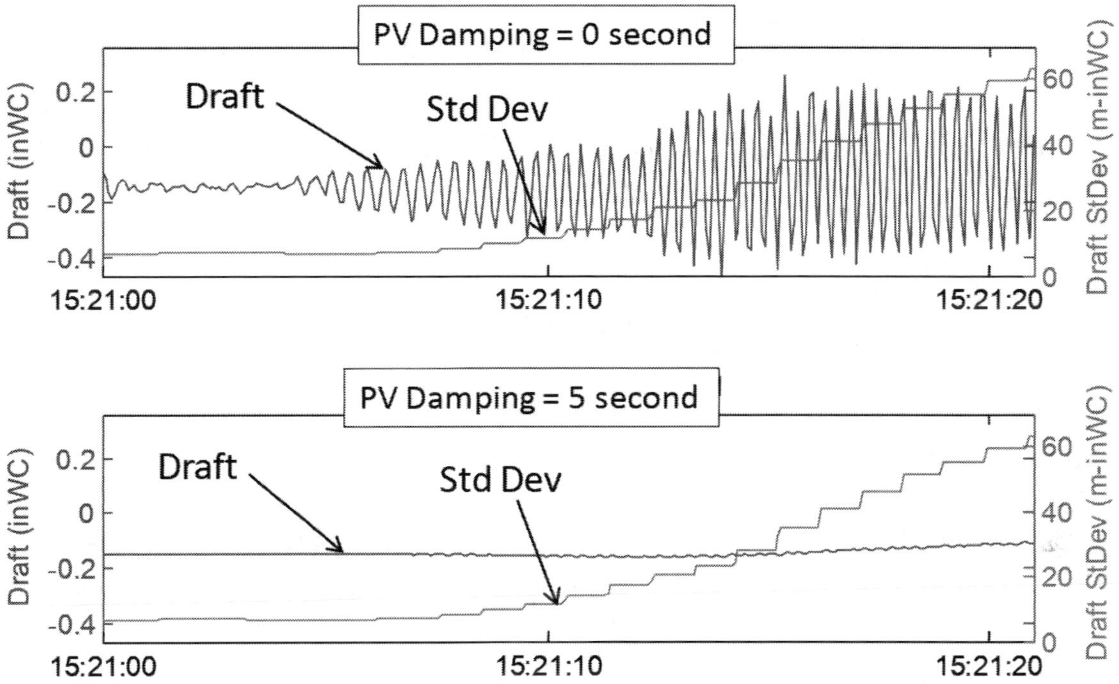

Figure 10 – SPM Standard Deviation is not affected by PV Damping

Implementing SPM technology to monitor an operating fired heater requires the following:

1. Measure draft using a draft transmitter with SPM
2. Trend SPM data into DCS process historian
3. Configure flame instability alarm

Item 1 requires simply replacing or upgrading the existing draft measurement with an advanced diagnostics draft transmitter with SPM. Referring back to Figure 3, the draft can be measured at any elevation on the heater. Previous work [13, 14, and 17] has shown that the SPM can detect flame instability regardless of the elevation of the draft measurement.

Item 2 requires first bringing the SPM data into the control system I/O via any of the interfaces shown in Figure 7, and then configuring the process historian to be trending the SPM data into the database. The SPM data may also be displayed to the heater operator so that they can use it together with other heater operational data.

Item 3 may take some additional effort. The flame instability alarm limit is the minimum value of the SPM standard deviation at which to notify a heater operator of combustion problems and/or flame instability. Obviously, in an operating fired heater, one cannot deliberately make the heater flame out, just to determine how much the standard deviation changes, in order to set the alarm limit. However, manufacturers of burners have test facilities that enable burners to be tested in a wide variety of conditions, including substoichiometric and flame instability conditions. It is possible that results from these tests at burner vendors may be utilized to gauge how much the SPM standard deviation changes under flame instability. The purpose of this joint effort between Rosemount and Valero was to evaluate the use of SPM technology for flame-instability detection, and to determine a method for setting the alarm limit on operating fired heaters.

Testing at Burner Vendors

Over two weeks in December of 2011, Valero tested five different burners at three different burner vendors. These tests were part of the standard Valero procedure for validating the suitability of each burner for various heater design or retrofit projects across the company. During these tests, each burner was evaluated on typical burner performance factors, such as range of firing rate and turn-down, flame size (length and diameter), emissions (CO, NOx, etc.), and acoustic noise levels. All five burners were the newest Ultra-low NOx burners made by each of the respective burner vendors. Four of the burners were round-flame burners, while one was flat-flame burner geometry.

Rosemount was also invited to participate in these burner tests to help evaluate the use of Advanced Pressure Diagnostics Statistical Process Monitoring technology for flame instability detection. On each test furnace, a Rosemount 3051S draft transmitter, with the Advanced Diagnostics (DA2 option), was installed. On all tests, the draft transmitter was installed at approximately the burner level (location 6 or 7 in Figure 3). Throughout the burner tests, both the un-damped draft measurement, and the draft standard deviation were recorded.

In addition to the regular burner tests, each burner was driven to various levels of substoichiometric combustion. Although burner vendors may have the capability of testing substoichiometric operation, this testing is usually not part of the standard burner test protocol, because heaters are not typically operated below stoichiometry. However, the refining industry recognizes that although they would never purposely operate a heater sub-stoked, unexpected events (e.g. a surge in the fuel gas composition) may occur periodically, and they want to be certain that their heaters will not flame out under such an event. Thus, substoichiometric testing at burner vendors is becoming more prevalent.

A typical substoichiometric / flame instability test would consist of the following:

1. Set the burner at a desired firing rate (e.g. minimum, normal, or design) and normal excess oxygen (e.g. 3% O_2, wet basis, in the flue gas)
2. While keeping the stack and burner air dampers at a constant position, gradually increase the firing rate, until the stoichiometric point (0% O_2 in flue gas) is reached. This normally coincides with the CO Break (the point where the concentration of carbon-monoxide in the flue gas exceeds 1000 parts per million).
3. While still keeping air dampers at a constant position, continue increasing the firing rate into the substoichiometric range. Because the % of stoichiometry cannot be measured directly, it must be estimated based on how much the fuel gas flow is increased.
4. Continue increasing the firing rate until either flame instability is observed (on a camera or through a test furnace viewport) or until the test furnace operators feel they cannot safety operate any further into the substoichiometric range.

Figure 11 shows an example of one of the flame-instability tests at a burner vendor. The top trend shows the un-damped draft measurement, while the bottom trend shows the draft standard deviation. Note that throughout this paper, the draft standard deviation is shown in units of thousandths inch of water column (m-inWC). For example 6 m-inWC = 0.006 inWC.

In this burner test, operation started at a baseline firing rate (3.6% O_2), and then the firing rate was gradually increased. The burner was still stable at both the stoichiometric point (3.15 MMBtu/hr) and at 90% of stoichiometry (3.47 MMBtu/hr). However, when trying to hold the burner at 85% of stoichiometry (3.6 MMBtu/hr) the burner flames would suddenly start becoming unstable. The flame instability could be seen both on a camera, visible in the control room, and in observing directly through a viewport on the test furnace. During flame instability, the sides of the burner flames could be described as "clapping hands together", and an audible "huffing" sound near the burner air intake.

Prior to the point of flame instability, the draft standard was approximately 4 to 7 m-inWC. However, at the point of flame instability, the standard deviation went up to between 60 and 70 m-inWC (about a 10 times increase). In all of the observed instances of flame instability, there was a similar large increase in standard deviation.

An alarm limit on the draft standard deviation could be defined for detecting flame instability. In this example, 30 m-inWC may be an appropriate alarm limit.

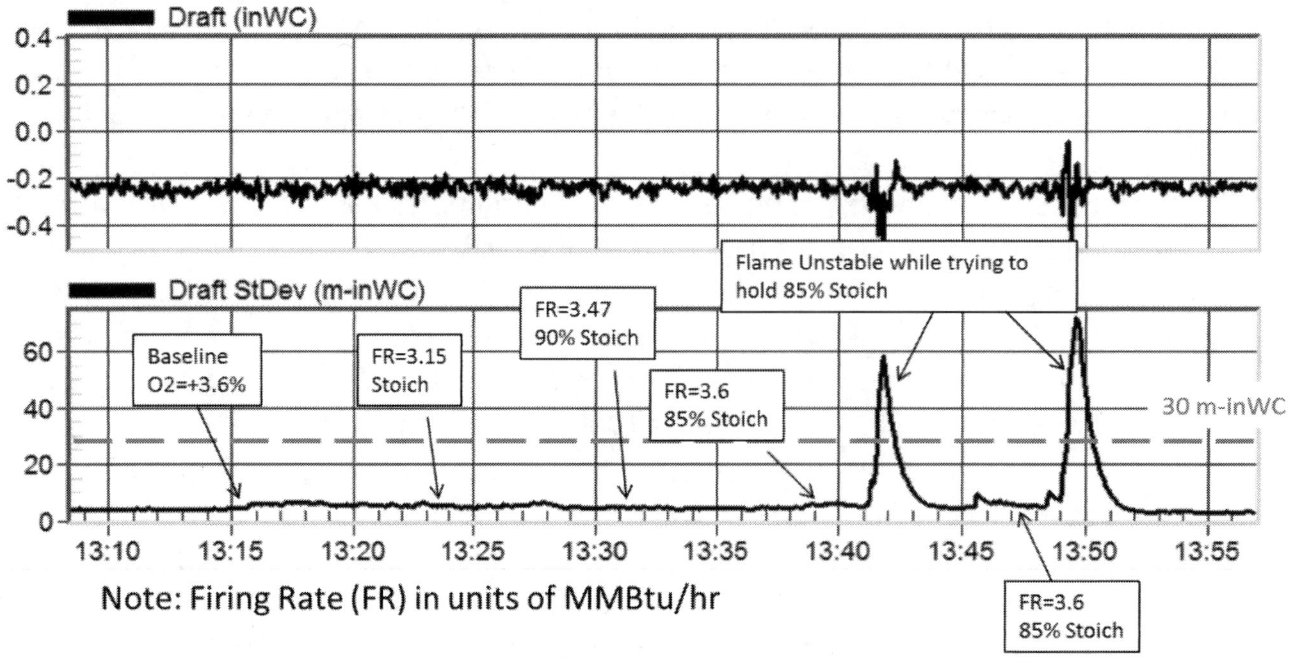

Figure 11 - Example Substoichiometric and Flame Instability Test at a Burner Vendor (Burner D-2 at Normal Firing Rate)

Table 1 shows a summary of all of the flame instability tests conducted at the burner vendors. Only burner vendor D was able to run the burners far enough into the sub-stoichiometric operating region to create flame instability. Burner vendors E and F showed their burners were stable down to 71% and 90% of stoichiometry respectively, but neither one had a test protocol that could take their burners to the point of flame instability.

Table 1 – Flame Instability Tests at Burner Vendors

Vendor	Burner	Geometry	Burner Firing Rate Conditions (MMBtu/hr)			Lowest % of Stoichiometry	Flame Instability in Burner	Flame Instability Detected with SPM
			Min	Norm	Max			
D	D-1	Round	1.12	5.05	6.72	57% (Norm FR) 70% (Max FR)	Yes *	Yes
	D-2	Flat	0.59	2.65	3.52	85% (Norm FR) 92% (Max FR)	Yes **	Yes
E	E-3	Round	2.29	7.96	9.15	71%	No	No
F	F-4	Round	1.96	6.26	7.83	90%	No	No
	F-5	Round	1.31	4.2	5.25	90%	No	No

* Burner D-1: When starting at the Normal Firing Rate, burner was stable down to 57% of Stoichiometry. Starting at the maximum firing rate, flame instability observed at 70% of Stoichiometry

** Burner D-2: Flame instability observed both when starting at the normal firing rate, and when starting at the maximum firing rate

It is interesting to note that of the two burners from vendor D, one of them was a round burner, while the other was a flat flame burner. The round flame burner (D-1) proved to be stable at as low as 57% of stoichiometry. (That is a, firing rate 43% higher than what would be needed to exactly match the available air.) However, the flat flame burner went unstable at as little as 92% of stoichiometry. Based on just a single test, one cannot make the conclusion that flat-flame burners are universally more unstable than round-flame burners. However, it appears that there may be a correlation between the geometry and stability of burners. It is recommended that more work be done in the industry to further explore this relationship.

Note in Table 1 that in the two burners where flame instability was created, it was detected by the draft transmitter with SPM. In the three burners that were not driven to the point of flame instability, the SPM standard deviation did not show the same large increase.

Figure 12 shows an example trend from a burner test at vendor F. This figure shows the values of draft and draft standard deviation that were observed over the full day of burner tests. Note that although there were a few periods when the burner operation was taken to the point of 90% of stoichiometry, the burner flames never went unstable. There were small changes in the draft standard deviation over the day of the tests, but it never went higher than 9 to 10 m-inWC. Using the same flame instability alarm limit of 30 m-inWC, we can see that this limit was not exceeded during the full day of the burner tests. Thus, the SPM did not give any false alarms.

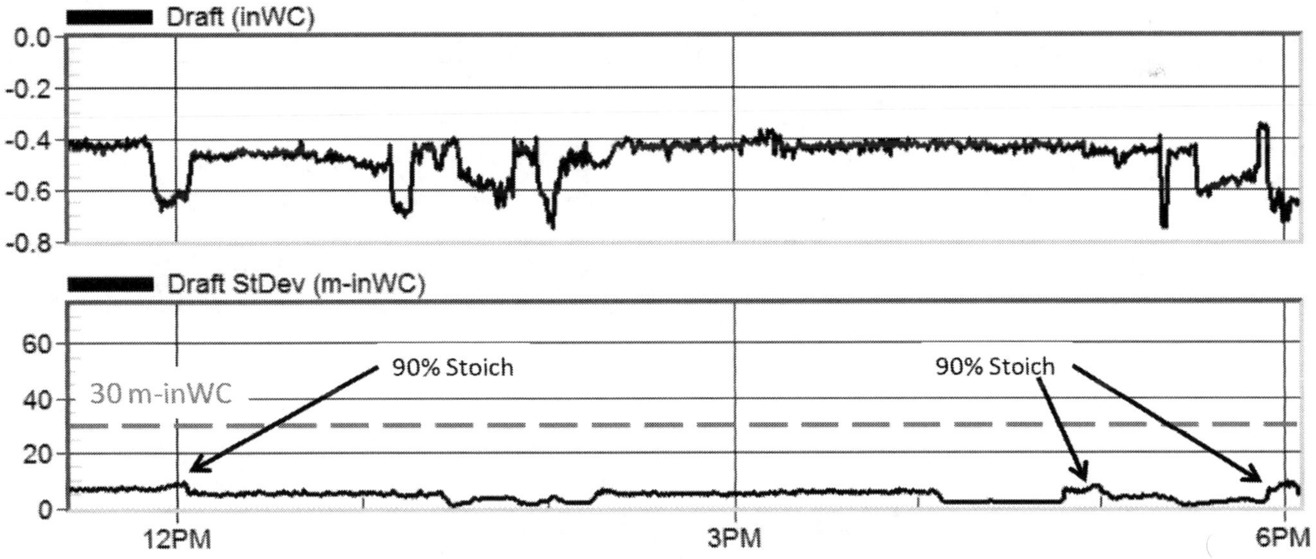

Figure 12 - Example Sub-stoichiometric test - Flame Instability Not Observed (Burner F-4)

Figure 13 shows one data point from a later test run on the same burner F-4. This data point was found rather opportunistically, as it occurred under a circumstance that typically doesn't happen in burner tests. In this case, the test furnace operators opened the stack damper to increase draft in the test furnace instead of keeping the damper partially closed and using steam injection to the base of the stack to increase draft in the firebox. Once the firebox was given line-of-sight to the sky, the burner flames started pulsating, and soon flamed out. A view of the trend of the draft standard deviation (Figure 13) shows that just prior to the flame out, there was a large increase in the standard deviation. In this case, the alarm limit of 30 m-inWC would be appropriate for detecting flame instability. There was also a point during light-off where the standard deviation exceeded 30 m-inWC. However, an implementation in an operating firing heater (such as in the Burner Management System) could have logic to ensure that a flame instability alarm is not generated during light-off.

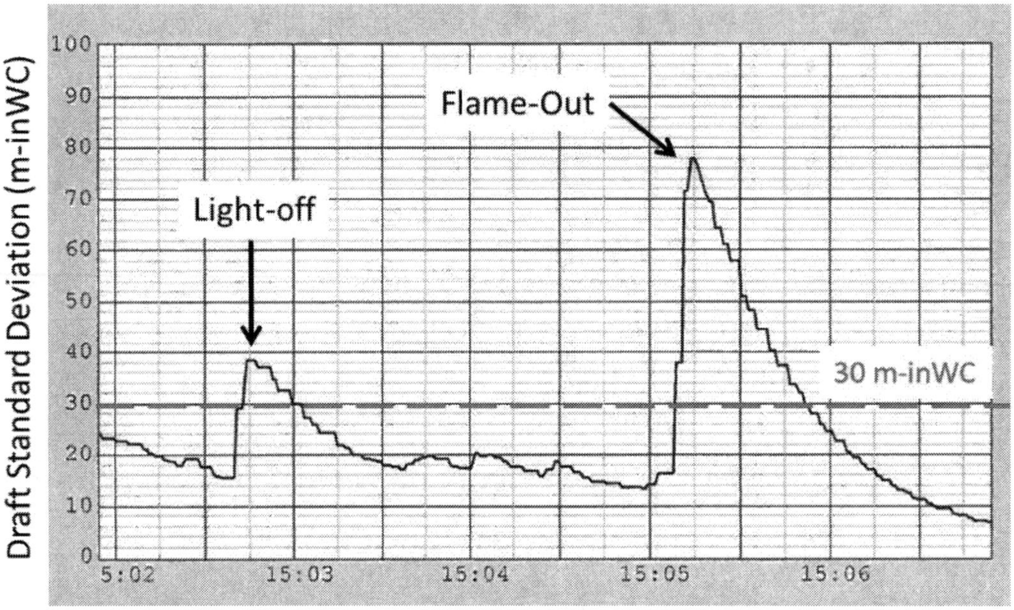

Figure 13 - Flame out Event Observed during a Later Test of Burner F-4

Refer to Appendix A for additional tends of data from the tests at the burner vendors.

Test on Refinery Fired Heater

As a follow-up to the burner vendor tests, in May of 2012, heater tests were conducted on the H-31 fired heater at the Valero Texas City Refinery. This is a radiant/convective vertical cylindrical heater, with a single row of tubes along the radiant axis. The heater has four burners, two on each side of the radiant tube bank. The burners in this heater were raw gas burners. This is an older model burner, not designed for the same low NOx levels as the burners tested at the vendors. Figure 14 shows a view looking from the bottom up into the radiant section of heater H-31.

Figure 14 – Inside the Radiant Section of Heater H-31

This test was a follow-up to the December 2011 burner vendor tests. The purpose of this test was to characterize changes in the draft standard deviation and to compare with the data that was taken at the burner vendors. Since this is an operating heater, the test team could not purposely flame out the heater, nor could the team create the same extreme substoichiometric conditions as were simulated in the test furnaces. However, it was determined that limited manipulation of firing rates, fuel pressures, and individual burners could be performed safely. During the tests, the team would be looking for characteristics of flame instability in the burner flames, and monitoring for any corresponding change in the SPM standard deviation.

Specifically, four tests were performed on this heater:

1. Low Burner Pressure on a single burner – Gradually close off the fuel supply valve to a single burner. See if as the burner is starved of fuel, it becomes "lazy" and affects the draft standard deviation.
2. High Fuel Gas Pressure – Increase the fuel gas pressure to feed more fuel to the burners (and reducing the O_2). See if burner flames go unstable.
3. Low O_2 – Increase fuel flow to drive the O_2 lower. Keep going until excess O_2 reaches 0.5%. (Typical O_2 on this heater is 3-4%). See if burner flames become unstable at low O_2.

4. Starve Single Burner – Close off the air register of a single burner to starve it of air. See if the burner flame goes unstable and affects the draft standard deviation.

Table 2 below summarizes the results from these heater tests. Flame instability could not be created in any of the tests. On the Low Burner Pressure test, the flame was completely extinguished before any change in draft standard deviation was observed. On both the High Fuel Gas Pressure and Low O_2 tests, there were small changes in standard deviation as the firing rate changed. However, these changes were significantly smaller compared to what is typically seen under flame instability. As seen looking through the furnace viewport, the burner flames remained in a stable, healthy combustion in both tests.

Table 2 - Heater Tests performed at Valero Texas City

Test	Lowest Recorded O_2	Flame Instability in Heater	Flame Instability Detected with SPM	Note
Low Burner Pressure on Single Burner	3.8%	No	No	Burner Completely Extinguished before any Change in Std Dev
High Fuel Gas Pressure	1.6%	No	No	Small increase in Std Dev with Firing Rate. Burner flames never went unstable.
Low O_2	0.3%	No	No	Small increase in Std Dev with Firing Rate. Lowest safe O_2 level reached. Burner flames never went unstable.
Starved Single Burner	2.1%	No	No	Single burner starved of air. Substoichiometric combustion observed through viewport.

The last test involved starving a single burner of air. There were small changes in the SPM standard deviation, but not of the magnitude that would indicate flame instability. A view of a starved burner through the heater viewport (Figure 15) showed very bad combustion (a bright orange flame that would jump up and down, as opposed to the stable blue flame of the normal operating burner). However, it was not the same as flame instability in the whole heater. Consequently, there was a small change in the draft standard deviation, but within the range of standard deviation changes that would be seen under normal heater operation.

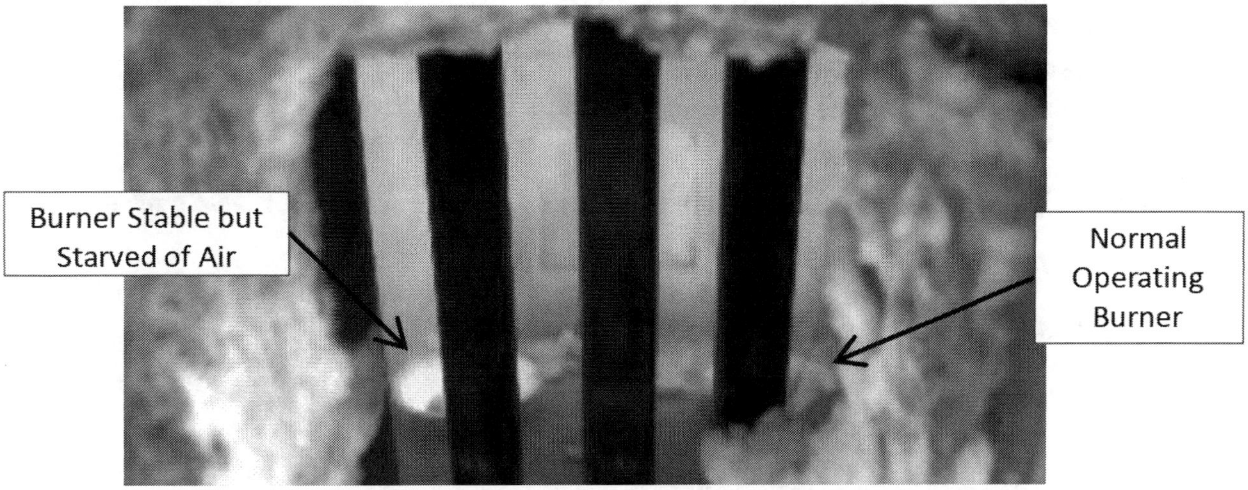

Figure 15 - Starved Burner seen in Heater Viewport

Figure 16 shows the trend of SPM Standard Deviation over the day of the heater tests. Although there were small changes in standard deviation with the different heater tests, these changes were all substantially less than what would be expected under flame instability. The alarm limit of 30 m-inWC was not tripped during the test.

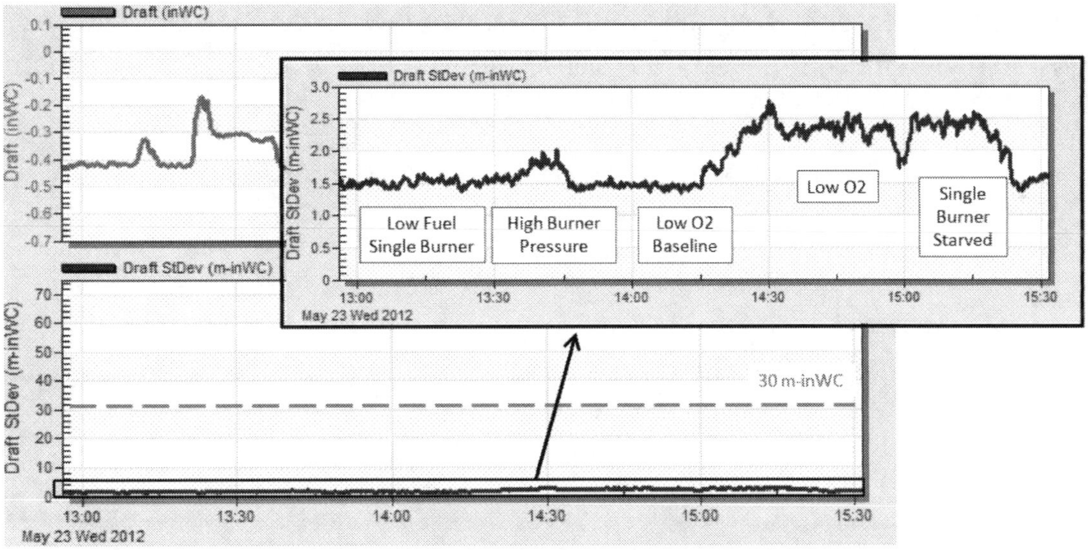

Figure 16 - Trend of Draft Standard Deviation during H-31 Heater Tests

Following the H-31 heater tests, the SPM standard deviation was monitored over five weeks of normal heater operation. During that time, the standard deviation did not go any higher than 2 m-inWC, always significantly less than the alarm detection limit of 30 m-inWC. Although no false alarm occurred during this time, we still don't know how high the standard deviation would go if flame instability and/or flame out actually did occur on this heater. It is possible that a much lower alarm limit (anywhere from 5 to 10 m-inWC) could be selected, which would not be tripped during normal heater operation. Additional heater operational experience is needed to confirm this.

Effect of Firing Rate and Comparison of Among Different Burners

During the burner and heater tests, it was observed that there was a correspondence between firing rate and draft standard deviation. In general, a higher firing rate of the burner(s) leads to a higher draft standard deviation. This makes sense because a higher firing rates means that there is more fuel and more combustion, thus more turbulence or combustion rumble. Figure 17 shows a comparison of the standard deviation values observed under normal operating conditions (e.g. minimum, normal, or maximum firing rate at 3% O_2), on each of the five burners tested at the vendors. The figure also shows standard deviation values observed during the H-31 heater tests.

For every burner, one can see that the standard deviation is the highest at the maximum firing rate, and the lowest at the minimum firing rate. This should be taken into account when setting the standard deviation alarm limit for detecting flame instability. The alarm limit must be chosen such that it would not be exceeded when the heater/burners are at the maximum firing rate. Note that the standard deviation values when burners D-1, D-2, and F-4 were unstable were in the range of 60 to 100 m-inWC and would be off this chart.

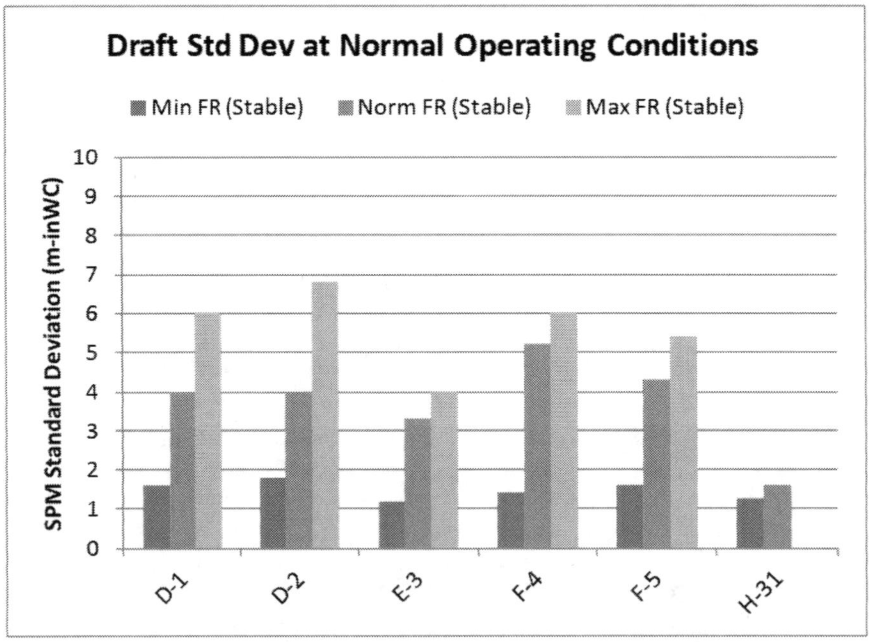

Figure 17 – Comparison of Draft Standard Deviation at Nominal Operating Conditions

Figure 18 shows a comparison of Total Firing Rate (MMBtu/hr) versus standard deviation for all five test furnace burners, and the four-burner heater H-31. All of these data points are from normal (stable) burner operation. For any single burner, there is the general relationship that standard deviation increases with firing rate. However, there is a large degree of separation between the different burners. In particular, note that on H-31, at higher firing rates, the draft standard deviation was substantially lower than that observed in all of the test furnaces.

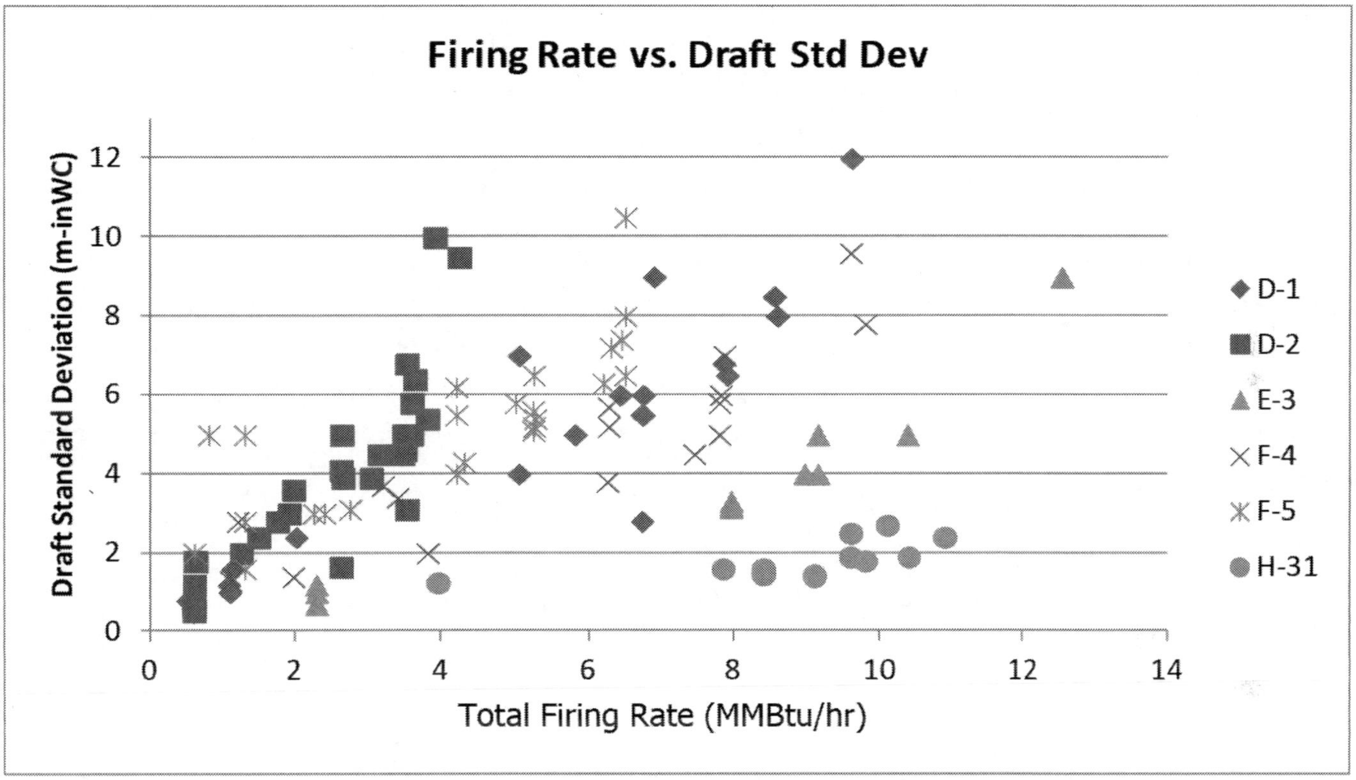

Figure 18 – Correlation of Firing Rate and Draft Standard Deviation

One factor that may explain the difference in draft standard deviation values among the different tests is the volumes of the fireboxes of the test furnaces and heater. A larger volume of the firebox means that the energy and turbulence of combustion is spread out more widely, and therefore the pulsations measured by the draft transmitter may not be as large.

Table 3 shows the approximate firebox volume on each of the test furnaces, and the calculated volumetric heat release at the Minimum, Normal, and Maximum firing rates.

Table 3 – Volumetric Heat Release for Each Test Furnace or Heater

Burner or Heater	# of Burners	Firebox Volume (ft^3)	Total Firing Rate (MMBtu/hr)			Volumetric Heat Release (Btu/hr/ft^3)		
			Min	Norm	Max	Min	Norm	Max
D-1	1	1215	1.12	5.05	6.72	922	4156	5531
D-2	1	792	0.59	2.65	3.52	745	3346	4444
E-3	1	1440	2.29	7.96	9.15	1590	5528	6354
F-4	1	2500	1.96	6.26	7.83	784	2504	3130
F-5	1	1000	1.31	4.2	5.25	1310	4200	5250
H-31	4	3049			12.5			3411

Figure 19 shows the same draft standard deviation values plotted against volumetric heat release. There appears to be a very definite correlation between draft standard deviation and volumetric heat release that is even consistent across different burners and fireboxes. Using volumetric heat release may be a way to predict the draft standard deviation values in a heater, using test data from a different furnace or heater. However, more work is required to validate this.

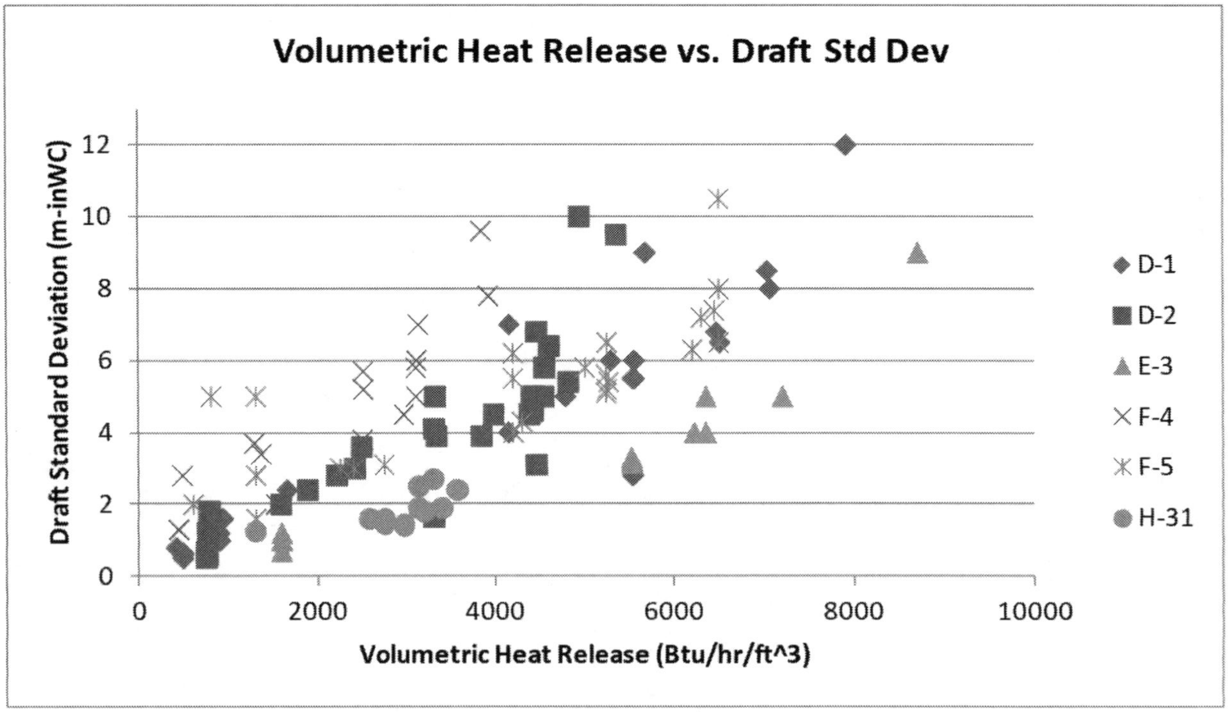

Figure 19 – Correlation of Volumetric Heat Release and Draft Standard Deviation

It may be possible to use operational data from a heater running in normal operation to set an alarm limit for flame instability detection. Figure 20 shows the draft standard deviation versus volumetric heat release for just the H-31 heater at the Texas City Refinery. Remember that during the tests on this heater, and in the weeks following, flame instability never occurred.

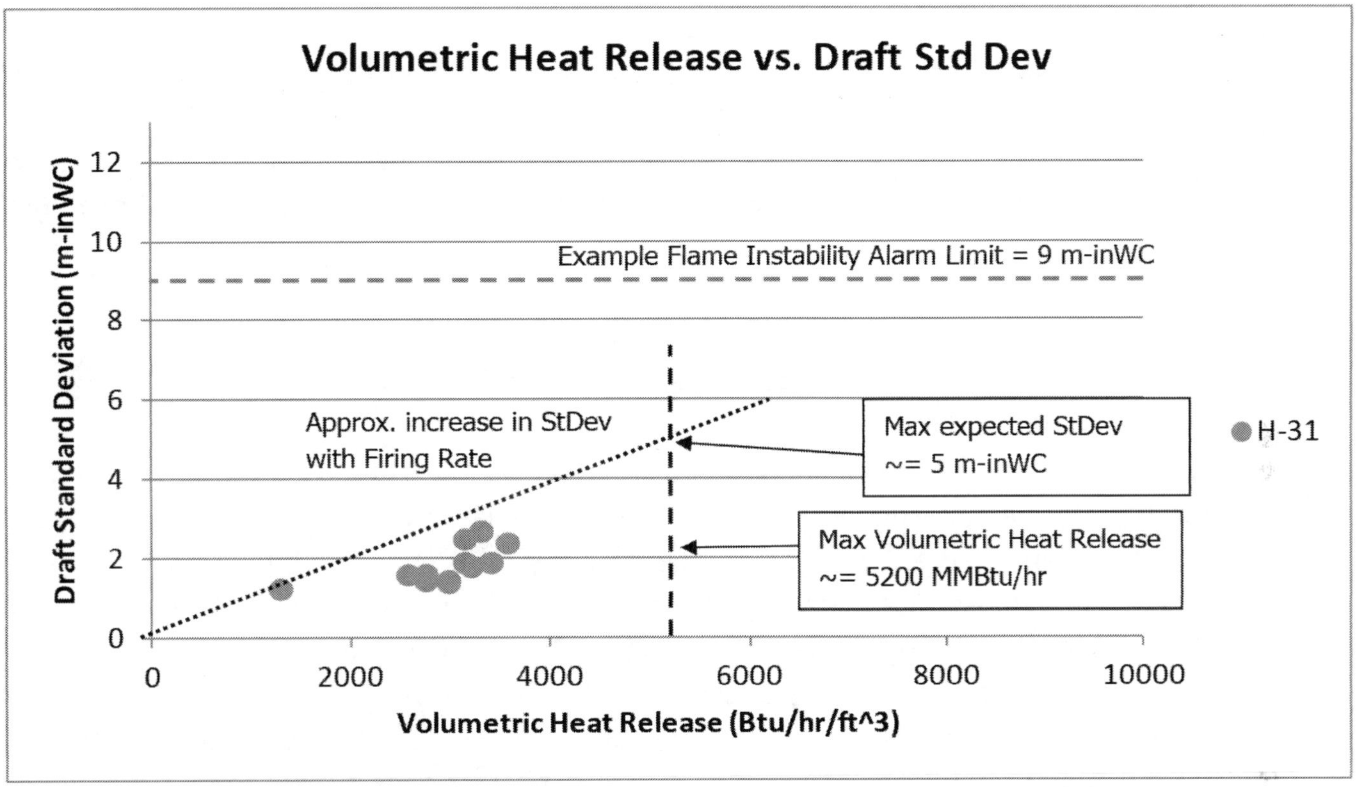

Figure 20 – Using Normal Heater Operational Data for Determining Flame Instability Alarm Limit

The H-31 heater has a maximum firing rate of 16 MMBtu/hr, corresponding to a maximum volumetric heat release of about 5200 Btu/hr/ft^3. This is marked with the vertical dashed line. The approximate increase in standard deviation with firing rate is marked by the diagonal dotted line. The point where these lines cross may correspond to the largest standard deviation value that would be expected in normal heater operation (approximately 5 m-inWC in this example). One could choose the flame instability alarm limit to be some multiple of this value. In the example, we have selected an alarm limit at 9 m-inWC. There would have to be substantial changes in the combustion conditions to make the draft standard deviation go higher than this limit.

The advantage to choosing a smaller alarm limit (e.g. as opposed to the 30 m-inWC from the burner vendor test data), is that, in this heater, we really don't know how high the standard deviation will go during substoichiometric combustion or flame instability. A smaller detection limit gives greater possibility that the alarm will occur under some future, but unknown magnitude, increase in draft oscillations.

The method shown in Figure 20 could be used to set the flame instability alarm limit on any operating fired heater. However, this method is based on only a limited set of test data. A linear relationship between standard deviation and firing rate may not necessarily exist on every burner/heater/furnace system. Furthermore, the actual magnitude of increase of the standard deviation during flame instability may vary between different heaters or burners. In addition to substoichiometric combustion, there may be other mechanisms that cause flame instability and/or flame out. As additional industry experience is gained using SPM technology for heater monitoring, it is expected these other factors will be better understood, and that new guidelines and best practices may be determined for setting the alarm limit, and integrating SPM technology with fired heater operation.

Considerations for Industry

Statistical Process Monitoring technology has been shown to have significant promise for monitoring heaters and detecting flame instability and flame out. However, more data is needed to fully evaluate the potential of this approach. Whenever adding a new draft transmitter to an oil refinery fired heater, owner/operators may wish to purchase a pressure transmitter with SPM technology. The SPM Standard Deviation could be integrated with the other heater operational data either via direct HART I/O, with a HART to analog converter, or with a wireless interface. Heater operators could monitor the Standard Deviation along with other heater measurement points, and use it as an indicator of combustion problems or flame out. An initial flame instability alarm limit could be set based on the standard deviation during normal heater operation and/or standard deviation values observed under flame instability in a test furnace. This alarm limit could be adjusted as experience is gained with a given heater.

SPM is a relatively new technology for monitoring fired heaters. If one chooses to implement SPM for monitoring heaters, the technology should be utilized in addition to (not as a replacement for) industry recommended best practices. Refer to API Recommended Practice 556 [18] for best practices in the instrumentation, control, and protective systems for fired heaters.

It is recommended that purchasers request burner vendors to monitor and trend the SPM standard deviation during burner tests. This data can be utilized during substoichiometric testing as an indicator of the stability of the burners. The SPM data can be used as a means to compare the stability of different burners at various levels of sub-stoichiometry. The SPM data can also be given to a heater designer or an end-user owner-operator, and used by the industry to further understand the application of SPM technology for heater monitoring.

Summary and Conclusion

Advanced pressure diagnostics Statistical Process Monitoring is a promising technology for detecting flame instability and/or flame out in fired heaters. The draft pulsations associated with flame instability can be detected by a draft transmitter with SPM. An increase in the draft standard deviation may be indicative of flame instability. A heater operator may be able to use the draft standard deviation together with other heater operational data to gain insight into the health of the burner combustion. An alarm may be configured to generate an indicator of flame instability when the draft standard deviation exceeds the alarm limit. Initial results show that for some heater/furnace/burner systems, an alarm limit of 30 m-inWC may be suitable for detecting flame instability. On other systems, a limit of 5 to 10 m-inWC may be appropriate. There is a general relationship between volumetric heat release and draft standard deviation. This relationship may be exploited to select an initial flame instability alarm limit on an operating fired heater. Although additional work may allow more precise determination of the optimal alarm limit for a given heater, current experience suggests that all fired heaters would benefit by utilizing pressure transmitters with SPM technology for the draft measurement.

References

1. F. Eisinger, R. Sullivan. *Evaluation of Non-Vibrating Utility Burner/Furnace Systems for Resistance to Thermoacoustic Oscillations.* PVP2006-ICPVT-11-93054. Proceedings of 2006 ASME Pressure Vessels and Piping Division Conference July 23-27, 2006, Vancoucer, BC, Canada
2. F. Eisinger, R. Sullivan. *Elimination of Thermoacoustic Furnace Vibration in a Gas-Fired Package Boiler--A Case Study.* PVP2008-61209. Proceedings of 2008 ASME Pressure Vessels and Piping Division Conference July 27-31, 2008, Chicago, Illinois, USA.
3. F. Eisinger, R. Sullivan. *Avoiding Thermoacoustic Vibration in Burner/Furnace Systems.* Transactions of the ASME, Journal of Pressure Vessel Technology, Vol. 124, November 2002, pp 418-424.
4. F. Eisinger, R. Sullivan. *Prediction of Thermoacoustic Vibration of Burner/Furnace Systems in Utility Boilers.* Paper No. PVP 2007-26090. ASME PVP 2007 Conference, July 22-26, 2007, San Antonio, TX.
5. Patent #7,950,919 – *Method and apparatus for monitoring and controlling the stability of a burner of a fired heater.* (G. Johnson, B. Hamilton). Filed: October 12, 2005. Issued: May 31, 2011.

6. D. Wehrs . *Detection of Plugged Impulse Lines using Statistical Process Monitoring Technology*. Whitepaper, Rosemount.com, December 2006.
7. R. Szanyi, M. Raterman, E. Eryurek. Diagnostics capabilities of FOUNDATION fieldbus pressure transmitters. Hydrocarbon Processing, April 2003, pp. 53-59.
8. D. Wehrs, A. Klosinski. *Entrained Gas Diagnostic with Intelligent Differential Pressure Transmitter*. Whitepaper, Rosemount.com, January 2008.
9. D. Wehrs, A. Klosinski. *Wet Gas Diagnostic with Intelligent Differential Pressure Transmitter*. Whitepaper, Rosemount.com, January 2008.
10. Rosemount Technical Note – *Distillation Column Flooding Diagnostics with Intelligent Differential Pressure Transmitter* – www.rosemount.com
11. R. Pihlaja. *Detecting Incipient Flooding in a Distillation Column* – Presented at 2010 AIChE Spring Meeting and 6th Global Congress on Process Safety – March 21-25, 2010 – San Antonio, TX, USA.
12. E. Eryurek, P. Sharpe, D. White. Abnormal Situation Prevention through Smart Field Devices. NPRA Hydrocarbon Processing, March 2006, pp. 41-48.
13. J. Miller - *Furnace Flame Instability Detection Using Pressure Measurement with Advanced Diagnostics* – Session 57 – Advances in Fired Heater Safety and Operations – 2010 AIChE Spring Meeting and 6th Global Congress on Process Safety – March 21-25, 2010 – San Antonio, TX, USA
14. Rosemount Technical Note 00840-0500-4801 – Furnace Flame Instability Detection with Advanced Pressure Diagnostics – J. Miller – July 2010
15. R. Stier, J. Miller – Session 6B-2216 *Improving Refinery Fired Heater Safety using Advanced Pressure Diagnostics* – presented at Emerson Global Users Exchange, October 8-12, 2012, Anaheim, CA.
16. US Patent #7,536,274 – System and method for detecting an abnormal situation associated with a heater (L. Heavner, J. Sharpe, E. Eryurek, K. Kavaklioglu) Filed: March 14, 2005. Issued: May 19, 2009
17. Patent Publication No: 2010/0151397 – System and Method for Controlling Fired Heater Operations. (J. Farrell, S. Chhotray, G. Dobbs, M. Alvarez, P. Schweitzer). Filed: December 14, 2009
18. API Recommend Practice 556 – Instrumentation, Control, and Protective Systems for Gas Fired Heaters, Second Edition, April 2011

Appendix A

This appendix shows the trends of draft and SPM Standard Deviation for all of the burner vendor tests that were performed in December 2011. All labels of Firing Rate (FR) are in units of MMBtu/hr.

Burner E-3

Vendor	Burner	Geometry	Burner Firing Rate Conditions (MMBtu/hr)			Lowest % of Stoichiometry	Flame Instability Observed
			Min	Norm	Max		
E	E-3	Round	2.29	7.96	9.15	71%	No

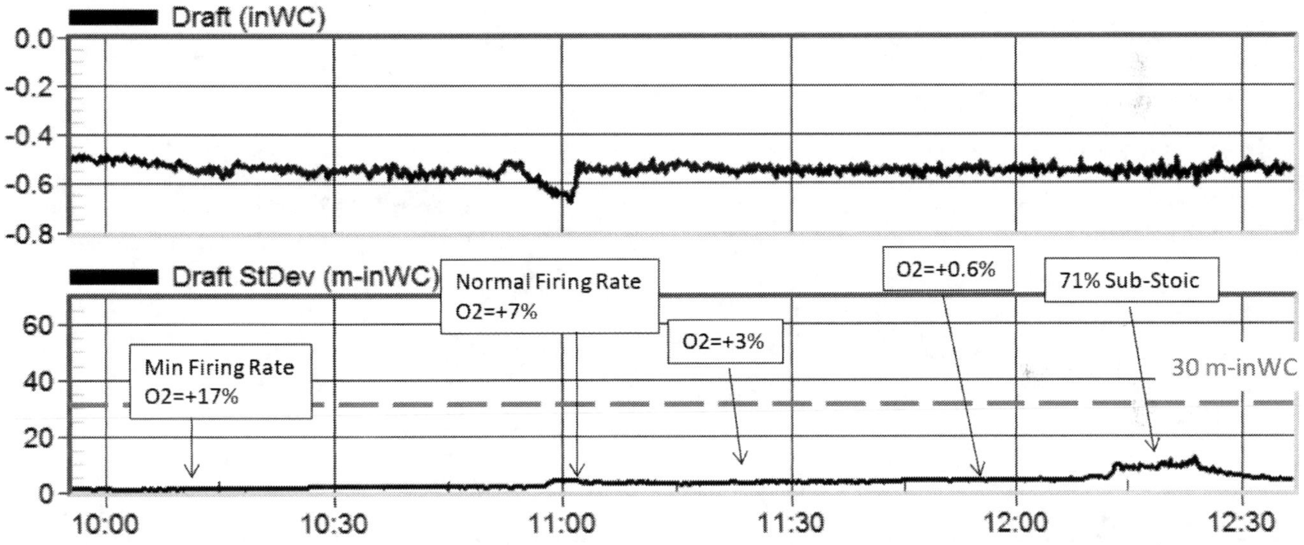

Burner D-1

Vendor	Burner	Geometry	Burner Firing Rate Conditions (MMBtu/hr)			Lowest % of Stoichiometry	Flame Instability Observed
			Min	Norm	Max		
D	D-1	Round	1.12	5.05	6.72	57% (Norm FR) 70% (Max FR)	Yes (Max FR only)

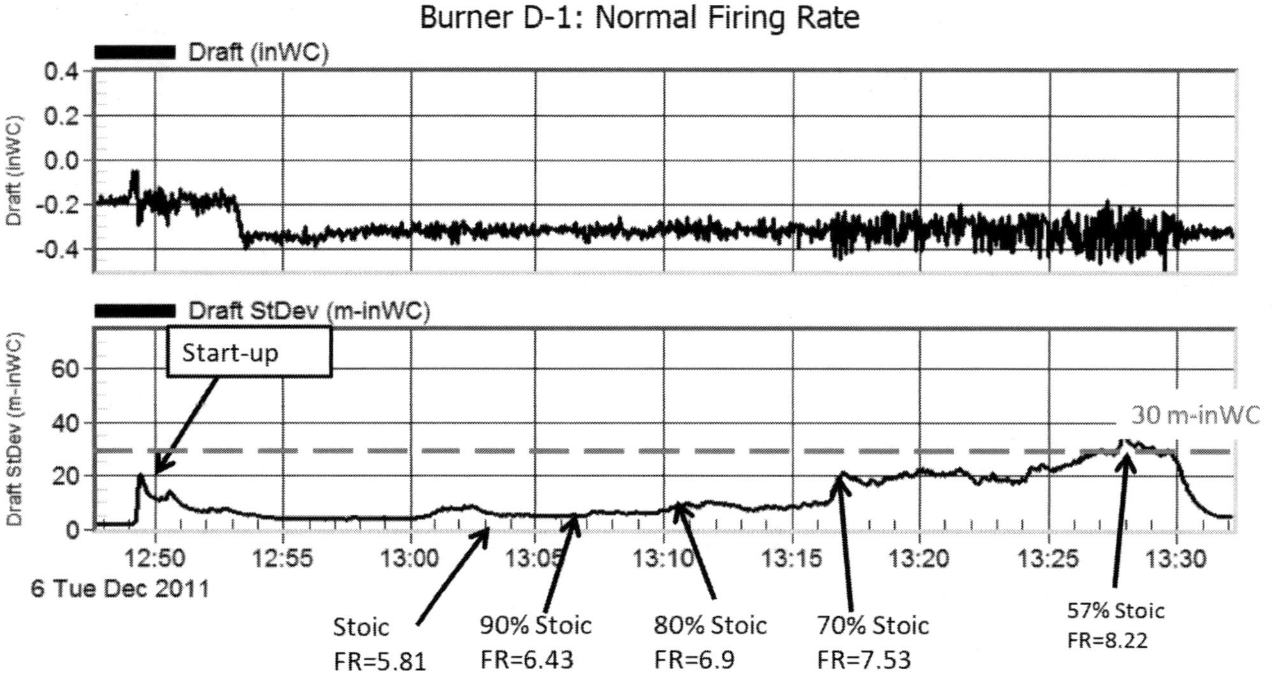

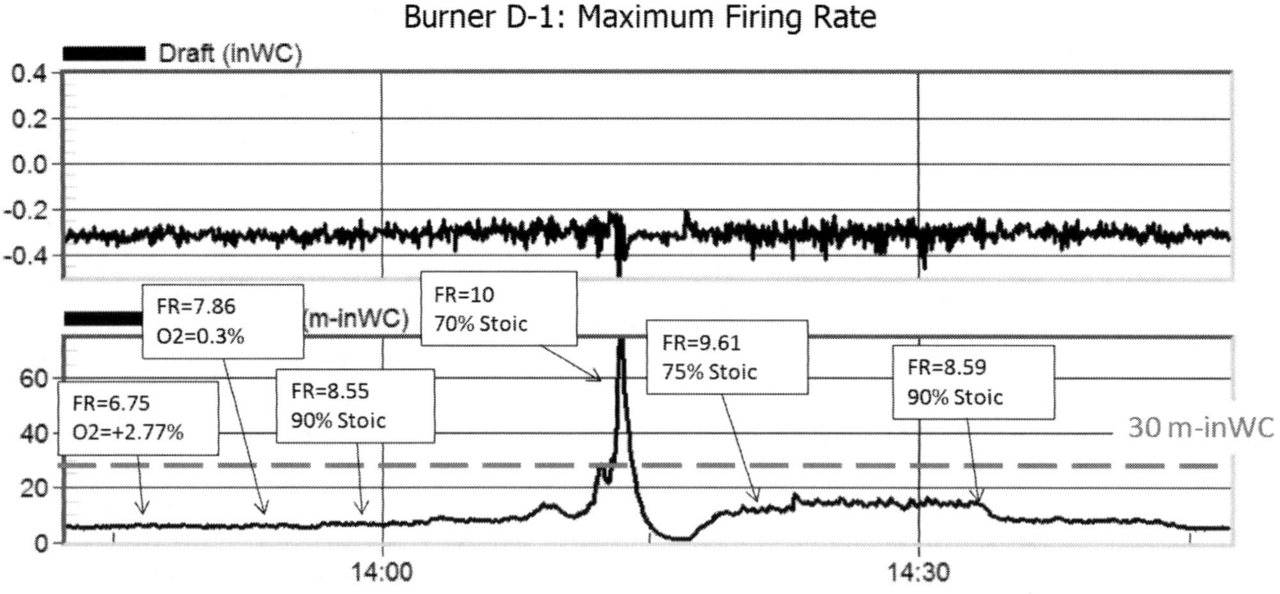

Burner D-2

Vendor	Burner	Geometry	Burner Firing Rate Conditions (MMBtu/hr)			Lowest % of Stoichiometry	Flame Instability Observed
			Min	Norm	Max		
D	D-2	Flat	0.59	2.65	3.52	85% (Norm FR) 92% (Max FR)	Yes

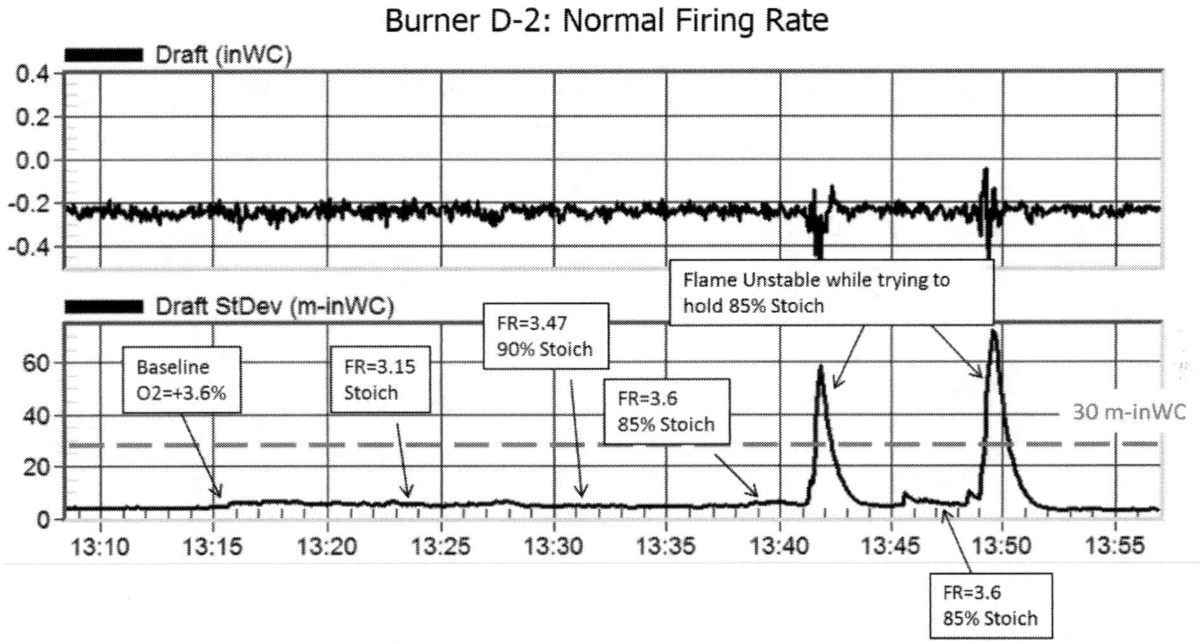

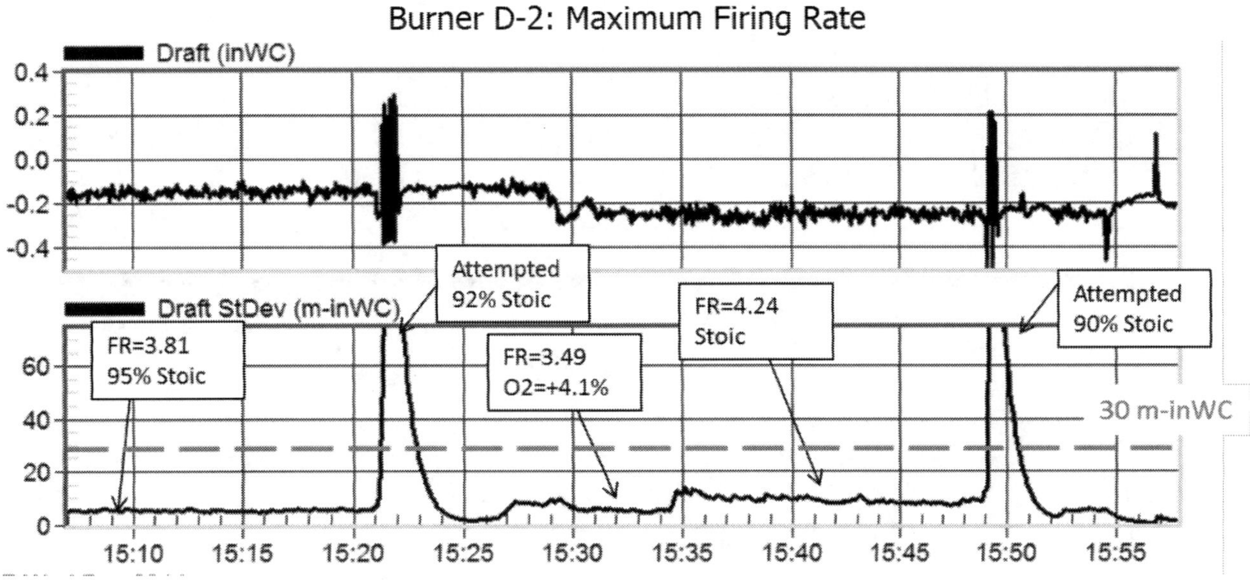

Burner F-4

Vendor	Burner	Geometry	Burner Firing Rate Conditions (MMBtu/hr)			Lowest % of Stoichiometry	Flame Instability Observed
			Min	Norm	Max		
F	F-4	Round	1.96	6.26	7.83	90%	No

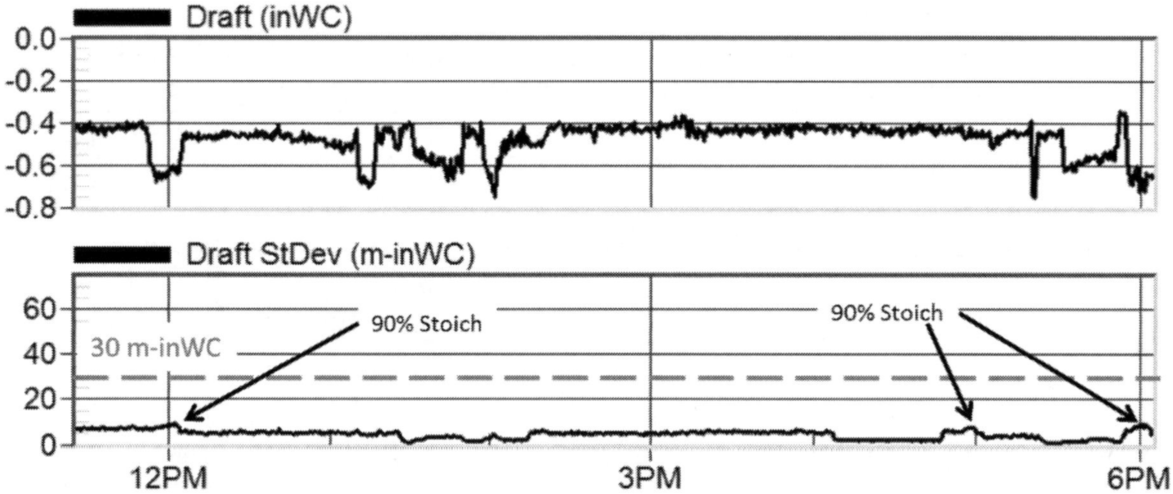

Burner F-5

Vendor	Burner	Geometry	Burner Firing Rate Conditions (MMBtu/hr)			Lowest % of Stoichiometry	Flame Instability Observed
			Min	Norm	Max		
F	F-5	Round	1.31	4.2	5.25	90%	No

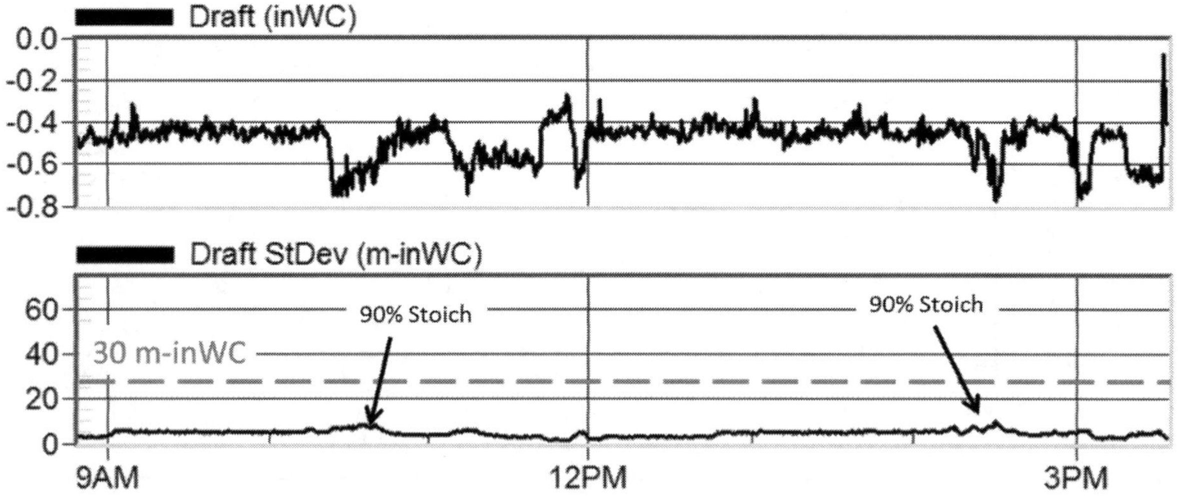

Abqaiq Plant Virtual Plant RTO Implementation

Dr. Kuangyao Brian Peng and Saleh H. AL-Musallam,

Saudi Aramco, Abqaiq, Saudi Arabia

Introduction

Abqaiq Plant is the largest worldwide crude stabilization plant with $200MM energy consumption per year. Abqaiq NGL plant shares 30% of total energy consumption. High NGL recovery rate has been the main objective because of undesired loss of hydrocarbon liquid condensate in the offgas pipeline. Recent efforts for process improvements including the commissioning of TEG dehydration unit and the compression train upgrades have been implemented to enhance NGL recovery and to improve the availability of desired cryogenic cooling for C2/C3 separation.

Abqaiq NGL Plant includes three major processes: compressing, high pressure stripping and deethanizing. Simultaneous optimization for all three major processes under the constraints of NGL product spec. requires fundamental understanding and a robust simulation tool. It has been a dream and a challenge for process and control engineers to develop such a complete model which can map the plant operating conditions in situ and perform a complete plant optimization online.

ROMeo (Rigorous Online Modeling with Equation-based Optimization), a widely used real time optimization (RTO) platform, can be used to achieve our goal. A real time optimizer can take the variations in upstream units such as changes in crude feed flow combinations and stabilization depth in real time, and calculates the most profitable mode of operating condition by using the rigorous plant model with energy consumption equations and process and product constraint equations. Then the calculated optimal set points are submitted to the DCS control room to ramp the plant process.

Abqaiq Plant NGL engineering department has successfully implemented and tested an in-house developed RTO NGL plant model. A potential increase of $200K to $400K revenue increase for each adjustment has been predicted under the current operating mode and unit operation constraints. In addition to the quantifiable revenue advantages, this presentation will also highlight other substantial advantages of the system such as measurement health monitoring, next generation process KPIs, and energy optimization.

Background of NGL process optimization initiatives

To meet the future energy demand and sustainability requirements in the Kingdom of Saudi Arabia, Abqaiq Plant starts several energy efficiency programs for all process units. A rough study with Saudi Aramco P&CSD for NGL recovery enhancement has shown great potentials with online optimization. As a result of this study, by simply adjusting the operating conditions of HP-Stripper columns could increase NGL recovery rate by 3 to 5 MBD. Further study with relaxing product spec. of individual columns and maintain the final product spec has shown an increase of NGL production up to 9 MBD.

A joint effort among corporate optimization specialists and plant operation engineers took off in 2010. A road map was proposed to implement ROMeo RTO at Abqaiq NGL plant by the end of year 2013. The road map includes three major milestones: rigorous model development, PI integration and offline optimization.

Process description

Continuous feed gas from oil stabilization units is fed into the NGL plant with a flow rate of 550MMSCFD at 2psig and 140°F. Feed gas is compressed with two stage compressors to 450psig and is cooled by cooling tower for the outlet temperature at 110-125 °F.

Feed compositions are studied with the available historical data. The major variation of feed compositions come from crude oil mix containing neat condensate, Arabia Light and Arabia Extra light. As a baseline model, average compositions over the past few years are used for model stability. Typical compositional data for hydrocarbon contains C3+ above 70%. Table 1 shows the average compositional data for a typical feed.

	N2	C1	CO2	C2	H2S	C3	IC4	NC4	IC5	NC5	C6	C7	C8
Overall	0.019706	0.0785301	0.028414	0.224106	0.043989	0.342631	0.056587	0.132501	0.032532	0.032833	0.006334	0.0018	3.4E-05

Table 1

Abqaiq NGL plant contains five major high pressure stripper columns and one deethanizer column. The average operation condition of high pressure column is 145 °F and 400 psig. Column efficiencies vary from 68% to 85% because of different tray designs. The five high pressure stripping columns were designed with 22 bubble trays for separation of C2 and C3+. Each column equips with steam kettle reboilers and temperature controller at the 7th stage of the column. The column temperature is the key handle for NGL product specification.

The deethanizer chiller is designed to operate at 65 °F and the reflux condenser is 60 °F . Currently, operation requires all columns product spec at 0.4 vol% which is a conservative approach. Overstripping of NGL product to offgas line would raise the liquid condensation in the pipeline, reduction of valuable NGL product recovery and waste of energy.

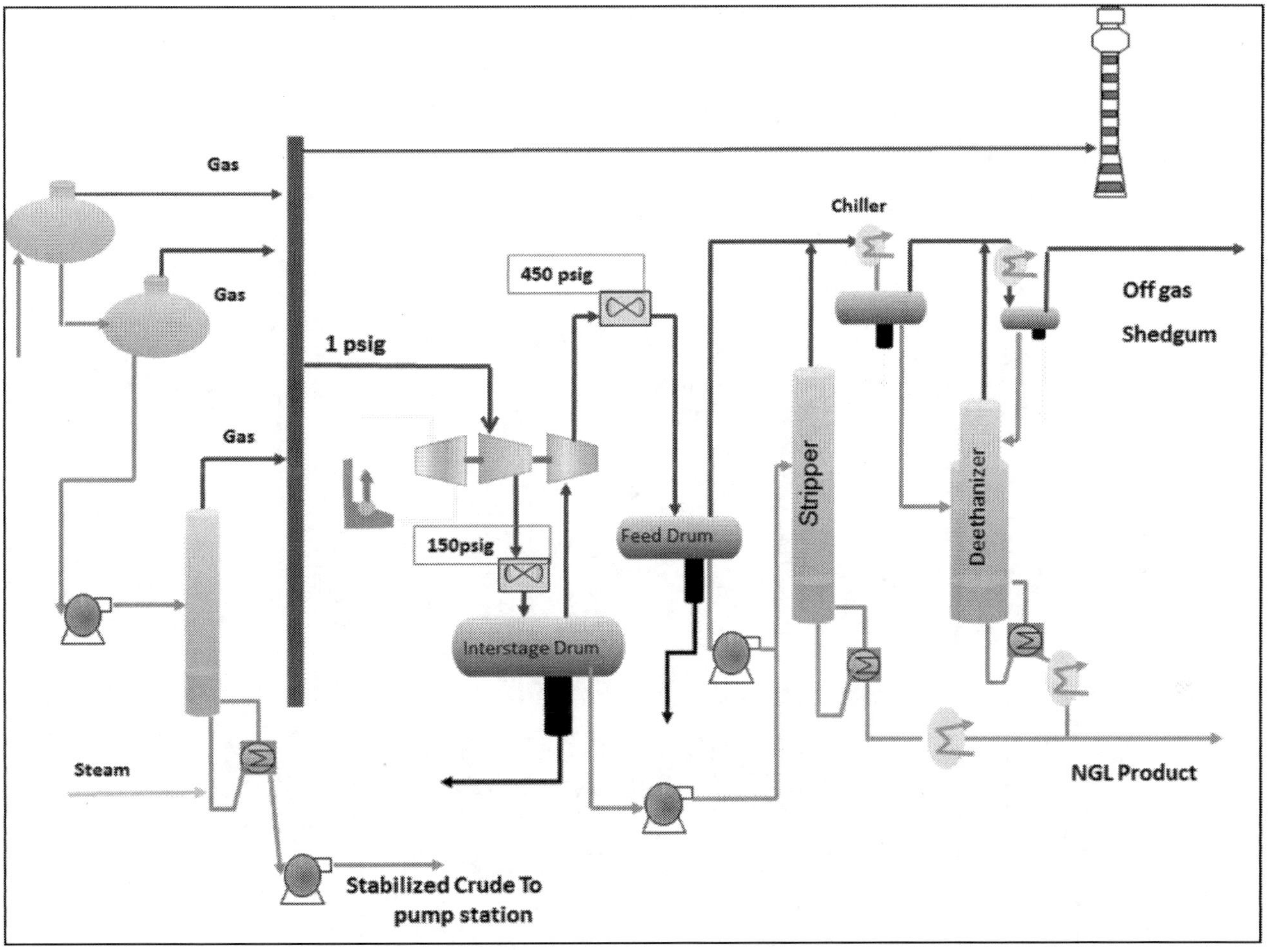

Figure 1. A simplified NGL process flow diagram.

Model development and implementation

Rigorous process model is developed based on the current process conditions in the field. Due to time constraint, the preliminary process model does not integrate with detail performance data of compressors. However, because phase I of this program is to maximize NGL recovery and to minimize the impact of liquid condensation, the performance functions are not required at this stage.

One of the challenges of the team is to identify all the piping routes of cross feed lines. The cross feed flow rates have to match the field instrumentations. The on-off setting of process units is relied on the macro programs that were developed in house with the engineering support and training from Invensys. A total number of 480 major PI-tags were linked to PI historian for data reconciliation.

Economic variables for the sources include 60# and 625# steam, propane and feed gas. For the product steams, two offgas streams and one liquid NGL stream are included. The environmental variables include the release of CO2 to atmosphere and the amount of sour water steam to downstream facility.

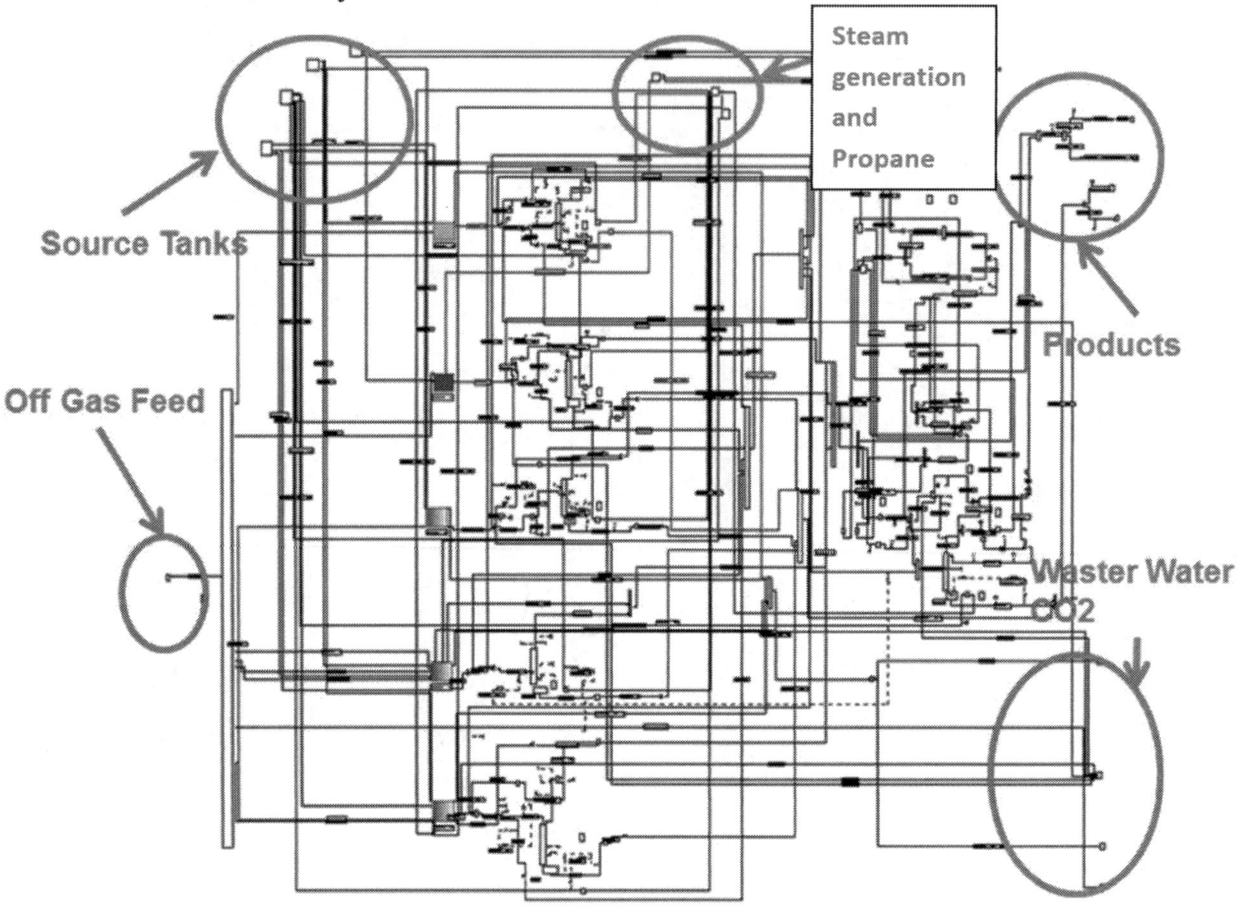

Figure 2

Figure 2 demonstrates the process models with eight compression trains, five high pressure stripping columns, one deethanizer column and a propane refrigeration system. All process units in this project are simulated with rigorous thermodynamic models.

Data Reconciliation

Process model has been integrated with PI historian for data reconciliation purpose. Macro routines with TCL scripts are developed to turn on and turn down major equipment under different operation modes. Data reconciliation results deliver the report for measurement system analysis that indicates major instrumentation issues. Operation, instrumentation engineering and maintenance are able to utilize this report to identify the issue and improve the situation.

Furthermore, accuracies of some major KPIs are improved after the improvement of measurement system.

Tags	Unit	Scan	Model	Bias	% ABS(Bias)	Comments
462-FI-310.PV	MMft3/day	114.851	91.047	-23.804	26.14474	
340-Total-Off-Gas.CAL	MMft3/day	393.3	297.296	-96.004	32.2924	
461-FIC-182.PV	MMft3/day	18.471	14.189	-4.282	30.17831	
499-TI-764AI1PV.CV	F	159.65	125.356	-34.294	27.35729	
462-FI-1147.PV	MMft3/day	79.345	61.404	-17.941	29.21797	
334-FI-1005.PV	lb/hr	38297.5	25444.8	-12852.7	50.51209	
461-FIC-12.PV	lb/hr	62396.6	29909.7	-32486.9	108.6166	

Table 2

Table 2 shows an example of data reconciliation results with problematic instrumentations related to the major KPIs. Engineering and operation was able to take immediate actions with the reported warnings and resolved the issue in time.

Online Open Loop Real Time Optimization

The online process model has been up running to provide advisory for operation engineers. The new process set points for high pressure stripping columns such as the feed flow rates and individual column specs are reported on daily basis. A typical adjustment could increase a few MBD of NGL product. When feed composition shift due to variation of oil feed combination and stabilization depth, the NGL recovery can be enhanced and lined out quickly with ROMeo RTO.

Table 3 shows an example of production increase with RTO. The long term benefit is estimated at least $3MM per year.

Optimization Summary

Variable	UOM	Value	Change
Net Objective	currency/day	5.628E+006	56752.500
Positive Contributors			
+ Products	currency/day	1.119E+007	54915.900
Negative Contributors			
- Steam Feeds	currency/day	5.563E+006	-1836.620

Table 3

ROMeo has also identified a potential increase of 3MBD recovery with simple process modification. As shown in the figure 3, under the optimization mode, ROMeo tends to divert all the overhead gas from deethanizer feed drum to bypass deethanizer overhead condenser. It has been the normal practice for NGL operation since the system was designed. Similar result was concluded from an independent energy optimization study report from corporate process optimization group. The online RTO demonstrates capability of evolving process by justifying the boundaries of current process design.

Other Monitoring System Development

A couple of monitoring indexes have been proposed for the NGL plant and pipeline operations. For NGL process, a fouling index is defined to capture the early stage of fouling. Instead of relying on the steam flow meter, a fouling index is defined with fundamental equations to remove the impact of temperature gradient across the heat exchanger. The fouling index can provide information for unplanned maintenance schedule.

Table 4 shows the examples of fouling indexes for the kettle reboilers. Out temperature at the cold side is assumed to be the bubble point of the hydrocarbon flow. Hot side temperature is assumed to be the dew point of the superheated steam flow. Temperature difference is calculated with the average temperature from both sides. Based on the engineering data, the heat transfer coefficient in red shows the inconsistent dimension which requires further validation during turn around activity.

Fouling Studies

Variable	UOM	Initial	Value
/340_E3_Duty_CU/e3ua	BTU/hr-F	337104.000	317283.000
/340_E3_Duty_CU/e3u	BTU/hr-ft2-F	67.421	63.457
/461_C_101Duty_CU/e103ua	BTU/hr-F	1.900E+006	1.789E+006
/461_C_101Duty_CU/e102u	BTU/hr-ft2-F	383.799	361.470
/462_stripper_duty_cu/e203ua	BTU/hr-F	2.037E+006	1.905E+006
/462_stripper_duty_cu/e203u	BTU/hr-ft2-F	59.916	56.036
/499_E3_Duty_CU/E3UA	BTU/hr-F	1.515E+006	1.515E+006
/499_E3_Duty_CU/E3U	BTU/hr-ft2-F	69.030	69.030

Table 4

Table 5 shows another environmental variable established for carbon dioxide release. The total amount of carbon dioxide can be tracked with online optimization model. With data collection from historian, justification for carbon dioxide capture can be established because supercritical carbon dioxide injection could be the potential technology for the depleted oil fields in the Kingdom.

Overview Stream Economics: CO2 Generation

Variable	UOM	Value	Change
Flow	MMft3/day	59.916	0.000

Table 5

Figure 3 shows an important process variable, the predicted hydrocarbon liquid condensate, to be monitored. The offgas pressure will drop along the pipeline to the gas plant for C1-C2 separation and sulfur removal. When the initial temperature is higher than the ambient temperature, hydrocarbon liquid condensate is expected in the pipeline. The amount of hydrocarbon liquid condensate can be predicted with online optimization model which has been updated on daily basis. The amount of accumulated liquid can be correlated to the back pressure incidents for future optimization of pipeline scraping frequency and plant design criteria. For instance, the NGL target recovery and frequency of pipeline scrapping shall be determined based on the total cost analysis.

Tracking of hydrocarbon liquid condensate can be achieved with an online model to mitigate back pressure and flaring of valuable products. This new performance index could potentially save the company millions of dollars.

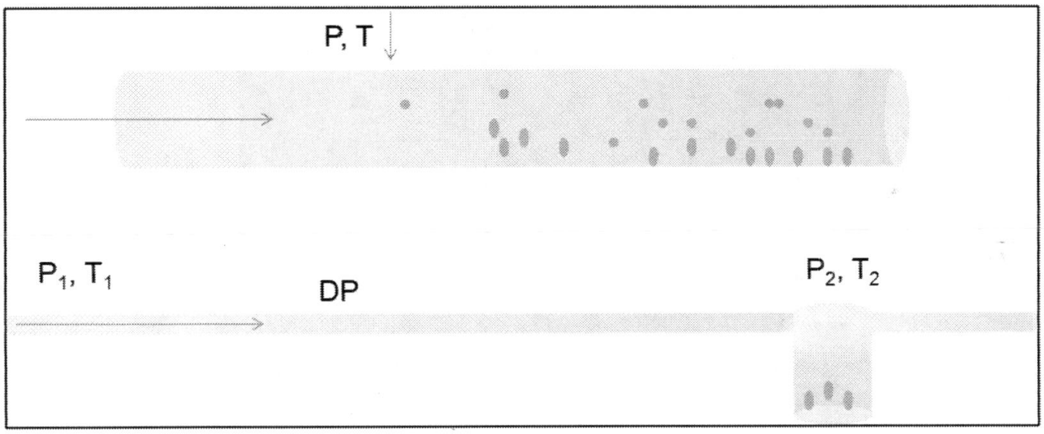

Figure 3

Conclusion

Abqaiq NGL Plant has successfully implemented and tested an in-house developed RTO NGL plant model. A potential increase of $200K to $400K revenue increase for each adjustment has been predicted under the current operating mode and unit operation constraints. In addition to the quantifiable revenue, this presentation also highlights other substantial advantages such as measurement health monitoring, next generation process KPIs, and energy optimization. Future development of this project will focus on supervisory PI Processbook for DCS operators and implementation of engineering details for major equipment.

Acknowledgement

The authors thanks for the supports from the managements of South Area Oil Production Department at Abqaiq and Process Simulation and Control System Department at Dhahran under Saudi Aramco Oil Company. This joint technology development has demonstrated a successful story for Saudi Aramco accelerated transformation program.

Economic Load Allocation Optimization of an Industrial Power Plant

Ravi Nath
Honeywell Process Solutions
Houston, TX

ABSTRACT

Operational Optimization of Industrial Power Plants (IPP) is an often ignored area. One reason for this complacency is that IPPs are usually cost centers in production plants and the cost of energy production is prorated amongst the various production units, so it suffers what is at times referred to as "the tragedy of the commons". Another reason for this complacency is that IPP structure is significantly different than other production units and traditional operations optimization techniques do not readily apply to IPP operations optimization.

This paper will discuss various aspects of IPP optimization; the solution will be illustrated with a recently completed project for a US refinery.

Unlike most refineries, this refinery purchases steam from a neighboring process facility to supplement their own steam production. The implemented solution is an on-line optimizer that optimizes purchased and produced steam to maximize the economic benefit as well as reliability. The implemented solution determines the optimum allocation of steam generators and adjustments to steam balance in the boiler house; furthermore, it also suggests changes to equipment line up in an opportunistic way. The project was completed last year and has been in operation since then, it has been instrumental in streamlining their operations and decreasing their IPP operating costs.

I. INTRODUCTION

A Process Plant comprises one or more process units and an Industrial Power Plant. The process units transform raw materials to products and in doing so consume energy (or utilities) in the form of steam, fuel, electricity, compressed air, cooling water, etc. The Industrial Power Plant (IPP), also referred to as the Plant Utility System, is responsible for providing utilities to the process units. Typically the IPP self generates some of the utilities and purchases others from external sources. Typically all of the cooling water and compressed air is self generated, much of the steam is self generated and at least some of the electrical power and much of the fuel is purchased. Electrical power is typically purchased from the local power grid. Fuel is purchased from one or more local fuel suppliers. Steam, if purchased, is usually from a neighboring process or cogeneration facility. External purchases are usually governed by legal contracts which tend to be complex with provisions for minimum "take or pay", "tier pricing", "time of use" rates etc. Figure 1 is a conceptual representation of a Process Plant showing the interactions between the Process Units and the IPP.

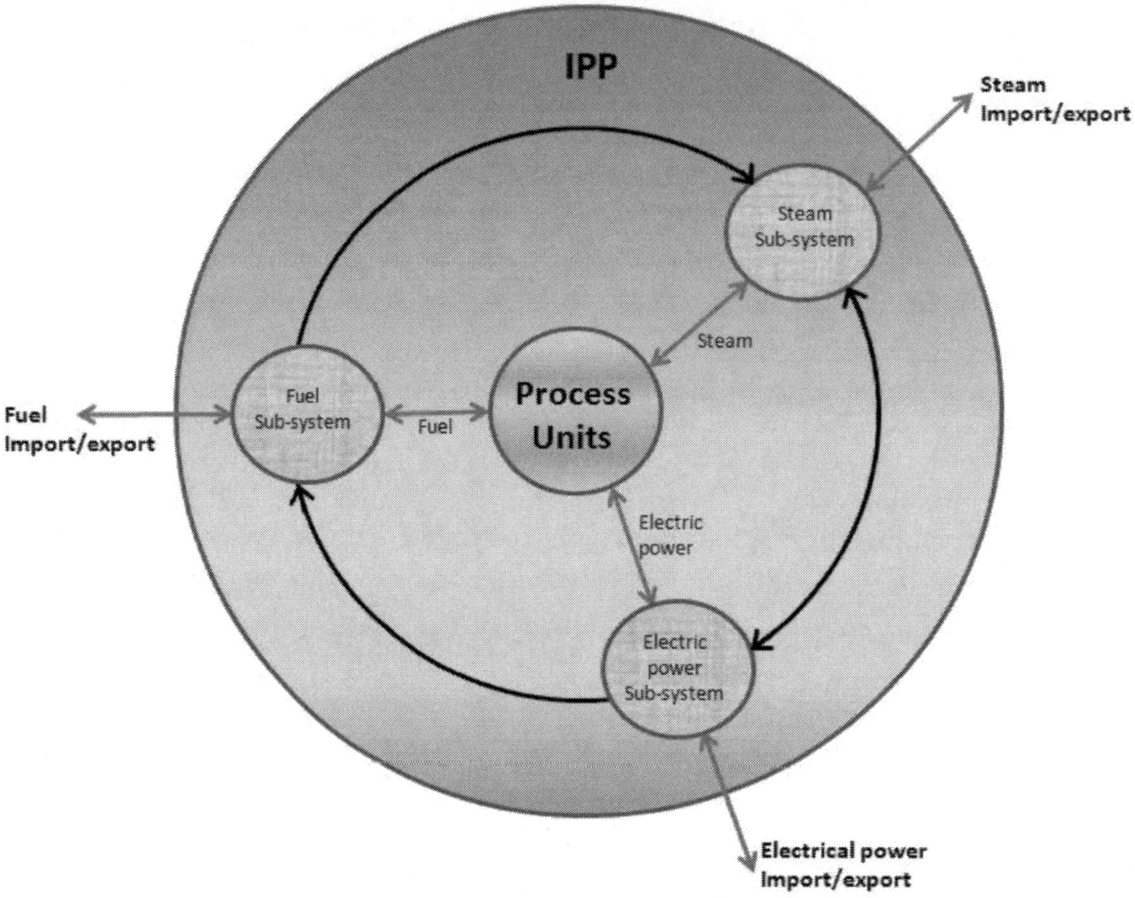

Figure 1: Conceptual Representation of an Industrial Process Plant

Depending on the process units' utilities needs and the location of the production site, there is a large variation in the structure of Industrial Power Plants; however they all share some common characteristics that will be discussed in section III.

As discussed in section II, there are many alternate ways to operate an Industrial Power Plant to satisfy the utility needs of the process units at any given instant. Each alternative however requires different mix of purchased energy and therefore has a different operating cost and different level of emissions to the environment. Because IPPs in large production sites tend to be complex, it is difficult to understand all tradeoffs; therefore there is a tendency to sub-optimize parts of the system rather than optimize the whole, doing so however could result in a significant missed opportunity.

This paper describes a methodology for operations optimization of Industrial Power Plants that has wide applicability and guarantees a global optimum. The proposed solution is based on Mixed Integer Linear Programming (MILP) formulation of the problem. An example based on a recently completed project is presented in section VI.

II. A TYPICAL INDUSTRIAL POWER PLANT

For illustrative purposes, we will consider a simplified IPP with three sub-systems: the steam subsystem, the fuel subsystem and the electric power subsystem. Each of the subsystems is briefly described below;

Steam Subsystem:

The steam subsystem comprises a network of headers that connect to the production units. As shown in Figure 2, there are three steam headers – HP, MP and LP; one boiler feed water (BFW) header and a steam condensate header. Steam is generated in multiple HP and MP boilers that supply to the HP and MP headers respectively. There are back-pressure and condensing steam turbo-generators that produce electrical power. There also are several steam turbines and electrical motors that drive various fans, pumps and compressors such as boiler fans, BFW pumps, cooling water pumps, plant air compressors etc. There are Pressure Reducing Stations that let down steam from higher to lower pressure and a steam vent to maintain header pressures. There is also a blow down flash unit that recovers LP steam from the boiler blow down streams. The BFW used in the boilers is generated in one or more deaerators that consume LP steam.

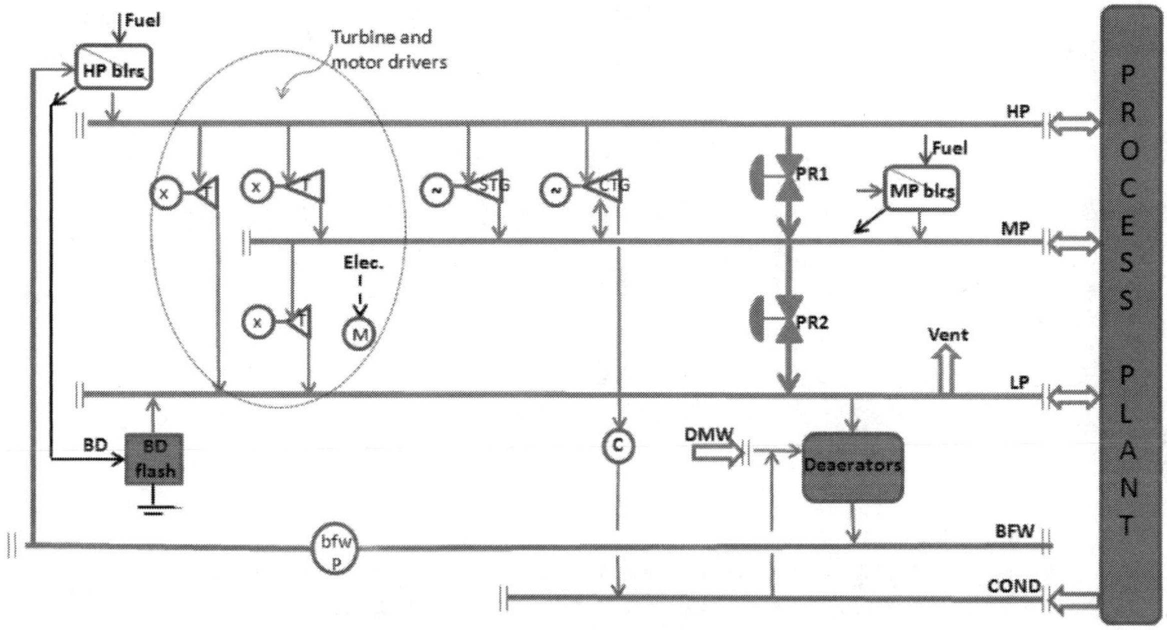

Figure 2: IPP: Steam subsystem

Fuel Subsystem:

Typically there is more than one fuel header at the site; for each fuel header there may be multiple suppliers. With each supplier there usually is a different contractual agreement; for example one may be a long term supplier with a minimum take or pay agreement, another may be short term supplier with a tiered pricing agreement and the third may be a spot supplier with price tied to a fuel price index. Fuel may be high pressure Methane, low pressure Methane or plant fuel which could be a mixture of Hydrogen and Methane. There are several consumers of fuel such as fired heaters, furnaces, steam generators, gas turbines etc. The fuel consumers may have certain restrictions, for example, some could only consume a single type of fuel, while others could consume any mix of fuels. Sometimes there are fuel compressors to uplift the fuel pressure; the fuel compressors could be driven by electrical motors, steam turbines or gas turbines. There are letdown stations and a flare to maintain fuel header pressures. One such fuel subsystem schematic is shown in Figure 3.

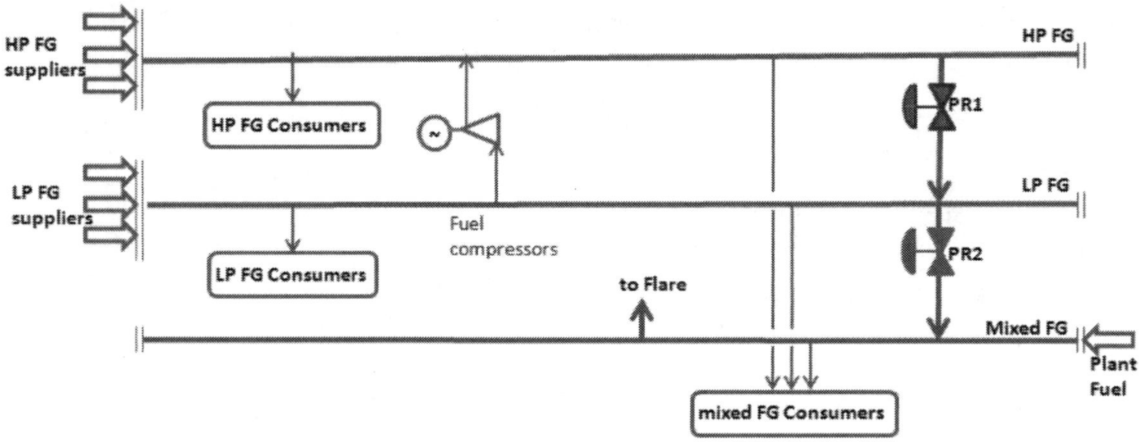

Figure 3: IPP: Fuel subsystem

Electrical Power Subsystem:

The electrical power subsystem is the simplest to conceptualize. Usually a simple balance model (IN = OUT) is sufficient, and it is usually not necessary to model the transformer stations and various voltage sub-headers unless they are constraining. One such electrical power subsystem schematic is shown in Figure 4.

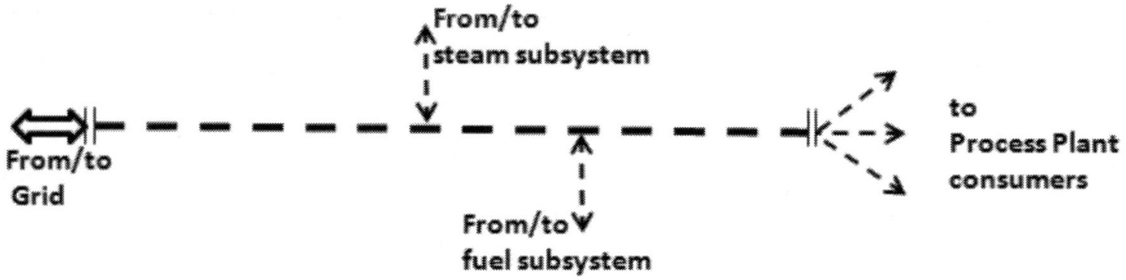

Figure 4: IPP: Electrical power subsystem

Sub-system Interactions:

The above mentioned subsystems interact with the process units by supplying the energy needs of the process. In addition, the subsystems interact amongst each other; for instance the fuel subsystem supplies the fuel required by the boilers in the steam subsystem and the electrical power subsystems accepts surplus power from the steam subsystem and supplies power to the fuel subsystem consumers. Such interactions are shown by arrowed lines in Figure 1.

III. CHARACTERISTICS OF INDUSTRIAL POWER PLANTS

There are several characteristics unique to IPP operations that are noteworthy;

- **Surplus generation capacity:** IPP typically has more generation capacity than required for normal operation. This is by design so that the IPP could accommodate the extremes in demand that are usually encountered during the process unit start ups and shut downs. The surplus generation capacity is manifested in terms of surplus boilers, pumps and compressors. For example, there may be four boilers that are required during start ups, but only two may be needed during normal operation and a third one may be in hot standby to provide additional steam on a short notice. During normal operation, the surplus boiler equipment represents additional degrees of freedom and creates an opportunity to select the most economic set of boilers that could reduce the operating cost. Similarly there would be more BFW pumps than normally needed. Since each pump has different operating characteristics, an opportunity exists to select a set of BFW pumps to reduce the operations cost.

- **Dual Drives:** Typically a boiler house has spare drivers for some of the larger fans, pumps and compressors. Typically one of the drives is steam turbine while the other one is an electric motor. Such situations create opportunities for optimization; this is, at times use of a particular drive may be more beneficial than the other drive.

- **Complex contracts with external suppliers**: Most plants are not self sufficient in their energy needs and have to purchase energy from external suppliers. For electric power, almost all plants are tied to the electrical grid for reliability reasons. And for fuel, most plants have agreements with one or more fuel supplier.

  For purchase of electrical power, there usually are charges both for the use of electrical energy as well for what is referred to as "electrical demand". Electrical demand is the instantaneous consumption peak that occurred over a specified period of time, usually the previous 12 to 24 months. This means creating a new peak in electrical demand will entail penalty for the next 12 to 24 months. Some IPPs also have "Time of Use" contracts in which case the price of electric power varies with the time of use, usually significantly higher during the "peak hours" in a day and during the "peak season" in a year.

  For purchase of fuel, typically there are multiple suppliers. Some with long term contracts to have a reliable supply for which one would usually pay a premium; for such supplier there usually would be a minimum "take or pay" type of agreement. For short term suppliers, depending on the local natural gas market, there could be volume discount or penalty for excess purchase.

- **IPP as Cost Center:** In an industrial setting, the IPP is usually a cost center and their cost is prorated amongst the various process units. Such an arrangement, many a times leads to short term thinking and discourages investment to improve the efficiency of IPP, which is unfortunate since energy costs in some industries are second only to raw material cost [1].

- **Flowsheet simplicity**: Compared to typical production processes, an IPP has a considerably simpler flowsheet. Other than steam and power generation, other unit operations could be modeled as simple mixers and splitters, there are no chemical reactors and distillation columns. In terms of number of components, we have just one in the steam subsystem.

- **Equipment performance curves**: Detailed, first principles modeling of steam and power generators (boilers, turbro-generators, GT/HRSG etc) could be complicated. However, for operations optimization of existing system, detailed modeling is not justified and instead the usual practice is to use equipment performance curves, such as boiler efficiency curves, turbine water ratios and Gas Turbine iso-charts to predict their behavior in the normal operating range.

In summary, characteristics of IPP are simplicity in flowsheet but complexity in purchased energy contracts and opportunities in equipment selection. These characteristics have profound implications on the design of an operations optimizer for an IPP.

IV. THE OPTMIZATION PROBLEM

The optimization problem comprises determining the most economical (usually least cost) operations policy that satisfies the energy demands of the process units and is consistent with external supplier contracts.

From an implementation perspective, the optimizer determines the set of desired values for the decision variables that gives the most economic operation. There are several decision variables in an Industrial Power Plant. These can be broadly classified in two groups;

- **Continuous decision variables**: are variables that could operate over a range, such as boiler and turbo-generator loadings. From an optimization point of view, continuous variables are simpler to handle as the traditional optimization methods apply to problems with continuous variables.

- **Discrete decision variables**: are variables that could take only discrete values such as the number of boilers in service. The simplest of the discrete variables are the "binary" variables that could be one of possible two values. Examples include ON/OFF status of a boiler and selection of turbine versus motor drive for a dual drive pump. From an optimization point of view, discrete decision variables are computationally demanding. Although systematic procedures for discrete optimization exist, each additional binary decision has the potential to double the computation time.

In most cases there will be many alternate ways to operate the Industrial Power Plant that will all satisfy the process energy demands and be consistent with other operating constraints. The optimization problem then is to find the operations policy that is the best amongst all such feasible alternatives.

The best operating mode may not always be obvious. Past experience could help but not always. To gain an appreciation of the problem, let us consider a simple scenario – with a steam subsystem that has only three steam headers, a HP boiler, a condensing turbo-generator CTG, two pressure reducing stations – PR1 and PR2 between steam headers, and a deaerator as shown in Figure 5.

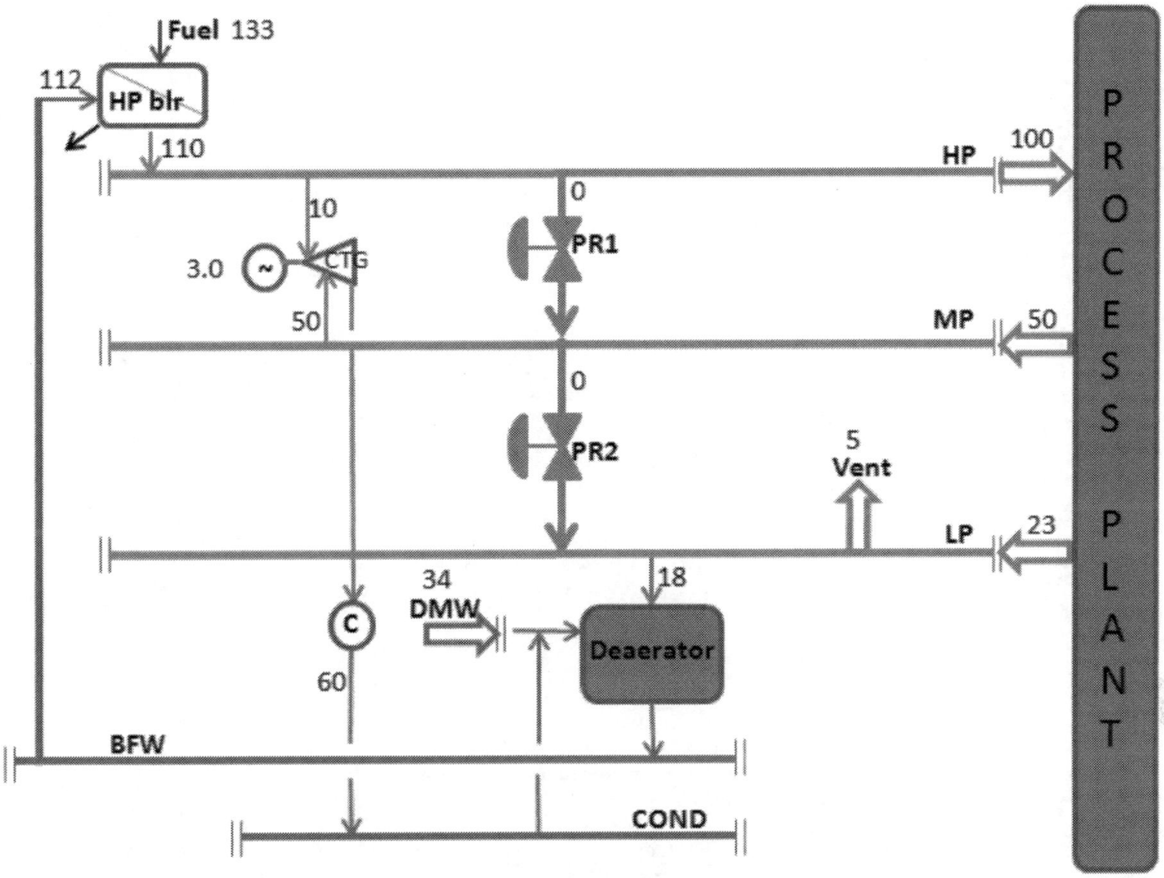

Figure 5: Simple steam subsystem operation: Base case

Note the following equipment constraints;

	Min	Max	EU
HP Boiler	100	300	MPPH
STG – HP inlet	10	85	MPPH
STG – MP inlet	20	50	MPPH
STG – generation	2	10	MW

And the process demands are as follows;

	Demand	EU
HP	100	MPPH
MP	-50	MPPH
LP	-23	MPPH

Base case

Figure 5 shows the base case operating conditions, the boiler produces 110 MPPH HP steam and the turbine throttles 10 MPPH HP steam and 50 MPPH MP steam from the process. Note that the boiler fuel consumption is 133 MMBtu/hr. and power generation is 3 MW. Note that there is steam vent of 5 MPPH, which is an obvious inefficiency and one may seek alternatives to prevent it.

Alternative A

Figure 6 shows an alternative to prevent steam venting by using the surplus LP steam to generate additional boiler feed water that is used to generate additional steam which is expanded through the condensing turbo-generator. Note that the boiler fuel consumption increases to 172 MMBtu/hr. and power generation increases to 6 MW. The venting of steam is stopped. Now, we are consuming little more fuel and producing little more power; this could be profitable if electrical power is a higher value commodity such as during peak hours. In such a case, one may wish to increase power generation further.

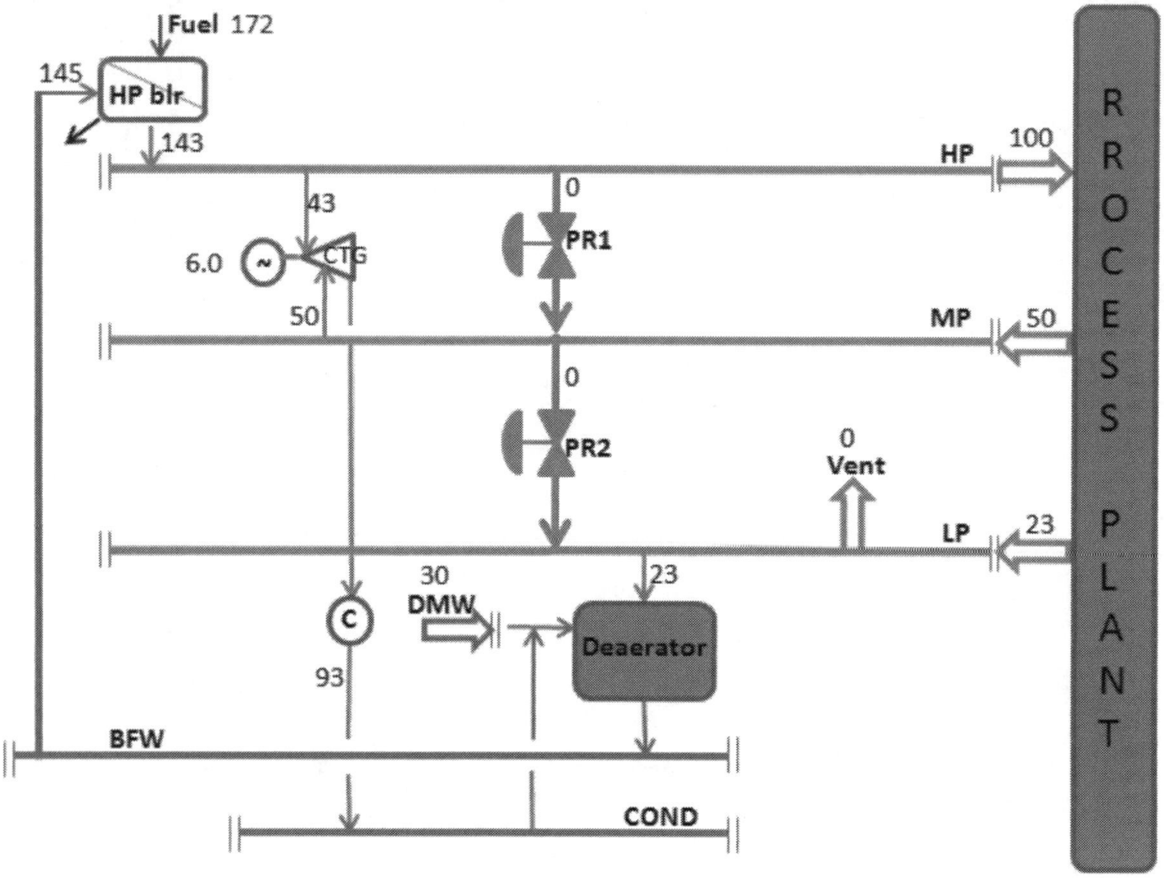

Figure 6: Simple steam subsystem operation: Alternative A

Alternative B

Figure 7 shows an alternative operation to increase power generation by generating additional steam and letting it down through the turbo-generator. Obviously the BFW demand and the deaerator LP steam demand will also increase which could be met by letting down 5 MPPH through PR2. Note that the boiler fuel consumption increases to 211 MMBtu/hr. and power generation increases to 8.6 MW. Depending on the purchased energy costs, one may wish to increase power generation further to the limits of the turbo-generator.

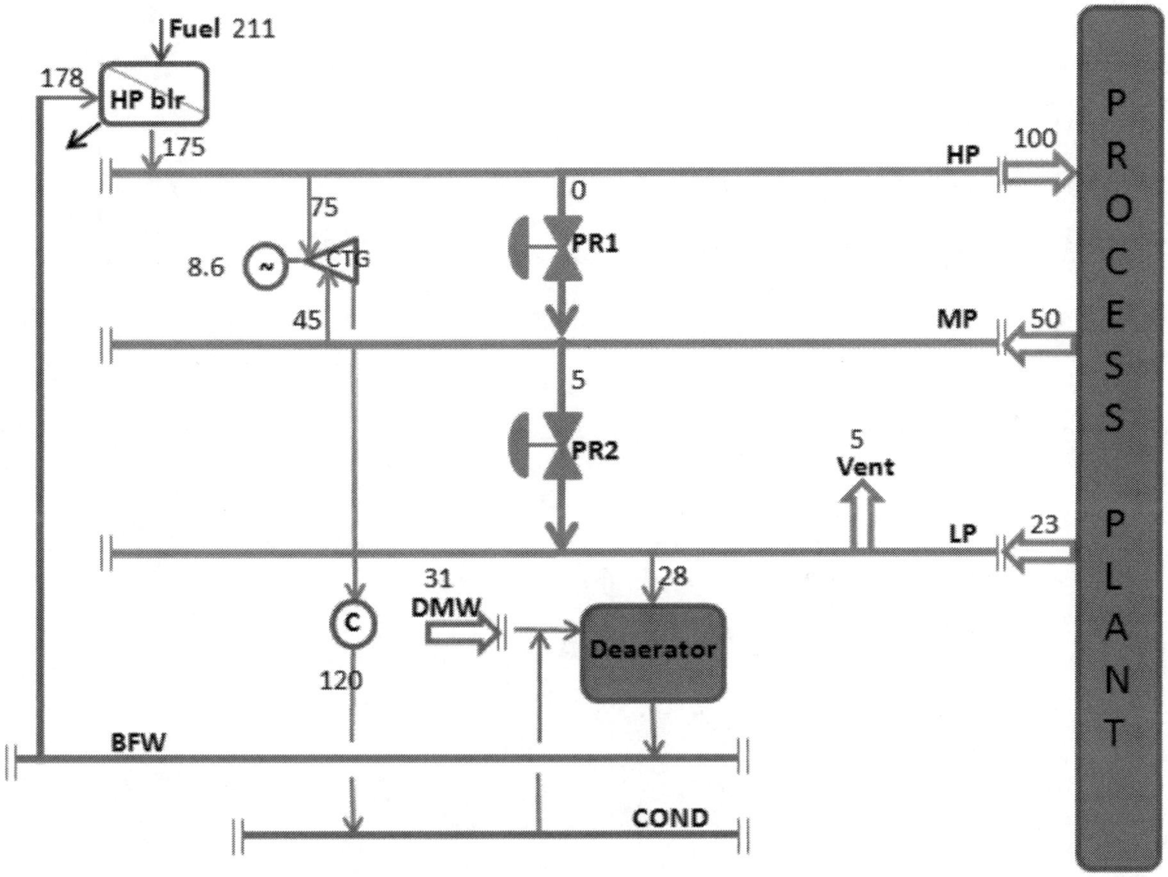

Figure 7: Simple steam subsystem operation: Alternative B

Alternative C

Figure 8 shows an alternative operation that maximizes power generation by generating additional steam and admitting it to the turbo-generator. Obviously the BFW demand and the deaerator steam LP steam demand will also increase which will now come by letting down 8 MPPH HP steam all the way down to LP. Note that the boiler fuel consumption increases to 233 MMBtu/hr. and power generation increases to 10 MW which is the power generation limit of the condensing turbo-generator.

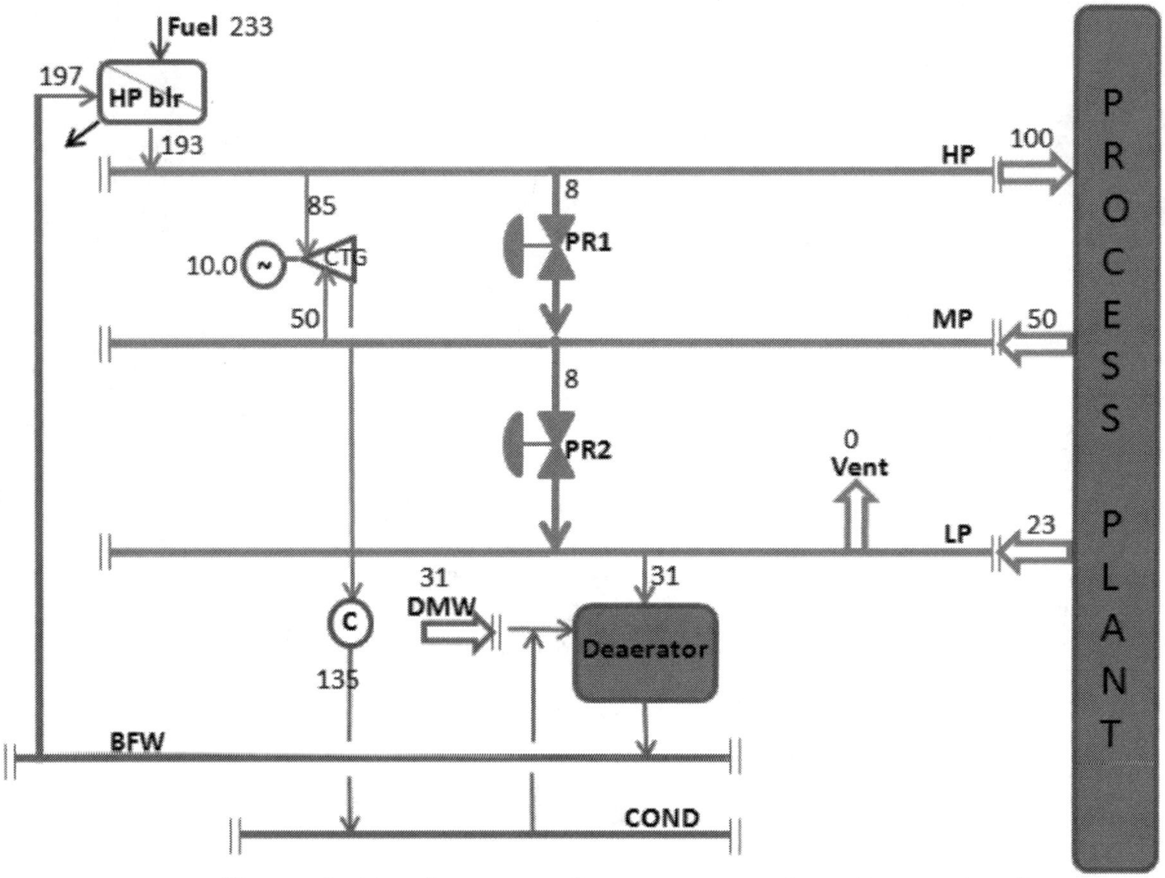

Figure 8: Simple steam subsystem operation: Alternative C

Note that each of these alternatives is feasible because the steam demand of process units is satisfied and the turbo-generator and the boiler are operating within their respective operating limits.

One may ask, "Which of these four operations policy is the best?" Answer to this question, is that any of these alternatives may be the best, it all depends on the relative cost of fuel and power. For relatively inexpensive power, the base case would be the best. As the relative cost of power increases (or fuel cost decreases), the alternative A will become the best. As the power cost increases further (or fuel cost decreases further) alternative B will become the optimum. And for even higher power cost (or lower fuel cost) the alternative C will become the optimum.

The example shown here has been simplified for clarity; it only has 2 degrees of freedom, has simple pricing for fuel and power and no surplus equipment. And even here the optimum is not obvious. For an industrial system, with multiple continuous degrees of freedom, complex contracts and multiple discrete degrees of freedom corresponding to surplus equipment, the problem can quickly become overwhelming and beyond manual optimization; a computerized solution will become a necessity to obtain a definitive solution in a reasonable amount of time.

V. SOLUTION TECHNOLOGY

In simple terms, optimization, is a search task. It is a search for the most desirable feasible operation amongst all feasible operations. Most desirable is usually defined in economic terms, for example for an Industrial Power Plant operation, the best is usually defined as the least expensive mode of operation. The constraints define as to what is operationally possible. Constraints could be as simple as the operating range of equipment or complex relationships that exist between variables, such as the heat and material balances for a boiler or a deaerator.

For most practical problems of modest complexity, the number of all possible solutions is still very large. For example, optimization with fixed equipment line up for an Industrial Power Plant with 4 boilers and two back-pressure turbo-generators there would be 6 continuous decision variables – corresponding to the loading of individual equipment. Mathematically it is a 6 dimensional problem. So optimization is a search in a 6 dimensional space, it is like finding a needle in a 6 dimensional hay stack – certainly a daunting task. However, if we now consider the flexibility in equipment lineup changes we would add another dozen or so discrete decision variables corresponding to 4 boilers, 2 turbo-generators, and perhaps half a dozen dual drives for various pumps and compressors in the IPP; thus creating a search for the needle in 4096 (2^{12}), 6-dimensional hay stacks, by no means a trivial task.

Fortunately, the mathematicians over the years have developed some very efficient search methods, called optimization algorithms [2]; that computer scientists and engineers have coded in computers for use by non-mathematicians.

Mathematically, an optimization problem is;

$$\text{MINIMIZE } F(X, Y) \qquad (1)$$

Subject to

$$G(X, Y) \gtreqless RHS \qquad (2)$$

$$Xmin \leq X \leq Xmax \qquad (3)$$

$$0 \leq Y \leq 1 \qquad (4)$$

Equation 1 is the objective function, F(X, Y), which specifies the cost function to be minimized; for example, it could be the total cost of purchased fuel and electrical power by the Industrial Power Plant. Equations 2 are the constraints, G(X, Y), that define all relevant relationships between the variables; for example, boiler and turbine performance curves, material balances on the steam and fuel headers, energy balances on the deaerators etc. Note that these relationships usually are equalities. Lastly, equations 3 and 4 are the bounds on variables, for example, the operating range of equipment, flow limits through pressure reducing stations etc.

Basically, there are two types of optimization problems

Linear Programming

An optimization model is linear if the objective function F and constraints G are all linear expressions, that is, all equations describing the various relationships are linear.

An expression is linear if each term in the expression is a constant times a variable. The following is an example of linear expression,

$$A1 * X1 + A2 * X2 + A3 * X3 + A4 * Y1 + \ldots$$

Linear optimization problems are also called Linear Programs (LP). The technology of solving LPs is very advanced and LPs can be solved very efficiently with guaranteed globally optimum solutions. Global optimum means that a better solution does not exist which is very reassuring. A variety of commercial solvers are available [3]. Such solvers are robust and many are fairly inexpensive.

Non Linear Programming

An optimization model that has at least one non-linear term in the objective function F or one or more of the constraints G is a non-linear optimization problem, also called a Non Linear Program (NLP).

An expression is non-linear if any term in the expression is anything other than a constant times a variable. Following are two examples of non-linear expressions;

$A1 * \sqrt{X1} + A2 * X2 + \ldots$
$A1 * X12 + A2 * X1 * X2 + A3 * \log(X3) + \ldots$

The technology of solving NLPs although has advanced significantly, is not as robust as that for LP. In general. Solution to NLPs is not as efficient and the solution is only a local optimum. Local Optimum means that is there is no guarantee that a better solution does not exist, which is not very reassuring. Although a variety of commercial solvers are available, they all suffer from the aforementioned difficulties.

Integer Variables

So far the optimization discussion in this section has been limited to continuous variables; that is the variables could take fractional values, such as 0.4 and 1.928. However as mentioned earlier IPPs have spare equipment so equipment selection variables need to be included in the optimization variables. Selection variables are the simplest integer variables as they could either be 0 or 1 corresponding to "on" and "off" states. Such variables are called Binary variables. General Integer variables that could take integers values over a range could be modeled using multiple Binaries. So we will limit the remainder of the discussion to binary integer variables. In the above formulation the binary variables are represented by symbol Y and are restricted to take value of either 0 or 1.

Binary variables add a new dimension to modeling and make the following possible;

- Equipment Selection (ON/OFF)
- Equipment operation with multiple states, such as ON, OFF or stand by
- Discontinuities in equipment operation, such as high range operation and low range operation.
- Minimum up time or downtime requirements, such as after start up a boiler must continue to operate for a minimum specified period
- Delay between events, such as start up and shut down time which is useful in multi-time period modeling, and
- Other operations logic

General discussion on the use of binary variables in modeling is beyond the scope of this paper. Note that the first three uses may be applicable to IPP modeling and will be discussed in the next section.

Although the binary variables add modeling capabilities that would be impossible with continuous variables alone; addition of binary variables to an optimization problem comes at a cost in terms of increased execution time. In the worst case, addition of a binary variable could potentially double the execution time as both options may have to be fully evaluated. So, one must be careful in the use of integer variables. However, properly formulated models, even those with large number of binary variables can be efficiently solved in a reasonable amount of time [4].

Use of binary variables in an otherwise LP renders the optimization problem to be a Mixed Integer Linear Program (MILP). Similarly use of integers in an otherwise NLP renders it a Mixed Integer Non Linear Program (MINLP).

Both LP and NLP solvers could be expanded to accommodate integer variables. Commercial software for both MILP and MINLP are available. Similar to their continuous counterparts, global optimum can be guaranteed for MILP, but not for an MINLP.

VI. CASE STUDY

Scope

This case study is based on a recently completed IPP optimization project for a US refinery. The scope of optimization was limited to the Boiler House and included the following;

- Energy purchase: steam, fuel and electricity
- HP and MP Boilers and auxiliaries
- Steam headers, fuel header and Electrical Bus
- Let down stations
- Blow down flash
- Deaerators
- BFW pumps, BFW pump turbines and motors
- Let down stations, vent and flare

The Boiler House includes three High Pressure (HP) and two Medium Pressure (MP) gas fired boilers. In addition, the plant purchases High Pressure steam from a neighboring Cogeneration plant. Steam from the Cogeneration plant is a little cheaper but it is also somewhat less reliable. In addition to the HP and the MP steam headers, there is also a Low Pressure (LP) header that supplies to the Process Plant. Most auxiliary pumps and fans in the Industrial Power Plant are turbine driven while a few are motor driven which offers some flexibility in balancing of steam headers. The process plant is usually a net importer of steam, however occasionally it exports LP steam to the Industrial Power Plant which can at times cause venting of LP steam in the steam subsystem.

All boilers are gas fired; the HP Boilers are tied to the purchased Natural Gas header while the MP Boilers are tied to the refinery gas and the purchased fuel headers. Overall, the site is net importer of fuel gas from a local Natural Gas Company. The site is also a net importer of electrical energy from the local grid.

Continuous blow down from the boilers is collected in two Blow Down drums that generate LP flash steam.

Steam balance is maintained by use of Pressure Reducing Stations that let down HP to MP and MP to LP steam headers.

Boiler Feed Water is prepared in LP deaerators. Process condensate return is fed to the deaerators. Make up water comes from the Demin system.

There are two groups of Boiler Feed Water pumps. Each group has an electric and two turbine driven pumps.

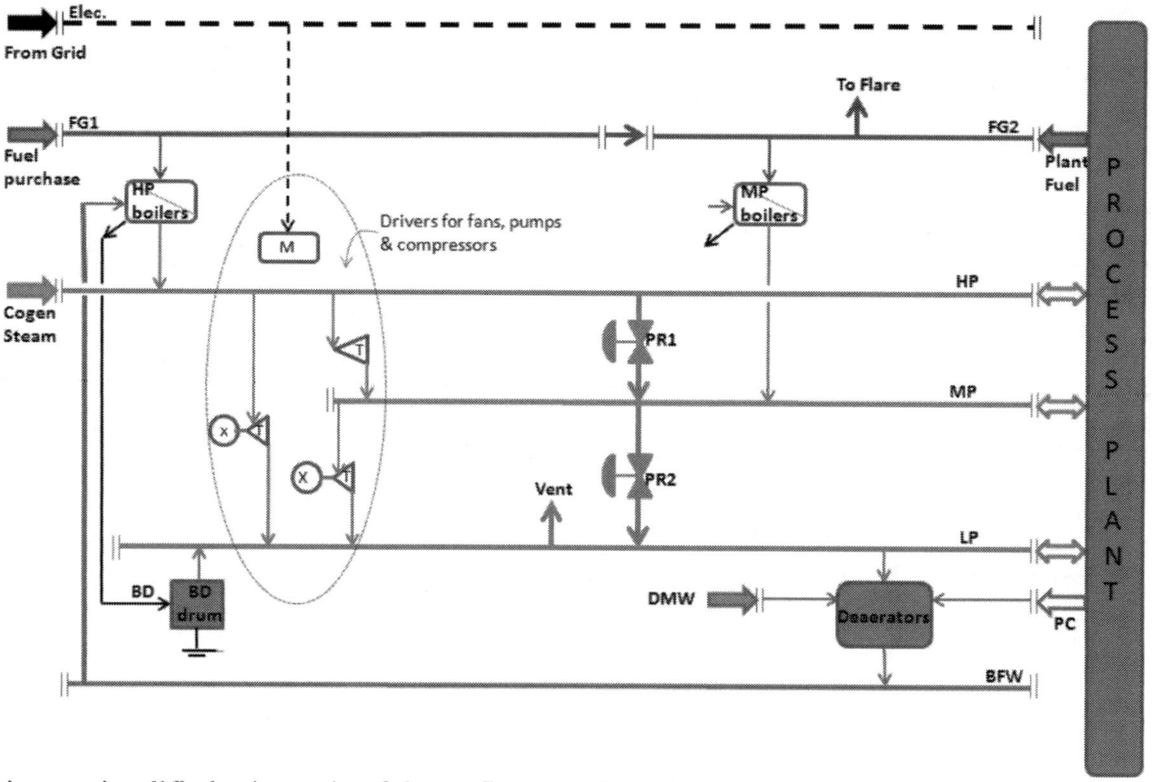

Figure gives a simplified schematic of the Refinery Boiler House.

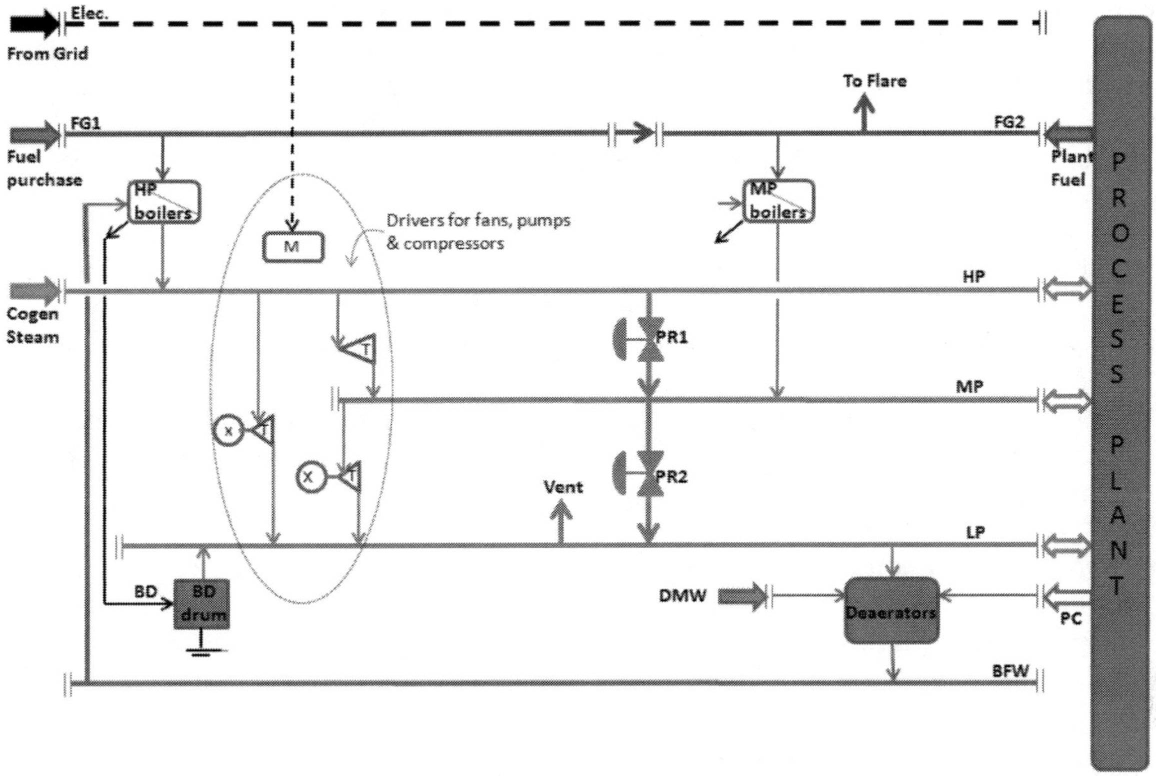

Figure 9: Case Study: IPP Boiler House – simplified schematic

The complexity and uncertainty associated with Cogeneration steam purchase, the inherent variability in process plant steam demand and flexibility in the operation of the auxiliaries, collectively create an environment where determination of minimum cost operation without additional risk is not straight forward. An automated optimization tool that could quickly analyze the situation and recommend optimal operations policy would be a valuable tool for more economical operation of the boiler house.

With this in mind, the Refinery had contracted Honeywell Process Solutions to design, configure and implement an Economic Load Allocation (ELA) Optimizer for the Boiler House. The ELA Optimizer has been designed for an on-line, open-loop execution that provides advisory information to the operations staff. Its execution is at a high frequency, once every two minutes, so the results of ELA Optimizer are always fresh. All data communication to and from the ELA application is via the Honeywell DCS. The Refinery operations staff is responsible for reviewing the ELA optimizer recommendations and making the ultimate operations changes.

The Refinery Boiler House operation usually runs with surplus steam generation capacity so as to be prepared for possible Cogeneration steam supply disruption.

Optimizer Modeling

Equipment operations data was collected and Mixed Integer Linear models for each of the equipment in the scope of the Optimizer were configured. An MILP model represented the operations quite well. In addition, several tuning constants were built in the model and were used to fit the model to actual operation in the first step of the execution cycle; this way the model maintains good fidelity.

One particular concern in this case was the reliability of steam import. This concern was mitigated by adding reliability constraints which ensured that the surplus steam production capacity was at least equal to the steam import plus a desired excess; a similar constraint was applied to boiler feed water capacity. The overall result of these constraints was that the optimum operation could gracefully handle any disruption in supply of imported steam by quickly making up for the loss by utilizing the surplus capacity in boiler feed water and steam generation.

Another aspect of optimization modeling was the recognition that making discrete changes to equipment lineup is a more deliberate endeavor which requires careful consideration of other factors. For this reason, the following three optimization problems were solved in each execution cycle;

- *COPT*: for optimization of only the continuous variables with current equipment line up fixed. This solution is suitable for immediate implementation manually by the operator or via automation.

- *POPT*: for optimization of continuous variables plus added flexibility to be able to change the pump lineup. The motivation behind this formulation is the recognition that implementing pump lineup change is easier then implementing boiler lineup changes. POPT recommendations need consideration only when the additional benefits beyond COPT are worth the additional trouble.

- *DOPT*: for optimization of all continuous variables as well as all discrete variables. DOPT case had the additional flexibility to change the boiler lineup. DOPT recommendations need consideration only when the additional benefits beyond POPT are worth the additional effort.

On-line Execution

In absolute terms, Industrial Power Plant optimization is a relatively small optimization problem and execution is extremely fast. The three Optimization cases mentioned earlier were solved every two minutes and the results of optimization were presented to the operators on the DCS. Figure 10 summarizes the optimization execution cycle.

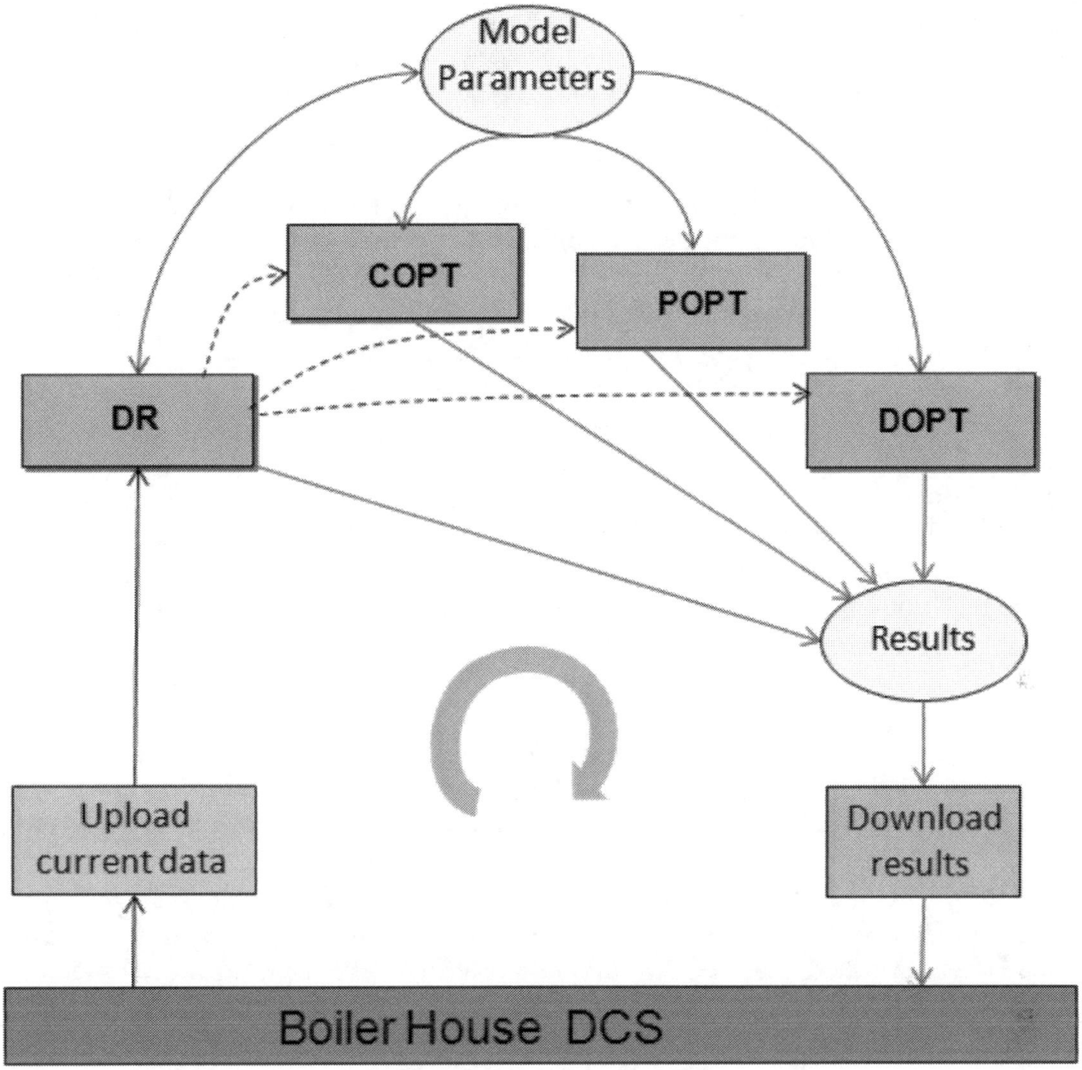

Figure 10: Case Study: IPP Boiler House – ELA Optimizer Execution Cycle

User Interface

All operator interactions with the Optimizer were via an Operator Station on the DCS. Custom graphics per client specification were configured that displayed both the inputs and the outputs. Economic parameters were also displayed on the input graphics that could be modified by the operator. The results of the three optimization runs (COPT, POPT and DOPT) were shown side by side along with the current operations so the operator could clearly see the optimization recommendations and make the necessary moves.

Summary and engineering reports are also produced on the engineering station that are accessible to the engineers from anywhere in the plant.

Optimizer Performance

Benefits of Optimization have been two folds;

- *One time Benefits*: During the modeling stage the current operating practices were thoroughly reviewed and examined. This led to a better understanding of the current operations and some modifications in operating philosophy were made which resulted in quick improvement in certain operations.

- *Continuous Benefits*: The optimizer runs every two minutes so it responds quickly to changes in operating conditions and identifies opportunities as they arise. It clearly summarizes the expected benefits and the way to realize them. The optimizer has been well received by the operators because of the ease of it's of use. It has been integrated in the daily work flow and has resulted in a more streamlined operation of the refinery boiler house. The economic benefits have been estimated to be about 1% reduction in the Industrial Power Plant operating cost.

Future Plans

The boiler house is in the midst of an upgrade; a new boiler is being commissioned and two of the older boilers are to be decommissioned. The optimizer model has been designed with these changes in mind and should continue to operate in the new configuration with minimal changes, if any. It is expected that the new configuration will have more optimization opportunities and then closing the loop with COPT results may be worthwhile.

VII. CONCLUSIONS

Operations optimization of Industrial Power Plants is often overlooked for a variety of reasons. However, for medium to large industrial production facilities, energy usage is the most significant cost after raw material cost, so operations optimization of the industrial power plants is well worth the effort. Non-traditional approach, such as MILP, is well suited for operations optimization of industrial power plants as has been demonstrated in a recently completed Refinery IPP Optimization project.

ACKNOWLEDGEMENTS

We would like to thank J. Schilling, J. Charr, N. Badgujar, D. Boudreax and P. Lynch for their support of this project. We would also like to thank the management of Honeywell for permission to publish this work.

ABBREVIATIONS

A	Constant coefficient
BD	Boiler Blow Down
BFW	Boiler Feed Water
COND	Condensate
COPT	Continuous Optimization
CTG	Condensing Turbo-Generator
DCS	Distributed Control System
DMW	De-Mineralized Water
DOPT	Discrete Optimization
DR	Data Reconciliation and parameter estimation
ELA	Economic Load Allocation

EU	Engineering Unit of measure
F	Objective Function
FG	Fuel Gas
G	Constraint equation
GT	Gs Turbine
HP	High Pressure
HRSG	Heat Recovery Steam Generator
IPP	Industrial Power Plant
LP	Low Pressure or Linear Programming
MILP	Mixed Integer Linear Programming
MINLP	Mixed Integer Non Linear Programming
MMBtu	Million (10^6) BTU
MPPH	Thousand (10^3) Pound Per Hour
MW	Mega (10^6) Watts
MP	Medium Pressure
NLP	Non Linear Programming
POPT	Pump Optimization
PR	Pressure Reducing station
RHS	Right Hand Side (of an equation)
STG	Steam Turbo-Generator
X	Continuous variable
Y	Binary variable

REFERENCES

1. http://en.wikipedia.org/wiki/Energy_in_the_United_States
2. Luenberger, D. G., "*Linear and Nonlinear Programming,*" Addison-Wesley (1984).
3. Anonymous, "*LINGO User's Guide*", Lindo Systems Inc, (2011).
4. Salbidegoitia, I. et. al, "*Operations Optimization of Concentrating Solar Power Plants*", paper presented at the AIChE Spring Meeting, Houston, TX (2013).

(24a) Crude Unit Overhead Systems: Process Configurations to Mitigate Fouling

Andrew W. Sloley, CH2M HILL, Bellingham, WA

In the crude distillation column up to 40% of the total heat removal may be in the tower overhead condenser. The quantity of duty available makes heat integration between the column overhead and the unit feed tempting. Approximately 70% of the current crude units recover some portion of this duty. However, many of these units suffer from persistent and serious corrosion problems in the overhead system. In extreme cases, the viability of the entire refinery site may be in doubt due to fouling and corrosion problems. Up to 50% of the refinery's entire corrosion expense may occur in the crude unit. The root of the fouling and corrosion often lie in basic unit configuration choices. Once the unit configuration is set, anti-fouling chemical addition and unit operating condition limits, while helpful, may never truly solve the unit operating problems. An overview of major factors and a case study of modifications for a specific unit are presented.

"IMPORTANCE OF FIELD TESTING OF LIQUID DISTRIBUTORS"

German Luna-Mejias, P. Eng.
ACM Facility Safety
Calgary, AB, Canada

Biography

German Luna-Mejias holds a Chemical Engineering Degree from Universidad Simon Bolivar, Venezuela (1985), and he is a Senior Member of AIChE. He has been working in the Oil & Gas industry for more than 27 years. His areas of expertise includes: crude distillation, hydrogen manufacturing (process and operations), H2S abatement, visbreakers and HF alkylation. Most recently he has been working in the Process Safety Management area.

Introduction

Good performance in a packed tower is strongly dependant on not only the selected packing, but on the proper design of internal distributors. Internal distributor selection will impact the efficiency as well as the capacity of the column. Proper liquid-vapour contact is key to maximizing the packing efficiency. Uneven distribution of liquids is possible in any column, so designers must determine the appropriate column internal device for the service to minimize this risk.

Liquid distributors are generally classified as pressure distributors or gravity distributors. The most common pressure distributors are spray nozzles. As a general rule, pressure distributors provide a greater area liquid-vapour contact area and are cheaper, lighter and easier to install. However, they have the disadvantage of higher operating costs, susceptibility to plugging and corrosion, entrainment, and if not properly designed the liquid distribution may be poor.

There are many types of gravity distributors, the most common being the orifice and weir distributor. This type of distributor has a broader capacity in terms of flow and good turndown ratio; however, they are extremely sensitive to levelling and surface liquid waves. Orifice distributors are susceptible to plugging, so they should not be use in dirty services. The "notched trough" weir distributor is the one recommended for dirty services.

This document analyses the use of gravity and pressure distributors (specifically spray nozzles) in fractionation columns, including an example of the effects of poor distribution and why it is so important to perform a series of checks and verifications on them before starting the unit. Typical testing procedures are also included for each type of distributor, as well as, some safety recommendations required for the safe execution of the procedure.

Spray Nozzles

Introduction

Spray nozzles utilize fluid pressure as an energy source to break liquids into droplets. This type of nozzle is less energy consuming than a gas atomized or twin-fluid spray nozzle. As the fluid pressure increases, the flow increases and the droplet size decreases. However, this leads to difficulties in selecting a droplet size to achieve a certain flow rate at a given pressure.

Spray nozzles are used for two primary functions: to increase the liquid surface area to enhance evaporation, or to distribute a liquid over an area. In this particular case the goal is to properly distribute a liquid over an area to assure the proper wetting of a surface (packing).

Spray nozzle distributors are mainly used for heat transfer and refinery applications where a uniform distribution pattern is critical due to short bed heights. They are also commonly used in scrubbers, wash zones, condensing zones, and pump around zones of crude oil vacuum towers.

Design Rules

Spray nozzles can be designed for liquid rates as low as 0.1 gpm/(ft^2· ft of tower area), and have been used in columns with diameters in excess of 30 feet.

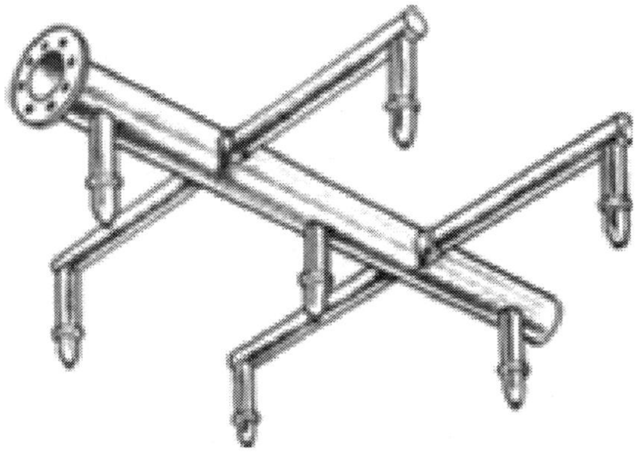

Figure 1. Typical arrangement of a spray nozzle distributor

The maximum turndown of a spray nozzle distributor is 2:1 at a typical pressure drop across the nozzles of 10-15 psi. In the past, several systems have been designed with a pressure drop up to 30 psi. The nozzle tips are typically 12" - 30" above the packing, depending on the spray nozzle layout, spray angle, flow rate, and column diameter [5, 6, 7, 8]. Some authors [1, 2] report the optimum distance above packing to be from 18" to 24". With full cone nozzles, the recommended spray angle is between 90° and 120°.

Spray nozzles have demonstrated their effectiveness for many years; however, there are many reports of failures due to improper installation & testing [1,3,4]. This is the driving force to write this paper: to give advice to the public of the importance of spray nozzle testing before any start-up.

Spray Nozzle Testing - General Considerations

In any fractionation column, checking of the spray nozzles before operation begins is a key to success. Poor or incomplete checking can lead to problems such as loose spray nozzles, leaking gaskets, or loose branches immediately after start up or later, during run-length.

Any of these failures will create an unexpected shut down of the unit for inspection or repair [1].

Due to the time-consuming effort required after plant maintenance is completed to perform a spray nozzle test, some refiners do not perform this type of testing. To guarantee a safe and productive operation, spray nozzle tests must be performed.

Spray Nozzle Testing

Spray nozzle testing consists of 4 steps: Manufacturer Testing, Workshop Testing, Piping Flushing and Internal Testing.

Manufacturer Testing

If the nozzles are constructed from special materials, it is beneficial to complete a process check in the manufacturing plant before sending them to the facility. Even if no special materials are used, it is recommended to send a witness for the factory test (also called FAT or Factory Acceptance Test).

The test in the manufacturing plant, if passed, will guarantee that the nozzles are ready for immediate use and that no debris or dirt is blocking the orifice. This test will not guarantee that the nozzles will work correctly once installed in the column.

Refinery Workshop Testing

Additional testing can be completed in the facility workshop before sending the nozzles to the plant for installation.

In order to perform a test in the workshop, a testing facility needs to be built. A schematic of a typical facility workshop testing facility is shown in Appendix 1.

A Process Engineer will design the testing facility based on the nozzles needed to be tested to ensure that the test covers all the operating range of the facility were the nozzles are going to be installed.

The workshop test should not be done for columns with diameters smaller than 25 feet, but in larger diameter columns, where it may not be possible to test the nozzles inside the column, this is a viable alternative.

Pipe Flushing

Once the nozzle check has been performed, either at the manufacturing plant, the refinery workshop or both, the next step before installation of the nozzles is the flushing of the piping.

Once the piping for the distributor is installed, it must be flushed with high quality water [1, 4] (e.g. BFW quality). This is to prevent any debris from potentially blocking the nozzles. Depending on the flow, it may also help to detect flange leaks and any debris in the line from the pump. Use as much of the current circuit as possible even if using an external water source (use own pump if possible).

Since not all P/A circuits are subject to maintenance during a turnaround, this flushing procedure is fundamental in the prevention of a short run length due to ineffective nozzle operation. A typical procedure for pipe flushing is included in Appendix 2.

Nozzle testing inside the column

This is the final and most important step for spray nozzle testing. The engineer must verify that all of the nozzles are the appropriate ones and that they have been correctly installed and tested.

The intent of this test is to detect: plugged nozzles, defective nozzles, wrong flow pattern of the nozzle, excessive pressure drop across the nozzle, gaskets or branch leaks, etc.

As a part of this test, it is important to have a method or "apparatus" to measure the angle and distribution of the water spray from the nozzle. An example of how to construct a suitable device to measure this is shown in Appendix 3. To save time during testing this device should be prepared ahead of time. A typical procedure for the internal nozzle testing test is described in appendix 4.

In some cases, for example with nozzles bigger than 2" diameter, this final test represents a real challenge due to the amount of water and the high flow rates required. In these particular cases, it may be necessary to install a scaffold above the distributor so the engineers checking the flow rate will be above the spray path and not below. Also in this situation it is helpful to evaluate the use of the draw-off trays of the tower as a "temporary storage" facility for the water.

Larger columns are definitively a challenge for the test of spray nozzles, a consideration on how to test the spray nozzles should be evaluated at the design stage of the system. This is normally overlooked at design stages leading to the impossibility of performing the test.

Gravity Distributors

Introduction

The most common type of gravity distributors are the orifice and weir. Both can handle large liquid loads, and in some aspects can be considered "problem-free" distributors because they are less affected by corrosion or dirty fluids. However, these types of distributors are very sensitive to level installation and liquid waves.

Special attention should be paid to areas directly underneath non-perforated troughs, vapour passages, support beams, support rings and near the column wall. (Contact manufacturer for support)

Liquid is introduced into the parting box, which properly distributes the liquid into the laterals. Generally, one parting box is required for towers up to 8 feet in diameter. Multiple parting boxes are used for larger diameter towers or when liquid rates are high.

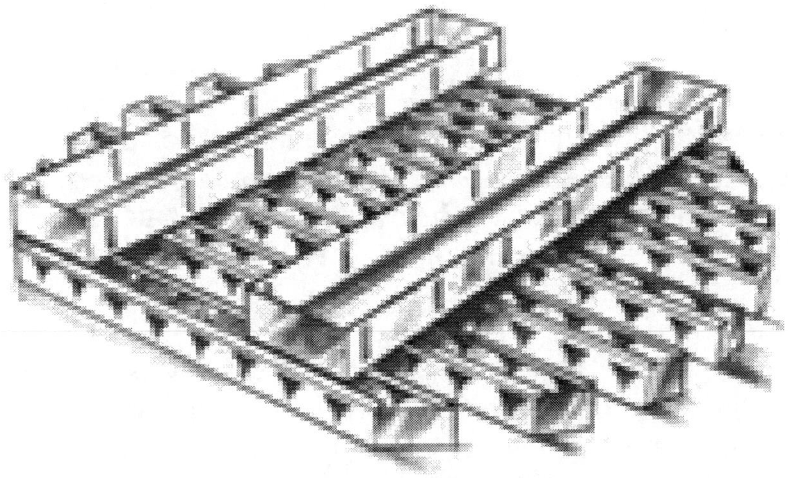

Figure 2. A typical gravity distributor

General Design Rules

A gravity distributor is generally used in fouling services or in towers with high liquid rates. Turndown characteristics are very good (at least 4:1) because of the tapered notch design. This is one advantage of gravity distributors over spray nozzles, which have a maximum turndown of 2:1.

The dripping rate is 4 drips/($ft^2 \cdot$ min), but in some cases can reach up to 20. This will depend on the packing height and liquid flow.

A minimum of 4 -6 inches above the packing should be allocated for vapour disengagement. (Consult manufacturer for any specifications)

Gravity Distributor testing - *General Considerations*

In any fractionation column, testing of the gravity distributor before operation begins is a key to success. Poor or incomplete testing can lead to problems such as dry zones in the packing, uneven distribution of liquid, leaking gaskets, loose collectors, or even problems with dripping devices immediately after start up or during run-length. Any of these failures will create an unexpected need to shut down the unit for inspection or repair.

Due to the time-consuming effort required after plant maintenance is completed to perform a gravity distributor test, at some facilities, this type of testing is usually not performed. To guarantee a safe and productive operation, gravity distributor tests must be performed.

Gravity Distributor Testing

Gravity distributor testing consists of 3 steps: Manufacturer Testing, Installation Verification (by owner) and Field Testing.

Manufacturer Testing

The test in the manufacturing plant, if passed, will guarantee that the distributor has been properly constructed, and if properly levelled during installation, there will be no problems with performance. The manufacturer test is performed as per the manufacturer's standard procedures (also called Factory Acceptance Test).

Installation Verification (by owner)

During the installation of the distributor tray, caution must be taken to ensure that all manufacturers recommended steps have been followed and the distributor has been properly levelled. Any small deviation in levelling will cause problems in the near future. A typical procedure for installation, verification and the required equipment is included in Appendix 5.

Field Testing

Once the piping for the distributor is installed, it must be flushed with high quality water [1, 4] (e.g. BFW quality). This is to prevent any debris from potentially blocking the dripping orifices. This test must be performed at the manufacturer-recommended flow. All dripping points must be checked to assure proper wetting. Depending on the flow, this may also help to detect flange leaks and any debris in the line from the pump. Use as much of the current circuit as possible even if using an external water source (use own pump if possible). A simplified procedure for gravity distributor field testing is included in Appendix 6.

Example of poor distribution impact

The poor performance of a spray nozzle or gravity distributor will impact the packing below. It is quite common to observe fouling/plugging of packing when an uneven distribution occurs.

Mal distribution of wash-oil is often blame for the coke formation in packing within the vacuum units; this phenomena is not always identified as during T/A little or no care is observed to inspect the packing as it is usually removed and disposed.

The fouling or plugging of packing can be observed by changes in the yield or quality of the product or increase of ΔP across the packing.

In an inspection of packing just after T/A it was found a localized coking of packing. After the observation of packing a test was performed in the spray nozzles above and was found a poor pattern in the area above the packing:

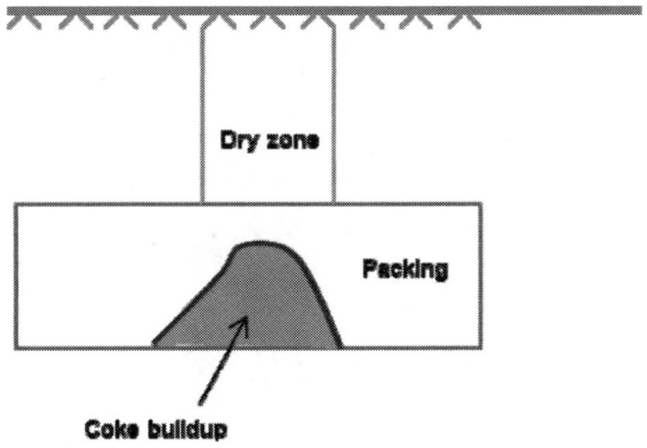

Figure 3. Coke buildup due to poor liquid distribution.

Depending on criticality of the system, this may have an enormous impact on production/yields and economy of the unit.

Conclusions

- A good designed distributor with plugged orifices or nozzles will create an uneven liquid distribution leading to poor product quality and yield or plugging of packing.
- Spray nozzle testing must include these 4 steps: Manufacturer Testing, Workshop Testing, Pipe Flushing and Internal (inside the column) Testing.
- Gravity distributor testing must include these 3 steps: Manufacturer Testing, Workshop Testing, Internal (inside the column) Testing.
- Field tests are expensive, but may detect a problem such as uneven distribution of liquid, leaking gaskets, loose collectors, plugging of dripping devices or nozzles that could lead to an unexpected shutdown.
- A Job Safety Analysis or equivalent must be performed before starting any testing inside the column (internal testing) to ensure a good and safe environment for the team performing the test.

Recommendations

- Larger columns are definitively a challenge for the test of spray nozzles, consideration on how to test the spray nozzles should be taken at the design stage of the system.

References

1. Henry Kister, Personal correspondence (2008)
2. German Luna, "Evaluacion de boquillas distribuidoras de una unidad de vacio" (Field Testing of Spray Nozzles for a Vacuum Column, in Spanish). 1997
3. Henry Kister, Distillation Operation (1990)
4. German Luna, "Mejoras a la Operacion de la Unidad de vacio" (Improvements to vacuum units operation, in Spanish). 1996
5. Koch-Glitsch: http://www.koch-glitsch.com/koch/products/liquid_distributors.asp
6. Sulzer Chemtech: Gas & liquid Separation Technology (www.sulzerchemtech.com)
7. Lechler USA: Industrial Process Bulletin – How to design spray nozzle headers. (www.Lechlerusa.com)
8. Spray Systems Co: Optimizing your spray system. (www.spray.com)
9. Norman Lieberman. Troubleshooting Process Operations (1985)

APPENDIX 1

Schematic of a Typical Refinery Workshop Nozzle Testing Facility

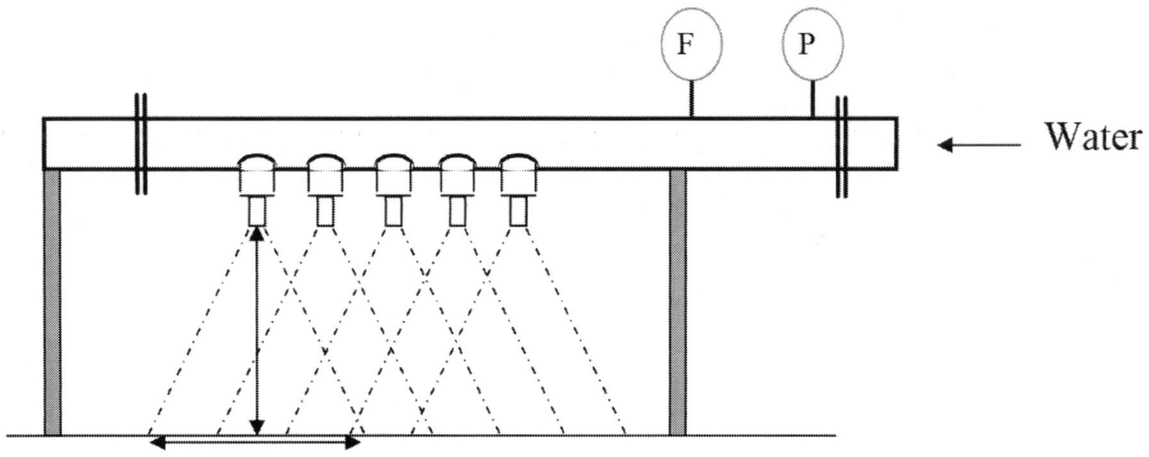

Figure **4**. A typical nozzle testing facility at a refinery workshop

Figure **5**. A typical full cone spray nozzle liquid path

APPENDIX 2

Guidelines for Distributor Flushing (Before Installation of the Spray Nozzles)

1. Obtain a set of PFDs and P&IDs that show the circuit to be tested.
2. Identify blinds that need to be removed or closed. Make a list and include this information with the P&IDs.
3. Discuss blinding and removal of blinds with Operations and Maintenance before proceeding.
4. Perform a Job Safety Analysis (JSA) before commencing the procedure.
5. If needed, install a ladder inside the column to allow engineers (testing team) to reach the distributor or packing.
6. Verify that all spray nozzles are removed.
7. Check the water quality as per Process Engineer request (BFW recommended)
8. Verify that the external water source is available and connected to the system.
9. Verify using the DCS, that all instruments associated with the circuit are available for the test. (e.g. operating ranges, alarms, etc)
10. Install a manometer (60-90# depending on test pressure) in the distributor entrance to the column.
11. Provide the testing team with the appropriate personal protection equipment (PPE).
12. Ask operations to verify that the circuit is lined up completely using the PFDs/PIDs.
13. The team performing the test must verify communication with the control room and pump operator before the start-up of the test.
14. A person must be ready to rescue the person(s) inside the column if a problem should occur. This person must be located at the manhole (This should be identified in the JSA).
15. A whistle test between the people inside the column and the rescue team must be performed before starting the test. (This should be identified in the JSA).
16. Once all previous steps have been completed, proceed with the flushing test.

17. Ask the operator to start the pump and adjust the control panel to normal flow.

18. During the test, all flanges, gaskets and any type of connections inside the distributor header should be checked for leaks; also flow paths (looks for irregular flow paths).

19. This test should be conducted for a minimum of 10 minutes; the length of the test will be defined by the inspector inside the column.

20. The test is complete when the people inside the column check that all branches in the distributor are clean and free of debris or hydrocarbons.

Note:

1. This procedure assumes that the column was previously flushed with hydrocarbons before become ready for maintenance.

APPENDIX 3

Use of a "Wired Cone" For Measuring the Water Angle & Distribution from a Spray Nozzle

When completing an internal nozzle test it can be difficult to verify whether or not the nozzles are producing the correct water flow path. The easiest way to test this is through the use of a "wired cone". This is simply a wire shaped into the same height and diameter as the desired spray path. This "wired cone" can be placed underneath any of the nozzles to confirm that the water flow is properly distributed.

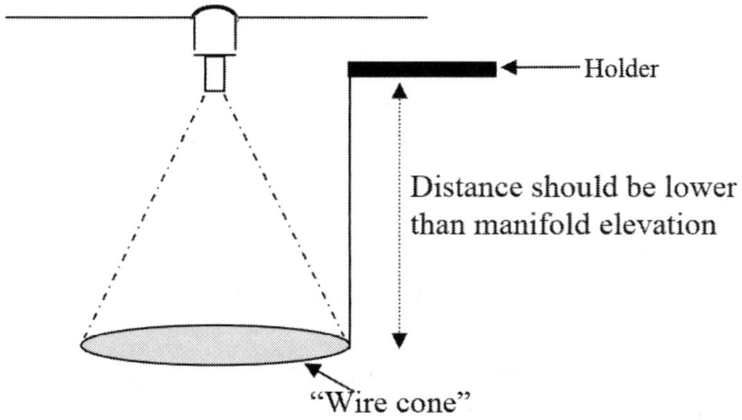

Figure **6**. A "wired cone" apparatus used for spray nozzle performance testing.

APPENDIX 4

Procedure for Internal (Inside column) Spray Nozzle Testing

Before beginning the installation of any spray nozzles, verify that the flushing process has been completed (refer to Appendix 2).

The recommended process crew to complete the internal nozzle testing consists of: 1 or 2 engineers inside the column, 1 engineer in the manhole (outside the column) and, if available, one engineer in the control room (could be the operations engineer). All personnel should remain in contact using radios set to the control room frequency.

It is strongly recommended to perform a JSA (Job safety Analysis) before start any of these procedures.

1. The process engineers should ensure that the nozzles to be installed are the correct ones. If all nozzles have the same diameter, it is very easy to install nozzles in the wrong location.

2. Maintenance personnel should start by installing the spray nozzles.

3. Once the spray nozzles have been installed, the process engineer(s) should check them one by one to be sure the installation has been done properly.

4. Nozzles should be hand tightened to ensure that none are loose.

5. The operator then checks that the system is properly lined up and that no changes have been made since the flushing test.

6. Perform all safety checks before starting the internal nozzle test (as per refinery standards).

7. Due to the amount of water to be sprayed, the process engineer should be located close to manhole to start the test

8. The test will start when the process engineer inside the tower asks the operators to start the pump.

9. The panel operator will adjust the flow to normal and inform the process engineer(s).

10. The process engineer(s) will check each nozzle to verify the spray angle and also to check for any leaking gaskets.

Figure **7**. Incorrect liquid distribution from a malfunctioning or missing nozzle

Figure **8**. Proper liquid distribution from a spray nozzle

11. The process engineer outside the column will check the pressure and flow (in the control room) once all nozzles have been checked, and instruct the operator to adjust the flow to turndown rates.

12. The checking inside the tower will be repeated.

13. A record of pressure and flow rate is required.

14. If any nozzle or gasket leak or malfunction is detected, the process engineer(s) inside the column will ask the panel operator to stop the test.

15. If an emergency occurs inside the column, the inside engineer(s) sound the whistle; the outside engineer will signal the operator to stop the test immediately.

16. This test should be done as fast as possible due to the large amount of water required.

17. Once the test is completed, again verify that all nozzles are tight. If any are not tight, ask maintenance to tack weld all nozzles. If any of the nozzles are not tight, a lot of trouble could be caused during normal operation because the nozzles will loosen.

Notes:

1. Use of a closed circuit allows for reuse of the water and saves time during the shutdown.

2. A flashlight is required inside the tower. If the tower is very large, the person(s) inside may require a self-breathing apparatus. The HSE department should analyze any risk before starting the test.

3. Special gloves may be necessary to avoid injury caused by packing. Consider the use of wood to protect the packing; however, this may make it more difficult to see the wetting performance.

4. The packing may get very slippery due to the amount of water used. Caution must be observed to avoid a fall.

5. If the tower diameter allows, a maximum of 2 people may be inside the tower. A 3rd process engineer will be at the manhole.

6. A whistle will be used only for emergencies if person(s) inside the tower are in trouble and the test needs to be immediately stopped.

APPENDIX 5

Installation of a Gravity Distributor

It is strongly recommended to perform a JSA (Job safety Analysis) before start any of these procedures.

1. The installation of a gravity distributor begins with installation of the packing below.

2. Once the packing bottom supporting grid is installed, proceed to check for levelness.

3. To check levelness use a liquid hose [9] as it is easier to use with water. This method is better than a standard leveling procedure due to the size of the column.

4. Proceed with the installation of the packing.

5. Once the packing is in place, install the packing upper supporting grid.

6. Level the upper supporting grid using the same method as the bottom support grid.

7. Install sub-collectors and check for levelness using the method recommended by Liebermann [9].

8. Adjust as needed to guarantee that the collectors are properly leveled.

9. Check all collector drip points to guarantee that they are clean and free of debris and dirt.

10. If installing a collector with partition boxes out of headers, check the levelness using a "hand level gauge". This section of the distributor is the most critical and therefore required more careful level checking.

11. Install the main collectors (depending on tower diameter may be more than one).

12. Once again, verify the levelness of the main collector.

13. Once all collectors have been installed and leveled, the water test can begin.

Notes:

1. If the distributor being installed is an old distributor that was removed for packing replacement, inspect every part of the distributor for damages or corrosion before proceeding with re-installation.

2. Partition boxes are very likely to be corroded or damaged. If any are found to be damaged, replace with a new one before installing the collector. Future operation of the tower will rely on this step.

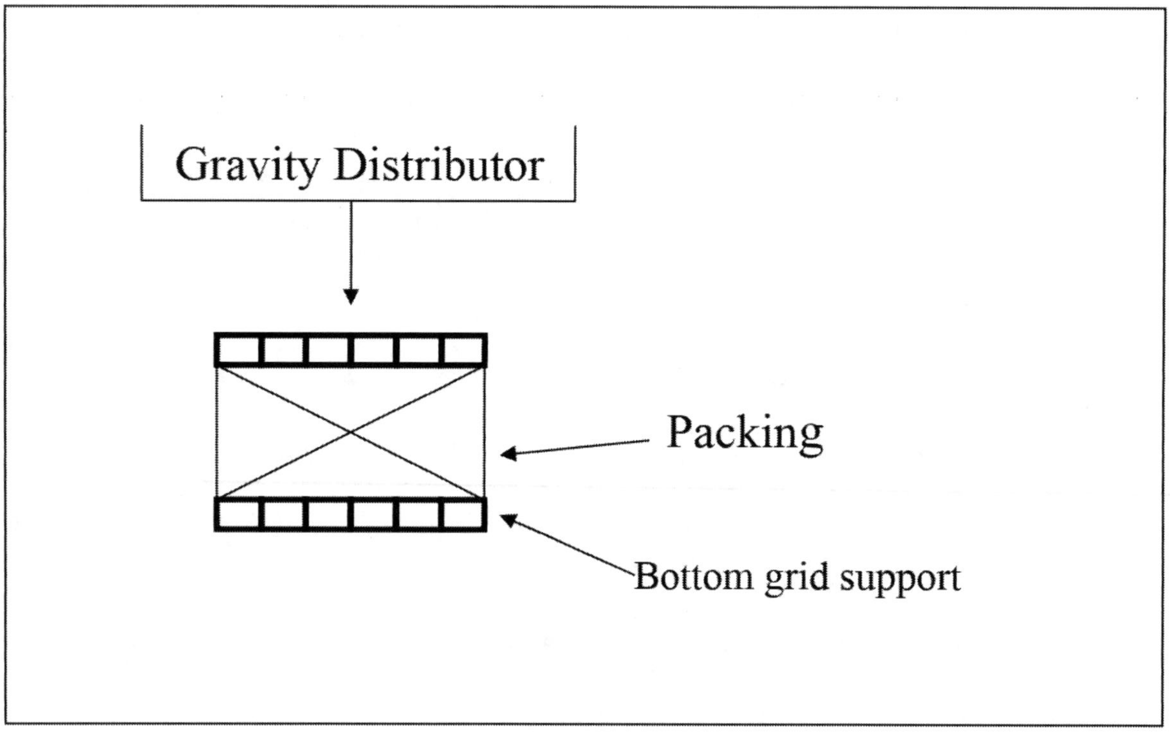

Figure **9**. Details of gravity distributor installation

APPENDIX 6

Guidelines for Gravity Distributor Testing

It is strongly recommended to perform a JSA (Job safety Analysis) before start any of these procedures.

1. Obtain a set of PFDs and P&IDs that show the circuit to be tested.
2. Identify blinds that need to be removed or closed. Make a list and include this information with the P&IDs.
3. Discuss blinding and removal of blinds with Operations and Maintenance before proceeding.
4. Before entering the column, verify that all safety features have been checked and are acceptable; this must be done by HSE personnel.
5. If needed, install a ladder inside the column to allow engineers to reach the distributor or packing.
6. Check the water quality as per Process Engineer request (BFW recommended).
7. Verify that the external water tank is connected to the system.
8. Verify, using the DCS, that all instruments associated with the circuit are available for the test.
9. Provide the testing team with the appropriate personal protection equipment (PPE).
10. Ask operations to verify that the circuit is lined up completely using the PFDs/PIDs
11. The test team must verify communication with the control room and pump operator.
12. A person must be ready to rescue the person(s) inside the column if a problem should occur. This person must be located at the manhole.
13. A whistle test between the people inside the column and the rescue team must be performed before starting the test.
14. Ask the operator to start the pump and adjust the control panel to normal flow.
15. People inside the column will verify the flushing process and will check for gasket leaks.

16. Review, one by one, all dripping points and partition boxes to check if water flow is correct.
17. Check for any overflow or water waves in the collector.
18. Ensure that the water coming from main pipe is not spilling outside the main collector.
19. Verify even distribution of the water in all sections of the distributor
20. Once all points have been checked and none are failing, the distributor is ready for startup.

Notes:

1. If any of the sections of the collector are not distributing the water in a smooth and even path, stop the test and correct where needed.
2. Not a single partition box or dripping point should be left incorrectly installed. This will dry out the packing in that section.
3. This procedure assumes that the column is ready for maintenance

(24c) Evaluating Individual Instrument Readings and Potential Errors in the Analysis of Monitoring Data

G. T. Polley, Chemical Engineering, University of Guanajuato, Guanajuato, Mexico, Elvis K. Tamakloe, Chemical Engineering, Universidad de Guanajuato, Guanajuato, Mexico and Yash-Vardhan Kapoor, TOTAL, GONFREVILLE, France

Evaluating Individual Instrument Readings and Potential Errors in the Analysis of Monitoring Data

E. Tamakloe[1] , G.T.Polley[1] & Y.V.Kapoor[2]

1. Dept of Chemical Engineering, University of Guanajuato, Mexico

2. CERT, Total s.a., France

Data reconciliation techniques have been applied to measurements obtained from the monitoring of pre-heat train performance for many years. In fact, the pre-heat train appears to be a standard problem addressed in text books on the subject of data reconciliation [1]. These techniques are usually based on statistical analysis.

In 2011 Ishiyama et al [2] introduced an alternative to data reconciliation based upon statistical analysis. This approach involved "data adjustment". It involved the application of a fouling model to the monitoring data in order to minimize differences in hot and cold heat balances, generate missing temperature information and the grouping of data in order to identify trends over specific operating periods.

The technique can be used to identify "best values" of missing measurements and to identify which measurements are likely to be unreliable and generate estimates of what the true values are likely to be.

In this paper the authors follow the lead of Ishiyama et al but look at how errors in individual instrument readings can be addressed.

Instrument errors

Instrument errors are assumed to be made up of two components: an off-set and a drift.

Off-set is an absolute temperature difference between actual and apparent reading. It can be positive or negative.

Drift is a percentage of the reading relating to difference between actual and apparent reading. Again, it can be positive or negative.

Identification of Onset of Instrument Failure

Experience in the analysis of monitoring data has shown that the behavior of temperature instruments deteriorates several days before it fails. It is therefore possible to detect the onset of failure and provide a warning to the plant operator.

Example will be described.

References

1. Narasimhan, S., Jordache, C., Data Reconciliation & Gross Error Detection, Gulf Publishing Company, Houston, TX, 2000.

2. Ishiyama E, Pugh S.J., Wilson D.I., Paterson W.R. & Polley G.T., "Importance of data reconciliation on improving performances of crude refinery preheat trains", AIChE Spring Meeting, 2011

HOW TO GET THE BEST FROM YOUR PREHEAT TRAIN – AN UK REFINERY BASED CASE STUDY

Edward M. Ishiyama, IHS Downstream Energy Research, London, UK
Simon J. Pugh, IHS Downstream Energy Research, London, UK
Graham T. Polley, University of Guanajuato, Mexico
James Kennedy, IHS, London UK
D. Ian Wilson, University of Cambridge, UK
Alison Ogden-Quin, Petroineos, Grangemouth, UK
Graham Birch, Petroineos, Grangemouth, UK

Abstract

Crude refinery preheat trains (PHTs) are network of heat exchangers that recover heat from product and pump-around streams to raise the crude temperature before it enters an atmospheric distillation column for fractional separation. Fouling is a long known problem in the PHTs; the ability of a refinery to process different crude blends or to increase its production capacity strongly depends on the thermal and hydraulic performance of the PHT under fouling conditions. This work describes a data reconciliation approach that utilizes historical plant data, consisting of missing information and uncertainty, to generate a so called 'dynamic fouling model'. Dynamic fouling models are used in a simulation methodology to predict plant performance under variable operating conditions. A discussion in the latest developments in modelling and simulation of PHTs subject to fouling is provided. A novel software tool, SmartPM, was successfully utilized to study heat recovery paths, cleaning schedules and furnace firing capacity on a UK based crude refinery PHT case study.

Introduction

Crude oil is a major source of global energy; its usage has increased during the past decades reaching a consumption of 73 MM bbl per day (Figure 1) [1]. In prior to undergoing fractional separation, crude oil passes through a network of heat exchangers known as the preheat train (PHT). PHTs recover up to 70% of the heat required for fractional separation from process streams, that require cooling [2]. Additional heat required to raise the crude to its distillation temperature is provided through a fired heater. Due to the high capacity of crude being processed, the effective performance of the PHT is vital to achieve desirable economic and environmental performance. Fouling is known to be responsible for reduced heat transfer and reduced production, which itself results in problems of higher energy consumption and increased furnace emission, lost of throughput margin and maintenance and cleaning issues which could result in plant shutdowns ([3–5]).

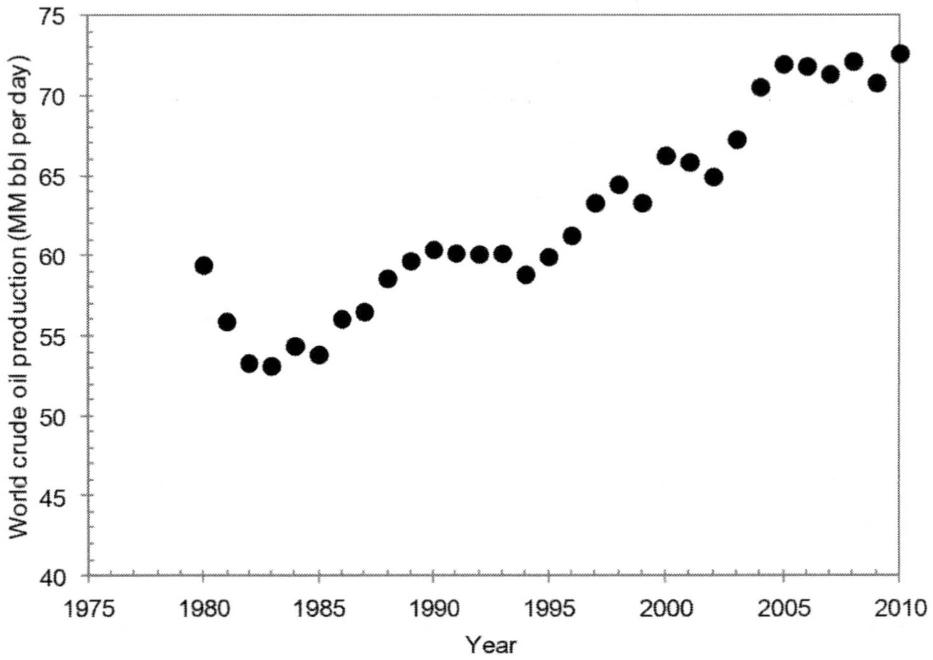

Figure 1: World crude oil production[1].

The composition of the crude being processed is complex and subject to frequent variation. Crude oil is prone to fouling and the fouling behaviour is directly linked to its operating condition [6]. Mitigating the effects of fouling on PHT is a widely researched area both within academia and industry due to its economic and environmental relevance [7–10].

A systematic approach to manage fouling in PHT was recommended in Ishiyama *et al.* (2013)[7], consisting of 5 main steps:

1. Data reconciliation
2. Fouling analysis
3. Prediction of PHT performance under fouling condition and identification of exchangers responsible for reduced PHT performance
4. Identification of retrofit and revamp options
5. Systematic scheduling of cleaning

The recommended guideline requires the evaluation of the PHT performance at various operating scenarios. Formulation of a simple analytical model for PHTs is not practically achievable; hence the thermo-hydraulic PHT simulation software, SmartPM was used in this work. The SmartPM software is built upon the combination of advanced heat exchanger network simulation and cleaning scheduling methodology and industrially accepted shell-and-tube heat exchanger design and rating technology. The network simulation code was developed at the University of Cambridge as part of the UK Government-fund CRude Oil Fouling (CROF) research program. The heat exchanger design technology was developed for the computer program EXPRESSplus and re-implemented in SmartPM. The techniques described in this manuscript, such as the key steps noted above, and other techniques are incorporated into SmartPM.

Model description

The thermal effect of fouling, the overall fouling resistance, R_f, is obtained through

$$R_f = \frac{1}{U} - \frac{1}{U_{cl}} \tag{1}$$

Here, U is the overall heat transfer coefficient. U_{cl} is the overall heat transfer coefficient, which is calculated when the exchanger is clean and operated under the same temperature and mass flow conditions pertaining to the calculation of U.

Heat exchangers in a PHT commonly consist of grouped shells (*i.e.* shells in series or shells in parallel arrangement).

A recommended method in modelling shells in series is not as individual units but as independent shells. The main issue arising through this approach is that most intermediate temperature between shells are not measured due to inability to access the stream and the isolation valves are only available to isolate all shells but not individually. The importance arises of monitoring individual shells, as the build-up of fouling is unequal in each shell due to different temperature and flow fields.

A methodology was introduced by Ishiyama *et al.* (2011, 2013)[7,12] to determine the intermediate temperatures based on the distribution of fouling resistances between shells. For a heat exchanger undergoing chemical reaction fouling, the rate of fouling, $\dot{R}_f$, was described by Polley (2010)[11] as

$$\dot{R}_f = \frac{a}{h_c} exp\left(\frac{-E_a}{RT_f}\right) P \tag{2}$$

Here, a is a dimensional constant, h_c, is the film transfer coefficient of the cold (crude) stream, R is the gas constant, T_f is the film temperature and P, an attachment probability based on surface shear stress. For a particular crude slate, a is constant.

Assuming all shells are clean, the ratios of the initial fouling rates of the shells, 1 to n is given by

$$\dot{R}_{f,1}:\dot{R}_{f,2}: \dots : \dot{R}_{f,n} = \frac{1}{h_{c,1}} exp\left(\frac{-E_a}{RT_{f,1}}\right): \frac{1}{h_{c,2}} exp\left(\frac{-E_a}{RT_{f,2}}\right): \dots : \frac{1}{h_{c,n}} exp\left(\frac{-E_a}{RT_{f,n}}\right) \tag{3}$$

The distribution of the fouling resistances between the shells at an operating instance could be allocated based on the ratios of initial fouling rates of the individual shells.

$$R_{f,1}:R_{f,2}: \dots : R_{f,n} = \dot{R}_{f,1}:\dot{R}_{f,2}: \dots : \dot{R}_{f,n} \tag{4}$$

This information is helpful, during a cleaning operation. Usually, shells in series require being isolated together due to the location of the isolation valves. Still it is not necessary to clean all shells during isolation, if the shells with higher fouling could be identified.

A UK based refinery PHT undergoing sever fouling was considered for methods to improve its operation.

Case study network

The case study PHT is part of the Petroineos refinery situated in Grangemouth, UK. The medium and hot section of the PHT is presented in Figure 2. The crude enters the desalter in an average of 110 °C and its temperature raised to ~ 155 °C via a set of heat exchangers. The crude temperature is further raised to ~ 205 °C, before entering the furnace. The current average capacity of the PHT is 52,000 bbl day^{-1}. Temperature variations for a 10 month period at the coil inlet are plotted as solid circles in Figure 3(a). During this period a turn-around has occurred where all units present in Figure 2 underwent cleaning. The coil inlet temperature dropped to a minimum of 170 °C. The throughput exhibits continuous fluctuation, with a gradually decreasing trend till the plant shutdown (solid circles in Figure 3(b)).

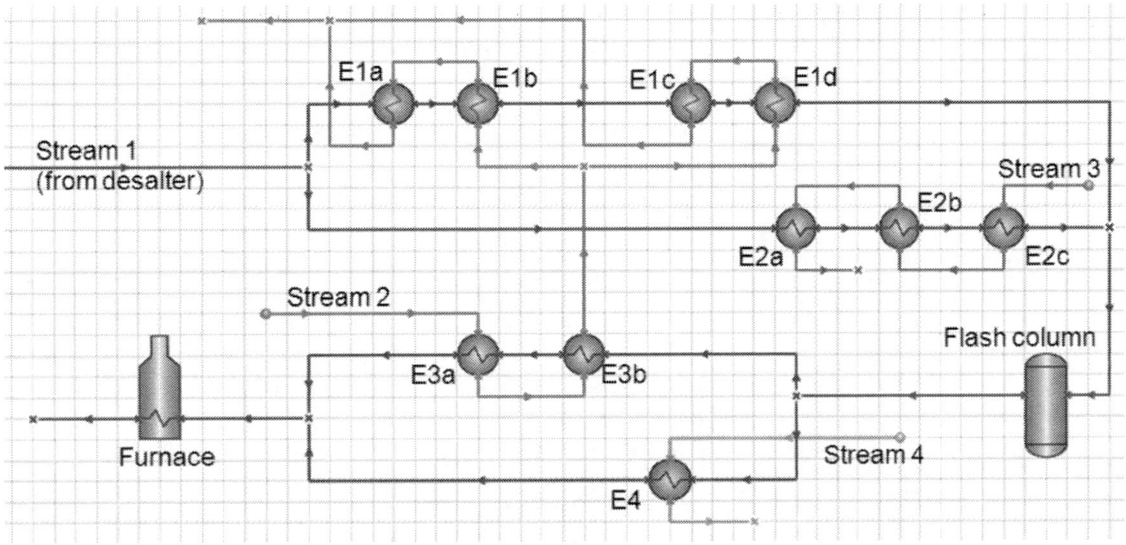

Figure 2: Hot end of the case study preheat train. Label 'E' denote heat exchangers.

(a)

(b)

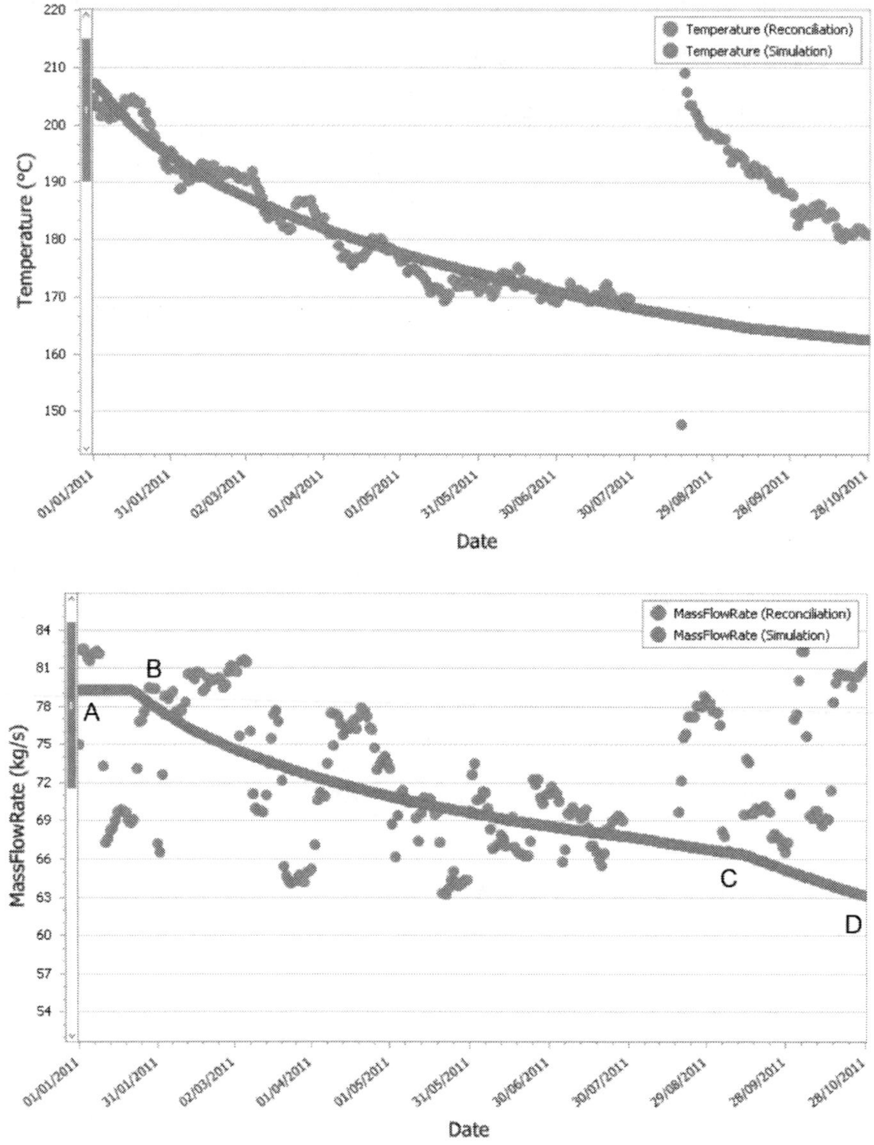

Figure 3: Variation in (a) coil inlet temperature and (b) throughput

Plant monitoring data for a period of 10 months was collected and data reconciliation performed for this period. The fouling resistance profile of unit E4 is given in Figure 4. A cleaning action was performed at the 7th month for all exchangers (this was a plant shut-down). The exchangers are returned online consisting of a minimum fouling resistance, which again increases with fouling. The observed minimum resistance is relatively high (0.001 $m^2K\ W^{-1}$ for E4), even after a cleaning action. This indicates that a higher degree of cleaning could be achieved through alternative cleaning methods (*e.g.* use of hydro drilling).

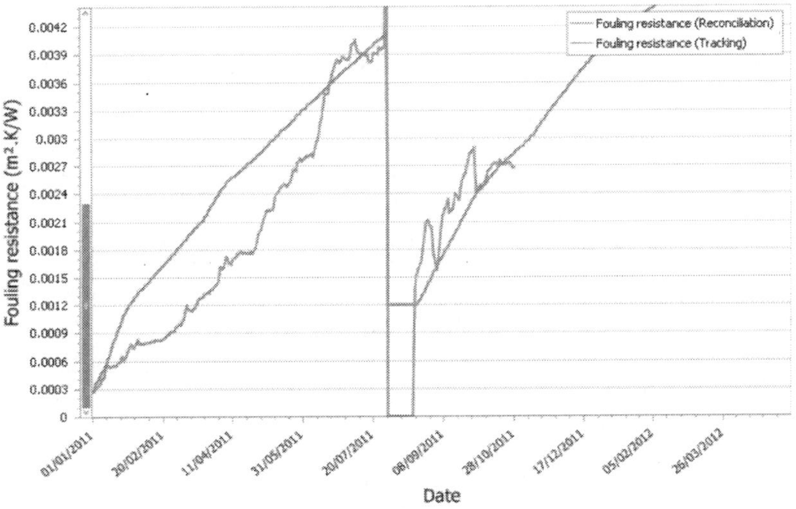

Figure 4: Variation in fouling resistance profile for E4.

The slopes of the fouling resistance profiles, *i.e.* the fouling rate, are a strong function of operating temperatures and flow. The region of stable operation is separated for each exchanger within the SmartPM program, generating a matrix of operating periods. The corresponding fouling rates and operating data are used to evaluate parameters for fouling models such as that described in equation (2)). This allows the modelling of fouling performances in periods where particular crude blends are processed, hence allowing a database of crude slate specific fouling model parameters to be built that can be used in simulation and cleaning calculations. Fouling analysis showed that when equations (2)) was used to evaluate the fouling behaviour of the crude at the period where the data were corrected, a fouling parameter of $a = 450\ h^{-1}$ would best fit the reconciled data.

During the data reconciliation period, the PHT was operated to achieve a target throughput. Due to the furnace reaching its maximum capacity, a gradual reduction in throughput has occurred. Taking the PHT operating parameters at the beginning of data reconciliation period as the starting point of simulation, the CIT and throughput profiles were simulated (solid line in Figure 3a,b). During the simulation the throughput was assumed ideally to be maintained at a target throughput of 85 kg s^{-1}. The cleaning actions were not taken into account. There are three distinct sections in the simulated throughput graph Figure 3(b). Section AB is a region of constant throughput, where the furnace is able to accommodate constant throughput with the reduction in CIT as the furnace has not reached its duty limit. Section BC presents a region where the reduction in throughput has occurred as a consequence of the furnace reaching its maximum duty with fouling. Section CD presents a region where the reduction in throughput is steeper compared to section BC. This is reflecting the boiling properties of the crude – the furnace is required to accommodate additional duty where a majority of the energy requirement is dominated via latent heat.

The fouling model together with the network simulation and furnace model was shown to reliably predict the historical plant behaviour. Future performance simulation studies are now performed to evaluate the best methodology to manage the PHT. An initial simulation study was performed to evaluate the scenario when the PHT is operated without any cleaning or modifications to the network. This was done by taking the data set after the turn-around as the starting point. A forward prediction for a period of 2.5 years was performed. At the beginning of the simulation, the furnace is operating at its maximum duty. Hence with the decrease in PHT thermal efficiency (solid line in Figure 5(a)), the furnace would not be able to provide the required heat duty to bring the crude to its desired outlet temperature under a constant throughput. A reduction in the throughput will occur with fouling to maintain the coil outlet temperature (solid line in Figure 5(b)).

The cleaning scheduling was formulated based on a heuristic algorithm detailed by Ishiyama *et al.* (2009)[9]. The objective function is to minimize the total economic impact during the operating period:

Total economic loss = Cost of energy + Cost of lost throughput margin + Cost of cleaning (5)

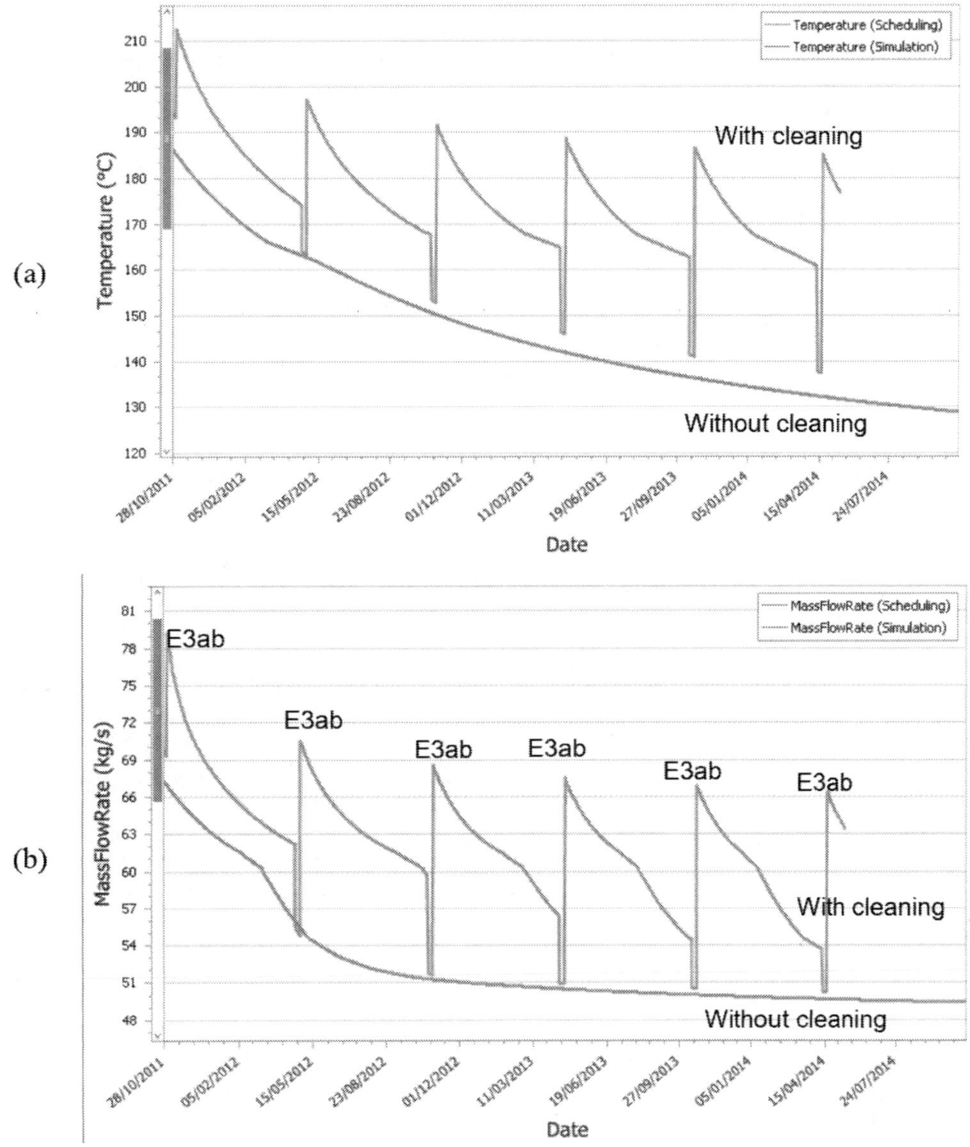

Figure 5: Variation in (a) coil inlet temperature and (b) throughput; for simulation with and without cleaning.

The economic loss is reflected through the lost throughput margin when the furnace is operated at its maximum duty. The optimum cleaning schedule generated for this network indicates that exchanger E3ab requires frequent cleaning. In this paper the retrofit and revamp analysis are identified first to reduce the rate of fouling and second to enhance energy recovery (which will consequently increase the throughput under limited furnace duty). The following opportunities were identified:

Case study 1: Use of tube inserts for E3ab: E3ab is a main heat duty contributor in the PHT and also subject to sever fouling. The tube insert is identified as an option to increase the tube-side wall shear stress, which would reduce the rate of fouling. The choice of tube-insert, twisted tape, was made based on the allowable pressure drop and the ease for installation. Twisted tapes could be fabricated on-site.

Case study 2: Use of tube inserts for E2abc: The hot stream outlet of E2abc is in the order of 190 °C. This unit was identified to be able to enhance heat recovery through tube-inserts.

Case study 3: Divert all crude flow rates to E3ab after the pre-flash: The exchanger, E4, appears to be one of the least energy contributors to the network. If there is sufficient cooling water duty available to cool Stream 4, the diversion of crudes to E3ab will further increase the wall shear stress to reduce fouling in a long term. The exchanger E4 could be utilized elsewhere in the system (e.g. in pre-desalter section to maintain desalter inlet temperature).

The overall economics during the 2.5 year period of simulation was evaluated through comparison of the case studies by taking Figure 5 (without cleaning) as the base case. As the furnace is operating at its maximum duty, any difference in the economic performance is directly related to the throughput processed.

$$Benefit = C_{lo} \int_{0}^{t_f} \left(m_j - m_{reference}\right) dt - total\ cleaning\ cost \qquad (6)$$

Here, m is the throughput, C_{lo} is the throughput margin, subscript 'j' denotes case studies 1 to 4, and subscript 'reference' denote case study 1 under no cleaning condition. t_f denote the time duration where the scheduling study was performed.

The economic comparison of the above three case studies are summarized in Table 1. With the current fouling rate, if no cleaning action were performed a throughput margin reduction amounting to over 10 MMUS$ would occur due to the furnace reaching its capacity. Installation of tube inserts in units E3ab and E2abc, shows long term benefit in terms of both reduced rate of fouling and enhance heat recovery. A network revamp option combined with cleaning activities gives the biggest economic benefit (case study 3).

Table 1: Comparison of the economic benefit for the case studies 1 – 4 over a period of 2.5 years.

	Benefit (calculated from equation (6)) MM US$	
	Without cleaning	With cleaning
Original network	0.00 (base case)	11.30
Case study 1	2.39	12.95
Case study 2	3.51	13.66
Case study 3	4.25	15.24

Conclusions

Data reconciliation and fouling analysis methodology discussed by Ishiyama *et al.* (2011) [12] was applied to a UK refinery through the use of a high definition thermo-hydraulic simulator, SmartPM. The fouling behaviour was successfully modelled by a dynamic fouling model, which enabled the identification of the reduction in production capacity due to the furnace reaching its firing capacity. A retrofit, revamp and cleaning scheduling analysis was combined to identify economic opportunities to recover 2 – 4 MMUS$ during a 2.5 year period through fouling mitigation.

Nomenclature

a	dimensional constant, s^{-1}
C_{lo}	throughput margin, US$ kg^{-1}
E_a	activation energy, J mol^{-1}
h	film transfer coefficient, W m^{-2} K^{-1}
m	mass flow rate, kg s^{-1}
P	attachment probability, –
R	gas constant, J K^{-1} mol^{-1}
R_f	fouling resistance, m^2K W^{-1}
$\dot{R}_f$	fouling rate, m^2K J^{-1}
T	temperature, K
U	overall heat transfer coefficient, W m^{-2} K^{-1}

Subscripts

c crude stream

cl clean condition

f film

j case study 1, 2 or 3

reference reference case

References

(1) Mundi World crude oil production by year, http://www.indexmundi.com/energy.aspx?product=oil&graph=production, visited on 19th, March **2013**.

(2) Panchal, C. B.; Huangfu, E.-P. *Heat Transfer Engineering* **2000**, *21*, 3–9.

(3) ESDU *exchanger fouling in the pre-heat train of a crude oil distillation unit*; IHS ESDU International plc., London, 2000.

(4) Ishiyama, E. M.; Paterson, W. R.; Wilson, D. I. *Heat Transfer Engineering* **2009**, *30*, 805–814.

(5) Ishiyama, E. M.; Paterson, W. R.; Wilson, D. I. *Chemical Engineering Science* **2008**, *63*, 3400–3410.

(6) Ebert, W. A.; Panchal, C. B. In *Fouling Mitigation of Industrial Heat Exchange Equipment*; Begell House: New York, 1997; pp. 451–460.

(7) Ishiyama, E. M.; Pugh, S. J.; Paterson, B.; Polley, G. T.; Kennedy, J.; Wilson, D. I. *Heat Transfer Engineering* **2013**, *34*, 692–701.

(8) Ishiyama, E. M.; Heins, A. V.; Paterson, W. R.; Spinelli, L.; Wilson, D. I. *Applied Thermal Engineering* **2010**, *30*, 1852–1862.

(9) Ishiyama, E. M.; Paterson, W. R.; Wilson, D. I. *Energy and Fuels* **2009**, *23*, 1323–1337.

(10) De Oliveira Filho, L. O.; Liporace, F. S.; Queiroz, E. M.; Costa, A. L. H. *Applied Thermal Engineering* **2009**, *29*, 3073–3080.

(11) Polley, G. T. Jersey City, 2010; pp. 13–17.

(12) Ishiyama, E. M.; Pugh, S. J.; Wilson, D. I.; Paterson, W. R.; Polley, G. T. In *11AIChE - 2011 AIChE Spring Meeting and 7th Global Congress on Process Safety, Conference Proceedings*; 2011.

(25a) Current Myths about LNG

Christopher Caswell[1], David Coyle[2], Charles Durr[3], Heinz Kotzot[3] and **Jeffery Gaspard**[2], (1)Gas Monetization, KBR, Houston, TX, (2)KBR, Houston, TX, (3)KBR, houston, TX

Since the startup of the first baseload LNG plant (Camel) in 1964, there have been several generations of LNG plant designs. Each generation can be defined by a noticeable step change in elements of the overall plant configuration. These steps may include developments in equipment, changes in cooling media, breakthroughs in train throughput, or combinations of changes leading to new industry standards or benchmarks. Since the first LNG projects with train capacities of less than 1 Mt/a, generations of technical innovations have led to operating trains of nearly 8 Mt/a, and front end engineering designs (FEED) exceeding this current benchmark.

Regardless of the state of the LNG industry, there are often misconceptions regarding the application of new technologies, determination of process configurations, or implementation of project execution strategies to reduce project costs and shorten schedules. These misconceptions or innocent mistakes, if perpetuated at conferences or reflected in FEED, can be perceived as reality. These "design and execution myths" will result in less accurate cost estimates or increased technical and execution risk for new projects.

This paper will review LNG technology and project execution myths that could derail the innovation necessary to develop the next generation of LNG projects. By openly discussing these myths, we can stimulate constructive debate on the challenges facing today's projects.

(25b) Boil-off Gas System Design for Integrated Liquefaction and Regasification Facilities

Bindupriya Bandari, Bechtel Oil, Gas & Chemicals, Inc., Houston, TX and Dipanjan Bhattacharya, Bechtel Oil and Gas, Houston, TX

BOG system design for integrated liquefaction and re-gasification LNG facilities

Dipanjan Bhattacharya

Senior Process Engineer

Bechtel Oil, Gas, and Chemicals, Inc.

Bindu Bandari

Process Engineering Specialist

Bechtel Oil, Gas, and Chemicals, Inc.

Designing boil-off gas (BOG) system for a LNG facility that offers both liquefaction and re-gasification capabilities is inherently complex and challenging. Steady state simulation tools are commonly employed to design the BOG system for independent liquefaction or regasification facilities. But their application is limited when working with a complex network of multiple jetties, multiple LNG storage tanks, multiple boil-off gas compressors and multiple modes of operation.

This topic discusses a case study based on a dynamic simulation model of an integrated facility that is converted from an existing LNG import terminal to a bi-directional facility. The design challenges are compounded by the fact that the model has to evaluate the hydraulic adequacy of the existing systems for an entirely new operating envelope.

A well-developed integrated dynamic simulation model can be used for hydraulic analysis, compressor selection, what-if-analysis and preliminary tuning of the controllers and sizing of control valves. The what-if-analysis can be performed to evaluate transient operating conditions, system performance and response; to various operating modes - ship loading mode, normal operation and upset scenarios such as compressor trip

This case study illustrates that a rigorous dynamic model is an effective means to analyze a complex BOG system for an integrated LNG facility.

(25d) Removal of Heavy Hydrocarbons from Lean Natural Gas

Yu-Nan Liu, Air Products & Chemicals, Allentown, PA and **Gowri Krishnamurthy**, Air Products and Chemicals, Inc., Allentown, PA

Removal of heavy hydrocarbons, such as C_6+ from natural gas is essential prior to cryogenic liquefaction in order to prevent freeze-out of these components in the liquefier. Traditionally, cryogenic methods such as passing feed gas through a scrub column or a front-end NGL extraction unit have been employed for this purpose. However, these conventional approaches may not be feasible for certain feed gas streams such as lean natural gas containing low amounts of C_2-C_4, yet significant amounts of C_6+. Lately, several sources of natural gas, such as domestic pipelines, shale gas, coal-bed methane, etc., report lean feed gas of this nature. Adsorption-based separation processes are an alternative method to strip trace heavy hydrocarbons from natural gas. In this paper, we discuss important considerations for evaluating such adsorption systems as well as possible integration with conventional cryogenic methods for the effective removal of heavy hydrocarbons from natural gas.

"Qatar: the largest LNG exporter in less than ten years"

By J C Kuo, Hao Dong, Marco Vaccaro and Dr. K H Wang.

University of Houston

Abstract

Natural gas is one of the most important energy sources. Due to its large reserves, there is enough supply to last over two hundred years for human consumption. Due to supply constraints and location, natural gas is often transported as liquefied natural gas (LNG). Qatar first started exporting LNG to Spain in 1997 and became the world's largest LNG exporter within a decade. This amazing growth is due to the development of the North Field which is the world's largest non-associated gas field. Qatar is now exporting LNG all over the world and represents over 30% of total LNG exporting. As the world energy demand expands, Qatar will continue to play a major role in LNG exports.

This paper will cover the following:

1) An introduction and background on Qatar
2) Specifics on Qatar's natural gas reserves and production
3) General information on Liquefied Natural Gas
4) Details on Liquefied Natural Gas in Qatar
5) Future and conclusions of Liquefied Natural Gas in Qatar

By J C Kuo, Hao Dong, Marco Vaccaro and Dr. K H Wang.

University of Houston

(1) Introduction to Qatar

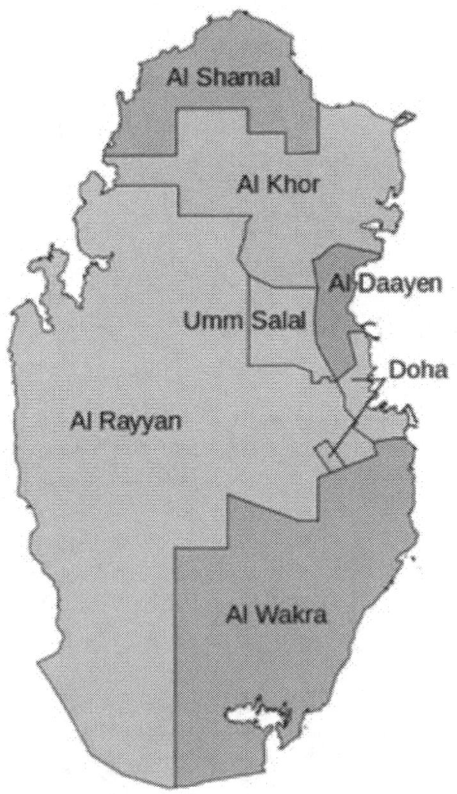

Figure 1 Qatar Municipality Map after 2004 [4]

Qatar is located in the Middle East. Qatar's sole land border is with Saudi Arabia, and the rest of the territory is surrounded by the Persian Gulf. The country spans 4,416 square miles with a population of 1,759,000. Although this population is small it is very diverse. The Qatari ethnic groups include: Arabic 40%, Indian 20%, Nepali 13%, Filipino 10%, Pakistani 7%, Sri Lankan 5%, and other 5%. Key languages spike in Qatar are: Arabic, English, Malayalam, Hindi, Tamil, Kannada, Tagalog, Urdu, and Punjabi. Three quarters of the country is Muslim while other religions make up the final 25%. The geography in Qatar is mostly desert, where water is scarce and agriculture is minimal. [2]

Qatar was once controlled by the Sheikhs of Bahrain until 1867. The country has been ruled by the Al Thani family since the mid-nineteenth century. Qatar was a British protectorate noted mainly for pearling and fishing in 1916. Once oil was discovered in the 1940's, it became one of the region's wealthiest states due to its enormous oil and natural gas production. Qatar became independent in 1971. Since then Qatar has built intimate military ties with the United States during the Persian Gulf War in 1992, and signed a defense pact with United States in 1994. [3]

Qatar now has commercial ties with the United States. Historical trade volumes have grown from $738 million in 2003 to $3.2 billion in 2009. Over the same period, the U.S has become one of the largest import partners for Qatar. Qatar attracted an estimated $100 billion in energy investment, approximately $60 to $70 billion coming from the United States.

Qatar's government is an elected, unicameral, federal republic. It is divided into seven municipalities, as shown by Figure 1. Qatar is a member of the Organization of the Petroleum Exporting Countries (OPEC), Gulf Cooperation Council (GCC), and the Oil & League. 60% of Qatar's gross domestic product is from oil and natural gas revenues. Qatar's GDP is $172.98 billion and has the world's largest GDP per capita at $109,000. In 2011, Qatar consumed approximately 147,000 bbl/d of oil equivalent, 79% of this total was attributable to natural gas. [4]

Qatar Petroleum (QP) controls all aspects of Qatar's oil sector, including exploration, production, refining, transport, and storage. It also holds the rights to all petroleum resources in the Qatari territories. QP often enlists foreign company involvement such from ExxonMobil, Chevron, Shell, and Total.

Qatar has the world's largest and second largest natural gas companies, Qatargas and Rasgas. Both are located in Doha, the capital of Qatar. Doha is a very large and developed city. Doha contains more than 60% of the Qatar's population. Due to the overcrowding in Doha, Qatar is building a new city nearby named Lusail.

(2) Qatar's natural gas reserves and production

The Persian Gulf's coastal areas are the world largest single source of crude oil. The Safaniyi Oil Field and the North field represent the world's largest offshore oilfield and non-associated natural gas field respectively..

Qatar has proved natural gas reserves of 896 Tcf, which represents 14% of the world's natural gas reserves. Qatar's natural gas reserves rank third behind Russia and Iran, Figure 2. Qatar produces 4.12 Tcf/y, of which it exports 3.35 Tcf/y.

Qatar is now exporting nature gas all over the world. This includes Asia, Europe, South and North America. Because of the development of the North Field, Qatar became the world's largest LNG exporter in 2006. Qatar exported over 80 MTPA of LNG in 2011, around 70% of which was LNG. Qatar currently exports 2 Bcf/d of natural gas to UAE and Oman through the Dolphin pipeline, which designed capacity, is 3.2 Bcf/d. [5]

Qatar also utilizes gas-to-liquid facilities for some of its excess gas production. Two world class GTL facilities include the Oryx GTL plant and the Pearl GTL project which is the largest GTL plant in the world . Qatar is one of the only three countries along with South Africa, Malaysia to have operational GTL facilities. Pearl is the largest GTL plant in the world and the first integrated GTL operation.

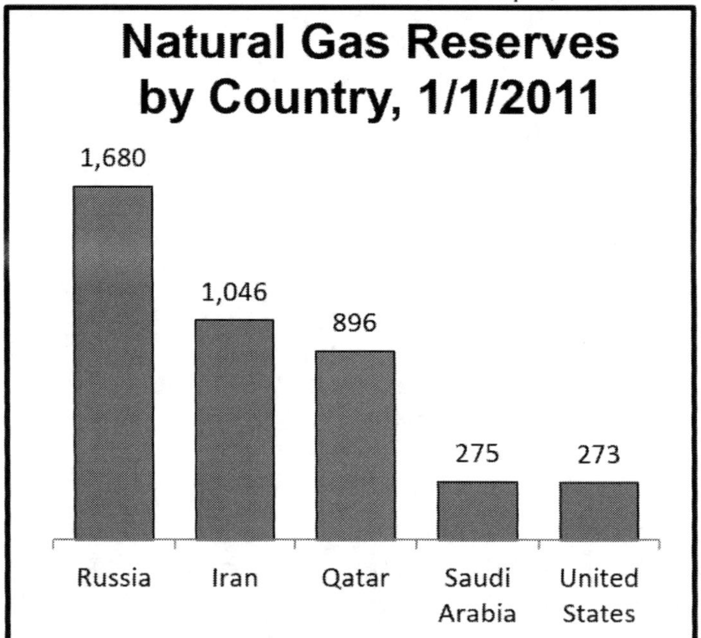

Figure 2 World Natural Gas Reserves TCF, by Country, 2011 [6]

(3) Introduction to Liquefied Natural Gas

The demand for natural gas has risen by 28% from 2001 to 2011.[7] However, this demand is not being met by supply in some remote parts of the world. This is mainly due to the difficulties of transporting natural gas through terrains such as mountains, and oceans. Liquefied Natural Gas provides a solution to this problem by significantly reducing the volume of gas and using tankers for transport. Along with volume reductions LNG provides a much purer gas, consistent of more methane than typical natural gas. LNG capital costs are significant, however shipping costs are very competitive compared to other forms of gas transport.

The LNG process includes liquefaction, transport and regasification. Liquefaction consists of cleaning upstream gas through field facilities, followed by cooling the gas to -160°C. Gas liquefaction is assisted with refrigerants that match the cooling curves of the gas. There are many different methods used for gas liquefaction. Today commercial liquefaction is dominated by only a few licensed technologies. The Propane-MR (PMR) and AP-X of Air Products (APCI) process, and its derivatives, accounts for about 75% of the market share. The Optimized Cascade (OC) process of

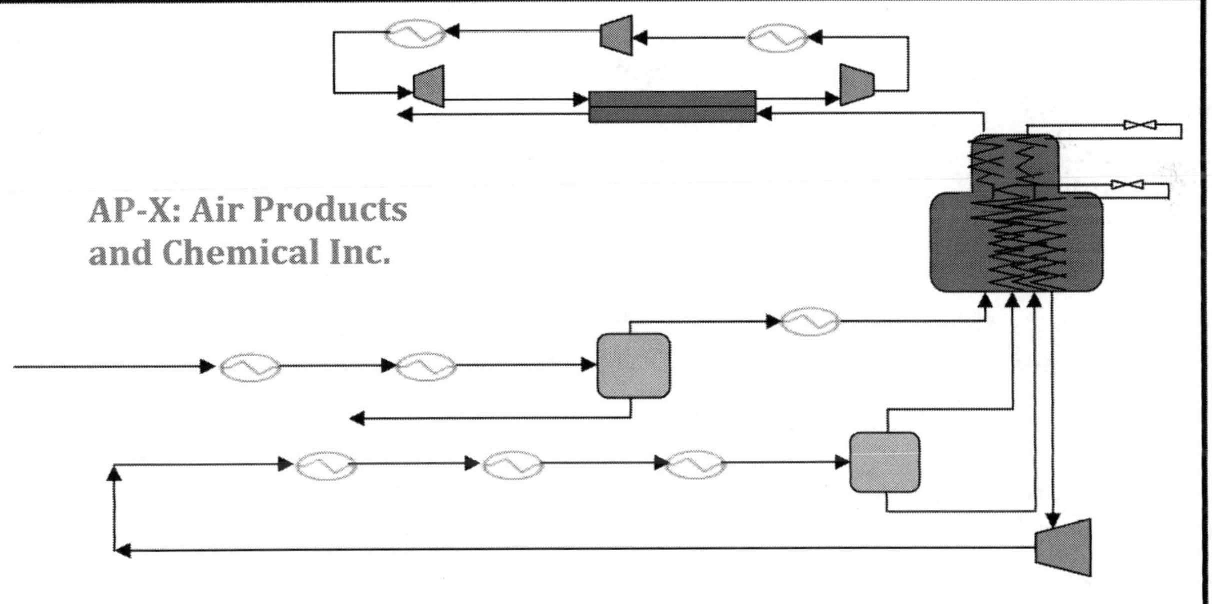

Figure 3 The AP-X Process of Air Products [17]

ConocoPhillips

(COP) represents another 20%. The other processes, such as the Multi-Fluid Cascade (MFC) of Linde and Single Mixed Refrigerant (SMR), make up the balance.[11] All of these processes reduce 54.6 Mscf of natural gas to one metric ton of LNG. This is a 600:1 volume reduction. The Optimized Cascade (OC) process of ConocoPhillips (COP) is shown in Figure 4. Mathematical examples of the cooling process are

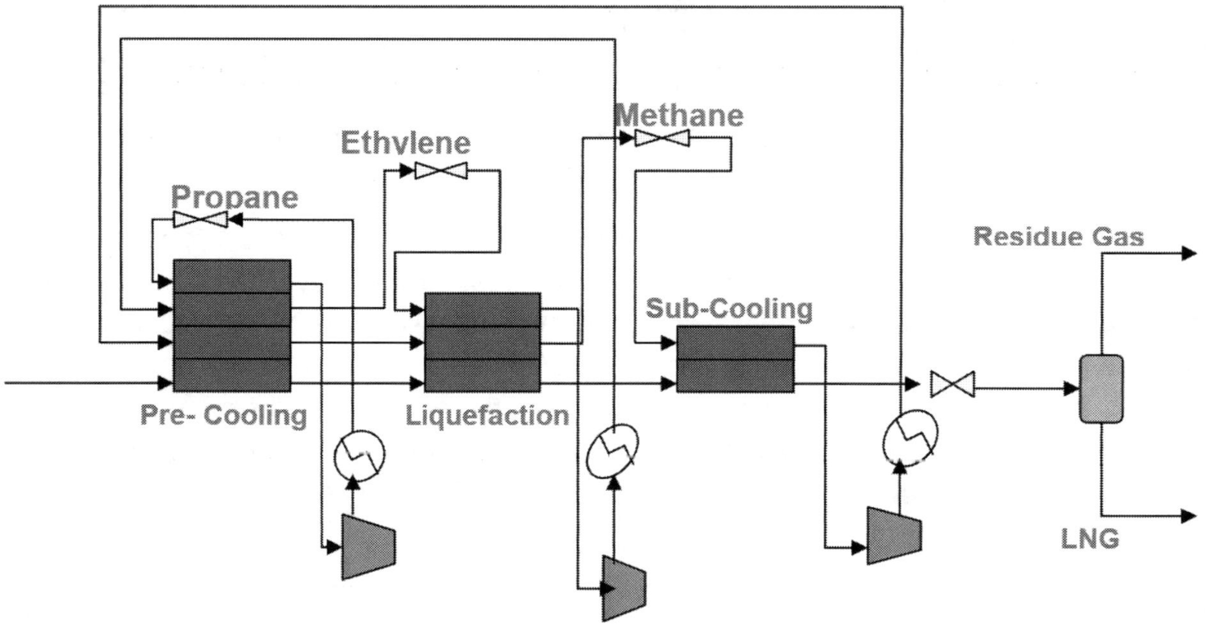

Figure 4 The Optimized Cascade (OC) Process of ConocoPhillips [17]

demonstrated in Wang, X., and Economides, M.J.: "Advanced Natural Gas Engineering".[1]

LNG train capacity has been increasing over the past thirty years in order to capture economies of scale. The capacity of an LNG plant is determined by many factors. The first consideration is the size of the gas reservoir which can support several large LNG trains for a reasonable time period. The second consideration is the technology limitations such as the size of gas turbine, compressors and piping. The third consideration is the marketability of the product.

LNG transport typically involves large carriers containing containment systems. As with the liquefaction process, there are many different LNG carrier designs. These carriers contain spherical or membrane type holding systems. Carriers allow for remote sea travel which eliminates geographical issues to natural gas transport.

Gas regasification begins once the LNG is offloaded from the carrier. Heat is added to the LNG to get an initial fraction of natural gas. Once all of the gas is in its natural form, it is odorized and sent to gas pipelines.

Natural gas liquefaction was developed by Michael Faraday in 1873 and by 1912 the first liquefied natural gas plant was built in West Virginia. The first LNG tanker was built from a converted World War II freighter in 1959. The United Kingdom became the first LNG importer when they started trading with Algeria in 1964.

As of 2011, there are 96 liquefaction trains and 89 regasification terminals around the world. This represents 279 MTPA in liquefaction capacity and 608 MTPA in regasification capacity. The global LNG carrier fleet currently consists of over 360 LNGC.[16] In total, 2011 LNG import/exports exceeded 240 MT. This was a 10% increase from 2010.[14] Recent increases in LNG demand were fueled by the severe earthquake and tsunami in Japan. A breakdown of the growth of LNG and exports by country is given below in figures 5, and 6.

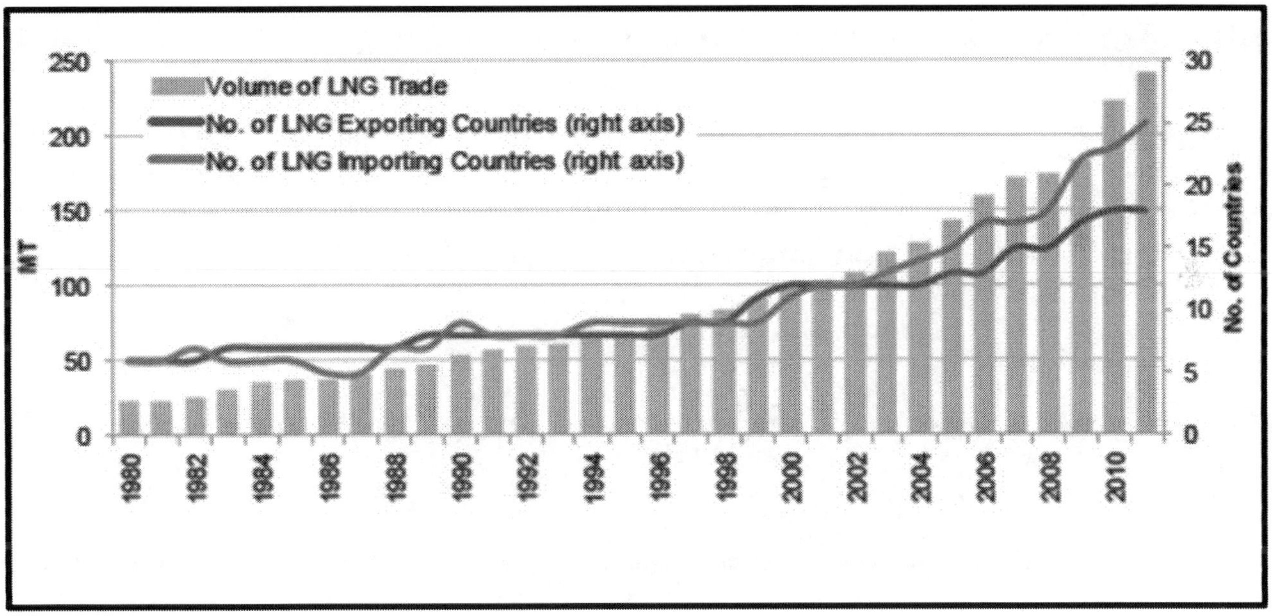

Australia will have signaficant growth in the near future due to the new projects such as Gorgon, Wheatstone and other LNG base load plants.

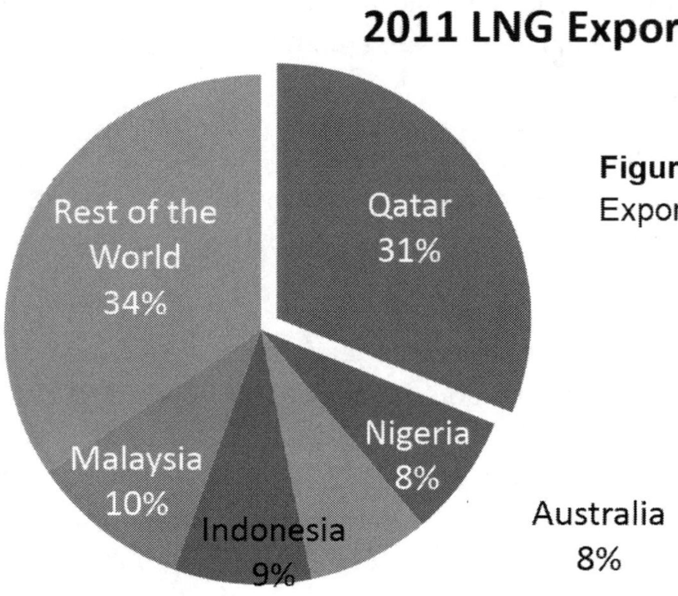

Figure 6 2011 LNG Exports by Country [7]

(4) Details on Liquefied Natural Gas in Qatar

LNG began in Qatar with the establishment of the Qatargas company in 1984. Qatargas was founded simply as an LNG company. By 1996, Qatargas produced and shipped its first supply of LNG to Japan. Qatargas has grown and into four separate

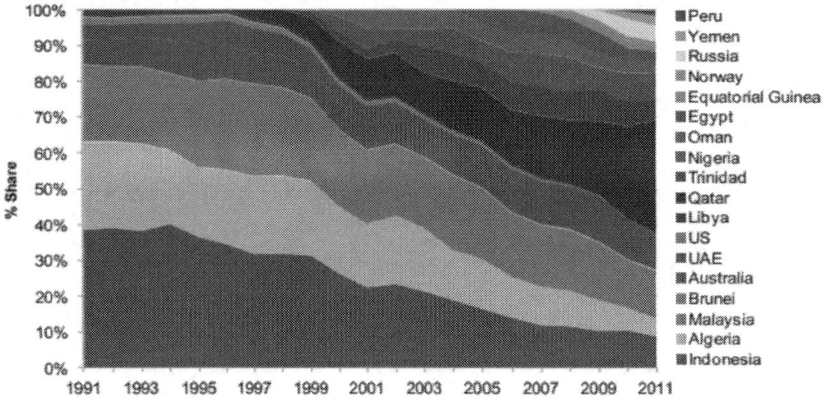

By J C Kuo, Hao Dong, Marco Vaccaro and Dr. K H Wang.

University of Houston

ventures since 1996 and now supplies 42 MTPA of LNG. Figure 7 shows a chart of the growth of LNG in Qatar and other countries since 1991. As you can see Qatar did not get into LNG until after other countries had shown success in the industry.
In Qatar, the typical LNG base load plant from feed gas to transportation is shown in Figure 8. Gas come from the offshore or onshore fields and then will go through Acid gas removal, Mercury removal, water removal/dehydration, heavy hydrocarbon removal, and liquefaction. Finally the LNG is put on a LNGC (LNG carrier) for ship transportation.

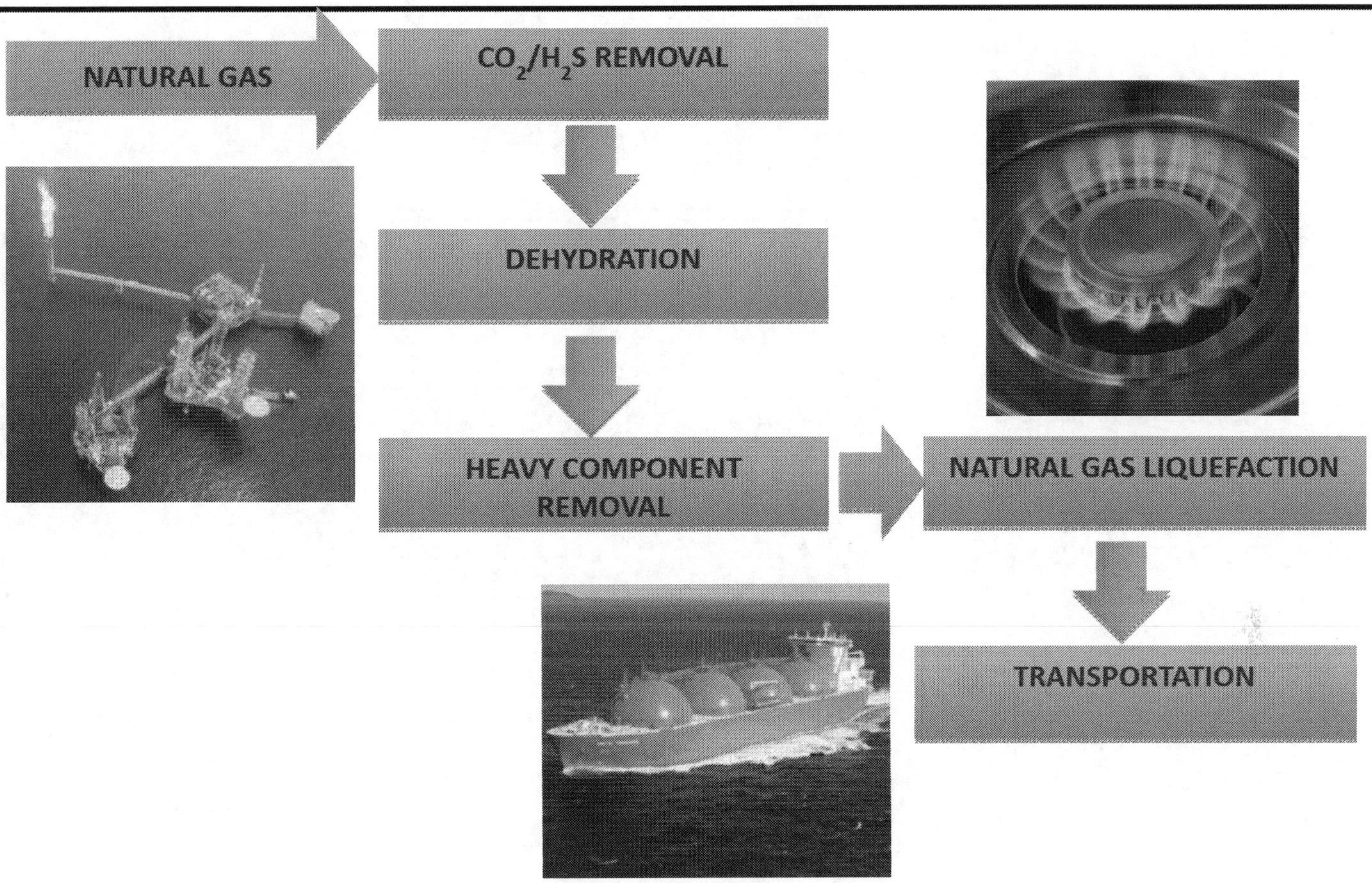

Figure 8 LNG from feed gas to transportation [17]

Qatar only uses APCI process trains. Qatar has 14 LNG trains, including six "mega-trains" that produce 7.8 MTPA. Trains 6 and 7 of Ras Laffan Liquefied Natural Gas Company (RasGas), and trains 4, 5, 6 and 7 of Qatar gas.

Qatar's LNG capacity has grown from zero to about 80 MTPA in less than 10 years. The offshore gas supply is the only bottle neck for increased production. Therefore, most of the LNG trains have pre-investment for more capacity. Debottlenecking has already been successful on Qatar gas trains 1, 2 and 3, where it has increased their combined capacity from 6.6 to 10.2 MTPA (54.5% increase).

Each mega-train will be able to give about 1.6 MTPA of additional capacity. Qatar could add 7.2 to 10.0+ MTPA at short notice due to their pre-investment in the early stage. Typically without major pre-investment, the maximum additional capacity is around 5-7+%.

Qatar's LNG Infrastructure, January 2011			
Unit	Liquefaction Capacity	Start-up	Primary Market(s)
RasGas Facilities			
Trains 1& 2	2 x 3.2 MMt (320 Bcf)	Aug. 1999	South Korea
Train 3	4.7 MMt (230 Bcf)	Feb. 2004	India
Train 4	4.7 MMt (230 Bcf)	Aug. 2005	Europe
Train 5	4.7 MMt (230 Bcf)	Nov. 2006	Europe & Asia
Train 6	7.8 MMt (380 Bcf)	Jul. 2009	Asia, Europe & N. America
Train 7	7.8 MMt (380 Bcf)	Feb. 2010	Asia, Europe & N. America
QatarGas Facilities			
Trains 1-3	3 x 3.2 MMt (480 Bcf)	Dec. 1996	Japan & Spain
Train 4	7.8 MMt (380 Bcf)	Apr. 2009	UK, Europe & Asia
Train 5	7.8 MMt (380 Bcf)	Sep. 2009	UK, Europe & Asia
Train 6	7.8 MMt (380 Bcf)	Nov. 2010	Asia & N. America
Train 7	7.8 MMt (380 Bcf)	Mar. 2011	Asia & N. America
Source: RasGas, Qatargas, media reports			

Table 1 LNG Trains in Qatar [4]

Before 2006, there were approximately 155 LNG carriers in operation with a total capacity of about 18 million cubic meters. In 2010, Qatar added another 54 ships. 9 of which are conventional LNGC (120 km3), 31 of these ships are Q-Flex (205 km3) and 14 are Q-Max (250 km3). This is a total added capacity of 11 million cubic meters.

The Q vessels are a new breed of LNGC that are being considered as an industry breakthrough in size and design. The tankers will enable more efficient transportation of natural gas throughout the world. The Q-Flex and the even larger Q-Max are a new generation of LNG mega ships with about 40% lower energy requirements and carbon emissions than conventional vessels due to the economies of scale created by their size and the efficiency of the engines.

Alongside excellent operational servicing, these vessels are also known for their excellent constructional features. Specifically, the usage of silicon ensures the protection and safety of the marine ecology. The price of a Q-Max increased from $300 million in 2005 to about $500 million currently because steel and raw material prices have risen.

(5) Future of LNG in Qatar and Conclusion

Qatar only has a population of 1.8 million including the expats that work there. This is only about 0.025% of the world's population. However, they have built enough LNG base load plants and LNG carriers to support about 7.2% of the world's total energy consumption. Furthermore, Qatar became the world's no. 1 LNG exporter in 2006. This was achieved in less than a 10 year time span.

Qatar's North Field, located in the Persian Gulf, is one of the largest non-associated natural gas fields in the world. This is a very good geographic petroleum location in terms of development and transportation. Qatar's natural gas reserves are no. 3 in the world and could supply LNG for about 225 years with current reserve and production. Russia's natural gas reserves are currently the largest in the world. However, many reservoirs are located in very severe climate conditions. Therefore, natural gas development cost is higher and scheduling takes longer. Iran has the second largest natural gas reserves. However, their current political standing and relationships with the United States and European countries is unsteady. The most dominate LNG technology providers, APCI (Air Products) and COP (ConocoPhillips), are both US companies.

Qatar has signed a defense pact with United States. They also have excellent relationship with the US and most industrialized countries. Furthermore they work with most of the major international oil companies. This includes ExxonMobil, Chevron, Shell, Total and ConocoPhillips in many projects. Typical engineering and construction of any major LNG based plant takes four to five years to complete. Additional trains will take one and a half to three years. Nevertheless, with excellent planning and the large oil companies' expertise, Qatar was able to build an LNG empire in a short period of time.

There are no plans to build new LNG facilities in the near-term future in Qatar. So any additional capacity increases will be the result of improvements to their existing facilities. Most LNG trains have pre-investment for additional capacity. Each mega-train will be able to give about 1.20 to 1.66+ MTPA of additional capacity. Qatar could add 7.2 to 10.0+ MTPA at short notice due to the pre-investment in the early stage.

Currently, Qatar exports LNG to numerous countries all over the world. This includes Spain, Turkey, Italy, U.S., France, South Korea, India, China, and the U.K. All of these countries share a strong relationship with Qatar. The European countries are dependent on LNG due to their location and their production of natural gas. The Asian countries are booming economically. This means that they will need lots of energy in the future.

As we can see from Table 2, Japan is the largest LNG importer in the world, but the country is much diversified in selection. Japan imports average 10% from each country such as Australia, Indonesia, Malaysia, Russia, and Qatar.

Korea is the second largest LNG importer in the world with more than 20% coming from Qatar. U.K., Spain, India, and Taiwan have a similar situation to Korea. They import a lot of LNG, but Qatar is already their major exporter.

China has grown by an average of ten percent per year over the past two decades, and became the second-largest economy and energy user in 2010 after the U.S. [18] China's bill for LNG has surged in the past four years as it feeds its booming economy and cuts reliance on coal for power generation. [19] China consumed 2,432 MT of energy in 2011 However, only 4% was from Natural gas compare to 26% in the U.S. Also, the nature gas price is high in China compared to the rest of the world. Average natural gas price in China is $282/T, which is about three times the U.S. price. Even though China is the fifth largest LNG importer, it has the most growth potential in the market. Qatar has all of the key essentials to meet this demand. This includes: reserves, management skill, relationships, hardware and track record.

Importer	LNG Import From Qatar, MT	Total LNG Import, MT	Percentage
Japan	7.91	78.8	10.04%
Korea	7.58	35.8	21.17%
UK	10.57	18.6	56.83%
Spain	4.20	17.1	24.56%
China	1.27	12.8	9.92%
India	8.05	12.7	63.39%
Taiwan	2.88	12.2	23.61%
France	1.77	10.7	16.54%
Italy	4.56	6.4	71.25%
US	0.94	5.9	15.93%
Turkey	1.46	4.6	31.74%
Belgium	4.51	4.5	100.22%
Argentina	0.18	3.2	5.63%
Mexico	0.72	2.9	24.83%
Chile	0.12	2.8	4.29%
Canada	0.18	2.4	7.50%
Kuwait	0.00	2.4	0.00%
Portugal	0.06	2.2	2.73%
UAE	0.12	1.2	10.00%
Greece	0.03	1	3.00%
Dom. Rep.	0.00	0.7	0.00%
Thailand	0.00	0.7	0.00%
Brazil	0.38	0.6	63.33%
Netherlands	0.00	0.6	0.00%
Puerto Rico	0.00	0.5	0.00%
Total	57.49	241.3	

Table 2 LNG Imports by Country, 2011 [9]

References

1. Wang, X., and Economides, M.J., "Advanced Natural Gas Engineering". Gulf Publishing Company (2009).
2. Central Intelligence Agency, World Wildlife Fund, US State Department, Qatar, January 8, 2012. <http://www.eoearth.org/article/Qatar?topic=49460>
3. Qatar, <http://www.infoplease.com/ipa/A0107901.html>
4. Wikipedia, Qatar, <http://en.wikipedia.org/wiki/Qatar>
5. EIA, Qatar Energy Data, Statistics and Analysis - Oil, Gas, Electricity, Coal, 2011
6. Oil & Gas Journal, Jan. 1, 2011 and EIA Natural Gas Navigator (U.S. only), Dec. 31, 2009
7. 2011 BP Statistical Study, BP, <http://www.bp.com/sectionbodycopy.do ?categoryId=7500&contentId=7068481>
8. Dominion, 2012, <https://www.dom.com/business/gas-transmission/cove-point/history-of-lng.jsp>
9. International Gas Union, PETRONAS, "World LNG Report 2011", (2012)
10. Saggas, <http://www.saggas.com/en/caracteristicas-tecnicas/proceso-de-regasificacion/>
11. Pek, J. J. B., van Driel, A., de Jong, E. C. J. N., Nagelvoort, R. K., "Large Capacity LNG Plant Development", paper presented at LNG 14, Doha, Qatar, March 21-24 (2004).
12. Avidan, A., Martinez, B., Varnell, W., "Study Evaluates Design Considerations of Larger, More Efficient Liquefaction Plants", Oil & Gas Journal, issue of August 23 (2003).
13. Bauer, H., "A Novel Concept", Hydrocarbon Engineering, issue of May (2002).
14. Dam, W., Ho, S.-M., "Unusual Design Considerations Driver Selection of Sakhalin LNG Plant Facilities", Oil & Gas Journal, issue of October 1 (2001).
15. Grootjans, H. F, Nagelvoort, K. N., Vink, K., J., "Liquefying a Stream enriched in Methane", U. S. Patent 6,370,910 B1 (2002).
16. Grootjans, H. F, "Compression Apparatus for Gaseous Refrigerant", U. S. Patent 6,637,238 B2 (2003).
17. Aduku, A., Harris, O., Rivera, V. Bagajewicz, M., "Evaluation of LNG Production Technologies", University of Oklahoma
18. Saltanat Berdikeeva, China Turns to Natural Gas to Fuel their Economic Growth, (2012) <http://oilprice.com/Energy/Natural-Gas/China-Turn-to-Natural-Gas-to-Fuel-their-Economic-Growth.html>

(26a) Sulphur Technology Tutorial – Conventional Wisdom and Innovative Frontiers

David Stevens, KPS Technology & Engineering, Overland Park, KS and **Angela Slavens**, WorleyParsons, Monrovia, CA

This tutorial will provide a brief overview of conventional sulphur recovery, tail gas treating and sulphur degassing technologies which are routinely employed to achieve environmental and operational requirements. The intent is to provide participants with a broad understanding of the types of processes which are available, and how they are typically applied, as a foundation for exploring possibilities for improvements upon these traditional practices.

Following the review of conventional technologies, an exploration of some recently-developed, innovative technologies will be undertaken. The technical and commercial merits of each new technology will be compared to conventional technologies, and the advantages / disadvantages will be explored. Examples of some of the new technologies to be presented include:

- SMARTSULF – this technology applies the sub-dewpoint principle, which is widely-known to increase sulphur recovery efficiencies in the SRU due to operation of the Claus reactors at lower temperatures, which increases the extent of the exothermic Claus process. The SMARTSULF technology further improves sulphur recovery through the use of internally-cooled reactors, which further reduce catalyst bed temperatures and therefore increase sulphur recovery efficiency.
- Low Temperature Tail Gas Oxidation Process – with this technology, the conventional fuel gas fired incineration process is omitted and indigenous hydrogen and carbon monoxide in SRU tail gas are consumed in a catalytic oxidation process, avoiding additional fuel consumption.
- In-Situ Sulfur Degassing Process – this technology removes the dissolved H2S present in Claus-produced sulfur within the sulfur condenser itself, as opposed to external, conventional degassing methods, which may require additional plot space, utility consumption and/or which may pose HSE concerns.
- RSR Production of Elemental Sulphur from SO2 – this technology converts post-combustion SO2 into an elemental sulphur product, rather requiring H2S as the feed to the sulphur production process. The technology combines two well-established processes: 1) reaction of CH4 and sulphur vapor to produce CS2, followed by 2) catalytic hydrolysis of CS2 to H2S, along with Claus reaction of H2S and SO2 to form elemental sulphur.

Successful Sulfur Unit Startups

Doug Cicerone
Cicerone & Associates, LLC
1302 Waugh Drive, Suite 248
Houston, TX 77019 USA
doug.cicerone@gmail.com

Prepared for Presentation at
American Institute of Chemical Engineers
2013 Spring Meeting
San Antonio, Texas
April 28-May 2, 2013

1. Introduction

The startup of any facility can have a big effect on the project or turnaround schedule and budget. As facilities become more integrated, the smooth startup of each unit within the facility becomes more important as it directly affects other units. Sulfur Recovery Units (SRUs) are often critical for the overall facility startup as their operation is a requirement from an environmental permitting standpoint.

This paper gives general guidelines and information for the safe and successful startup of an SRU. These guidelines do not supersede any site or vendor procedures or guidelines. Specific startup procedures must be developed for each facility prior to startup.

2. Sulfur Recovery

Although many sulfur recovery and tail gas cleaning technologies exist, only a standard straight-through Claus process will be discussed. (See reference [1] for more details on the sulfur recovery process.)

2.1 Process Chemistry

The modern Modified Claus process for the conversion of hydrogen sulfide (H_2S) to elemental sulfur (S) was first used in 1936. The process is divided into a high temperature thermal conversion section and a lower temperature catalytic conversion section. Although many reactions take place in the thermal section, the majority of the thermal conversion occurs according to the following two reactions:

$$\mathbf{H_2S + 3/2\ O_2 \rightarrow SO_2 + H_2O} \text{ [Eq. 1]}$$

$$\mathbf{2H_2S + SO_2 \rightarrow 3/x\ S_x + 2H_2O} \text{ [Eq. 2]}$$

In the catalytic section, the conversion of hydrogen sulfide occurs according to Equation 2 (also known as the Claus reaction) over a catalyst that is typically alumina or titania based.

2.2 Process Description

Figure 1 below shows a simplified process flow diagram for a typical sulfur recovery unit with two catalyst beds. The H_2S bearing feed gas (acid gas) flows to a Reaction Furnace where it is combusted with air supplied by a Combustion Air Blower. The air is metered such that one-third of the H_2S is burned to sulfur dioxide (SO_2). The unburned H_2S reacts with the SO_2 according to Equation 2 and forms elemental sulfur. Temperatures in the Reaction Furnace range from 1,500°F to 2,500°F depending on acid gas composition. Heat is recovered in the Waste Heat Boiler, typically by generating high pressure steam.

Figure 1: Sulfur Recovery Unit Simplified Process Flow Diagram.

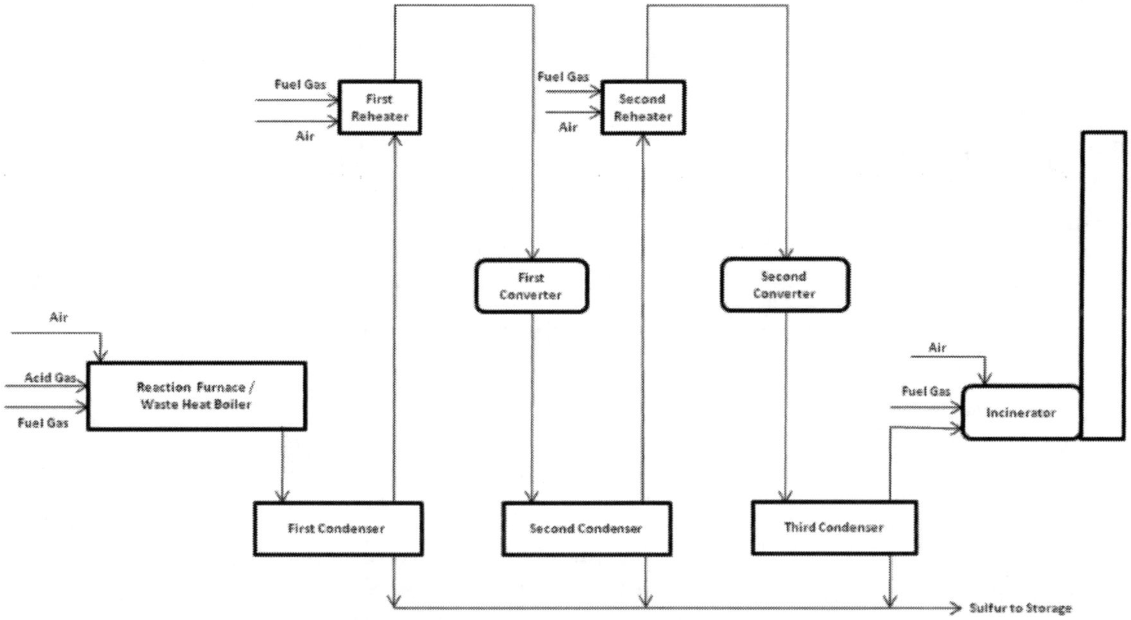

The process gas flows to a Condenser where the sulfur vapor is condensed to liquid sulfur, which is sent to storage. Heat is recovered in the Condenser, typically by generating low pressure steam. The process gas exits the condenser at its sulfur dewpoint and passes through a Reheater. This Reheater raises the process gas temperature above the dewpoint to prevent sulfur from condensing on the downstream catalyst. Sulfur condensation causes a reduction in conversion by blocking active catalyst sites. The heat for the Reheater is typically supplied by high pressure steam or by the direct firing of fuel gas or acid gas.

The heated process gas flows to the Converter where Equation 2 takes place across a catalyst bed. Temperatures in the Converter are typically in the range of 525°F to 625°F. The sulfur vapor is condensed in the downstream Condenser, and the Reheat, Conversion and Condensing steps are repeated.

The process gas from the Final Condenser (tail gas) flows to the Incinerator where any remaining H_2S (along with any other reduced sulfur compounds) is oxidized to SO_2 before exiting to the atmosphere through a stack.

2.3 Process Safety

Note that this section is not in any way meant to be a discussion of all SRU safety concerns, and is only meant to give the background necessary for the remainder of the paper.

The principle safety concern in a sulfur recovery unit is personnel exposure to and environmental release of H_2S. The Threshold Limit Value (TLV) for H_2S is 10 ppm, and it is lethal in concentrations of 1,000 ppm and above. Although sulfur recovery units operate at low pressures (5 to 15 psig), H_2S concentrations are very high. Typical concentrations in the front end of an SRU are in the range of 50% to 90%, and in the back end of the unit they range from 1% to 2%. Consequently, any loss of containment of the process gas has the potential to cause injury or death.

Wind socks should be placed at high points throughout the unit to indicate wind direction. All personnel in an operating unit must be aware of the wind direction. If an H_2S release occurs, the plant evacuation should proceed upwind or crosswind. Stationary and personal H_2S monitors should also be used.

Burns are another safety concern as liquid sulfur is in the temperature range of 300°F to 350°F. Special care must be taken to prevent burns when checking the sulfur lookboxes.

3. Startup Planning

As early as possible during the detail design phase of a project, a Startup Plan should be developed. This plan should be refined as the project progresses through construction and approaches startup. The Startup Plan becomes the road map for turning the plant over from Construction to Operations. It should address issues such as:

- Organizational Chart
- Matrix of Responsibilities
- Overall Schedule
- Budget
- Spare Parts
- Acceptance criteria for mechanical and electrical equipment
- Order in which utility and process systems will be completed mechanically
- Specific procedures for line flushing, steam bows, equipment run-ins, refractory dryout, etc.

This plan must be agreed upon by all parties involved prior to Startup activities. See reference [2] for detailed information regarding startup planning.

4. Precommissioning

Precommissioning is the project phase that occurs between construction and startup. During this period the equipment and piping are cleaned of the greases, dirt and debris that accumulate during manufacturing and construction.

The overall smoothness and success of a startup is almost always a direct result of the care taken by the commissioning team during precommissioning. While it can be very tempting to relax standards and cut corners in order to compress the schedule in the short term, these compromises will often have the opposite effect on the schedule in the long term.

For example, a proper instrument loop check includes the testing of all aspects of the loop, including controller action, valve stroke and DCS display. It is easier and takes less time to check only the point-to-point continuity of the control loop, which only ensures that the right instruments are wired correctly in the loop. This leaves the bulk of the instrument troubleshooting to occur during startup which can cause delays.

The following sections list items to consider with regards to the precommissioning of equipment, piping and instrumentation. Typically experienced operations, maintenance and engineering personnel are required to sign off on all items prior to the plant being deemed "ready for startup."

4.1 Piping & Equipment

General Considerations:

- Insides of all vessels cleaned of rust, dirt and debris.
- All vessel internals installed.
- Proper gaskets installed at flanges (and spare gaskets in warehouse).
- All fired equipment refractory installed and dust removed.
- All vessel slide plates free of debris and anchor bolts are loose or tight as per mechanical drawings.
- All motors greased.
- All motors checked for proper rotation.
- All seals installed with proper seal fluid.
- All lubricating systems full of oil.
- All temporary and/or permanent strainers installed.
- All compressor and pump alignments checked.
- Filter elements installed (and spares in warehouse).

Special Considerations:

- Ensure the manufacturer's recommended spare parts for all rotating and fired equipment are in good condition and available in the warehouse. This is particularly important if the site is in a remote location.
- When flushing or blowing pipe and equipment, ensure control valves and regulators are protected by removing or isolating. Orifice plates should also be removed when flushing.
- All utility and process piping should be blown with air or nitrogen to remove dirt and debris. Blinds should be used to prevent the transfer of dirt from one section to another or to equipment.
- For the boiler feedwater and cooling water systems, water can be used to flush the piping.
- For the condensate systems, condensate should be back-flushed to the downstream side of the steam traps.
- For the steam systems, steam should be used to blow the lines (and special steam blow procedures may be necessary if the system includes steam turbines).
- The catalyst support grids and screens in the Converter vessels must be installed properly such that catalyst cannot slip between the grid/screen and vessel wall.
- The sulfur pit should be cleaned of all debris. Any standing water should be pumped out.

4.2 *Electrical & Instrumentation*

General Considerations:

- All switchgears and breakers tested.
- Heater settings verified for motors.
- Acceleration times set for motors.
- All transmitters calibrated.
- All control valves stroked.
- All orifice plates installed in proper orientation.
- All PSVs tested and installed with block valves car sealed or locked as necessary.
- All instrument loops properly checked.

Special Considerations:

- Machinery monitoring and anti-surge systems for air blowers and compressors must be set up, typically by the vendor.
- The Burner Management System for each piece of fired equipment must be functionally checked prior to the introduction of fuel gas. This means verifying each line of the Cause & Effect Diagram with all automated valves moving to their proper positions. Additionally, any communication between PLC systems and DCS systems must be fully checked.
- The flame scanners on the Reaction Furnace, Incinerator and any other fired equipment may require calibration as the burners are fired on fuel gas. They may need additional calibration when acid gas is introduced. Ensure the calibration procedures are understood and/or vendor support is available.

4.3 *Refractory Dryout & Boilout*

The Reaction Furnace, Incinerator and any fired heaters will contain refractory to protect the vessel shell from the heat of the flame. For small units, the refractory may be installed and dried in the shop. For larger units, the refractory is installed and dried in the field.

The dryout can be done with the Reaction Furnace burner or by a third party dryout company. The third parties use their own burners and normally exhaust out of a manway on the furnace or boiler. The advantage to using the Reaction Furnace burner is that it will allow for run time on the burner and Burner Management System before startup. The advantage to using the third party is that they can normally provide tighter temperature control.

The dryout procedure includes periods of heating up slowly, followed by periods of holding temperatures. These heat-up and hold periods allow water contained within the refractory to boil off without damaging the refractory. Heating up too quickly, or shortening the hold times, can cause refractory damage due to the rapid escape of the water.

The refractory vendor will supply the specific procedure for drying out their refractory. A typical procedure increases the temperature at 50°F per hour with hold points at 250°F, 500°F and 1,000°F, holding each temperature for one hour per inch of refractory. Cool down rates are typically 50°F to 100°F per hour. For example, a Reaction Furnace with 14 inches of refractory, the dryout will take about three days for the heat-up and hold points, and another three or four days to cool down to ambient. The cool down proceeds at the prescribed rate to the discharge temperature of the Combustion Air Blower, after which it cools down to ambient much more slowly. It is advisable to enter the vessels and inspect all of the refractory after dryout. This will allow for repairs if necessary.

If the Reaction Furnace burner is used for the refractory dryout, it is common to perform a boilout of the Waste Heat Boiler and Condensers at the same time. Otherwise the boilout is done just before startup. The boiler vendor will provide a detailed boilout procedure. It typically calls for a boilout solution of detergent and alkaline cleaning agents to be added to the boiler feedwater in the boilers. This boilout solution helps to clean any greases or oils that have accumulated during manufacturing and construction, and ensures the heat transfer surfaces are clean. This solution is flushed out with boiler feedwater during the boilout. It typically takes three to five volume change-outs to flush out the solution.

The refractory dryout of the Incinerator and fired Reheaters will follow the same basic procedure, although the hold and end temperatures will vary based on the normal operating temperature. Often the Converters have several inches of refractory for fire protection. This can be dried out as the Reaction Furnace is dried out, or as part of the startup.

4.4 Catalyst Loading

Catalyst is typically loaded as the last step of precommissioning in order to protect it from becoming fouled with dirt and debris from line blowing, and soot from the refractory dryout. There are usually at least two layers of catalyst, with a larger diameter support layer on the bottom, and smaller diameter catalyst balls on top of the support. Some catalyst systems use two types of catalyst, and some include a larger diameter hold-down layer on top. Prior to loading, markings should be made on the Converter vessel walls to show the heights of the layers. These markings are used when loading.

The catalyst vendor will supply a loading procedure. The procedures typically call for the use of a hopper and sock. The catalyst is dumped into the hopper, and flows through the sock into the Converter vessel. The goal is to limit the free-fall of the catalyst balls to prevent them from breaking apart.

Periodically during the loading process, the catalyst should be leveled inside the converter. Catalyst loading can create a lot of dust, so the vessel needs to be ventilated and workers in the vessel must wear proper respirators. Plywood can be used to facilitate walking on the catalyst. When the loading is complete, the catalyst bed should be a uniform level across the length and width of the vessel.

When the catalyst loading is complete, the beds will still contain dust generated during the loading process. This dust should be blown out of the unit prior to startup. This is normally done using the Reaction Furnace Air Blower to blow air through the unit. It is preferable to expel the dust from each converter upstream of its condenser. If this is not possible, then the dust can be blown through each condenser and exhausted through the rundown line. In this case, the sulfur seals must be blinded to ensure the dust does not enter them.

5. Startup

Once the piping and equipment cleaning, instrument checks, refractory dryout and catalyst loading have been completed, the unit is "ready for startup." At this point there are several final preparations that are necessary before the SRU can be heated and acid gas admitted.

5.1 Final Preparation

The following are items to consider while preparing the unit for Acid Gas.

- Ensure all firefighting and safety equipment (including breathing air and/or rescue packs) is in place and in good working condition.
- Heat up all steam lines draining condensate from boots and low points. Do not line up steam traps until the condensate drains clear.
- Warm up the Sulfur Pit steam coils 24 to 48 hours before startup in order to boil off any water that has collected in the pit.
- Warm up all steam jackets and steam tracing at least 24 hours before startup, and check all steam traps for proper operation.
- The fuel gas and acid gas systems should be purged up to the final block valves and have less than 0.5% oxygen (typically nitrogen is used to purge).
- The acid gas system from the battery limit to the Reaction Furnace should be leak checked at normal process pressure (normally around 15 psig). All flanges and instrument connections should be taped and checked with soapy water solution. The same leak check should be performed on the process gas pipework and equipment from the Reaction Furnace to the Incinerator.
- If seal pots are used, it is recommended that they be filled with sulfur prior to startup. Solid sulfur pellets can be used, which then melt when the seal pot jacket steam is turned on.
- Ensure all boilers are full of boiler feedwater and boiler feedwater treatment systems are ready for operation.
- Ensure the Burner Management System and all other Safety Systems are fully functional and all bypasses have been removed.
- If possible, analyze samples of the fuel gas and acid gas to determine the stoichiometric air ratios for each. For fuel gas the target is 95% to 98% of stoichiometric. For acid gas, the target is to burn 1/3 of the H_2S and all of the hydrocarbons (if any).
- Follow the vendor instructions for the startup of the Air Demand Analyzer (also known as the Tail Gas Analyzer). The analyzer sample systems are typically isolated until the unit operation has stabilized on acid gas.

5.2 Unit Heat-Up

Most SRU licensor and operating companies heat their SRUs above the acid dewpoint and sulfur melting temperature before admitting acid gas to the unit (all temperatures above 250°F is the normal target). The Incinerator is lit and heated according to the refractory vendor's schedule. Then the Reaction Furnace is fired with fuel gas to heat the furnace and to provide flow through the unit. The refractory vendor should supply a heat-up curve for normal startups after the refractory has been dried, and this curve should be followed. Heating the refractory too quickly can cause the inner layer(s) to expand more quickly than the outer layer(s) which can cause refractory damage.

Note that excess air can be used in the Reaction Furnace and fired Reheaters to control temperature ONLY during the initial startup or after a shutdown when the unit has been freed of sulfur and the equipment open to atmosphere. Once acid gas has been admitted to the unit, a layer of iron sulfide forms on the piping and equipment. Iron sulfide is pyrophoric and will easily catch fire if excess oxygen is present downstream of the Reaction Furnace.

As the Reaction Furnace heats up, the Reheaters are used to heat the catalyst beds. A heat-up rate of 50°F per hour is typical. All process temperatures should be brought to their normal operating temperature based on the Heat and Material Balance.

The site ports of all the fired equipment should be checked regularly during startup. The flame should be checked for stability and soot (which normally appears as white-hot specs in the flame), and the refractory within the viewing areas should be checked for cracks.

Most Reaction Furnaces have a tempering steam (or sometimes nitrogen) connection to help control the furnace temperature. Once all downstream process temperatures are above 250°F, the tempering steam should be started. Admitting steam when the rest of the piping and equipment is cold can cause corrosion from the condensation of the steam.

As the unit heats up, the level controls on the boilers can be put in automatic, along with the temperature controls of the Reheaters. The Reaction Furnace and Incinerator burner controls should also be put in automatic.

The final step before admitting acid gas is to perform a stoichiometric fire test in the Reaction Furnace. During this test, the air flow is adjusted until the process gas at the First Condenser outlet contains no oxygen. The resulting air to fuel gas ratio is used on subsequent startups to ensure oxygen does not escape the Reaction Furnace to cause fires in the downstream equipment. A portable oxygen analyzer can be used at the Condenser outlet to check the oxygen concentrations. Note that as the furnace approaches stoichiometric conditions, the furnace temperature will increase and tempering steam will be necessary to control the temperature to the normal operating range. Be sure the temperatures do not increase above the design temperature of the refractory.

Once all of the process temperatures are within their normal operating range, the Reaction Furnace is firing at a stoichiometric ratio and there is no oxygen measured downstream, the unit is "ready for acid gas."

5.3 *Admitting Acid Gas*

As Acid Gas is admitted to the plant, the air requirements in the furnace will change. If an acid gas sample has been analyzed, use the calculated air ratio. If not, use the Heat and Material Balance ratio. It can be helpful to create a chart with the fuel gas flow on axis, the acid gas flow on the other axis, and the calculated air flow required for the combined acid gas and fuel gas flows in the body of the chart.

Slowly admit acid gas to the Reaction Furnace, and increase the air flow by a corresponding amount, based on the air-to-acid gas ratio. Once the acid gas is above its minimum flow (above 50% of design flow is preferred), the fuel gas flow can be decreased and the air flow decreased correspondingly based on the air-to-fuel gas ratio. When the fuel gas is below its minimum flow, it can be stopped and the fuel gas system isolated. The tempering steam flow should be stopped at the same time.

Monitor all furnace and catalyst temperatures as the operation is switched from fuel gas to acid gas. Rapid increases in catalyst temperatures can indicate oxygen breakthrough from the furnace. In this case, reduce the air flow.

The Reaction Furnace flame will switch from blue or clear to a blue-green color when acid gas is admitted. The flame should be checked regularly to monitor its stability and to check for soot.

5.4 *Initial Operations*

Once the Reaction Furnace is operating on acid gas only and operation has stabilized, the process temperatures should be closely watched. The temperature differences in the catalyst beds should approach their normal operating ranges. The temperature rise in the catalyst beds is a direct indication of the amount of reaction that is taking place.

When the unit operation has stabilized, place the Air Demand Analyzer on-line and the remainder of the Reaction Furnace controls in automatic or cascade. Most SRUs run with a tail gas H_2S:SO_2 ratio of 2:1 or 4:1. A ratio of 2:1 maximizes the efficiency of the SRU, while 4:1 provides operating stability for a downstream Tail Gas Unit.

Watch the process pressure measured upstream of the Reaction Furnace, typically on the air line just upstream of the burner. If this backpressure starts to rise while the feed flows are constant, it is an indication that sulfur is collecting in at least one of the condensers. Check all of the steam traps associated with the liquid sulfur lines to ensure they are hot. Often the plug will happen at the last Condenser where sulfur flows are the lowest.

Other items to consider during initial operations:

- As the catalyst temperatures increase, sulfur should start flowing in the look boxes (or site glasses). These flows should be checked regularly.
- Some sulfur pit pumps have a minimum sulfur level requirement for operation. Be sure the sulfur pit is at or above this level before starting the pumps.

- Check all steam traps regularly, including any steam traps associated with Reheaters.
- Check all fired equipment site ports regularly.
- Blow down the boilers regularly.
- If catalyst temperatures rise quickly, indicating a sulfur fire, do not shut down the Reaction Furnace. Reduce the Reaction Furnace air flow until the temperatures start to decrease.

5.5 Restarts

There are two types of SRU restarts: hot and cold. A hot restart typically occurs soon after a shutdown, when Reaction Furnace temperatures are above 1,500°F and all downstream process temperatures are above 250°F. In this case, the Reaction Furnace is typically lit with fuel gas and acid gas is reintroduced as soon as possible.

Cold restarts happen after the unit has been shut down for at least several days, and temperatures have cooled to below the hot restart criteria. In this case the Reaction Furnace should be heated using the same schedule followed in section 5.2. Unlike the initial startup, however, excess air cannot be used during a cold restart because the piping and equipment will contain sulfur and iron sulfide. The stoichiometric air-to-fuel gas ratio determined during the initial startup must be used while firing fuel gas during a cold restart. Tempering steam is used to control furnace temperatures.

5.6 Hot Standby

Hot Standby is the practice of firing fuel gas in the Reaction Furnace to keep the furnace and downstream equipment hot and ready for acid gas. This can be done relatively safely if the fuel gas composition is stable. If the fuel gas composition changes frequently, it can be more challenging to maintain a stoichiometric air-to-fuel gas ratio.

In either case it is not recommended to keep the unit in Hot Standby for long periods of time because of the potential for the following to happen:

- Temperature spikes in the Reaction Furnace due to changes in fuel gas composition, air-to-fuel gas ratio or tempering steam flow.
- Fires in the catalyst bed due to oxygen breakthrough from the Reaction Furnace.
- Soot deposition on the catalyst from firing the Reaction Furnace substoichiometric, leading to catalyst deactivation.
- Corrosion in the Final Condenser from the condensation of the water from combustion in the Reaction Furnace. The Final Condenser is the coldest part of the plant, and the potential for condensation and corrosion is the greatest when the mass flow through the unit is low, such as during Hot Standby operation.

6. Recommendations

The following are some final recommendations for a smooth and successful startup:

- Take care during precommissioning to ensure all pipework and equipment is clean.
- Ensure all spare parts are available and in good working condition.
- Follow the refractory vendor's guidelines for refractory dryout and do not try to shorten the schedule.
- Use excess air in the Reaction Furnace only during the initial startup or after the unit has been swept of sulfur. Use the stoichiometric air-to-fuel gas ratio any other time fuel gas is fired in the Reaction Furnace with tempering steam to control the temperature.
- Monitor catalyst temperatures closely during the startup and initial operations. Reduce the air flow to the Reaction Furnace if the temperatures rise quickly.
- Check all steam traps regularly, particularly those associated with liquid sulfur lines.
- Minimize periods of Hot Standby operation in the Reaction Furnace.

7. References

[1] Engineering Data Book, 10th Edition, Gas Processors Suppliers Association, Tulsa, OK, 1994.

[2] Planning for Startup, Implementation Resource 121-2, Construction Industry Institute, Austin, TX, 1998.

Sulfur Pit Explosion Protection by Deflagration Venting

WorleyParsons

Bill DeWees, Angie Slavens, Shadi Al Adel, Vincent Yih, Dan Samani

Disclaimer

The authors have prepared this paper for the purpose of presenting available information on the subject matter and discussing and evaluating the advantages and possible disadvantages of utilizing deflagration vents on sulfur pits. Neither the authors nor WorleyParsons make any recommendations or representations as to the merits of deflagration vents, and hereby disclaim any responsibility or liability with respect to decisions with respect to use or non-use of such vents or any other use of the information contained in this paper.

Contents

Introduction

Liquid sulfur produced by the Claus process contains dissolved H_2S, which when liberated into the vapor space of its container can form a flammable mixture with air. Concrete, underground storage pits have been the industry standard for the collection, degassing and storage of elemental sulfur, produced in the Claus process, for many years. Most sulfur pits include features designed to prevent deflagrations using the principle of combustible concentration reduction, whereby the concentration of H_2S is maintained below the lower flammability limit (LFL) by dilution with sweep air drawn through the vapor space of the pit using a steam ejector. Backup ejectors and natural draft "chimneys" (steam jacketed vents to atmosphere), together with safety interlocks monitoring H_2S concentration, sweep air and/or motive steam flow, are commonly used features designed to ensure that sweep air keeps flowing at an adequate rate, and to alert operators if there is a deficiency.

Should all of these protective features fail, many sulfur pits also include deflagration vents (or "explosion hatches") designed to relieve pressure resulting from a deflagration, thereby preventing a full-blown explosion event that could compromise the structural integrity of the pit and injure personnel. NFPA 68 – Standard on Explosion Protection by Deflagration Venting, as well as some European standards, contain guidelines for designing deflagration vents. However, these standards can be challenging to interpret and apply. Something as seemingly straightforward as determining the strength of a concrete enclosure can be difficult. There are also different types of venting devices to choose from, ranging from explosion doors or hatches to rupture diaphragms, each with different efficiency, costs, operability and maintenance concerns.

Complicating matters even further is the fact that owners, operators and engineers don't always agree on the need for deflagration vents and often ask, "are these really required?" This straightforward question is not always simple to answer, as some consider the sweep air and monitoring features to be sufficient, while others consider deflagration vents to be essential, similar to a relief valve on a pressure vessel.

This paper discusses some of the arguments for and against deflagration vents. It also provides interpretation of the guidelines and design methods included in the applicable NFPA standards and reviews the different types of vent designs commonly used. Alternative sulfur storage designs that can eliminate the need for deflagration vents altogether are also explored.

Sulfur industry literature is noticeably silent on this subject. This paper seeks to raise the profile of this important issue within the sulfur industry.

Typical Sulfur Storage Pit Design and Safety Features

Liquid sulfur produced by the Claus process typically contains about 300 ppmw of dissolved hydrogen sulfide (H_2S) and hydrogen polysulfides (H_2S_x) [1]. Over time, polysulfides decompose to form H_2S and sulfur by Reaction 1 below. As the sulfur cools, either from heat lost to the surroundings or with the aid of a sulfur cooler, the equilibrium of Reaction 1 shifts to the right. H_2S is physically desorbed into the gas phase by Reaction 2, aided by any agitation that may take place in the rundown container.

$$H_2S_x \leftrightarrow H_2S(l) + S_{x-1} \qquad (1)$$

$$H_2S(l) \leftrightarrow H_2S(g) \qquad (2)$$

Degassing technologies of various types may be used to speed up the process by increasing the level of agitation, improving the efficiency of vapor/liquid contact and/or via the addition of a catalyst to promote Reaction 1. However, even without a special degassing process, H_2S will be released into the vapor space of the rundown container and, if not removed can accumulate and potentially form a flammable mixture with air. The lower and upper flammability limits (LFL and UFL) of H_2S in air at 135 °C (275 °F) are 3.3 mol% and 50.4 mol%, respectively [11, 12]. Therefore, the rundown storage system must be designed to avoid this potentially flammable mixture, as well as possible ignition sources, during normal operation and maintenance activities.

NFPA 69 – Standard on Explosion Prevention Systems, describes systems for the prevention of explosions in enclosures that contain flammable concentrations of gases, vapors, mists, dusts or hybrid mixtures. These systems fall into six general prevention or control methods:

1. **Deflagration prevention by oxidant concentration reduction**
2. **Deflagration prevention by combustible concentration reduction**
3. Deflagration prevention by hot particle detection and intervention systems
4. Deflagration control by suppression
5. Deflagration control by isolation
6. **Deflagration control by pressure containment**

As explained above, it is possible for the vapor space of a molten sulfur enclosure to contain an explosive mixture of H_2S and air; therefore, the prevention methods described in NFPA 69 should be applied. While methods 1, 2 and 6 are commonly employed in this service, the authors are not aware of any sulfur storage systems using methods 3, 4 or 5.

Method 1, deflagration prevention by oxidant concentration reduction, involves maintaining the oxidant concentration in the system low enough to prevent deflagration [NFPA 69 5.1.1]. This method has been employed in some sulfur plants using fuel gas, nitrogen or other inert gas as the purge gas [1] to maintain the oxidant concentration below the *Limiting Oxidant Concentration (LOC)*, below which a deflagration cannot occur [NFPA 69 3.3.23]. The primary drawback to this method is that iron sulfide is allowed to form and accumulate on steel surfaces when air is not continually present to oxidize it. Upon eventual contact with air (during maintenance, for example) the built-up, pyrophoric iron sulfide can spontaneously ignite, causing serious safety concerns [1].

Most sulfur pits utilize **method 2**, combustible concentration reduction, for the prevention of explosive mixtures within the vapor space. In this method, the concentration of H_2S is maintained below the LFL by dilution with ambient air. By maintaining an oxidizing atmosphere, the buildup of iron sulfide is prevented.

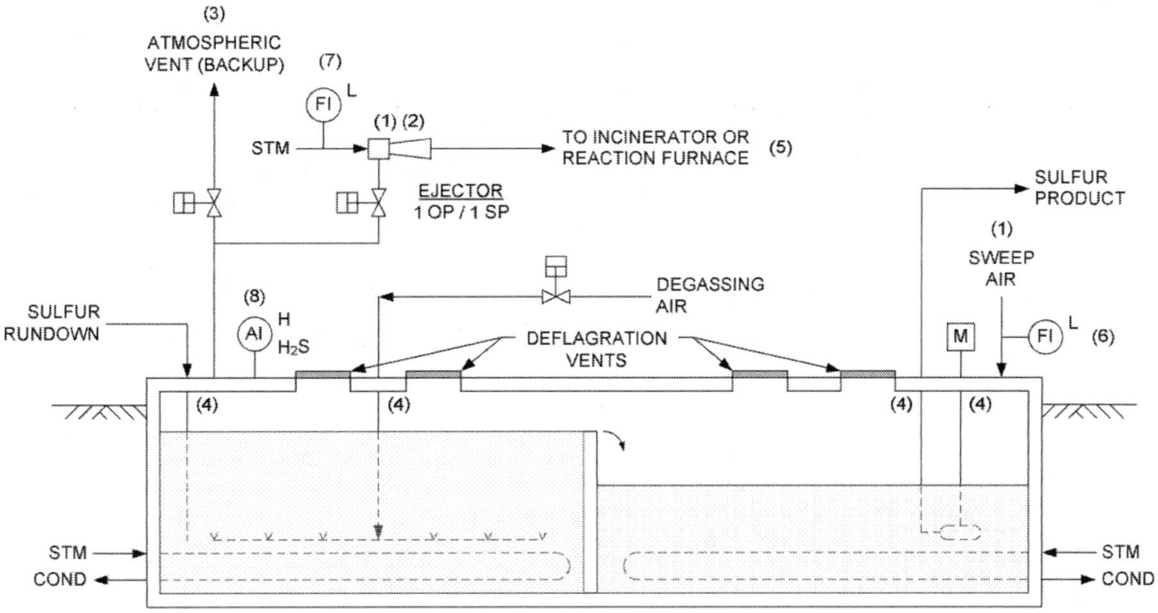

Figure 1 – Typical Sulfur Pit with Sweep Air

Figure 1 illustrates a typical scheme of this type, including the following features:

(1) A steam ejector located at one end of the pit is used to draw atmospheric air through air intakes at opposite corners of the pit, sweeping the entire vapor space and ensuring that no stagnant zones exist.

(2) A backup steam ejector is often installed which can either be started manually or automatically, in the event that the normally-operating ejector fails.

(3) A backup natural draft vent to atmosphere that can either be opened manually or automatically.

(4) Well sealed pit roof penetrations to prevent air ingress through these connections rather than through the dedicated air-intakes .

(5) In most systems, the sulfur pit ejector discharges the vent gas directly to the incinerator; however, tightening environmental regulations may force some systems to discharge the pit vent to the SRU Reaction Furnace for recovery of the H_2S and sulfur vapor in the stream. Implementation of a pit vent recycle scheme to the SRU Reaction Furnace requires additional controls and safeguards to prevent the backflow of hot furnace gas. Sulfur pit deflagration vents would not typically be designed to protect against this scenario because the carbon steel piping between the pit ejector and the reaction furnace would most likely fail before the furnace gases ever made it to the sulfur pit.

(6) The flow rate of sweep air is monitored, typically at the air inlet(s) rather than at the ejector suction or discharge. Sweep air flow measurement at the air inlet

location is selected for two reasons: 1) to ensure that sufficient air is entering through the air intake line rather than through pit penetration leaks, creating the opportunity for stagnant zones and inadequate sweeping despite achieving what appears to be an adequate pit sweep flow rate through the ejector, and 2) to avoid plugging the instrument taps with sulfur. A low air intake flow alarm alerts operators or trigger an automated response to start the backup ejector or open the backup atmospheric vent.

(7) The flow rate of ejector steam is monitored and a low flow alarm alerts operators or triggers an automated response similar to the low sweep air flow response.

(8) An H_2S analyzer monitors the vapor space and a high alarm alerts operators or triggers an automated response to shut-off the degassing air, if present, start the backup ejector or open the backup atmospheric vent. NFPA provides guidelines on acceptable H_2S concentration limits. The combustible gas concentration shall be maintained at or below 25% of the LFL (up to 60% of LFL with automatic safety interlocks) [NFPA 69 6.3.1]. Operations shall be discontinued whenever instruments show a combustible gas concentration of 35% or more of the LFL and should not be resumed until the instruments indicate a concentration of 15% or less of the LFL [NFPA 655 5.2.3 and 5.2.4].

The above safety features are aimed at preventing the possibility of forming an explosive mixture. For additional precaution, the following safety features are designed to prevent a source of ignition:

- The liquid sulfur inlets use a dip tube to deposit the sulfur below the liquid level and adequate grounding is provided to avoid static charge.
- The oxygen-rich environment prevents the buildup of pyrophoric iron sulfide.
- All instruments use explosion-proof housings in accordance with the electrical area classification.

Other typical safety features, such as monitoring the vapor space temperature to detect a fire, or providing snuffing steam to extinguish a fire, while important, do not specifically prevent deflagrations.

Protection against overpressure due to phenomena other than internal deflagrations (e.g. water leaks from heating coils or a sulfur cooler which may result in overpressure due to the rapid vaporization of water) are not addressed by NFPA 68.

Are Deflagration Vents Required?

If all of the protective features designed to prevent a deflagration from occurring should fail, many sulfur pits also include deflagration vents (or "explosion hatches") designed to relieve pressure resulting from a deflagration, thereby preventing a full-blown explosion event that could compromise the structural integrity of the pit and injure personnel.

Owners, operators and engineers don't always agree on the need for deflagration vents and often ask, "are these really required?" This straightforward question is not always simple to answer, as some consider the sweep air and monitoring features to be sufficient, while others consider deflagration vents to be essential, similar to a relief valve on a pressure vessel.

Sulfur Industry Literature Sources

While a great deal of information related to deflagration venting is available in general literature, it is surprising that sulfur industry literature is noticeably silent on this subject. Many papers have been written on safe sulfur storage practices for sulfur pits [1, 2, 3], but few mention the role of explosion hatches in the overall safety scheme of the sulfur pit. Those that do, mention them only briefly and don't discuss the specific details and challenges associated with their implementation in this service.

A literature search including Brimstone and Laurence Reid conference archives, ASRL Quarterly Bulletins, and other sulfur industry resources revealed only two references to explosion hatches specifically for sulfur storage. One was a slide presentation about safe sulfur storage in which a single bullet on a single slide said, "Include explosion hatches in design" [2]. The second was a Q & A style article about safe sulfur storage in which a question was asked, "Does anyone use explosion doors to minimize damage?" The answer was simply, "We have never used blow-off doors on pits" [3].

NFPA Standards

The National Fire Protection Association (NFPA) develops codes and standards intended to minimize the possibility and effects of fire and electrical risks in buildings, processing facilities and other industrial applications in the United States, as well as many other countries. These codes & standards may be adopted into law or adopted as design standards by private companies. NFPA is the recognized authority in the US for fire and explosion prevention.

The main NFPA standards applicable to deflagration prevention and control within sulfur storage enclosures are described below.

- NFPA 655 – *Standard for Prevention of Sulfur Fires and Explosions*, 2012 Edition
- NFPA 69 – *Standard on Explosion Prevention Systems*, 2002 Edition
- NFPA 68 – *Standard on Explosion Protection by Deflagration Venting*, 2007 Edition

NFPA 655

NFPA 655 applies to the handling of sulfur in any form, liquid or solid [NFPA 655 1.1.1]. Chapter 5, *Handling of Liquid Sulfur at Normal Handling Temperatures (246 °F to 309 °F)*, covers the "typical" deflagration prevention features of a sulfur pit, such as the need for an H_2S vent system, H_2S monitoring requirements, safe concentration limits of H_2S in the vapor space, avoiding static buildup/discharge, avoiding exposure of heating coils to air, plus other safety features unrelated to deflagration prevention. However, Chapter 5 does not require, or even mention, deflagration vents.

Chapter 6, *Handling of Liquid Sulfur and Sulfur Vapor at Temperatures above 309 °F*, explicitly requires that process equipment *shall be* provided with heated deflagration venting [NFPA 655 6.2.2.1]. This is perhaps because at temperatures above 309 °F, sulfur is nearer the *flash point* - the minimum temperature at which a liquid or solid emits vapor sufficient to form an ignitable mixture with air near the surface [NFPA 68 3.3.14]. Reported flash points of sulfur vary but are consistently greater than 309 °F. However, since most sulfur pits operate well below 309 °F, NFPA 655 does not provide a clear mandate for deflagration vents in normal service.

NFPA 69

NFPA 69 covers the design of the H_2S vent systems, which are required by NFPA 655 Chapter 5. As discussed previously, the most commonly used deflagration prevention method for sulfur pits is combustible concentration reduction with sweep air.

NFPA 69 also covers deflagration control by **method 6**, pressure containment, which may be an alternative to deflagration venting. It is only practical where a pressure vessel is used instead of a concrete pit. However, there is no mandate in NFPA 69 that deflagration vents or containment be provided in addition to a sweep air system.

NFPA 68

NFPA 68 applies to the design, location, installation, maintenance and use of devices and systems that vent the combustion gases and pressures resulting from a deflagration within an enclosure so that structural and mechanical damage is minimized [NFPA 68 1.1].

NFPA 68 only applies where the need for deflagration venting has already been established [NFPA 68 1.3]. It does not itself establish the need for deflagration vents. That responsibility is left to the authority having jurisdiction (AHJ), which is a phrase used throughout NFPA documents to refer to the organization, office, or individual responsible for approving equipment, materials, an installation, or a procedure. The AHJ could be a government agency, insurance agency or the property owner [NFPA 68 A.3.2.2].

European Standards

NFPA is not the only organization which addresses electrical and fire safety. The European Committee for Standardization (CEN) is a significant contributor to the development and implementation of European standards and technical specifications. The CEN's 33 national members work together to develop voluntary European Standards (ENs).

The main European standards that can be interpreted as being applicable to deflagration prevention and control within sulfur storage enclosures are:

- EN 1127-1:2007 – Explosive atmospheres - Explosion prevention and protection
- EN 14994:2007 – Gas Explosion Venting Protective Systems
- EN 14460:2006 – Explosion Resistant Equipment
- EN 14797:2006 – Explosion Venting Devices

The concepts in the European standards are similar to those in the NFPA standards but the specific requirements for their implementation and the formulas used for sizing, location, etc. are sometimes different. Nevertheless, one key consistency between the NFPA and CEN standards is that mandates for the installation of deflagration vents on sulfur pits are noticeably absent in each, despite that guidelines for their sizing and placement are provided.

In summary, there is no industry consensus with respect to the need for deflagration vents on sulfur storage enclosures, nor does there appear to be any definitive obligation to provide them in published safety standards. Therefore, the responsibility for deciding whether they are required belongs to the authority having jurisdiction, which is likely the facility owner in this case.

Making an Informed Decision

In order for the owner to make an informed decision about the need for deflagration vents, the topic should be investigated in-depth during the design phase of a project as part of the HAZOP review. However, it is possible that deflagration may not be properly considered during these safety reviews, as the vent devices may not always be shown prominently on P&IDs, which provide the primary basis for the HAZOP. They may be shown as combination maintenance/explosion hatch devices, which masks their importance as safety features. Furthermore, deflagration vent devices are rarely assigned tag numbers, possibly because they are considered an integral part of the concrete pit and so are designed by civil/structural engineers, who do not typically participate in HAZOP meetings.

The front-end designer typically does not have access to the details that will be developed in later phases of a project, such as sulfur pit detailed design calculations and construction techniques. Additionally, as mentioned previously, the owner's operating, maintenance and safety philosophies/procedures are also very important in determining whether the installation of deflagration vents should be considered, yet the front-end designer may not have full access to the specific owner's vigilance with respect to implementing and maintaining such practices. In the absence of these important details, in order to ensure a conservatively safe design, during the Basic Design stage of a project, deflagration vents may be suggested with a brief note on the sulfur pit data sheet; something like "Provide maintenance/explosion hatch. Size and number to be determined during detail engineering."

Some of the detailed design and/or operational factors that may make it necessary to consider sulfur pit deflagration vents include:

- Backup protection against deflagrations is needed due to inadequate safeguards to ensure the reliability of the sweep air system.
- Lower than normal sulfur pit strength due to design, construction materials, field installation procedures and/or soil conditions.

Some of the reasons for attempting to avoid installing sulfur pit deflagration vents include:

- In the event of a deflagration relief, toxic gas must be vented into the atmosphere, possibly putting personnel at risk of exposure to flame, heat and/or SO_2.

- The strength of the concrete pit may weaken over time to lower than the relieving pressure that the vents were originally sized for.
- Heated vents may be required to prevent plugging and/or maloperation.
- Onerous inspection and maintenance, of both the pit and the relieving devices, is required to ensure adequate protection as the facility ages.
- Depending on the strength of the pit, there may be insufficient area available on the pit roof for adequate venting.
- A sulfur pit equipped with deflagration vents may be more expensive, in terms of engineering and construction, than a sulfur pit without them.

Designing Deflagration Vents

If, in fact, it is concluded that deflagration vents are required, NFPA 68 provides guidelines for their design and implementation. The following sections discuss the basics of deflagration vent design, as described in NFPA 68. Many specific NFPA citations are provided, with accompanying interpretation which is specific to sulfur pit overpressure protection applications. The discussion and examples are intended to provide a general overview and point out some of the nuances that are specific to sulfur pits, but should not be considered a substitute for NFPA 68.

Fundamental Properties of Combustible Gases

The equations provided in NFPA 68 for sizing deflagration vents rely on two fundamental measured properties of the combustible gas: The *Deflagration Index for Gases*, K_G, and the *Fundamental Burning Velocity*, S_u.

The *Deflagration Index for Gases*, K_G, is determined by the maximum rate of pressure rise attained by combustion in a closed test vessel with volume, *V*, and is defined by the following equation:

$$K_G = \left(\frac{dP}{dt}\right)_{\max} \cdot V^{1/3}$$ **NFPA 68 Equation 6.1.1**

The value of K_G for H_2S is provided in NFPA 68, Annex E, Table E.1. The higher the value of K_G, the higher the rate of pressure rise upon deflagration of the particular gas. The value of K_G for H_2S in Table E.1 is 45 bar-m/sec. The gas with the next closest value of K_G in Table E.1 is CH_4 at 55 bar-m/sec.

The *Fundamental Burning Velocity*, S_u, is the rate of flame propagation, relative to the velocity of the unburned gas ahead of it, of a laminar flame under stated conditions of composition, temperature, and pressure of the unburned gas. The values of S_u for many gases have been measured and published in NFPA 68, Annex D, Table D.1(a). The value of S_u for H_2S is not provided in this table; however, the value of S_u for one flammable gas can be approximated from that of another flammable gas by the following equation:

$$(S_u)_2 = (S_u)_1 \cdot \frac{(K_G)_2}{(K_G)_1}$$ **NFPA 68 Equation E.1**

Equation E.1 produces the most accurate values where the two flammable gases have similar values of K_G [NFPA 68 E.1]; thus, CH_4 properties can be utilized to derive the value of S_u for H_2S. The value of S_u for CH_4 in Table D.1(a) is 40 cm/sec. Therefore, the value of S_u for H_2S calculated by equation E.1 is 33 cm/sec.

Gas Mixtures and Sulfur Mist

For the purpose of this paper, the flammable/explosive mixture is considered to be H_2S in air. The concentration of sulfur vapor in the vapor space is assumed to be negligible compared to the concentration of H_2S. Sulfur mist in the vapor space is also considered negligible, as would be the case without sulfur degassing inside the sulfur pit and/or when in-pit degassing is limited to air sparging from the bottom of the pit, below the liquid sulfur level. However, where a significant quantity of sulfur mist is present, which may be the case for some degassing processes where liquid sulfur is sprayed into the vapor space through nozzles, a different approach may be required. For combustible mists, the K_G for propane of 100 bar-m/sec or the equivalent S_u for propane of 46 cm/sec shall be used unless specific test data are available [NFPA 68 7.1.3]. It is observed that these values are substantially higher than those for H_2S, resulting in greater consequences, should a deflagration occur.

It may also be necessary, in some circumstances, to consider the case where a sulfur seal fails and process gas from the sulfur condenser flows into the pit. In this case, the combustible gas might be a mixture of H_2S, SO_2, CO_2, CO, COS, CS_2, H_2, N_2, H_2O, sulfur vapor and air. The mixture may or may not be flammable depending on the dilution effect of inert species and quantity of oxygen available.

Another scenario that may be considered is the presence of contaminants, such as unburned hydrocarbons, dissolved in the liquid sulfur, which may be subsequently released into the pit vapor space. Each credible scenario should be evaluated and the controlling case chosen for the design of the deflagration vents.

Where the hazard consists of a flammable gas mixture, the vent size shall be based on the K_G or S_u of the mixture [NFPA 68 6.2.1.1]. Where the gas mixture composition is not certain, the vent size shall be based on the component having the highest K_G or fundamental burning velocity [NFPA 6.2.1.2]. Hydrogen may present a serious concern in this case because it is a fast-burning gas with K_G of 550 bar-m/sec and S_u of 312 cm/sec. Fast-burning gas deflagrations can readily undergo transition to detonation, in which case NFPA recommends using alternative measures discussed in NFPA 69 [NPFA 68 A.7.1.1].

Effect of Turbulence

If the gas system is initially turbulent, or if internal structures promote turbulence, the rate of deflagration increases [NFPA A.7.5]. In most instances, the vapor space of the sulfur pit can reasonably be considered quiescent, or non-turbulent. However, for some upset scenarios that may be considered, such as failure of sulfur seals, turbulence may exist, hence requiring consideration in the deflagration calculations. To account for the effects of initial turbulence and internal structures for enclosures with initial pressures near atmospheric, the K_G for hydrogen (550 bar-m/sec) shall be used for gases that have values, in the quiescent state, that are close to or less than that of propane (100 bar-m/sec) [NFPA 7.5]. Thus, similar to the scenario where the presence of sulfur mist is considered, greater deflagration consequences would also be expected if turbulence was present in the sulfur pit vapor space.

Enclosure Design Pressure

NFPA 68 defines several terms related to the pressure inside on an enclosure, as follows:

- *Maximum Pressure* (P_{max}) is the maximum pressure developed in a contained deflagration of an optimum mixture.
- *Reduced Pressure* (P_{red}) is the maximum pressure developed in a vented enclosure during a vented deflagration.
- *Static Activation Pressure* (P_{stat}) is the pressure that activates a vent closure [NFPA 68 3.3.27].
- *Enclosure Strength* (P_{es}) is the static pressure that results in either the deformation (yield strength) or failure (ultimate strength) of the weakest structural component of an enclosure [NFPA 68 6.3.2].

The amount by which P_{red} exceeds P_{stat} is a complicated function of the rate of pressure development within the enclosure, vent size, and mass of the vent closure device. These relationships are illustrated in Figure 2. Where deflagration vent area to enclosure volume ratio is large, P_{red} approaches P_{stat} (C). As the vent area is reduced, P_{red} increases (B) and approaches P_{max} as the vent area goes to zero (A) [NFPA 68 A.6.3.1.1].

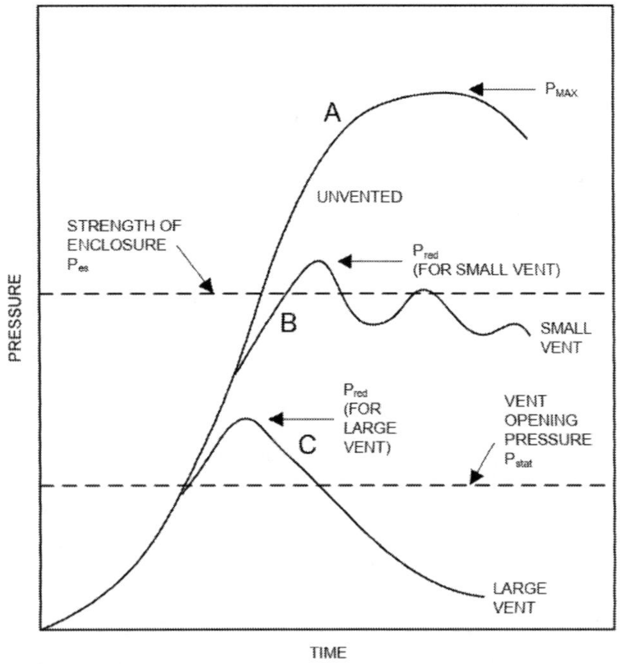

Figure 2 – Pressure-time characteristics of vented and unvented deflagrations from initially closed vessels illustrating the relationship of P_{max}, P_{red}, P_{stat}, and P_{es}.
(Adapted from [13])

As illustrated in Figure 2, the pressure from a vented deflagration rises to a maximum then decreases over a short time period after the vent opens. The load on the enclosure is therefore dynamic rather than static, causing the peak load to be amplified. The degree of amplification is called the *Dynamic Load Factor (DLF)* and depends on factors such as the duration of the load and the natural frequency of the system. In the absence of detailed structural response analysis, it shall be permitted to assume a worst-case value of $DLF = 1.5$ [NFPA 68 6.3.2.1]. Therefore, P_{red} shall not exceed two-thirds of the ultimate strength of the vented enclosure where deformation of the equipment can be tolerated, or yield strength where deformation cannot be tolerated [NFPA 68 6.3.1]. For a concrete sulfur pit, deformation cannot be tolerated, and therefore yield strength would apply.

$$P_{red} \leq \frac{P_{es}}{DLF}$$ **NFPA 68 6.3.2**

The structural design of the concrete pit and the design of deflagration vents are inseparable because the strength and geometry of the pit impacts the size and location of the vents. The following sections provide the methodology for determining the strength of the pit and corresponding venting requirements.

NFPA 68 provides different methods for sizing deflagration vents depending on the reduced pressure, P_{red}. Low-Strength Enclosures are defined as those that are capable of withstanding reduced pressures of not more than 0.1 barg (1.5 psig). High-Strength Enclosures are defined as those that are capable of withstanding reduced pressures from 0.1 barg (1.5 psig) up to 2 barg (29 psig).

In general, the higher the reduced pressure, the smaller the vent area required for both high-strength and low-strength enclosures. Therefore, there is incentive to design the pit for a high strength. However, at some point, the cost of the extra concrete and reinforcing steel that is required to reinforce the pit will eventually outweigh the benefits of smaller vent area.

Figure 3 plots reduced pressure, P_{red} on the x-axis against required deflagration vent area, A_v, on the y-axis for two sections of a hypothetical sulfur pit: a 30 ft wide x 30 ft long x 10 ft deep degassing section followed by a 30 ft wide x 80 ft long x 10 ft deep storage section. The plotted values were determined by calculation methodology provided in NFPA 68, which will be discussed in detail later in the paper. Figure 3 illustrates that for low strength enclosures, required vent area increases exponentially, as reduced pressure decreases. At the upper pressure limit of 1.5 psig for a low-strength enclosure, the deflagration vent area extends across 30-35% of the roof area. Increasing the reduced pressure to 3 psig (perhaps a practical maximum for a concrete sulfur pit), upgrades the pit to a high-strength enclosure, and allows the vent area to be reduced to about 12-17%.

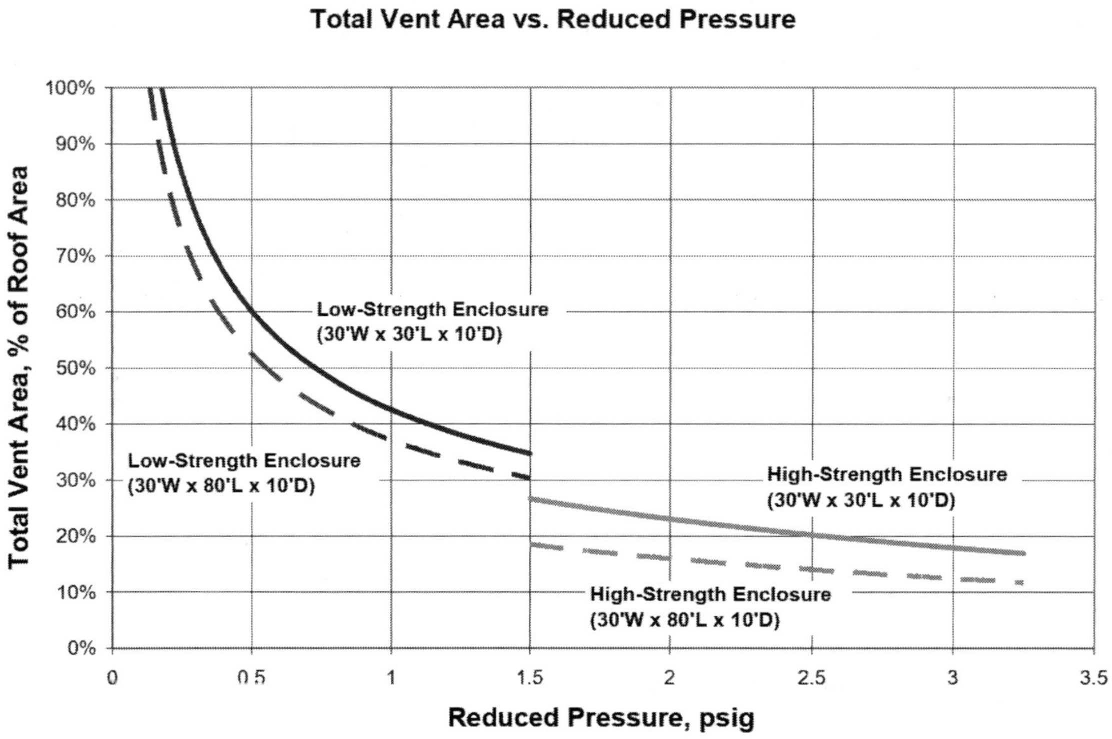

Figure 3 – Vent Area vs. Reduced Pressure

Figure 3 begs the question, "can the pit be designed to withstand a higher pressure without increasing the cost significantly?" The answer to this question is not simple because the strength depends on many factors. Pit geometry, design of the roof, number and size of roof penetrations, the type of roof-to-wall joint, air and soil temperatures and the internal reduced pressure, all contribute to the ultimate pit design strength.

In most cases, if a low-strength design is chosen, based on the geometry and configuration of the enclosure, the sizing of the walls and slabs are minimally impacted by the internal design pressure. This is due to the fact that the thermal stress caused by the temperature gradient between the hot molten sulfur side of the concrete and the colder soil side will govern the design. Referring back to Figure 3, recall that this option will require at least 30-35% of the roof area to be apportioned for deflagration venting devices. With so much open area, along with other roof penetration requirements, it is challenging to incorporate adequate support which will ensure high structural integrity. On the other hand, if a high-strength design is selected, the higher internal design pressure (in excess of 1.5 psig) will reduce the required venting area, but may have a significant impact on the sizing of the walls and slabs, and may result in additional concrete and reinforcing steel.

Unlike a vessel that is fabricated and tested in a shop, field fabrication of sulfur pits results in many uncertainties, such as quality, storage and preparation of materials, installation skill level of laborers, and weather conditions, to name a few, that can detrimentally impact the actual strength of the pit, regardless of its design strength. With so many variables in play, it is impossible to determine the optimum enclosure strength that works for every situation. Thus, specific calculations should be conducted for each project during detailed design. And, if necessary, calculations should be revised following construction, if it is suspected that field fabrication conditions may have contributed to a reduction in design pressure below the intended design value.

The calculation methodology which was used to derive the data included in Figure 3 is described in detail in the sections that follow.

Enclosure Length-to-Diameter Ratio (L/D)

The L/D ratio of the enclosure is an important consideration, especially for high-strength enclosures where it can have an impact on the vent area sizing calculation.

$$L/D = \frac{H}{D_{he}}$$ **NFPA 68 6.4.3.7**

where:

H = Maximum flame length (in meters) along which a flame can travel based on the maximum distance, taken along the central axis, from the farthest end of the enclosure to the opposite end of the vent [NFPA 68 6.4.3.2]. The effective volume, V_{eff}, is the volume of that part of the enclosure defined by H.

D_{he} = Effective hydraulic diameter (in meters) and is calculated by following equation:

$$D_{he} = 4 \cdot \frac{A_{eff}}{p}$$ **NFPA 68 6.4.3.6**

Hydraulic diameter is used for noncircular cross-sections, where A_{eff} is the cross-sectional area normal to the longitudinal axis of the space and p is the perimeter of the cross-section [NFPA 68 3.3.18]. See Figures 4, 5 & 6.

The most conservative method for calculating L/D is to determine H and V_{eff} based on the total enclosure, irrespective of vent location [NFPA 68 6.4.3.4], as illustrated in Figure 4. However, this method may result in a larger vent area than is actually required if L/D greater than 2.

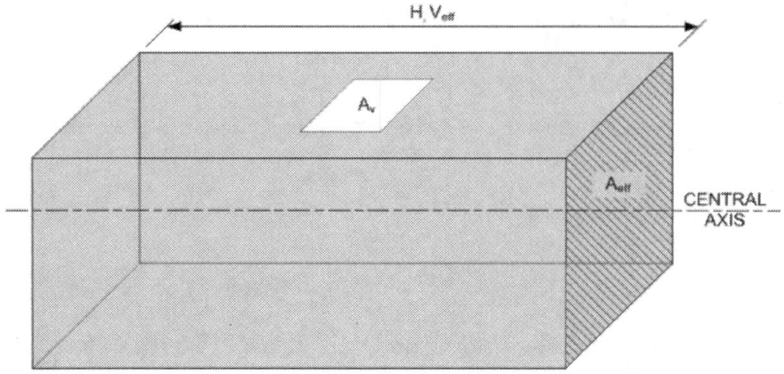

Figure 4 – *L/D* Method 1

When multiple vents are provided, a single value of *H*, and *L/D*, shall be permitted to be determined for the enclosure based on the farthest vent [NFPA 68 6.4.3.2.1], as shown in Figure 5.

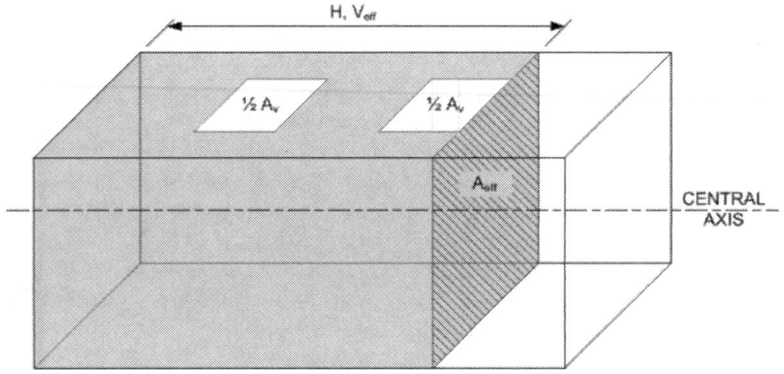

Figure 5 – *L/D* Method 2

The value of *H*, and *L/D*, can also be determined for each section using the maximum distance from the closest end of one vent to the opposite end of the next vent [NFPA 68 6.4.3.2.2], as illustrated in Figure 6.

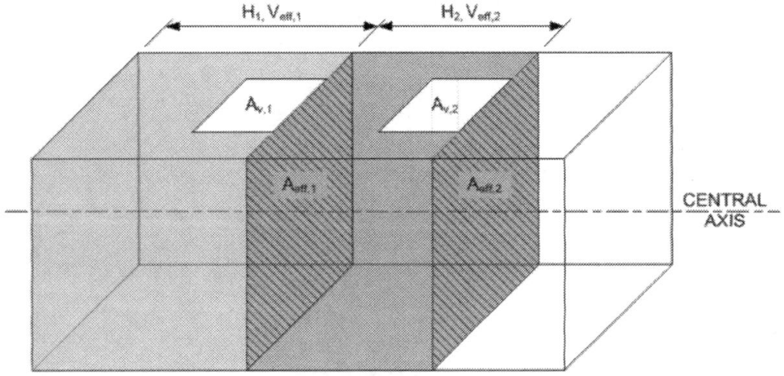

Figure 6 – *L/D* Method 3

Enclosure Internal Surface Area (A_s)

Internal surface area, A_s, should include the areas of the roof, floor and walls [NFPA 68 7.2.4.1]. Since the liquid sulfur in the bottom of the pit is incompressible, it acts like a "floor;" thus, the level will be a factor in determining the effective wall area. The choice of liquid level to use for the worst case scenario depends on the circumstances and should consider startup and shutdown as well as normal operation. The most conservative approach is to base the wall area on the empty pit dimensions [10].

Enclosure Compartments

If the pit is subdivided into separate compartments, each compartment shall be provided with its own vent(s) [NFPA 68 7.2.6.2]. This may be the case where separate degassing and storage compartments are provided, as shown in Figure 1. The wall between the two compartments extends from the pit floor to within 1-2 feet of the roof. While the vapor space of the two compartments is contiguous, the relatively small opening acts as a restriction to movement of expanding gases during a deflagration. NFPA 68 is not entirely clear if the two compartments must be treated separately in this specific case, but it seems prudent to do so. One reason is that this geometry may lead to a phenomenon called *pressure piling*, which can occur when a deflagration takes place in a vessel that is interconnected, through a pipe or duct, to another vessel, as illustrated in Figure 7. The pressure front accelerates through the restriction, causing it to move faster than the reaction front, which increases the initial pressure, P_o, ahead of it by about 8 x P_o. When the reaction front arrives, the final pressure becomes 8 x 8 x P_o = 64 P_o. The restriction also increases the combustion-generated turbulence, which amplifies the combustion properties of the gas. These phenomena may lead to a *deflagration-to-detonation transition* (DDT) if enough energy accumulates in the pressure wave so that the adiabatic compression of the gas leads to autoignition and initiation of a detonation [12]. Given these possibilities, it is recommended that each compartment be treated as a separate enclosure for the purpose of sizing the deflagration vents in that particular compartment.

When carrying out vent area calculations for an individual compartment, the designer must keep in mind that some of the calculations depend upon the longest dimension of the enclosure or the L/D of the entire pit. Thus, while the overall pit may be rectangular with the longest dimension in the direction of the x-axis, for example, an individual compartment may have its longest dimension in the direction of the y-axis.

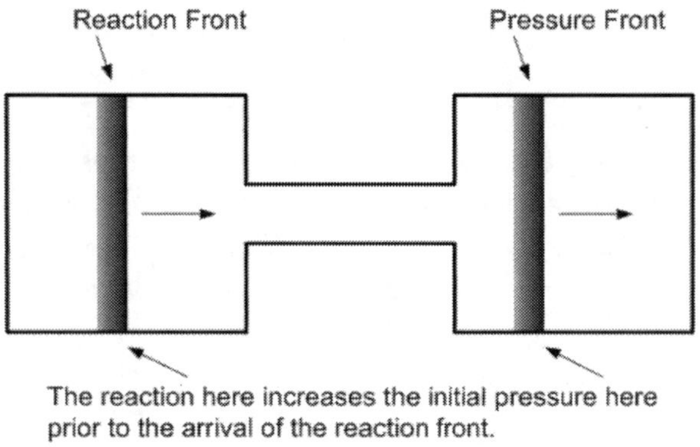

Figure 7 – Pressure Piling in Interconnected Vessels
(Adapted from [12])

Venting Area Requirement for Low-Strength Enclosures

The following equation determines the minimum venting area required for a low-strength enclosure [NFPA 68]:

$$A_v = \frac{C \cdot A_s}{P_{red}^{1/2}}$$ **NFPA 68 Equation 7.2.2**

where:

A_v = vent area (m^2)
C = venting parameter ($barg^{1/2}$)
A_S = internal surface area of enclosure (m^2)
P_{red} = maximum venting pressure (barg)

The venting parameter, *C,* is determined by the following equation for a fundamental burning velocity, S_u, less than 60 cm/s [NFPA 68]:

$$C = 1.57x10^{-5} \cdot (S_u)^2 + 1.57x10^{-4} \cdot (S_u) + 0.0109$$ **NFPA 68 Equation 7.2.2.1a**

Where:

S_u = fundamental burning velocity (cm/sec)

Note that for low-strength enclosures there are no dimensional restrictions on the enclosure or the placement of the vents as long as the vents are distributed as evenly as possible with respect to the longest dimension. If the vents are restricted to one end of the pit, the *L/D* ratio should not exceed 3 [NFPA 68 7.2.3.1].

It is observed that Equation 7.2.2 in NFPA 68 for venting gas or mist deflagration in low-strength enclosures is the same one used in EN 14491:2006 for venting dust deflagrations in buildings.

Venting Area Requirement for High-Strength Enclosures

For high-strength enclosures, the *L/D* ratio of the enclosure, and of the sections between vents, can have an impact on the vent sizing.

If *L/D* is 2 or less, the following equation determines the minimum venting area, A_v required:

$$A_v = \{(0.127\log_{10}(K_G) - 0.0567)\cdot P_{red}^{-0.582} + 0.175P_{red}^{-0.572}(P_{stat} - 0.1)\}\cdot V^{2/3}$$

NFPA 68 Equation 7.3.3.2

where:

K_G $\leq$ 550 bar-m/sec
P_{red} $\leq$ 2 barg and at least 0.05 barg > P_{stat}
P_{stat} $\leq$ 0.5 barg
V $\leq$ 1000 m^3
Initial pressure before ignition $\leq$ 0.2 barg

If *L/D* is 2-5, A_v shall be increased by adding more venting area, ΔA, which is calculated using the following equation:

$$\Delta A = \frac{A_v \cdot K_G \cdot (L/D - 2)^2}{750}$$

NFPA 68 Equation 7.3.3.3.1

Notice that the larger *L/D* becomes, the larger ΔA becomes, up to a maximum of 54% additional vent area for *L/D* = 5, in the case of H_2S. Therefore, when locating the vents along the major axis of the pit, as shown in Figure 6, it is best to limit the spacing so that the *L/D* ratio of each section is 2 or less, if possible, to avoid the need for extra vent area.

Effects of Vent Closure Panel Inertia

The free area of a vent does not become fully effective in relieving pressure until the vent closure moves completely out of the way of the vent opening. Until this occurs, the closure obstructs the combustion gases that are issuing from the vent [NFPA 68 A.6.7.4]. For a given vent area, the reduced pressure increases with the specific mass (kg/m^2) of the vent closure panel because heavy panels take longer to clear the vent opening than light panels do [NFPA A.6.7.2].

The specific mass of a vent panel, *M*, is the total mass of the closure (including counterweights or insulation) divided by the area of the vent opening [NFPA 68 6.7.1, 6.7.3 and EN 14797:2006 3.10]. The threshold mass, M_T, is the specific mass above which a correction to the vent opening area is required to account for inertia effects. Where *M* is less than M_T, no correction is required to the vent sizing equations given in NFPA 68 [NFPA 68 6.7.3]. Where M is greater than M_T, but less than 40 kg/m^2, the venting area must be increased by ΔA as shown below.

$$M_T = \left[6.67 \cdot \left(P_{red}^{0.2}\right) \cdot \left(n^{0.3}\right) \cdot \left(\frac{V}{K_G^{0.5}} \right) \right]^{1.67}$$ **NFPA 68 Equation 7.2.2.5.2**

where:

M_T = threshold mass (kg/m^2)
P_{red} is in barg
n = number of panels
V > 1 m^3

$$\Delta A = A_v \cdot (0.0075) \cdot M^{0.6} \cdot \frac{K_G^{0.5}}{n^{0.3} \cdot V \cdot P_{red}^{0.2}}$$ **NFPA 68 Equation 7.2.2.6**

where:

ΔA = additional vent area required (m^2)
M = mass of vent panel (kg/m^2)
A_v = vent area (m^2) calculated by Equation 7.2.2 or 7.3.3.2

The value of K_G used in equation 7.2.2.6 shall be 75 bar-m/sec, if K_G for a certain component is less than 75 bar-m/sec.

Where M is greater than 40 kg/m^2, it is necessary to perform testing to determine the efficiency of the vent closure or apply alternate explosion protection methods per NFPA 69, Standard on Explosion Prevention Systems [NFPA 68 A.7.2.2.5.2].

European Standards differ from NFPA 68 in the limits imposed on the value of M before it becomes necessary to perform testing.

$$\Delta A = A_v \cdot \left(\frac{1}{E_f} - 1\right)$$ EN 14994:2007 5.2 (2)

$$E_f = \frac{A_E}{A_d}$$ EN 14797:2006 7.3.4

where:

E_f is the dimensionless venting efficiency of the venting device
A_E is the effective vent area of the device, m^2
A_d is the actual vent area of the device, m^2

- E_f = 1 for explosion venting devices with M less than 0.5 kg/m^2
- E_f = 1 for explosion venting devices with M greater than 0.5 kg/m^2 and smaller or equal to 10 kg/m^2 provided $A_v/V^{0.753}$ < 0.07, where A_V is the vent area and V the vessel volume. This is valid for $P_{stat} \leq 0.1$ barg and 0.1 barg < P_{red} < 2 barg

- for all other conditions and for explosion venting devices with an M greater than 10 kg/m^2 the efficiency E_f has to be determined by tests.

In summary, NFPA says that vent closures with $M > 40$ kg/m^2 require testing to determine their efficiency while European Standards set the limit much lower at $M > 10$ kg/m^2.

Hazard Zone

NFPA 68 discusses the hazard zone around the discharge of deflagration vents in the context of protecting personnel from the hot gases and material that may be ejected. It does not address the potentially toxic cloud that may result from the combustion of H_2S and/or other sulfur species, in the case of a sulfur pit deflagration. The determination of the hazard zone area is important because it may have some bearing on the location of deflagration vents on the sulfur pit roof and/or may present a requirement for cordoning off certain areas around the pit to prevent regular personnel access.

The fireball from a vented gas or dust deflagration presents a hazard to personnel who may be in the vicinity. People caught in the flame itself will be at obvious risk from burns, but those who are outside the flame area can be at risk from thermal radiation effects. The heat flux produced by the fireball, the exposure time, and the distance from the fireball are important variables to determine the hazard [NFPA 68 A.7.6.4].

The hazard zone from a vented gas deflagration is calculated by the following equation:

$$D=3.1\cdot\left(\frac{V}{n}\right)^{0.402}$$ **NFPA 68 Equation 7.6.4.1**

where:

D = axial distance (front-centerline) from vent (m)
V = volume of vented enclosure (m^3)
n = number of evenly distributed vents

The hazard zone measured radially (to the sides, measured from the centerline of the vent) shall be calculated as one-half D [NFPA 68 7.6.4.2]. Ducts should be carefully designed because they can impact the vent area sizing [NFPA 68 7.4].

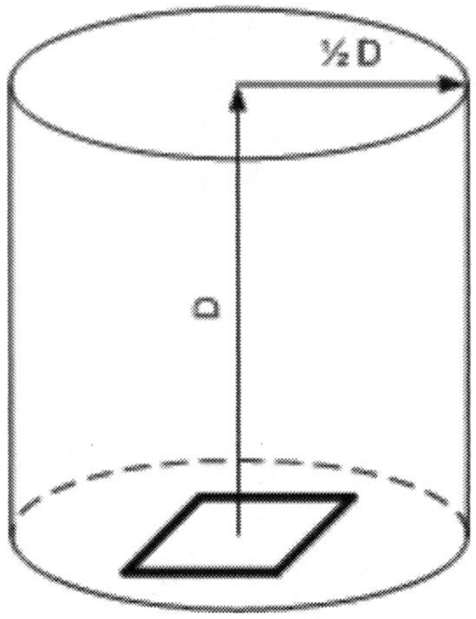

Figure 8 – Hazard Zone

Vent Closure Static Activation Pressure (P_{stat})

Similar to API limitations for relief valve overpressure, NFPA specifies permissible limits for deflagration vent activation pressure. Low strength and high strength enclosures each have their own specific requirements.

- For low strength enclosures, P_{red} shall exceed P_{stat} by at least 0.024 barg (0.35 psig) [NFPA 68 7.2.6.1].
- For high strength enclosures, P_{red} shall exceed P_{stat} by at least 0.05 barg (0.73 psig) and P_{stat} shall be less than 0.5 barg (7.3 psig) [NFPA 68 7.3.3.2].

Deflagration vent closures should release at a P_{stat} value that is as low as practical, yet should remain in place when subjected to external wind forces that produce negative pressures, to prevent vents from being pulled off. In most cases, a P_{stat} value of 0.01 barg (0.14 psig) is acceptable. In areas subject to severe windstorms, release pressures up to 0.015 barg (0.21 psig) are used [NFPA 68 A.7.6.2.1].

Vent Closure Types

Similar to a relief valve, a vent closure is a device that is installed on a sulfur pit vent opening that is designed to open at a pre-determined pressure, P_{stat}, to prevent the pressure in the enclosure from exceeding P_{red}. Vent closures should withstand pressure fluctuations that are below P_{stat} [NFPA 68 6.5.5]. For sulfur pits, this would include any slight vacuum potentially created by the sweep ejector.

Vent closures may have non-reusable elements or reusable elements. Those with reusable elements may automatically reclose following a deflagration or may require manual repositioning. Some common types of vent closures are described in the following sections.

Rupture Diaphragms

Rupture diaphragms (also called rupture or bursting panels, disks or membranes) are non-resealing, one-time use devices that rupture at a pre-determined differential pressure (P_{stat}). They can be round, square or rectangular and are typically mounted in a support frame which is attached to the enclosure (Figures 9-13). Individual panels are available in a range of sizes up to a few square meters in area. Flat panels can be used for vacuum service if backed by a support grid (Figure 10). Domed panels can resist vacuum conditions on their own without a support grid (Figure 11).

Thermally insulated vent panels are available (Figure 13) to reduce heat loss to the atmosphere, which can lead to sulfur condensation and/or solidification on and around the device, potentially hindering the relief device's performance. Heat loss and sulfur condensation/solidification concerns associated with sulfur pit deflagration vents are specifically discussed in a subsequent section of this paper.

Rupture diaphragms are almost inertia-free, with a specific mass less than 0.5 kg/m^2, and do not impede the venting process. Therefore, they have a venting efficiency, $E_f = 1$. [EN 14797:2006 7.3.3.1].

Some benefits of rupture diaphragms include:

- They are cost effective when infrequent activations are expected. For example, indicative cost of materials for a total vent area of 70 m^2 is about $58,000 for rupture diaphragms, compared to $525,000 for explosion doors [15].
- They have low inertia and therefore no penalty on vent area.

Some drawbacks of rupture diaphragms include:

- Their service life may be negatively impacted by pulsation or vibration.
- They must be replaced after each incident.

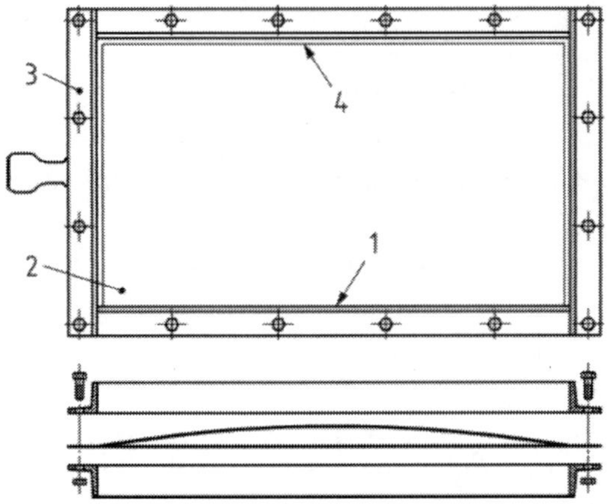

Key

1 fixed side

2 area of the explosion venting device

3 supporting frame

4 sides yielding during venting process

Figure 9 – Rectangular Rupture Diagram
(Source: EN 14797:2006)

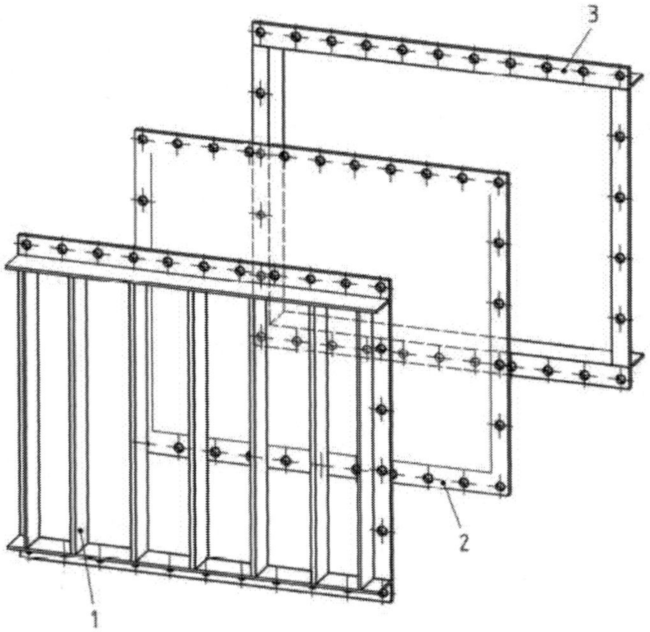

Key

1 back pressure support

2 vent panel

3 supporting frame

Figure 10 – Back Pressure Support
(Source: EN 14797:2006)

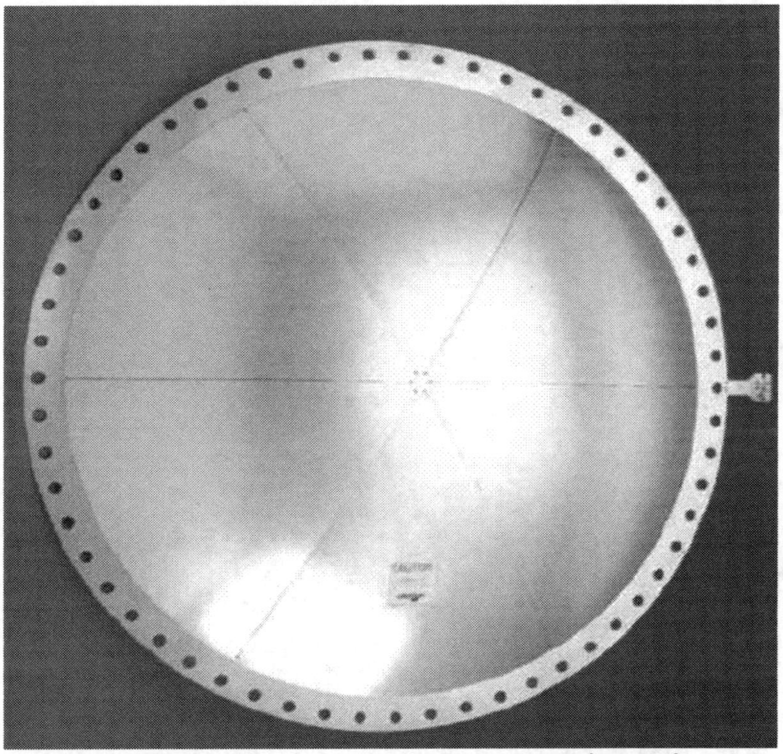

Figure 11 – Typical Domed Rupture Diaphragm (Round)
(Photo provided by BS&B Safety Systems)

Figure 12 – Typical Rupture Diaphragm (Square)
(Photo provided by BS&B Safety Systems)

Figure 13 – Thermally Insulated Rupture Diaphragm
(Photo provided by BS&B Safety Systems)

Pop-Out Panels

Pop-out panels come in many forms; a few examples are shown in Figures 14 and 15. They may be semi-rigid panels, secured at the edges to a frame, designed to bend and become detached at the desired P_{stat}. They may also use rigid panels secured to the frame at the edges with clamping devices, designed to break or release at the desired P_{stat}. Pop-out panels may be of single-wall or double-wall construction with insulating material between, so called "sandwich panels".

Pop-out panels may be installed in the vertical or horizontal orientation. If installed in the horizontal position, provisions to prevent the accumulation of snow, ice, rainwater or debris must be included in the design.

Pop-out panels have restraining devices (chain, wire, etc.) to prevent them from becoming projectiles. Restraining devices shall not impede the operation of the vent of vent closure device [NFPA 68 6.5.3].

The primary benefit of pop-out panels is that they may be reusable following an incident. The primary drawback is that broken clamps must be replaced after each incident.

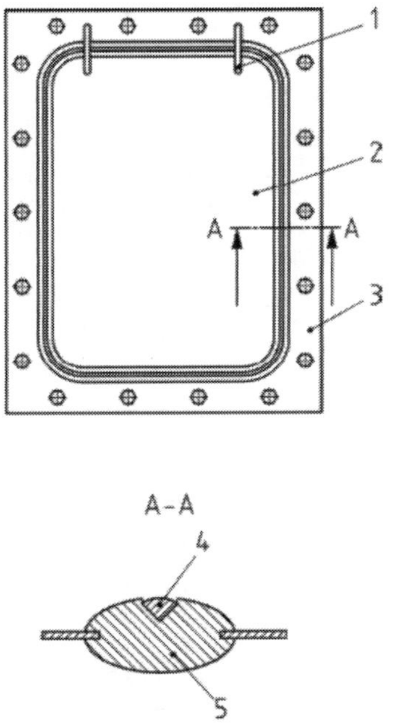

Key

1 restraining element
2 venting element
3 supporting frame
4 clamped device
5 plastic or rubber seal

Figure 14 – Restrained Pop-out Panel
(Source: EN 14797:2006)

Figure 15 – ATEX Reflex Pop-out Panel
(Photo provided by ATEX Explosion Protection L.P.)

Explosion Doors

Explosion doors are hinged devices, typically held shut with spring-loaded, magnetic or friction latches; some examples are shown in Figures 16 and 17. Hinged closures shall be permitted to be used, provided the following conditions are met [NFPA 68 6.7.4]:

(1) There are no obstructions in the path of the closure that prevent it from opening.
(2) Operation of the closure is not restrained by corrosion, sticky process materials, or paint.

Some benefits of explosion doors include:

- They are cost effective when multiple activations are expected or when the replacement of vents after opening is very difficult and/or time consuming.
- They can be made self-closing, which can limit oxygen supply and minimize the damage from fire.
- They can be field-calibrated for easy setup and maintenance testing.

Some drawbacks of explosion doors include:

- Explosion doors have greater inertia than rupture diaphragms and so may require additional vent area.
- Self-closing doors may create a vacuum in the pit resulting from cooling of the hot combustion gases following a deflagration.
- Explosion doors may be more challenging to heat, if required (this is discussed in more detail in a subsequent section of the paper).
- Maintenance of the hinge and latch mechanism is required.

Figure 16 – Explosion Doors
(Photo provided by ATEX Explosion Protection L.P.)

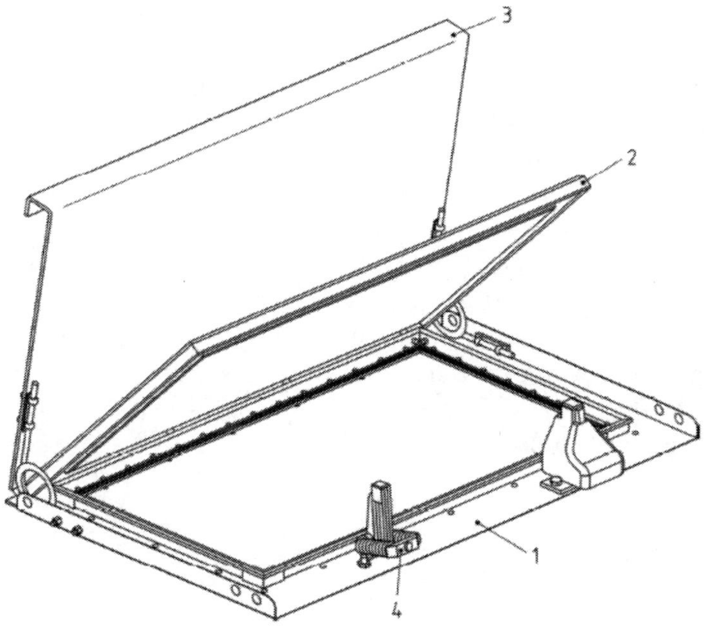

Key

1 supporting frame
2 venting element
3 baffle
4 retaining elements

Figure 17 – Explosion Door
(Source: EN 14797:2006)

Vent Closure Inspection and Maintenance Requirements

Vent closures shall be inspected at least annually [NFPA 11.4.1]. The inspection frequency depends on the environmental and service conditions to which the devices are exposed, and on documented operating experience [A.11.4.2]. The inspector shall verify [NFPA 68 11.4.4]:

- The opening is free and clear of any obstructions
- The closure is not corroded or mechanically damaged
- The closure has no buildup of deposits on the inside surfaces or between layers of the vent
- The closure shows no fatigue and has not released
- Records of inspections shall be retained for a minimum of 3 years [NFPA 68 11.7.2].
- If process material has a tendency to adhere to the vent closure, the vent closure shall be cleaned periodically to maintain vent efficiency [NFPA 68 11.9.3]

Vent Closure Heating

Vent closure operation should not be hindered by snow, ice, corrosion, debris or buildup of deposits on the inside surfaces [NFPA 68 6.5.2]. For sulfur pits, this might require heating and insulating the vent closures to keep the inside surface temperature above 250 °F to prevent sulfur from freezing on the panel.

When designing the heating system, many variables need to be considered, including ambient air temperature, wind speed, sweep air temperature (heating the incoming sweep air may be required), size of the panel, insulation thickness, etc. ControTrace heating elements may be well suited to this type of application because they can be adapted to a variety of panel mounting frame shapes and styles. Heating elements should not be placed directly on top of the vent panel (Figure 18) if they interfere with the operation or increase the panel inertia. However, heating elements may be placed on the surrounding frame to which the panel is mounted, which allows heat to be transferred to the vent panel by conduction (Figure 19). For large vent panels, the convective heat loss to the surroundings may be too substantial to keep the center of the panel above 250 °F, even with insulation [14]. Adding a vent duct to block the wind (Figure 20) may resolve this. However, vent ducts should be carefully designed because, similar to the discharge piping on a relief valve, the additional backpressure imposed by a vent duct can increase the vent area required [NFPA 68 7.4].

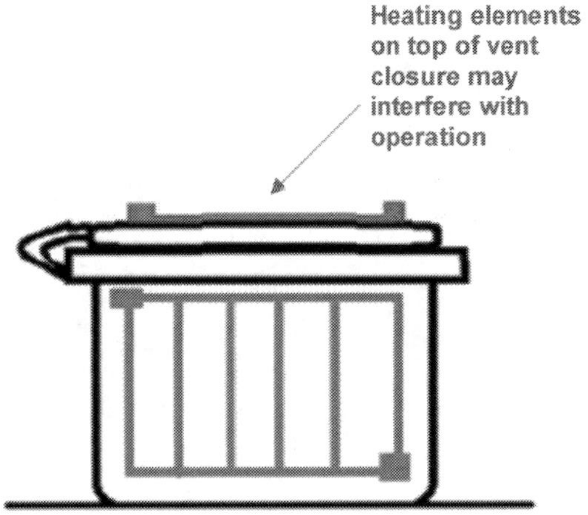

Figure 18 – Concept for Heated Explosion Door
(Figure provided by Controls Southeast Inc.)

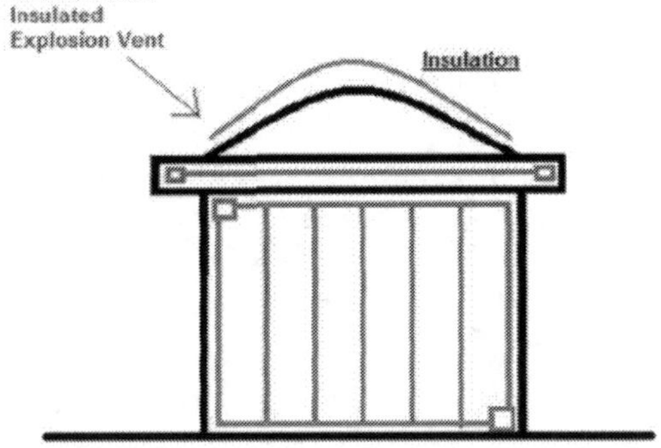

Figure 19 – Concept for Heated Rupture Membrane
(Figure provided by Controls Southeast Inc.)

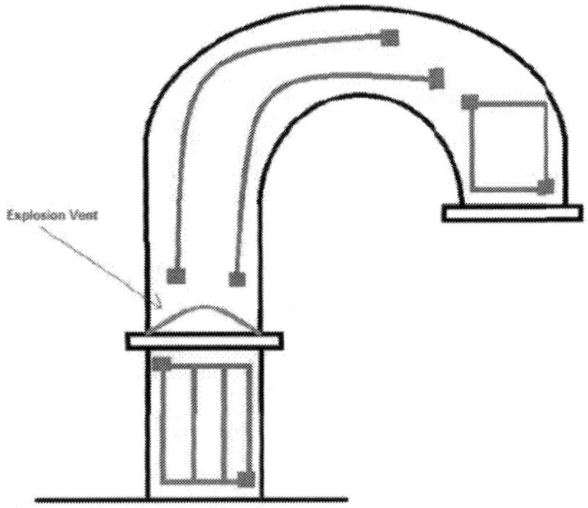

Figure 20 – Concept for Heated Rupture Membrane with Vent Duct
(Figure provided by Controls Southeast Inc.)

Conclusions

The primary conclusions derived from this study can be summarized as follows:

- There is no clear consensus within the industry on the requirement for deflagration vents on sulfur pits.

- Owners need to make deliberate, informed decisions about the need for sulfur pit deflagration vents during the design phase HAZOP, in consultation with their detailed engineering contractor, while also considering site-specific operation procedures and safety practices.

- Sulfur pit deflagration vents are challenging to implement.

- If it is deemed that other safety devices, methods and procedures cannot adequately mitigate the possibility of forming explosive conditions within the vapor space of the sulfur pit without the installation of deflagration vents, the owner may elect to install a pressure vessel rated to withstand deflagration pressure, rather than installing a concrete pit with protection devices that are difficult to design and maintain.

References

1. Johnson, J., Hatcher, N., *Hazards of Molten Sulfur Storage & Handling*, Laurance Reid Gas Conditioning Conference, February 2003.
2. SulfurExperts, *Liquid Sulfur Handling, Treating and Storage*, Sulfur Recovery Seminar, September 2005.
3. Amine Best Practices Group, *Are You Handling Sulfur Safely?*, Hydrocarbon Processing, July 2003.
4. NFPA 68, *Standard on Explosion Protection by Deflagration Venting*, 2007 Edition
5. NFPA 69, *Standard on Explosion Prevention Systems*, 2002 Edition
6. NFPA 655, *Standard for Prevention of Sulfur Fires and Explosions*, 2012 Edition
7. BS EN 14797:2006, *Explosion Venting Devices*, December 2006
8. BS EN 14994:2007, Gas Explosion Venting Protective Systems, February 2007
9. BS EN 14460:2006, Explosion Resistant Equipment, May 2006
10. B. Chase, Telephone interview. August 17, 2012. Mr. Chase is a technical expert with NFPA who provided informal interpretation of some sections of NFPA 68.
11. N. Dowling, *Explosion Limits for $H_2S/CO_2/Air$ and $H_2S/N_2/Air$ Mixtures*, ASRL Quarterly Bulletin 148, Vol. XLV No. 4, Jan-Mar 2009, pp. 1-2
12. D. Crowl, *Understanding Explosions*, American Institute of Chemical Engineers, New York, 2003
13. Center for Chemical Process Safety, *Guidelines for Engineering Design for Process Safety, Center for Chemical Process Safety*, American Institute of Chemical Engineers, 1993
14. D. Campbell, Interview. August 29, 2012. Mr. Campbell represents Controls Southeast Inc., makers of ControTrace heating elements, who conducted heat transfer calculations for several conceptual vent closure configurations.
15. D. Guaricci, e-mail. August 28, 2012. Mr. Guaricci represents ATEX Explosion Protection L.P., who provided photos and indicative price quotes for rupture diaphragms and explosion doors for the example sulfur pit.

Considerations of Temperature Measurement in the Claus Thermal Reaction Furnace

Steve Croom

Delta Controls Corporation

It is necessary that the temperature in the Claus Furnace be monitored to protect it from damage, minimize maintenance, avoid costly shutdowns, and improve safety and efficiency. There are two ways to do this: the Thermocouple and the Infrared Pyrometer. The question of which is better is irrelevant, since both offer advantages as well as limitations.

For decades, thermocouples were generally thought to be unreliable in Claus furnaces, which they were. They had a tendency to fail and when they did, they could not be easily repaired or replaced until the unit was shut down. So, the thermocouple gave way to the Infrared device, which did not succumb fatally to improper installation or the harsh conditions of the furnace. And when they did malfunction, they could be serviced, by the use of an isolation valve, while the unit remained in operation.

However, the tendency for the Infrared Sensors (IR) to drift over time and the routine maintenance requirements of the IR units created a demand for something better. So, special thermocouples and infrared pyrometers were designed specifically for the Claus furnace that addressed the operating issues.

Careful examination of Claus furnace thermocouple failures revealed that problems we not so much in the design, but in the installation technique, or more accurately, the lack thereof. All thermocouple failures in Claus service are due to corrosion of the thermocouple elements. Careful examination revealed the reasons for the failures; they

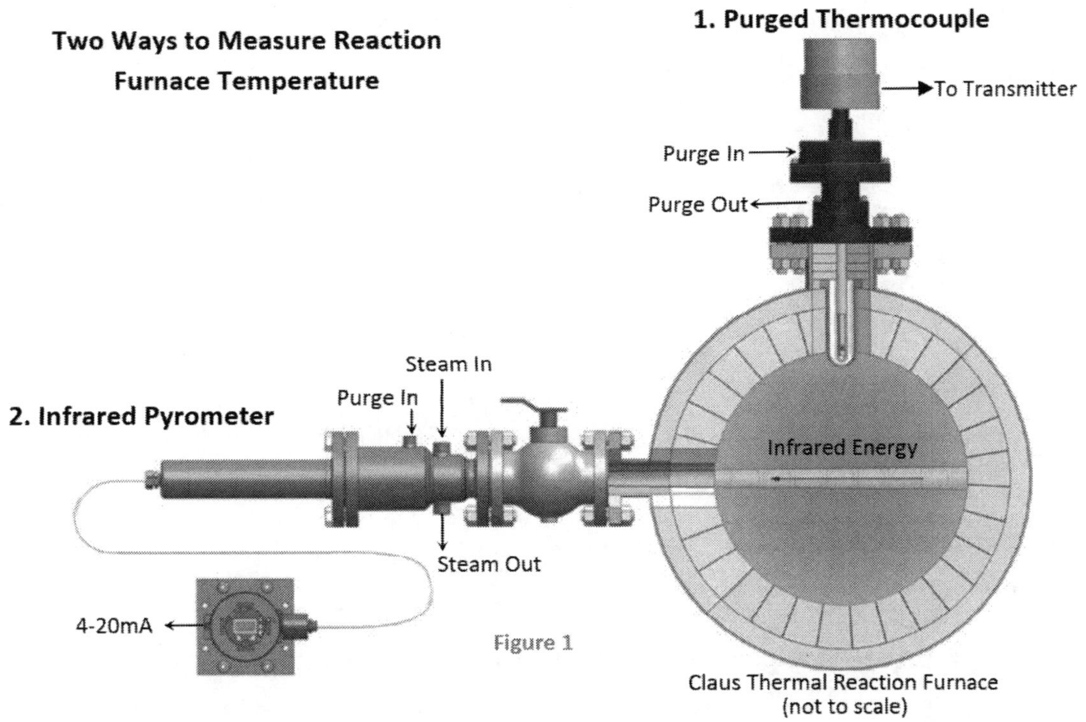

Figure 1

often were of a mechanical nature and not because of the process conditions, severe as they are. Failure was generally breakage of the thermowell, usually because of improper handling and installation, which exposed the elements to corrosion. In a couple of instances, it was observed that the thermocouple was broken during installation resulting in failure by contamination a few months after startup.

So, the solution was to design a thermocouple that did not succumb to the stresses of refractory movement and to design installation tools and fixtures that make installation simple, safe and foolproof. In addition, installation tools and proper installation technique ensure the thermocouple is positioned properly so as to provide an accurate reading of temperature at the refractory hot face.

So, why bother with the thermocouple and its apparent vulnerabilities? The reason is that it is the most accurate device to measure temperature; it is the world standard.

Let's compare them:

Thermocouple advantages:

- Relatively low cost,
- Little or no maintenance,
- High accuracy,
- Unaffected by changing gas composition.

Thermocouple drawbacks:

- May be damaged by severe thermal shock, extreme high temperature, shifting refractory, and improper installation.
- Must be continuously purged to specified pressure and flow.
- Cannot be repaired until furnace shutdown.

Infrared Pyrometer advantages:

- Mechanical isolation using the nozzle valve, permitting service or replacement while the furnace remains in service.
- Withstand severe system upsets. Normally not harmed by moderate over-temperature conditions or shifting refractory.

Infrared Pyrometer drawbacks:

- Vulnerability to sulfur condensing, building up in the nozzle bore
- Sulfur coating the viewport glass window (see Figure 2)

  - (Both conditions cause an attenuation of the Infrared energy reaching the sensor, resulting in an ever-increasing inaccurate low measurement).
- Drifting sensor / electronics requires frequent recalibration. Some IR pyrometers incorporate a mechanical chopper mechanism that reduces long term reliability.

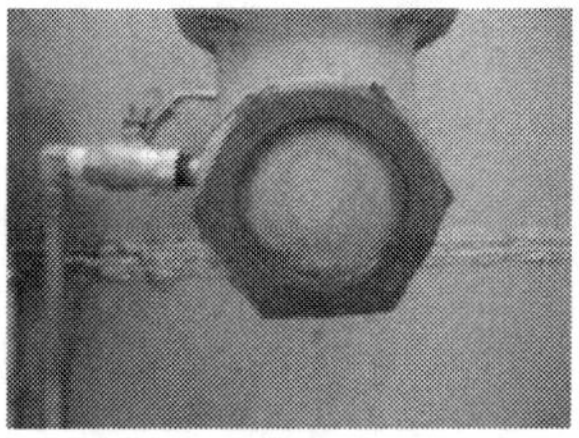

Figure 2 Sulfur coating a viewport window

So, what specifically has been done to make both of these two technologies more suitable for Claus furnaces?

For the thermocouple…

1. Installation of a secondary large outer protective thermowell that rests on the refractory and moves with it relative to the vessel shell. It is not attached to the thermocouple assembly, but fits loosely in a large hole bored through the refractory. This permits some movement and shifting of the firebrick relative to the insulating refractory layer as well as shifting insulating refractory relative to the vessel shell. In addition, this secondary well provides protection of the inner element thermowell from thermal shock.

2. Thermocouple installation fixtures provide simple and easy installation minimizing the risk of breakage. A refractory drilling kit ensures that the correct bore is created in the refractory, on center and perpendicular to the mounting flange. This ensures that the proper clearances are established permitting refractory movement without damage. Another valuable tool is an installation mounting bar kit that permits safe and easy installation in other than the recommended vertical top centerline nozzle location. In these non-vertical locations, it is difficult for the installer to insert a heavy thermocouple into the nozzle and simultaneously keep the assembly perfectly centered to avoid any side loads on the ceramic thermowell. If the installer relaxes his grip on the unit during horizontal installation, it will become broken due to a combination of gravity and the heavy flanged headworks. The mounting bar kit attaches to the vessel nozzle flange and supports the full weight of the unit, keeping the thermowell centered during insertion until it is solidly bolted in place.

3. Thermocouples must be purged to prevent contamination over the long term. No materials exist that are "gas-tight" in extremely high temperatures. And in the Claus furnace, the use of ceramic materials is necessary and at the high temperatures, certain elements of the reaction gases migrate though the wall of the ceramic thermowell. Without a purge, these corrosive elements will, in time, collect inside the thermowell and begin to corrode the precious metal thermocouple elements.

Figure 3 - Purge Panel

To prevent this, a continual purge flow at elevated pressure is maintained throughout the thermocouple assembly. Typical flow rates of approximately 11 LPH (~ 0.4 SCFH) are sufficient to protect the thermocouple without cooling the thermocouple elements any more that 1-2 degrees. Thermocouples can eventually fail if the purge and pressure are not maintained over the long term. To eliminate the purge issues, simple, easy to install purge control panels (Figure 3) are provided. In addition, the purge controls also serve as a diagnostic tool to determine the causes of any thermocouple accuracy issues.

Nitrogen is the recommended purge gas; clean dry instrument air may be suitable, but only if it contains *no* hydrocarbons; trace quantities of oil in the purge air will collect and burn inside the thermowell, leaving deposits that eventually damage the thermocouple. (see Figure 4)

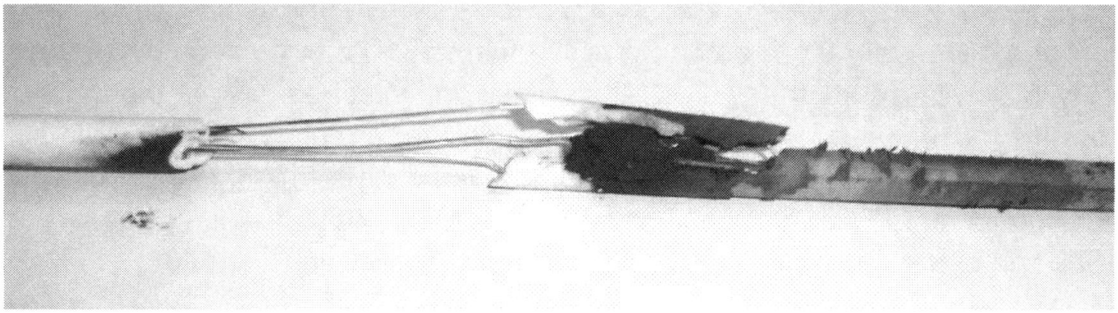

Figure 4- Damage to thermocouple element support due to burning of trace quantities of oil in the purge gas.

For the Infrared Pyrometer...

1. Implementation of a heated and insulated lens assembly eliminates the problem of sulfur buildup in the refractory bore, nozzle and block valve. The heated lens housing generates sufficient heat that when combined with the furnace heat, prevents sulfur from solidifying and obstructing the IR energy pathway to the sensor. High purge flow into the nozzle (a historic industry standard practice) should not be employed as the cooler air encourages the solidification of sulfur and corrosion of the nozzle. To prevent accumulation of dust or ash on the lens window surface, new designs incorporate a very small purge flow that is directed onto the lens window in a swirl path fashion. In addition, the low flow purge is pre-heated to avoid any cooling effect that would encourage sulfur to condense on the lens glass.

2. Early IR instruments suitable for the rigors of Claus service incorporated sensors that were subject to aging and drift, resulting in the need for frequent recalibration. This technology also employed the use of a mechanical "chopper" motor which compromised long term reliability of the instrument.

Modern Claus pyrometers use high stability semiconductor sensors that do not exhibit the aging effects of early designs. These sensors provide long term reliability and freedom from frequent calibration or maintenance. Because semiconductor sensors cannot tolerate the high temperature of the vessel nozzle, a high temperature fiber optic cable is used to connect the lens to the sensor.

Thermocouple Installation Criteria

The thermocouple performs most reliably when installed on the vessel top centerline in the vertical down position. This is because it is the easiest position to make the physical installation and minimizes the risk of breakage during insertion. Also, movement or shifting of the refractory at the top centerline is normally limited to separation of the firebrick from the castable materials and occasional separation of insulating brick from the vessel shell; both are easily tolerated by the thermocouple. There is little shear motion at the top and any that occurs is along the longitudinal axis. Mounting at the side positions of the furnace, although acceptable, makes installation more difficult and, in addition, refractory shifting may also be of some radial shear motion. This refractory movement, if severe, can cause breakage, particularly if any refractory dust/chips settle around the outer refractory thermowell and fill the space needed to permit refractory movement.

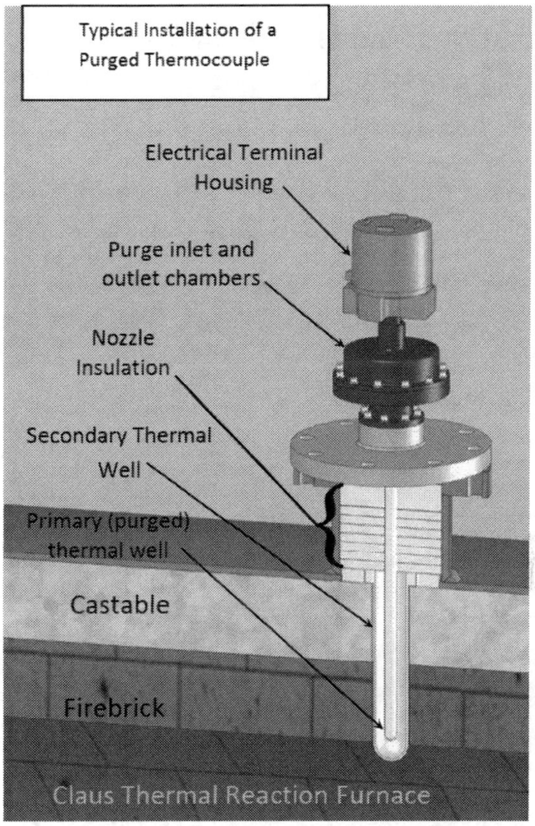

Nozzle sizes of 4" or 6" permit easy installation of nozzle insulating materials that minimize the buildup of sulfur in the nozzle, which could place stresses on the inner "gas-tight" element thermowell, again causing damage. Also larger nozzles permit refractory movement without causing the outer refractory thermowell well to press against the ID of the nozzle.

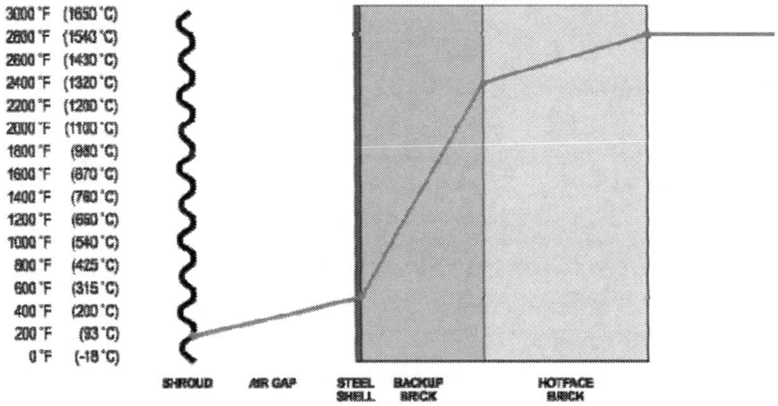

Figure 5 Temperature Profile of Refractory, Shell, and Shroud
Illustration courtesy of Thorpe Specialty Services

Nozzles should be as short as practical. Shorter nozzles maintain heat closer to that of the vessel shell, minimizing corrosion (See Figure 5) illustrating the temperature profile of refractory, shell and shroud). Taller nozzles are cooler, encouraging corrosion and the solidification of sulfur. In addition, shorter nozzles reduce the thermocouple insertion length, lowering the cost of the thermocouple. Shorter thermocouples are also easier to install and minimize the chances of physical damage during handing. It is advantageous to have the thermocouple mounting flange positioned beneath the rain shield/shroud. Nozzles should not be purged as this needlessly cools the nozzle, encouraging corrosion and the solidification of sulfur, destroying the protective capability of the specially designed nozzle insulating materials. Weld neck flanges make ideal nozzles since they have no weld beads in the ID that restrict insertion of the solid insulating rings.

The most critical detail of thermocouple installation is the creation of the bore hole through the refractory. For maximum reliability, this hole must be of the proper diameter, centered, and be perpendicular to the nozzle mounting flange. Although this seems simple, logistical issues often result in incorrect bore hole creation. A bore hole that does not meet the specification causes undue needless expense to repair the refractory for another drilling exercise, and takes valuable time. It is important to protect against castable material entering the base of the nozzle; should this occur, the nozzle depth dimension will be altered, preventing installation of the secondary outer refractory thermowell that rests on the cold face of the refractory. If material is found to have entered the nozzle above the inside surface of the vessel shell, it must be removed leaving a level surface on the refractory and flush with the inside surface of the vessel shell. The easiest and simplest means to protect against this is to use the refractory stops supplied by the thermocouple manufacturer. But, someone must take responsibility of assuring that these stops are fitted to the nozzles prior to the beginning of refractory installation.

Installing a stop on the nozzle prior to refractory installation prevents castable from entering the nozzle.

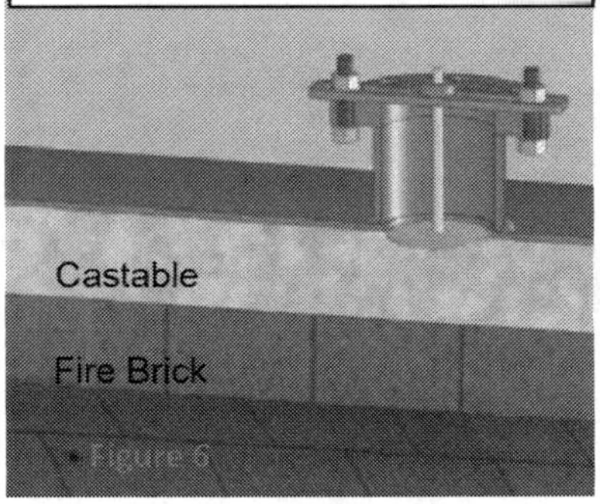

A thermocouple installed in a nozzle / refractory bore that does not meet specification may fail during startup due to shifting refractory and the failure will not become evident until the furnace has been in service for weeks. The drilling of the refractory cannot be performed "free style" without the proper guide fixtures; doing so usually results in failure. The easiest, quickest, safest and simplest means is to utilize the drilling kit furnished with the thermocouple.

Infrared Pyrometer Installation Criteria

The pyrometer, contrary to the thermocouple, is usually installed in a side position on the vessel. This is to provide for easy access, since it can be adjusted or serviced while the furnace remains in operation; however, keeping the nozzle location above the 9:00 / 3:00 o'clock position prevents the accumulation of particulate and molten sulfur, which could eventually block the IR energy path to the sensor. Gravity and vibration provide cleaning assistance as long as there is some downward angle of the nozzle toward the vessel centerline. Remote electronics type units may be installed in a vertical downward position, but consideration must be given to the limited length of interconnecting fiber optic cable from the lens to the electronics.

As with the thermocouple, the nozzle should be short. Again, this is to maintain heat that helps prevent corrosion and the buildup of sulfur. For IR units featuring remote electronics and steam jacketed lens assemblies, no nozzle purge should be used and the nozzle, block valve and lens should be insulated to maintain higher temperature, closer to that of the vessel shell (see Figure 7). This minimizes corrosion and discourages the deposition of solid sulfur and eliminates coating of the lens window glass, preventing the need for routine maintenance. Remote electronics type IR units require that 50 - 100# steam be fed to the steam jacketed lens housing in order to realize the benefits of immunity to sulfur build up and corrosion.

Figure 7 Pyrometer with Insulation

With integral IR units, whereby the electronics and lens are mounted directly to the nozzle, consideration must be given to the manufacturer's installation instructions regarding nozzle purging and any cooling requirements to protect the instrument. No insulation would normally be recommended for these type units.

Nozzles are usually 2" or 3". It is important to assure that the bore hole through the refractory is clear and in alignment with the lens. The bore hole should be a minimum 1.5" ID, centered and perpendicular to the mounting nozzle flange to assure a clear pathway to the interior of the vessel. Although not normally necessary, the creation of this bore hole through the refractory is most easily performed using the drilling kit furnished with the IR pyrometer.

Longitudinal Location of Temperature Measurement Devices

There is no standard recommended longitudinal placement of temperature measurement mounting nozzles in a Claus Reaction Furnace. The ideal placement is usually determined by the burner manufacturer using CFD (Computational Fluid Dynamics) to select the most appropriate locations. Consideration must also be given to physical obstructions, access to the instruments and of the refractory at the point of penetration. With

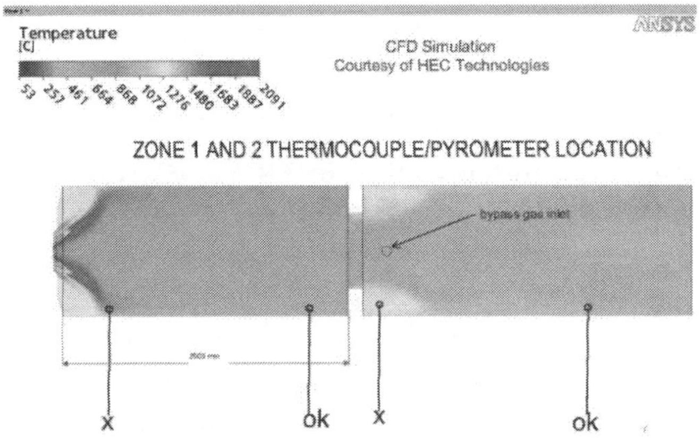

Figure 8 Furnace Temperature Profile – Uniform Combustion

no recommendation by the burner manufacturer or the benefit of CFD, placement is usually at a point about two-thirds the distance from the burner to a checker wall / choke ring and another about two-thirds the distance from the checker wall / choke ring to the WHB tube sheet. (See Figure 8). Additionally, CFD profiles may reveal other possible problem areas, such as burner design, feed streams or choke ring / checker wall effectiveness, any of which could result in premature failure, shut down and costly remediation (See Figure 9).

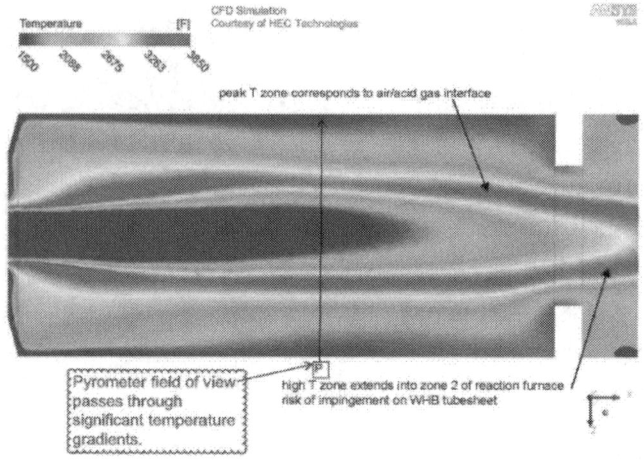

Figure 9 CFD Simulation of poor burner design. Pyrometer or T/C sees relatively cool refractory even though hot gases impinge on WHB tube sheet.

Refractory Dry-Out Temperature Measurement

The curing of refractory is critical to long term refractory performance. Failure to adhere to the refractory contractor's recommendations may cause premature refractory failure and invalidate warranties. Accurate measurement of the critical and relatively low temperatures during dry-out is a challenge. Generally, IR pyrometers and thermocouples that are suitable for measuring reaction temperatures are unable to accurately measure the critical low temperatures. But, there are a couple of ways to accomplish this.

One is to select the proper thermocouple element type. Type "S" or "R" are generally suitable for Claus service to a maximum temperature of 1500°C and provide usable accuracy to as low as 100°C, useful in measuring the refractory dryout profile. To maximize immunity from damage from excessive temperatures, Type "B" thermocouple elements are employed (the most commonly used in Claus service) and will perform to 1800°C without damage. However, the Type "B" is not suitable for measuring the low "dryout" temperatures. To utilize the Type "B" and also have low temperature measurement capability, an auxiliary element (usually a Type "K" element) can be fitted as a "third" element making the unit suitable for accurate measurement of the low end dryout temperature. Although this is a convenient method, the single drawback is that when the reaction temperature exceeds that of the dry-out thermocouple element's max temperature (about 1400°C), it will melt away, making it a "one use only" "dry-out" thermocouple. So, if it becomes necessary to shut down for additional refractory work and subsequent dryout, this "third" thermocouple would no longer be available.

A more suitable method of dryout would entail the use of the insertable "calibration" thermocouple that is an option offered with Infrared Pyrometers. This provides an excellent means to monitor this dry out profile. However, it requires that the IR unit remain out of service until the refractory dry-out cycle is complete, but the IR pyrometer normally is of little use at low temperatures anyway. Upon reaching the end of the dry out cycle and before going on line, the insertable thermocouple is withdrawn and the IR device returned to its operating position and placed in service.

A point of interest… some refineries have opted to prohibit opening of any Claus Furnace port during operation, including that of utilizing thermocouples and cleaning devices fitted with a mounting gland and an operable block valve.

Pyrometers installed to monitor the reaction gas temperature in lieu of the refractory hot face may not effectively read the gas temperature of fuel gas or natural gas during the dry-out phase.

By far, the most effective means to obtain ideal refractory monitoring is to install multiple "K" thermocouple elements at critical locations inside the furnace. This minimizes the chance of improper readings due to a single element placement, which could be in an unsuitable location

Use of Both IR and Thermocouple

Use of both technologies provides redundancy and protection from common-cause failures; i.e., conditions that affect one technology will likely not affect the other technology.

Since thermocouples are the accuracy standard, they provide a constant means for verification of IR units in actual service.

Thermocouples succumb to extreme over-temperature conditions (1825°C melting point of "B" type), loss of long term purge, severe shifting refractory or severe thermal shock. None of these conditions would normally have any detrimental effect on IR units.

IR units, as mentioned above, may fail because of other problematic issues. Setting aside the newer remote electronic units designed to eliminate the problems of blockage of the IR sight path due to build up of sulfur and coating of the lens glass, these are common problems. Although these issues may be corrected by opening the nozzles and cleaning them out, it is a costly procedure. With IR units that suffer these problems or that require frequent calibration due to sensor drift, the thermocouple can be used to determine when IR maintenance should be performed.

This cleaning maintenance work on the IR can only be performed however, when the nozzle block valve is operable and not frozen in the open position. Means to prevent this are the selection of a proper fireproof valve and keeping it at elevated temperature. Also, periodically operating the valve handle may prevent it from freezing.

Differences in Measurement Reading between IR Pyrometers and Thermocouples

It has been frequently observed that there is a discrepancy between the temperatures reported by thermocouples versus pyrometers. In these instances, there are three possible causes:

1. The instruments are reporting the actual temperatures which are different.
2. The thermocouple is wrong.
3. The pyrometer is wrong.

Of course, discrepancies can also be caused by any combination of the above.

Actual Temperature Differences

First, it is important to understand that no two points in the furnace are at exactly the same temperature. Reaction gases start out at near ambient temperature at the burner and rise in temperature as they react while traveling down the furnace. At the same time, heat is being lost through the wall of the furnace. High gas velocities, turbulence, stratification and radiant heating tend to smooth out the temperature gradients somewhat, but temperature differences of several hundred degrees at different points along the furnace length are not uncommon. These differences can change with gas flow rate and gas composition.

It is also important to understand what is being measured. A thermocouple reports the actual temperature at the thermocouple junction. An infrared pyrometer measures the infrared energy emanating from a small region on the refractory wall; about 20% of this energy is radiated by the wall that the pyrometer is focused on, the other 80% of the energy is reflected energy that was radiated by other parts of the furnace. Thus the infrared measurement is a kind of weighted average of the temperature in that region of the furnace. (Those familiar with infrared temperature measurement may be tempted to think that the emissivity adjustment on the pyrometer can allow the instrument to read only the temperature of the wall at the focus point, but that concept only applies where a hot object is surrounded by a cooler environment – in that case there is virtually no reflected energy. Inside a reaction furnace, the reflected energy is abundant and cannot be distinguished from radiated energy.) The result is that if there are significant temperature gradients within a furnace, then a perfectly calibrated pyrometer could read a different temperature than a thermocouple, even if the pyrometer were focused directly on the thermocouple's thermowell.

In practice, there is usually less than 25°C difference between two instruments located close together. If the measurements are within this range, it is probably about as good as can be reasonably expected.

Thermocouple Errors

Thermocouples cannot read erroneously high. If the thermocouple is reporting a higher than expected temperature, strong consideration must be given to the possibility that expectations, rather than the thermocouple, are in error. One reason for a high temperature reading is a direct flame impingement on the thermocouple. In this case, the thermocouple is not in error, but is reporting the high temperature to which it is exposed.

Thermocouples can become inaccurate by reading lower than the actual temperature. This is usually caused by contamination or corrosion. Contamination is caused by lack of proper purging of the thermocouple, either due to a problem in the purge system, or else due to breakage of the ceramic thermocouple element well.

Most Claus furnace thermocouples are equipped with a second 'reference' thermocouple that is internally embedded in ceramic and not directly exposed to the purge gas inside the thermowell. If these two separate thermocouple elements in a given thermocouple unit are consistently reading within approx 25°C of each other, it may be assumed that they are reading accurately.

Thermocouples can also read low if the purge flow rate is significantly higher than the recommended purge rate (approx 0.4 scfh or 11 LPH). This low flow rate cools the thermocouple by only about a degrees or two, but cooling rapidly increases as the purge rate is increased. Excess purge cooling can easily be tested by temporarily turning off the purge and observing a rise in the indicated temperature. (Briefly turning off the purge will not harm the thermocouple.)

Of course, as with any instrument, the temperature transmitter connected to the thermocouple may be improperly programmed or spanned.

Infrared Pyrometer Errors

IR pyrometers are factory calibrated by pointing them into a 'black-body' calibration environment. This type of device very precisely models the IR emission of a furnace at uniform temperature. As discussed above, actual Claus reaction furnaces are not completely uniform in temperature, so temperature reported by a pyrometer will always be a weighted average of the furnace temperature. The measured wavelengths are carefully chosen to be such that the reaction and reactant gases are transparent, so normally the gas composition and flame temperature will have little effect on the IR measurement.

The most common problem with pyrometers has to do with material coming between the furnace and the sensor and occluding the infrared energy. Usually, pyrometer errors come on gradually whereby the IR pyrometer slowly decreases its signal relative to the thermocouple(s). Usually, the cause is from a very thin layer of sulfur accumulating on the lens viewport glass. As this becomes more and more opaque, the IR energy reaching the sensor is attenuated, causing a lower and lower reading over time. Another may be sulfur building up in the base of the nozzle or in the bore through the refractory, again, partly blocking the view path or the IR. This problem is minimized and often prevented by keeping the viewport at a temperature above the freezing point of sulfur. This is done with the use of heated lens assemblies, pre-heated low flow lens purge (only). And by insulating the nozzle, block valve and lens assembly, buildup of sulfur in the optical path will be eliminated. High flow purging of the nozzle cools the nozzle and encourages the solidification of sulfur; it is not recommended.

Older IR designs for Claus service often experience drifting of the accuracy of the sensor / electronics, requiring frequent recalibration. This is a historic issue with lead-based sensors; modern semiconductor sensors usually perform for many years without calibration.

Although uncommon, vibration or refractory shifting as the furnace heats rises, particularly upon initial start up, can cause a misalignment of refractory to the sight path of the IR lens, partially obscuring the "view". Most modern pyrometers provide for visually checking of the boresight of the instrument to detect misalignment.

Mitigating Measurement Discrepancies

If the difference in measurements between two nearby instruments is relatively small, (perhaps 30°C, although that number is somewhat arbitrary) things are probably doing as well as can be expected. Larger differences may require some diagnostics as mentioned above to determine which instrument(s) may be in error. Discrepancies that increase over time are a particular warning that something is amiss.

After verifying that the optical path is not occluded, the operator may determine that the pyrometer needs to be recalibrated. The best method to calibrate a pyrometer is with another pyrometer. The process pyrometer is temporarily removed and replaced with a portable pyrometer that has matching optical characteristics. After reading the furnace temperature, the process pyrometer is placed back in service and adjusted to match the portable pyrometer reading. Note that the portable pyrometer should be one that is specifically designed for the job. A general-purpose IR gun typically has too wide of a viewing angle to accurately see down a valve, nozzle, and borehole into the furnace. And its calibration does not take into account any attenuation in the viewing window.

An easier, but less accurate method of calibration is to set the IR device to match the reported reading of the nearest permanent thermocouple, once it is established that the thermocouple(s) are operating properly.

The use of insertion thermocouples at the IR port does not necessarily make an accurate calibration, although it is the best method if the above methods are not available. When using the insertion thermocouple, it is wise to turn off any high flow nozzle purge, as this has a cooling effect. And, the insertion thermocouple will likely be destroyed if introduced to a temperature higher than 1400°C, which would melt the "K" type element. Higher temperatures would also cause physical damage to the insertion thermocouple's stainless steel sheath.

Recommendations:

- Use both thermocouple and IR technology to reduce susceptibility to common-cause failures.
- Install both in different regions of the furnace
- Use thermocouples designed specifically for Claus service and ensure that proper purge specifications are carefully followed.
- Consider IR pyrometer designs that are designed for Claus service and which are heated to protect against sulfur build-up in the optical path.
- Follow the details of installation instructions and use the proper tools and fixtures for installation.
- Span Instruments to read higher than the maximum expected operating temperatures.

UNDERSTANDING TODAY'S LIMITING FACTORS ON

REACTION FURNACE LININGS

by

Thorpe Specialty Services Corporation
www.ThorpeSSC.com

Andy Piper – VP/Sales & Marketing Manager
Jeff Proctor, P.E. – Technical Services Manager

Prepared For Presentation at the 2013 AIChE Spring National Meeting
April 28 – May 2, 2013
16th Topical Conference on Refinery Processing
Sulfur Management Session
San Antonio, Texas

ABSTRACT

SRU operators are experiencing increasing demands for improved reliability due to significant changes in emission regulations and penalties, more severe operating conditions, high cost of repairs and the consequential impact a reaction furnace failure can have on other operating units in refineries. Management is finally paying more attention to these issues, however, before the SRU problems can be properly addressed, they first must be fully understood.

Our paper outlines the common problems plaguing SRU linings today, including a lack of understanding of refractories in general and the unique requirements for SRU applications, as well as the consequences of the technical and commercial decisions made by licensing companies, project teams and owner/operating companies. We will show examples of common failures throughout the industry and then illustrate some of the complex technical aspects that need to be understood and tradeoffs that must be considered in the design process. The ultimate goal of the design/implementation process is to "balance" many competing factors in order to maximize the performance, reliability and life of the lining system; not for theoretical steady-state operating conditions, but for the real world conditions that are imposed on the refractory linings.

INTRODUCTION

If you are reading this paper then the chances are fairly high that you or your company has suffered the pain common to refractory issues in reaction furnaces. This is not a pleasant or an inexpensive experience; nor is it isolated to only you or your company. We see the same issues regarding premature refractory failure repeated time and again all over the world. The painful impact to the plant may include the following:

- Premature &/or frequent maintenance repairs or lining replacements required between or during planned outages
- Longer downtimes due to extra refractory work requirements
- Extended startup times or even failures due to dryout issues
- Shell breeches and the safety & environmental issues associated with this extreme failure
- Loss of production capability of related units during unexpected or extended reaction furnace outages

Typical Extreme Refractory Failures

In other words - less runtime, less reliability, higher costs and worse, less profit. Many operating companies today are only too aware of these issues but are unaware of how to correct them.

There are reasons the same failures occur repeatedly and there are ways to significantly improve these situations. As with most problems, the devil is in the details. In this case, the devil is in even understanding what details. This is especially true for refractories as they are normally considered a commodity that anyone can design, supply &/or install; making refractories almost an afterthought.

In actuality, SRU linings are complex structures that require careful consideration. They are not only the first line of defense to protect the vessel shells, but are the <u>only</u> defense. In our opinion, these units operate under more severe service conditions than any other unit in a refinery. The consequences of "getting it wrong" up front then brings in the added costs and frustrations of dealing with substandard design, materials of construction, or workmanship for the entire working life of the SRU or until lining improvements are implemented.

We recognize that under certain conditions, all thermal reactors are capable of melting any refractory made today. Having in the past supplied these units on a turnkey basis and having performed countless emergency repairs for customers due to unexpected failures, we base our lining recommendations on an <u>Optimized</u> lining approach. Typically we find that due to cost pressures &/or lack of awareness of the operating potential, linings are designed to the theoretical normal operating conditions. We are aware of the particular difficulties with these units due to startups and shutdowns, natural gas firing, O_2 enrichment, hydrocarbon slugs, operational errors, thermal striations or complete unit overheating from poor burner mixing or damaged burner components. We consider these to be the real world operating conditions for thermal reactor service; it is not a matter of "if" but "when" a thermal event stresses the lining. Every unit should be evaluated for its size, geometry, thermal profiles, etc. in an effort to optimize the lining to better endure these events.

We will begin with the commercial or project structure issues involved in most new projects and then outline some of the important technical issues that need to be considered to optimize the refractory performance.

PROJECT CONSIDERATIONS

Many of the problems with today's refractory lining systems actually begin in the project phase. While it is sometimes politically expedient or convenient to identify only one thing as the root cause of a problem, we find in refractory that there are normally numerous contributory issues combining to eventually result in a failure or need for repair, sooner than is necessary. Below we will discuss some of the typical problems we find in different projects which we believe have contributed to various refractory issues.

Project Structure – How a project is structured and how the team is rewarded can have a significant impact on how decisions are made and therefore on the final end product and its performance. The general refractory guidelines are normally established by the licensor early in a project and are used to determine vessel and duct sizing, etc. Unfortunately, due to the extreme variations in size, geometry, and operating requirements for these units, such guidelines are often inadequate. Even worse, some specifications require spending excessive dollars to satisfy unnecessary requirements that bring no true value to the lining, while totally disregarding more important considerations. Getting refractory knowledgeable people involved early in the project is dependent on how the project is structured and is essential in mitigating problems that can develop "down the road". If an EPC is awarded the full turnkey project, the customer normally will not challenge the design basis, believing the EPC has the refractory knowledge as well as being reluctant to impose any requirement that could void a performance warranty. If the Project has broken the refractory components into pieces between a licensor, mechanical erector(s), various equipment OEM's, vessel fabricators and local installers, the communication path and responsibility trails are often too long to effectively assure optimal lining designs and installations. Too often inappropriate companies are involved in the refractory process.

Problems associated with the project structure are typically the result of the team underappreciating the complexity and importance of the refractory system as they focus primarily on a low cost or convenient solution. From the standpoint of assuring the success and reliability of the lining, we believe that the refractory design, supply and installation should not be fragmented and broken apart, and the refractory company should be involved at the very beginning of the project as the vessels and ducts are sized.

Lack of Refractory Knowledge – Several factors contribute to the lack of an industry wide, thorough understanding of the diverse and complex issues associated with sulfur reaction furnaces today. First of all, it is important to recognize that owners, manufacturers, designers and installers have become more and more dependent upon improved monolithic refractory lining systems in recent decades. The significant improvement over the last 25 years in monolithic product formulation, installation knowledge and equipment, has contributed to the replacement of many brick applications with monolithic linings throughout the industrial markets. There has been a corresponding loss of understanding of brick systems. This is particularly true in refineries where either monolithic materials or ceramic fiber based products are used in most other units outside the SRU. Since it is generally accepted (and we agree) that brick linings are the best choice for thermal reactors there has arisen a "disconnect". We have a preferred product form that has experienced a declining user base.

With the loss of brick design and installation skills, it is easy to understand why the EPC/Licensor/OEM's, for which refractories are only a small component to the entire SRU process, are not aware of the details involved in this critical, specialized part of the unit. While we have not worked or visited with every EPC/Licensor/OEM, we have worked or visited with many. As yet, we have not found one with a true refractory design engineer on staff. We normally find an engineer assigned to follow that company's refractory use and specifications. The lack of proper refractory design knowledge combined with their companies having to compete for projects on a price basis, results in a lack of desire to consider improved methods. This lack of internal knowledge leads to the use of outdated specifications ultimately allowing the use of suboptimal materials and systems. Most of the specifications focus primarily on the individual materials to be used without sufficient regard for how those materials can be successfully integrated for the particular application to meet performance expectations. It is the systems' performance that is critical, not the individual materials.

Lack of Customer Feedback – In an attempt to defend the EPC/Licensor/OEM's from some of our comments above, it should be noted that many are very good, quality conscious companies and individuals operating on the firm belief they are providing the best linings possible. Part of the problem between belief and reality is that most times customers do not give proper feedback to the original project participants. We have seen numerous occasions where failures occurred well within the warranty period but the plant proceeded to perform repairs without contacting the original responsible parties. This makes for a more economical repair compared to getting the original parties notified, mobilized to inspect, time to evaluate the failure, etc…all before beginning the actual repairs. Plants typically take this approach as lost production is far more expensive than just performing the repairs and getting the unit back up. The reasons for not contacting the responsible parties may change, but the results do not change; a continuation of poor designs or processes from project to project by the original suppliers who are unaware they had a problem.

WHAT NEEDS TO CHANGE?

Higher Priority on the Refractory Process – To optimize lining life and performance it is strongly suggested to think about refractories sooner in the project rather than later. If reliability is important, the refractory linings should be considered as an integrated system requiring careful consideration through all phases of the entire project instead of passing this important function down the project food chain. Ideally the lining design, material supply and installation would be of single source supply and responsibility in order to prevent miscommunications and misunderstandings. As these functions are normally broken apart and each step bid out independently, it may require restructuring the project to allow for a single-source evaluation and award. Overall, since these linings should be brick construction, the evaluation process should focus on companies that emphasize and are active in industries and units involving complicated brick construction, but also including active participation in brick design and construction in SRU linings. The demanding nature of these units makes it imperative that awards be based on the knowledge and ability of the refractory company rather than the low dollar proposal or to a "Good Ole Boy" network system.

Better Understanding of Reality – First of all, again, the burner systems in these units, whether for low temperature gas plant operations or for refinery related units with or without O2 enrichment, can (and do) melt any refractory made today. Designing for only the theoretical operating range should not be acceptable. Optimizing the system to better survive those upset conditions is the only way to get the best reliability and longest refractory life with the lowest life cycle cost.

Demand an Engineered System – One of the biggest misconceptions in the marketplace is that if it is on a drawing, then the lining is engineered. If so, one can definitely argue the quality and depth of the engineering on some of the drawings utilized. Many of these should be labeled "Bill of Material" or "Refractory Placement" drawings which would be more accurate. Material science does apply to refractories just as it does to the steel components. There are physical material properties that when understood against the operational demands of these systems, can be applied to thermally and mechanically balance the system to optimize the lining performance. An engineered system will address and control many aspects of the system which goes through many changes on its way to a potential of say 3000°F. Much of the true engineering of a self-supported brick system is performed behind the scene and is not obvious to the untrained person looking at the drawing. To illustrate the point, we will discuss a few of the concerns that need to be addressed to optimize a lining system.

- Thermal Balance Of The Entire System – While the refractory lining system is designed to protect the steel shell from high temperature corrosion and metal failure, it is the exterior weather shroud (often inappropriately called a rain shroud) that protects the shell from low temperature corrosion. Both systems, in unison, are designed to protect the vessel shell from high and low temperature corrosion. To those unaware, it seems natural to separate these two items (the refractory and the external shroud) in the initial construction. Many times the two systems are designed by completely different companies (assuming an external shroud was even used). However, the external shroud can have a significant effect not only on the vessel shell temperature, but also on the thermal profile through the refractory lining. This thermal profile can have a significant effect on the ability of the refractory lining components to perform under the extreme stresses and temperatures to which they are exposed. These systems should be designed by the same engineer familiar with the dynamics of both systems and the interaction between these systems.

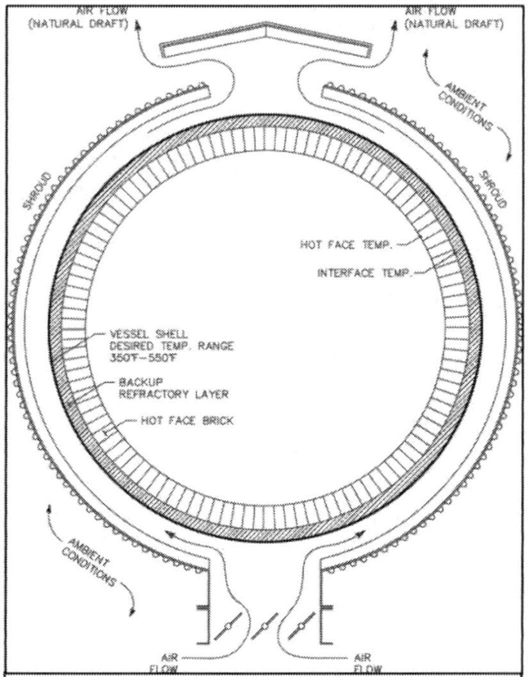

Full Weather Protection "Shroud"

- Proper Lining Thickness – Determining the ideal lining thickness to optimize performance is a complicated process. The importance of this step is often overlooked. Many times, generalized rules of thumb or a similar sized unit is used from a past job that appears to have been successful and as long as a simple thermal calculation seems to show a proper shell temperature; this process stops. However, lining thickness is a function of many factors, some of which are; vessel diameter, thermal profile thru the system, mortar quality and design, brick shape, physical properties of lining materials, etc. Determining proper lining thickness is usually an iterative process as the design engineer balances the effects of the complete thermal system to stresses in the refractory lining and how to control or moderate those effects within the properties of the materials utilized.

- Hot Load Strengths – For unsupported (unanchored) brick linings, consideration of the hot load strength of the refractory products is critical and is the beginning point for designing these systems. Generally the unrestrained thermal expansion of the lining will be approximately 3 times as much as the vessel shell. When heat is applied, the lining goes into severe compression. If the thermal profile thru the lining is excessive, it will plastically deform under the stress resulting in what is referred to as "Hot Load Deformation". So essentially, hot load deformation occurs when the temperature of the material exceeds the point at which it will be able to resist the compressive stress to which it is subjected. While there can be other contributory issues, the evidence of hot load deformation is seen when units are inspected after a service campaign in the form of open joints overhead, sagging brick, or complete collapse of rings overhead.

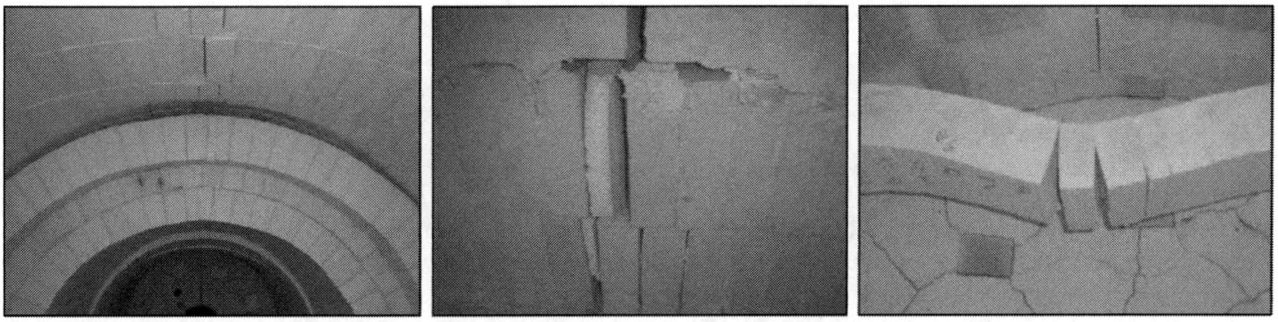

Classic examples of high temperature deformation

Minimizing deformation is a critical component of an overall design. Most specifications do not even address these issues or are misdirected by attempting to make references to specific manufacturers' brand names (or equals) or other physical properties that may have no bearing on the design. One example is specifications for alumina content. The error is assuming that higher alumina content automatically translates into higher hot load strengths. This is sometimes a result of looking at the melting temperature of the individual brick product and not understanding that materials weaken before they melt; i.e., these properties are not directly related to each other. A severe, high temperature, long duration testing regimen clearly shows there are significant differences in these supposedly competitive high alumina bricks and also show, for example, that the highest alumina (99%) bricks are not good for typical usage in reaction furnaces. Field experience has also proven this to be true. Direct testing for the needed properties should be required.

- Brick Keying Action – Due to lining movement as a result of thermal expansion/contraction, it is essential that the individual brick shapes have the necessary "keying action" to keep them from slipping out of position. One must remember that gravity is pulling down on the overhead brick 24/7. If there is not enough keying action or taper on a brick, it will slip out. Should this happen, the entire ring will become unkeyed and the likely result is more brick falling down creating a larger hot spot and typically a shell breech.

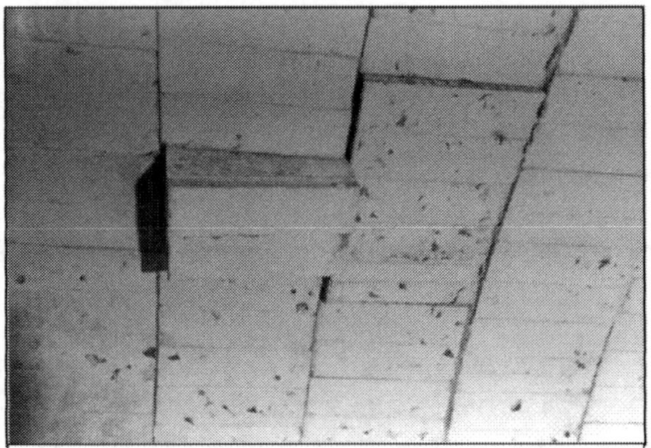

Example of brick slippage due to insufficient keying action.

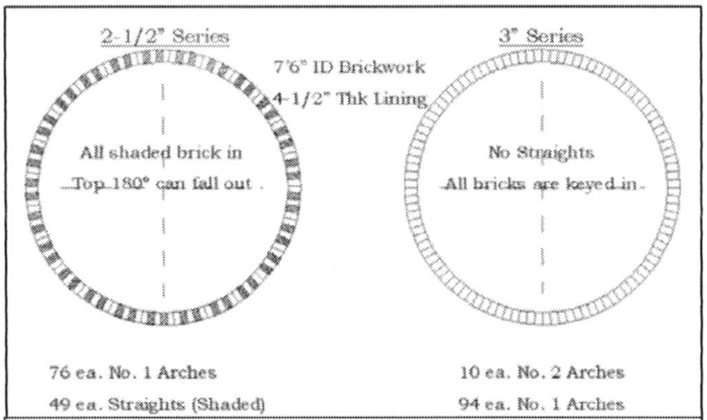

Two sample ring combinations from standard industry manufacturers' guides. The tables do not indicate what combinations are appropriate for the applications.

Utilizing standard brick ring tables will provide a design engineer many options to simply turn a circle, but will not tell them which combinations are best for the application. Choosing the correct brick shapes with the proper amount of keying action requires an experienced designer familiar with the brick shape options, their benefits and disadvantages combined with the need of the specific application.

- Controlling Brick Movement – We have already described the significant thermal movement expected with these products. Unrestrained, brick will continue to move as the unit is thermally cycled with the end result being large open joints, hot spots or collapse or failure of the immediate area of the lining. On the other hand, an excessive amount of restraint will overstress the lining and the vessel shell resulting in possible damage to both. Controlling brick movement requires knowledge and experience to utilize the correct balance between restraining brick movement by use of the selective brick shapes in the correct places while relieving the expected stresses with properly placed and designed expansion relief.

Drawing on left illustrates improperly oriented brick on the dished head and the result of unrestrained brick on the bullnose ring (center picture). Drawing on right illustrates properly oriented head brick resulting in an anchored bullnose ring with expansion relief designed into the outlet throat.

Summary

Improved reliability and performance is possible for these units, even for full time oxygen enriched, high temperature operations. First one must realize the importance of the refractory linings and then raise and maintain the visibility (the coordination) of all aspects involving refractory. Once this is accomplished we believe that by following three simple keys, reliable linings are possible for these severe operating units. Detail attention at every step of the project, start to finish. We call these the **Three Keys to Reliability**:

1. **Engineered Lining Design** – Qualifying a Designer cannot be based on how old they are or how many drawings they have generated. Careful consideration must be given to:
    - Knowledge and familiarity of the operational atmosphere and capabilities of a thermal reactor; normal and upset; real world conditions
    - Understanding the needs of the refractory system to survive the normal and upset conditions, as much as possible.
    - Must be completely familiar with complicated brick construction specifically for SRU's.
    - Able to perform the complicated thermal calculations resulting from the shroud effect and operational swings under different outside ambient conditions
    - Familiar with the information important to the installers of the lining system
    - Clear understanding of the critical material properties needed for the specific lining design

2. **Proper Material Selection** – This should not include brand named products (or equals) as the only criteria. The appropriate material properties critical to meeting the needs of the system must be properly tested for before shipment. For the hotface brick, this should include a regimen of high temperature testing for extended periods of time to verify suitability for service per the design. Additional standard quality control testing should be performed and related to those of the testing regimen to verify consistency for the balance of the manufactured brick. For the other backup and/or monolithic materials, specific critical properties should be determined and included in the manufacturer's requirements.

3. **Experienced Installation Crews** – Not all refractory installers are knowledgeable and skilled at complicated brick construction for SRU reaction furnaces. In today's world of high monolithic refractory use, there are fewer skilled craftsmen than in the past. Careful attention should be paid to the skill level of the installers, the use of appropriate equipment and installation procedures to assure that only experienced crews are involved in the installation of these linings.

(45a) Pilot and Full-Scale Application of Technologies for Mercury Control

Noah Meeks, Research & Environmental Affairs, Southern Company Services, Inc., Birmingham, AL

With the final MATS rule going into effect in 2015, electric utilities must now consider mercury control as a needed technology in their emissions train. This talk wiill cover recent full-scale injections of activated carbon and alkali injection, as well as pilot-scale studies of non-halogenated carbons or non-carbon sorbents. We will also discuss recent mixed matrix membrane pilot studies. **(Most recent data will be included in the abstract closer to the meeting)

(45b) Acid Tunneling Vs Acidizing : A Case Study of Rudatain Oilfield-Kuwait

Adel Elsharkawy, Petroleum Engineering, Kuwait University, Kuwait, Kuwait and Mohamed Al-Hadher, Pet Eng, Kuwait Oil Company, Safat, Kuwait

Acid Tunneling Vs Acidizing : A Case Study of Rudatain Oilfield-Kuwait

By Mohammed Al-Hadher and Adel Elsharkawy

Kuwait University, College of Engineering & Petroleum

Po Box 5969 Safat 13060 Kuwait

Abstract

Acidizing is one of the most important techniques used in oil industry in order to enhance the reservoirs and increase oil production. There are three types of acidizing techniques which are matrix acid stimulation, acid fracturing, and acid tunneling. In this paper a comparison these types of acidizing is discussed. The case study of RA-0007 acid tunneling completed in Kuwait Oil Company is presented. It represents the first dendritic with extensive worm holes created by acid tunneling for maximum reservoir contact. The whole effect is aptly demonstrated by the change in PI of the well before and after the acid tunneling as mentioned before. Analysis of the well designed and implemented job showed that two tunnels were created successfully in MA-C formation through a two days rig-less operation. The results indicated an improvement in the well productivity index (PI) which had been increased from 1.3 to 3.87 bpd/Psi.

(45e) Sustained Casing Pressure Modeling and Well Integrity

Tony Rocha-Valadez, Mary Kay O'Connor Process Safety Center, Artie McFerrin Department of Chemical Engineering, Texas A&M University, College Station, TX, A. Rashid Hasan, Harold Vance Department of Petroleum Engineering, Texas A&M University, College Station, TX, C. Shah Kabir, EPT, Hess Corporation, Houston, TX and Sam M. Mannan, Artie McFerrin Department of Chemical Engineering, Mary Kay O'Connor Process Safety Center, Texas A&M University, College Station, TX

SUSTAINED CASING PRESSURE MODELING AND WELL INTEGRITY

T. Rocha-Valadez[1*], A.R. Hasan[2], C.S. Kabir [3] and S.M. Mannan[1]

[1]Mary Kay O'Connor Process Safety Center, Artie McFerrin Department of Chemical Engineering, Texas A&M University, College Station, Texas 77843-3122, US., email: *tonyrochav@tamu.edu, mannan@tamu.edu.

[2] Harold Vance Department of Petroleum Engineering, Texas A&M University, College Station, Texas 77843-3122, US, email: rashid.hasan@pe.tamu.edu.

[3]Hess Corporation, 1501 McKinney Street, Houston, TX 77010, US, email: skabir@hess.com.

With an increasing demand for hydrocarbons and other sources of energy, the methods for delivering such energy needs have become complex. In the case of the oil and gas industry, exploration and production have gone to deeper waters and deeper horizons than ever before, resulting in drilling several miles into the earth's core. Deeper well penetration means encountering increasingly high formation pressures and temperatures. For example, in deepwater drilling, well construction and production management deal with significant heat transfer between the produced fluids and its surroundings. The high-rate production and the attendant heat transfer demand close attention in terms of operational safety and potential environmental impact. One phenomenon that occurs in producing wells is sustained casing pressure (SCP), which presents a risk to both process safety and the environment.

In oilfield terminology, the casing is a steel pipe cemented to the surrounding formation. Multiple concentric casings are involved in a typical well construction. The casing immediately surrounding the production tubing is known as the production casing, which is most vulnerable to thermal and mechanical stresses and fluid leakage. The cement sheaths provide zonal isolation, support the casing, and protect it from corrosive fluids. Ideally, pressures monitored at the wellhead for multiple casings should change minimally after being bled off through a needle valve at the surface. When the pressure rebuilds after being bled off, that casing is said to exhibit SCP [1].

SCP occurs because of stresses that the system experiences owing to fluids production up the tubing. Besides the normal source of SCP, a poor cement job may allow hydrocarbon migration from the producing or other horizons and accumulation into one or more of the casings. The risk of not attaining well integrity can range from the activation of rupture discs to the blowout of hydrocarbons. Because of the safety and environmental dangers of SCP, regulations require remedial treatments for a casing string that exhibits significant SCP problems. Unfortunately, available methodologies to test for SCP are often qualitative and empirical in nature [2, 3], rendering them inapplicable in many situations. No practical systematic testing procedure has emerged to quantify SCP.

In this paper, we present a model for SCP that is rooted in the transport processes of the system. The model assumes gas leaking through the cement barrier, and then migrating up the annular-liquid column. It accounts for the fluid-temperature profile along the wellbore and liquid compressibility. Some robust assumptions about the transport processes allow us to write the governing equation as a first-order, linear-differential equation that is solved with appropriate boundary conditions. The resulting algebraic expression for the casing pressure rise as a function of time paves the way for all applications. A comparative study of the model's performance with field data suggests excellent agreement for wells where the casing pressure increase is owes gas influx into the annulus. The proposed model could form the basis for standardization of SCP testing. This study also offers some test procedures to assess SCP-related safety concerns.

References

1. Bourgoyne Jr., A.T.J., S.L. Scott, and W. Manowski, ***A Review of Sustained Casing Pressure(SCP) Ocurring on the OCS***, 2000, Final Report Submiteted to MMS.

2. Xu, R. and A.K. Wojtanowicz, ***Diagnosis of Sustained Casing Pressure from Bleed-off/Buildup Testing Patterns.*** SPE, 2001: p. Productions and Operations Symposium.

3. Wojtanowicz, A.K., S. Nishikawa, and R. Xu, ***Diagnosis and Remediation of Sustained Casing Pressure in Wells***, July 2001, MMS.

(45f) Evaluation of Particle Growth in an Electrochemical Fluidised Bed Reactor

Emmanuel Ehirim, Rivers State University of Science and Technology, Port Harcourt, Nigeria

The process of electrodeposition within electrochemical reactor results in the increase of particle sizes, which increases the volume of the bed. This work entails a mathematical model to describe the effect of this phenomenon on the behaviour of a three-dimensional fluidised bed electrode operating under the condition of electron transfer. The model incorporates the pseudo-stationary behaviour in the reactor since the bed increases in volume due to the particle size growth. The Orthogonal Collocation method was used to solve the developed differential equations. The overpotential distribution and the reaction rate within the reactor have been calculated. The principal results are the effects of the electron transfer operation on the superficial displacement with time. The effects of total applied current, electrode geometry and the bed porosity on the behaviour of the system have been analysed. The result shows a very gradual increase in the bed height.

(45g) Debottlenecking Gas Dryers and Desiccant Adsorbers

Benjamin A. Schmitt, Hydrocarbon Processing Industry (HPI) - Business Unit, Johnson Screens, Inc., New Brighton, MN and Michael Ekholm, Engineering, Johnson Screens, Inc., New Brighton, MN

Debottlenecking Gas Dryers and Desiccant Adsorbers

Though Maximization of the Useable Bed Volume with Johnson Screens' SSG^{TM}

Benjamin Schmitt, PE

Senior Mechanical Engineer, Technology

Michael Ekholm, PE

Applications and Product Design Manager

Johnson Screens Inc. St. Paul, MN

AIChE Spring Meeting, April 2013

13th Topical Conference

San Antonio, TX

Introduction:

The following presentation of the Shaped Support Grid (SSG^{TM}) releases additional supporting data and updates to the initial presentation given by Robert G. Norell at the 2010 AIChE Spring Conference.

Limitations on ethylene dryers and desiccant adsorbers can create bottlenecks in the production process which can lead to a need for increased dryer and adsorber capacity. Capital intensive investments may be necessary in order to eliminate the bottleneck which may include the addition of an additional vessel. With limits on the plot plan, this may not be a workable solution. Adding to the problem is the amount of time consumed in engineering, planning and construction, which can have negative effects on the operation of the ethylene unit. Increased desiccant volume and operational efficiency can increase the unit's production with minimal capital expenditures. Johnson Screens', patent pending, SSG^{TM} can add the increased process efficiency and cycle life necessary to extend the operating capability of the unit.

Extending the operation, though debottlenecking, can be achieved by maximizing the time before breakthrough of the media bed. There are three primary functions that maximize the breakthrough time:

Proper gas distribution in both normal operation and regeneration conditions which maximizes the bed utilization.

Increased molecular sieve or desiccant bed volume are achieved through the elimination of traditional bed supports comprised of support grids and beams. The support interface for the screen configuration also allows for thermal expansion typically present in vessels with a thermal cycle as part of the process. The design allows for thermal cycling without the risks of seal or structural failure inherent in other solutions.

Minimizing the bed volume consumed by the distribution and support system increases the net desiccant volume lost when utilizing alternative bed supports.

Johnson Screens ***SSG*** is designed to provide the industry with an alternative, cost effective, approach to increasing capacity. Supporting data and preliminary case studies will be depicted to quantify the benefits that are achievable through the implementation of the ***SSG***.

Conclusions:

The use of a Johnson Screens ***SSG*** provides a reliable immediate solution to reduce capacity bottlenecks. The ***SSG*** increases the bed volume, when compared to traditional bed support arrangements comprised of grids and beams. The increase in desiccant volume will allow for increases in feed flow rate and/or adsorbtion cycle time. The ***SSG*** maximizes the flow distribution during desiccant regeneration, maximizing the utilization of the complete cross section of the vessel. This solution can be implemented in both new and existing vessels, providing consistency in design between planned increases in performance of current vessels and future expansions.

The modular design of the ***SSG*** allows for easy installation and maintenance. Connection of the ***SSG*** to the vessel is performed within the outlet nozzle, eliminating all welding on the pressure vessel and eliminating any need for recertification of the pressure vessel after installation. The design allows for ease of dismantling for inspection of the vessel head during routine maintenance and inspections. The low profile will also reduce interference with any required media exchanges. The design package of the ***SSG*** provides the end user with a holistic solution that fulfills debottlenecking needs with limited capital expenditures.

Biography of Speaker:

Benjamin Schmitt is a Senior Mechanical Engineer in Johnson Screens' Hydrocarbon Processing Industry Technology Group. Ben graduated from North Dakota State University in 2004 with a Bachelor of Science degree in Mechanical Engineering. He currently working toward his Master's in Business Administration from Hamline University. Ben is member of AIChE and is a registered professional engineer in the state of Minnesota.

AMACS
PROCESS TOWER INTERNALS

AIChE 2013 Spring Meeting Poster board
Design Considerations for effective liquid removal In Suction Drums

Presented By:
Kanti Patel, PE

Compressor down again?

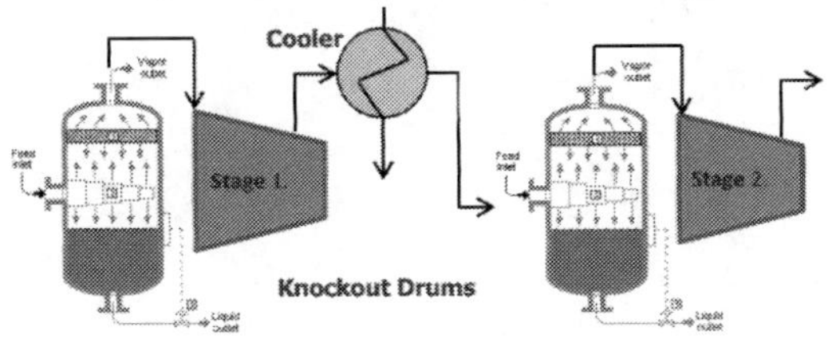

Is liquid carryover the culprit?

<u>We have heard it all before</u>:

-Premature compressor shutdown.

-Production time lost.

-High yearly compressor maintenance costs.

-Impeller damage due to pitting corrosion.

-Corrosion due to salt deposits.

AMACS
PROCESS TOWER INTERNALS

Common Design Problems with Suction Drums which promotes liquid carryover

1. Choosing the correct mist eliminator (M.E. equipment)
2. Knowing where to install the M.E. in a suction drum.
3. Consider a high mist loading in the gas when selecting your M.E.
4. Fouling considerations.
5. M.E. disengaging space, bad gas distribution & liquid slug in the feed.
6. One micron and great droplets, versus submicron mist removal.
7. High viscosity liquids.
8. Dislocated M.E. due to pressure surges.
9. Through put exceeds original design capacity.
10. Capacity range too large for suction drum.

AMACS
PROCESS TOWER INTERNALS

Choosing the correct Mist eliminator equipment

- Proper sizing

+ Use Souders-Brown equation to find optimum velocity. $$Vd = K \cdot \sqrt{\frac{Pl}{Pg} - 1}$$
+ Select the correct vapor velocity factor K. based on the operating pressure, and type of equipment.
  - Lower K-factor 0.01 for every 100 psi
  - K factor to select when using:

Mesh Pad	K= 0.1 - 0.55
Chevron Vane	K= 0.45 - 0.65
MultiPocket® Vane	K= 0.75 – 1.0

AMACS
PROCESS TOWER INTERNALS

Choosing the correct Mist eliminator equipment cont.

- Selection based on droplet size.

Table 2. Droplet sizes (water in air) typically captured with 99.9% efficiency by mist eliminator elements of various types

Element type	Size range microns
Fiber candles or panels	0.1 and larger
Mesh with co-knit yarn	2.0 μm and larger
0.006-inch knitted mesh	5.0 μm and larger
0.011-inch knitted mesh	10 μm and larger
AMACS MultiPocket ® Vane	10 μm and larger
Conventional vane arrays	15 μm and larger

AMACS
PROCESS TOWER INTERNALS

Where to install the M.E. in the suction drum

- Improperly installed M.E. can cause severe liquid carryover problems.
- General accepted spacing guidelines.

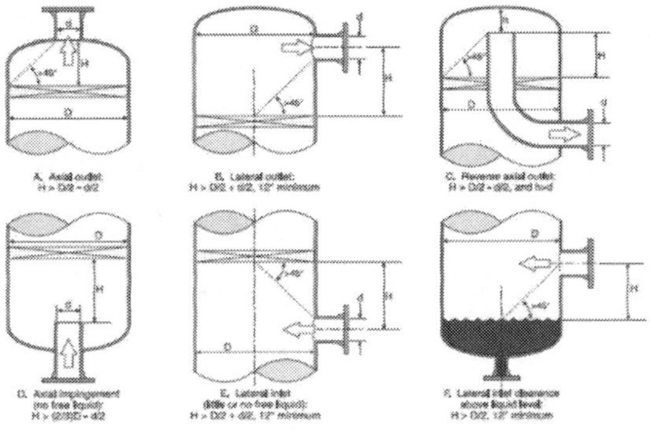

AMACS
PROCESS TOWER INTERNALS

Consider high mist loading when selecting your M.E. equipment

+ High mist loading in the gas stream can flood your M.E. equipment and cause liquid carryover.

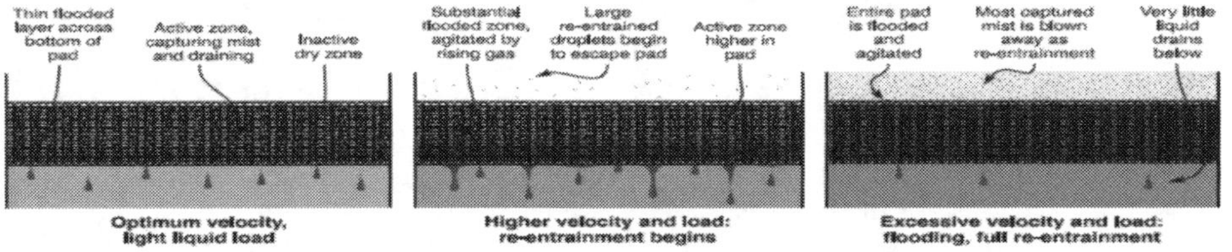

Figure 4. Envisioning mesh pad performance degradation as liquid and gas loads increase (vertical cross-sections)

AMACS
PROCESS TOWER INTERNALS

Consider high mist loading when selecting your M.E. equipment cont.

+ Limited drainage capabilities
  - mesh M.E. <0.5 gal/FT2/min
  - Vane type M.E. <2 gal/FT2/min

+ Select a combination of vane & mesh mist eliminators

AMACS
PROCESS TOWER INTERNALS

Fouling Applications

+ Fouling of mesh mist eliminator will cause liquid carryover.

+ For mild fouling install a water spray system below the mist eliminator for occasional inline cleaning (based on Delta-P)

+ For consistent fouling problems, install a Vane mist eliminator upstream of the mesh mist eliminator.
    - Vanes remove solids and bulk liquids therefore, protecting the downstream mesh mist eliminator

AMACS
PROCESS TOWER INTERNALS

Fouling Applications cont.

+ For consistent fouling problems where the solids are "sticky" install a water spray system below the vane mist eliminator.

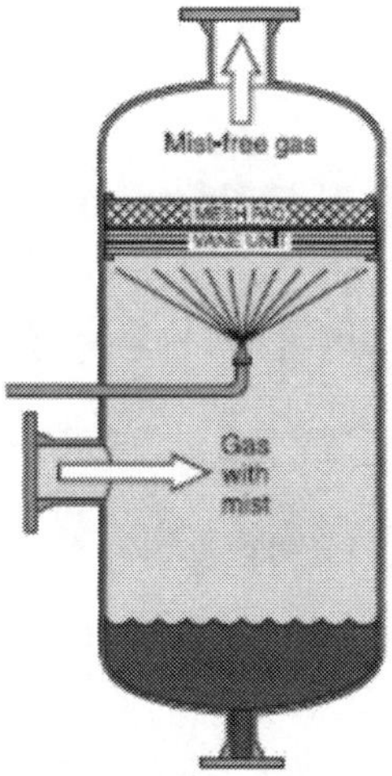

AMACS
PROCESS TOWER INTERNALS

Bad Gas distribution, limited disengaging space and liquid slugs.

+ Bad gas distribution, limited disengaging space in a suction drum can cause severe liquid carryover.

+ CFD analysis can assist in locating high velocity areas.

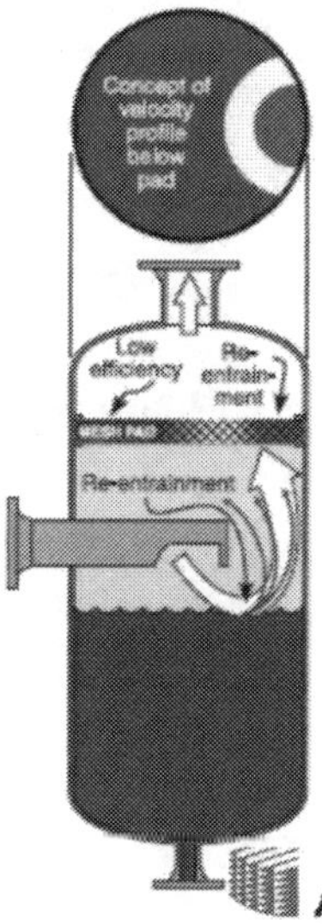

AMACS
PROCESS TOWER INTERNALS

Bad Gas distribution, limited disengaging space and liquid slugs cont.

+ Use inlet diffuser to force uniform gas distribution i suction drum.

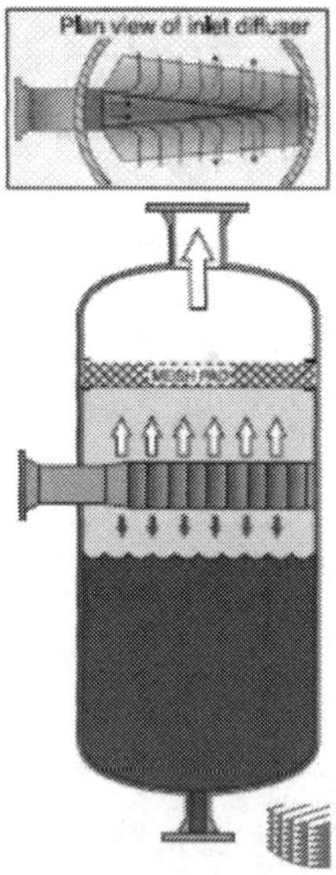

AMACS
PROCESS TOWER INTERNALS

Inlet Feed Device

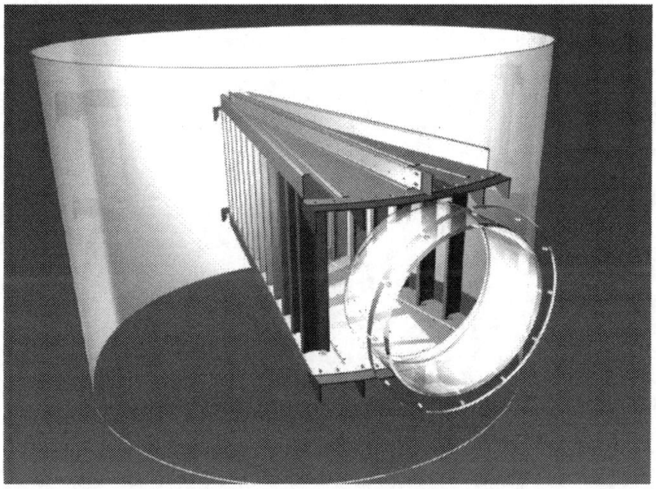

AMACS
PROCESS TOWER INTERNALS

Inlet Feed Device

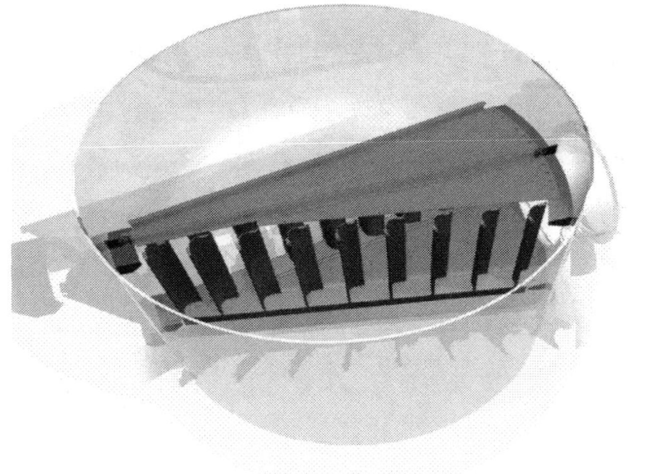

AMACS
PROCESS TOWER INTERNALS

Bad Gas distribution, limited disengaging space and liquid slugs cont.

+ Use a perforated distribution plate downstream of the M.E. for additional distribution assistance.

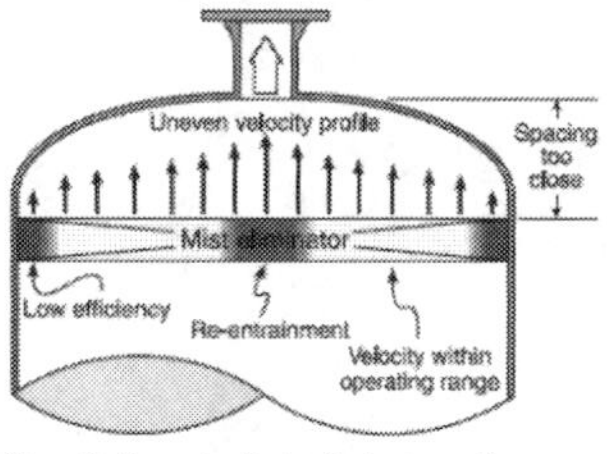

Figure 7. Example of mist eliminator performance degradation due to uneven velocity profile

+ Inlet diffusers also handle incoming liquid slug which otherwise can cause flooding of the Mist eliminator.

AMACS
PROCESS TOWER INTERNALS

Know what mist size droplets You are dealing with

+ Know if your mist is > 1 micron or submicron.
+ When gas travels from A to B, condensation in the pipeline will cause a fine mist (fog or submicron droplets)

  Examples:
  - Incoming gas from an offsite location
  - Fuel gas to furnaces/heaters with low NOx burners

AMACS
PROCESS TOWER INTERNALS

Know what mist size droplets You are dealing with cont.

+ Submicron mist is not captured by conventional M.E.

+ Submicron mist requires a coalescing filter, typically
installed in a two stage filter separator.

AMACS
PROCESS TOWER INTERNALS

Know what mist size droplets You are dealing with cont.

<u>1^{c} stage;</u> submicron droplets are coalesced to > 10 micron droplets.

<u>2^{c} stage;</u> greater than > 10 micron droplets are ejected from the coalescing filter and captured by a vane, which captures droplets > 8 micron.

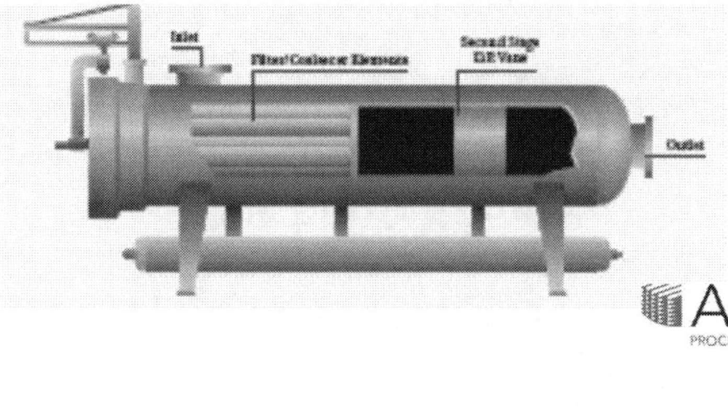

AMACS
PROCESS TOWER INTERNALS

High viscosity liquid as mist

+ Viscous liquids drain slower out of the M.E. and could cause localized flooding.
+ Use a mesh mist eliminator with an open drainage layer on the bottom to improve liquid drainage.
+ Use a lower K- value (lower face viscosity) for the M.E. To allow additional drainage time.
+ Use a vane type M.E. upstream of the mesh mist eliminator, to remove the bulk of the viscous liquid, so the M.E. can effectively handle the residual liquids.

Dislocated M.E. due to System pressure surges

+ M.E. sections can be dislodged into the downstream equipment, which then creates massive carryover.
+ Standard M.E. equipment often gets installed and secured to the support ring with tie wire.

Dislocated M.E. due to System pressure surges cont.

+ Where system pressure surges are common, there are more secure ways to install the M.E.
  - dual support rings
  - through bolts
  - install a pressure relief door

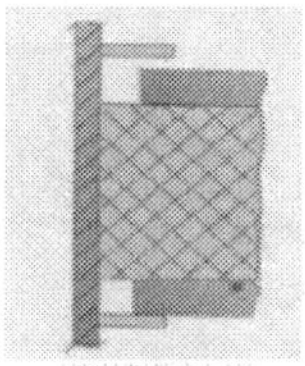

Dual support rings

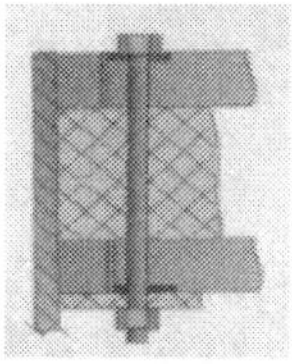

Through bolts

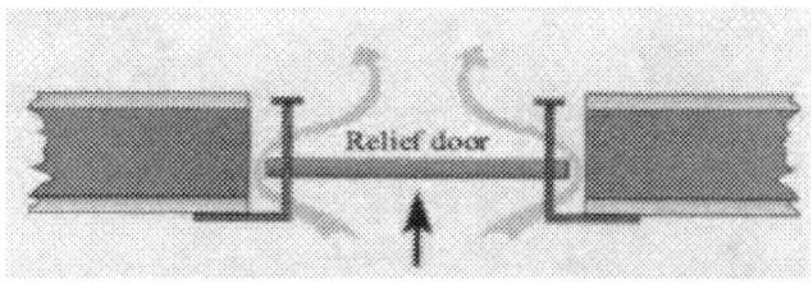

AMACS
PROCESS TOWER INTERNALS

When throughput exceeds original capacity

+ There are various solutions to increase capacity in an existing vessel.
  - Patented High capacit'
    mist eliminator 0% - 1(

AMACS
PROCESS TOWER INTERNALS

When throughput exceeds original capacity cont.

-Patented MISTFIX™ cylindrical mist eliminator 0% - 25%

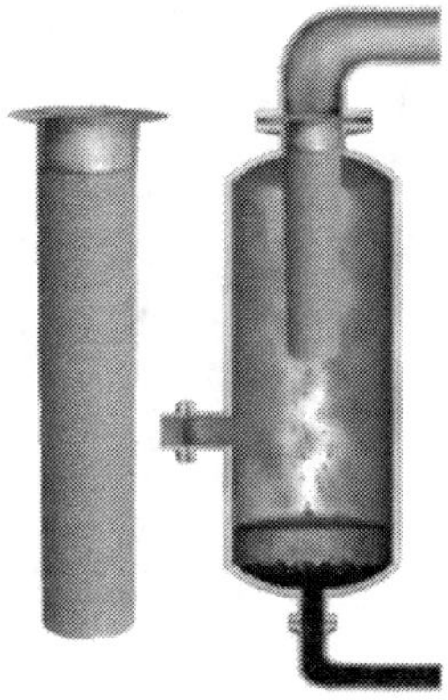

When throughput exceeds original capacity cont.

- High capacity Patented MultiPocket® vane. 0%-50%

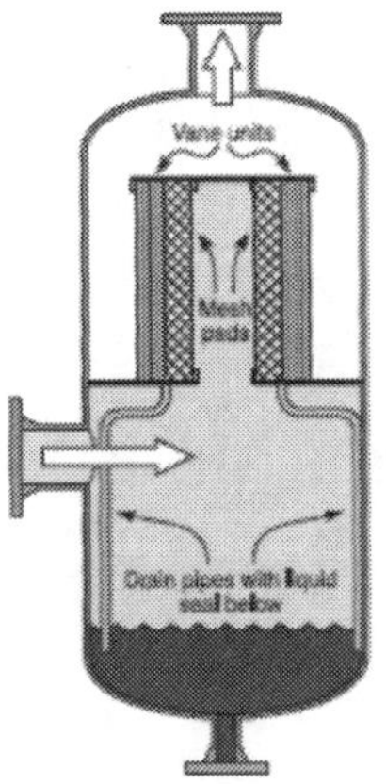

AMACS
PROCESS TOWER INTERNALS

When throughput exceeds original capacity cont.

+ MistFix™ is installed through the gas outlet, all other M.E. equipment comes in sections, installed through the manway.
+ With increased capacity comes increased bad gas distribution.
  - Use an inlet diffuser.
+ To avoid welding new support rings and housing for the high capacity new equipment, use double expansion rings, a compression fit.

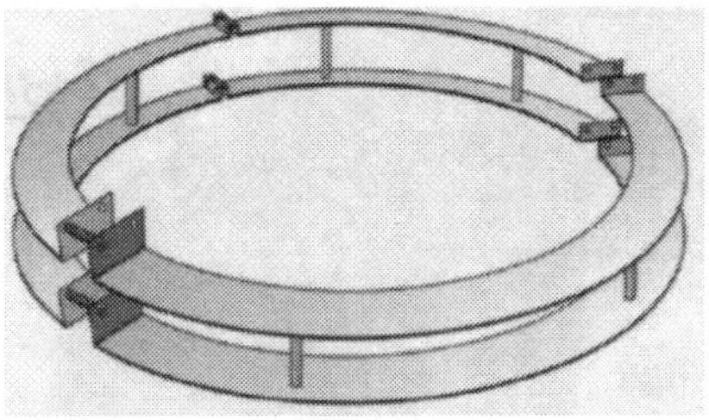

AMACS
PROCESS TOWER INTERNALS

Support Rings

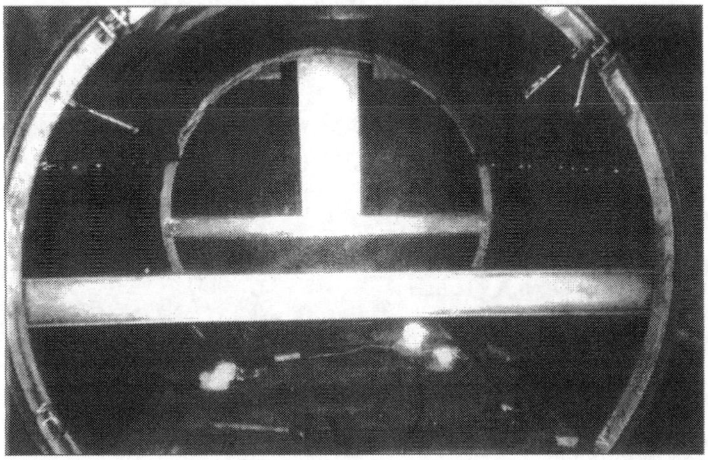

AMACS
PROCESS TOWER INTERNALS

Support Rings

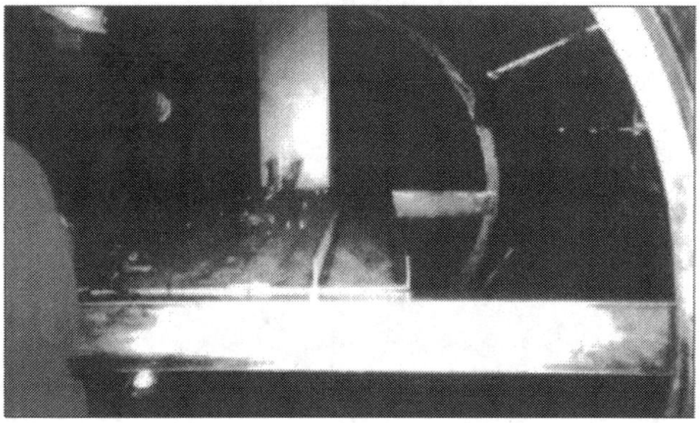

AMACS
PROCESS TOWER INTERNALS

When the capacity range is too large for a suction drum

+ Typical M.E. equipment will operate between

  30% – 110% capacity.

+ When a suction drum often operates below 30% liquid carryover may occur.

AMACS
PROCESS TOWER INTERNALS

When the capacity range is too large for a suction drum Cont.

To facilitate a capacity range of 10% - 110% consider a 2 phase M.E. in the same drum.

- one that operates 30% - 1
- second that operates between 10% - 30%

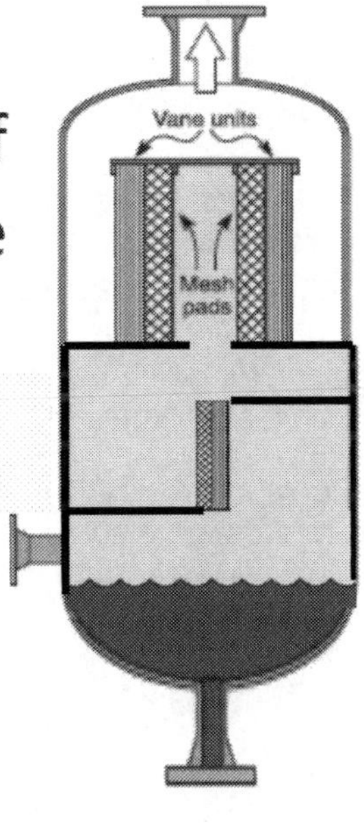

(45i) Renewable Feedstock for Steam Crackers: Catalytic Upgrading of Crude Tall Oil (CTO) Into Bio-Naphtha

Jinto Anthonykutty, Juha Linnekoski, Antero Laitinen and Ali Harlin, VTT Technical Research Centre of Finland, Espoo, Finland

Renewable feedstocks from bio resources are better alternatives to petroleum derived naphtha for the production of olefins and aromatics by steam cracking or reforming.[1, 2] Mostly, these renewable feedstocks (bio-naphtha) rich in n-alkanes are produced by catalytic upgrading methods via hydrotreating/hydroprocessing; in which oxygenates are removed from bio-oil by various deoxygenation processes. At present, hydrotreated vegetable (HVO) oil is produced in many countries on industrial scale using existing petroleum refinery methods[3]. However, employing higher priced vegetable oils could be a major concern in these processes and the main challenge is therefore finding an inexpensive and non-food chain affecting feedstock. Tall oil, obtained as a by-product from Kraft pulping process, is abundant and cheap;and hence, meets the criteria of a low cost starting material for upgrading[4]. Moreover, the utilization of tall oil in conventional oil refineries could serve as an eco-friendly method to convert bio-wastes into bio-naphtha[1].

In the present research approach, catalytic upgrading (hydrotreating) of crude tall oil (CTO) into bio-naphtha suitable for a steam cracker is discussed. CTO, which contains mainly fatty acids (42.9 wt. %) and resin acids (25.7 wt. %) was obtained from Stora-Enso Kaukopää (Finland) mill and used as received. A commercial NiMo catalyst was employed and the experiments were conducted in a continuous, down flow, fixed bed reactor at different process conditions (WHSV 1-3h^{-1}, Pressure 5 MPa and Temperature 350 - 450 ^{o}C). After the stabilization period (6 h.), the liquid sample was collected and separated into organic and aqueous phases.The product distribution in organic phase was analyzed with a Shimadzu GC/MS equipped with a Zebron MS 30m × 0.25 mm × 0.25μm column.

This current approach reports that CTO produces mainly paraffins, cycloalkanes and aromatics during upgrading process. Paraffinic hydrocarbons mainly comprise n-octadecane (20-25 wt. %) and n-heptadecane (15-20 wt. %), are obtained in maximum yield (55- 60 wt. %) in low temperature reactions. The yield of paraffinic hydrocarbons increased as a function of increasing space time in low temperature reactions. 18-Norabietane and other mono-, di- and tri- cycloalkanes are obtained from resin acids in CTO, and these cyclic structures are converted to aromatics such as naphthalene and phenanthrene in high temperature reactions (>400 ^{o}C). Stigmastane, apparently produced from sterols in CTO are also present as a major three ring component in the product stream. Additionally, gaseous products obtained and the yield increased as a function of temperature. The trend observed with the distribution of gaseous products reveal that decarboxylation and cracking reactions are favored at higher temperatures, while hydrogenation and hydrodeoxygenation (HDO) are suppressed.

The elemental composition of CTO shows that it contains relatively high sulfur content (1800 mg/kg), could be useful to keep the catalyst's activity with time on stream (TOS). However, the considerable amount (87 mg/kg) of alkali metals (Na, K, Ca and P) in CTO might be critical for the catalyst activity with TOS, will be discussed in this contribution.

The product composition obtained with CTO is assessed in comparison with compositions and characteristics of refinery naphtha feeds, and will be discussed in detail. Furthermore, this contribution will discuss the achieved deoxygenation rate and success criteria for the product composition of upgraded CTO.

References

1. Gornay, J.; Coniglio, L.; Billaud, F.; Wild, G. Steam Cracking and Steam Reforming of Waste Cooking Oil in a Tubular Stainless Steel Reactor with Wall Effects. ***energy&fuels***, **2009**, 23(11), 5663-5676.
2. Pyl, S.P.; Schietekat, C.M.; Reyniers, M.F.; Abhari, R.; Marin, G.B.; Van Geem, K.M. Biomass to olefins: Cracking of renewable naphtha. ***Chemical Engineering Journal***, **2011**, 176-177, 178-187.
3. Sunde, K.; Brekke, A.; Solberg, B. Environmental Impacts and Costs of Hydrotreated Vegetable Oils, Transesterified Lipids and Woody BTL- A Review. ***Energies***, **2011**, 4(6), 845-877.
4. Coll, R.; Udas, S.; Jacoby, W.A. Conversion of the Rosin Acid Fraction of Crude Tall Oil into Fuels and Chemicals. ***energy&fuels***, **2001**, 15, 1166-1172.

AIChE Paper Number 45j
High Efficiency Ethylene Cracking Furnaces

Rajaram Ramesh

Product Development Manager - Ethylene
Technip
Netherlands

Marco W. M. van Goethem, PhD

Product Development Manager – Modeling and Testing
Technip
Netherlands

25th Ethylene Producers Conference
AIChE 2013 Spring National Meeting
San Antonio, Texas, April 29 – May 3, 2013

Abstract

Environmental and economic pressures generate a driving force to produce products more efficiently. A direct effect of more efficient processes is that the emission of green house gases, particularly carbon dioxide decreases. In households nowadays "high efficiency furnaces" are available with efficiencies in the range of 94-97%. This paper describes the outcome of a feasibility study on applying the high efficiency furnace concept to steam cracking furnaces. The high efficiency is achieved by enhanced heat recovery from the flue gases before they are released to the atmosphere at ca. 65°C.

Technip studied the application of this concept for cracking furnaces based on gas as well as liquid feedstocks. Natural gas was used as basis for the energy price. Since energy prices have a strong impact on project economics, the evaluations were done for an EU average natural gas price and a US natural gas price.

Two alternative heat sinks have been considered: preheating of combustion air and preheating of cold hydrocarbon feed. Both are viable for liquid cracking furnaces. In the case of gas cracking furnaces feed preheating results in a pinch and this makes it not viable.

Capital costs are based on Technip's cost database and quotations from air preheater vendors. Special attention has been paid to the material selection of the affected convection section banks.

This study shows that high efficiency ethylene cracking furnaces are economically viable. For Europe the study indicates an attractive payback period. In the US the payback period is longer due to the lower energy prices. If energy prices increase in the future, the concept's economic viability will become better. Inclusion of CO2 credits will make the payback period more attractive.

Furnace Efficiency

In this paper furnace efficiency has been calculated as net thermal efficiency as interpreted by API560[1)]. Furnace efficiency is determined by drawing an envelope around the furnace and including in it the radiant section, convection section and any heat exchanger equipment that recover heat from the flue gas (air preheater, economizer, steam superheater, etc.). The efficiency is the ratio of the useful heat absorbed to the fired heat of the fuel and any other external heat inputs. Ambient temperature is a reference temperature for efficiency calculations and any heat inflow or outflow above this reference is taken into consideration as a heat input or heat loss.
Furnace efficiency can be determined as follows:

$$\frac{\textit{heat absorption in radiant section} + \textit{convection section}}{(\textit{firing rate} \times \textit{fuel heating value}) + \textit{sensible heat of fuel and air above ambient temp.} + \textit{other external heat inputs}}$$

Alternately, Furnace efficiency $= 100\% -$ Total heat loss, where:

$$\text{Total heat loss} = \frac{\textit{stack loss} + \textit{radiation heat loss}}{(\textit{firing rate} \times \textit{fuel heating value}) + \textit{sensible heat of fuel and air above ambient temp.} + \textit{any external heat inputs}}$$

In a furnace where fuel and combustion air enter the furnace at ambient temperature and there are no other external heat inputs, the first expression can be simplified to the following:

$$\textit{Furnace efficiency} = \frac{\textit{heat absorption in radiant section} + \textit{convection section}}{\textit{fuel firing rate} \times \textit{fuel heating value}}$$

Please note that net thermal efficiency, as defined by API 560, is based on the lower heating value of the fuel.

Study objectives

Technip undertook this study with the goal to reduce fuel firing in cracking furnaces by at least two percent. Although this figure does not look very ambitious, it must be remembered that since efficiencies in current designs are already in the mid nineties (upper part of the technology S-curve), further improvements will not be easy and can only be incremental.

Since stack loss is the main source of inefficiency (radiation heat loss is comparatively smaller and more difficult / expensive to reduce) it was decided to make stack loss reduction as the main focus of the study.

By a relatively straight forward calculation it can be shown that the two percent efficiency rise target would require the flue gas to be cooled to approximately sixty or seventy degrees Celsius.

The main challenge in reducing stack temperature to this level is finding an appropriate heat sink for the recovered heat. In a cracking furnace there are two suitable heat sinks, combustion air and hydrocarbon feed, and both were studied. The following sections discuss the results for liquid furnaces and gas furnaces.

Basis of Study

We have done the study on a naphtha furnace and an ethane furnace. We have chosen these two feedstocks since a large number of furnaces operate on these two feeds and we can draw conclusions that can be applied approximately to other liquid and gas feeds. The study has been performed on the following bases:

- The reference liquid and gas furnace designs used in this study are based on actual furnaces from Technip's past projects. The furnaces were simulated using Technip's proprietary tool SPYRO which is used in the design and simulation of cracking furnaces.
- The budgetary capital cost need to achieve the target furnace efficiency was calculated as an incremental cost. It was estimated from vendor quotes and Technip's internal cost database.
- The cost estimates are for a new project, not a revamp, and are for a West European location.
- Operating costs are based on cost of base utilities and derived utilities. Base utilities are natural gas and demineralized water. These are taken from market data.
- Since natural gas prices are very different in different parts of the world, the operating costs and economic analysis have been done for European and United States gas prices.
- EU natural gas price was fixed at €25.1 /MWh of higher heating value (HHV) which is equivalent to €28.0 /MWh of lower heating value (LHV). The prices are based on EUROSTAT data for 1st half of 2010 and were obtained from the publications of the Department of Energy & Climate Change in the UK. They correspond to the EU-15 average for Large Industrial Consumers (i.e. 2.13 pence/kWh HHV).
- US Natural gas prices were obtained from the US Department of Energy. The average value was 6 $/MM BTU (HHV) or 17 €/MWh (LHV) for 1st half of 2010. The average HHV used is 1029 BTU per cubic foot or 1.029 MM BTU per thousand cubic feet.
- Cost of demineralized water was assumed at € 2.00 /ton.
- The prices of other utilities (boiler feed water, steam and electricity) have been calculated from those of the two base utilities. Electricity was priced assuming 35% overall efficiency of power generation.
- The economic calculations do not include any CO2 credit for the fuel saved.
- Assumed exchange rate: 1 US dollar = 0.75 euros
- The plant is assumed to operate 8400 hours per year

The resultant utility prices used in the study are summarized below.

UTILITY COSTS

		EU basis	US basis	
FUEL as NATURAL GAS	Base	**28.0**	**17.0**	€ / MWh
DEMINERALISED WATER	Base	2.0	1.5	€ / ton
VHP STEAM (net value)	Derived	23.7	14.4	€ / ton
MP STEAM (net value)	Derived	13.9	8.5	€ / ton
LP STEAM	Derived	9.7	5.9	€ / ton
ELECTRICAL POWER	Derived	80.1	48.6	€ / MWh

Study Results

Naphtha Furnace using Air Preheat

Combustion air can be preheated in a plate type air preheater (APH). These are cross flow type exchangers and have low pressure drops. Combustion air is at a lower temperature than any of the process or utilities streams entering the convection section; therefore, from the standpoint of efficiency, APHs are ideally suited to be the last heat recovery section before the stack.

Fig 1 shows the Base Case Naphtha furnace while Fig 3 shows the same furnace with an APH. Please also refer to Table 1 for a performance comparison of the APH design with the Base Case naphtha furnace. The following points are of interest:

1) The overall efficiency of the furnace will improve from 94.4 to 97.0% for the APH option. This is mainly due to the lower stack temperature but also partly due to the smaller excess air, which is explained below.

2) This option will require a forced draft combustion air fan to overcome the pressure drop in the APH. As a result the furnace operation will change from induced draft mode to balanced draft. The availability of air in forced draft will enable more accurate control of the excess air and it will be possible to operate the furnace with lower excess air (5% instead of 8%). This leads to a lower flue gas flow, a lower stack loss and therefore better efficiency.

3) There will be a drop of 4.8% in fuel firing which is greater than the efficiency increase. This is because of the lower absorbed heat. The furnace is designed and controlled in such a way as to supply all the required process heat and to absorb the surplus heat into utilities to the extent possible. When the furnace operates more efficiently and the heat demanded by the process media remains unchanged, as in the APH case, there will be less surplus heat available for VHP steam. As a result the VHP steam generation will go down from 79.7 to 76.0 t/h.

4) The heat transfer surface in some parts of the APH will be below the water dew point of the flue gas (around 60°C). This can lead to water condensation and create two possible issues: 1) accumulation of water in the APH and 2) corrosion due to the presence of liquid water, oxygen and carbon dioxide.

   Two measures have been included in the design to counter these problems:
   - Placement of unit such that flue gas flows downward to aid easy draining
   - Use of corten steel to resist corrosion

5) We have evaluated the impact of the air preheat option on NOx emissions using API 535[2)]. The rise in air temperature and the drop in excess air will have opposing effects on the NOx production. The net effect is that the NOx concentration will be close to the Base Case.

   If a DeNOx unit is present, as is often the case in the US, the APH option may help to slightly extend the life of the DeNOx catalyst. This is because the total NOx load will be smaller than the Base Case due to the smaller flue gas flow.

Table 2 summarizes the results of the economic analysis of this option. We have used payback period to compare different options. Payback period is the ratio of the incremental capital cost above the Base Case CAPEX divided by the savings in yearly operating cost.

1) The capital cost (CAPEX) of the APH option includes the cost of the APH, the combustion air fan, extra ducting and structural steel. The operating cost (OPEX) savings is the savings in furnace fuel but this must be offset against the value of the lost VHP steam production (which notionally must be made up in a utility boiler elsewhere) and the additional electricity for the two fans.

2) It can be seen that the payback period is three years based on European gas prices and five years based on American gas prices.

Naphtha Furnace using Feed preheat

Naphtha is normally stored at ambient temperature and is preheated to 60°C using quench water before it is routed to the furnaces. Most cracking furnaces are equipped with a Feed Preheater (FPH) bank in the convection section. If flue gas is utilized to preheat naphtha from ambient conditions the stack temperature can be lowered to the desired range of sixty to seventy degrees Celsius. Till now designers and operators have not taken full advantage of this mainly because of the higher CAPEX and fear of dew point corrosion.

One of the consequences of this feature will be that the quench water duty previously absorbed by the naphtha will now shift to the air coolers in the quench water circuit. The additional electrical power that the air cooler fans will require will be about 10 KW per furnace.

Please refer to Table 3 for an overview of the performance of a naphtha furnace with and without an extended feed preheater. Please note that the Base Case naphtha furnace is the same as the one considered in the APH option.

1) The overall efficiency of the furnace can be improved from 94.4 to 96.0% for this option.

2) Unlike the APH option a combustion air fan will **not** be required. The furnace will continue to operate in induced draft mode.

3) The fuel firing and the VHP steam generation will drop marginally.

4) The upper tube rows of the FPH will be below the water dew point of the flue gas and some condensation will occur. As a precaution we have considered that the top four rows will be of stainless steel.

5) The flue gas NOx concentration will be unaffected by this option. However the total hourly NOx emissions and load on the DeNOx unit, if any, will decrease to the same extent as the fuel firing, which is around 0.6%.

Table 4 summarizes the results of the economic analysis of this option.

1) The capital cost (CAPEX) of this option covers the incremental cost of the extended FPH, extra ducting and structural steel. The OPEX savings is the reduction in furnace fuel after adjusting for the reduction in VHP steam production and the additional electricity for the ID fan and the Quench Water Air Cooler fans.

2) In the Base Case the naphtha feed piping between the external feed preheater and the convection section is insulated for heat conservation. In the FPH option naphtha will be fed at ambient temperature to the furnace and so these lines do not have to be insulated. This will yield some capital cost saving but this has not been taken into account in our analysis.

3) The payback period excluding CO2 credit will be three and a half years with European natural gas and five and a half years with US natural gas.

Ethane Furnace using Air Preheat

Fig 2 shows the Base Case Ethane furnace that was used in the study. The APH arrangement on an ethane furnace will be quite similar to the one on the naphtha furnace. Table 5 shows the impact that the APH would have on the furnace performance. The main points are as follows:

1) The overall efficiency of the furnace will improve from 91.7 to 96.5% for the APH option. This is mainly due to the lower stack temperature but also partly due to the smaller excess air.

2) In this furnace also the APH will require a forced draft combustion air fan. This will allow the excess air to be reduced from 8% to 5% which will result in lower stack losses and a better efficiency.

3) The fuel firing will go down by 11.0%, from 8.2 to 7.3 t/h. The absorbed heat will drop from 116.8 to 110.6 MW. As a result of these the VHP steam generation will go down from 97.0 to 88.8 t/h.

4) A portion of the APH will operate below the flue gas water dew point. Consequently the unit must be designed to prevent water accumulation and corrosion.

5) The impact on NOx was evaluated here also. The rise in air temperature will have a greater impact on NOx than the fall in excess air. We estimate that as a result the flue gas NOx concentration will increase slightly (around two percent). This can be a cause for concern if the furnace does not have a DeNOx unit and is operating close to the limits imposed by authorities. However since the total firing will go down, the hourly NOx production will drop. Plants can consider discussing with authorities to obtain a higher allowable NOx limit,

   If the furnace has a DeNOx unit, the chemical load on the unit will reduce, which should help to obtain a longer catalyst life.

The operating cost savings and payback for this option have been summarized in Table 6.

1) The capital cost includes the cost of the APH, the combustion air fan, extra ducting and structural steel. The operating cost savings is the savings in fuel firing with an adjustment for the value of the reduced VHP steam generation and the extra electricity for the ID fan and FD fan.

2) The payback period with natural gas priced at European rates will be one and a half years. The corresponding figure for US gas prices will be three years.

Ethane Furnace using Feed Preheat

Using feed at ambient temperature does not significantly reduce the stack temperature in a gas furnace. This is because the feed preheat curve has a steeper slope than the flue gas cooling curve, causing a pinch on the hot side of the Feed Preheater. Fig 4 shows the heating and cooling curves of ethane, boiler feed water and flue gas in the convection section. The hatched shape on the plot shows the maximum additional heat duty that can be recovered into feed, over and above the Base Case. The plot clearly shows that when the ethane feed temperature is lowered from the base value of 55°C to 30°C, the stack temperature will drop from 170°C to about 150°C. The flue gas cannot be cooled any more than this modest amount.

For this reason the feed preheat option was not investigated any further.

Conclusions and Summary

The key results from the different study options are listed together below.

Cracking furnace feed	Naphtha		Naphtha		Ethane	
Cold sink	Combustion air		Hydrocarbon feed		Combustion air	
	EU nat. gas	US nat. gas	EU nat. gas	US nat. gas	EU nat. gas	US nat. gas
OPEX savings	€ 0.60 mill	€ 0.36 mill	€ 0.17 mill	€ 0.10 mill	€ 1.27 mill	€ 0.77 mill
Payback time	3 years	5 years	3.5 years	5.5 years	1.5 years	3 years
Salient features of each option	Small number of additional equipment		No additional equipment		Small number of additional equipment	
	Higher CAPEX		Lower CAPEX		Higher CAPEX	
	More OPEX savings		Less OPEX savings		Best OPEX savings	
	Slight effect on NOx concentration		No effect on NOx		Around 2% higher NOx concentration	

All three scenarios have attractive payback periods and will be worth considering in high energy cost countries such as Europe and Asia. In US and other energy rich nations, the economic incentive will be much lower. In the US, where ethane cracking is common and increasing even more, improvement of furnace efficiency with air preheat can be worth considering as the payback time is less than 3 years.

If the options are compared against each other, in our opinion the Air preheating option has more potential, irrespective of feedstock.

The Feed Preheating option in liquid furnaces is cheaper but it also gives proportionally modest savings. Moreover, as earlier mentioned, the feed preheat duty is being shifted from one waste heat source (quench water) to another (flue gas). This option will not lead to a net improvement or cash savings in the plant as a whole unless there is a shortage of quench water and there are other, better uses for the saved quench water duty

Acknowledgements

The authors wish to express their thanks to their colleagues Mr Ashok Laghate, Ms Susan Cooper, Mr Sachin Singare and others in Technip Benelux who contributed to this study.

References

1) API 560: Fired Heaters for General Refinery Service, 4th edition

2) API 535: Burners for Fired Heaters in General Refinery Service, 2nd edition

Table 1
Naphtha Furnace with Combustion Air Preheating – Performance Summary

		Base Case	Air Preheat Case
Naphtha feed	t/h	60.8	60.8
Ambient conditions	-	15°C, 85% R.H.	15°C, 85% R.H.
Air temp. to burners	°C	15	77
Stack temperature	°C	108	61
Excess air	%	8	5
Absorbed heat	MW	113.2	110.4
Overall efficiency	%	94.4	97.0
Fuel flow	t/h	8.4	8.0
Fired heat	MW	119.9	113.9
VHP steam production	t/h	79.66	76.01
NOx concentration		Base	Close to Base

Table 2
Economics of Naphtha Furnace with Combustion Air Preheating

		Air Preheat Case (EU location)	Air Preheat Case (US location)
Savings in fuel firing	million €/yr	1.42	0.86
Cost of lost VHP steam capacity	million €/yr	0.73	0.44
Cost of additional electricity	million €/yr	0.10	0.06
Net OPEX saving	million €/yr	0.60	0.36
Payback	Years	3	5

Table 3
Naphtha Furnace with Feed Preheating – Performance Summary

		Base Case	Feed Preheat Case
Naphtha feed	t/h	60.8	60.8
Feed temperature	-	60°C	30°C
Combustion air temp.	°C	15	15
Stack temperature	°C	108	71
Excess air	%	8	8
Absorbed heat	MW	113.2	114.3
Overall efficiency	%	94.4	96.0
Fuel flow	t/h	8.4	8.35
Fired heat	MW	119.9	119.1
VHP steam production	t/h	79.66	79.63
NOx concentration		Base	Base

Table 4
Economics of Naphtha Furnace with Feed Preheating

		Feed Preheat Case (EU location)	Feed Preheat Case (US location)
Savings in fuel firing	million €/yr	0.19	0.12
Cost of lost VHP steam capacity	million €/yr	< 0.01	< 0.01
Cost of additional electricity	million €/yr	0.01	< 0.01
Net OPEX saving	million €/yr	0.17	0.10
Payback	Years	3.5	5.5

Table 5
Ethane Furnace with Air Preheating – Performance Summary

		Base Case	Air Preheat Case
Ethane feed	t/h	51.9	51.9
Ambient conditions	-	15°C, 85% R.H.	15°C, 85% R.H.
Air temp. to burners	°C	15	129
Stack temperature	°C	169	64
Excess air	%	10	5
Absorbed heat	MW	116.8	110.6
Overall efficiency	%	91.7	96.5
Fuel flow	t/h	8.2	7.3
Fired heat	MW	127.3	114.5
VHP steam production	t/h	97.0	88.75
NOx concentration		Base	Base + 2%

Table 6
Economics of Ethane Furnace with Air Preheating

		Air Preheat Case (EU location)	Air Preheat Case (US location)
Savings in fuel firing	million €/yr	3.01	1.83
Cost of lost VHP steam capacity	million €/yr	1,64	1.00
Cost of additional electricity	million €/yr	0.10	0.06
Net OPEX saving	million €/yr	1.27	0.77
Payback	Years	1.5	3

Fig 1
Naphtha cracking furnace

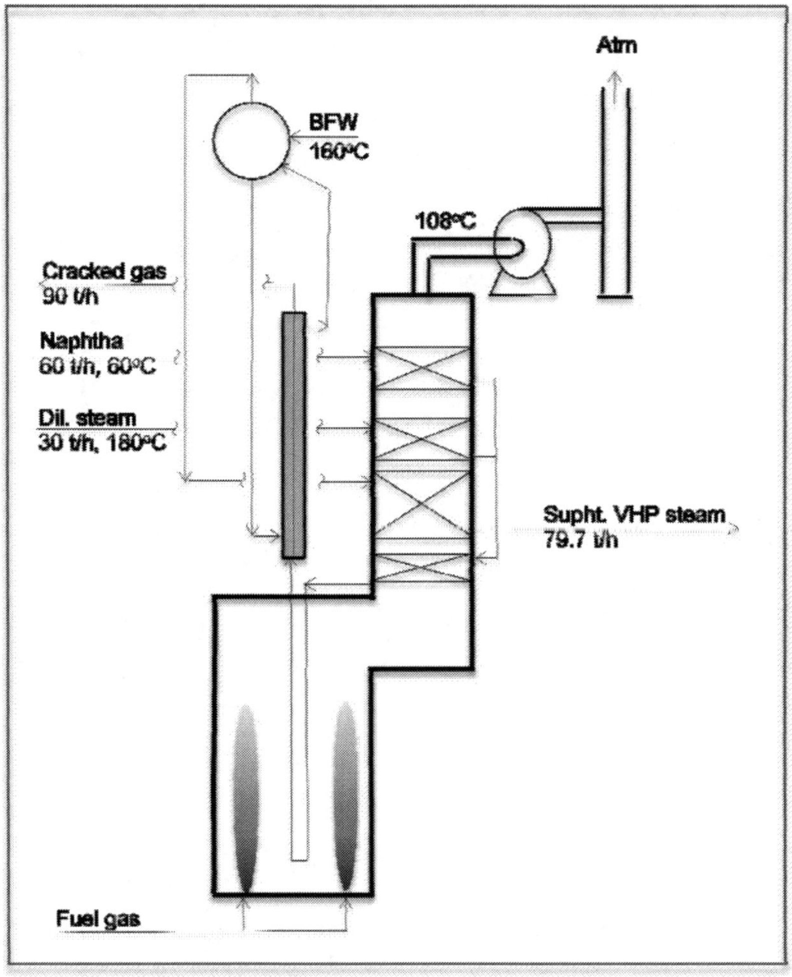

Fig 2
Ethane cracking furnace

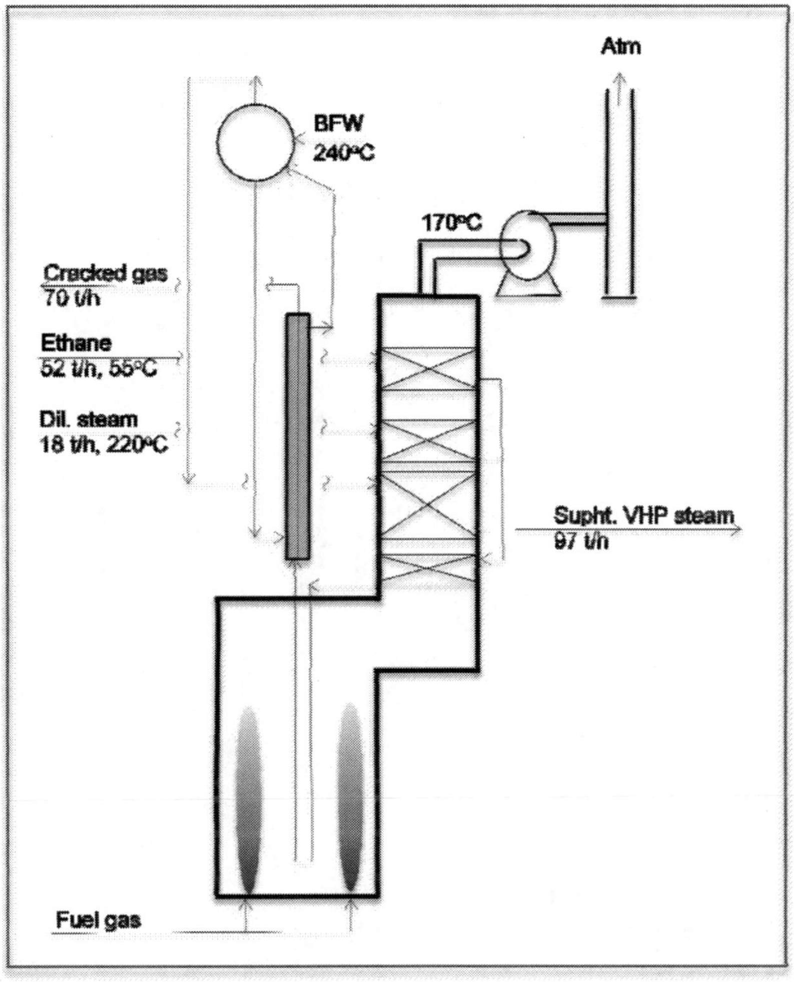

Fig 3
Naphtha cracking furnace – Air Preheater option

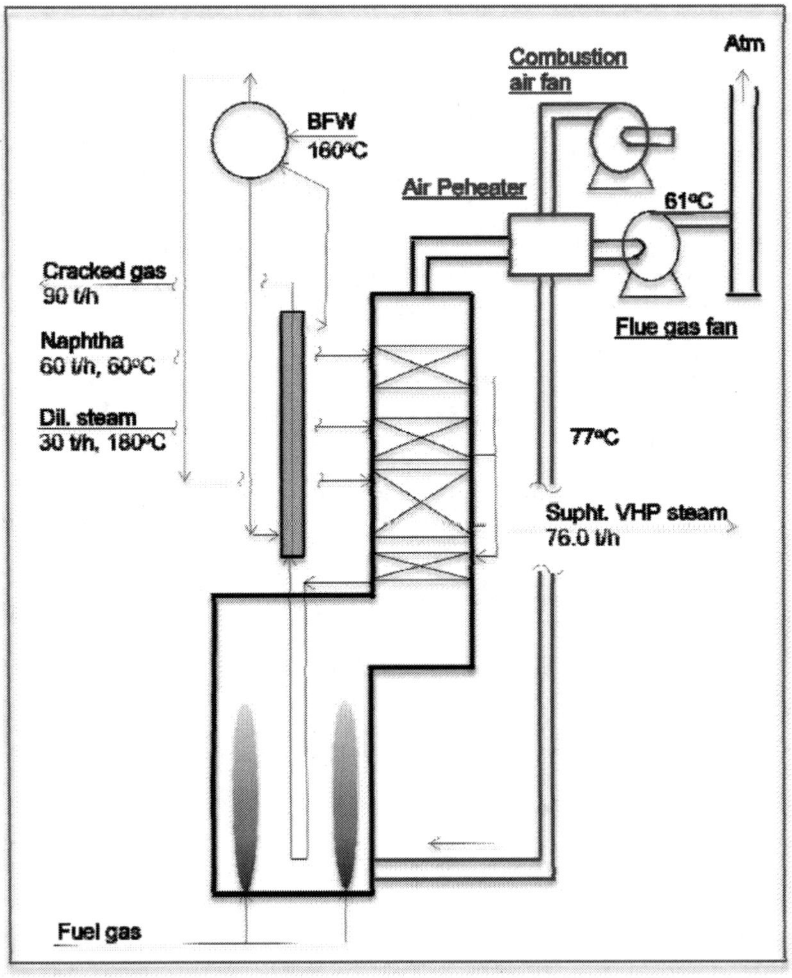

Fig 4

Ethane cracking furnace – Feed Preheater option

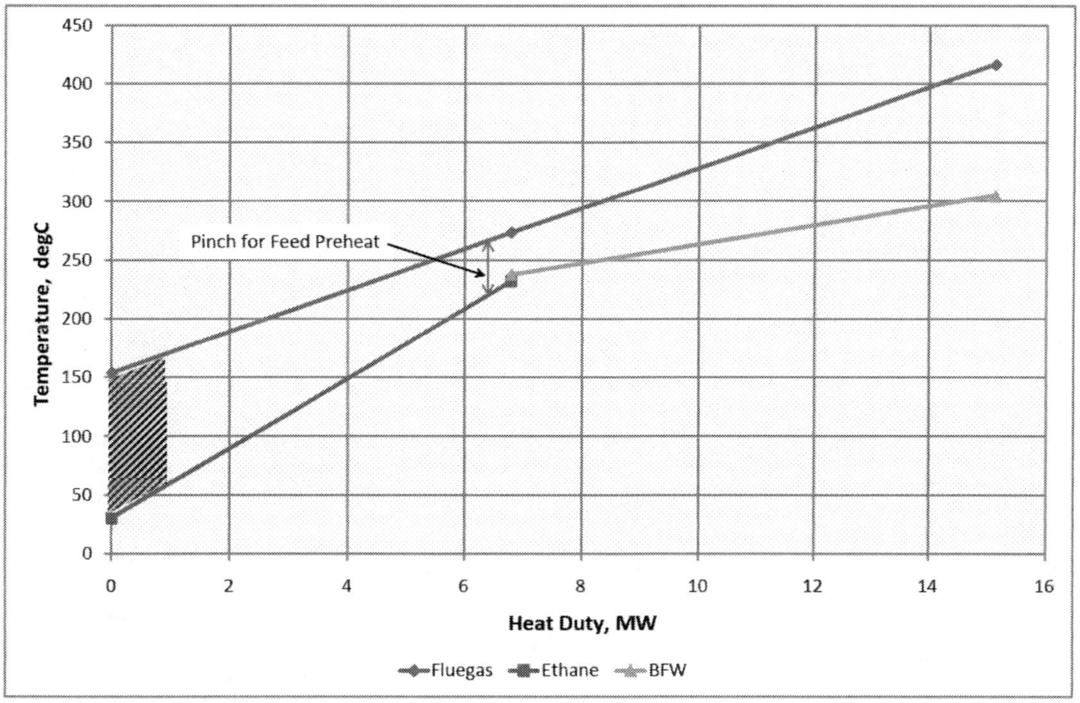

Modernize Your Control Functionality
Easy Steps to Start Process Optimization

13 AIChE
Spring Meeting, San Antonio, TX

Presenters

- Laurie Ben
- James Beall

EMERSON.
Process Management

Agenda

- Introduction
- Examples – Best Practices & Lessons Learned
  - Split Range Duty
  - Smith Predictor
  - Gain Scheduling/Adaptive Control
  - Signal Characterization
- Summary

Introduction

- Modernizing Control System – Decision Done Check!
- 'Do I' or 'Don't I' take advantage of new technology
  - Implemented older cumbersome techniques?
  - Complex custom strategies that handle automation challenges?
  - Hard to implement & maintain?
- Standard features that are field proven, practical, and easy to implement
  - Easier to implement & maintain
  - More efficient and effective control strategies
  - Achieve improvements in process performance
  - Lays the foundation for optimization

Agenda

- Introduction
- Examples – Best Practices & Lessons Learned
  - Split Range Duty
    - Control Challenge
    - Legacy Solution
    - Newer Technology Solution
    - Summary
  - Smith Predictor
  - Gain Scheduling/Adaptive Control
  - Signal Characterization
- Summary

Split Range Duty – Control Challenge

- Use steam and hot oil as energy source for distillation column for temperature controller.
- Hot oil is "free energy" and should be maximized
- Control Scenarios
  - Maximize hot oil and supplement with steam
  - Base load steam (if meter bad) and supplement with hot oil
  - Base load hot oil flow (if bad meter) and supplement with steam
- Constraints
  - Minimum and maximum hot oil valve position and flow rate
  - Minimum steam flow rate

Split Range Duty – Legacy Solution

– Old legacy system implemented this feature using a velocity PID with incremental output and a "micro expert rules system"

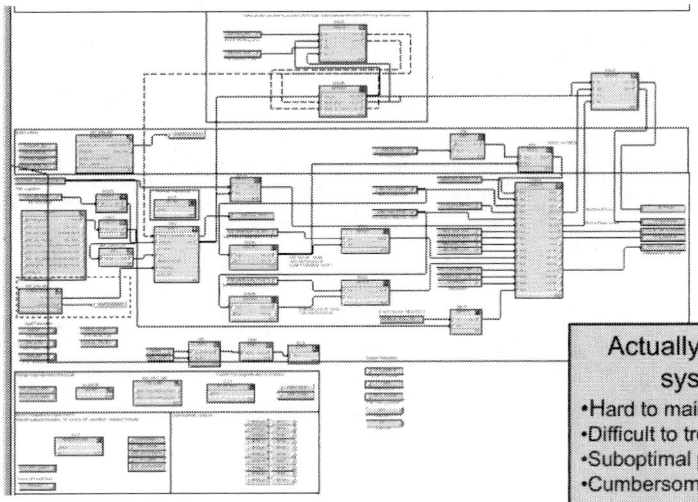

Actually Duplicated in New system, but still…
- •Hard to maintain and modify
- •Difficult to troubleshoot
- •Suboptimal performance
- •Cumbersome/Kludge/Ugly

Split Range Duty – New Technology Solution

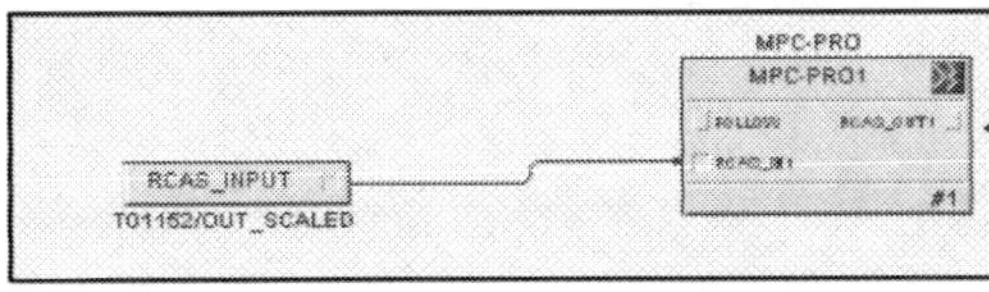

Embedded MPC Function Block provides all control functionality! EASY!!!

Simple calculations to calculate limits

Split Range Duty – New Technology Solution

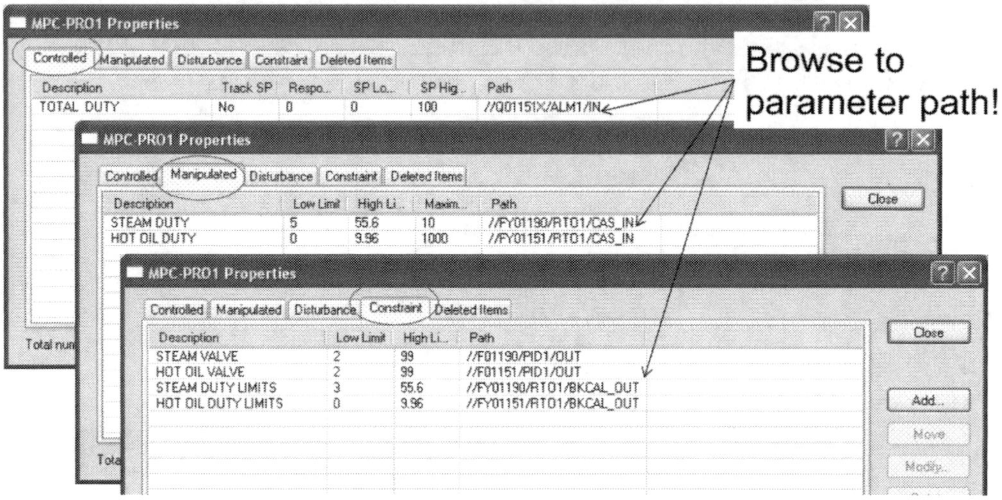

Split Range Duty – New Technology Solution

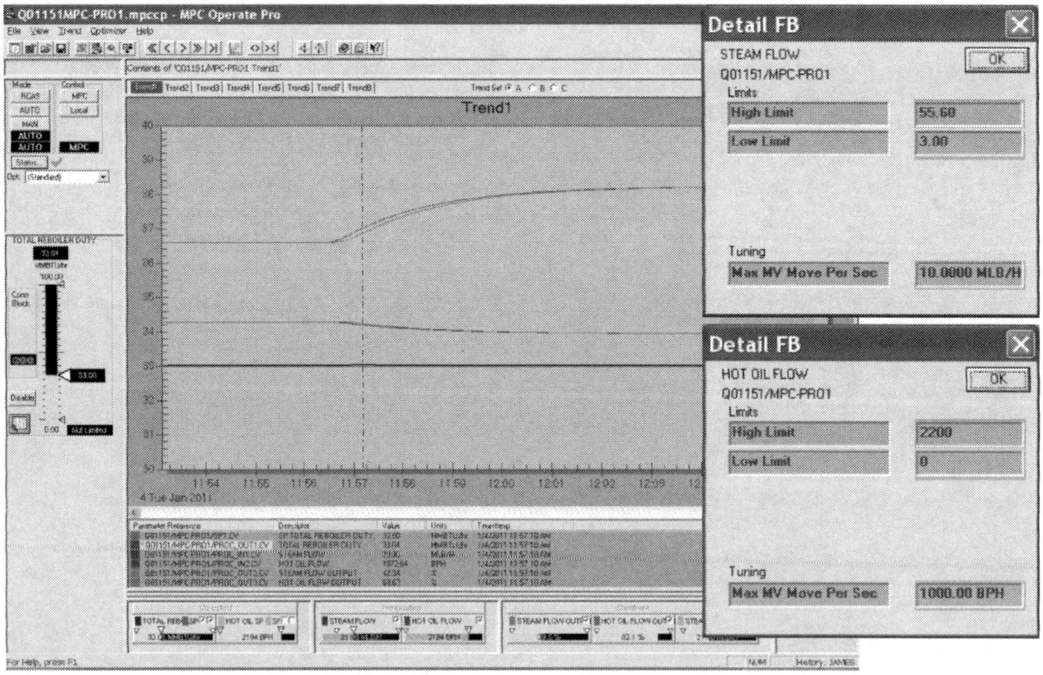

Summary Split Range Duty Control

Embedded MPC (New Technology) vs Legacy Solution

- Better performing solution
  - Bumpless transfers and constraint control
- Easier to Maintain
  - Change limits and settings
  - Simpler implementation
  - Easy to understand, train others, document
- Easily expands to handle future constraints
- Reduce energy costs!

Agenda

- Introduction
- Examples – Best Practices & Lessons Learned
  - Split Range Duty
  - Smith Predictor
    - Control Challenge
    - Legacy Solution
    - Newer Technology Solution
    - Summary
  - Gain Scheduling/Adaptive Control
  - Signal Characterization
- Summary

Smith-Predictor PID – Control Challenge

- Control of processes that are "dead time dominant" (dead time > time constant)
- Typically custom configured
  - some legacy control systems have preconfigured options, e.g. PROVOX
- Algorithm uses a process model
  - (gain, dead time and time constant)
- Challenge: as process dynamics change from model used, performance degrades rapidly

Smith-Predictor PID – Legacy Solution

- Traditional Smith-Predictor

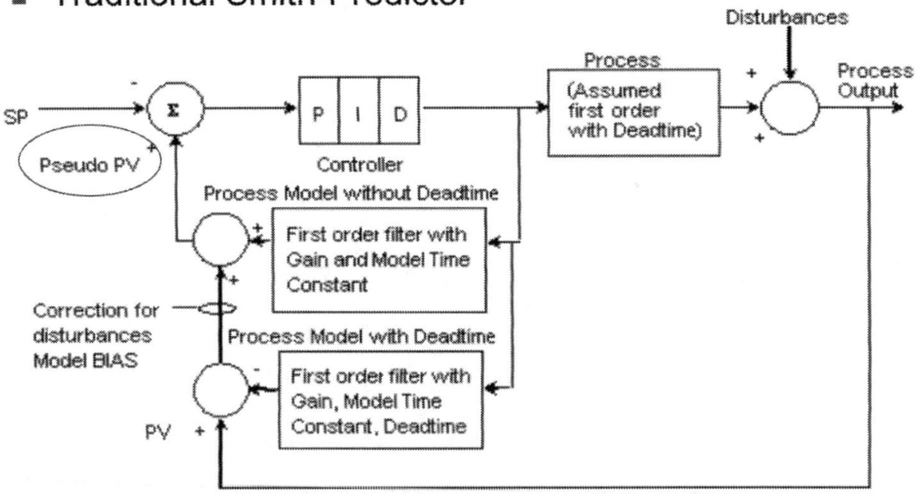

•High maintenance when process dynamics change
•Performance degrades rapidly as process dynamics change from configured model

Smith-Predictor PID – New Technology Solution #1

Easy to implement the Smith-Predictor but…

- High Maintenance when process dynamics change
- Performance degrades rapidly as process dynamics change from configured model

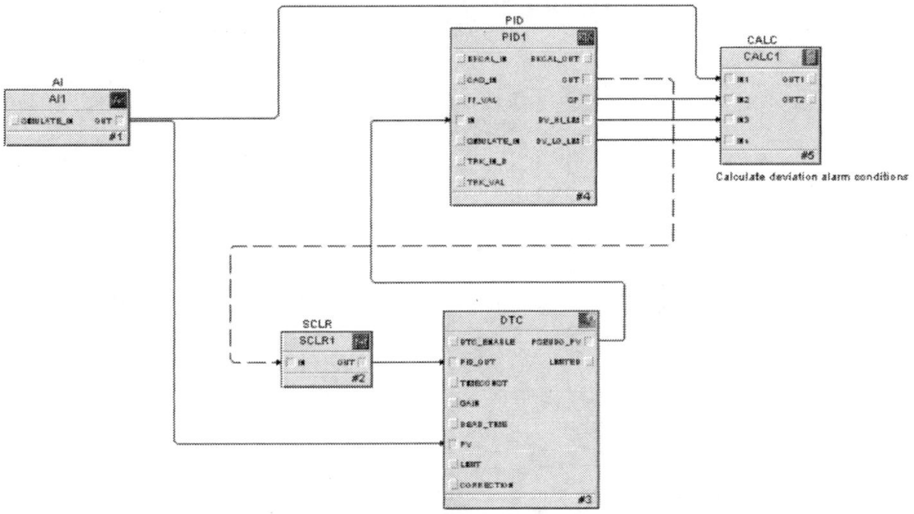

Smith-Predictor PID – New Technology Solution #2

- Embedded MPC Solution

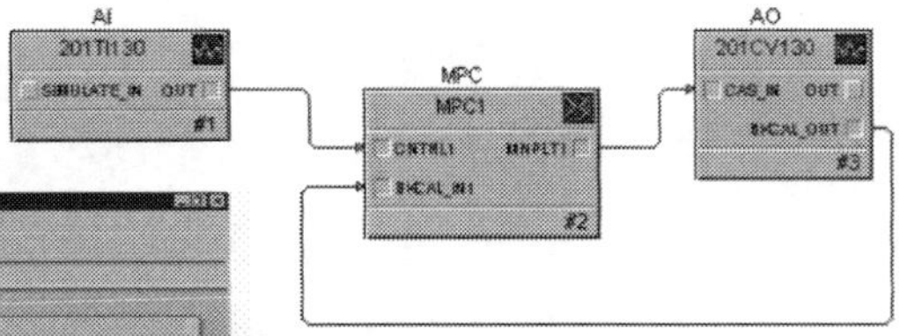

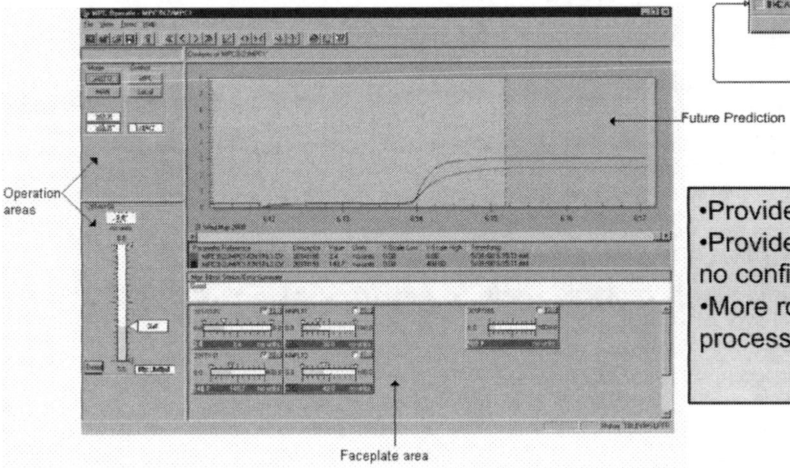

- Provides Future Prediction
- Provides MPC Faceplate – no configuration needed
- More robust under changing process dynamics

Smith-Predictor PID – New Technology Solution #2

- Embedded MPC Solution

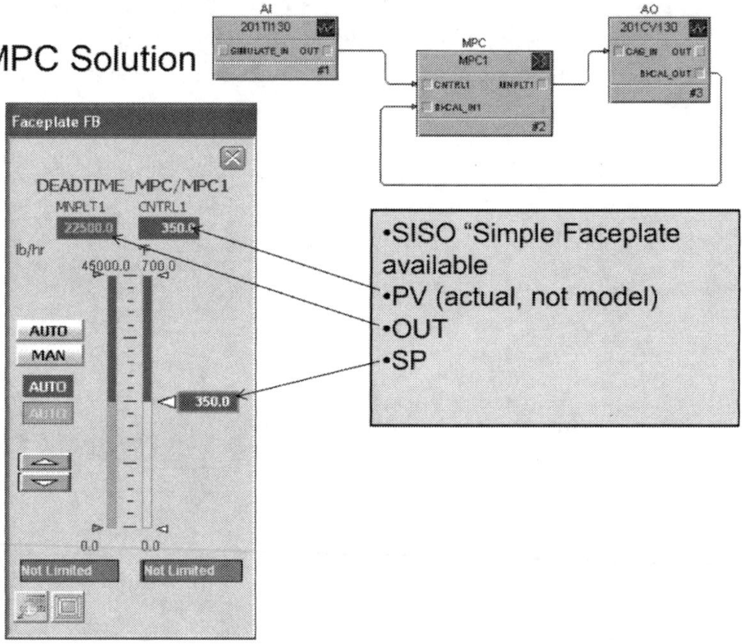

Summary Smith-Predictor PID

New Technology Solutions

- Standard Smith Predictor (#1) and Embedded MPC (#2)
- Both easy to implement and understand!

New Technology Solution #2 (Embedded MPC)

- More robust solution under a changing process model
- Provides visual display of the "future" PV
- Save money through better control performance

Agenda

- Introduction
- Examples – Best Practices & Lessons Learned
  - Split Range Duty
  - Smith Predictor
  - Gain Scheduling/Adaptive Control
    - Control Challenge
    - Legacy Solution
    - New Technology Solution
    - Summary
  - Signal Characterization
- Summary

Gain Scheduling – Control Challenge

- Legacy systems have limited capability to modify tuning based on process conditions
- Others had add-on application software, e.g. PROVOX "Gain Scheduler" system provided 3 sets of Gain, Reset, and Rate applied in 3 separate regions of a "state variable"
  - State variable = PV, OUT or "other"
  - Transition between tuning sets interpolated over "dead band"
- Beware: some suspend loop execution when tuning parameters change, so don't change the tuning every scan

Gain Scheduling – Legacy Solution

- Custom code
- Multiple points, registers, displays
- Restricted (slower) update rate

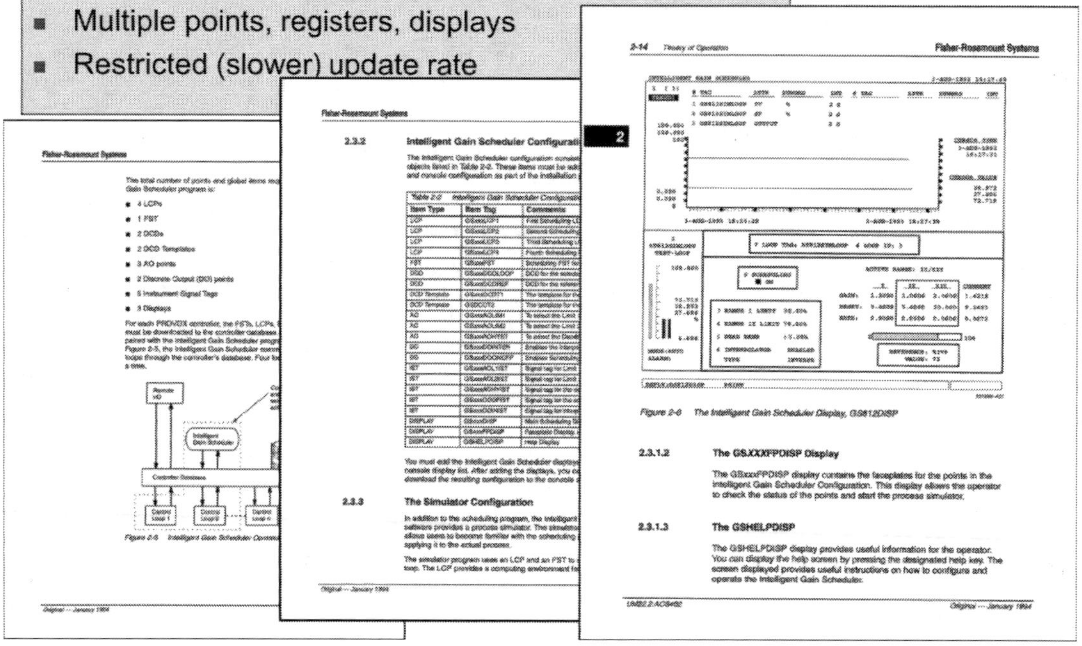

Gain Scheduling – New Technology Solution #1

Standard Modules
- Standard configuration
- Preconfigured faceplate & detailed display
- Easy to implement
- Tuning updates at loop execution

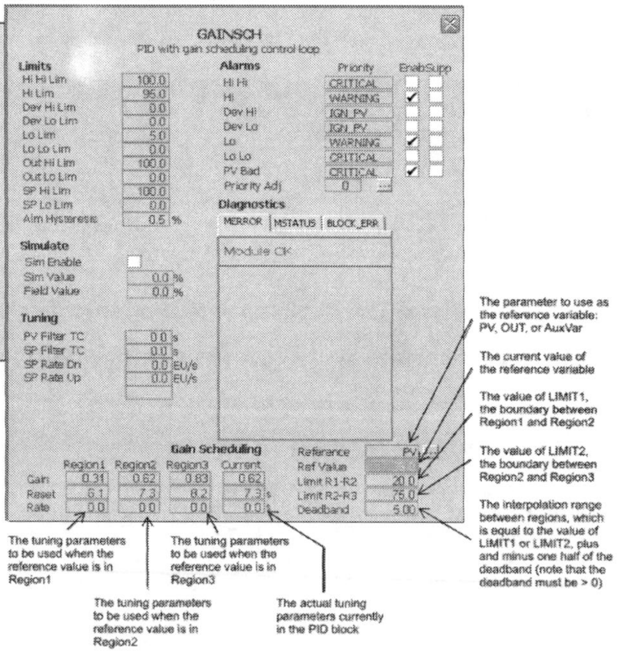

Gain Scheduling – New Technology Solution #2

New Technology "Adaptive Control"

- "Learns" process models from normal operating moves
- Allows up to 5 tuning regions of the "state variable"
- Partial Adaptive Control
  - Use "approved model" only & selected tuning rule in each region
  - Lambda PI or IMC PID; "Closed loop time constant" or "Lambda Factor".
- Optional Full Adaptive Control
  - continuous "Learning" of process model in each region
  - updates tuning while in each region per selected tuning rule
  - set allowable deviation (%) from approved model in each region

New Technology – Adaptive Control

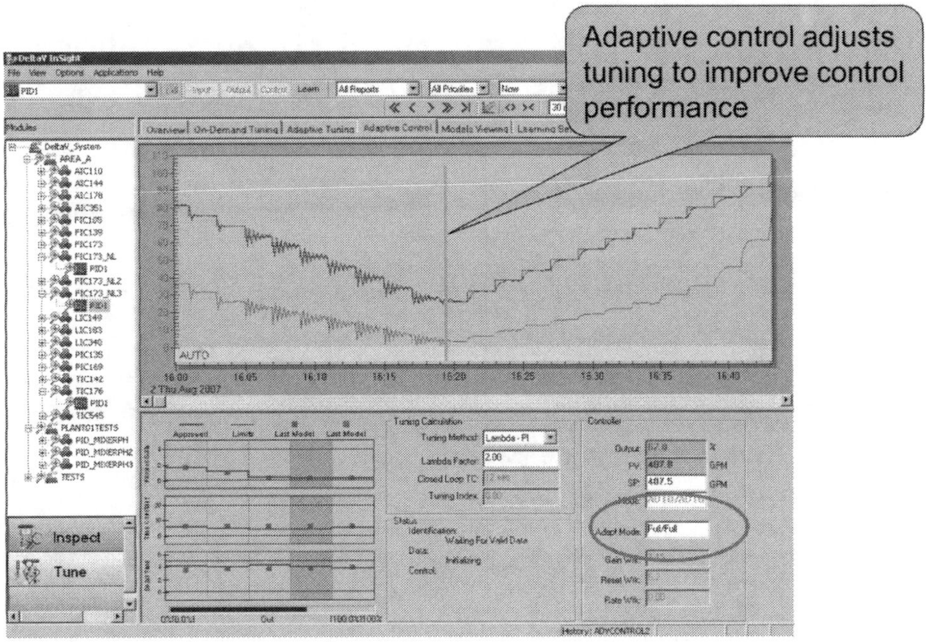

Summary Gain Scheduling

New Technology Solution(s) vs. Legacy Solution

- Gain Scheduling easy to implement and understand!
  - Standard configuration
  - Preconfigured faceplate & detailed display
  - Tuning updates at loop execution
- Adaptive Control provides additional benefit and ease of implementation
  - Learns models during normal operation – NO STEP TESTs!
  - More Regions – Up to 5
  - Automatic tuning updates when process model changes
- Improved control performance saves money!

Summary Gain Scheduling

New Technology Solution(s) vs. Legacy Solution

- Gain Scheduling easy to implement and understand!
  - Standard configuration
  - Preconfigured faceplate & detailed display
  - Tuning updates at loop execution
- Adaptive Control provides additional benefit and ease of implementation
  - Learns models during normal operation – NO STEP TESTs!
  - More Regions – Up to 5
  - Automatic tuning updates when process model changes
- Improved control performance saves money!

Agenda

- Introduction
- Examples – Best Practices & Lessons Learned
  - Split Range Duty
  - Smith Predictor
  - Gain Scheduling/Adaptive Control
  - Signal Characterization
    - Control Challenge
    - Legacy Solution
    - New Technology Solution
    - Summary
- Summary

Signal Characterization - Control Challenge

- Input Characterization (AI)
  - Combustion control – characterize fuel and air
  - Level – characterize pv (e.g. level to volume)
- Output characterization (AO)
  - modify installed flow characteristics of valves, dampers or other final control elements
  - Use to modify loop gain, e.g. output path to slave loop
- Legacy System Challenges
  - Lack of resolution (few points)
  - External black box characterizers may shift, lose documentation
  - Potential operator/maintenance confusion on signal going to final control element

Signal Characterization – Legacy Solution

- Legacy system using external analog characterizer

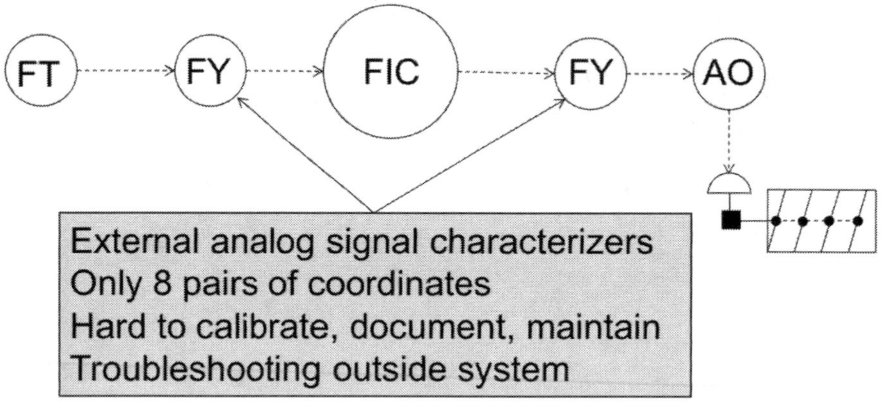

Signal Characterization – New Technology Solution #1

- Standard Module implementation provides 21 point signal characterizer function block (SGCR)

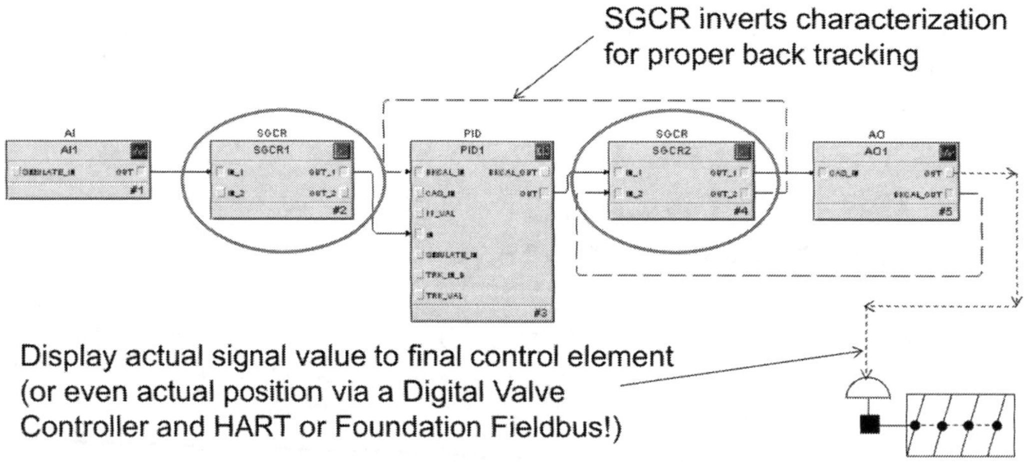

Signal Characterization – New Technology Solution #2

- Adaptive Control
  - Adjust tuning to compensate for final control element characteristics

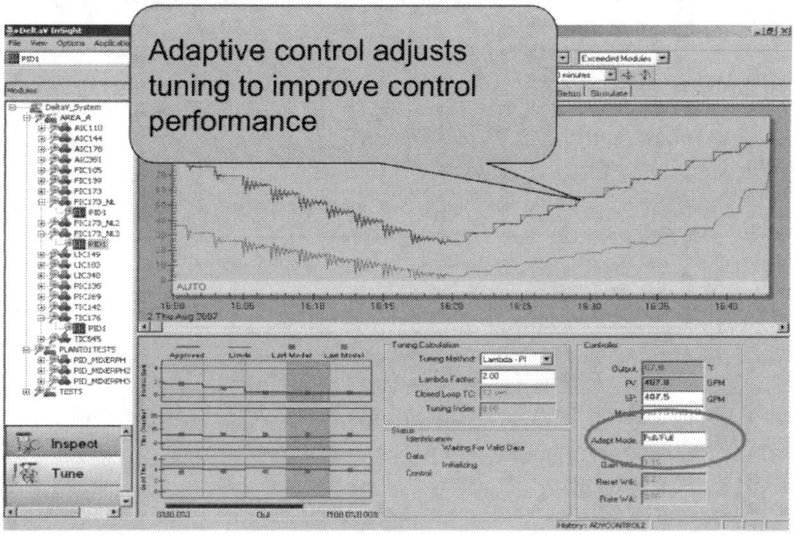

Summary Signal Characterization

- Newer Technology Signal Characterization function block
  - Better resolution (21 pairs of coordinates)
  - Self documenting (exists in control system)
  - No calibration required vs. 'external black box'
  - Provides "inverse" characterization for analog output characterizations
  - Reduce confusion: display signal being sent to final control element
- Applying Adaptive Tuning may provide a superior solution for AO characterization of final control element

Agenda

- Introduction
- Examples – Best Practices & Lessons Learned
  - Split Range Duty
  - Smith Predictor
  - Gain Scheduling/Adaptive Control
  - Controller Characterization
- Summary

Summary

- Practical application of control features that deal with common challenges
- Highlighted best practices & lessons learned
  - Split Range Duty
  - Smith Predictor
  - Gain Scheduling/Adaptive Control
  - Signal Characterization
- Do look at new functionality available when modernizing
  - Allows jump start to process optimization
  - Leads to improved process performance
  - Easier to implement and maintain over the life investment

Contacts & Questions?

James.Beall@Emerson.com
903-235-7935

Laurie.Ben@Emerson.com
512-834-7237

Paralleled Double-Effect Distillation: Simulative Case Studies

*Pei Wang, Leilei Dai, Ming Shi , Fan Wang, Hong Gao, Jinsheng Sun**
School of Chemical Engineering and Technology, Tianjin University, No. 92 Wei Jin Road, Tianjin 300072, P. R. China

Abstract

The increasing scarcity and growing demand for energy have led to an increasingly attention to the energy-efficient methods. Double-effect distillation is a common method to achieve remarkable energy conservation, which has three kinds of typical configurations including the feed splitting (FS), the light split/forward (LSF) and the light split/reverse (LSR). In spite of a good pathway for energy saving, better options are expected to cut down energy expenditure by a bigger margin than double-effect distillation does.

In this paper, case studies of paralleled or combined double-effect distillation schemes are proposed on the basis of current double-effect distillation. In order to verify the feasibility of this method, the improved methanol distillation scheme and organosilicon distillation scheme are optimized under this principle as examples. Exergy losses and the energy consumptions are measured as a tool to compare the two prevalent schemes with their corresponding optimized counterparts, respectively. The results of the total exergy loss and the energy consumption of these schemes show that the optimized schemes can cut the corner of the exergy loss by 22.8% and 40.41% and the energy consumption by 15.61% and 44.28%, respectively, in comparison to their popular cousins.

Keywords: double-effect distillation, paralleled double-effect distillation, methanol distillation, organosilicon monomer distillation, distillation process simulation, energy saving, exergy analysis

1. Introduction

Distillation, in chemical process and related industries, is by far the most widely applied for the separation of liquid mixture. In spite of its well-known benefit and widespread application, the major drawback is its high energy consumption. In classical distillation process, heat is transferred to the reboiler of distillation column. Along with the distillation column to separate the mixture, a large number of heat are directly discharge into the surrounding environment and cooling water. Araujo AB [1] has done some research to achieve that distillation has a relatively low thermodynamic efficiency and it requires the input of high quality energy in the reboiler to push the separation of liquid mixture. Therefore, the focus of energy conservation is to consider how to recover or reuse the discharged heat, or how to reduce the quantity of transferred heat to the column. In the early 20th century, many scholars have devoted to the study of saving energy in distillation, and now most of correlative techniques are widely used in distillation process [2,3]. These energy saving methods involve multi-effect distillation, heat pump distillation, heat integrated distillation and etc [3-8]. But without extra rotating machines by electricity or other power producers, multi-effect distillation has been more favored in research and application [9].

Many researches confirmed multi-effect distillation to be one of the effective methods to reduce the energy consumption [10-12]. The energy saving effect of those schemes used multi-effect distillation is associated with the number of the effects and the characteristics of separation system. Generally speaking, those schemes can offer energy savings around 30% with respect to the original or classical multicomponent distillation scheme [23, 27, 28]. Considering the interaction between investment of plants and energy saving effect, the double-effect distillation has been the most commonly used in not only research but also field application.

The use of the double-effect distillation technology for optimizing the conventional distillation schemes can cut a large amount of the energy requirement in comparison with the correspondingly prevalent schemes [13]. Although the system is very difficult to control in the sense that it is nonlinear, multivariable and interacting [14]. More and more published studies [15-19] have been focused on further questing the way of energy saving on the basis of the double-effect distillation schemes. However, the overhead vapor of one high pressured column only can integrate with the reboiler of one low pressured column in those published studies. In view of this, in this paper, the paralleled or combined double-effect distillation is proposed to broaden the thought for the application of the double-effect distillation, in which the overhead vapor from one of the high pressured column is divided into two or more streams to successively integrate with the reboiler of the other columns [9].

2. *Theory*

2.1 Double-effect or heat integrated distillation system

A double-effect distillation system is made up of two adjacent distillation columns running in different pressure, which can lead to a temperature difference between the top of the high pressured column and the bottom of the low pressured column. Through the thermodynamic connection, the reboiler of the low pressure column plays a role of the condenser for the high pressure column. According to the direction of the mass flow and the heat integration, the double-effect distillation can fall into three kinds of configuration: the feed splitting (FS), the light split/forward (LSF) and the light split/reverse (LSR), as showed in Figure. 1.

One of the configurations with a high thermodynamic efficiency is the feed splitting (FS), in which the feed is divided roughly equal between the two columns running in different pressure, and the products can be obtained from both of the top and bottom of the two columns, as illustrated in figure. 1 (a). The overhead vapor form the high pressured column is used to heat the reboiler from the low pressured column. There is only energy exchange between the two columns, so the heat duties of the feed splitting (FS) are approximately equal to the heat duty for single column with the same conditions divided by the number of effect.

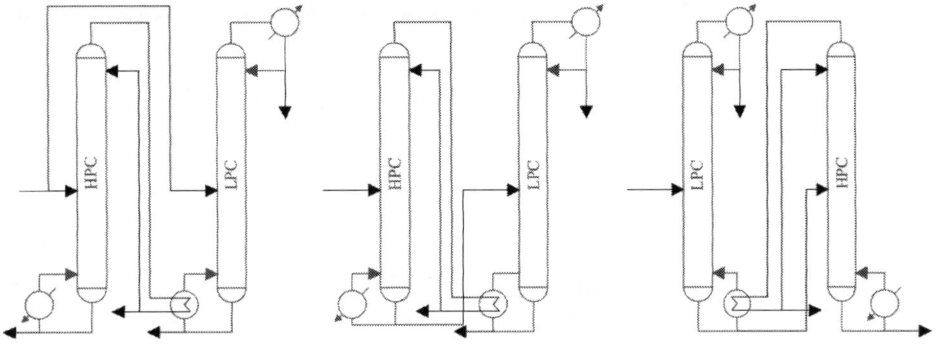

Fig. 1. Three configurations of double-effect distillation

Figure. 1 (b) and (c) respectively shows the configuration of the light split/forward heat integration (LSF) and the light split/reverse heat integration (LSR). Different from the feed splitting (FS), the total feed stream was sent to one column in the other two configurations. In the light split/forward (LSF), the total feed is pumped to the high-pressure column (HPC) and the bottom stream of HPC flows into the low-pressure column (LPC). In this way, all of the heavy components are from LPC and the light components from the top both of the two columns. Figure. 1 (b) indicates the direction of mass flow and heat integration is the same. The last configuration is the light split/reverse (LSR), as shown in figure. 1 (c), all of the feed is sent to the low-pressure column (LPC). It can remove the two kinds of light components from the top both of the two columns and one kind of heavy components from the bottom of HPC. Moreover, the direction of heat integration is in the reverse with the mass flow. These two kinds of configurations have not only energy exchange but also substantial exchange, therefore, when the operating parameters of one column are changed, the other column will also change.

Theoretically, the three kinds of configurations are empirically 30% less energy consumption than the single column distillation. As for choosing what kind of the configurations of the double-effect distillation, it depends on the properties of the multi-component mixture, the pressure of saturated steam, the temperature of cooling water, structure parameters of existing distillation column and so on.

2.2 Exergy analysis of the distillation process

According to the First Law of Thermodynamics, also called the Law of Energetic Balances, no matter what happens, the total amount of energy has not yet changed. However, when the various types of energy generate the mutual conversion, the quality of the energy is always decreased. The separation process in distillation is irreversible. In order to separate the mixture in distillation column, it needs the driving force which must consume a certain amount of external work, by this means, the decline of the quality of energy is inevitable. Therefore, we should choose a reasonable driving force in distillation process to avoid unnecessary loss of the quality of energy. It defined the concept of exergy (E_x) to compare the quality of the different types of energy. Exergy, which is a measure for evaluating the extent of available energy or the quality of energy, is based on the Second Law of Thermodynamics. Exergy analysis reveals the essence of the energy, and can be used as an important tool to evaluate the effect of energy saving. And it usually combines with process optimization studies, and as its guidance, to identify cost saving opportunities and amend the existing scheme.

Szargut [29] has summarized a comprehensive definition of the exergy, that is: 'Exergy is the amount of work obtainable when some matter is brought to a state thermodynamic equilibrium with the common components of the natural surroundings by means of reversible processes, involving interaction only with the above mentioned components of nature'. The state of one stream recorded as 1, according to the definition of exergy, the exergy of the stream can be expressed as:

$$E_x = (H_1 - H_0) - T_0(S_1 - S_0) \tag{2}$$

where, H_1 and S_1 is the enthalpy and the entropy of the stream 1, and T_0 is the temperature of the surrounding and H_0 and S_0 represent the enthalpy and the entropy of the surrounding.

The state of another stream recorded as 2, when the system stream change from state 1 (T_1, P_1) to state 2 (T_2, P_2), the variation of exergy (ΔE_x) is given by

$$\Delta E_x = E_{x,1} - E_{x,2} = \Delta H - T_0 \Delta S \tag{3}$$

In the conventional distillation process, it must input the heat or energy to drive the occurrence of separation process. The bottom of the column input the heat (Q_R) at this temperature (T_R), and the top of the column was sent to the heat (Q_C) at this temperature (T_C). The net work consumption (W) in distillation is

$$W = Q_R(1 - \frac{T_0}{T_R}) - Q_C(1 - \frac{T_0}{T_C}) \quad (4)$$

When the enthalpy difference between the product and the feed is very small, $Q_R=Q_C=Q$ can be approximately gotten, then Equation (4) is rewritten to be:

$$W = QT_0(\frac{1}{T_C} - \frac{1}{T_R}) \quad (5)$$

Then, the total exergy loss (Ex_{loss}) of the distillation column can be calculated by

$$Ex_{Loss} = W - \Delta E_x \quad (6)$$

3. *Prototype case: the prevalent distillation schemes*

3.1 The improved four-column methanol distillation scheme

In the methanol industry, it has gradually appeared two-column, three-column, four-column and five-column distillation scheme [20-23]. With the increase of the number of distillation column, the corresponding distillation scheme can not only increase the efficiency of the process, but also cut the energy consumption. At present, the four-column scheme has a dominant status to produce the purified methanol. However, the improved four-column distillation scheme can cut some more energy requirement with respect to the traditional four-column distillation scheme. The improved four-column distillation scheme is seen as the comparative reference in the paper.

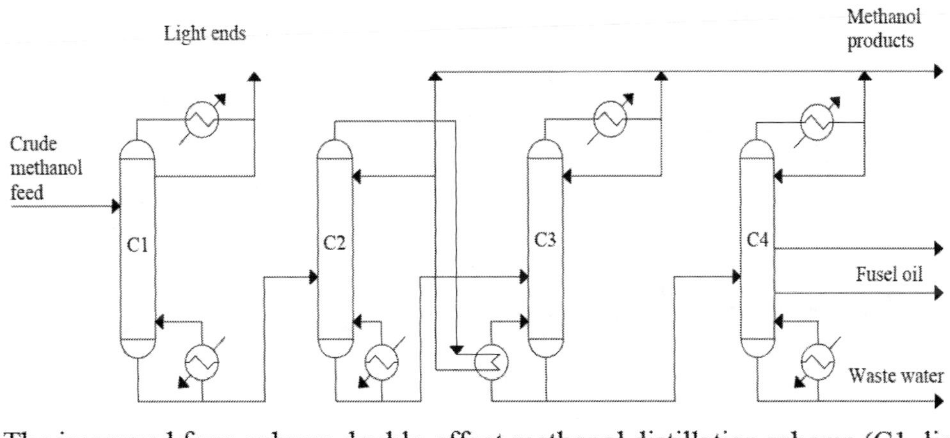

Fig. 2. The improved four-column double-effect methanol distillation scheme (C1, light ends column; C2, pressured column; C3, atmospheric column; C4, water column)

In the improved four-column distillation scheme, as shown in figure. 2 [24], the crude methanol flows into the light ends column (C1) in 87.5 ℃ for removing the light ends and the remnants of dissolved gas from the top of C1. Besides, in order to raise the methanol recovery rates, the light ends column (C1) designs a two-stage condenser system. The crude methanol was distillated successively from the pressured column (C2) and the atmospheric column (C3), and most of the methanol products can be received at the top both of C2 and C3. The bottom stream of C1, C2 and C3 is pumped respectively as the feed into C2, C3 and the water column (C4). Residual methanol, fusel oil and waste water are extracted from the corresponding position of the top, side stream and the bottom of the water column (C4).

In the improved scheme, the operation pressure of C2 is 0.9MPa, and the other three columns run under the atmospheric pressure. Due to the different of operating pressure, the overhead vapor of C2 can undertake thermal coupling connection with the reboiler of C3.

3.2 The current organosilicon monomer distillation scheme

According to the survey, there are several kinds of organosilicon distillation schemes. However, the 9-column organosilicon distillation scheme [9], in China, has an important role in the organosilicon industrial, as illustrated in Figure. 3, it is used as the base case.

Figure. 3 shows that the crude monomer is delivered to the heavies removal column (C1') and the bottom stream of C1' has a lot of the unsteady substances of high boiling points. These unsteady substances of high boiling points are pumped to the high boiling point column (C9') as the feed, and the products from the top of C9' continue to serve as the crude monomer by circumfluence. By this way, it can accomplish the separation between the unsteady substances of high boiling points and the useful substances at a relatively low temperature in the high boiling point column (C9') and avoid the unsteady substances of high boiling points from thermal cracking to generate many unknown substances which would greatly affect the subsequent separation of product. The low removal column (C2') provides trimethylchlorosilane (Me3), methyldichlorosilane (MeH), and other light components for lights removal column (C3'), MeH column (C4'), azeotrope removal column (C5') and Me3 column (C6'), which can finally achieve two products (MeH and Me3) and three wastes. It also provides methyltrichlorosilane (Me1) and dimethylchlorosilane (Me2) and other heavy components for Me1 column (C7') and Me2 column (C8'), which can finally get two products (Me1 and Me2).

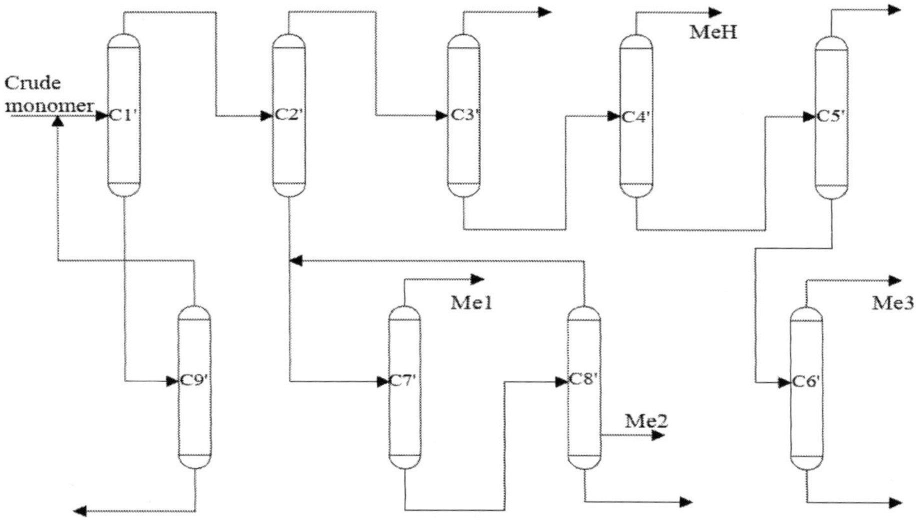

Fig. 3. The prevailing 9-column distillation scheme (C1', heavies removal column; C2', low removal column; C3', lights removal column; C4', MeH column; C5', azeotrope removal column; C6', Me3 column; C7', Me1 column; C8', Me2 column; C9', high boiling point column)

In this scheme, the operating pressure of all nine columns is relatively low, and the highest operating pressure is 260 kPa in the lights removal column (C3'). Although it has a different operating pressure between C3' and the other columns, but the temperature difference is not enough to make them use double-effect or heat integrated distillation.

4. *Simulation and analysis both of the prevalent schemes and the optimized schemes*

4.1 The novel five-column heat integrated methanol distillation scheme

Although the energy saving of the improved four-column methanol distillation scheme can reach 10% compared with the traditional four-column scheme, it still exist several shortages, which will be illustrated by the results of computer simulation to the improved four-column methanol distillation scheme. The flow rate of the crude methanol is 239230 kg/hr, and the temperature is 87.5 ℃, and the composition of the crude methanol is exhibited in table 1. The inlet and outlet temperature of the cooling water are 30 ℃ and 40 ℃.The methanol content is specified as 99.99 wt% in the methanol products and 10ppm in the wastewater of the water column. Table 2 shows the operating parameters of each column in the improved four-column scheme.

Table 1. Crude methanol feed data

Component	Mass fraction%	Component	Mass fraction%
H_2	0.0100	Pentanol	0.0338
N_2	0.0710	Formic acid	0.0030
Methane	0.0100	Acetic acid	0.0030
CO	0.0360	Methyl formate	0.0092
CO_2	0.6640	Acetone	0.0193
Methanol	95.0000	Methyl ethyl ketone	0.0030
Dimethyl ether	0.0309	Diisopropylether	0.0056
Trimethylamine	0.0010	Water	3.7100
Ethanol	0.0900	i-Pentane	0.0562
i-Propanol	0.0024	n-Pentane	0.0169
Propanol	0.0264	n-Hexane	0.0487
i-Butanol	0.0338	n-Heptane	0.0009
Butanol	0.0185	n-Octane	0.0050
3-Pentanol	0.1407	n-Nonane	0.0050

Table 2. The operating parameters of improved four-column scheme

The operating parameters	C1	C2	C3	C4
Top pressure (kPa)	150	900	110	110
Top temperature (℃)	81.0	132.6	67.7	67.7
Bottom temperature (℃)	83.8	139.1	96.5	104.5
Reflux ratio	0.58	2.44	1.68	11.57
Number of theoretical stage	33	52	58	61

The results of simulation indicate that the hot utility is still in a high level because the reboilers of C1, C2 and (C4) still need the saturated steam to transmit the heat. Table 2 shows the reflux ratio of C4 is 11.57 and this large reflux ratio can cause the increasing of the energy requirement of the water column (C4). Besides, according to the flowsheet of this scheme, the diameter of column is so large that the emergence of many urgent cases.

In view of above these defects in the improved four-column scheme, the novel five-column heat integrated methanol distillation scheme (shown as Fig. 4) is designed in this paper to reduce the impact of these factors as far as possible by using the paralleled or combined double-effect distillation. Juntao Zhang [23] and Jinsheng Sun [25] are also proposed a five-column methanol distillation scheme by using triple-effect distillation and double-effect distillation respectively, and the effect of energy saving of the two schemes is outstanding.

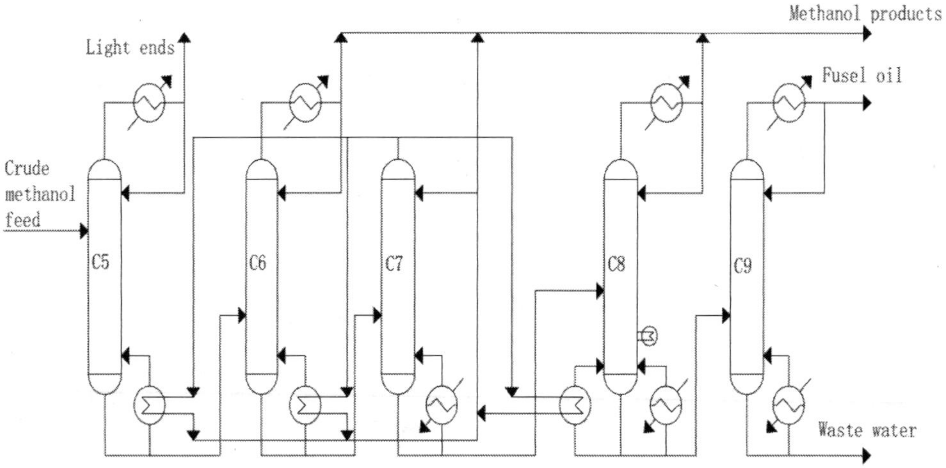

Fig. 4. The five-column heat integrated methanol distillation scheme (C5, light ends column; C6, prepositive atmospheric column; C7, pressured column; C8, postpositive atmospheric column; C9, water column)

According to the principle of the paralleled double-effect distillation, the four columns are unable to constitute the heat integration network in the improved four-column scheme. Through repeated verification, the prepositive atmospheric column (C6) is added after the light ends column (C5), meanwhile, the operating pressure of the pressured column (C7) is reduced to increase its throughput. In this way, the overhead vapor of C7 is split into three streams to respectively integrate with the light ends column (C5), the prepositive atmospheric column (C6) and the postpositive atmospheric column (C8). Due to use the paralleled double-effect distillation, the duties of the hot utility have decreased in a great margin. However, there are still part of heat from the background process is absorbed by the cooling water. So an inter-reboiler with a low temperature level is designed in the postpositive atmospheric column (C8) and it can share the heat with the background process streams, besides, we have proved the feasibility by using the column grand composite curve analysis of columns. Figure. 4 shows that the product of the top of the water column (C9) is fusel oil instead of purified methanol, which can not only reduce the reflux ratio of C9, but also make the diameter of column smaller to reduce the difficulty of operation.

Due to produce the fusel oil instead of purified methanol in the water column (C9) in the novel five-column distillation scheme, the methanol content of the feed for the water column have to limit at 0.016 wt%. In the five-column scheme, the operating pressure and the number of theoretical stage for each column are determined through the existing process, the onsite data and the requirement of the heat coupling. The proportion of the methanol products from the C6, C7 and C8 is determined by their own double-effect or heat integrated distillation system. The simulation results of the novel five-column scheme are listed in table 3.

Table 3. The operating parameters of five-column scheme

Column	C5	C6	C7	C8	C9
Top pressure (kPa)	160	110	400	110	110
Top temperature (℃)	83	66.6	104	66.6	86.5
Bottom temperature (℃)	85.6	84	118.8	104.5	103.2
Reflux ratio	1.46	1.49	2.23	1.87	4
Number of theoretical stage	33	54	54	65	29

4.2 Energy consumption and exergy loss analyses for the two methanol schemes

In the improved four-column scheme, the pressured column and the atmospheric column constitute the distillation system of heat coupling, there are three columns (C1, C2, C4) need the hot utility, and the entire duty of reboiler is 145.4×10^3 kw. As described above, the light ends column (C5) and the prepositive atmospheric column (C6) don't need the saturated steam in the five-column scheme and the total duty of reboiler is reduced to 122.7×10^3 kw. The detailed data of the reboiler duty as shown in table 4.

Table 4. The duty of reboiler in the two schemes

Column	C1	C2	C4	C7	C8	C9
Duty of reboiler (10^4 kw)	3.30	10.04	1.20	8.97	2.86	0.44

Obviously, the novel five-column scheme is much better than the improved four-column scheme in energy saving and the energy consumption of the novel scheme can be reduced by 15.61%. However, the recovery rate of methanol of the novel scheme is approximately equal with the improved four-column scheme.

Only the energy consumption analysis is not revealed the energy saving essence of the novel five-column scheme. Exergy analyses are used to further compare the two schemes, and the exergy flow rates are calculated through using the computer simulation data. The activity coefficient of the liquid is calculated using the Alcohol Package thermodynamic model. The calculation of the net work consumption of distillation column should pay more attention to the temperature of the cooling water. When the cooling water exchanges heat with other streams, the temperature of the cooling water is momentarily changed, so it is replaced the temperature of cooling water with the arithmetic average of inlet and outlet temperature of cooling water. Figure. 5 show the overall exergy loss of each column in the two methanol distillation schemes, in addition to the overall exergy loss of C6. Due to optimize the four-column scheme, the prepositive atmospheric column (C6) is added in the transformed scheme, and the overall exergy loss of C6 is 9157.6 kw. The exergy loss of each column, besides the pressured column (C7), in the novel five-column scheme is obviously less than their counterparts in the improved four-column scheme.

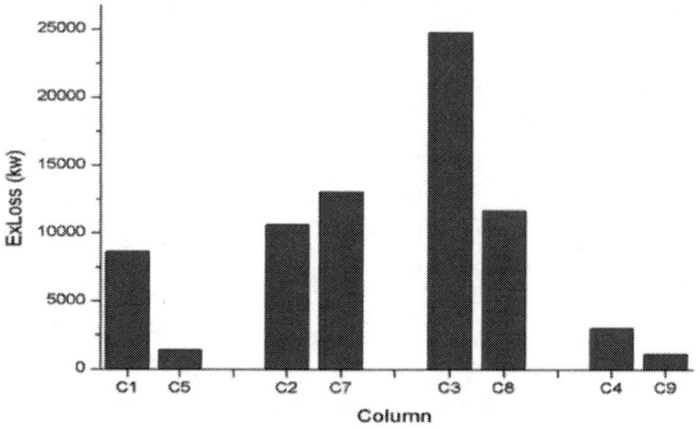

Fig. 5. The overall exergy loss analysis of each column in the two schemes

The total exergy loss of the improved four-column scheme and the transformed scheme is, respectively, 47100.1 kw and 36355.1 kw. The total exergy loss of the optimized scheme is dropped by 22.8%. Through the comparison between the exergy loss and energy consumption of the two schemes, it is indicated that the novel five-column scheme should be applied to produce the purified methanol in industry.

4.3 The novel heat coupling organosilicon distillation scheme

The primary source of the crude organosilicon monomers is the catalytic reaction products between chloromethane and silicon at high temperature and the table 5 displays the normal boiling points and mass percentage of the main components of the crude organosilicon monomers. Methyltrichlorosilane (Me1), dimethylchlorosilane (Me2), trimethylchlorosilane (Me3) and methyldichlorosilane (MeH) are the most valuable components in the crude organosilicon monomers. The purification requirements on organosilicon of these materials are shown in table 6.

Table 5. The crude organosilicon monomer feed data

Component	Normal boiling point at 1 atm/℃	Mass percentage %
LB		1
Methyldichlorosilane	41.8	4
Silicon tetrachloride	57.6	0.5
Trimethylchlorosilane	57.9	2.5
Methyltrichlorosilane	66.4	11
Dimethyldichlorosilane	70.2	80
HB		1
Temperature (℃)	40	
Pressure（kPa）	100	
Flow rate (kg/h)	26200	

Table 6. Purification demand on organosilicon

Monomer	Purification demand
MeH	99.5%
Me3	99.5%
Me1	99.5%
Me2	99.95%

The conventional nine-column scheme is simulated with the aid of existing process and onsite data to achieve the operating parameters of each column, as illustrated in table 7. The bottom temperature of the heavies removal column (C1') and the high boiling point column (C9') are obviously much bigger than that of the other seven columns in table 7. According to the principle of the double-effect distillation, the top temperature of the high pressured column is relatively high in comparison with the bottom temperature of the low pressured column, and the high pressured column can integrate with the low pressured column. Although the top temperature of C1' is less than the bottom temperature of the other seven columns, the overhead vapor of C1' still has a great opportunity to conduct heat with the other seven columns with the increasing of its top temperature.

Table 7. The simulation results of prevailing 9-column

Column	C1'	C2'	C3'	C4'	C5'	C6'	C7'	C8'	C9'
Top temperature (℃)	70.8	65.3	57.3	66.7	72.1	70.6	71.2	74.9	72.7
Bottom temperature (℃)	124.1	89.1	80.4	85.2	74.9	75.4	75.3	75.3	126.5
Reboiler duty (MW)	11.67	5.61	1.13	0.51	1.19	0.46	12.27	16.08	0.02
Condenser duty (MW)	-12.47	-5.53	-1.08	-0.53	-1.20	-0.47	-12.69	-16.12	-0.02
Top pressure (kPa)	110	180	260	220	170	150	110	110	110

According to the way of the paralleled double-effect distillation, the increasing operating pressure in the heavies removal column (C1') can make the top temperature of C1' ascending to integrate with the other seven columns. The idea is still in study to overcome some difficulties. However, we have got another feasible method, in which the operating pressure of C1' and the Me1 column (C7') is raised to 360 kPa and 280 kPa, respectively. In this way, the top temperature of C1' and C7' is increased to ensure they can integrate with the other columns, besides the high boiling point column, and the parallel double-effect or heat integrated organosilicon monomer distillation scheme is designed as shown in figure. 6 [26]. The feasibility of this heat coupling scheme has confirmed through three x-y diagrams including the heavies component and Me1, the heavies component and Me2, Me1 and Me2 [9].

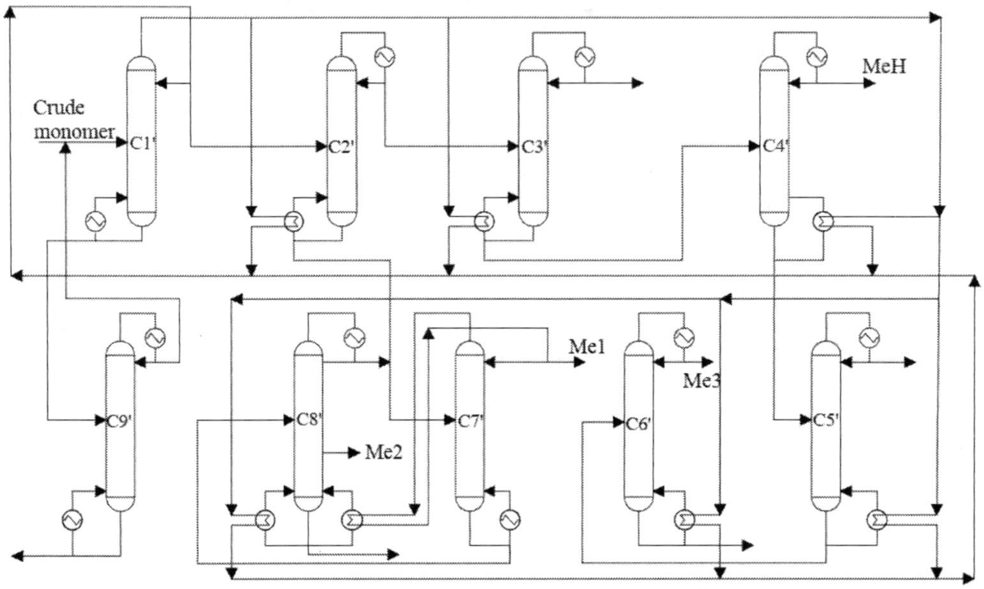

Fig. 6. The heat coupling 9-column scheme (C1', heavies removal column; C2', low removal column; C3', lights removal column; C4', MeH column; C5', azeotrope removal column; C6', Me3 column; C7', Me1 column; C8', Me2 column; C9', high boiling point column)

Figure. 6 shows the heavies removal column (C1') make the heat coupling with the low removal column (C2'), the lights removal column (C3'), the MeH column (C4'), the azeotrope removal column (C5'), the Me3 column (C6') and the Me2 column (C8'), successively. In order to gain a better energy saving effect, C8' is designed with two reboilers, the one integrated with C1' and the other integrated with the Me1 column (C7'). The operating parameters and the duty of reboiler and condenser of each column in the heat integrated 9-column are enumerated in table 8.

Table 8. The simulation results of heat integrated 9-column

Column	C1'	C2'	C3'	C4'	C5'	C6'	C7'	C8'	C9'
Top temperature (℃)	114.3	65.3	57.3	66.7	72.1	70.6	105.1	74.9	72.7
Bottom temperature (℃)	124.9	89.9	78.8	84.5	75.4	75.4	110.9	76.6	121.4
Reboiler duty (MW)	13.06	5.62	1.13	0.51	1.19	0.46	13.83	14.12	0.38
Condenser duty (MW)	-13.72	-5.54	-1.08	-0.53	-1.20	-0.47	-14.02	-14.43	-0.41
Top pressure (kPa)	360	180	260	220	170	150	280	110	110

4.4 Energy consumption and exergy loss analyses for the two organosilicon schemes

Figure. 4 indicate that each column of the prevalent nine-column scheme is not integrated with the other columns and the reboilers of all of these columns are required the saturated steam (600 kPa). According to the data from the table 7, the total duty of reboiler is 48.94 MW for the prevailing 9-column scheme. In the new nine-column scheme, C1' integrate with C2', C3', C4', C5', C6' and C8' successively and C7' integrate with C8', so C2', C3', C4', C5', C6' and C8' don't consume the saturated steam. The total duty of reboiler is 27.27 MW. Obviously, the heat integrated 9-column scheme has an absolute advantage over the prevailing 9-column scheme in energy saving, and the energy savings can reach 44.28%.

As the same with the methanol distillation scheme, the inlet and outlet temperature of cooling water are 30 ℃ and 40 ℃, and the total exergy loss is calculated through the results of computer simulation of the two schemes. The overall exergy loss comparison of each column in the two schemes is illustrated in figure. 7. In the normal scheme, the throughput of C9' is so small that the overall exergy loss of C9' is very near to zero, so it only have the exergy loss of C9' of the new scheme in figure. 7. The overall exergy loss of each column in the optimized scheme is declined in comparison with that of the corresponding column in the normal scheme except the high boiling point column (C9'). Because the crude organosilicon monomers contains some unstable components, in order to avoid these components happening thermal cracking in the heavies removal column (C1') in the novel scheme, more liquid from the bottom of C1' is pumped to C9' as the feed. That can cause the ascent of the throughput and the overall exergy loss of C9'.

The total exergy loss is 14035.5 kw and 8363.2 kw, correspondingly, in the conventional scheme and the novel scheme, and the exergy loss of the new scheme can be reduced by 40.41%. Through the above analysis and comparison, the heat integrated nine-column organosilicon distillation scheme should be a substitute for the prevalent nine-column scheme.

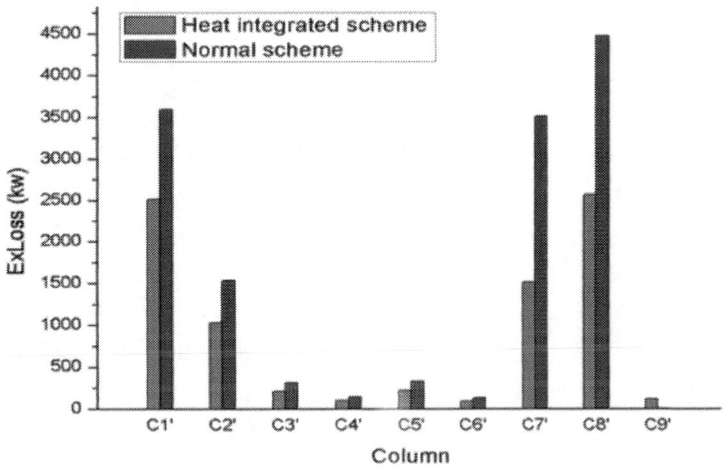

Fig. 7. The exergy loss comparison between the heat integrated scheme and the normal scheme

5. Conclusions

Paralleled double-effect or heat integrated distillation for energy saving is proposed, exampled by heat integrated five-column methanol distillation scheme and the heat coupling 9-column organosilicon monomer distillation scheme, with the aid of computer simulation. Besides, the optimized methanol distillation scheme and the transformed organosilicon scheme are comprehensively analyzed through the energy consumption and the exergy loss in comparison with the correspondingly prevalent scheme.

Adding a prepositive atmospheric column and decreasing the operating pressure of the pressured column in the novel five-column scheme can make the pressured column to simultaneously integrate with the light ends column, the prepositive atmospheric column and the postpositive atmospheric column. Besides, an inter-reboiler is added to integrate with the background process streams. The energy consumption and the exergy loss of the new scheme are 15.61% and 22.8% lower than the improved scheme.

Increasing the operating pressure of the heavies removal column and the Me1 column can make them to integrate with the other columns, in addition to the high boiling point column, to constitute a paralleled double-effect distillation system in the optimized nine-column scheme. The energy consumption and the exergy loss of the novel scheme are 44.28% and 40.41% less than the normal scheme.

The two cases expound that the key of paralleled double-effect distillation process lies in the reasonable adjustment of the energy distribution, the operating parameters and the configuration of some columns. It not only reduces the quantity of energy consumption, but also avoids the decrease of the quality of energy. For factories with a large numbers of distillation schemes of high energy requirement, the paralleled schemes lead to improve economic benefit.

References

[1] Araujo AB, Brito RP, Vasconcelos LS, "Exergetic analysis of distillation processes-a case study," Energy 2007;32:1185-93.

[2] Michele Mascia, Francesca Ferrara, Annalisa Vacca, Giuseppe Tola, Massimiliano Errico, "Design of heat integrated distillation systems for a light ends separation plant," Applied Thermal Engineering 2007;27:1205-1211.

[3] Amiya K. Jana, "Heat integrated distillation operation," Applied Energy 2010;87:1477-1494.

[4] Hilde K. Engelien, Sigurd Skogestad, "Multi-effect distillation applied to an industrial case study," Chemical Engineering and Processing 2005;44:819-826.

[5] James G. Gebbie, Michael K. Jensen, Piotr A. Domanski, "Experimental transient performance of a heat pump equipped with a distillation column," International Journal of Refrigeration 2007;30:499-505.

[6] Massimiliano Errico,Giuseppe Tola, Michele Mascia, "Energy saving in a crude distillation unit by a preflash implementation," Applied Thermal Engineering2009;29:1642-1647.

[7] Massimiliano Erricoa, Giuseppe Tola, Ben-Guang Rong, Daniele Demurtas, Ilkka Turunen, "Energy saving and capital cost evaluation in distillation column sequences with a divided wall column," chemical engineering research and design 2009;87:1649-1657.

[8] Ivar J. Halvorsen, Sigurd Skogestad, "Energy efficient distillation," Journal of Natural Gas Science and Engineering 2011;3:571-580.

[9] Jinsheng Sun, Fan Wang, Tingting Ma, Hong Gao, Yanzhen Liu, Fang Cai, "Exergy analysis of a parallel double-effect organosilicon monomer distillation scheme," Energy 2012;47:498-504.

[10] Hilde K. Engelien, Sigurd Skogestad, "Multi-effect distillation applied to an industrial case study," Chemical Engineering and Processing 2005;44:819-826.

[11] Kazuo Matsuda, Kenichi Kawazuishi, Yasuki Kansha, Chihiro Fushimi, Masaki Nagao, Hiroshi Kunikiyo, Fusao Masuda, Atsushi Tsutsumi, "Advanced energy saving in distillation process with self-heat recuperation technology," Energy 2011;36:4640-4645.

[12] M.A. Sharaf, A.S. Nafey, Lourdes García-Rodríguez, "Exergy and thermo-economic analyses of a combined solar organic cycle with multi-effect distillation (MED) desalination process," Desalination2011;272:135-147.

[13] Weiwen Wang, Wenjian Wang, "Energy Saving in Methylchlorosilane Distillation," International Conference on Computer Distributed Control and Intelligent Enviromental Monitoring 2012:290-293.

[14] Myungwan Han, Sunwon Park, "Multivariable control of double-effect distillation

configurations," J. Proc. Cont. 1996:247-253.
[15] Hilde K. Engelien, Sigurd Skogestad, "Multi-effect distillation applied to an industrial case study," Chemical Engineering and Processing2005;44:819-826.
[16] Debadrita Maiti, Amiya K. Jana, Amar Nath Samanta, "A novel heat integrated batch distillation scheme," Applied Energy2011;88:5221-5225.
[17] Anton A. Kiss, Servando J. Flores Landaeta, Carlos A. Infante Ferreira, "Towards energy efficient distillation technologies-Making the right choice," Energy2012;47:531-542.
[18] Diego C. Alarcón-Padilla, Lourdes Garcí a-Rodrí guez, Julián Blanco-Gálvez, "Experimental assessment of connection of an absorption heat pump to a multi-effect distillation unit," Desalination2010;250:500-505.
[19] Diego C. Alarcón-Padilla, Lourdes Garcí a-Rodrí guez, Julián Blanco-Gálvez, "Design recommendations for a multi-effect distillation plant connected to a double-effect absorption heat pump: A solar desalination case study," Desalination2010;262:11-14.
[20] Fiedler E, Grossman G. Methanol, "Ullmann's Encyclopedia of Industrial Chemistry," Weinheim : Wiley-VCH Verlag/Gmbh & Co; 2002.
[21] Meyers RA, "Handbook of synfuels Technology," New York : McGraw-Hill; 1984.
[22] Chu YZ, Qin LP, "Optimum design of methanol distillation process and columns," Chem Ind Eng Prog 2008;(27):1659-62.
[23] Juntao Zhang, Shengrong Liang, Xiao Feng, "A novel multi-effect methanol distillation process," Chem Eng Pro 2010;(49):1031-37.
[24] Jinsheng Sun, Fan Wang, Tingting Ma, Hong Gao, Peng Wu, Lili Liu, "Energy and exergy analysis of a five-column methanol distillation scheme," Energy2012;45:696-703.
[25] J. S. Sun, P. Wu, Y. H. Wang, H. Gao, L.L. Liu.CN Patent102039058 A. 2011.
[26] J. S. Sun, P. Wu, Y. H. Wang, Y. X. Su, H. Gao, J. Li, D. Yan. CN Patent 101798322 A. 2010.
[27] Michele Mascia, Francesca Ferrara, Annalisa Vacca, Giuseppe Tola, Massimiliano Errico, "Design of heat integrated distillation systems for a light ends separation plant," Applied Thermal Engineering 2007;27:1205-1211.
[28] Olga A. Flores, J. Carlos Ca´ rdenas, Salvador Herna´ ndez, Vicente Rico-Ramı´rez, "Thermodynamic Analysis of Thermally Coupled Distillation Sequences," Ind. Eng. Chem. Res. 2003;42:5940-5945.
[29] Szargut J, "Exergy Analysis of Thermal, Chemical and Metallurgical Processes," New York: Hemisphere Publishing Corporation;1988:55-152.

(45n) In Situ Transesterification of Jatropha Curcas Seed Particles with Alkaline Phase Transfer Catalyst

Subbarao Duvvuri[1], Shuhaimi Mahadzir[2] and Sintayehu Hailegiorgis[2], (1)Department of Chemical Engineering, Universiti Teknologi PETRONAS, Tronoh, Malaysia, (2)Chemical Engineering, Universiti Teknologi PETRONAS, Tronoh, Malaysia

Economic viability of biodiesel technology is at stake due to the high cost of edible oil feedstock and slow reaction rates due to limited solubility of oil in methanol. In situ transesterification of non edible oil seed particles can minimize the cost of feedstock. Microwave pretreatment of seeds can make oil molecules more reactive. Use of phase transfer catalysts can enhance the reaction rates. In the present work, these three concepts are utilized together to investigate alkaline in situ transesterification of microwave pretreated jatropha curcas seed particles in presence of benzyltrimethylammonium hydroxide (BTMAOH) as a phase transfer catalyst (PTC). It was observed that addition of PTC to alkaline catalyst has improved conversion and reduced reaction time. Microwave pretreatment of jatropha seed particles reduced reaction time drastically. The quality of biodiesel produced was within the limits of the international standard requirements of ATME D6751 and EN-14214. Thus, combined use of microwave pretreatment of seed particles, phase transfer catalysis in the presence of alkaline catalyst can increase the rate of in situ transesterification and product yield.

(45o) Catalyzed-Assisted Manufacture of Olefins (CAMOL): Updated for Use in Naphtha Service

Maximilian Walter, Petrochemicals, BASF SE, Ludwigshafen, Germany and **Suzanne Rech**, BASF Qtech, BASF Corporation, Iselin, NJ

Since the commercialization of the olefins plant, operators have searched for ways to be more competitive. One area always targeted includes keeping their furnaces in operation longer between decoking steps. Over the years, many anti-coking methods/technologies have become commercially available using different approaches to prevent coils from coking. One of the latest available technologies is the novel CAMOL™ (Catalyzed-assisted Manufacture of Olefins) Technology from BASF Qtech. As reported in the introduction of the technology to the EPC in 2008 & 2010, the CAMOL™technology differs from other available anti-coking technologies in that the technology not only eliminates the catalytic (or filamentous) coke make through a physical barrier, but also gasifies the thermal (or amorphous) coke for an even greater coking resistance. The coating process has generated a coating with almost identical thermo-mechanical properties as the base metal substrate (alloy). This allows bonding to the base metal similar to being welded with a minimum risk of spallation when going through thermal cycling. Commercially, the CAMOL Technology has been available since 2006 and is now in operation in furnaces of various designs.

With CAMOL™, operators have experienced extended run lengths, reduced energy requirement and GHG emissions, lower average TMT operating temperatures, reduced DMDS consumption, and less carburization. In addition, the technology has exhibited high operating temperature stability (>1130°C/ 2065°F).

At BASF SE in Ludwigshafen, Germany, the technology has been in operation since early 2010 in mostly naphtha service. After overcoming some initial challenges associated with a catalytic surface technology, such as reactivating the catalytic species after decoking, the coil is now successfully operating with run lengths significantly higher than uncoated coils. This paper will report on the results and learnings from the experiences gathered over the past three years along with the advancements made since the last report in 2010.

(45p) Extending Furnace Run Length and Coil Life by Optimizing Decoking

Vinod Mishra, Ingenero, Mumbai, India and **James Brigman**, Ingenero, Houston, TX

Abstract:

In the Ethylene manufacturing process, coke deposited in the radiant coils and Transfer Line Exchangers (TLEs) during normal operation is removed periodically by decoking the furnaces using a mixture of steam and air. There is a significant variation in operating parameters adopted by different plants for decoking. These are often either too conservative or aggressive, leading to production loss from increased off-stream time, TLE limitation and deterioration in coil life. This paper describes how the decoking procedure was modified based on furnace and TLE operating conditions and the decoking steps changed to obtain the best possible Ethylene yield, without adversely impacting coil life and TLE performance. This optimization has now been done in several major ethylene facilities and furnace run length has more than doubled and production has significantly increased.

SOLVING OPERATIONAL PROBLEMS IN A NAPHTHA SPLITTER

Flávio Weissheimer, GE Water & Process Technologies do Brasil Ltda, São Paulo/Brasil
Giorgio Franceschetti, Braskem, Santo André/Brasil
Rafael Henrique Camparin, GE Water & Process Technologies do Brasil Ltda, São Paulo/Brasil
Patrick Lucas, GE Water & Process Technologies, The Woodlands/US

Abstract

This paper sets out how the polymerization process of Reboilers of the (A-100) Crude Naphtha Fractionation Unit of Braskem UNIB 3 ABC can be controlled by applying polymerization inhibitors developed by GE Water & Process Technologies Ltda and Braskem´s engineering solutions.

These heat exchangers have a high degree of polymerization because of the content of olefinic compounds present in the feed composition which, when in contact with high temperatures in the heat exchange areas of reboilers, tend to polymerize. The joint solution of GE technology and Braskem engineering solutions applied to this system ensures regular operation of such equipment for periods of time greater than those hitherto recorded without manifesting loss of heat transfer efficiency.

In the 2012 shutdown while the opening of these reboilers was being inspected, it was observed that pipes had not been clogged nor was surface fouling present, which is a very positive state when compared to the historical records of previous campaigns.

The strategy adopted by GE/Braskem to ensure a longer campaign for these pieces of equipment was based on the quality control of feeding by adjusting dosages and monitoring performance using statistical analysis tools.

The use of statistical models based on Multivariable Regression Analysis (MRA) to evaluate performance, which is validated by inspecting reboilers internally, has been an efficient way to determine the degree to which the system has been fouled and also to schedule shut-downs for cleaning the mechanical parts.

Description of the Process

The (A-100) Naphta Fractionation Unit includes two distillation columns, the objective of which is to separate the light (C_6), medium (C_6 - C_8) and heavy (C_{9+}) cuts, included in the feed supply of crude naphtha.

The light cut and a portion of the heavy cut are sent to pyrolysis furnaces so as to produce ethylene. The medium cut is sent to the Hydrotreating and Catalytic Reforming Unit, with the objective of increasing the aromaticity of the system so as to produce BTX.

The A-100 processes 100 tons/h of crude naphtha, and approximately 35% of this current represents the average cut. Figure 1 below shows the schematic flow diagram of this fractionation process.

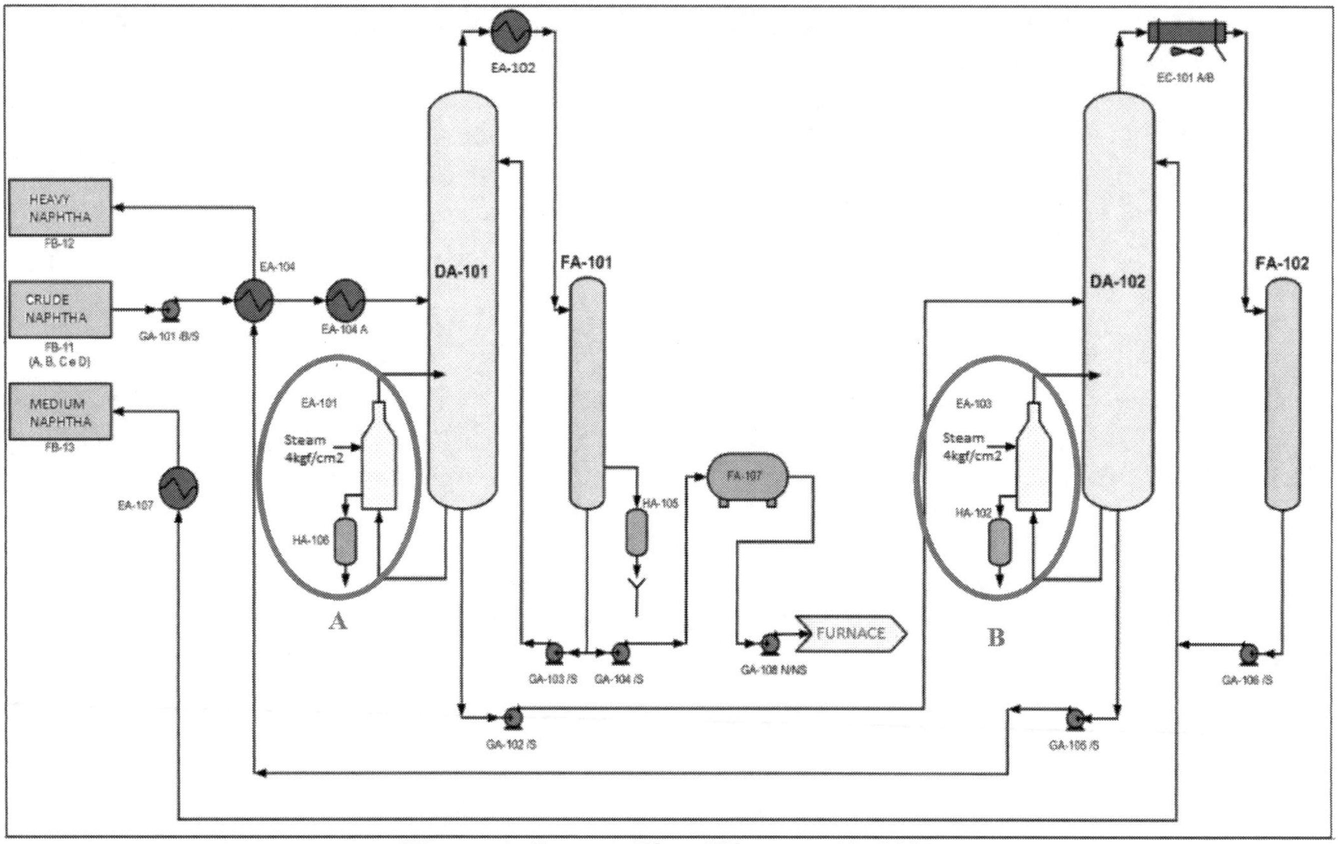

Figure 1. Process Flow Diagram (A-100)

The Reboilers, indicated by the letters A and B in Figure 1 above, operate with an injection of low steam (4kgf/cm2) to promote vaporization of the bottom stream of the A-100 distillation columns. These devices operate over a temperature range of about 150°C, which is sufficient to promote the polymerization of certain olefinic compounds present in the feed of crude naphtha.

Poor performance of these Reboilers can lead to production stoppages for maintenance and cleaning, which very often affects up to five days of production. The possibility of raising the times of campaigns and predicting the degree of fouling in downstream units helps to minimize production losses due to unscheduled stoppages.

This study will discuss the results of Reboiler A, which is the most critical equipment, with a view to synthesizing data, but gains in equipment B were also significant.

Evaluation System

Historical Record of the Campaign

Reboiler A was installed in 1972 but examination of the historical records did not reveal that the sheaf of tubes had ever been replaced. This exchanger has currently 11 tubes plugged into it.

The Campaign times of this equipment are shown in Table 1, with the last campaign being highlighted in blue so as to set it apart from the other ones. Table 1 gives a summary in the order of the timings of the campaigns, the average load processed in the period, the total load processed, the reason for operational stoppages etc.

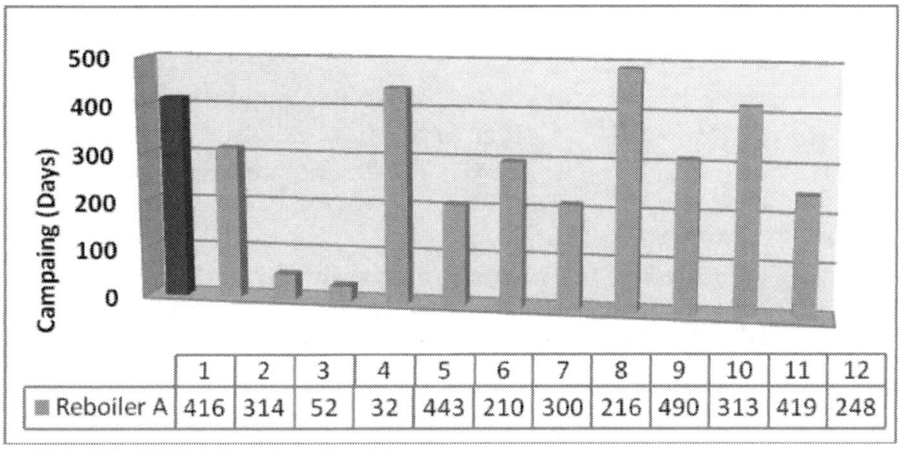

Figure 2. Campaign of the Reboiler A

Table 1. Historical Record of Reboiler A Campaigns

Date	Campaign *(days)*	Capacity *(ton/h)*	Capacity *(ton/campaign)*	Occurrence	Tubes Clogged *(%)*
13/06/12	416.0			Programmed Shutdown	0
17/03/11	314.0	93.1	701601.6	Heat exchange performance	30
03/05/10	52.0	90.8	113318.4	Stuck tube	-
11/03/10	32.0	71.1	54604.8	Heat exchange performance	50
09/02/10	443.0	72.0	765716.6	Heat exchange performance	20
16/11/08	210.0	98.1	494575.2	Heat exchange performance	-
16/04/08	300.0	104.0	748800.0	Programmed Shutdown	-
16/06/07	216.0	92.8	481075.2	Programmed Shutdown	-
10/11/06	490.0	-	-	Programmed Shutdown	-
30/06/05	313.0	-	-	Stuck tube	-
17/08/04	419.0	-	-	-	-
18/06/03	248.0	-	-	-	-
10/10/02	-	-	-	-	-

Conditions of Opening

In this review, a comparison is presented between two operational shut-downs of A-100, which occurred in March 2011 and May 2012, in order to observe the improvement in performance from the last campaign associated with the change in the GE Treatment strategy.

Campaign I: Production shutdown on 03/2011

Table 2 below summarizes the features observed during the inspection of the state of the inlet of this piece of this equipment during the shutdown of March 2011:

Table 2. Internal inspection of Reboiler A in 03/2011

Bottom Cover	Surface residue
Bottom Mirror	No deposits or internal blockage of the tubes were observed.
Top Mirror	Approximately 30% of the tubes were clogged, especially in the region of the vapor inlet, which has higher temperatures. Magnetic and non-magnetic deposits.

May/2011

Figure 3. The mirror of Reboiler A before it was cleaned.

May/2011

Figure 4. The mirror of Reboiler A at the vapor inlet before it was cleaned.

Note: It is difficult to carry out internal cleansing by Hydro jetting (pump pressure during cleaning is 20,000 psi). This reboiler has 11 tubes plugged into it, but none of them are plugged in during this shutdown.

Campaign II: Production shutdown on 03/2011

Table 3 below summarizes the features observed during the inspection of the state of the inlet of this piece of equipment during the shutdown in May 2012:

Table 3. Internal inspection of Reboiler A in 05/2012

Bottom Cover	Surface residue
Bottom Mirror	No deposits or internal blockage of the tubes were observed.
Top Mirror	Clogged tubes were not identified, not even in the region of the inlet vapor. There was only a thin layer of deposit that was easy to remove and which was evenly distributed over the upper surface of the mirror.
Spool	A thin deposit on the inner surface of the caisson.

May/2012

Figure 5. The mirror of Reboiler A before it was cleaned.

May/2012

Figure 6. The mirror of Reboiler A before it was cleaned.

Note: Hydro jetting was done quickly, without any obstructions due to deposits having been identified. This equipment has 11 tubes plugged into it, but none of them were during this shutdown.

Analysis of Deposits

The characterization of the deposits found in Reboiler A showed an 86.0% loss on ignition, thus indicating the strong presence of organic compounds in their composition, with the functional groups of carboxylic acid and ester standing out. The inorganic analysis of the sample indicated a high concentration of iron oxides and also sulfur oxides in the form of sulfides on the cover and spool of this reboiler. This is summarized in Table 4 below:

Table 4. Analysis of the Deposits of Reboiler A

Organic Compounds - LOI (%)	86.0
Water of hydration	Smaller
Sulfur compounds	Smaller
Carboxylic Acid and Ester	Greater
Dichloromethane soluble (%)	2.0
Elemental Sulfur	Greater
Dichloromethane insoluble (%)	98.0
Water of hydration	Smaller
Inorganics	Smaller
Carboxylic Acid and Ester	Greater
Composition of Dichloromethane insolubles (%)	
Carbon	56.0
Hydrogen	5.0
Nitrogen	3.0
Sulfur	8.0
Inorganic Compounds (%)	
Iron, $Fe_2O_3 + Fe_3O_4$	82.0
Sulfide, S^{2-}	16.0
Phosphate, P_2O_5	1.0
Sodium, Na_2O	1.0

Evaluation of Performance

The evaluation of performance of Reboiler A was prepared using the MRA tool, which is a GE good practice to predict fouling in critical systems. This statistical tool enables a dependent variable to be determined, in this case the percentage of opening the steam valve, by means of the numerical regression of the other operational variables of the column. This method generates a mathematical equation to predict the percentage of inlet steam when the equipment is clean, given that the equation obtained by regression includes data of this condition.

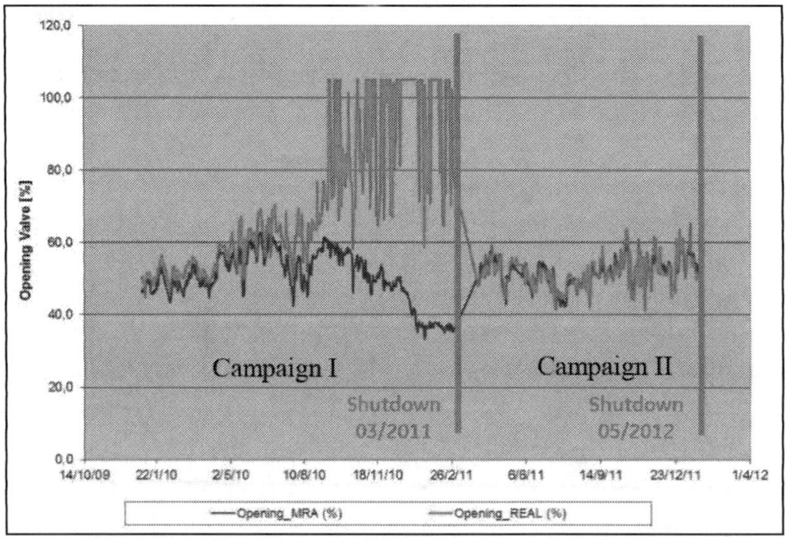

Figure 7. Performance Evaluation of Reboiler A

In Figure 7 the dark blue line represents the percentage of inlet steam when the equipment is clean, as per the equation defined by the MRA. The pink line represents the actual percentage of inlet steam. The sharp upward shift evidenced in the period between November 2010 and March 2011 shows the fouling of the system during Campaign I (prior to the 2011 operational shutdown).

After the 2011 shutdown, it can be seen that the two curves follow the same trend, without a shift, indicating that the unit has the equivalent performance of a clean piece of equipment. The change in the GE treatment strategy, taking the control of the dosages of the product Petroflo 20Y25 into account, was seen to have been effective in increasing the campaign time of this equipment.

Conclusion

The change of strategy in applying dispersant and controlling the dosages after the March 2011 shutdown, presented positive results, given the states of the inlet seen during the inspection of reboilers A and B in May 2012..

The low degree of fouling of these reboilers evidenced in the 2012 shutdown led to the production process being considered reliable, as the A-100 reboilers are no longer the pieces of equipment responsible for the loading limit of the unit. The downtime during the cleaning of the mechanical parts of the reboilers was about 10 times less than that observed in the previous shutdown, thus reducing Braskem´s maintenance costs.

In addition, the MRA statistical tool was able to produce a reliable representation of the degree of fouling of the system. This tool can be effective for scheduling shutdowns more than 03 months in advance.

References

1. Setal Lummus, "Manual de Operação da Unidade de Processamento de Nafta Bruta (Área 100)", Santo André, Brazil.
2. Ponce de Leon, Carolina; Weissheimer, Flavio and Camparin, Rafael H. (2011), "Proposta Técnica de Fornecimento de Especialidades Químicas, Equipamentos e Serviços para Tratamento Químico para a Unidade de Processamento de Nafta Bruta (Área 100)", Santo André, Brazil.
3. Kleinbaum, David G., Kupper, Lawrence L., Muller, Keith E. (2008) "Applied Regression Analysis and Other Multivariable Methods", Belmont, USA.

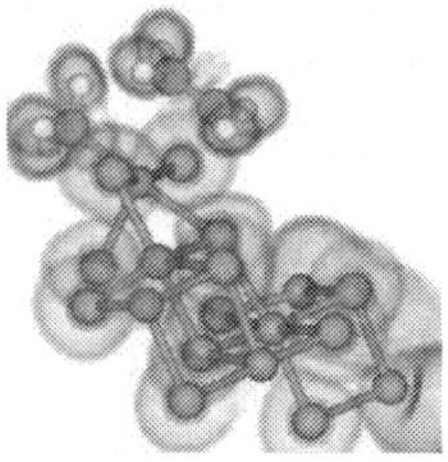

13AIChE
Spring Meeting, San Antonio, TX

SmartProcess® Application Strategy Achieves Faster and More Efficient Control of an Ethylene Complex

Patrick Truesdale
Principal Consultant
Industry Solution Group
Emerson Process Management

April 29, 2013
San Antonio, TX

EMERSON.
Process Management

Speaker Introduction

- Principal Consultant
  Emerson Process Systems & Solutions
- BS Electrical Eng, PE
- 40+ Years experience with Refinery & Petrochem Process Control
- Joined Emerson in 2007
- Previous employers: duPont, Bonner & Moore, SETPOINT, AspenTech, Comcept Technology

Patrick Truesdale

E-mail: patrick.truesdale@emerson.com

13AIChE
Spring Meeting, San Antonio, TX

EMERSON.
Process Management
Page 2

Agenda

- Performance Package Review
- SmartProcess Ethylene Overview
  - Furnaces
  - Distillation
- Benefits

SmartProcess Ethylene Packages

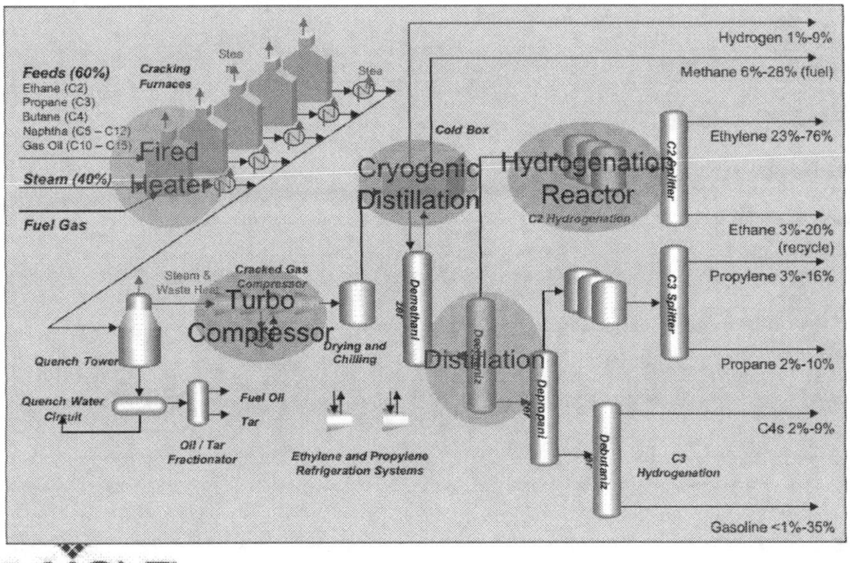

SmartProcess Scope

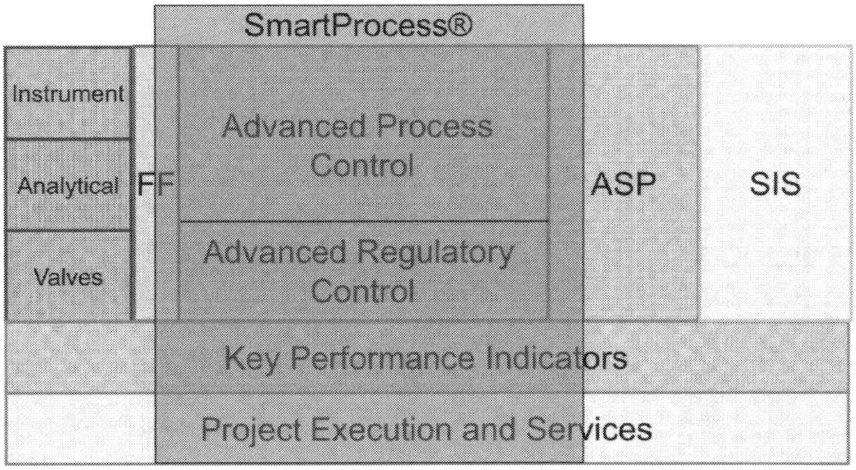

Basic Ethylene Furnace Variables

Key Operating Objectives:

- **Constant severity**
- **Maintain HC/Stm ratios**
- **Min excess air**
- **Maintain within constraints**
- **Max charge (if needed)**

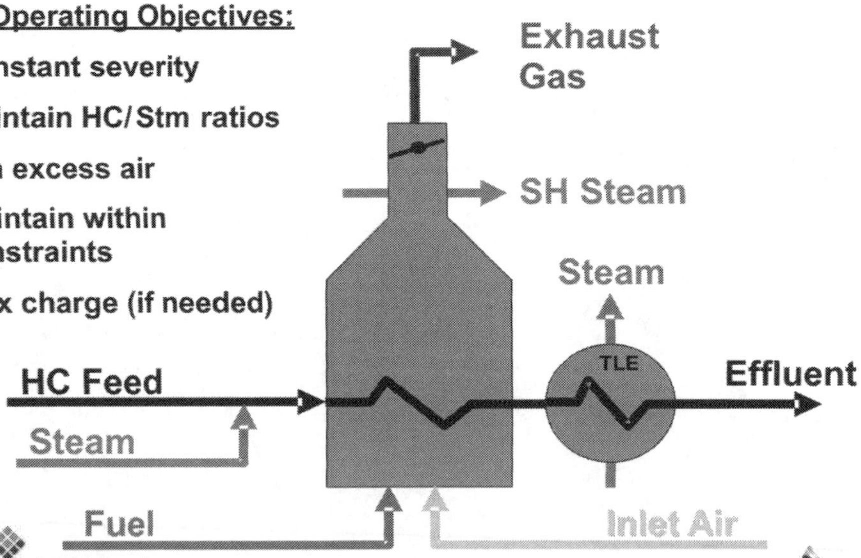

SmartProcess Furnace Module

- Design allows for different furnace configurations
  - Process Zones
  - Firing Zones
  - Multiple TLE analyzers
- Advanced Regulatory
  - Charge Control
  - Steam Control
  - Severity Calculation
  - Combustion Control
  - KPI's
- Advanced
  - Pass Balancing
  - Heat Balancing
  - Severity Control
  - O2 Control

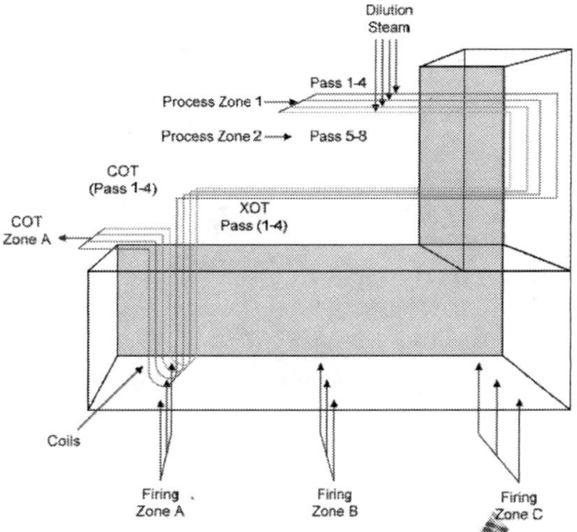

Standard Furnace Module Design

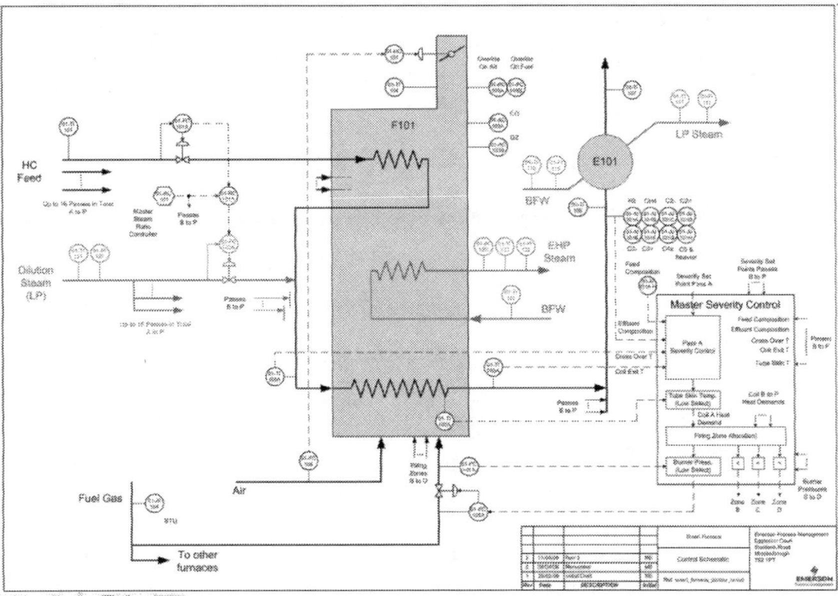

Charge Rate Control

Functions:
- Pass Balancing
- Constraints
- Feed Maximization

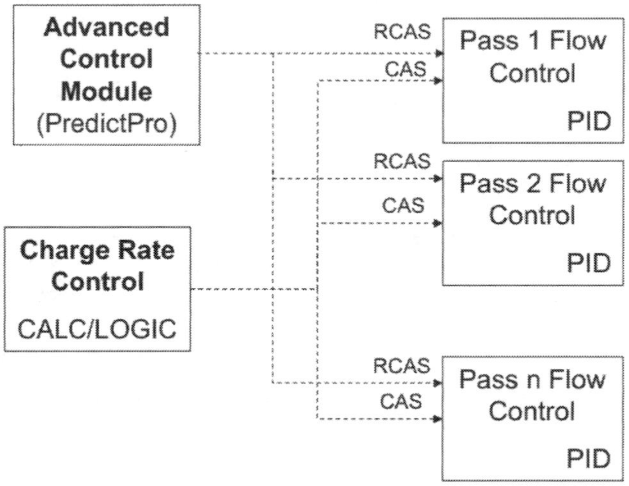

Functions:
- Ramping
- Limits & Alarms
- Multiple Feed Types
- Furnace Status
- Pass Offsets

13 AIChE
Spring Meeting, San Antonio, TX

EMERSON. Process Management

Page 9

Steam Ratio Control

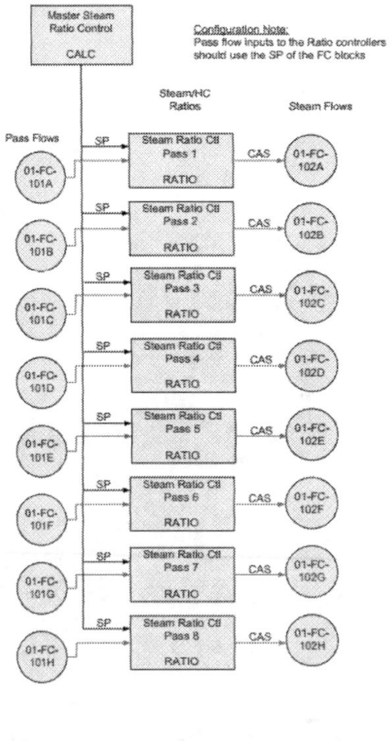

13 AIChE
Spring Meeting, San Antonio, TX

EMERSON. Process Management

Page 10

Severity Control

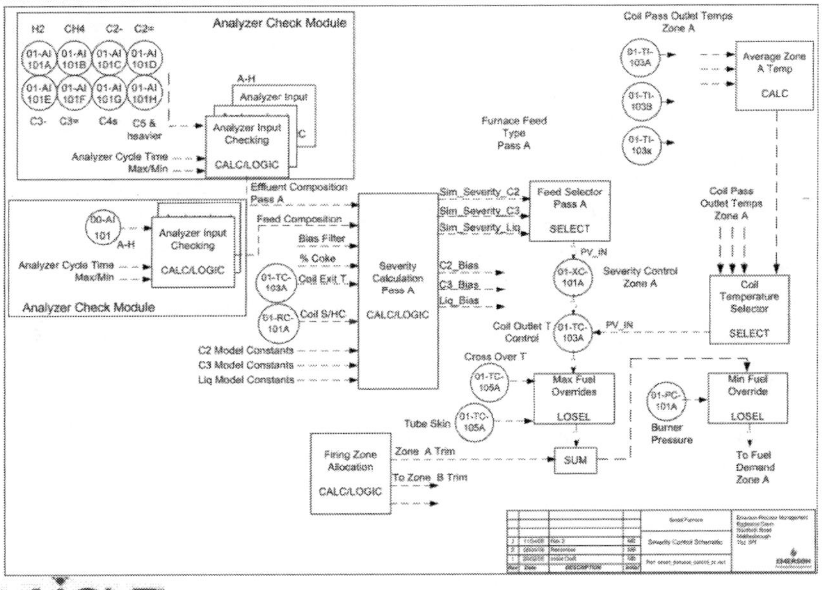

Automated Furnace Sequences

- Standard module templates to be used for Furnace Startup, Shutdown and Decoking sequences.
- Utilizes preconfigured Sequence Function Charts (SFC) running in DeltaV controllers or App Station
- Standard sequence functions can be customized to match site requirements
- Manual operator steps with confirmation can be mixed with automated steps
- Sequence execution control graphics and faceplate templates

Key Performance Calculations

- Furnace efficiency
  - Heat loss method
- Furnace Duty
- TLE Duty, Fouling factor
- Stack gas estimated composition
- Stack gas daily total tons exhaust for CO, CO2, S, N, etc.
- Stack gas NOx emissions (optional – based on CEMS)

13 AIChE
Spring Meeting, San Antonio, TX

EMERSON. Process Management

Distillation - Classic Multivariable Problem

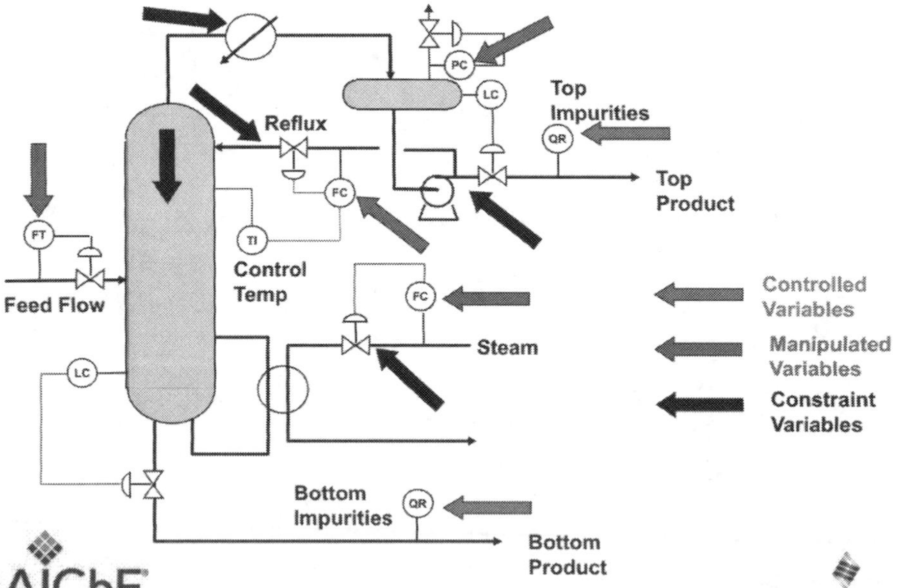

13 AIChE
Spring Meeting, San Antonio, TX

EMERSON. Process Management

Distillation Control Module

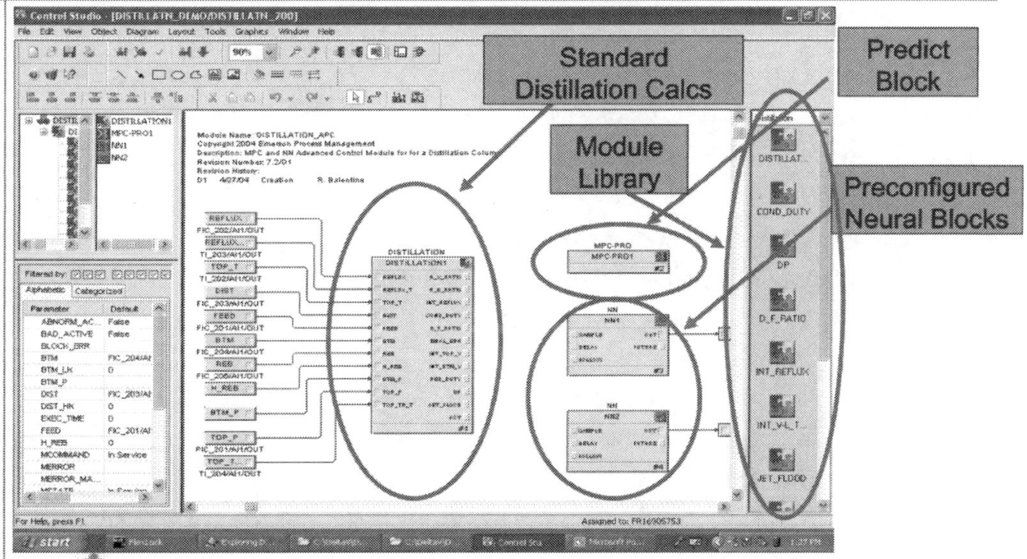

13AIChE
Spring Meeting, San Antonio, TX

EMERSON. Process Management

SmartProcess Distillation

- DeltaV PredictPro with up to 4 MV's
- DeltaV Neural for up to 2 product qualities
- Standard set of calculations & constraints
  - Internal reflux
  - Pressure compensated temperatures
  - Jet flooding
  - Reboiler & condenser duty
- Engineering Services to design, implement and commission

13AIChE
Spring Meeting, San Antonio, TX

EMERSON. Process Management

SmartProcess Applications

- Licensed industry applications for common processes that are:
  - Pre-engineered, tested, documented
  - Designed by APC Consultants with extensive practical experience
  - Supported & enhanced
- Bundled combination of:
  - DeltaV libraries: composite blocks, modules & templates
  - DeltaV simulation and demonstration example
  - APC tools (e.g. Predict, Neural)
  - Application code (e.g. VB, .Net)
  - Other databases and apps (e.g. SQL*Server, Excel, TankMaster)
  - Sample DeltaV Graphics
  - Documentation
- Templates to be used to configure for specific site requirements
- May be implemented by Emerson or Customer

Benefits

Ethylene Furnace High Temperature Imaging Cameras to Monitor Burner Flames ,Tube Temperature and Tube Failure Detection

Thomas M. Canty, P.E.
President
J.M. Canty Inc
Buffalo, NY USA

Abstract

The development of explosion proof, high temperature imaging cameras which operate in ethylene furnaces without cooling air now allows monitoring of ethylene furnaces to improve energy/operating efficiency along with providing important safety benefits. Imaging cameras work off the visual wavelength which provides 2 dimensional view and temperature measurement of the tube temperature which is more accurate than thermocouples and IR probes. In addition visual images of the furnace burners lets you detect flame detachment, smoking burners and flame shape in order to increase burner efficiency and save energy. The safety aspects of the Camera imaging systems is significant as it allows visual detection of pin hole leaks and tube failure in the furnaces.

Introduction

Ethylene furnaces, or any type of cracking unit, rely on thermal management in order to make product efficiently and to operate in a safe manner. Having 'eyes' in the furnace can help to identify problems in their early stages of development. High temperature camera technology is able to survive the environment and give a wide angle view of the tubes from which temperature and tube blockage may be determined. Achievable benefits from a camera system include:

1. Temperature analysis capacity for all areas of the image, not just single point. Multiple zone capability for user customization of analysis.
2. Flame detection and size measurement
3. Detection of tube movement and ability for software to automatically move measurement zones as tubes distorts.

Imaging Fundamentals

Basics of an imaging system include a high resolution sensor, proper illumination, appropriate hardware for the process and a robust software analysis package. For this particular application illumination is created by the process and need only be managed optimally for best results. A high density pixel array yields resolution of the image and measurement as well. The hardware and software are the more challenging aspects of the application and will be discussed in some detail.

Hardware

The design of the camera system is unique in that it's internally exposed lens assemblies are **designed to survive without cooling air**. When installed properly, the combination of system materials and heat transfer coefficients at the camera tip limit heat transfer to the system and allow it to run hot without chance of damage. This feature, when system is installed in a clean, gas burning environment, does not disrupt the temperature or air balances of the furnace because no cooling air is being introduced to the furnace. In addition, the design includes a disposable, quarts shield which protects the process exposed lenses from direct contact with the internal atmosphere. Should any particulate contact the system, the quartz shield is there to protect the optical lenses. Replacement of the protective quartz shield is inexpensive. Replacement of damaged lenses is very expensive and so the Canty design guards against that possibility.

Internally, the system is a two piece assembly consisting of the internal insertion lens and the external camera package. As mentioned earlier in this document, Canty employs a fused glass window as part of the electronic housing package. This has a dual purpose; Firstly, the camera housing assembly is a separate entity which carries WP and EXP ratings. Secondly, the camera housing bolts to the internal lens system and transmits the optical rays through the fused glass window to the CCD. This mounting also creates a hermetic barrier between the process and the electronic housing thereby preventing migration of hot, process gasses outside the pressure boundary of the system.

Figure 1. High Temp Camera Assembly

The camera electronics come with a full array of controls that allow it to be optimized to the software function to achieve accurate, consistent data output.

As previously mentioned, the installation is an important feature of the design. When installed correctly, the camera sits inside the refractory (see Figure 2) where its exposure to the process environment is minimized thus minimizing heat transfer into the internal lens.

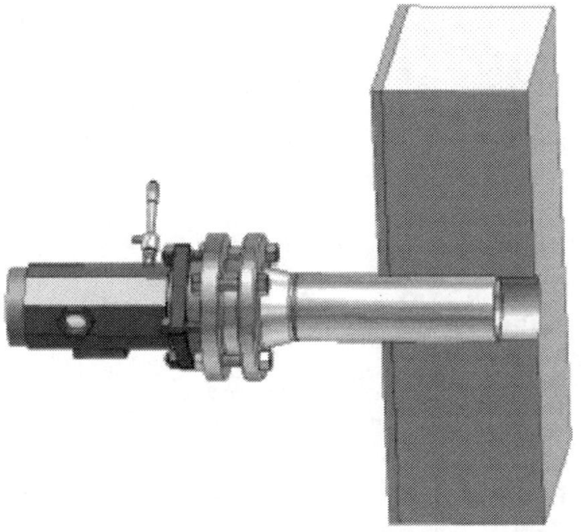

Figure 2. Camera Installation Schematic

Software

With respect to software design, the Canty software is fully flexible in analyzing any portion of the captured image. Multiple analysis zones can be drawn to monitor discreet areas of the view and each zone can run its own particular algorithm depending on need. The benefit here is that multiple tubes can be continually surveyed for changes and trends. The analysis data is also not single point, but is comprised of all the data points in the analysis zone. By choosing proper software filters various information can be obtained from the full array of data such as hot or cold spots. This data can also be archived in total or as snap shots at determined increments of time. The on screen data output can also be updated in the short or near term and can thus be used to judge what is happening in the furnace over time with just a glance at the screen. For instance, using the histogram feature with a 1x daily output a long term trend can be presented on screen in real time.

The basic theory of temperature measurement is based on the radiative emissions of a body at temperature – as the body temperature increases, the radiation curve shifts towards the shorter wavelengths and that is why as a body heats up the eye cannot detect it until it reaches 1000F which is the approximate visible point where emitted radiation is dark red in color. As the temperature increases, the color graduates to bright red, then orange then toward white. A pyrometer functions by measuring a narrow band of wavelength intensity and comparing it to the intensity of a black body to determine the temperature. From the graph below it can be seen that if a single wavelength is monitored the intensity of that measurement changes as the body temperature changes (curve shifts) which then corresponds to a different temperature on the calibration scale.

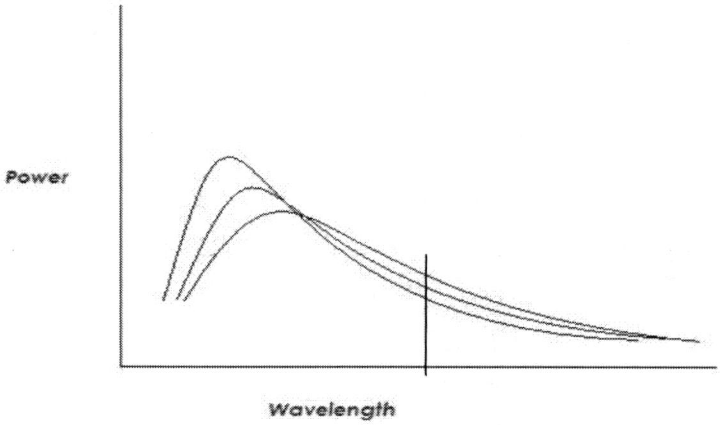

Figure 3. Power Curve at Varying Temperatures

The same is true of a camera system measuring the increasing intensity of an object as it radiates from approx 1000 F (dim, dark red color) on up the temperature scale to 3000+ F for instance which would be a bright white. From the graph above one can see that as color goes from red toward white the peak intensity is always increasing and it is the measure of this intensity which indicates the object temperature. Using an RGB camera the measurement can also be made by monitoring the individual color components of the signal impinging on the chip. For industrial uses, the measurement of intensity usually suffices.

Standard instrumentation used in industry include thermocouple type devices (contact) and thermal imaging and pyrometer systems (non contact). In severe temperatures and atmospheres, the contact devices deteriorate quickly and provide gradually less reliable information leaving the operator running blind. They also create a discontinuity on the surface they are welded to which, in the case of a furnace tube, can lead to premature tube failure.

Non contact systems have difficulties with calibration drift and emissivity adjustments to account for the fact that many items being measured are not black body emitters. In addition, these technologies are single point measurement systems and are best used as hand held devices which can be difficult to use and provide subjective measurement results based on operator preferences for targeting.

Vision technology offers several benefits these other technologies do not. The visual aspect of the system allows the operator to view the process and confirm the readings received. It also allows flexibility to set up measurement zones on screen at any time to assess any or all areas of the process in view should a change be required. Since vision systems are generally fixed installations, an added benefit is that the information is taken from the same process points and features within a given measurement zone at all times which means there is no error in sighting of the instrument.

The vision system is also unique in that it does not suffer calibration drift as almost all instruments do. Once the calibration is set the readings repeat themselves within a couple degrees F. The calibration is based on thermocouple or pyrometer readings initially, however after that calibration is finished it does not have to be redone in the future as long as the system remains in its location.

Process history and performance can also be best documented by a vision system. First off, the camera can analyze any area of process that is within the systems view. This affords the operator the capability of viewing many points of interest instead of just single point temperature readings. As an example, if the system is monitoring a tube bank it can assess the temperature of the tubes for the full length of view. If there is a local flow problem in a tube it will become overheated in one section. The camera can locate that section and indicate its temperature rise over the rest of the tube or adjacent tubes. This information allows the tube location to be identified as well as the position in the tube that has encountered the upset condition. Over time this historical data, recorded in video or still image format by the control software, may be significant in improving the process. In addition to the ability to monitor the full area of the view, the full area of view can also be segmented into multiple zones of measurement, i.e. each tube having its own measurement zone for automated process control capabilities. This capability allows the tube to be operated at optimum efficiency since the addition of excess heat can be avoided thereby reducing energy draw to operate the furnace.

Thermal Analysis

Tube Temperature

The following figures illustrate the temperature analysis capability for tubes in a live test situation. Measurement zones are placed over the tubes (figure 4) as the furnace is heated and cooled. The temperature plots are shown in figures 5 and 6. The results are as expected and uniform from tube to tube, although a small number are included in this demonstration.

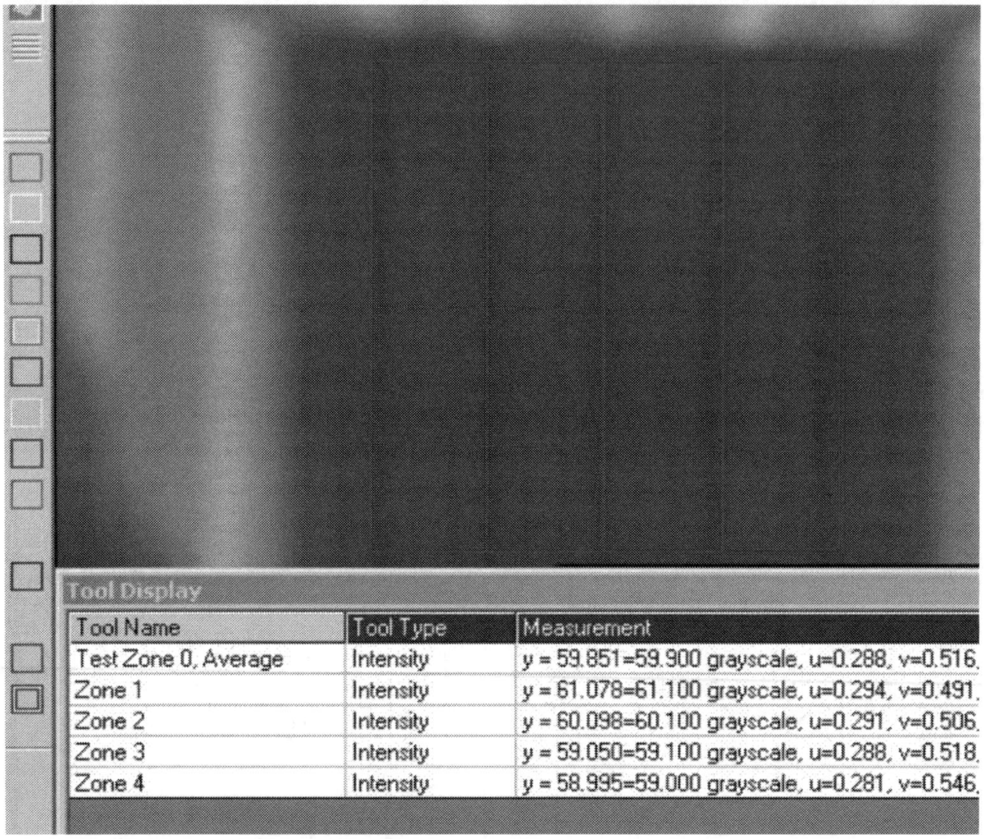

Figure 4. Software Screen Shot

As thermal loads are fluctuated up and down the following graphs track those changes. The cooling cycles are left to natural convection and thus are not linear in nature.

For the following curves, the vertical scale is gray scale intensity which is measured 0 – 255 (pitch black to bright white). The hotter the tube, the higher, or more intense, the gray scale reading. This value in process would be calibrated to the temperature. The horizontal scale is the image number, or time.

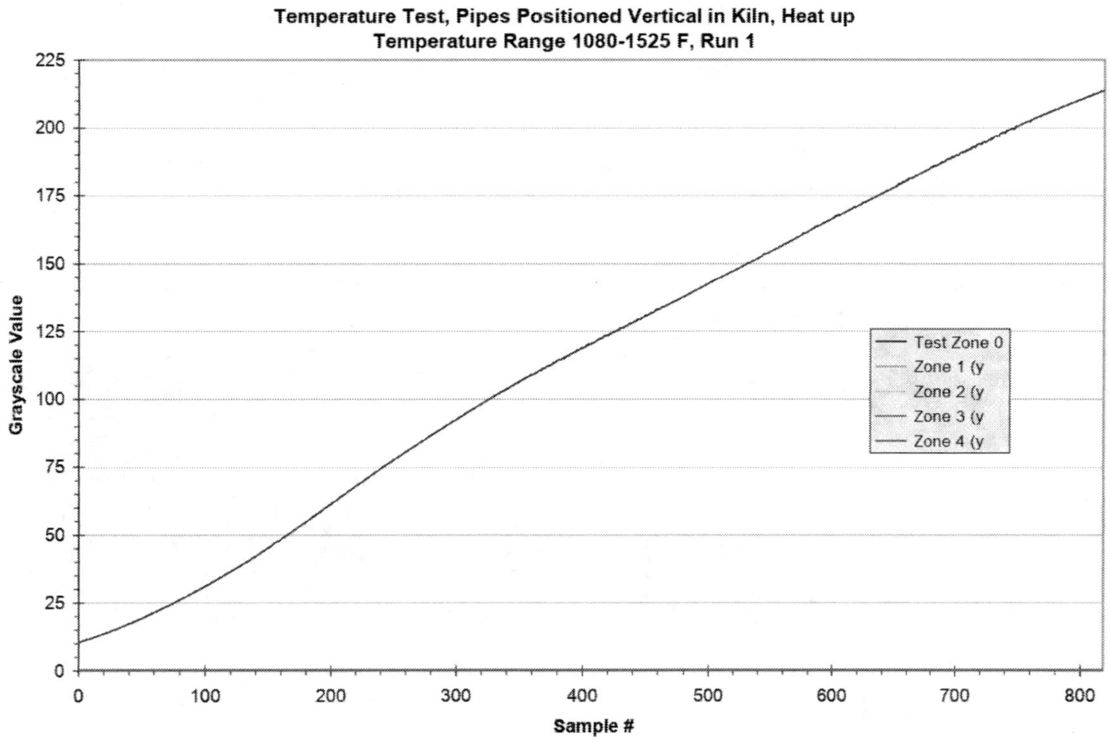

Figure 5. Heating Cycle

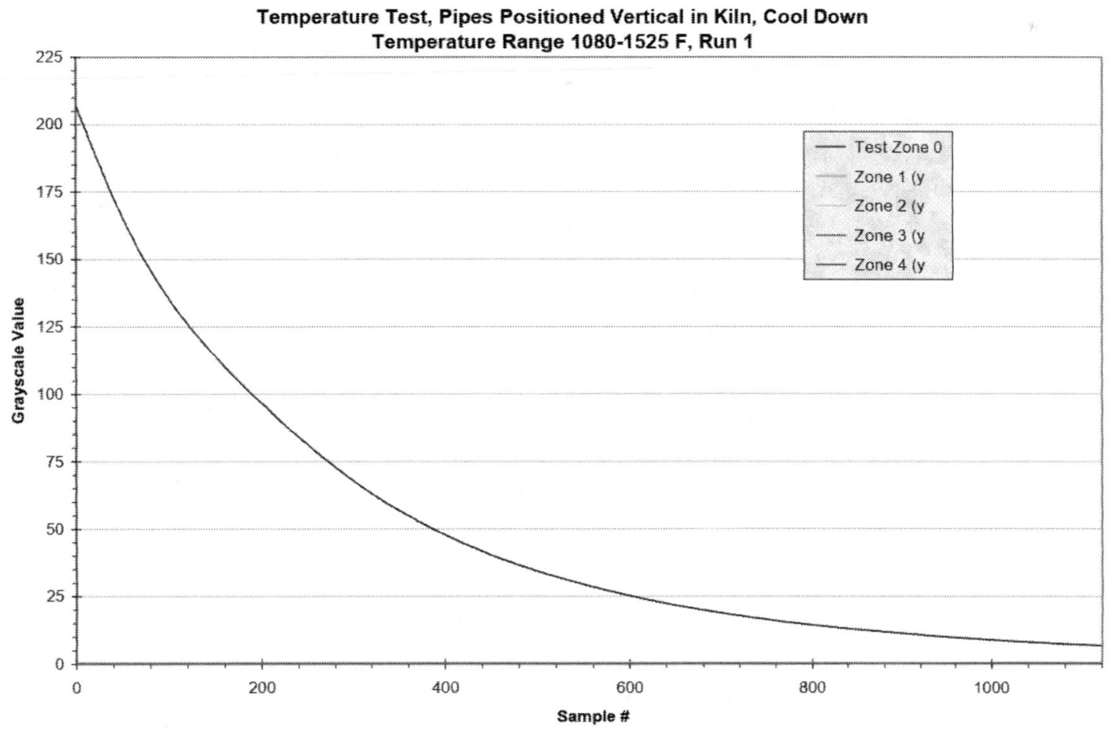

Figure 6. Cooling Cycle

Flame Measurement
Flame temperature can be monitored in the same way as described previously, however the flame size, or presence, can also be monitored. Figure 7 shows a live flame and Figure 8 shows a digital capture. Computing the flame area, or shape, can be done to track its changes over time.

Figure 7. Kiln Flame (Live)

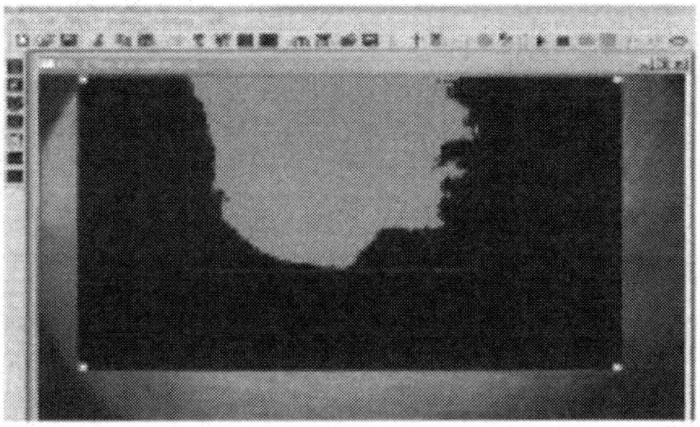

Figure 8. Kiln Flame (Digital)

Conclusion

Vision Technology presents an opportunity to increase effectiveness of furnace monitoring through hardware and software capabilities. The thermal design at the core of the hardware enables the system to provide the operator a view into the furnace and feed back on its thermal performance over time. This data can assist in detecting building upset conditions, but also provide a record for study to assist in improving the operational method of the furnace.

(45v) Straighten up! Non-Linear Compensation Using Feed Forward Control

Barbara Hamilton, Process Systems and Solutions, Emerson Process Management, Hampden , ME

Non Linear processes occur throughout the petrochemical plant. While the PID loop is surprisingly robust for most situations, there are cases where compensation of these non-linearity's is critical for stable control. Similarly, disturbances can cause major disruptions in control stability.

A method of feed forward control, using data based models, can provide robust control throughout the process range. Historical data are used to build models to drive the output toward the solution with PID control providing the final drive to zero error.

The model is also used to develop dynamic tuning. The result is a simple to configure, robust solution to manage non-linear loop and/or disturbance variations.

(45w) Adsorption of CO2 by Alkaline Activated Ca-Al2O3-SiO2 Material

Naim Faqir, King Fahd University of Petroleum & Minerals, Dhahran, Saudi Arabia and Reyad Shawabkeh, Chemical Engineering, King Fahd University of Petroleum & Minerals, Dhahran, Saudi Arabia

Natural gas contains a slight level of carbon dioxide which leads to several industrial and domestic heating problems. In addition, it is one of main greenhouse gases that contribute for global warming, where its pollution is a wide-spread problem arising from a wide range of industries, leading to complex contaminant in atmosphere. Given the increasing pressures worldwide to minimize the emission of greenhouse gases into atmosphere, CO_2 must be remediated efficiently and cost effectively.

In this short abstract a chemically treated limestone and alumino-silicate material were utilized for removal of CO_2 from natural gas using a combined adsorption-chemical reaction processes. This functionalized Ca-Al_2O_3-SiO_2 material was characterized according to its crystalline phase, surface area, pore size distribution and surface morphology. Then it was tested against different concentrations of CO_2 in CO_2-N_2 gas streams.

Pressure Gauges: Hidden Danger or Visible Safety

by Jeff Placek, WIKA Instrument, LP, Lawrenceville, GA

Introduction

In recent years, refining plants, petrochemical plants and chemical plants have been required to increase capacity, but have been provided with very few upgrades to existing equipment to accommodate the increased capacity requirements. This has caused plants to have increased stress and strain on their systems and operating units. For example, higher pulsating pressures and heavy mechanically-induced vibrations are translating to all areas of the processing equipment, especially pressure instrumentation.

Pressure gauges are one of the most important, yet most overlooked or ignored keys to ensuring the continued safety, productivity and profitability of process plants. Process pressure gauges are used to read pressure on transfer lines, discharge pumps, and other places where it is important to be able to read pressures.

Problems Associated with Misapplied, Faulty or Ignored Gauges

Incorrect, misapplied, damaged, faulty or ignored gauges can lead to a number of issues, including safety problems, downtime, and costly regulatory agency fines and other penalties.

Process Safety:

One group of problems relates to process safety issues, such as loss of containment that leads to fugitive emissions, fires and explosions. Bourdon tubes in malfunctioning gauges can rupture, causing the release of media, such as gasoline, oil or gas. Because this media is often volatile and corrosive, it can cause fires, explosions, or other serious safety issues, such as a toxic gas releases or acids spraying onto workers.

Media releases are common results of misapplied, faulty or ignored gauges. Fires and explosions are less common but they do occur. When reviewing plant instrumentation, findings have revealed instances where the wrong gauges have been installed in locations on pumps. In one instance, a gauge vibrated so severely when the pump came out of alignment that it caused the Bourdon tube in the gauge to crack. The media coming out of the tube leaked onto the pump, which was excessively hot because of age and misalignment, causing a fire.

Expensive Downtime:

A second group of problems relates to expensive downtime, which can result from the breakdown of process safety just mentioned, or simply from having to shut down a process in order to replace incorrect, misapplied, broken or otherwise damaged gauges.

1 - Safety-Related Downtime:

As a result of the problems identified above (emissions, fires, explosions, etc.), there can be any number of causes of subsequent downtime, including fire suppression activities, released media containment, line shutdown, personnel evacuation, and then eventually having to rebuild or repair the lines, replace equipment, etc. In addition, if there is a media release with no injuries or fatalities, the plant is required to notify the EPA. However, if there is a media release that does cause injuries or deaths, the plant is required to notify OSHA. These can lead to very long process shutdowns, anywhere from a day to a month or more. (More on OSHA and the EPA below.)

2 - Non-Safety-Related Downtime:

When a line fails, causing a gauge to fail, even if there is no release or safety incident, downtime can still occur as a result of the process needing to be shut down in order to replace the gauge. Gauge failure rarely causes other parts of the process to fail, although gauge failure is usually an alert that there is a problem. In most cases, a plant worker will simply replace the pressure gauge and move on. When another failure occurs, the worker will do the same, until the root cause of the failure, such as a misaligned pump, causes the plant to shut down the line, resulting in more costly, unplanned downtime and a major maintenance time.

Regulatory Agency Fines and Penalties:

A third group of problems relates to the potential for significant fines and other penalties from federal regulatory agencies (primarily OSHA and the EPA), as well as from state and local jurisdictions, that result from compromised process safety.

1 - OSHA:

OSHA addresses many process-related safety issues in 1910 Subpart H, "Hazardous Materials." More specific is 1910.119, "Process safety management of highly hazardous chemicals." In its Process Safety Management section, OSHA notes that, "Unexpected releases of toxic, reactive, or flammable liquids and gases in processes involving highly hazardous chemicals have been reported for many years in various industries that use chemicals with such properties. Regardless of the industry that uses these hazardous chemicals, there is a potential for an accidental release at times they are not properly controlled, creating the possibility of disaster."

OSHA goes on to note that: "To help ensure safe and healthful workforces, OSHA has issued the Process Safety Management of Highly Hazardous Chemicals standard (29 CFR 1910.119), which contains requirements for the management of hazards associated with processes using highly hazardous chemicals."

OSHA's standard emphasizes the management of hazards associated with highly hazardous chemicals and establishes a comprehensive management program that integrates technologies, procedures, and management practices.

While OSHA does not monitor gauges in specifically, they will enforce guidelines for worker safety. OSHA investigates injuries and deaths, and may levy fines or other penalties.

2 - EPA:

According to Greg Neneman, director of business development solutions for Valin Corporation (San Jose, Calif.), fugitive emission regulations from the Federal level have tightened up over the past several years. In many states, such as California, Oregon and Washington, the regulations are even tighter.

"One challenge companies face is ensuring compliance with emission regulations when they upgrade or expand their processes," points out Neneman. "They are not allowed to produce additional emissions beyond the original limit set by regulations."

From the standpoint of federal EPA regulations, each company that produces or emits volatile organic compounds (VOCs) is required to come up with a leak detection and repair program. "In some areas, county and local requirements are even tighter than federal requirements," continues Neneman.

In certain areas, companies that produce or emit VOCs are required by regulation to do audits on valves, flanges, gauges, and all other process connections on a quarterly basis. "And in some jurisdictions, the regulatory agencies can give plants 24 hours notice before they descend on that plant to check all of their records and conduct other investigations," adds Neneman.

In May 2012, the U.S. Environmental Protection Agency (EPA) levied a $1 million civil penalty on a plastics company for alleged violations of the Clean Air Act at two of its chemical manufacturing plants. In addition to paying the penalty, the company agreed to implement a comprehensive program to reduce fugitive emissions of hazardous air pollutants (HAPs) from leaking equipment such as valves and pumps. The settlement also required the company to implement enhanced work practices, including more frequent leak monitoring, better repair practices, and new efforts designed to prevent leaks. After being fined by the EPA, the company realized that instrumentation was one of the areas where they having fugitive emissions. Until that time, workers had been replacing gauges left and right, because they weren't aware of the problems that could occur. The gauges weren't a big expense to them, so they considered the replacement activity to be more of a nuisance than anything. It's when they had the rupture that they realized there was a real problem.

An audit of the plant, which involved reviewing at all of the instrumentation, found that in many applications, they were using the wrong instrumentation. To correct the issues, the right instrumentation was specified for them, and the correct gauges were ordered. The plant now has its instrumentation under control.

While conforming to EPA regulations has always been a challenge for process facilities, it is going to become even more challenging in the future as a result of EPA's recent introduction of new requirements related to fugitive emissions and leak detection and repair (LDAR) strategies. On April 17, 2012, the EPA Administrator signed a suite of final air quality regulations that affect both upstream and midstream oil and gas sectors - the production/processing and transmission/storage sectors of the oil and natural gas industry.

For the first time, EPA will regulate air emissions from natural gas wells that are hydraulically fractured, as well as other emission sources associated with exploration, production, processing, and transportation of oil and natural gas.

Promulgation of the final rules will result in new control equipment requirements, as well as increased requirements for monitoring, recordkeeping, and reporting for owners, operators, contractors and vendors.

The final rules require the use of emission controls, as well as work practices, through two different air regulatory programs. One is the New Source Performance Standards (NSPS). The other is the National Emissions Standards for Hazardous Air Pollutants (NESHAP).

EPA finalized a new NSPS at 40 CFR Part 60, Subpart OOOO, which regulates emissions of volatile organic compounds (VOC) and sulfur dioxide (SO_2) from a variety of oil and gas exploration, production, processing, and transportation facilities, including natural gas wells, certain compressors and pneumatic controllers, storage vessels, and sweetening units at onshore natural gas processing plants. Subpart OOOO also includes startup, shutdown and malfunction provisions.

The EPA has also made it clear that oil and gas production and processing sources will be the focus of its National Enforcement Initiative (NEI) for fiscal years 2012 and 2013. With the new NSPS and revised NESHAPs in place, EPA now has the tools it needs to enforce in a comprehensive manner.

As a result of these new regulations, plants should look at everything more closely, including gauges.

How Problems Can Occur With Gauges

Technical problems can occur with process pressure gauges in a number of ways, which can be conveniently categorized into seven broad categories with specialties or subsets. The most common reason for gauge problems is that workers often install gauges without knowing how they should be installed or accessorized. As a result, the gauges can fail because they were never designed to function in those specific situations. They can also fail because they were installed incorrectly, or because they were not properly maintained.

One way to eliminate or minimize these problems is to consider specific pressure gauge accessories. These can prove to be the extra step needed to ensure a successful, proactive preventive maintenance program, and for extending the life of the gauges. Accessories can provide additional protection for the gauges during their lifetimes.

1 - Mechanical Vibration.

This is oscillating, reciprocating or other periodic motion of a rigid or elastic body or medium forced from a position or state of equilibrium. Typically, vibration is the main cause of pressure gauge failure in process plants. Pump applications produce most of this vibration.

Gauges that aren't designed to withstand vibration may be installed in applications where vibration occurs, causing the gauges to come out of accuracy. Even more serious, a gauge can vibrate so much that the Bourdon tube inside it can crack, releasing media."

One way to prevent vibration-related problems is to install liquid-filled gauges, which help to dampen the vibration and also protect the internal components. The liquid case filling also lubricates the internals and prevents corrosion.

2 - Pulsation.

When media rapidly cycles through the pressure system of the gauge, pulsation occurs. If the pointer pulsation increments are greater than 5 percent of the full scale value, you must intervene to prevent damage to the gauge.

A restrictor or throttle screw can solve most pulsation problems. It will slow down the media by reducing the size of the orifice and averaging out pressure fluctuations. For extreme pulsation, such as spiking and surging, a snubber works best. It allows more control over the amount of snubbing and the flow of media into the pressure system.

3 - Temperature:

If a gauge is exposed to a media with very high temperature, the metal alloy in the gauge may not be able to handle the temperature, causing the Bourdon tube to fail and release media.

Long pipes that separate the pressure gauge from the process, or small cooling elements with radiator fins, protect the pressure instrument from high process temperatures by allowing ambient air to flow over the surface area. In environments with steam, using a siphon with the pressure gauge cools the steam down and drives it back to its liquid state before it gets to the gauge. With very high-temperature media, a diaphragm seal with a cooling element will isolate and prevent the hot media from reaching the gauge.

4 - Spikes or Overpressure.

In some cases, process media may be under extreme pressure. Pumps also can cause spikes and over-pressure, both of which shorten gauge life. Gauges can fail as a result of these pressure spikes. When workers turn pumps on or off, or open or close valves, a surge of media or oil goes through the pipe and hits the pressure gauge, which causes a huge spike. For example, if a line is pumping 100 psi of oil, someone may install a 100 psi gauge, not realizing that this leaves no allowance for any overpressure, surging, or spiking. When pressure does increase, it causes overpressure on the gauge, sometimes to the point of where the Bourdon tube ruptures and leaks media.

One way to prevent this problem is to use a different type of gauge, such as 200 psi gauge for a 100 psi line. Another option is to accessorize the gauge with an over-pressure protector, which would shut off the pressure if there were a surge or spike. Overpressure protectors are designed to avert spikes and the resulting overpressure damage. These protectors provide protection for pressure gauges against the effects of pressures exceeding their maximum pressure ratings.

5 - Corrosion.

Corrosion problems can occur with gauges if the process media is not compatible with the pressure gauges' wetted parts material. Media can sometimes be so corrosive that it corrodes through the Bourdon tube and leaks onto the ground.

When the media is not compatible with the wetted parts material of the pressure gauge, a diaphragm seal is required to isolate the gauge from the media. The benefit of a diaphragm seal is that it can be made of special alloys that are designed to withstand corrosion. It is essential to select a diaphragm material that is compatible to the process media.

6 - Clogging.

In some processes, the media may eventually crystallize or may otherwise harden. If a standard gauge is used in such an application, it will eventually get clogged and will not be able to read the pressure accurately. For example, if a gauge gets clogged and ends up reading zero pressure, when in fact it is under tremendous pressure, there will be an unexpected and dangerous release of media if and when the piping system is opened. Diaphragm seals protect the pressure gauge from solids, particles and viscous media by creating a barrier between the process and the pressure inlet of the gauge. There is also an option to equip the diaphragm seal with flushing ports to constantly flush material out from the diaphragm surface.

7 - Mishandling and Misapplication.

Gauges can also fail prematurely if they are misapplied. For example, a gauge may have been ideal for the application when it was installed. However, if there are equipment modifications or more components subsequently added, this can change process parameters, resulting in the gauge not longer being right for that application. Misapplication can lead to serious problems associated with failure and subsequent media release.

It is also possible to accidentally damage gauges during installation. Gauges should be installed using the proper sized wrench on the flats. Gauges should never be installed by turning the case. After installation, gauges can also fail if they are mishandled or used improperly.

Why Problems Can Occur With Gauges

In a typical process plant, about 25 percent of pressure gauges have failed or are about to fail.[1]
A lack of management focus and worker education and experience are key reasons plants experience issues with gauges.

Management Focus.

Gauges are often viewed as inexpensive commodities, with management failing to realize that they are very critical parts of the processes. Often, plants will replace failed gauges with the same replacement gauges, which will also fail, only to have to be replaced again. Eventually, their luck may run out and a serious failure may occur because the incorrect gauge has been installed.

Why do gauges tend to receive such little attention from management? Management tends to focus on the expensive equipment - $100,000 or million dollars items. They often don't understand the importance of focusing attention on $100 gauges. This is not the fault of management. The problem is that no one has explained to them just how important gauges are.

As noted earlier, though, ignoring instrumentation can cause serious safety and financial problems. Plant process gauges can be thought of in the same way as vehicle gauges. If the temperature gauge on an automobile is defective, it is possible to destroy the engine as a result of overheating that is not registered by the defective temperature gauge. Similarly, if the oil pressure gauge on an automobile is defective, it is again possible to destroy the engine as a result of seized components resulting from loss of oil that is not indicated by the defective oil pressure gauge.

Worker Education and Experience.

A lack of worker education and experience also contributes to gauge issues. In many instances, when maintenance technicians look at gauges, they don't pay much attention to them unless they leaking or look as if they are going to explode. They don't realize all of the information that that gauge can provide. In addition, technicians may not realize that a gauge has failed or is about to fail. Again, as long as it is not leaking media, they will often just keep walking. This is rarely the fault of the technicians. Many have not been properly trained to understand process pressure gauge performance.

Along the same lines, the industry has been experiencing a "brain drain" for several years, with large numbers of experienced technical workers and engineers retiring and being replaced by younger people who lack the education and experience to understand what gauges are telling them, or, worse, not being replaced at all.

According to research conducted by Cambridge Energy Research Associates in 2007, more than half of all oil-field professionals will reach retirement age during the next decade. The Interstate Oil & Gas Compact Commission found that the average age of production and service company employees is 46 to 49 years, and the average retirement age in the oil and gas industry is 55 years. [2]

Historically, plants had specialists whose jobs focused exclusively on instrumentation. They spec'd the gauges, maintained them, and made sure they were operating correctly. However, as plants got leaner in recent years, they started letting these people go. Plants first shifted gauge responsibility to the pipefitter whose main responsibility was to make sure the pipes were put together correctly and not leaking.

[1]WIKA data aggregated from more than 250 instrument audits.
[2]Peter Peek, Jocelyn Fenard, Patrick Gantes, Christine Theiler. December 2008, "Skills Shortage in the Global Oil and Gas Industry: How to close the gap" Part I. Pages 14-15.

Then, as plants began letting pipefitters go, they shifted the responsibility to the unit operators or technicians, who now install the gauges. These people usually don't have any experience or knowledge of how to install instrumentation. They usually just replace old gauges with similar gauges. The problem is that the gauges are usually the wrong ones to begin with. The result is that, these days, few plants have people who have much if any education and experience with instrumentation.

Taking Stock of Process Pressure Gauges

When it comes to improving the safety, productivity and profitability of process plants through an effective gauge management program, companies can take a couple of steps. First, an understanding of the vital importance of gauges must be established. Second, workers must understand what kind of gauges will be effective.

Focus on Gauges:

World-class process plants maintain their assets proactively. In such plants, pressure gauges receive as much attention as pumps and control valves. Management understands the importance of proactively maintaining their assets, and they realize that one of the best ways to maintain their most expensive assets is to properly source and maintain the gauges that protect these assets. That is, they understand the value of spending $100 for a gauge that can subsequently prevent damage or destruction to a $100,000 or $500,000 piece of equipment and the resulting negative safety, downtime and regulatory implications.

There is an interest in predictive maintenance these days in process plants, which is a good thing. A lot of this, though, involves expensive software. In addition, when something is out of kilter, the system may be set up such that technicians may have to wait for it to get far enough out of the normal range before it will generate an alert on the computer. However, properly-maintained pressure gauges, studied by people who know what to look for, can be very inexpensive but reliable and useful mechanical predictive maintenance tools.

Gauge Procurement Considerations:

Over time, as noted earlier, the status of pressure gauges has unfortunately diminished to that of commodities in many process plants. In order to ensure process plant safety, productivity and profitability, it is essential that plant professionals understand the importance of selecting the right gauges for applications.

It is also important that these be high-quality gauges, not "commodity" gauges. That is, not all process gauges are created equal. The so-called "lowest price" gauges appear attractive at first, but are rarely the "lowest cost" in the long run, because they usually have to be replaced more often, leading to excessive downtime and inconvenience. Worse, they may fail unexpectedly, leading to catastrophic results.

Consider these 10 steps to select the right gauges that will help to ensure safe, productive and profitable plant processes.

1 - Know what is in the pipe. This way, you can select a gauge alloy that will be compatible with the process media.

2 - Next, know the temperature of the media. If it is very hot, for example, it will be unsafe to use a standard pressure gauge. A long pipe, siphon or diaphragm seal assembly may be needed to cool the media sufficiently before it reaches the gauge internals.

3 – Third, know the ambient conditions surrounding the gauge. For example, if it is corrosive around the gauge, you need to make sure that the external components of the gauge such as the case and window can handle the environmental conditions.

4 - Next, know the pressure in the piping that the gauge is meant to measure. A best practice is using a gauge that is designed for two times the normal operating pressure. For example, if normal pressure is 100 psi, use a 200 psi gauge.

5 – Know the conditions that affect gauge wear. These could include mechanical vibration, pulsation, pressure spikes, etc. For example, if vibration exists, a gauge with a liquid-filled case is recommended. If there is pulsation, use a throttle plug or restrictor accessory.

6 - Select gauges that are designed for a particular service. Process type gauges are designed with strong Bourdon tubes and other heavy duty components. If a gauge is misapplied, the internals of the gauge may not be rugged enough to resist failure and subsequent media release in these applications.

7 - The gauge case must be rugged and have the correct safety features. In some case designs, if there is a rupture of the Bourdon tube, media will be released out the front of the gauge, which can spray workers. Select a gauge where media would release out from the back of the gauge.

8 - Accuracy is important. In most process plants, +/-0.5% accuracy is the norm, and +/-1% is acceptable.

9 - Size is also important. The gauge dial needs to be large enough that operators can read it from a distance – four inches or larger.

10 - Finally, the gauge should be mounted properly. A gauge is usually mounted by its stem, but it can also be mounted in a panel or against a flat surface.

Looking Outside for Assistance

As noted earlier, it can be difficult for plant management to devote enough time and attention to an effective process pressure gauge program. Similarly, it can be difficult for technical personnel to get up-to-speed in terms of education and experience related to process pressure gauges.

There has been a growing trend over the last 30 years in virtually all industries for companies to outsource services that are not considered "core competencies." That is, companies have come to realize that it makes sense to focus their attention on their missions and what they do best, and leave secondary and peripheral activities to third parties that specialize in these other activities and services. Becoming experts in pressure gauges is certainly not a core competency for refining plants, petrochemical, and chemical plants, so this is an area that can easily be outsourced to an experienced third-party audit team.

If a plant elects not to gain the experience that it needs on its own, free assistance is available. To ensure that gauges are working properly and are able to protect the valuable assets that they were designed to protect, a five-phase instrument review is recommended. A best practices review identifies gauges with unwanted conditions and recommends corrective actions to ensure optimum performance. The review identifies issues that affect reliability, longevity and safety. Such a review can uncover evidence of pulsation, vibration, pressure spikes, excess process or ambient temperature, installation damage, clogging, chemical incompatibility, and other problems that can lead to serious problems.

Phase 1 - Interview. This involves discussions between the third-party audit team and plant personnel. During the interview, plant personnel learn the purpose of the instrument audit, how it will be conducted, and how they will benefit from the audit. The interview also helps determine the best way to match the audit to the needs of that specific plant.

Phase 2 - Pre-Meeting. This involves determining the units or sections of the plant to audit, as well as outline roles and responsibilities. It also identifies prohibited and exempt areas of the plant. The third-party team reviews problematic instrument installations and discusses safety procedures with plant personnel.

Phase 3 - Instrument Application Audit. This involves having key plant personnel lead the third-party team on an actual plant walk-through. The audit team reviews the target gauge installations, identifies gauges that demonstrate unwanted conditions, as well as those that demonstrate installation best practices.

Phase 4 - Post-Meeting. This involves explaining the instrument audit results to key plant personnel, specifically noting gauge installations that pose immediate safety concerns. Together, the audit team and plant personnel determine which installations to investigate immediately and which to block out. The audit team and plant personnel then set schedules for the follow-up meeting.

Phase 5 - Follow-up Meeting and Presentation. The third-party team presents its general findings and the executive summary resulting from the instrument audit. The team also presents a spreadsheet that lists details about the gauges from the instrument review, color-coded by severity. The team also provides recommendations for improvements based on experience and best practices. Together, the audit team and plant personnel develop a strategy for moving forward.

The plant should also consider leveraging the audit team in other ways. For example, over time a plant's storeroom inventory can get out of hand. A qualified third-party audit team can help plant personnel rationalize, consolidate and manage their inventory of pressure gauges. For example, if a plant stocks spares for 10 or more ranges, the audit team may be able to help reduce that number to five or six ranges.

Essentially, when used properly, gauges play a significant role in plant safety, productivity and profitability. They act like silent alarms, indicating when something is seriously wrong or helping to diagnose deeper issues within systems and processes that can eventually lead to serious safety issues. To ensure safety and reliability at your plant, make certain the proper gauges are in place by reviewing your processes, gauges and your inventory. Everything is at stake.

REFERENCES

1. WIKA data aggregated from more than 250 instrument audits.

2. Peter Peek, Jocelyn Fenard, Patrick Gantes, Christine Theiler. December 2008, "Skills Shortage in the Global Oil and Gas Industry: How to close the gap" Part I. Pages 14-15.

CFD simulation of laminar liquid film condensation on the entrance region of vertical plate

Bo Sun, Chun Jiang Liu*, Bo Tan Liu

State Key Laboratory of Chemical Engineering, Tianjin University, Tianjin 300072, China

Abstract:

A computational fluid dynamics model based on volume of fluid method is developed to investigate the laminar film condensation accompanied by mixed convection of humid air at the entrance region of a vertical plate. The effects of vapor in different concentration ranges are taken into consideration in this research. It is observed that the condensation rate and heat transfer coefficient is higher in larger concentration gradient of water vapor. Latent heat transfer is dominant in the process of the interface heat transfer process, and is about 100 times larger than sensible heat transfer. Due to the heat generated by condensation, the liquid film is heated and the temperature can be higher than that of the gas phase in some position.

Key words: Condensation, Liquid film, Computational fluid dynamics (CFD)

1 Introduction

Condensation, associated with laminar condensate, is an important thermal-hydraulic phenomenon, which has been widely applied into desalination plants, refrigeration and air-conditioning equipments [1]. Several experiments related to the condensation of vapor-gas mixtures in vertical plates and tubes have been performed to develop new condensation plants with high mass and heat transfer efficiency. Many empirical correlations and mechanistic models have also been proposed to provide a better prediction for the heat and mass transfer coefficients.

* Correspondence: Prof. C. J. Liu (author to whom correspondence should be addressed, cjliu@tju.edu.cn)

Koh [2] is the first that simulated the condensation process of pure vapor. The variable in the governing equations was simplified to balance mass and heat transfer process in the vapor-liquid interface. Minkowyca and Sparrow [3,4] presented theoretical investigations of laminar falling film with condensation by natural and forced convection on an isothermal vertical plate. They were the earliest researchers that paid their attentions to the effect of non-condensable gas. After that, the condensation of vapor from vapor-gas mixture has been studied by many scholars [5,6]. Chin et al. [7] applied the finite volume numerical method into making clear of the condensation heat transfer process of gases mixture on inclined plate, and concluded the effect of inclination angle on the heat transfer. Dehbi et al. [8] and No et al. [9] developed a simplified mechanistic model to make an easier way to obtain the temperature and pressure in the interface. Diffusion layer model was applied by Revankar et al. [10] to investigate the relation between radical velocity, temperature, non-condensable concentration, liquid film thickness and condensation rate. Moreover, they also found that the changing trend of heat transfer coefficient with gas phase Reynolds number in the presence of non-condensable gas is opposite to that in the condition of pure vapor steam. Martín-valdepeňas [11] compared four kinds of condensation models with CFX code. Siow et al. [12,13] investigated the heat transfer during condensation of steam-air mixture on vertical and inclined plates with numerical method. The influence of inlet flow rate was considered to be vital in the condensation process. Lan et al. [14] focused on the dropwise condensation, and developed a model with the consideration of interfacial interaction between the condensate and condensing surface. Paoli et al. [15] investigated the effect of turbulent mixing on the droplet condensation via direct numerical simulation. Meanwhile, Mimouni et al. [16] also modeled the dropwise condensation by CFD approach to study the heat and mass transfer coefficients on droplet in the core of the flow and condensation/evaporation phenomena on the wall. In addition to the numerical investigations, condensation in the presence of non-condensable gas has also been experimentally discussed because of its important applications in various industrial systems [17-21].

In view of the above literatures, it is evident that the mixed convection with laminar condensate at the entrance region of the vertical wall has not been sufficiently studied, and the heat transfer characteristics in the region are important for the condensation plants. Based on CFD method, the authors aim to study the convective film-wise condensation of vapor and air mixture flowing downward on a vertical plate with a short distance, and to further obtain the effect of water vapor concentration on the condensation heat transfer process in different concentration ranges.

2 CFD modeling

2.1 Physical model

According to the entrance region distance of the condensation plants, a two-dimensional model with a flow field of 120 × 7 mm is applied in the paper. The detailed description of the geometry of the physical model is shown in Figure 1, and the boundary condition of the physical model is illustrated in Table 1. The wall can be subjected to a uniform heat flux q_w and is wetted by a water film flowing downward with inlet temperature T_{l0} and inlet mass flow rate L_{l0}. The vapor and air mixture enters into the channel concurrently with water film with gas temperature T_{g0}, vapor concentration w_{v0} and gas velocity u_{g0}.

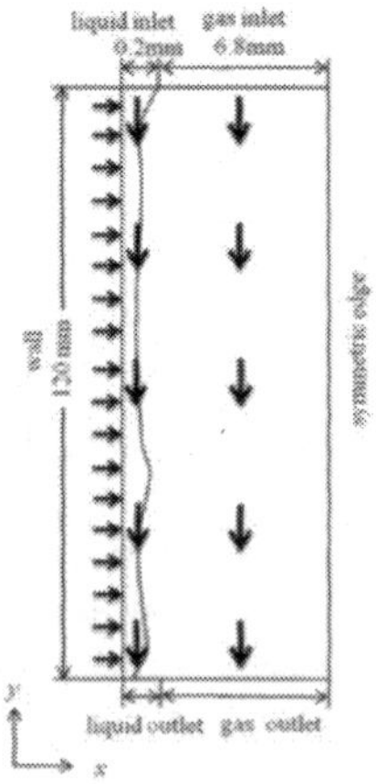

Figure 1. Physical model

Table 1 Boundary condition of the physical model

	Boundary condition
Liquid inlet	Velocity inlet
Gas inlet	Velocity inlet
Liquid outlet	Pressure outlet
Gas outlet	Pressure outlet
Wall	No slip wall
Symmetric edge	Symmetry

For the mathematical formulation of the problem, the following simplifying assumptions are taken into account [1,10,22]:

a) Liquid and gas phase are laminar and two dimensional.

b) The condensate film is impermeable to non-condensable gas.

c) Humid air is an ideal mixture of vapor and dry air, and gas-liquid interface is in the thermodynamic equilibrium.

d) The surface tension, radiation heat transfer, and viscous dissipation are neglected.

2.2 CFD modeling

The commercial CFD code, FLUENT, is employed to simulate the laminar film condensation of vapor-air mixture based on the volume of fluid (VOF) method. In the VOF model, the liquid phase is specified as the primary phase, thus the volume fraction of the vapor phase is solved. It follows that (a) when the volume fraction is 1, the computational cell is filled with liquid; (b) when the volume fraction is 0, it is filled with vapor; (c) when the volume fraction is between 0 and 1, the computational cell is partially filled with liquid. The VOF model adopts a single set of momentum equations to model two immiscible fluids, which means the two phases share a common velocity and temperature field.

The continuity equation takes the following form:

$$\frac{\partial \rho}{\partial t}+\nabla\cdot(\rho\vec{u})=|m_i| \tag{1}$$

The source term in continuity equation m_i, which describe the volume change for gas and liquid phase in the condensation process, is defined as [1,23]:

$$m_{g,i} = -m_{l,i} = \left[\frac{\rho_g D}{1-w_v}\right]\left(\frac{\partial w}{\partial x}\right)_{g,i} \cdot \alpha_e \tag{2}$$

Where α_e is effective interfacial area per unit volume, and is defined as [24]:

$$\alpha_e = |\nabla \alpha_g| = |\nabla \alpha_l| \tag{3}$$

Where $\nabla \alpha_g$, and $\nabla \alpha_l$ is the volume fraction gradient of the gas and liquid phase, respectively.

The momentum equation is:

$$\frac{\partial}{\partial t}(\rho\vec{u}) + \nabla\cdot(\rho\vec{u}\vec{u}) = -\nabla P + \nabla\cdot\left[\mu\left(\nabla\vec{u} + \vec{u}^T\right)\right] + \rho\vec{g} + F_i \tag{4}$$

The drag force term F_i between two phases can be described by the friction pressure drop model [25], and it is defined as follows:

$$F_i = \alpha_e f_i \rho_g \left(u_g - u_{l,i}\right)\left|u_g - u_{l,i}\right| \tag{5}$$

Where f_i is the drag force factor, proposed by Wallis [26] and it can be formulated as:

$$f_i = 0.005\left[1 + 300\left(\delta^*/D^*\right)\right] \tag{6}$$

Where δ^* and D^* are dimensionless liquid film thickness and hydraulic diameter, and they are given by Eqs. (7).

$$D^* = D/LL \qquad \delta^* = \delta/LL \tag{7}$$

LL is Laplace length, and can be calculated by Eqs. (8):

$$LL = \left[\sigma/(\rho_l - \rho_g)g\right]^{0.5} \tag{8}$$

The energy equation is:

$$\frac{\partial}{\partial t}(\rho E) + \nabla\left(\vec{u}(\rho E + p)\right) = \nabla(k_{eff}\nabla T) + Q_i \tag{9}$$

The heat transfer source term is based on the $m_{g,i}$ mentioned above, and it can be described as:

$$Q_i = -m_{i,g} \cdot H = -\left[\frac{\rho_g D}{1-w_i}\right]\left(\frac{\partial w}{\partial x}\right)_{g,i} \cdot \alpha_e \cdot H \tag{10}$$

Species equation:

$$\frac{\partial(\rho w)}{\partial t} + \nabla\cdot\left(\rho\vec{u}w\right) = \nabla\cdot(\rho D\nabla w) + m_i \tag{11}$$

2.3 Mesh strategy

The number and quality of the meshes used in numerical investigation are crucial to the simulation result. Therefore, quadrilateral mesh with high quantity is applied and mesh independence test is conducted by changing the mesh sizes for the region of falling liquid film flow. The range of each mesh sizes considered are as follows: 0.05 mm $\leq \Delta Y \leq$ 1 mm, 0.02 mm $\leq \Delta X \leq$ 0.04 mm. The results of vapor concentrations at gas outlet for different mesh strategies are shown in Figure 2. It is found that the vapor concentration at gas outlet decreases with the decline of the mesh sizes, which indicates a higher condensation rate in smaller mesh sizes. Especially for the cases of $\Delta Y/\Delta X<5$, the vapor concentration at outlet is obviously smaller than the concentration value in the condition of $\Delta Y/\Delta X>5$. By considering of the accuracy of simulation results and the computer performances, the reference mesh size is selected as 0.02 mm wide, 0.1 mm long, and the difference of the results between the reference mesh size and the finest mesh size is within 0.2%. The detailed description of the mesh strategy applied is shown in Table 2.

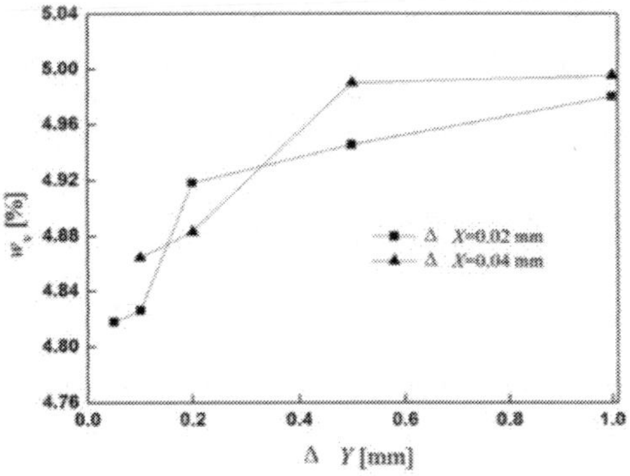

Figure 2. The effect of mesh sizes on the vapor concentration at gas outlet
(L_{l0}=0.002kg/(m.s), T_{g0}=50□, T_{l0}=25□, u_{g0}=1m/s, w_{v0}=0.05, q_w=0)

Table 2. Detailed description of the mesh strategy applied

x			*y*	Total number
0-0.2 mm	0.2-0.4 mm	0.4-7 mm		
▯ X=0.02 mm	▯ X=0.025 mm	0.04 mm≤▯ X≤0.8 mm	▯ Y=0.1 mm	69600

2.4 Simulation scheme

Since the small liquid flow rate, and the low velocity of co-current gas phase, the laminar model was applied in the paper. The geo-reconstruct method was used to precisely trace the gas-liquid interface. The time step size was set as 5×10^{-5} s. A first order upwind differencing was chosen as the solution of momentum equation. The body force weighted method was used for pressure discretization, and Pressure Implicit with Splitting of Operators scheme was adopted for pressure-velocity coupling. The momentum, heat and continuum equation source term was implemented by user defined functions into each mesh at the gas-liquid interface. The scholar transport equation (Eqs.11) was solved as a user defined scalar equation in FLUENT. Unsteady simulation was applied, and the calculation is terminated when it reached a quasi-stable state.

3 Result and Discussion

3.1 Validation of the CFD model

In order to test the reliability of the CFD model applied in this work, the numerical result is compared with the experimental data published [27] under the same conditions, and the results are shown in Figure 3.

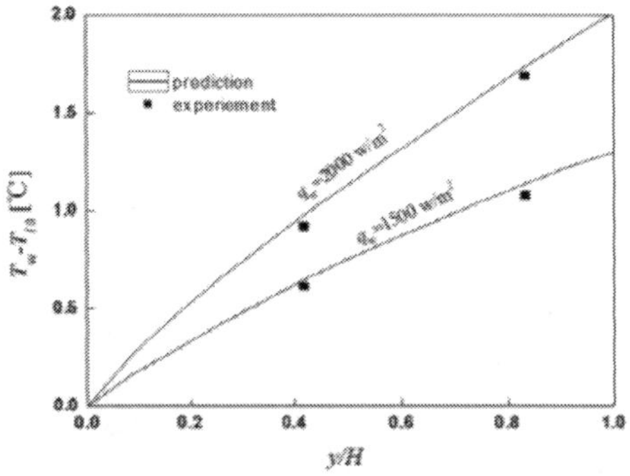

Figure 3. The comparison between numerical and experimental results [27]
(L_{l0}=0.02 kg/(m.s), T_{g0}=T_{l0}=30 □, u_{g0}=2 m/s, w_{v0}=0.013)

With the wall subjected to a uniform heat flux, the liquid film that flows downwards along the plate is heated, and the evaporation occurs at the gas-liquid interface. It can be considered as the inverse process of condensation. The good agreements of the numerical and experimental results indicate the high veracity of the CFD model, which is taken in the simulation of evaporation and condensation process.

3.2 The effect of low vapor concentration

The effect of vapor concentration (w_{v0}) on the condensation rate (m_{cond}) in low vapor concentration range is illustrated in Figure 4. Since the simulation results fluctuate with the waves on the liquid film, the average value within a distance of 10 mm along the vertical plate is adopted. Same method is also applied in the treatment of the following simulation results. It is observed in Figure 1 that the m_{cond} is enhanced for higher value of w_{v0}, and decreases with the down-flow of liquid film, which results from the larger vapor concentration gradient between the gas phase and gas-liquid interface for larger w_{v0} in the position near the inlet. Similarly, due to the heat generated by condensation, the liquid film temperature increases, and the temperature and vapor concentration gradient decrease with the liquid film flowing downwards, which leads to a decreased m_{cond}.

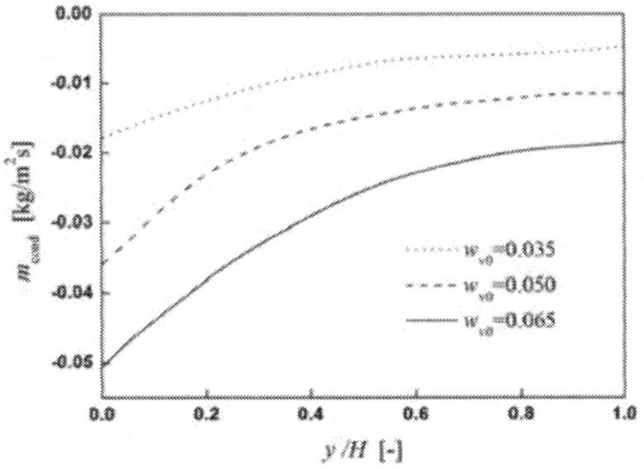

Figure 4 Axial distribution of condensation rate at low vapor concentrations
(L_{l0}=0.006 kg/(m.s), T_{l0}=25 □, T_{g0}=50 □, u_{g0}=1 m/s)

Figure 5 presents the axial evolution of the condensation heat transfer coefficient (h_{cond}), and it is defined as $h_{cond} = Q_{cond}/(T_g - T_l)$. Where Q_{cond} is the condensing heat,

T_g and T_l are the average temperature of gas and liquid phase in the same horizontal position. Compared with the effect of w_{v0} on m_{cond}, it is clear that the change tendency of h_{cond} is similar to that of m_{cond}. Higher vapor concentration can result in a lager h_{cond}, and smaller h_{cond} is found near the outlet.

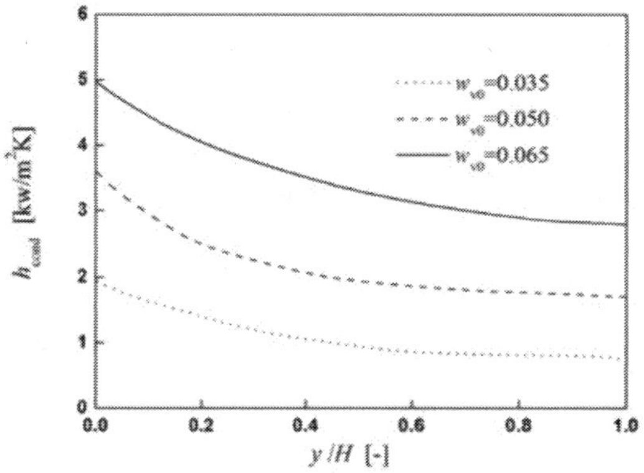

Figure 5 Axial distribution of the condensation heat transfer coefficient at low vapor concentrations

(L_{l0}=0.006 kg/(m.s), T_{l0}=25 □, T_{g0}=50 □, u_{g0}=1 m/s)

Figure 6 gives the axial evolution of the latent and sensible heat near the gas-liquid interface. It is seen that the value of Q_l and Q_s at inlet are evidently larger than that at outlet, which indicates a higher condensation and sensible heat transfer rate exist near the inlet. Moreover, Q_l is about 100 times larger than Q_s. The large value of Q_l / Q_s can be explained by the increased temperature of the interface, which results from the heat generated in the condensation process. The increased interface temperature decreases the vapor concentration gradient between the liquid and gas phase near the interface as well as the temperature gradient, and it limits the sensible heat transfer process, which leads to a large value of Q_l / Q_s.

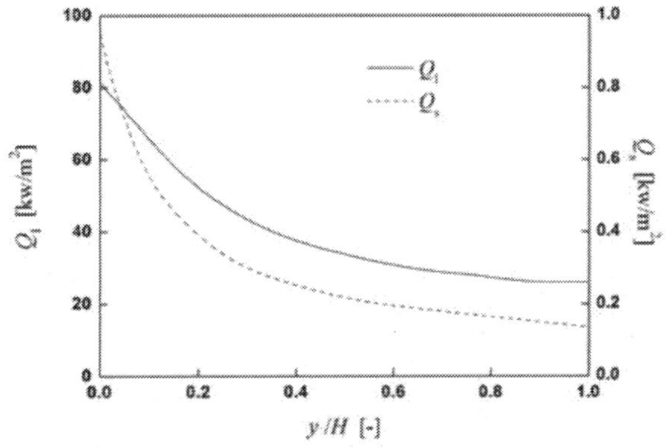

Figure 6 Axial distribution of the latent and sensible heat at low vapor concentrations
(L_{l0}=0.006 kg/(m.s), T_{l0}=25 □, T_{g0}=50 □, u_{g0}=1 m/s, w_{v0}=0.05)

3.3 The effect of high vapor concentration

Figure 7 illustrates the effect of w_{v0} on m_{cond} in high vapor concentration range. It is observed that the higher w_{v0} leads to a larger m_{cond} in the upstream half of the plate. However, the larger m_{cond} decreases greatly along the plate, and in the lower plate part, m_{cond} is smaller at higher w_{v0}. This is the reason that more heat generated by condensation with higher vapor concentration greatly increases the interface temperature in the lower part. It enhances the saturated concentration of vapor at interface, and reduces the concentration gradient between the interface and gas bulk.

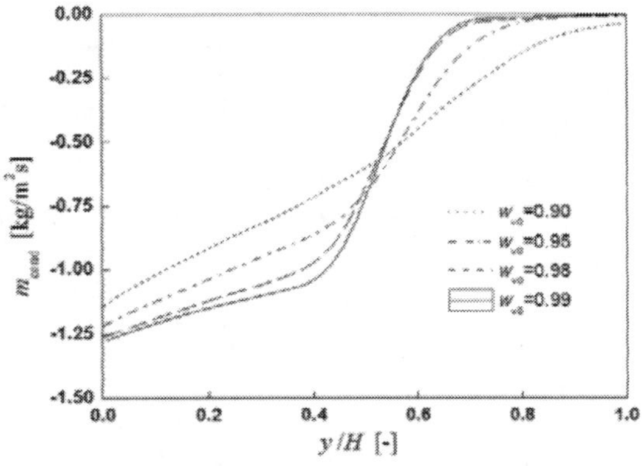

Figure 7 Axial distribution of the condensation rate at high vapor concentration
(L_{l0}=0.006 kg/(m.s), T_{l0}=25 □, T_{g0}=100 □, u_{g0}=1 m/s)

Figure 8 presents the effect of w_{v0} in high vapor concentration range on the temperature of the gas mixture and liquid phase at the middle and outlet of the channel. It can be found that the gas mixture temperature decreases firstly and then increases with the increase of the distance between the interface and gas phase. The heat generated by condensation increases the interface temperature, and the interface temperature is higher than that of the liquid film and the gas phase nearby, which can be evidently found at the middle of the channel (y/H=0.5). In addition, it can be found that, with the water flowing down, the liquid temperature increases continuously, and the temperature difference between liquid film and gas phase reduces.

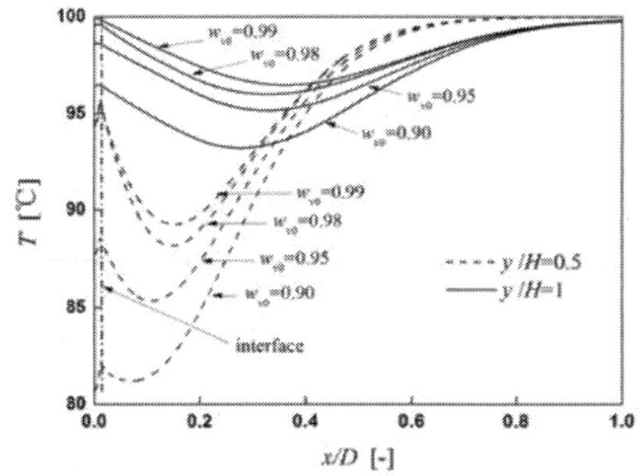

Figure 8 Transverse distribution of temperature at high vapor concentration (L_{l0}=0.006 kg/(m.s), T_{l0}=25 □, T_{g0}=100 □, u_{g0}=1 m/s)

Figure 9 illustrates the axial distribution of the condensation heat transfer coefficients in high vapor concentrations range. For all the cases, h_{cond} increases to an extreme value, and then decreases rapidly to a small value near the outlet. As the the liquid film temperature increases and condensation process occurs continuously, when the liquid temperature is similar to the temperature of gas phase, h_{cond} increases significantly. Therefore, the extreme value of h_{cond} can be attributed to the marginal temperature difference and the existing condensation heat induced by the great concentration gradient between the gas-liquid phases. The decrease of h_{cond} in the downstream of the plate results not only from the decreased condensation rate, but the increased temperature difference. Therefore, it can be concluded that, the liquid film temperature can be higher than that of the gas phase at specific position. Moreover, it can also be found that the location of the extreme value of h_{cond} approaches to the outlet with the decrease of w_{V0}. It results from the decreased temperature rise speed of liquid film, and the liquid film needs more time to be heated to the temperature similar to that of the gas phase.

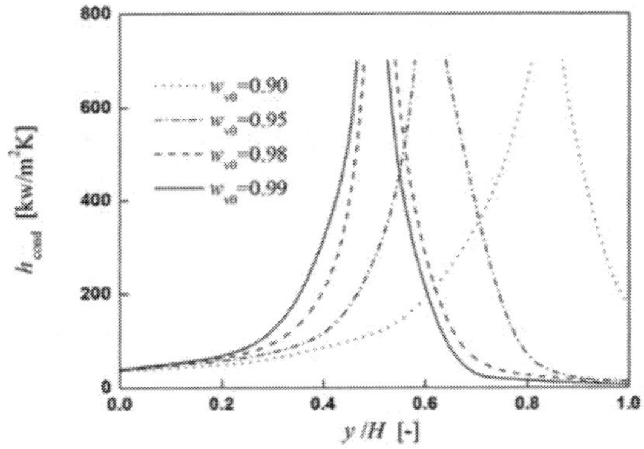

Figure 9. Axial distribution of the condensation heat transfer coefficient at high vapor concentration

(L_{l0}=0.006 kg/(m.s), T_{l0}=25 □, T_{g0}=100 □, u_{g0}=1 m/s)

The distribution of the latent and sensible heat at high vapor concentration is shown in Figure 10. It is observed that Q_s decreases rapidly near the inlet, and the tendency is similar to that at low vapor concentration. However, when $y/H > 0.4$, $Q_s < 0$, and Q_s increases slowly with the down-flow of liquid film. The negative value of Q_s indicates a higher temperature of liquid film than that of the gas phase, and the heat begins to transfer from the liquid film to the gas buck. For Q_l, a higher value is obtained at half part of the plate, and decreases greatly to a small value. The value of Q_l is still significantly larger than Q_s. Therefore, according to the results in Figure 6 and 10, it can be concluded that latent heat transfer is the dominant form in the interface heat transfer process. In addition, both Q_l and Q_s decrease to a small value in a distance of 100 mm.

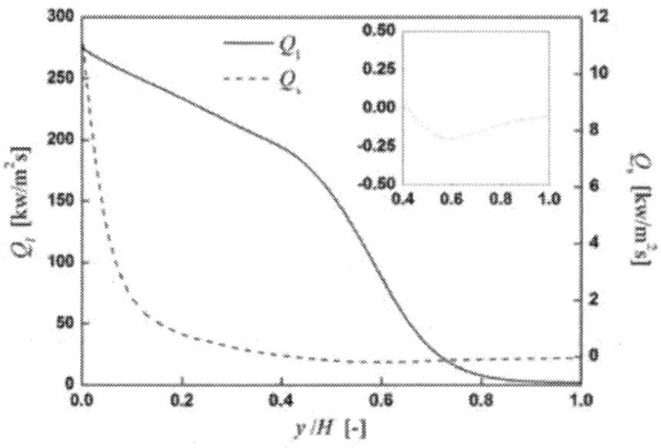

Figure 10 Axial distribution of the latent and sensible heat at high vapor concentration (L_{l0}=0.006 kg/(m.s), T_{l0}=25 ℃, T_{g0}=100 ℃, u_{g0}=1 m/s, w_{v0}=0.95)

4 Conclusions

A numerical analysis has been carried out to explore the condensation heat transfer process associated with laminar condensate at the entrance region of a vertical plate. The influence of vapor concentration in different ranges on the heat transfer is investigated in detail. It is demonstrated that the condensation rate and heat transfer coefficient depend greatly on the vapor concentration, and they vary significantly at the short distance near the inlet, which indicate the important significance of the entrance region.

In low vapor concentration range, m_{cond} and h_{cond} increase with the increasing of w_{v0}, while, in high vapor concentration range, m_{cond} with higher w_{v0} is larger at the upstream half of the plate, and is smaller than that with lower w_{v0} near the outlet. In addition, the extreme value of h_{cond} can be observed in specific position in the high vapor concentration range.

The latent heat transfer occurs in the gas-liquid interface is the dominant form in the heat transfer process. The heat generated by condensation enhances the liquid film temperature, which limits the sensible heat transfer. Moreover, the condensation heat makes the interface temperature higher than that of the liquid film and gas phase nearby, and the temperature of the liquid film can be higher than that of the gas phase in the condition of high vapor concentration.

Symbols used

D	[m^2/s]	Diffusion coefficient
D^*	[-]	Dimensionless hydraulic diameter
F	[N/m^3]	Momentum source term
f	[-]	Drag force factor
H	[kJ/kg^1]	Vaporization enthalpy
h_{cond}	[$kw/K.m^2$]	Heat transfer coefficient
LL	[m]	Laplace length
L	[kg/m.s]	Mass flow rate
T	[□]	Temperature
m	[kg/m^3 s]	Source term in continuity equation
m_{cond}	[kg/m^2s]	Condensation rate
Q	[w/m^3]	Heat transfer source term,
Q_l	[kw/m^2 K]	Latent heat
Q_s	[kw/m^2 K]	Sensible heat
q	[w/m^2]	Heat flux,
u	[m/s^1]	Velocity,
w	[-]	Mass fraction

Greek symbols

α_e	[m^2/m^3]	Effective interfacial area per unit volume
α	[-]	Volume fraction
ρ	[kg/m^3]	Density
δ	[m]	Thickness of liquid film
σ	[N/m]	Surface tension
μ	[kg/m s]	Viscosity

Superscripts and subscripts

g	Gas phase
i	interface
l	Liquid phase
w	wall
0	initiate state

References

[1] A. B. Mohamed, J. Orfi, C. Debissi, S. B. Nasrallah. *Desalination*, **2007**, *204 (1-3, 5)*, 471-481.

[2] J. C. Y. Koh, E. M. Sparrow, J. P. Hartnett. *Int. J. Heat. Mass. Transfer*. **1961**, *2 (1-2)*, 69-82.

[3] W. J. Minkowycz, E. M. Sparrow. *Int. J. Heat. Mass. Transfer.* **1966**, *9 (10)*, 1125-1144.

[4] E. M. Sparrow, W. J. Minkowycz. *Int, J. Heat Mass Transfer*, **1967**, *10 (12)*, 1829-1845.

[5] W. D. Baasel, J. C. Smith. *AIChE J.* **1962**, *9 (6)*, 826-830.

[6] C. Y. Wang, C. J. Tu. *Int. J. Heat Mass Transerf.* **1988**, *31 (11)*, 2339-2345.

[7] Y. S. Chin, S. J. Ormiston, H. M. Soliman. *Heat. Mass. Transfer*. **1998**, 34 (4), 271-277.

[8] A. Dehbi, S. Guentay. *Nuc. Eng. Design*. **1997**,*177 (1-3)*, 41-52.

[9] Hee Chen No, Hyun Sik Park. *Int, J. Heat Mass Transfer.* **2002**, *45 (4)*, 845-854.

[10] S. T. Revankar, D. Pollock. *Applied Mathematical Modelling*. **2005**, *29 (4)*, 341-359.

[11] J. M. Martín-valdepeñas, M. A. Jiménez, F. Martín-Fuertes, J. A. Fernández Benítez. *Heat. Mass. Transfer*. **2005**, *41 (11)*, 961-976.

[12] E. C. Siow, S. J. Ormiston, H. M. Soliman. *Heat. Mass. Transfer.* **2004**, *40 (5)*, 365-375.

[13] E. C. Siow, S. J. Ormiston, H. M. Soliman. *Int. J. Therm. Sci*. **2007**, *46 (5)*, 458-466.

[14] Z. Lan, W. H. Ma, X. D. Zhou, M. Z. Wang. *Enhanced Heat Transfer*. **2009**, *16*, 61-71.

[15] R. Paoli, K. Shariff. *J. Atmos. Sci*. **2009**, *66*, 723-740.

[16] S. Mimouni, A. Foissac, J. Lavieville. *Nucl. Eng. Design*. **2011**, *241 (11)*, 4445-4455.

[17] S. Z. Kuhn, V. E. Schrock, P. E. Peterson. *Nuc. Eng. Design*. **1997**, *177 (1-3)*, 53-69.

[18] J. Orfi, M. Laplante, H. Marmouch, N. Galanis, B. Benhamou, S. Ben Nasrallah, C.T. Nguyen. *Desalination*, **2004**, *168 (15)*, 151-159.

[19] R. H. Xiong, S. C. Wang, Z. Wang, L. X. Xie, B. L. Li, A. M. Zhu. *Chinese. J. Chem. Eng.* **2005**, *13 (3)*, 324-328.

[20] J. Liu, H. Aizawa, H. Yoshino. *Building and Environment*. **2004**, *39 (8)*, 905-911. [21] B. J. Chung, M. C. Kim, M. Ahmadinejad. J. Mech. Sci. Technol. **2008**, *22 (1)*, 127-133.

[22] J. Orfi, C. Debbissi, A. Belhaj Mohamed, S.B. Nasrallah. *Desalination*. **2004**, *168 (15)*, 161-168.

[23] W. M. Yan, D. Lin. *Int. J. Heat. Mass. Transfer.* **2001**, *44 (6)*, 1143-1151.

[24] Y. Y. Xu, S. Paschke, J. U. Repke, J. Q. Yuan, C. Wozny. *Chem. Eng. Technol.* **2009**, 32 (8), 1227-1235.

[25] G. F. Woerlee, J. Berends, Z. Olujic, J. de Graauw. *Chem. Eng. J.* **2001**, *84 (3)*, 367-379..

[26] G. B. Wallis. *One-Dimensional Two-Phase Flow*. McGraw-Hill, New York **1969**.

[27] Y. L. Tsay, T. F. Lin. *Experimental Thermal and Fluid Science*. **1995**, *11 (1)*, 61-71..

Improving Fired Heaters using CFD

Ashutosh Garg
Furnace Improvements
www.heatflux.com

Fired Heaters

- Important equipment in refineries and chemical plants
- High temperature heat transfer
- Fuel gas fired
- High tube metal problem
- Coking
- Shutdown

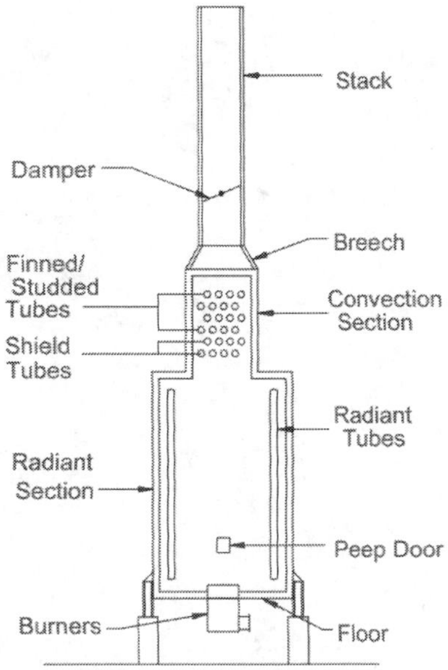

CFD Modeling

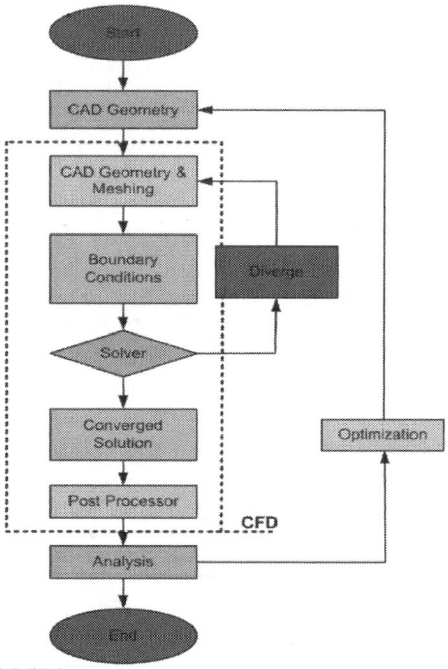

- Computational Fluid Dynamics
- Uses numerical methods and algorithms to study fluid flow
- Flow patterns in fired heaters, burners, fireboxes
- 3D models of fired equipment are created

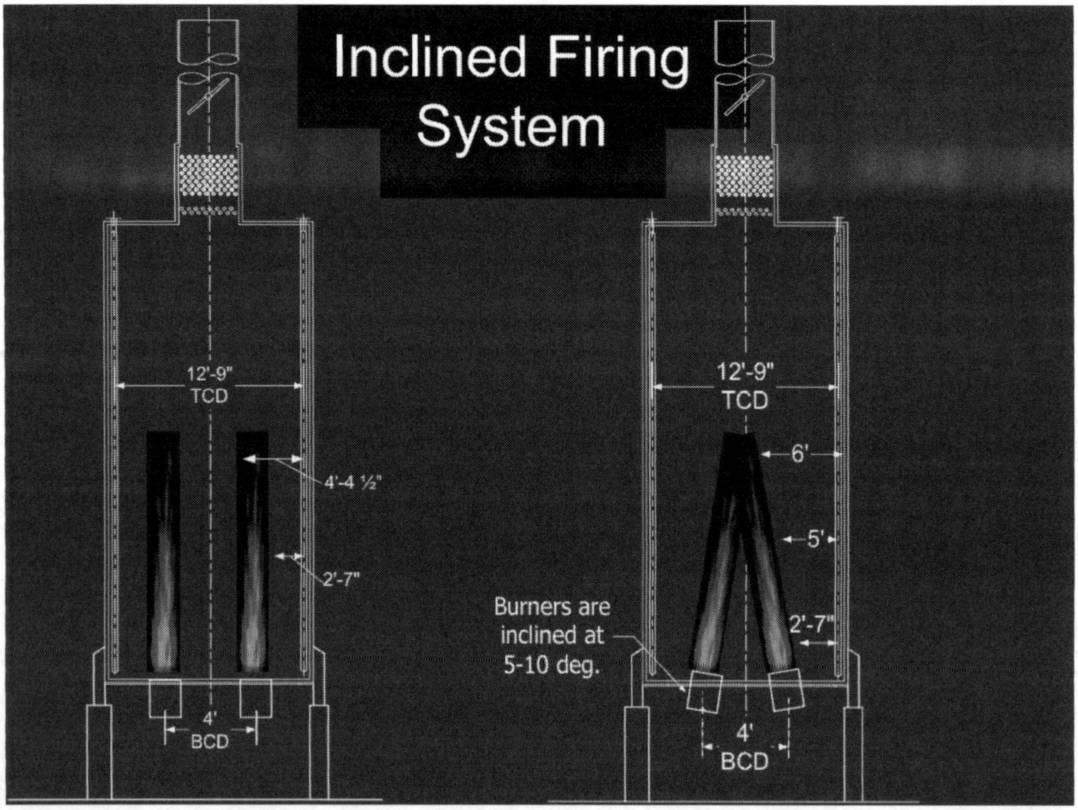

Coker Heater

Furnace Improvements Services

- High tube metal temperature issues
- Fuel gas fired low Nox burners

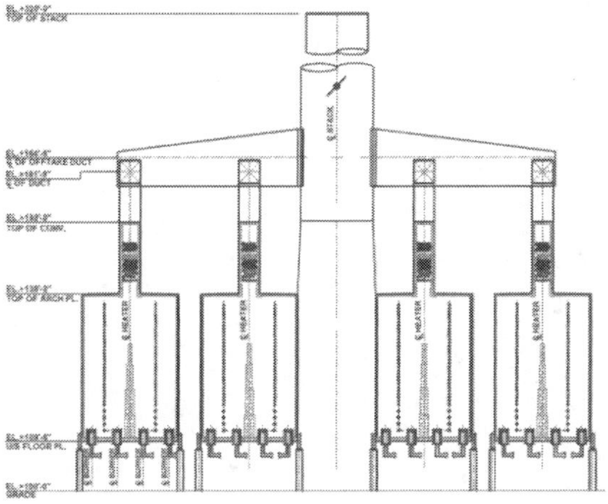

Heater Geometry

Furnace Improvements Services

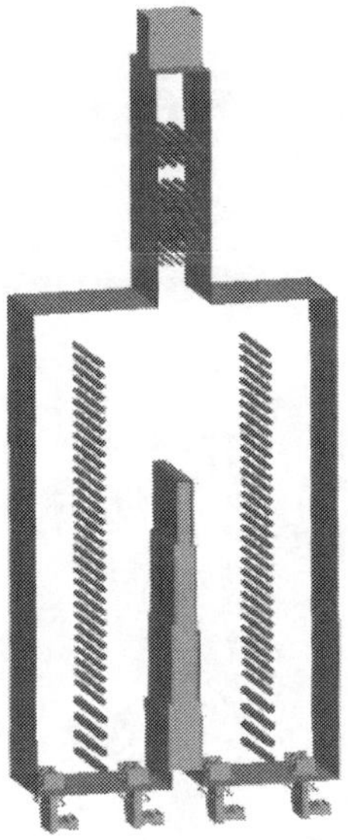

Heater Geometry

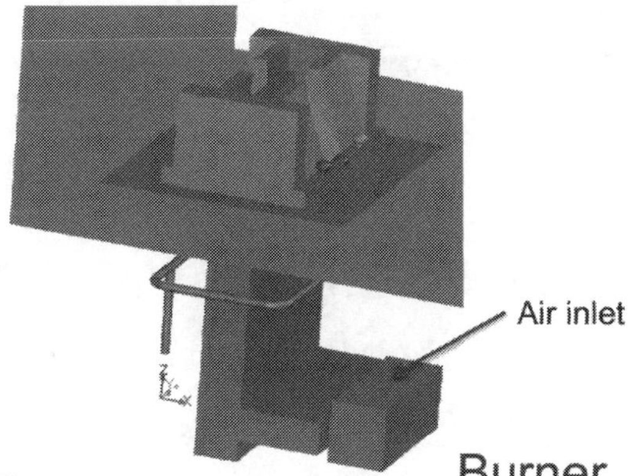

Burner Geometry

Burner Tip Details

Furnace Improvements Services

Primary Tip – Stamped B

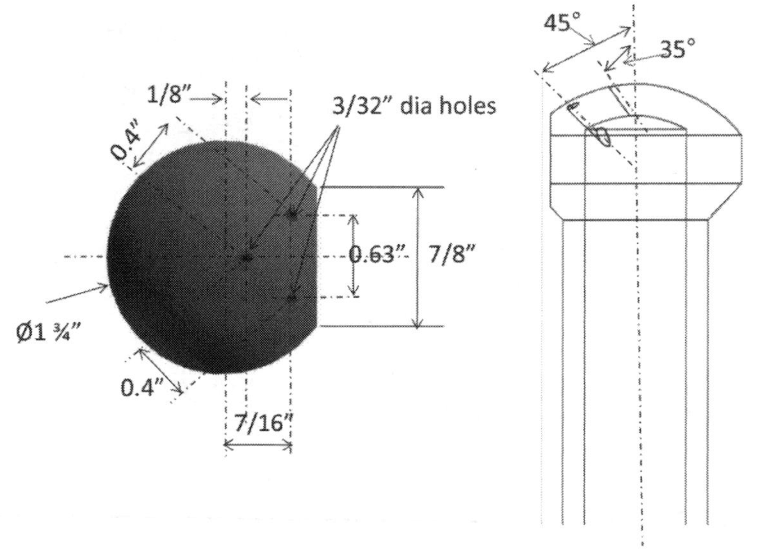

March 8, 2013 www.heatflux.com 7

Furnace Improvements Services

Results

- The flow pattern of flue gases is shown using
  - Velocity Vectors
  - Path Lines
- The temperature profile in the heater is shown using
  - Temperature Contours
- The flame profile is shown using
  - CO envelope of 2000 ppm

March 8, 2013 www.heatflux.com 8

Planes Used for Results

- Planes passing through primary and secondary tips are considered to plot
  - Temperature
  - Velocity

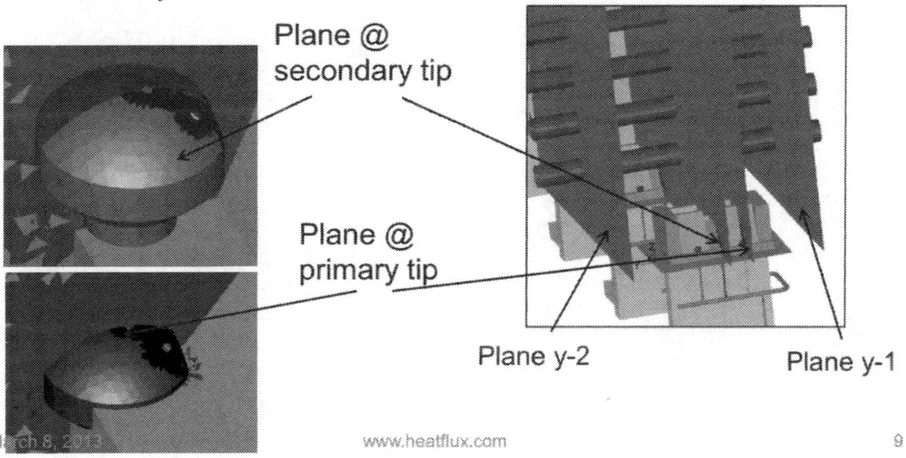

Planes Used for Results

- Planes along the center of burner
- Planes along the height of heater

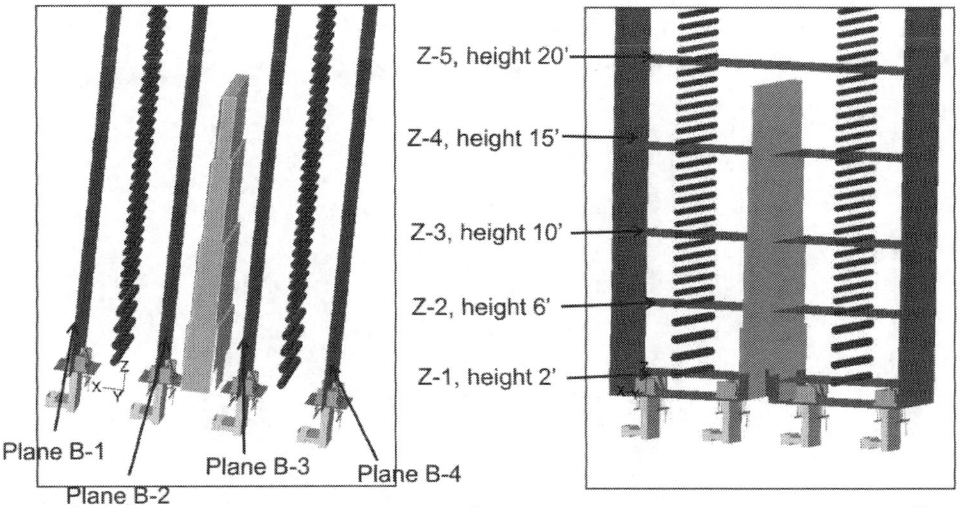

Velocity Vectors

Furnace Improvements Services

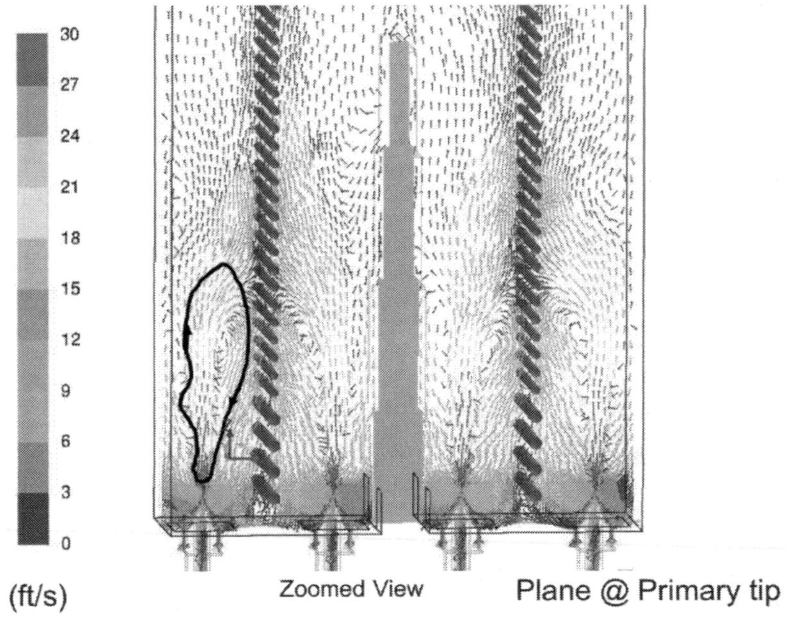

March 8, 2013 www.heatflux.com 11

Velocity Vectors

Furnace Improvements Services

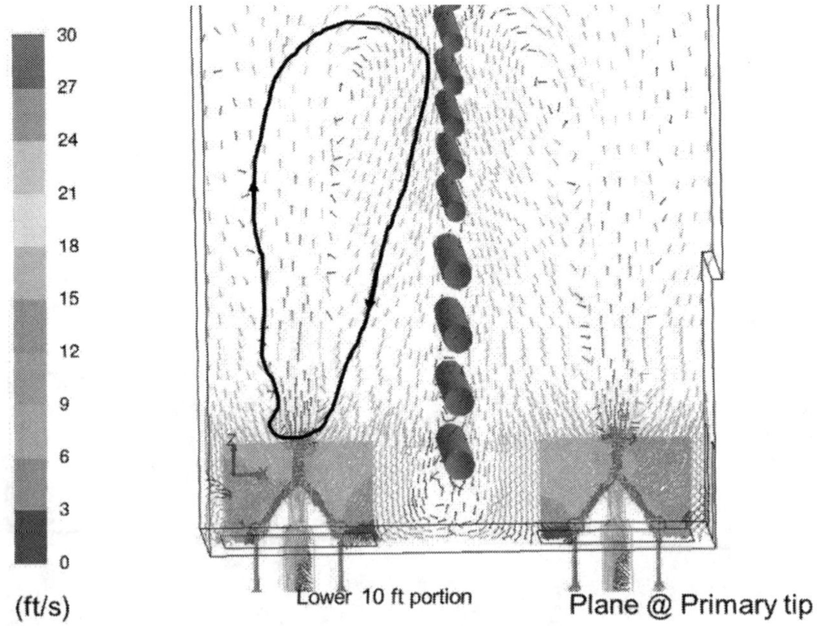

March 8, 2013 www.heatflux.com 12

Velocity Vectors

Furnace Improvements Services

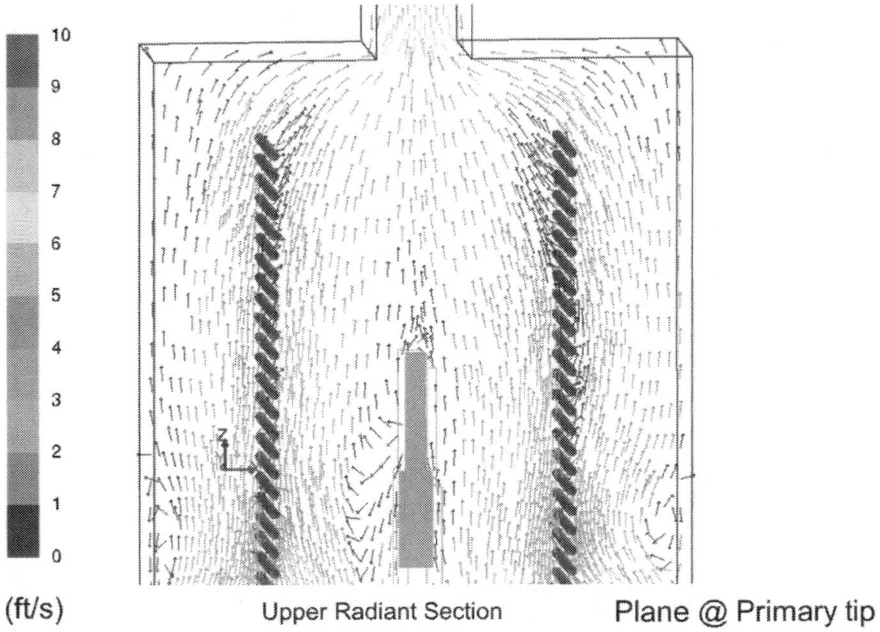

Velocity Vectors (by temperature)

Furnace Improvements Services

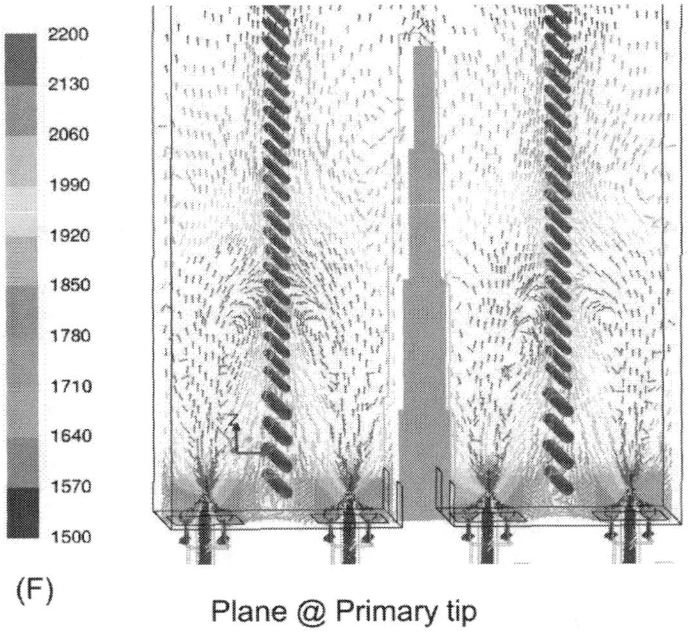

Path Lines

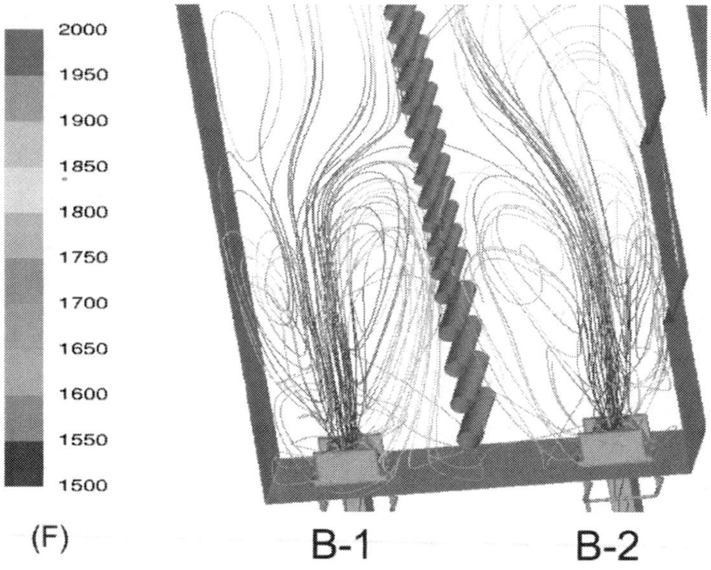

Temperature Contours

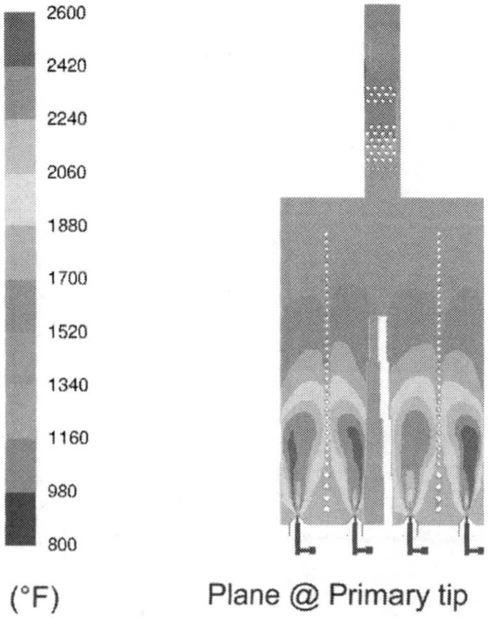

Plane @ Primary tip

Temperature Contours

Furnace Improvements Services

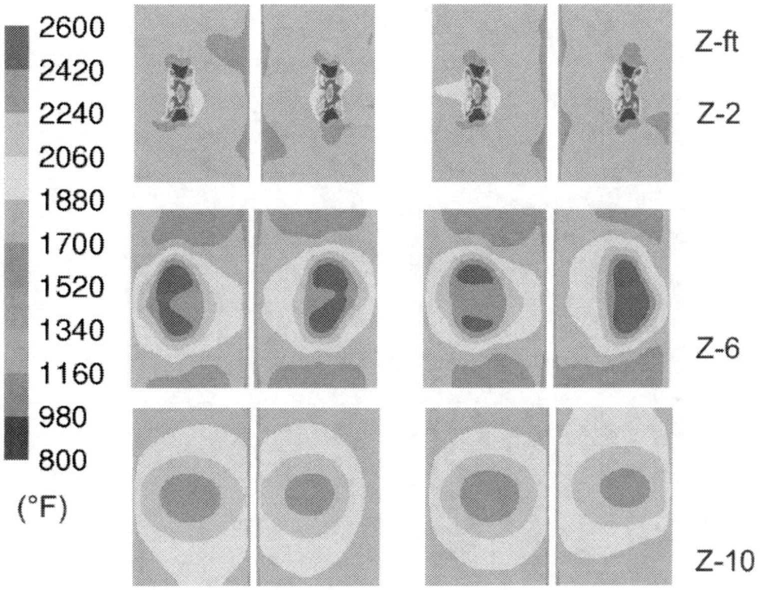

March 8, 2013 www.heatflux.com 17

Tube Metal Temperature Contours

Furnace Improvements Services

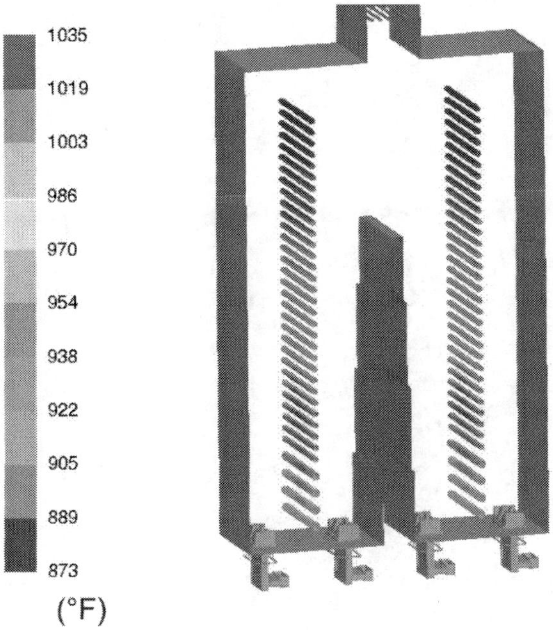

March 8, 2013 www.heatflux.com 18

Refractory Temperature Contours

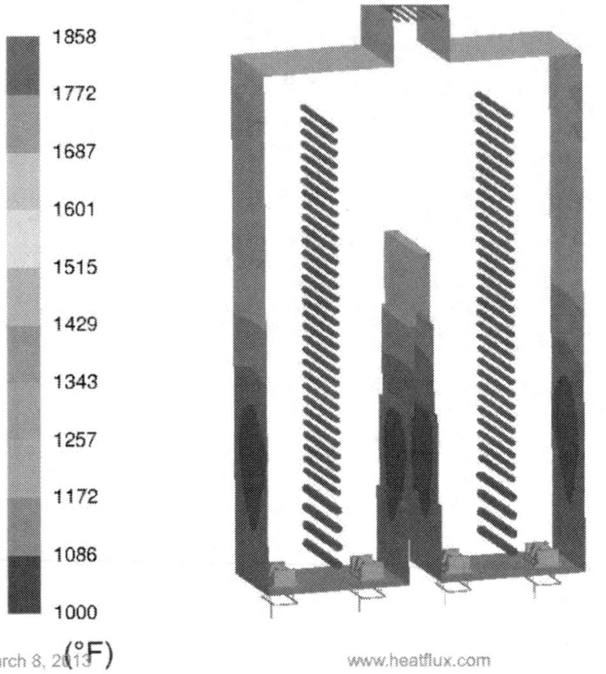

Flames Shape (Iso-Surface of CO 2000 ppm)

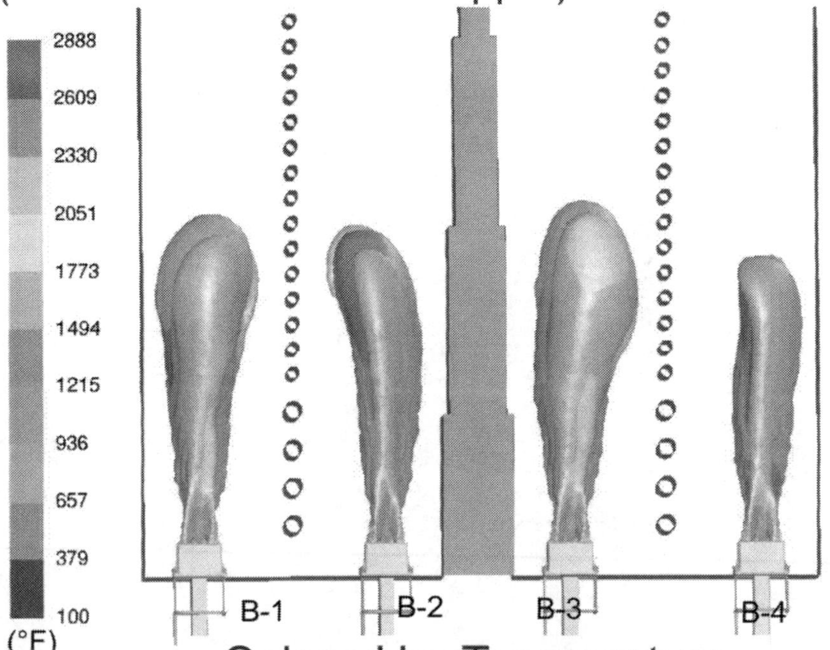

Colored by Temperature

Inclined Firing

Furnace Improvements Services

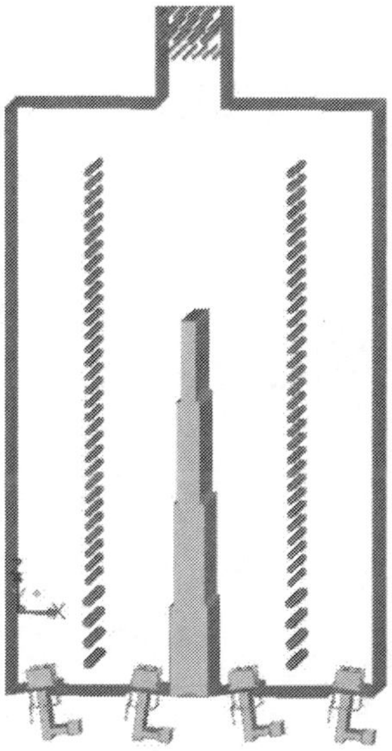

Heater Geometry

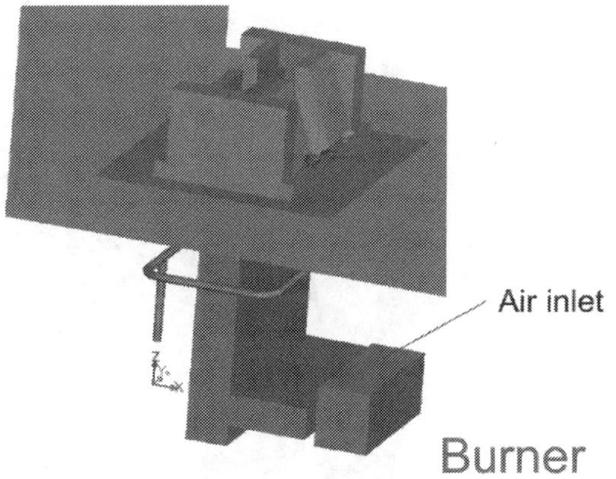

Burner Geometry

Burners are inclined by 7.5 deg towards the walls

Velocity Vectors

Furnace Improvements Services

Inclined Firing

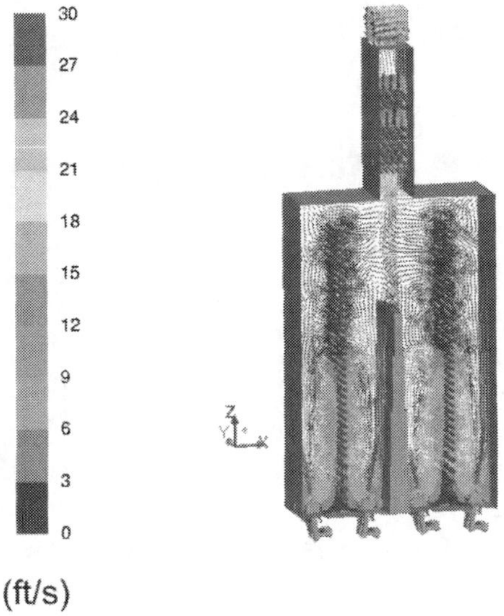

Plane @ Primary tip

Velocity Vectors

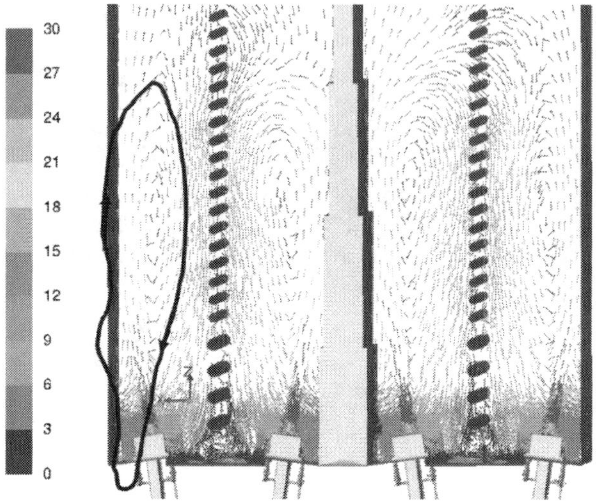

Velocity Vectors

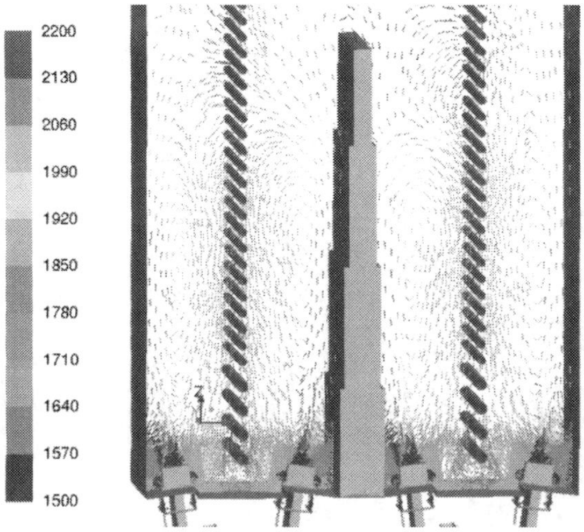

Temperature Contours

Inclined Firing

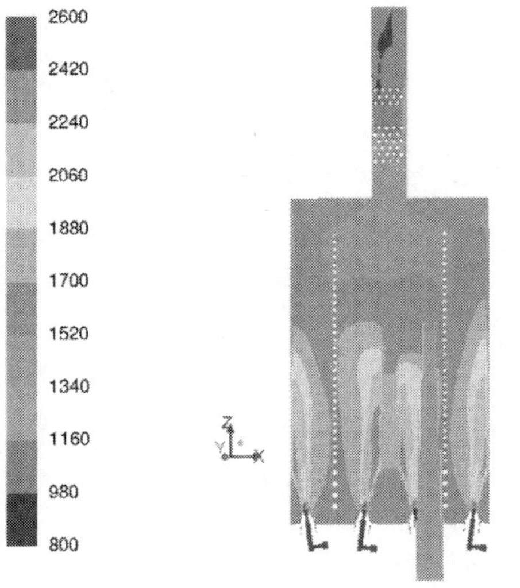

(°F) Plane @ Primary tip

Tube Metal Temperature Contours

Inclined Firing

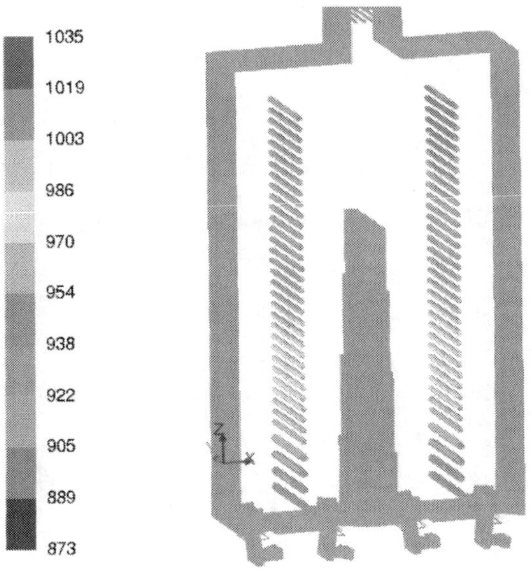

(°F)

Iso-Surface of CO 2000 ppm

Furnace Improvements Services

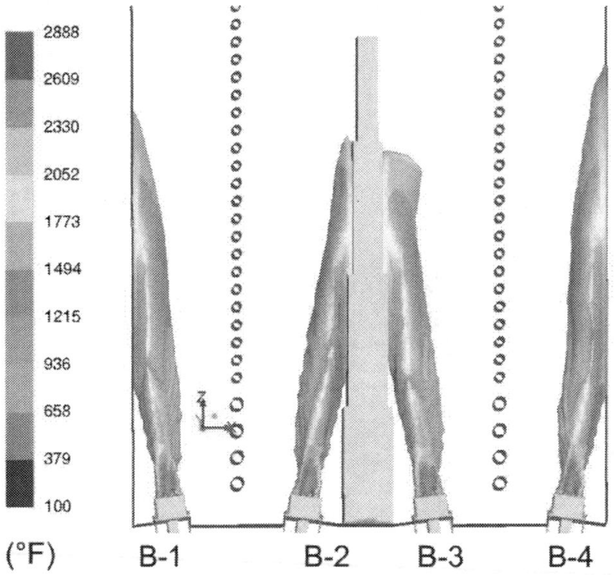

Inclined Firing

Colored by Temperature

Vertical Up Firing VS. Inclined Firing

Furnace Improvements Services

- Flow Profiles

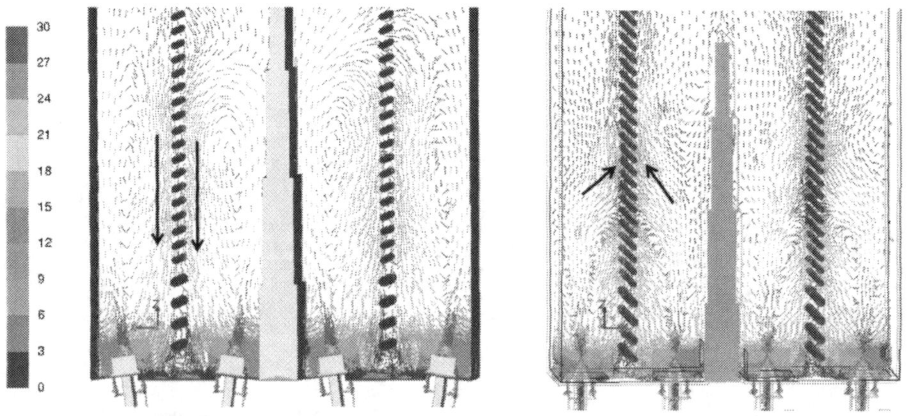

Vertical Up Firing VS. Inclined Firing

CHEM INNOVATIONS 2012 CONFERENCE & EXPO

- Temperature

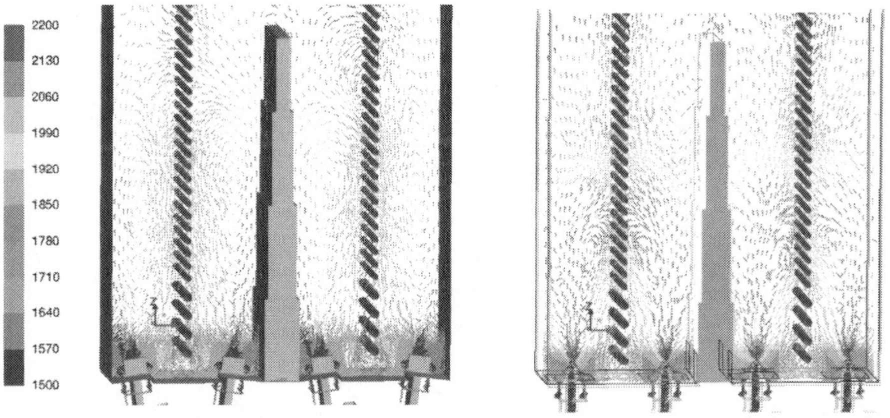

Vertical Up Firing VS. Inclined Firing

Furnace Improvements Services

- Flame Profile

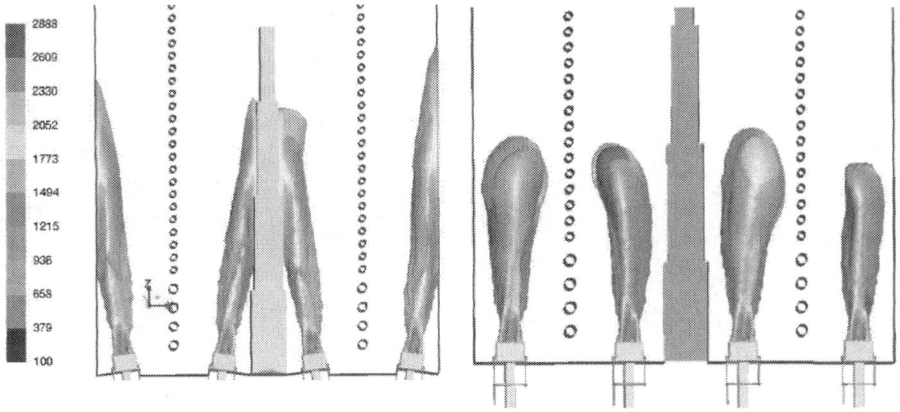

Conclusion

- Inclined firing resulted in reduction of tube metal temperatures by about 40-50 F
- Elimination of flame impingement
- Longer life of tubes
- Longer heater run lengths

Case Study-2-Crude Heater

- Client facing flame impingement and high tube metal temperatures in radiant section
- FIS recommended inclined burner firing system to eliminate flame impingement
- Carry out CFD Modeling of the Burners and Radiant Section to demonstrate the benefit of inclined burner firing system

Furnace Improvements Services

Existing Burner Arrangement

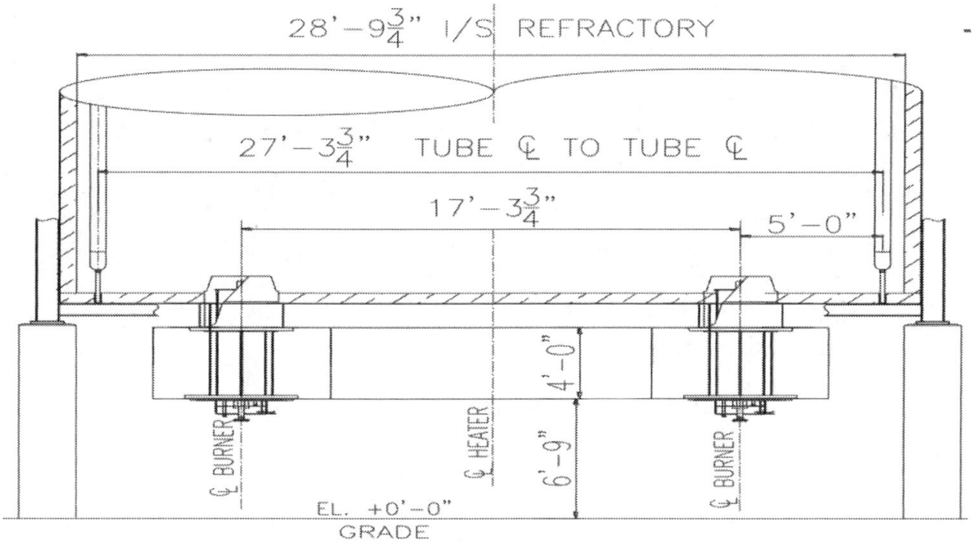

Furnace Improvements Services

Inclined firing system Burner Arrangement

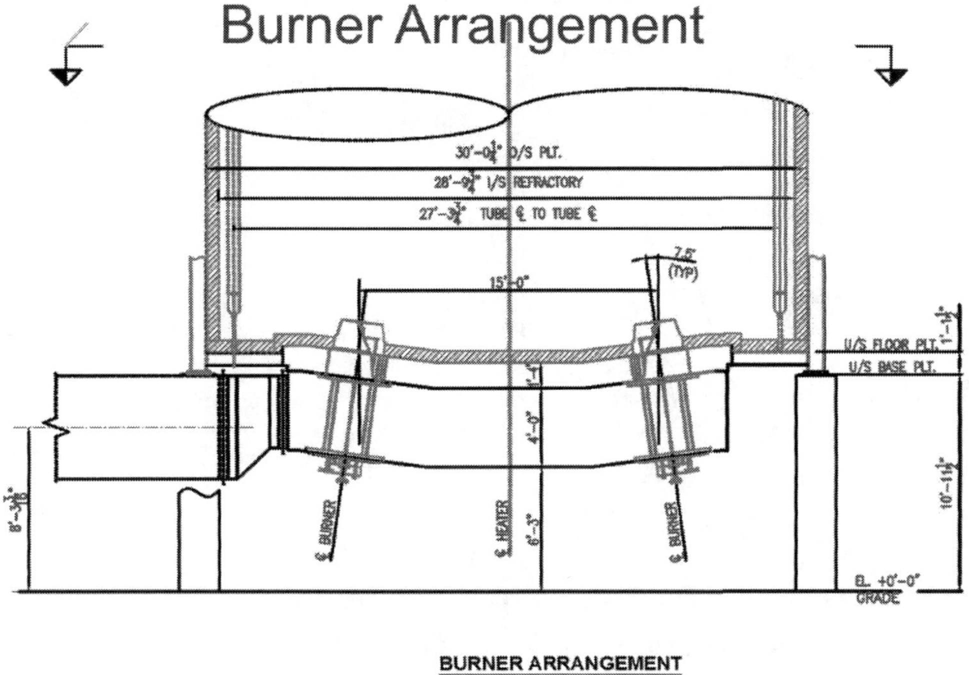

BURNER ARRANGEMENT

12 NOS. JOHN ZINK BURNERS

Burner Geometry

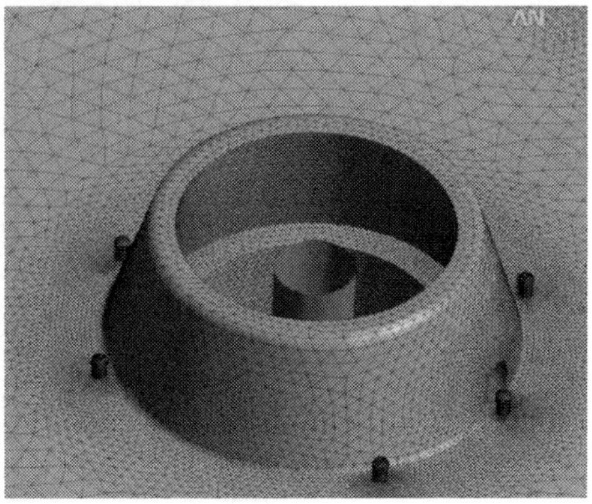

Burner tiles

Recirculation hole

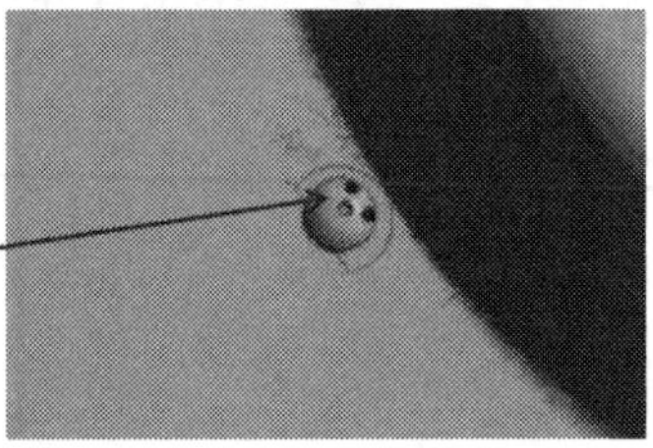

Staged tip with 3 holes

Furnace Improvements Services

Velocity vectors (ft/s)

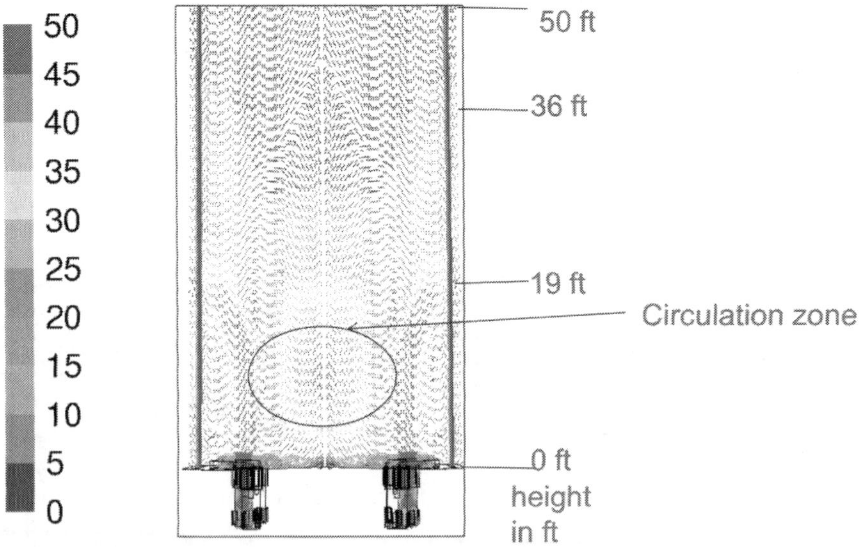

- Velocity vectors shows that it is moving away from center due to circulation at center

Velocity vectors (ft/s) Inclined Firing System

Furnace Improvements Services

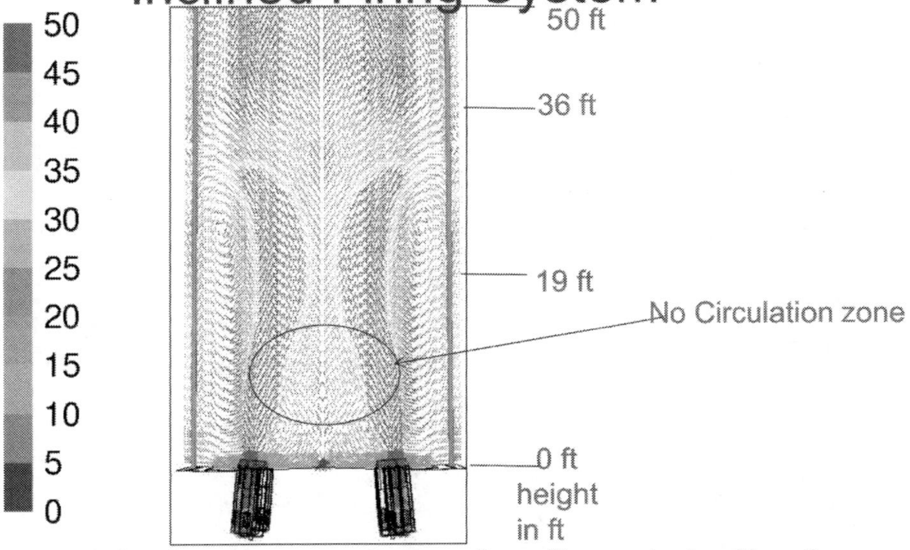

- Velocity vectors shows that flame is inclined towards center, this is due to no recirculation formed at the center of heater

Velocity vectors (ft/s) ° Base case vs. Inclined Firing System

Furnace Improvements Services

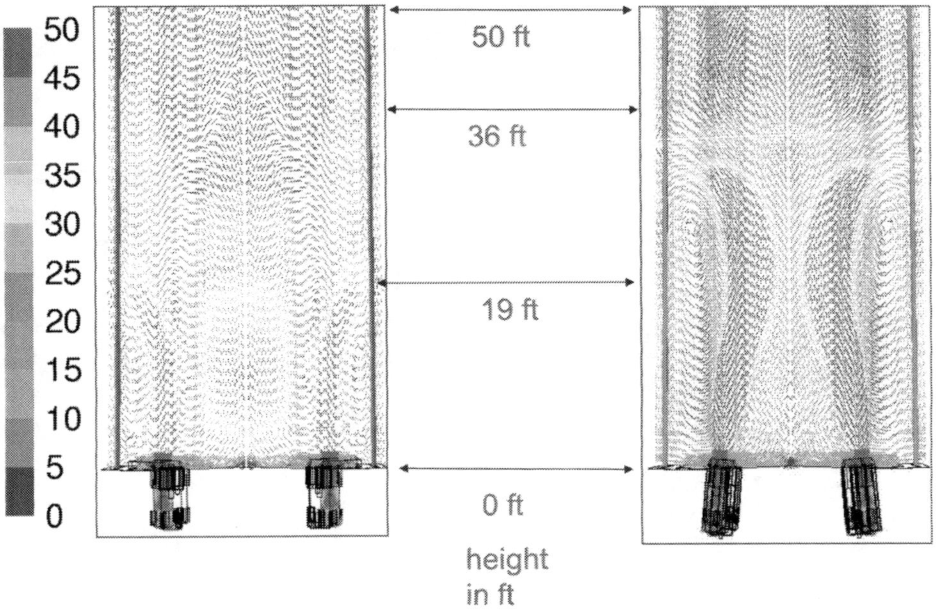

Furnace Improvements Services

Temperature Contours Base Case VS. inclined firing system

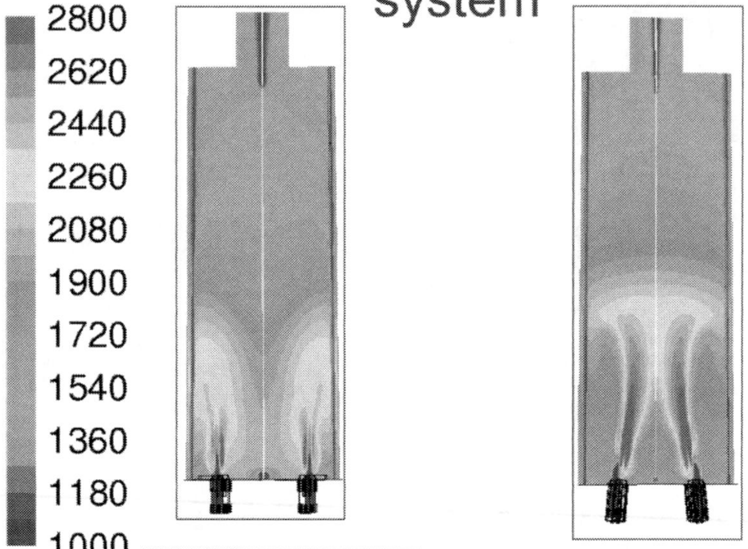

Base case

inclined firing system

Furnace Improvements Services

Temperature contours(°F) at 10 ft section- Base Case VS. inclined firing system

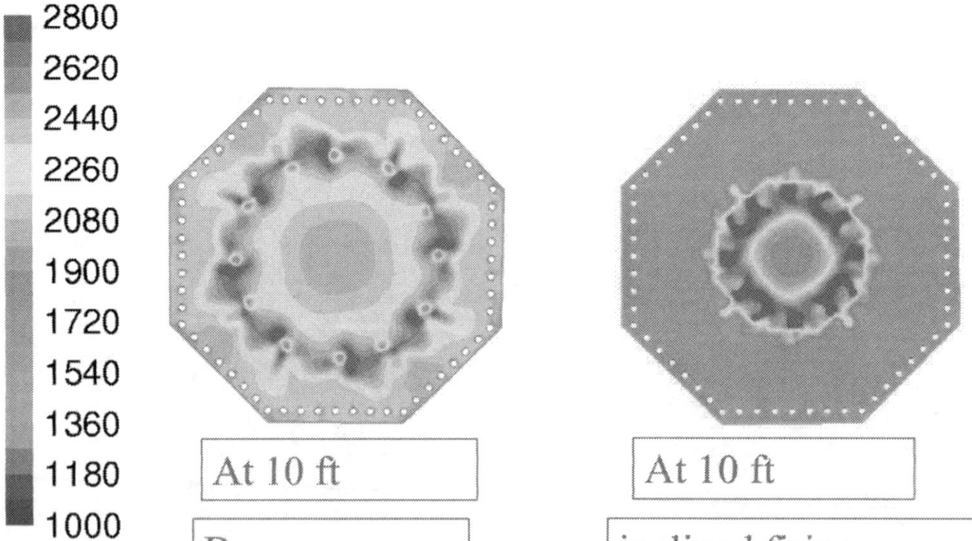

Base case

inclined firing system

Temperature contours (°F)
at 15 ft section- Base Case VS. inclined firing system

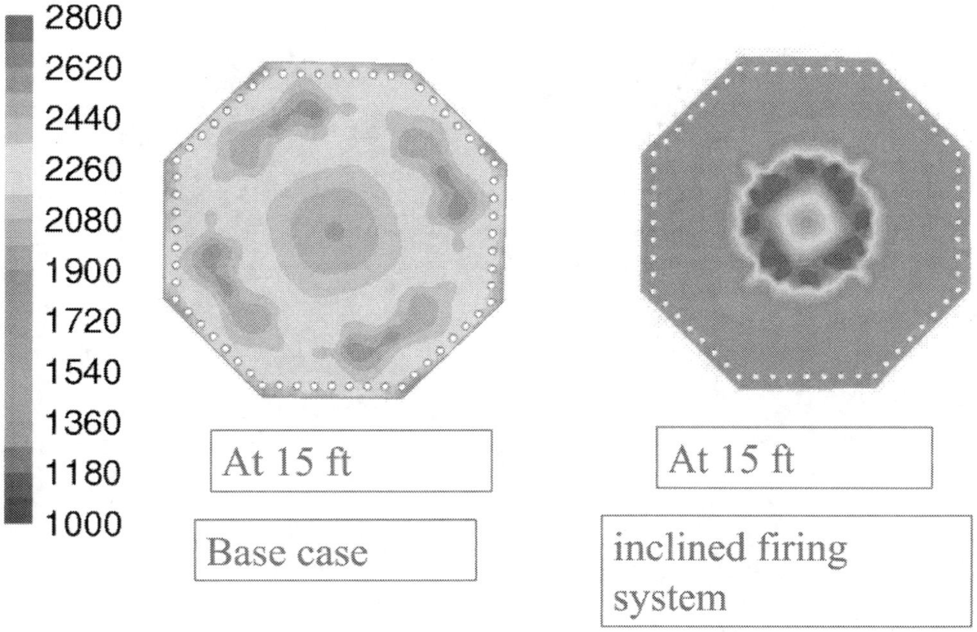

Furnace Improvements Services

CO mass fraction contours (PPM)
at 10 ft section- Base Case VS. inclined firing system

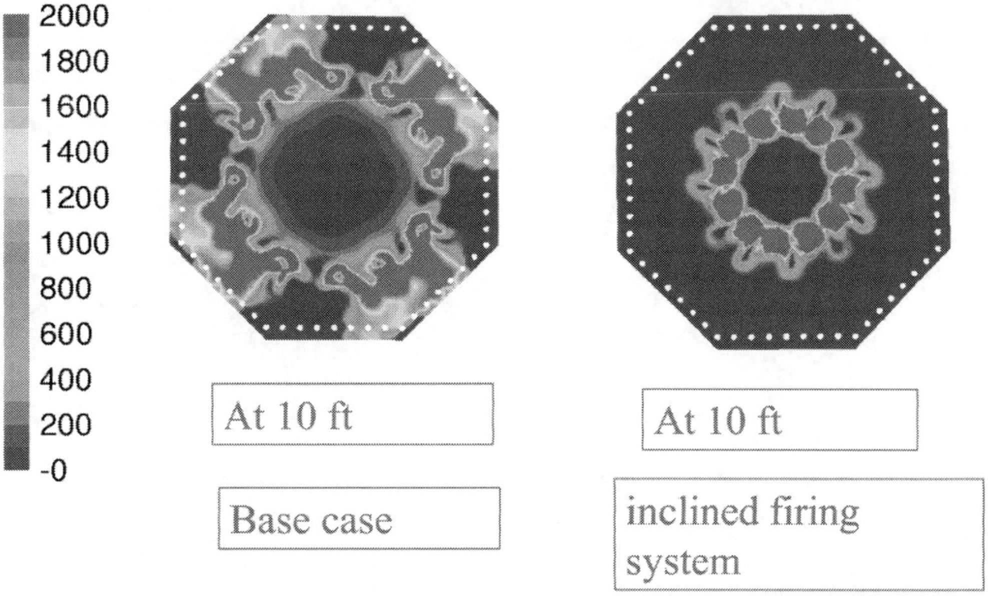

CO mass fraction contours (PPM) at 15 ft section- Base Case VS. inclined firing system

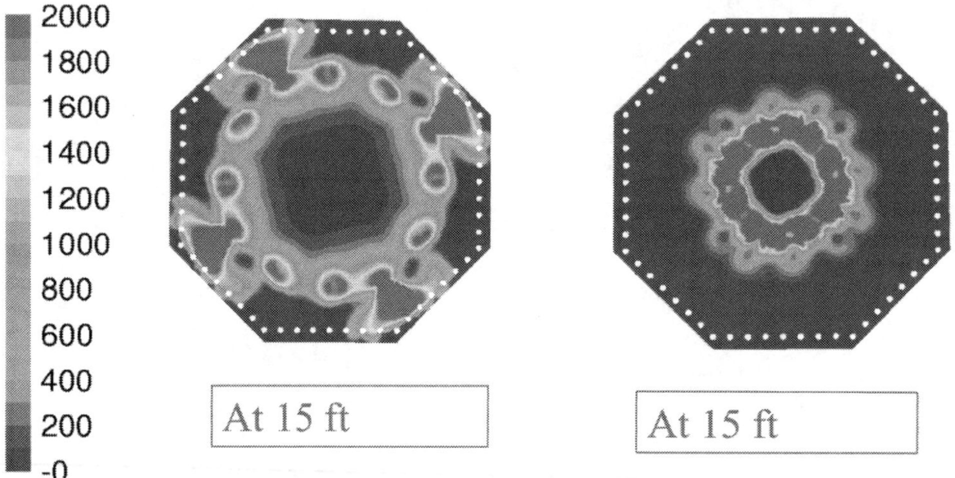

At 15 ft

At 15 ft

Base case

inclined firing system

Furnace Improvements Services

Base Case VS. inclined firing system

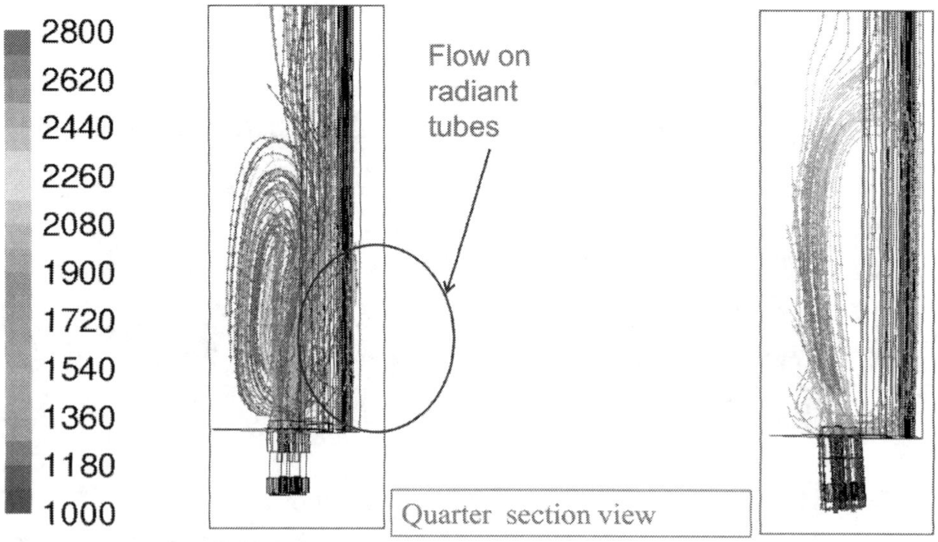

- Pathlines shows the flow towards radiant section for the base case at entrance of heater
- However for inclined firing system flow does not interact to radiant tubes at entrance of heater

Iso surface of CO mf 2000 PPM colored by temperatures (° F) Base Case VS. inclined firing system

Furnace Improvements Services

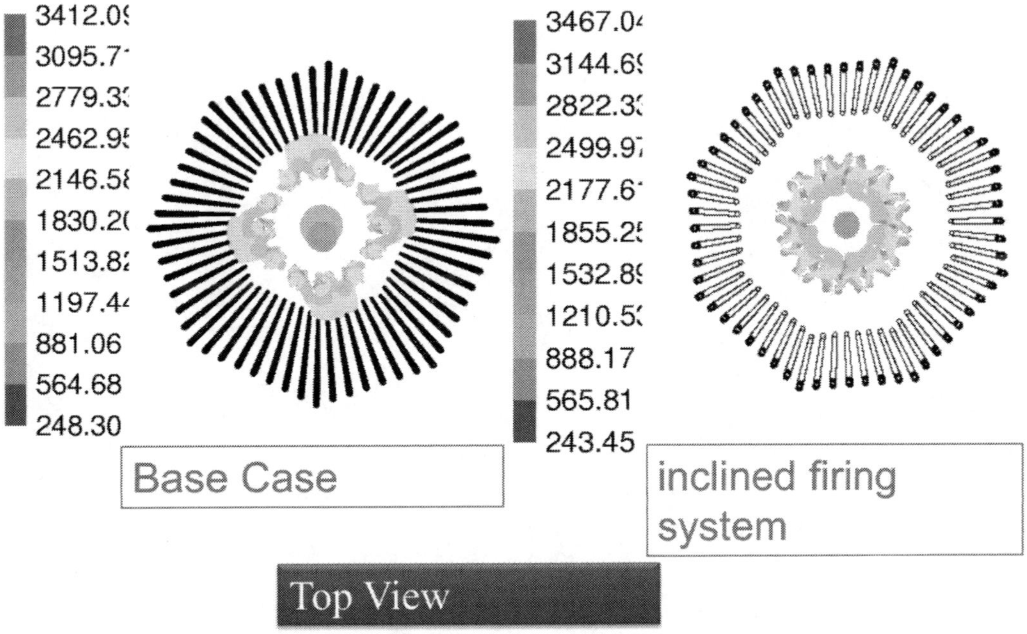

Iso surface of CO mass fraction 2000 PPM colored by flame height (ft)

Furnace Improvements Services

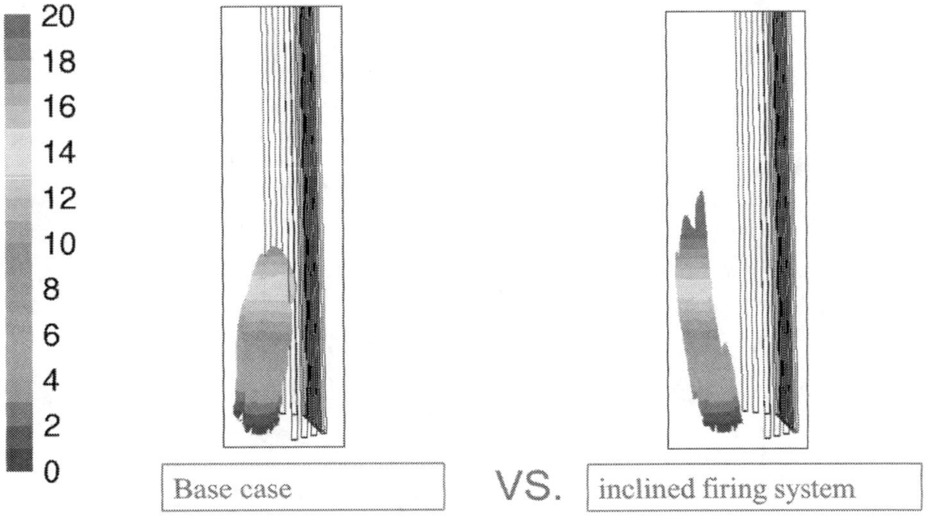

- Height of flame is increased from 20 to 24 ft for inclined firing system due to non existence of recirculation pattern

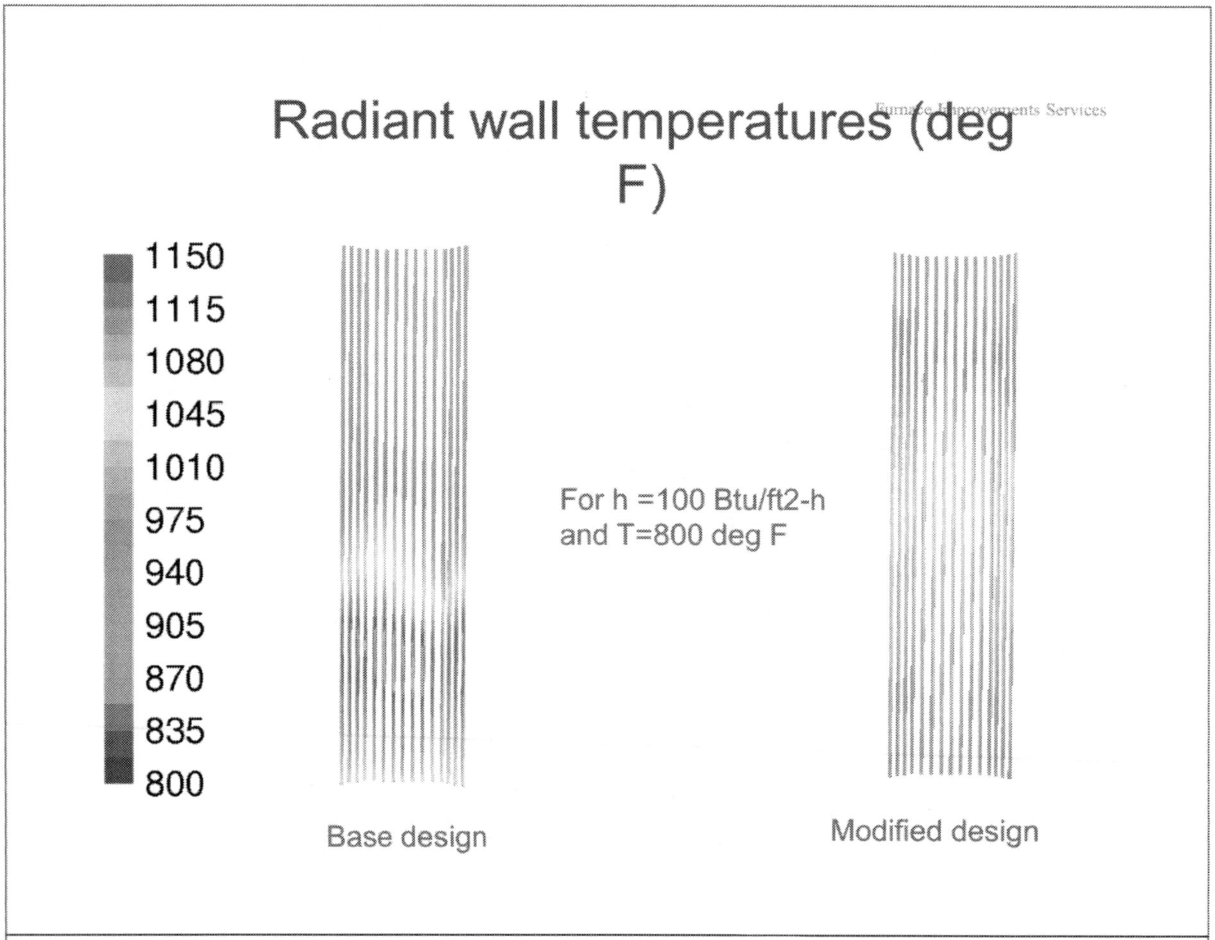
Radiant wall temperatures (deg F)
Furnace Improvements Services
1150
1115
1080
1045
1010
975
940
905
870
835
800
For h =100 Btu/ft2-h
and T=800 deg F
Base design
Modified design

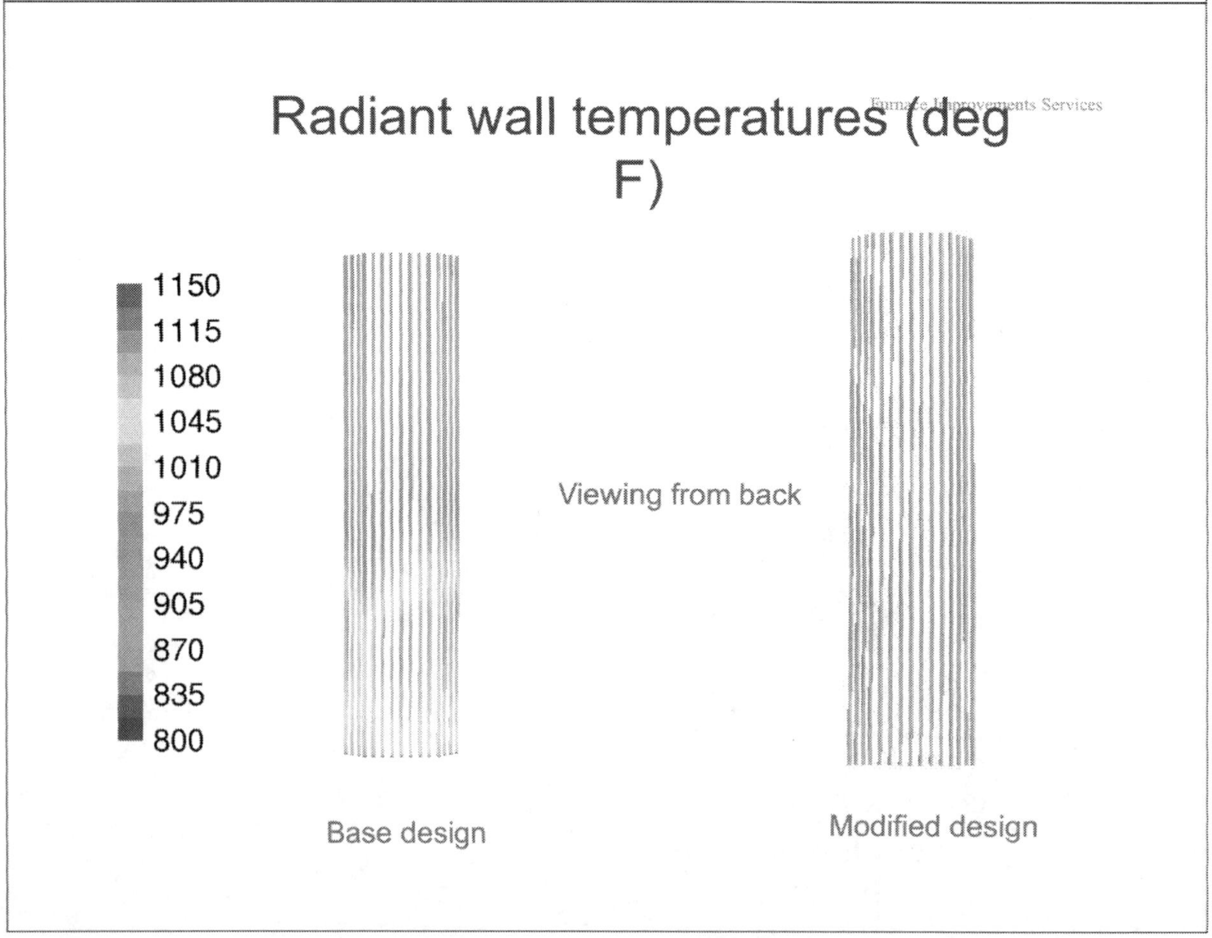
Radiant wall temperatures (deg F)
Furnace Improvements Services
1150
1115
1080
1045
1010
975
940
905
870
835
800
Viewing from back
Base design
Modified design

Furnace Improvements Services

Summary

- CFD Modeling helped in identifying the problems of high tube metal temperatures
- CFD Modeling helped in demonstrating the benefit in performance
- CFD Modeling is very effective in identifying the problems and solutions if done correctly.

Furnace Improvements Services

We hope you will find our presentation helpful and informative.

(55c) Assesment of an Eulerian-Eulerian Multiphase CFD Framework for A Three Dimensional Modeling of A Trickle BED Reactor in Counter-Current Operation

Sebastian Moreno Cardenas[1], Vivian Passos de Souza[2], Adriana de Souza Ferreira[2], Carlos Alberto Araujo Monteiro[2], José Roberto Nunhez[3] and E. M. Matos[4], (1)UNICAMP, Campinas, Brazil, (2)PETROBRAS, Rio de Janeiro, Brazil, (3)Chemical Process Department, UNICAMP, Campinas, Brazil, (4)DPQ, UNICAMP, CAMPINAS, Brazil

Commercial HDT processes usually operate in a trickle-bed regime, with co-current downward flow of gas and liquid over a randomly fixed bed of catalyst particles where reactions take place. It is well known that sulfur removal is strongly inhibited by the competitive adsorption effect of H_2S at the sulfided active sites of the catalyst. As a consequence, it is important in these process to maintain the concentration of H2S as low as possible during the reaction to achieve a low sulfur content product at the reactor outlet. According to Ancheyta et al. (2007), a more convenient profile of H2S concentration can be provided by operating the reactor in counter-current mode, for instance, introducing the feed at the top and H_2 at the bottom of the reactor. Thus, in the lower section of the reactor, the H_2S concentration is lower and H_2 is greater, promoting higher reaction rates. The main problems on the counter-current operation are a lower efficiency on the gas-liquid contact; a difficulty in preventing flooding conditions and a higher pressure drop. In order to investigate the above processes in more detail, a three-dimensional fluid-dynamic study of a counter-current trickle bed reactor was carried out using CFD (Computational Fluid Dynamics) techniques with emphasis on determining the pressure drop and the distribution of liquid hold-up in the bed. A CFD tridimensional model based on an Eulerian – Eulerian multiphase approach was used to model the hydrodynamics of the pseudo two-phase flow in a trickle bed reactor (TBR) in counter-current operation. The closure terms for the phase interactions have been based on the fluid-fluid interfacial force balance concept (Attou et al. 1999), which was used and validated to simulate co-current HDT reactors by Gunjal e Ranade (2007). Radial variation of porosity was estimated using the Klerk (2003) distribution. The above set of model equations were implemented inside the commercial software ANSYS-CFX Release 13. The model provided information about phase distribution, velocity profiles, pressure drop liquid hold up for a isothermal trickle bed reactor at laboratory scale in counter-current operation.

(55d) Use of Cold Flow Modeling and CFD for New Product Development

Zhanping Xu, UOP LLC, A Honeywell Company, Des Plaines, IL and Pengfei Chen, UOP LLC, a Honeywell Company, Des Plaines, IL

Internals are installed in reactors and distillation columns for fluid mixing, distribution and separation of hydrocarbon mixtures under high pressure and high temperature conditions. In the development of these internals, it is unlikely to use the process fluid under similar conditions for conducting the testing of the internals to verifying their performance. CFD and cold flow modeling can be valuable tools in the development.

This paper will discuss the development of reactor and distillation internals using CFD and cold flow modeling. Examples and methods will be shown on using CFD and cold flow modeling in new product developments. Guidance will be provided on what CFD and cold flow modeling can do as well as their limitations.

(55e) Evaluation of the Dispersion Coefficient and Residence Time Distribution in Packed Bed Using Computational Fluid Dynamics

Hadjira Iddir and Quan Yuan, UOP a Honeywell Company, Des Plaines, IL

Packed beds are widely used in the chemical and petroleum industries. The processes often involve a single fluid or mixture of fluids that pass through a packed bed for processing or purifying the fluids. The fluid flow distribution inside the packed bed plays an important role affecting the process efficiency, yield and sometimes the lifetime of the particles. Flow in the packed bed is greatly affected by the design of the vessel, internals and bed loading. The dispersion coefficient and Residence Time Distribution (RTD) are two important parameters that are widely used to quantify the flow characteristics in packed bed. The measurement of the two parameters in large-scale industrial units is expensive, time consuming and sometimes not feasible. As an alternative, Computational Fluid Dynamics (CFD) can be used to calculate the RTD and dispersion coefficient.

This paper presents a few different methods on how to calculate the dispersion coefficient and RTD using CFD. The results will be compared to experimental data reported in literature. The issues and potentials of this technique are then discussed, along with several examples from the UOP technologies.

(56a) Selection of Optimized Gas Monetization Option with FLNG or Fixed FLNG

Joseph H. Cho, SK E&C, Houston, TX

In general, as the gas E&P technology advances, gas fields development costs with the same conditions decreases, results in allowing development of gas fields in remote areas or deep water field. Upon adoption of advanced technologies and widen their practical applications to gas monetization, those of stranded gas field developments are feasible and economical.

Even several FLNG project are under construction stage with overcoming technical limitations and commercial challenges; however, it may be said that project developers prefer to follow the conventional development options, which have been proven with commercial operation by others. It is mainly because the conventional development option is relatively easier to identify potential risks with required costs associated with mitigation methodologies than newly developed options, such as FLNG and Fixed FLNG options.

Main parameters that affect selection of development options are 1) proven technology, 2) potential risks and mitigation, 3) economic returns, and 4) gas reserve (location, size, reservoir characteristics). It is clear if the economic returns of a development option cannot absorb potential risks, a project developer would not take risks in the selected gas monetization option.

This paper discusses the cost analysis, cost savings, potential risks in gas monetization with FLNG concept compared to the green field development option with conventional technology. This paper also addresses potential benefits and risks of FLNG and Fixed-FLNG concept with a virtual project development case as a reference. Advantages and limitations of the Fixed-FLNG concept have also been presented. This includes cost benefit, application flexibility, permit issue, environmental impact, construction requirements, etc.

(56b) Smart Cargo Containment System Advanced (SCA) for LNG Carrier

Dai Gil Lee[1], Munkuen Ha[2] and Haeki Jang[2], (1)KAIST (Korea Advanced Institute of Science and Technology), Daecheon Si, South Korea, (2)Samsung Heavy Industries Co., Ltd, Gyeongsangnam Do, South Korea

Membrane type cargo containment system has been evolved in terms of sloshing resistance and low boil off aspect. Membrane type cargo containment system, so called Smart containment system Advanced, has been developed by Samsung Heavy Industries to achieve less barred range of filling and lower boil off rate.

Smart containment system Advanced can achieve less barred range of filling 15% to 70% of tank height and lower boil off rate, 0.12 % /day. It has been verified and approved by LR and ABS and now ready for real LNG carrier application. This paper discusses features of Smart containment system Advanced.

LNG FSRU and FPSO Pump Considerations for Ship Motion

Gregory P. Wood, P.E.
Engineering Manager
Ebara International Corporation
Sparks, Nevada USA

Scope

LNG Floating Storage and Regasification (FSRU) and Floating Processing and Send Out (FPSO) vessels are a growing component of the LNG infrastructure. These marine vessels use in-tank mounted vertical centrifugal submerged motor cryogenic pumps. Unlike land based pumps, marine pumps installed in FSRU and FPSO applications are exposed to three degrees of ship movement as a result of wave motion. There is concern that adding ship motion to a traditional stationary pump technology can cause issues such as unseating, free spinning, and bearing damage during non-operation. This paper reviews the technical impact of the ship motion and the effectiveness of current and proposed solutions.

Pump and Installation Description

The in-tank mounted vertical centrifugal submerged motor cryogenic pump used in the FSRU/FPSO service consists of a pump and motor assembly, suction valve, column, retraction system, head plate, and power system (see Figure 1). The vertical centrifugal pump uses a Thrust Equalizing Mechanism™ (TEM) (see Figure 3) to balance the axial thrust loads and weight forces during pump operation, eliminating all axial loads on the main ball bearing. During non-operation, the main ball bearing carries the axial weight of the rotating components. The pump is installed into a spring loaded suction valve that seals the column from the tank when the pump is not installed. The pump fluid discharge goes directly into the column and is separated from the tank inlet by a static face type seal between the pump and the suction valve. During non-operation, gravity secures the pump onto the suction valve seal and the rotating elements are allowed to rotate freely. During pump operation, the column fluid pressure applies additional force to the pump weight that is applied to the suction valve seal.

Ship Motion Conditions

The FSRU and FPSO vessels are built in shipyards, transferred to their final location and moored during operation. They may also be relocated to a new location periodically. FSRU and FPSO vessels sometimes contain their own propulsion system, but often rely on auxiliary vessels to provide propulsion. During transit the non operating FSRU and FPSO vessels are exposed to sea conditions. For this analysis, ship motion is described in linear and rotational terms, and is based on 50 to 80 foot sinusoidal wave motion over typical 8 to 20 seconds producing 0.13 to 0.05 Hz frequencies.

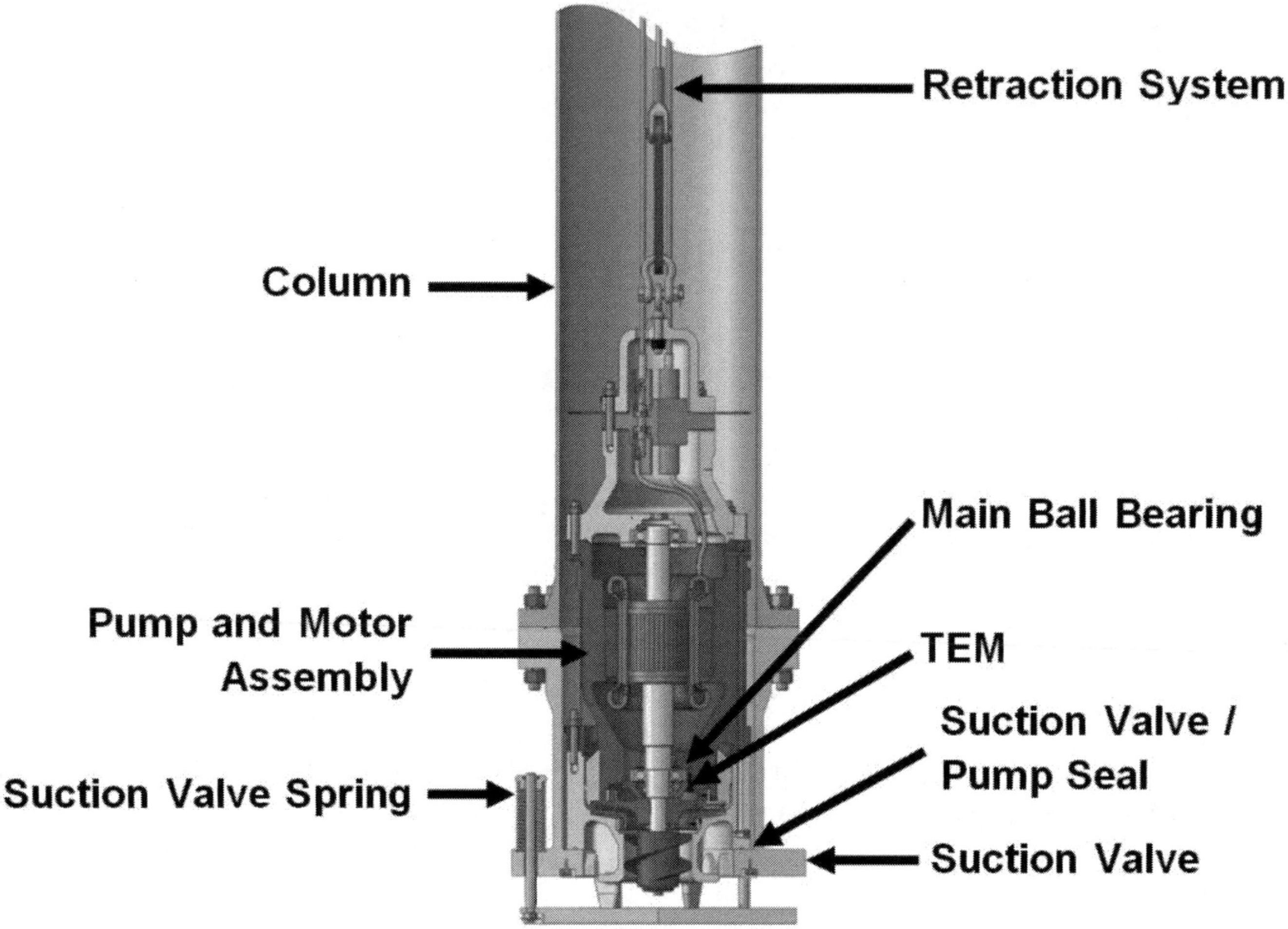

Figure 1: In-tank mounted vertical centrifugal submerged motor cryogenic pump installation

Ship Motion Description

Ship motion is described in linear and rotational terms (see Figure 2). The linear motions are heave (up/down), sway (side-to-side), and surge (front/back). The rotational motions are roll, the rotation of a vessel about its longitudinal (front/back) axis, pitch, the rotation of a vessel about its transverse (side-to-side) axis, and yaw, the rotation of a vessel about its vertical axis. Each type of ship motion introduces its own force and direction impact on the vertical in-tank pumps. The location of the pump in the vessel will also result in different force amplitudes and direction. The analysis going forward will focus on the worst case combination of ship motion and pump location.

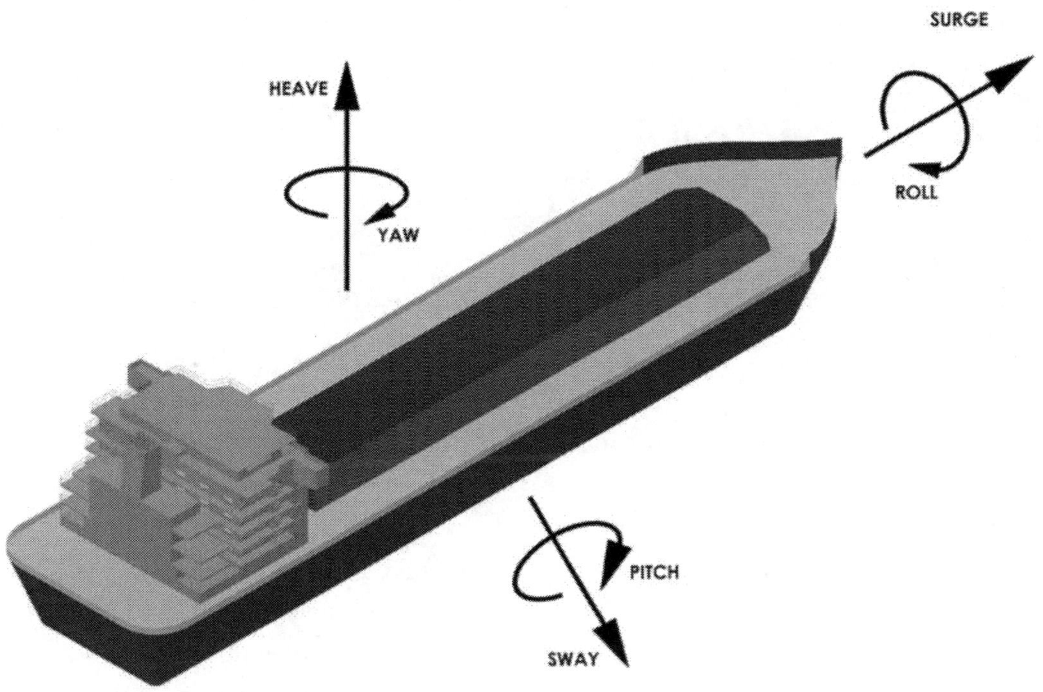

Figure 2: Linear and rotational ship motion

Ship motion will be based on sinusoidal waves. Although a ship will typically dampen the effects of sea wave motion, this analysis will use the actual sinusoidal wave motion as a worst case scenario to be conservative. The sinusoidal wave condition is based on 50 to 80 foot wave motion over typical 8 to 20 seconds producing 0.13 to 0.05 Hz frequencies (see Figure 4). Longer wave frequencies are not considered since they lessen the impact on the pump.

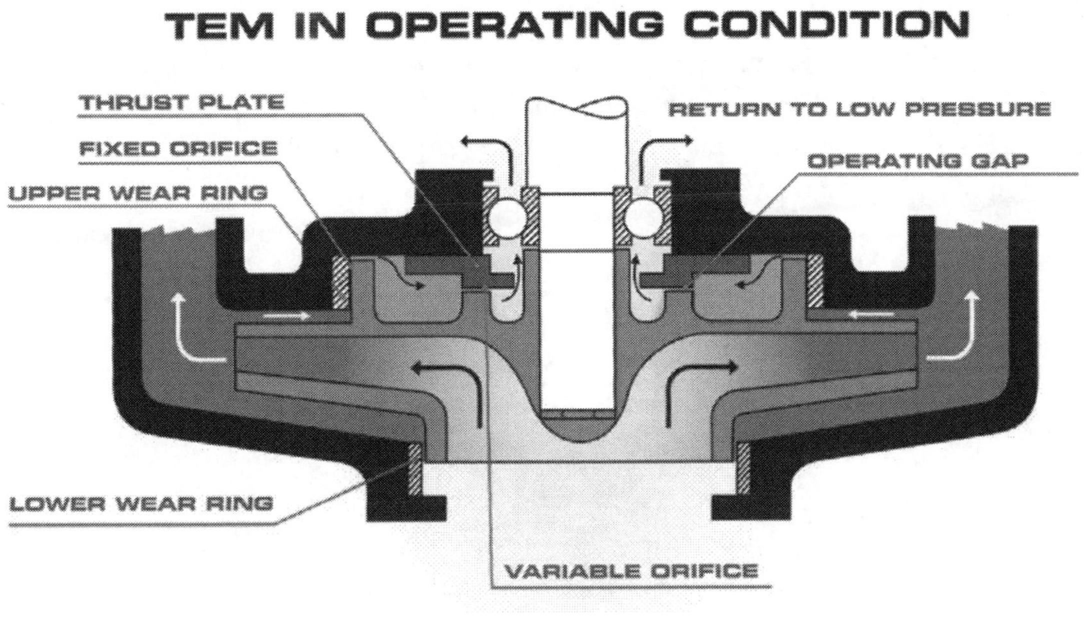

Figure 3: Thrust Equalizing Mechanism™

Period (Seconds)	8	12	16	20
Frequency (Hz)	0.13	0.08	0.06	0.05
Minimum Wave Height (Feet)	5	5	5	5
Maximum Wave Height (Feet)	50	80	80	80

Figure 4: Wave period (frequency) and height range

Ship Motion Impact on Non Operating Pumps

Operating pumps are not subject to ship motion damage. During pump operation there is a large amount of force on the suction valve seal from the pump discharge fluid pressure. This force is many times the pump weight, overcoming any tendency for the pump to unseat or tilt with regards to the suction valve seal. The Thrust Equalizing Mechanism™ balancing forces during pump operation are many times greater than the weight of the rotating element, eliminating any possibility of loading the main ball bearing as a result of ship motion.

During non-operation the pump is seated on the suction valve seal plate using pump weight and the rotating elements are resting on the main ball bearing. The concern is that adding ship motion to this traditional land based pump technology can cause issues such as unseating and/or tilting from the suction valve, free spinning of the rotating elements, and bearing damage due to excessive axial loading. Each of these potential impacts on the in-tank mounted vertical centrifugal submerged motor cryogenic pumps will be analyzed individually.

Pump Unseating Potential

Pump unseating from the suction valve would be the result of ship heave motion. The pump rests on the suction valve seal with a force equal to the weight of the pump minus the upward force of the suction valve springs. The typical spring force is equal to 25% of the pump weight. Therefore, the vertical acceleration required to unseat the pump from the suction valve is:

Equation 1: $a = (1 - f_s)g$

a = acceleration required to unseat pump

f_s = spring force as a percent of pump weight

g = gravitational constant (32.2 ft/sec^2)

$a = (1 - 0.25)\ X\ 32.2 = 24.2$ ft/sec^2

The peak acceleration for sinusoidal wave motion based on sea conditions must exceed 24.2 ft/sec^2 to unseat the pump from the suction valve. The sinusoidal wave characteristics are shown in Figure 5.

Equation 2: $a_{max} = 2\pi^2F^2\text{D}$

a_{max}= maximum acceleration based on sinusoidal wave

F = wave frequency (Hz)

D = wave height (feet)

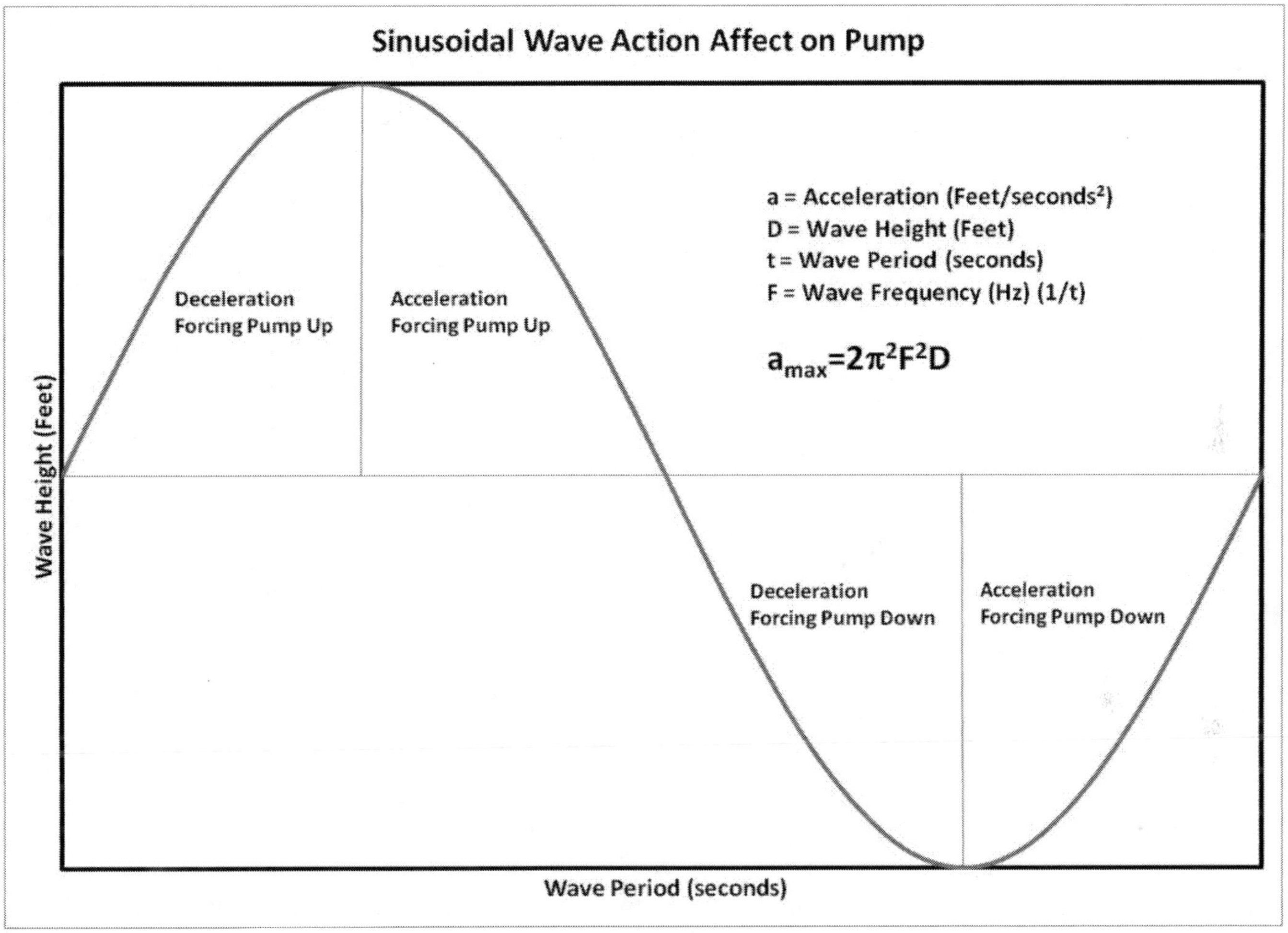

Figure 5: Sinusoidal wave motion

The results of calculating the maximum vertical acceleration derived from a sinusoidal wave form derived from sea conditions is 15.4 ft/sec^2, only 64% of the required 24.2 ft/sec^2 to unseat the pump from the suction valve (see Figure 6). Therefore, there is no potential to unseat the pump in the suction valve.

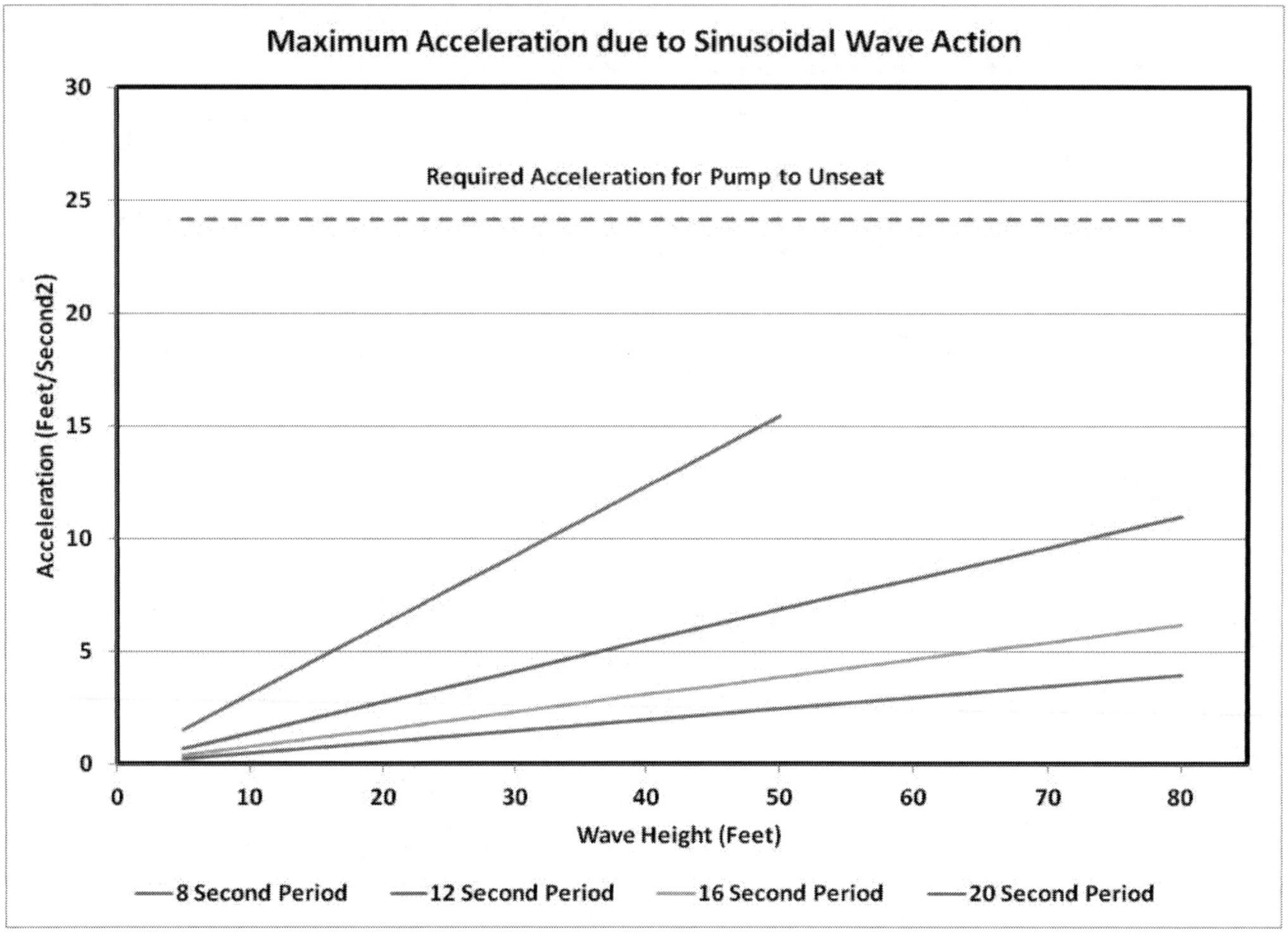

Figure 6: Sinusoidal wave maximum acceleration

Pump Tilting Potential

Pump tilting is based on ship roll movement. The pump is loose in a column and the pivot point is the suction valve seal. The center of gravity (CG) for typical in-tank vertical pumps is about 40% of the total pump length away from the seal plate pivot point (see Figure 7). Based on this CG location, when the ship rolls to some angle and combined with the angular momentum, the pump will tilt. The tilt stops when the guide pads on the side of the pump make contact with the column wall, approximately 0.25 inches. Although this is a small movement, it would cause the pump to suction valve seal to lose contact. Losing seal contact does not cause any immediate issues since the column is not sealed from the tank while the pump is installed. The seal would have to be damaged, however unlikely, to compromise pump performance to a noticeable degree. Even though this does not represent a high risk for premature pump failure, it is not a desirable condition.

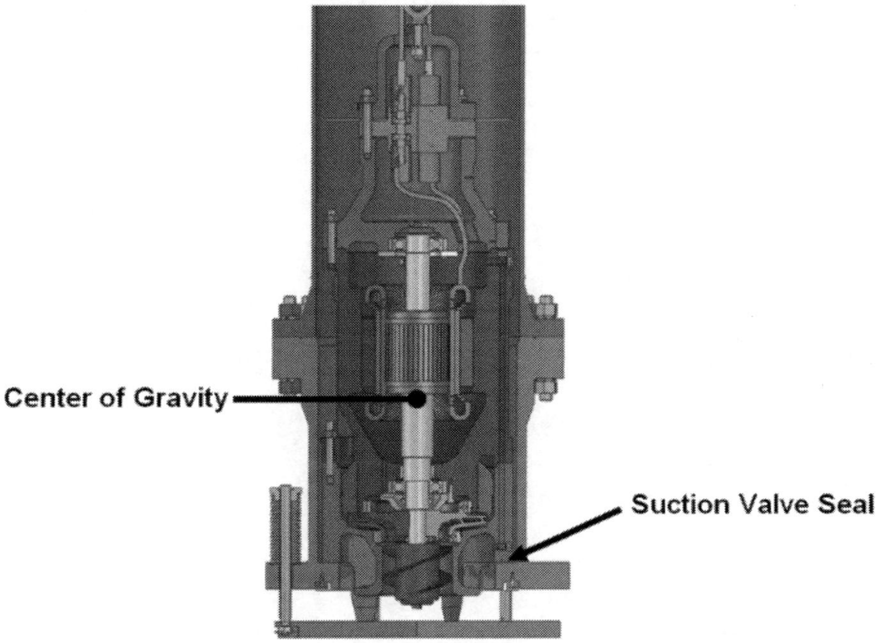

Figure 7: Typical Pump Center of Gravity in relationship to the suction valve seal

Main Bearing Damage Potential

Pump main bearing axial loading that could cause brinelling is based on ship heave motion. The vertical acceleration forces from the ship movement must be greater than one gravitational force to allow the rotating elements to rise above the main bearing. The rotating elements must then impact the bearing with enough force to cause brinelling. The typical main bearing is a deep groove ball bearing size 6320 constructed of 440C stainless steel. The axial force required to brinell the races of this 6320 bearing is 23,000 pounds. Based on this, a typical in-tank vertical centrifugal submerged motor cryogenic pump rotating element requires more than 12 G's to cause bearing brinelling. The maximum vertical force the pump would be exposed to (Figure 6) is less than 1 G. There is no possibility of brinelling the main bearing of an in-tank vertical centrifugal submerged motor cryogenic pump. Another bearing failure mode to consider is fretting wear false brinelling. This is caused by minute high frequency oscillations of the bearing balls displacing grease resulting in surface oxidation that creates indentations over time that look like brinelling. The pumps use stainless steel bearings, are exposed to low frequency vibration and typically are in an oxygen free environment eliminating this possibility.

Free Spin Potential

There is some thought that ship motion can potentially cause a pump to free spin. This spin has to occur around a vertical axis or yaw rotation motion, the only ship motion that rotates around a vertical axis. For a pump to see the full effects from ship yaw motion, it would have to be located close enough to the ship pivot axis to absorb most of the yaw motion. The yaw motion must produce acceleration large enough to overcome the calculated starting torque of the rotating elements that accounts for the ball bearing friction (see Equation 3). The friction coefficient for a deep groove ball bearing is 0.0010 to

0.0015. The typical yaw movement for large marine vessels is small with low acceleration. The yaw motion is considered to be a sine wave type motion.

Equation 3: $M = \mu Pd/2$

$M =$ friction moment (start torque)

$\mu =$ friction coefficient

$P =$ load on bearing (pounds)

$d =$ bearing bore diameter (inches)

Minimum Pump Rotation Start Torque: $M = .0010\ X\ 1244\ X \frac{3.937}{2} = 2.45$ in-lb

Equation 4: $T = I\alpha$

$T =$ torque on pump rotating element

$I =$ mass moment of inertia of pump rotating element

$\alpha =$ angular acceleration from ship yaw motion at the pivot point

Assuming the yaw rate correlates with the wave period as a worst case, the maximum angular acceleration is relatively low. To ensure the pump rotating element does not spin, $M \geq T$ must be true (see Figure 8). Based on this analysis, the pump rotating elements will not begin to rotate on the main bearing unless the pump is located at the yaw pivot point and the yaw motion exceeds 6 degrees in 8 seconds minimum, beyond any expected yaw motion. There is no free spin in a non operating pump.

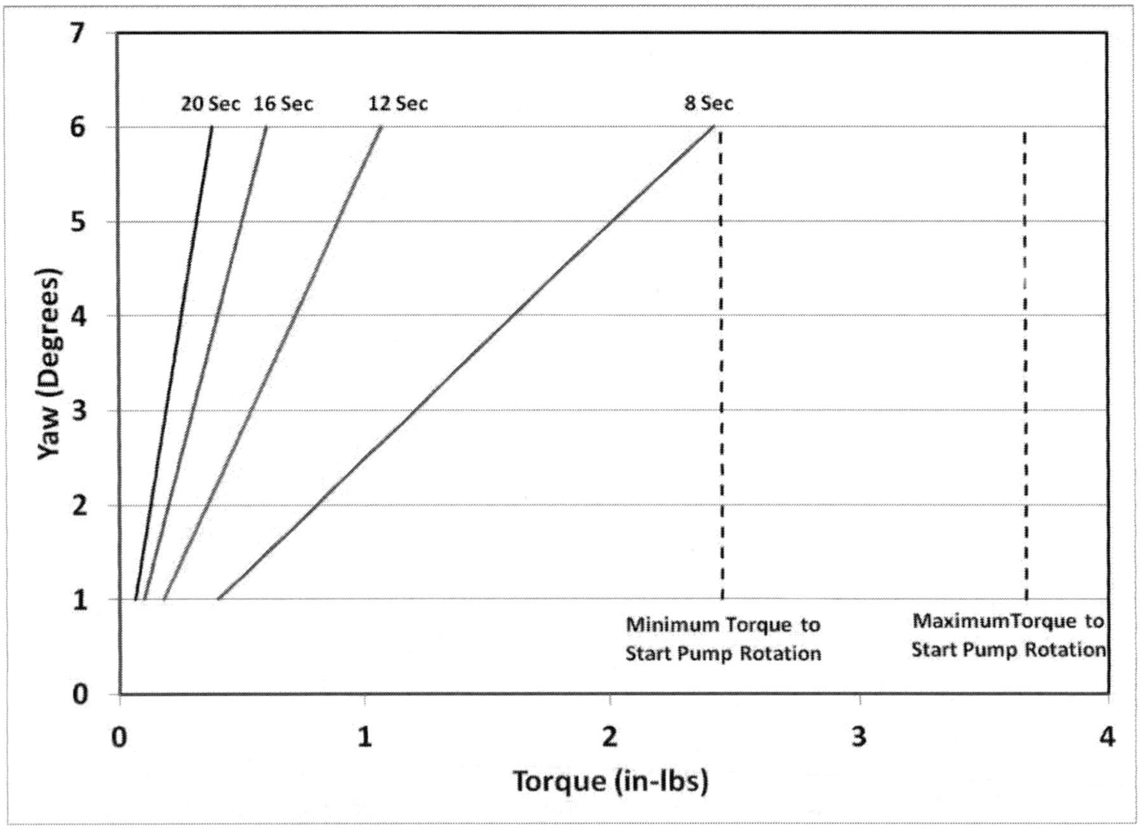

Figure 8: Correlation of torque from ship yaw motion to actual torque required to rotate pump based on 8 to 20 second wave period

Experience with Installed Marine Cargo Pumps

There are thousands of LNG marine cargo pumps in active service with some exceeding 30 years of active service. These pumps are used specifically for offloading LNG from cargo ships. Therefore, they do not operate during sea transit between ports. The pumps are continuously exposed to wave motion conditions during the transit times.

The marine cargo pump designs are essentially the same as a land based type pump (see Figure 9). They are rigidly mounted on a mast eliminating the suction valve which eliminates any potential pump unseating possibilities. There are no additional protection systems. There have been no reported bearing failures or degradations observed during periodic dry dock service. No cargo pump has failed to operate normally between dry dock services. There are thousands of marine cargo pumps in active service with no bearing failures or degradation. The service data and supporting calculations leads to the conclusion that marine vertical centrifugal submerged motor cryogenic pump bearings are not subject to damage during non-operation exposure to wave motion.

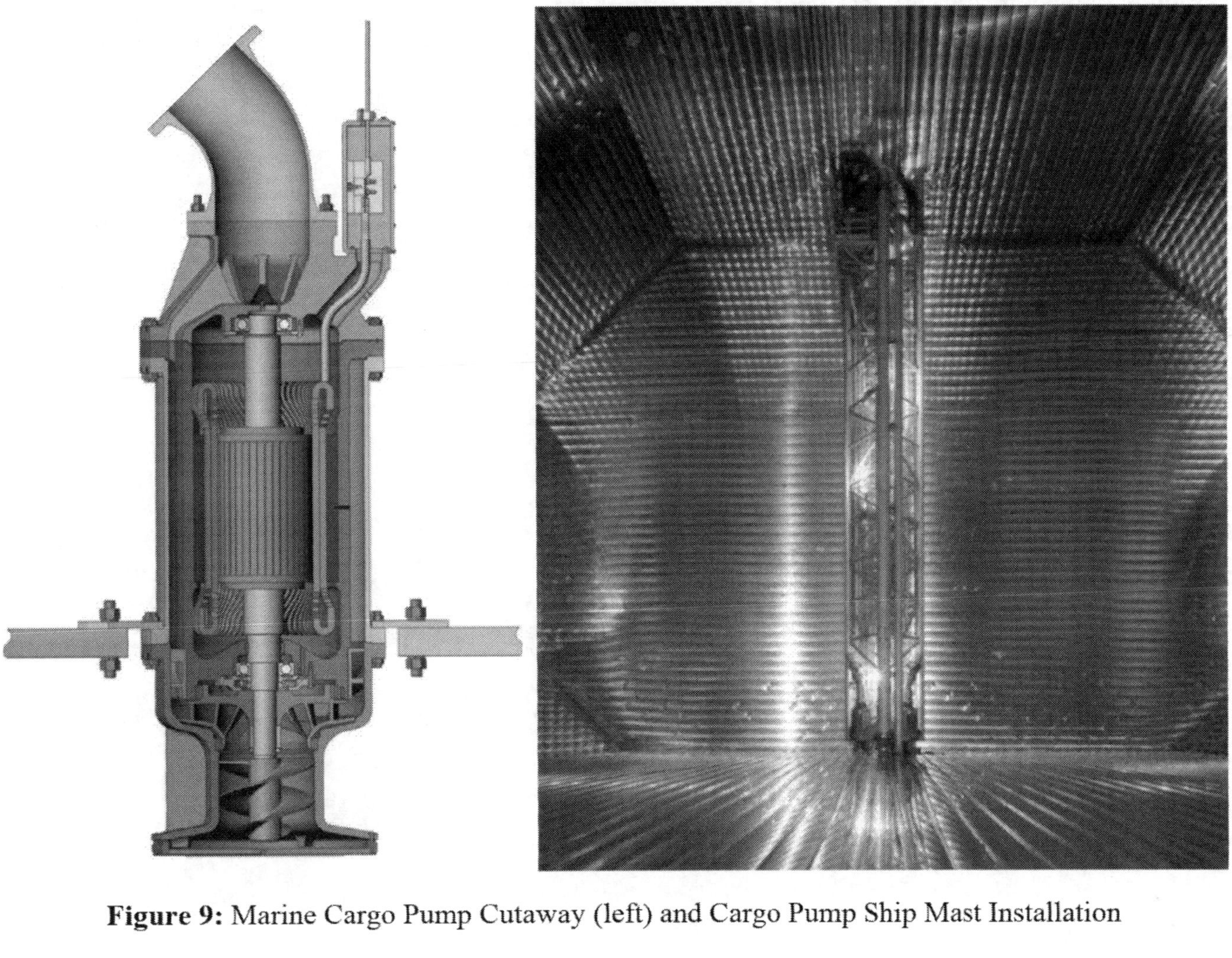

Figure 9: Marine Cargo Pump Cutaway (left) and Cargo Pump Ship Mast Installation

Pump Protection

Before any pump protection design is implemented, it would be prudent to determine if protection is necessary for the targeted failure mode. Ship motion analysis concludes that there is no pump unseating or bearing damage potential. A protection system can add weight, cost and complexity. If the decision is made to add protection to a pump from ship motion, several options may be considered.

Rigid Retraction System

A Rigid Retraction System is designed to eliminate any axial pump movement such as unseating or tilting. It is a retraction system constructed of rigid pipe sections instead of cables (see Figure 10). The rigid pipe sections transcend from the pump to the head plate. A constant down force is applied to the rigid pipe from springs located on the head plate eliminating any need for external control. In addition to a down force, the springs allow for coefficient of thermal expansion differences between the aluminum pump and 304 stainless steel piping and column. The coefficient of thermal expansion generates 0.008 inch per foot of pump length change at -160 degree C temperature (see Figure 11). The down force is recommended to be approximately equal to the pump weight. This will provide 1.75 G total down force, including the counter effect of the suction valve springs.

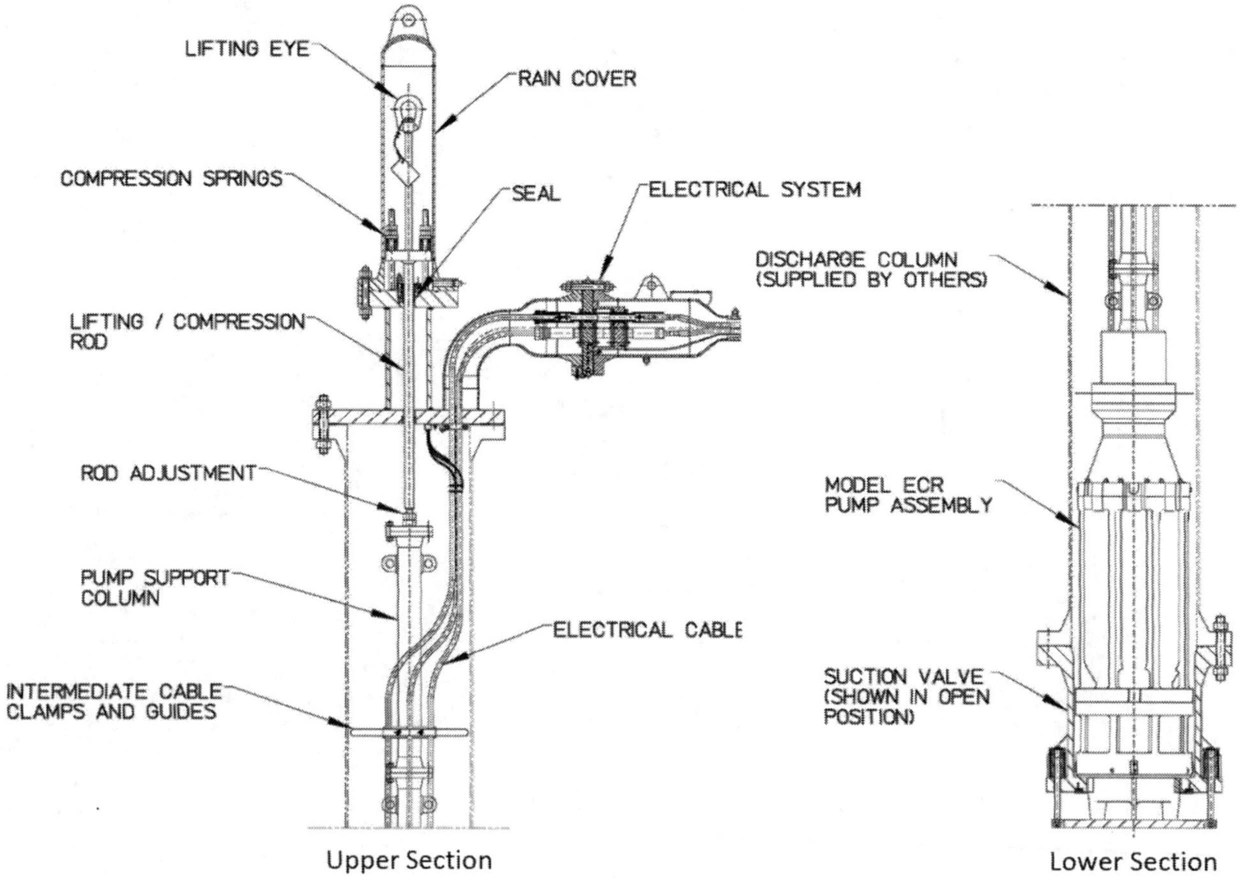

Figure 10: Rigid Retraction System using springs to provide down force

The spring type Rigid Retraction System provides continuous down force on the in-tank pump that ensures the pump will remain seated and in the upright position. However, the calculations show that wave motion will not cause an unseated condition. Another solution is to lower the pump Center of Gravity (CG) with respect to the suction valve eliminating the Rigid Retraction System while preventing pump tilting. This low CG system is presented in the next section.

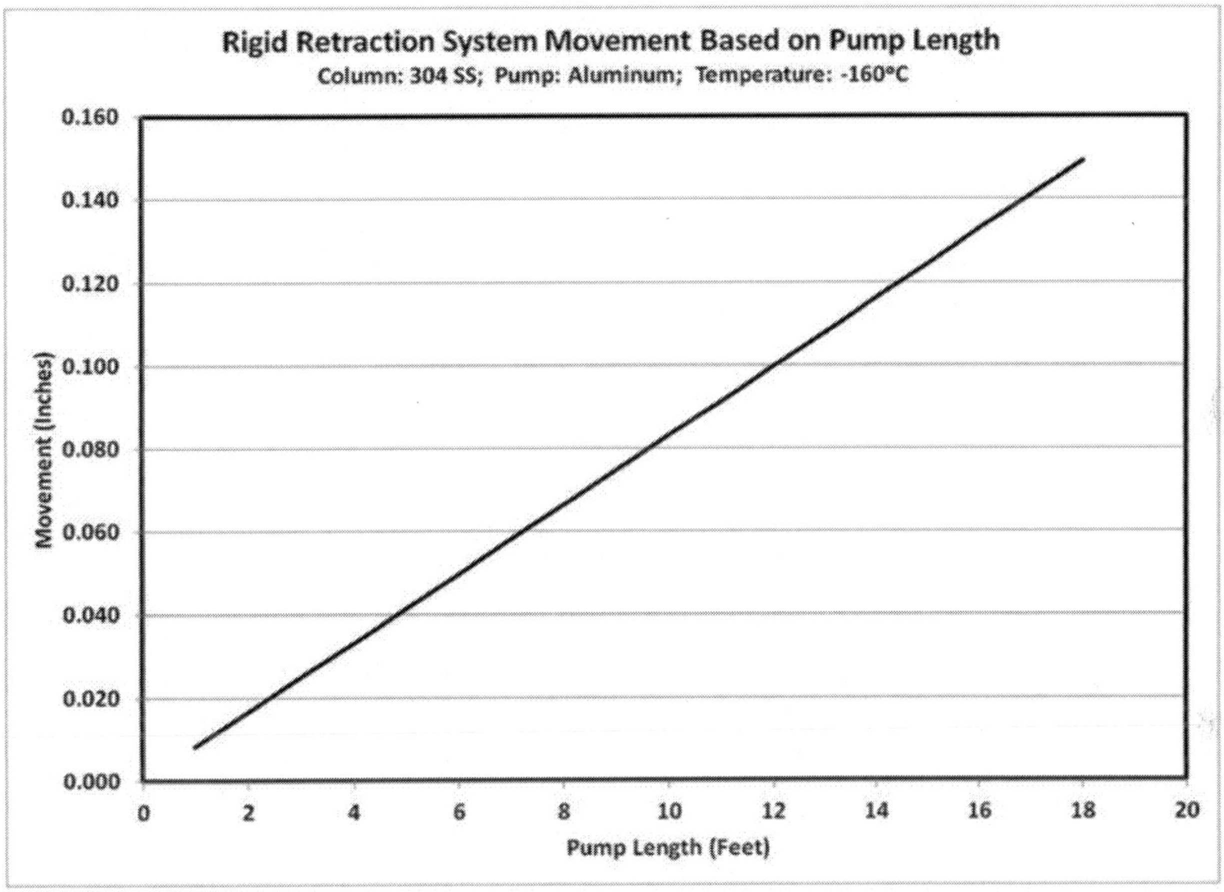

Figure 11: Rigid Retraction System differential length change at LNG temperature conditions

Low Center of Gravity Pump Seating

Since vertical pump unseating is eliminated as a possibility (see Pump Unseating Potential section), lowering the center of gravity (CG) relative to the suction valve sealing plate can now restrain the pump from tilting as a result of ship pitch and roll. The CG is lowered by lengthening the suction valve to reduce the pump sealing plate to pump CG distance (see Figure 12). The low CG pump seating effectively combines the pump weight and friction factors to ensure the pump remains seated. The suction valve seal is PTFE with a low coefficient of friction. If the ship were to tilt enough, the pump will slide until the bumpers make contact with the column and the pump makes contact with the suction valve. The column is 304 stainless steel and the pump, suction valve and bumpers are aluminum. The coefficient of friction between 304 stainless steel and aluminum is 0.45 and aluminum to aluminum is greater than 1.0. The pump weight, suction valve spring force, roll angle, and coefficient of friction are used to calculate the unseating. Since the seal is constructed of PTFE with a friction coefficient of 0.2, the pump will slide on the seal plate before it will tilt starting at 12 degrees. Once the pump bumpers make contact with the column, the friction forces between the column and bumpers keep the pump from unseating through 90 degrees of roll. The forces used for the calculations include pump weight, friction, and suction valve spring. The pump holding force is calculated based on the roll angle and the resultant friction forces, weight on springs, and spring force. The friction forces between the column and bumpers are enough to keep the pump from unseating all the way to a 90 degree angle (see Figure 13). The low center of gravity pump seating system ensures positive pump seating during ship motion eliminating the need for a more complex rigid retraction type system.

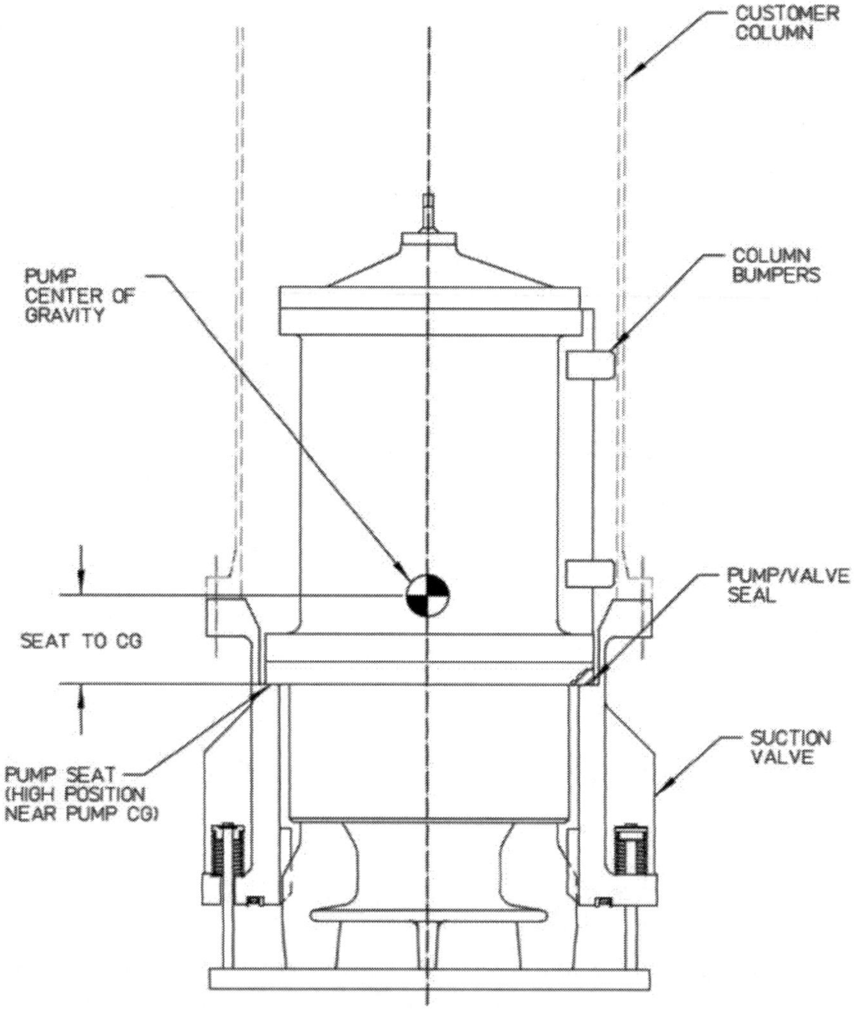

Figure 12: Low Center of Gravity System

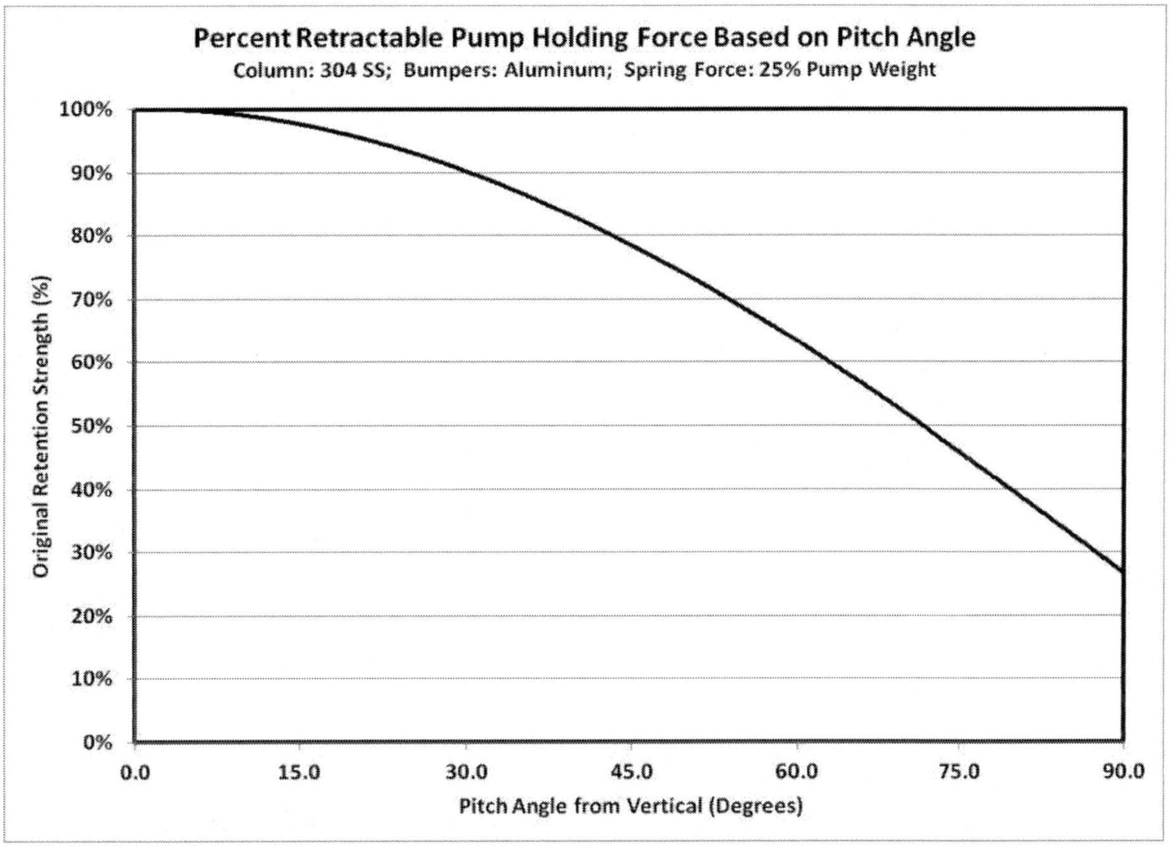

Figure 13: Retractable pump holding force compared to roll angle

Shaft Brake System

There is some thought that ship yaw and heave motions can potentially cause the rotating elements to spin, and load the main bearing during non operating conditions. A brake system can prevent shaft rotation and axial bearing loading. There are two types of shaft brake systems. One system presses a pad against the shaft to keep it from rotating (see Figure 14). The other lifts the shaft off the main bearing and presses against the shaft to keep the shaft from rotating and moving vertically (see Figure 15). Each system uses nitrogen actuated pistons which are actuated after the pump comes to a complete stop.

There is solid information that supports the idea that a brake system is not required to protect pumps against ship motion during non operating conditions. A typical in-tank retractable and rigid mounted cargo pump uses a 6320 type ball bearing that requires 23,000 pound axial load to cause brinelling. This would require more than 12 G's to cause brinelling in a typical pump. There are several thousand rigid mounted marine cargo pumps in service, some with over 30 years marine service. These pumps are non-operational during ocean transit conditions in both loaded and unloaded LNG cargo conditions. There have been no bearing failures or degradation.

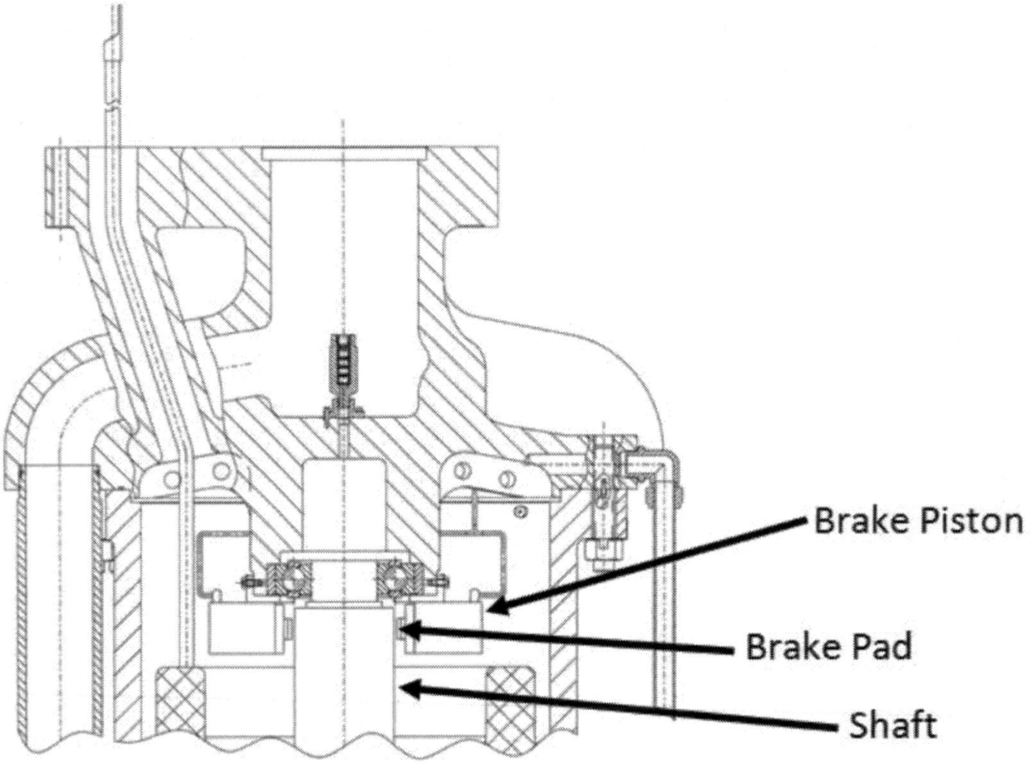

Figure 14: Shaft brake system that protects against shaft rotation movement

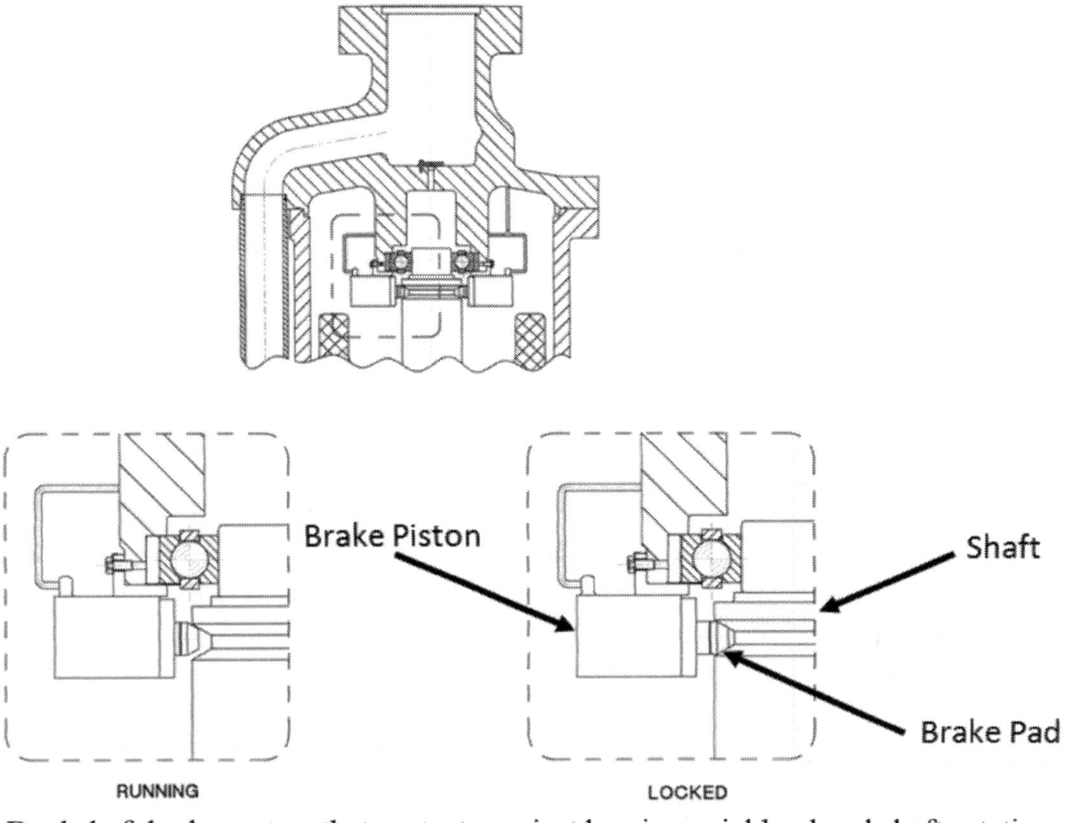

Figure 15: Dual shaft brake system that protects against bearing axial load and shaft rotation movement

Conclusion

LNG FSRU and FPSO vessels use in-tank mounted vertical centrifugal submerged motor cryogenic pumps that are exposed to three degrees of ship movement as a result of wave motion. The only recommended change to pumps used on LNG FSRU and FPSO vessels to provide protection against ship motion is a low CG suction valve. Ship motion does not cause free spinning and bearing damage during pump non-operation, which is confirmed by calculations and thousands of marine cargo pumps in actual service. Therefore, the pump life expectancy is not impacted. Adding additional equipment such as a rigid retraction and brake system to protect the in-tank pumps from free spinning and axial load bearing damage only adds complexity, cost, and weight. Ship roll motion may have the potential to cause a pump to tilt on the seal plate, which could potentially damage the seal. The least complex and most economical solution is to use a low CG suction valve design, moving the CG close to the pump/suction valve seal eliminating the possibility of pump tilting. The low CG suction valve design allows the ship to roll 90 degrees while maintaining the pump seated condition.

Study on determination of optimal operating conditions of single mixed refrigerant process for LNG FPSO

Yeonmi Suh, Taeyeong Lee, Gyeongho Han, Kiil Nam, and T.J. Kim, Hyundai Heavy Industries, Ulsan, Korea

Introduction

The demand and consumption of natural gas alike have been increasing all over the world. It is estimated that the global needs of natural gas will grow up to 345 million tons between 2010 and 2020, accordingly an average of 4.7% increase every year. Because 20% of the world gas reserves are in the offshore fields, LNG FPSO (Liquefied Natural Gas-Floating, Production, Storage and Offloading unit) has become to be economical for exploring natural gas well in this area and has been recently spotlighted.

The LNG FPSO is largely composed of two parts, i.e. hull and topside. While the LNG FPSO is moored steadily above the well, the unprocessed natural gas is pumped from well to the LNG FPSO's topside via riser. When the gas reaches the topside, it is processed to separate natural gas, LPG and natural gas condensate. The processed feed gas is treated to remove impurities and liquefied through the liquefaction process prior to being stored in the hull. When the liquefaction process turns the pretreated natural gas into LNG, the volume of natural gas shrinks to six hundredth. It is known that the liquefaction process accounts for 30~40% of overall capital cost as shown in Figure 1 [1]. Therefore, it should be regarded as top priority among all topside systems on the LNG FPSO.

The LNG liquefaction processes can be categorized according to the number of cycles. The number of cycles means the number of working fluids when the natural gas is pre-cooled, liquefied and sub-cooled. If the number of cycles increases, the LNG production efficiency will be enhanced accordingly. But, as the process gets more complex, consequently the capital cost increases. The single mixed refrigerant (SMR) process uses only one working fluid for pre-cooling, liquefaction and sub-cooling of the natural gas and is known for having the advantage that the system is relatively simple. In case of the large scale LNG FPSO and onshore plant, the DMR (dual mixed refrigerant) process, which uses two working fluids, can be applied because the space constraint is not a big problem. But in case of the small and medium scale LNG FPSO, the compact and light LNG liquefaction system is essentially required due to the limited space. For these reasons, the SMR process is generally suggested to an alternative of the LNG liquefaction process that catches up with the modern trend of LNG FPSO.

To build more economical liquefaction process, it is vital to optimize the liquefaction system. So, there have been a lot of research studies about the optimization of LNG liquefaction processes. Lee et al. [2] proposed a novel method for the selection of refrigerant composition using the nonlinear programming (NLP) techniques. Jensen [3] modeled and optimized the PRICO LNG from the perspective of process operation and control. Chang et al. [4], Wang et al. [5, 6] and Lee et al. [7] performed the optimization based on in-depth thermodynamic analysis. Nogal et al. [8], Alabdulkarem et al. [9] and Hwang et al. [10] used the stochastic optimization (genetic algorithm, GA) to overcome the local optima. Most of studies above dealt with onshore processes which were relatively large scale but didn't correspond to the offshore process.

In this study, the method was proposed to determine the optimal operating conditions of the SMR process on the LNG FPSO. The complicated configuration of real application process sufficiently reflected on this model by coupling the rigorous process simulation program (Aspen HYSYS®) with GA in MATLAB.

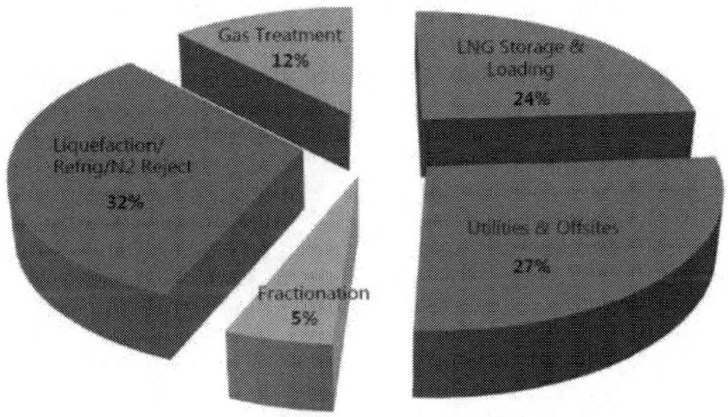

Figure 1 Ratio of capital cost for LNG FPSO

Optimization of SMR Process

Description of studied SMR Process

The SMR process consists of two compressors, two coolers, two phase separators, two mixers, three valves and three heat exchangers as shown in Figure 2. The mixed refrigerant is separated into two streams by means of molecular weight at phase separator 1. After entering into heat exchanger 1, respectively, the relatively heavy refrigerant is used for pre-cooling the natural gas, but the lightweight one is cooled by the heavier refrigerant. And then the cooled lighter refrigerant is separated into two fractions again at phase separator 2. After that, the heavier one is used for liquefaction in heat exchanger 2 and the lighter one for sub-cooling of the natural gas in heat exchanger 3. The mixed refrigerant is composed of methane, ethane, propane, n-butane, i-butane, n-pentane, i-pentane and nitrogen.

Modeling Conditions of SMR Process

The centrifugal type compressors are used for the SMR process. It is assumed that the compressors operate at 75% of adiabatic efficiency. If the inlet stream of the compressor is liquid phase, the compressor will be damaged. Therefore the inlet temperature of the compressor must be higher than the saturated temperature of the inlet stream. In cooler 1 and 2 located after compressors, seawater is used as cooling medium and the pressure drop effect is not considered. The temperature of the refrigerant has to be above seawater temperature of 25°C through the seawater cooler. The refrigerant is separated into vapor and liquid in the phase separator after compression. The pressure drop and the heat loss at each phase separator can be neglected because they are so small. Thus, the pressure and temperature of the outlet stream remain the same as those of the inlet stream in the phase separator. In heat exchangers 1, 2 and 3, the refrigerant exchanges the thermal energy with the natural gas. The pressure drop through the heat exchanger is ignored for simplicity. It is assumed that the outlet temperatures of the hot streams in the heat exchangers have the same value with the effectiveness of 100% for the heat exchangers [11]. All expansion valves are operated at an isenthalpic condition. To prevent back flow effects in the mixer, all inlet streams have to be the same pressure values.

With the above-mentioned conditions, the rigorous process simulation for the SMR process is implemented by Aspen HYSYS®, a commercial process simulation program. In this simulation, the Peng-Robinson equation of state is selected for modeling the property of substances. As the result of the simulation, the sixteen independent variables of the SMR process are defined and listed in Table 1.

Table 1 List of independent variables for SMR process

Variables	Baseline values
Heat exchanger 1 outlet temperature of natural gas (℃), T_{23}	-35.22
Heat exchanger 2 outlet temperature of natural gas (℃), T_{24}	-109.3
Cooler 1 outlet temperature (℃), T_2	27.1
Cooler 2 outlet temperature (℃), T_8	29.6
Compressor 1 outlet pressure (bar), P_1	26.8
Compressor 2 outlet pressure (bar), P_7	48
Valve 1 outlet pressure (bar), P_5	6.5
Mole fraction of mixed refrigerant composition (N2/C1/C2/C3/nC4/iC4/nC5/iC5)	0.0671/0.3115/0.3612/0.0619/0.1176/ 0.0580/0.0056/0.0170
Molar flow of mixed refrigerant (kgmole/s), m	14.05

Objective Function

In general, the required total power of the LNG liquefaction is summation of the required compressor power and the produced expander power. Because there is no expander in the SMR process, the objective function of this model is to minimize the total compressor shaft power required to liquefy natural gas.

Optimization Framework

Most of previous studies formulated their own mathematical equations for the thermodynamic properties, mass and heat balances [8, 10] to simulate and optimize their liquefaction processes. These equations are so complicated that it takes a lot of efforts to construct these equations and solve them. There are some assumptions to simplify the problems in the previous studies. It is difficult to utilize these approaches in the industrial company which do not have enough time and resources to construct such complicated models. To avoid this problem, we use Aspen HYSYS® which is a widely used flowsheet program. By using Aspen HYSYS®, we can easily construct the rigorous process models considering complicated mass and energy balances on all of the process units. Aspen HYSYS® has its own optimization program, HYSYS Optimizer. However, it is not powerful enough to solve the optimization problems of the SMR process which include the various compositions of streams as independent variables. To solve these problems, the SMR process is simulated with Aspen HYSYS® and then optimized with GA code in MATLAB Global Optimization Toolbox. In addition, a MATLAB code is created to make the objective function and to connect the process flowsheet in Aspen HYSYS® and GA code in MATLAB Global Optimization Toolbox. The optimization framework used to solve the optimization problem of the SMR process is shown in Figure 3.

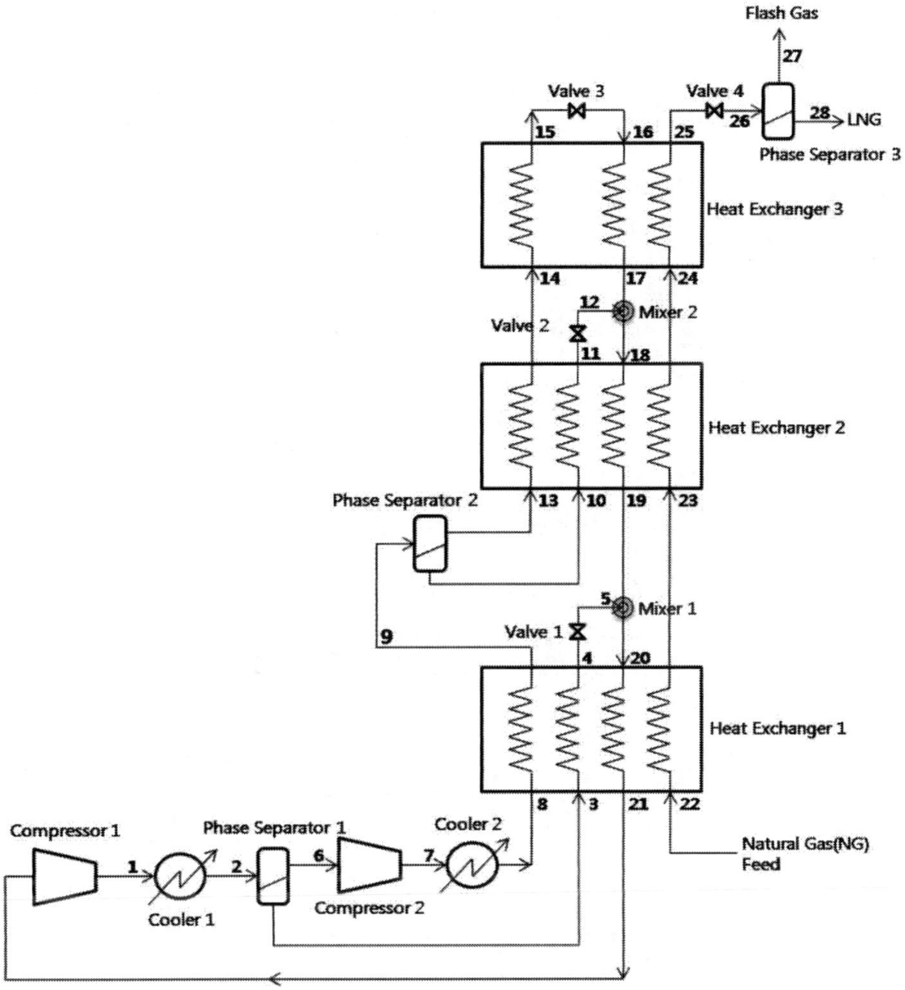

Figure 2 Schematic flow diagram of SMR process

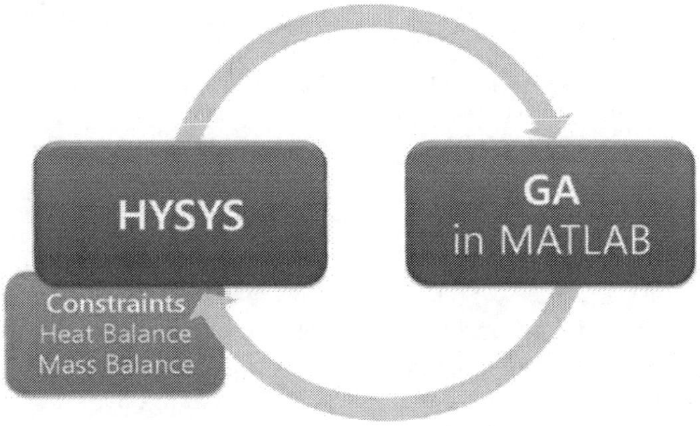

Figure 3 Optimization framework

Results and Discussions

As displayed in Figure 4, the SMR process is simulated by using Aspen HYSYS®. The composition of feed gas is listed in Table 2. The design specification of the SMR process is based on an engineering research project data and assumptions mentioned previously. The simulation result is only regarded as the baseline case. The baseline case is optimized by the proposed optimization framework.

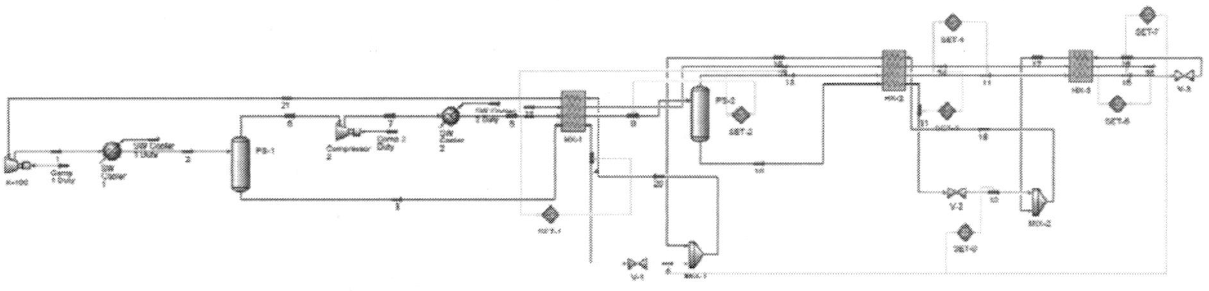

Figure 4 SMR process used in Aspen HYSYS®

Table 2 Typical composition of feed gas

Component	Mole fraction (%)
Nitrogen	1.5679
Methane	84.9954
Ethane	7.1809
Propane	3.5942
n-Butane	0.8727
i-Butane	0.7752
n-Pentane	0.2884
i-Pentane	0.3861
Hexane	0.3392

The model of the SMR process has sixteen independent variables as listed in Table 1. The objective function is to minimize the total power consumption of the SMR process, which is the sum of power consumption of all compressors. The optimal values of independent variables and the objective function values of each case are listed in Table 3. In baseline case of Table 3, the data are the simulation results to which no optimization tool is applied. The optimization results using HYSYS Optimizer program are shown as the HYSYS Optimizer case. The GA case means the results from optimizing the SMR process by utilizing GA in MATLAB Global Optimization. The GA tuning parameters used in the optimization are listed in Table 4. As shown in Table 3, the GA case has less molar flow of mixed refrigerant and lower overall compression consumption rate than any other cases. The mole fraction of methane, propane and i-pentane increase while the mole fraction of ethane and isobutene decrease by the optimization of the mixed refrigerant's composition. Thus the power consumption of the GA case is 15.9% less than that of the baseline case and is 6.52% less than that of the HYSYS Optimizer case.

The temperature profiles of the streams in the heat exchangers of the SMR process are shown in Figure 5. The efficiency of the heat exchanger can be maximized if the temperature of the hot stream (e.g. feed gas) approaches that of the cold stream (e.g. mixed refrigerant) along its entire length because the heat exchanger approaches to reversible operating condition under such circumstances [12]. The increased efficiency of the heat exchangers reduces the power consumption of the SMR process. The heat exchangers 1 and 2 in the GA case show small gaps between the temperature profile of the hot and cold components than any other case, while the heat exchanger 3 indicates larger gap. On the whole, the heat exchangers in the GA case show better efficiency than any other cases.

Table 3 Results of optimized SMR process

Case	Variables								
	T23	T24	T2	T8	P1	P7	P16	m	
Baseline	-35.22	-109.30	27.10	29.60	26.80	48.00	6.50	14.05	
HYSYS Optimizer	-10.00	-106.60	29.63	26.94	23.50	39.57	6.89	14.05	
GA	-25.06	-93.65	25.00	25.01	18.09	39.79	6.46	13.00	
Case	Variables								Objective function
	N2	C1	C2	C3	nC4	iC4	nC5	iC5	Power consumption (MW)
Baseline	0.0671	0.3115	0.3612	0.0619	0.1176	0.0580	0.0056	0.0170	82.27
HYSYS Optimizer	0.0671	0.3115	0.3612	0.0619	0.1176	0.0580	0.0056	0.0170	74.03 (Reduced by 10.02%)
GA	0.0714	0.3298	0.3148	0.0936	0.1191	0.0243	0.0045	0.0425	69.20 (Reduced by 15.88%)

Table 4 Typical GA tuning parameters

Tuning parameters	Values
Population size	100
Reproduction count	2
Maximum number of generations	500
Crossover fraction	0.8
Selection method	Stochastic uniform
Fitness scaling method	Rank
Mutation method	Adaptive feasible

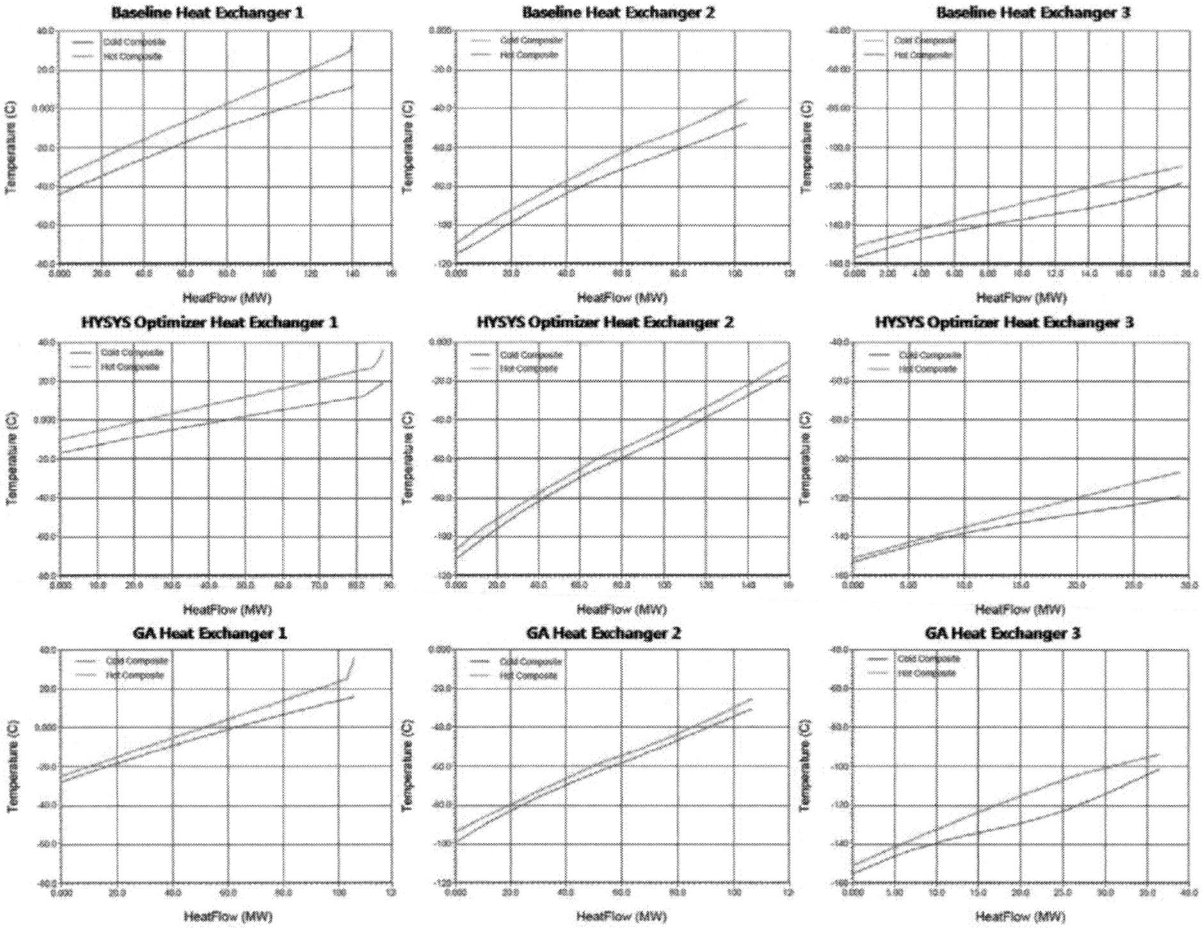

Figure 5 Temperature profiles of baseline, HYSYS Optimizer and GA cases

Usually, the heat exchanger of LNG liquefaction process is designed for minimum approach temperature to be within 1~3 °C [11]. To examine the effect of the minimum approach temperature, the additional cases, GA 1, 2 and 3, are optimized with each minimum approach temperature of 1, 2 and 3°C. As shown in Figure 6, the power consumption of the process lessens with the decrease in the minimum approach temperature. The saving in the power consumption from the heat exchangers with 1°C minimum approach temperature is 8.83% lower than that of the heat exchangers with 3°C minimum approach temperature.

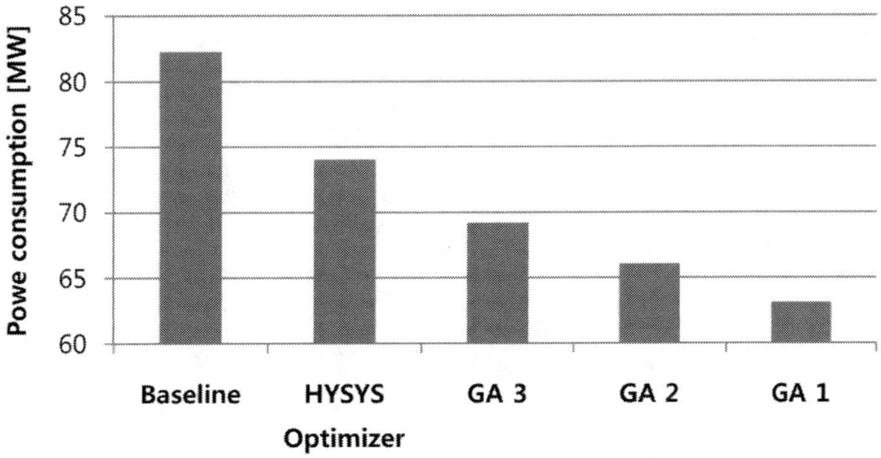

Figure 6 Comparison of power consumptions of baseline and optimization results

Summary

In this study, the SMR process installed on the LNG FPSO was optimized by connection Aspen HYSYS® with GA in MATLAB Global Optimization Toolbox. The result shows the power consumption of optimized process can be reduced by 15.9% when compared with the baseline case. It is derived that there is room for lessening the power consumption additionally by adjusting minimum approach temperature. In the future, the existing SMR process will be consistently developed and optimized by considering more realistic constraints.

References

1. Shukri, T. (2004). LNG technology selection. Hydrocarbon Engineering, 9(2).
2. Lee, G. C.; Smith, R.; Zhu, X. X. (2002), "Optimal synthesis of mixed refrigerant systems for low-temperature processes," *Industrial and Engineering Chemistry Research*, 41, pp. 5016–5028.
3. Jensen, J. B. (2008), "Optimal operation of refrigeration cycles," Ph. D. Thesis, Norweglan University of Science and Technology.
4. Chang, H. M., Chung, M. J., Kim, M. J., & Park, S. B. (2009), "Thermodynamic design of methane liquefaction system based on reversed-Brayton cycle," *Cryogenics*, 49(6), pp. 226–234.
5. Meiqian Wang, Jian Zhang, Qiang Wu, Kuyen Li. (2011), "Thermodynamic-Analysis-Based Energy Consumption Minimization for Natural Gas Liquefaction," *Industrial and Engineering Chemistry Research*, 50, pp. 12630-12640.
6. Meiqian Wang, Jian Zhang, Qiang Wu. (2012), "Optimal design and operation of a C3MR refrigeration system for natural gas liquefaction," *Computer and Chemical Engineering*, 39, pp. 84-95.
7. Sanggyu Lee, Nguyen Van Duc Long, Moonyong Lee. (2012) "Design and Optimization of Natural Gas Liquefaction and Recovery Processes for Offshore Floating Liquefied Natural Gas Plants," *Industrial and Engineering Chemistry Research*, 51, pp. 10021-10030.

8. Nogal, F. D., Kim, J. K., Perry, S., & Smith, R. (2008), "Optimal design of mixed refrigerant cycles," *Industrial and Engineering Chemistry Research*, 47, pp. 8724–8740.
9. Alabdulkarem, A., Mortazavi, A., Hwang, Y., Radermacher, R., & Rogers, P. (2011), "Optimization of propane pre-cooled mixed refrigerant LNG plant," *Applied Thermal Engineering*, 31, pp. 1091–1098.
10. Ji-Hyun Hwang, Myung-Il Roh, Kyu-Yeul Lee. (2013) "Determination of the optimal operating conditions of the dual mixed refrigerant cycle for the LNG FPSO topside liquefaction process," *Computer and Chemical Engineering*, 49, pp. 25-36.
11. Douglas, J. M. (1988). Conceptual design of chemical processes. New York, NY, USA: McGraw-Hill.
12. Nexant, Inc. (2007) "LNG: The Expanding Horizons of Liquefaction Technology and Project Execution Strategies."

Constructability Issues of FLNG for Design and Project

Moon, Herbert
Technical Director, Project Development & Technology

Kim, Ingu
Engineering Manager, Project Development & Technology

DAEWOO Shipbuilding and Marine Engineering LTD, Korea

AIChE Spring Meeting, April 2013
13th Topical Conference on Natural Gas Utilization
San Antonio, April 28 – May 2, 2013

Introduction

Constructability issues shall be reviewed and incorporated in the Project Development Plan and Design of the FLNG. This paper addresses important features on the FLNG Constructability, and recommends significant aspects from the Pre FEED to Detail Engineering Stages.

FLNG Projects Status and Brief Comparisons between FLNG and Oil FPSO.

Several FLNG projects are under the investigation and study, and some of them are in construction phases. Overall view of the Current FLNG Projects and comparison between the FLNG and Oil FPSO is presented.

Overall review of the construction schedule and man-hour will be done, and several recommendations will be also described.

DSME's Yard Capacity Overview

The brief summary of DSME's Okpo Construction Facilities is presented, and brief plans for construction and commissioning will be presented.

Also, new facility for FLNG construction, test and commission will be explained.

Trends of Hull Dimensions, Weights of FLNG, and Limitations

The hull of FLNG is bigger and heavier than that of Oil FPSO, and the limitations for Construction will be presented. Also, recommendations will be also identified.

Trends of Topside Module Dimension and Weights and Topside Module Integration Methodology

The Topside Module Dimensions and Weights are much larger and heavier. It is recommended that new facility will be mobilized for efficient topside module integration.

Summary of Major Utility Consumptions and its Impacts on the Hull Main Deck Layouts and Arrangements

Major utility consumptions are reviewed and the aspects of the piping, electrical layouts and arrangements are discussed and presented. Comparisons between FLNG and Oil FPSOs will be made, and several recommendations will be also proposed.

Identification of Long Lead Items and Design and Construction Issues

The delivery durations of major equipments in FLNG are expected to be more than one year, and may be more than two years for some equipment. It is required that most of the equipments shall be delivered minimum 6 months prior to module fabrication.

However, if the equipment delivery is later than construction plans, several measures and mitigations will be presented.

Construction Issues of Different Cargo Containment Systems for FLNG

For FLNG, membrane tanks and SPB are normally applied. For each Cargo Containment System, the features considering the construction plans will be presented

Review of Electrical Installations and Recommendations

Pre-Commissioning, Commissioning Methodology and Recommendations

Two major methodologies for commissioning works will be presented and compared. And, differences in the conceptual and basic design designs will be presented.

Also, several recommendations will be proposed to perform the commission works efficiently.

Methodology of Gas Trial and Near Shore Commissioning

Unlike the LNGCs, the methodology of Gas Trial and near shore commissioning for FLNGs are not finalized yet. Several restrictions on these issues will be presented, and several future plans for maximizing the tests and near commissioning activities will be presented.

Conclusions

Constructability is one of the key considerations for FLNG Project. From the conceptual stage, it must be reviewed and implemented in the design and development plan.

References Cited

Not Applicable

Biography of Speaker

Herbert Moon is a Technical Director and in charge of Technical Proposals for FLNG Projects in DSME. He has been working with DSME for over 26 years and has been involved in many offshore projects. He was graduated from Seoul National University and received BS and MS in Naval Architect.

(57a) Challenges in Syngas Preparation for Different Applications

Vk Arora, Kinetics Process Improvements, Houston, TX

The syngas composition & purity requirements differ significantly for different applications. Depending on the end application of the syngas, the required configurations could be from a simple scheme to a very complex and capital intensive one. As the size of the syngas derivative units are getting much larger in scale- there is all the more the need to minimize the complexity & the cost of producing syngas- which is major contributor in the overall economics.

This presentation will examine the challenges in design and relative economics of various syngas configurations for different applications including world scale Methanol, Ammonia, Oxo-syngas, CO & GTL plants. It will also include key measures in energy savings including co-production of Methanol & Ammonia & their benefits.

(57b) Topsoe Tigas Technology for Conversion of Abundant Low Cost Resources Into High Value Synthetic Fuels

Livia Sierra Llorens, Haldor Topsoe, Houston, TX

In times with increasing oil prices it becomes very attractive to consider alternative sources than crude oil for transportation fuel. The Topsoe Improved Gasoline (TIGAS^{TM}) technology is designed to produce high quality and valuable gasoline from natural gas, shale gas, coal, petcoke or biomass.

The process converts natural gas to gasoline, and the gasoline product properties are such that it can readily be used as a drop-in fuel or used as a perfect blend stock in the gasoline pool. Plant capacities can range from as small as 1,500 bbl/d to above 30,000 bbl/d and can easily be matched to the available gas resources.

A 23 bbl/d TIGAS^{TM} demonstration plant, DOE-funded project, will start-up early 2013. Data and results from the initial operation will be presented.

(57c) Value Additiion by Integrating C3 Derivatives with PDH Units

Vk Arora, Kinetics Process Improvements, Houston, TX

With rapid development of Ethane based Ethylene production in Middle East and switch to lighter feedstock in US crackers have been mainly responsible for shortfall in Propylene and C4s. To fill this gap in Propylene supply, a number of new on-purpose Propylene production units (PDH) are being built in US, China & other parts of the world. With so many PDH units planned in US, an excess of propylene is already being speculated while most of the Propylene is used up in polypropylene units. The projected scenario of excess propylene will impact the long term profitability of propylene producers and will likely promote and justify propane export – which will not be in the best interest of US producers using propane feedstock. To sustain the long term profitability of on-purpose propylene producers, downstream integration with the relevant propylene derivatives is called for. Based on the increasing global market demand of those derivatives with better returns- such integration units are already being planned.

This presentation will examine the relative economics of various propylene derivative combinations- including Acrylic Acid & its derivatives as well as Oxo-Alcohols/aldehydes & its derivatives. Other propylene derivatives (like Acrylonitrile, PO, Cumene/Phenol will be briefly included but not for the economic comparison). A brief relevance of the market will also be included in the presentation.

(57d) Opportunities for Modular Gtl in North America

Jeff McDaniel, Velocys, Inc., Plain City, OH

Modular Gas to Liquids (GTL) plants can bring stranded, associated or undervalued gas to market and reduce the flaring of associated gas. Of particular interest is the opportunity for using small scale distributed GTL plants located near the feedstock resource to monetize shale gas fields in North America. Velocys patented technology for the Fischer-Tropsch process, the central step of GTL, has been commercially demonstrated in the field. The unique combination of microchannel reactors and super-active catalysts provides superior economics, especially in the range 1,000-15,000 barrels per day, or 10-150 mmscfd of gas. The modular process produces clean-burning, premium diesel or jet fuel: valuable products which are easy to transport and serve a global market.

The presentation will include the latest results from an integrated GTL pilot plant that includes a steam methane reformer and microchannel FT reactor. The pilot plant is designed to provide operational data to support a variety of process configurations and conditions that may be found in commercial GTL facilities.

(57e) Compact Heat Exchange Reactors in Pilot Plants for Fischer-Tropsch Synthesis and Mixed Alcohol Production

Hani Gadalla, Steven J. Vallee and Zhijun Jia, Chart Energy & Chemicals, La Crosse, WI

Chart's Compact Heat Exchange Reactor (CHER) is a brazed metal reactor with integrated heat exchange capabilities. Because of its high metal surface area and short distance for the heat to travel, the reactor can control highly exothermic and endothermic reactions that are otherwise hard to control with conventional reactors. CHER can be constructed out of a variety of metals, including aluminum and stainless steel. The choice of material is based on the operational conditions that the reactor will experience. Recent data from two pilot plants will be presented to demonstrate the effectiveness of aluminum and stainless steel CHERs to control the reactions.

The first pilot plant is a 1bpd Waste-to-Energy plant where the CHER is used for the highly exothermic Fischer-Tropsch (FT) reaction. The reactor is constructed out of an aluminum core that is housed in a steel vessel (based on Chart's Core-in-Kettle patented design). The second pilot plant uses a stainless steel CHER for the production of mixed alcohols from syngas. The alcohol production reaction is also highly exothermic and is operated at higher temperatures and pressures than FT.

A comparison between aluminum and stainless steel reactors will be reviewed. Data for the reactions will be presented and the performance of the reactors will be analyzed. Scalability to commercial plants will also be discussed.

Performance Enhancement and Profitability Improvements of Refinery Operations using 3-D Flow Modeling Techniques

Praveen Kumar, Amarvir Chilka and Damo Vedapuri, Ph.D.
Tridiagonal Solutions Inc.

Abstract

Efficient operations of refinery processes rely on optimal flow distribution, fluid mixing and phase separation. Poor mixing or separation result in process bottlenecks, reduced efficiency and frequent shut downs. Optimizing flow can de-bottleneck a process, lead to an enhancement of the process performance and reduce the frequency of shut downs. This leads to improved profitability of the refinery operations. Advancements in simulation technology have enabled reliable computational models of refinery process units to be built. In the current presentation, the use of computational fluid dynamics (CFD) techniques to improve refinery operations is discussed with the aid of three case studies:

- Analysis & improvement of crude blending operations
- Design improvements to vapor horns for efficient removal of entrained liquids in a vapor fractionator
- Retrofit of a flare knock out drum to increase capacity

The case studies illustrate the opportunity to improve performance and avoid the need to purchase new equipment through the use of flow modeling techniques.

Introduction

Advancements in simulation technology have enabled reliable computational models of refinery process units to be built. Computational Fluid Dynamics (CFD) is one of the preferred engineering simulation techniques for improving flow performance. CFD is a technique based on the physical laws that govern the flow of fluids under various conditions. It can be used for the diagnosis of operational issues, improving the performance of existing processes and the design of new process equipment and components. It has been successfully applied to various applications involving blending, flow distribution, heat transfer, mixing, separation, erosion, reactions and particulate flows of refinery operations. CFD has enabled engineers to go beyond the rules-of-thumb, empirical correlations and reduced the number of expensive experiments that are conventionally employed for the design and optimization of processes.

A CFD based design procedure typically involves the following steps:

1. Developing computational models that closely predict the unit operational performance when compared to the observed plant performance
2. Conceptualize multiple design options and assess their effect on performance
3. Evaluate the design options and identify the most suitable option

A few examples that demonstrate the above design procedure are presented below.

Case Study 1: Crude Oil Blending

Crude oil blending is a process in which oils of different grades are blended in different proportions in order to produce consistent crude of desired properties. The consistency in the properties of the crude is vital for downstream refinery operations. Blending thus becomes a very crucial operation. An inappropriate blend configuration could lead to a long blending time and inconsistent quality of the blend.

Large tanks are used for storage and blending crudes of different grades. CFD based simulations were used to evaluate the mixing for an existing configuration of mixing nozzles. Multiple nozzle configurations were evaluated to arrive at an improved nozzle configuration which reduced the blending time considerably.

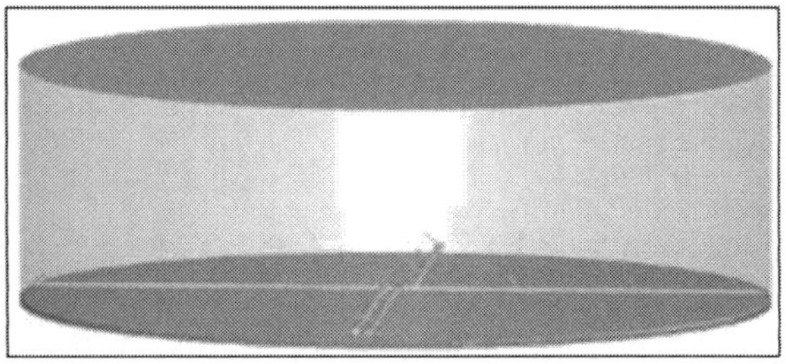

Figure 1: Tank geometry

Figure 1 shows the image of the large storage tank with the nozzle configuration used to blend the crudes. To ensure effective mixing, it is important to have uniform flow distribution across all the nozzles. The nozzles are installed with a single inlet connected to a pump for circulation, so the layout and distribution of the nozzles is critical to achieving this.

The existing nozzle configuration had three nozzles equally spaced on four arms as shown in Figure 2. Detailed flow and mixing analysis showed a very significant variation of mass flow across the nozzles. The deviation in mass flow was as high as 30% of the mean flow for a few of the nozzles.

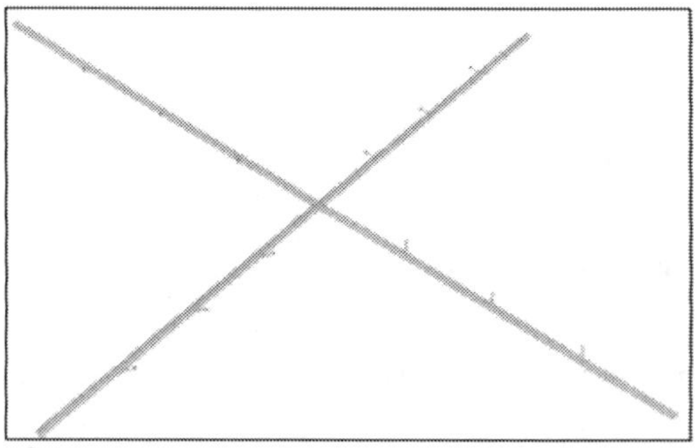

Figure 2: Existing nozzle configuration

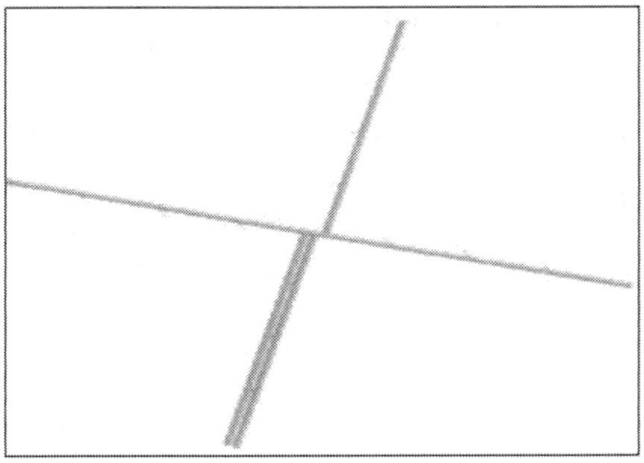

Figure 3: Proposed nozzle configuration

Multiple variations of nozzle configurations were evaluated to achieve a uniform mass flow distribution across all the nozzles. The nozzle configuration shown in Figure 3, improved the mass flow distribution substantially. The deviation in mass flow distribution across the nozzles was within ±2% of the mean value.

Path lines released from inlet were analysed to evaluate the strength of jet through all the nozzles. The jets from all the nozzles were found to be symmetrical as shown in Figure 4. Detailed blending simulations were performed using the existing and proposed nozzle configuration.

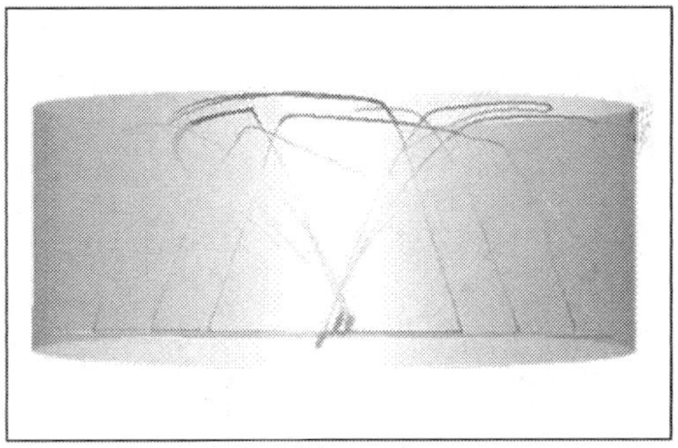

Figure 4: Particle tracks

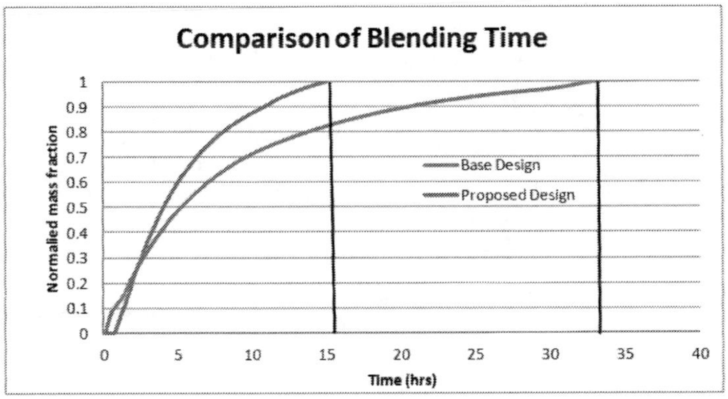

Figure 5: Comparison of blend time

Figure 5 shows the comparison of blending time based on the normalized minimum mass fraction within the tank for existing and proposed nozzle configurations. Blending time for the proposed design is considerably reduced as compared to the existing nozzle configuration.

Case Study 2: Enhancing Performance of a Vapor Horn

Most of the vacuum towers for the refinery fractionators have a two phase vapor-liquid mixture entering the unit at a very high vapor flow rate. It is essential to have a very high pre-separation of any free liquids and a vapor horn is a widely used inlet device for these applications. An efficient vapor horn design will ensure a high separation of the vapor and reduce the liquid fed into the unit. A base case simulation is developed for the tower based on the installed design of vapor horn. The detailed flow analysis is carried out to understand the flow characteristics within the tower. Simulation of the vapor phase flow coupled with Lagrangian particle tracking to study the separation dynamics is performed.

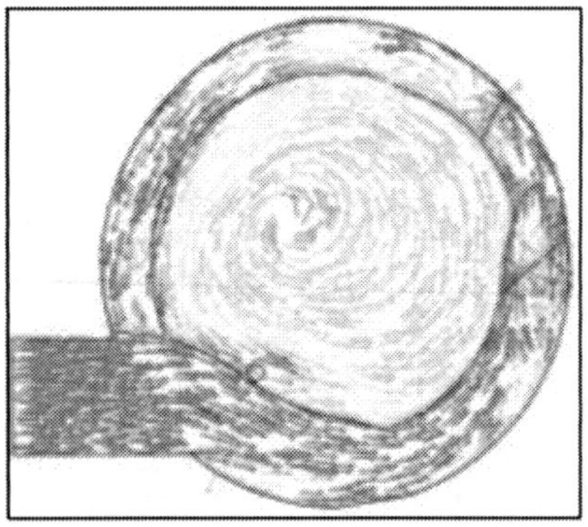

Figure 6: Flow pattern on horizontal plane – existing design

Figure 6 shows the flow pattern on a horizontal plane in the tower. A strong circulation pattern is observed at the periphery. Such a pattern is desirable for good separation. However at the outlet end, high velocity is observed which causes entrainment of smaller droplets.

The flow pattern on the same plane for the proposed design modifications is shown in Figure 7. The high velocity region at the outlet end is reduced, improving the separation of smaller droplets.

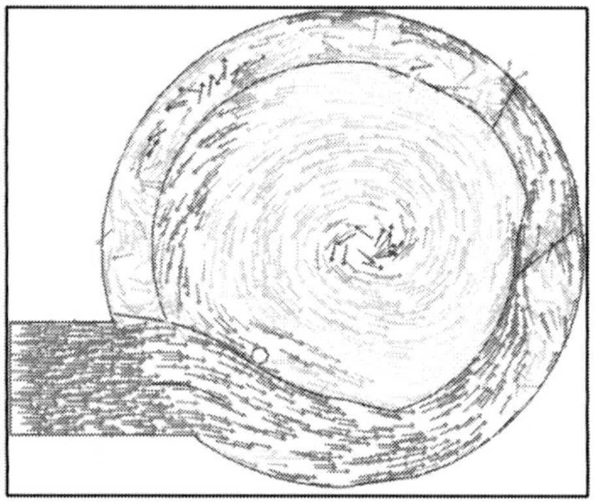

Figure 7: Flow pattern on horizontal plane – proposed design

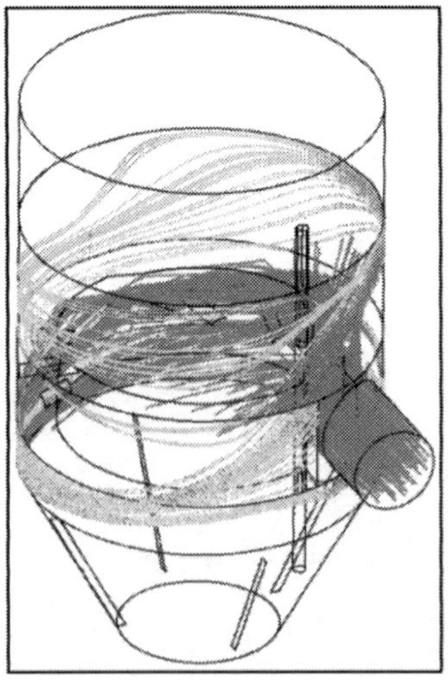

Figure 8: Trajectory of droplets in the tower for existing Design

Droplets were released from inlet and tracked in the tower to evaluate the separation efficiency. It shows some of the droplets are being carried up with the vapor flow and not being separated in the tower as shown in Figure 8.

Droplets trajectory for the proposed vapour horn design showed considerable improvement in the separation efficiency. Droplets of various sizes were tracked to evaluate the separation efficiency over a wide range of droplet size distribution. The proposed vapour horn design achieved improved separation with no carryover for droplets above 200 microns as shown in Figure 9.

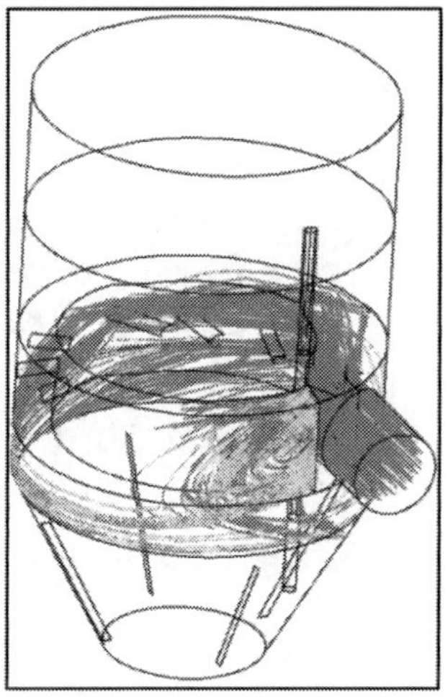

Figure 9: Trajectory of droplets in the tower for proposed Design

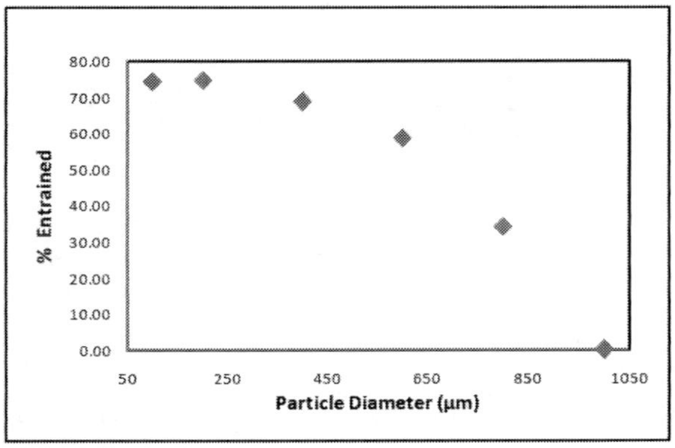

Figure 10: Percentage of droplets escaped vs. Particle diameter - Existing Design

Figure 10 shows the percentage of droplet carry over for the existing design over a range of droplet diameters. Droplets above certain diameters are being separated efficiently. Smaller droplets have a considerable carry over in the tower.

The proposed design achieved a significant improvement in separation efficiency. Droplets above 200 microns diameter are being completed separated with no entrainment as shown in Figure 11.

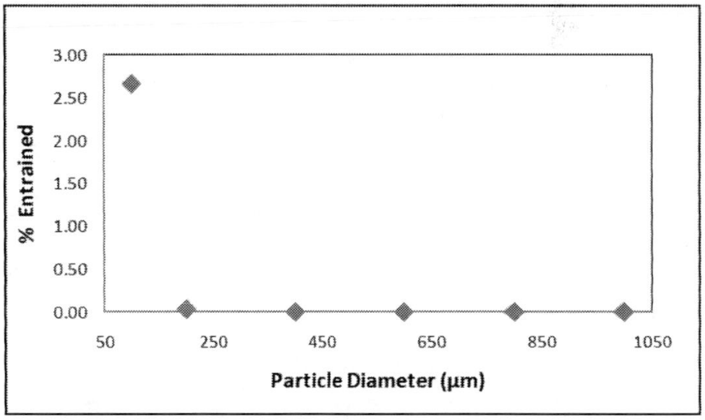

Figure 11: Percentage of droplets escaped vs. Particle diameter - Proposed Design

Case Study 3: Retrofit of a flare knockout drum

A Knock Out Drum (KOD) is crucial equipment in a flare system. The KOD is expected to efficiently separate entrained liquid from a vapor stream and prevent the liquid from being transported to the flare. The operation of a knockout drum is ideal when a plug flow pattern is established in the drum. Due to the high turn down ratios, the flow pattern in the knockout drum could be very different from the assumed plug flow pattern. CFD based simulations were performed to evaluate and retrofit a knockout drum operating at a very high turn down ratio.

Tangential inlet develop a swirling flow pattern within the drum, multiple cross sectional planes along the length of drum are used to plot the contours of velocity magnitude as shown in Figure 12. This flow pattern will generate centrifugal forces that will allow droplets to separate in the drum.

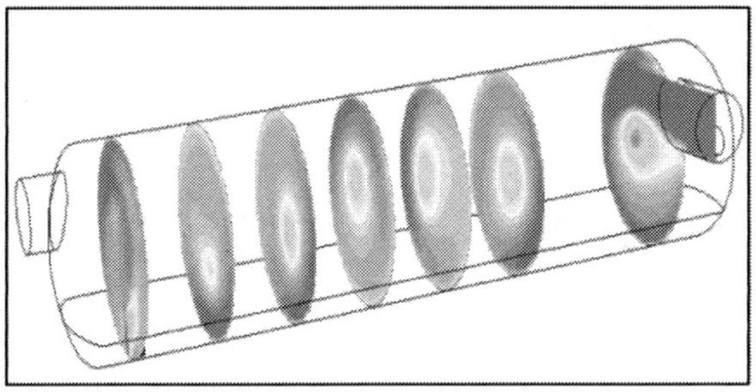

Figure 12: Velocity flow profile in existing design

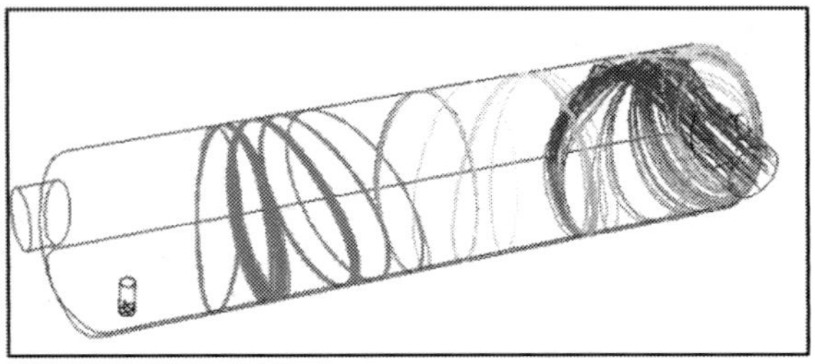

Figure 13: Droplet Tracks for existing design colored by residence time

The droplet trajectories for the existing design are given in Figure 13. Due to the strong circulation flow pattern the droplets circulate near the wall and are being separated. However there is a possibility of potential carryover of separated liquid due to strong swirling flow.

A multiphase simulation was performed considering a layer of separated liquid at the bottom of the drum. The objective was to study the impact of swirling flow pattern and the re-entrainment of droplets from the liquid layer. Figure 14 shows the liquid volume fraction in the drum. It shows there is a substantial carryover of liquid droplets in the drum. This suggests that the existing drum is not suitable for the high turn down ratio process conditions.

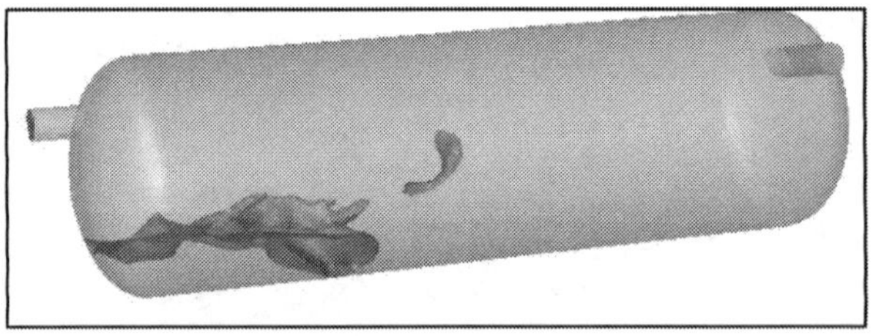

Figure 14: Entrained liquid in existing design

Various arrangements of internal baffles were evaluated to suppress the circulation flow pattern and maintain high separation efficiency. The most feasible design is put forth in the preceding section.

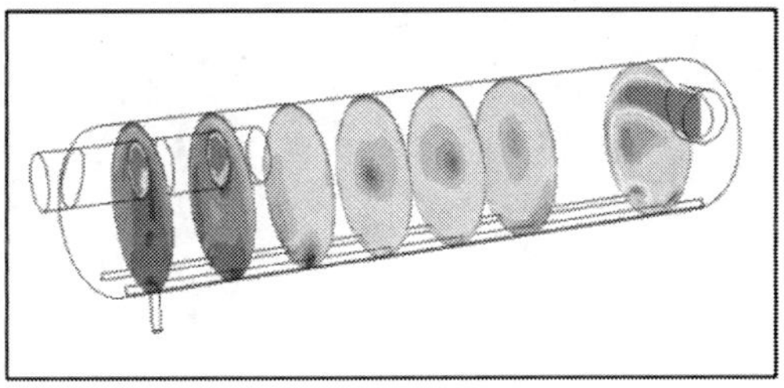

Figure 5: Velocity flow profile in proposed design

Placement of internal baffles breaks the strong circulation flow pattern generated by the tangential inlet. Figure 15 shows the contours of velocity magnitude on cross sectional planes at different length of the drum. It could be seen that after the inlet plane velocity drops down significantly suppressing the circulation flow pattern.

Droplets of various sizes were tracked to evaluate the separation efficiency of the proposed design. Droplets trajectories shown in Figure 16, showed that all the particles are being captured suppressing the circular flow pattern.

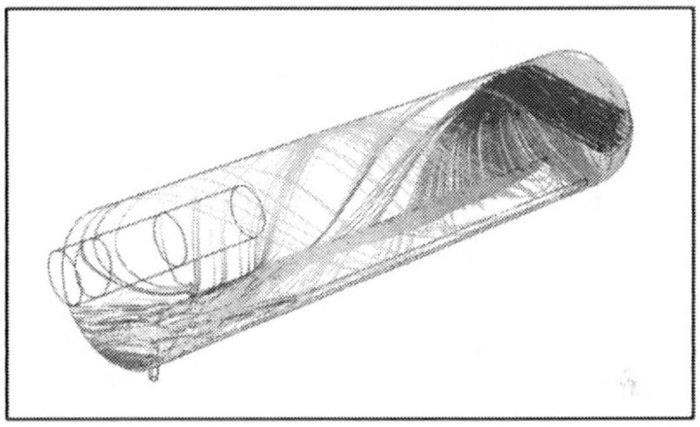

Figure 16: Droplet Tracks for proposed design colored by residence time

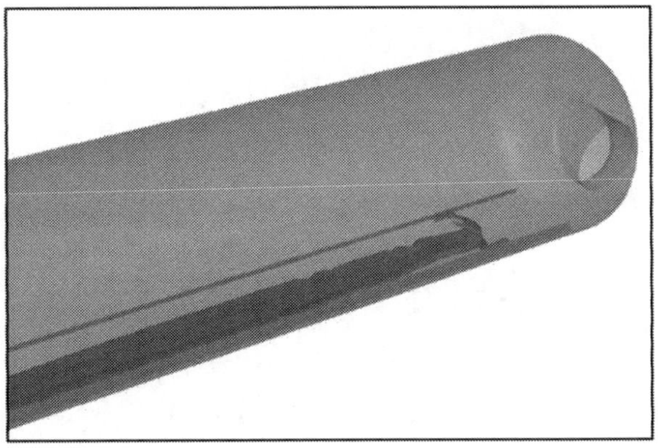

Figure 17: Entrained liquid for proposed design

Multiphase simulation was performed for the proposed design to evaluate the re-entrainment of the separated liquid. Figure 14 shows the liquid volume fraction for the proposed design. Due to the placement of internal baffles, carryover of liquid is reduced. This design provides high separation efficiency and avoids the potential of separated liquid being carried over along with the vapor stream. The retrofit design could be used for higher turn down ratios avoiding the need to change the complete drum for higher turn down process conditions.

Conclusions

The examples in this paper show how CFD based modeling provides the ability to capture the flow features and fluid dynamics with great detail. This technology helps to understand the performance of a particular system and the impact of different process parameters on the operation. Potential design variations were evaluated to determine which would enhance the performance of the unit. The availability of complex models allowed the relevant physics to be captured in the model to provide a simulation that behaves close to plant observations. Simulations help to develop a deeper insight into the operation of a unit and based on this, modifications to de-bottleneck the unit or enhance the performance can be achieved.

References

ANSYS Fluent 13.0 User's Guide, Ansys Inc

API Recommended Practice – 521, 1997

Edward L.; Atiemo-Obeng, Victor A.; Kresta, Suzanne M Paul,(Editors) ,Handbook of Industrial Mixing - Science and Practice , John Wiley & Sons, 2004

PROFITABLY RECOVER REFINERY BY-PRODUCT NGLS, HYDROGEN AND ETHYLENE/PROPYLENE

Guang Lee, Jie Yu, Sudhir Golikeri and Benjamin Klein

Bechtel Hydrocarbon Technology Solutions, Inc.
Houston, TX

Summary

Refineries have long been dealing with constantly changing market conditions. Currently, reduced domestic gasoline consumption, required ethanol blending, improved vehicle miles per gallon and more stringent fuel quality specifications have exacerbated the situation. Capturing opportunities to improve margins will be crucial for survival and sustained profitability.

Over decades of operation, refinery fuel gas systems have become the repository for purge gases from Fluidized Catalytic Cracking, Delayed Coking, Catalytic Reforming, Hydrotreating, Hydrocracking and other process units. Figure 1 illustrates the various sources of hydrocarbon offgas and hydrogen purge streams in a typical refinery. Flare gas recovery systems also contribute to the build-up of non-methane components in refinery fuel gas. Hydrogen concentrations of 50 volume per cent (and higher) are not unusual. Segregating fuel gas blending streams and extracting hydrogen, NGLs and ethylene/propylene provides a moderate capital cost route to improved profitability, triggered by the "perfect storm" conditions of low natural gas cost (to replace the heating value of extracted gases) and higher values for recovered components. Hydrogen recovery and purification avoids costs of onsite hydrogen manufacture or import from area pipelines. Capital costs vary widely, dependent on refinery size / complexity and scope of targeted component recoveries. While there is no "typical" recovery project, case study has shown such projects have attractive economics with payback from 1 to 3 years. Capital costs ranged from less than $5 million to approximately $100 million.

Technology strategies for extraction and purification include cryogenic, pressure swing adsorption (PSA), semi-permeable membranes, compression and gas treating. BHTS has developed optimized configurations utilizing a combination of technologies to meet a refiner's specific recovery and purity targets.

An initial screening study provides in-depth understanding of the current operation and identifies opportunities for segregation, extraction, treating and purification of hydrogen, NGLs and olefin fractions of various streams connected to the fuel gas system. This is followed by a comprehensive analysis which includes capital & operating cost estimating, equipment layout / constructability analysis, piping changes and utility / offsites impacts.

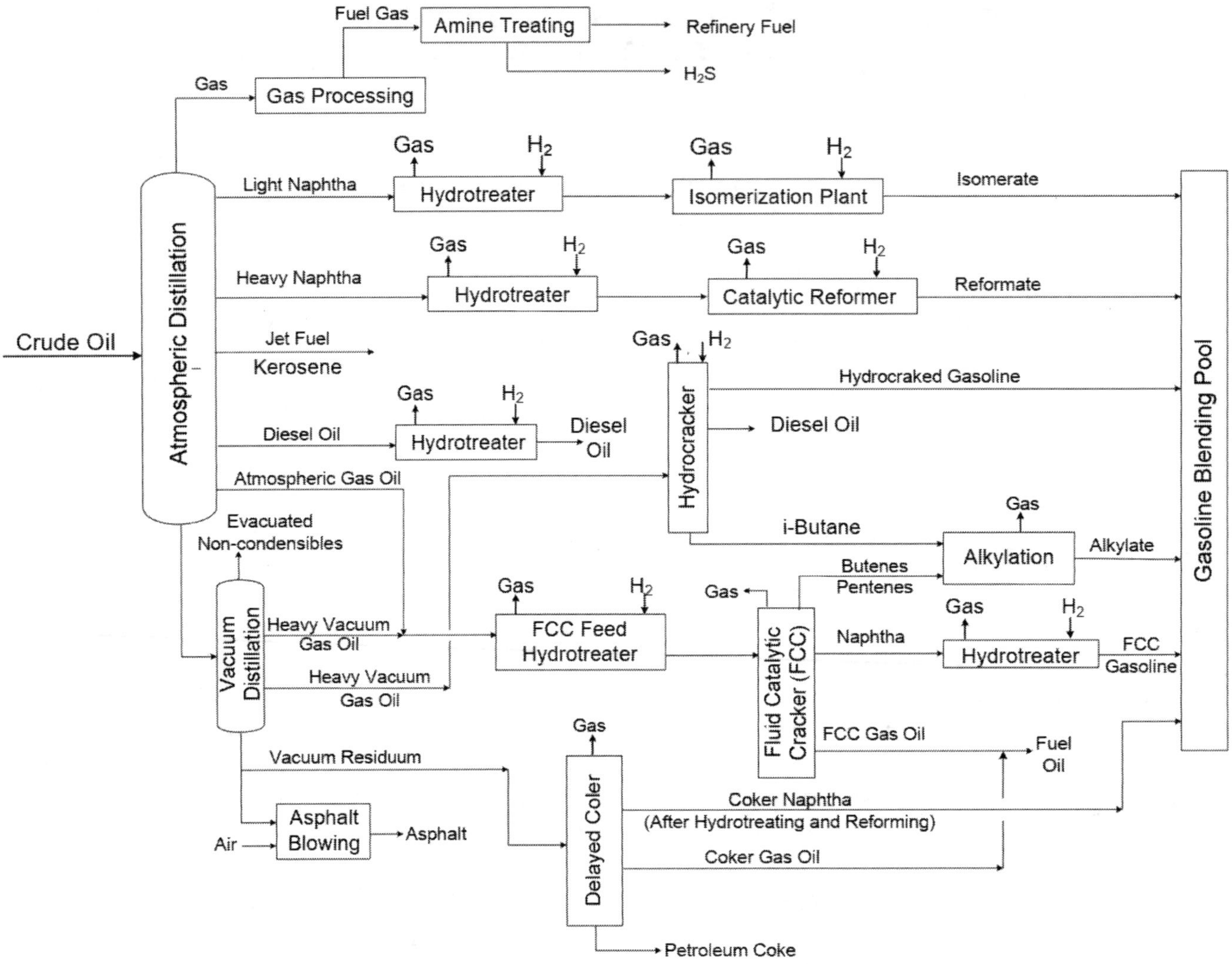

Figure 1. Sources of refinery offgas and hydrogen purge streams.

Background

While most refineries have added conversion and treating units and undergone significant capacity creep over several decades, fuel gas systems have generally received little attention. Refinery fuel gas contains significant concentrations of hydrogen, ethylene, propylene, ethane, propane and heavier paraffinic components. Today's fuel gas composition is significantly different from the original design based on natural gas.

The current outlook for refinery gas processing in U.S.is bullish for several reasons. The production from newly discovered abundant reserves of shale gas has sharply driven down the natural gas price to the \$2-\$3.50/MMBTU range. Industry experts project the price to remain around \$4/MMBTU over the next decade[1]. At these prices, recovering natural gas liquids (NGL) as feedstocks for petrochemicals and other uses is becoming attractive. Refinery fuel gas is valued the same as natural gas. Thus, the NGLs and olefins in the fuel gas are becoming increasingly more valuable. Naphtha derived from crude oil as well as NGLs are the alternative routes to produce ethylene, propylene and other basic chemicals. Since commodity chemical prices are globally set by naphtha-based producers, cheaper natural gas is thus giving U.S. chemical companies an edge over their global rivals. That enables U.S. companies to earn higher margins from using less expensive natural gas. This dynamic justifies investments in gas processing projects for liquids recovery. Also, hydrogen recovery incentives derive from increasing demand to produce cleaner transportation fuels, to process heavy and sour crudes, from increased hydroprocessing capacity and also the need to abate refinery CO2 emissions.

How to Perform a Refinery Gas Processing Project

In order to assess the attractiveness of the liquids and hydrogen recovery project, one must have an in-depth understanding of current operation and focus on the following considerations:

- What process, equipment and piping changes are needed to segregate and reconfigure individual blending streams?
- Can the hydrogen recovery significantly reduce purchases or internal production by SMR?
- Will this be a brown field or green field installation?
- Is there a local market or existing product pipeline(s) for liquid products?
- What are the capital and operating costs?
- What are the product prices?
- Are the utilities and support services available?

A refinery by-product recovery project will follow the standard stage-gate process as shown in Figure 2.

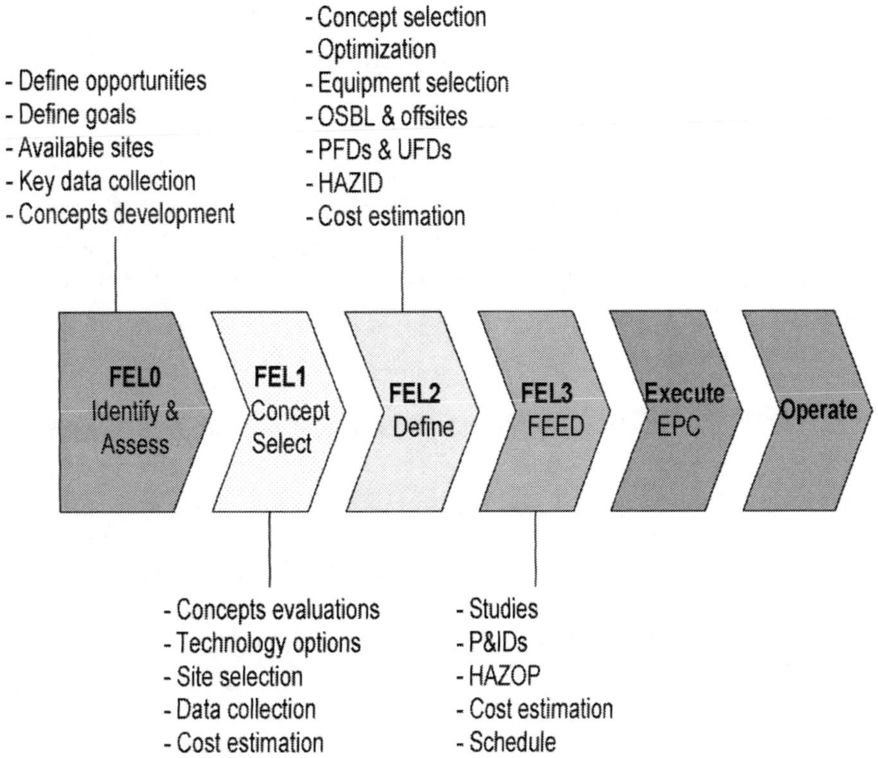

Figure 2. Typical stage-gate process for project development & execution.

To assess if a particular refinery is a candidate for potential hydrogen and liquids recovery project, we work closely with the client to identify opportunities and goals through brainstorming sessions. Once opportunities are identified, various technical concepts are developed and the following information is collected:

1. Tabulation of all streams (vents, purges) entering the fuel gas system with their compositions including trace contaminants, average/peak flow rates, operating pressure and temperature at its source.
2. Existing fuel gas distribution flow schematic
3. Hydrogen system flow schematic and stream composition data – hydrogen generation & purification, compression, make-up H2 streams to individual hydroprocessing units, hydrogen purges from individual hydroprocessing units, catalytic reforming unit(s) net hydrogen production, other hydrogen recovery / purification (PSA / membrane)
4. Existing NGL recovery (if any) – flow schematic and stream composition data
5. Existing propylene and / or ethylene recovery (if any) – flow schematic and stream composition data

In FEL1 study, each concept is simulated for evaluation. A systematic evaluation of all competing technologies for each recovery section is vital to eventually arrive at the best combined process scheme. Site visit with piping/layout, construction and cost estimating professionals is essential to determine suitable location as well as to improve constructability, maintainability and the accuracy of cost estimate. Cost estimate can involve multiple project execution strategies, such as a single project for all recovery sections for best cost efficiency or phased implementation to spread the investment due to funding or construction intensity constraints.

During FEL 2, it is customary to develop in more detail and optimize the best concept by developing, PFDs, UFDs, equipment data sheets, etc. In this phase, more accurate cost estimate is performed based on vendor quotes for major equipment, site preparation, project schedule, and any utilities, OSBL and support services requirement. Risk assessment review such as HAZID should be performed in this phase.

Case Study for hydrogen, NGL and olefins recovery

Figure 3 shows a scenario where multiple offgas streams with significant concentrations of hydrogen, olefin and NGLs are fed to the fuel gas system. The first step to evaluate a liquids/hydrogen recovery project would be to collect stream data, understand current operation/configuration, and to establish recovery goals for hydrogen, NGL and refinery grade propylene. Product and feed pricing information will be necessary to determine the project economics.

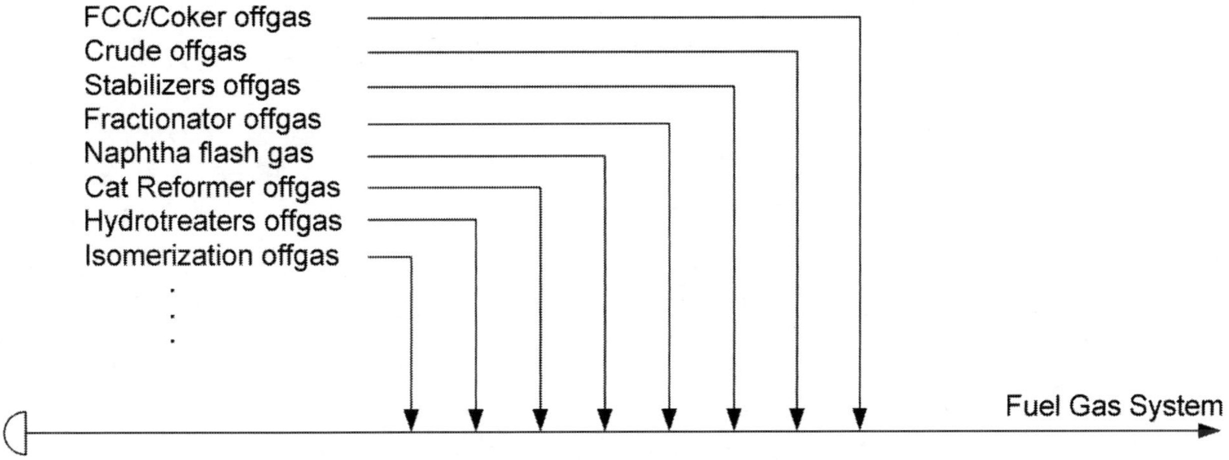

Figure 3. Streams feeding into the fuel gas system.

Table 1 lists several refinery offgas streams and their typical compositions. Significant quantities of valuable components such as ethane, propane, ethylene, propylene and hydrogen are present in these streams. The goals are to:

1. Increase hydrogen recovery to reduce steam methane reformer (SMR) firing.
2. Recover Ethane+ components as NGL product.
3. Recover refinery grade propylene (RGP) as product.

Table 1. Several offgas sources and their typical compositions.

Offgas source	Crude Distillation	Hydrotreaters	FCC/Coker
Component	mol%	mol%	mol%
Hydrogen	9.00	70.00	10.50
Nitrogen	2.50	0.00	5.00
Methane	5.50	15.00	40.50
Ethane	10.00	7.50	12.00
Ethylene	0.00	0.01	7.50
Propane	22.00	3.60	4.50
Propylene	0.00	0.06	6.50
C4's	20.50	2.50	0.75
C5's	12.00	0.75	1.90
C6+	4.50	0.55	1.65
Oxyegn	0.00	0.00	0.20
CO	0.00	0.00	0.50
CO2	4.00	0.01	2.00
H2S	10.00	0.02	6.50

The next step is to evaluate alternatives to segregate and process these streams to meet the above goals. Figure 4 shows a potential process configuration and its overall material balance.

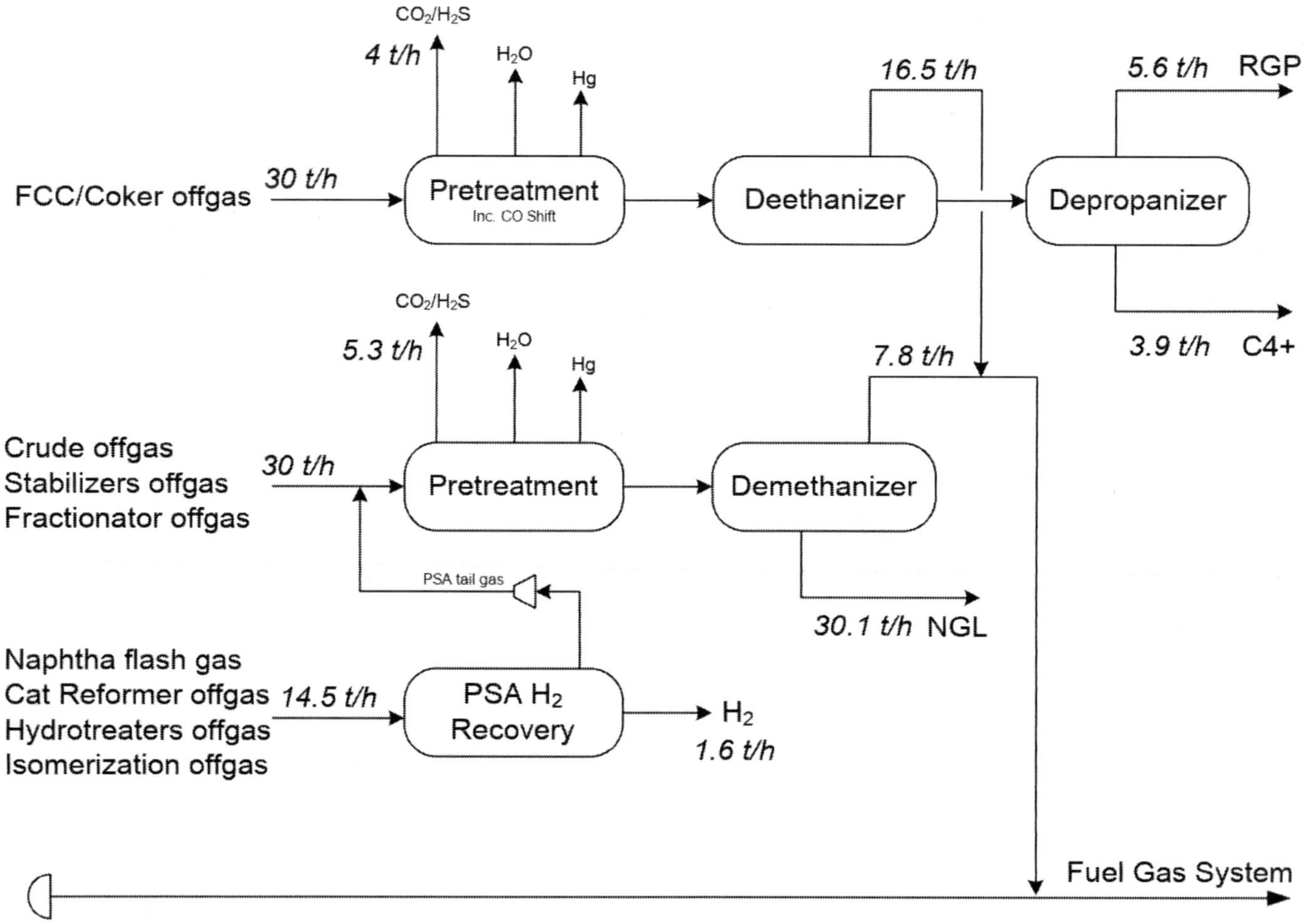

Figure 4. Stream segregation and flow configuration for Hydrogen/NGL/RGP recovery.

The FCC/Coker offgas stream, containing unsaturated hydrocarbons, is segregated from the saturated hydrocarbons. After pretreatment to remove CO, CO_2/H_2S, mercury and moisture, the dried stream is sent to a deethanizer to recover C3+ component, while the residue gas is sent to the fuel gas system. The pretreatment unit may also contain feed compression. The recovered liquid stream is further fractionated in a depropanizer to achieve 95% recovery of RGP from the overhead. The remaining C4+ stream is sent to existing fractionation units.

Streams with high hydrogen content, such as the catalytic reformer offgas and hydrotreaters offgas, are segregated from other lower hydrogen content streams. The combined high hydrogen stream is routed to a PSA unit to recover at least 80% of the hydrogen at 99.9 mol% purity. The hydrogen lean tail gas is compressed and combined with other lower hydrogen content streams. The combined stream undergoes pretreatment to remove CO_2/H_2S, mercury and moisture. The pretreatment unit may also contain feed compression. The dried stream is sent to a demethanizer to recover about 80% of the C2+ component as NGL product. The residue gas is sent to the fuel gas system.

Key Project Considerations

The following two tables summarize simple economic evaluation. Table 2 contains results for the NGL and hydrogen recovery and Table 3 provides the results for RGP recovery. The evaluations are based on the use of PSA for hydrogen recovery and cryogenic turboexpander technology with supplemental propane refrigeration for NGL and RGP recovery.

Table 2. Simple economic evaluation for NGL and hydrogen recovery.

NGL price Mont Belvieu Feb-19-2013	43.64	$/BBL
Net NGL production	9230	BBL/d
Additional revenue*	132.9	$million
Total shrinkage cost due to additional H2 and NGL production= (assuming $3 per MMBtu)	20.7	$million
NGL production margin per year	112.2	$million
Payback time	1 to 2 years	

*: Assume 330 days operation/year

Table 3. Simple economic evaluation for RGP recovery.

RGP (65-75% propylene content) price	0.40	$/lb
Net RGP production	1650	BBL/day
Additional revenue*	39.3	$million
Total shrinkage cost due to additional RGP production= (assuming $3 per MMBtu)	11.3	$million
RGP production margin per year	28.0	$million
Payback time	2 to 3 years	

*: Assume 330 days operation/year

For the NGL and hydrogen recovery project, the payback is estimated to be 1 to 2 years based only on NGL recovery. In case the refinery currently purchases hydrogen, the economics will be even better since the additional hydrogen recovery reduces these purchases.

For the RGP recovery project, the payback time is estimated to be less than 3 years. The economics for both projects can significantly improve with sufficient availability of utilities and by sharing a common refrigeration system.

With simultaneous execution of NGL, RGP and hydrogen recovery projects, the payback time below 2 years is possible.

Technology Options

Hydrogen Recovery Options

Commercially proven hydrogen recovery technologies include Pressure Swing Adsorption (PSA), Membranes, and Cryogenic systems. Each technology has its own unique merits and constraints. The following table provides a comparison among those technologies:

Table 4. Comparison of Hydrogen Recovery Technologies[2]

Features	Adsorption	Membranes	Cryogenics
H2 Purity	99.9%+	90 - 98%	90 - 96%
H2 Recovery	75 - 92%	85 - 95%	90 - 98%
Feed Pressure	10 - 40 barg	20 - 160 barg	> 5 -75 barg
Feed H2 Content	> 40%	>25 - 50%	> 10%
H2 Product Pressure	Feed Pressure	<< Feed Pressure	Feed/Low Pressure
H2 Capacity	1 - 200 MMSCFD	< 50 MMSCFD	5 - 60 MMSCFD
Pretreatment Requirements	None	Minimum	CO_2, H_2O Removal
Multiple Products	No	No	Liquid HC's

Adsorption (PSA system) is the best choice when ultra-high purity, 99.9%+, hydrogen product is required. This is a cyclical process based on pressure swing adsorption that employs multiple beds in parallel. The impurities in the hydrogen rich feed as in a refinery offgas (ROG) are adsorbed at moderate pressure and subsequently desorbed at low pressure. PSA produces hydrogen at essentially feed pressure. Hydrogen recoveries (typically about 80%) are dependent on feed conditions, hydrogen product specifications, and system configuration. The tail gas consisting of impurities in the PSA feed and unrecovered H2 is delivered at low pressure and is available for fuel. Since the tail gas is rich in NGL, it can be compressed and sent for NGL recovery. The ability of a PSA system to produce hydrogen at high purity can provide benefits to the refiner through longer catalyst life, and reduction of inerts introduced into the hydroprocessing unit.

Membranes could provide a low cost solution for high pressure ROG feeds, especially when integration with existing hydrogen compression is possible. Membrane systems are bulk separators that produce refinery-quality hydrogen purities of 90 to 98% at high recoveries (85% or better). Since pressure drives the separation, they are best suited for high-pressure feeds (>20 barg or 300 psig) such as hydroprocessing purge gases. A common disadvantage of membranes is that the hydrogen product stream is produced at pressures much lower than the feed pressure, and may need recompression. The hydrogen product may be sent to existing make-up or recycle compression in hydroprocessing loops. An important consideration for ROG applications is that H_2S is also a "fast gas" through the membrane polymer and will end up in the hydrogen product.

However, concerns of fouling, plasticization, and condensation of hydrocarbon vapors on the membrane surface need to be properly addressed when considering application of membranes in refineries.

Cryogenics is another route to recover 95 mol% hydrogen from ROG feeds when by-product liquid hydrocarbons can subsidize the higher capital expenditure. Temperatures in the range of -250°F are required. Proper pretreatment is necessary to avoid freezing of moisture, CO_2 and aromatic components such as benzene, toluene, ethylbenzne and xylenes (BTEX). Other contaminants in the feed such as mercury and NOx must be monitored and removed to avoid corrosion and ensure safe operation of the commonly employed brazed aluminum exchangers. Without liquids recovery to offset the higher capital costs, cryogenic hydrogen separation systems cannot generally be justified over the other separation technologies.

NGL/Olefins Recovery Options

There are several commercially proven technologies for NGL recovery:

1. J-T expansion
2. Conventional mechanical refrigeration
3. Cryogenic turboexpander
4. Lean Oil Absorption
5. Membranes

J-T expansion process uses the Joule-Thompson effect to reduce the gas temperature upon expansion from a pressure-reducing (J-T) valve. With appropriate heat exchange and sufficient pressure differential across the J-T valve, sufficient temperature reduction can be achieved to recover NGL. The J-T expansion process offers advantages such as simplicity in design and operation. However, it is only suitable for dew point control or when low NGL recovery is needed.

Conventional Refrigeration generally maintains the gas pressure through the process. The gas is heat exchanged and then cooled by the refrigeration chiller to a specified temperature. Liquid is separated from a cold separator. The straight refrigeration process is most suitable for dew point control or for modest (<50%) level of ethane recovery.

Cryogenic turboexpander process has been predominantly used for high ethane recovery. The process expands the feed gas through a turboexpander to produce needed refrigeration. Energy recovered from this gas expansion is typically used to drive a linked compressor to compress the residue gas or the feed gas. External refrigeration may be needed when processing feeds rich in NGL components and/or high ethane recovery is targeted. The focus of this design is to provide sufficient reflux to the demethanizer to attain low overhead temperature and high ethane recovery. Because of the cryogenic temperatures, proper metallurgy (stainless steel and aluminum) is needed. Typically brazed aluminum exchangers are employed which result in compact layout. Care must be taken to provide proper pretreatment to remove potential contaminants in the feed that are prone to freeze at cryogenic temperatures. A properly designed cryogenic turboexpander process can provide highly reliable and efficient performance.

Lean oil (Solvent) absorption is a process in which the NGLs are extracted by contacting the ROG with a liquid hydrocarbon solvent (lean oil) in an absorber column. The rich oil with absorbed NGLs is then distilled to recover the NGLs and return the regenerated lean oil to the absorber column. A refrigerated system enhances the recovery of lighter hydrocarbon components such as ethane and propane. The lean oil is usually a mixture of paraffinic compounds, frequently extracted from the feed, with a molecular weight between 100 and 200. Lean oil absorption processes have the advantage that the absorber can operate at essentially feed gas pressure with minimal loss of pressure for the exiting gas stream. Further, lean oil plants, whether ambient or refrigerated, are constructed of carbon steel. An additional benefit of the refrigerated lean oil process is it is less susceptible to contaminants in feed gas such as CO_2, H_2S, moisture, mercury, arsenic, etc. The lean oil process usually requires larger processing equipment with higher energy requirements. An improved version of the lean oil (solvent) extraction technology is offered by Advanced Extraction Technology, Inc.

Membrane technology can be used for propane or heavier components recovery[4,5]. After compression and air cooling, a portion of the water and C3+ hydrocarbons is condensed and recovered. The off-gas from the condenser is then passed across a silicone rubber membrane that preferentially permeates C3+ hydrocarbons. It should be noted that ethane is not recovered in the permeate.

The table below compares the two potential technology options for higher (>80%) ethane and NGL recovery:

Table 5. Comparison of high NGL recovery technologies

	Turboexpander	Refrigerated Solvent Extraction
Ethane Recovery	>80%	>80%
Coldest Temperatures	-100 to -125 deg F	-50 deg F
Heat Exchangers	Brazed Aluminum	Shell & Tube
Number of Towers	one	two larger diameter towers
Materials of Construction	Stainless & Aluminum	Carbon Steel Construction
Capex	Lower	
Utilities Requirement	Significantly Lower	

Potential Contaminants in Refinery Fuel Gas System

A refinery fuel gas system can contain a host of impurities and trace contaminants as listed in Table 6 that can have adverse impact on operation and/or product purities. The recovery processes need to take these into account carefully and design appropriate removal methods[6,7].

Table 6. Potential Contaminants in Refinery Fuel Gas System and removal methods

Contaminant	Adverse Effect	Removal Method
Water	Hydrate foramtion	TEG, Molecular sieve
CO2 + H2S	Ice formation, catalyst inhibitor, impurity	Amine, Caustic
COS	Catalyst inhibitor, impurity	Alumina adsorption, COS hydrolysis
Acetylene	Promote coking, C2H4 product impurity	Selective Hydrogenation
Nox	Formation of hazardous material Nitrous Gum and salt, explosion risk	Selective Hydrogenation
Oxygen	Corrosion, fouling, safety critical	Selective Hydrogenation
Butadiene	Coke/fouling precursor, NOx gum formation	Selective Hydrogenation
Ammonia	Catalyst inhibitor, formation of hazardous material with NOx	Water scrubbing (water wash section of amine)
Mercury	Poision, Amalgamation with aluminum	Guard bed, sacrificial catalyst bed
Arsine	Product impurity, catalyst inhibitor	Guard bed, sacrificial catalyst bed
Chlorides	Catalyst poison, corrosion	Water scrubbing (water wash section of amine); Adsorption bed
CO	Formation of Ni-carbonyls, highly hazardous	CO shift conversion

Design for Ethane Recovery/Rejection Flexibility

Among the NGLs, ethane prices are most volatile. Units designed for high ethane recovery operate profitably when ethane prices are high. However, from time to time when ethane prices drop, there is little incentive for high recovery. Thus, the units may need to consider the flexibility to switch between ethane recovery and ethane rejection modes to adapt to the swings in ethane prices.

Process technology such as BHTS' proprietary Enhanced NGL Recovery Process^{SM} (US. Patent 5,992,175 and US Patent 6,116,050) has the built-in flexibility to switch between the ethane recovery/rejection modes. Figure 5 depicts how a gas processing plant can accommodate the two operating modes by using this technology.

- Valve positions of normally open (NO) and normally closed (NC) correspond to the Ethane Rejection mode.

- In the Ethane recovery mode, NC valves will be open and NO valves will be close.
- Plants can be designed to recover 95% propane in ethane rejection mode, and to recover 85% ethane in ethane recovery mode.

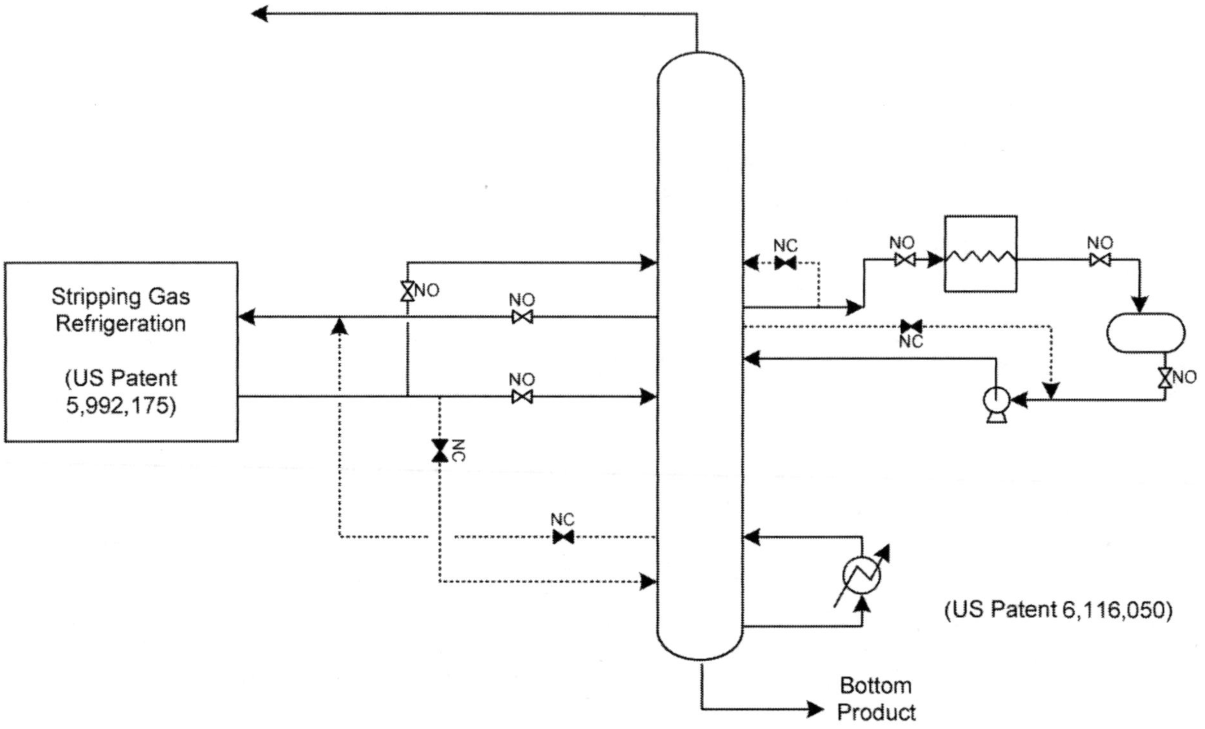

Figure 5. Flexibility of switching between ethane recovery and rejection modes.

Additional Benefits for Fired Heaters

Fuel gas quality changes create formidable challenges for burner management in process and utility fired heaters. Due to recent environmental regulations, most fired heaters have been retrofitted with low NOx burners that use gas tip orifices that are significantly smaller than older technology units. The fired heaters provide best performance with fuel of steady heating value and in composition similar to that used for the burner design. Burner control problems occur when there is sudden change in heating value due to intermittent streams of vastly different composition entering the fuel gas system . Fuel with high hydrogen burns with higher flame temperature and produces higher NOx. Fuel with heavy hydrocarbons has a higher dew point and is prone to liquid condensation and carryover to the burners resulting in coking and plugging. Other contaminants include pipe scales, catalyst fines, water and hydrocarbon aerosols. Of course, these problems can be mitigated by proper pretreatment such as filtration and providing superheat. Fuel gas recovery projects would eliminate some of the operating problems and could result in improved and more reliable operation.

Conclusions

Over decades of operation, refinery fuel gas systems have become the repository for purge gases from Fluidized Catalytic Cracking, Delayed Coking, Catalytic Reforming, Hydrotreating, Hydrocracking and other process units. Flare gas recovery systems also contribute to the build-up of non-methane components in refinery fuel gas. Valuable components such as hydrogen, olefins and NGLs are present in significant amount in the fuel gas system.

Triggered by the "perfect storm" conditions of low natural gas cost and higher values for hydrogen, NGLs and ethylene/propylene, recovering them from the fuel gas system often provides a moderate capital cost route to improved profitability. Hydrogen recovery and purification avoids costs of onsite hydrogen generation or import from area pipelines. Capital costs vary widely, dependent on refinery size / complexity and scope of targeted component recoveries. While there is no "typical" recovery project, case study has shown such projects have very attractive economics with payback time from 1 to 3 years. Capital costs have ranged from less than $5 million to approximately $100 million.

A successful project requires thorough understanding of the specific refinery configurations, operation, market and economic goals. The study team needs to be able to efficiently evaluate and select best combination of technologies and process scheme from all potential options. Usually an initial survey of a refinery-wide analysis of fuel gas blending streams (flow rate, source pressure, and composition) would help determine if the refinery is a candidate for potential hydrogen and liquids recovery project.

References

1. Annual Energy Outlook 2013 (AEO2013) Early Release Overview, U.S. Energy Information Administration.
2. Patel, N., Baade, B., and Khurana, V. (2006), "Creating Value through Refinery Hydrogen Management", ARTC in Singapore.
3. Yuv R. Mehra and Ali H. Al-Abdulal (2005) "Hydrogen Purification in Hydroprocessing (HPH^{sm} Technology)", presented at the 103rd NPRA Annual Meeting.
4. Baker, R. (2002), "Future Directions of Membrane Gas Separation Technology", Ind. Eng. Chem. Res., 41, 1393-1411.
5. D. Alvarado and D.E. Gottschlich, "Membrane Matrimony," Hydrocarbon Engineering (September 2005).
6. Chuang, Sheng Y., Nanda, Rajeev and Rizopoulos, John (2008), "Refinery Off Gas Treatment and Recovery – Challenges of Designing ROG Units", NPRA Annual Meeting, March 9-11, 2008, San Diego, CA.
7. Malik, Zaheer I and Slack, John W (2009) "How to Treat Refinery Gases And Make Them Suitable to Recover Valuable Liquids", 88th GPA Annual Convention, San Antonio, TX.

Exploiting Hydrogen Plant to Improve Refinery Margins

Sanjiv Ratan,
DVP, H2 Product-line, Technip Stone & Webster Process Technologies

---

Abstract

In addressing the global drive for cleaner transport fuels against increasing proportion of heavier and sourer crudes, refiners are faced with the challenge to satisfy their current and future hydrogen demand economically. Hydrogen, with no direct revenue potential of its own, carries a substantial cost burden in a high complexity deep conversion refinery and constitutes a sizeable portion of the refinery processing and operational costs. Hence there is a strong drive for improving refinery profitability or margins through added-value options in order to lower the unit cost of hydrogen (UCH), and thereby improving overall economics of a hydrogen facility whether new or existing or under the 'make' or 'buy' scenarios, which eventually improves refinery margins.

These options include advanced H2 management along with integrated utilization of refinery off-gas (ROG), augmentation of existing hydrogen capacity, economies of scale and strategic steam-power synergy. Furthermore, most of these options carry inherent added merits of improved availability and reduced environmental impact.

The paper outlines some of these proven value-enhancement options and case histories providing potential reduction in net unit cost of (on-purpose) hydrogen and thus enhancing refinery's margins as well as its extrinsic economics.

Introduction

Modern day high complexity and deep conversion refining complexes with the challenging task of meeting the clean fuels product slate against lower grade (opportunity) crude mix, involve growing intensity of hydro-processing, thus making hydrogen supply and its costs increasingly critical in establishing any refinery's overall margins and profitability. Though hydrogen carries substantial cost of its own both in terms of Opex as well as Capex, with no direct revenue potential of its own, it has been gaining the status of an 'asset' in refineries beyond being a critical utility in refinery operations, based on the added value it renders, based on the widening "oil-gas price gap" in recent years.

With the advent of unconventional gas revolution, especially in the US, deeper hydro-processing and bottom-of-the-barrel strategies are finding more and more justification for rendering more of the premium clean fuels products through higher specific usage of hydrogen (scf/bbl) produced from relatively cheaper gas, compounded by the application of various smart concepts or optimization options for lowering the unit cost of hydrogen.

For enhancing the cost-effectiveness of refinery hydrogen, few strategic options have been explored and practiced with the underlying aim to create an 'added value' component for lowering the net Opex against optimized capital investment. Strategic reduction of hydrogen costs, by incorporating various state-of-the-art developments and latest advances aimed to optimize the so called unit cost of hydrogen (UCH) based on the fixed as well as variable costs, beyond lower priced feedstocks, play an important role in further improving the overall refinery margins, while also satisfying the needs for higher operational reliability and HSE compliance.

Several proven options have been employed to improve capital efficiency of a hydrogen facility, both new and existing. Modern hydrogen plant designs keep pace with state-of-the-art technological advancements and developments in catalysts, adsorbents, high temperature metallurgy, automation, process control and safety, as well as the advanced unit operations which enhance energy efficiency, reduce capital costs and minimize emissions. Their effect on plant economics but also on its ease of operation, flexibility, required maintenance and the continuous on-stream time, in order to provide best value to the user.

Hydrogen Management

Hydrogen is typically supplied and balanced through a "network" fed by hydrogen derived from various off gas streams from the processing units and largely supplemented by make-up from the on-purpose hydrogen generation. Advanced and creative solutions are being sought for ensuring and enhancing refining profitability based on not merely balancing the hydrogen but through innovative and smart hydrogen management methodology for optimizing the refinery H_2 networks.

Advanced refinery H2 management technique as developed by Technip and called Hydrogen Network Design Tool **(HyN·DT^{TM})** contains models for the various refinery operations and fall back modes to identify the optimized on purpose hydrogen generation capacity based on the feed, product, hydrogen and fuel network optimization as well as the steam system and energy philosophy, along with consideration of the overall refinery CO_2 emission.

Case History: Advanced optimization of H2 recovery cluster

For a grass root refinery, the HyN·DT^{TM} program was applied, which also established the optimum level of H2 recovery from refinery off gases (ROG), thereby also minimizing the net capacity of on-purpose hydrogen generation.. The results are represented in Figure-1 showing the relationship between the cost of H2 Recovery v/s curtailed Generation.

The benefit from incremental lowering of hydrogen generation capacity should be exceeding the cost for the incremental H2 recovery. However beyond a stage of ~ 60% cluster H2 recovery, further increase in recovery, also taking into account the cost of compression of low pressure purge gas with relatively low hydrogen content, is not justified since the total cost for recovery exceed the contribution benefit.

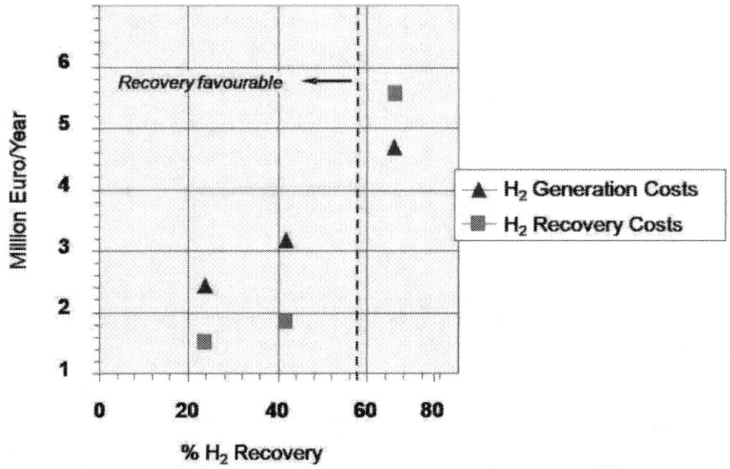

Figure-1: Cost of H2 Recovery versus savings on H2 generation

Refinery Offgas Integration with H2 Generation

For capturing the full potential of refinery off-gases (ROG), there has been an increasing trend of their 'integrated' utilization for on-purpose hydrogen generation. The underlying criteria being that any hydrogen burned as fuel is a loss of its 'asset' value above its heating value. Accordingly, by integrating ROG in a hydrogen generation plant, not only does most of its hydrogen content get recovered, but also the price attached to the ROG is often lower than the premium hydrocarbon feedstocks for hydrogen production.

Thus it offers enough incentives to identify potential ROG streams in the refinery, which can be integrated cost-effectively in the hydrogen plant to enhance its economics. Utilization of ROG through integration with the hydrogen plant can be done in basically 3 routes (as represented in Figure-2):

1. Hydrogen recovery by mixing with the process gas upstream (common) PSA
2. Direct use as (part) feedstock for H_2 generation
3. Dedicated recovery PSA with integration of its purge gas as feed for H_2 generation

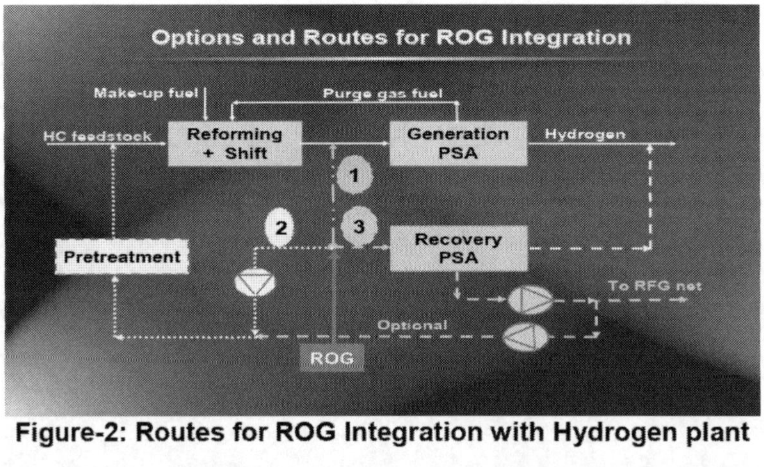

Figure-2: Routes for ROG Integration with Hydrogen plant

Typical reduction in net hydrogen cost can be between 2-10% depending upon the relative pricing of ROG versus the base feed, available quantity of ROG stream(s), hydrogen (or hydrocarbon) fraction, available pressure and level of impurities.

A case history is described below for utilizing ROG (with average 40 vol % H_2) as primary feed for hydrogen generation based on medium level contribution. The comparative economics is presented in Table-1. Though the pricing of ROG was similar to NG in terms of LHV basis, the operating cost benefit due to potential (contained) hydrogen contribution was substantial (~190 $/h or ~7% of unit cost of hydrogen).

In case NG feed compression was needed; the payout for ROG integration would have become even more attractive. Other inherent benefits for operating with ROG feed mix included relaxation on recycle hydrogen, easier start-up and longer reformer tube life.

Table-1 ROG Integration as SMR feed (45 mmscfd H2)

		Case-1 ROG as partial feed	**Case-2 NG feed only**
Operating cost savings[1)]	US$/h	188	base
Additional TIC investment [2)]	MM $	4.1	base
Pay-out	Years	2.5	base

1) Pricing data

ROG :	260 $/t (avg. 40 vol % H_2)	Steam	8 $/t
NG	180 $/t	Power	60 $/MW

2) For compression and pre-treatment

Though the pricing of ROG was similar to NG in terms of LHV basis, the operating cost benefit due to potential (contained) hydrogen contribution was substantial (in the order of 7%). The payout of additional investment was less than 3 years, without downsizing the reformer and downstream section, which were still sized for NG feed as the controlling design case. Other inherent benefits for operating with ROG feed mix included relaxation on recycle hydrogen; NG feed flexibility, easier start-up and longer reformer tube life. Various ROG integration schemes have been implemented and proven, also providing better refinery margins.

Capacity Revamp of Existing H2 Plants

Often the refiners are faced with hydrogen shortfall in addressing the changes in crude mix versus the market based clean fuels product slate, which is large enough to impact overall operation and refinery profitability but may not be large enough to justify a new dedicated hydrogen plant.

Achieving additional hydrogen through revamping of existing plants can prove to be attractive by providing reduction in the unit cost of hydrogen through cost-effective retrofitting. The actual economics however depends upon the degree of uprate, available design margins, condition of existing equipment and the level of modifications required. Normally a capacity revamp not only provides additional hydrogen at lower cost, but also the benefits of shorter time schedule based on the interfacing-facilities/ off-sites being already in place, and can also entail feedstock change or flexibility; improved efficiency and environmental performance.

There are various proven options for cost-effective augmentation of hydrogen capacity up to 30% and each case needs proper identification of the bottlenecks and assessment for selecting the most appropriate option.

In cases where the target capacity increase is substantial (>15%) with major limitations on the reforming section, one of the effective solutions without overloading the reformer is regenerative reforming based Technip Parallel Reformer (TPR). The underlying concept is to use high-level heat of the reformed gas to reform additional feed through convective heat exchange, also known as post reforming. Furthermore, this option also provides 10-15% reduction in emission levels of CO_2, NOx and SOx per unit hydrogen.

In the expected additional H2 capacity range of 2 and 25 mmscfd, TPR-based retrofit investment largely depends upon the % increment of hydrogen capacity, design conditions and available design margins in the existing plant and level of modifications required in the plant. Generally, when compared with a new hydrogen plant for the same additional capacity, together with the necessary offsite / utilities and auxiliaries, such investment can prove very attractive. (see Table-2)

Table-2 : Hydrogen Plant Capacity Revamp

Option	Typical incremental hydrogen (%)[1)]	Level of investment [2)]
Reformer Upgrade	5 - 15	Medium
Regenerative reformer (TPR) integration	15 - 30	High

1) Typical additional hydrogen between 2 and 25 mmscfd
2) Based on the typical range of $ 0.5-2 MM per mmscfd H2 depending upon % increase and design specific factors

Economies of Scale and Energy Efficiency Enhancement

On-purpose hydrogen generation plants are capital intensive due to high temperature catalytic processing and customary gas phase purification, and the total investment can vary considerably depending upon site-specific factors like location, feedstock, export steam conditions, degree of utility integration and reliability needs.

For larger hydrogen plants, the variable costs linked to specific energy consumption costs are more governing the unit cost of hydrogen. Table-3 provides an overview of the sensitivity of variable and fixed costs to plant capacity. An 1% reduction in the energy costs of a hydrogen plant of 100 MM scfd can result in approximately $ 750,000 savings per year based on $5.0 / MMBtu. Thus, for larger hydrogen plants, there is also a bigger incentive for incremental investment in terms of extended heat recovery as well as optimization of the flowsheet and operating conditions on a typical payback period of 2-4 years.

The Capex is though relatively less critical for larger plants; their reduction becomes quite vital for lowering of the unit cost of hydrogen. Based on economies of scale, the cost of hydrogen produced by a single-train large capacity hydrogen plant can be appreciably lowered by the capital exponent contribution.

Table-3: Hydrogen Generation Cost Split

	Large (> 60 mmscfd)	Medium (20-60 mmscfd)	Small (< 20 mmscfd)
Variable costs %	60 - 80	50 - 70	30 - 50
Fixed costs [1)] %	20 - 40	30 - 50	50 - 70

Figure-3 illustrates the economy of scale based reduction in the unit cost of hydrogen from a 50 MM scfd to a 250 MM scfd hydrogen plant based on US gulf-coast economics covering average of natural gas pricing range of $4-7 /MM Btu.

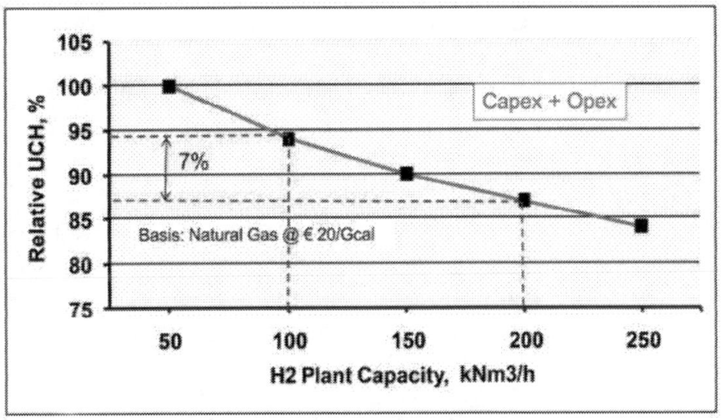

Figure-3 : Economy of Scale for Hydrogen Generation Plant

Though economies of scale favor larger plants, there is of certainly a capacity limit above which a single train plant starts becoming cumbersome and requires detailed evaluation to establish the break-point for two or more trains. Physical size, weight and transportable limits on the equipment, valves and piping as well construction facilities must be taken into consideration. Such limits have progressively increased from 100 mmscfd up to a recent project size of 220 mmscfd based upon compacting of equipment, advanced equipment design, piping modeling and modular construction concepts.

Energy Efficiency and Design Optimization

Based on the case-specific data, a sensitivity analysis on a variety of variables like Feed/Fuel price ratio, Fuel/Steam price ratio is conducted for optimization of Opex against incremental investment for Capex analysis to ensure overall economics within the given evaluation criteria.

Table-4
Hydrogen Process Optimization :
Typical Relative Price Ratios

	Higher Reformer outlet temperature (> 870 °C)*	Higher S/C ratio (> 3.0)	Higher level of shift conversion (> HT)*	Higher PSA H_2 recovery (> 86%)*
Operating costs				
■ Feed / Fuel	> 1.2	> 1.3	> 1.3	> 1.1
■ Steam / Fuel	> 1.1	< 0.9	> 1.2	-
■ Feed / Steam	-	> 1.4	-	-
Investment	>	>>	>>	>

* For large plants, lower price ratios could be applicable

A typical energy balance of a hydrogen plant is illustrated in Figure-4. The main thrust for improving the thermal efficiency of a hydrogen plant thus wrests in reducing the eventual heat loss thru the flue gas to stack and thru process cooling to cold utility (air / cooling water).

Further, as also explained above, the energy or opex related variable costs become more governing for optimizing larger H2 plant designs in order to lower the UCH, which in turn also allow higher capital investment for applying enhanced and advanced heat recovery below the heat pinch by extending the so-called "cold composite", as depicted in Figure-5

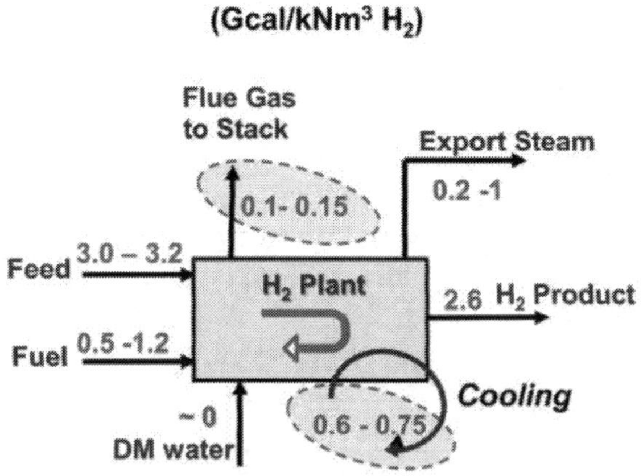

Figure-4 : Energy Balance of a Hydrogen Plant

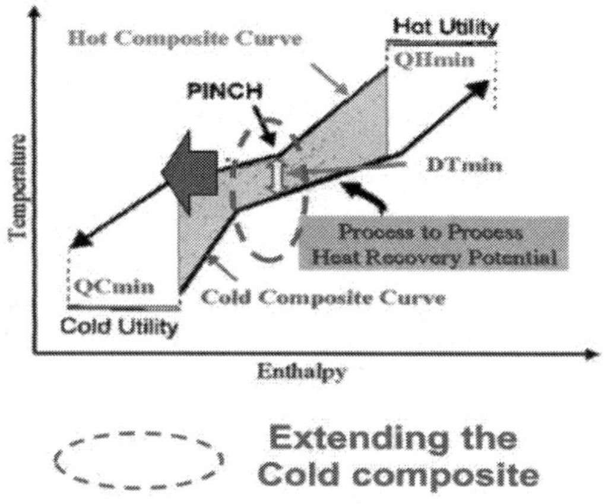

Figure 5: Enhanced Heat Recovery

Steam-Power Synergy

A hydrogen plant is typically a net exporter of steam based on the high temperature processing and related threshold waste heat recovery. The amount of export steam can be adapted to a reasonably large extent by different design options to match the site-specific steam philosophy or the potential steam-power integration, especially in case of low-credit or excess refinery fuel gas.

In a refinery landscape, the export steam from the hydrogen can be a major optimization parameter based on case-specific and site-wide requirements. The availability of low level heat from the process gas, in combination with overall furnace efficiencies well in excess of 90%, makes the hydrogen plant an efficient steam producer, producing typically 10-15% more steam for a given heat liberation in the furnace. For a 100 mmscfd H2 plant, export steam can be tailored between virtually zero to 180 t/h and upto 300 t/h with auxiliary firing (see Figure-6A).

Concerning Capex, "natural steam make" H2 plants with no or low air preheat, the investment cost is usually the lowest and considered as baseline for comparative assessment. Maximizing export steam via auxiliary firing will increase capital investment, based on additional equipment and/or larger steam generation system, whereas an increase of the heat integration would also increase the investment cost due to the utilization of more expensive high temperature materials and reduced temperature difference on the heat recovery equipment.

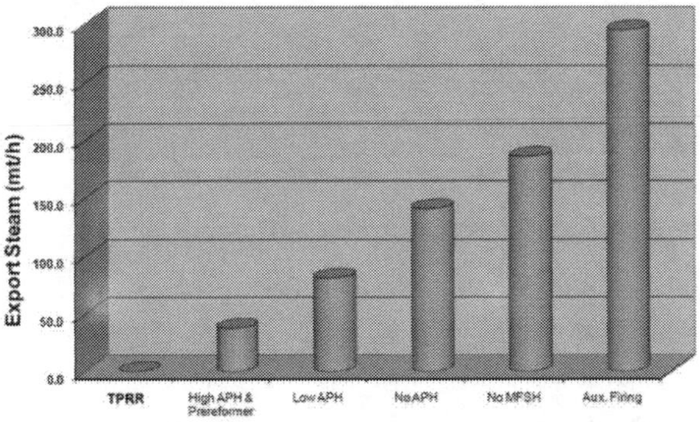

Figure 6A: Export steam flexibility

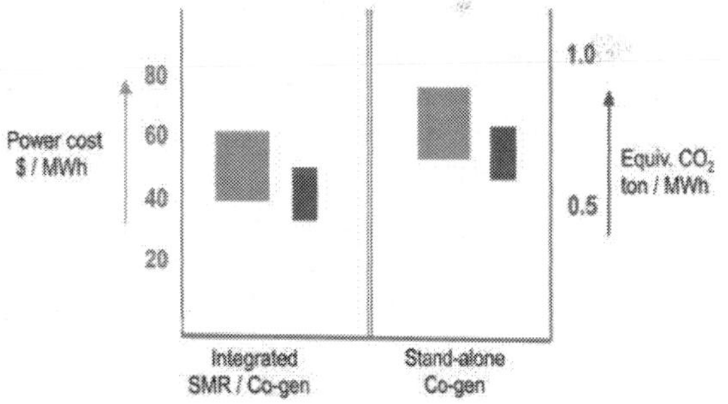

Figure 6B : Cogen with 100 mmscfd H2 plant

Steam-power synergy by integrating H2 plant export steam with cogeneration or combined cycle in refineries, especially with larger hydrogen plants, say above 70 mmscfd, can prove quite attractive (see Figure-6B). Especially under certain fuel-power scenario and for locations having insufficient or unreliable grid power supply, such synergy not only offers site-wide power self-sufficiency and reliable availability, apart from lowering the specific cost of hydrogen. Also, it can provide few additional advantages like integrated project execution, better energy efficiency and emissions compliance.

The steam turbine can be implemented as a topping / back-pressure turbine or as an extraction / condensing turbine, with case-specific match for being a direct drive on large machines or coupled to an alternator for power generation. Further integration can be done with Gas turbines in a combined-cycle cogeneration scheme, with extended possibility of using gas turbine exhaust as combustion air in the hydrogen plant reformer.

The different options and schemes for steam-power integration carry a distinct balance between their (annual) capital cost and the heat to power efficiency (heat rate). Table-5 presents the summary of the steam-power integration economics in a large hydrogen plant of 80 mmscfd based on natural steam generation.

However, having such integration, the operating hydrogen capacity dictates the 'steam dispatch', which tends to limit the ability to independently vary steam or power output. Thus it needs to be studied how the export steam demand from the hydrogen plant affects its end-user operation.

Table-5 : Economics of Steam-Power Integration

80 mmscfd H_2 plant; Natural steam make

Integration Option →		Base Case (No synergy)	Case A Topping Turbine		Case B Condensing Turbine
Steam to Turbine [1] / Export	t/h	-- / 130 [2]	210 / 130 [3]		210 / 0
Power (-) Import / Export	MW	(-) 2.6	12		37
Operating cost credit [4]	$/h	910	1510		1850
Savings	$/h	base	600		940
	MM$/y [5]	base	5.1		8.0
Budgetary capital charges [6]	MM$/y	base	2.0 [7]	1.6 [8]	3.7
Net annual revenue	MM$/y	base	3.1	3.5	4.3
Reduction in H_2 price [9]	%	base	~ 4		~ 5

Notes:

1) 1500 psig and 950 F
2) 600 psig and 750 F
3) 80 t/h process steam extraction at 600 psig; 220 psig back pressure MPS
4) Steam price : $8/t (600 psig); $7/t (220 psig); Power = $ 50/MW
5) 8500 hours per year
6) Based on 20% / year
7) Based on STG + Alternator for power generation / export
8) Based on Direct drive(s) within refinery, with cost adder for individual turbines
9) Based on $1200/t H_2, with added benefit of captive power reliability

Conclusions

The various attributes and options summarized above illustrate that there are several ways to optimize the hydrogen capacity as well as lower the Unit cost of hydrogen in a refinery complex, thereby also effectively contributing towards improving the overall refinery margins, based on its substantial cost contribution, apart from higher premiums through optimization of hydrogen usage in processing of lower cost crude mix and enhancing the product slate. Additional underlying benefits could also include improved environmental performance and operational reliability.

(60e) Optimized Combustion of Hydrocarbon and Hydrogen Fuel Mixtures Using Coriolis-Based Measurements

Andrew J. Verdouw, ISG, Emerson Process Management, Georgetown, IN

Process-derived hydrogen gas is often utilized by powerhouse boilers and fired heaters for steam production and process heating. The extremely high heating value of hydrogen gas coupled with its water vapor combustion product makes it a desirable primary or supplemental fuel. In addition, process hydrogen can easily help to offset natural gas or other fossil fuel consumption, thus lowering operating costs. Hydrogen is often mixed together with natural gas and sent to the burners as a single fuel source. Efficient combustion of this mixture over a wide range of hydrogen concentrations can prove problematic for conventional combustion controls. This is due chiefly to the very different stoichiometric air-to-fuel ratio requirements of hydrogen versus the alkane series. The typical method used to ensure safe combustion is to fire the boiler or heater with excessive air. The consequence of this practice is reduced thermal efficiency which translates to increase purchased fuel use. Using a cost-effective method to measure and determine the hydrogen concentration in real time can allow precise calculation of the required stoichiometric and excess combustion air by the control system. The result is optimized combustion over a very wide range of hydrogen concentration.

This paper will discuss a technique for determining the hydrogen concentration in a hydrocarbon fuel (namely natural gas) using coriolis-based devices. The principles of measurement will be explained as well as how the process data obtained is translated into a specific fuel-to-air ratio that can be employed by an advanced combustion control system. The principles of fuel and air cross-limiting and the calculation of excess combustion air using this method will also be examined. It will be illustrated how real-time knowledge of a varying fuel-to-air ratio allows a boiler or fired heater combustion process to be safety controlled and thermally optimized across all loads.

(60f) Setting the Revamp Basis: What Is Current Operation?

Andrew W. Sloley, CH2M HILL, Bellingham, WA

A revamp aims to modify a plant to meet some new combination of capacity, feed, product, safety or regulatory requirements. The starting point on any revamp is the current plant operation. Often, this is poorly understood. Plant mythology may have grown up over years. This mythology hides both problems and opportunities. Without understanding a plants current operation, revamps fail due to not fixing the hidden problems. They also cost too much by missing the hidden capabilities of the existing asset. Full understanding of the needs and potential of a plant requires a careful test to find the hidden flaws and unexpected capabilities of the facility.

A rigorous test run requires planning. Objectives must be clearly outlined, personnel assembled, and data gathered. All required data must be gathered and kept. A successful test run creates the starting point for identifying needed modifications. The objective is to define all needed modifications necessary to meet goals – and avoid overinvestment at the same time. An outline of necessary steps, applied to an example test run is shown.

13 AIChE
Spring Meeting, San Antonio, TX

Design & Revamp of Cooling Water Networks

E. K. Tamakloe, G. T. Polley & M. Picón Núñez

University of Guanajuato,

Mexico

Appear Complex

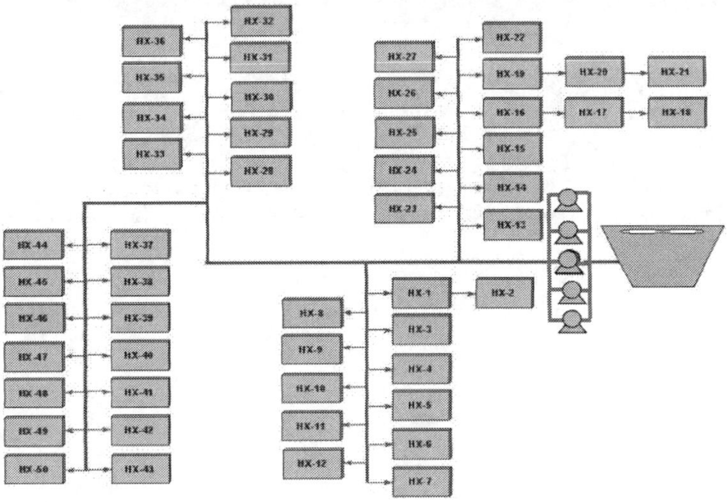

13 AIChE
Spring Meeting, San Antonio, TX

Reality

- Simulations can be set up quickly
- Analysis quite simple
- Significant Benefits

13 AIChE

Design of New Networks

13 AIChE

Topics

- Design Principles
- Flow Simulation
- Fouling Analysis
- Blowdown & Make-up

13AIChE

Design Principles

13AIChE

Cooling Towers

- Individual Plants – Plant Regions – Centralized System

- Packaged Units – Specialist Suppliers

- Selected with future expansion in mind

13AIChE

Cooling Towers

- Minimum Wetting Rates apply: so, system may have flow set by cooling tower rather than process plant

- Maximum water return temperatures apply

  (function of water chemistry)

- Thermal Behaviour : Effectivess-Ntu Model

  Tested, Reliable

13AIChE

Piping

- Dominates Capital Cost
- Set with allowance for plant expansion
- Equations for economic pipe size
- Capital Cost : related to diameter to power 3

  (exception being very long straight runs)
- Installation: again diameter to power 3

13 AIChE

Piping

- Function of Plant Layout – not use of temperature driving force

13 AIChE

Heat Exchanger Sizing

- Must be undertaken in "system context"
- Cannot arbitrarily set pressure drop or cooling water temperature rise

13 AIChE

Flow Simulation

13 AIChE

Components

- Gravitational head
- Frictional losses
- Momentum changes

13 AIChE

Special Feature

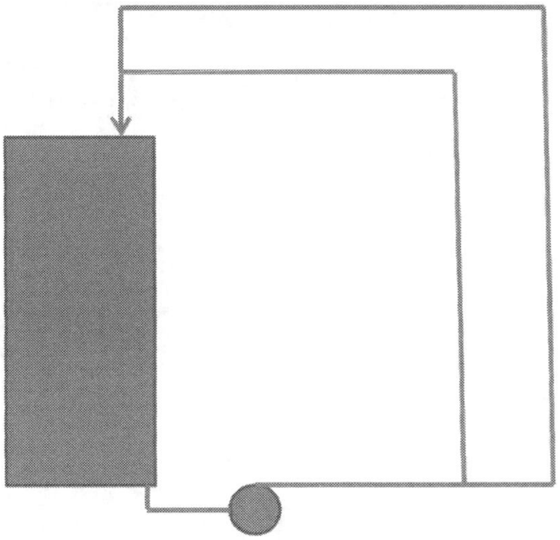

CLOSED SYSTEM

13 AIChE

IMPLICATIONS

- MOMENTUM EFFECTS INDEPENDENT OF FLOW PATH
- GRAVITATIONAL EFFECTS INDEPENDENT OF FLOW PATH (Provide flow initiated)

13AIChE

Gravitation

- Initiation of flow may require provision of back pressure – restriction at cooling tower return
- Vacuum can be pulled in heat exchangers – checks for vapour blocking are required

13AIChE

Friction Equation

- Great advantage in using equation based on Volumetric Flow Rate rather than velocity
- Velocity can vary along specific flow path – but volume flow remains constant

13AIChE

Pipe

$$\Delta P_f = 2f\left(\frac{l}{d}\right)\left(\frac{\rho u^2}{g_c}\right)$$

13AIChE

Valves & Fittings

$$K_{valve} = \left(\frac{8\rho}{\pi^2 d_{pip}^4} \right) k_{valve}$$

13 AIChE

Exchanger Pressure Drop

Function of Square of volumetric throughput

$$\Delta P = f(V^2)$$

13 AIChE

Components are Additive

Pipe + Valve + Heat Exchanger + Valve

$$K_{leg} = 2K_{valve} + K_{exchanger} + K_{pipe}$$

13 AIChE

Branch

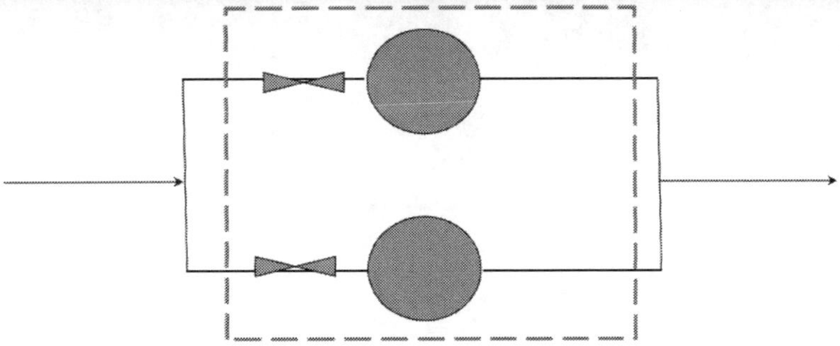

$$K_{branch} = \frac{1}{\left(\left(\frac{1}{K_{branch\,1}}\right)^{0.5} + \left(\frac{1}{K_{branch\,2}}\right)^{0.5}\right)^{2}}$$

13 AIChE

K for Whole System

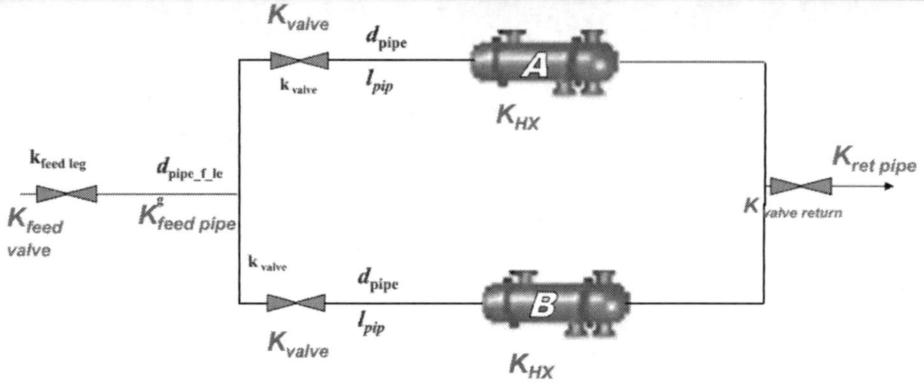

$$K_{sys} = K_{feed-valve} + K_{feed-pipe} + K_{branch} + K_{ret-valve} + K_{ret-pipe}$$

13 AIChE

Flow Through Network

$$V_T = \left(\frac{\Delta P_T}{K_{sys}} \right)^{0.5}$$

Solved simultaneously with Pump Curve Equation

13 AIChE

Distribution Through Network

$$\frac{v_j}{V_T} = \frac{\sqrt{K_j}}{\sum_k \sqrt{K_k}}$$

13 AIChE

Heat Exchanger Sizing

- Relationship between heat ransfer coefficient and pressure drop

$$\Delta P = \frac{bA}{V} \alpha^m = KV^2$$

V = Volumetric flowrate, A = Area

b = constant (physical properties, characteristic dimension – e.g. Tube i.d.

m = exponent (tubes = 3.5) based on exchanger type (available for tubes, shells, plate-and-frame)

- Solve together with basic exchanger design equation

13 AIChE

Fouling

13AIChE

Calcium Sulphate

- Najibi Kraus (1997) developed a model for fouling due to crystallization of calcium sulphate.
- The rate of fouling is defined by two consecutive processes:
  - First, transport of dissolved salts to the convective heat transfer surface as defined by the mass transfer coefficient and concentration gradient between the flow and the surface
  - Second, the reaction and precipitation of the salt ions in the flow to the heat transfer surface to form deposits. This reaction is defined by the Arrhenius relationship.

13AIChE

Predictions of Model

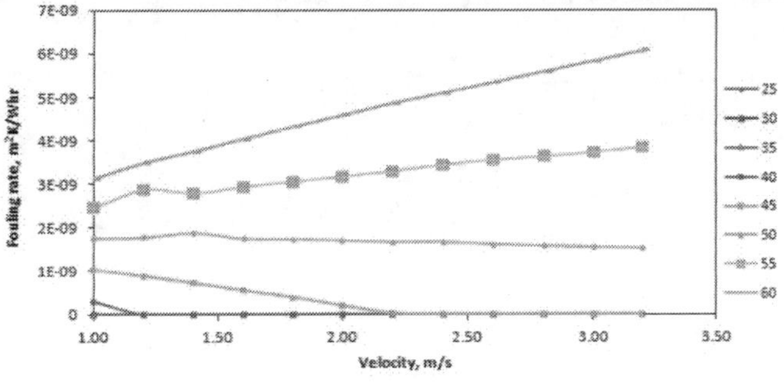

High water temperatures: fouling rate increases as velocity increases

Moderate temperatures: fouling rate decreases as velocity increases

A fouling "threshold" can be observed

Fouling Thresholds

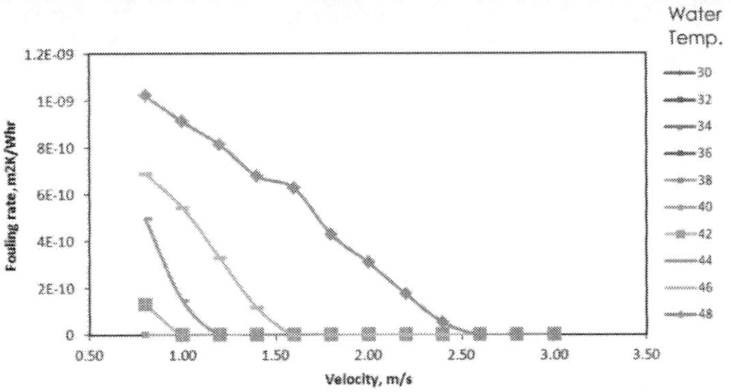

Since water temperature (rather than "film temperature" has been used, plots are function of process side h.t.c.

Blowdown & Make-up

13 AIChE

Governing Equations

- Concentration of calcium sulphate in make-up water is known.
- Concentration in recycled water needs to be controlled.

Water balance:

Make-up = Evaporation + Blowdown

Salt balance:

Make-up(feed conc) = Blowdown (system conc)

Approximate Heat balance:

Evaporation = Tower Load/Latent Heat of water

13 AIChE

Typical Results

Blowdown: kg/s

		Feed (ppm)			
		10	50	100	200
System (ppm)	1000	0.03	0.14	0.3	0.68
	1200	0.02	0.12	0.25	0.55
	1500	0.02	0.09	0.19	0.42
	1800	0.02	0.08	0.16	0.34
	2000	0.01	0.07	0.14	0.3

Tower Heat Load: 6 MW

Vaporisation: 2.73 kg/s

13AIChE

Revamping Networks

13AIChE

Two cases considered

1. Reduction of Load on Tower

2. Installation of New Cooler

Other cases exist: e.g. elimination of pumps

Result of Improved Heat Recovery

- Load on coolers is reduced
- **BUT** Process streams enter and leave coolers at lower temperatures

Case study

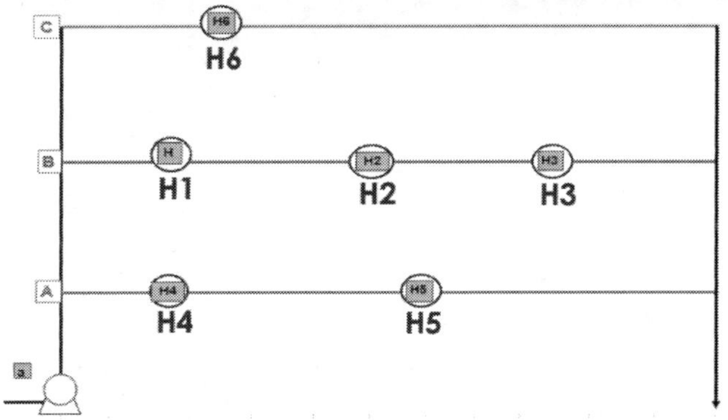

- 3 Branches in parrallel
- 6 Heat exchangers

13 AIChE

Current Performance

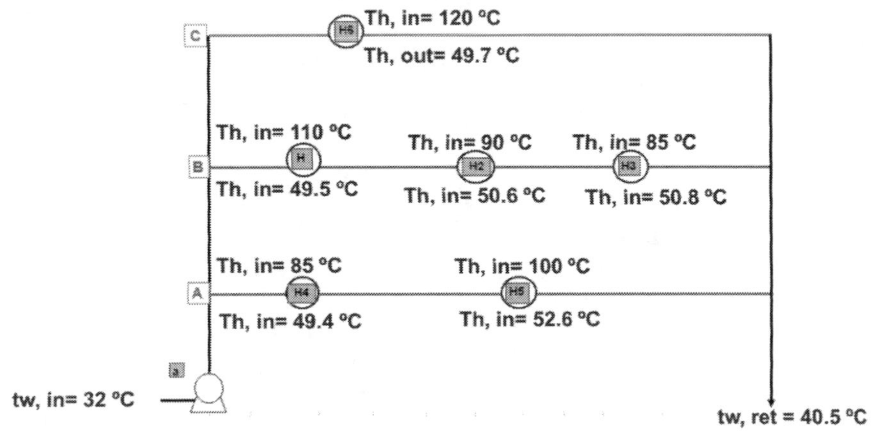

13 AIChE

Coolers

Coolers		
Unit	Tin	Tout
H1	110	49.59
H2	90	50.6
H3	85	50.8
H4	85	49.49
H5	100	52.6
H6	120	49.7
Load on CT = 9.58 MW		
T_{Wret} = 40.5° C		

13 AIChE

Result of improved Heat recovery

- Reduced entrance temperatures
- "Leads to reduced outlet temperatures"

13 AIChE

After energy savings

After energy savings			
Unit	Tin	***Tout***	Tout (current)
H1	70	***40.57***	49.59
H2	65	***41.99***	50.6
H3	65	***42.49***	50.8
H4	70	***44.54***	49.49
H5	75	***45.29***	52.6
H6	75	***40.68***	49.7
Load on CT = 5.27 MW T_{Wret} = 36.7° C			

13AIChE

Observations

- Streams go to storage at lower temperatures
- "Over cooled"
- Load on cooling tower higher than necessary

13AIChE

Options for revamp of coolers

1. Reduce cooling water flow
   - Throttling
   - Bypass on cold side

2. Bypass on hot side

3. Reduce area – plug tubes

13 AIChE

Effects of bypass, plugged fraction and flow reduction

- Hot bypass fraction:

  $h_h \downarrow$

- Flow reduction (Cold bypass fraction):

  $F\downarrow \quad h_c \downarrow$

- Plugged fraction area (plugged tubes):

  $A \downarrow \; Q \downarrow$

13 AIChE

Restore Cold Outlet Temperatures

One option

Unit	Bypass, cold	Plug. Tubes	Bypass, hot	Tout
H1	0	0.4768	0	50
H2	0	0.5299	0	50
H3	0	0.6	0	49.97
H4	0	0.2943	0	50
H5	0	0.2611	0	49.99
H6	0	0.423	0	49.99
Load on CT = 3.81MW				
T_{Wret} = 35.5° C				

13AIChE

Summary

- Current operation 9.58 MW
- After improved Heat recovery 5.27 MW
  ΔQ = 4.31 MW
- After revamp of coolers 3.81 MW

ΔQ_{saved} = 5.77 MW **75% through heat recovery additional 25% through cooler modification**

13AIChE

Identifying ALL Options

Procedure Producing Matrix – Each Unit

Cold Bypass

Hot Bypass

	0	0.02051952	0.041039	0.061559	0.082078	0.1026	0.12312	0.164156142	0.16415614	0.1846757
0	0.476744938	0.47454218	0.4722194	0.469767	0.467173	0.46443	0.46151	0.458415376	0.45511867	0.4516022
0.04737	0.457881777	0.45558658	0.4531654	0.450608	0.447902	0.44504	0.44199	0.438757875	0.43531245	0.4316355
0.09474	0.435785526	0.43337403	0.4308291	0.42814	0.425293	0.42228	0.41907	0.415663636	0.41203165	0.4081534
0.14211	0.409378219	0.40681567	0.4041101	0.401249	0.39822	0.39501	0.39159	0.387961156	0.38408721	0.3799479
0.18947	0.376985097	0.37421817	0.3712952	0.368203	0.364927	0.36145	0.35775	0.353815384	0.34961382	0.3451209
0.23684	0.335816061	0.3327568	0.3295228	0.326099	0.322469	0.31861	0.31451	0.310135588	0.30546364	0.3004627
0.28421	0.280744013	0.27723108	0.2735139	0.269575	0.265393	0.26095	0.25621	0.251153062	0.24574479	0.2399472
0.37895	0.200827021	0.19650992	0.1919345	0.187077	0.181912	0.17641	0.17053	0.164249423	0.15751146	0.1502701
0.37895	0.065656457	0.05948539	0.0529206	0.045924	0.038451	0.03045	0.02187	0.012644932	0.00269392	-0.0080679
0.42632	-0.291636584	-0.3079721	-0.325769	-0.345244	-0.366662	-0.39035	-0.4167	-0.44628658	-0.4797441	-0.5180001

Assumes correct inlet condition
Evaluates exchanger behaviour

Fraction of plugged tubes

Adding a Cooler

13 AIChE

Case Study 2

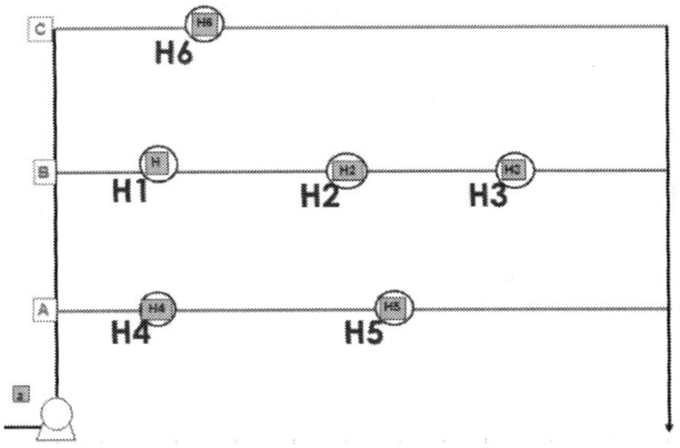

New Cooler to be added. Geographical location close to branch A.

13 AIChE

Options

- New parallel branch
- New branch plus pump revamp/replacement
- In series – after exchanger H5
- In parallel with exchanger H5

13 AIChE

Parameters for New Unit

- Thermal behaviour: Load of New Unit
- Hydraulic behaviour: K value
  (variable to be explored)
- Fouling behaviour: model used for new systems

13 AIChE

Design of New Unit

- Simulation gives flow and temperature field
- Procedure described for new design applies

13AIChE

Option 1: New Parallel Branch

Unit	Change in Load MW	Change in Hot Outlet Temp.
1	-0.105	+2.3
2	-0.081	+3.2
3	-0.203	+3.8
4	-0.047	+1.2
5	-0.105	+2.6
6	-0.044	+2.1

13AIChE

Change in Fouling

Unit	Original		Revised	
	Inlet	Outlet	Inlet	Outlet
1	0	0.0008	0	0.00145
2	0.0002	0.0023	0.0014	0.00307
3	0.0013	0.0034	0.001	0.0035
4	0	0	0	0
5	0	0.0009	0	0.00154
6	0	0	0	0

13 AIChE

Option 2: Parallel Branch + Pump Modified

Unit	Change in Load MW	Change in Hot Outlet Temp.
1	+ 0.04	- 0.9
2	+ 0.014	- 0.6
3	+ 0.013	- 0.3
4	+ 0.038	- 0.9
5	+ 0.025	- 0.65
6	+ 0.04	- 0.55

13 AIChE

Change in Fouling

Unit	Original		Revised	
	Inlet	Outlet	Inlet	Outlet
1	0	0.0008	0	0.0006
2	0.0002	0.0023	0	0.002
3	0.0013	0.0034	0.0012	0.0033
4	0	0	0	0
5	0	0.0009	0	0.0007
6	0	0	0	0

13AIChE

Option 3: New Unit in Series with H5

Unit	Change in Load MW	Change in Hot Outlet Temp.
1	-0.006	+0.1
2	+0.003	-0.2
3	-0.048	-0.2
4	-0.045	+1.2
5	-0.067	+1.7
6	-0.013	+0.3

13AIChE

Option 4: New Unit in Parallel with H5

Unit	Change in Load MW	Change in Hot Outlet Temp.
1	-0.03	+0.6
2	-0.018	+0.7
3	-0.008	+0.7
4	+0.007	-0.2
5	-0.092	+2.3
6	-0.02	+0.36

13 AIChE

Conclusion

BETTER OPTIONS ARE:

NEW PARALLEL BRANCH PLUS PUMP MODIFICATION

NEW UNIT PLACED IN BRANCH A IN SERIES WITH H5

13 AIChE

Overall Conclusions

- Thermo-hydraulic Simulation of Cooling Water Networks is not difficult
- Simulation includes consideration of tower behaviour and pump characteristics
- Fouling models can easily be incorporated into the analysis
- Design principles for new systems have been established

13 AIChE

Overall Conclusions

- In revamping the effect that improved heat recovery has on the performance of the existing system has to be evaluated if true reduction in tower load is to be achieved
- Existing exchangers can be modified in order to achieve reduction in tower load – full range of possible options can be identified
- Care must be exercised when installing new coolers – options easily evaluated
- All coolers need to be designed in "systems context" – procedures established

13 AIChE

(61b) Water / Wastewater Optimization Methods for Industrial Systems

Joseph W Guida, Process Technology, Fluor, Sugar Land, TX

Efficient management of water and wastewater systems in large industrial complexes is important for minimizing costs and optimizing natural resources. This paper provides an overview of methods used to analyze water use efficiency such as water pinch analysis and water cascade analysis as well as other visual techniques. Discussion will also include techniques for water recovery and recycle such as zero liquid discharge and an explanation of when recycling makes sense and when it doesn't.

(61c) Balancing Act - Simple Techniques for Steam Header Controls

Barbara Hamilton, Process Systems and Solutions, Emerson Process Management, Hampden , ME

An industrial steam distribution system with multiple headers, turbines, boilers, and reducing valves is managed with header control strategies that mitigate upsets, increase power generation, and reduce clean condensate makeup. Effective use of override, anti-windup, and TG governor interface strategies insure header stability. Upsets are managed without cascading into multiple equipment failure. An overall coordination scheme is built from optimum use of pressure reducing valves that is easily grasped by operating personnel.

Coordinating interconnected steam headers is impossible without automatic control. A major upset on one header can cascade throughout the system causing multiple equipment failures and eventual plant outage. A control strategy that coordinates turbines, boilers, and reducing valves can balance a multiple header system during normal operation and prevent a catastrophic failure during a major upset.

Coordinated control of the pressure reducing valves (PRVs) is the most effective too for stable header control. Proper configuration of override controls and limiters keep the valves closed until needed. Interfaces with the Woodward TG governor systems coordinate the PRVs with TG extraction to allow PRVs to take over when the TGs are unavailable.

Thermal Fluids: the Key to Overall Site Integration

E. K. Tamakloe, G. T. Polley & M. Picón Núñez

University of Guanajuato,

Mexico

Topics

- Key Message
- Advantages of Thermal Fluids
- Integration between adjacent plant sections – example an FCC Plant
- Total Site Analysis
- Linkage with Fired Heaters

Key Message

Large number of factors MUST be considered simultaneously when undertaking systems design:

- Energy efficiency
- Heat exchanger behaviour
- Capital costs
- Control
- Flexibility

13AIChE

Advantages of Thermal Fluids

- Major capital cost advantage – piping costs – dominated by valves and fittings – vary with diameter raised to a power of around 3.
- Pipes carrying steam very much larger than those transporting liquid
- Steam temperature set by mains pressure – becomes superheated when pressure is let down
- Heat can be extracted from thermal fluids over significant temperature range

13AIChE

Integration between Adjacent Plant Sections – Example Revamp of an FCC Plant

13AIChE

Process Flow Diagram

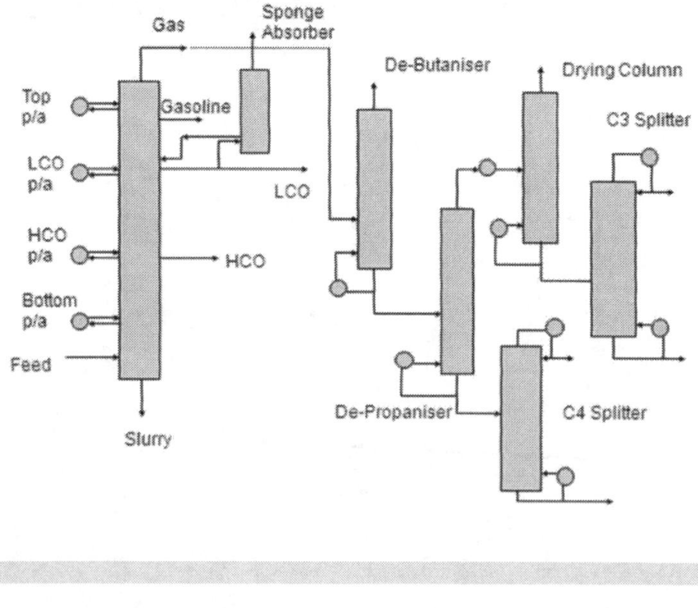

13AIChE

Geographical arrangement

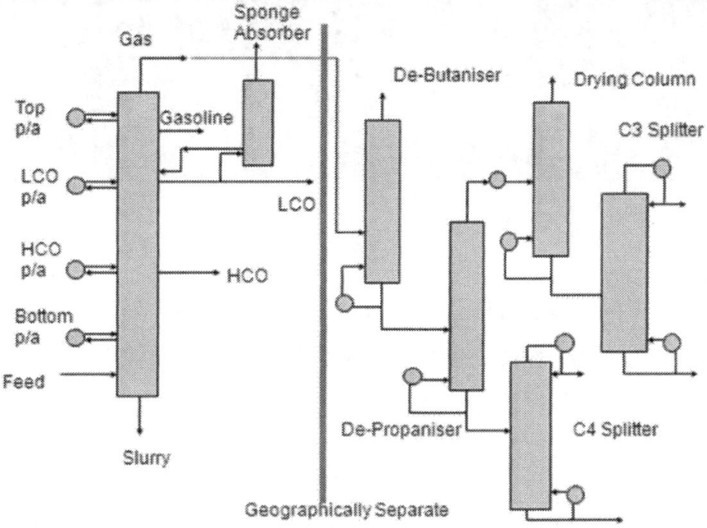

13AIChE

Energy Analysis

- Full Problem Energy 36 MW
  T.A.C. 6.04 M$/annum
- Column Energy 3.2 MW
  T.A.C. 2.5 M$/annum
- Gas Processing Energy 74.1 MW
  T.A.C. 7.5 M$/annum

13AIChE

Energy Efficiency Requires Transfer of Heat from Column Region to Gas Processing Region

Problem: Re-boilers in Gas Processing Region operate at low temperature

13 AIChE

Column Zone

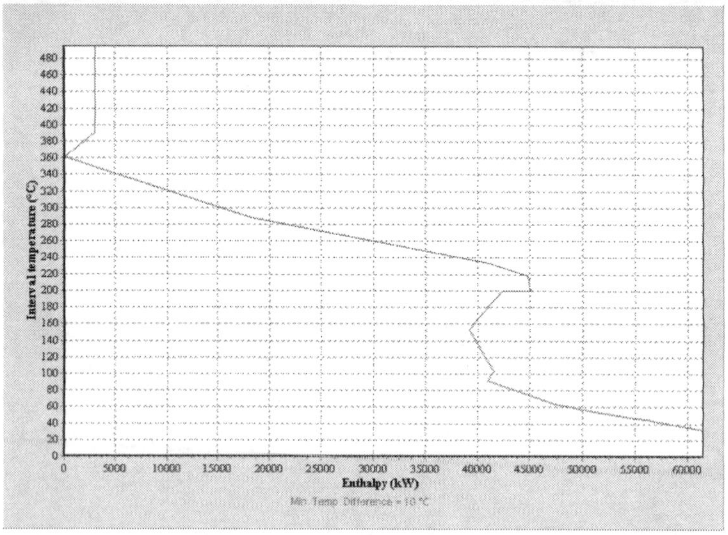

13 AIChE

Gas Processing zone

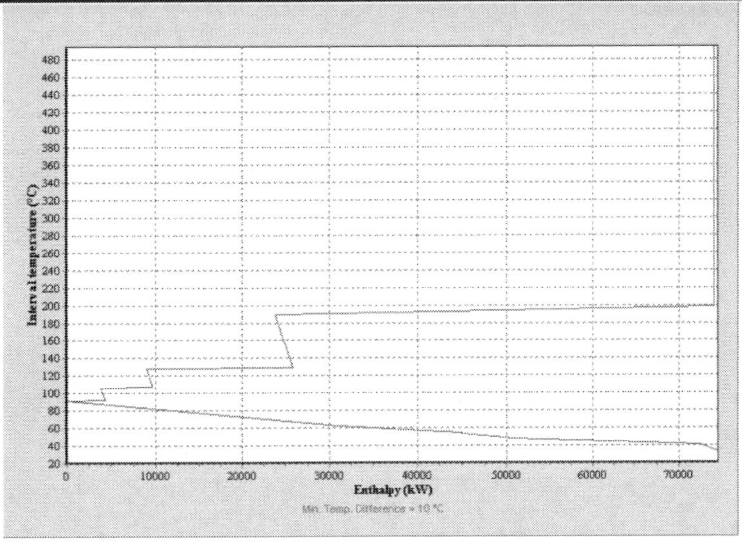

13 AIChE

Column Zone

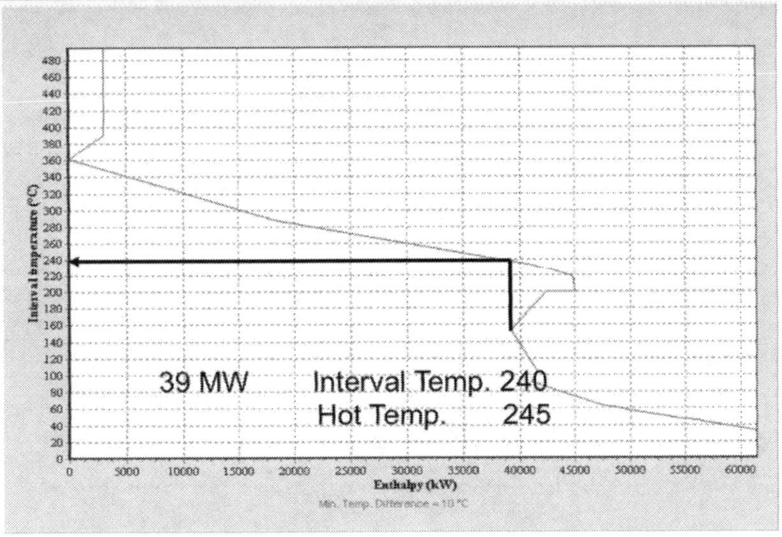

13 AIChE

Gas Processing Zone: Option 1

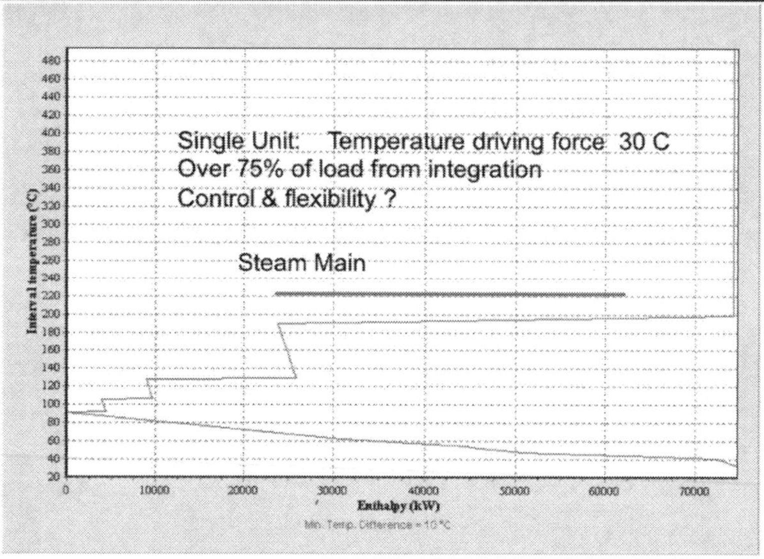

Gas Processing Zone: Option 2

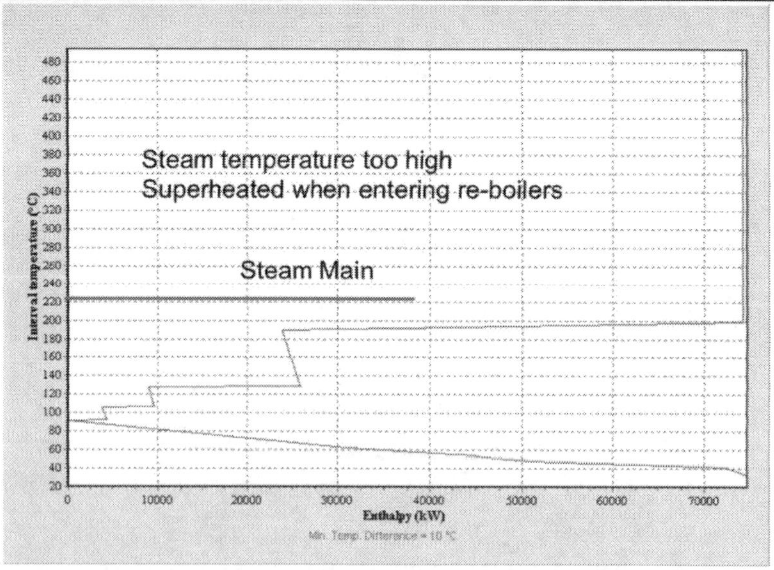

Column Zone: Used to Heat Thermal Fluid

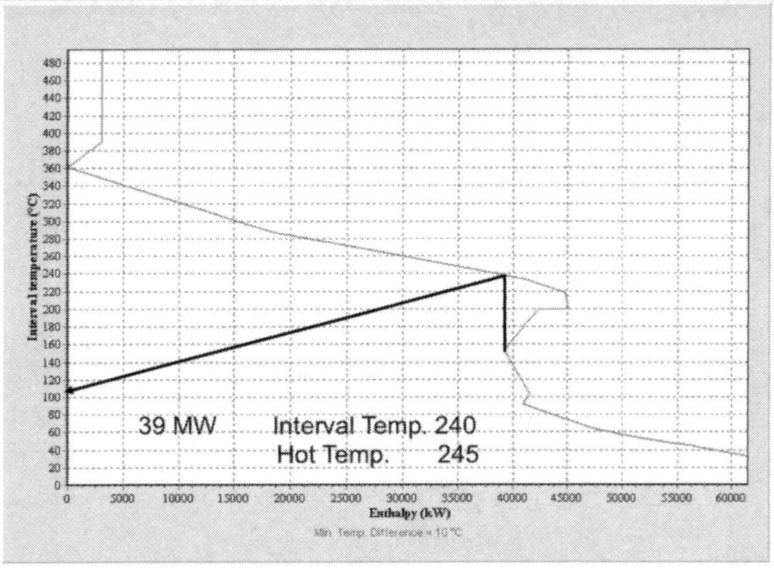

13 AIChE

Liquid Re-circulation: Determined from Other Zone

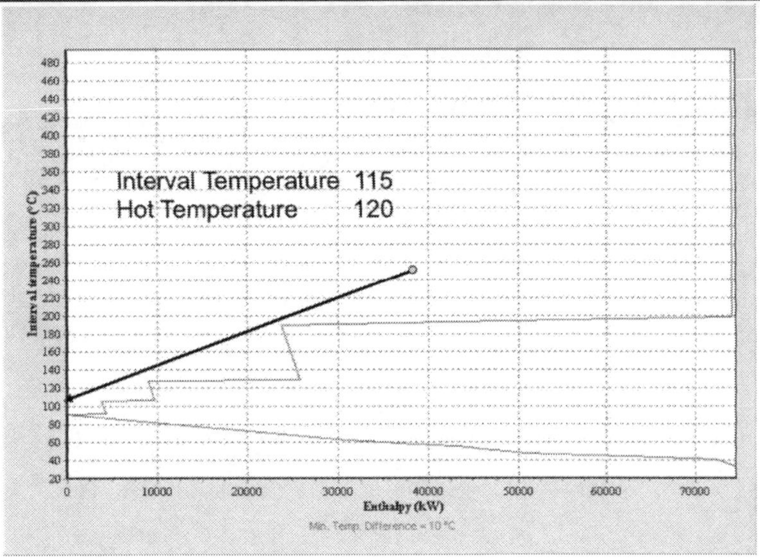

13 AIChE

Observations

- Hot oil used to drive 4 individual re-boilers
- Ring-main can be employed
- Piping costs relatively low.
- Temperature at which oil is used consistent with re-boiler duty
- Control very easy – rate at which oil is extracted from ring main
- Supplementary oil heater provides easy start up and good flexibility

13 AIChE

Total Site Analysis

13 AIChE

Overall Production Site

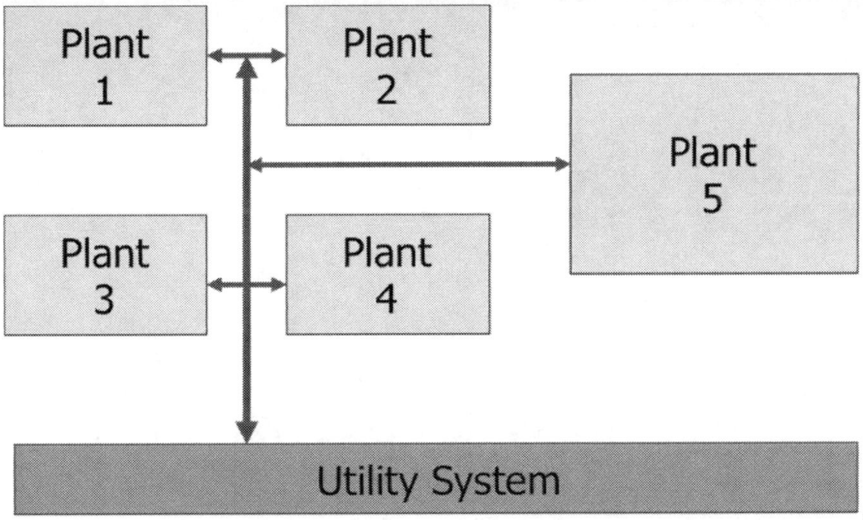

Individual GCC's

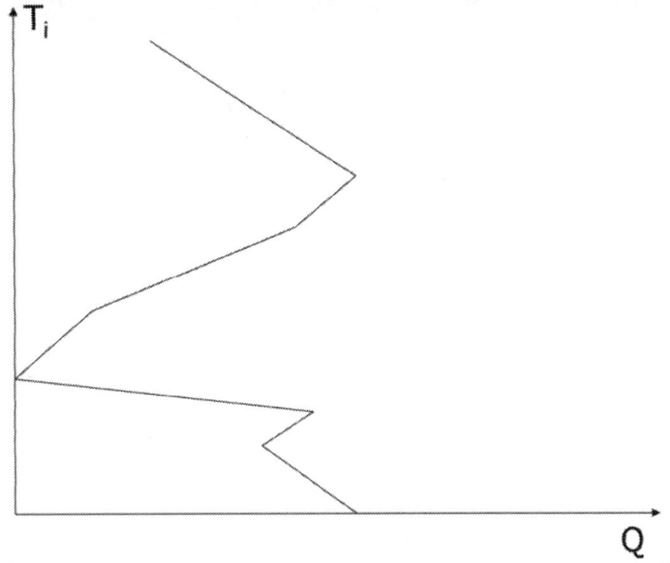

Demands

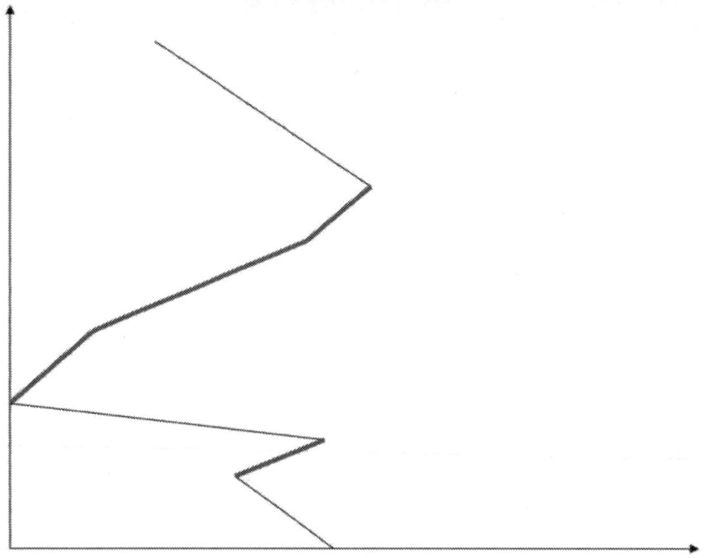

Supplies

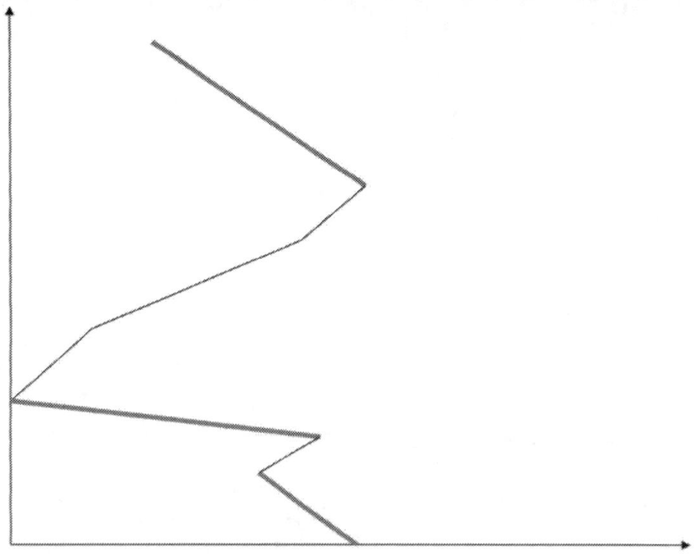

Site Composites

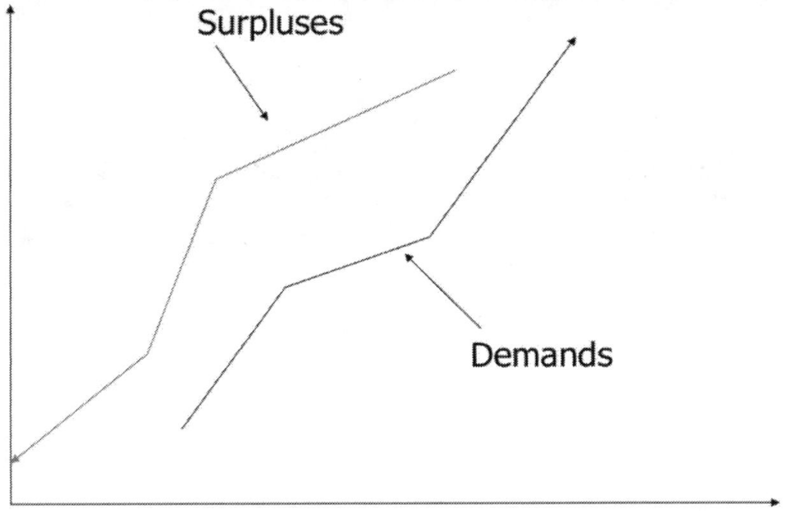

Site Composites

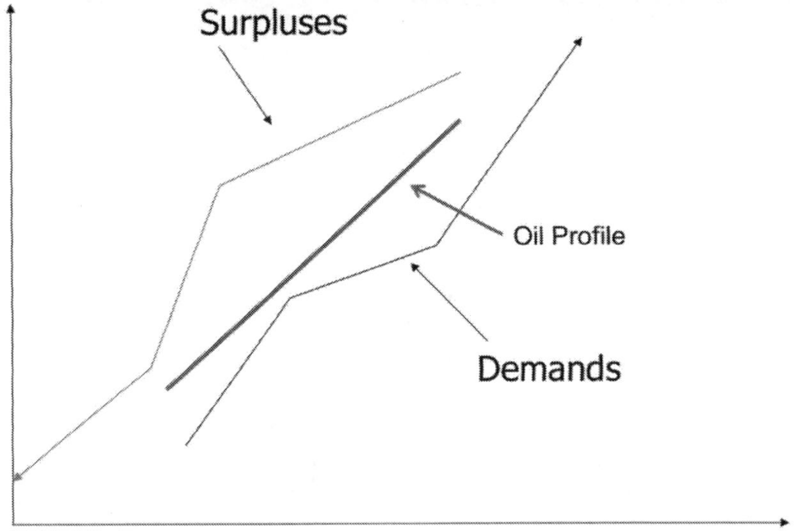

Linkage with Fired Heaters

13AIChE

Typical Furnace

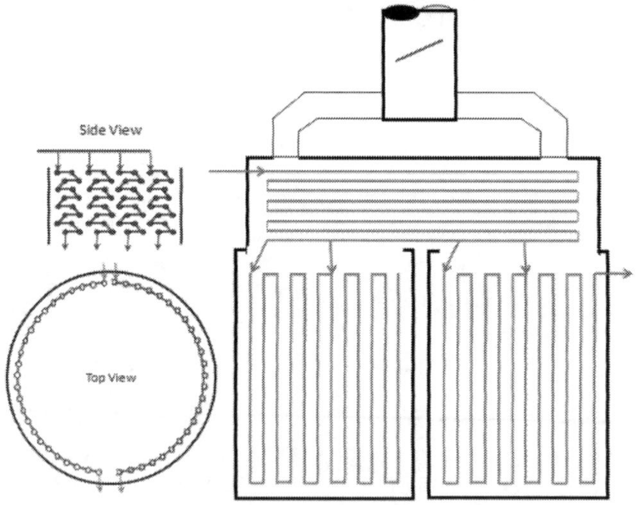

13AIChE

Reducing Fouling Rates in Fired Heaters

- Fouling rates in convective section of a fired heater are affected by the flow conditions on the crude side

- If crude enters as a single phase liquid fouling rates are high

Improved Efficiency

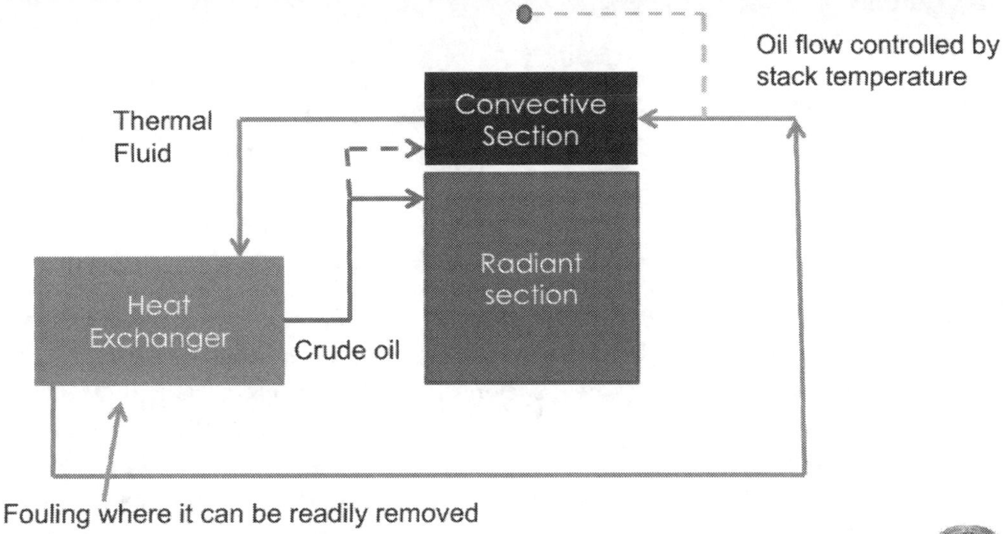

Solid State Carbon Sequestration by Utilization of Agro-wastes to Produce Value-added Products

Soh Fong Lim[1], Mohammad Omar Abdullah[1], Andrew Ragai Henry Rigit[2]

[1] Department of Chemical Engineering and Energy Sustainability, University Malaysia Sarawak, 94300 Kota Samarahan, Malaysia

[2] Department of Mechanical and Manufacturing System Engineering, University Malaysia Sarawak, 94300 Kota Samarahan, Malaysia

Abstract

Solid state carbon sequestration via utilization of agro-wastes to produce value added products such as biofertilizer and biosorbent is studied in this paper. The carbon sequestration is a sustainable development process which can be used to curb the negative impacts of human activities by reducing the emission of CO_2 via locking the carbon in solid form. Analytical analysis has found that both of the products contain relatively high percentage of organic carbon compare to other organic material such as soil. These might be due to the locking of carbon in solid form during the application of these products.

Keywords: agro-wastes, biofertilizers, biosorbents, solid state carbon sequestration.

I. INTRODUCTION

Around the globe, waste is piling up at an alarming rate and we need an effective waste disposal schemes very urgently while at the same time making use of the unwanted waste to provide clean energy or new materials via smart and efficient composting or recycling methods. The second important issue is linked to the goal towards global sustainability in terms of reducing greenhouse gas emissions, in particular CO_2 (Abdullah, 2012).

Solid state carbon sequestration is used to reduce carbon footprint and to minimize waste. The carbon footprint is formed by the existence of carbon dioxide (CO_2) gas in the atmosphere. CO_2 which is well known for driving global warming and climate change is the most important greenhouse gas emitted by industries and human activities. In this instance, agricultural activities are increasingly being required to account for its contribution to the release of CO_2. Thus, sustainability in the agricultural activities is important to conserve the natural environment and resources for our future generation. The sustainability should also extend beyond the economic well-being of the current generation and reflects the ability of future generations to meet their needs. These could be achieved by steering the agricultural activities along a more sustainable path using green technology or sustainable technologies.

Carbon sequestration is a sustainable development process that removes carbon footprint. This process can be used to curb the negative impacts of human activities by reducing the emission of CO_2 via locking the carbon in solid form such as in trees, soils, and other value-added solid products (eg. bio-fertilizer, bio-sorbent, etc.). There are numerous agro-wastes being produced in plantation and processing such as skins/peels, cores, pits, leaves, brunches, and pericarp. Agro-wastes are recognized as having hidden economic values which could be benefited by transforming the agro-wastes into useful products such as biofertilizer and biosorbent (Fortuna et al., 2003; El-Haggar, 2007; Kukal et al., 2009). Rivera-Cruz et al. (2008) reported that banana wastes are effective biofertilizer carriers for promoting plant growth and soil sustainability. Sud et al. (2008) reviewed the potential of agricultural waste materials in producing biosorbent for sequestering heavy metal ions from aqueous solutions. These agro-wastes largely compose of hemicelluloses, lignin, extractives, lipids, proteins, simple sugars, water hydrocarbon, starch, and etc. (Sud et al., 2008). Thus, these agro-wastes will be targeted for potential green technology application and sustainable development/practices.

2. MATERIALS AND METHODOLOGY

2.1 Materials

Banana peels (*Musa Acuminata*) that were used in this study were agro-wastes that were produced from local industry.

2.2 Biofertilizer

Composting method was used to produce biofertilizer from banana peels. 500 kg of banana peels were placed in a blender with 200 ml of water. Then, the blended banana peels were exposed to the sunray for 2 days (dehydration process). The dried banana peels were then put into a bin together with 50 g of sawdust, 250 g of lignite, and 25 ml of nipah palm sugar. The mixture was stirred for 5 minutes. After that, the bin was lid tightly and kept for 3 weeks for composting process purposes. Within the 3 weeks of composting process, compost pile was stirred frequently to allow the mixture to be well mixed and reach optimum temperatures as well as not overheat. After 3 weeks of the composting process, all the compost materials were removed from compost bin and the dehydration process was repeated for 4 hours. This is to ensure that the biofertilizer are in dry solid state (Matu, 2010; Muhammad, 2011).

2.3 Biosorbent

The banana peels were washed thoroughly with tap water to remove the adhering dirt and dried in a laboratory oven at 60°C for 58 hours. Then, the dried banana peels were crushed into a powder form using a laboratory grinder. The powdered banana peels were sieved to obtain a uniform size of the crushed banana peel. Finally, the powdered banana peels were thoroughly washed with deionized water to remove physically adsorbed contamination and dried in a laboratory oven at 80°C for a period of 24 hours (Sugumaran, 2010; Meikle, 2011).

3. RESULTS AND DISCUSSION

3.1 Value-added Products

Agro-wastes from banana industry mainly banana skins are used to produce two types of useful products; biosorbent and biofertilizer (Figure 1). These products are fabricated using simple and cost-effective methods. Both of the products have shown significant results in their applications. The biosorbent is able to remove crystal violet dye from aqueous solution. The maximum uptake of the crystal voilet dye is 245.74 mg/g. Removal of the dye is very fast, as it reached equilibrium in 3 hours of contact time. It is also found that the removal of the dye is highly pH dependent. Within the tested pH range (pH 2 to 12), the maximum removal of the dye is between pH 6 and pH 8. On the other hand, the biofertilizer which was fabricated using solid state fermentation method is able to increase the yield of crop which was proven by the plantation of the mustard and water spinach vegetables.

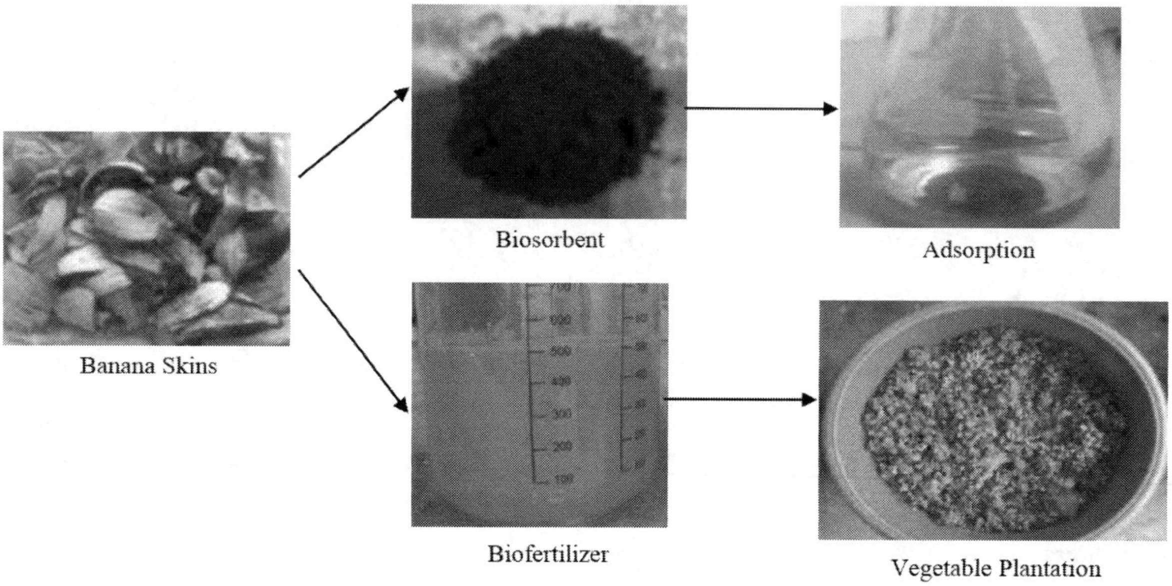

Figure 1 Flow diagram for value-added products from banana skins

3.2 Carbon analysis

Samples from the biosorbent and biofertilizer were analyzed to determine the amount of carbon content using Walkley-Black method (Saarani, 2011). Walkley-Black method is chosen based on its simplicity and rapidity (Islam and Weil, 1998; Schumacher, 2002; Krishan *et al.*, 2009). Walkley-Black method is wet combustion method where the amount of carbon is determined by the reaction of $Cr_2O_7^{2-}$ and H_2SO_4 without external heating. In this method, the amount of organic carbon content in the sample is determined from the difference of amount $FeSO_4$ used in the blank titration with the amount of $FeSO_4$ used for reaction with the remaining un-reacted part of $K_2Cr_2O_7$. The chemical reaction that involved in Walkley-Black methods is shown in equation (1).

$$2\ Cr_2O_7^{2-} + 16H^+ + 3\ C \rightarrow 4\ Cr^{3+} + 8\ H_2O + 3\ CO_2 \quad (1)$$

Desired amount of sample (W g) was placed in the 250mL Erlenmeyer flask. 5.0mL of 1.0*N* $K_2Cr_2O_7$ solution is added into the flask that containing the sample and the flask is swirled to mix the sample with $K_2Cr_2O_7$ solution. Then, 10.0mL of concentrated H_2SO_4 is added rapidly to the mixture and followed by adding 100mL of deionized water. The suspension is filtered out and ferroin indicator solution is added into the flask containing the solution. Then the solution is titrated with 0.5N $FeSO_4$ from 50 mL burette. As the endpoint approach, the solution takes on green cast and change into dark blue green. Then additional 0.2 mL of indicator is used to sharpen the endpoint. At this point, the $FeSO_4$ solution is added drop by drop until the colour change from blue green into brownish (maroon in reflected light) and the amount of ferrous sulfate solution used is recorded (*T* mL). The experiment is run again with blank and the amount of FeSO4 solution used is recorded as (*B* mL). The amount of

$FeSO_4$ solution in the back titration which is obtained from the analytical procedure is then be used to calculate the amount of carbon in the sample in percentage using equation (2).

Where,

B	=	the amount of $FeSO_4$ solution used to titrate blank
T	=	the amount of $FeSO_4$ solution used to titrate sample
N	=	the normality of $FeSO_4$ solution
W	=	weight of the sample in mg
3	=	equivalent weight of carbon in mg.
y	=	oxidation factor

Carbon analysis is performed on both of the value-added products. Table 1 demonstrates the percentage of carbon in biofertilizer using Walkley-Black analytical method. Based on Table 1, the amount of carbon in the biofertilizer is approximately 30.09 – 35.10 wt%.

Table 1 Amount of $FeSO_4$ used in back titration and percentage of carbon in biofertilizers

SAMPLE	Total amount of $FeSO_4$ used in back titration (mL)	Percentage of carbon (wt%)	
		y = 1.14	y = 1.33
Biofertilizer 1	5.60	31.16	36.35
Biofertilizer 2	2.58	29.01	33.85
Average		30.09	35.10

Figure 2 shows an increase of carbon content in biosorbent at different crystal voilet (CV) dye concentration. For both oxidation factors, the amount of carbon in the biosorbent was significantly increased as the dye concentration increased until it reached saturated state which is at CV dye concentration of 120 mg/L and above. This saturated state shows that the biosorbent is not able to absorb the CV dye anymore. Thus, the adsorbed carbon which is in CV dye structure remains constant when the adsorptions reach saturated state (Andersen *et al.*, 2009). Manocha *et al.* (2011) also stated that the formation of sample sizes and the interconnectivity between adsorbate and adsorbents can affect the rate of adsorption capacity. The weight of the biosorbents also can affect the percentage of absorption of the dye where as the weight of biosorbents increased, the absorption rate also increased (Raymundo, 2010).

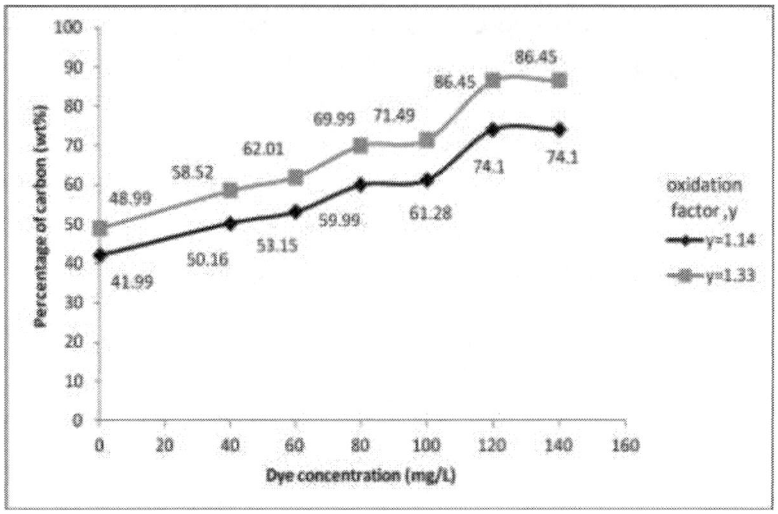

Figure 2 Percentage of carbon versus dye concentration in (mg/L)

Thus, analytical analysis has found that both of the useful products contain relatively high percentage of organic carbon compare to other organic material such as soil (Kim, 2005). Thus, the carbon is locked in solid form which could reduce the emission of CO2. This process is called carbon sequestration which is a sustainable development process to curb the negative impacts of human activities.

ACKNOWLEDGEMENT

The authors would like to thank Universiti Malaysia Sarawak for the use of laboratory facilities and equipments. This project is funded by Fundamental Research Grant Scheme (FRGS/02(17)/738/2010(24)) from the Ministry of Higher Education, Malaysia.

REFERENCES

A. Fortuna, R. Harwood, K. Kizilkaya, E. A. Paul, "Optimizing nutrient availability and potential carbon sequestration in an agroecosystem." *Soil Biology and Biochemistry*, Vol. 35, 1005-1013, 2003.

A. S. Raymundo, R. Zanarotto, M. Belisário, and M. de Godoi, "Evaluation of Sugar-Cane Bagasse as Bioadsorbent in the Textile Wastewater Treatment Contaminated with Carcinogenic Congo Red Dye," *Brazilian Archives of Biology and Technology* ,vol. 53, no.4, pp. 931-938, 2010.

B. A. Schumacher, *Methods For Determination of Total Organic Carbon (TOC) in Soils and Sediments* , pp. 1-18, 2002.

D. Sud, G. Mahajan, and M. P. Kaur, "Agricultural Waste Material as potential adsorbent for sequestering heavy metal ions from aquenous solution – A review." *Bioresource Technology*, vol. 99, no. 14, pp. 6017-6027, 2008.

F. S. Saarani, "Elucidation of Carbon Content via Laboratory Analysis," Bachelor of Mechanical and Manufacturing Engineering Thesis, Department of Mechanical and Manufacturing Engineering, Faculty of Engineering, Universiti Malaysia Sarawak, 2011.

G. Krishan, S. K. Srivastav, S. Kumar, S. K. Saha, and V. K. Dadhwa, "Quantifying the underestimation of soil organic carbon by the Walkley and Black technique – examples from Himalayan and Central Indian soils," *Current Science* , vol. 96, 2009.

H.T. Kim, *Soil Sampling, Preparation, and Analysis* (2nd Edition ed.). Boca Raton: Taylor & Francis Group, 2005.

K. R. Islam and R. R. Weil, *A Rapid Microwave Digestion Method for Colometric Measurement of Soil Organic Carbon*, 1998, pp. 2269-2284.

M. C. Rivera-Cruz, A. T. Narcía, G. C. Ballona, J. Kohler, F. Caravaca and A. Roldán, "Poultry manure and banana waste are effective biofertilizer carriers for promoting plant growth and soil sustainability in banana crops." *Soil Biology and Biochemistry*, Vol. 40, pp. 3092-3095, 2008.

M.O. Abdullah, *Applied Energy: an introduction.* New York: CRC Press (Taylor & Francis Group), 2012.

N. Muhammad, "Evaluation of Bio-Fertilizer from Agricultural Waste," Bachelor of Mechanical and Manufacturing Engineering Thesis, Department of Mechanical and Manufacturing Engineering, Faculty of Engineering, Universiti Malaysia Sarawak, 2011.

S. Manocha, J. Bhagat, M. Patel, N. Patel, and L.M. Manocha, Adsorption Behaviour of Carbons from Bio-Mass, pp. 727-731, 2011.

S. M. El-Haggar, "Sustainable Industrial Design and Waste Management." Academic Press, Chapter 7, pp223-260, 2007.

S. U. Matu, "Utilization of Agro-waste to Produce Biofertilizer," Bachelor of Mechanical and Manufacturing Engineering Thesis, Department of Mechanical and Manufacturing Engineering, Faculty of Engineering, Universiti Malaysia Sarawak, 2010.

S. S. Kukal, Rehana-Rasool, D. K. Benbi, "Soil organic carbon sequestration in relation to organic and inorganic fertilization in rice–wheat and maize–wheat systems." *Soil and Tillage Research*, Vol. 102, pp. 87-92, 2009.

T. A. Sugumaran, "Utilization of Agro Waste to Produce Biofertilizer," Bachelor of Mechanical and Manufacturing Engineering Thesis, Department of Mechanical and Manufacturing Engineering, Faculty of Engineering, Universiti Malaysia Sarawak, 2010.

V. Meikle, "Removal of Contaminants Using Biosorbent," Bachelor of Mechanical and Manufacturing Engineering Thesis, Department of Mechanical and Manufacturing Engineering, Faculty of Engineering, Universiti Malaysia Sarawak, 2011.

W. C. Andersen, S. B. Turnipseed, C. M. Karbiwnyk, R. H. Lee, S. B. Clark, W. D. Rowe, M. R. Madson, K. E. Miller, Quantitative and Confirmatory Analyses of Crystal Violet (Gentian Violet) and Brilliant Green in Fish. *U.S. Food and Drug Administration, Animal Drugs Research Center,* 2009.

(105b) Ni-CaO-ZrO2 Nanocomposite Catalyst for CO2 Reforming of CH4

Qingjun Chen[1], Jun Zhang[1], **Tiejun Zhao**[1] and Yuhan Sun[2], (1)Low Carbon Conversion Center, Shanghai Advanced Research Institute, CAS, Shanghai, China, (2)State Key Laboratory of Coal Conversion, Institute of Coal Chemistry, Chinese Academy of Sciences, Taiyuan, China

1. **1.** Introduction

Recently, utilization of natural gas to fuels and chemicals receives more attention due to its abundance over the world and less CO_2 emission. The production of syngas by CO_2 reforming of methane is one of the attractive routes to reach this goal. Compared with conventional steam reforming, CO_2 reforming reaction tends to easier coking formation and needs extra 20% heat to obtain similar conversion. Nickel-based catalysts have shown an excellent behavior in this reaction with a comparable activity to noble metal catalysts [1]. However, coke formation and sintering of nickel particles at a high temperature condition are often observed on Ni catalysts [2]. Developing a high-temperature stable catalyst is highly interesting for this reaction by controlled synthesis of catalyst. In this contribution, a stable Ni-CaO-ZrO_2 nano-composite catalyst for CO_2 reforming of CH_4 is prepared by a coprecipitation method. The effect of preparation condition is carefully investigated. The relationships between catalyst structures and activity of nanocomposite catalysts are discussed.

1. **2.** Experimental

Ni-CaO-ZrO_2 catalysts were prepared by a coprecipitation method with a reflux-digestion process. After precipitation, the obtained slurry was washed and refluxed in de-ionized water, then filtered, dried and calcined at 600 °C for 5 h. The catalysts prepared at different reflux-digestion condition were assigned as NC-X-Y, where X and Y are the reflux temperature and time, respectively. The catalyst without reflux was assigned as NC-0. The reforming reaction was conducted at 850-900 °C in a quartz fixed-bed reactor at atmospheric pressure. Obtained catalysts are tested by feeding the reaction mixture of CO_2 and CH_4 (molar ratio 1.2: 1) without dilution and a gas hourly space velocity (GHSV) of 79 000 ml/(hg) . In addition, comparison of calciantion procedure and coking quantity was performed at different operating conditions.

1. **3.** Results and discussion

3.1 **Physical and chemical properties of the Ni-CaO-ZrO_2 catalysts**

The pore structure parameters are collected in Table 1. The catalyst without reflux-digestration has the smallest specific surface area (63 m^2/g) and pore volume (0.09 cm^3/g). With the increase of the reflux time, the specific surface area and pore volume of the catalysts increased. NC-100°C-24h exhibits the largest surface area of 245 m^2/g. Reflux temperatures also have a significant effect on the pore structures of the catalysts. At the same reflux time, the specific surface areas of the catalysts increased with increasing the reflux temperature.

Table 1. Pore structure parameters of the catalysts

Samples	S_{BET} m^2/g	V_t cm^3/g	D_m nm

NC-0	63	0.09	5.58
NC-100°C-6h	152	0.18	4.73
NC-100°C-12h	176	0.21	4.85
NC-100°C-24h	245	0.30	4.95
NC-60 °C-24h	92	0.13	5.83
NC-80 °C-24h	160	0.21	5.31

The XRD patterns of the catalysts prepared at different reflux-digestion condition are shown in Figure 1. The catalyst without reflux has strong diffusion peaks of ZrO_2 and NiO. As the increase of the reflux time and temperature, diffusion peaks assigned to ZrO_2 and NiO became weaker and wider. This result indicates that reflux digestion improves the dispersion of NiO and decreases the size of ZrO_2 crystals in the catalysts.

TPR profiles of Ni-CaO-ZrO_2 catalysts are illustrated in Figure 2. The peak or shoulder at about 400 °C is attributed to the reduction of large NiO particles, while the peaks at 600 °C and 700 °C may belong to the reduction of NiO that have the weak and strong interactions with ZrO_2, respectively. For the catalyst without reflux, NiO mostly has the weak interaction with the carrier or is in the form of bulk particles. At the reflux temperature of 100 °C, the interaction between NiO and ZrO_2 became stronger with the increase of reflux times (Figure 2 a). However, when the reflux temperature is lower than 100 °C, the interaction between NiO and carrier was not significantly enhanced (Figure 2 b). So the high reflux temperature and long reflux time are key factors to prepare the catalysts with strong interaction between metal and the carrier.

Figure 1. XRD patterns of the catalysts   Figure 2. TPR profiles of the catalysts

3.2 **Activity of the Ni-CaO-ZrO_2 catalysts for CO_2 reforming of methane**

Results of the catalytic tests at 850 °C with a high GHSV (CO_2/CH_4=1.2:1) of 79, 000 ml/(h·g) are shown in Figure 3(a). The catalyst without reflux has a low activity with the CH_4 conversion of about 70 %. In addition, CH_4 conversion of this catalyst decreased with time and only 55% (conversion) is achieved after reaction on stream for 150 h. A higher conversion of CH_4 and stability (very close to equilibrium) is observed on the catalyst refluxing at 100°C for 24 h. Therefore, prolonging the reflux time will increase activity of Ni-CaO-ZrO_2 catalysts for CO_2 reforming of methane. The catalyst with 24 h reflux-digestion (NC-100 °C-24h) exhibits a highly stable activity over 1000 h without obvious deactivation at a GHSV of 79, 000 ml/(hg) at850 °C (Data are not shown) .

Figure 3. Catalytic performance of catalysts for CO_2 reforming of methane

(a) Effect of reflux temperature and time; (b) Effect of calcination temperature; (c) Effect of calcination procedure

Furthermore, the calcination temperature effect on the catalyst activity is carried out at similar operating conditions. It is found that the activity decreases when the sample is calcined at static air at a higher temperature. The NC-100-24h sample calcined at 900°C presents a very low activity (Fig.3b) . XRD examination indicates that monoclinic ZrO_2 is gradually formed and particle size of ZrO_2 becomes larger. TPR measurement indicates no obvious peaks for NiO reduction for the sample calcined at 900°C. These observations indicate that the sintering occurs at high temperatures. However, if the catalyst is directly reduced in the mixtures of H_2 and N_2 at 850°C or 900°C for 3 h, the conversion of CH_4 still can maintain a high CH_4 conversion (Fig. 3c), implying that the calcination procedure has a profound impact on the activity of catalyst. Interestingly, the catalyst performing at 900°C gives a lower carbon deposition than 850 and 800°C after time on stream for 200 h. The full operating window will be given by a more systematic study afterwards.

1. **4. Conclusions**

A high-temperature stable Ni-CaO-ZrO_2catalyst is developed by the coprecipitation method with a reflux-digestion process. Reflux digestion can improve the dispersion of NiO and enhance the strong interaction between metal and support. The obtained Ni-CaO-ZrO_2 catalyst can be further improved by modifying the calciantion procedure.

References

[1] J. R. Rostrup-Nielsen, J. H. B. Hansen, ***Journal of Catalysis*** 144 (1998).

[2] C. J.Liu, J. Y. He, J. J. Jiang, Y. X.Pan, ***ChemCatChem*** 3 (2011)

(105c) Gas Cleanup Using Elastic Layered Metal-Organic Framework Adsorbents

Francisco Sotomayor and **Christian M. Lastoskie**, Civil & Environmental Engineering, University of Michigan, Ann Arbor, MI

The suitability of elastic layered metal-organic frameworks (ELM) for the removal of carbon dioxide from methane-rich gas mixtures was evaluated using molecular modeling and experimental measurements for two adsorbents, ELM-11 [$Cu(BF_4)_2(bpy)_2$] and ELM-12 [$Cu(OTf)_2(bpy)_2$]. These materials are promising absorbents for CO_2 separation from biogas and natural gas mixtures, due to their reversible gated isotherm features and their high selectivity adsorption of CO_2 over CH_4 (S = 50-100 at 298 K) from gas mixtures with CH_4 and CO_2 compositions representative of biogas / natural gas extraction streams.

Grand canonical Monte Carlo (GCMC) simulation was used to calculate the framework adsorption capacity and CO_2 selectivity from binary CH_4 / CO_2 mixtures. Simulated CO_2 capacities above the gating pressure and CO_2 isosteric heats of adsorption agree well with ELM-11 and ELM-12 experimental data. Ideal adsorbed solution theory (IAST) estimates of CO_2 selectivity are generally in good agreement with the GCMC simulations, although modifications of the IAST methodology are needed in order to take into account the co-adsorption condition for CH_4 uptake by CO_2 gate-opening. Density functional theory (DFT) calculations are also presented for CO_2 and CH_4 interactions with the ELM-11 and ELM-12 framework structures. Upon the adsorption of CO_2 in the ELM framework, the metal-anion bond of the framework was extended, whereas the same bond length is virtually unchanged upon CH_4 adsorption. DFT analysis of the adsorption binding energies of other trace components in natural gas and biogas production (e.g. H_2S, H_2O, N_2) and CH_4 will also be presented and discussed.

(105d) Effect of Preparation Parameters On Mixed Amine Functionalized Silica Sorbents for CO2 Capture

Adefemi Egbebi[1], Sonia Hammache[1] and McMahan L. Gray[2], (1)URS-NETL, Pittsburgh, PA, (2)U.S. Department of Energy, National Energy Technology Laboratory, Pittsburgh, PA

Effect of preparation parameters on mixed amine functionalized silica sorbents for CO_2 capture

Adefemi Egbebi[1, 2], Sonia Hammache[1, 2], and McMahan L. Gray[1]

[1]US DOE-National Energy Technology Laboratory, Pittsburgh, PA

[2]URS, Pittsburgh, PA

Global climate change has been linked to growing atmospheric concentrations of carbon dioxide. Therefore the development of practical methods to remove, reduce or prevent the release of anthropogenic CO_2 is prudent to slow this increase. The most mature technology for the separation of CO_2 from flue gas streams from power plants is via liquid amine scrubbing which is considered expensive and energy-intensive. The potential therefore exists for the development of alternative CO_2 removal technologies, including the use of solid amine-based sorbents among others, capable of less energy-intensive CO_2separation.

Previous work in our laboratory investigated co-impregnated mixed amine (Polyethyleneimine (PEI) + 3-aminopropyltriethoxysilane (APTES) / Silica) solid sorbents with the goal of improving sorbent performance in terms of its CO_2 uptake capacity, thermal and hydrolytic stability as well as the kinetics during sorption and regeneration cycles. Results showed improved CO_2capture capacity and stable performance of the mixed amine sorbent relative to a single amine (PEI) sorbent.

Current work investigates the effects of synthesis parameters/variables such as bath temperature, solvent type and impregnation sequence on sorbent performance. Modified Pressure Swing Adsorption (PSA) tests were conducted with a thermogravimetric analyzer (TGA) on two sorbents with same composition of 15 wt. % PEI: 15% APTES: 70 % SiO_2 but with different solvents used during their syntheses - methanol for one and methanol-hexane mixture for the other. Adsorption was performed using dry 10% CO_2 /N_2 feed gas and desorption was conducted with 100% N_2 feed gas while temperature was kept the same at 100°C. Preliminary results show negligible differences between the sorbents in terms of CO_2uptake capacity and thermal stability.

The effect of various preparation parameters will be elucidated from the results of TGA, fixed bed reactor and other characterization experiments on sorbents and will be presented.

CO2 Reduction Efforts for Canadian Oil Sands Industry

Ravishankara Subraya, Oscar Aguilar

The Quest project is a joint venture of the Athabasca Oils Sands Project partners (Shell Canada, Chevron Canada Limited, and Marathon Oil Canada Corporation) to fully integrate CCS technology to capture, transport, inject, store, and monitor CO_2 from the Scotford Upgrader in the Industrial Heartland Region near Fort Saskatchewan, Alberta. CCS is an important part of both Canada's and Alberta's CO_2 reduction plans. Project is currently under execute phase. This facility will sequester 1.08 Mtpa of CO_2 on completion in Q1 2015. The technical presentation will cover technology selection, process & utility integration challenges and key learning from the project.

(76a) Modeling Dataset Selection in Multivariate Statistical Process Control

Zheng Li, Department of Chemical Engineering, Texas Tech University, Lubbock, TX and **M. Nazmul Karim**, Chemical Engineering Department, Texas A&M University, College Station, TX

Using methods from multivariate statistical analysis, statistical process control has found wide applications in different industrial processes. Owing to the data-based nature of statistical process monitoring, it is relatively easy to apply to real processes with nonlinear and time varying behavior, in comparison with other methods based on systems theory or rigorous process methods. In a typical industrial application of statistical process control, the purpose is diagnose an evolving batch as normal or not, and to assist the trouble shooting during batch process. For the application of batch process in biochemical process, especially process with limited number of total runs, knowing of minimum number of historical data is really important. In this paper, a simple approach to calculating the minimum number modeling dataset for a validate model is developed. In addition, the proposed approach effectively estimates the extent of capturing meaningful variations and analyzes the stability of the partial least square (PLS) model. The method is illustrated by several industrial applications.

(76b) Continuous Data Mining to Sustain Advanced Process Control and Process Capability

John Antanies, Envoy Development, Phoenix, AZ

Multiple factors can affect the performance of advanced controls. These may be due to changes associated with the instrumentation, regulatory controls, or the process itself. If not detected and diagnosed, the effect can be a partial loss of benefits, or in the worst case, the complete loss of benefits due to the controls being switched off. It is therefore important to monitor not just control-specific metrics, but also process measurements and process metrics (KPIs) that may not directly be used by the advanced control system. This presentation will describe a "big data small footprint" approach to statistically analyzing large amounts of historian data to determine when significant changes occur, and identifying candidate variables to facilitate cause and effect determination. Plots and reports are automatically generated for control engineers, process engineers, and instrumentation technicians. A benefit of the approach is bringing these groups together to solve problems and ensure process objectives are met. Examples are presented to highlight the approaches and statistical techniques used.

(76c) An Advanced Control Platform for Smaller Plants

John Keenan, Optima Powerware, Houston, TX

The cost of implementing advanced control can be prohibitive for certain plants in the processing industry. These tend to be smaller plants where incremental advanced control benefits do not translate into the large savings that are typically used to justify advanced control in large facilities. However, there may still be opportunities for achieving significant benefits in these smaller plants. Beyond the cost issues, many smaller plants also face challenges: Maintaining the skill set to implement advanced functions at the DCS or PLC level; Implementing on the various hardware that may be in use; and operating with minimal or no control room operators.

To meet the needs of smaller plants, an integrated control platform called AutoPilot has been developed and applied to a range of processes over the last 15 years. In the following, the key components - MPC, fuzzy logic, inferred values, external calculation, and optimization - are described and specific technical features that have proven useful highlighted. Examples are presented which show the capabilities and application of the product in midstream, chemical, and mining industries.

(76d) A Case Study of the Use of Integrators in Model Predictive Control

Chad Segura, Advanced Control Solutions, Houston, TX

In the practice of model predictive control (MPC), for long time to steady state processes, it can be advantageous to use an integrator for some controlled variables to reduce the controller time-to- steady-state. In the terminology for one commercial model predictive control package, DMCPlus®, the controlled variable would be set up as a pseudoramp. Several examples are outlined which will give the audience information on when to use or not use integrators within the most used commercial MPC packages.

(76e) Non-Invasive Closed Loop Step Testing Technology for MPC Applications

Quinn Zheng, Michael Harmse, Hong Zhao and John Campbell, Aspen Technology, Inc., Houston, TX

Multivariable Predictive Control (MPC) is the most widely used advanced process control technology in the process industries. For instance, there are more than 6,000 DMCplus based applications currently in service worldwide. A common and challenging problem is that MPC control performance degrades over time due to inevitable changes in the underlying process unit, due to fouling, catalyst deactivation and mechanical wear. Among all possible causes of control performance degradation, poor model accuracy is the primary factor in most cases. To sustain good control performance, the model needs to be periodically recalibrated.

However, it is a technically challenging and resource-intensive task to pinpoint a problematic model curve and re-identify a new model curve for replacement in a MPC application. In a large scale MPC application, over a hundred variables may be involved, and there are often more than 500 model curves in a typical MPC model matrix. Conducting a re-test and re-identification by a conventional approach may take an experienced engineer several weeks of intensive work and cause significant interruption to the normal operation. The process industry has been looking for an automated, safe and less invasive approach to conducting plant testing for many years. Over the past ten years, some technologies and tools have emerged to address this need; technologies such as step testing in an automated way with the process safeguarded by a specially designed multivariable controller. Further the automated test can simultaneously perturb multiple input variables instead of the traditional single variable test, greatly increasing the average signal-noise ratio, leading to superb model accuracy. Even using these tools, the plant test will inevitably have a negative impact on normal process operation: the process may have to operate in an economically unfavorable condition for one to two weeks at a time. Therefore, the available technology in this field is still considered too invasive for the process industry.

In this paper, an innovative approach for non-invasive closed loop step testing is presented. It can be configured and executed to generate the required test data for model quality assessment and model re-identification, while also minimizing any negative impact to the underlying process operation by using a tunable trade-off between optimal process operation and process perturbation. More importantly, this technology can tolerate high model uncertainty in that it can keep an otherwise inoperable controller running smoothly with certain trade-offs in economic optimization. The new technology provides the process industry an opportunity to run a small amplitude background step test over a relativelive long period that is suitable for model identification purposes while can still keep the controller runing. Such a test will not move too far away from optimal control targets therefore minimizing the interruption to the normal operations.

Multiple industry applications and simulation results are used to illustrate the advantages of the new technology, including: maintaining robusteness when the model is significantly degraded, keeping the process operating inside a sub-optimal range specified by maximum allowed economic give-away, and generating step testing signals maximizing the signal-noise ratio subject to both the process constraints and operational optimization requirements.

AN AFFORDABLE NON-LINEAR APC TECHNOLOGY FOR DISTILLATION COLUMNS USING FIRST PRINCIPLE MODELS

Bert Pluymers, IPCOS NV, Leuven, Belgium
Edwin M.C.J. Weetink, IPCOS BV, Boxtel, The Netherlands

Introduction

This paper discusses a novel APC technology[1] that has been developed specifically for control and optimization of distillation columns. Traditional APC solutions make use of linear, empirical models extracted from plant test data to control certain key variables around the desired operating point. The disadvantage of this traditional approach is the cost of model building (especially for very slow processes) and the limited operating range within which acceptable performance can be obtained.

The new approach discussed in this paper makes use of first principle distillation column models that are directly integrated in the APC controller. This results in higher-performance APC controllers that can deliver larger benefits over wider operating ranges. At the same time, no dedicated plant tests need to be performed, reducing the cost of obtaining the required models. Finally, additional process insights can be obtained due to the first principle nature of the models.

After a detailed discussion of this technology, which is marketed as INCA MPC4Distillation [1], the results of a real-life application are presented in the last part of this paper. These results clearly show the benefits of this new APC technology.

Advanced process Control

Advanced Process Control (APC) is a technique that has been used for several decades already to control and optimize production facilities, among which refining, (petro-)chemical and related plants. An APC controller makes use of a *dynamic process model* and acts as a supervisory layer on top of the base layer (PID) controllers implemented in the DCS and makes sure the plant is operated more tightly and more optimally. There are three key features that enable the benefits that an APC controller can deliver:

- ***Predictive***: by using the process model, an APC controller can predict future effects of external disturbances and past control actions and can therefore pro-actively compensate for these effects.
    - E.g. increase steam flow to a distillation column in response to a load increase, to prevent the product composition to go out of spec.

- ***Multiple-Input, Multiple-Output***: as opposed to a PID controller, which can only control one key process variable by manipulation one other process variable (i.e. single-

[1] INCA MPC4Distillation, where INCA stands for IPCOS Novel Control Architecture. [2]

input, single-output control), an APC controller can control and manipulate multiple process variables simultaneously. This allows the controller to take into account all the interactions that might be present in the plant, which is something a set of independently operating PID controllers cannot achieve.

- E.g. accurately balance steam and reflux flows in a distillation column in order to change one product composition without changing other compositions.

- ***Optimizing***: While base layer controllers will typically try to keep its controller process variable close to a specified target, APC controllers can actually try to push certain key variables towards more optimal values, within the operational constraints that are imposed.
  - E.g. maximize throughput in a distillation column, while keeping product compositions within specifications and keeping the top cooling periphery in control.

By means of these three main features, APC controllers can use a *squeeze and shift* strategy in order to deliver benefits: due to its predictive and multiple-input, multiple-output capabilities an APC controller will be able to significantly reduce the process variance (*squeeze*). This can then be exploited to move (*shift*) the process to a more optimal operating point, without increasing the risk of exceeding the operational constraints of the process.

It goes without saying that the performance of such a controller highly depends on the accuracy of the embedded process model. On the other hand, the engineering effort spent in obtaining an accurate process model is a significant part of the overall design cost of an APC controller. Therefore during the design of an APC controller and especially when designing new APC technology special attention needs to be paid to this area of tension.

This paper discusses a new APC technology that makes it possible to design with limited effort (and hence cost) non-linear APC controllers for distillation columns, using non-linear first principle models.

Empirical Models for APC

In this section a brief overview is given of the empirical modeling techniques that are typically used for APC, and their advantages and disadvantages are highlighted. Empirical modeling means that the dynamical models needed for APC are built using sufficiently rich process data. This typically involves *plant testing*, where the process is disturbed by applying step changes to the most important process inputs that will be manipulated by the APC controller (MVs), while continuously monitoring and logging the key process variables that need to be controlled/optimized (CVs). This can vary from single step changes that are applied sequentially to the different process inputs to entire (random or predefined) step sequences that are applied simultaneously.

The way in which such data is obtained, as described above, already shows one of the disadvantages of pure empirical modeling. Since such plant tests typically span large multiples of the time-to-steady-state of the process, testing can take days to weeks. Over this period of time the process needs additional supervision and regular production might be disturbed, both of which can incur a significant cost.

Traditional Linear Models

Traditional linear models express the behavior (as a function of time) of each CV when a unit step change would be applied to each MV. These relationships are often referred to as *model curves*. Essentially, such curves describe:

- The static gain (final change in the CV value for each unit change in the MV value)
- Dynamical information
  - Initial delay
  - Time to steady state
  - Rise time
  - Overshoot and/or any initial inverse response
  - Amount of oscillation/damping present after the initial response

All the above aspects are assumed to be invariant in time, irrespective of the operating point (load, feed composition, column pressure, product specifications). For many columns (high-purity columns, difficult separations, some multi-component or azeotropic separations) this assumption is a stretch, meaning that acceptable APC performance cannot be achieved over the envisioned operational range (load range, product spec. range, feed composition range) due to the fact that the models are not accurate throughout that operational range.

Static Transformation

In order to overcome the above problems, static non-linear transformations can be added to the above described linear empirical models. To some extent this can overcome changes in the static gain of the process, but typically dynamics also depend on the operating point. Furthermore, interactions between key variables (e.g. a change in gain from steam flow to bottom composition, due to a change in the top pressure) cannot be taken into account either.

Gain Scheduling

One of the classic approaches of dealing with non-linear processes is the gain scheduling or multi-model approach. In this approach multiple linear models are built for different operating points. The APC controller will switch (or interpolate) on-line between these models, depending on the operating point. It is obvious that this approach exacerbates the disadvantages of performing long plant tests, since now a multitude of such tests need to be performed, for each of the different operating regions.

Full Non-Linear Empirical Modeling

Finally, another alternative would be to try to build a completely non-linear black-box model using e.g. neural networks. However, also here the excessive data requirements directly make this approach economically unviable.

First Principle Models for APC

A modeling approach that can yield superior results to empirical models is the use of first principle models. Such models rely on physical, chemical, thermodynamic, etc. relationships rather than purely on process data, to establish the relationships that the APC controller needs to control the process. Also within this class of models, different approaches are possible.

Rigorous Modeling

One approach is to completely rigorously model all phenomena that are known to take place in the distillation column under consideration. This includes detailed tray hydraulics, rigorous pressure model, mass balance of all components, detailed energy balance, etc… Similar approaches are taken in two related contexts:

- **Column design**: In typical software packages a rigorous approach is taken to build static models with the aim of obtaining a suitable column design prior to the actual building of the column. This context, however, induces significantly different technology requirements:
  - There is no need to make sure that, when the column is actually operational, the models are able to predict future (next several hours/days) behavior of the column. As a result, no functionality is needed to keep the model aligned with the actual process (cfr. observer functionality, see later)
  - No process data is available, so it is not possible to fit certain physical parameters to this data. This means that all physical properties, tray/packing design specifications, etc… need to be known in detail.
  - Dynamic process behavior does not need to be captured.

- **Real-Time Optimization**: In real-time optimization applications a static, highly detailed process model is typically used to perform regular optimization runs that provide optimal steady-state targets to an APC controller (which uses empirical models). Due to their relatively large cost such solutions are limited to a select few processes for which an RTO application can generate sufficiently high benefits. For most APC applications such a high cost would be excessive and would prohibit actual implementation.

Simplified First Principle Modeling

In the specific context of first principle modeling for APC, several beneficial circumstances can be exploited to make this approach economically viable:

- **Historical plant data**: the advantage – compared to the column design context – is that actual plant data is already available. This removes the need for many of the hard-to-obtain physical, chemical and thermodynamic parameters. Examples are:

  - ***Tray efficiency***: in order to predict tray efficiency, detailed low-level information on the mixture, flow regime, tray design, etc… is needed. This information is often hard to obtain and can also change over time and depending on the actual operating point. However, when lab samples of product stream compositions are available, efficiency can be extracted from actual plant operational data.

  - ***Pressure drop***: also for this property, typically very low-level information on the tray or packing design is needed, which is often not available. Also, as trays or packings foul, pressure drop will typically also evolve. However, when measurements are available, the model can be calibrated to these measurements, without the need for such detailed information.

- **Required level of detail:** for control purposes, different requirements exist in terms of necessary model accuracy. Due to the fact that APC controllers have both feed forward (i.e. prediction) and feedback (i.e. adapt to incoming information) capabilities, a slight model mismatch is often acceptable. Note that due to the non-linear character of the first principle model, the accuracy is already significantly better than "classical" linear empirical models. On top of that, several phenomena (e.g. phenomena related to column startup) are less relevant for control purposes and hence do not need to be present in an APC model.

- **Adaptivity:** when working in an on-line context, where e.g. every minute new process measurements are obtained, new data can be used to continuously adapt certain model parameters to make sure the model is kept in line with the real process. This again reduces the need for full rigorous modeling, since many (small) mismatches can be compensated for in this way.

It is this approach of simplified first principle modeling that is used in the INCA MPC4Distillation technology discussed in this paper. The next section will first elaborate on how this new modeling approach was designed. Afterwards, we discuss how these models are integrated in the actual APC controller. Finally, some real-life results are discussed.

Novel First-Principle Distillation Modeling Approach

Modular modeling approach

First of all, in order to maximally enable reuse of existing models or model structures, a modular modeling framework was set up, where models of distillation columns can be constructed from basic building blocks that each represent one part of the column. Using the following pre-defined model blocks a wide range of column topologies can be constructed:

- ***Section***: part of a column (trayed or packed) without external flows.
- ***Feed/sidedraw tray***: single tray with (in- and/or outgoing) external flows.
- ***'Reboiler'***: bottom part of a column, including the reboiler
- ***'Condenser'***: top part of a column, including condenser

This is illustrated in Figure 1.

These pre-defined model blocks represent the dynamic behavior (flows, temperatures, compositions, pressure) of one part of the column and are based on simplified first-principles relationships. Essentially, all these blocks contain a simplified mass and energy balance model over the different (theoretical) trays that are modeled as illustrated in Figure 2.

Each block contains some "free" parameters that are column-specific and need to be set/identified by the user. Free parameters were designed such that they match directly with process information that is typically available (e.g. number of trays, length of packing, volume of condenser vessel) or that can easily be identified from available process data (e.g. pressure drop coefficients, mean tray efficiency, etc.).

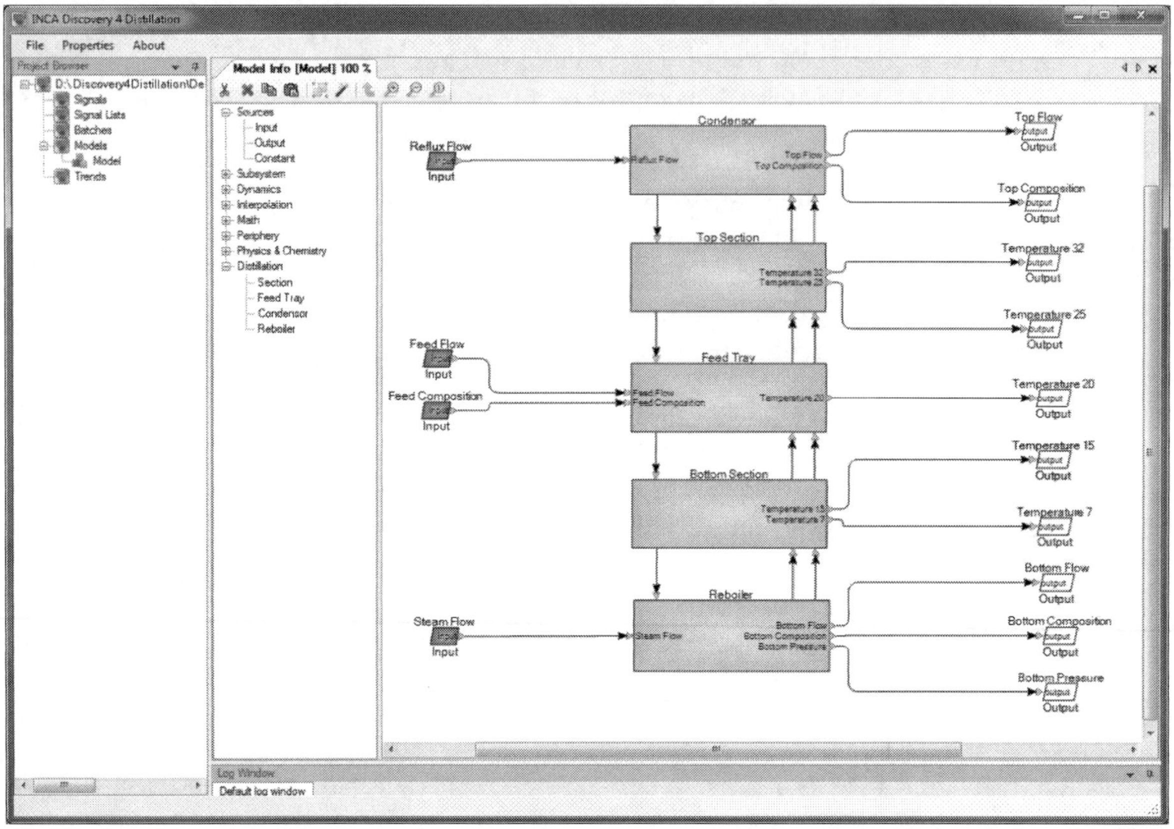

Figure 1. Modeling tool enabling easy construction of various distillation column topologies, illustrated here with the most basic column topology (for clarity reasons). Multiple feed streams, side-draws, etc. are possible with this modular approach.

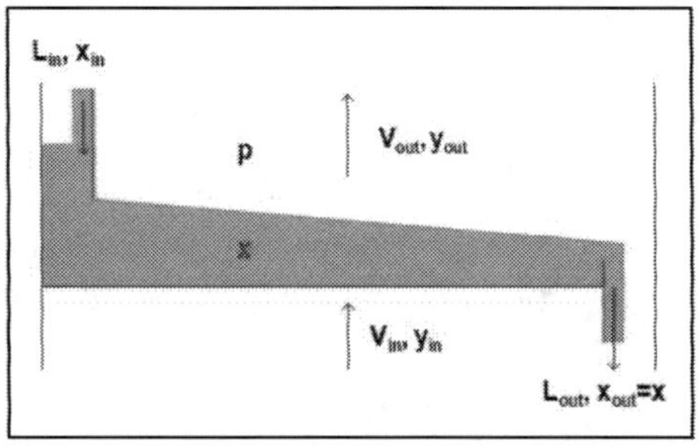

Figure 2. Illustration of the mass balance over a single (theoretical) tray.

Finally, a description of the mixture that is being separated in the column also needs to be specified. Flexible models for describing vapor-liquid equilibria (VLE) of such mixtures and physical properties of the components in the mixture are provided within the technology. A wide range of mixtures (binary, multi-component, (a)zeotropic) can be described, an example of which is illustrated in Figure 3.

Modeling workflow

The workflow of building a non-linear first-principle model for APC using the INCA MPC4Distillation technology, can be summarized as follows:

1. Define mixture properties and its components
2. Build column topology using pre-defined model blocks and set known parameters (#trays etc.)
3. Estimate remaining free parameters using historical process data:
    - Fit parameters of pressure drop model using available pressure measurements
    - Estimate mean tray efficiency and related static parameters using available temperature and composition measurements
    - Estimate dynamic parameters using historical process data (column holdup, liquid dynamics, etc.)
4. Build custom models (using available generic model blocks) for the existing periphery that is relevant for APC (e.g. cooling equipment related to the condenser that might be throughput- or pressure-limiting)

With this approach building a first-principle model for APC could be completed in a matter of hours for the relatively simple columns.

Process insight and inferentials

One of the major advantages of first principle modeling is the additional process insight that is obtained. Due to the fact that on each (theoretical) tray of the column the energy and mass balance is computed, a detailed profile of the most important column variables (liquid composition, temperature, pressure, liquid flow, ...) can be visualized. This is illustrated in Figure 4.

An additional advantage that is directly linked to the above observation is the fact that inferential sensors (i.e. models that predict certain key process variables that are difficult to measure directly and/or frequently) can very easily be added to the model. Since e.g. compositions are computed at each position in the column anyway, adding a soft sensor for one such composition simply involves adding a model output in the modeling tool and calibrating it to the available (lab) measurements.

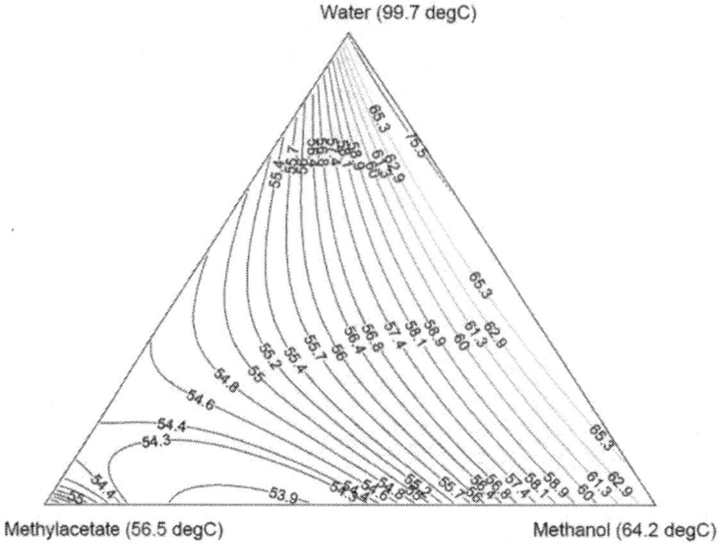

Figure 3. Illustration of the boiling temperature of a ternary azeotropic mixture as described by the INCA MPC4Distillation modeling technology using a Wilson activity model.

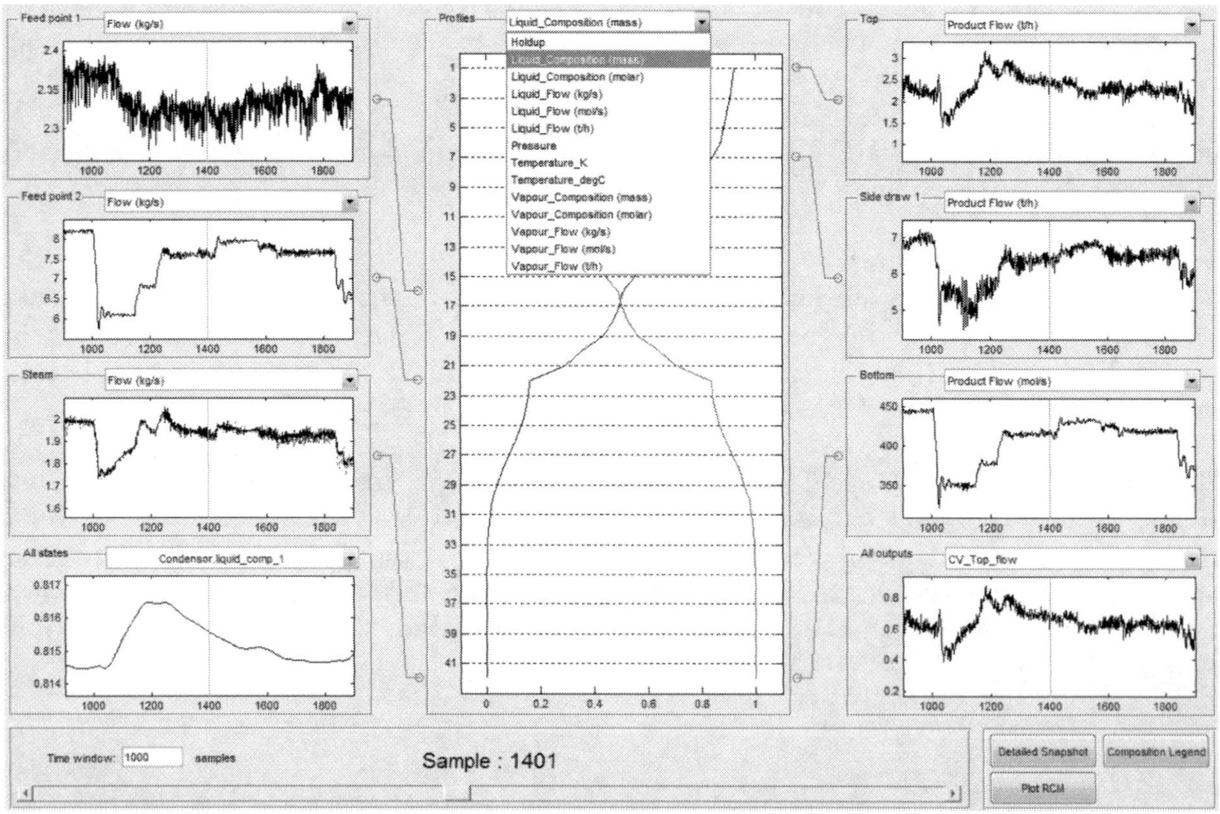

Figure 4. Visualization of a column profile using an INCA MPC4Distillation first principle model of a distillation column with multiple feed points and multiple product streams. The drop-down box at the top shows the different properties that can be visualized with a simple mouse click. Using this information, accurate inferentials for these properties can very easily be added to the model.

Integration with INCA APC Controller

Tight Integration With Existing INCA Infrastructure

One of the challenges of first-principle models or non-linear models in general is coupling these models with or integrating these models in the actual APC controller. Many APC technologies are not well (or not at all) suited to deal with non-linear models. Over the past decade the INCA APC platform has been extended [3] to be able to transparently handle both linear and non-linear models.

The INCA MPC4Distillation models have been natively embedded in the on-line INCA APC infrastructure, and are used in a very similar way as classical linear models, resulting in a truly non-linear APC controller, with prioritized constraints and targets, weighted static and dynamic optimization, hard and soft constraints, etc.

Observer

One very important technology component that needs to be available, is functionality to keep the non-linear model aligned with the actual process, i.e. an observer. The observer of the INCA MPC4Distillation technology is EKF[2]-based and is used to modify certain carefully chosen parameters that influence the model behavior. These parameters are chosen in order to be able to compensate for:

- Mismatches in the pressure model
- Mismatches in tray efficiency (fouling, different flow regimes, etc.)
- Mismatches in the VLE
- Mismatches in the energy balance (mismatch in vaporization enthalpies, heat loss through the column wall, …)

Real-Life Case Study

Problem Description

In this section we describe a real-life industrial case where INCA MPC4Distillation technology was applied. The column under consideration is a binary, high-purity distillation column, part of a distillation train aimed at separating A and B products coming from a reactor section. The column under consideration is a big steam consumer at the production site and therefore was considered a primary candidate for the application of APC with steam consumption reduction as the main goal. Steam flow, reflux flow and top pressure were selected as primary control variables, with the aim of minimizing steam consumption, while respecting top and bottom product impurity constraints.

In 2000 IPCOS installed a linear APC controller using INCA technology, resulting in a reduction of several percent in steam consumption, which was very significant. The application was kept in good use for many years. However, due to increased flexibility requirements in 2009, the column started to be operated under much more varying conditions, straining the APC controller. While pre-2009 operations would aim at 250-350ppm bottom impurity and 95%-100% column load, operation in 2009 would vary between 50ppm and 500ppm bottom impurity and 60%-100% column load. Due to the limited validity range of the originally identified linear black-box models, the controller could only achieve 25% up-time in 2009, which reduced the benefits obtained from the APC application. Therefore in 2010 it was decided to evaluate the INCA MPC4Distillation technology to

---

[2] Extended Kalman Filter [3]

improve the APC controller performance, although at that time the technology was only in a prototype phase. The goals of "upgrading" the existing APC application to an INCA MPC4Distillation application were three-fold:

1. Increase the APC operating range in terms of bottom impurity to 50ppm-500ppm.
2. Increase the APC operating range in terms of column load to 60%-100%.
3. Obtain equal or better variance reduction with respect to the existing linear APC solution.

Results

As a first test the APC controller was configured to follow a sequence of target bottom impurity levels, ranging from 50ppm to 350ppm, the results of which can be seen in Figure 5. It can clearly be seen that the controller is able to control the column in a stable way over the entire operating range, which indicates that the INCA MPC4Distillation model is able to sufficiently accurately capture the non-linearities of the actual column behavior.

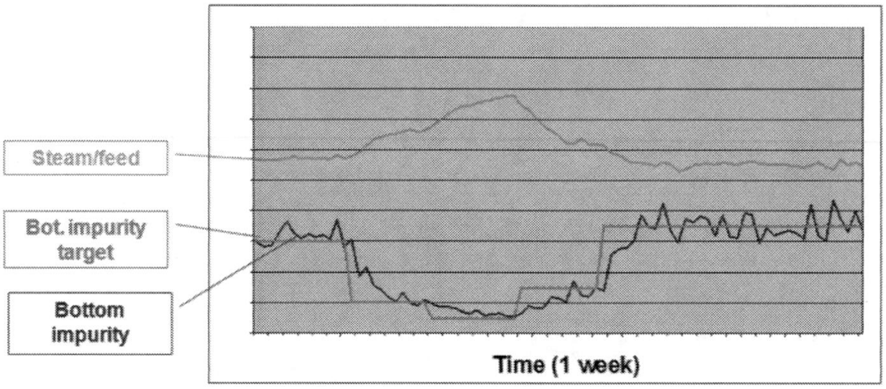

Figure 5. Illustration of stable controller operation over a wide range of bottom impurities.

As a second test, the column was operated over a wide range of column loads, i.e. a wide range of total feed flow. The column was brought from 100% to 75% load, while the APC controller needed to make sure the top and bottom compositions would still remain within the imposed specifications. The results are illustrated in Figure 6.

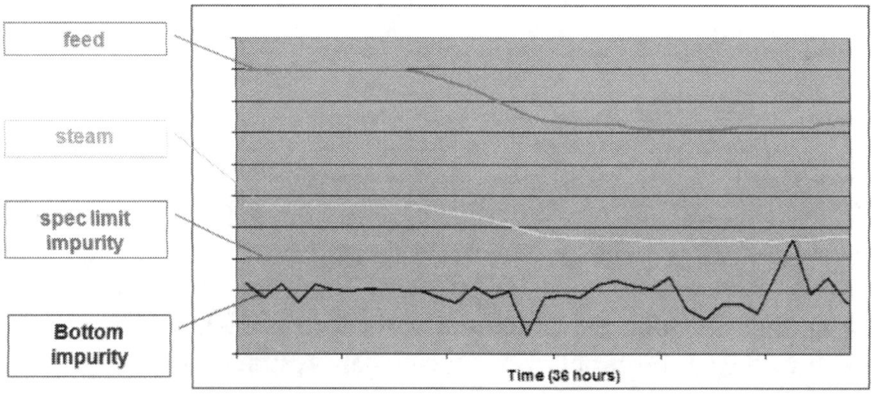

Figure 6. Illustration of stable controller operation during a load change from 100% to 75%.

Finally, an analysis was performed on data collected over a period of 50 days in order to assess the amount of variance reduction that the new APC controller was able to establish. It could be observed that the standard deviation of the bottom impurity was further reduced with 40% compared to the original linear APC controller. This allowed the operators to increase the target value of the bottom impurity from 350ppm to 400ppm, which can be observed in Figure 7.

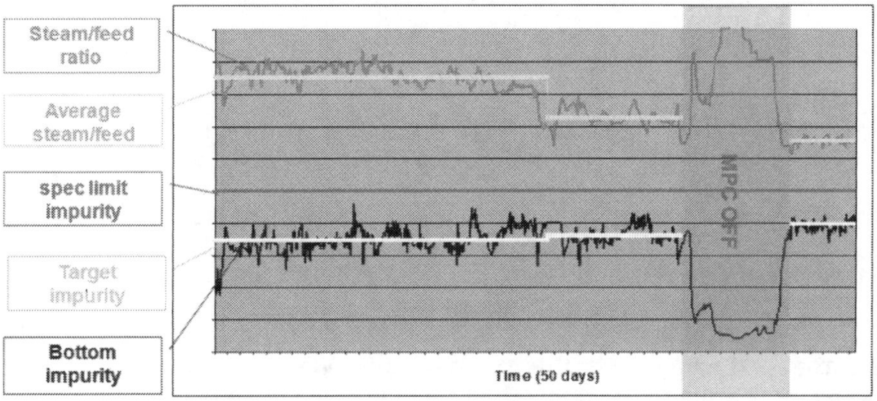

Figure 7. Illustration of a further increase of the target impurity without violating the specification, which was made possible by a further decrease in the variance. Note that when the APC controller is turned off, steam consumption increases significantly, due to the more conservative manual operation.

Obviously, since all tests were clearly successful, the case can be considered a technical success. Also from a commercial point of view this project was evaluated positively due to the increased benefits that were created, compared to the original linear MPC implementation, even though the plant is operated over a significantly wider range.

Conclusion

This paper discusses a novel modeling and control methodology, INCA MPC4Distillation, specifically targeted at distillation columns. The technology is aimed at obtaining accurate non-linear APC models in significantly shorter amount of time than what would be required to obtain linear black-box models using classical technology and also much cheaper than what other first-principles modeling approaches would allow. These affordable, accurate, non-linear models enable the resulting non-linear APC controller to achieve higher benefits over a wider operating range than what would be achievable with classic technology.

The resulting APC technology was applied on an industrial binary high-purity distillation column with results that are superior to the results achieved with a classical, linear APC controller.

References

1. INCA MPC4Distillation website (http://www.ipcos.com/en/mpc4distillation)
2. Ludlage, J. and Backx, A. (1999). "*A new generation model predictive control technology.*" In Proceedings of the 5th International Seminar on Mathematical Simulation.
3. Simon, D. (2006). "*Optimal State Estimation.*" Hoboken, NJ: John Wiley & Sons.
4. Vandecraen, B., Espinosa, J., Pluymers, B., Vinson, D., Ludlage, J., and Brempt, W.V. (2007). "*An industrial approach for efficient modeling and advanced control of chemical batch processes.*" In Proceedings of DYCOPS'07.

(77a) Hydrocracking of Aromatics and Naphtheno-Aromatics

Pedro Rojas, Artie McFerrin Dept of Chemical Engineering, Texas A&M University, College Station, TX and Gilbert F. Froment, Department of Chemical Engineering, Texas A & M University, College Station, TX

Advanced kinetic models for the development and simulation of hydrocarbon processes require detailed information on the reaction network for a variety of potential catalysts. The present work deals with the hydrocracking of aromatics and naphtheno-aromatics on a PtPd/US Y zeolite

Experiments on the hydrocracking of respectively tetralin and 1-methylnaphthalene, diluted with n-heptane, were conducted in a fixed bed catalytic reactor in the temperature range 275°C to 350°C, at pressures between 28 and 35 bar, space times between 21 to 69 gcat h/mol of reactant and with hydrogen/hydrocarbon molar ratios ranging from 100 to 340.

With tetralin feed as many as 10 components were detected and identified in the reactor effluent, with 1-methylnaphthalene up to 30. Following the rules of carbenium ion chemistry very detailed networks of elementary steps were generated for each feed. The network for tetralin contains a total of 96 elementary steps, that for 1-methylnaphthalene 230. Typical pathways to a given product include dehydrogenation/hydrogenation of ring structures on the metal sites and protonation/deprotonation, ring contraction/expansion, PCP branching of side chains, and endocyclic β-scission on the acid sites. The presence of Pt in the catalyst also led to side chain and ring-opening hydrogenolysis.

A significant decay of the conversion of the feed component with process time was observed, more pronounced with 1-methylnaphthalene. It is attributed to irreversible adsorption of the basic feed species on the strongest acid sites. For the kinetic modeling the data were extrapolated to zero process time.

(77b) Modeling and Simulation of Cyclic Operations Applied to Trickle Bed Hydrotreaters

Jinwen Chen, CanmetENERGY, Natural Resources Canada, Devon, AB, Canada, Mohsen Hamidipour, CanmetENERGY, Devon, AB, Canada and Faical Larachi, Chemical Engineering Department, Laval University, Quebec, QC, Canada

CanmetENERGY scientists in collaboration with Laval University are conducting research with the goals of improving operations and intensifying process performance of petroleum trickle bed hydroprocessing reactors, both of which complement intensive research that have been conducted over the past several decades. At CanmetENERGY the research focuses on exploration of cyclic operations by periodically alternating the flow rates of gas or liquid, or both phases to enhance mass and heat transfer, and catalyst efficiency, and to reduce/eliminate solid fines deposition on catalyst surface.

To achieve the overarching research goals, this study utilizes computational fluid dynamics (CFD) modeling and simulation of a trickle bed reactor under gas, liquid, and gas/liquid alternating cyclic operations. The commercial CFD software FLUENT was used under an unsteady multiple-Euler framework. A number of modeling and simulation factors, such as reactor geometry meshing, numerical schemes, drag forces, gas and liquid properties and catalyst particle size, were examined to investigate their effect and sensitivity on simulation results. To validate the computation results, electrical capacitance tomography (ECT) imaging was applied to a lab-scale trickle bed reactor to track the liquid holdup waves along the bed.

Researchers at CanmetENERGY and Laval University concluded that a one-dimensional multiple-Euler CFD model with first order numerical scheme was sufficient to accurately capture the non-steady state hydrodynamic behaviours of trickle bed reactors under different type of cyclic operations. Moreover, the model was able to predict accurately the morphological characteristics of a liquid wave inside the bed for all the examined cyclic operations, including base-peak and ON-OFF modes. In addition, simulation results were in excellent agreement with experimental data in terms of both value and fluctuating behaviour of overall bed pressure drop. Finally, researchers also concluded that the liquid-solid drag force is the main cause for wave attenuation along the catalyst bed.

AUTHOR INDEX

AUTHOR INDEX

AUTHOR INDEX

AUTHOR INDEX

Fuels and Petrochemicals Division 2013

Core Programming Topic at the 2013 AIChE Spring Meeting & 9th Global Congress on Process Safety

San Antonio, Texas, USA
28 April – 2 May 2013

Volume 2 of 2

ISBN: 978-1-62748-464-0

Printed from e-media with permission by:

Curran Associates, Inc.
57 Morehouse Lane
Red Hook, NY 12571

CURRAN ASSOCIATES INC. proceedings.com

Some format issues inherent in the e-media version may also appear in this print version.

Printed by Curran Associates, Inc. (2013)

For permission requests, please contact AIChE
at the address below.

AIChE
3 Park Avenue
New York, NY 10016-5991

Phone: (203) 702-7660
Fax: (203) 775-5177

www.aiche.org

Additional copies of this publication are available from:

Curran Associates, Inc.
57 Morehouse Lane
Red Hook, NY 12571 USA
Phone: 845-758-0400
Fax: 845-758-2634
Email: curran@proceedings.com
Web: www.proceedings.com

TABLE OF CONTENTS

VOLUME 1

VOLUME 2

Activation of Hydroprocessing Catalysts: An in depth understanding of Dimethyldisulfide (DMDS) decomposition chemistry on Hydroprocessing Catalysts during their activation via Sulfiding

Francis Humblot
R&D Manager, Thiochemical Division, Arkema, France

Vijay Srinivas
Principal Research Scientist, Thiochemicals Division, Arkema Inc., USA

Paper # 293309

Abstract: The progress in the improvement of hydroprocessing catalysts has been impressive since the discovery of Type-II acid sites. Strict regulations on the sulphur and nitrogen levels in finished transportation fuels are being enforced globally, and this has promulgated the need to capture all the available activity of these new catalysts. The new generations of CoMo and NiMo catalysts have to be carefully activated via sulfiding to accomplish this task. Dimethyl disulfide (DMDS) has been and continues to be the agent of choice to sulphide "*in-situ*", and obtain optimal activity from these catalysts. The decomposition of DMDS, while verified under laboratory conditions, has not been clearly understood under actual refinery unit conditions. The DMDS decomposition chemistry under a range of commercial unit conditions has been investigated. Experimental data obtained have been used to build a kinetic model based on the fundamental decomposition mechanism. The data can be used beneficially to fine tune existing sulfiding procedures established by Catalyst Manufacturers and Refining Companies so as to make them optimal for and applicable to different types of units in the refinery. Additionally, this information can be used to ascertain essentially complete use of the 'sulfur' introduced via DMDS and prevent SO_2 excursions due to inadvertent loss of sulphur intermediates to the flare or flue gas, during the catalyst sulfiding and activation. Being the leader in DMDS manufacture, Arkema has recently introduced a new formulation, *"DMDS Evolution® E2"* that has an improved user-friendly odor. Details of this formulation and of the *"Carelflex® Service"* that Arkema provides to sulfide these catalysts in-situ will be described.

1. Introduction

For more than 20 years, the refining industry has greatly improved the refining process to be able to convert the bottom-of-the-barrel to valuable fuels and to drastically decrease or exclude the deleterious impurities present[1,2]. Thanks to this effort and particularly the reduction of sulfur content in gasoline and gasoil, the acid rain phenomena observed in Europe and United-States in the 90's has just about disappeared. Furthermore, the deactivating effect of sulfur compounds in gasoline and diesel on the noble metal catalysts used in catalytic converters on cars to reduce hydrocarbon and NOx in tail pipe emissions prompted Europe and the United States to promulgate regulations targeting 10-15 PPM sulfur in these fuels. This trend is now progressively being adopted all around the world.

To meet this challenge for cleaner fuels, the refining industry and catalysts manufacturers have worked diligently hand-in-hand on developing new hydroprocessing technologies and new generations of catalysts[3]. While the metals used in todays hydroprocessing (HDP) catalysts are still the same as those used in the 1990's (bimetallic Cobalt-Molybdenum or Nickel-Molybdenum, Nickel-Tungsten) their relative activities have been drastically improved. These improvements have been made possible by the development in sophisticated catalyst characterization techniques that have resulted in a deeper understanding of the catalyst structure and the nature of their "active sites". Invoking mixed sulfide structures such as "CoMoS"[4,5,6] and their interaction with the carrier or support, leading to the concept of 'Type-I' and 'Type-II' active sites[7,8] has allowed catalyst manufacturers to introduce the current line of highly active catalysts with defined specificity via the use of templates provided by organic molecules such as glycol ethers[9,10,11].

---

1 H. Topsoe, B.S. Clausen, F.E. Massoth, "*Hydrotreating catalysts Science and Technology*", Ed. Springer, 1996

2 World wide fuels regulations: United States, Canada, Europe, Japan, India... (http://www.dieselnet.com/standards/)

3 A. Stanislaus, A. Marafi, M.S. Rana, "Recent advances in science and technology of ultra low sulfur diesel production", Catal. Toady, 153, 1-68 (2010)

4 H. Topsøe, B.S. Clausen, "Importance of cobalt-molybdenum-sulfur type structures in hydrodesulfurization", Catal. Rev.-Sci. Eng., 26, 395 (1984)

5 P. Ratnasamy, S. Sivasanker, "*Structural chemistry of cobalt-molybdenum-alumina catalysts*", Catal. Rev.-Sci. Eng., 22, 401 (1980)

6 B.S. Clausen, H. Topsøe, R. Candia, J.V. Villadsen, B. Lengeler, J.Als. Nielsen, F. Christensen, "*Extended X-ray Absorption Fine Structure Study of Co-Mo Hydrodesulfurization Catalysts*", J. Phys. Chem., 85, 3868 (1982)

7 B. Hinnemann, J.K. Norskov, H. Topsøe, "*A density functional study of the chemical differences between type I and type II MoS2-based structures in hydrotreating catalysts*", J. Phys. Chem. B, 109, 2245, (2005)

8 E.J.M. Hensen, V.H.J. de Beer, J.A.R. van Veen, R.A. van Santen, "*A refinement on the notion of type I and type II (Co)MoS phases in hydrotreating catalysts*", Catal. Letters, 84, vol. 1-2, 59 (2002)

9 S. Eijsbouts, L.C.A. van den Oetelaar, R.R. van Puijenbroek, "*MoS2 morphology and promoter segregation in a commercial type II Ni-Mo/Al2O3 and Co-Mo/Al2O3 hydroprocessing catalysts*", J. Catal., 229, 352 (2005)

10 D. Nicosai, R. Prins, "*The effect of phosphate and glycol on the sulfidation mechanism of CoMo/Al2O3 hydrotreating catalysts: an in situ QEXAFS study*", J. Catal., 231, 259 (2005)

Hydroprocessing catalysts are manufactured as a mixture of active metal oxides on a support, typically alumina. These have to be converted to their corresponding active metal sulfides prior to use. The metal oxides are fully converted to metal sulfides, which are dispersed on the carrier surface to create nano-scale crystals by treating them with a sulfiding agent. This activation is generally performed on site (*in-situ* sulfiding), in the catalytic reactor to avoid the handling of metallic sulfides, which are known to be pyrophorric. Sulfiding is an exothermic process thus needing careful monitoring to control the temperature and avoid deleterious side reactions such as metallic oxide reduction by hydrogen and coke formation that reduce the final catalyst activity. These side reactions are minimized if the sulfiding process is started in the temperature range (200^{0}C–240°C; 392^{0}F-464^{0}F) and completed at temperatures greater than 300°C or 572^{0}F. Metal oxides react readily with hydrogen sulfide (H_2S) to form the corresponding sulfides, however the use of H_2S on a large scale poses a high safety risk due to its toxicity. Hence most refiners and catalyst manufacturers around the world prefer to activate their catalysts using a sulfiding agent that efficiently generates and delivers H_2S in-situ.

For several decades, dimethyldisulfide (DMDS) has been recognized as an efficient, relatively safe and convenient sulfiding agent for HDP catalysts[12,13]. It is a liquid that contains 68% sulfur, which efficiently converts to H_2S on the HDP catalysts in the presence of hydrogen at relative low temperatures (180^{0}C - 200°C) thus being optimally suited for in-situ sulfiding of HDP catalysts[14,15,16]. Produced for more than 30 years by Arkema and available worldwide, it has become the reference standard for the sulfiding agent market. Until recently, DMDS decomposition data on HDP catalysts was known under ambient pressure conditions only. There was a need to understand the DMDS decomposition chemistry under a range of commercial unit operating conditions. Arkema embarked on an investigation to obtain such data and establish a mechanism of DMDS decomposition supported by a kinetic model as well. This data and information can now be used beneficially to fine tune existing sulfiding guidelines established by catalysts manufacturers as needed and maximize sulfur transfer from DMDS to the catalyst.

---

[11] T.S. Nguyen, S. Loridant, C. Lorentz, T. Cholley, C. Geantet, "*Effet of gloycol on the formation of active species and sulfidation mechanism of CoMoP/Al2O3 hydrotreating catalysts*", Appl. Catal B: Environmental, 107, 59 (2011)

[12] H. Hallie, "*Experience reveals best presulfiding techniques for HDS and HDN catalysts*", Oil and Gas J., pp 69-74, (Dec. 20, 1982)

[13] A. Tuxen, H. Gøbel, B. Hinnemann, Z. Li, K.G. Knudsen, H. Topsøe, J. Lauritsen, F. Besenbacher, "*An atomic-scale investigation of carbon in MoS2 hydrotreating catalysts sulfided by organosulfur compounds*", J. Catal., 281, 345 (2011)

[14] C. Brun, T. Saint-Pierre, "Thiochemicals as sulfiding agents for hydrotreatment catalysts", Hydrocarbon Eng., pp26-30, Jan-Feb. 1997

[15] S. Texier, G. Berthault, G. Pérot, V. Harlé, F. Diehl, "Activation of alumina-supported hydrotreating catalysts by organosulfides: comparison with H2S and effect of differents solvents", J. Catal., 223, 404 (2004)

[16] F. Humblot, "Agent of change: choose of sulfiding agents for hydrotreating catalysts", Hydrocarbon Eng., pp41-46, Aug. 2007

2. Experimental

2.1. Catalysts and Chemicals

Latest generation CoMo and NiMo commercial catalysts were used for this study (trilobe extrudates). The metal loadings are confidential, however, the catalyst manufacturer indicated that the sulfur uptake expected for these catalysts after sulfiding is the following:

- CoMo/alumina catalyst : 11.0% sulfur
- NiMo/alumina catalyst : 10.5% sulfur

Synthetic liquid feed (n-dodecane, from Merck, purity > 99.0%) was used for this study as a surrogate for straight run gasoline/gasoil (SRG/SRGO) typically employed for industrial sulfiding. The n-dodecane used was sulfur-free and allowed for a more accurate determination of the sulfur species present in the gas and liquid exiting the reactor to perform a sulfur balance. Additionally, the boiling point (417–424°F) of n-dodecane falls within the boiling point range for gas oil.

Pure hydrogen (purity > 99.995%) was supplied by Linde.

Dimethyldisulfide was manufactured by Arkema, at Lacq in France (Purity 99.82-99.85%).

Silicon Carbide powder (diameter 1.68 mm) was supplied by VWR.

2.2. Hydrotreating Pilot Unit Description

This pilot was designed by Vinci technology (see Figure-1). It is capable of handling pressures of hydrogen up to 100 bars (1450 psig). The liquid feed is injected in the up-flow mode into a 100 cm3 reactor (diameter 19 mm, 524 mm of total height). The Catalyst temperature is measurement with 3 thermocouples located at different heights in the center of the reactor. Silicon carbide is placed at the bottom and at the top of the catalyst bed so that catalyst bed is located in the isothermal zone of the reactor. The pressure is monitored by a control valve after the catalytic reactor and before the Gas – liquid separator.

2.3. Procedure

In order to study the impact of Liquid Hourly Space Velocity (LHSV) on the DMDS decomposition chemistry, different catalyst volumes were loaded in the reactor:

- 0.5 h^{-1} LHSV : 80 cm^3 catalyst volume
- 4.0 h^{-1} LHSV : 10 cm^3 catalyst volume

Catalyst was always loaded in the middle of the reactor, between 2 beds of silicon carbide. During the study, the ratio of hydrogen / n-dodecane flow rates was maintained at 250. The hydrogen pressure ranged from 217 psig to 870 psig. The following steps were taken:

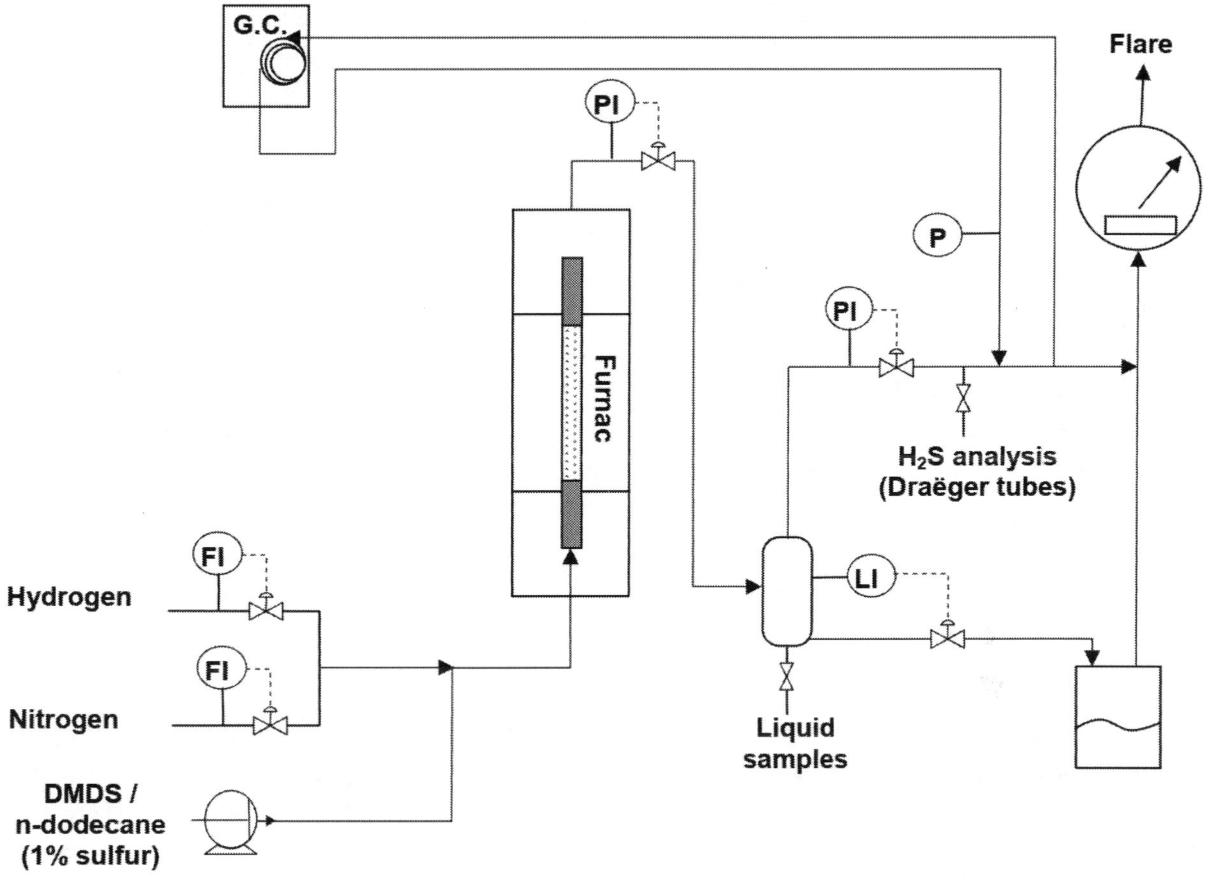

Figure 1 : Hydrotreating Pilot used to study the decomposition of DMDS during the sulfiding of CoMo or NiMo catalyst

1. Catalyst drying was done at 300°F (150°C) with a down flow of nitrogen at low pressure (30 psig)
2. Catalyst wetting with n-dodecane was performed at a low pressure (2 bars) of nitrogen, until it was observed in the gas – liquid separator.
3. The nitrogen was slowly displaced with hydrogen flowing at 10 L to 20L/hr. The pressure was allowed to build up to the value that needed to be tested.
4. Once pressure was stabilized and the temperature of the catalyst bed was fixed at 300°F, a solution of DMDS (1.47% w/w) in n-dodecane was injected.
5. After about 2 hours, the outlet hydrogen was analysed on-line for 2 hours by gas chromatography. The H_2S concentration in the hydrogen can also be measured using Draëger tubes. Once a constant composition in species was observed (H_2S, methane, etc...), a liquid sample was taken at the bottom of the gas–liquid

separator and analyzed by gas chromatography. Liquid and gas compositions were then used to established DMDS conversion and selectivity.

6. The reactor temperature was then increased by 18°F (10°C) and steps #5 and #6 were repeated. Hence, DMDS decomposition is described step by step at 300°F, 318°F... up to the full conversion of DMDS into H_2S and methane

Sulphur balance (sulphur introduced with DMDS compared to the sum of the sulphur species analyse in gas and liquid) was done for each test at each temperature. The results were found to be inside +/- 10% of the expected value showing that all the compounds were detected and appropriately quantified by gas chromatography.

2.4. Kinetic Model

Specific experiments in gas phase only (without n-dodecane) have been done for the kinetic estimation in order to avoid the liquid phase in the reactor. The hypothesis to establish the model were the following :

1. The reactor is isothermal
2. The reactions take place only in gas phase
3. The flow is not dispersive

Each reaction was modelled as a Power law and the chemical equilibrium was treated as two reversible reactions. The software gPROMS® was used to develop the reactor model and estimate the kinetic parameters. The methodology to define the kinetic of the reactions was the following:

1. The low temperature of the specific experiments (248-284°F) were used to estimate the parameter of the first reaction
2. To define the kinetics of the equilibrium, other experiments were done by injecting a mix of H_2S and DMS under Helium pressure to avoid the two last reactions. The middle range temperature (284-338°F) of the gas phase experiment were also used to estimate those kinetic parameters
3. The two last reactions (Methane and H_2S formation) were estimated by using the results from high temperature (354-428°F) gas phase experiments

The results from the kinetic model and the parameters established with gas phase experimental tests successfully matched the tests performed in presence of n-dodecane.

3. Results and discussion

The sulfiding reactions of molybdenum oxide and tungsten oxide have been established for a long time and can be described bythe following equations:

(1) $MoO_3 + 2\,H_2S + H_2 \longrightarrow MoS_2 + 3\,H_2O$

(2) $WO_3 + 2 H_2S + H_2 \longrightarrow WS_2 + 3 H_2O$

The exact stoichiometry of the cobalt and nickel sulfides formed during sulfiding has not been well established. However, equations (3) and (4) below are generally accepted because they adequately correlate to the observed sulfur uptake of the catalyst.

(3) $3 NiO + 2 H_2S + H_2 \longrightarrow Ni_3S_2 + 3 H_2O$

(4) $9 CoO + 8 H_2S + H_2 \longrightarrow Co_9S_8 + 9 H_2O$

The transformation of metallic oxides to the corresponding sulfides is facile and temperature dependent. The reaction is relatively slow below 350°F and above 570°F, essentially complete conversion is obtained. Hence, most sulfiding procedures and guidelines for commonly used HDP catalysts recommend starting the injection of sulfiding agent at about 390°F. Therefore we started our investigation of the DMDS decomposition around 350^0F.

Starting at 356°F and at each subsequent temperature thereafter, we obtained stable concentrations of all sulphur species in both gas (H_2) and liquid (n-dodecane) phases within 2 hours of stabilization. The composition and concentration of the sulphur species changed with each temperature used in the study. The figures 2 and 3 show the DMDS conversion to the various sulphur species on CoMo and NiMo respectively, at a pressure of 362 psig of hydrogen. One important observation was that between 430^0F and 450^0F, the DMDS is completely converted to methane and hydrogen sulphide exclusively on both CoMO and NiMo catalysts, as depicted by the following equation:

(5) $CH_3\text{-}S\text{-}S\text{-}CH_3 + 3 H_2 \longrightarrow 2 CH_4 + 2 H_2S$

Most catalysts manufacturers recommend performing the primary sulfiding in the 430^0F to 450^0F range and wait for the breakthrough of H_2S. Most HDP catalysts are about 55% to 60% sulfided when H_2S breakthrough occurs. This renders the catalyst robust enough for the temperature to be further increased at a prescribed rate while DMDS injection is continued, albeit at a reduced rate. Presence of an active form of sulphur such as H_2S in sufficient amounts ensures the appropriate oxidation state of the active metal to maintain and avoid reduction by hydrogen present in the system. Some of the more active and selective catalysts are known to be stable to reduction up to a temperature of 480^0F, thus the catalyst is never at this temperature without at least 3000-5000 ppm of H_2S being present in the recirculating hydrogen gas.

The results obtained in our study confirm that DMDS/E/E2 decomposes completely and exclusively to H_2S and CH_4 at these temperatures, thus providing excellent protection for these catalysts while activating them at the same time. Some of the intermediates formed at lower temperatures are good sulfiding agents in their own right and thus

DMDS/E/E2 provides active sulphur species in a wide range of temperatures spanning both the primary and secondary stages of sulfiding.

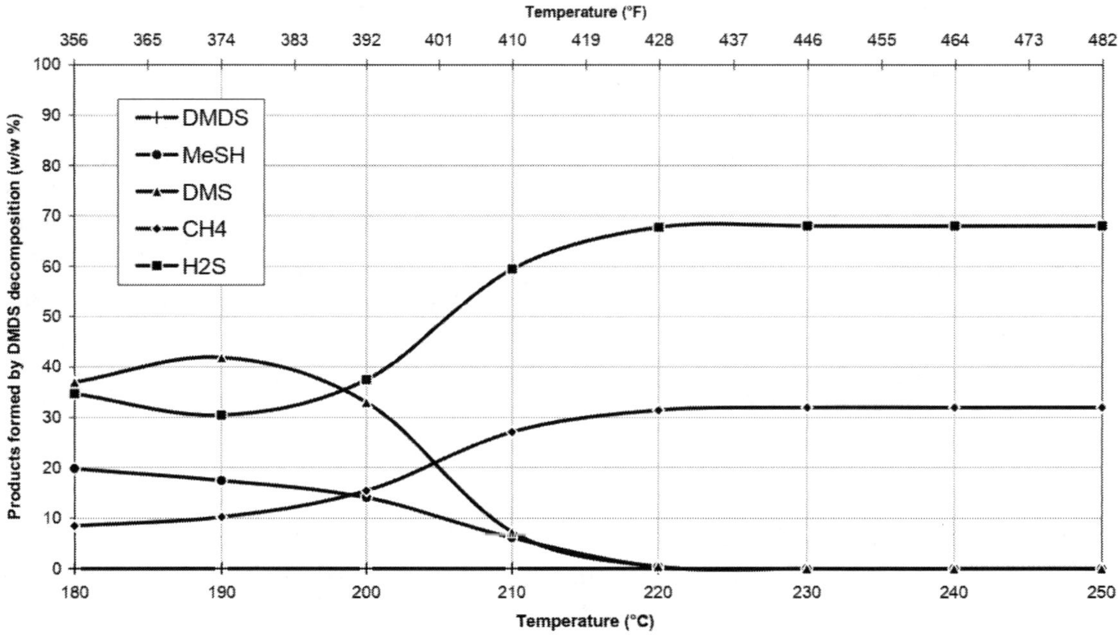

Figure 2: DMDS decomposition on CoMo/Alumina catalyst under 362 psig of hydrogen and in presence of n-Dodecane (LHSV=0.5 h^{-1})

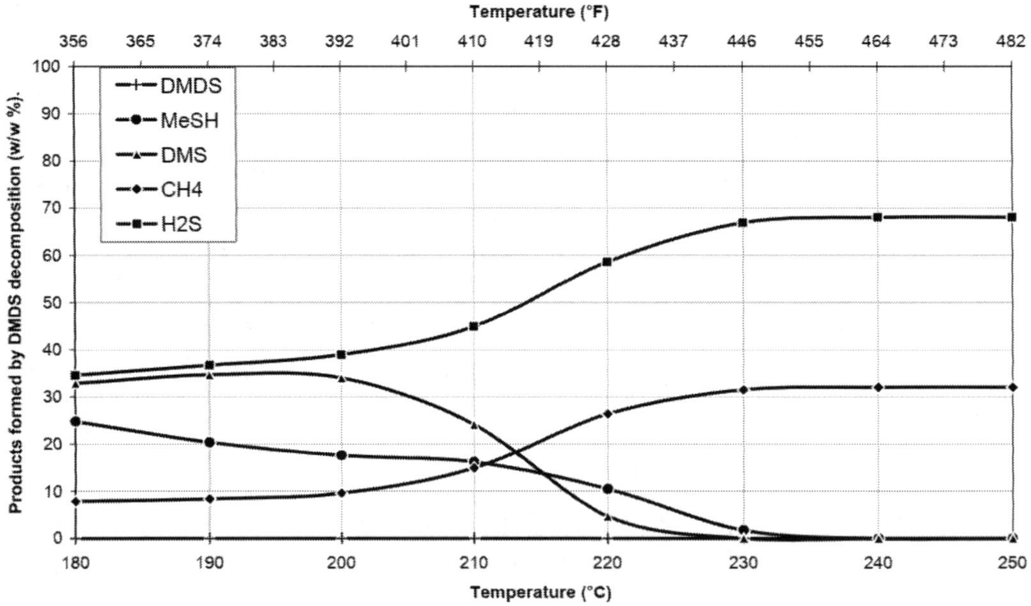

Figure 3: DMDS decomposition on NiMo/Alumina catalyst under 362 psig of hydrogen and in presence of n-Dodecane (LHSV=1h^{-1})

At temperatures as low as 356°F, DMDS is completely decomposed. A mixture of methylmercaptan, dimethyl sulphide, hydrogen sulphide and methane are formed. Methyl mercaptan decreases steadily as the temperature is increased whereas dimethylsulfide (DMS) reaches a maximum at around 380°F before disappearing completely at 430°F. Methane and hydrogen sulphide are formed by the hydrogenolysis of methylmercaptan. Dimethylsulfide, known to be more stable needs a higher temperature to be converted to methane and hydrogen sulphide via sulfhydrolysis.

In order to establish the onset temperature of DMDS decomposition, the reaction conditions were changed to a space velocity of 2 hr^{-1} and a temperature of 300°F. Stabilization of the reactor effluent composition took a longer time (6 to 12 hours) due to an overall lower reaction rate. Results of this experiment are shown in Figure-4. Even at 302°F, the conversion of DMDS was 80% and the primary product was methylmercaptan, formed by the hydrogenolysis of the sulphur-sulphur bond in DMDS. Dimethylsuphide is a secondary product obtained from methylmercaptan[17]. This reaction is described as equilibrium, catalyzed by the acid sites on the catalyst:

(6) $CH_3\text{-}S\text{-}CH_3 + H_2S \leftrightarrow 2\, CH_3SH$

Internal experience corroborates this hypothesis: DMS is a by product in the manufacture of methyl mercaptan.

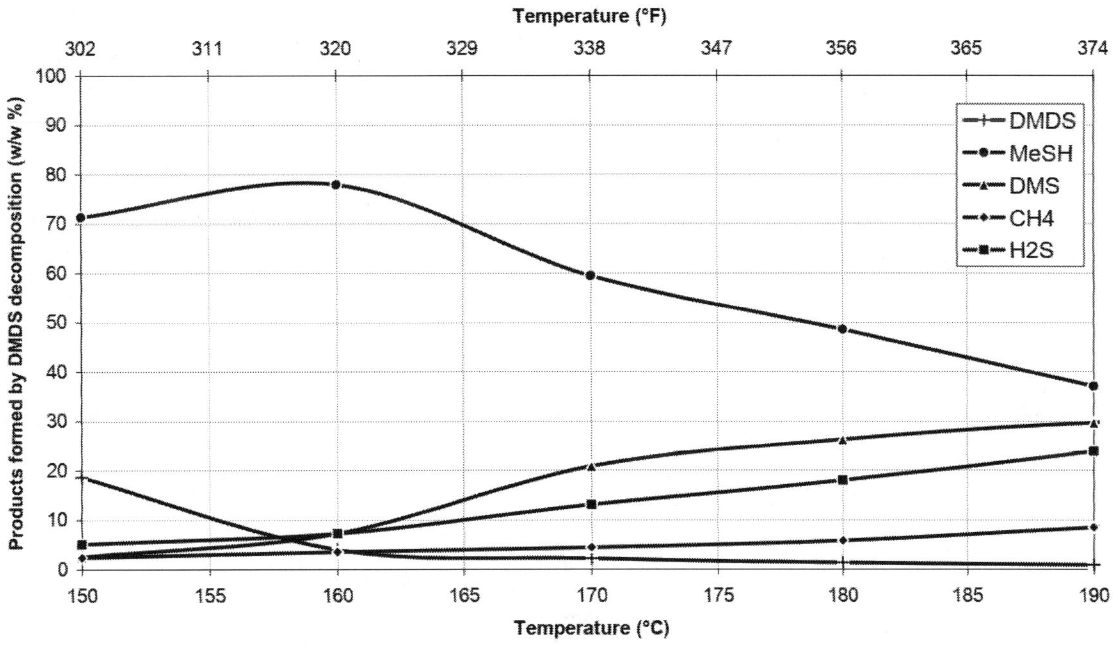

[17] V.Yu. Mashkin an S.N. Koshelev, "*Kinetic studies of methylmercaptan production from dimethyl sulphide and H_2S*", React. Kinet. Catal. Lett., 46, N°1, 187-192 (1992)

Figure 4: DMDS decomposition on CoMo/Alumina catalyst under 507 psig of hydrogen in presence of n-dodecane (LHSV=2h^{-1}) and at low temperatures.

We continued this study by extending the range of hydrogen pressure and space velocity used to the following:

- Hydrogen pressure: from 217 and 870 psig (see figure 5)
- Space velocity: 0.5 to 2.0 hr^{-1} (see figure 6)

Increasing the hydrogen pressure from 218 to 870 psig., increases the DMDS conversion, as expected and similar amounts of hydrogen sulphide, and methane are formed at 20^{0}F lower temperatures at 870 psig compared to that at 217 psig. It is believed that the catalyst surface gets saturated with adsorbed hydrogen even at 217 psig, thus increasing pressure further has only a limited impact on the hydrogenolysis of DMDS.

The effect of varying the space velocity, however, is more pronounced. Decreasing it from 2 hr^{-1} to 0.5 hr^{-1} has the same effect as increasing the temperature 30^{0}F. In these tests, hydrogen to n-dodecane volumetric flow ratio was kept constant. As a consequence, an increase of Liquid Hour Space Velocity - LHSV (n-dodecane flow rate) is accompanied by an increase of hydrogen flow rate, leading to a significant change in residence time thus affecting the hydrogenolysis of DMDS. It was however observed that DMDS was totally converted to hydrogen sulphide/methane at about 450°F, irrespective of the space velocity.

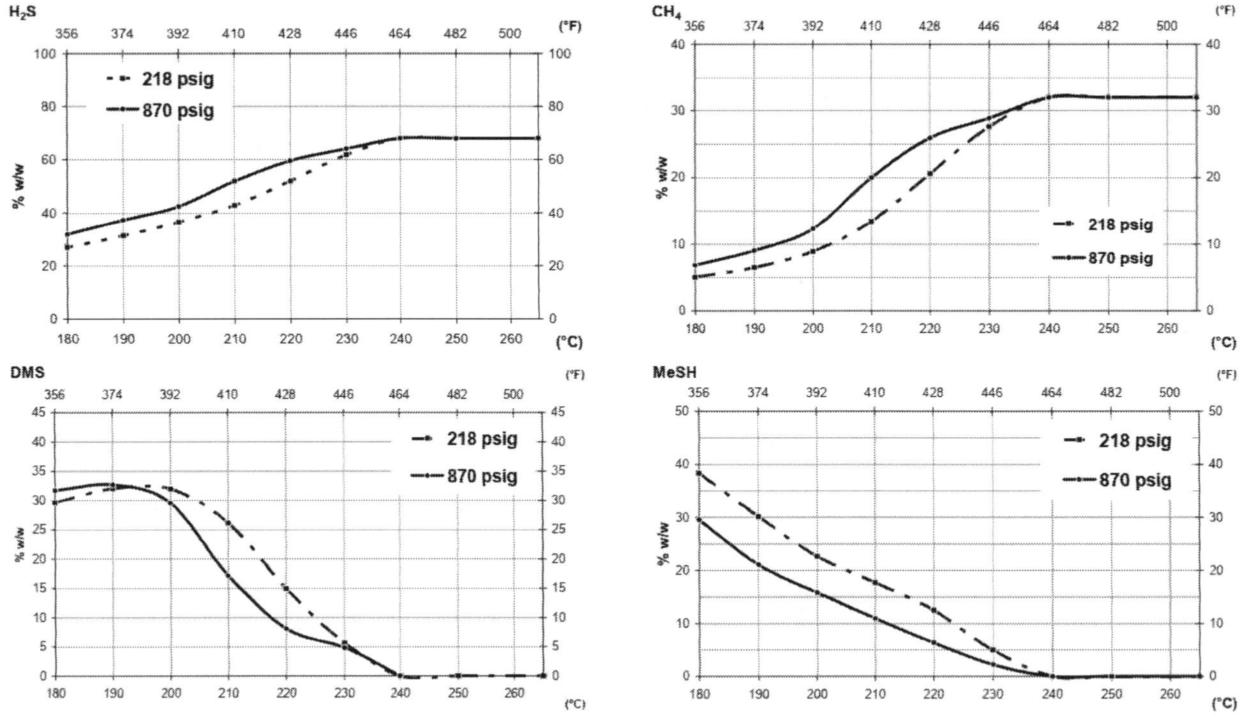

Figure 5: DMDS decomposition on CoMo/Alumina catalyst in presence of n-dodecane (LHSV=1h^{-1}), between 218 and 870 psig of hydrogen.

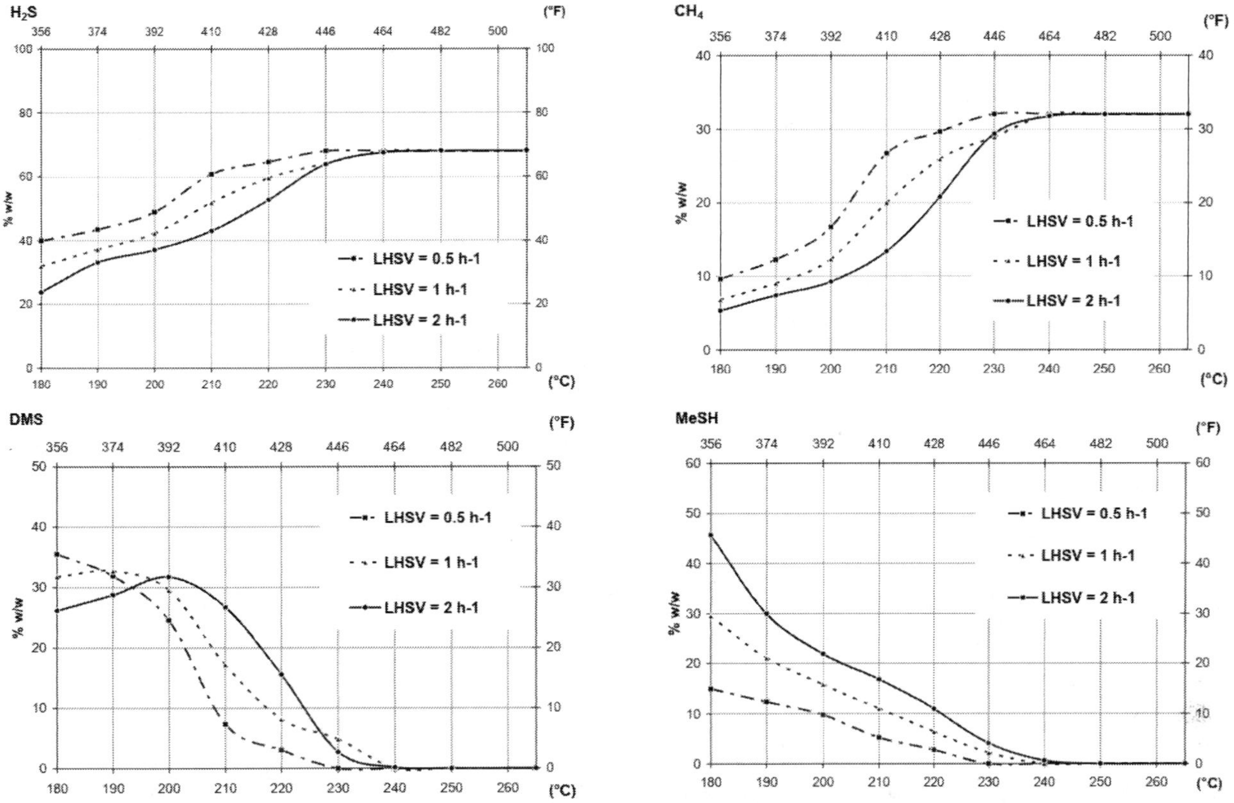

Figure 6: DMDS decomposition on CoMo/Alumina catalyst under 60 barg of hydrogen and in presence of various n-dodecane (LHSV 0.5 hr^{-1} to 2 hr^{-1}).

The proposed hydrogenolysis mechanism for the decomposition of DMDS is supported by the data presented in Figure 4. The low temperature experiments clearly indicate that the hydrogenolysis of sulphur-sulphur bonds as the primary reaction. DMDS is converted at low temperature (below 300°F) to methylmercaptan and at 340°F, the presence of methane proves that methylmercaptan also start to get hydrogenolyzed to hydrogen sulphide and methane. However, at 340°F, another side reaction is observed, where methylmercaptan gets converted partially to generate dimethylsulfide and hydrogen sulphide. Finally, at 360°F, which is generally the temperature where some refiners start to inject DMDS, about half of the introduced sulphur is converted to hydrogen sulphide from both the partial hydrogenolysis of dimethyldisulfide and from the methylmercaptan – dimethylsulfide equilibrium. At this temperature, this hydrogen sulphide reacts rapidly with hydrotreating catalyst and initiates its conversion to the active metal sulfides. As the temperature is increased, at 430-450°F (see Figures 2 and 3), no more methylmercaptan or dimethylsulphide are formed, with methane and hydrogen sulphide being exclusively produced from DMDS. All these reactions from dimethyldisulfide are summarized in the Figure 7 below.

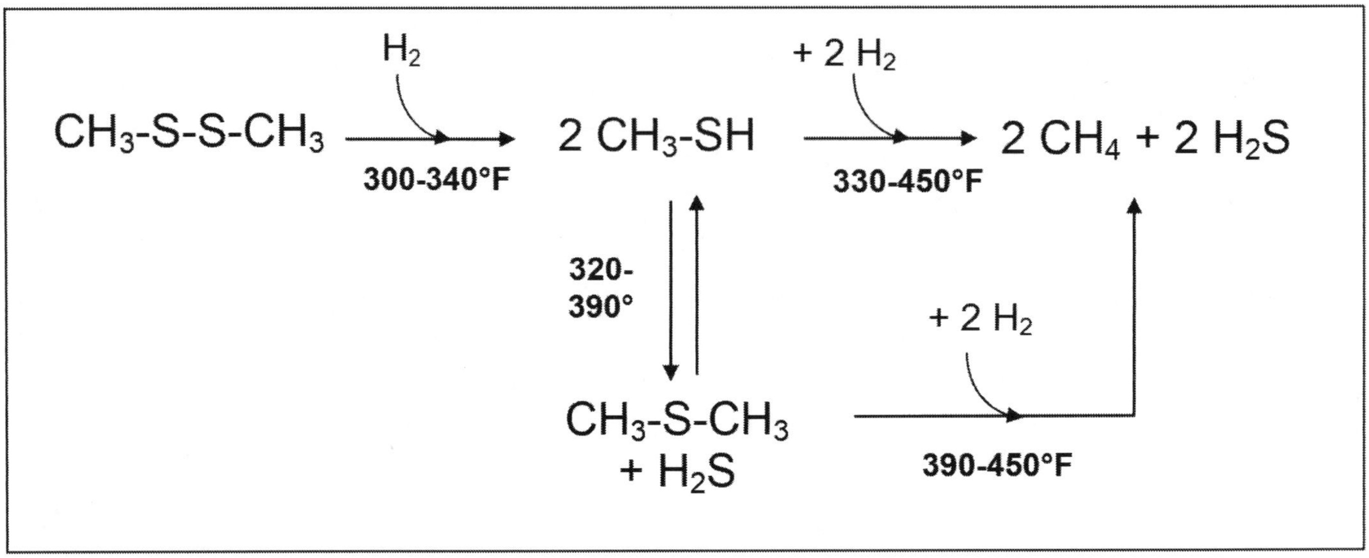

Figure 7: Mechanism of DMDS decomposition on CoMo or NiMo/Alumina catalyst with hydrogen (200 – 900 psig).

In order to provide a theoretical basis for our proposed mechanism (Figure 7) of DMDS decomposition under hydroprocessing conditions via hydrogenolysis on CoMo and NiMo catalysts, we attempted to build a kinetic model. Pre-exponential factor, partial orders and activation energy were estimated for each reaction shown by obtaining extensive data at different temperatures. Some specific data had to be obtained and evaluated for the DMS-methylmercaptan equilibrium on partially sulfided HDP catalysts in order to more precisely get the rate of reaction. Figure 8 depicts a check of the accuracy of our best fit model on one set of experiments on CoMo catalyst, under conditions of 290 psig of hydrogen and 1 hr^{-1} LHSV. Methane and hydrogen sulphide were particularly well estimated. Minor differences remain for the methylmercaptan and dimethylsuphide, especially when their concentrations are low. At least partially, these differences could be due to analytical limitations (detection limit). The kinetic model should however, predict better estimations of the concentrations of these species.

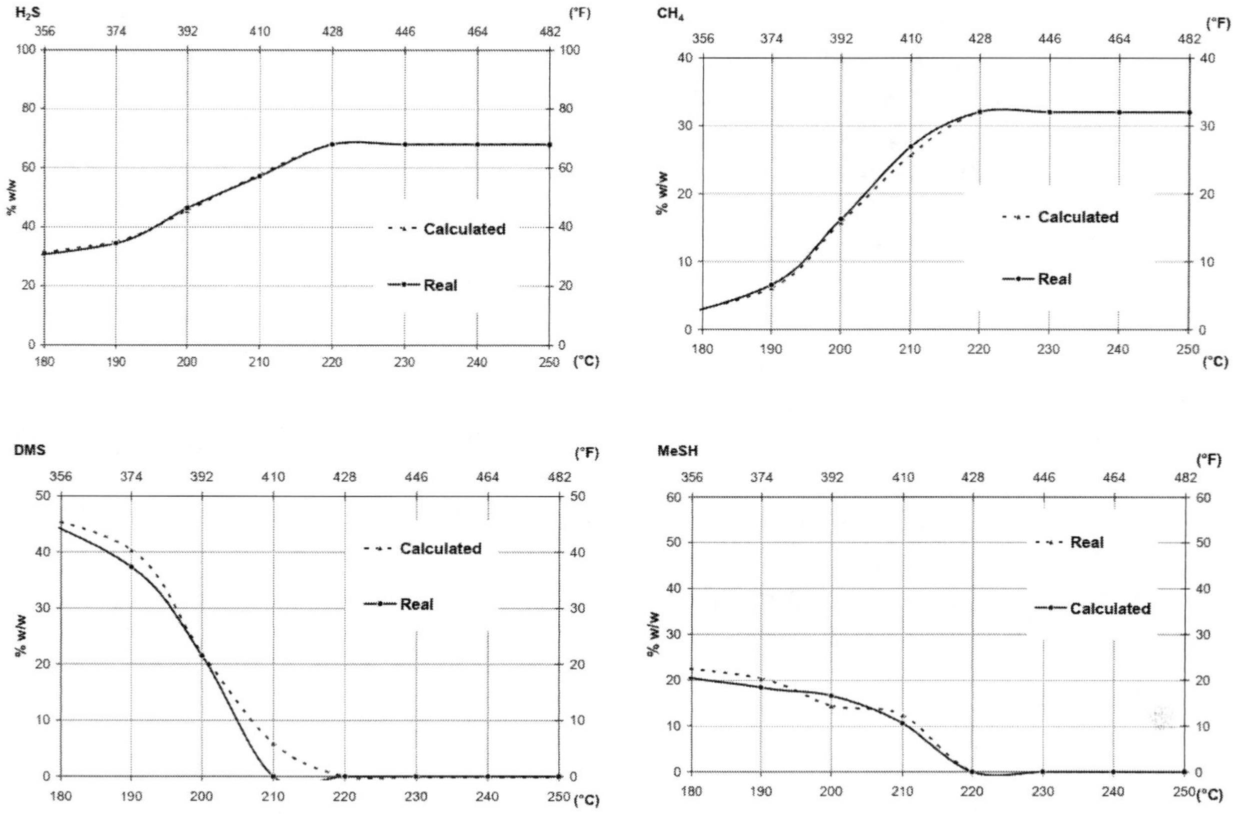

Figure 8: Example of kinetic model results, in comparison with experimental data of DMDS decomposition on CoMo/Alumina under 290 psig of hydrogen and 1 h^{-1} LHSV.

4. Summary and conclusion

DMDS is one of the most commonly used thiochemical for sulfiding HDP catalysts insitu. DMDS belongs at the top of the list of sulfiding agents recommended by all catalyst manufacturers. The results from our study clearly indicate that DMDS is an excellent surrogate for H_2S in the activation of hydroprocessing catalysts via sulfiding, from both the safety and efficiency points of view. The sulfur contained in DMDS is all "efficient sulfur". All of this sulfur is transferred on to the catalyst during the process of insitu sulfiding. DMDS decomposes via hydrogenolysis, first to methyl mercaptan (CH_3SH) and then as the temperature increases, the methyl mercaptan is converted to methane and H_2S, a small portion of it via DMS. The catalyst manufacturers have innovatively improved the intrinsic activity of the HDP catalysts steadily in the last decade. All of these catalysts can be activated to exhibit the expected optimum activity via sulfiding with DMDS. The decomposition of DMDS can be controlled by the conditions employed and thus eliminates unplanned exotherms and excursions that could compromise safety.

Refineries, especially those located close to residential neighborhoods and public places like schools etc are under greater pressure to keep the odors to a minimum. Innovation by Arkema to incorporate a new fragrance in DMDS to impart a user-friendly odor to DMDS in their current offering DMDS-E2, has essentially helped refiners avoid unpleasant odor outages. In fact the original version, DMDS-E has been doing so for well over a decade. The Carelflex® service offered by Arkema assists in delivering the DMDS-E2 to the refinery in a safe and reliable manner and safely introducing it into the HDP reactor. This service has now been expanded to inject DMDS at elevated pressures.

Increasing hydrogen pressure during sulfiding and decreasing the LHSV, decreases the temperature at which DMDS decomposition to the intermediates and eventually to H_2S takes place. The more pronounced effect seems to be that of decreasing the LHSV. The corroboration of results between the Kinetic Model and the Pilot Tests further goes to show that DMDS is perfectly suitable for both liquid and gas phase sulfiding of HDP catalysts. The low temperatures at which DMDS starts to generate H_2S allows it to protect the catalyst from undergoing reduction by hydrogen. Since maintaining a healthy concentration of H_2S in the recycle hydrogen gas at relatively low temperatures, is seen as the key to obtaining an optimally active catalyst, DMDS is well positioned to satisfy this need in an efficient and controlled manner.

Hydrotreating Reaction Optimisation Using Carbon Nanotubes Supported NiMo And Mo Based Catalysts In Refining Crude Oil

Aman A. Dhanani, Pandit Deendayal Petroleum University, Gandhinagar, India

1. Introduction:

Sulfide transition metal catalysts have been widely used for petroleum refining & hydroprocessing such as hydrodesulfurization (HDS) and hydrodenitrogenation (HDN). The traditional hydrotreating catalysts usually contain cobalt or nickel and molybdenum or tungsten with the support commonly used is alumina.

This type of hydrotreating catalysts have been used successfully and commercially used, there are difficulties in these traditional catalysts like difficulty in sulfurization and in reduction of Co-Mo and lower valence active phases. When alumina supported (single-layer) molybdenum oxide was heated in oxygen and H_2S/H_2 mixture upto 825K, a strong oxide-support interaction was observed, leading to highly dispersed molybdenum phases on alumina support as was studied by Hayden[21].

Moreover, for the Hydrodenitrogenation, adsorbed nitrogen and metal heteroatoms are most often the contaminants responsible for downstream catalyst deactivation. The basic nature of these compounds causes them to adsorb onto Lewis acid sites on the catalyst surface, inhibiting the availability of the sites. This poisoning may be reversible or irreversible, depending on the hydroprocessing conditions (Furimsky and Massoth, 1999). High concentrations of organic nitrogen compounds can cause significant deactivation for reforming, cracking, hydrotreating, or any other type of hydroprocessing catalysts.

In the present area of nano-science and nano-technology the CNT supported catalysts have attracted much more attention in this field because of certain reasons like the CNT supports of the hydrotreating catalysts show a high activity, excel in lower coking deposition, easily recoverable from the waste catalysts by burning of the carbonaceous supports. Being a novel material, CNTs have high mechanical and unique electronic properties similar to hollow graphite fibres with seamless tube like graphitic walls with a perfect structure, sp^2 carbon-carbon atoms, surface properties(e.g. high specific surface areas maintaining the large pore diameters(>50nm)) which can be easily modified and excellent electron-transporting capability. Moreover the CNTs are available cheaply with a good quality.

In the first study, some oxide state Mo, Co–Mo and sulphide state Mo supported on CNT catalysts were prepared, while the corresponding Al_2O_3 supported catalysts were prepared as comparison. Laser Raman Spectroscopy (LRS) and X-Ray Diffraction (XRD) techniques were adopted to characterize these CNTs and γ-Al_2O_3 supported catalysts. In the case of oxide state catalysts, the reducibility of active species was monitored by Temperature Programmed Reduction (TPR) technique. The catalytic performances of CNTs or γ-Al_2O_3 supported Co–Mo HDS catalysts were evaluated with dibenzothiophene (DBT) as model compound. For comparison γ-Al_2O_3 supported Co–Mo catalysts were also tested under the same conditions.

In the second study Multi-Walled Carbon Nanotubes (MWCNTs) were synthesised with specific pore diameter by chemical vapour deposition (CVD) of a carbon source on anodic aluminium oxide (AAO) template. AAO templates consist of pore channels in a uniform hexagonal arrangement that run parallel to surface of the film. AAO pore sizes are altered by creating an electrolysis cell (40V potential difference) with passive aluminium anode with (0.4M) oxalic acid as an electrolyte, thus enabling us to regulate the MWCNT pore size. Several grades of NiMo/MWCNT sulphide catalysts were prepared to find optimum pore size and their characterisation was done by Transmission Electron Microscopy (TEM), CO chemisorption, N_2 adsorption and H_2 Temperature Programmed Reduction (TPR).

2. Results and Discussion:

The Fig.1 shows XRD pattern of the CNT support with oxidised Mo catalysts with different MoO_3 loadings with a strong peak at 26.05° showing the graphitic basal plane (002) in CNTs in all samples with or without catalysts loading. A relatively weak and broader peak at 43° due to diffraction of (100) plane showing graphitisation degree of CNTs. XRD pattern of 4%m MoO_3/CNT catalyst are similar to that without catalyst loading implying well dispersion of Mo oxides but XRD pattern of 6%m MoO_3/CNT and 8%m MoO_3/CNT catalyst depict the broad and weak peaks of diffraction due to amorphous MoO_2 active species in the catalyst. For 8%m MoO_3/CNT catalyst diffraction peaks at 36.9°, 53.49°, 60.52° of MoO_2 are obtained 8%m is the maximum loading beyond which bulk phase is lower.

The reason for this hydrophobic nature of CNT support results due to weak interaction between support and active phase. Mo oxide is reduced by CNT support during heat treatment process at 500 °C and active phase of HDS activity is MoS_2 so oxides needs to be transformed to sulphide before use.

The Fig.2 shows XRD patterns of S-MoO_3/CNT and S-Co–Mo/CNT catalysts. For of S-MoO_3/CNT(10%m MoO_3) two broad diffraction peaks are observed characteristic of Mo_3S_4 with low valence. For S-Co–Mo/CNT new weak and very broad peaks at 14.4° and 32.71° got due to Mo_3S_4 crystallite and at 29° due to Co-MoS phase at low valence like Co-$Mo_{3.13}$ and Co-$MoS_{2.17}$. The active species are well dispersed on CNT support shows that introduction of Co species and sulfiding process help in active phase dispersion on CNT. It should be noted that a small amount of MoO_2 still remains unsulfided even in the sulfide state catalysts, as evidenced by the XRD patterns.

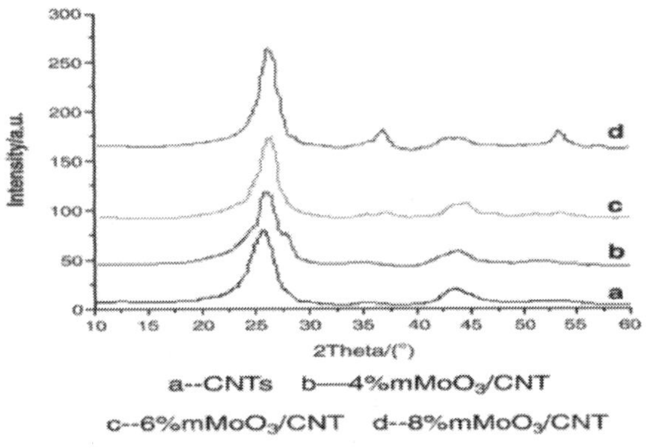

Fig.1 XRD patterns of oxide MoO_3/CNT catalysts[22]

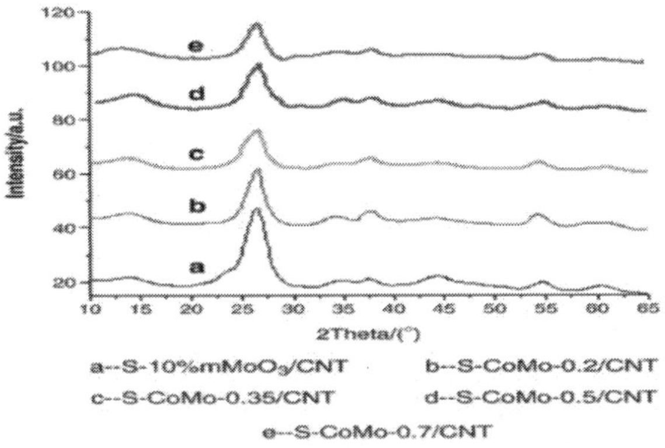

Fig.2 XRD patterns of sulfide state catalysts with different MoO_3 loadings.[22]

The Raman Spectra of MoO_3/CNT catalyst and the CNT are shown in Fig.3 accords the XRD pattern showing low graphitisation of CNT by peaks at 1346.64cm^{-1} and at 1585.36cm^{-1}. Rest peaks are due to C-H vibration. In case of the catalyst two new weak bands in low frequency region show vibrations of Mo=O bond in MoO_2 proving the XRD result.

The Raman Spectra of the sulfided Mo-based bimetallic catalysts with Co as promoter depict a peak at 353cm^{-1} due to Mo=S bond vibration. As no peak at 696cm^{-1} and at 820cm^{-1} found it concludes no bulk Co oxides or MoO_3 in catalysts.

For the Co–Mo-0.7/CNT catalyst, it is believed that there exist three kinds of active species because there are three hydrogen consumption peaks in its TPR curve. In the TPR curve of Co–Mo-0.5/CNT catalyst, two reduction peaks appear. And taking the TPR property of Co–Mo/γ-Al_2O_3 into account, the reduction of well-dispersed MoO_3 species takes place at about 900 °C while the reduction of bulk MoO_3 starts at about 500 °C. This results show that active species in Co-Mo/CNT are easily are reduced than those in Co–Mo/γ-Al_2O_3, indicating CNT supports facilitates reduction of active phases and Co/Mo atomic ratio has great influence on number of active species that existed in catalysts and also the reducibility of active phases. In addition, the electron donor effect and the semiconductor property of CNT may also play a role in the reduction of active species. Moreover it can also be concluded that at least 2 kinds of active species are reduced by Co-Mo-0.7/CNT more easily that those with Co/Mo atomic ratio of 0.2,0.35 and 0.5 resp. Moreover it is interpreted that the all the Co-Mo/CNT catalysts show high selectivity and activity than the γ-Al_2O_3 supported Co-Mo catalyst. Fig 5,6,7,8 also show that the highest activity is obtained for the catalyst with Co/Mo ratio 0.7 confirm the Bouwens report that the maximum in HDS activity of a kind of activated carbon supported Co–Mo catalyst is situated at a Co/Mo atomic ratio of about 0.73.

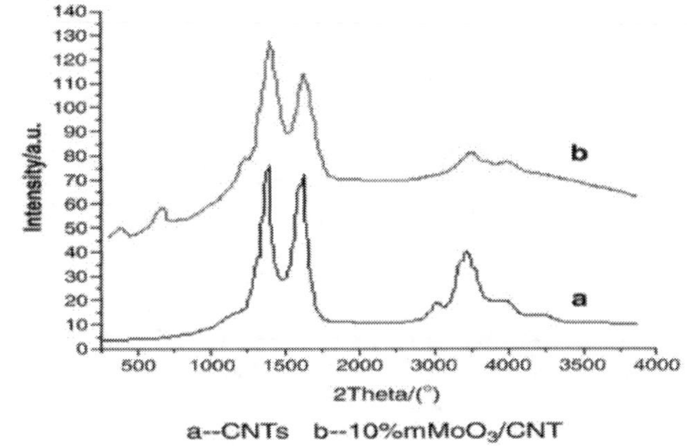

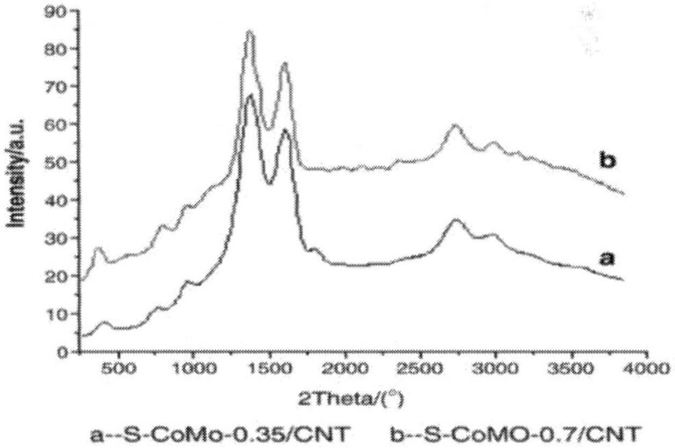

Fig.3 Laser Raman spectra of oxide catalysts.[5] Fig. 4: Laser Raman spectra of sulfide catalysts[5]

Now in the second study discussion the results of the following factors in detail as follows:

1. Variation of multi-walled carbon nanotube inner diameters
2. The effect of MWCNT pore diameter variation on catalyst performance
3. Variation of CVD parameters

Table 1. Selectivity and HDS conversion ratio for different catalysts shown below:

Catalysts	Selectivity	HDS ratio/%m
CoMo-0.35/γ-Al_2O_3-500	4.35	83
CoMo-0.7/CNT-500	10.26	99.6
CoMo-1.0/CNT-500	11.1	91.24
CoMo-0.2/CNT-500	12	94.22
CoMo-0.5/CNT-500	12.4	95.99
CoMo0.35/CNT-500	16.48	96.71

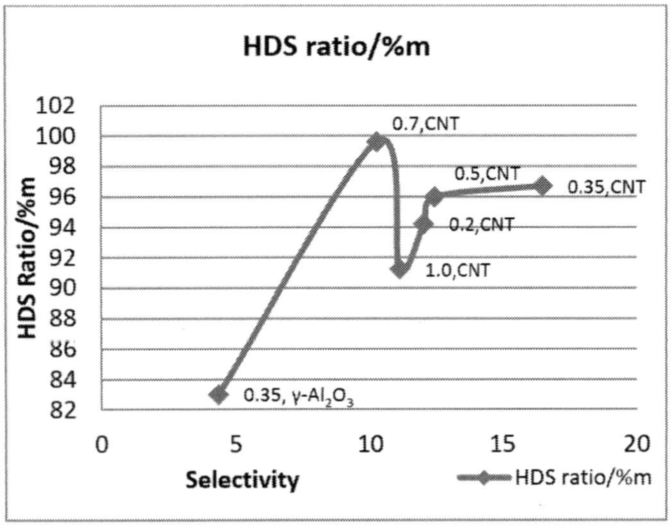

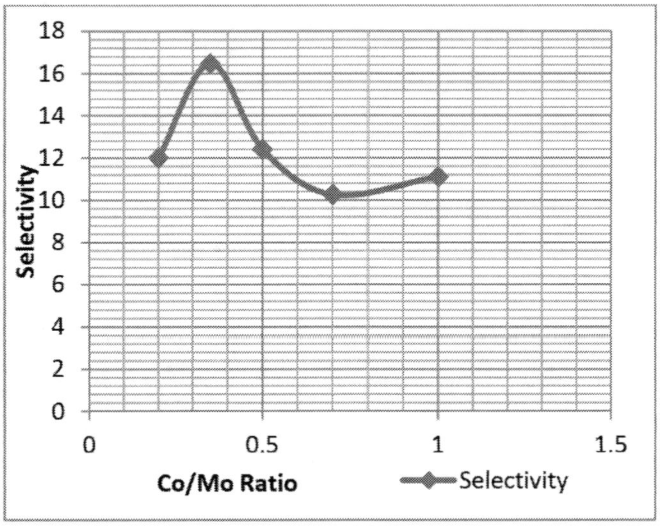

Fig.5: HDS conversion vs. Catalyst Selectivity

Fig.6 Catalyst selectivity affected by Co/Mo ratio

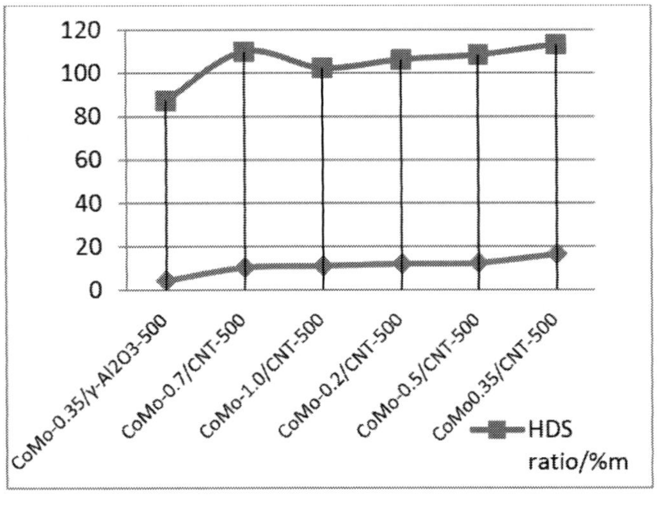

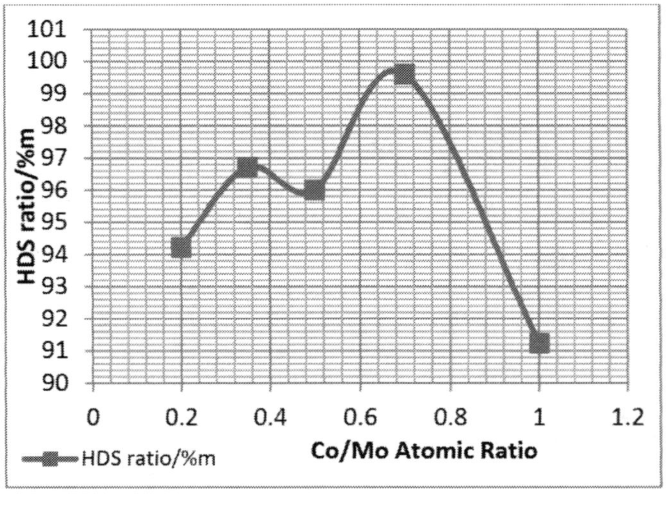

Fig.7 Variation of selectivity & HDS ratio/%m for different CNT and Alumina catalysts

Fig.8 Variation of HDS conversion ratio with with Co/Mo atomic ratio for varied CNT loadings

2.1 Variation in MWCNT pore diameter:

2.2 TEM Image of Functionalised MWCNTs:

The significant CNT thickness indicated a mutli-walled morphology and majority of the CNTs were quite linear. As the pore channel diameters increased within the AAO templates the Y-branched CNTs appeared to increase in number but CNT got by 0.5 mol/L oxalic acid concentration and Maximum annodizing voltage of 60V (fig 10).

2.3 Surface Characterization by N_2 Adsorption/Desorption:

The Fig.shows the isotherm exhibited at 77K was consistent with the one defined by BET classification. There is marked increase in the adsorption at high relative pressures between an adsorbate and macroporous materials (>50nm). The surface area (A) was found by implementing the BET method, while the total pore volume (V) was determined by the single point method at a relative pressure of ~0.98 and for pore diameters less than ~110 nm. The average pore diameter (d) was then determined by assuming perfectly cylindrical pore spaces (i.e. $d = 4V/A$), a reasonable assumption for the nanotubes produced, with high reproducibility for each run. The pore diameters estimated after this test were far less than TEM images as CNTs remained close-ended after HNO_3 treatment also affecting the pore volume due to Y-branching of the CNTs. Thus from fig.11, optimum value of 3 surface parameters. Also, Smaller AAO pore channels would then result in smaller CNT pore diameters, while larger AAO pore channels would produce more detrimental Y-branched CNTs.

2.4 Raman Spectroscopy:

The vibration characteristics for the different samples of the MWCNTs is shown in Fig.12, where 2 prime intense bands are common in each spectrum characteristic peaks of graphitic materials.

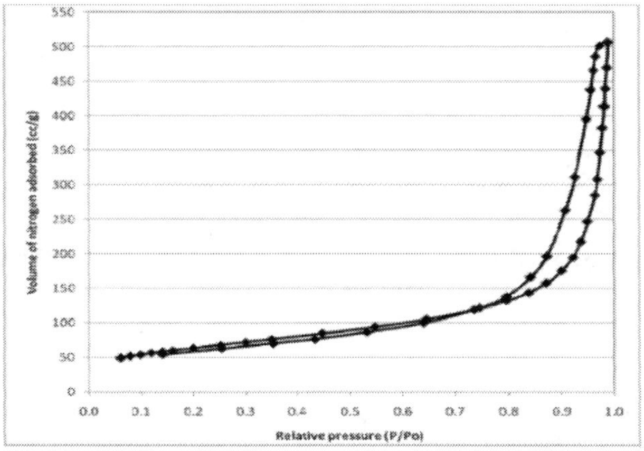

Fig.9 N_2 adsorption/desorption isotherm exhibited for condition 4 CNTs & similarly for all HNO_3 functionalized CNTs.[1]

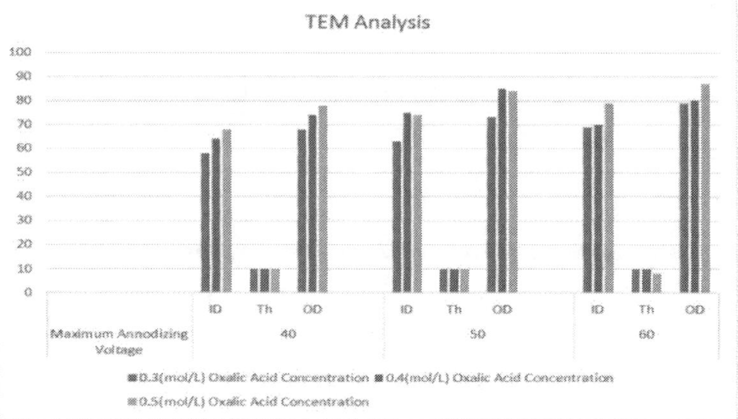

Fig.10 TEM analysis shows pore diameters remain constant at different voltages and oxalic acid concentrations (mol/L)

The band that occurs between wavelengths of 1580 to 1600 cm^{-1} corresponds to the stretching mode of sp^2 graphite bonds and is commonly referred to as the G-band (Eklund et al., 1995; Lefrant, 2002; Dresselhaus et al., 2005; Delhaes et al., 2006). The band appearing at wavelengths between 1330 and 1350 cm^{-1} is referred to as the D-band and is interpreted as the extent of imperfections and disorder in the graphite sheets (Eklund et al., 1995; Lefrant, 2002; Dresselhaus et al., 2005; Delhaes et al., 2006). The ratio of the intensity of these bands ($R_I = I_D/I_G$) is often used to represent the quality of the MWCNT morphology.

R_I values of the MWNCTs was relatively low(average of 0.97) while for commercial grade of MWCNTs with smaller pore diameters had R_I value of 1.57. Pore diameter variation don't affect the R_I values of the graphite layers

2.5 Thermogravimetric Analysis:

The TGA results for all the samples was similar as in as highest rate of CNT oxidation was found to occur at 630°C.Variation in the CNT pore diameter by variation of the AAO synthesis parameters has no effect on the thermal stability of the CNTs.

2.6 Effect of the MWCNT Pore Diameter on Catalyst Performance:

Specific grades of functionalized MWCNTs were chosen for screening as catalyst supports for HDS and HDN of coker light gas oil derived from Athabasca bitumen. The catalysts prepared by incipient wetness co-impregnation method, were denoted according to their approximate inner diameters: cat-60, cat-65, cat-70, and cat-75.

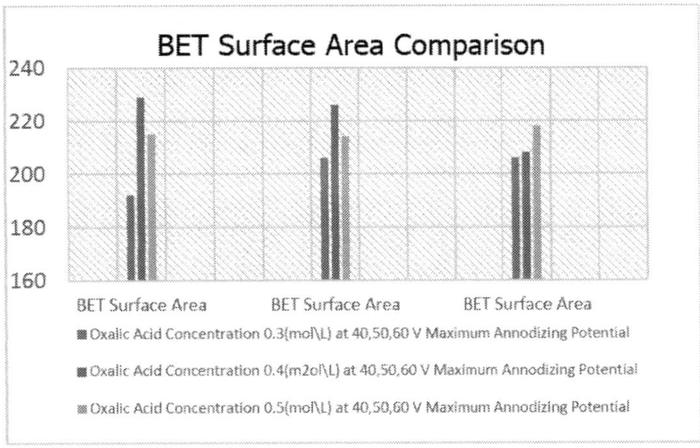

Fig.11 BET surface Area Comparison

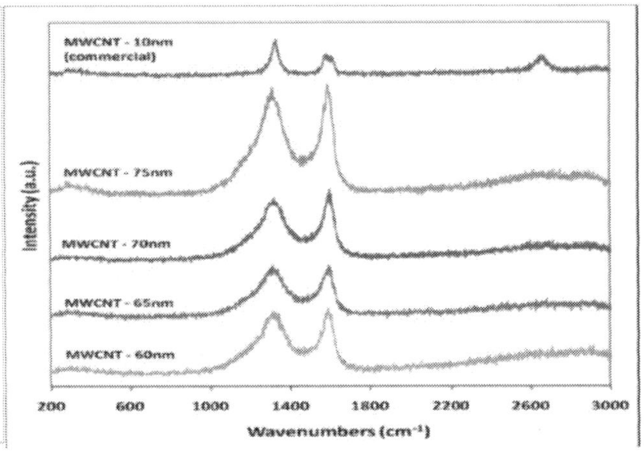

Fig.12 Results by Raman Spectroscopy[1]

2.7 HAADF-STEM Images of Functionalised MWCNTs:

High angle annular dark field scanning transmission electron microscopy (HAADF-STEM) images shows that the catalyst particles were chemisorbed on inner and outer walls of the MWCNTs and their sizes varied from 1 to 5nm in diameter. Catalyst sample image after the hydrotreating operation show that coke filaments branched of the metal catalysts (one of the reason for catalyst deactivation).

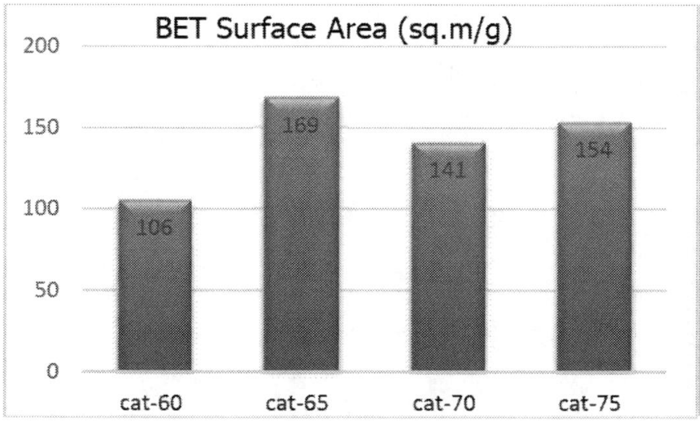

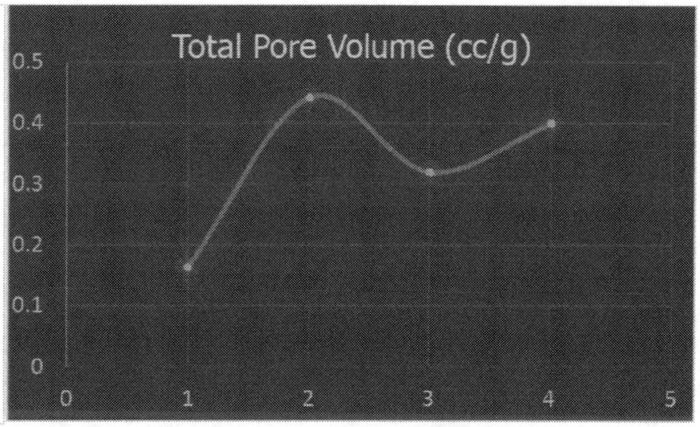

Fig.13 BET surface Area for different samples Fig.14 Total pore volume for different samples

2.8 Characterization by N_2 Adsorption/Desorption:

The analysis performed on the pelletized catalysts show give the result that for cat-60 surface characteristics decrease significantly due pore blockage and support compression. But cat-65 has most desirable surface characteristics as per fig. Structural characteristics of the prepared NiMo/MWCNT catalysts (~2.5 wt.% Ni, ~13 wt.% Mo), wherein the thickness of each of the catalyst was found to be same[1].

2.9 CO Chemisorption for Varied NiMo/CNT Pore Diameters:

This analysis gave the results as follows with cat-65 had the highest amount of CO uptake; 1.43 cc/g for a system pressure of 400 mmHg. The analysis as per fig. 16(red column: estimated metal exposure & blue column: CO uptake) showed that as surface area increased large amount of CO was chemisorbed and catalyst performance increased. Low exposure of the metals show large number of catalyst atoms were left inaccessible (here ~1.5%).

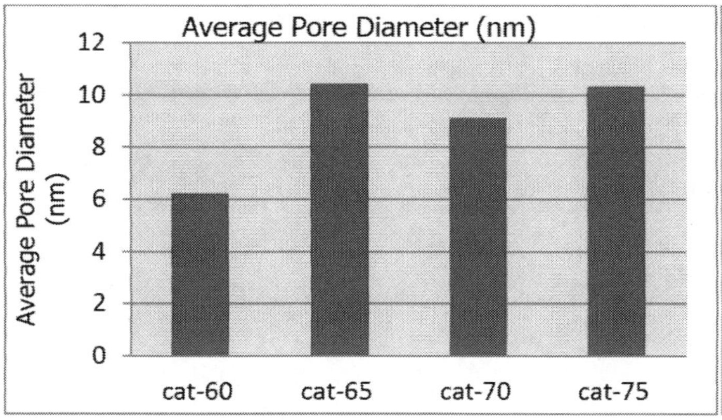

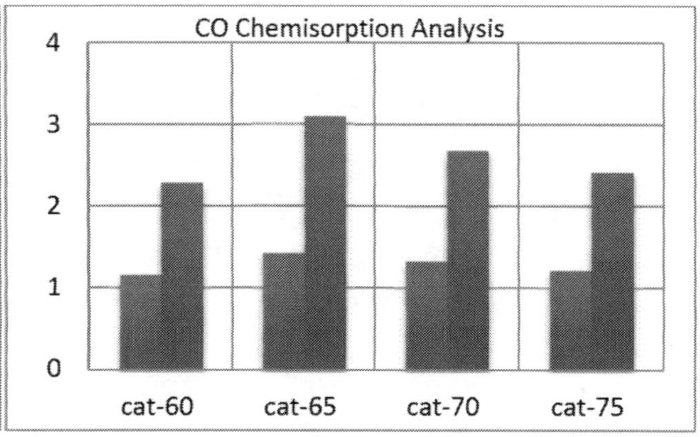

Fig. 15 Average Pore Diameter variation Fig. 16 CO chemisorption analysis

2.10 Temperature Programmed Reduction with H_2:

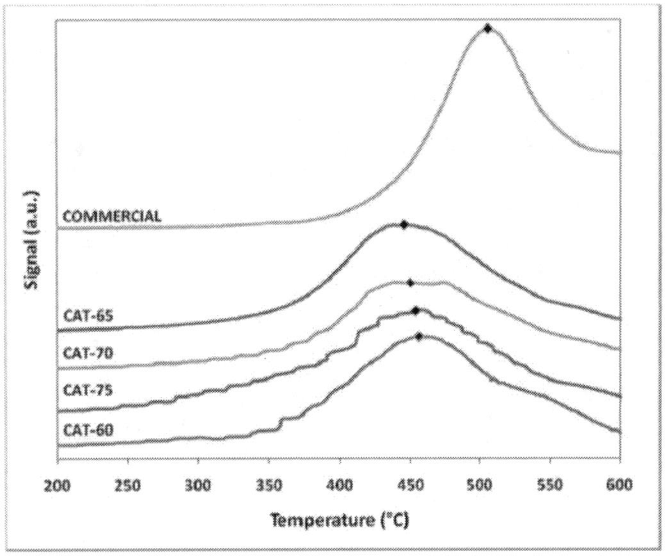

Fig.17 TPR using H2 of NiMo/MWCNT catalysts with varied pore diameters[1]

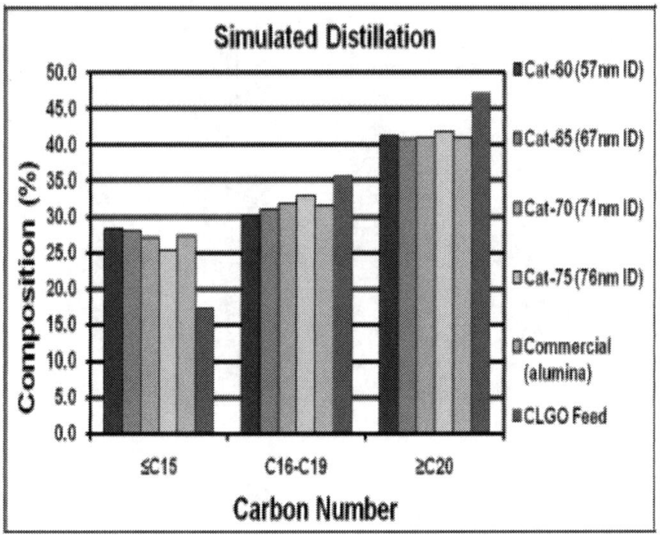

Fig.18 Carbon number composition of the steady-state coker light gas oil product[1]

The reduction peaks for the catalysts with varied pore diameters can be found in Fig.17. The hydrogen consumption peaks, which occurred within a domain of 445-456°C for the MWCNT-supported catalysts and at 506°C for the commercial γ–alumina catalyst, are the result of Mo reduction from the amorphous Mo^{6+} (octahedral) oxide phase to the ordered Mo^{4+} (tetrahedral) oxide phase (Leyva et al., 2008; Qu et al., 2003; Yu et al., 2008). The lower reduction temperatures of the MWCNT catalysts indicate less interaction with metal catalyst compared to γ-alumina thus making it easier for transition to sulphide phase for MWCNT catalysts in hydrotreating application. The TPR temperature being nearly similar for cat-70 and cat-75 showed that two supports were similar enough in surface features. The specific consumption peak temperatures for each catalyst were 456°C for cat-60, 453°C for cat-75, 450°C for cat-70, and 445°C for cat-65.

2.11 Boiling Point Distribution of CLGO Feed and Products:

The results of the hydrotreating the CLGO feed samples for commercial catalyst and prepared samples show that average pore diameter lowered from 75 to 60 nm, the percentage of carbon numbers ≤C15 increased steadily from 25.4% to 28.4%. The inverse trend was found for carbon numbers ranging from C16 to C19 and ≥C20, as the percentage of molecules within this range decreased steadily from 32.9% to 30.3% and 41.7% to 40.9% as the average pore diameter of the catalysts dropped from 75 to 60 nm resp. Hence, by making the pores of the MWCNT supports more confined leads to more extensive cracking of CLGO (coker liquid gas oil) molecules during the hydrotreating process. Refer Fig.18.

2.11 Catalyst Performance Based on N and S Conversion:

HDS & HDN steady-state activities of each catalyst from the CLGO feed can be found numerically in Fig. 19 and Fig. 20 for 4 MWCNT-based catalyst with a commercial catalyst with γ-alumina support under equivalent loading condition to evaluate the applicability of the NiMo/CNT sulphide catalyst.

Here the steady-state activity is obtained by averaging the conversions found for all the samples at steady-state (got after 24 hours) at 370°C 350°C and 330°C. The catalyst displayed sulfur conversions of 77.0%, 61.0%, and 47.4% and similarly for nitrogen, steady-state conversions of 37.5%, 26.7%, and 19.3% were found at respective temperatures of 370°C, 350°C, and 330°C for cat-65 rep. Under the same reaction conditions and at equal mass loading, the commercial γ–Al_2O_3 catalyst performed steady-state sulfur conversions of 51.1%, 37.8%, and 30.7%, along with nitrogen conversions of 26.6%, 16.2%, and 8.7%, for the three descending reaction temperatures. The cat-60, cat-70 and cat-75 had decreasing order of activities and long term detrimental effects was due to precooling taking place for the smaller pores for cat-60 and large inner diameters of cat-70 and cat-75 resulted in less dispersion of catalyst metals or direct mass transport of gas oil through catalyst without any reaction. Cat-65 had sufficient mass transfer of the reactant liquids and gases though maintaining high surface areas necessary for sulphide metal dispersion.

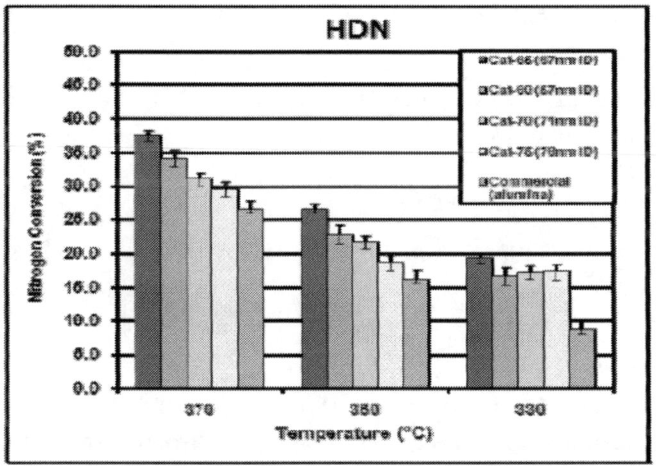

Fig.19 HDN activities in steady-state[1]

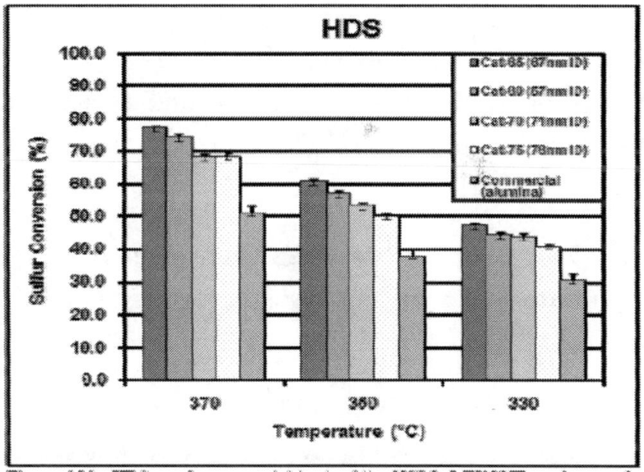

Fig.20 HDS activities in steady-state1

2.12 Variation of CVD Parameters:

Carbon yields from C_2H_2 achieved for nine chemical vapor deposition conditions and 0.250 grams of AAO template per run. Depict that highest yield of CNT was obtained at 650°C and yield also decreased with the temperature of the reaction as shown in Fig. 21 and Fig.22 .

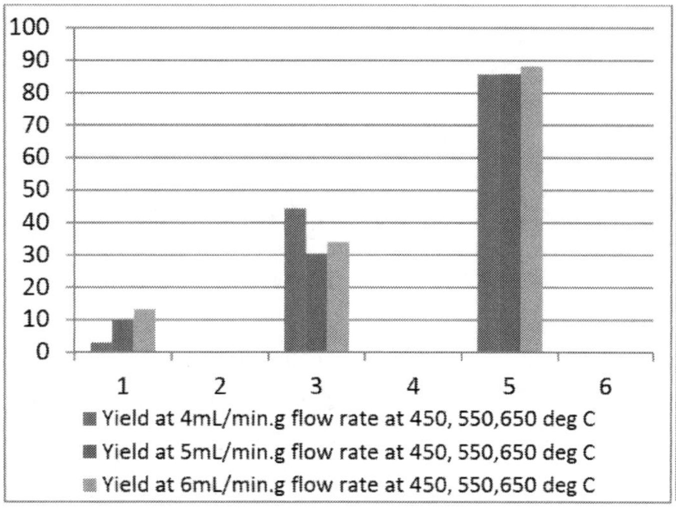

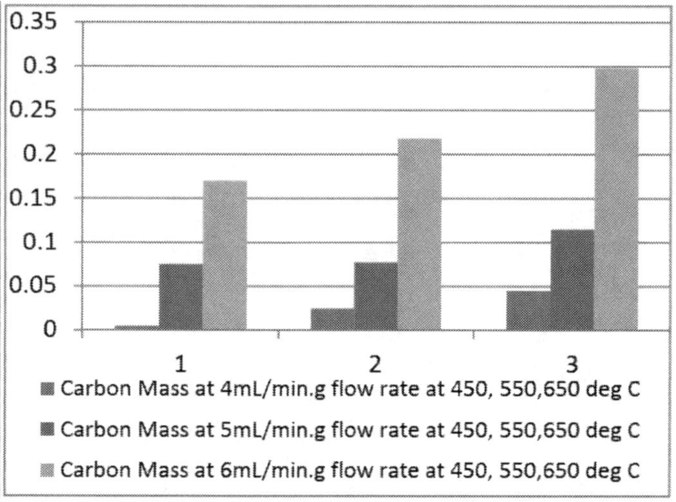

Fig.21 MWCNT yield at different temperatures Fig.22 Carbon mass deposited during CVD

3. Conclusions:

From the first study it can be concluded that, the pore diameter variation of the AAO film influenced the preparation of the CNTs with desirable structural characteristics (at 40V potential and 0.4M oxalic acid concentration). The CNTs have average pore diameter of 10-11.5 nm, 229 m2/g BET surface area and a 0.658 cc/g single-point pore volume.

The MWCNTs with approximate inner diameters of 65-67 nm were found to provide the optimum HDS and HDN activities for a Ni (2.5 wt.%) Mo (13.0 wt.%) catalyst. The catalyst displayed sulfur conversions of 77.0%, 61.0%, and 47.4% and nitrogen conversions of 37.5%, 26.7%, and 19.3% at reaction temperatures of 370°C, 350°C, and 330°C, respectively which is relatively high compared to commercial alumina catalysts.

Moreover for the chemical vapour deposition process operation conditions at 650°C reaction temperature produced highest carbon yield of 85% MWCNT product.

Now from the second study, the conclusions can be drawn as follows, the TPR results show that active species of Co-Mo/CNT catalysts can be reduced at relatively lower temperature compared to that of Co–Mo/γ-Al_2O_3. Moreover the CNTs favour s the reduction of the active species than the γ-Al_2O_3. The Co/Mo ratio greatly affected the reducibility of catalysts where catalyst with Co/Mo ratio 0.7 having the highest reducibility and the reducibility decreasing for the 0.2, 0.35 or 0.5 resp.

The HDS experiments prove that the Co-Mo/CNT catalysts have higher activity and selectivity for the hydrogenation compared to Co–Mo/γ-Al_2O_3 where the Co/Mo atomic ratio of 0.7 has highest activity and 0.35 ratio having the highest selectivity for the Co-Mo/CNT catalyst. The weak interaction between the active phases of CNT result in greater dispersion, surface atomic concentration of Co, Mo and S and chemical environment for Mo atom compared to γ-Al_2O_3 catalyst. Thus it can also be concluded that there is relation between the reducibility and HDS activity of the catalysts.

4. Refernces:

1. Stefan Kasey Sigurdson "Hydrotreating of Light Gas Oil Using Carbon Nanotube Supported NiMoS Catalysts: Influence of Pore Diameters" December,2009
2. Ancheyta J., Speight J.G., Hydroprocessing of Heavy Oils and Residua, Boca Raton, FL: CRC Press, 2007
3. Chen Y., Ciuparu D., Lim S., Haller G., Pfefferle L.D., "The effect of the cobalt loading on the growth of single wall carbon nanotubes by CO disproportionation on Co-MCM-41 catalysts", Carbon, 67-78, 44, 2006
4. Dai H., "Carbon nanotubes: opportunities and challenges", Surface Science, 218-241, 500, 2002
5. Diaz-Real R.A., Mann R.S., Sambi I.S., "Hydrotreatment of Athabasca bitumen derived gas oil over Ni-Mo, Ni-W, and Co-Mo catalysts", Industrial and Engineering Chemistry Research, 1354-1358, 32, 1993
6. Dong K., Ma X., Zhang H., Lin G., "Novel MWCNT-support for Co-Mo sulfide catalyst in HDS of thiophene and HDN of pyrrole", Journal of Natural Gas Chemistry, 28-37, 15, 2006
7. Dresselhaus M.S., Dresselhaus G., Saito R., Jorio A., "Raman spectroscopy of carbon nanotubes", Physics Reports, 47-99, 409, 2005
8. Eklund P.C., Holden J.M., Jishi R.A., "Vibrational modes of carbon nanotubes; Spectroscopy and theory", Carbon, 959-972, 33, 1995
9. Eswaramoorthi I., Sundaramurthy V., Das N., Dalai A.K., Adjaye J., "Application of multi-walled carbon nanotubes as efficient support to NiMo hydrotreating catalyst", Applied Catalysis A: General, 187-195, 339, 2008
10. Furimsky E., "Role of MoS2 and WS2 in hydrodesulfurization", Catalysis Reviews - Science and Engineering, 371-400, 22, 1980
11. Gras R., Duvail J.L., Minea T., Dubosc M., Tessier P.Y., Cagnon L., Coronel P., Torres J., "Template synthesis of carbon nanotubes from porous alumina matrix on silicon", Microelectronic Engineering, 2432-2436, 83, 2006
12. Gruia A., "Chapter 8: Hydrotreating", Handbook of Petroleum Processing, 321-354, Dordrecht: Springer, 2006
13. Lee O.J., Hwang S.K., Jeong S.H., Lee P.S., Lee K.H., "Synthesis of carbon nanotubes with identical dimensions using an anodic aluminum oxide template on a silicon wafer", Synthetic Metals, 263-266, 148, 2005
14. Sigurdson S., Sundaramurthy V., Dalai A.K., Adjaye J., "Effect of anodic alumina pore diameter variation on template-initiated synthesis of carbon nanotube catalyst supports", Journal of Molecular Catalysis A: Chemical, 23-32, 306, 2009
15. 207 (2001) 421 from Fuel Processing Technology, M.C. Abello, M.F. Gomez, O. Ferretti, Appl. Catal., A Gen.
16. 187 (1999) 213 from Fuel Processing Technology, A Gen.
17. 105(1987) 299 from Fuel Processing Technology, F.T.J. Hayden, A. Dumesic, R.D. Sherwood, R.T.K. Baker, J. Catal.

Overview of P.E. Licensing Process in United States

William R. Parrish, P.E. and Cory D. Jensen, P.E.

Professional engineering licensure was established in the United States to protect the health, safety and welfare of the public from unscrupulous or incompetent decisions of engineers. Licensure provides a proven and accepted assessment of experience, education and ethical standards.

This presentation gives an overview of the benefits of being licensed, the key players involved in becoming licensed and the path to licensure. The overview presents the background information to help put the other session presentations in perspective. More details about licensure are available in the March issue of *Chemical Engineering Progress (CEP).*

Value of being licensed

Being licensed shows other that you have a strong commitment to high technical and ethical standards. It can provide you with better career opportunities and provide a competitive edge when seeking employment. Many employers state that they would choose a licensed engineer over an unlicensed engineer, given both are otherwise equally qualified. Also, companies often offer incentives to encourage their employees to become licensed.

Some engineers think that licensure is important only in consulting or expert testimony. However, licensure is often required in firms doing design, construction and startup of processes. State laws also may include those teaching "advanced" engineering subjects be licensed.

Some states allow an "industrial exemption" for large companies who have the financial resources to cover liabilities arising from employee decisions. Most companies encourage licensure but many oppose eliminating the industrial exemption if they have operations in multiple states because it requires their engineers to be licensed in multiple states. There is a push to eliminate the industrial exemption. This important topic is discussed in another presentation.

No one can predict the exact future course of their career. However, being licensed may provide unexpected rewards and opportunities.

Key players in licensure

Becoming a P.E. involves education, experience and examination. Besides your coworkers, who can help guide your job experience, there are two important players in becoming licensed. The first is your local engineering board that establishes the standards for education and experience and ability to take the examinations. You will work with the board throughout the licensure process.

The second important player is the National Council of Examiners for Engineering and Surveying (NCEES) (www.ncees.org) . This is a nonprofit organization whose members are the various engineering licensing boards. The NCEES was created to provide some uniformity of standards of education, experience and examination. The membership establishes Model Law to offer uniformity. However, Model Law is a set of suggested best practices and no board follows Model Law exactly. All exams are administered and graded by the NCEES.

The licensure process

Obtaining a P.E. involves the following steps:

1. Having the appropriate formal education
2. Passing the Fundamentals of Engineering (FE) exam
3. Obtaining the appropriate experience (engineering internship)
4. Providing required documentation to board
5. Passing the Chemical Professional Engineering (PE) exam

While the general requirements are uniform across all jurisdictions, exact details vary. It is critical to work with your local board so that you understand and meet the exact requirements. A brief synopsis of each step is given below. More details are provided in the *CEP* article and in other presentations in this session.

1. Formal education: All jurisdictions require BS engineering degree, or equivalent, from an EAC/ABET-accredited four-year engineering program. Foreign degrees and other degrees are considered but must be approved by the board.
   Current Model Law states that as of 2020, engineers taking the PE exam must have a Masters or equivalent (MOE). This requirement is strongly opposed by AIChE and many other organizations. More details are presented in another presentation in this session and in the *CEP* article.

2. FE Exam: This exam tests your fundamental knowledge of engineering. It is easiest to pass when you are first eligible to take it (usually senior year or shortly after graduation). Another presentation and the *CEP* article provide more details of the exam and the upcoming changes in the exam.

3. Practical experience: Your work experience must be appropriate engineering practice. Typically four years are required although some boards permit fewer. In some jurisdictions advanced degrees can reduce the number of years required.

4. Documentation: The exact format of documentation and what is required varies between jurisdictions so it critical to work with your local board on this. However, all will require proof of education, passing the FE exam and work experience. The work experience will be supported by several references, who will also vouch for your ethical behavior.

5. PE Exam: This is the final step in becoming a licensed professional engineer. The exam is designed to be passed by a "minimally competent engineer with four years of practical experience". This exam tests the practical application of your engineering knowledge. Both a session presentation and the *CEP* article provide more details on the exam.

Summary

The AIChE recommends that its members become licensed to help ensure a competent and ethical chemical engineering workforce. While the process may seem daunting at first, working with your fellow employees and your local board will make the effort much easier. Each year about 400 chemical engineers become P.E's. They have made a significant investment in themselves and possibly in their future career.

(78b) Licensure--An Academic Perspective

S. Ranil Wickramasinghe, Department of Chemical Engineering, University of Arkansas, Fayetteville, AR

Most Chemical Engineering faculty members are not licensed. Why? Is it important for faculty to be licensed? Most undergraduates take the Fundamentals of Engineering exam but do not take the Professional Engineering exam. Why? Should faculty be licensed? Is licensure important for faculty who are conducting research? Should we encourage faculty to obtain their PE license? In this presentation these questions will be discussed.

TESTING FOR COMPETENCY - THE PROFESSIONAL ENGINEERING EXAMINATION

Anne Bertelsmann, PhD, PE, Galveston Bay Refinery, Marathon Petroleum Company LP, Texas City, TX
And
Denise Chastain-Knight, PE, CFSE, CH2M HILL, Augusta, GA

Introduction

The decision to pursue engineering licensure is a personal one, and the road includes many hurdles to overcome. Engineers take this road for a variety of reasons, with different motivations and experiences along the way, and finish the trek with the PE Exam. The Principals and Practice of Engineering Examination (PE Exam) is a nationwide exam used to ensure that licensed PE's are "minimally competent". The exam must include test content applicable to the competency being measured, effectively differentiate between competent and not competent candidates, and provide a consistent competency measure from administration to administration. This paper will define minimal competency, discuss the purpose of the Professional Engineering Exam, exam format using the Chemical Engineering Exam as the example, development methodology, item quality metrics and offer suggestions for test taking preparation. It will also describe the function of the NCEES Exam Development Committee and their role in developing a fair, effective exam, and explain why it takes weeks to get a result for a computer scored exam.

Path to Licensure

Whether an engineer undertakes the path to licensure early or late in their career, each must successfully negotiate the same hurdles. Each step in the process has a specific goal to either develop or to measure the engineer's capabilities toward professional competency. Figure 1 on the next page illustrates the path to licensure.

The quest to licensure begins with obtaining a Chemical Engineering degree. The next step is to pass an exam that measures competency on engineering basics. This first exam is the Fundamentals of Engineering Exam (FE) – also sometimes called the EIT (Engineer in Training) Exam. The FE exam is typically taken either during the senior year in college or shortly after graduation. The types of items resemble "textbook" questions and are designed to test understanding of the engineering fundamentals as taught in a typical engineering degree program. Currently, the FE examination consists of a general morning session (common to all disciplines) and a choice of seven discipline specific afternoon modules. Starting in the beginning of 2014, the FE examination will convert to a computer based test, which is discipline specific (i.e. no general module). For more information on the FE exam content and the conversion to computer based testing, refer to the NCEES website (http://ncees.org).

Figure 1: Path to PE Licensure

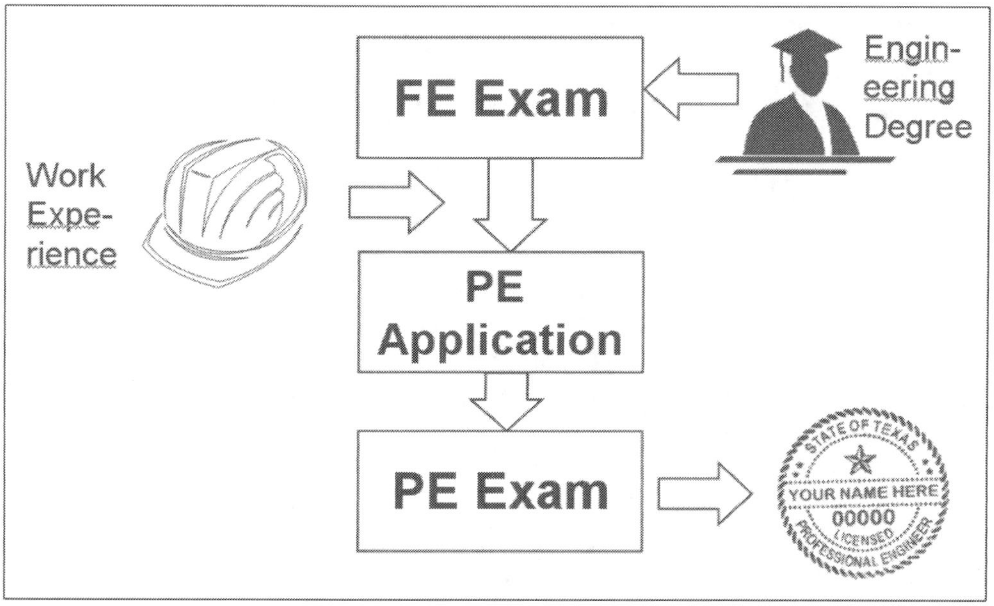

Following graduation, engineers enter the practice phase of their careers where fundamental knowledge is applied to real world application. In this phase of their careers, engineers use the theoretical principals learned from their education to solve problems that ultimately deliver physical products to market. Most states require a minimum of 4 years engineering related work experience before allowing candidates to sit for the PE exam. In order to become licensed, the candidate needs to submit an application to the state licensing board, describing education and experience, documenting passing of FE exam, and providing character and skills performance references. Format of the application and requirements for references, transcripts etc. vary from state to state. Once the application is accepted by the board, the candidate can register for the PE exam. After passing the PE examination, the engineer is licensed and can obtain the PE stamp. It is through this process that the state licensing boards ensure that only qualified candidates become licensed.

PE Exam Purpose

The PE exam is the final step in the road to licensure and its purpose is to ensure that only competent engineers become licensed. This helps to uphold the high standards of licensure and protects the health and welfare of the public. Since the exam is the same nationwide, it ensures consistency between states. Thus, one of the main requirements for a PE examination is that it can distinguish between competence and incompetence and that it can demonstrate this reliably and consistently. It is important to remember that "competent" should not be equated with "expert". The purpose of licensure is to identify individuals who have at least the minimum knowledge at which a person can practice professional engineering in such a manner that will safeguard life, health and property and promote the public welfare. That is why the term "minimally competent" is used.

A minimally competent engineer:

- should be able to perform identified tasks
- should meet minimum education and experience requirements
- is beginning to specialize in a sub-discipline
- may safely practice in another sub-discipline even though he/she does not practice in this area on a daily basis

The definition of "minimally competent" has implications on how the questions on the exam are written. The PE exam differs from the FE exam in that it addresses both the principles and the practice of engineering. So, while the FE questions are typically textbook related, questions on the PE exam are predominantly practice related. The questions build upon the experience that an engineer has gained during his or her first few years of working and need to recognize that chemical engineering experience can be gained in a variety of areas. The difficulty level of the questions is geared towards an engineer with about four years of experience.

PE Exam Content

The content of the examination is determined through a survey of practicing chemical engineers (PAKS study = Professional Activities, Knowledge and Skills), which is repeated about every 6-8 years to ensure that the examination stays current. The PAKS process includes conducting a technical content survey with input from practicing professionals. This survey is conducted over a period of weeks to ensure that the respondent pool size is adequate and is sufficiently diverse to be representative of the overall profession. Survey results are analyzed by a team of licensed Chemical Engineers, most with no prior exposure to PE exam development work, under the direction of a psychometrician who ensures that proper procedures are followed. The PAKS team will determine both the domain (subject area) content and item weight of the subdomains. PE exam content is established based on the PAKS study, and is published in the format of the PE exam specification available from the NCEES website (http://ncees.org/exams/pe-exam/) and is summarized in Table 1 on the next page.

The ChE exam specification was recently updated and became effective for the April 2011 exam administration. It includes six major domains and fifteen sub-domains that investigate theoretical and practical knowledge. Exam content is distributed within the major and subdomains according to the percentages shown in Table 1 to ensure a consistent content basis for every exam. The specification also includes guidelines for topical content within each sub-domain. For example the exam will include 4% of the items from sub-domain VI. C. Operation. Figure 2 includes example content from the 2011 exam specification.

Table 1: April 2011 ChE PE Exam Specification Summary

Knowledge Area	Approximate Percentage of Examination
I. Mass/Energy Balances and Thermodynamics	**23%**
A. Mass Balances	10%
B. Energy Balances and Thermodynamics	13%
II. Heat Transfer	**16%**
A. Mechanisms	9%
B. Applications	7%
III. Kinetics	**11%**
A. Reaction Parameters	6%
B. Reactors	5%
IV. Fluids	**16%**
A. Mechanical-Energy Balance	12%
B. Flow and Pressure Measurement Techniques	4%
V. Mass Transfer	**14%**
A. Phase Equilibria	5%
B. Continuous Vapor-Liquid Contactors	8%
C. Miscellaneous Mass Transfer Processes	1%
VI. Plant Design and Operation	**20%**
A. Economic Considerations	1%
B. Design	10%
C. Operation	4%
D. Safety, Health, and Environment	5%

Figure 2: Sample Exam Specification Sub-domain Content

C. Operation 4%
1. Process and equipment reliability (e.g., testing, preventive maintenance, startup/shutdown procedures, robustness)
2. Process improvement and troubleshooting (e.g., debottlenecking, experimental design and evaluation, optimization)

PE Exam Format:

The PE examination is given twice a year - once in April and once in October. It is an eight hour exam consisting of a four hour morning session and a four hour afternoon session with a lunch break in between. There are a total of 80 multiple choice questions (40 in the morning, 40 in the afternoon). Each question has four answer choices of which one and only one is the correct answer. Each question is stand-alone, i.e. it is independent of the other questions on the exam and can be solved with the information given in the problem, sometimes with the help of standard references. Some questions involve calculations while others are qualitative. All questions count equally and there is no penalty for guessing.

Each of the 80 multiple choice questions on the exam is of the same general format. In exam development, questions are referred to as "items" rather than questions because not every item has a question mark (like the one in the example below). Figure 3 illustrates the anatomy of an item, i.e. its general format and identifies the different parts of the item.

Figure 3: Anatomy of an Item

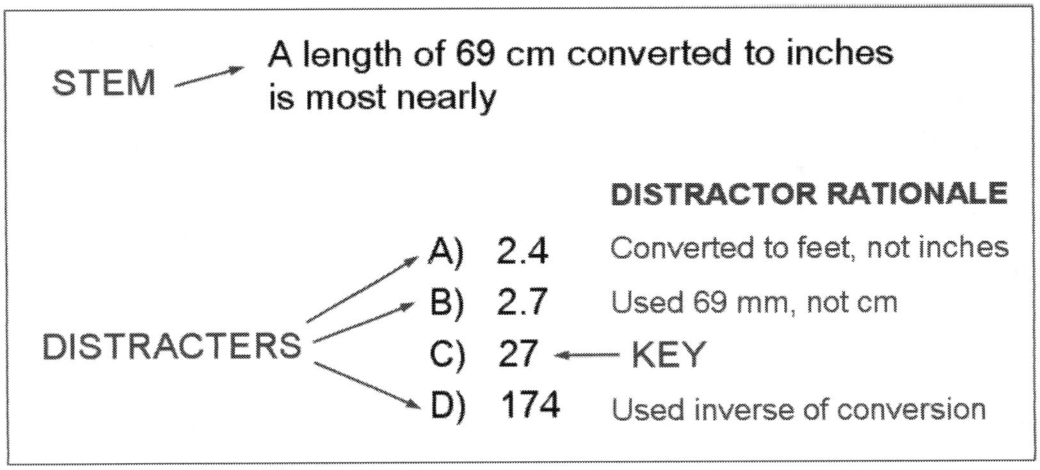

Each item consists of a stem and four answer choices. The stem (or problem statement) identifies what calculations are needed. It includes the information that is provided to solve the problem (sometimes even in table format) and states what is required for an answer. Beneath the stem are the four answer choices. Of these four answer choices, one (and only one) is the correct answer – the key. The other three are so-called distracters. They are incorrect answers – but they are not random numbers.

Each of the wrong choices represents a typical error that a less than minimally competent engineer (or someone not familiar with this specific topic) may make. Each distracter has to have a "rationale", i.e. an explanation which error was made. So, when a candidate is taking the exam and calculating a number that is exactly one of the given answer choices it does not necessarily mean that the problem was solved correctly. Maybe the candidate just made one of the "expected likely errors"...

The distracters can have a significant impact on the statistics of an item. If one of the distracters is obviously wrong (such as a negative absolute pressure), then even less than competent candidates will disregard it. That reduces the feasible choices, improves the "guessing" performance, and increases the chance that a not minimally competent candidate gets the question right. The Exam Development Committee strives to develop items that include the key, and three equally likely errors as distracters.

Exam Development

The PE Examination (as well as the FE, and the Surveying Exams) is a service that NCEES provides to the licensing boards. As such, NCEES provides the framework and the oversight for the exams, but the responsibility for the content rests with each exam committee. The committee writes the items, assembles the exams and reviews the statistics. NCEES has established rules and procedures that the exam committees follow to ensure consistency. Committees report back to NCEES with their exam quality measures. NCEES also provides the logistical support (psychometric consultant, publishing and printing, administrative support, maintaining records and the item bank, administering the exam, scoring, etc) and provides all security functions. For the larger exams, NCEES also pays and arranges for volunteer travel. For the smaller exams, this is the responsibility of the sponsoring society.

The Chemical Exam Committee is the group that actually writes and reviews the Chemical PE Exam. To be on the committee, one must be a licensed PE and express interest through the NCEES website or through a committee member. The committee has about 50 active members and meets twice a year in Clemson, South Carolina at the NCEES headquarters. Not everyone comes to every meeting, of course, so typically 30 – 35 committee members present at any given meeting. During the meeting items are written and reviewed by committee members who work with one another to critique and improve the items. Task teams also work to assemble, review and pretest exams.

Before an exam is administered, each item will have no less than six documented formal reviews and likely have been assessed by more than 20 people. Committee members are asked to complete detail oriented and often challenging engineering work and can earn PDH (professional development hours) for this activity. Developing the PE exam helps committee members to keep their engineering skills current and to give back to the profession. Without the work of this committee, the PE exam and with it the licensing wouldn't happen.

Preparing an exam that effectively and consistently measures minimal competency is a challenging task that succeeds or fails based on the ability of the committee to predict item performance. Each committee relies on a structured process and employees a variety of quality tools and techniques to improve the opportunity for success. Figure 4 is a workflow of the exam development activities. Items are written by a PE and submitted for review. Many items are prepared by the committee members at each meeting, while others are submitted by PEs who do not attend the meetings. The individuals who write and review items are referred to as Subject Matter Experts (SME). The author and two reviewers will independently work the item and develop the rationale for each distracter. Through this process the item is often revised and edited to improve its presentation and quality. This step is considered one of the most critical activities of the committee because items are the building blocks of the exam and each must meet high quality standards. Each committee member is responsible for item quality and is granted authority to modify draft items as needed to meet quality objectives, or reject items that fail criteria and cannot be edited to usable items. Finally the item is given a final check by the committee chair (or authorized designee) to verify that documentation is complete, both reviewers agree to the item content, and all quality checks have been completed, and approves the item for acceptance into the item bank. The item bank is a repository for items that are ready to be placed on exams.

Figure 4: Exam Development Process

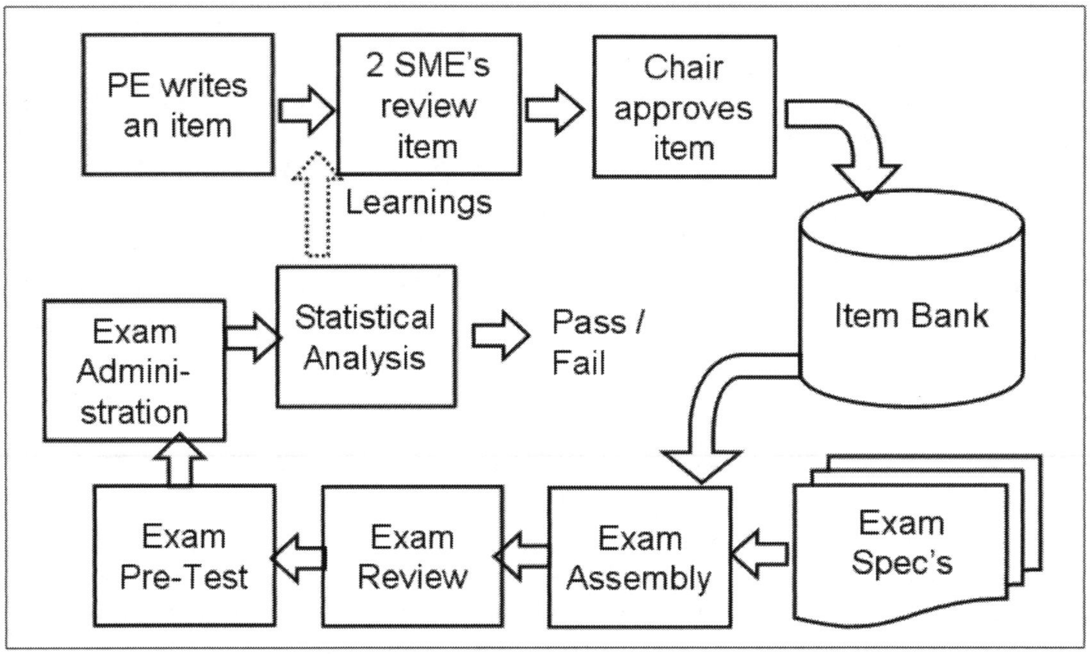

Exams are built, reviewed, revised, reviewed, quality tested and reviewed over a series of meetings. Each exam will have three independent team reviews, a pretest and a quality assessment. There are specific checks required at each workflow gate with task teams assigned to maximize the number of eyes that evaluate any given exam. An exam is compiled using items from the bank following the content distribution of the specification and sequenced in the specification order. The build team will then conduct a complete review of the exam, looking at each exam item for individual quality and checking for overall content flow and replacing items if necessary. The exam is then reviewed in two more meetings by independent teams who provide a "fresh set of eyes" and continue to improve the content to ensure that each item on the exam and the overall exam meet quality criteria. They also verify that the documentation is complete in accordance with the NCEES exam development requirements. Each exam must also have a pre-test where a group of committee members take the exam in as if they were actual test takers. After working all the items on the exam, they answer a series of questions about the items designed to predict item performance, then perform a distracter evaluation. Comments from the pre-testers are incorporated into the exam. In the final meeting before administration, a quality team conducts one last review of the exam verifying that all preceding checks (and associated documentation) are complete, all comments have been addressed, specification distribution is correct, and that author diversity limits and other NCEES criteria are met.

When the exam is administered and scored, data is compiled that can be used to evaluate the quality of the exam as a whole, and individual items. The data is analyzed and used to help the committee to learn what makes items perform effectively, and improve our performance prediction skills. Authors are provided a 'report card' so they know how items they wrote performed and have the opportunity to see how those items evolved in the review process. At each meeting, the committee will review items that had great statistics and try to find out what made them perform well. Similarly, we will review items that didn't perform well and identify traits that are undesirable in future items.

Item Statistics

Item quality is ultimately measured by the statistics generated from the candidate responses after an exam administration. Before discussing the statistical analysis of the examinee responses, one must first consider what makes an item a "good" item. The definition of a good item is one that is capable of distinguishing between "minimally competent" and not "minimally competent" candidates, does not discriminate against any particular group of test takers, and performs consistently in multiple exam administrations. To achieve those goals, it must be relevant, not confusing, and measure competency.

Most importantly, the content of the item must be relevant. Relevance is defined as the appropriateness of the topic and the difficulty of the item. The topic of the item must test knowledge or skills that are directly related to the exam specifications. For example: a question on high voltage power distribution may be relevant for an electrical engineer, but will not provide information on engineering competency of a *chemical* engineering candidate – no matter how well written it is. If the item does not relate to the exam specifications, it does not belong on the exam. Second, the difficulty level and complexity of the item must be such that a "minimally competent" engineer with 4 years experience can solve it in about 6 minutes. When writing items, authors often think about the interesting and challenging problems that are encountered in practice; however, if an engineer with 20 years experience requires several hours or days to solve the problem, what are the chances that a junior engineer will get the answer in 6 minutes? These type of problems, as interesting as they may be, do not make good exam items.

Another important principle to consider is that items should not be confusing to people who know the answer. When an item is confusing, a competent engineer who understands the concept being tested may not select the key, thus the question does not help to distinguish between competent and not competent. The so called "trick" question does not evaluate a candidate's competence and is considered a "bad" item to be weeded out in the exam development process.

It is equally undesirable to use items that give the answer away to people who don't know the answer. If the item includes clues to the key, then an incompetent engineer can choose the right answer without understanding the subject matter; thus the item does not measure competency. Clues can take many different forms:

- If some of the answer choices can be ruled out as impossible (like negative absolute pressures), the chance of guessing the right answer goes up.
- Grammar can give clues when one of the answer choices "stands out", i.e. is more detailed or does / does not match the grammatical structure of the item statement

For example, an item that asks about the 19th president of the United States, where three of the four answer choices are well known pop stars, does not require the candidate to know anything about US history to pick the correct answer.

Following an exam administration, a psychometric analysis is performed (psycho – metric = measuring the mind) on examinee responses. The analysis is solely based on information provided on the marked answer sheets without regard to the actual content of each item, but quite a bit of information about an item can be observed just from examinee responses. The psychometric analysis will evaluate a number of things. Of course, the simplest analysis is to determine how many candidates selected the key. Statistically, there is a 25% chance of guessing the correct answer because there are four choices. An item where less than 25% of the candidates selected the key may be flawed because it may have directed candidates towards preferentially selecting a distracter over the key. The next question is to ask *who* got the question right. Ideally, the candidates who got the question right were the same people who performed well on the exam overall. The analysis looks at the distribution of answers between the distracters and who selected them. This analysis also considers if all distracters were selected equally or if there was one distracter that was particularly attractive; maybe it even got selected more frequently than the key.

The objective of this analysis is to confirm the validity of each item as a tool capable of measuring engineering competence. An item that follows the basic three criteria (relevance, no confusion and no clues) can be expected to distinguish between competence and incompetence; however, sometimes clues, confusion or complexity can creep into an item unintentionally and be missed during the review before the exam is given. Statistics are then used to identify potential flaws. One such flaw is a mis-key. Mis-key is when the committee identified for example option A as the correct right answer, but option B is actually correct. Another flaw is an ambiguous item where more than one of the options could be correct. Despite the committee's efforts, such flawed items occasionally make it through the review process and are placed on an exam. The rigor of review process prior to exam administration minimizes the likelihood of a flawed item; however, it cannot totally eliminate them. The statistics after the exam will reflect item imperfections and provide an opportunity to compensate for item flaws in exam scoring so that candidates are not negatively impacted by flawed items.

Figure 5 is an example of the statistical analysis report for an item prepared by the psychometric consultant. One such sheet is prepared for each of the 80 items on an exam. The top left corner shows the "number of cases", i.e. the number of examinees who took the exam in this administration. In this example, there were 2910 candidates. So, this example is not from any Chemical PE exam where typically about 300 candidates sit per exam. The number below the case number (highlighted in yellow) is the Mean (or the P+ value). This number shows what percentage of the candidates selected the key; in this case, 65%. P+ alone indicates that the question is of appropriate difficulty for the average test taker; not too easy and not too hard. Statistically, the higher the P+, the easier the item.

Figure 5: Item Statistics

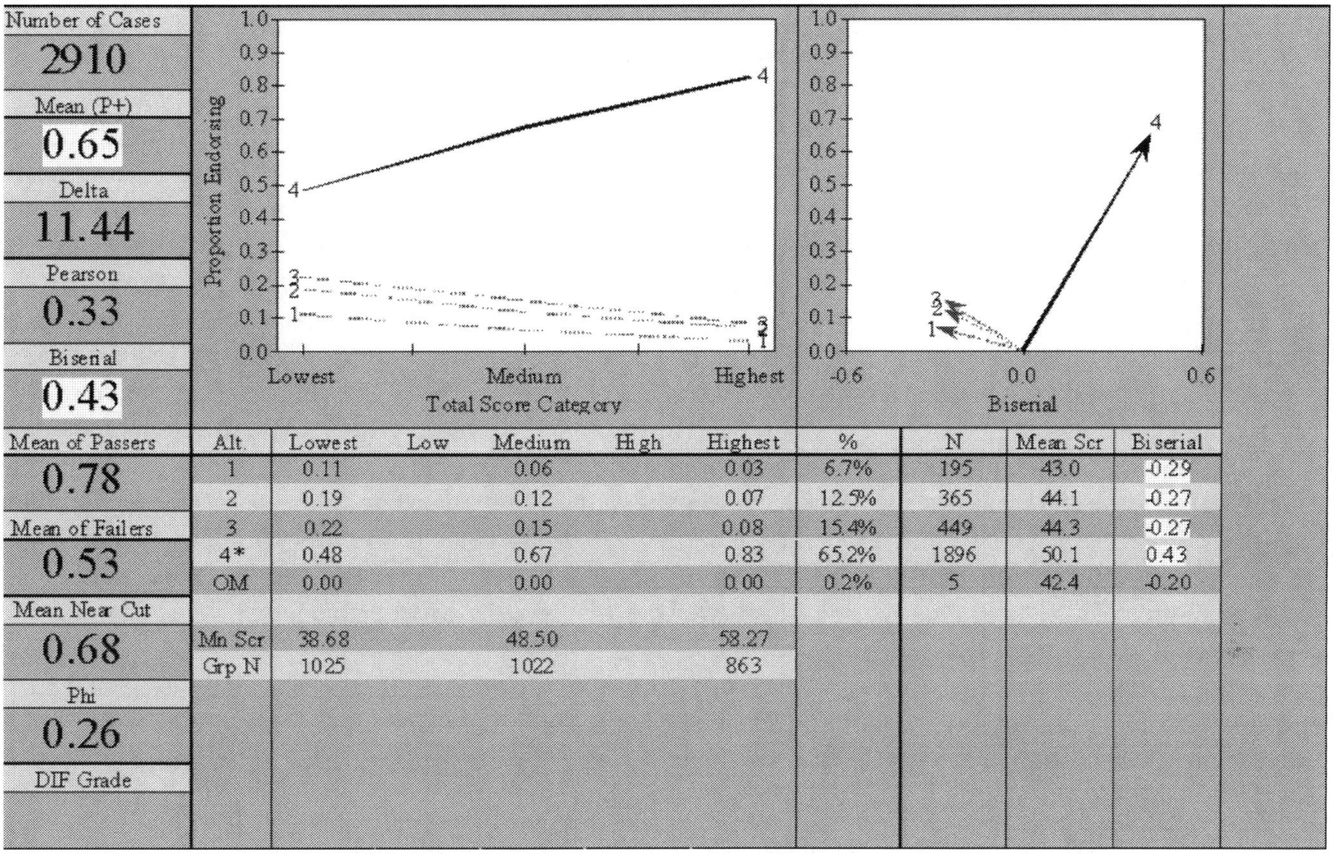

Alt.	Lowest	Low	Medium	High	Highest	%	N	Mean Scr	Biserial
1	0.11		0.06		0.03	6.7%	195	43.0	-0.29
2	0.19		0.12		0.07	12.5%	365	44.1	-0.27
3	0.22		0.15		0.08	15.4%	449	44.3	-0.27
4*	0.48		0.67		0.83	65.2%	1896	50.1	0.43
OM	0.00		0.00		0.00	0.2%	5	42.4	-0.20
Mn Scr	38.68		48.50		58.27				
Grp N	1025		1022		863				

The statistics not only evaluate how many candidates selected the key, but how selecting the key on this item correlates to overall exam scores. Test takers are divided into three groups of roughly equal size based on overall score on the exam; the low scoring third, the medium scoring third and the high scoring third. For each of these groups, the report in the table under the left graph shows how many selected which answer choice. For example on this question, "4" (or "D") was the key as indicated by the asterisk. Of the low scoring third, only 48% got the right answer, while 83% of the high performers selected the key. Thus, someone who performs well on the exam overall (which is an indication of competence) is significantly more likely to select the key than someone who scored low (may indicate lack of minimal competence). This spread shows that the question is capable of testing competency.

Figure 5 also illustrates that the incorrect answers were relatively evenly distributed between the distracters, indicating that none of the distracters appears to be overly attractive or easily ruled out. Another way of looking at the same data is presented on the right of the page. The table shows average overall scores for the people who selected each of the answer choices. In this case, the people who selected the key had an average overall score of 50, while the people who selected the distracters had overall scores in the range of 43 to 44. Five candidates with an average score of 42 did not answer the item. This clearly demonstrates that the people who selected the key were more likely to do well on the exam overall, and indicates that the item can effectively measure competency.

The data is evaluated in yet another way by the cumulative statistics in the lower left: Mean of Passers indicates that of those candidates who passed the exam, 78% selected the key. Mean of Failers shows that only 53% of the people who failed the exam selected the key. The larger the spread between these two numbers, the better the item is at measuring competency.

A handy way of summarizing the "who got the question right" analysis is the R-Biserial number. Basically, this number is related to the slope of the lines in the graph on the left. For a good item, this slope for the key should be positive and the larger the better because it expresses the difference in performance of the low and the high scorers. Theoretical range for the R-Biserial number is from -1 to +1, positive values indicate that the candidates who selected the key also scored higher on the overall exam. A negative R-Biserial number indicates that higher scoring candidates selected one of the distracters, which could be an indication of a mis-key or ambiguous item. The R-Biserial number is highlighted in yellow on the left and shown in the picture on the right. The R-Biserial number is important because it evaluates how well an item meets the three basic criteria. We previously observed that the sample item is of appropriate difficulty based on the percentage of correct answers (P+ value). The R-Biserial number shows that we did not confuse the competent candidates (a high percentage of high performers got it right) and apparently it did not give any clues to the less competent candidates (. It also appears to be relevant with regard to the topic since the high performers (those with better understanding of chemical engineering) did better on this than the low performers.

To further illustrate the concept of the R-Biserial number, consider an example of an item that might produce undesirable statistics. The item is asked on the final exam for a graduating class of chemical engineers: "Which bar in town has the most beers on tap?" One would expect a certain percentage of students to select the correct answer, but are those candidates the most knowledgeable students who spend a lot of time in the lab, or the "party crowd"? This question would likely have a negative R-Biserial number, indicating the overall least knowledgeable students are more likely to select the key, which means this item is not an effective measure of a graduating chemical engineer competency, nor is it relevant to chemical engineering knowledge and skills.

Determining Competency

The individual item statistics are the building blocks for determining the overall exam performance. To make the move from the individual item statistics to determining competency of the candidate, a passing score, also called cut score, needs to be established. Figure 6 on the next page illustrates the philosophy behind relating the score distribution to the cut point, or passing score. Total score is simply the number of items answered correctly. For the Chemical Engineering PE Exam, the total score has a theoretical range of 0 to 80. Most candidates have at least some knowledge and get lucky guessing on some questions, so scores on the very low end are uncommon.

If it were known upfront which candidates are competent and which ones are not, the scores of the two groups could be plotted as shown in Figure 6. The at least minimally competent engineers have higher scores, the not minimally competent engineers have lower scores. However, there will be some overlap between the two groups. The better individual items differentiate between competency and incompetency, the smaller the overlap. However, because there is a range of competency and because a person's performance can vary, there will always be some overlap and thus the cut point decision is not clear cut.

Figure 6: Score Distribution

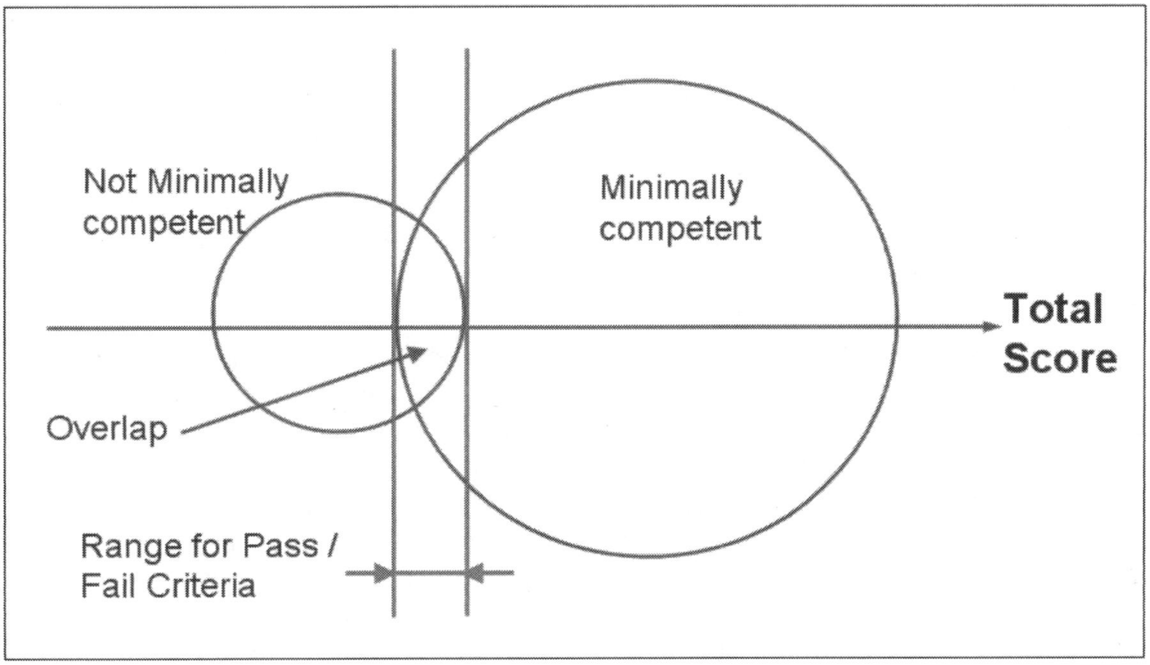

The overlap represents the range where the pass/fail decision should be made. All the at least minimally competent candidates on the right side should pass, all the not minimally competent candidates on the left should fail. With any decision point within that overlap range, there is a potential for competent engineers to fail and/or for less than minimally competent engineers to pass. Failing a competent engineer means that they will need to re-take the exam and includes the risk that they may give up on licensure. Passing a less than minimally competent engineer may allow an under-qualified person into public practice. Since the second type of error is more consequential, there is a tendency towards raising the bar when determining the pass/fail criteria.

Without knowing upfront which candidates are "minimally competent' and which ones are not (after all, it is the purpose of this exam to establish this), statistical measure are used to determine the pass/fail criteria, i.e. establish the "cut score". The cut score is the number of correct answers that a candidate needs to have in order to be considered "minimally competent". One way of determining the cut score is through panel of licensed PE's. This process is fairly involved, since it requires about 15-20 PE's who can volunteer 2-3 days of their time and travel to attend to cut score session. The panel must have good diversity (age, gender, area of practice, geographic location, demographics, etc…) and the cut score study must be completed before the results of the exam can be reported back to the state boards and ultimately to the candidates who are eagerly waiting for the scores. For these reasons, a cut score panel is usually only done when there was a change in exam specifications where the exam results cannot be directly compared to recent prior exams. The first exam where the format or specification is changed is called the "anchor exam" and cut point is established by a formal study.

For a cut score study the diverse panel of volunteers (all licensed PE's) meet and spend about a day to work the exam in a timed environment just like the actual administration. They can use any references, cannot consult with one another and they don't have access to the exam solution. The next day, they receive their score, and review and discuss each item as a group. They look at the statistics similar to those described above, but lacking the pass/fail metrics, review the solution, the distracters and discuss the topic and difficulty of each item. On the basis of that discussion, the panel members make a judgment call on how many "minimally competent" engineers out of 100 would answer this question correctly. A psychometrician guides the session and collects these estimates from each panel member. This data is then used in statistical analysis, generating a proposed passing score. The proposed passing score can then be adjusted (for example, to err on the "safe side"). This then becomes the basis for the pass/fail decision for the anchor exam.

Such a cut score study is not practical for every exam at every administration. So as long as there is no change in the exam specification, it is possible to "equate" the current exam to a previous exam. The equating process, heavily supported by a psychometrician, compares examinee performance on the current exam with examinee performance on previous exams. From this comparison it can be determined if the current exam is harder or easier than the previous exams, and cut point adjusted accordingly. If the data indicates that the exam is easier, the cut score is adjusted upward and a candidate will need to answer more questions correctly in order to pass. Conversely, candidates are not penalized if they take a harder exam because the cut point is revised downward and fewer questions must be answered correctly to pass. Equating is NOT the same as grading on a curve. When grading on a curve, the passing percentage would be fixed and the cut score adjusted to achieve that percentage. With equating, the competency of a candidate is measured on personal performance, and not dependent on the individual candidate's position within the test taking population, nor on the difficulty of the exam.

The psychometricians calculate other measures from the data. Coefficient Alpha measures the internal consistency or statistical homogeneity of the data, thus basically expresses the reliability of the exam. A measure called "Subkoviak" provides an estimate of the consistency of the pass/fail decision or an estimate of the probability that an individual would receive the same pass/fail decision on multiple administrations of the exam. All these measures are regularly reviewed to ensure that the exams can consistently measure competency.

After an exam is administered and all the answer sheets are scored, the statistical item analysis is performed. Exam Committees need to review potentially flawed items in case the key needs to be adjusted and the exam rescored. Then the cut point is determined, either through a cut score study with a panel or through equating. Only then can the pass / fail decision be made. State boards are notified with the results, perform state specific pass/fail assessments and then notify candidates of the results. NCEES recommends that only pass/fail should be reported to the candidates, though some states provide the actual score (or a "scaled percentage" based on a fixed passing percentage) to candidates. Candidates who fail get a diagnostics report so that they know how they performed in the various subject areas. NCEES recommends to only report the pass/fail decision for two reasons: The cut score can vary from exam to exam, so a candidate may pass one year and another candidate taking the exam another year may fail with the same score. The other reason is that the exam is designed to make a pass/fail decision on minimum competency. In that regard, the PE exam is different from an exam like the SAT, which is designed to measure "aptitude" over a wide range of abilities. The PE exam is not designed to measure ability accurately at the far ends of the scale. The PE exam does not distinguish between the "outstandingly competent" and the merely "extraordinarily competent engineer". Similarly, it cannot distinguish between the "totally incompetent" and the "absolutely clueless". Thus, a candidate who received a score of 75 out of 80 is not necessarily a better engineer than a candidate with a score of 72 out of 80 – both are "minimally competent" and pass.

The skeptical reader asks; "Does this really work?" Can an exam such as this really measure competency as the statistical "number magic" appears to indicate? The answer is found in the historical exam pass rates. NCEES does not report an overall pass rate for the exam, but breaks it up into a pass rate for first time takers and for repeat takers. Table 2 includes pass rates for exams given under the current Chemical PE Specification. The most recent pass rates for other exams can be found on the NCEES website.

One specific item to note about the data in Table 2 is that there is a gap between the pass rate for first time takers and for repeat takers. The pass rate for first time takers is typically rather high; in the 70 – 80% range for the Chemical PE exam. This high pass rate is because the examinee pool is already a pre-selected group. In order to be admitted to the exam, a candidate had to pass the FE exam, had to prepare an application with documentation of engineering experience, and had to be approved by the state board, a process that weeds out some of the less qualified potential candidates. Also, especially in chemical engineering where there are many jobs that come under the industrial exemption and do not require licensure, the people who decide to pursue licensure are already a non-random subset of all engineering graduates. This contributes to the relatively high pass rates for first time takers. The portion of the candidate population that passes on a repeat attempt includes in candidates that are minimally competent but may have been inadequately prepare in their first exam sitting.

Table 2: Chemical PE Exam Pass Rates

Chemical PE Exam Pass Rates	April 2011	October 2011	April 2012	October 2012
1st Time Pass Rate	78%	75%	74%	71%
Repeat Pass Rate	31%	29%	31%	20%

How to Prepare for the Exam

All the fascinating information presented in this paper aside, chances are the reader really wants to know how to prepare for the exam. Preparation philosophy will differ for a young candidate near the 4 year experience mark from that for a veteran engineer. One theory is that any minimally competent engineer can pass the examination without any preparation if he or she had unlimited time - time to look up references, read up on concepts and topics outside one's familiar area of practice, find the right equations and conversion factors, etc. But there is a time limit, and 6 minutes per question is not a lot of time.

Exam preparation should focus on building speed and efficiency for solving items. Through study and preparation, a candidate should become familiar with reference content, placing tabs to help find frequently used tables or compile a personal reference of frequently used information. Common equations and frequently used conversion factors should be memorized. Candidates should reactivate the habit of using math: solving equations, working with variables and using a scientific calculator instead of a computer. Speaking of calculators, it is strongly recommended to prepare for the exam using one of the approved models. Precious time can be wasted searching for the logarithm button on an unfamiliar calculator. Approved calculator models can be found on the NCEES website.

Exam readiness is achieved through lots and lots of practice. Preparation classes are useful for refreshing material basics, but are no substitute for rehearsal. A candidate should prepare by solving as many sample problems as one can find. By solving many, many problems, one will automatically remember the frequently used equations and conversion factors, enhance math skills and remember where to find information. The process also develops test day confidence because the items encountered on the exam have been encountered in similar form during exam preparation. Candidates should conduct one or more mock exam day(s) to practice pacing and practice maintaining focus for 8 hours. NCEES publishes a sample exam for each specification that has been developed by the exam committees in the same manner as a "real" exam. One author of this paper frequently recommends to potential exam candidates that they practice with readily available materials, complete a timed trial with the NCEES exam, then continue practicing with more materials.

Finally each candidate should prepare for the test environment and stage to be at peak performance on test day. Exams are administered in finite locations within a state and often require candidates to travel. Plan a travel itinerary that will allow arrival at the test site early, rested and unstressed. Have a satisfying but not heavy breakfast and make a plan for a relaxing lunch. Consider the test room conditions, dress comfortably and in layers. Be aware of the materials that are allowed in the test hall, and prepare accordingly. Candidates will be turned away without a valid government photo ID. Cell phones and other forbidden materials may be confiscated or admission refused. The exam is worked in a printed booklet, which contains sufficient space for working the questions and pencils are provided. Candidates need to bring an approved calculator, unapproved calculators will be rejected. The exam is open book, so candidates also need to bring any references that they want to use during the exam. References must be bound (no loose paper), but can be marked with flags. Review the test day policies on the NCEES website and plan accordingly (http://ncees.org/exams/exam-day-policies/).

Closing Comments

Professional Engineer Licensure is a personal decision and requires one to undertake a dedicated and challenging journey. Candidate should embark on this journey knowing that the exam development process is designed to ensure each exam measures for minimal competency in a consistent manner, and items are statistically validated. Finally, the factors influencing exam score processing timelines are understood. The authors of this paper encourage engineers to pursue licensure. We speak for ourselves and our fellow PEs when we say that nobody ever regrets the resources and effort spent obtaining a PE, and we appreciate the industry approval and instant expert acceptance granted by those two little letters following our names. P.E. Professional Engineer.

(78d) The Current Hot Button Issues in Domestic Professional Licensing

Amos Holt, Southwest Research Institute, San Antonio, TX

Amos Holt, a past president of ASME and current VP for Environmental, Safety and Quality Systems will review the issues currently receiving the greatest attention by the PE community which includes all of the various state and territorial licensing boards, the National Council for Examiners of Engineers and Surveyors as well as the numerous professional society stakeholders and educational organizations and societies. The discussuion will touch on the Model Law and its currect sections on the Masters or Equivalent (MOE) requirements as well as Industrial Exemptions. The various positions of key players covered including the position of a consortium of professional societies called "Licensing that Works" or LTW will be summarized. This group includes both ASME and AIChE and represents about 240,000 engineers thtroughout the world. Amos will tell you what you will likely be hearing more about in the next few years and how it might affect both current PE licensees and prospective PE licensees.

The California Divide – And Why You Need To Engage Your Legislators

Diane Spencer, PE, Lawrence Livermore National Laboratory, CA
and Emmett Miller, PE, consultant and AIChE Fellow, CA

Abstract

California is crippled by a great divide that has nothing to do with tectonic plates. California's professional engineering statute gives civil engineers free reign, allows electrical and mechanical engineers to work in their disciplines so long as that does not conflict with civil engineering, and acknowledges by title only that other engineering disciplines exist.

This division presents problems for engineers whether they have a professional engineering license or not. For example, Spill Prevention Control and Countermeasure (SPCC) plans are mandated by federal regulations. Such plans are prepared by chemical engineers all over the country, but NOT in California. SPCC plans are part of the exclusive domain of civil engineers.

A fire protection engineer was once told he could not design a fire suppression system because that was the exclusive domain of mechanical engineers. A chemical engineer was once told he could not submit a Risk Management Plan (RMP) (required by Cal EPA) without a civil engineer stamp because even the hazard analysis on a facility that uses chemicals is the civil engineer's domain. These are just a few examples where engineers are not allowed to work on projects that similarly trained engineers are expected to do in the rest of the country and the rest of the world.

So how do engineers in California navigate this regulatory nightmare? And how do we influence the legislators, engineering board, and others to nudge the statute in the right direction?

California's answer included forming a coalition of professional engineering societies that work together to present a united voice. The California Legislative Council of Professional Engineers (CLCPE) not only represents the interests of engineering professions to the legislature, but serves as a pro bono resource for the legislature on technical issues as well.

This paper will discuss: (1) our lessons learned from working with legislators; (2) California's professional engineering statute and why it is written as it is; and (3) why one needs to be engaged in the legislative process. Professional engineers are entrusted with protecting the health, safety, and welfare of the public. Getting engaged has the benefit of letting legislators understand the impact of their decisions and allows engineers to serve the communities where they live, and provide a safer, better community for everyone.

California's Problem

While most other states utilize some form of a generic registration model for engineers, California's model is discipline-specific, with civil engineering held on a pedestal. The definition of "civil engineering" in the Professional Engineers Act (California Business and Professions Code Sections 6700-6799.) is so broad that it encompasses nearly the entire spectrum of engineering. This overly broad definition, coupled with a mandate that only civil engineers can do civil engineering, means there is little room for those in other engineering disciplines to legally work in their own professions. The law allows overlap from civil engineering into all other disciplines, but other disciplines cannot overlap into civil engineering. So, not only are chemical engineers impacted, the other engineering disciplines are impacted as well.

Section 6731 states the following, with parts underlined for emphasis:

> "Civil engineering embraces the following studies or activities in connection with fixed works for irrigation, drainage, waterpower, water supply, flood control, inland waterways, harbors, municipal improvements, railroads, highways, tunnels, airports and airways, purification of water, sewerage, refuse disposal, foundations, grading, framed and homogeneous structures, buildings, or bridges:
>
> (a) The economics of, the use and design of, materials of construction and the determination of their physical qualities.
>
> (b) The supervision of the construction of engineering structures.
>
> (c) The investigation of the laws, phenomena and forces of nature.
>
> (d) Appraisals or valuations.
>
> (e) The preparation or submission of designs, plans and specifications and engineering reports.
>
> (f) Coordination of the work of professional, technical, or special consultants.
>
> (g) Creation, preparation, or modification of electronic or computerized data in the performance of the activities described in subdivisions (a) through (f)."

For many years the conventional wisdom has been that the term "fixed works," as used in the statute, (see Section 6731 above) somehow limits the scope of civil engineering, leaving room for other engineers to practice their craft. In order to clarify where that boundary lies, Senator Mimi Walters asked the California Legislative Counsel Bureau for a legal opinion (2011). The findings were shocking. "Fixed works" includes not only the structures, but also all the attachments to the structure. So, if a mechanical engineer designs a pumping system or a chemical engineer designs a fractionating column, and that hardware and the hardware is secured to a foundation, those engineers have crossed into the exclusive domain of the civil engineer. Other engineers are similarly impacted.

When a reasonable person reviews this list of studies or activities, it becomes apparent that the list precludes even scientists from legally studying the phenomena and forces of nature. And in most jobs, engineers are expected to produce engineering reports or to coordinate the work of other professionals.

How Did California Get Into This PE Mess?

The perverse nature of the current Professional Engineers Act in California is a result of good intentions gone awry.

In March 1928, a dam northeast of Castaic, CA, broke and flooded the valley below, wiping out the town of Santa Paula and flooding other towns downstream. Hundreds of people died and millions of dollars' worth of property was damaged.

The Professional Engineers Act was then created to ensure that only qualified people (civil engineer PEs) could work on dams and other fixed works. The legislators made the law as inclusive as possible to make sure there were no loopholes. Later, the Legislature expanded the Act to include Chemical, Electrical, Mechanical, Petroleum, Industrial and Metallurgical Engineering as Title Act Disciplines. With the passage of time, the Board for Professional Engineers, Land Surveyors, and Geologists (BPELSG, aka "Engineering Board") was empowered by the Legislature to create Title Disciplines on its own authority, which it did, eventually increasing the count to 15 title protected disciplines (Title Act disciplines), These are defined by Board Rules (16CCR 400-476) and have minimal stature in law. Along the line, Mechanical and Electrical Engineering were converted by statute to practice protection (Practice Act discipline)[1].

This patch job to add mechanical engineering and electrical engineering to the Professional Engineers Act has been ineffective (per Legislative Counsel's review). Inclusion of Title Act disciplines constitutes an even lower tier where those engineers are not able to take responsible charge of work within their field of expertise. In summary, the field is far from level.

The problem with the Professional Engineers Act has been known for many years. The difference now is the Legislative Counsel's legal opinion that was issued in July 2011 (www.CLCPE.org). According to that opinion, companies that rely on engineering services are on notice that the only engineering discipline allowed to conduct engineering work (not just supervise) in California is civil engineering. If a project requires a chemical engineer, a dual-licensed engineer will be needed to meet the competency requirements of both a chemical engineer and the legal requirements of a civil engineer. A dual license is needed for the other engineering disciplines as well.

---

[1] Four of the title disciplines (Corrosion, Manufacturing, Quality and Safety) were dropped ~1999 for lack of an NCEES exam.

The Industrial Exemption

Common to most states, California has an Industrial Exemption that allows companies engaged in manufacturing to use unlicensed engineers in their businesses— except for civil engineering work. The justification for the Industrial Exemption has been that such employers are capable of and competent in evaluating their own engineering talent, that they do not need interference from the state in doing so, and that failure that will impact the public is subject to strict liability.

There are exemptions for people who perform some specific and narrow civil engineering services. Architects performing architecture and building contractors following building codes are examples. But no general exemption exists for engineers who are not civil engineers if they perform "civil engineering." The Industrial Exemption even states that it does not apply to civil engineering.

With the recent legal opinion, however, all engineering practice in connection with facilities that are geographically fixed is now regulated as civil engineering, and the Industrial Exemption is therefore meaningless.

Engaging the Legislature

In the late 1940's, engineering societies in California organized the California Legislative Council for Professional Engineers (CLCPE) in order to work together as a resource to assist the California legislature, support laws that benefit the public and profession, and help keep society members informed about legislation that might affect them. Application of the Professional Engineers Act was not a problem to engineers in the state so long as the Engineering Board and regulators maintained a liberal interpretation. Likewise, CLCPE did not intensify its effort to change the law until the literal interpretation ultimately reinforced by the Legislative Counsel ruling came into focus.

CLCPE's efforts to change the law began in the 1980s. Our members helped draft a bill and found a senator to carry it. In California, the legislator who carries the bill is the "author" and organizations that support the bill, such as CLCPE, are "sponsors" regardless of who actually drafts the bill. Drafting a bill and finding "authors" is a process CLCPE repeated about a dozen times.

Let's reframe the process we used in California in terms that may help others take advantage of our lessons learned.

1) Coalition. We had a consortium of engineering societies working together as a coalition. This provided more credibility in that we were able to speak with one voice instead of individual organizations promoting separate proposals. A coalition also allows a sponsoring organization such as CLCPE to represent a larger portion of a legislator's constituency. AIChE local section leaders met with a senator to promote the one bill that chemical engineers sponsored without going through CLCPE. The senator listened intently about the impact of current law, asked a few clarifying questions, and then asked about how many chemical engineers were in the state. She was not satisfied with our answer, stood up and thanked us for our time.

Creating a coalition of several engineering societies enabled us to afford a lobbyist. AIChE and local sections are a 501(c)(3) nonprofit organization, which restricts how much money can be spent on lobbying. CLCPE however, was organized as a 501(c)(6) nonprofit, which allows for lobbying, but since the societies and CLCPE are all volunteer organizations, we do not have an intense political presence, and that restricts campaigning for or donating to candidates or political parties. Unfortunately, the opposition we face in California has major campaign funding and they openly support candidates that vote as the opposition wants them to vote.

Another advantage of working within a coalition of engineering societies goes beyond promoting legislation. The conversations we had when we met allowed us to gain insight into the problems each of the different disciplines were facing and we heard personal testimony about those various issues. This perspective has been invaluable.

2) Hiring a lobbyist. Over its lifetime, CLCPE has successively employed three lobbyists who helped with efforts to change the Professional Engineers Act, in addition to supporting the overall mission. A good lobbyist helps focus the discussion and facilitate consensus on language for a given bill. They also know how to navigate through the legislature, explain the bill or other proposals to legislators, legislative staff, and to others, and to find legislators who will author a bill. Typically engineers have full-time jobs, and working a bill is very time-consuming. The first of CLCPE's lobbyists was an engineer who later became Executive Officer of the Engineering Board. The other two were professional lobbyists with legal backgrounds.

3) Language for the bill. Earlier this paper discussed drafting language for a bill and then entering that bill into the legislative process (i.e., finding an author). "Entering" is the key word here. In the late 1990s, our bill attracted attention from the Professional and Technical Consultants Association (PATCA). In our quest to expand the law to allow engineers to practice their own profession, we needed to be careful not to adversely impact scientists (e.g., chemists) and other technical professionals. After several meetings with PATCA, we were able to work out compatible language. Going forward on subsequent bills, we deliberately engaged with the chemists (PATCA) early in the process.

4) Bill's "author". Over the decades we have been pursuing change, our bills have been authored by different senators and assemblymen. In several cases, the legislator had been an engineer in his/her former career. One such senator, who had been an aeronautical engineer, had commented that it took multiple disciplines to design an aircraft. A legislator with a technical or science background readily grasped the issue we were dealing with. One of the first committees to hear a bill dealing with engineering (i.e., the bills we were promoting) was either the Business and

Professions Committee (B&PC) in the Senate or the B&PC in the Assembly. When seeking an author we often approached members of these committees as well. The intent was to get early support and move the bill through the legislative process.

5) Finding support. In addition to sponsors, it helps to find other organizations that will support a bill through (1) letters, (2) testimonies, and (3) lobbying efforts. We collected binders of letters from individuals, municipalities, regulators, and corporations. These included Chevron, the Chemical Industry Council of California, and Genentech. The letters provided a lot of anecdotes on how this perverse law affected people. The volume of the letters attracted the attention of legislators.

Once a bill is scheduled for a vote, testimonies at the hearing can be very compelling. The city of San Diego, California Fire Prevention Officers Association, a division of the California Fire Chief Association, along with various individual department fire chiefs and fire marshals, individuals impacted by current law, and others have testified at legislative hearings over the years. The Dean of Engineering at University of California—Davis testified on behalf of the University of California at the hearing in January 2012, stating that the state is educating engineers who cannot work in their own professions. The dean was shocked to find out that the legislators still would not vote to support changing the current law.[2]

The third level of support is when companies/ organizations involve their lobbyists. However, we found it very difficult to convince companies that had lobbyists in Sacramento to get engaged at this higher level.

6) Other curve balls. We experienced numerous unexpected challenges. One such curve ball happened in 1997, when chemical engineers tried independently (without CLCPE) to change the law. The Engineering Board required that we provide a Sunrise Package[3]. Essentially converting engineers from Title Act (defined in by BPELSG Board Rules) to Practice Act (defined in the Professional Engineers Act) is viewed as creating a new engineering discipline. We were given one week to prepare this. Fortunately, CLCPE had prepared something similar 10 to 15 years earlier, which we were able to mine and then create the required Sunrise Package. The Engineering Board still did not support our bill. Note: Keeping a good archive of material produced in supporting legislation could have unexpected value in the future. The key is being able to find the needed archived information when it is needed. Another advantage is that many of the same people remained engaged in CLCPE for long periods.

---

[2] Incidentally, the western region of ASCE sent their lobbyist to testify in opposition at the same hearing where the UCD dean testified.

[3] Sunrise is a process under which an occupation or profession wishing to receive state certification or licensure must propose the components of the legislation, along with cost and benefit estimates of the proposed regulation

What We Proposed and Why

Our approach to changing legislation evolved over the years. As mentioned earlier, chemical engineers made an attempt in 1997 to raise chemical engineering to a Practice Act (defined by statute). Our opposition protested that our attempt was part of a turf battle among the engineering disciplines. Legislators do not want to be drawn into controversies (e.g., infighting). Our next attempt involved CLCPE, with the intent to define all the Title Act disciplines as Practice Act disciplines. Because we had a coalition involving more than 15 engineering societies, we were able to combat this accusation. Still we did not succeed.

We also shifted our approach and proposed legislation that was closer in alignment with generic engineering. Specifically, we proposed that engineers should be allowed to do work based on their education, experience, or testing. In talking to legislators and to our lobbyist, we learned that we needed to approach any change in engineering statute with baby steps. Adopting generic licensure language (similar to other states or similar to Model Law proposed by the National Society of Professional Engineers (NSPE) was too much at one time. Because of the heavy influence by civil engineering companies in the state, we drafted bills that would continue all the advantages afforded to civil engineering. Unfortunately, our opposition would not negotiate.

In 2000, CLCPE had caught the attention of the legislature sufficient enough that they commissioned a study of the two-tier approach (Title Act/Practice Act) used in California. The study considered natural overlap between the professions, whether there was enough distinction between the engineering disciplines to support this unique regulatory structure, and whether the three Practice Act disciplines (civil, mechanical, and electrical engineering) posed more of a risk to the public (relative to the other engineering disciplines) and thus justified the existing two-tier structure. (See **www.CLCPE.org** for study by the Institute for Social Research.)

The study strongly supported changing the existing regulatory structure. The task of making this change fell to the Engineering Board. The bill it sponsored was opposed by American Consulting Engineers Council (ACEC) - California (a contractors' association) and Professional Engineers in California Government (PECG, a public employee union). Both were well-funded with strong ties to legislators — and the bill failed. [PECG is allowed to donate directly to candidates and its 13,000 members individually campaign for legislators. That has a big influence on those legislators who might support our position.]

The study marked a shift in how we approach changing legislation; CLCPE did not sponsor bills after the study. Instead, CLCPE and our individual engineering societies supported a succession of bills that were sponsored by the Engineering Board or by companies and organizations that use engineering services.

The disadvantage of this approach is that we had less influence on making amendments to the language in the bills. So while sponsoring our own bills had not been successful, this other approach allowed us to continue to pursue fixing the Professional Engineers Act.

Another challenge we had was finding companies and organizations that would commit their lobbyists. Many of these companies were very willing to prepare letters of support, but were not ready to involve their lobbyists.

End Results

This paper provides numerous lessons learned from decades of experience in trying to change California law to allow all engineers to practice in their own professions without finding a civil (typically) or other Practice Act engineer to take responsible charge of the work. Despite years of effort, we have not succeeded in changing the Professional Engineers Act. We are debating our next steps. It is our hope that the lessons we learned will be helpful to others who find the need to engage their state legislators.

(80a) Utilization of Low-Grade Heat and Process Integration in Oxy-Combustion Coal Based Power Plants

R. Soundararajan, Department of Energy and Process Engineering, Norwegian University of Science and Technology, Trondheim, Norway and **Truls Gundersen**, Energy and Process Engineering, Norwegian University of Science and Technology, Trondheim, Norway

CO_2 Capture and Storage (CCS) is a technology that collects and concentrates the CO_2 emitted from large point sources, transports it to a suitable storage location and stores it away from the atmosphere for a long time. CCS is expected to play a pivotal role in stabilizing the atmospheric greenhouse gas levels within acceptable limits. It has been estimated that the average contribution of CCS in total emission reduction would range from 15% to 54% for stabilization targets of 750 ppmv to 450 ppmv CO_2 respectively[1]. The deployment of CCS could help bring down the overall cost of mitigation of climate change in the longer run[2].

Oxy-combustion is one of the three main capture routes that is considered to be promising and in the demonstration phase. This particular method that burns the fuel in oxygen instead of air has several environmental and economic advantages[3, 4]. The upstream production of oxygen and the downstream purification and compression (CPU) are responsible for an efficiency penalty of around 10 percentage points[4]. Novel concepts such as pressurized coal combustion improves the overall efficiency of an oxy-combustion power plant with CO_2 capture and helps reduce the capture penalty[5]. Even with pressurized coal combustion, the efficiency penalty is still considerable and needs to be reduced in order to make the oxy-combustion technology commercially viable in the near future.

Operating the boiler at a higher pressure also helps recover more latent heat from the flue gases before it enters the CPU for final purification and compression. This low grade heat is usually recovered by employing special types of heat exchangers that are resistant to acid corrosion. They are made of highly inert materials such as Teflon and are called acid condensers or plastic heat exchangers. The most common way of utilizing this heat is to heat the boiler feed water in the steam cycle. This reduces the amount of steam extracted from the turbines and hence increases the gross power produced in the steam turbines. Apart from this method, low temperature power cycles such as Organic Rankine Cycles or CO_2 Brayton cycles can be used to improve the overall efficiency of the power plant[6]. The rising cost of electricity could justify the additional capital investment required for this method.

Both methods mentioned above for utilization of the low grade heat are designed independently of the steam cycle and the integration with the rest of the power plant is not optimized. Thus, development of a more tightly integrated power plant using mathematical modelling and optimization principles could yield better solutions at a much lower capital cost[7]. In this context, a systematic approach is required to achieve optimal utilization of the available low-grade heat. Simulation of an oxy-combustion coal based power plant with CO_2capture is presented here along with various integration options and their efficiency improvement potential. Also, the current work based on mathematical modelling is discussed in brief.

References:

[1] T. Morita, N. Nakicenovic, J. Robinson, Overview of mitigation scenarios for global climate stabilization based on new IPCC emissions scenarios, Environmental Economics and Policy Studies, 3 (2000) 65-88.

[2] J.A. Edmonds, P. Freund, J.J. Dooley, The role of carbon management technologies in addressing atmospheric stabilization of greenhouse gases., in: Proceedings of the Fifth International Conference on Greenhouse Gas Control Technologies, 2000.

[3] F. Petrakopoulou, G. Tsatsaronis, A. Boyano, T. Morosuk, Exergoeconomic and Exergoenvironmental evaluation of power plants including CO2 capture, Chemical Engineering Research and Design, 89 (2011) 1461-1469.

[4] M. Kanniche, R. Gros-Bonnivard, P. Jaud, J. Valle-Marcos, J.-M. Amann, C. Bouallou, Pre-combustion, post-combustion and oxy-combustion in thermal power plant for CO2 capture, Applied Thermal Engineering, 30 (2010) 53-62.

[5] R. Soundararajan, T. Gundersen, Coal based power plants using oxy-combustion for CO2 capture: Pressurized coal combustion to reduce capture penalty, in: Chemical Engineering Transactions, Vol. 29, 2012, pp. 187-192.

[6] J.P. Roy, M.K. Mishra, A. Misra, Parametric optimization and performance analysis of a waste heat recovery system using Organic Rankine Cycle, Energy, 35 (2010) 5049-5062.

[7] X. Luo, B. Zhang, Y. Chen, S. Mo, Heat integration of regenerative rankine cycle and process surplus heat through graphical targeting and mathematical modeling technique, Energy, 45 (2012) 556-569.

HEAT EXCHANGER NETWORK (HEN) RETROFIT FOR ENERGY EFFICIENCY OPTIMIZATION IN CRUDE OIL REFINING

Abdulaziz Najjar, Zeeshan Farooq & Mahmoud Bahy Noureldin

ABSTRACT

There are many initiatives for reducing energy consumption in a given HEN in oil refining. However, in practice, implementation of such initiatives is very difficult due to constraints like topology, safety, downtime, boundary of retrofit and so on. There are initiatives which are very attractive at one stage but become hindrance for further retrofit in future because most of the time not all initiatives can be implemented in a single go. In this paper HEN "Retrofitability" concept through a case study for integrated atmospheric and vacuum distillation units (ADU/VDU) in an oil refinery, is presented.

INTRODUCTION

The economics of industrial production, limitation of global energy supply, and the realities of environmental conservation are an enduring concern for all industries. Wherever you turn, there's another entreaty to save energy, reduce carbon emissions and protect the environment for posterity.

Industrial plant's heat exchanger networks (HEN) shall always be ready to face high level of operational changes along its lifetime. These changes are short-term, such as process disturbances; uncertainty in feed stock conditions and product demand, and long-term such as the need to process more raw materials that warrants the debottlenecking of the facility including the HEN to increase its capacity. In present scenario, the retrofit of heat exchanger network (HEN) is important way of improving energy efficiency in process industries. An industrial plant may need to be retrofitted several times throughout its lifetime to improve energy efficiency and/or to meet the increased production rate. There are various strategies to achieve energy savings in a retrofit study, for example, reducing the use of utilities, modifying appropriate network topology, upgrading heat transfer units, installing additional heat transfer area, re-piping streams and re-assigning heat recovery matches. The retrofit objective is to identify a cost effective HEN, subject to any design and operating constraints. However, implementing aforementioned retrofit strategies in practice may be difficult, due to constraints related to the topology, safety and maintenance which often exist in a complex HEN.

Besides, the capital cost is usually high because of considerable piping and civil works required for the retrofit and potential production losses during process modification. Nowadays and since late seventies of the last century, another important frequent change is happening due to the continuous escalation in energy prices in a rate that is higher than plant equipment cost. It also warrants continual modification of the facility's

HEN to enhance the plant energy efficiency along its life time (sometimes reaching up to 30 years). In such cases, the HEN retrofit objective/task is to produce a practically implementable cost effective HEN design modification that satisfies the new process objective and its new operating constraints. There are many possible modifications for an existing HEN to retrofit the original design to the new objective. It can include a combination of all possible process operating and design condition modifications, existing HEN topological/structural modifications, and existing HEN unit design modifications and parametric modifications (such as heat transfer enhancement to improve U) as well.

In huge industrial facilities such problems can be solved using pinch technology or MINLP, where we have an objective to minimize total cost of the existing HEN modifications and the rest of the plant's process units design and operation modifications minus the realized energy cost saving. There are some known drawbacks in the existing methods. In one hand the most widely used to date pinch method and its modifications is non-systematic and involves heuristic that produce solutions whose quality depends entirely on the designer experience in applying heuristics. For instance, determining the location of pinch point for an assumed best global ΔT_min, (which global ΔT_min should be used) is a little arbitrary. As the value of assumed best global ΔT_min changes, the location of the pinch point may change and therefore the assessment of which unit to transfer heat across the pinch also changes. The same also applies upon process changes that result in same output. Furthermore, the retrofit is a one package solution that has no consideration for yearly step-by –step modifications and practically for future retrofit projects. The mathematical programming- based methods on the other hand have the drawback of providing little or almost no scope for user interaction. Nowadays, it is possible to solve larger problems than before but also not on the scale needed in many industrial mega sites. The problem of the HEN retrofit is NP-hard, and including all possible process design modifications, different types and configurations of heat exchanger units and so on superstructure is impractical and will make the problem harder, intractable and impossible to solve. In addition, MINLP model cost objective functions need comprehensive data-base for the model solution to be realistic with information/data that is normally not available during conceptual phase of the retrofit project, especially in de-centralized engineering environment where projects cost control depends entirely on rigorous cost calculation that always comes after the retrofit project basic engineering and is carried out at different departments in most of the companies. Besides neither practical nor logical to assume with high fidelity all possible retrofit solutions that can be implemented for a given plot plan of a facility and rigorously develop cost correlation to all of it before running the mathematical program especially when there is always more than one option to carry out a designated modification such as the increase of the network surface area.
Neither one of these methods (pinch technology, network pinch and mathematical programming) can rigorously assess the existing network; rigorously target for the solution before the go ahead of the retrofit project for practically attainable waste heat recovery and/or minimum number of units that need to be added to reach desired

energy consumption targets or part of it, and to systematically renders solutions that can be implemented on phases with a guarantee that the solution implemented today in phase one will not become an obstacle or need to be demolished, for a tomorrow retrofit to be implemented at later time in the future on phases two, three and so on along the plant's lifetime. This paper illustrates through a case study plants' lifetime retrofitability concept in a typical oil refining ADU/VDU application.

HEN RETROFITABILITY

Many design methods for the retrofit of heat exchanger network (HEN) have been proposed and currently used, but none of them had addressed the HEN retrofitability from the implementation point of view. The new, patented, method for HEN retrofit-with-retrofit-in-mind focuses on finding different routes for attaining HEN energy consumption targets and goes further to explores the way to supplement the implementation of these retrofit initiatives. Among all possible routes for achieving the targets, the route which seems most economical and complimenting the phased implementation by each phase implementation step for the following phases is selected to attain the ultimate target. This approach for the plant's HEN lifetime retrofitability eliminates any possible contradiction between now and any in the future of HEN retrofit projects.

CASE STUDY

The objective of this case study is to find the most cost-effective retrofit projects via arranging each initiative on phases for implementation using the existing topology and available capital investment. Each energy saving' initiative is implemented in a way that does not hinder the future initiatives.

Crude oil refining facilities are normally integrating atmospheric and vacuum distillation units to produce different products such as LPG, Naphtha, Motor Gasoline, Kerosene, Jet Fuel, Diesel, Fuel Oil and Asphalt.

In this oil refinery case study pinch analysis shows that the current refinery ΔT_min used for ADU/VDU is 86 F corresponding to the existing hot and cold utility load of 383 MMBtu/hr and 212 MMBtu/hr respectively. The cold streams to be heated and hot streams to be cooled used in this study along with hot and cold Composite Curves and Grand Composites Curves are shown below.

Cold streams					Hot streams				
Stream	HX	Tin (°F)	Tout (°F)	Duty (MMBtu/hr)	Stream	HX	Tin (°F)	Tout (°F)	Duty (MMBtu/hr)
Total feed	E-101	87.6	162	145	OH of DC	E-101	264	168	145
Feed1 to D	E-261	162	251	94		E-102	162	130	27.3
	E-131	251	285	37	VC top SC6	E-261	382	214	94
Feed2 to D	E-142	162	188	26.6		E-262	214	192	27.3
	E-121	188	227	40.1	SC3	E-131	486	280	37.3
	E-116	227	290	70		E-133	280	66	31.8
Feed1 to F	E-283	284	305	22	SC2	E-121	445	219	40.57
	E-271	305	346	42.5		E-122	219	72	22
	E-292	346	429	91.8	Top RECY	E-115	361	325	57
	E-281	429	534	126.2		E-116	325	278	70
Feed 2 to F	E-115	284	332	57.1			278	271	1.1
	E-141	332	379	56.2	SC4	E-141	565	344	56.2
	E-282	379	424	57.4		E-142	344	224	26.6
	E-145	424	513	119.7		E-143	223	102	23.7
	E-291	513	560	66.5	Bottom recycle	E-145	581	501	119
Furnace	F-100	548	679	383		E-305	475	435	56.3
Re-boiler forC3	E-305	340	364	56.3		E-146	435	423	16.1
Feed to C3	E-301	130	295	41.4	Vacuum bottom	E-291	676	603	66.5
						E-292	603	489	91.8
					Visbreaker feed	E-293	489	365	51.2
					SC7	E-271	499	324	42
						E-272	324	317	1.7
					SC8	E-281	628	518	126
						E-282	518	465	54.2
					SC8-1	E-283	465	392	22
					Bottom C3	E-301	350	170	56

TABLE1: Cold and Hot Streams for Pinch Analysis

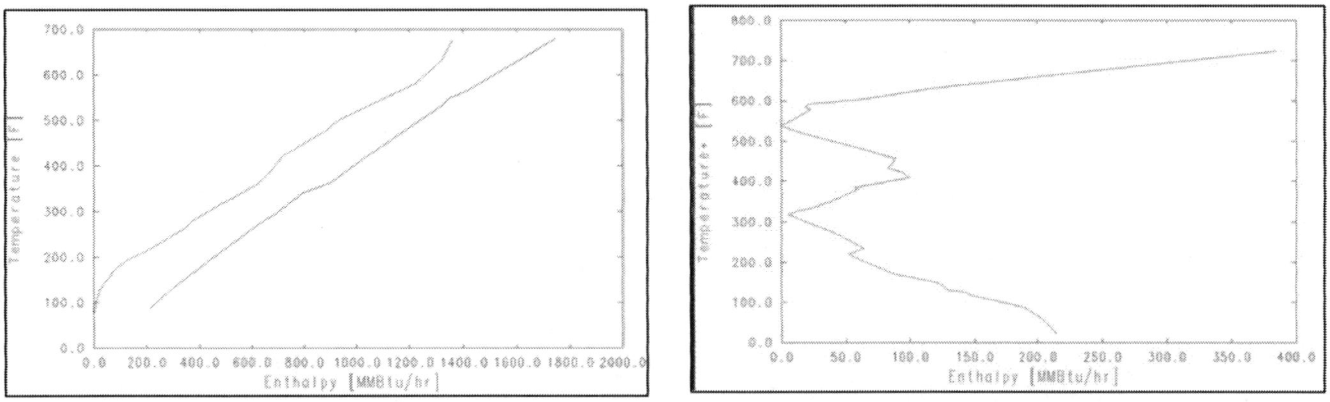

FIGURE 1: Composite Curve and Grand Composite Curve for Existing Case

From the plant lifetime perspective ΔT_min of 30 F is considered for long term plant HEN retrofit. The new minimum approach temperature results in huge potential for energy saving equal to about 133 MMBtu/hr.

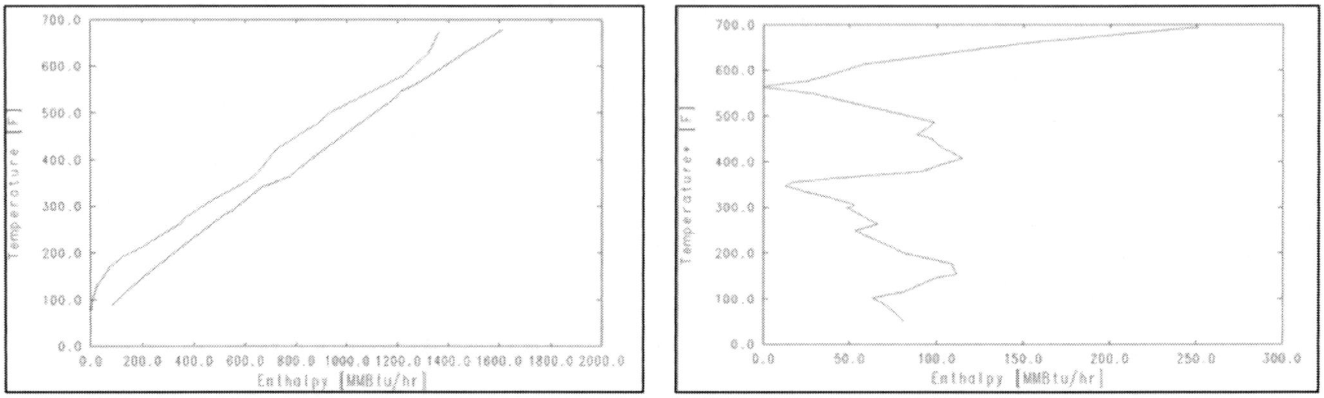

FIGURE 2: Composite Curve and Grand Composite Curve for ΔT_min of 300F

The pictorial representation for the ADU/VDU HEN after pre-flash tower before retrofit is shown below.

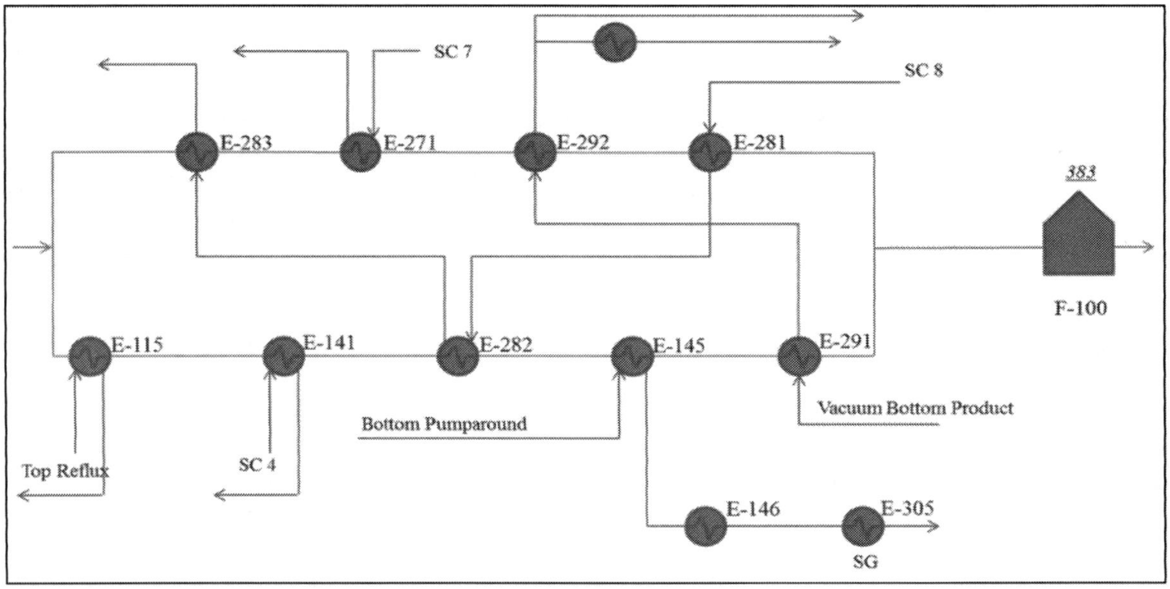

FIGURE 3: ADU/VDU HEN Pictorial Representation for Existing Case

To achieve the desired energy target based on a ΔT_min of 30 F many initiatives can be identified. Our focus will be on a few of them with a plan to implement them on phases. We got three practical routes accepted by the plant owners to implement the initiatives to enable them attain the desired energy saving targets.

<u>ROUTE 1</u>

It consists of the following three initiatives to reach desired energy target:

(1) Vacuum Bottom Product Stream Waste Heat Recovery Modification

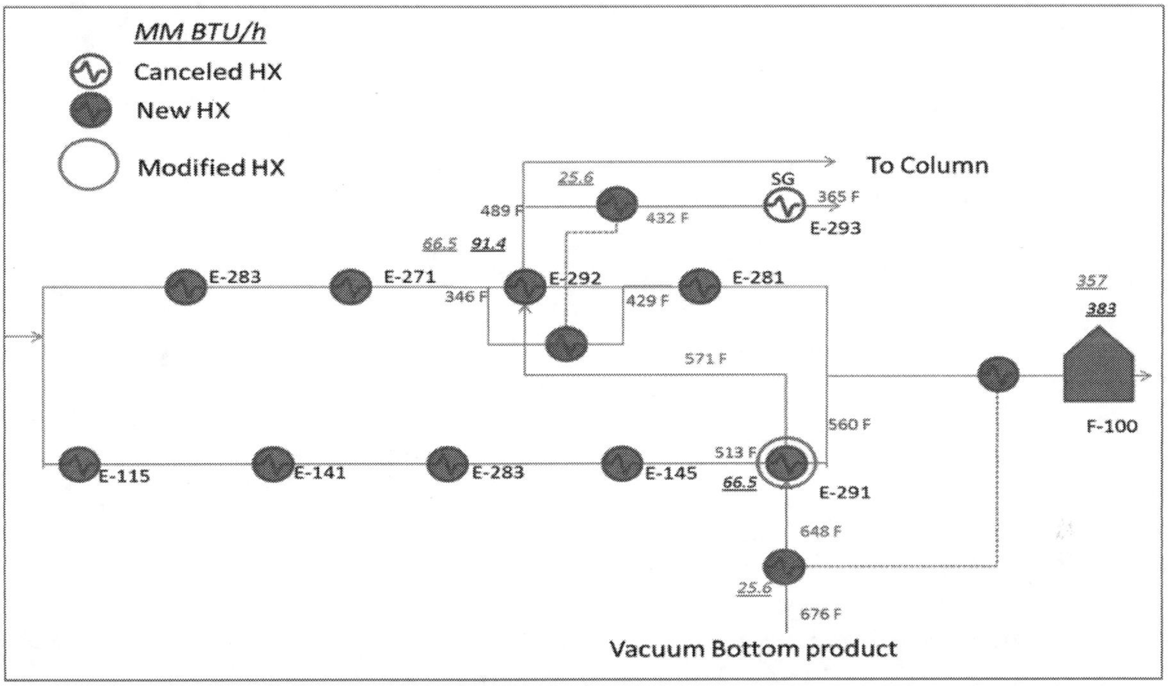

FIGURE 4: Pictorial Representation for Vacuum Stream Waste Heat Recovery Modification

	NHX1		E-291	
	Vac. product	0.2 Crude oil	Vac. product	Crude oil
Temp in (°F)	676	548	648	513
Temp out (°F)	648	590	571	560
Duty (MMBtu/hr)		25.6		66.5
	E-292		NHX2 (parallel)	
	Vac. product	Crude oil	Vac. product	Crude oil
Temp in (°F)	571	346	489	346
Temp out (°F)	489	429	365	429
Duty (MMBtu/hr)		66.5		25.6

TABLE2: Estimated Energy Saving for Vacuum Stream Waste Heat Recovery Modification

(2) Bottom Pump-Around Stream Waste Heat Recovery Modification

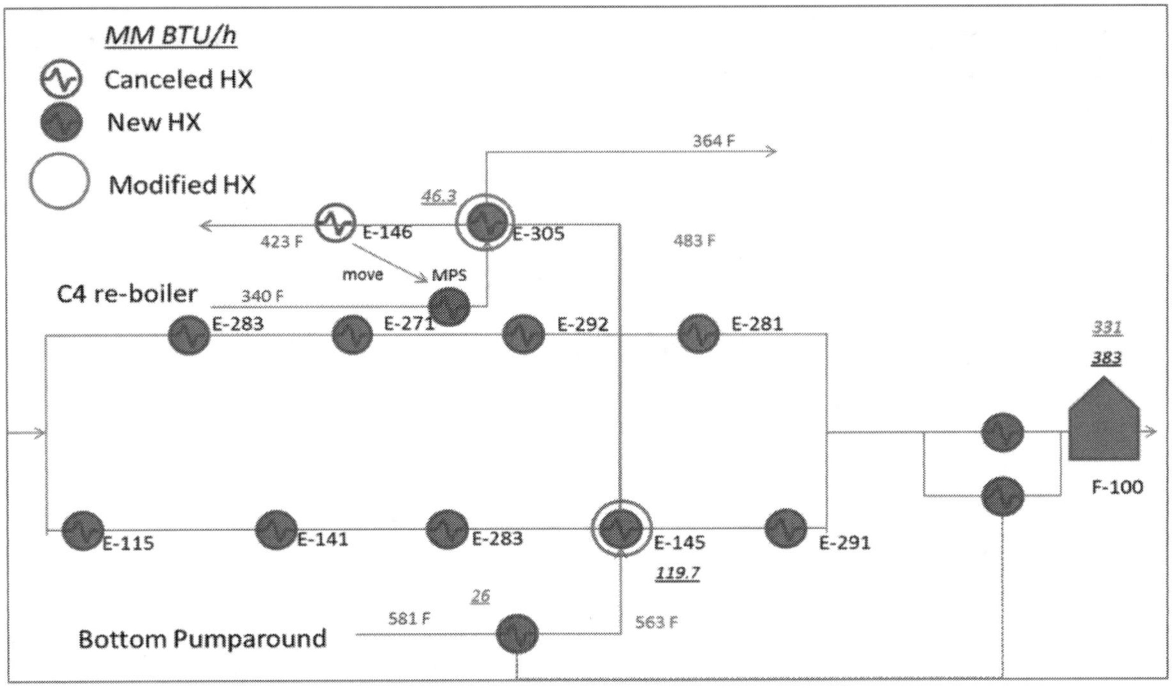

FIGURE 5: Pictorial Representation for Bottom Pump-Around Stream Waste Heat Recovery Modification

	NHX		E-145	
	BPAF	0.5 Crude oil	BPAS	Crude oil
Temp in (°F)	581	548	563	424
Temp out (°F)	563	566	483	513
Duty (MMBtu/hr)	26.1		119.7	
	E-305		E-146	
	BPAS	C4		C4
Temp in (°F)	483	340	M.P.S	340
Temp out (°F)	423	364		364
Duty (MMBtu/hr)	46.3		10	

TABLE 3: Estimated Energy Saving for Bottom Pump-Around Stream Waste Heat Recovery Modification

(3) Side Cut-8 Stream waste heat Utilization

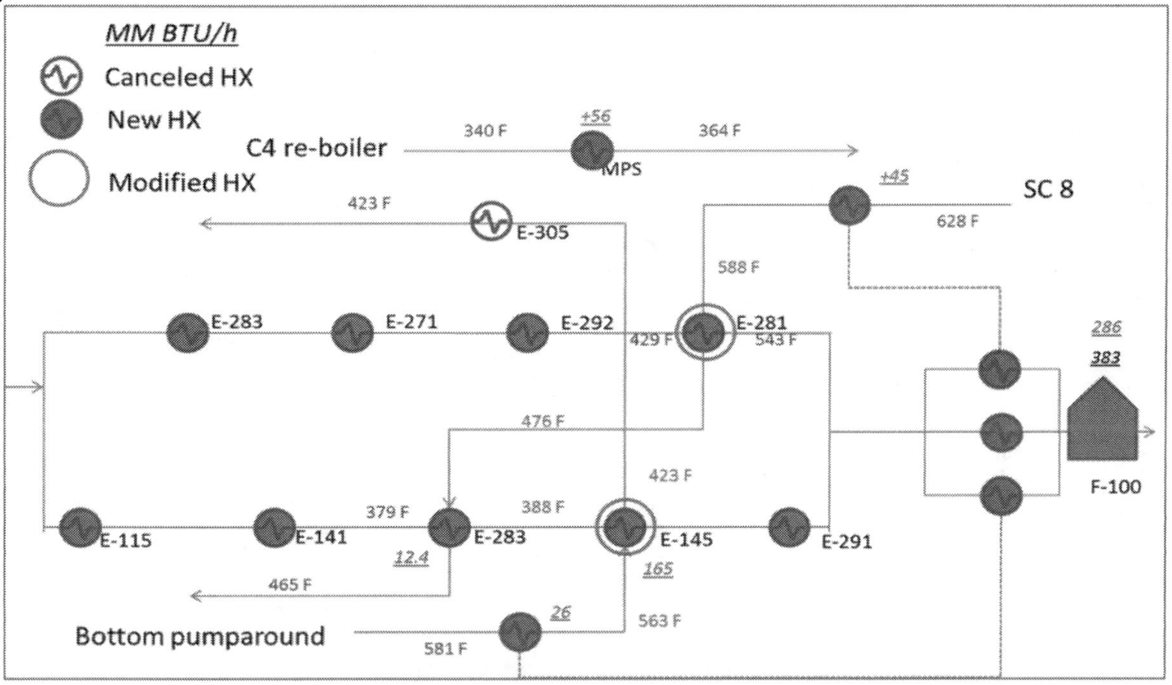

FIGURE 6: Pictorial Representation by Side Cut-8 Stream Utilization

	E-145		E-305	E-146 (M.P.S)	
	BPAS	Crude oil	Cancel	M.P.S	C4
Temp in (°F)	563	388			340
Temp out (°F)	423	513			364
Duty (MMBtu/hr)	165		0	56	

	NHX		E-281		E-283	
	SC8	Crude oil	SC8	Crude oil	SC8	Crude oil
Temp in (°F)	628	548	588	429	476	379
Temp out (°F)	588	599	476	543	465	388
Duty (MMBtu/hr)	45		126		12.4	

TABLE 4: Estimated Energy Saving by Side Cut-8 Stream Utilization

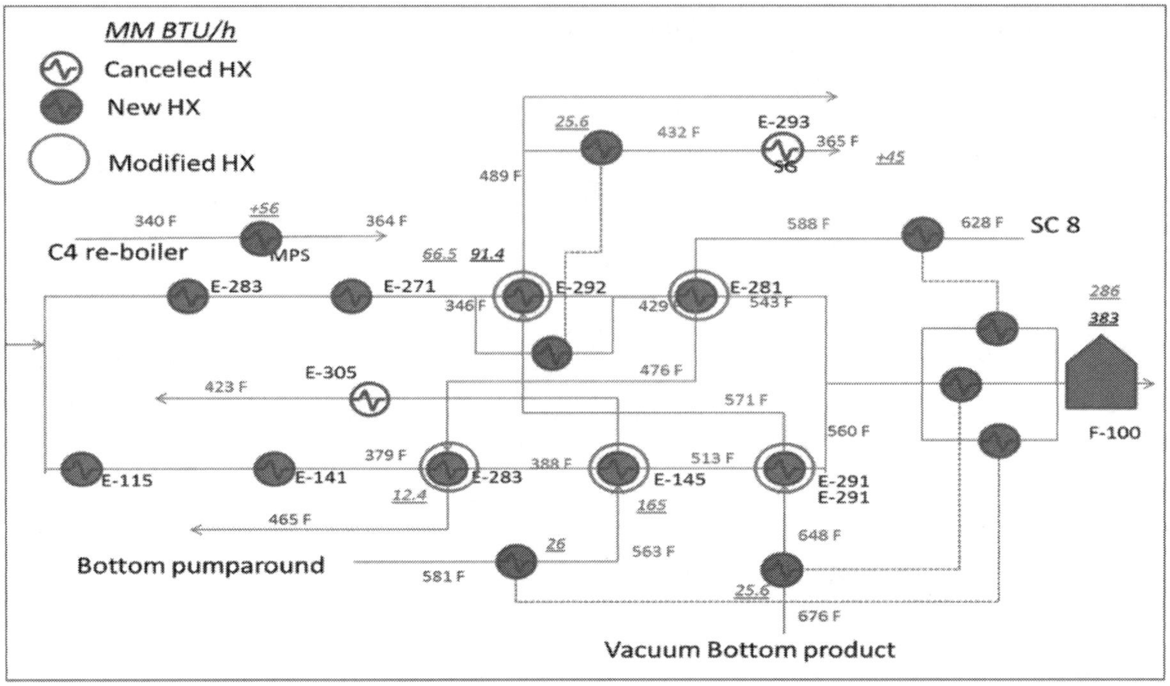

FIGURE 7: Pictorial Representation for Route 1

ROUTE 2

In the second route, the first initiative (Vacuum bottom product utilization) will be implemented as is, and the following changes in initiatives (2) & (3) will be implemented.

Bottom Pump-Around & Side Cut-8 Streams Waste Heat Recovery

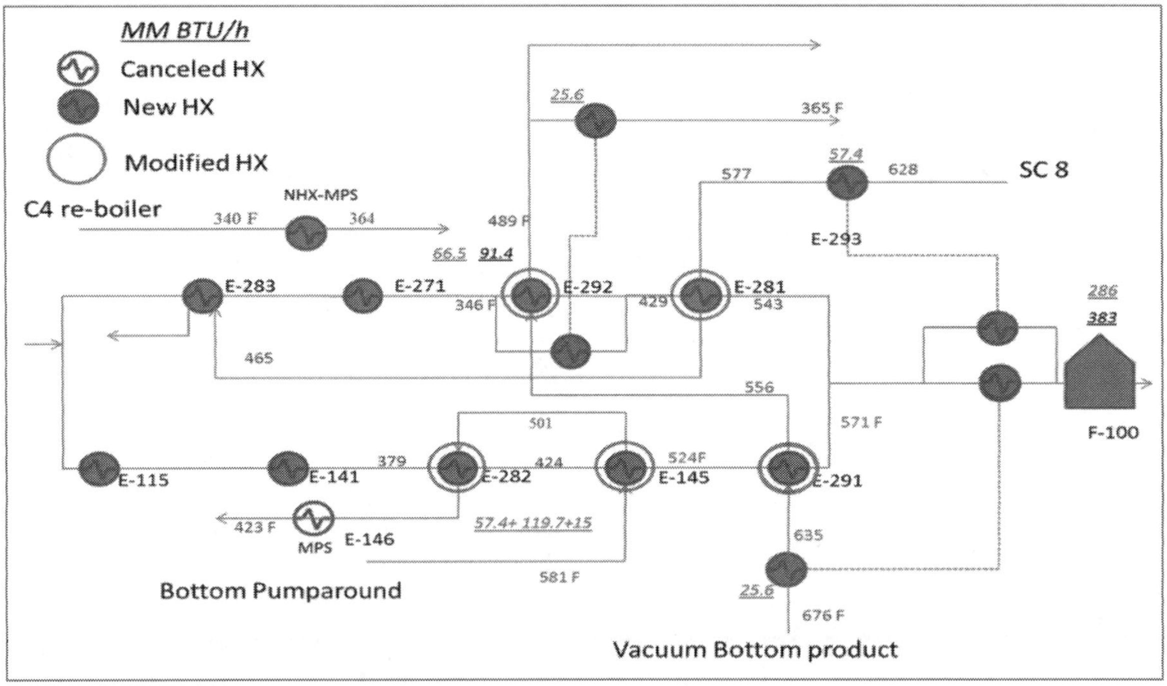

FIGURE 8: Pictorial Representation for Route 2

	NHX		E-145& E-282	
	BPAS	Crude oil	BPAS	Crude oil
Temp in (°F)	628	548	581	379
Temp out (°F)	577	599	501	524
Duty (MMBtu/hr)		57.4	119.7+ 57.6+15	

	E-281		E-146
	SC8	Crude oil	
Temp in (°F)	626	524	cancel
Temp out (°F)	547	571	
Duty (MMBtu/hr)		126	

TABLE 5: Estimated Energy Saving by Side Cut-8 Stream Utilization

ROUTE 3

In the third route, which is mainly a modification in route (2), we are only shifting duties in the system to reduce the number of capital projects.

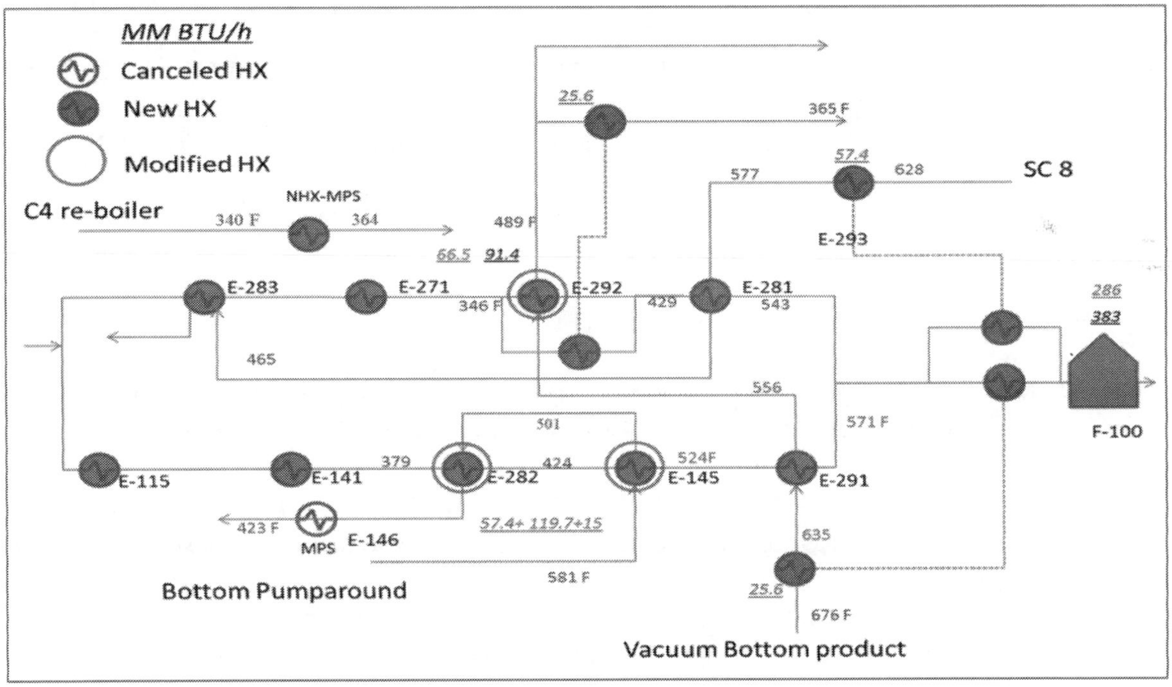

FIGURE 9: Pictorial Representation for Route 2

RESULTS

The table below shows that route (3) is the most attractive one.

ROUTE 1			
Initiatives	Extra unit	Modification	Total saving
Vacuum Bottom Product	2	1	25
Bottom Pump Around	1	2	26
SC 8	2	2	49
Total	5	5	97
NPV	$97 MM		
ROUTE 2			
Initiatives	Extra unit	Modification	Total saving
Vacuum Bottom Product	2	1	25
Bottom Pump around and SC8	2	4	73
Total	4	5	97
NPV	$103 MM		
ROUTE 3			
Initiatives	Extra unit	Modification	Total saving
Vacuum Bottom Product	2	1	25
Bottom Pump around and SC8	2	2	73
Total	4	3	97
NPV	$110 MM		

TABLE 5: Results Summary for Different Routes of Initiatives

Route 1 uses 5 new units and 5 modifications while route 2 uses 4 new units and 5 modifications. Moreover, Route 3 uses 4 new units and 3 modifications. The first initiative in route 3 is independent from the other two initiatives. Hence, it can be implemented in retrofit project (1). The second and third initiatives in route 3 are related. Hence, they have to be implemented together in retrofit project (2).

CONCLUSION

The availability of several routes to achieve desired energy targets in any plant's HEN retrofit mandates the need to find the best project(s) route that maximizes the plants' HEN retrofit Net Present Value (NPV) along its lifetime retrofit projects. Plants' HEN "Retrofitability" concept is an important new key performance index (KPI) for heat exchanger network retrofit projects' evaluation for mega size energy intensive facilities. The adaptation of HEN "Retrofitability"concept opens the door for more aggressive energy saving and more reduction in energy-based GHG emissions in any industrial facility, as shown in this case study.

REFERENCES

Linnhoff, B., and Hindmarsh, E., "The pinch design method for heat exchanger networks" Chemical Engineering Science, Vol. 38, PP 745-763, (1983).

Floudas, C.A., Ciric, A.R., and Grossmann, I.E., "Automatic synthesis of optimum heat exchanger network configurations" AICHEJ., Vol. 32, 276-290 (1986)

Gundersen,T., and Naess, L., "The synthesis of cost optimal heat exchanger networks: An industrial review of the state of the art" Computer and Chemical Engineering, Vol. 12, PP 503-530, (1988)

Ciric, A.R., and Floudas, C.A., "Heat exchanger network synthesis without decomposition" Computers and Chemical Engineering, Vol. 15, PP 385-396, (1991)

Linnhoff, B., "Pinch analysis-state of the art overview" Transactions Institution Chemical Engineering Research and Design, Vol. 71(A5), PP 503-522, (1993)

Biegler, L.T., Grossmann, I.E., Westerberg, A.W., "Systematic Methods of chemical process Design" Prentice Hall, New Jersey (1997)

Noureldin, M.B., "Methods for Heat Exchanger Network Energy Efficiency Assessment and Lifetime Retrofit", USP 8,150,560, (2012)

Noureldin, M.B., "Systems and Program Product for Heat Exchanger Network Energy Efficiency Assessment and Lifetime Retrofit", USP 8,150,559, (2012)

J.L.B. Ven Reisen, G.T. Polley, P.J.T. Verheijen, Structural targeting for heat integration retrofit, Applied Thermal Engineering 18 (5) (1998) 283-294

N.D.K. Asanta, X.X. Zhu, An automated approach for heat exchanger network retrofit featuring minimal topology modification, Computer & Chemical Engineering 20 (1996) 7-12

B.A. Al Riyami, J. Klemes, S. Perry, Heat integration retrofit analysis of a heat exchanger network of a fluid catalytic cracking plant, Applied Thermal Engineering 21 (2001) 1449-1489

G.T. Polley, J.D. Kumana, Energy saving retrofit of FCC plant, in: NPRA Annual Meeting (March 2007). San Antonio, Texas.

B.L. Yeap, D.I. Wilson, G.T. Polley, S.J. Pugh, Mitigation of crude oil refinery heat exchanger fouling through retrofits based on thermohydraulic fouling models, ChED 82A (2004) 53-71

(80c) Adiabatic Compression As an Open Heat Pump for Process Integration within Oxy-Combustion Coal-Based Power Plants

Chao Fu, Department of Energy and Process Engineering, Norwegian University of Science and Technology, Trondheim, Norway and **Truls Gundersen**, Energy and Process Engineering, Norwegian University of Science and Technology, Trondheim, Norway

Coal was responsible for 41% of global energy related CO_2 emissions in 2005 and is expected to be responsible for 44% in 2030 [1]. Power generation is the largest application of coal and amounts to more than 40% of the annual coal consumption in the world [1]. Thus coal based power plants are extremely interesting candidates when Carbon Capture and Storage (CCS) is going to be implemented. For coal-based power plants, oxy-combustion is a promising technology to enable the capture of CO_2, since the reduction in power efficiency and the additional investment cost are less than (or at least comparable to) other CO_2 capture technologies. The possibility of co-capturing SOx and NOx also makes oxy-combustion competitive. In addition, some other environmental impacts (rather than the global warming impact) related to CCS are smaller for oxy-combustion than pre- and post-combustion.

The core concept of oxy-combustion is to use high purity oxygen instead of air for the combustion process so that the flue gas is composed mainly of CO_2 and H_2O. The CO_2 can be separated by condensing the H_2O and then purified by chilling. The oxy-combustion technology is ready for demonstration in both new power plants and for retrofitting existing facilities [2]. The main challenge for implementing this technology is the large power penalty caused by the air separation unit (ASU) and the CO_2 compression and purification unit (CPU). A 500 MWe power plant will consume around 10,000 tons of O_2 per day. Cryogenic air separation through distillation is the only commercially available technology for large volume O_2 production. When a traditional double-column distillation cycle is applied for the supply of O_2 in a coal based oxy-combustion power plant, the total thermal efficiency penalty related to CO_2 capture is around 10.3% points (based on the higher heating value) [3]. The ASU and the CPU contribute 6.6 and 3.7 percentage points respectively. It is expected that the total efficiency penalty will be reduced to less than 6% points in the future [4], while the theoretical minimum is 3.4% points when all the unit operations in the ASU and the CPU are assumed reversible [5]. Process integration is expected to play an important role in reducing this power efficiency penalty.

Both the ASU and the CPU are subambient separation processes. Mechanical energy (work) is consumed by compressors in the refrigeration processes. The compression heat is normally removed by cooling water to lower the operating temperature, and thus reduce the work consumption in the compressors. An alternative option for the compression heat from the ASU and the CPU is to be integrated with the steam cycle by partially preheating the boiler feedwater, so that less steam is extracted from the power cycle. Adiabatic compression may be more favorable than multi-stage compression with interstage cooling when the compression heat is utilized for preheating the boiler feedwater. In this case, compression heat is upgraded (the temperature is lifted) through adiabatic compression, which acts as an open heat pump. As a result, the boiler feedwater is preheated to a higher temperature. The extraction of steam from the higher pressure levels can be reduced, resulting in more power generation from the steam turbines. On the other hand, however, more work is consumed in the adiabatic compression case.

This paper presents a detailed study on the integration of the compression heat with the steam cycle. The integration potential is first investigated in the case when interstage cooling is applied in the compression processes. In this case, the compression work in the ASU and the CPU is not influenced by the integration. Then fully and/or partially adiabatic compression is applied instead of interstage cooling, so that the temperature of the compression heat is lifted. Such heat integration has been investigated by several case studies. The features of the open heat pumps will be addressed. The results show that adiabatic compression may be a more favorable compression scheme for the ASU. The thermal efficiency of the power plant increases by 0.78% points when the compression heat from the ASU and the CPU is integrated.

References

[1] EIA, International Energy Outlook 2008.

[2] IEAGHG, Greenhouse News, 2011: NO. 104.

[3] C. Fu. Process integration in coal based oxy-combustion power plants with CO_2 capture. PhD thesis, Norwegian University of Science and Technology, 2012.

[4] J.P. Tranier, R. Dubettier, N. Perrin. Air separation unit for oxy-coal combustion systems. 1st International Oxyfuel Combustion Conference, Cottbus, Germany, 7-11 September 2009.

[5] C. Fu, T. Gundersen. Exergy analysis of an oxy-combustion process for coal-fired power plants with CO_2 capture. Submitted to ***Fuel*** (under review), June 2011.

13AIChE
Spring Meeting, San Antonio, TX

Identifying Heat Recovery Paths to be Used in Improving the Energy Efficiency of Pre-Heat Trains

E. K. Tamakloe, G. T. Polley & M. Picón Núñez

University of Guanajuato, Mexico

Topics

- Overview of Pre-Heat Train Revamp Procedure
- Significance of Fixed Loads
- Case Study

13AIChE
Spring Meeting, San Antonio, TX

Overview of Pre-Heat Train Revamp Procedure

13 AIChE

Sources of Additional Heat

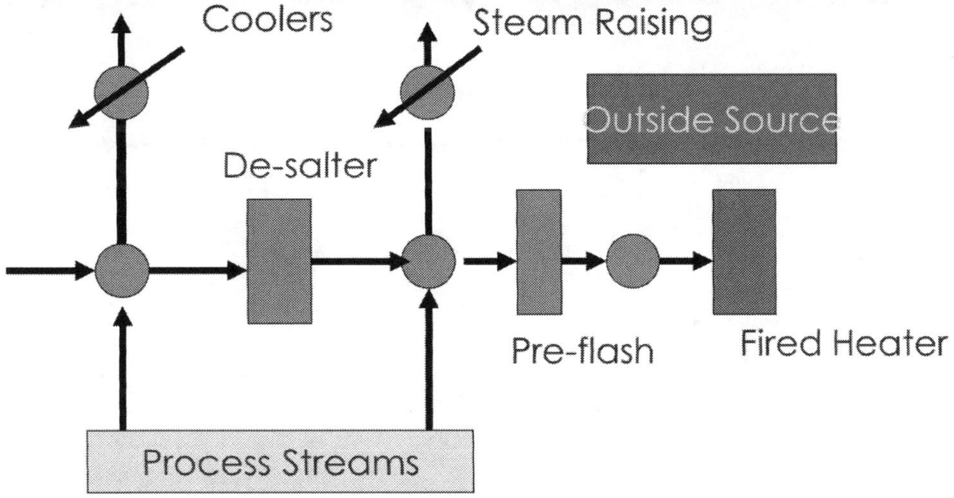

13 AIChE

To Recovery Heat need Path

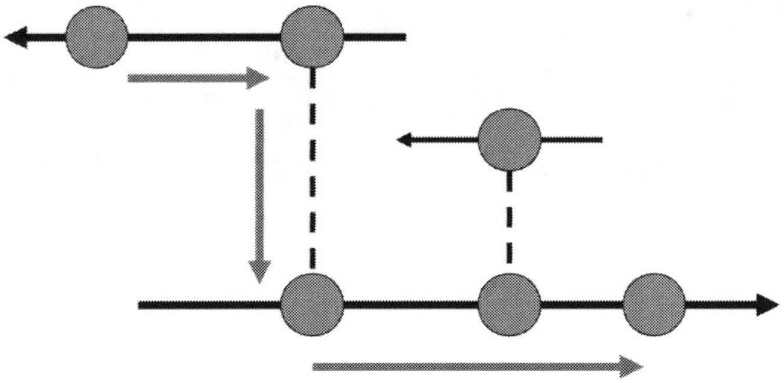

13 AIChE

Heat Balances

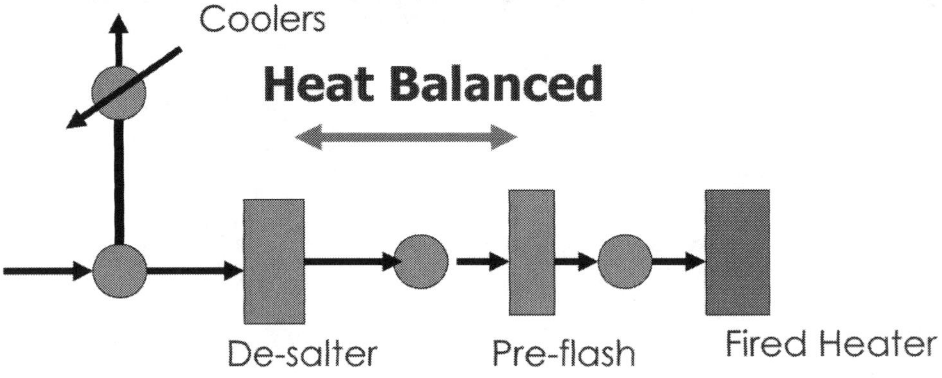

13 AIChE

Required Structural Features

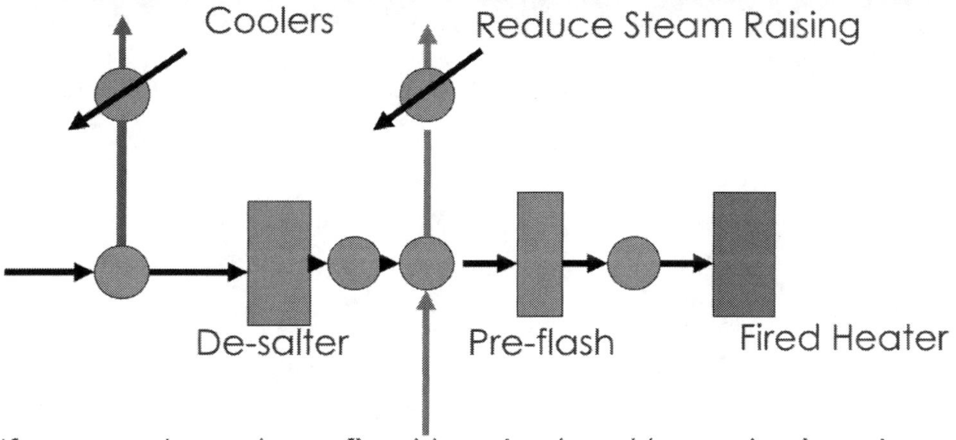

If we are to reduce fired heater load by reducing steam raising what structural feature do we need?

Required Structural Features

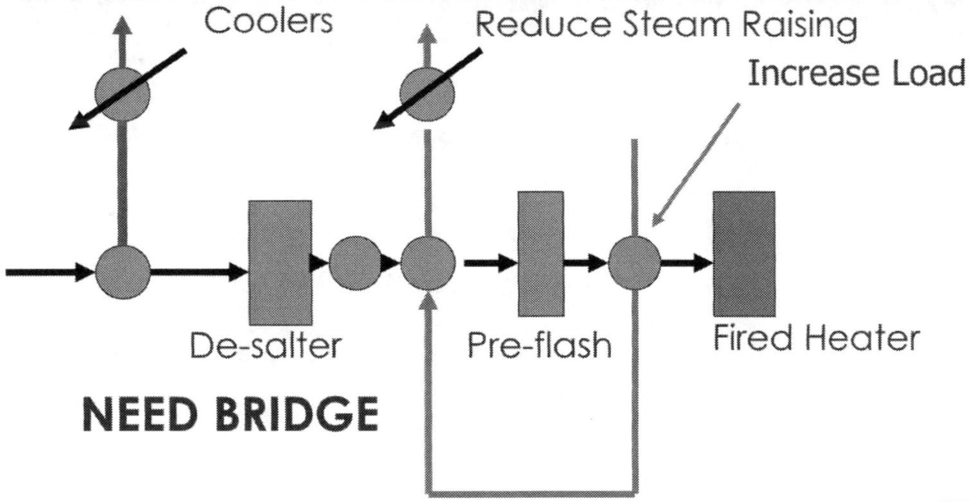

Bridge Does Not Need to be on Source Stream

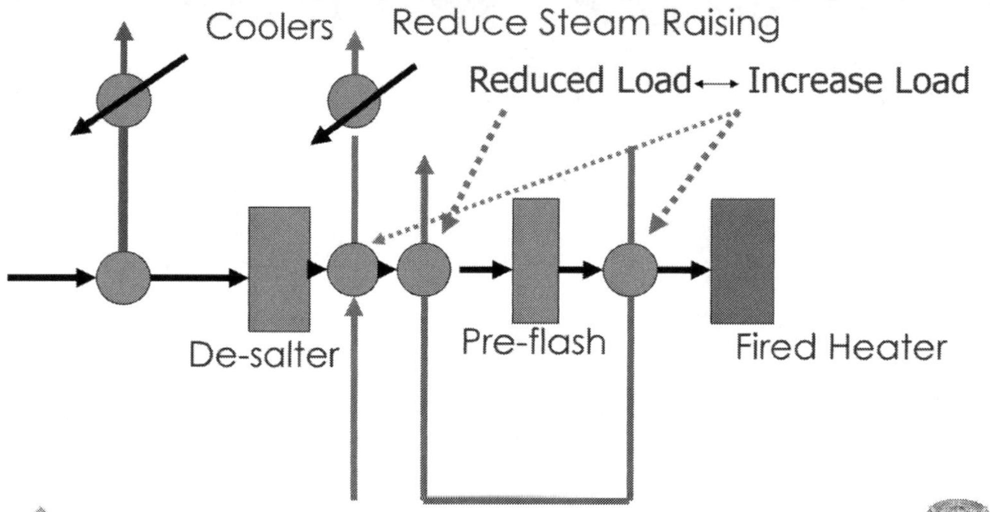

Heat Balances

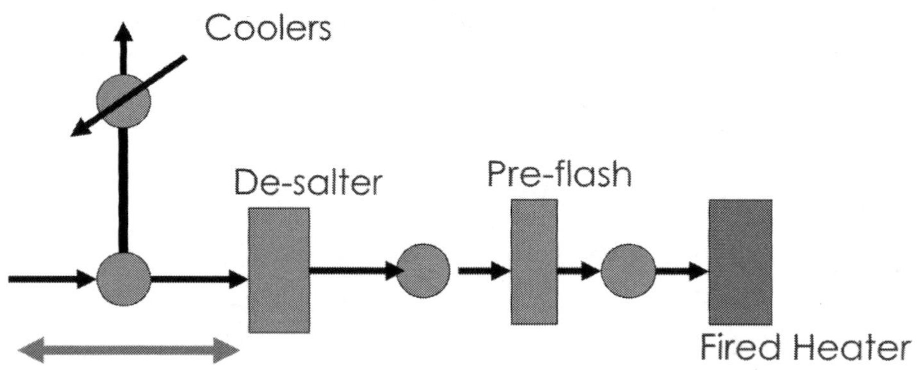

Heat Load Fixed

Required Structural Features

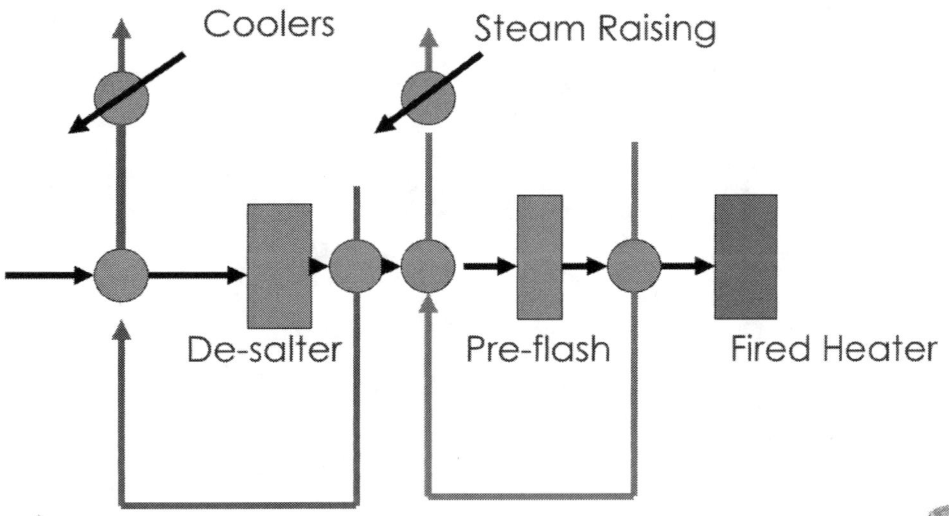

13 AIChE

Capacity of 'Bridge'

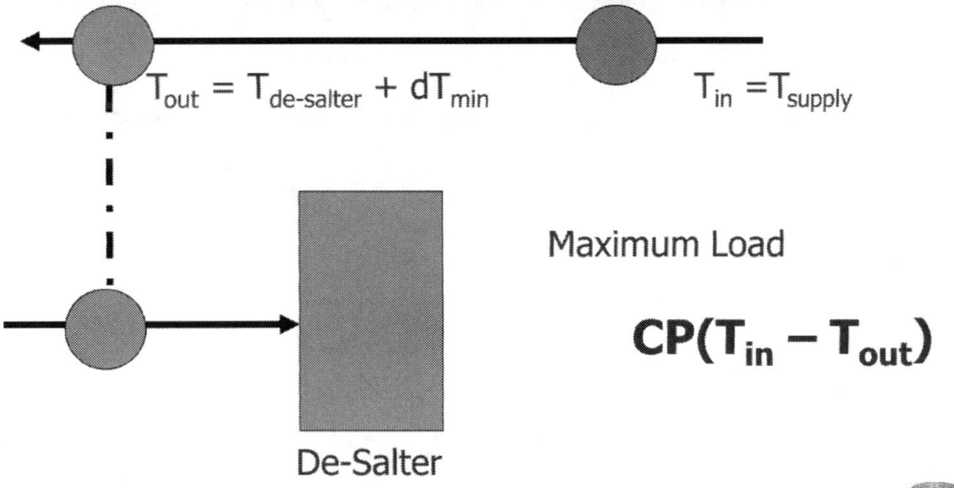

13 AIChE

"Block" on Heat Recovery

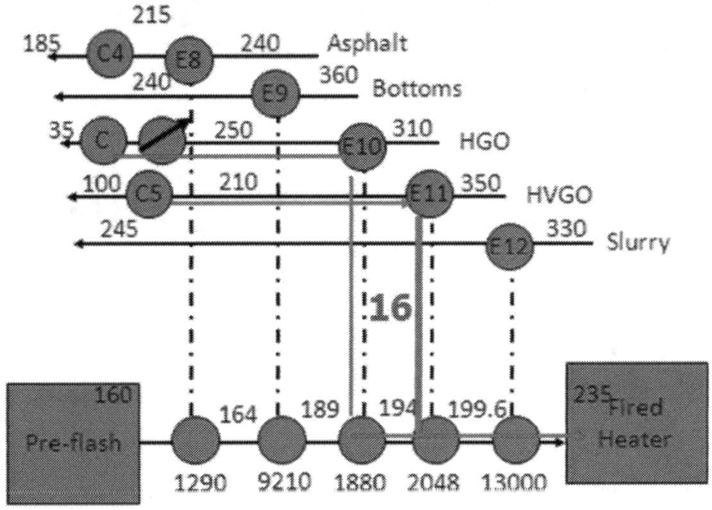

Driving Force Plot

Exchangers Arranged in Series

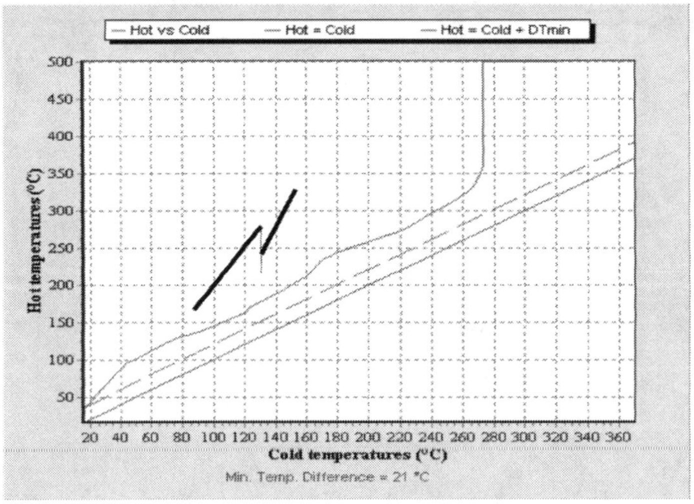

Open up Minimum Temperatures

Exchangers Arranged in Parallel

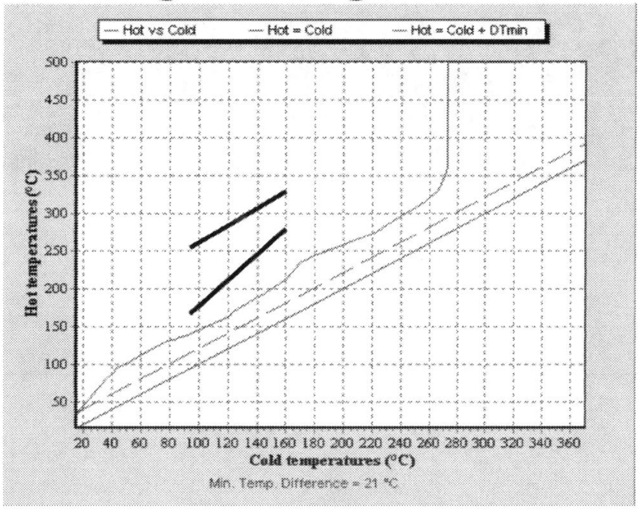

13 AIChE

"Fixed Loads"

Example:

Heat extracted from distillation columns through use of pump around streams

KEY to identifying paths that should be used – in terms of both adding new heat recovery surface and in mitigating fouling

13 AIChE

Case Study

13 AIChE

Process Operations

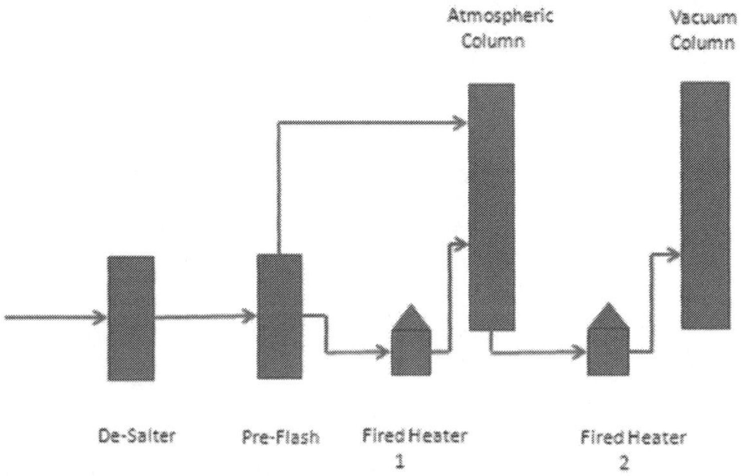

13 AIChE

Step 1.
Identify Heat Recovery Units and Heat Sources

13 AIChE

Identify Heat Recovery Exchangers
Between Storage & De-salter

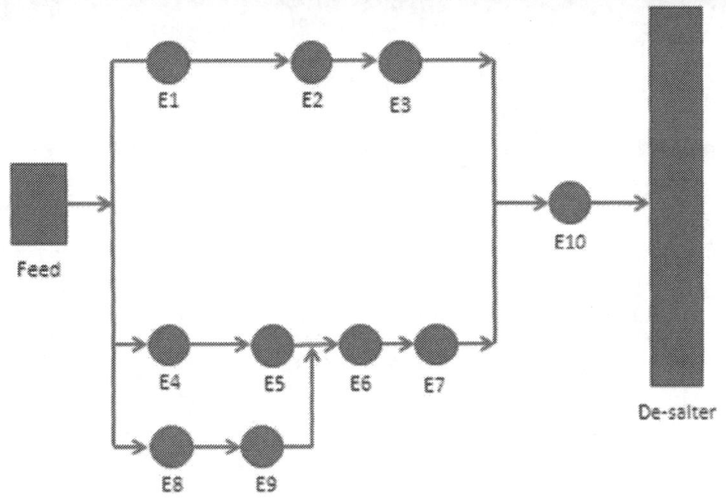

13 AIChE

Identify Heat Sources
Coolers

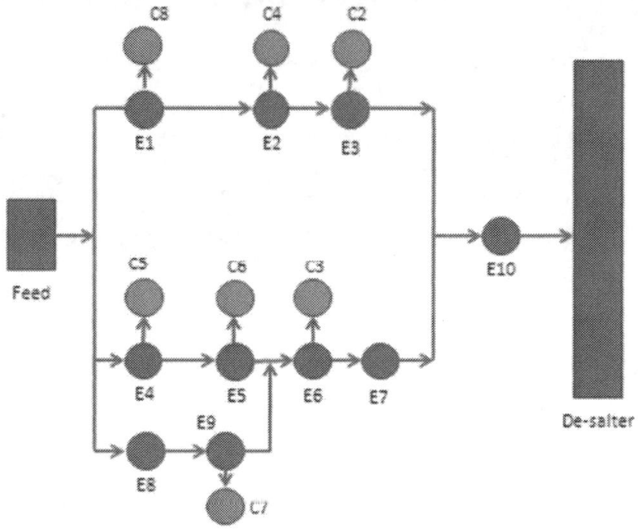

13 AIChE

Identify Heat Recovery Exchangers
Between De-salter & Pre-flash

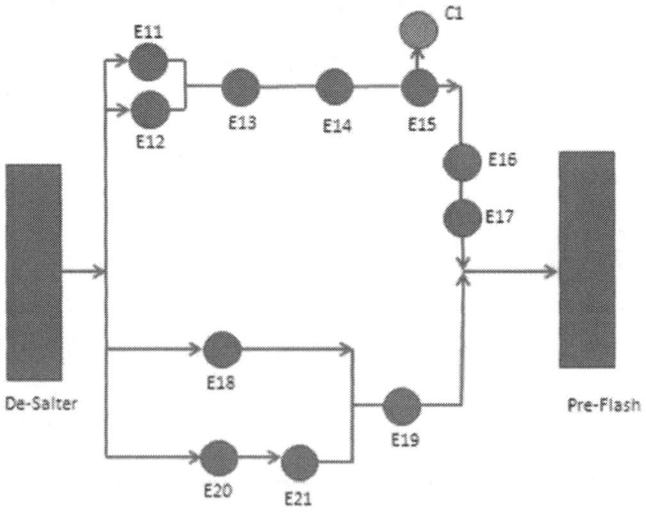

13 AIChE

Identify Heat Recovery Exchangers
After Pre-flash

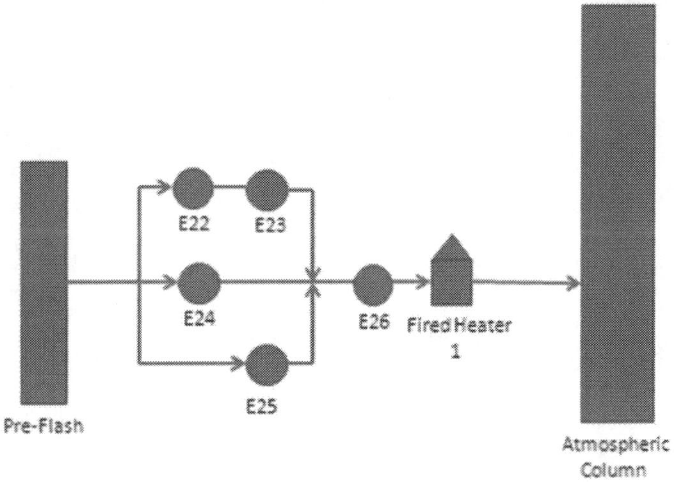

13 AIChE

Step 2.
Identify Fixed Loads

13 AIChE

Fixed Loads
Below the De-Salter

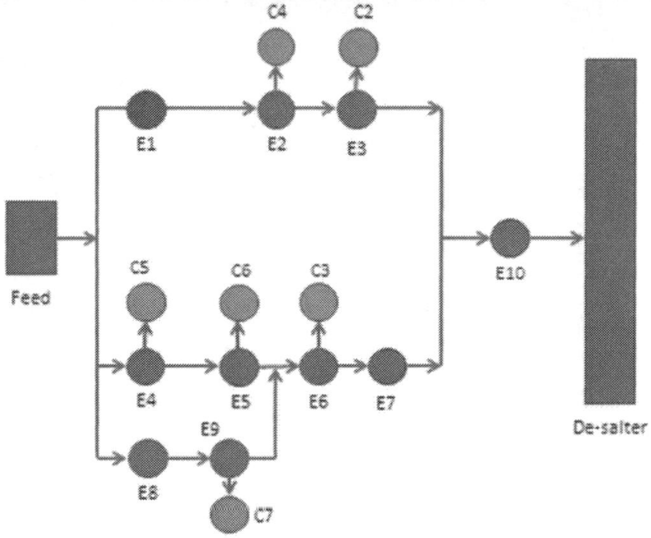

Fixed Loads
De-Salter to Pre-Flash

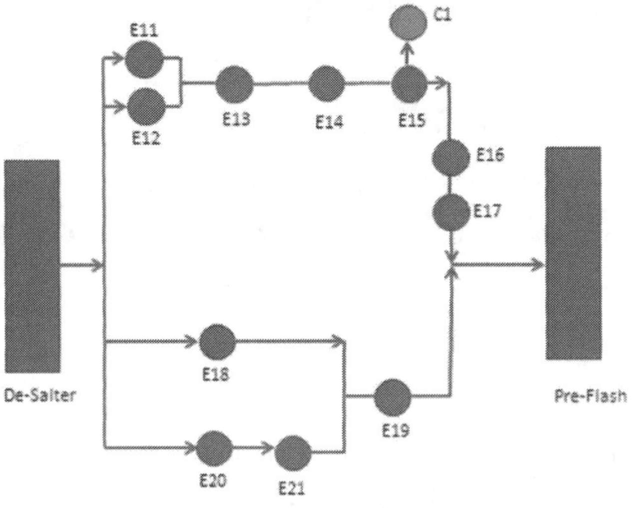

Fixed Loads

After Pre-Flash

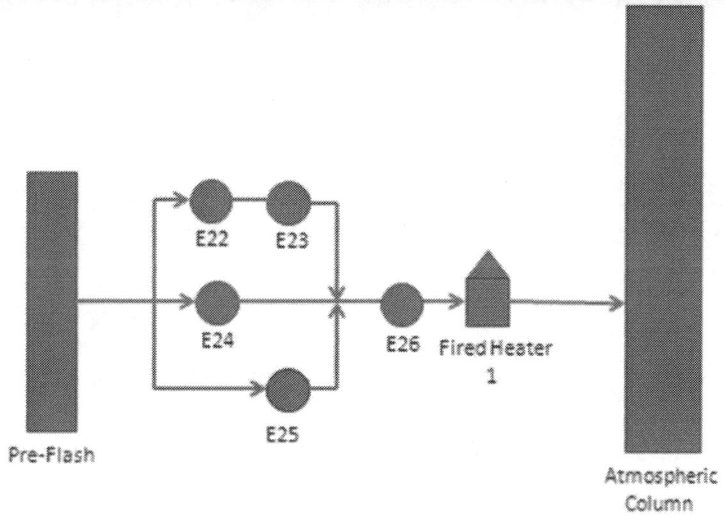

13 AIChE

Step 3.

Identify Bridges

13 AIChE

Bridge Points

Below the De-Salter

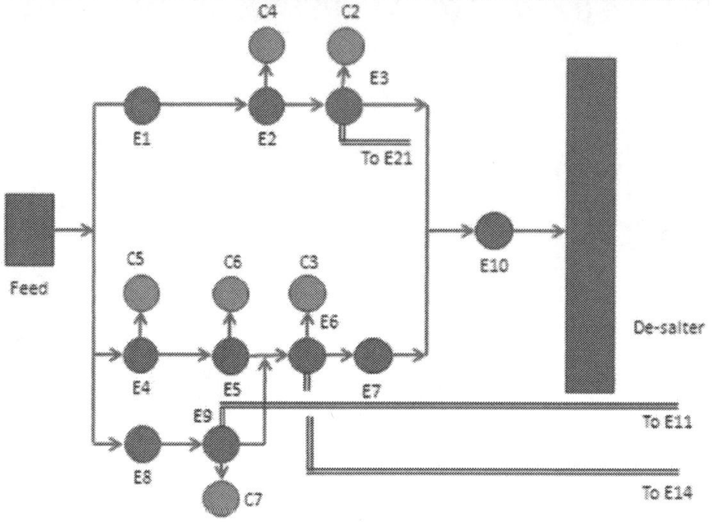

13 AIChE

Bridge Points

Between De-Salter and Pre-Flash

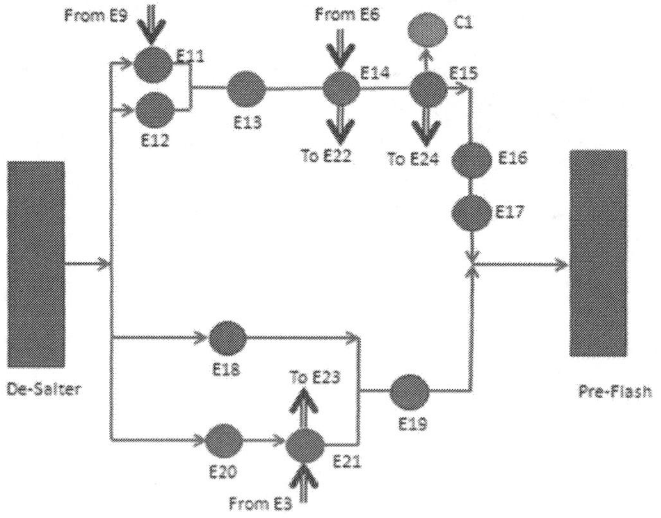

13 AIChE

Bridge Points

After Pre-flash

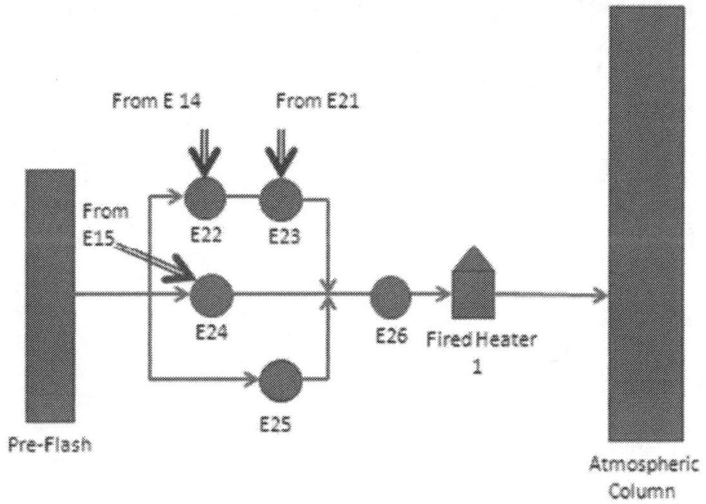

Step 4.

Identify Path Options

Possible Paths

- PATH 1. Use of heat rejected through C2 and C4 can only be undertaken using the bridge provided by E3 & E21. This heat can be transferred across the pre-flash using the bridge provided by E21 & E23. However immediately down stream of E21 is a fixed load unit (E19). Consequently, the use of this path would require careful control of performance of E19.

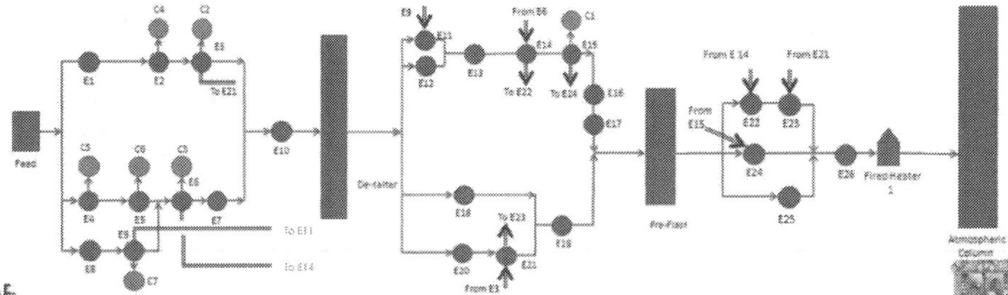

13AIChE

Possible Paths

- PATH 3. Use of heat rejected through C7 can also be passed along PATH 2.

- PATH 4. Use of heat rejected through C7 can be undertaken using the bridge formed by E15 & E24.

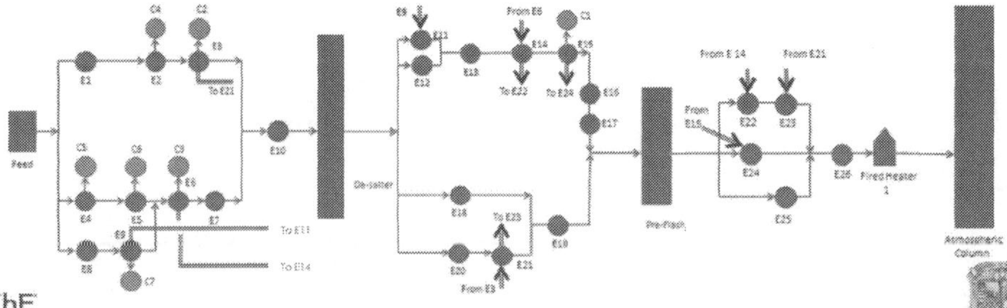

13AIChE

Favored Paths

- PATH 2.
- PATH 4.
- PATH 1 can only be used if careful control of E19 is instigated.

13AIChE

Conclusions

- The improvement in heat recovery – through improved integration, fouling mitigation, or both – involves the transfer of heat along easily identifiable "paths"
- Units having a "fixed load" have a strong influence on the viability of a path
- A systematic procedure for the identification of options has been developed.

13AIChE

(80e) Energy and Capital Savings in DME Process Design and Integration

Mansour Emtir, Industrial Department, National Oil corporation, Tripoli, Libya and Elmahboub Edreder, Industrial Department, National Oil Corporation, Tripoli, Libya

In recent years, dimethyl ether (DME) is widely used as a feed for the synthesis of chemicals including acetic acid and its esters and as a hydrogen source for fuel cells. The major process for the production of DME on an industrial scale comprises dehydration of methanol using an acidic dehydration catalyst in a fixed bed reactor followed by rectification of the product stream to recover high purity DME. An important motivation for studying integrated distillation processes is to reduce the energy consumption. Basically, distillation (batch or continuous) is an energy intensive process and consumes large amount of thermal energy (via steam), so even small improvements which become widely used, can save huge amounts of energy. The main focus in this work is to apply the heat-integrated and reactive distillation designs on DME process using the steady state approach. Here, the energy and capital cost saving of each design and the environmental impacts have been considered and compared with that by the conventional process.

An observation result using heat-integrated design of the process shows that a significant operating cost savings of 50% compared to the conventional process. The reactive DME distillation column design results show that the highest savings in the capital cost of 69 % and operating cost of 84 % can be achieved compared with the conventional base case. Fundamentally, the saving in capital and operating costs will be directly saving the total annual cost and this saving shows the benefit of using the process integration rather than conventional process. Beside the cost saving, the flue gas emissions have been reduced significantly. The total emissions have been reduced by 51% and 86% using the heat-integrated design and the reactive distillation column design respectively.

Development of Heat Exchangers for Use as Replacement Plates In Distillation Columns: Proof of Concept and Collection of Design Information

[2]Graham Thomas Polley, [1]Edgar Enrique Vázquez Ramírez [1]Antonio Alberto Aguilar Moreno, [1]Jose Manuel Riesco Ávila, [1]Carlos Omar Ríos Orozco.

[1]Departamento de Ingeniería Mecánica, Universidad de Guanajuato,
Carretera Salamanca–Valle de Santiago km. 3.5 + 1.8 km, Comunidad de Palo Blanco, Salamanca, Gto., México
Teléfono: 01 464 64 79940 ext. 2306

[2] Departamento de Ingeniería Química, Universidad de Guanajuato

Introduction

Separation of chemical species by distillation is an energy intensive process. Whilst it has been recognized for many years that the extraction of heat from intermediate stages within a distillation column can result in significant energy savings the equipment required to do this is generally expensive. One option is the use of side strippers [1]. These are often employed in columns used for the distillation of crude oil.

In more recent years the option of actually undertaking the distillation inside heat exchangers has been examined [2]. However, this usually results in significant reduction in the concentration driving forces used in the separation with significant effect upon column size and cost.

Work being undertaken at FIMEE is directed at the examination of an inexpensive form of heat exchanger (fabricated from parts manufactured for use in the motor industry) that can be installed as a replacement for a plate within an existing distillation column.

Using such an exchanger positioned above the feed plate it becomes possible to extract heat from the distillation column at a temperature that is higher than that present in the column overheads. This heat can then be used to satisfy a suitable heat demand elsewhere within the chemical process.

Using an exchanger positioned below the feed plate it becomes possible to provide part of the heat needed to drive the distillation column at a lower temperature than that present in the re-boiler at the base of the column. It may prove possible to provide such heat at through a heat source available elsewhere in the chemical process.

By placing heat exchanger either side of the feed plate it becomes possible to extract heat from above the feed plate and use a heat pump to provide boil up below the feed plate.

Exchanger Design

One possible design for a heat exchanger installed within a column was proposed by in 1993 [2]. A distillation column consisting of a number of sieve plates is illustrated in Figure 1. A sieve plate is simply a flat plate through which a number of small diameter holes have been drilled. The vapor traveling upward through the column has sufficient velocity to prevent liquid flowing through these holes. As a consequence, a two phase mixture is produced on top of the plate. It is maintained on the plate by a weir. Whilst the vapor velocity is sufficient to prevent liquid flowing through the holes it is not high enough to carry liquid to upward through the column. Consequently, liquid and vapor separate above the reservoir formed by the weir with the liquid flowing down the plate down-comer that feeds the next plate down the column.

The new exchanger fits in the space of two existing plates. The upper outlet then acts as a sieve plate. The liquid over flows the weir and feeds the next plate down the column.

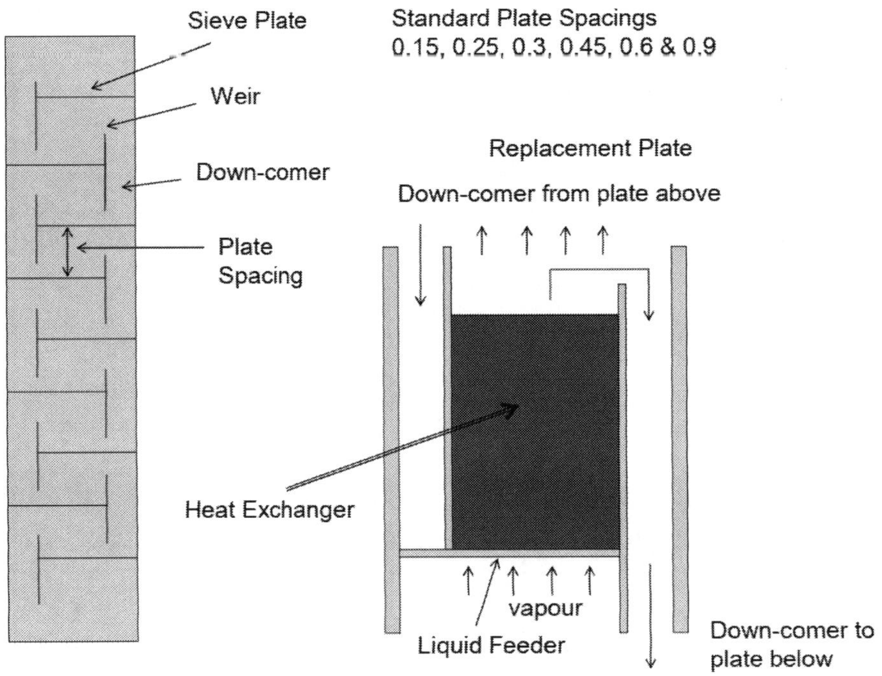

Figure 1. Distillation Column Arrangement and New Exchanger

The proposed design has two significant features. The first is a liquid feeder plate that distributes the liquid from the feed down-comer across the heat exchanger. This is illustrated in Figure 2.

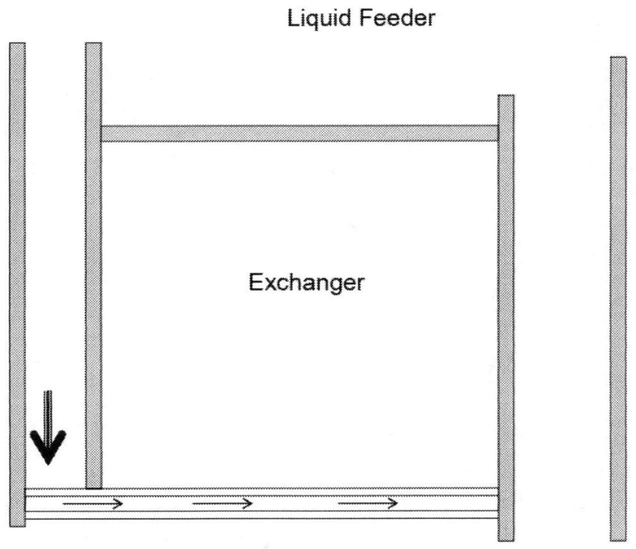

Figure 2. Liquid Feeder

This base of this plate is sealed. The upper plate is perforated.

The second feature is the "flow inducer" and is illustrated in Figure 3.

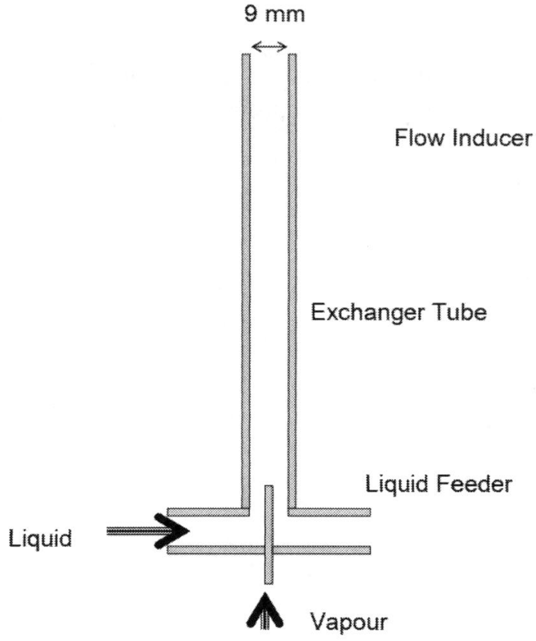

Figure 3, Flow Inducer

Vapor flowing from the sieve plate below the exchanger passes through tubes installed in the base of the liquid feed plate. These tubes protrude into the individual heat exchanger tubes. Liquid from the feed down-comer is then entrained by this vapor flow and drawn through the heat exchanger.

Experimental Apparatus

The apparatus used in this study only considers hydraulic performance of the heat exchanger. It consist consists of a single transparent tube (in order that flow patterns can be observed using high speed photography) that represents a tube within the heat exchanger that is fitted with flow inducer (as described above - Figure 3).

The rig (both schematic and photograph) is shown in Figure 4.

This tube sits inside a large diameter tube that provides a reservoir of liquid. The reservoir has gauge marks that are used to measure the height of liquid above the flow inducer.

Liquid is fed to the feed reservoir from tank A. The flow being controlled by a valve positioned in the feed leg.

Air is injected into the tube at a controlled and measured rate (measured using rotameters).

The two-phase mixture leaving the exchanger tube is separated in a tank (tank C), with liquid flowing to a collection vessel (tank D). This vessel is used to measure the rate of flow of liquid through the exchanger tube (through measuring the time to collect a given quantity of liquid). On completion of the measurement this liquid is emptied into tank B. Tanks A and B operate on a constant recycle in order to minimize pressure variation in the feed leg and provide good flow control (confirmed by differential height calculations during collection in tank D).

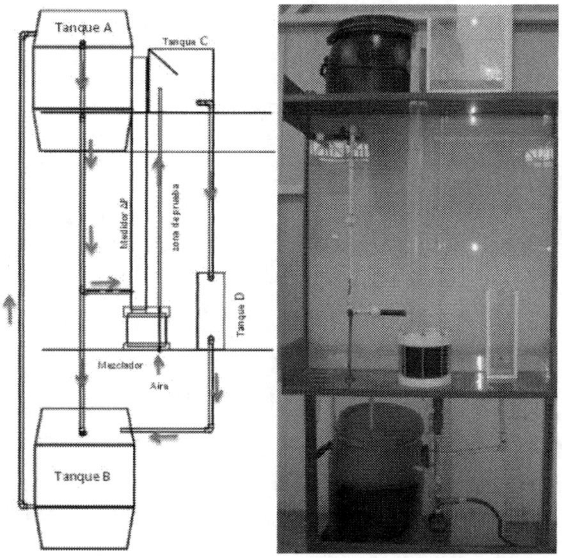

Figure 4. Experimental Rig

Analysis of Experimental Data

Data obtained in the experiments are given in Table 1.

The head required to provide the recirculation is assumed to consist of three terms: gravitational head (a function of the voidage within the tube), frictional losses and a kinetic energy term. This kinetic energy ter mis the difference between that of the two phase flow leaving the tube and that of the liquid and vapor flow entering the tube. The kinetic energy of the vapor entering the tube is calculated from the measured air flow. The kinetic energy of the liquid entering the tube is assumed to be zero.

A "homogeneous" model is found to provide good correlation of the measurements for conditions in which frictional lossses dominate. Accuracy of prediction falls as the % of overall pressure drop due to kinetic effects increases. This suggest that assumption that the incoming liquid has zero kinetic energy is incorrect.

Further experimentation is necessary to determine the significance of kinetic energy effects within the wáter injection system.

Flux másico aire kg/s m²	*Flux másico agua kg/s m²*	*Variación* **ΔP experimental** *vs Δ calculada %*	**% de ΔP Total por Fricción**	**% de ΔP Total Energía Cinética**	**% de ΔP Total Energía Potencial**
3.22	701.9	8	48.5	29.9	21.6
3.22	609.7	11	48.6	29.1	22.3
3.22	516.2	13	48.8	28.3	22.9
3.22	429.1	16	49.0	27.5	23.5
3.22	333.5	21	49.4	26.5	24.1
1.93	728.0	18	39.4	24.0	36.6
1.93	631.0	20	38.9	23.0	38.1
1.93	520.8	25	38.4	21.8	39.8
1.93	430.1	27	38.0	20.7	41.3
1.93	316.0	34	37.6	19.3	43.1
0.64	651.5	23	19.9	11.6	68.5
0.64	497.5	29	17.2	9.4	73.3
0.64	366.0	36	14.9	7.6	77.5
0.64	251.8	44	13.0	6.1	81.0
0.64	158.9	54	8.9	5.0	86.1

Table 1. Experimental Data and Predictions of Homogeneous Flow Model

Evaluation of Thermal Performance

Heat transfer within the tubes of the exchanger will be controlled by the balance of two mechanisms: two-phase convective heat transfer and nucleate boiling heat transfer.

The convective heat transfer coefficient is given by [4,5]:

$$\alpha_{fc} = \alpha_L \left(\frac{\Delta P_{TP}}{\Delta P_L} \right)^{0.45}$$

The predictions of this equation have been compared with a large data base of measured heat transfer coefficients during upward flow boiling within circular tubes. The data are well correlated.

This equation has been used to determine how much heat can be added through the internal exchanger.

The principle problem that has been identified is that under many circumstances the liquid Reynolds Number falls in the laminar flow region. This places uncertainty on the validity of the calculation.

However, it has also led to the identification of two potential improvements in the exchanger design. These are the provision of a liquid recycle and the use of tube inserts to promote heat transfer within the exchanger.

Linking Column Design, Exchanger Design and Integration

The vapor flow thru the column is set by the column duty.

The maximum cross section of the heat exchanger is dependent on the area taken up by the down-comers. This cross section can be related to the number of tubes installed in the heat exchanger allowing calculation of the minimum mass flux of vapor flowing through an exchanger handling the specified column duty. Higher mass fluxes are obtained by reducing the number of tubes in the exchanger.

The mass flux of liquid flowing through the exchanger can be set independently by adjusting the positioning of recycle vanes installed in the liquid down-comer.

So, it is possible (using a quite simple spreadsheet) to determine the heat transfer coefficient achieved in the heat exchanger for different combinations of vapor and liquid mass fluxes. The homogeneous pressure drop model will then allow calculation of minimum tube length for the heat exchanger.

Knowing the overall heat transfer coefficient for the heat exchanger the heat load obtained for a given temperature difference between the column and the fluid with which it is integrated.

Case Study: Methanol-Water Separation

An overview of column integration is provided in Appendix 1.

The first step in the analysis is an evaluation of exchanger performance as a function of vapor and liquid mass flex through the unit. This is presented in Table 2.

It is found that the vapor mass flux should be at least 50 kg/m^2she higher the liquid mass flux the higher is the quantity of heat that can be transferred through the tubes.

The exchanger has a square face and the tube length equates with the plate spacing. The use of a square face provides space for the provision of liquid recycle lanes.

The column diameter is assumed to be 1 metre. The vapor mass flux within the column is assumed to be 1.8 kg/m^2. Therefore, to obtain the vapor mass flux necessary for efficient heat transfer in the exchanger the exchanger dimensions need to be 0.25 m x 0.25 m x 0.9 m (yielding a vapor mass flux of 57.3 kg/m^{2}s).

The next step is examination of Table 2 to determine the heat duty that can be obtained as a function of liquid mass flux.

mass flux vapour	mass flow	mass flux liquid	x	rho-tp	vis-tp	Re-L	f-L	Re-tp	f-tp	dP-tp	dP-L	multiplier	Nu	h-l	htc	Q/dT per tube
23	0.001463	50	0.315068	3.858492905	3.25541E-05	933.9975	0.017131	20181.82	0.006628	1830.8174	11.45098	9.81095	3.7	77.7	762.3108	19.40096
50	0.003181	100	0.333333	3.648099781	3.09375E-05	1818.182	0.0088	43636.36	0.005466	6742.3417	23.52941	12.75674	3.7	77.7	991.1991	25.22621
50	0.003181	200	0.2	6.060461162	4.85294E-05	4363.636	0.00972	46363.64	0.005384	11104.194	103.9569	8.182484	42.57413	894.0568	7315.605	186.1836
50	0.003181	400	0.111111	10.83857686	7.81579E-05	9696.97	0.007961	51818.18	0.005236	19565.426	340.5779	6.189745	80.64469	1693.539	10482.57	266.7836
100	0.006362	100	0.5	2.436026801	2.12903E-05	1363.636	0.011733	84545.45	0.004633	15214.667	31.37255	16.16481	3.7	77.7	1256.006	31.9656
100	0.006362	200	0.333333	3.648099781	3.09375E-05	3636.364	0.010173	87272.73	0.004596	22678.444	108.805	11.05437	36.79602	772.7164	8541.894	217.3929
100	0.006362	400	0.2	6.060461162	4.85294E-05	8727.273	0.008173	92727.27	0.004527	37349.909	349.668	8.182484	74.12587	1556.643	12737.21	324.1645
100	0.006362	800	0.111111	10.83857686	7.81579E-05	19393.94	0.006694	103636.4	0.004403	65809.987	1145.563	6.189745	140.4106	2948.622	18251.22	464.4972
50	0.003181	200	0.2	6.060461162	4.85294E-05	4363.636	0.00972	46363.64	0.005384	11104.194	103.9569	8.182484	42.57413	894.0568	7315.605	186.1836
50	0.003181	400	0.111111	10.83857686	7.81579E-05	9696.97	0.007961	51818.18	0.005236	19565.426	340.5779	6.189745	80.64469	1693.539	10482.57	266.7836
50	0.003181	800	0.058824	20.21252866	0.000121957	20534.76	0.006599	62727.27	0.004992	35687.012	1129.31	4.730028	146.9801	3086.583	14599.62	371.5633
50	0.003181	1200	0.04	29.3510704	0.000152778	31418.18	0.005934	73636.36	0.004796	51059.915	2284.666	4.047304	206.5436	4337.416	17554.84	446.7742

Table 2. Exchanger Performance

The results of these calculations are given in Table 3.

Liq. Mass Flux	dQ/dT Each tube	Exchanger Duty MW	dx/dT
200	186.2	0.092	0.032
400	266.8	0.132	0.047
800	371.6	0.184	0.066
1200	446.8	0.221	0.079

It is concluded that given a heat source that is 10 degrees greater than local column temperature 0.9 MW of heat can be exchanged. 32% of recycled liquid would be vaporised.

The final stage of the design would be the identification and validation of the integration options (see Appendix 1).

Conclusions

The concept of an integrated internal heat exchanger that takes position previously occupied by distillation plates has been demonstrated using a simple apparatus.

It has been found that the hydraulic behavior of such an exchanger is well modeled by a homogeneous flow model.

There is significant scope to improve the heat transfer performance of such exchangers by installing tube inserts.

The flexibility of such exchangers can be improved by installing vanes in the liquid down-comers in order to provide liquid recycle and increased liquid flow through the exchanger.

A procedure that allows the engineer to link column duty, heat exchanger design and process integration opportunity has been developed.

REFERENCES

[1] S. Bandyopadhyay, Thermal Integration of a Distillation Column Through Side-Exchangers, Chemical Engineering Research and Design, Vol. 85, No.1, 2007, pp. 155-166.

[2] M.Nakaiwa, K.Huang, A.Endo,T.Ohmori,T.Takamatsu, "Internally heat integrated columns: A review", ChERD, 81, 162-177, 2003

[3] G.T.Polley, Heat Exchanger "Design and Process Integration", The Chemical Engineer, 8 April 1993

[4] J. C. Chen "Correlations for boiling heat transfer to saturated fluids in convective flow", Ind. Eng. Chem. Process Design Development, Vol. 5, no. 3 (1966) pp. 322-329

[5] E.E. Vázquez Ramírez, G.T Polley , J.M. Riesco Avila "Engineering equations for the estimation of heat transfer in pool boiling and flow boiling during upward flow in tubes", AMIDIQ XXX, (2009), México.

Appendix 1

Integration of Simple Columns

The scope for integrating a single simple distillation column into a process is easily identified by superimposing the temperature/heat load lines representing the column reboiler and condenser on the Grand Composite Curve for the background process (example given in Figure 1).

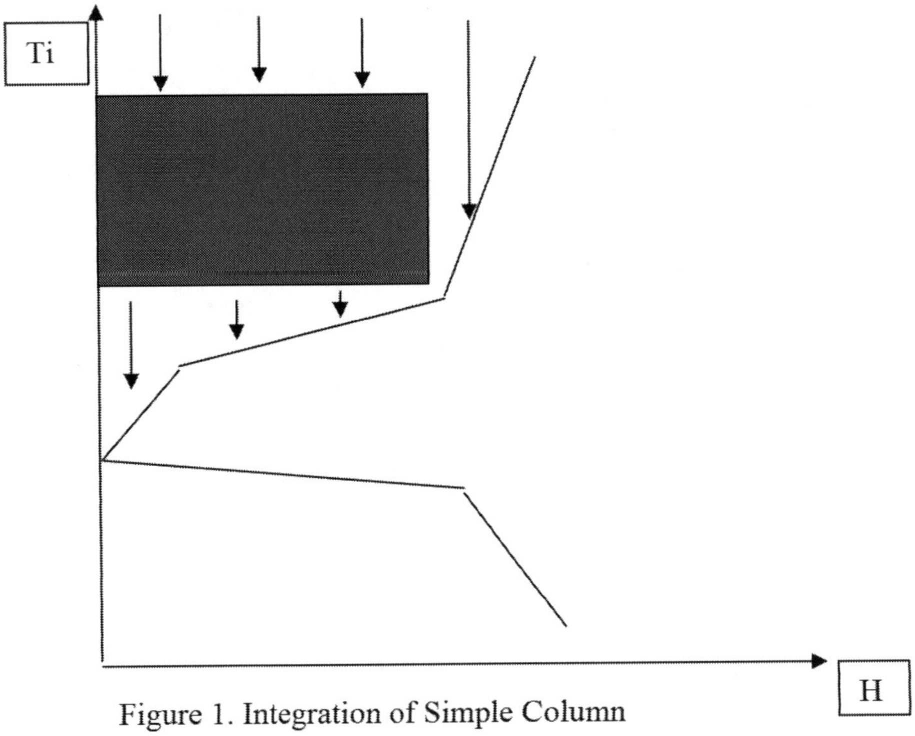

Figure 1. Integration of Simple Column

This lines can be considered to make up a 'block'. If the 'block' fits well, integration is straight forward. A column positioned above the Pinch would usually be driven by hot utility and the heat released in the condenser used for other process heating duties. The column could be controlled using the boil-up rate.

A column positioned below the Pinch would usually be driven by heat from the process with the heat given up in the condenser being absorbed by cold utility. The column could be controlled using the reflux.

On occasions, the 'block' may fit partially (on rare occasions totally) into a 'pocket' (Figure 2) in the Grand Composite Curve. In the example shown, the column would be partially driven by heat recovery and partially driven by hot utility. The heat released through the condenser would be recovered within the process. Such an arrangement could provide control and operability challenges. These would have to be examined and an acceptable control strategy identified before the integration scheme could be accepted.

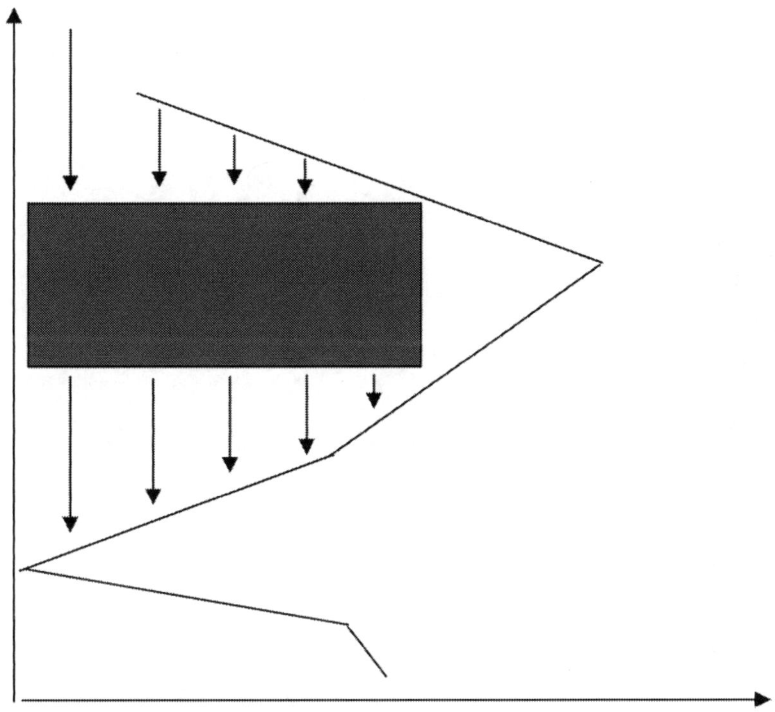

Figure 2. Integration into a 'Pocket'

Improving Integration

It is sometimes found that a simple column does not easily fit into the Grand Composite Curve. For instance, consider the situation illustrated in Figure 3.
Here a heat process demand positioned at a temperature level intermediate between the column re-boiler and the condenser is restricting the ability of the process to absorb the full condenser heat load.

In this situation the designer has a number of options. The choice may be to only re-absorb part of the column heat duty (i.e. accept the energy penalty). Alternatively, the designer may seek to develop a process modification (e.g. raise the column pressure and hence the temperature at which the condenser operates, or modify the background process in order to remove the restriction placed upon heat recovery from the column). Another option would be to introduce an intermediate condenser such that some of the column heat is rejected at a higher temperature level than the main overhead condenser (Figure 4).

The scope for identifying the possibility of using an intermediate condenser (or, re-boiler) cannot be determined through arbitrarily drawing lines to form new 'blocks' representing the distillation column. An examination of the thermodynamics of the distillation process is required.

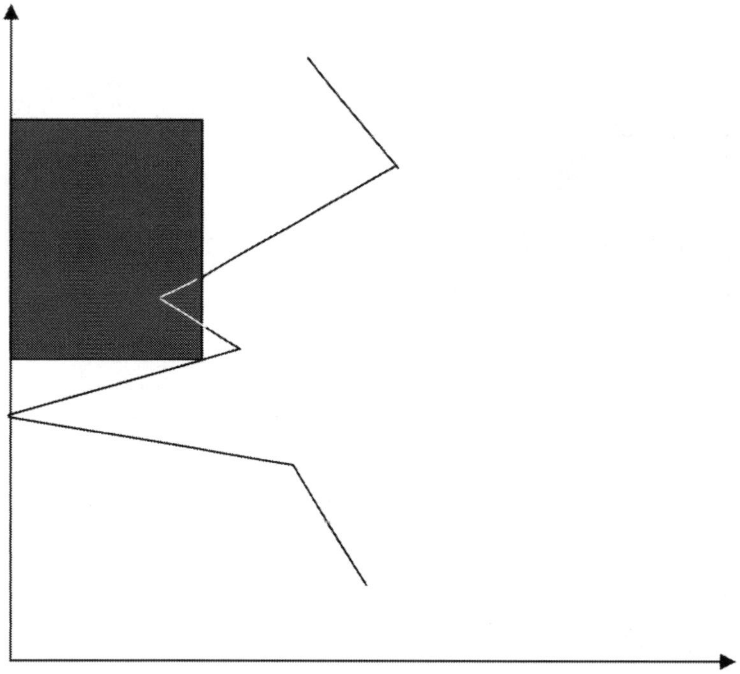

Figure 3. Process Restricts Column Integration

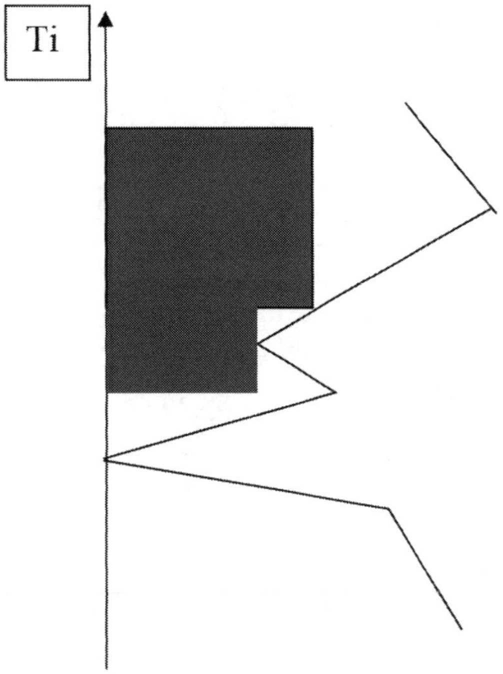

Figure 4. Use of Intermediate Condenser?

Distillation Column Thermodynamics

The thermodynamics of the distillation process have been appreciated for many years (see, for example, Benedict [1947]). A good description has been given by Flower & Jackson [1964].

An appreciation can be gained by first considering the concept of 'minimum reflux ratio'. Using a McCabe-Thiele plot, this can be determined from the slope of line connecting the required overhead composition to a point at which the feed q-line cuts the equilibrium line (or, if this point cannot be reached without cutting the equilibrium line, a line from this point tangential to the equilibrium line). This is the limiting 'operating line' for the rectification section.

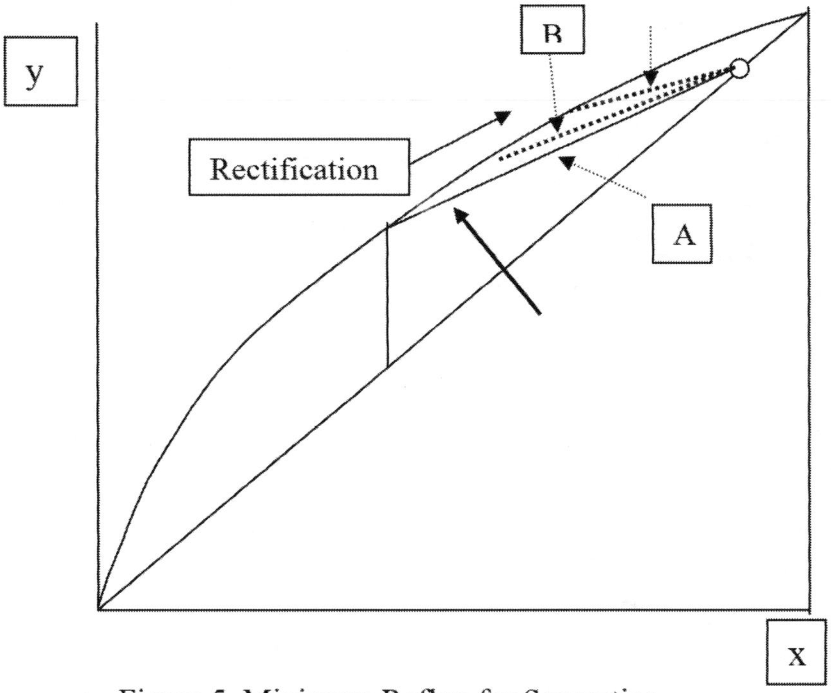

Figure 5. Minimum Reflux for Separation

The operating line of minimum reflux is illustrated in Figure 5. As the rectification takes place the concentration of more volatile component in both liquid and vapour phases increase. The distillation process moves up the equilibrium line. The amount of reflux required to complete the separation therefore changes. For the equilibrium line illustrated, it is observed that the minimum reflux required to complete the separation falls as the rectification process continues (moving up the curve from line A to line B. to line C etc.). Consequently, not all of the heat that needs to be removed from the column has to extracted through the top condenser (which otherwise provides the full reflux down the column). Some heat can be extracted via an intermediate condenser. The quantity of heat that can be extracted in this way can be determined (as a function of temperature) from the differences in minimum reflux ratio.

A similar analysis can be conducted for the stripping section of the column. The minimum boil-up required for a separation can be determined from the slope of a line joining the required bottom concentration with the point at which the feed q-line cuts the equilibrium line (Figure 6). It is found that some heat can be provided through the re-boiler positioned part way up the column (using an intermediate re-boiler). Possible intermediate load against temperature level can be determined in a manner similar to that use for intermediate condensers.

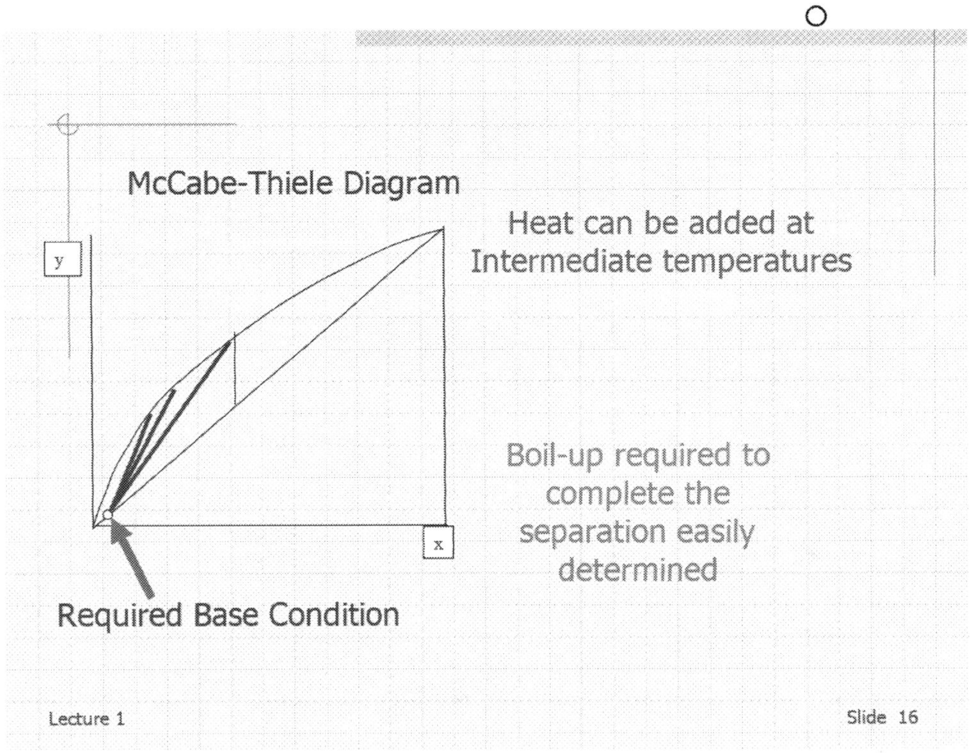

Figure 6. Column Boil-up Requirements

The Q-Curve

Flowers & Jackson proposed that the heat rejection profile for the rectification part of the column and the heat absorption profile for the stripping section be brought together to form a single curve, that they named the Q curve.

They showed that for the rectification ***the quantity by which the heat rejection from the column between the feed plate and the top condenser exceeds that needed to condense the top product*** is given by:

$$\Delta Q = D\left(\frac{x_D - x^*}{x^* - x}\right)\Delta h_v$$

where:

$$D = molal.fowrate.of.top.product$$
$$x_D = concentration.of.top.product$$
$$x = local.liquid.concentration$$
$$x^* = local.equilibrium.concentration$$
$$\Delta h_v = molal.heat.of.vaporisation$$

For the stripping section ***the quantity of heat that must be added to the column between the feed plate and a local point*** is given by:

$$\Delta Q = B\left(\frac{x - x_B}{x^* - x}\right)\Delta h_v$$

The Q Curve is a plot of excess against local temperature. It can be used to identify how heat can be added and rejected at different temperature levels. A typical example is shown in Figure 7.

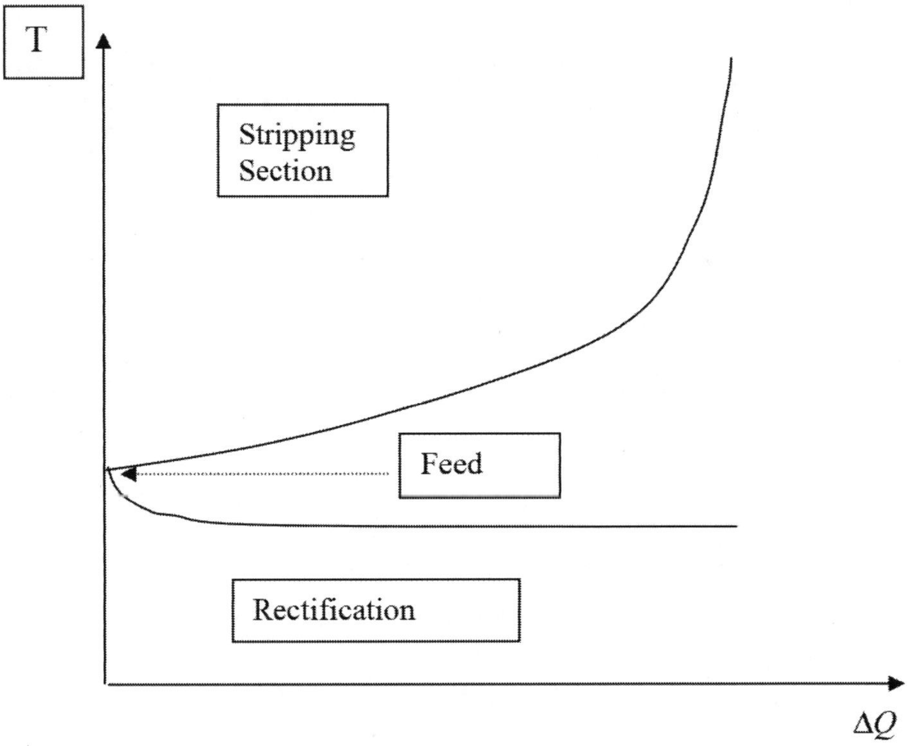

Figure 7. Q Curve

Development of T-Q Plot from Column Simulation Results

The McCabe-Thiele construction is based on the assumption that the molal latent heat does not vary with composition. For many systems that is not a good approximation. For these systems the Q-curve is better developed using the Enthalpy-Concentration Diagram for the separation. The construction (developed by Flower & Jackson) for the rectification section is shown in Figure 8. That for the stripping section is shown in Figure 9.

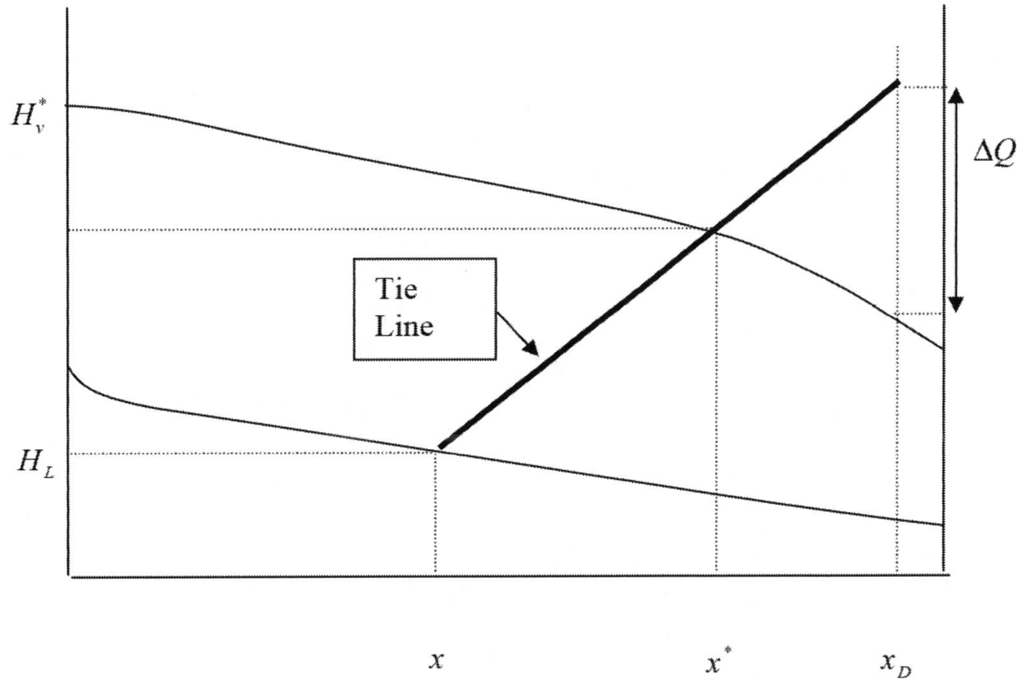

Figure 8. Rectification: Excess Heat Profile

For the rectification section, the following equation gives the excess heat profile:

$$\Delta Q = D\left(H_L + \left(H_v^* - H_L\right)\left(\frac{x_D - x}{x^* - x}\right) - H_c - \Delta h_{vD}\right)$$

For the stripping section the excess heat can be calculated from:

$$\Delta Q = B\{\left[H_v^* - H_{vF}^*\right] - \left[H_v^* - H_L\right]\left[\frac{x^* - x_B}{x^* - x}\right] + \left[H_{vF}^* - H_{Lf}\right]\left[\frac{x_f^* - x_B}{x_f^* - x_f}\right]\}$$

Prior to the development of distillation column simulators and physical property packages, this more accurate approach was limited by lack of Enthalpy-Concentration Diagrams. Now, as discussed by Dhole [1991], the information required for the solution of these equations is to be found in the standard output from commercial process simulators.

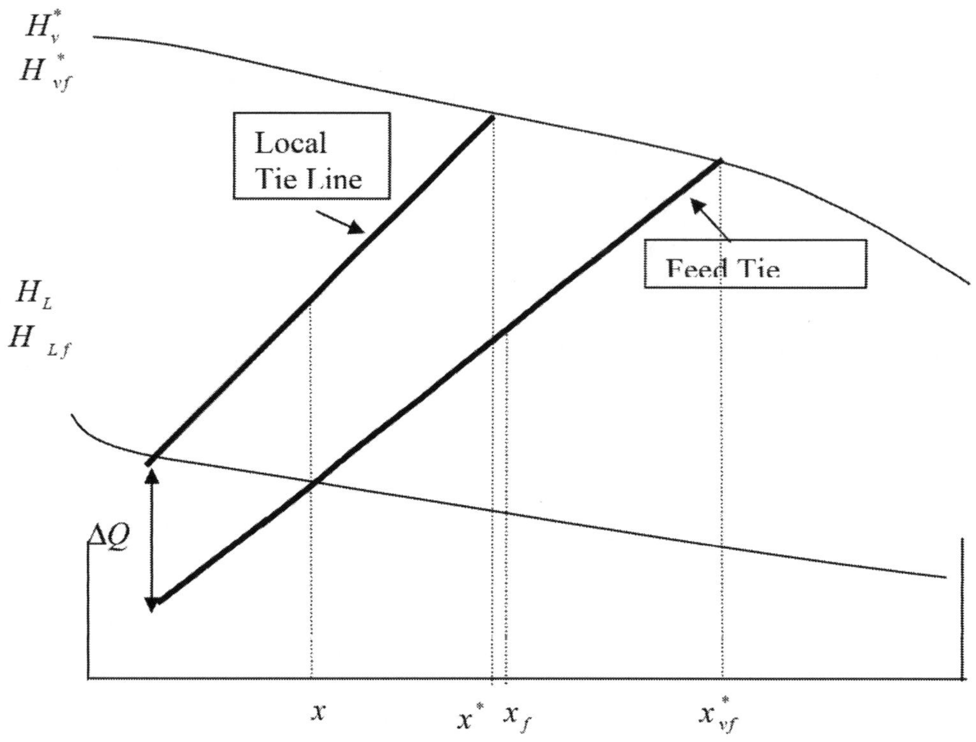

Figure 9. Heat Input Through Stripping Section

Most significantly, Dhole has further suggested that, as a first approximation, the analysis can be extended to cover multi-component as well as binary separations. All that is needed is a converged column simulation.

Dhole also pointed out that in addition to the Q curve, a curve relating heat requirements to plate position can be directly derived for the converged simulation.

Interpretation of Q Curve

The Q curve (also known as the Column Grand Composite Curve, CGCC) indicates how heat can be added and extracted from the column at intermediate points. Columns have to operate at reflux ratios above the minimum value in order to complete the separation with a finite number of trays. Typically economic reflux ratios fall in the range 1.05 to 1.25 times the minimum value. Consequently a heat content block for a distillation column would consist of the heat load associated with the excess reflux plus the Q curve profile.

In the case shown in Figure 10 a single re-boiler at the base of the column (Figure 10a) could be used. Alternatively, re-boiler at the base and a number of intermediate re-boilers (Figure 10b) could be used. Since, the satisfaction of overall heat requirements using the intermediate addition of heat effects the internal reflux ratio, the use of intermediate reboilers increases the number of plates required within the column. The effect upon column size and cost needs to be determined before a decision to use intermediate units is taken.

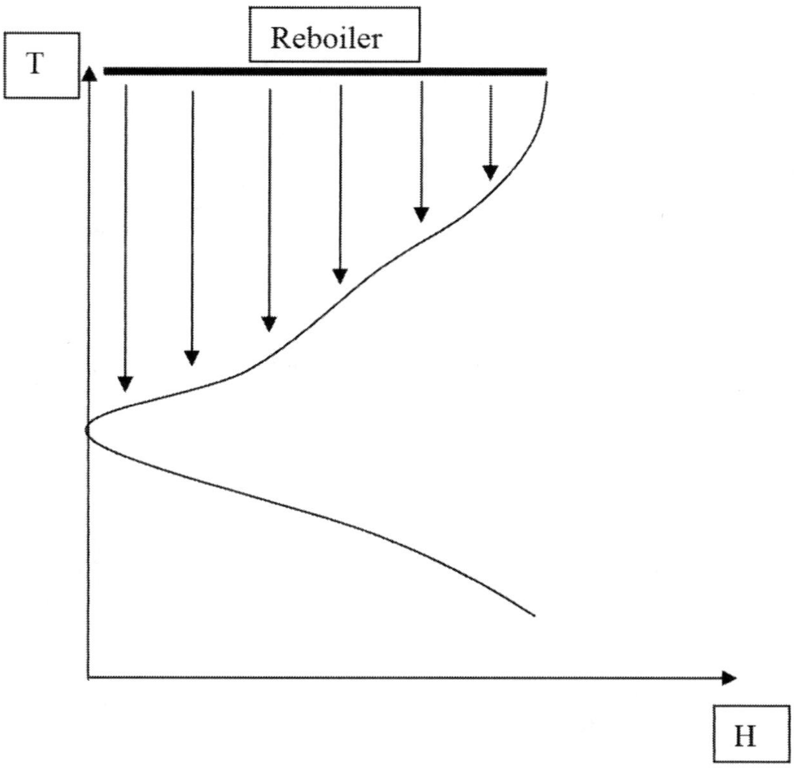

Figure 10a. The Use of a Single Reboiler

The Q Curve also provides us with guidance on how the column feed can be conditioned. If the curve shows a sharp increase in enthalpy immediately below the feed plate (decreasing temperature), cooling of the feed could be beneficial (Figure 11). So, in an integrated plant context, the feed to the column could be a valuable heat source.

If the curve shows a marked increase in enthalpy immediately above the feed plate (Figure 12), partial or complete vaporization of a saturated feed would result in reduced re-boiler load. In an integrated context the feed stream would be treated as a heat sink in the overall process.

Vaporization of the feed changes the temperature at which it enters the column. A different thermodynamic profile exists and a new Q Curve should be developed (Figure 13).

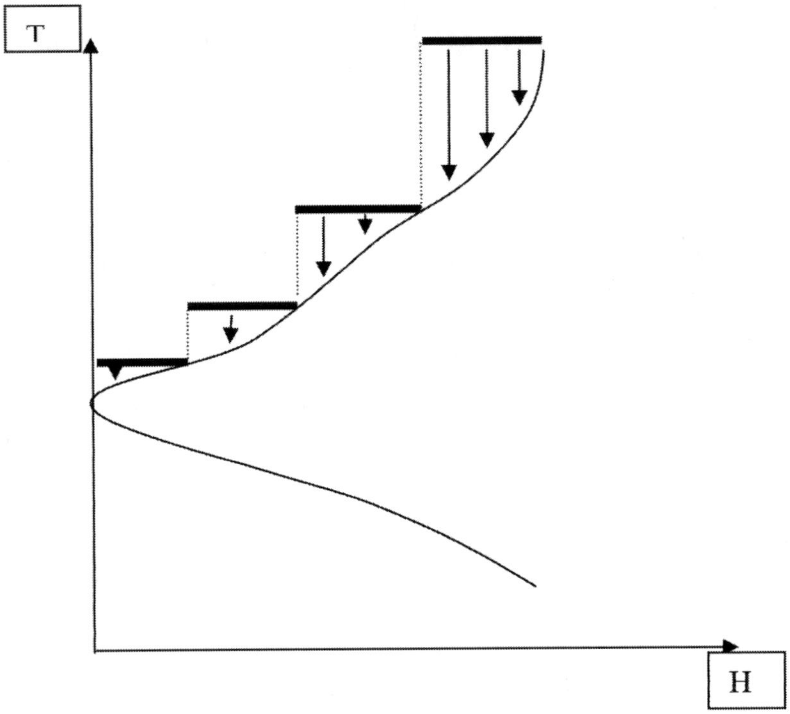

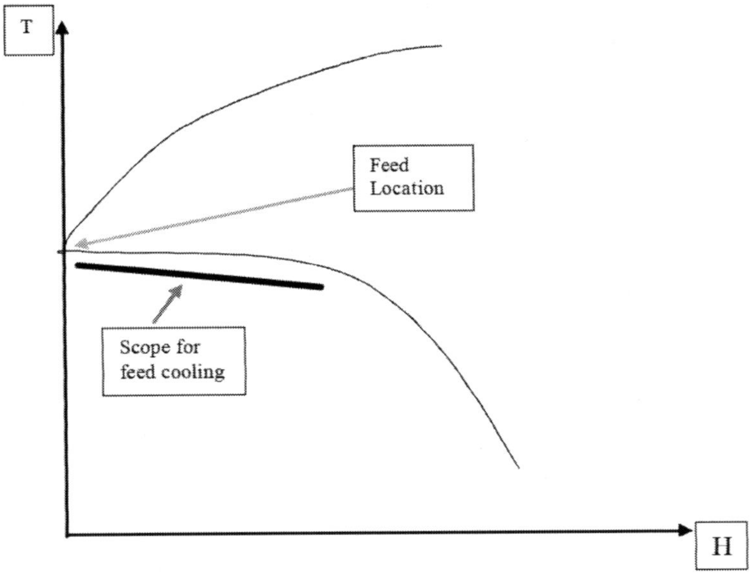

Figure 11. Profile Indicating feed cooling

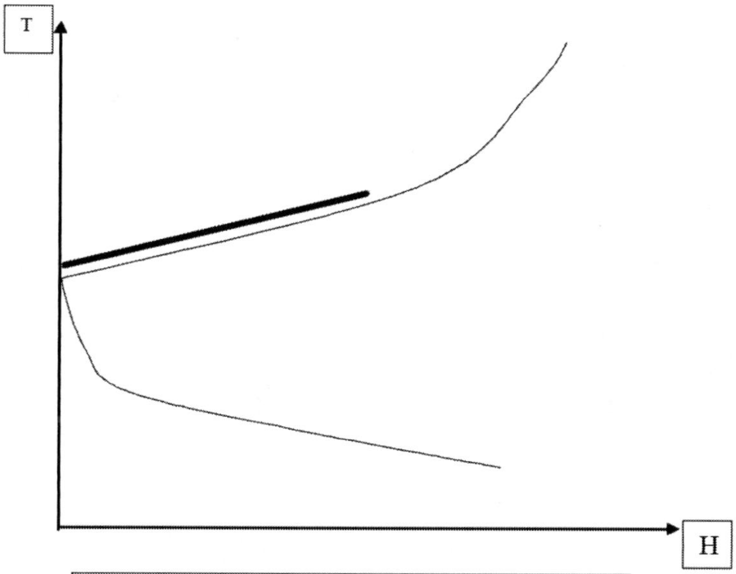

Figure 12. Profile indicating that feed pre-heating would be advantageous

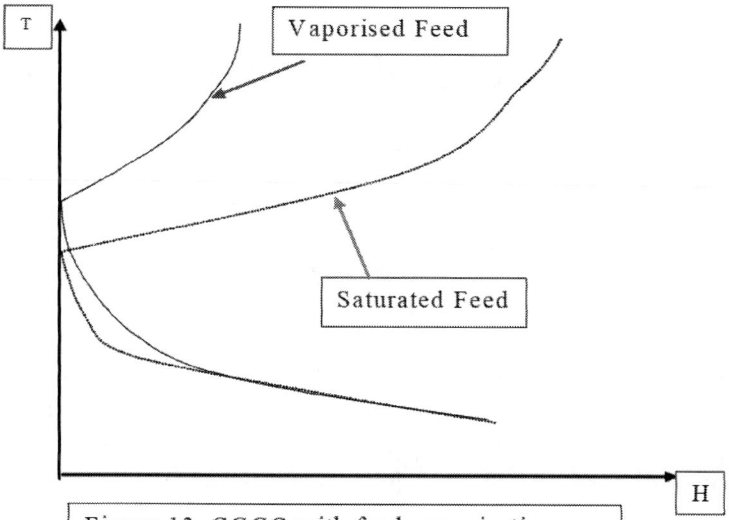

Figure 13. CGCC with feed vaporisation.

The Q Curve is particularly useful in identifying where heat pumping across the feed plate is an option (see Figure 14).

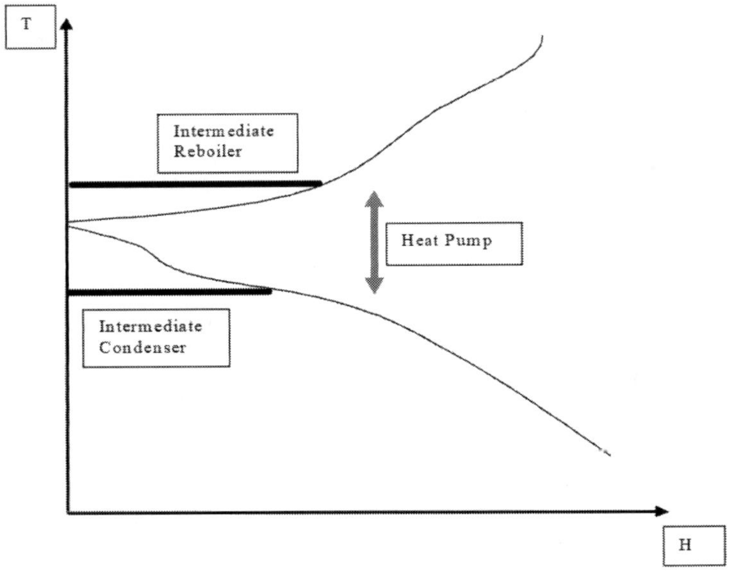

Figure 14. Application of Heat Pump

Benedict M. 'Multi-stage Separation Processes', Trans.A.I.Chem.E. 1947, 43(2), 41-60

Flower J.R. & Jackson R. 'Energy requirements in the separation of mixtures by distillation', Trans.I.Chem.E. 1964, 42,249-258

Dhole V.R. Ph.D.Thesis, UMIST, 1991

Dhole V.R. & Linnhoff B. 'Distillation Column Targets', Paper presented at ESCAPE-1 Conf., Denmark, May 1992

Rajah W. & Polley G.T. 'Synthesis of Practical Distillation Schemes', ChERD, 73A,953-966, 1995

(81a) Simulation and Optimization of Tubular Fixed-Bed Fischer-Tropsch Synthesis Reactor with Recycle

Bo Liao Sr.[1], Luhaibo Zhao[1], Xiaoquan Wang[2], Xiaohao Liu[1] and Yuhan Sun[3], (1)Low Carbon Conversion Center, Shanghai Advanced Research Institute, Shanghai, China, (2)Low Carbon Conversion Center, Shanghai Advanced Research Institute,Chinese Academy of Sciences, Shanghai, China, (3)State Key Laboratory of Coal Conversion, Institute of Coal Chemistry, Chinese Academy of Sciences, Taiyuan, China

Theincreasing demand of world's energy and the fast depletion of liquid fossil fuel reserves have made natural gas, coal and biomass increasingly attractive as feedstocks for synthetic liquid fuels. For the last two decades, Fischer–Tropsch synthesis (FTS) has gained renewed interest for the conversion of syngas to liquid hydrocarbons.

In industrial FT reactors, tubular fixed-bed reactor with recycle is widely used. Although many fixed-bed models for FT process have been reported in literature, research on concentration/temperature distributions in fixed-bed reactor with recycle is very limited. In this work, a 2D heterogeneous fixed-bed reactor model with recycle is developed and the simulation results are validated by using the industrial data. The effects of operating conditions on concentration/temperature distributions in FT reactor, such as inlet temperature, pressure, recycle ratio and feed syngas composition, are investigated. The results show that inert content in feed syngas and recycle gas, such as CH_4, N_2, etc., have significant effect on the conversion of CO and temperature distributions in fixed-bed FT reactor. The increase in recycle ratio results in the decrease in the CO conversion for inert gas rich syngas, and the inert gas content in syngas plays an important role to determine the hotspot location and temperature, which can be easily applied to biomass conversion process. Finally, the model is used for multiobjective optimization to maximize CO conversion and minimize the operation cost for fixed-bed FT process with recycle.

(81b) Fischer-Tropsch Cobalt Catalyst Improvements with the Presence of

Jennifer L. Klettlinger, Combustion Branch, NASA Glenn Research Center, Cleveland, OH

The objective of this study was to evaluate the effect of titanium oxide, lanthanum oxide, and zirconium oxide on alumina supported cobalt catalysts. The hypothesis was that the presence of lanthanum oxide, titanium oxide, and zirconium oxide would reduce the interaction between cobalt and the alumina support. This was of interest because an optimized weakened interaction could lead to the most advantageous cobalt dispersion, particle size, and reducibility. The presence of these oxides on the support were investigated using a wide range of characterization techniques such as SEM, nitrogen adsorption, x-ray diffraction (XRD), temperature programmed reduction (TPR), temperature programmed reduction after reduction (TPR-AR), and hydrogen chemisorptions/pulse reoxidation. Results indicated that both La2O3 and TiO2 doped supports facilitated the reduction of cobalt oxide species in reference to pure alumina supported cobalt catalysts, however further investigation is needed to determine the effect of ZrO2 on the reduction profile. Results showed an increased corrected cluster size for all three doped supported catalysts in comparison to their reference catalysts. The increase in reduction and an increase in the cluster size led to the conclusion that the support-metal interaction weakened by the addition of TiO2 and La2O3. It is also likely that the interaction decreased upon presence of ZrO2 on the alumina, but iii
further research is necessary. Preliminary results have indicated that the alumina- supported catalysts with titanium oxide and lanthanum oxide present are of interest because of the weakened cobalt support interaction. These catalysts showed an increased extent of reduction, therefore more metallic cobalt is present on the support. However, whether or not there is more cobalt available to participate in the Fischer-Tropsch synthesis reaction (cobalt surface atoms) depends also on the cluster size. On one hand, increasing cluster size alone tends to decrease the active site density; on the other hand, by increasing the size of the cobalt clusters, there is less likelihood of forming oxidized cobalt complexes (cobalt aluminate) during Fischer-Tropsch synthesis. Thus, from the standpoint of stability, improving the extent of reduction while increasing the particle size slightly may be beneficial for maintaining the sites, even if there is a slight decrease in overall initial active site density.

(81d) Enhanced Fischer-Tropsch Catalysts for Conversion of Biomass and Coal to Transportation Fuels

Gokhan Alptekin, Ambalavanan Jayaraman, Douwe Bruinsma, Ewa Muteba and Alex Wickham, TDA Research Inc., Wheat Ridge, CO

The United States needs viable alternatives to crude oil to moderate the effect of oil price hikes and provide an interim bridge until other fuel sources can commercially supplant petroleum-based fuels. The candidate resources that can be used to produce liquid fuels include biomass, oil sands, oil shale and coal. Coal is by far the largest and least expensive resource, with over 250 billion tons of known domestic reserves. Co-gasification of coal and biomass will reduce the carbon footprint of the CBTL processes while Fischer-Tropsch (FT) synthesis is the most economically viable option to produce transportation fuels from syngas. Although the FT process is run on a commercial scale in both South Africa and Malaysia, several challenges face the production of liquid fuels from coal, including the attrition resistance of the catalyst and downstream hydrocarbon separation. The FT process could also benefit from a highly selective catalyst to produce desired range of hydrocarbon products.

TDA Research, Inc. (TDA) is developing a unique catalytic material to address three technical challenges: FT activity, catalyst selectivity and better heat management. TDA's FT catalyst will produce the desired (gasoline or diesel) range hydrocarbons (rather than producing heavy wax and then cracking it into lower molecular weight hydrocarbons). A key advantage of our approach is the reduction in the number of processing steps and process severity in the conversion of FT products into the respective transportations fuels (gasoline or diesel). In a conventional FT process, without cracker processing of the product stream, only 16% of the plant output falls into the range of a gasoline hydrocarbon cut. Even with further processing, this production rate only increases to 26% of the plant output, with a loss of 9% of the available production during the downstream processing. TDA's novel size-selective FT catalyst has achieved 68% selectivity in the naphtha range in preliminary bench scale testing experiments. Naphtha is used in gasoline blending and this cut represents the highest value product that can be produced from the FT process. The results from our FT catalyst development will be presented in the meeting.

(81e) Aqueous-Phase Fischer-Tropsch Synthesis by Using Ruthenium Nanoparticle Catalyst

Venkat Ramana Rao Pendyala, Wilson D. Shafer and Burtron H. Davis, Center for Applied Energy Research, University of Kentucky, Lexington, KY

Fischer-Tropsch synthesis has long been known as an alternative process to produce long-chain hydrocarbons for their use in transportation fuels. In the Fischer–Tropsch reaction, synthesis gas produced from natural gas reforming or gasification of coal or biomass is transformed in the presence of cobalt, ruthenium, and iron catalysts. The product distribution is highly dependent on the process conditions and the type of catalyst used. Commercially FTS is performed in the gas or slurry phase. Recent studies demonstrated that FTS can be performed in aqueous phase too [1, 2]. The present work aims to investigate the effect of reaction temperature and reaction time on the performance of ruthenium nanoparticle catalyst during aqueous-phase Fischer-Tropsch synthesis by using a stirred tank batch reactor.

(86a) Explosion and Fire At Caribbean Petroleum Tank Terminal in Bayamòn, Puerto Rico

Vidisha Parasram and Maria Mazzocchi, Incident Screening, US Chemical Safety Board, Washington, DC

On October 23, 2009, normal tank filling operation became catastrophic when a 5,000,000 gallon atmospheric tank overflowed for approximately 30 minutes at the Caribbean Petroleum terminal in Bayamòn, Puerto Rico. The developing vapor cloud enveloped part of the tank farm and reached the waste water treatment area where it ignited. A flash fire and explosion ensued and the contents of the tank burned for 2.5 days. Neighboring communities were evacuated and oil contaminated the nearby wetlands. This incident revealed a number of safety system deficiencies at the facility and gaps in the regulatory requirements and industry standards addressing overfill prevention and emergency response. This incident also raises questions regarding the adequacy of safety management systems in place at tank terminals and the need for automated overfill prevention alarms including high, high high level alarm and independent alarms to detect and shut down flow into a tank during filling operations. In addition, due to the close proximity of the community to the Caribbean Petroleum tank terminal, this incident further demonstrates the need for a risk assessment process that addresses high hazard facilities. This presentation will provide an overview of the US Chemical Safety Board's Caribbean Petroleum investigation and discuss recommendations made by the CSB to prevent a similar incident.

(86b) Use Dynamic Simulation to Reduce Startup Flare Emissions for Ethylene Plants

Jian Zhang, Meiqian Wang, Jie Fu, Chuanyu Zhao, Ha Dinh, Shujing Zhang, Qiang Xu and Kuyen Li, Dan F. Smith Department of Chemical Engineering, Lamar University, Beaumont, TX

Ethylene plants play an important role in petrochemical industry. The products of an ethylene plant, such as ethylene, propylene, butadiene, benzene, toluene and other co-products, are all basic feedstocks in petrochemical industry. Ethylene production process is one of the biggest and most complicated processes in petrochemical industry. Startup of an ethylene plant is also very critical. A typical startup of an ethylene plant could flare 400,000 - 1,000,000 lbs of materials, and generate 200,000 lbs of CO and VOCs (volatile organic compounds). Reduction flare emissions will help reducing environmental impact and saving materials.

In this paper, rigorous steady-state and dynamic models of an ethylene plant was built. The models include CGC (charge gas compressor), chilling train, demethanizer, deethanizer, depropanizer, C2 splitter and C3 splitter sections. Both steady-state and dynamic models have many details so that they are more close to startup procedure. For example, performance curves were used in CGC section, so that the section can simulate scenarios on different feed and bypass flowrate. In chilling train section, information of equipment size and heat capacity is considered for simulating precooling process. For distillation columns, rigorous distillation model and tray sizing calculation are employed. Both steady-state and dynamic models are tuned and validated to make them close to real plant operations.

The validated dynamic model was used to simulate startup scenarios. A startup procedure has dozen of steps, and some steps have several flare reduction options. Through dynamic simulation, all major steps and options can be simulated and evaluated, and the flare emissions can be calculated. For example, CGC startup is the most critical step and has largest amount of flare vent, because charge gas has to be continuously produced through furnaces and sent to flare to keep inlet pressure of compressors. We simulated several scenario cases with charge gas of different flowrate, methane and nitrogen, and found the best scenario, which could largely reduce flare emissions of CGC section. Another case study is regarding the precooling of chilling train. A chilling train should not be chilled very fast, otherwise it could be damaged by thermal expansion. On the other hand, chilling train cannot work properly before chilled to certain temperature. Through dynamic simulation, several chilling scenarios were simulated. For each scenario, control strategies are employed to control the chilling rate within a certain range. Simulation results showed some scenarios could even cool down chilling train without any flare emissions.

Flare calculation showed that the total flare emissions could be greatly reduced through dynamic simulation. The final scenarios have been acknowledged by ethylene plants.

(86c) Development of Novel Reducing Agents for NOx Reduction by Sncr Method

Maoqi Feng, Div. of Chemistry and Chemical Engineering, Southwest Research Institute, San Antonio, TX and Reggie Zhan, Emissions and Vehicle Research Division, Southwest Research Institute, San Antonio, TX

Nitric oxides (NO_x), mainly including nitric oxide (NO) and nitrogen dioxide (NO_2), are of the most toxic pollutants formed during combustion and refining processes. Post-combustion treatment methods include selective catalytic reduction (SCR), selective noncatalytic reduction (SNCR), reburning, and the combination of these aftertreatment technologies. Among the post-combustion technologies, the SNCR technology is considered to be an effective approach to reduce NO_x, with reasonable capital investment and operation cost. To reduce NO_x in a lean environment, a reductant is needed. For SNCR, the most commonly used reductant is ammonia or urea water solution as an alternative to NH_3. One of the disadvantages of ammonia is the cost of the injection system. Different reducing agents for NOx removal by SNCR method will be evaluated in this paper, including carbon monoxide, hydrogen, char, hydrocarbons, and oxygenates, etc.

Rotating Zigzag Bed Industrial Application Research on Ammonia Stripping

Zhi-chao Xu, Xiao-hua Li, Guang-quan Wang, Jian-bing Ji
(College of Chemical Engineering and Materials Science, Zhejiang University of Technology,
Zhejiang Province Key Laboratory of Biofuel, Hangzhou, Zhejiang, China 310032)

Abstract: About 400 units of rotating zigzag bed (RZB) have been used in industrial distillation process. These successful applications fully demonstrated its advantages such as high gas-liquid mass transfer rate and small size. There will also be broad prospects of RZB applied in waste water treatment process. In this paper a preliminary attempt was made to study ammonia nitrogen removal from industrial waste water by RZB before emission. Firstly, stripping experiments were carried out to remove ammonia nitrogen from the waste water containing ammonia and ammonium sulfate in a small single-rotor RZB, obtaining ammonia solution with a proper concentration. Secondly, a unit of industrial-scale three-rotor RZB with outer diameter of 750mm was designed based on experimental data and used to treat ammonium sulfate solution with 2% ammonia nitrogen. The industrial results showed that the feed capacity was greater than 1500 kg/hr, ammonia concentration in the bottom was less than 30 mg/L, and aqua ammonia concentration in the top was greater than 25%.

With the improvement of human life, there is growing public concern about environment problems. Wastewater emission is one of the important factors contributing to environment deterioration. In the wastewater emissions, ammonia nitrogen concentration is a very important target which should be strictly controlled because excessive ammonia in the river will cause water eutrophication. The ammonia wastewater has a wide variety of sources, such as refining, fertilizer, inorganic chemicals, pesticide, ferroalloys, glass and food industries. Therefore, studies on removal ammonia from the wastewater and how to recover ammonia to reuse is of great significance. Rotating zigzag bed (abbreviated to RZB) has been successfully applied in about 400 cases of industrial distillation[1-5]. Due to its higher mass transfer rate it was expected to have a very wide application in wastewater treatment. Some traditional enterprises have demand for retrofit and optimization of the production process without occupied space increase because of limited land area. RZB can exactly meet this demand owing to its very small volume. In this paper lab-scale experiment of stripping ammonia from wastewater with RZB was conducted and ammonia solution with a certain concentration was obtained. Then, the experimental results were scaled up to design a unit of RZB successfully used in industrial production.

1.Lab RZB device and flowsheet

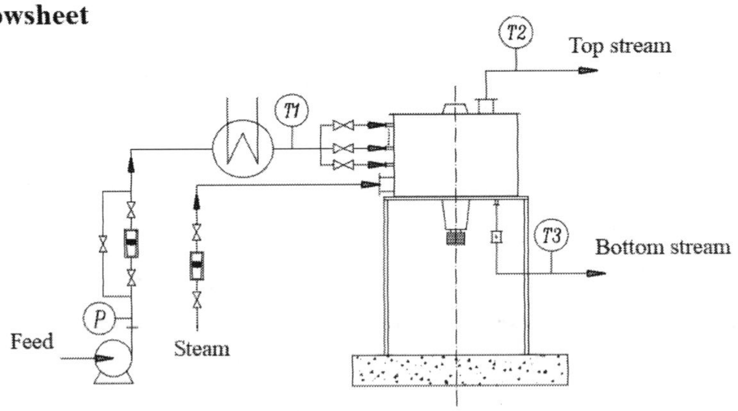

Fig. 1 Lab RZB device and flowsheet

Lab RZB device was set with multiple liquid feed inlets, which were used in variety of processes with different number of theoretical stages. For feeds with different concentration of ammonia, their optimal operation and equipment parameters can be determined by experiments.

In experimental flowsheet shown as Fig.1, the liquid feed was measured with a rotormeter and pumped into RZB device with multiple inlets after heating by a heat exchange. The steam with higher temperature enters into vapor inlet after flow measurement. The treated water was drawn from liquid outlet on the bottom of RZB. Vapor phase containing ammonia withdrawing from the top of RZB was condensed and absorbed in the next steps. T1-T3 were thermometers to measure and record the temperatures of several key positions.

2. Experiments and results analysis

In the experiments, the concentration of the prepared feed was close to that of industrial stuff, which was ammonium sulfate solution with about two percent ammonium. The feed position can be determined through the preliminary experiments. For this study, the flow ratio of feed to steam was varied and analyzed the ammonium compositions of bottom and top outlet stream as well as the power consumption of the motor. These data provided the basis for scale-up design of RZB.

Feed preparation: Ammonium sulfate solution with concentration of 15.5%, which pH was adjusted to about 13 by dripping NaOH solution, was prepared by adding ammonium sulfate powder into water. Ammonium sulfate solution has ammonium concentration of 3.3% and was chosen as the feed.

In the experiments, originally the top outlet vapor stream entered into an absorber but not a condenser. The system was maintained under vacuum condition by adjusting the relief valve at the end of absorber. The pressure of the system was 72kPa and the temperature of the top outlet vapor phase was about 80℃. The typical process and operation parameters in the experiments were shown in Table 1:

Table 1 Typical process and operation parameters

feed				steam		Bottom stream	Top stream
$(NH_4)_2SO_4$ %	Ammonium concentration %	pH	flowrate L/hr	flowrate kg/hr	pressure MPa	Ammonium concentration mg/L	Ammonium concentration(vapor) %
15.5	3.3	13	1000	120	0.35	<100	>48

If the top outlet vapor stream was connected to a condenser, solution with ammonium concentration higher than 20% can be obtained.

3. Industrial application of RZB in ammonium stripping

Fig. 2 Industrial equipment

After the results of industrial field tests were accepted by the customer, scale-up design was made for RZB. The section between the feed position and the top, corresponding to the rectifying section of distillation column, has a separation ability of 4.5 theoretical plates. This section can effectively prevent entrainment. On the other hand, if the actual ammonium concentration was lower than designed, partial reflux can be introduced to increase the ammonium concentration in the top outlet stream. The type of RZB is BZ750-3P, which means the largest one in the three rotors has a diameter of 750mm. Its appearance was shown in Figure 2.

The RZB was used in Taiwan, with industrial process flowsheet shown in Figure 3. The feed and alkali solution for adjusting the pH value, which flowrate were measured by flowmeter respectively, were mixed in a fixed proportion. The mixed stream was heated by the bottom outlet liquid and entered into the RZB. The steam was introduced by vapor inlet. The outlet vapor entered into absorption system which performs cooling and absorption steps. The absorption systems consist of four sections: pure water (absorbent), low concentrated ammonium recycle, high concentrated ammonium recycle and finished ammonium. The flow order is pure water section → low concentrated ammonia section → high concentrated ammonia section → finished ammonia section. The finished ammonia left from the bottom of the absorption system.

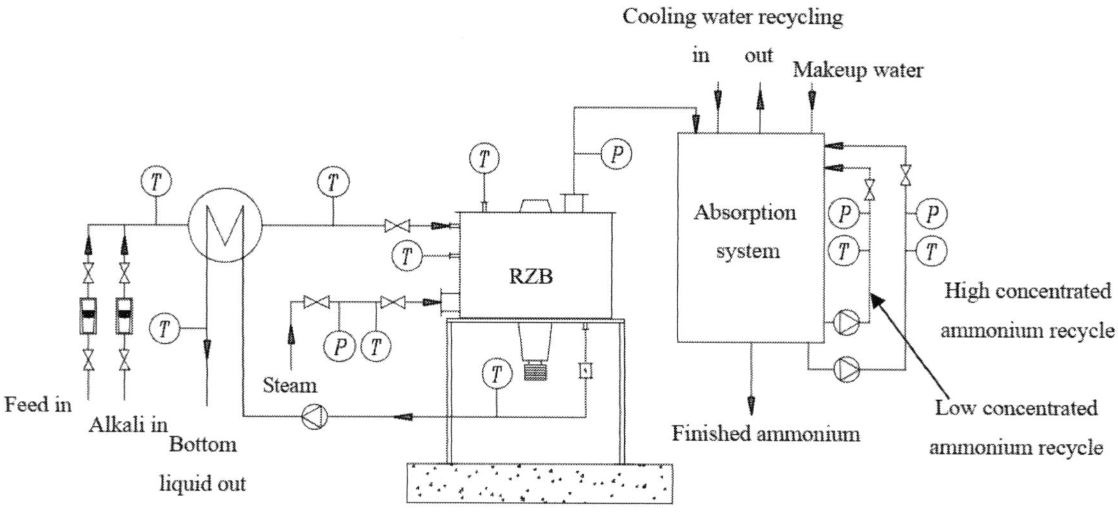

Fig. 3 flowsheet of industrial application of RZB

-Pump (P)-Manometer (T)-Thermometer

Actual operating parameters are as follows: the feed rate 1500kg/hr, ammonium concentration 18g/L, pH value 11.5-12, outlet vapor temperature about 80℃, no reflux, finished ammonia concentration > 25%, ammonia in outlet liquid<30mg/L, unit consumption of steam 90kg/ton feed, bottom temperature about 98℃. the temperature approach between inlet steam and heated feed was about 10℃. The customer was very satisfied with the results.

4. Conclusion

About 400 units of RZB used in a variety of industrial distillation processes demonstrated its higher separation ability. Its successful application in stripping ammonium from the wastewater was another good example.

RZB has a small volume with a height of lower than 1m. For traditional enterprises with small unused space in the factory, this kind of device can find its place here.

In this system, feed was heated by bottom outlet stream which sensible heat was recovered. So the system shows energy-saving effect.

5. Prospects

Modification of existing RZB device can realize liquid cut off between its rotors, which can extend its applications in recovery of ammonium from the wastewater. For RZB with more than three rotors, its top and bottom section perform absorption and stripping respectively, thus these two processes can be integrated in single equipment.

Reference

1 Ramshaw, C. (1983)., "Higee distillation—An example of process intensification, " Chem. Eng., February, pp. 13–14

2 Rao, D.P., Bhowal, A. and Goswami, P.S. (2004), "Process intensification inrotating packed beds(HIGEE): An appraisal," Ind. Eng. Chem. Res., pp. 1150-1162.

3 Li, X.P., Liu, Y.Z. and Li, Z.Q. (2008), "Continues distillation experiment with rotating packed bed," Chin. J. Chem. Eng., pp. 656-662.

4 Wang, G.Q., Xu, Z.C., Yu, Y.L.and Ji, J.B. (2008), "Performance of a rotating zigzag bed—A new HIGEE," Chem Eng Process, November, pp. 2131–2139.

5 Ji, J.B. , etc. (2004), "Equipment of zigzag high-gravity rotating beds," CN Pat., 01134321.4.

56e

World Scale Boil-Off Gas Reliquefaction

Heinz C. Bauer, Linde Engineering, Pullach, Germany
Hans Mattsson, Siemens, Finspång, Sweden
Sven-Erik Brink, Siemens, Duisburg, Germany

Abstract

Boil-off gas reliquefaction has become an industry standard for many LNG carriers, which no longer use gas as fuel for steam boilers, but tap others energy sources to generate power for propulsion. Well known examples are the ships of the Q-Flex and Q-Max class. Most of these vessels use simple nitrogen expander cycles with relatively high energy consumption (typically 800 kWh/t) and a moderate capacity of about 200 t/d LNG liquefaction.

Increasing LNG demand in Asia, especially in Japan after the nuclear catastrophe triggered innovative ideas to increase the LNG output of existing liquefaction plants. Boil-off gases and ship return gases from LNG tank farms have been identified quickly as promising source for additional LNG, as these gases are perfectly suitable for cryogenic processing. Obviously, dehydration and sweetening is not required. However, low pressure, low temperature and elevated nitrogen content pose challenges to an efficient design.

Linde Engineering has developed an innovative concept for a world scale boil-off gas reliquefaction unit with 0.3 to 1.0 mtpa capacity, which includes the following features:

- Boil-off gas compression starting from cryogenic conditions
- Mixed refrigerant cycle with two heat exchanger bundles only
- Load balancing between feed gas and refrigerant cycle compressors
- Nitrogen rejection into the gas turbine fuel gas with a double flash process
- Use of identical gas turbines as mechanical drive

Introduction

Large LNG export terminals include an LNG tank farm with large LNG tanks providing storage capacity of several hundred thousand meters cubed. Low pressure gases at tank pressure (about 100 to 200 mbarg) originate from several sources like

- Boil-off gas caused by heat ingress through the tank walls and LNG pumps
- Flash gas generated in the run-down line between liquefaction and storage
- Ship return gas / displacement gas during ship loading
- Flash gas during cool-down of warm send-out systems

In total the overall LP gas (hereinafter called boil-off gas or BOG) flow rate may accumulate to about 1,000 t/d for a 10 mtpa LNG facility (this figure can be scaled almost linearly for larger plant capacities). The quality of the BOG may raise concerns as it will be enriched with nitrogen and may contain contaminants from ship return gases. The distance between the LNG liquefaction trains and the LNG storage may add up to a few kilometers depending on the site conditions. Thus, returning the BOG to the liquefaction trains and merging it with the end-flash gas may not be the optimum solution for every site.

The decision to install a dedicated BOG reliquefaction system to the entire LNG facility will depend very much on the specific site conditions including the fuel gas balance, the capacity of the existing refrigeration systems, availability of natural gas as feed stock, distance between liquefaction trains and jetty/tank area, and flare load especially during peak BOG generation as it happens for example during ship loading. Under all circumstances, however, BOG reliquefaction will generate a high value product starting from very favorable feed stock conditions, as the absence of water, sour gas components and heavy hydrocarbons in BOG simplifies cryogenic processing significantly.

Concept Selection

A BOG reliquefaction unit with a capacity between 0.3 and 1.0 mtpa LNG can be compared with mid-scale LNG plants. All land based LNG plants in this capacity range are using a single mixed refrigerant (SMR) cycle. This choice is based on its good efficiency (typically 350 kWh/t[1]), minimum count of rotating equipment and good safety records. Considerations to use nitrogen expander cycles for floating LNG concepts are not applicable for a land based BOG reliquefaction facility, mostly as compact design and readily available make-up supply are not likewise important.

Even if there is some discussion about the use of brazed aluminum plate-fin heat exchangers (PFHE) for SMR based mid-scale LNG plants, experienced operators prefer coil wound heat exchangers (CWHE) in this service, as especially load changes during ship loading will cause extra stress to the heat exchangers. All experience shows that CWHEs are more tolerant against frequent load cycles than PFHEs. The marginal cost adder to go with CWHEs is more than offset by the peace of mind, if downtime caused by heat exchanger repair is taken into account.

BOG reliquefaction requires large compression systems for the refrigeration cycle and also for BOG compression, as liquefaction under low pressure would be very inefficient. A typical SMR process with a specific energy consumption of 350 kWh/t needs about 15 MW for a feed gas flow rate of 1,000 t/d under a pressure of at least 50 bar. Feed gas compression starting from tank pressure to at least 50 bar requires almost the same power. For compressor trains with more than 10 MW each gas turbines have been selected as driver in the past, if fuel gas is available under economically attractive conditions. Using electric motors instead of gas turbines seems to be preferable only, if and when electric power can be sourced reliably at low prices.

Process Optimization

World scale BOG reliquefaction units have not been built so far. Thus, there is no proven template for a cost effective and robust process design. It is fair, however, to start with building blocks as they are known from mid-scale or world scale LNG plants.

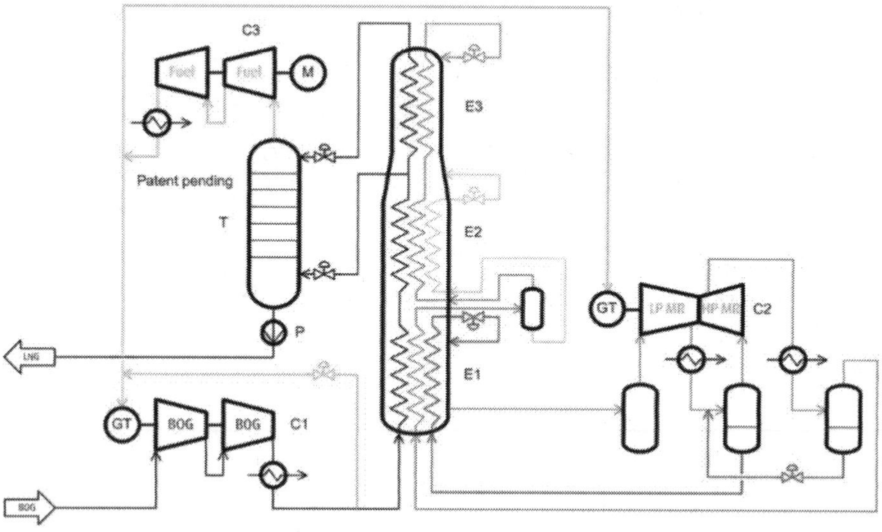

Fig. 1 Base concept with N2 stripper

Fig. 1 shows a conventional design for a BOG reliquefaction system. BOG from the LNG storage tanks has to be compressed first. The low molecular weight of the BOG and the high pressure ratio require two compressor casings for the single shaft, centrifugal compressor C1. In the absence of a need for sour gas removal and dehydration the compressed BOG will be sent directly to the cryogenic heat exchanger, which consists of a pre-cooling bundle E1, a liquefaction bundle E2, and a sub-cooling bundle E3. In most of the cases the nitrogen content of BOG will exceed the standard specification for LNG (<1 vol%) so that nitrogen rejection is required.

At elevated nitrogen levels in the BOG a nitrogen stripper T operated at low pressure will be the right solution for this purpose. An elegant manner to provide strip gas[2] for the stripper is the use of 'warm' LNG after the liquefier E2, which will generate a sufficient quantity of gas, when it is flashed into the column. On-spec LNG will be sent by pump P via run-down lines to the storage tanks. The nitrogen rich stripper overhead gas has to be recompressed in compressor C3 to the required fuel gas pressure of the selected gas turbines. The chosen single mixed refrigerant cycle is the LIMUM® process, which has been developed and patented[3] by Linde Engineering and is successfully in operation since many years in different locations. It is built around a single casing, barrel type compressor C2 with two compression stages.

This conventional design was scrutinized with value engineering methods with the following findings:

- Shaft power of C2 is about 150% of shaft power of C1, identical duties are desirable
- LP suction conditions of C3 result in an expensive multi casing design
- Stable operation of stripper T under frequent load changes is hard to achieve

The result of the subsequent process optimization is shown in Fig. 2. The MRC compressor C2 is unloaded by shifting part of its duty to the BOG compressor C1. This can be achieved by replacing the sub-cooling refrigeration duty of the MRC, which was used in E3, by an open methane cycle.

Instead of sub-cooling LNG to such a low temperature that only the desired fuel gas stream is generated, a larger methane rich flash gas rate is deliberately accepted, which is recycled to the suction side of the BOG compressor C1. Thus, a very good match between the shaft power of C1 and C2 could be achieved[4].

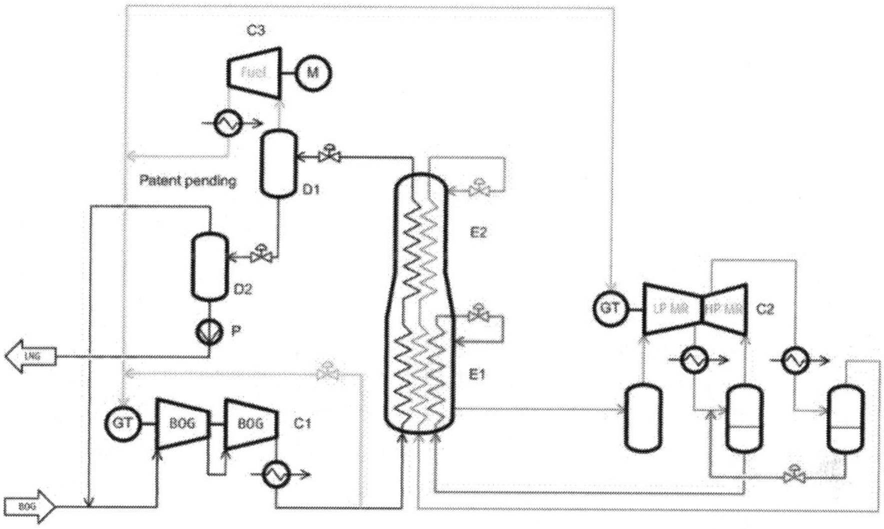

Fig. 2 Final concept with double flash system

In a second step the nitrogen stripper T was replaced by a medium pressure flash drum D1 and a low pressure flash drum D2. The operating pressure of D1 can be selected such that the fuel gas can be compressed in a single casing, single stage compressor C3, which can be operated optionally at ambient temperature suction conditions, as the elevated suction pressure allows for a warm-up of the fuel gas before it is compressed.

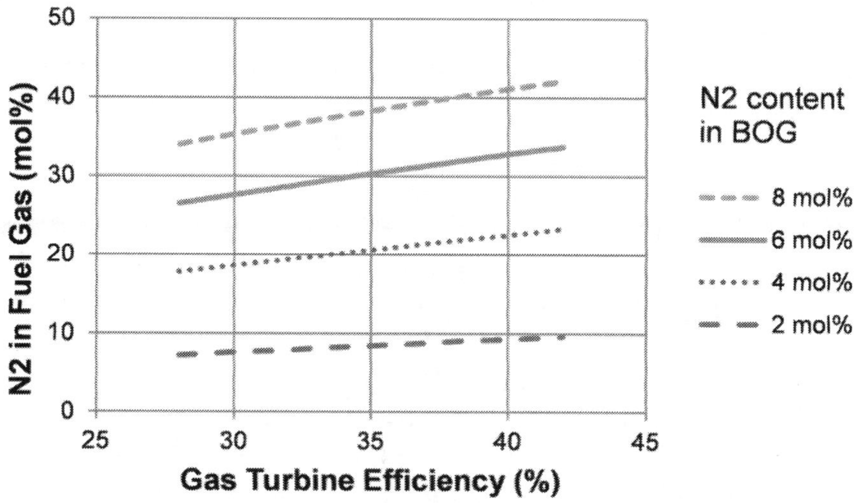

Fig. 3 Correlation between GT efficiency and fuel gas quality

In total the value engineering exercise helped to eliminate the sub-cooling bundle E3 and the low temperature LP casing of the fuel gas compressor C3. The stripper column T was replaced by two simple flash drums D1 and D2 and the main compressors C1 and C2 can be hooked up to drivers of identical rating.

After a successful streamlining of the process the suitable equipment had to be selected. One very complex decision was the selection of the best fitting gas turbine. In a situation, where the BOG is worked up into only two products, LNG and fuel gas, obviously most of the nitrogen has to go with the fuel gas, as LNG can accept only a very small amount (max. 1 vol% N2). As a result highly efficient gas turbines with a low heat rate have to be operable with higher nitrogen content in the fuel gas and vice versa. This correlation is plotted in Fig. 3 for different BOG compositions.

Low BTU gas turbines

The combination of DLE (Dry Low Emission) requirements with a low heating value of the fuel gas requires certain measures to guarantee the flame stability. Fig. 4 shows a suitable concept as it is used e.g. in the Siemens industrial gas turbine family SGT-600/700/800 being equipped with eighteen 3G fuel nozzles, also called burners, in an annular combustor. The burners use pilot gas, which is fed at the burner outlet and is burned around the main flame for flame stabilization. Typical for DLE combustion, which is used by Siemens for more than 20 years, fuel is burned with twice the amount of air in comparison to stoichiometric combustion. The large amount of air keeps the flame temperature low reducing emissions of nitrous oxides, but on the other hand increasing the risk of combustion instabilities and incomplete combustion leading to emissions of CO.

At load levels exceeding 50 % most of the gas is burned in the main flame, which has a low ignition temperature producing low concentrations of CO and NO_x. At lower power output, more gas is burned in the pilot flames, which operate at a higher temperature and thus create more NO_x. At all levels of power output the amount of pilot flame gas is limited to the amount needed to ensure stable operation of the turbine.

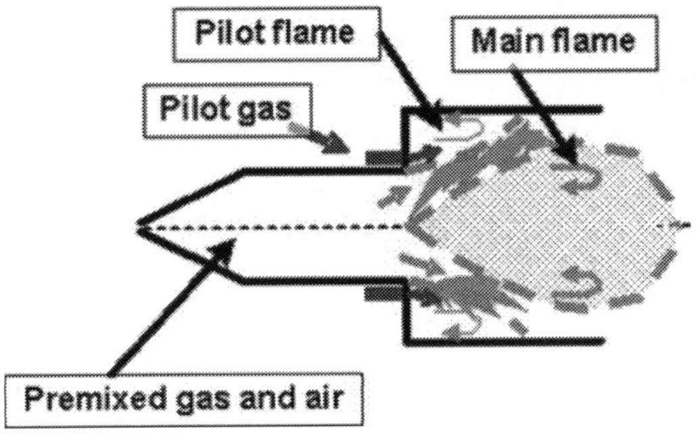

Fig. 4 3G burner schematic

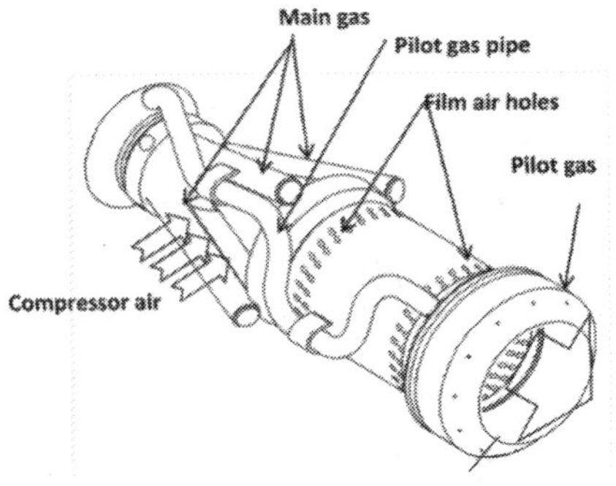

Fig. 5 3G DLE burner details

The unique fuel nozzles (see Fig. 5) of the Siemens industrial gas turbines have been proven to be very flexible in terms of the heating value of the fuel, respectively the Wobbe index. A high N_2 content in the fuel acts just as ballast since N_2 does not contribute to the combustion. Results from test beds and from actual operation in LNG plants[5] show that up to 40% N_2 in fuel gas can be accepted without changes in the standard configuration. This feature comes with the full capability to accept substantial changes in Wobbe index. However the starting gas needs to be correlated with the control system for safe start-up, but once in operation a very reliable operation can be achieved with this system.

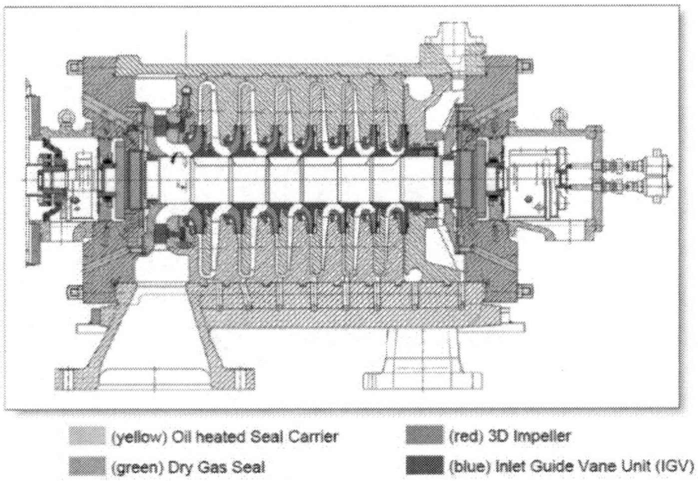

Fig. 6 Turbo compressor cross sectional drawing

BOG compressor

Among all turbo compressor applications the cryogenic BOG compressor service is one of the most demanding applications. Due to the nature of the BOG service the compressor trains are periodically exposed to large fluctuations regarding volume flow and suction temperatures. The proprietary Siemens design features (see Fig. 6) include variable Inlet Guide Vane (IGV) control (blue) in front of the impeller (red), protected dry gas seals (green) in a heated and cooled seal carrier (yellow), extensive expertise in material selection, compressor design, manufacturing and testing.

The IGV unit (Fig. 7) is an aerodynamic device and induces a vortex onto the flow increasing the compressor head respectively adjusting the angle of attack (incidence) to the impeller blading. The IGV concept improves the turbo compressor performance control and increases the operating range (turndown). The IGVs provide a map of individual performance curves within the IGV setting range, where the volume/head adjustment is quick and smooth. Main advantages of IGV control for BOG applications are

- Direct Online (DOL) start-up possible (start-up without any prior static cool-down, therefore no gas needs to be flared)
- Compact design, less space required, less weight, minimum equipment count, less spare parts, minimized footprint
- The IGV concept for cryogenic BOG compressor applications is well backed-up by references
- Largest turndown (highest process flexibility)
- Lowest maintenance

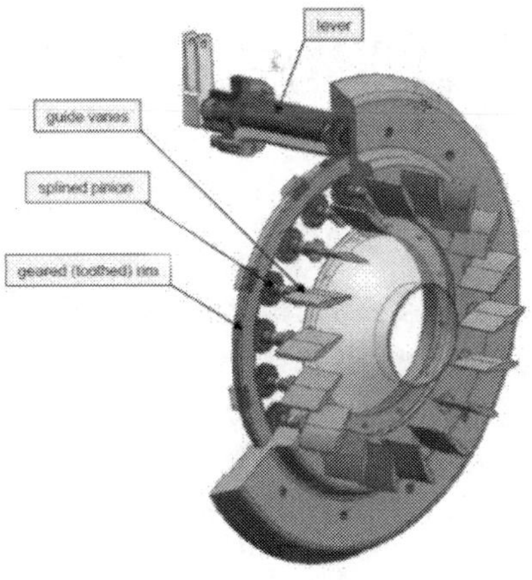

Fig. 7 variable IGV unit

This design was first executed in 1972 for the first single shaft compressor train supplied for BOG service (Brunei) and in 1992 the first time also for cryogenic (-160 degree C) temperatures (Qatar). Since then, this design has become industrial standard and today all new BOG single shaft centrifugal compressor trains have being equipped with this advanced technology.

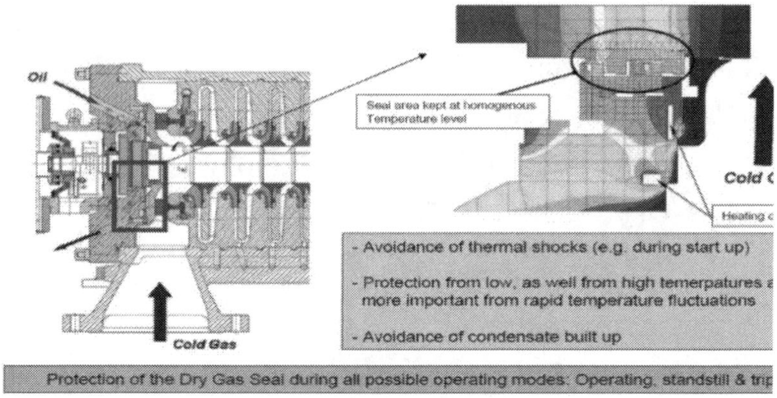

Fig. 8 heated/cooled dry gas seal carrier

Due to the nature of the BOG service the turbo compressor train is periodically exposed to very low suction temperatures (even down to -170 degree C) as well as to ambient and even high discharge temperatures during start up, compressor run in, or performance test. In order to eliminate or minimize thermal tension/stress for the dry gas seals caused by the extremely low suction temperature and periodically occurring temperature fluctuations (e.g. start/stop) resulting out of the different operation modes it is essential to keep the seal area at an almost constant temperature level during all possible operating modes. The required constant temperature level is achieved by providing oil heated/cooled seal gas carriers (see Fig. 8, the oil heated/cooled seal gas carrier is marked in yellow colour on the left hand side drawing).

The heating oil is split from the regular lube oil flow and the fluid is circulating around the dry gas seal cartridge in seal-welded channels inside the compressor seal carrier (casing cover) forming a thermal shield. This design was first executed in 1991 for the first single shaft compressor train supplied with dry gas seals for cryogenic temperature BOG service. Since that time all new single shaft BOG centrifugal compressors provided with dry gas seals are equipped with oil heated/cooled seal carriers (in 2013 total 83 compressor casings) ensuring highest reliability (safety) and start-up availability, long-life service periods and lowest operation, inspection and maintenance cost.

Conclusions

World-scale BOG reliquefaction has become reality. Linde Engineering has successfully developed a new product for the LNG industry. Siemens contributed with innovative compressor and gas turbine designs, which are proven in service and are highly respected worldwide. The first plant, which is based on this novel concept is under execution for Malaysia LNG and is scheduled for start-up in less than two years[6].

References

[1] Tariq Shukri, Michael Barclay, Single mixed refrigerant process has appeal for growing offshore market, LNG Journal, June 2007, pp.35-37

[2] Patent pending

[3] e.g. US patent 6,334,334

[4] Patent pending

[5] Development of Medium-sized Gas Turbines for LNG Applications, Gudmundsson B., Larfeldt J. , Troger C., Cox G. 16th International Conference & Exhibition on Liquefied Natural Gas, 18-21 April 2010, Oran, Algeria

[6] http://www.the-linde-group.com/en/news_and_media/press_releases/news_121126.html

(89b) Tuning and Optimization of Compressor Controls Using Dynamic Simulation

Sanjay Ganjam and Mohit Thakur, Process Engineering, Bechtel OG&C, Inc, Houston, TX

Compressor systems are critical and integral part of an LNG plant for safe, productive and cost effective operations. Robust control design and strategies are essential for stable operation and for protection of machines during abnormal operations. With advancements in dynamic modeling and its use in designing, testing and troubleshooting of plant operations it is important to apply it in designing and testing of control strategies of compressor systems.

In this paper we present and discuss a case study based on a high fidelity dynamic simulation model integrated with CCC emulator for centrifugal compressors. Integration of dynamic model with CCC controls eliminates emulated controllers in the model and brings the actual controller functionality. The integrated dynamic model augments the capability and accuracy of results in dynamic simulations.

Placement of SCL is done by balancing two different objectives – placed to the right of Surge Line for better controllability and as close to Surge Line as possible for production. Location of the SCL is usually determined at site during startup. Purpose of this work is to demonstrate how to achieve this at the EPC phase of the project using dynamic simulation model.

A high fidelity dynamic model gives a good representation of the plant leading to pre-determination of Surge Control Line (SCL), adequacy of SCL for compressor control, pre-tuning of the compressor controllers, check the adequacy of the limit controls.

Opportunity of compressor control tuning at site is less due to risk and complication of compressor surge. Therefore the final controls are usually sub-optimal. Main benefit of performing this activity at EPC stage of the project is to achieve optimal compressor control with minimal compressor surge tests at site during startup.

AIChE 2013 Spring Conference

Topical 6: 13th Topical Conference on Gas Utilization – LNG Simulation and Control

Considerations in Designing a Closed Loop Heating Medium System

Benjamin Gross
Process Engineering
Bechtel Oil, Gas & Chemicals, Inc.

Debby Sielegar, P.E.
Process Engineering
Bechtel Oil, Gas & Chemicals, Inc.

Abstract

A closed loop heating medium system using synthetic heat transfer fluids, such as hot oil, has become a desirable alternative to the conventional boiler feed water / steam in the oil, gas and petrochemical industry. Whether it is for onshore or offshore applications, such systems are reliable, low-maintenance and can be implemented at relatively low cost.

Closed loop heating medium systems can be designed using various configurations depending on the process heat demand, availability of waste heat recovery, desired process control schematic and system reliability, and other operational preferences. In an LNG plant, several factors can affect the required process heat input requirement, such as the feed gas sources and quality, operating conditions, and operating objectives. Waste heat recovery improves the overall plant thermal efficiency by utilizing the waste heat from the exhaust of refrigeration gas turbines or power generation units. While waste heat recovery reduces fuel consumption, increases LNG production and environmental emission benefits, and therefore the NPV of the plant, other factors such as plot space, local environmental regulation differences, and operational preferences may lead to other simple alternative closed loop heating medium system utilizing conventional fired heater.

These various system configurations along with equipment / mechanical design, process control and reliability, and relief system design challenges for a closed loop heating medium system integrated with or without waste heat recovery are discussed in this paper.

1. Introduction

At the dawn of macro-scale natural gas liquefaction, it was not uncommon to utilize direct fired heating to meet the plant's heat requirement. However, process and equipment technology has evolved, design capabilities have improved, and environmental regulations regarding emissions have become more stringent and concerned. For all of these reasons, along with our desire to maximize overall plant thermal efficiency, and minimize capital and operating costs, the methods by which heating demands are met in an LNG plant have become increasingly important.

The logical next step was to implement a robust heating system that could satisfy both the needs of a more efficient process and stricter environmental regulations. A closed-loop heating medium system integrated with a fired heater and/or a Waste Heat Recovery Unit (WHRU) installed from the exhaust of refrigeration gas turbines or power generation units is often chosen to cater to the needs of increasingly efficient processes. Such systems are stable and flexible, and thus respond favorably to process upsets and turndown scenarios.

So we turned to the conventional boiler feed water / steam system, a widely and commonly used system for all industrial applications. A typical boiler feed water / steam system integrated with a waste heat recovery unit is shown in Figure 1 below.

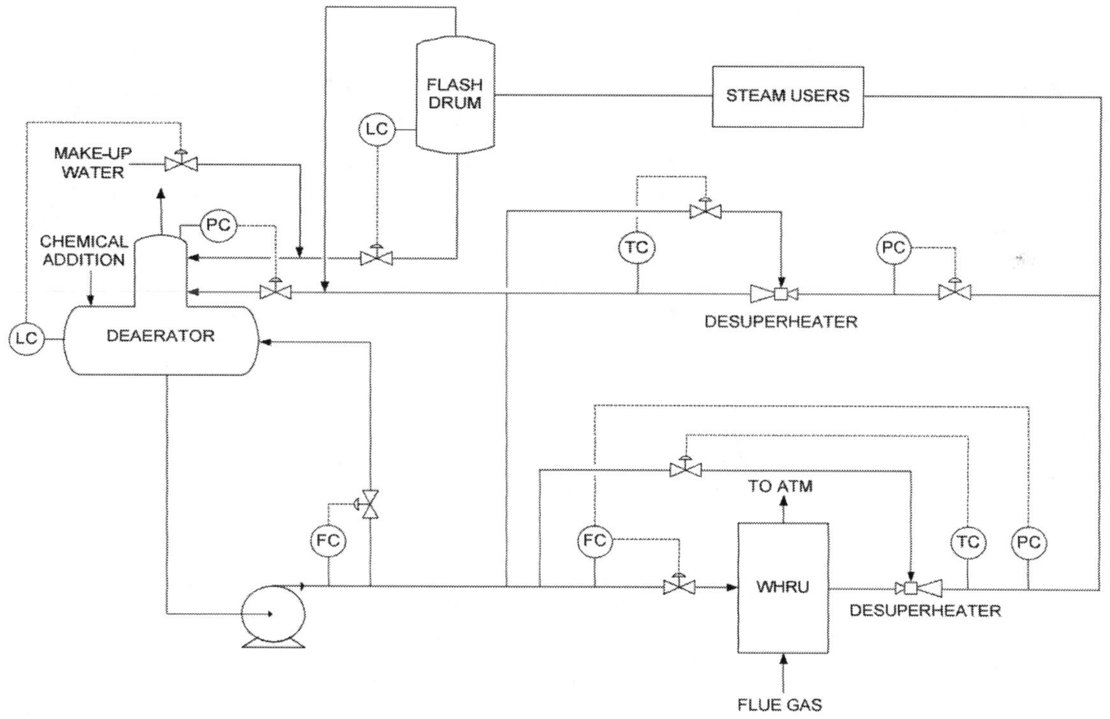

Figure 1. Closed Loop Boiler Feed Water / Steam System Integrated with Waste Heat Recovery Unit

Boiler feed water from a deaerator enters the Waste Heat Recovery Unit (WHRU) coils to generate MP steam. To maintain MP steam supply temperature and pressure, temperature control utilizing a desuperheater and pressure control regulating incoming boiler feed water flowrate are normally applied. LP steam from MP steam let-down and condensate return from MP steam users are routed back to the deaerator. The LP steam is used to strip the condensate return to prevent build-up of impurities.

As a heat transfer medium, steam is ideal. It has a relatively high heat transfer coefficient, unlimited fluid life, and is economical and readily available. However, the boiler feed water / steam system is anything but ideal. The excerpt below from "Waste Heat Recovery Considerations in LNG Plants" by Wiles, D., Kwan, S., and Rajkumar, V. [2] summarizes the disadvantages of such systems best:

> *"The preparation and production of steam requires significant monitoring and treatment of the boiler feed water to remove minerals which could cause deposition and therefore fouling in heat exchangers and pipes. In addition, steam systems with cogeneration require piping and equipment rated for high pressures. In addition, there are high maintenance costs for the removal of scale, the repair of steam traps and leaks."*

The industry has since answered the call for an alternative heating medium, such as hot oil, a synthetic hydrocarbon-based heat transfer fluid, lube-oil based heat transfer fluids, glycol solution, and many others. This paper primarily focuses on hot oil system.

While hot oil has a lower heat transfer coefficient and is more hazardous than steam, and is subject to thermal degradation, it has several very favorable properties: low vapor pressure, high boiling point, low pour point, and is non-compressible. Furthermore, a hot oil system operates at a relatively low pressure and runs at relatively lower operating and maintenance costs. And because hot oil system does not rely on latent heat transfer by condensation, such a system is efficient and reliable with proper design and temperature applications; thus, hot oil is becoming a more desirable heating medium in the oil, gas, and petrochemicals industry.

Factors to be considered in designing a heating medium system may include system configuration selection, equipment / mechanical design, process control and reliability, and relief system design. These factors with focus on LNG plant application are to be discussed in this paper. In an LNG plant, typical hot oil users may include feed gas preheater, amine regenerator reboiler, fuel gas heater, heavy removal reboiler(s), and many others.

2. System Configurations

There are various closed loop hot oil system configurations. Each system configuration is designed to fit the plant requirement and conditions, such as process heat demand, availability of waste heat recovery, desired process control schematic and system reliability, and other operational preferences. Each of these system configurations are outlined as the following:

2. 1. Closed Loop Hot Oil System Integrated with Fired Heating

Fired heating is inherently unsafe, and with regard to fuel consumption and high system design temperatures, it can be expensive given the number of services in an LNG plant (i.e. feed gas heater, molecular sieve dehydrator regeneration gas, amine regenerator reboiler, fuel gas heater, fractionation reboilers, etc.). However, a hot oil system still requires a primary source of heat, and a fired heater is a suitable source. Figure 2 below shows a closed-loop hot oil system integrated with fired heating.

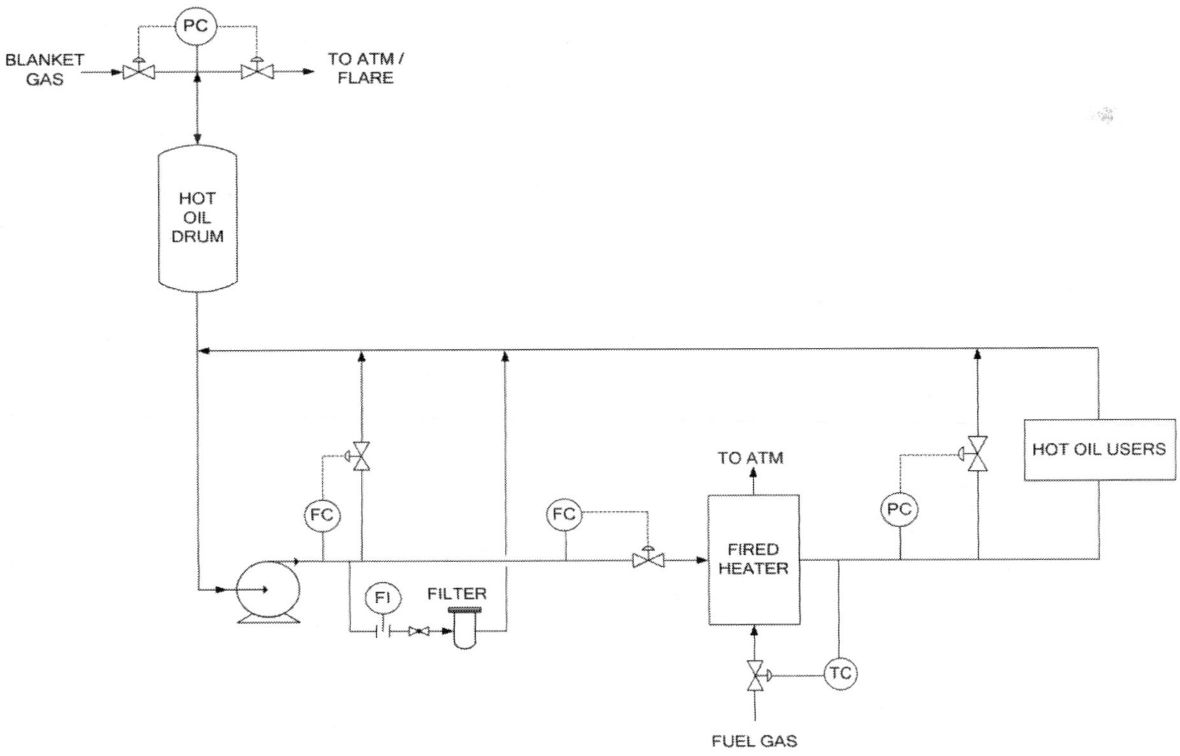

Figure 2. Closed Loop Hot Oil System Integrated with Fired Heating

If it looks like a closed-loop hot oil system is an oversimplified boiler feed water / steam system, that's exactly what it is. The advantageous physical and chemical properties of hot oil – no chemical treatment and no phase changes – allow for this simplicity.

The hot oil supply temperature is directly controlled by regulating the fuel gas flow rate to the fired heater. Note that fuel oil can also be used as a source of ignition for the fired heater. The bypass pressure control is used to provide system balance. If multiple fired heaters are installed in parallel, hydraulics must be symmetrical, and the hot oil flow rate through each unit must be individually controlled to ensure even distribution of the heating medium. The flow control shall have a minimum set point to protect heaters.

Fired heaters typically come equipped with O_2 and combustibles (i.e. CO, H_2) analyzers, which are used to monitor the completeness of combustion taking place inside the heater.

A hot oil system shall include a filter, which is implemented to remove contaminants and/or scales that result from fluid thermal degradation. Normally, these filters are designed to remove 98% of scales greater than 5-10 microns. There are two types of filter installations, which are in-line and side stream. In-line filters are used only with positive displacement pumps, while side stream filters are used with centrifugal pumps. Considering hot oil recirculation rate is high, side-stream filters are normally used. Flowrate through the side-stream filter is typically in the range of 3 – 10% of the recirculation design flowrate. [3]

Also, a hot oil drum is normally located at the highest point in the system. It serves as a main venting point of the system and provides adequate volume for thermal expansion. This vessel must be blanketed by a *dry* inert gas such as nitrogen, fuel gas, or natural gas in order to prevent the fluid from coming into contact with oxygen or moisture. Oxygen tends to accelerate heat transfer fluid degradation, while moisture ingress may lead to sudden pressure surges upon vaporization of the water. Therefore, if nitrogen is used, it is imperative that oxygen and moisture either be removed or reduced to a negligible concentration. The hot oil drum can be a horizontal or a vertical vessel. Hot oil return line can be routed back to the drum or to the suction of the hot oil pumps. Also, a tank, make-up unit, or sump drum is usually provided as part of the hot oil system for provision of filling at least one heat transfer service during pre-commissioning after maintenance shutdown.

2. 2. Closed Loop Hot Oil System Integrated with Waste Heat Recovery and Fired Heating

Whether taken from the front-end or the back-end, fuel gas consumption is an additional operating cost, and a hot oil system integrated with fired heating does not necessarily consume less fuel than an LNG facility that uses fired heating only. So how can we reduce fuel gas consumption, thereby increasing overall plant thermal efficiency? The answer is waste heat recovery.

Waste heat recovery is not a new concept. It has been used in steel mills, power plants, and refineries for decades. In an LNG plant, heat from refrigeration gas turbine and/or power generation exhaust gas can be recovered via coils in the turbine exhaust stacks that service hot oil and even regeneration gas. Figure 3 below shows a hot oil system integrated with both fired heating and waste heat recovery.

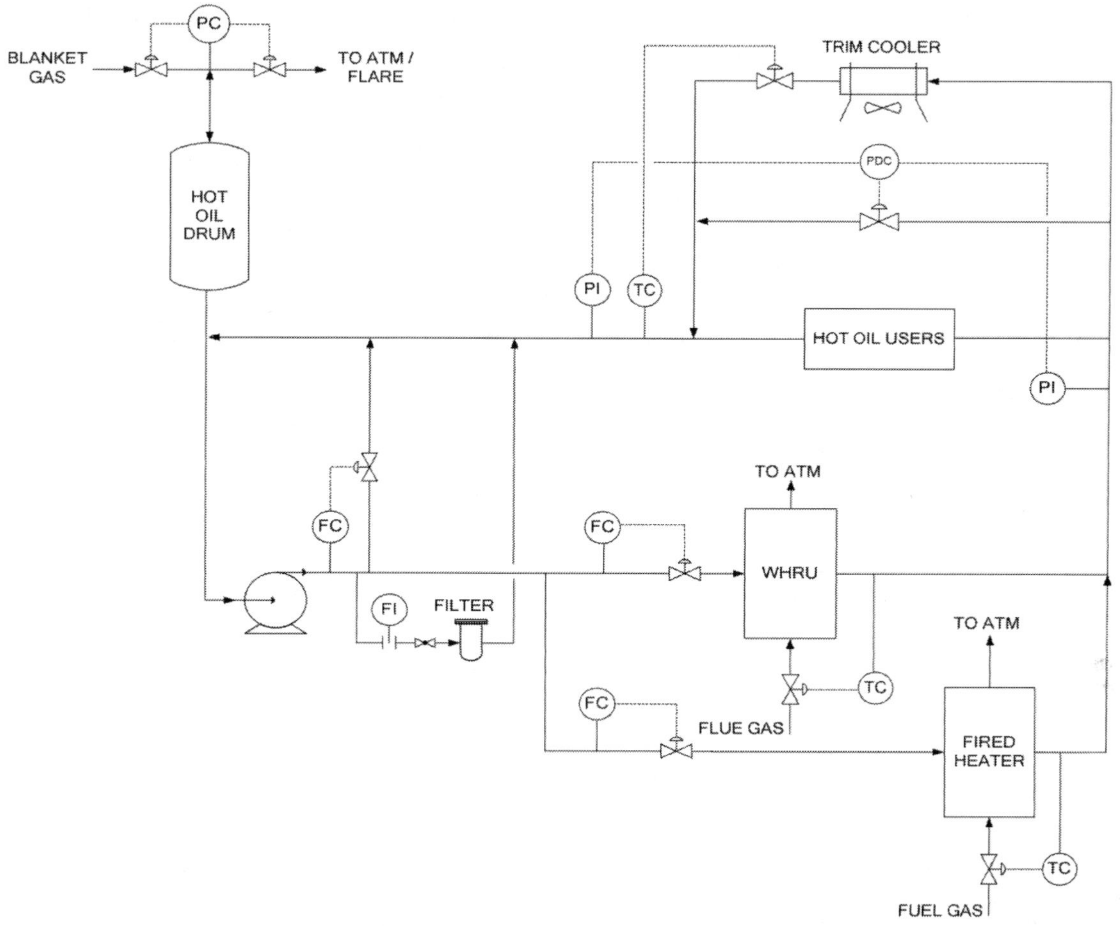

Figure 3. Closed-loop Hot Oil System Integrated with Waste Heat Recovery and Fired Heating

In this scheme, fired heating is typically only required for regeneration gas and during start-up and stand-by, when refrigeration compressors are offline. However, fired heating can be used to supplement the process heat demand if there is insufficient net recoverable heat.

The Waste Heat Recovery Unit (WHRU) process outlet temperature is normally controlled by regulating the flue gas flow rate between the main and bypass exhaust stacks via a damper valve. As damper valves are not as reliable as regular control valves, a certain hot oil bypass flow around the waste heat recovery unit may also be established under temperature control to provide flexibility and robustness in controlling the hot oil supply temperature, considering variability in the gas turbine and power generation turbine exhaust. Again, if the system requires that multiple WHRUs be installed in parallel, each unit must be hydraulically balanced.

Another important consideration in WHRU design is acid gas condensation, which may lead to corrosion in the exhaust system. The WHRU vendor shall consider the stack gas dew point when designing the coils taking into account all operating cases and apply a temperature approach well above the dew point for the worst case. Nevertheless, it is typical to monitor stack gas temperature in the waste heat recovery unit. For environmental regulation reasons, exhaust sample ports are provided on both the WHRU stack and bypass stack, from which exhaust gas can be tested.

In addition, turndown requirements and/or process upsets are a key consideration. The other major difference between this scheme and the one depicted in Figure 2 – aside from the heat source obviously – is the addition of a trim cooler, which helps to maintain a constant hot oil return temperature, especially during turndown. Furthermore, variance in hot oil supply and return pressures associated with fluctuation in process heat users and with turndown are managed by using pressure differential control. Finally, the trim cooler requires a minimum continuous hot oil flow rate in order to permit rapid ramp-up without risk of thermal stress damage. Note that air cooler fans may be turned off completely or turned down to achieve the desired minimum flow, when required.

2. 3. Closed Loop Hot Oil System Integrated with Waste Heat Recovery Unit (WHRU)

A hot oil system integrated with waste heat recovery only is the final stage in the evolution of this closed-loop heating medium system. Figure 4 below shows a closed-loop hot oil system integrated with waste heat recovery only.

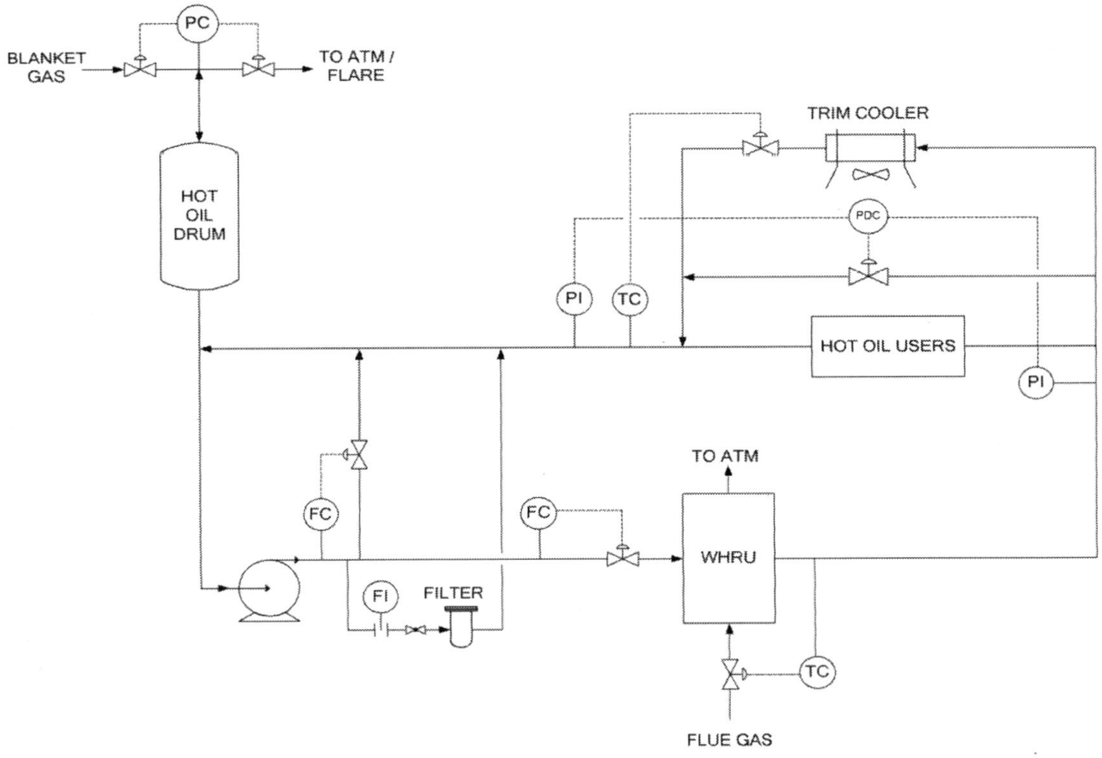

Figure 4. Closed-loop Hot Oil System Integrated with Waste Heat Recovery

The differences between this schematic and the one illustrated in Figure 3 – aside from the absence of a start-up fired heater – are not obvious. In both cases the waste heat recovery unit supplies 100% of the process heat requirement *during normal operation*, and the control strategy is generally the same. But removal of the start-up heater has introduced a new dynamic: If there are no power generation units, how do we heat the heating medium that is required for acid gas removal that must take place prior to liquefaction, which utilizes gas turbines whose exhaust heats the heating medium? It is important to consider the start-up procedure when applying this particular process scheme.

Inclusion of the regeneration gas coil in the hot oil WHRU has been considered. Regeneration gas heating is an intermittent service, and as such, it will introduce fluctuation in the WHRU duty requirement. But with proper mechanical design and coil module arrangement, this strategy further improves plant overall thermal efficiency and reduces installation costs associated with multiple waste heat ducts.

It is also important to evaluate the process heat demand for various design and rating cases, to ensure that there is sufficient recoverable heat and appropriate design approaches. Factors impacting the net recoverable heat include turbine engine degradation, fouling, erosion, radiation heat loss, and gas leakage.

2. 4. Direct Heating

When a heater is only required by a single unit operation, direct heating may be applied instead of a closed loop heating medium system due to its simplicity. For example, assuming only the amine regenerator reboiler requires heat input and there is no source of waste heat (i.e. all refrigeration compressors are electric driven), direct heating is often chosen as a preferred heating method for the reboiler. In this case, closed loop heating medium system is no longer required and a fired heater can be installed. Another example is molecular sieve dehydration regeneration gas heater application. It has been a common practice that the regeneration gas is directly heated via a separate coil in the waste heat recovery unit or via an independent fired heater, as heating medium supply temperature is typically lower than the required regeneration gas temperature.

3. Waste Heat Recovery Unit versus Fired Heating

Waste heat recovery might seem like the optimal choice when considering the reduction in environmental emissions and *potential* improvement in plant overall thermal efficiency; however, additional consideration and evaluation are needed during conceptual design phase. For instance, withdrawing fuel gas from the back-end can "...provide a purge for unwanted inert gases, such as nitrogen." [2] Nitrogen can be present in the feed gas at various concentrations. It accumulates by entering the refrigeration circuit via flash and boil-off gas and thus "...consume[s] unnecessary compression power, reducing overall plant thermal efficiency." [2] LNG product and fuel gas specifications may require that a Nitrogen Removal Unit (NRU) be installed. Also, if there are no or limited exhaust sources – all rotating equipment is electrically powered and all electricity comes from the grid – then waste heat recovery would not be an option.

Also, the extensive ductwork associated with the waste heat recovery unit along with the additional equipment (i.e. trim cooler) and piping require an higher installation cost and more plot space (this is an especially important consideration during a retrofit). Finally, longer exhaust ducts means more head loss, which means increased turbine heat generation and therefore less power available. [2]

Fired heating may be simpler from a controllability standpoint, but that controllability comes with increased environmental emissions and loss of feed and/or product. As discussed above, fired heating control scheme does not require constant return temperature to the hot oil drum, thus trim cooler is not required. Fired heater's fuel gas control valve is more robust in controlling supply temperature to the hot oil users than waste heat recovery unit's damper valve.

4. Relief System Design Challenges

The relief scenarios to consider when designing a closed-loop hot oil system are for the most part standard (i.e. pump shut-off, abnormal heat input, fire, etc.) with one exception: exchanger tube rupture. A tube rupture that causes leakage of hot oil to the process side may lead to loss of hot oil inventory. On the other hand, if an exchanger is servicing a high pressure vapor (i.e. feed gas or high pressure fuel gas) on the tube-side, and a tube ruptures, the result, which is analogous to water hammer, can be catastrophic. Subject to the piping component design, transient analysis can be used to determine which mitigation strategy is appropriate in conjunction with overpressure protection. For example, provision of fast acting valves at the inlet and outlet of the exchanger can be used to isolate the exchanger in the event of a rupture, preventing translation of a pressure wave to downstream equipment and piping.

When the hot oil recirculation is routed directly back to the drum, the drum should be sized adequately for vapor disengagement in the event of an exchanger tube rupture at one of the hot oil users. On the other hand, when the hot oil recirculation is routed to the hot oil pumps suction, provision of a high pressure interlock can be used to shut down the hot oil system immediately, thereby protecting the pumps from cavitation.

5. Conclusions

Just as with any process in the oil, gas, and petrochemicals industry, safety, cost, control, maintainability, and operability are the factors which should be considered for the design of a closed-loop heating medium system. And just as with any process, there is no compromise on safety and operability.

Fired heating requires fuel consumption that is ultimately a loss of revenue – lost feed, product, energy, and potential carbon tax benefits – but fuel consumption can help reject inert gases that have accumulated in the liquefaction system, which may require removal by other means (i.e. NRU). Waste heat recovery *can* increase overall plant thermal efficiency and the NPV of the facility, but this comes with increased installation cost, loss of turbine power, and a larger plot space requirement.

Fired heating can be dangerous but has simple, robust control. Waste heat recovery does not come with the dangers associated with an open flame, but the system configuration, control, and start-up are as a result more complex.

The owner's priorities, environmental emission regulations, economic evaluations, and ongoing turbine driver and waste heat recovery developments must be considered in the design of a closed-loop heating medium system for an LNG facility.

6. Acknowledgement

The authors would like to thank Stanley Kwan for his constructive inputs and David Messersmith for reviewing the paper.

7. References

1. Therminol, Liquid Phase Systems Design Guide, Pub. No. 7239128D, 2008
2. Wiles, D., Kwan, S., Rajkumar, V., Waste Heat Recovery Considerations in LNG Plants, AIChE Spring Conference, April 2012
3. MultiTherm Heat Transfer Fluids, TechTeam Report Volume 2 Issue 1, March 2005

8. Biography

Benjamin Gross joined Bechtel Oil, Gas, and Chemicals, Inc. as a Process Engineer in January of 2012 subsequent to completing his Bachelors of Science in Chemical Engineering from the University of Texas at Austin. As a Process Engineer at Bechtel OG&C, Inc., he has completed assignments in utility system design, lube oil, and LNG technology.

Debby Sielegar, P.E. is a licensed engineer in the state of Texas. She completed her bachelor's degree in Chemical Engineering from Purdue University in 2007 prior to joining Bechtel OG&C, Inc. While working for Bechtel as a process engineer, she pursued her Master's degree in Chemical Engineering from University of Houston and completed the degree in 2011. In Bechtel, she has satisfactorily completed assignments on various projects such as lube oil refinery, LNG liquefaction, offshore FPSO crude production, and delayed coking.

(89d) How Glycols Affect the Acid Gas Removal Process

Georg Sieder, BASF SE, Ludwigshafen, Germany, Torsten Katz, Gas Treatment Process Technology, BASF SE, Ludwigshafen,, Germany, Justin Hearn, BASF, Ludwigshafen, Germany and **Jay Habayeb**, BASF Corporation, Houston, TX

In some natural gas applications glycols, such as MEG, DEG or TEG, are added into the pipeline or the gas conditioning process, either as hydrate inhibitors or for dehydration purposes to protect downstream pipelines. These substances have high boiling points and are not supposed to be carried over into the downstream process. Proper separator design shall avoid carry-over of these substances into the Acid Gas Removal plant.

Some amine plants however report glycol build-up in the amine solution over time, which can lead to performance losses and ultimately to unexpected plant shutdowns. This paper deals with the sources and consequences of glycol build-up in the amine unit and delivers some understanding of how the glycol content affects the acid gas capture performance. It provides recommendations on how to control the full acid gas capture performance over time and how to avoid operational surprises.

13 AIChE
Spring Meeting, San Antonio, TX

CONDENSATE STABILIZATION: HOW TO GET THE MOST FOR YOUR MONEY

Mahdi Nouri, John Rizopoulos
Process Engineering & Technology
CH2M HILL Energy & Chemicals
Houston, TX

Prepared for Presentation at
American Institute of Chemical Engineers
2013 Spring Meeting
9th Global Congress on Process Safety
San Antonio, Texas
April 28 – May 1, 2013

CONDENSATE STABILIZATION: HOW TO GET THE MOST FOR YOUR MONEY

Mahdi Nouri, John Rizopoulos
Process Engineering & Technology
CH2M HILL Energy & Chemicals
Houston, TX

ABSTRACT

Condensate stabilization is one of the most common operations in the gas processing world. The principal considerations affecting the performance of the plant when debottlenecking or designing new condensate stabilization units are the configurations, product quality, energy saving, and equipment size. Sound process design can significantly reduce capital investments and operating costs. The purpose of this paper is to review and illustrate the fundamentals of this process, including flow-sheet configuration and key operating parameters, and to show how they affect the plant performance, including energy consumption and economics.

INTRODUCTION

The incoming gas from offshore or onshore pipelines carries condensates generated during transmission by heat transfer and pressure drop. The two phases are typically separated in the slug catcher or inlet separator. The gas stream is treated before feeding to the cryogenic system for extracting NGL. It is first routed to the acid gas removal unit, then dried in a Glycol unit, and polished in a molecular sieve dehydration unit, (Figure 1). The recovered hydrocarbon condensate is sent to the condensate stabilization unit for vapor-pressure reduction. Stabilized condensate can be stored in floating roof storage tanks, and safely transmitted in pipelines. The stabilized condensate, typically C5+, is sold as final product or as petrochemical feed-stock to olefins and aromatic plants. Figure 1 illustrates how the vapor from a condensate stabilization unit is compressed and recycled to the inlet feed separator gas stream before entering the gas treating system. The presence of contaminants (water, salts and suspended solids) in the slug catcher can cause problems on the condensate stabilization unit performance, including long-term maintenance issues. Water carryover from the slug catcher and hydrocarbon/water separator can lead to off-spec stabilized condensate.

This study (1) evaluates six alternative process configurations and key operating parameters for meeting the stabilized condensate specifications and saving energy, (2) develops optimized utility consumption for the alternatives, (3) develops comparative capital and operating cost appraisals, and (4) presents the optimum alternative in terms of process efficiency and economic analysis. An appropriate process simulator with the right thermodynamic package, optimum flow sheet configuration and equipment sizing is a very important tool for achieving the desired results.

Figure 2 illustrates the integrated engineering and economics work flow. In this paper, HYSYS and Process Economic Analyzer (APEA) is used for simulation and economic analysis. This Aspen Technology, Inc., software is very effective in carrying out comparative evaluation of process designs. Based on a HYSYS flow-sheet simulations, APEA determines the capital expenditures, operating costs, and finally the profitability of the proposed designs. APEA has a system which links process simulation programs to the estimating tool, though a "process engineer's window" to the "Icarus Estimating engine". It is used to (1) extend the results of process simulation, (2) generate sizing and cost estimates for process equipment, (3) perform preliminary mechanical designs, and (4) estimate materials and labor costs, indirect costs, total capital investment, and (5) the EPC schedule. [1]

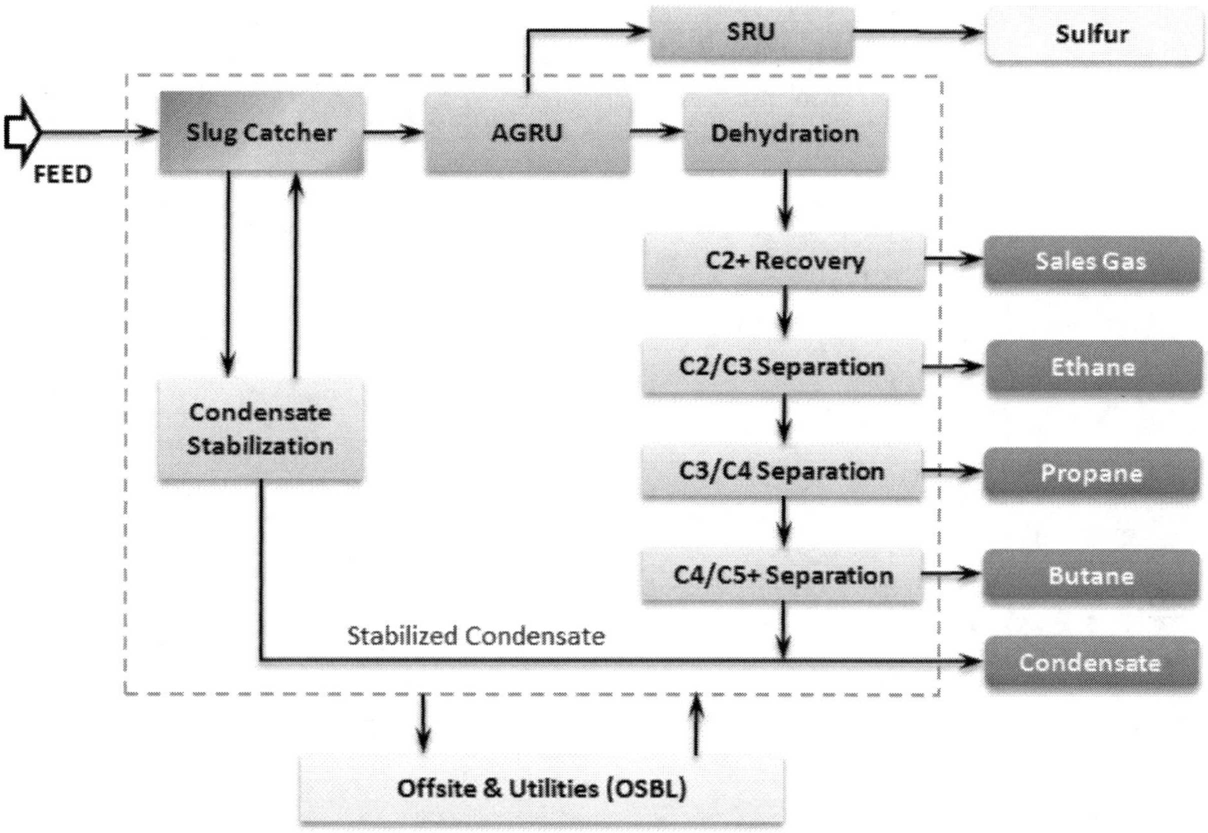

Figure 1 – Typical Block Flow Diagram for a NGL Recovery Plant

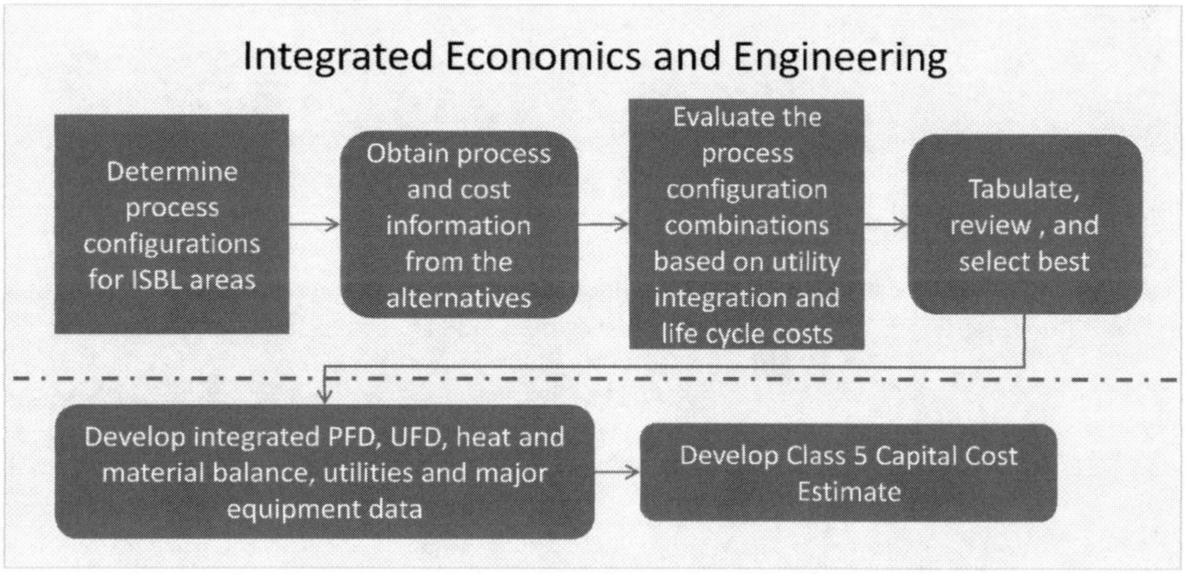

Figure 2 – Process Configuration Screening Work Flow Process

BASIS of DESIGN

Feed characteristics

The single-phase liquid feed properties used for the case studies are shown in Table 1.

Table 1 – Unstabilized Condensate Characteristics

Component	Molar Flows (lbmole/hr)
H_2O	1449.7
N_2	16.1
CO_2	63.5
H_2S	0.0
C1	1317.2
C2	271.3
C3	218.2
i-C4	99.6
n-C4	165.6
i-C5	146.8
n-C5	126.4
C6+	2330.0
Total	6204.3
Pressure, Psig	600.0
Temperature, °F	90.0
RVP, Psia	250.0

Products specifications

The products of condensate stabilization unit are:

- Stabilized condensate meeting the required specifications (Table 2),
- Off-gas compressed and recycled to the gas treating system unit.

Table 2 – Stabilized Condensate Specifications

Component	Specification	Test Method
RVP, Psia	<8.5	ASTM D-323
Color	Not Less than 25 Saybolt	ASTM D-156
Water Content	Nil	

Battery limit conditions

The operating conditions of the major streams to the condensate stabilization unit are:

Table 3 – Battery Limit Conditions

		Operating Conditions	
	Description	Pressure, Psig	Temperature, °F
Feed	Liquids from slug catcher	600	90
Products			
Stabilized condensates	To floating roof storage tank	Atm.	130
Off-gas	2nd stage aftercooler outlet	1100	130

Turndown ratio

The Condensate Stabilization unit is designed to operate 40-110% of the normal continuous flow rate.

CASE STUDIES

The objective of this case study is to evaluate different condensate stabilizer configuration alternatives and operating parameters. Table 4 describes the six process configurations compared in this paper. In all alternatives, the stabilizer is operated at pressure of 145 Psig except for Alternative 4, which is operated at 87 Psig. In all studied cases, 10 theoretical-stages of valve trays are used as additional stages appear to offer no further benefit. The following assumptions are made for the economic analysis based on U.S. Gulf Coast:

- Currency Conversion Rate = U.S. Dollar
- Stabilized condensate = $ 102 per barrel
- Gas Price = $ 3.30/1000ft^3
- Steam cost @ 400 Psia = $ 5/klb
- Electricity cost = $ 0.07/KWh
- Plant availability = 355 days per year
- Annual rate of return = 20%
- Plant life = 25 years

Table 4 – Process Configuration Alternatives for Comparison

	Refluxed Column	Non-Refluxed Column	Non-Refluxed LP Column	Feed Preheating	Feed Splitting	Side Reboiler
Alternative 1		X				
Alternative 2		X		X		
Alternative 3		X		X	X	
Alternative 4		X	X	X	X	
Alternative 5		X		X	X	X
Alternative 6	X			X		

Alternative 1: non-refluxed column, no feed preheating, no split feed & no side reboiler

The condensate from the slug catcher is flashed again at lower pressure in the hydrocarbon/water separator. All hydrocarbon condensate leaving the condensate/water separator is sent to the top of the column. The stabilized condensate leaving the column is cooled before storage, which increases the bottom condensate cooler duty. Stabilizer overhead vapor is compressed by a two-stage reciprocating motor-driven compressor. The first-stage discharge gas is cooled down to 130 °F in an air cooler and is then sent to the three-phase separator for removing any condensate and water. Overhead gas is then mixed with gas from the hydrocarbon/water separator prior to feeding the 2nd stage overhead compressor suction drum. Gas is further compressed by the second stage, cooled in the after-cooler, and sent to the acid gas removal unit for further processing. Figure 9 shows the relationship between the inter-stage compressor pressure and the compressor power.

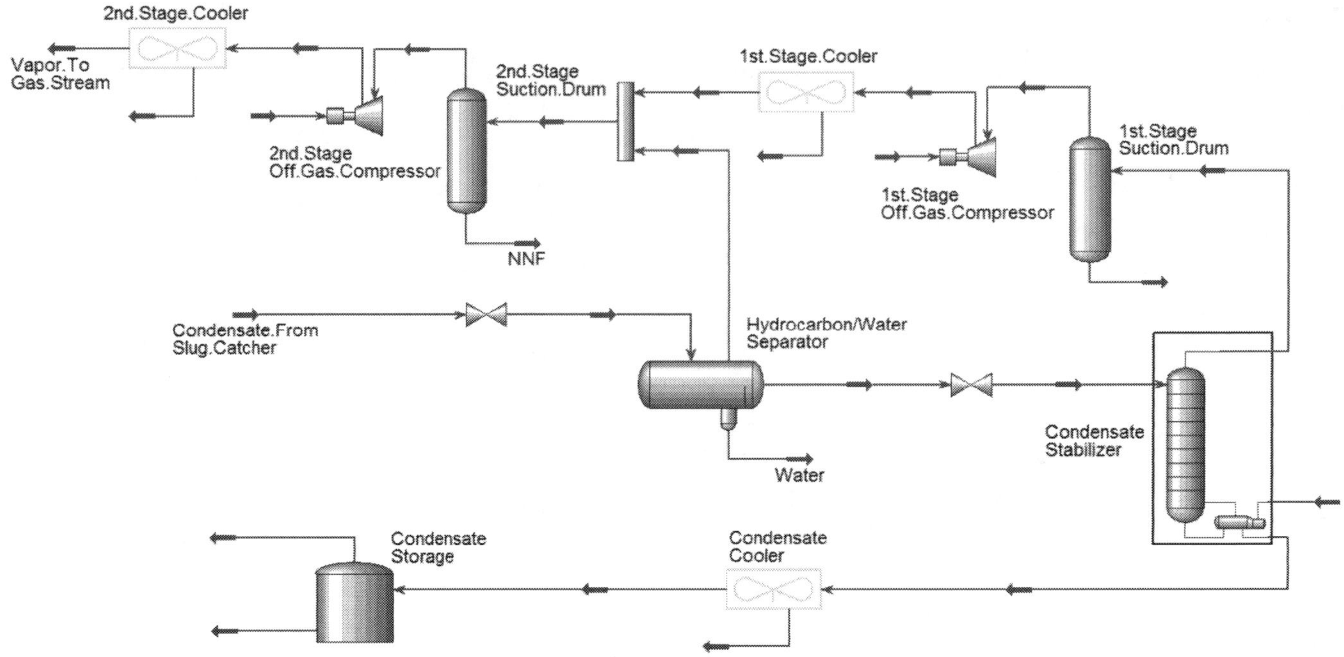

Figure 3 – Alternative 1 Process Flowsheet

Alternative 2: non-refluxed column with feed preheating, no split feed, and no side reboiler

This scheme is similar to Alternative 1. The difference is the presence of a feed/bottom exchanger as a preheating system. The entire hydrocarbon condensate stream from the hydrocarbon/water separator flows through the feed/bottom heat exchanger and enters to the top of the column. The vapor recovery compressors have relatively high power requirements, and reboiler has higher thermal duties. In addition, the bottoms from the 2nd stage suction drum are routed directly to the hydrocarbon/water separator to remove water.

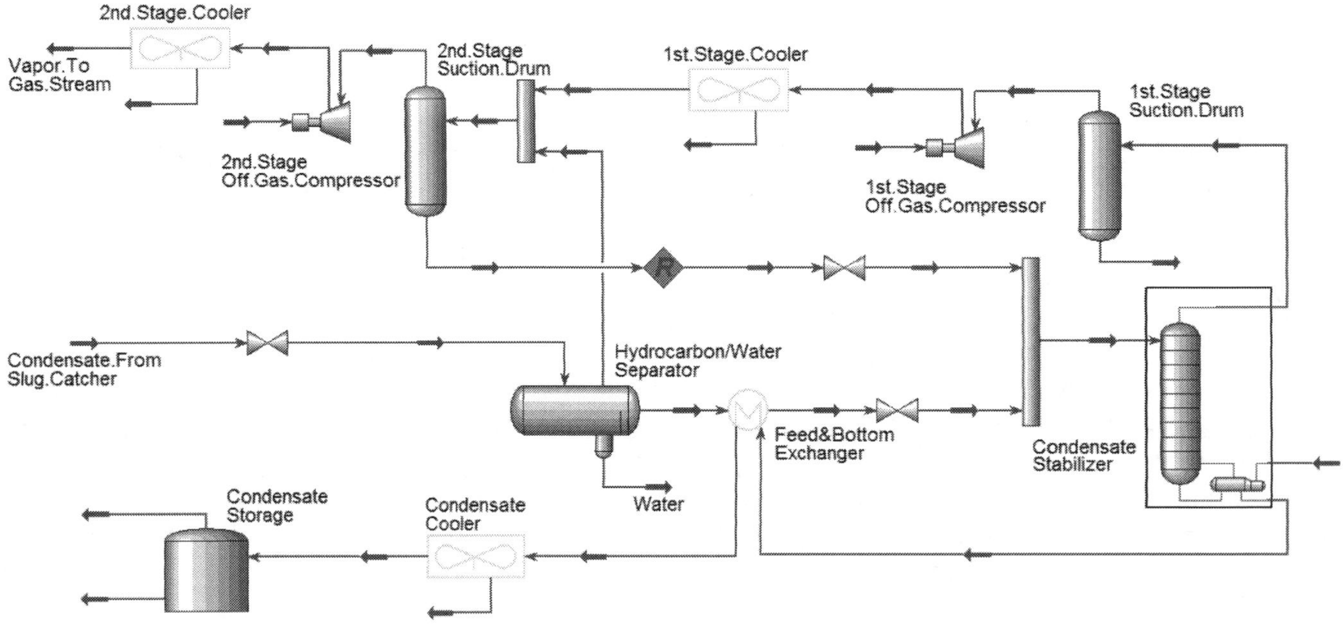

Figure 4 – Alternative 2 Process Flowsheet

Alternative 3: non-refluxed column with feed preheating, split feed, and no side reboiler

The process scheme detailed in Alternative 2 is modified so that the feed is split, with 17% sent cold to the top of the column. The residual feed passes to the feed/bottom exchanger before entering the column at the 4th theoretical stage. Figure 10 illustrates the impact of splitting feed and feed stage location on reboiler duty.

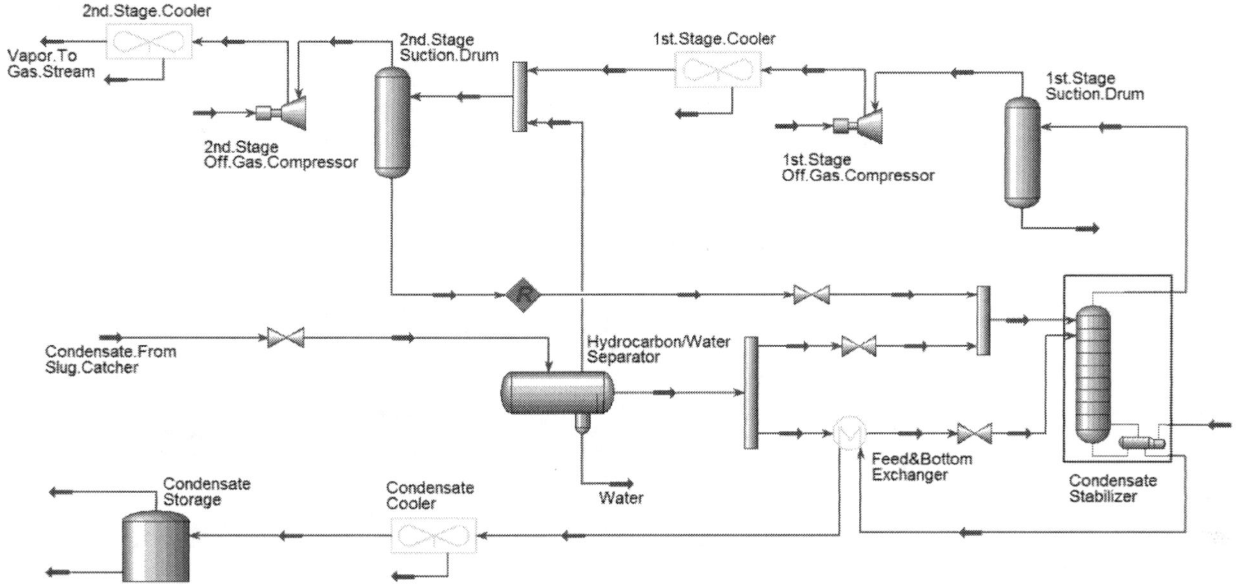

Figure 5 – Alternative 3 Process Flowsheet

Alternative 4: non-refluxed low pressure column with feed preheating & split feed & no side reboiler

This configuration is similar to Alternative 3, except that the stabilizer is operated at lower pressure (87 Psig), which requires an additional recompression stage. In this scheme, the feed is split, with 11% sent cold to the top of the column.

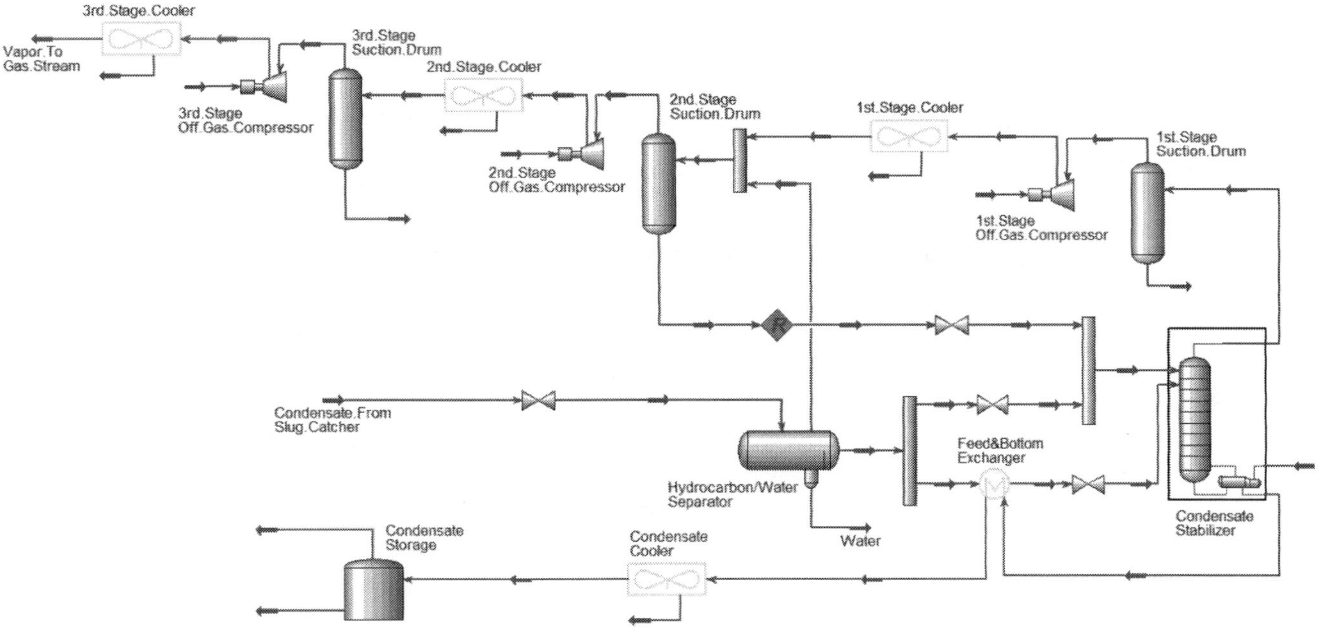

Figure 6 – Alternative 4 Process Flowsheet

Alternative 5: non-refluxed column with feed preheating, split feed, and side reboiler

This scheme of Alternative 4 is modified by the addition of a side reboiler at the 7th theoretical stage. The feed is split, with 10% sent cold to the top of the column. The residual feed is routed through the feed/bottom heat exchanger before entering the column at the 4th theoretical stage.

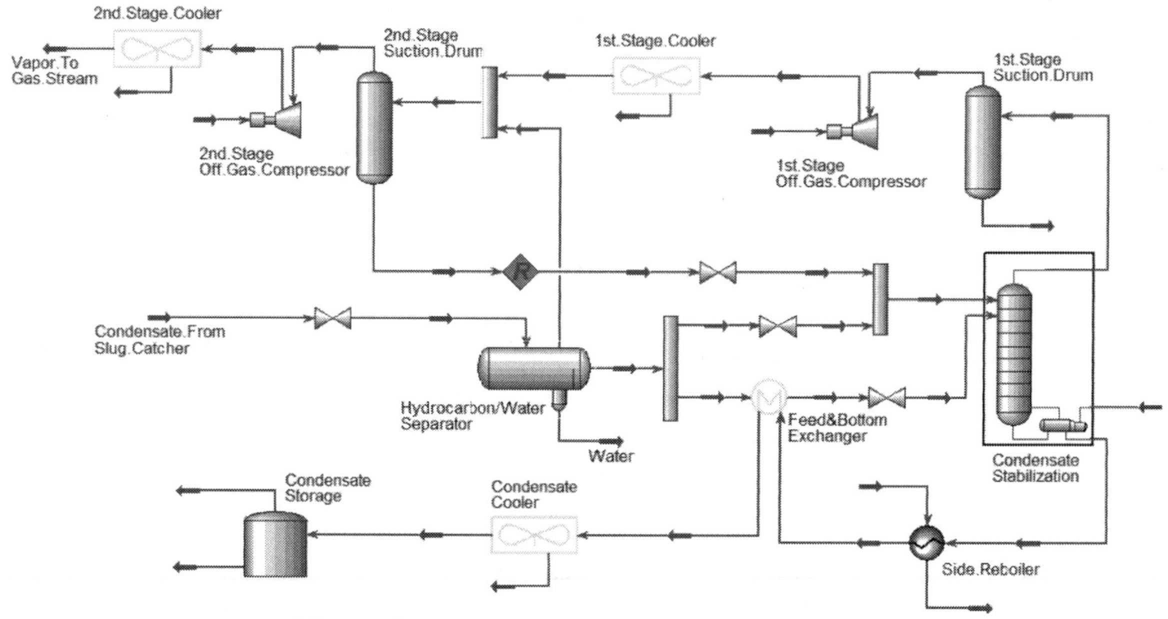

Figure 7 – Alternative 5 Process Flowsheet

Alternative 6: refluxed column with feed preheating, no side reboiler

The condensate leaving the hydrocarbon/water separator is heated using the feed/bottom heat exchanger, and enters the 4th theoretical stage of a conventional distillation column.

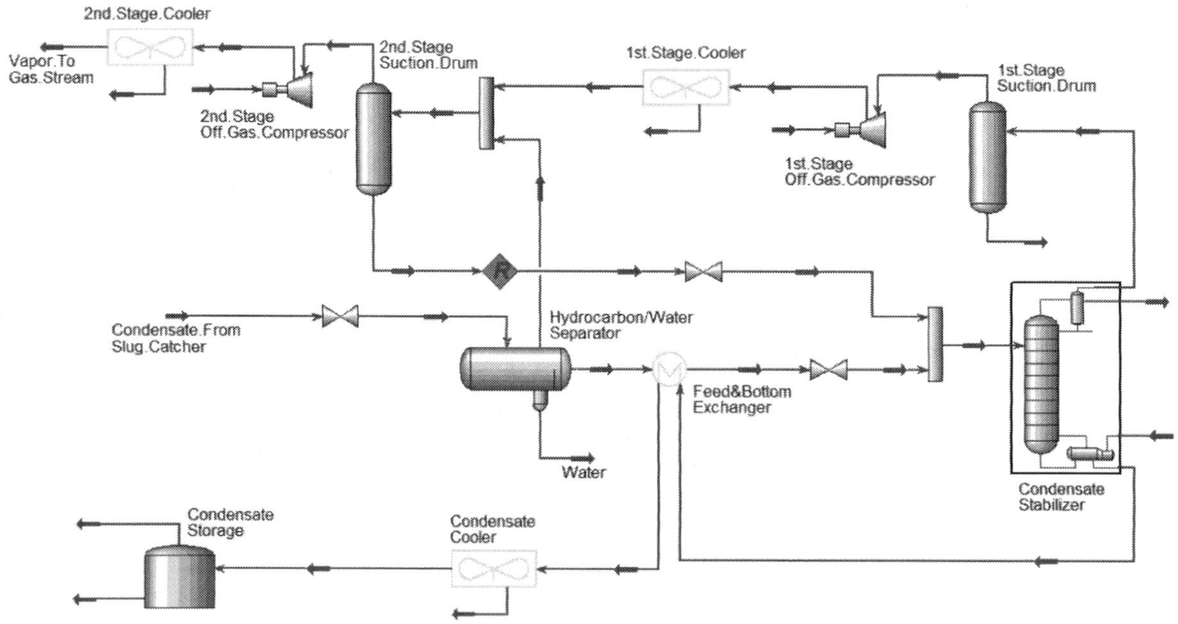

Figure 8 – Alternative 6 Process Flowsheet

RESULTS

Plant designers consider major design parameters for increasing process efficiency and saving energy. The following discussion is limited to the findings of the sensitivity analysis. Sensitivity analyses are performed by investigating and focusing upon various operating parameters, process configurations, and robustness of columns, by observing differences from simulation outputs, and economic analysis outputs. Table 5 presents the optimum alternative in terms of CAPEX and OPEX. Figure 11 illustrates that Alternatives 2 to 6 produce essentially the same stabilized condensate. The provided capital costs are a rough order of magnitude on a U.S. Gulf Coast basis. These estimated costs have been calculated by APEA, and cover engineering, procurement, and installation of process and utility equipment; storage, metering, and loading facilities; plus, construction indirect, spares, engineering, and commissioning. Items such as OSBL costs, owner's costs, land costs, escalation, and contingency are excluded.

Alternative 1 does not have the heat recovery system, and the hydrocarbon condensate with feed operating temperature is sent to the stabilizer. Operational experiences show that the product specifications are achieved with difficulty in terms of water content, because water separation from hydrocarbons is not easy at low temperatures. Figure 11 shows stabilized condensate flow rate, however, the highest reboiler, and condensate cooler duties in comparison to the other cases as shown Figure 13. Figure 9 shows the minimum off-gas compressors power at the lower operating temperature. This process configuration has a lowest capital cost because of the simplest design, fewest pieces of equipment, and less severe operating conditions, however, the operating and utilities costs are the highest, as reported in Table 5.

Alternative 2 can meet the product specifications with an efficient column temperature profile. The reboiler duty is significantly decreased by heat recovery (Figure 13).

Alternative 3 features a traditional design approach in terms of meeting product specifications, and operating cost optimization. Figure 13 shows that the reboiler duty is reduced by feed splitting in comparison to Alternative 2. Figure 10 shows the relationship between feed splitting, location of feed tray, and operating costs, including steam consumption and electricity usage.

Alternative 4 requires an additional compression stage because of the lower column pressure condition. Table 5 reports the highest capital investment; however, the reboiler duty is reduced in comparison to Alternative 3.

Alternative 5 shows an optimum design approach in terms of meeting product specifications, and operating cost optimization. The operating and utilities costs are minimized by the presence of the side reboiler exchanger for further heat recovery; however, the capital cost is slightly increased in comparison to Alternative 3 (Table 5).

Alternative 6 is a conventional distillation column. This case is the most complex system. CAPEX, OPEX and utilities cost are more expensive than Alternative 3 and 5 because of the additional reflux system, reported in Table 5.

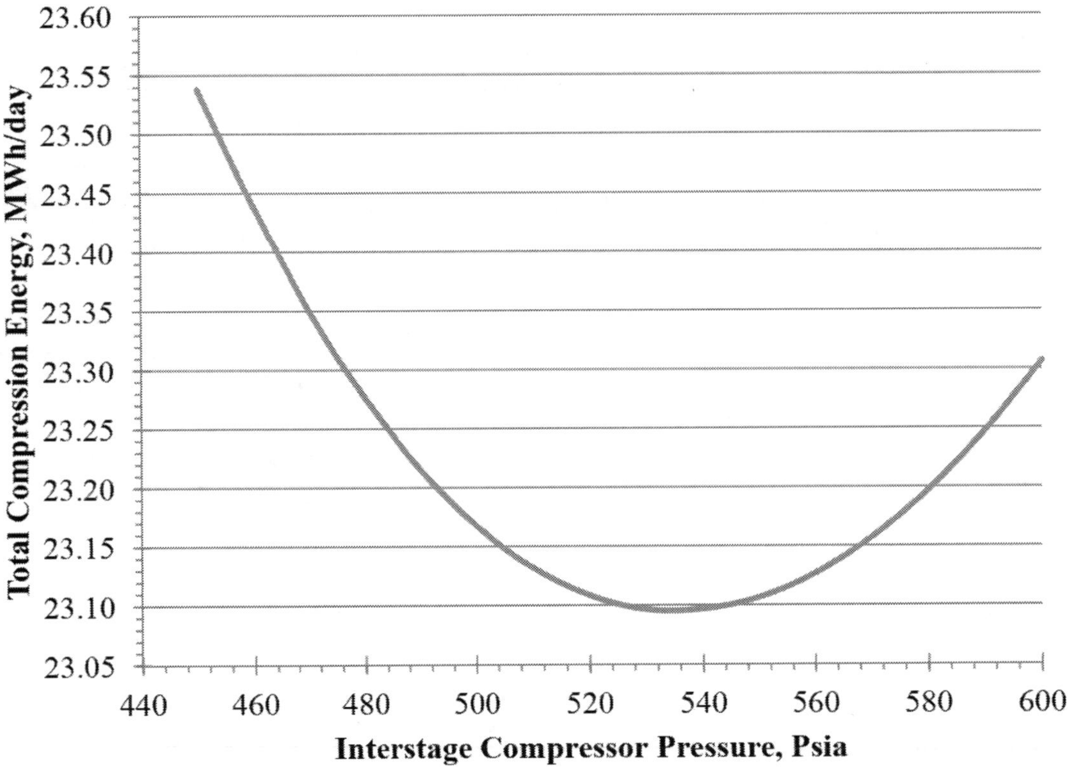

Figure 9 – Off Gas Compressor Power Optimization for Alternative 1

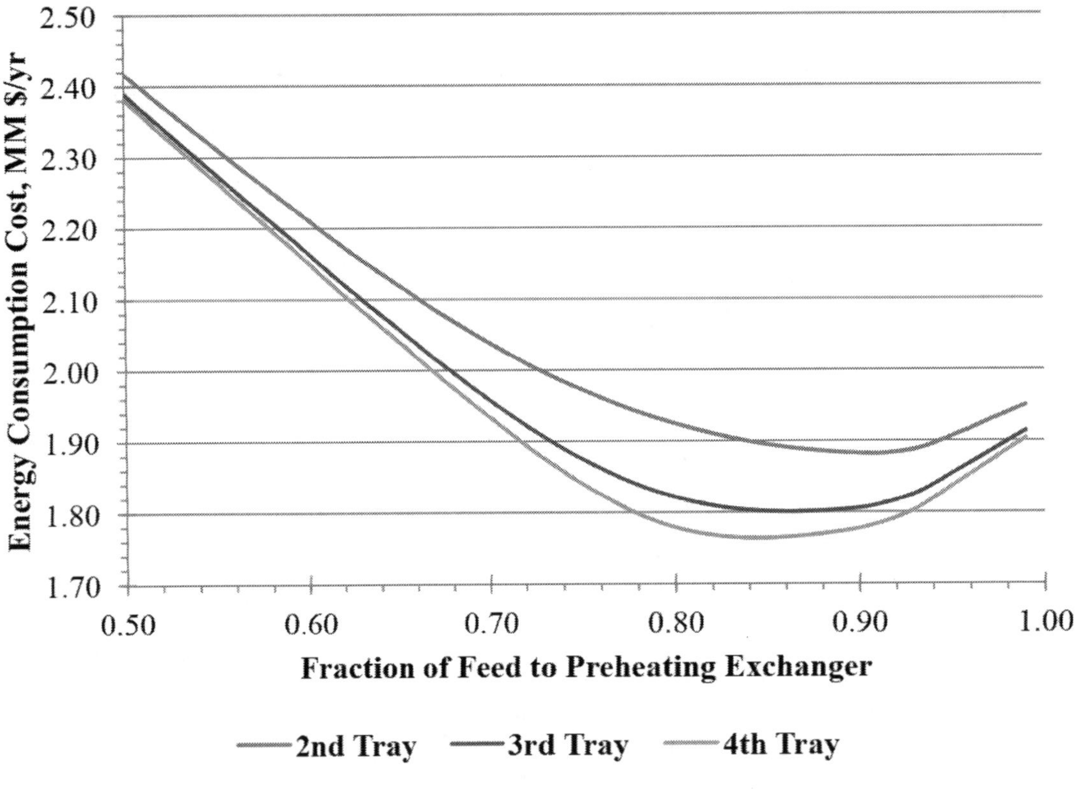

Figure 10 – Reboiler Duty Optimization by Splitting Feed & Feed Stage Sensitivity for Alternative 3

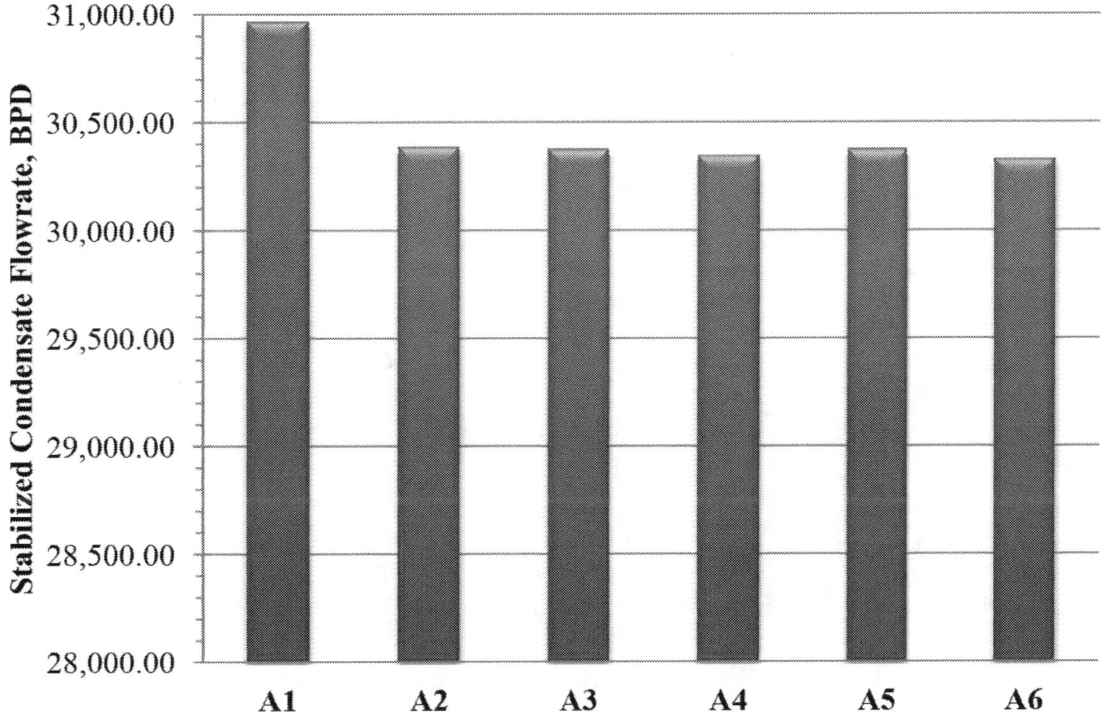

Figure 11 – Produced Stabilized Condensate for Each Process Configuration

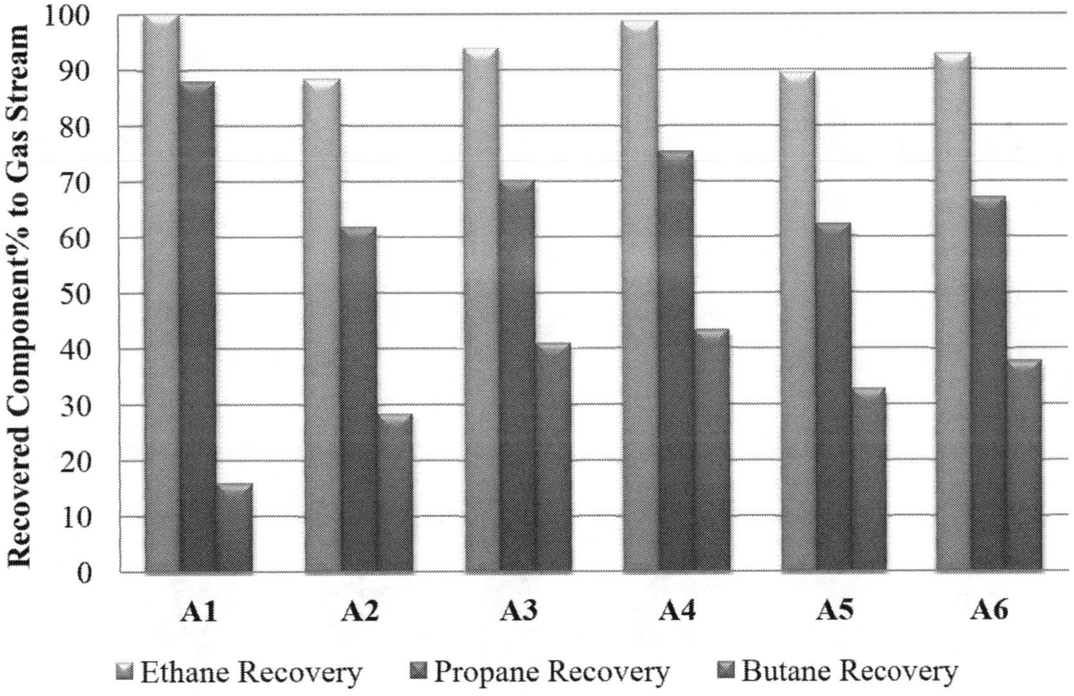

Figure 12 – Products Recovery for Each Process Configuration

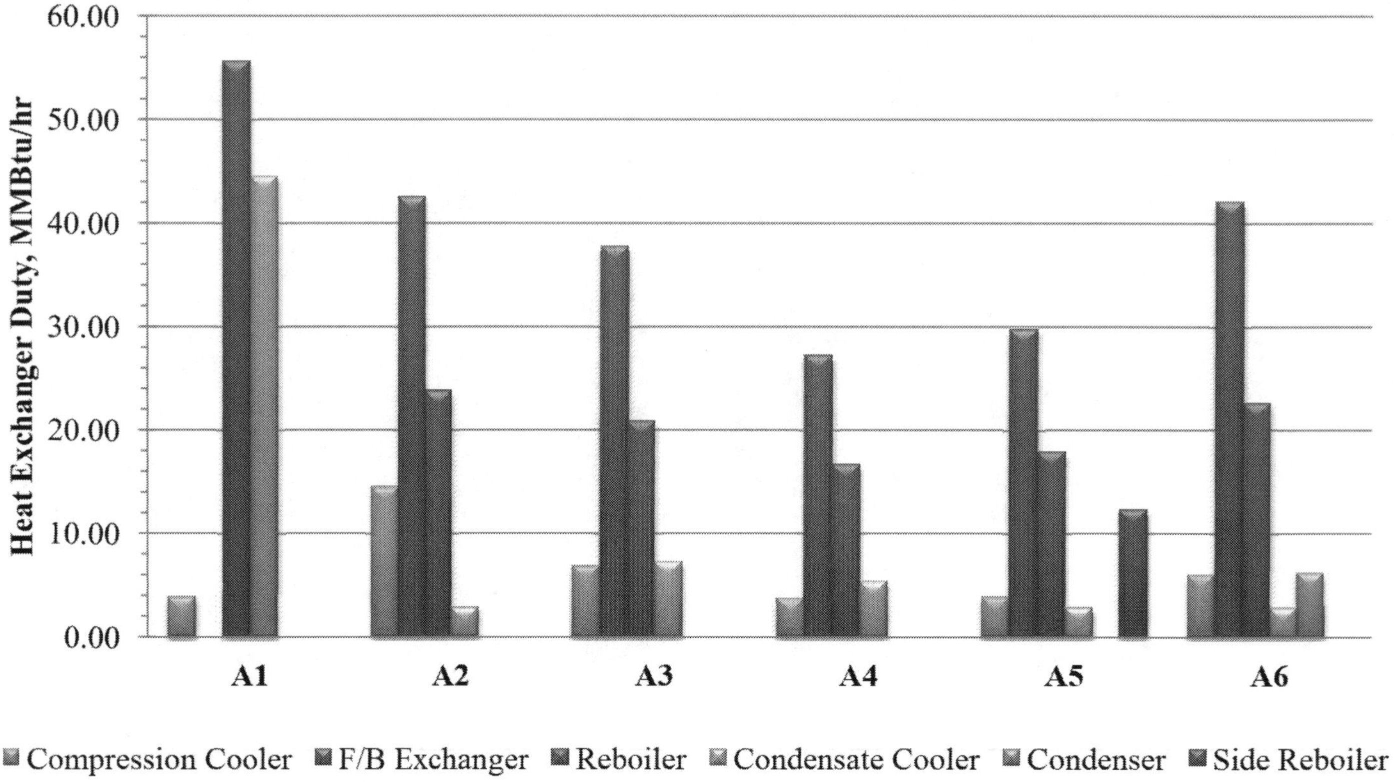

Figure 13 – Heat Exchangers Duty for Each Process Configuration

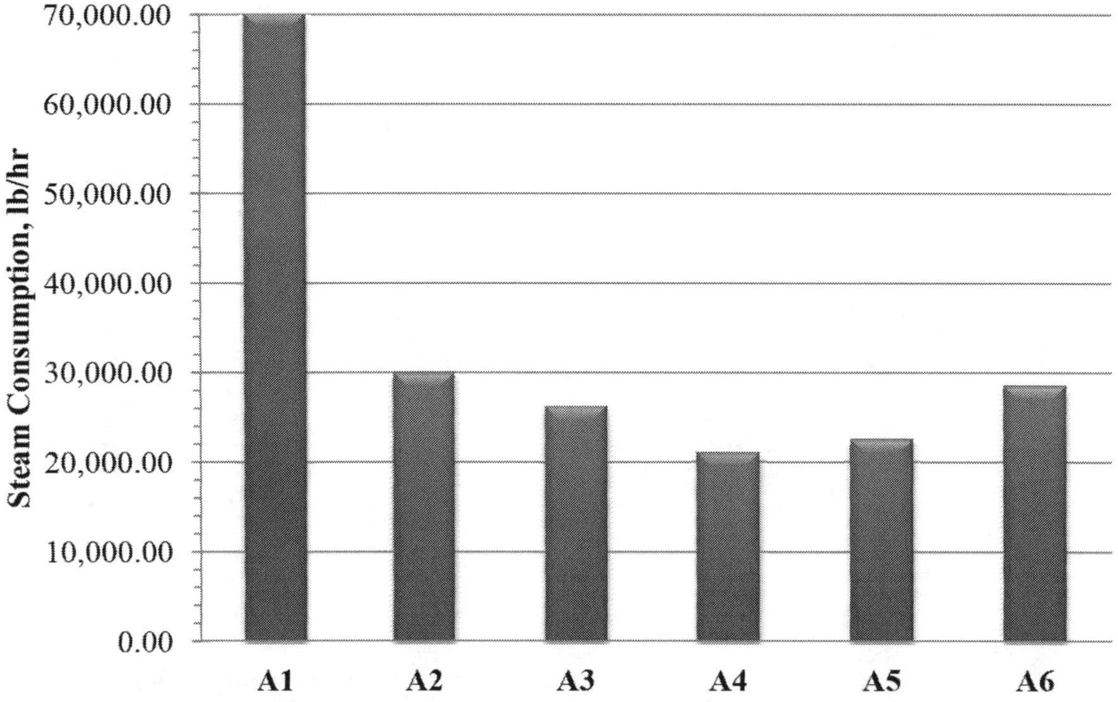

Figure 14 – Steam Consumption for Each Process Configuration

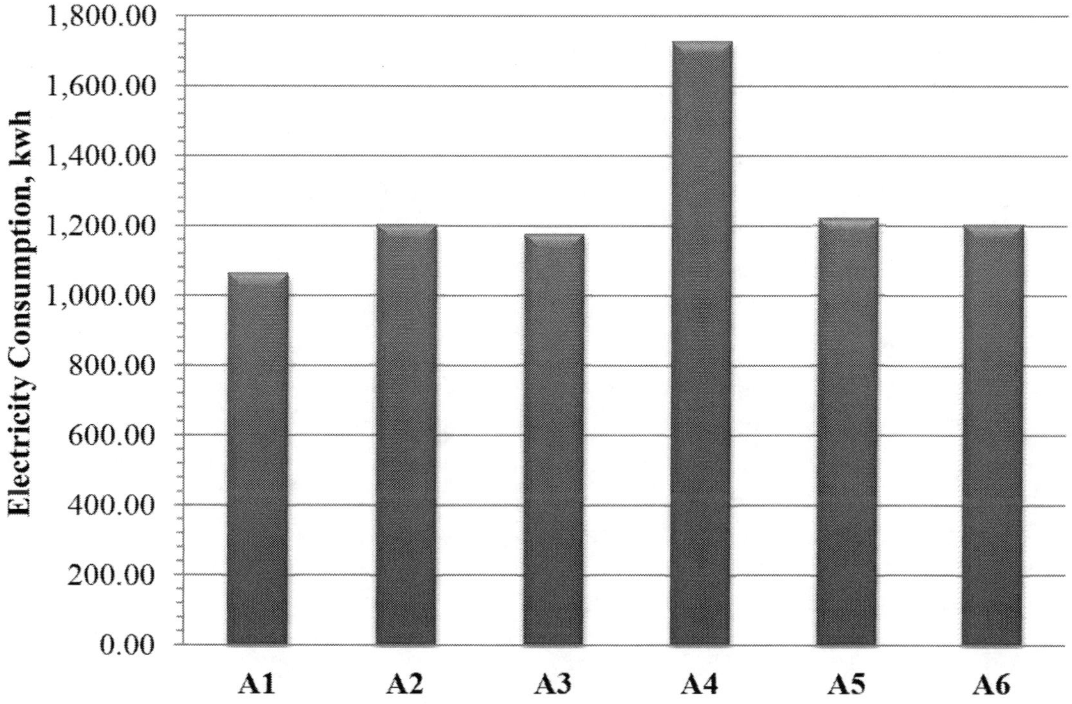

Figure 15 – Electricity Consumption for Each Process Configuration

Table 5 – Cost Comparison for Alternatives

	Capital Cost	Operating Cost [1]	Utilities Cost [2]	Product Sales
Alternative 1 [3]	0.89	1.68	2.16	1.02
Alternative 2	0.89	1.07	1.09	1.00
Alternative 3	0.97	1.06	1.08	1.00
Alternative 4	1.24	1.18	1.07	1.00
Alternative 5	1.00	1.00	1.00	1.00
Alternative 6	1.07	1.08	1.14	1.00

(1) The operating costs cover the raw materials, operating labor costs, maintenance, plant overhead, utility costs, electricity, fuel gas, potable water, and instrument air.

(2) The utilities costs cover the heating and cooling.

(3) Operational experiences show that the stabilized condensate specifications are achieved with difficulty with higher water content.

SUMMARY & CONCLUSIONS

Condensate stabilization units can be optimized by selecting the best process configuration and operating conditions depending on feed characteristics and required product specifications. A number of optimization runs are required using a steady-state simulator in order to determine the appropriate column pressure, column temperature, feed split ratio, and the feed tray location in terms of energy optimization. It is recommended that appropriate equipment design margins and sizing be used in order to ensure acceptable operational flexibility (particularly for water separation from hydrocarbon condensate). Table 5 summarizes the results of this study. Alternative 5 was chosen over the other process configurations through screening following on the following criteria:

- Acceptable capital cost, within the limits of study accuracy
- Lowest operating cost in terms of steam and electricity consumption
- Relatively low process complexity, which should lead to higher availability and lower maintenance costs in the long term, yielding the highest potential return to plant stakeholders.

Figures 9 – 14 illustrate sensitivities of various operating parameters. Based on this analysis and alternatives comparison, the following conclusions can drawn through the selection of the optimum process configuration:

- Optimizing of feed splitting and feed tray location:
  a. Reduces reboiler duty
  b. Reduces tower size
- Off-gas compressor power is optimized by adjusting the interstage pressure
- Off-gas after cooler duty is optimized by adjusting the interstage pressure
- Adding a bypass to the preheat exchanger and changing the feed tray location are obvious zero cost options.

REFERENCES

[1] Warren D. Seider, "Product and Process Design Principles" Second Edition, John Wiley, 2002

(91b) Simulation and Optimization of Internal-Loop Airlift Slurry Reactor Design Using Computational Fluid Dynamics

Luhaibo Zhao and Bo Liao Sr., Low Carbon Conversion Center, Shanghai Advanced Research Institute, Shanghai, China

Internal-loop airlift reactors (ILAR) are widely used in biotechnology and chemical industry, because they are suitable for processes with good mixing and close contact between the phases. However, the fundamental hydrodynamic behaviors in ILAR, which are essential for reactor scale-up and design, are still not fully understood. In the paper, the three-dimensional computational fluid dynamics (CFD) modeling is used to study the performance of three-phase ILAR with multi-layer ring gas distributor. The global flow characteristics, such as gas holdup, liquid velocity and solid particles distribution, are simulated, and the results agree well with the experimental data. Then, the effect of the sparger structure of gas distributor, including the number of layers, orifice size and angle, porosity, on hydrodynamic behaviors in ILAR is investigated based on CFD model. The optimal configuration of the multi-layer ring distributor is obtained and verified by cold model experiment, which results in even gas distribution and promotes the gas-liquid-solid mixing in ILASR, thus improves its yield and productivity.

(58a) Tunable Polyimide and Resulting Carbon Molecular Sieve Membranes for Gas Separation

Wulin Qiu, School of Chemical and Biomolecular Engineering, Georgia Institute of Technology, Atlanta, GA and William J. Koros, School of Chemical & Biomolecular Engineering, Georgia Institute of Technology, Atlanta, GA

Polymer membranes have been used for natural gas and olefin-paraffin gases separations successfully. However, the separation performance is limited by a productivity and efficiency trade-off upper bound. Therefore, advanced carbon molecular sieves (CMS) materials were developed. These materials show separation performance that exceeds the polymer "upper bound" for gas separations, while potentially maintaining economically attractive manufacturing. CMS membranes can be formed by thermal decomposition of polymer precursors, and contain "micropores" of 6–20 Å connected by smaller so-called "ultramicropores". The combination of ultramicropores and micropores provides the molecular sieving function and high permeability characteristics of these unusual CMS membranes.

We synthesized diverse but systematic sets of co-polyimides, containing symmetric and asymmetric polymer backbone structure, to explore effects of matrix rigidity and packing of precursor polyimides on final CMS properties. We also explored the interaction of pyrolysis conditions on final CMS properties. Permeation was tested using both pure gases (He, O_2, N_2, CO_2, CH_4,C_2H_4, C_2H_6, C_3H_6, C_3H_8) and mixed gases for both the polyimide precursor membranes as well as the resulting CMS membranes. We correlated separation performance of precursor polyimides and formation conditions with the CMS membranes' gas separation performance. This research provides an improved fundamental framework to guide the systematic understanding and fabrication of CMS membranes for important energy intensive separations.

(58b) Post-Spinning Infusion of Poly(ethyleneimine) Into Polymer/Silica Hollow Fiber Sorbents for CO2/N2 Gas Separation

Ying Labreche[1], Ryan P. Lively[2], Fateme Rezaei[3], Grace Chen[4], David S. Sholl[5], Christopher W. Jones[5] and William Koros[2], (1)School of Chemical and Biomolecular Engineering, Georgia Tech, Atlanta, GA, (2)Chemical & Biomolecular Engineering, Georgia Institute of Technology, Atlanta, GA, (3)School of Chemical & Biomolecular Engineering, Georgia Institude of Technology, Atlanta, GA, (4)Georgia Tech, Atlanta, GA, (5)School of Chemical & Biomolecular Engineering, Georgia Institute of Technology, Atlanta, GA

Post-spinning infusion of poly(ethyleneimine) into polymer/silica hollow fiber sorbents for CO_2/N_2 gas separation

Ying Labreche[1], Ryan P. Lively[2], Fateme Rezaei[1], Grace Chen[1], David S. Sholl[1], Christopher W. Jones[1,*], William J. Koros[1,*]

[1] Georgia Institute of Technology, 311 Ferst Drive NW, Atlanta, GA 30332-0100, USA

[2] Algenol Biofuels, 28100 Bonita Grande Drive, Bonita Springs, Florida 34315, United States

Abstract

Amine-loaded hollow fiber sorbents for post-combustion CO_2recovery were created by utilizing a novel post-spinning infusion technique. This technique infused poly(ethyleneimine) (PEI) into cellulose acetate/mesoporous silica hollow fiber sorbents during the solvent exchange after spinning. A suitable post-spinning infusing solution was found to be 10% PEI in methanol with an infusion time of 4 h.

After amine infusion, the 51 wt% silica hollow fiber sorbents were found to have a nitrogen loading of 0.52 mmol/g-fiber and a CO_2 uptake of 1.2 mmol/g-fiber at equilibrium. The amine-loaded fibers were formed into a shell-and-tube module and exposed on the shell side at 1 atm and 35 °C to simulated flue gas with an inert tracer (10 mol% CO_2, 80 mol% N_2 and 10 mol% He at 100% R.H.). The fibers were found to have a breakthrough CO_2 capacity of 0.58 mmol/g-fiber, with a mass transfer propagation front velocity of 1.12 cm/s. The CO_2 "equilibrium" uptake at 1 atm and 35 °C is 0.92 mmol/g-fiber. The sorption enthalpy release was monitored and the thermal front propagation velocity was 0.75 cm/s. Finally, the water uptake was found to be 3.2 mmol/g-fiber. The results indicate the post-spinning infusion method provides a new platform for ultimate scale-up of practical fiber sorbents for flue gas application.

(58c) Strategies for Acid Gas Enrichment and Tail Gas Treatment:

Tofik K. Khanmamedov, TKK COMPANY, Sugar Land, TX

Strategies for Acid Gas Enrichment and Tail Gas Treatment:

Special Solvents or Better to Stick With Generic Solvents

Tofik K. Khanmamedov

TKK COMPANY, Sugar Land / Houston, Texas

Abstract

The quality of the acid gas feed stream to sulfur recovery units is crucial for proper operation. In this regard, acid gas streams with low H_2S concentration should be enriched with H_2S. This is also true for the amine section of tail gas treating units. There are different approaches for acid gas enrichment. One approach is based on using generic N-Methyl-Diethanol Amine (MDEA) with the family of patented HIGHSULF™ processes. Other approaches use more selective and very expensive solvents.

Also there are claims that some specialty solvents allow designing units using relatively smaller equipment than in case of MDEA, there are limited advantages in long term to say the least.

Due to depreciation of the unit in most cases the capital cost is not the primary concern of the clients. Even operating cost based on the use of local sources of power is not the major issue. However, they do seriously care about the costs of the solvents, that they have to purchase annually from the single foreign vendors and this factor limits their freedom of the choice. While this approach might reduce initial capital cost, it could significantly increase the operating costs over time due to the necessity to continue to purchase the expensive specialty solvents from the single vendor.

In most cases, switching from special solvents to generic, less expensive, solvents in the future would require a reduction of the overall unit capacity. Units designed for MDEA based HIGHSULF™ technologies can be easily switched to other specialty solvents within certain capacity constraints.

This paper focuses on advantages of the MDEA based family of HIGHSULF™ processes that require no additional equipment.

(58d) Synthesis of Ion-Exchanged SAPO-34 Zeolite Membrane for CO2/CH4 Separation

Yanfeng Zhang, Meng Li and Yuhan Sun, Low Carbon Conversion Center, Shanghai Advanced Research Institute, Chinese Academy of Sciences, Shanghai, China

Synthesis of ion-exchanged SAPO-34 zeolite membrane for CO_2/CH_4 separation

***Yanfeng Zhang*[1],*Meng Li*[1], *Yuhan Sun*[1]**

[1]Shanghai Advanced Research Institute, Chinese Academy of Sciences, 100 Haike Rd, Pudong New District, Shanghai 201203, P.R China

Introduction

Natural gas usually is contaminated with CO_2, which must be removed to meet pipeline requirement. The presence of CO_2 in natural gas decreases the energy content of the gas, and causes pipeline corrosion in the presence of water. Amine adsorption is a mature technology for CO_2 removal from natural gas, but it suffers from high energy cost for solvent regeneration, especially for natural gas with high CO_2 content. Membrane separation has many advantages, such as low energy cost, low capital investment, flexible size, etc, and polymer membranes have been used for natural gas sweetening at commercial scale. However, polymer membranes cannot treat natural gas with high CO_2 content (>20%), since plasticization effect decreases membrane selectivity significantly. Zeolite membranes have great chemical and thermal stability, and their uniform pore size. Zeolite membranes[1¨C4] have potential for CO_2/CH_4 separation owing to their great thermal, mechanical, and chemical stability, and stability at high CO_2 pressures. SAPO-34, a silicoaluminophosphate with chabazite (CHA) type framework having pore diameter 0.38 nm, is perfect for CO_2/CH_4 separation because CO_2 (0.33 nm kinetic diameter) and CH_4 (0.38 nm) can be separated by molecular sieving mechanism. The selective adsorption of CO_2 in SAPO-34 pores also favours the separation of CO_2. Many factors affect the performance of SAPO-34 membrane, including feed pressure [1,2], Si/Al ratios [3,4], seed size [5], template types [5], membrane thickness [4], cation forms [6], CO_2/CH_4 feed ratio, support properties [7], concentration polarization [8] and template residue [9]. Mei et al prepared ion-exchanged SAPO-34 zeolite membrane and CO_2/CH_4 selectivity was increased by 30%. Here we report a simple one step method of preparing ion-exchanged SAPO-34 membrane by combining template removal and ion-exchange in to one step. Obtained Na, K and Li-SAPO-34 membranes have improved separation performance.

Experimental

Detailed information about synthesis, calcination and gas separation test of SAPO-34 membrane can be found in ref 9. Uncalcined SAPO-34 membrane was soaked in 1wt% $NaNO_3$, KNO_3 or $LiNO_3$ solution for 5 min and dried at 383K for 2 h. Then the SAPO-34 membrane was calcined in vacuum at 673 K for 4 h with 1K/min heating and cooling rates. For comparison, blank SAPO-34 membranes were also calcined in vacuum at the same condition.

Results and discussion

Fig 1 shows the SEM images of SAPO-34 membrane. Typical cubic crystals with perfect intergrowth were found in Fig 1a and membrane thickness is ~ 5 μm (Fig 1b). SAPO-34 crystals with thin plate morphology, as shown in Fig 1c, were used as seed for membrane synthesis. XRD patterns indicate that the obtained crystals and membranes have CHA structure (not shown here).

(a) (b)

(c)

Fig 1 SEM images of SAPO-34 membrane and SAPO-34 seed, (a)top view and (b)cross section view and (c)SAPO-34 seed.

Table 1. Separation performance of SAPO-34 membranes (50/50 molar CO_2/CH_4 mixtures, room temperature, 4.0 MPa feed pressure).

Membrane treatment	**CO_2 Permeance mol/(Pa m^2 s)**	**CO_2/CH_4 Selectivity**
1wt% $NaNO_3$ in H_2O	2.1°Á10^{-7}	75
1wt% $LiNO_3$ in Acetone	3.9°Á10^{-7}	70
1 wt% $NaNO_3$ in H_2O	2.6°Á10^{-7}	81
1 wt% KNO_3 in H_2O	2.8°Á10^{-7}	73
	-7	

Eight membranes calcined directly	3.0°Á10	52

Eight SAPO-34 membranes were prepared and calcined directly in vacuum to remove the template. The average CO_2 permeance and CO_2/CH_4 selectivity are 2.8°Á10^{-7} mol/(Pa m^2 s) and 52 respectively (feed pressure 4.0MPa and 50/50 CO_2-CH_4 mixture). Four membranes were soaked in 1wt% salt solution for 5 min and dried at 383K. Then these membranes were calcined at 673K in vacuum to remove template. The melting points of $NaNO_3$, KNO_3 and $LiNO_3$ are 581, 607 and 528K respectively, all lower than calcination temperature of 673K. This indicates that the deposited salt on the membrane surface will be melted during calcination. After template removal, the as prepared H-SAPO-34 membrane will be exchanged to Na, K and Li-SAPO-34 membranes. All four membranes exhibits 40 to 60% increase in CO_2/CH_4 selectivity, compared with control experiment. Ion-exchange of H-SAPO-34 increases basicity of SAPO-34 crystals, which favours the adsorption of CO_2. Ion-exchange may also reduce pore size of SAPO-34 crystals, which might limit the diffusion of CH_4. This result is consistent with Mei's result [6].

Conclusions

A simple method was developed to prepare ion-exchanged SAPO-34 membranes in one step. By combining template removal and ion-exchange into one step, Na, K and Li-SAPO-34 membranes were prepared. The obtained SAPO-34 membranes have much higher CO_2/CH_4 selectivity, which is the result of increased basicity from ion-exchange. The combination of template removal and ion-exchange eliminate one step and greatly reduce synthesis cost. This method can also be applied to other zeolite membranes.

REFERENCES

[1] S. Li, J.G. Martinek, J.L. Falconer, R.D. Noble and T.Q. Gardner, Ind. Eng. Chem. Res., 44 (2005) 3220.

[2] M. A. Carreon, S. Li, J.L. Falconer and R.D. Noble, J. Am. Chem. Soc., 130 (2008) 5412.

[3] S. Li, J.L. Falconer and R.D. Noble, Microporous Mesoporous Mater., 110 (2008) 310.

[4] S. Li, J. L. Falconer and R. D. Noble, Adv. Mater., 18 (2006) 2601.

[5] M. A. Carreon, S. Li, J.L. Falconer and R.D. Noble, Adv. Mater., 20 (2006) 729.

[6] M. Hong, S. Li, H. F. Funke, J.L. Falconer and R.D. Noble, Microporous Mesoporous Mater., 106 (2007) 140.

[7] S. Li, J.L. Falconer and R.D. Noble, J. Membr.Sci., 241 (2004) 121.

[8] A. M. Avila, H. H. Funke, Y. Zhang, J.L. Falconer and R.D. Noble, J. Membr. Sci., 335 (2009) 32.

[9] Y. Zhang, A. M. Avila, H. H. Funke, J.L. Falconer and R.D. Noble, J. Membr. Sci., 363 (2007) 29.

(58e) An Experimental Study of Effect of Cyclone Separator On Transmission of Pressure Fluctuation in a CFB

Chen Yong, Zhang Qiong, Yan Chaoyu and **Wei Yaodong**, China University of Petroleum-Beijing, Beijing, China

Abstract: In large scale circulating fluidized bed device, in order to analyze the mechanism and regulation which cyclone separator effect on transmission of pressure fluctuation,dynamic pressure of before and after cyclone separator and different locations in cyclone separator was measured under many kinds of conditions according to change the solid circulating rate at the entrance of the cyclone separator. The particles used were FCC catalyst. The experimental results show that in the axial direction of the cyclone separator, with the same solid circulating flow rate, pressure fluctuations of the cyclone side walls are same as pressure fluctuation of the top of the dipleg but different with the inlet's. Pressure fluctuations of the cyclone center have differences with both the top of the dipleg's and the inlet's; In radial direction, from the side walls to the center, the differences of pressure fluctuations between the cyclone and the inlet and the top of dipleg are gradually increasing, and the tendency increased with increasing particle circulating flow rate; In circumferential direction, There are a slight differences of pressure fluctuations in different angles near the cyclone center. Near the side walls, when the solid circulating rate is low, pressure fluctuations in certain areas are different with other areas'. While when the solid circulating rate is high, pressure fluctuations have no difference substantially.

(103a) Dynamic Real-Time Optimization of an Alkylation Unit

James Jones, Manufacturing IT, Chevron Downstream and Chemicals, Houston, TX

Chevron has recently completed a number of Closed-loop, Dynamic, Real-time Optimization projects. The approach taken differs from conventional RTO projects based on a rigorous steady-state model of the process. Instead, the model includes both steady-state and dynamic elements and the model scope and rigor is adapted to the specific needs of the application. The motivation for this approach is presented, along with a description of the technology used. Some results from a recent Alkylation unit application will be discussed.

Nonlinear APC Solution for Polypropylene Plants

Abstract

A Nonlinear Model Predictive Control Solution has been implemented on the two Polypropylene Lines at Saudi Polyolefins Company (TASNEE) complex in Jubail, Saudi Arabia. The project scope included both pre-audit and post-audit studies to evaluate the application performance before and after implementation of the solution. The acceptance criteria for the project included both increased production and improved Cpk (product quality improvements) for all four reactors. The collaborative project, involving TASNEE and the APC solution provider, was completed in seven months and all performance criteria were exceeded. The ROI for this project was calculated at less than three months.

The installed APC solution has been online for over 30 months and is well accepted by the TASNEE operations team. The average utilization is over 93%. The solution is maintained by TASNEE personnel so that it will continue to deliver the expected results. Before implementation of this solution TASNEE was focused on improving their lab QC performance and significant improvements were achieved. Due to this effort and implementation of APC Solutions, TASNEE has been able to reduce the number of lab samples required. This has reduced the lab workload resulting in more consistent quality of results and improved production Quality Control at the same time.

Introduction

Saudi Polyolefins Company (TASNEE Petrochemical) operates two world-scale Polypropylene units based on Novolen Technology, which include polymerization and extrusion sections, in Jubail, Saudi Arabia. The polymerization section of each line consists of two reactors operating in parallel mode for homo polymer production. Hydrogen is added continuously in a fixed ratio to propylene feed to control the molecular weight of the polypropylene. The molecular weight is expressed as the melt flow rate (MFR). Each reactor train has its own degassing section, intermediate product storage and extruder. The total capacity of the plant is 720 KTA.

In early 2009, TASNEE decided to deploy an Advanced Process Control (APC) Solution for its recently debottlenecked Novolen Polypropylene plant in Al-Jubail KSA. An extensive set of plant operating data was supplied to selected APC vendors to allow them to perform a preliminary benefits analysis.

The scope includes; Reactor Composition Controller, Virtual on-line Analyser (VOA®) for melt flow rate (MFR) and xylene soluble (XS) and Quality Controller as well as Automated Transition for each of the four reactors.

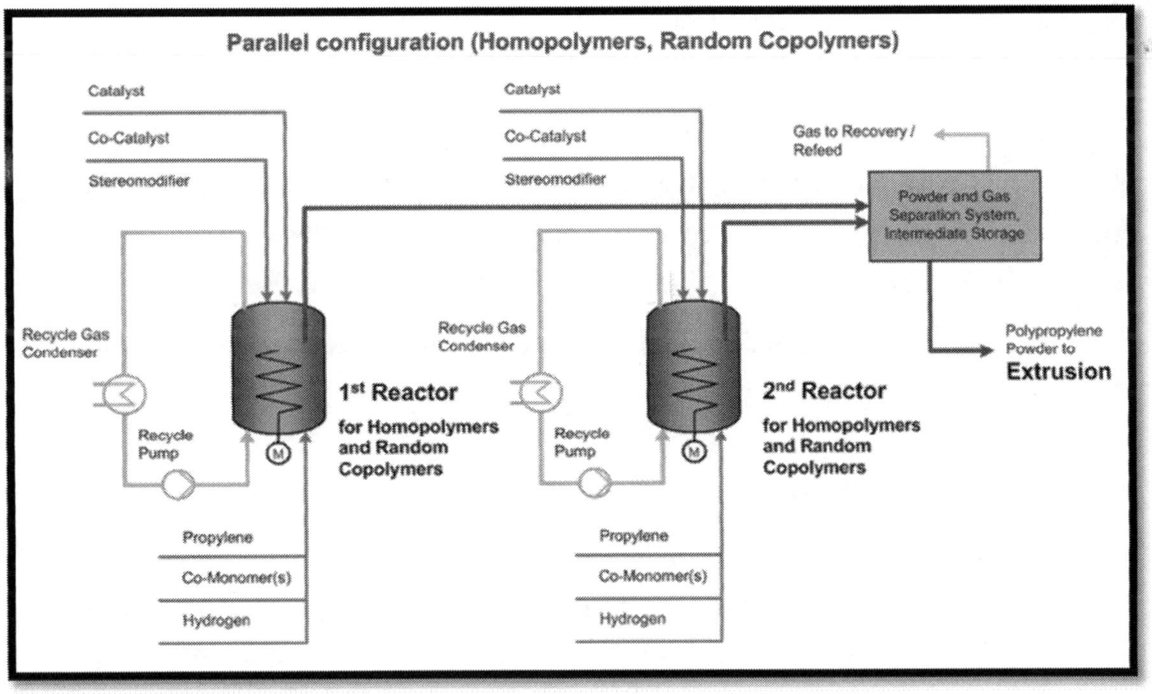

[1]Figure 1– Typical Process flow diagram with APC implementation

Benefit Study - Project Base Line

In order to estimate the potential benefits related to the implementation of an APC solution, a detailed benefits study was carried out in 2009 for the two Polypropylene lines at the Jubail complex.

The benefits related to implementation of *APC* for polymer plants were achieved mainly due to measurable improvements in the process stability, the ability to accurately predict key process variables which are hard to measure online (such as MFR & XS) and improvements in consistency of operation of the process (transitions execution). Another key benefit is that APC will push the process closer to its constraints without violating them thereby increasing the plant production rate.

This benefits study focused on three primary benefit areas; increased production, reduced off-spec product and reduction of operating costs. Results of the initial benefits study showed less than six months return on investment (ROI) can be achievable (i.e. increase production by over 2% and improve Cpk by over 50%).

Selection of the APC Technology

The TASNEE project team invested significant time and effort for vendor and technology selection. Factors such as technology capability, polymer experience, commitment to success, and cost were considered in the selection process. The vendor selection process took approximately 6 months. Each vendor was given the opportunity to present its benefits analysis and its technical solution. The TASNEE team paid special attention to the APC solution that provided remedies for chronic operational issues.

Project implementation

Within the first three months of the project kick-off, the majority of the APC solution including the VOA®'s, and the Quality Controllers were in place and commissioned. The operations team reported improvements in the process stability and product quality shortly after commissioning began.

To improve reactor stability, the reactor pressure controller was commissioned along with the polymer production rate controller. For the next four months, the project team focused on fine tuning the controller and implementation and testing of the automated transitions. The implementation was completed within seven months after the kick off meeting. The performance test and audit took an additional three months. The project exceeded all performance criteria outlined in the project objectives.

Project Challenges

This polymer APC project had to overcome typical and difficult issues related to the process and other project objectives. These included;

1. The process makes multiple products with varying catalyst activities due to varying hydrogen concentration in the reaction loop. This introduces a high degree of non-linearity in the gains and the dynamics.
2. The process with respect to some of the controlled variables is open loop unstable. This required that the controller respond in a quick and sufficient manner so as not to introduce further instabilities.
3. The process is sensitive to impurities in the feedstock and to those generated internally. The impurities tend to affect the catalyst activity and thus the response of the process.
4. The steady-state location of process variables for a product will vary from run to run.
5. Some of the final product properties that are important to the end user are not characterized by the parameters that are measured by the lab or controlled in the process.
6. Technology transfer to the operations and technical groups for maintenance and support.
7. The analyzer performance issues and lab result variability were flagged early in the project. The project team was able to offer solid technical solutions.

Results Achieved

The main project objectives were to increase production rate, improve process stability (reduce off-spec) and improve product quality or Cpk. Typically APC Solutions for polymers achieved these objectives by implementation of a nonlinear multivariable control solution.
An important aspect of reactor stability is the ability to control critical process parameters such as reactor pressure and reactor temperature. Figure 2 illustrates the pressure control improvement for one of the polymer lines.

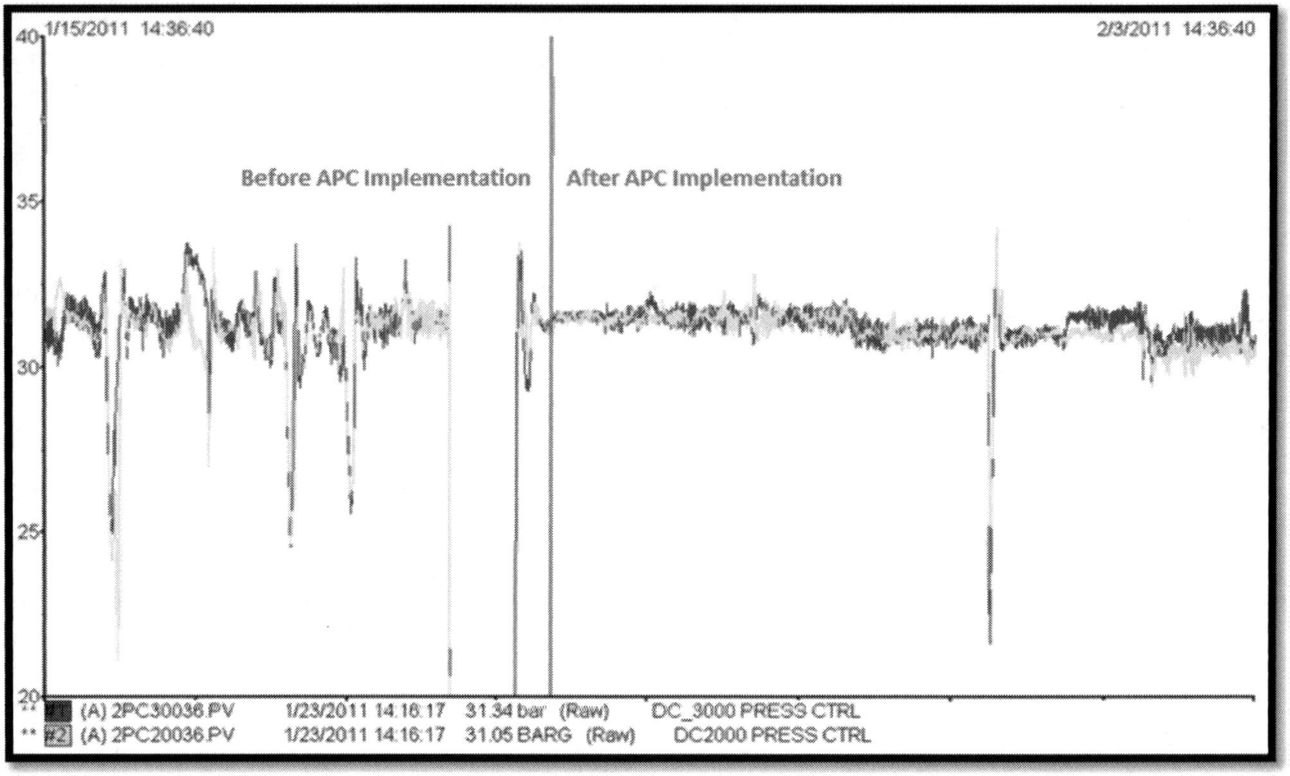

Figure 2 – Reactors Pressure Control before and after APC implementation

The pressure control solution was multivariable and included both the monomer and catalyst feeds as manipulated variables. This allowed the pressure and rate controller to work in conjunction with each other. This controller was used for start-up, rates changes and at steady state. Figure 3 shows the performance of the pressure control from "sleep" mode (no catalyst and minimum monomer) to full rates and pressures.

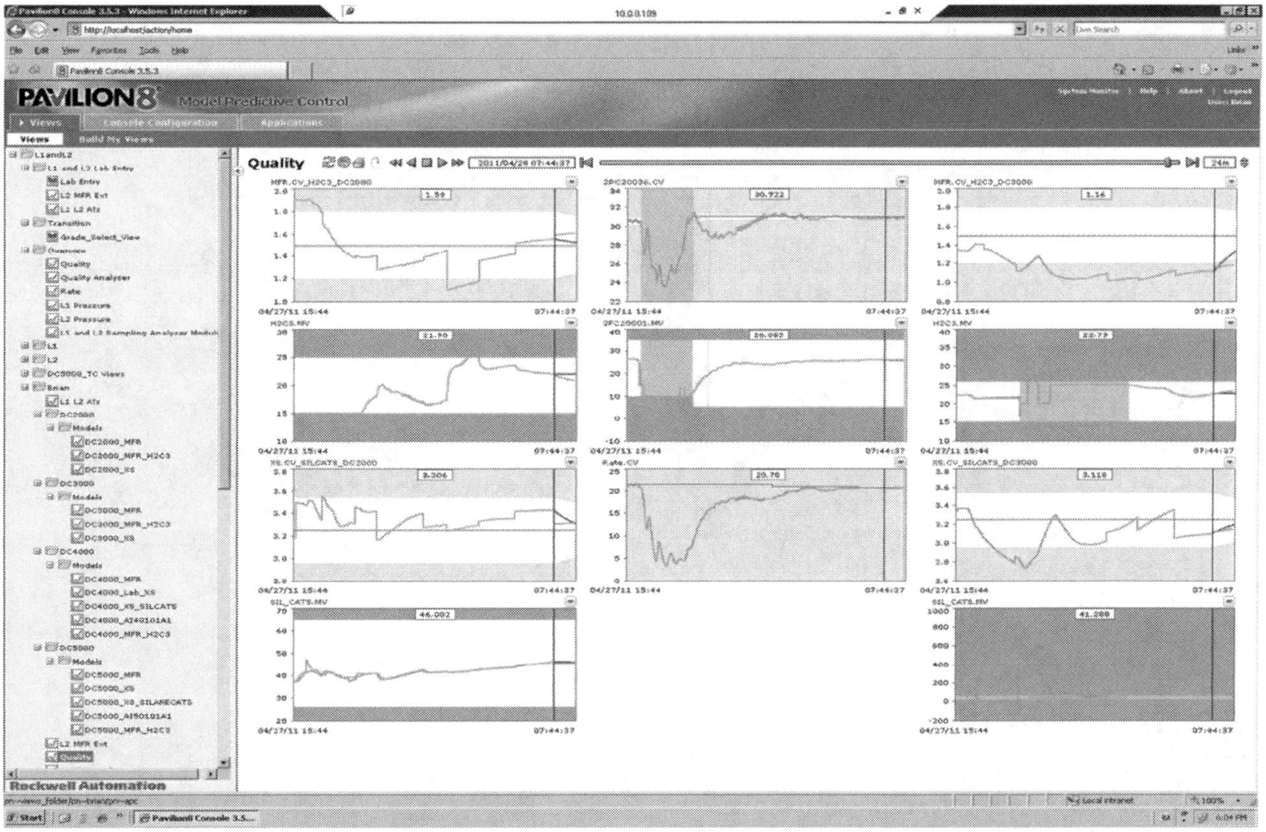

Figure 3 – Reactors Pressure Control Performance from "Sleep" Mode – Center plots

Figure 3 also illustrates the benefits of implementing a non-linear controller. The quality controlled variable is MFR. The prediction and control model is non-linear with respect to gains and dynamics. The left hand plots show the MFR on-line during sleep mode while the plot on the right shows MFR off-line. Even under the abnormal conditions of minimum rates the on-line reactor could maintain the quality with in specifications while the off-line MFR went off-specification.

Tables 1 and 2 show the improvements in process stability relative to the baseline during the 3 month performance period.

Table 1 & 2 – Improved Process Stability Train 1 and 2

Description Train-1	Standard Deviation Baseline	Standard Deviation After APC	Improvement
Reactor # 1 Pressure	0.74	0.31	58%
Reactor # 2 Pressure	0.68	0.27	61%
Reactor # 1 Temperature	0.93	0.47	50%
Reactor # 2 Temperature	0.93	0.68	27%

Description Train-2	Standard Deviation Baseline	Standard Deviation After APC	Improvement
Reactor # 3 Pressure	0.63	0.25	60%
Reactor # 4 Pressure	0.71	0.20	72%
Reactor # 3 Temperature	0.52	0.34	35%
Reactor # 4 Temperature	0.66	0.34	48%

The project team was able to achieve and exceed all project objectives.

Table 3 below shows the performance guarantee objectives of the project and table 4 shows the actual results for the 3 month performance period.

Table 3 – Performance Guarantee

Performance Guarantee Terms	Train # 1 RX1	Train # 1 RX2	Train # 2 RX3	Train # 2 RX4
Production Increase	2.0%	2.5%	2.5%	2.5%
Quality Improvement for MFR - (Cpk)	50%	50%	50%	50%

Table 4 – Performance Achievement

Results Achieved	Train #1 RX1	Train #1 RX2	Train #2 RX3	Train #2 RX4
Production Increase	3.5%	2.8%	2.8%	2.85%
Quality Improvement for MFR - (Cpk)	>50%	>50%	>50%	>50%
Quality Improvement for XS - (Cpk)	>50%	>50%	>50%	>50%

In addition to reactor stability, improvement in the product quality was an important objective. The product quality improvement is measured by the Cpk. This is a statistical measure of the plant's capability to operate between product limits and near the designated product target. For Tasnee Cpk for MFR was a project objective. For all products the Cpk objective was exceeded.

Improvements in other product properties including Xylene Solubles (XS) were also observed.

Figure 3 shows the Quality improvement for one of the product grades.

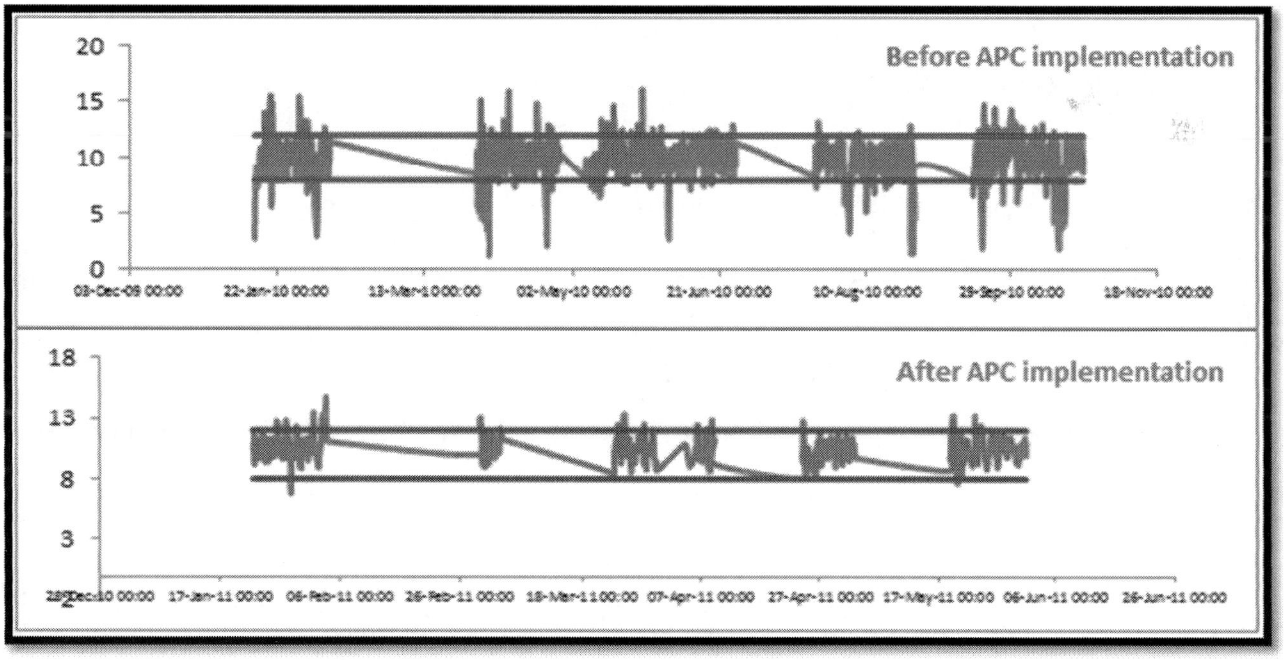

Figure 3 – Quality Control Performance Before & After APC implementation

In addition to improved process stability, increased production and better quality control, the solution was also able to improve product transitions by automation of the transitions. The product transitions are now faster and more consistent. Transitions times were reduced from 50% to 80% depending on the direction of the transitions and the magnitude of the change.

This translates to making less off-spec material due to products transitions. A typical MFR transition is illustrated Figure 4.

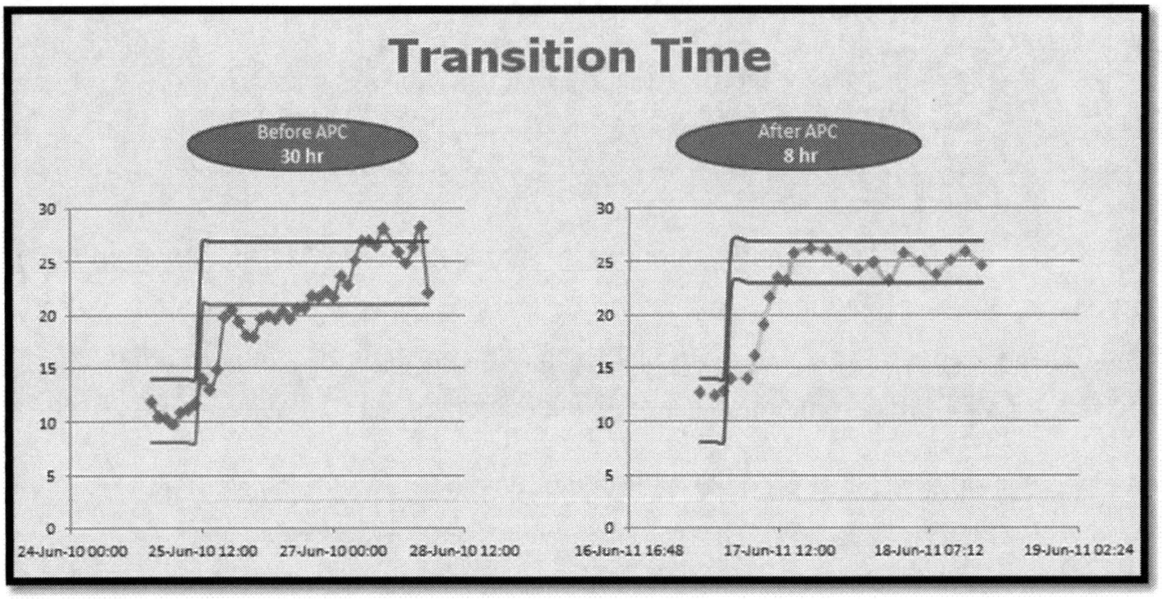

Figure 4 – Product Transitions before & after APC implementation

Results and Conclusions

The project was completed in seven months after the formal kick-off meeting. The total benefits achieved shows that the return on investment (ROI) was less than 3 months.

It is important to note a few important factors which contributed to the success of this project. The initial work on project baseline, scope definition and project preparation were essential elements. Also close collaboration between TASNEE and the APC vendor team, ongoing training, technology transfer to the TASNEE team and effective project management were critical.

The application performance was monitored during the evaluation period (3 months after commissioning) and the post-audit report shows that all performance criteria were exceeded. The total benefits achieved shows that the ROI was less than three months.

Subsequent internal audits have shown the overall benefits are exceeding those reported during the performance period.

Literature Cited:

[1] "Parallel configuration – Homopolymers and Random copolymers" (Available at www.cbi.com/images/uploads/Config_Parallel_New_3_788x450.gif)

(103c) Two-Level Algorithm for Gasoline Blend Planning Using Inventory Pinch Points to Reduce Number of Blend Recipes

Pedro Castillo Castillo, Chemical Egineering, McMaster University, Hamilton, ON, Canada and Vladimir Mahalec, Chemical Engineering, McMaster University, Hamilton, ON, Canada

We consider the problem of multiple blend recipes arising from multi-period blend models. Experience shows that the multi-period model solutions vary not only from one period to another, but also (for the same set of data) from one solver to another. Moreover, some solvers produce different solutions. Such solutions are awkward to apply in practice, since practitioners know intuitively that the same grade of gasoline can be blended by the same recipe for extended lengths of time.

Above behavior has been observed even with linear and nonlinear quality constraints. Recently introduced complex EPA formulations for gasoline quality constraints complicate the picture further, since they are highly nonlinear and require highly specialized algorithms to be solved. Solving multi-period MINLP model with these constraints remains in the domain of very specialized experts.

This work presents a two-level composite algorithm to solve the gasoline blend planning optimization problem while reducing the number of blend recipes. The first level deals with optimization of the blend recipes and in the second level production volumes and inventory profiles are computed. The first level is formulated as a multi-period NLP and the second level as a multi-period MILP; the time grids of both models are partially synchronized. In this way, after computing the lowest cost blend recipes, according to the problem instance and without the assumption of linear blending properties, a production plan that uses these optimal recipes is calculated. The algorithm (i) computes the blend plan that keeps the blend recipes as constant as possible along the blending horizon, (ii) avoids the need to use multi-period MINLP models, and (iii) computes the solutions that have the same cost as the multi-period MINLP optimum.

A pinch point on the cumulative total demand (CTD) curve is a point where cumulative average production (CAP) curve is tangent to the CTD curve and if we extrapolate the CAP curve from this point onward it will not cross the CTD curve. In other words, the inventory pinch points are defined by peaks on the cumulative total demand profile and they indicate time intervals where one recipe can be used without incurring in inventory infeasibilities. The inventory infeasibilities are defined as the missing volume to fulfill the demand of one or more products by using only the fixed recipes and the available amount of blending components. The inventory pinch points delineate the time periods in the recipe optimization problem. The planning horizon in the second level is divided in a fixed number of time periods and an integer number of these are contained in each one of the time periods of the recipe optimization level. The blend recipes and target inventories from the first level are fixed in their corresponding time periods in the second level.

Sometimes, inventory infeasibilities may still appear in the planning level when using the blend recipes from the first level. These infeasibilities are encountered because: (i) the recipe optimization problem has multiple global optimal solutions and not all of them can be used in the planning level, (ii) some logistic constraints (e.g., specific production sequences) may reduce the degrees of freedom to compute a production plan with a fixed recipe, or (iii) the presence of large variations on the blending components supply profiles. An iterative procedure based on the inventory infeasibilities and the time when they occur is used to generate a feasible plan. Slack variables are introduced in the model to detect the inventory infeasibilities, as well as in the objective function to minimize, multiplied by penalty coefficients which are larger than the blending components costs. In order to extend the use of a blend recipe, the penalty coefficients decrease with time, moving the infeasibilities as late as possible in the planning horizon. When inventory infeasibilities are found in the second level, the corresponding first-level time periods are divided at the same point in time where the first infeasibility appears. Then, new blend recipes are computed and used to calculate the production plan. The procedure repeats until no infeasibilities are found, or if infeasibilities appear in the first level which means that there are not enough blending components available to blend according to the demand and product specifications.

The results obtained with the two-level algorithm are very close to those provided by the corresponding multi-period MINLP model, the difference is less than 0.01%. In addition, the number of different blend recipes computed by our algorithm is less than the number given by the MINLP solution. The number of iterations is usually small and the algorithm is not affected by the blending components supply profile.

By decomposing the problem into two levels and using the inventory pinch point concept, complex rigorous nonlinear models can be used to compute optimal blend recipes without being solved at each time period of the production planning or scheduling horizon, avoiding the use of MINLP models and reducing the computational requirements to solve the overall problem without incurring in a great deviation from the optimal solution.

(103d) Hydrotreater Optimization by Direct Online Control of the Accumulated Product Tank Sulphur

Bhaskar V Iyer[1], **John MacGowan**[2], Huang Shu[3], Vimal Doraj[3] and Pradeep Appasani[4], (1)Honeywell Process Solutions, Honeywell International, Houston, TX, (2)Syncrude Canada Ltd, Fort Mcmurray, AB, Canada, (3)Syncrude Canada Ltd., Fort Macmurray, AB, Canada, (4)Syncrude Canada Ltd, Fort Macmurray, AB, Canada

Syncrude is one of the largest producers of crude oil from Canada's oil sands. The bitumen extracted from oil sands is processed in the upgrading section to produce value-added sweet crude oil. Hydrotreater units are used at the tail end of the upgrading section for Sulphur and Nitrogen treatment (and upgrading) of freshly distilled and re-run Heavy Gas Oil (HGO), Light Gas Oil (LGO) and Naphtha to produce marketable synthetic crude oil. Treated HGO product is sent to the storage tanks from where it is drawn down to the pipeline. Honeywell's multivariable predictive controllers are implemented on the HGO Hydrotreaters to control and maximize the rundown sulphur to the target.

Tank sulphur indication and control aids in running closer to the sulphur specification leading to substantial benefits by reducing over-treating requirement, reduced hydrogen consumption and extended catalyst run-length. This paper will present a method used by Syncrude for online indication and control of the Tank sulphur. The selected method uses the product analyzer readings to calculate the accumulated product tank sulphur and make correction to the hydro-treating reactor operation for online control of the Tank sulphur concentration.

In the present application, an online calculation has been implemented for predicting the current and future sulphur concentration in the HGO tank based on the Hydrotreater rundown flow and sulphur readings. A feedback control scheme has been implemented for controlling the accumulated product tank sulphur concentration by manipulating the hydrotreater rundown sulphur target in the linear multivariable controller. This paper will discuss some of the design and implementation issues.

(103e) Real-Time Monitoring of Trace Gas Concentrations in Syngas

Gerald Sprachmann[1], Rene Gutmann[2], Jens Herbig[2], Klaus Winkler[2] and Armin Hansel[3], (1)Shell Global Solutions International BV, Amsterdam, Netherlands, (2)Ionimed Analytik, Innsbruck, Austria, (3)Ion Physics and Applied Physics, University of Innsbruck, Innsbruck, Austria

A Proton Transfer Reaction Mass Spectrometer (PTR-MS) was implemented for real-time monitoring of volatile trace compounds in syngas in an industrial Fischer-Tropsch process. In industrial gas processes the knowledge about the exact gas composition is closely monitored and controlled. Beside the main constituents which typically appear in percent concentrations, also trace gas impurities at much lower concentrations can play a crucial role.

A poisoning effect of some impurities is a major problem in many catalytic reactions, especially hydrogen reaction such as methanation of coal synthesis gas or reforming of naphthas. Furthermore, when large amounts of off-gas are being released, also trace concentrations of compounds of extreme toxicity induce environmental risk.

A PTR-MS is a gas analytical device for the sensitive detection of volatile compounds. These devices are well established in several fields of research where their capability to analyze samples in real-time yields insight into system dynamics. Prime examples are the environmental monitoring, food and flavour science, medical and biotechnology research.

We demonstrate the extension of the PTR-technique to the monitoring of trace compounds in syngas. Together with a multiplexer this allows for real-time monitoring of the contaminations at different stages of a Fischer Tropsch process. Several volatile compounds, such as HCN, H_2S, RSH, carbonyls, acids, alcohols and others) have been measured in syngas down to parts per trillion levels. The potential of this novel approach is exemplified by a filter break-through.

We describe the setup to monitor syngas using PTR-MS. We summarize the result of this proof-of-principle project and outline the great potential for process monitoring.

Real-time Optimization of a Gasoline Run-down Header Blending Operation

Paper Number 103f

Shashi Mistry, Craig Mangan, Jim Eshpeter
Suncor Energy
Edmonton, Canada

Sanjay Sharma
Honeywell International
Houston, Texas

ABSTRACT

Refiners around the world are continually seeking ways to maximize margins on gasoline production. Optimization of various unit operations and gasoline blending using online blending packages are commonly employed. However, an area of opportunity that is less commonly exploited is the optimization of gasoline run-down headers. The economic benefits of gasoline run-down header optimization are enormous, but there are many challenges.

Run-down header optimization is challenging since many process units feed into the run-down header. Multiple control rooms, process operators and control systems are involved. This poses a significant challenge for a single application to optimize the entire system in real-time. Secondly, many of the process streams entering the run-down headers have multiple dispositions. Thirdly, the gasoline blending operation is highly non-linear. Lastly an on-line stream property measurement system that is sufficiently fast for on-line control and is able to provide a reliable property measurement over the entire blending range of components is a challenge.

Suncor Energy at Edmonton Refinery has been active in real-time optimization of the refinery operation for the last several years. While a majority of process units are being optimized using Advanced Process Control (APC), gasoline blending operation was, until recently, optimized using offline LP tools.

The gasoline blending operation at Suncor comprises Gasoline Blending operation and Gasoline Run-down Headers operation. An innovative project to optimize the gasoline run-down blending operation in real-time was recently completed and is the subject of this paper.

INTRODUCTION

Suncor Energy Corporation is one of the largest independent energy companies in the world. It is the fifth largest North American energy company. It is an integrated energy company with traditional exploration and production and refining operations, in addition to Wind Power and Biofuels operations. Suncor's second largest refinery is based in Edmonton, Alberta, Canada. This refinery processes 135,000 barrels per day (BPD) of the oil-sands based feedstock.

Suncor's Edmonton refinery is an integrated refinery and most of the process units in this refinery use Advanced Process Control (APC) for real-time optimization and control. Suncor management saw a need to minimize the quality giveaways in the gasoline product. A need was identified to optimize the gasoline blending operation.

Suncor's Edmonton refinery has standardized on Honeywell's Profit Suite to meet their Advanced Process (APC) applications needs. Therefore, Honeywell was selected to work with Suncor project team for the gasoline blending optimization. The gasoline run-down header optimization is a part of this project and is the subject of this paper.

The traditional approach for gasoline blending is to carry out the blending operation from run-down component tanks and the blend optimization is carried out as a part of this blending operation. There have been several documented approaches to optimize this operation and several commercial solutions are available. However, carrying out the real-time optimization on the run-down blending operation provides some key advantages over this traditional approach. The benefits of the real-time optimization on run-down headers are summarized below:

a. It minimizes the number of blend stock component tanks;
b. It provides a much faster response to dynamic changes in the refinery operations;
c. It provides a greater flexibility to produce multiple grades of gasoline at the lowest cost;
d. It minimizes product quality giveaways at the blend header level;
e. And finally, it provides a greater opportunity to optimize the blending and the refinery operation.

The project has resulted in over 5% increase in finished gasoline production while virtually eliminating the product quality giveaways. The project pay-back period is around two to three months. The controller was handed over to Operations in May 2012 and it has been in service since then with 100% uptime.

Benefits from the project are substantiated by the Edmonton Refinery Vice President as follows:

"By implementing a NIR based gasoline blending controller at the Suncor Energy Edmonton Refinery we have generated a number of benefits. True blending optimization is now confirmed through the capture of thousands of data points on tank properties, rather than a 'once per day' capture. This technology has enabled the refinery to:

- ***be proactive and blend gasoline in real-time with optimization***
- ***be able to complete target changes seamlessly while minimizing quality risks***
- ***achieve record maximum gasoline production volumes in 2012***
- ***manage gasoline blending to maximum profit or maximum volume, depending on the particular refinery constraints at that time***

The benefits and flexibility achieved have exceeded our initial expectations identified when this work was first undertaken.

John Prusakowski,
VP Edmonton Refinery "

The real-time gasoline run-down header blending optimization has some key challenges. First, the process and the geographic scope of the run-down header are large. It includes multiple process unit operators and multiple control rooms. Thus a control system infrastructure that provides communication to all associated process unit control system is necessary. Secondly, blending operation is non-linear. Therefore, a control system that can provide non-linear real-time optimization and control is necessary. This paper presents how these challenges were addressed. In addition, this paper also presents the benefits of implementing a real-time blending optimization on gasoline run-down headers.

PROCESS OVERVIEW

The Suncor Edmonton refinery processes 135,000 BPD of oil-sands based feedstock. The refinery is a complex and integrated refinery. Gasoline run-down components produced from different process units are sent to either one of the two gasoline run-down headers (Header A or Header B) or to a storage tank or other process units.

The property of each run-down component stream is managed by the individual unit operator. The run-down component stream flow rate into each header or to storage is set by planning. The run-down component from each process unit mixes with other gasoline run-down component at the process unit battery limit (Figure 1). The final blended stream is sent to its corresponding pre-blend tank.

The material in the pre-blend tank is then trimmed to the final gasoline specifications. This final blending operation is referred to as "trimming" in this paper. The trimming operation is carried out to make the desired gasoline product (BOB, RUL, SUL etc.).

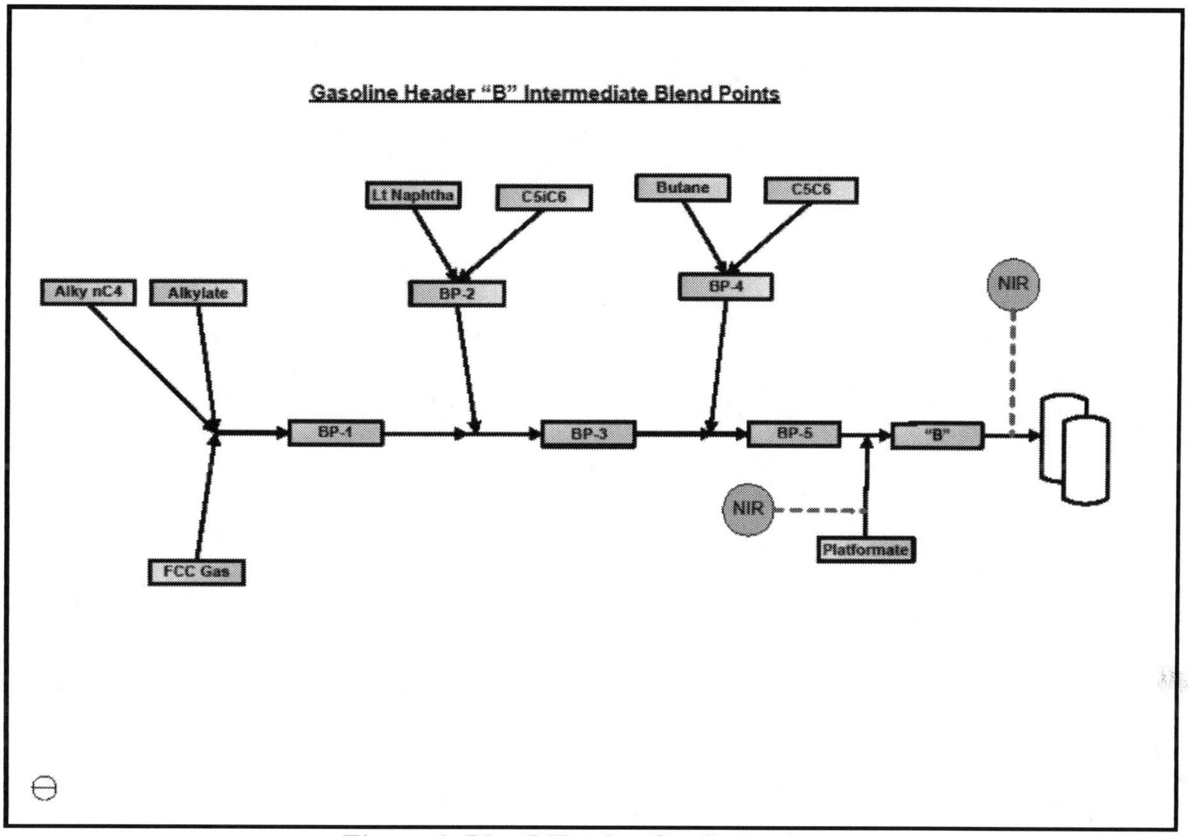

Figure 1 Blend Header Configuration

Figure 2, below, shows a simplified process flow diagram of the gasoline run-down headers.

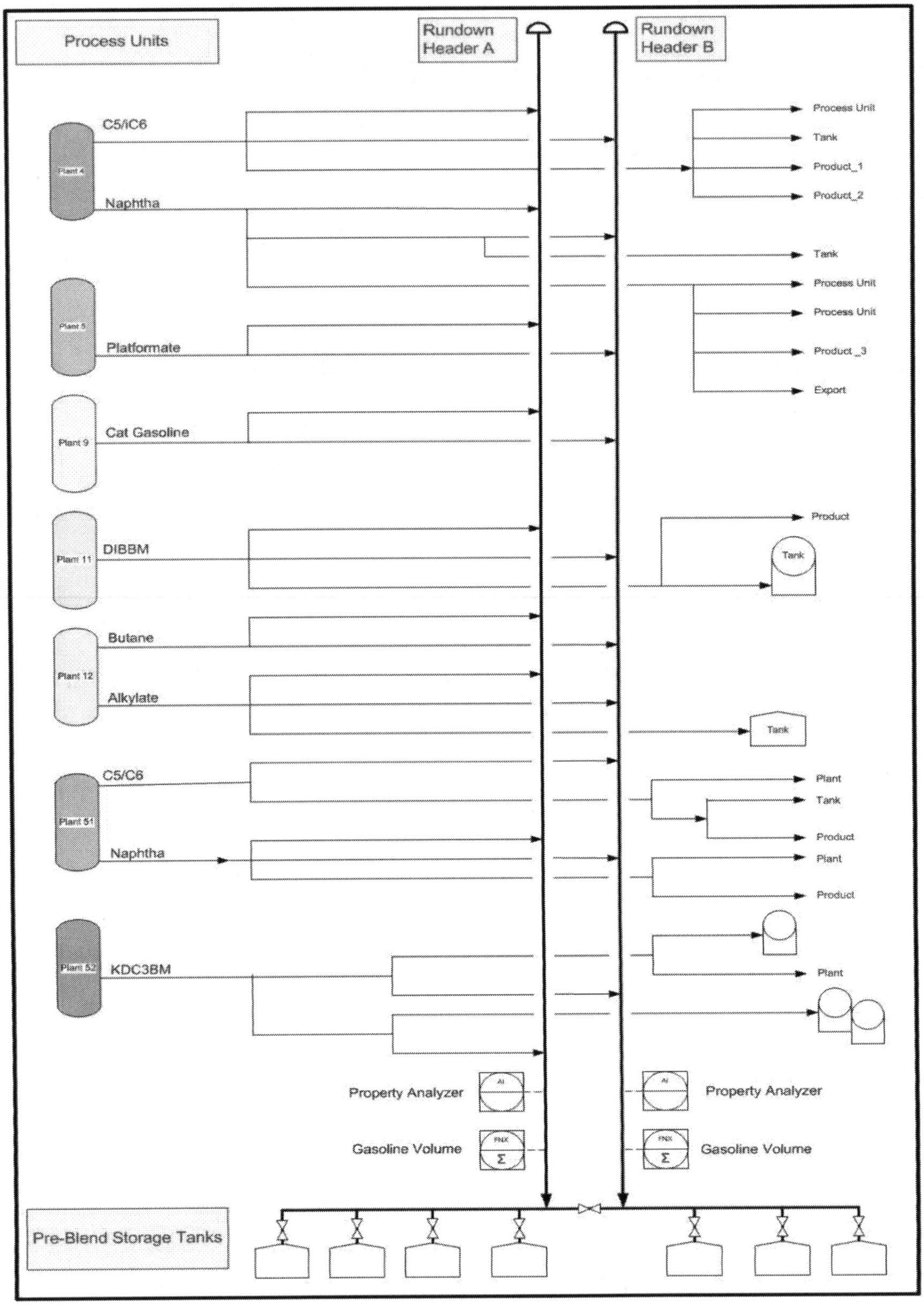

Figure 2 Run-down Headers Simplified Process Flow Diagram

OPERATING CHARACTERISTICS

Gasoline run-down header operation is different from a typical process unit in many ways which force the APC design to be unique. Some of the unique characteristics of the run-down header operation are:

Blend-stock quality variations

Multiple process streams from various process units feed into the run-down headers and each stream's quality varies dynamically. Hence, the run-down header stream's blended properties vary as individual input stream quality changes.

Blend-stock Availability variation

On an average, roughly ten process streams from various process units enter into each run-down header. The amount of available run-down material changes dynamically. In addition, some of the run-down streams are available only intermittently. This changes the available degrees of freedom to meet the operating objectives dynamically.

Secondly, because so many process units impact the operation of the run-down header, the run-down header operation is susceptible to upsets in any one of the process units.

As an example, some of the butane streams that enter the run-down header can change from zero (0) to five (5) percent of the total flow to the header in a matter of minutes. This has a very large impact on the RVP of the header.

Multiple product quality specifications

The refinery produces gasoline products with various specifications. Some of these include: Blend stock for Oxygenate Blends (BOB), Regular Unleaded (RUL) and Super Unleaded (SUL). The specifications of these products not only change with season, but sometimes change to meet a short term local demand.

Multiple control systems

As stated earlier, on the average, roughly ten process streams enter in each run-down header. Therefore, multiple units' control systems are involved. The real-time optimization applications have to be able to interact with multiple control systems.

No single "owner" of process

The run-down header operation involves multiple process units. Each process unit operator is responsible for his or her process unit and the process stream(s) that enter the run-down header. There is no single operator who has the visibility of the entire run-down header and thus can manage the run-down header as a whole. The refinery Planning and Economics department staff, the Planner, manages the run-down header on a daily basis or on an as needed basis wherever there is a significant change in the refinery operation. On a dynamic basis, there is no "owner" of the run-down header who has the visibility and the control of the run-down header property and volume.

Significant non-linear process

Gasoline blending is a highly non-linear operation. Both, the run-down component property and the flow changes, impact the blend header's quality in a non-linear fashion. The magnitude of the change can result in under or over specification of product quality that may lead to large economic penalties.

Additionally, run-down stream availability changes, product demand variations, product quality variations, process disturbances and the non-linearties of the blending operation results in a process that is highly dynamic and difficult to control and optimize in real-time. However, the real-time control and optimization of this process has significant economic benefit. Given that the dynamic nature of process operation is a very important consideration in the design of a real-time control and optimization solution, the run-down header control and optimization solution favors a solution with the following features:

- Robust MPC solution that handles model mismatch well;
- A robust real-time blending model that predicts the blend properties well;
- Robust MPC solution that has the ability to read the process model in real-time.

HIGHLIGHTS OF TECHNOLOGY USED

Honeywell's Profit Controller, Profit Controller Gain Mapper, Unified Real-time framework software development kit (URT/SDK) and non-linear blending model technologies available to Suncor are used in the project. These technologies are briefly described below.

Unified Real-time Software Development Kit (URT/SDK)

Honeywell's Profit Suite is based on a Unified Real Time Framework (URT) that enables the Profit Suite applications to execute in real-time and interact with other Profit Suite applications and other non-Profit Suite applications in real-time. One of the components of the Profit Suite is the Software Development Kit (SDK) that allows the user to create real-time applications using higher level languages such as Visual Basic.Net, C++ or C#. These user-created applications can seamlessly interact with the other Profit Suite applications and other non-Profit Suite applications in real-time. URT/SDK was used to build the non-linear blending model that executes in real-time and provides the updated model gains for the controller.

Profit Controller (RMPCT)

Profit Controller is Honeywell's model based, robust multivariable predictive controller. The controller executes in real-time under URT framework. This controller has been well documented by many authors [Lu, 1997; Nath, 2001]. Some of the key features of Profit Controller that are relevant to this implementation include:

a) Robustness – the ability to handle significant model mismatch including the ability to handle gain sign reversal in real-time;
b) Real-time gain updating;
c) Ability to optimize based on MV and CV values.

Gain Mapper

The Gain Mapper is a Profit Suite component. It provides the capability to seamlessly update the gains in Profit Controller in real-time. It includes the capability to validate the gains generated by an external model prior to updating the gains. The Gain Mapper application extracts the model gains from the non-linear blend model application and populates them in the Profit Controller in real-time.

Blending Model

The gasoline blending operation is highly non-linear. So, a reliable and accurate gasoline blending model was necessary to generate the real-time model gains for Profit Controller. The blending model uses each input stream's flow rate and various properties to calculate the blended property. The model calculates the following blended properties for each header:

- Research Octane Number (RON);
- Motor Octane Number (MON);
- Reid Vapor Pressure (RVP);
- ASTM Distillation 10% Point
- ASTM Distillation 50% Point;
- ASTM Distillation 90% Point;
- E200/E300
- Benzene
- Total Aromatics

For the run-down gasoline blending at Suncor, the final quality standard is the laboratory results. Periodic laboratory samples are collected from each header and the NIR analyzer results from the corresponding sample time are compared. In addition, the pre-blend tank samples are collected and validated with the laboratory results.

The model results are compared against each header's NIR reading periodically. The blending model is of sufficient fidelity that no process feedback in real-time was deemed necessary. In any case, since only process derivatives are required for gain updating, real-time bias update or rigorous parameter fitting is not required.

Figure 3 below compares the models' road octane calculation results and Figure 4 compares the RVP calculation results against the NIR readings for one of the headers.

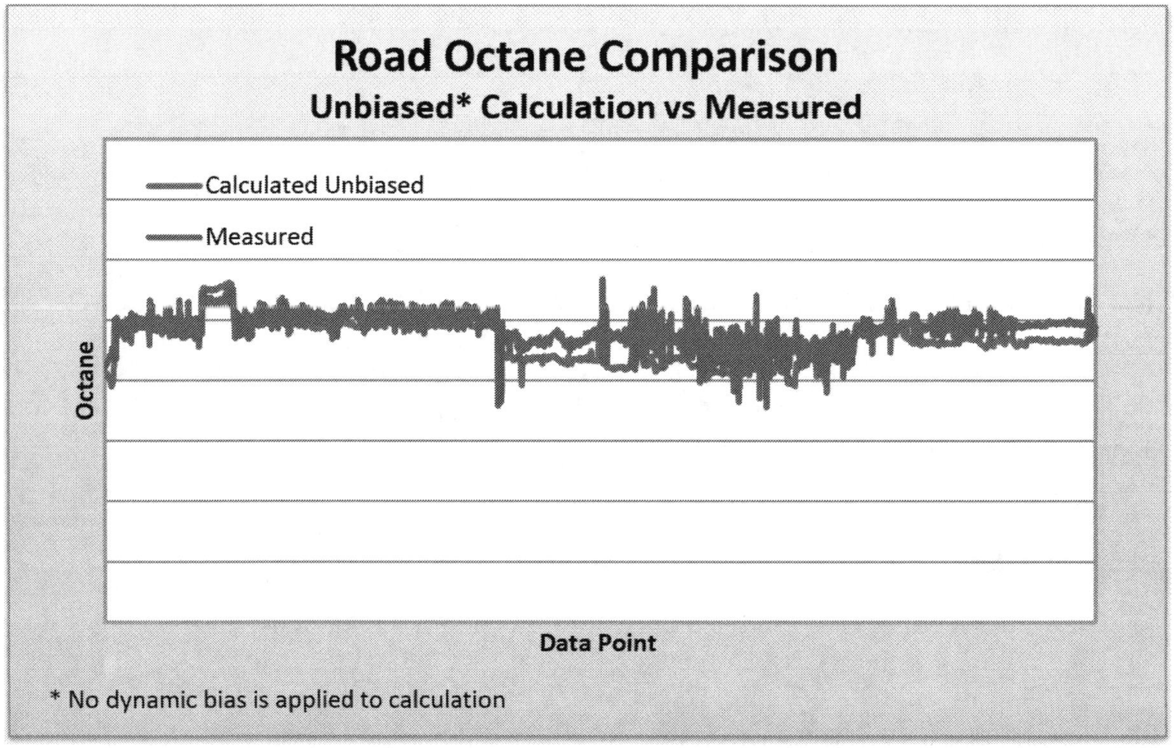

Figure 3 - Model vs. NIR Octane Comparison

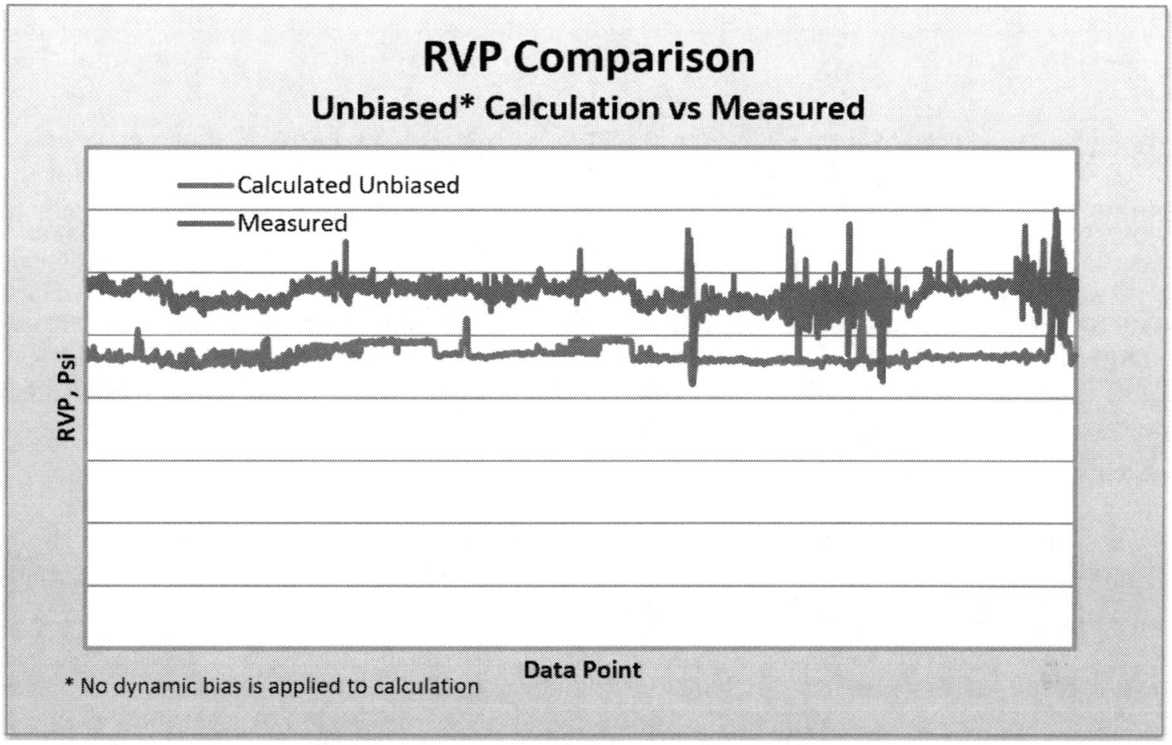

Figure 4 Model vs. NIR RVP Comparison

Goals and Requirements

From the planning department's point of view, the goals of gasoline blending can be simply stated is to maximize profit at the refinery level. From gasoline production perspective, generally, this translates into producing maximum amount of on-specification gasoline at minimum cost. Thus, the goal is to produce on-specification gasoline, at minimum cost, from the run-down headers while maximizing header volume.

PROJECT DETAILS

The gasoline run-down header optimization is a part of a larger gasoline blending optimization project. This paper focuses on the run-down header optimization part of the project.

One of the basic requirements for such type of project, that is generally unusual for a typical APC project, is the integrated control system infrastructure. The three basic challenges for this type of application are control system infrastructure, measurement and modeling system and ownership scope of the applications. Suncor had envisioned the gasoline run-down header optimization some time ago and had laid the foundation for this work. The run-down header APC implementation phase of the project commenced in January 2012.

Multiple process units impact the gasoline run-down headers. Additionally many operators in different control rooms impact the operation of the gasoline run-down headers. So, a control system infrastructure that provides the ability for the Advanced Process Control (APC) applications to work with all of the process units involved is necessary. Suncor has standardized on Honeywell's Experion PKS control system for the Edmonton refinery. Therefore, Experion PKS's DSA architecture was used to communicate with all of the process units involved in real-time.

Single Profit Controller application was implemented to control both header's properties and volume rate. In addition, run-down stream spillover flow rates to storage tanks were also included in the controller. One NIR analyzer on each header provides the blended stream's property to the Profit Controller. The controller's economic objective function was set to maximize profit.

A non-linear blending model was implemented to calculate the model gains for the Profit Controller at the current operating point. A Gain Mapper application then validated these model gains and subsequently populated them in the Profit Controller.

As a result of the project, the gasoline blending optimization workflow was revised. Prior to the project, the Refinery Planner provided daily targets to each operator for each run-down stream disposition. Although the Planner had the visibility of each gasoline run-down header's properties and volumetric flow rate in real-time, generally, new targets to each individual unit operators were sent only when a significant deviation from the plan were noticed.

At the beginning of the project, it was decided that the Planner should be the "owner" of the gasoline blend header optimization. The individual operators had control on the availability of their unit's controller while the planner provided the control targets for these controllers and other variables in the APC applications. The Planner then has the flexibility to adjust the blending operation in real-time as the economic or a unit operating conditions changed. Figure 5 below shows the work flow after the project implementation.

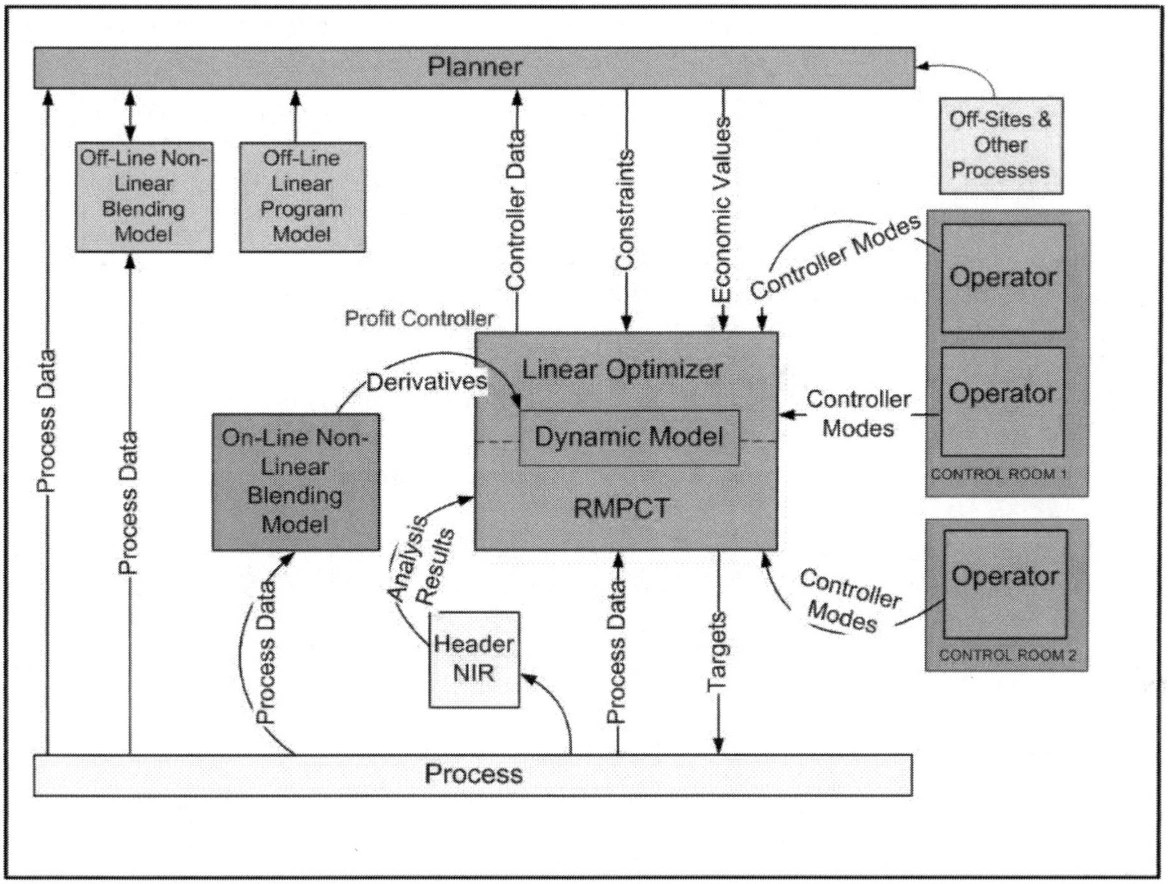

Figure 5 - New Workflow

The Planner has the ability to adjust the product specification of each header, the volumetric flow rate from each header and the volumetric flow rate of each run-down stream into the header in real-time. The Planner generally provides these values in terms of high and low limits. As an example, the Planner generally specifies a high and a low limit for road octane (RDON) for each header.

In addition, the Planner specifies an economic value of each stream entering into the header and a product value of each header flow rate. The APC application then seeks a feasible solution that maximizes profit, while honoring the constraints specified by the Planner.

PROJECT RESULTS

The economic benefits from the gasoline run-down header blending optimization are derived from the following outcomes of the APC application:

a) By improving the control of the header properties and volumes (minimize quality giveaways);
b) By increasing product value upgrade;
c) By increasing flexibility to adjust blending operation quickly (Logistics);
d) Operating constraints to be corrected at the plant level (Unit Operations).

The ability to control each header's property and volumetric production rate provides greater flexibility to the refinery planning personnel to respond to changing conditions. As unit operating conditions or economic conditions change, the Planner is able to adjust the blending operation quickly to adjust or to take advantage of these changes. This flexibility provides him or her with the capability to execute and confirm the refinery wide optimization plan in real-time. As seen in Figures 6, 7 and 8, the Planner is able to adjust the operating targets in real-time to increase profitability.

Figures 6 and 7 provide an example of one of the header's Road Octane (RDON) and RVP control before, during and after the commissioning time. As commissioning activities were completed, the controller was able to provide much tighter control on one of the header's properties (Octane, RVP etc.) while meeting the volumetric requirements.

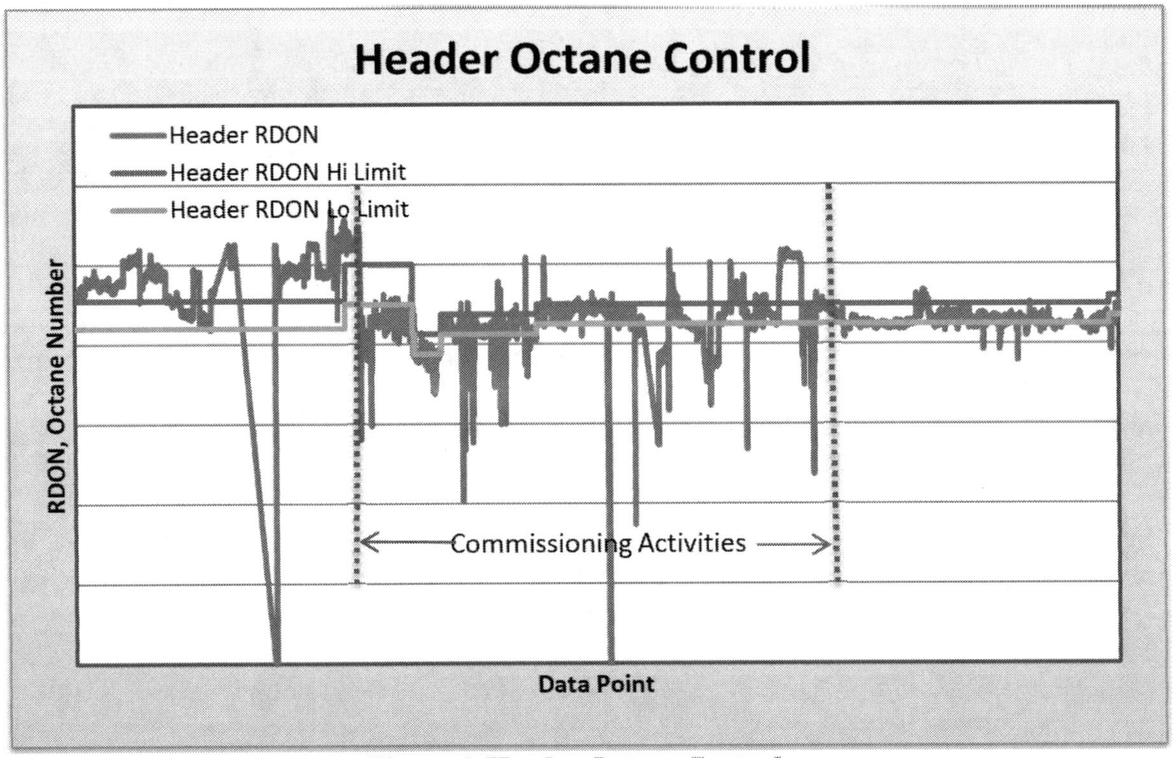

Figure 6 Header Octane Control

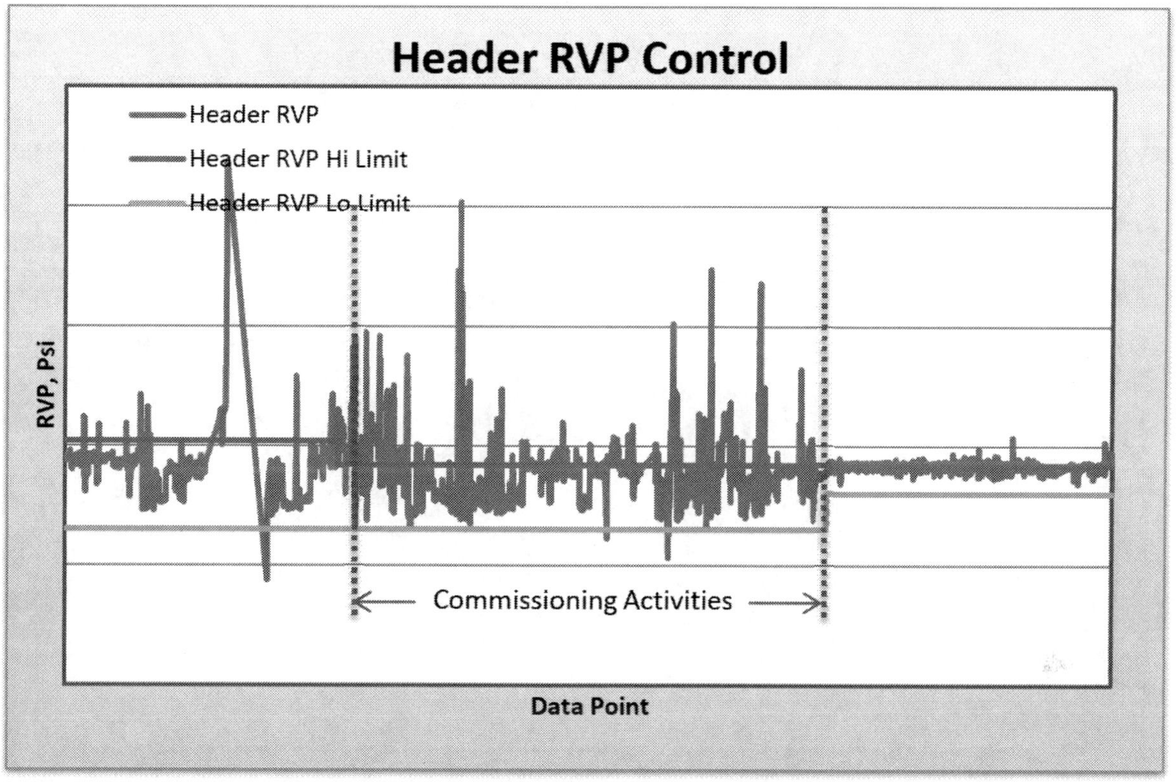

Figure 7 Header RVP Control

The second benefit from the APC application is the ability to carry out product value upgrade. The APC application is configured to maximize the profit, which entails maximization of the use of low cost streams, and generally maximization of gasoline production (the application will not increase gasoline production if the marginal cost is higher than the marginal value of the product.). In the case of the gasoline blending, this also results in increase in gasoline volume production (Figure 9).

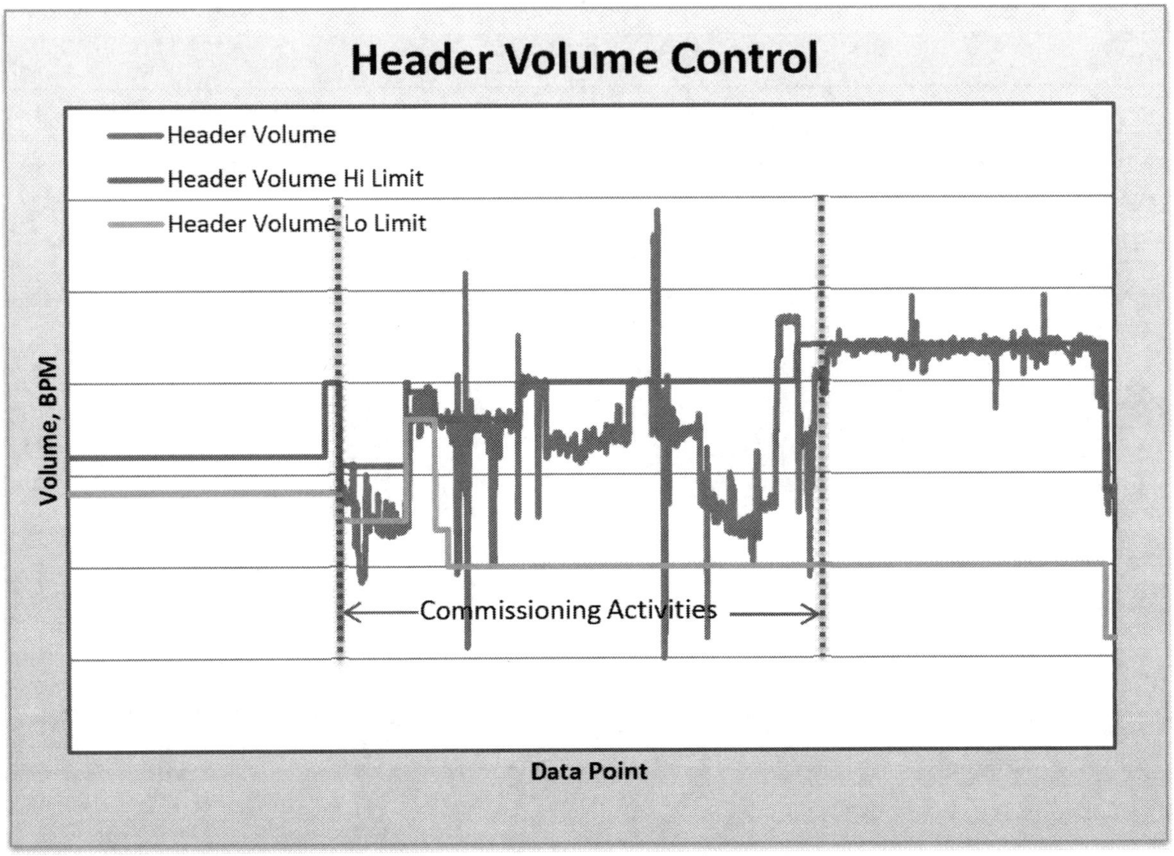

Figure 8 Header Volume Control

The overall, normalized gasoline production has increased by 5% (after accounting for factors such as crude rate changes etc.). The refinery has been able to set a new record in gasoline production.

This increase in gasoline production is due to increased usage of low cost blend components that otherwise would have been sent to storage. Because of piping infrastructure, once sent to storage, these components generally cannot be added back to the blend pool and are generally sold in the open market at a lower value than gasoline. Thus, APC application is able to help increase the marginal value of these components by blending them into the gasoline product.

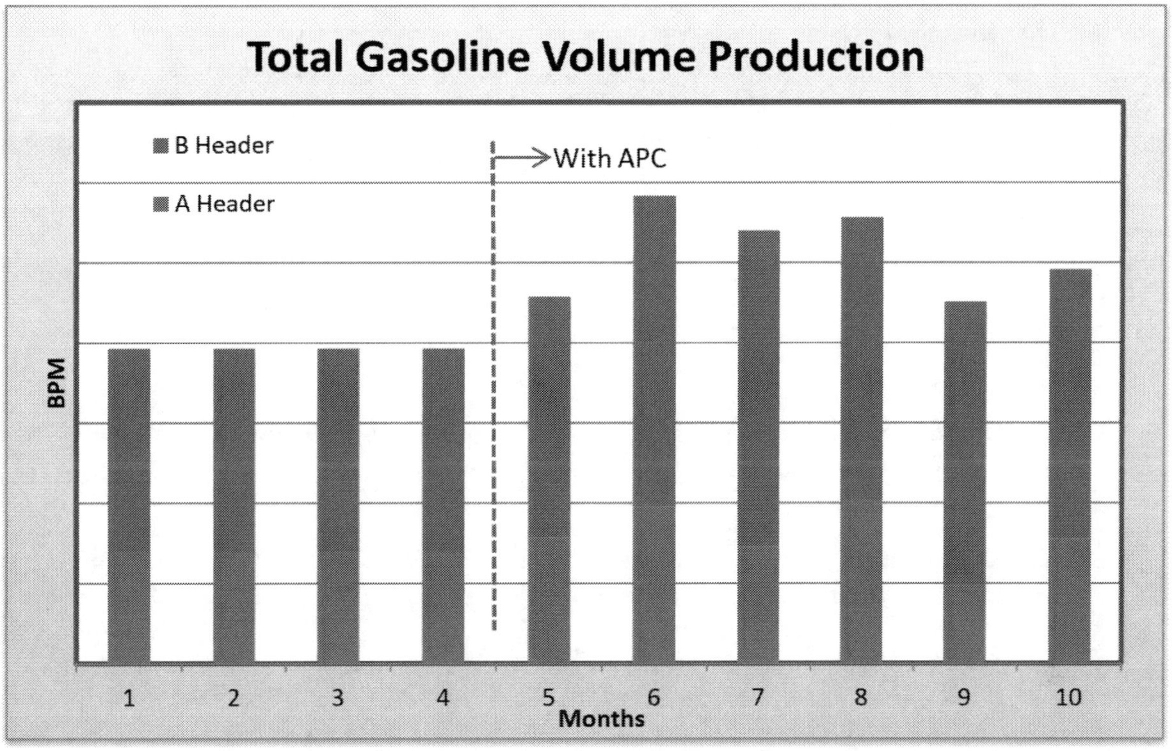

Figure 9 Total Gasoline Volume Production

Another benefit of maximizing the usage of all gasoline blend components in the run-down header is that it reduces the blend-stock tank inventory and provides maximum inventory of pre-blend gasoline. This pre-blend gasoline then can readily be used to make the desired final product.

CONCLUSION

Gasoline run-down blend header operation is very challenging and real-time optimization of the run-down header is, therefore, quite challenging as well. An innovative solution was implemented at the Suncor Refinery in Edmonton to optimize gasoline blending operation on the gasoline run-down headers. The APC solution has enabled Suncor to maximize the product value upgrade, by upgrading low cost blend components into higher value gasoline product. This has resulted in an increase of finished gasoline production by 5%. As a result, Suncor has set new records in gasoline production.

ACKNOWLEDGEMENT

We would like to thank the management of Suncor Energy and Honeywell for permission to publish this work. In addition, we thank Mr. Don Yeomans of Suncor Energy for his invaluable contribution this project.

REFERENCES

1. Anonymous (2008), "Advanced Process Control: Profit Controller Concepts Reference Guide". Profit Controller Documentation.

2. Anonymous (2008), "Advanced Process Control: Gain Mapper for Profit Controller Users Guide". Profit Controller Documentation.

3. Lu, Z. J., John A. Escarcega, "RMPCT: Robust MPC Technology Simplifies APC" Paper presented at AiChe Spring Meeting (1997).

4. Nath, Ravi, Zak Alzein, Rogier Pouwer, Mario Lesieur, "On-line Dynamic Optimization of an Ethylene Plant using Profit® Optimizer", Paper presented at NPRA Computer Conference (1999).

5. Anonymous (2010-12), "Gasoline Blending Technology Development for On-line Applications", IPCC Inc.

6. Anonymous (2012), Profit suite Product Documentation.

ACRONYM DICTIONARY

Table 1: Acronyms used in the paper

NO.	ACRONYM	DEFINITION
1.	APC	Advanced Process Control. This is generally achieved through the use of multivariable predictive control technologies.
2.	BOB	Blend stock for Oxygenate Blending
3.	BPM	Barrels Per Minute
4.	CV	Controlled Variable
5.	LP	Linear Programming
6.	MON	Motor Octane Number
7.	MPC	Multivariable Predictive Controller
8.	MV	Manipulated Variable
9.	NIR	Near Infra-Red Analyzer
10.	RDON	Road Octane Number
11.	RON	Research Octane Number
12.	RUL	Regular UnLeaded gasoline
13.	RVP	Reid Vapor Pressure
14.	SUL	Super UnLeaded Gasoline

(106a) Challenges and Opportunities in Oil Fields Produced Water Treatment

Rafique Janjua, Process, Fluor Enterprise Inc, Sugar Land, TX

Produced water treatment composition varies all over the map and requires specific unit processes and unit operations with each and every site. Produced water from Canadian Oil Sands is codendate from high pressure steam flooging/injection into Bitumen layer It have heavy oil emulsion, high Silica and dissolved solids as compared to tradional oil fields with more free oils, very high salinity and heavy metals. Produced or surface water used for Enhanced Oil Recovery have differnt reqiurements, such as suspended solids removal, corrosion and scale protection, complete Oxygen scavanging, selective Sulfate removal and Biocide injection which will require totally different treatment process. There are different challanges and opportunities for cost effective and reliable sytems for each application. This paper will discuss some real life examples how to achieve the desired results.

(106c) Sustainability in Industrial Water Treatment System

Vina Arjomandnia, Process Technology, Fluor, Sugar Land, TX

Sustainability in industrial water treatment system

Water is as important in the oil industry as the crude feed itself.

Proper water treatment is essential for reliable and efficient refinery operation as well as compliance with social and environmental regulations.

Sustainable water treatment system in industry should provide adequate water quantity and quality for general and specific needs such as high pressure steam generation. Wastewater streams selective source segregation and treatment enhances maximum sustainability and recovery for reuse as make up water in cooling towers etc. Non contaminated rain water along with treated contaminated rain runoff can provide a sustainable water supply for industry.

This paper will discuss designs where these methods could be implemented to ensure the sustainability of the industrial water demands, on domestic and international projects.

(106d) Overview of Boiler Water Treatment and Associated Wastes

William Moore, Industrial, WesTech Engineering, Salt Lake City, UT

Overview of Boiler Water Treatment and Associated Wastes

Boiler water treatment consists of steps to remove contaminants that will corrode or cause scale on the boiler tubes. As the boiler pressure increases, more contaminants must be removed. There are also various chemicals added to the boiler water. These chemicals and the wastes from the water treatment systems can produce an interesting soup for discharge. This paper discusses the systems used to produce boiler feed water and the ensuing discharge chemistry.

Genetic Algorithm Approach for Optimal Control of Air Emissions from an Oil Refinery

Mohammed S. Ba-Shammakh *
Chemical Engineering Dept., King Fahd University of Petroleum and Minerals,
P.O Box 5050 (31261), Dhahran, Saudi Arabia

E-mail: shammakh@kfupm.edu.sa

Abstract

Air pollution is a major issue facing the world nowadays. Many human activities and several industries have contributed to this problem. Oil refinery is an example of that and different air emissions are emitted. Examples of these emissions are; SO_2, NO_x and CO. Various studies have been conducted for air pollution control from previous decades, but no researches were extended about selecting air emissions control methods for oil refinery units considering, simultaneously, cost and efficiency of the control options. The aim of this study is to determine suitable control technique for oil refinery with two objective functions including minimization of cost and air emissions according to environmental standards. Genetic Algorithm approach is used to select an optimal control technology for each air emissions considered. The model is applied into a case study to show its application.

Key words: Air emissions, Genetic algorithm, Oil refinery

Introduction

The refining industry plays a very important role in international economics and in our daily life. Environmental regulations are pressing the refinery industry to minimize its air emissions. Petroleum refining converts crude oil into useable products, such as gasoline, asphalt, and jet fuel. This process heats crude oil to separate its components into several final products. By altering the temperature and the specific gravity of the crude oil, refineries may alter the overall composition of their final products.

The petroleum refining process results in the release of a number of air emissions from different sources or processes, including but not limited to SO_2, NO_x, and CO. In addition to the previous mentioned emissions, many greenhouse gasses (GHGs) are released from this industry but will not be the focus of this paper.

Sulfur dioxide (SO_2) itself can cause adverse effects on respiratory systems of humans and animals, and damage to vegetation. When dissolved by water vapor in air to form acids (acid gas) it can again have adverse effects on the respiratory systems of humans and animals, and it can cause damage to vegetation, buildings and materials, and contribute to acidification of aquatic and terrestrial ecosystems. When transformed into sulphate particles that are subsequently deposited on aquatic and terrestrial ecosystems, acidification can result, and when sulphate is combined with other compounds in the atmosphere, such as ammonia, it becomes an important contributor to the secondary formation of particulate matters (PM). Main sources of the SO_2 emissions in a petroleum refinery include combustion sources, thermal or catalytic cracking processes, reforming process and sulfur recovery plant.

Nitrogen oxide (NO_x) is another main air pollutant from a petroleum refining and can be defined as a generic term for the various nitrogen oxides produced during combustion process. They are believed to aggravate asthmatic conditions, react with the oxygen in the air to produce ozone, which is also an irritant and eventually form nitric acid when dissolved in water. The main sources of this pollutant in a petroleum refining are combustion processes.
Another major pollutant from a petroleum refining is carbon monoxide (CO). It is a colorless, odorless, tasteless and poisonous gas. It is a product of incomplete combustion of hydrocarbon-based fuels. CO can have a significant impact on human health. It enters the bloodstream through the lungs and forms carboxyl hemoglobin, a compound that inhibits the blood's capacity to carry oxygen to organs and tissues. The main source of this pollutant in a refinery is combustion process.

Since combustion is the major source for SO_2 emissions in an oil refinery, most of the research dealing with SO_2 control technologies, in an oil refinery, focused on either reducing the output from a unit that emits more SO_2 or control the emissions after it is emitted. These may include switching to run with fuel that has low sulfur contents or to control SO_2 emissions by installation of different control technologies. Each of these technologies has its own cost and it will defiantly affect the profit of a given refinery if it is installed or implemented.

Emissions control measures leading to reductions in NO_x is expected to reduce population exposures to all gaseous NO_x. Many technologies have been invented for NO_X reduction. Selective Catalytic Reduction (SCR), Selective Non-catalytic Reduction (SNCR) and low NOx burner are common technologies in the USA for oil refinery NO_X reduction.

Table 1 lists some common technologies that can be used for each pollutant under consideration.

Table 1: Air emissions with common technologies

Technology	NO_x	SO_2	CO
SNCR	X	-	-
SCR	X	-	-
Low-NO_x Burner	X	-	-
Thermal incinerator	-	-	X
Catalytic incinerator	-	-	X
Wet Scrubber	-	X	-
Spray Dry Scrubber	-	X	-

Many studies have been conducted in the field of air pollution control considering cost and efficiency [1-3]. The issue is very effective on point sources pollution control especially stacks and efforts have been conducted by several researchers [3-6]. To select effective control options with least cost, optimization approach should be used meeting air emissions standards. Optimization methods, heuristic or stochastic, are inherent for choosing suitable control technologies for different air emissions within an oil refinery [7].

Talking about modeling approach, the current available dispersion models are either analytical or stochastic. Analytical models, such as Box model, Plume Gaussian and Multiple-cell model, are based on mass transfer in air [8] whereas stochastic models such as Artificial Neural Network (ANN) or Fuzzy Logic (FL) are based on trained data [9]. Dispersion depends on the weather condition and how much air is needed to transfer and dilute pollutants. Dispersion of pollutants in air depends on different factors including wind velocity, wind direction and ambient air temperature. Multiple-cell model was used for displaying these factors.

The mathematical description of these processes leads to a system of partial differential equations as follows:

$$\frac{\partial C}{\partial t} = -\frac{\partial(uC)}{\partial x} - \frac{\partial(vC)}{\partial y} - \frac{\partial(wC)}{\partial z} + \frac{\partial}{\partial x}(K_x \frac{\partial C}{\partial x}) + \frac{\partial}{\partial y}(K_y \frac{\partial C}{\partial y}) + \frac{\partial}{\partial z}(K_z \frac{\partial C}{\partial z}) + E_s - (k_1^S + k_2^S)C + Q_s(c_1, c_2, \ldots, c_q),\ s = 1,2,\ldots,q \quad (1)$$

C : Concentration of the air emissions involved in model

U : Wind velocity,

K_x, K_y and K_z : diffusion coefficients,

E_S : Emission source,

k_1^s and k_2^s Deposition coefficients (for the dry deposition and wet deposition respectively)

Q_s for chemical reaction.

The assumptions are:

1. Steady state
2. U_y=U_z=0
3. (k_x=0) since transport by bulk movement in the x-direction surpasses diffusion
4. No reaction (Q=0)
5. No deposition in the system ($k_1^s = k_2^s$=0)

By considering the assumptions, equation (1) reduces to:

$$\frac{\partial(uC)}{\partial x} = \frac{\partial}{\partial y}(K_y \frac{\partial C}{\partial y}) + \frac{\partial}{\partial z}(K_z \frac{\partial C}{\partial z}) + E_s \quad (2)$$

The finite difference method is employed for solving the above equation and the following boundary and initial conditions are used:

at x =0, $C(0,j,k) = 0$

at y =0, $$\frac{\partial C}{\partial y} = 0$$

at y =W, $$\frac{\partial C}{\partial y} = 0$$

at z =0, $$\frac{\partial C}{\partial z} = 0$$

$\frac{\partial C}{\partial z} = 0$ at z = mixing height ,

After determining boundary and initial conditions, Eddy diffusivity, wind velocity profile and others are taken from literature [10].

Genetic Algorithm Approach

Genetic Algorithm (GA) was developed by John Holland in the 1960s and it is a stochastic optimization method for discrete variables [11]. The algorithm starts with randomly generating population of chromosomes and utilizing genetic operators that include selection, crossover and mutation operators [12]. The selection operator chooses chromosomes from the current population to produce new population. The crossover operator makes two new chromosomes from two existing chromosomes by cutting them at a random position and exchanging the parts of the following cuts. The mutation changes the chromosome genes randomly. These operators simultaneously guide random search method, which can yield an optimum set of function variables.

Main steps for carrying out GA are as follows:

1. **[Start]** Generate random population of *n* chromosomes (suitable solutions for the problem)
2. **[Fitness]** Evaluate the fitness *f(x)* of each chromosome *x* in the population
3. **[New population]** Create a new population by repeating following steps until the new population is complete
    1. **[Selection]** Select two parent chromosomes from a population according to their fitness (the better fitness, the bigger chance to be selected)
    2. **[Crossover]** With a crossover probability cross over the parents to form a new offspring (children). If no crossover was performed, offspring is an exact copy of parents.
    3. **[Mutation]** With a mutation probability mutate new offspring at each locus (position in chromosome).
    4. **[Accepting]** Place new offspring in a new population

4. **[Replace]** Use new generated population for a further run of algorithm
5. **[Test]** If the end condition is satisfied, **stop**, and return the best solution in current population
6. **[Loop]** Go to step **2**

Case Study

Oil refinery consists of different units and sources of air emissions. A typical simple refinery layout was taken as an example and ten different sources were identified and their emissions were reported as in the table below.

Table 2: Sources with air emissions

Source	NO_x (ppm)	SO_2 (ppm)	CO (ppm)
1	128	287	0
2	116	260	0
3	96	260	0
4	45	12	30
5	52	8	22
6	50	2	6
7	45	8	16
8	42	3	3
9	55	5	8
10	40	107	6

GA is applied targeting two objective functions dealing with minimizing the air emissions below the standard and the cost. For this problem, the length of chromosomes is the number of refinery's air emission sources which consists of individual genes. Each gene yields different controllers that can be used for related source. The initial population of the chromosomes was generated randomly. The fitness of each chromosome could be obtained by calculating the weighted value of cost and air emission concentration. The objective function presented in this paper is shown in Equation 3.

$$fitness(p) = W_{cost}\frac{\log_{10}(FC(p))}{\log_{10}(Cost_{\max})} + W_{pollution}\frac{\log_{10}(FP(p))}{\log_{10}(V_{CO}^2 + V_{NO_x}^2 + V_{SO_2}^2)} \quad (3)$$

p :	evaluated chromosome,
$FC(p)$:	controller's cost of chromosome p,
$FP(p)$:	weighted value of pollutants concentration on the ground
W_{cost} :	cost weight
$W_{pollution}$:	pollutants weight.

$W_{pollution}$ has more impact value than W_{cost} because air pollution control is more important than cost reduction. The value of these weight functions was settled by calculating average and variance of objective function for each value illustrated in Table 2. So the best cost and air pollutant weight were 0.2 and 0.8 respectively to have minimum value.

The amount of emissions (SO_2, NO_x, and CO) on the ground level after considering proposed controllers by utilizing the model developed was calculated and then *FP* was found based on equation 4. If each of the pollutant's concentration is more than its standard, penalty value is calculated for related pollutant by utilizing equations (a) to (c) below.

$$FP(p) = (V_{CO} + P_{CO}) + (V_{NO_x} + P_{NO_x}) + (V_{SO_2} + P_{SO_2}) \qquad (4)$$

Where V_{CO}, V_{SO_2} and V_{NO_x} are the pollutants volume on the ground level and P_{NO_x}, P_{SO_x}, P_{CO} are penalties of each pollutant which were computed by the following equations.

$$P_{CO} = (V_{CO} - V_{CO\_Standard})^2 \qquad (a)$$

$$P_{SO_2} = (V_{SO_2} - V_{SO_2\_Standard})^2 \qquad (b)$$

$$P_{NO_x} = (V_{NO_x} - V_{NO_x\_Standard})^2 \qquad (c)$$

After computing objective function for all existing chromosomes in each generation, the genetic algorithm was utilized and the chromosomes were sorted in order of increasing objective function. The best two chromosomes were transferred to new population. The couple was selected from mating pool by tournament selection and offspring produced by two point crossover technique. Mutation is accomplished by deliberately changing one gene in a random fashion. The whole process was repeated for predefined number of generations.

Results and Discussion

Air emissions control within an oil refinery is considered to be a challenging issue since the aim is to minimize cost and at the same time, minimize air emissions. An optimization approach is carried out in this study using genetic algorithm for a typical simple oil refinery layout. Table 3 summarizes the control option for each source mentioned earlier. The model does not allow applying more than a control technology on each source.

Table 3: Control option for each source

Unit.	Control
1	SNCR
2	Wet Scrubber
3	Wet scrubber
4	Thermal incinerator
5	No control
6	Low-NO_x Burner
7	No control
8	Thermal incinerator
9	No control
10	No control

The fitness function is an important factor for convergence and stability of genetic algorithm. Table 4 shows average and variance of objective function for various weight functions for the cost objective.

W_{cost}	Variance of fitness function	Average of fitness function
0.3	10.2E-4	0.47
0.2	8.5E-4	0.45
0.1	9.9E-4	0.46

The maximum concentration of SO_2, NO_X, and CO on the ground level before and after considering control options is calculated. The concentration of pollutants was reduced over 90 % and all of them are below standard value. Figure 1 shows the results for SO_2 as an example before and after application of control options.

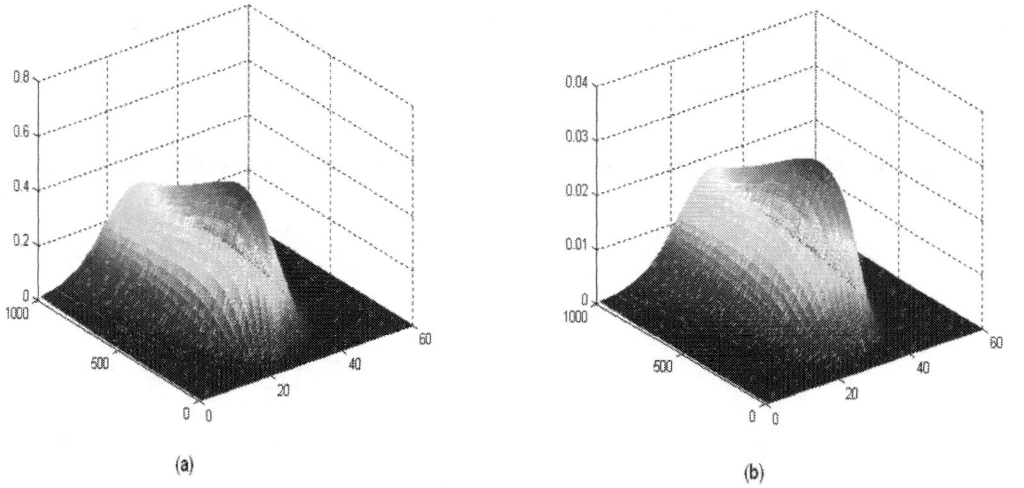

Figure 1: SO_2 emissions before (a) and after (b) considering control options

Conclusion

A genetic algorithm approach was carried out to find the suitable air emission control option for different sources within a simple typical refinery. Three emissions were considered in this study and the results showed that some sources that emit a certain emission in a quantity more than the standard needs to be controlled by applying the appropriate control option in terms of cost and efficiency. A significant decrease in the emissions was noticed when control options were implemented.

Acknowledgment

The author would like to thank KFUPM for their support.

References

[1] R. J. Heinsohn, R. L. Kabel, Sources and Control of Air Pollution. Prentice Hall, New Jersey, 1999.

[2] EPA, Air Pollution Control Cost Manual, sixth Ed, Environmental Protection Agency, EPA 452/B-02-001, Research Triangle Park. NC, USA, 2002.

[3] C.D. Cooper, Air Pollution Control a Design Approach, Waveland Press, Long Grove, Illinois, USA, 2002.

[4] S. Salvador, Y. Kara, J.M. Commandre, Factors influencing pollutant gas emissions of VOC recuperative incinerators—Large-scale parametric study, Applied Thermal Eng. 26 (2006) 1640–1651.

[5] Wikipedia, the free encyclopedia, 2010. Available at: http://en.wikipedia.org/wiki/Air_pollution, this page was last modified on 18 February 2010.

[6] A. Elkamel, E. Fatehifar, M. Taheri, M.S. Al-Rashidi, A. Lohi, A heuristic optimization approach for Air Quality Monitoring Network design with the simultaneous consideration of multiple pollutants, J. of Environmental Management. 88 (2008) 507–516.

[7] G.V. Reklaitis, A. Ravindran, K.M. Ragsdell, Engineering Optimization: Methods and Application. John Willy, New York, 1983.

[8] N.D. Nevers, Air Pollution Control Engineering. McGraw-Hill, New York, 2000.

[9] J.R. Jang, C.T. Sun, E. Hizutani, Neuro-Fuzzy and soft computing. Prentice Hall, New Jersey, 1997.

[10] H. S. Peavy, S.R. Rowe, G. Tchobanoglou, Environmental Engineering. McGraw-Hill, New York, 1985.

[11] R.L. Haupt, S.E. Haupt, Practical Genetic Algorithms. John Wiley & Sons, New York, 1998.

[12] M. Melanie, An Introduction to Genetic Algorithms. Massachusetts Institute of Technology, London, England, 1996.

Pretreatment of high concentration caustic waste wastewater from catalytic gasoline process unit in refinery

*Qiu Li-ping** and Luo De-chun, (College of Environmental Science and Engineering, Chang' an University, Xi' an, Shaanxi, China)

Abstract: The high concentration caustic waste wastewater of constituents complexity、heavy pollution and poor biodegradability of catalytic gasoline process unit in refinery were studied adopting the pretreatment of desulfurization (self-developed desulfurization agent), acidification(to recovery phenol) and the oxidation-coagulation of hydrogen peroxide- ferrous sulfate (H_2O_2-$FeSO_4$, Fenton reagent) . The results of experiment showed that the degree of contamination of caustic waste wastewater was debased greatly by the process. The removal efficiencies of chemical oxygen demand (CODcr)、sulfide、phenol and chromaticity reached 86.88%, 99.54%, 98.10% and 97.14% respectively. The biodegradability was improved significantly and B / C ratio was increased from 0.24 to 0.51 when desulfurization occurred for 48 hour and oxidation-coagulation occurred for 60 minutes at pH=6-7.Therefore, the CODcr loading rate in downstream biological treatment system was reduced greatly for incoming caustic waste wastewater.

Key words: Caustic waste wastewater. Pretreat- ment. Desulfurization agent. Fenton reagent. Oxidation-coagulation process

1 Introduction

The caustic waste wastewater with strong alkaline discharged from catalytic gasoline process unit in refinery generally have many pollutants of hard biochemical degradation and some of them such as COD_{cr}, sulfide and phenol in the stream account for 40%-50% (A Kolhatkar et al. 1995; Wujing 2007) of total amounts of pollutants. In addition, also contains a lot of naphthenic acid and other organic matter (Ye Zong-jun et al. 2006). If those pollutants are treated unreasonably or directly released to wastewater treatment plant (WWTP), they would result in the failure of biochemical treatment systems, reduce the removal efficiencies and increase the operating costs, and cause the adversely impact on the performance of WWTP. Considering the high toxicity of caustic waste wastewater, the WWTP would be vulnerable to the impact of high strength stream and reduced effluent quality if certain wastewater is directed into treatment units. Therefore, the most efficient methods should be considered to pre-treat the caustic waste wastewater prior to be discharged from refinery. At present, caustic waste wastewater treatment methods commonly used are neutralization, oxidation and biological treatment.

Neutralization methods mainly have neutralization of carbon dioxide and Sulfuric acid (Yu Wei 2001; Bai Yao-huai 1999; Li Shuchun et al. 2004). Carbon dioxide method use the waste carbon dioxide, which reflects the optimization of resource use, but the method is large in one-time investment of equipment, and not suitable to treat small batch processing of the catalytic gasoline caustic waste wastewater discharged periodically, and sodium carbonate produced by the technology is lower-purity and bad sales , in particular, there is the serious pollution of the exhaust gas of the spray-dried sodium carbonate in the production process, which lead to these devices often being half lay-off status and affect the treatment of caustic waste wastewater. Sulfuric acid neutralization method acidify generally caustic waste wastewater from 2 to 4 of pH to recovery phenols substances (Song

Yongxin. 2006), which greatly reduce the phenol and COD in wastewater (Ni Binghua et al. 2006), but the Na_2SO_4 produced after acidification easy form crystal and seriously affect subsequent processing, moreover the caustic waste wastewater containing high sulfide will lead to a large number of H_2S gas, which is very adverse for the subsequent processing of sulfides in caustic waste wastewater. Most of the pollution substances can be removed after the pretreatment by acidification of wastewater, but phenol effluent concentrations are still very high, up to thousands of to tens of thousands of mg/L(Zhao Hua et al. 2009)

Oxidation process is divided into air oxidation and chemical oxidation. Caustic waste wastewater oxidized is again pretreated to reduce COD load by sequencing batch activated sludge process (SBR),then wastewater after pretreatment is treated in WWTP (Zhang Xiheng 1996; Liu Yogsong 1990; Ma Fengrong et al. 2006) , which can effectively oxidize the catalytic gasoline waste lye , liquid hydrocarbon waste lye and other high concentrations wastewater. But air oxidation need large energy consumption and chemical oxidation needed high reagents cost, so direct the oxidation treatment of caustic waste wastewater containing a large number of organic compounds obviously is no reasonable in economic.

Biological treatment method is the process that the organic matter and other pollutants in wastewater are degraded and translated into harmless substances by the microbial metabolism (Ding Jiuliang 2001; Li fei 2006; Cai Qingling 2007), and have many advantages in treating caustic waste wastewater in refinery such as low cost , mild conditions, low energy consumption, low investment, no requiring a catalyst, no producing chemical sludge, small footprint, high processing efficiency, operation of stable and safe, etc., so, the method have broad application prospects. But the biological treatment of caustic waste wastewater in refinery still have many problems to be solved, on one hand, conventional biological treatment process is more difficult due to the toxic effects of the high concentrations sulfide and phenol pollutants on microorganisms in the conventional biological treatment system of wastewater. On the other hand, high efficient bacteria screening, microbial community structure and pollutant degradation mechanism etc. still need further studied (Chao Qun-fang et al. 2009). In this study, caustic waste wastewater was first desulfurized by self-developed desulfurization agent, then phenol of wastewater was recovered by acidification, finally, wastewater was pretreated by the oxidation-coagulation of H_2O_2-$FeSO_4$ (Fenton reagent). Based on the above experiments, the desulfurization principles、the effects of acidification recovering phenol and the biodegradability (BOD_5 / CODcr, abbreviated as B / C)of after wastewater treatment were explored, and determined the optimum pretreatment conditions that facilitate effective implementation of the follow-up processing.

2 Experimental

2.1 Materials and instrument

Concentrated sulfuric acid (H_2SO_4, mass percentage 98.3%, analytically pure), high efficient compound desulfurization agent (self-developed), hydrogen peroxide (H_2O_2, mass percentage 30%, analytically pure). Ferrous sulfate heptahydrate ($FeSO_4 7H_2O$, analytically pure, make it to the solution of mass fraction 10% when be used). Polyacrylamide (PAM, analytically pure, average relative molecular mass as $1.5\times10^7 \sim 2.0\times10^7$,make it to the solution of mass fraction 1‰ when be used).

SHIMADZU UV vensus spectrophotometer (UV-2001). ORION pH meter, a set of color-comparison tube(50ml), reflux units for CODcr, a set of distillation devices for volatile phenol, titration table,IR3-6 blender for coagulation, aeration pump, HH-4 isothermal water-bath pot etc..

2.2 Sampling

The caustic waste wastewater (pH>13, chromaticity =2100 times) used in this study were obtained from catalytic gasoline process unit in refinery, in which the major pollutants were CODcr (234,400mg/L),sulfide (12,022mg/l), and phenol(53,202mg/L).

2.3 Test Methods

The precipitate produced by adding concentration of 2-3 g/L desulfurization agent in caustic waste wastewater was separated from wastewater at 40℃ and micro aeration of 48h, from which desulfurization agent recovered could be recycling. The wastewater after desulfurization was immediately stratified by sulfuric acid acidifying to pH=6-7, sitting for a while, the phenol of upper organic phase was recovered in the recovery device, and the under aqueous phase was treated by oxidation–coagulation of Fenton reagent (P K Malik et al. 2003; Deng Xiaohui et al.2007) to investigate the effects of the dosages of H_2O_2 and $FeSO_4.7H_2O$ and pH value and other factors on removal of CODcr, sulfide, phenol, color etc., and to explore the appropriate process conditions and parameters of Fenton oxidation-coagulation method (Kang N et al. 2002).

Analysis methods were presented as follows: CODcr and sulfide were respectively determined by potassium dichromate method and iodimetry. Phenol was measured by using 4-aap and BOD_5 was determined by five-day biochemical cultivation. Chromaticity was measured by the method of dilution multiple.

3 Results and discussions

3.1 The Effect of Desulfurization and Acidification on Caustic Waste Wastewater

3.1.1 The Principles for Desulfurization

High-efficient compound desulfurization agent is a kind of polyelectrolyte coagulants made from multiple high polymers through complex physical and chemical reactions，which is characterized by superior ability of oxidization、coagulation and desulfurization. The sulfide in caustic waste wastewater, existing as the salt form of most NH_4^+ and Na^+, react chemically first with high valence metal ions (M^{x+}, x：represents valence) produced by hydrolysis of desulfurization agent. The reaction equation is presented as follows:

$$2M^{x+}+xS^{2-}\rightarrow M_2S_x\downarrow \qquad ①$$

At the same time，desulfurization agent also produce a variety of high-valence metal hydroxyl ions ($M(OH)^{(x-1)}$ 、$M(OH)_2^{(x-2)}$ ...) by hydrolysis and the causing turbidity substances of oil-soluble activity, which occur coagulation reaction with sulfide in wastewater ,and form polymer chain under the role of adsorption and bridging , which form floccules of large particle by intercross each other, then floccules are precipitated and separated，finally achieve the purpose of removing sulfide.

In addition，a small part of sulfide directly exposed in the air are oxidized into $S_2O_3^-$，then $S_2O_3^-$ are further oxidized into SO_4^{2-}. The reaction equations are presented as follows:

$$2HS^-+2O_2\rightarrow S_2O_3^{2-}+H_2O \qquad ②$$

$$2S^{2-}+2O_2+H_2O\rightarrow S_2O_3^{2-}+2OH^- \qquad ③$$

$$S_2O_3^{2-}+2O_2+2OH^-\rightarrow SO_4^{2-}+H_2O \qquad ④$$

Disulfide (RSSR), produced after thiols organic substances are oxidized, can be separated from wastewater. The reaction equation is presented as follows:

$$2RSNa+2R^1SNa+O_2+2H_2O \rightarrow 2RSSR^1+4NaOH \quad ⑤$$

3.1.2 Recover phenol by acidification

There are a large number of sodium phenolate in catalytic gasoline waste wastewater, when micro-aerated, phenol produced after acidification of wastewater was recovered. The reaction equation is presented as follows:

$$Ar\text{-}ONa + H^+ \rightarrow Ar\text{-}OH + Na^+ \quad ⑥$$

The remaining $S_2O_3^{2-}$ in the wastewater would react as follows:

$$S_2O_3^{2-} + 2H^+ \rightarrow H_2S_2O_3 \quad ⑦$$

$$H2S2O3 \rightarrow H2SO3 + S\downarrow \quad ⑧$$

$$H_2SO_3 \rightarrow SO_3^{2-} + 2H^+ \quad ⑨$$

$$2SO32- + O2 \rightarrow 2SO42- \quad ⑩$$

3.1.3 The Effect of Desulfurization and Acidification

Analysis data for desulfurization and acidification of wastewater were presented in table 1:

Table 1 The effect of desulfurization and acidification on removal efficiencies of pollutants

Parameters	pH	CODcr(mg/L)	Sulfide(mg/L)	Phenol(mg/L)	BOD_5(mg/L)	chromaticity (multiples)	B / C
Before desulfurization and acidification	> 13	234400	12022	53202		2100	
After desulfurization and acidification	6-7	65854	338	2172	15805	1050	0.24
Removal (%)		71.91	92.6	83.55		50	

From Table 1, it could been observed that the removal efficiencies for CODcr, sulfide, phonel and chromaticity were satisfactory by From Table 1, it could been observed that the removal efficiencies for CODcr, sulfide, phonel and chromaticity were satisfactory by desulfurization under micro aeration, for the reasons that, on one hand, micro-aeration would moderately agitate wastewater so as to make desulfurization agent more readily mixed with wastewater. On the other hand, plenty of sulfides and some small molecule compounds of high concentration such as ammonia and hydrogen sulfide, etc. decomposed by organic compounds containing heteroatom organics in wastewater (Zhong Huawen et al. 2003) would affect the desulfurization rate, but a small portion of oxygen provided by micro-aeration could accelerate the oxidation and desulfurization of sulfides.

3.1.4 The Biodegradability of Caustic Waste Wastewater

The pollution loads of wastewater after desulfurization and acidification(to recovery phenol) were largely reduced, but biodegradability (BOD/CODcr=0.24) was still low and could not be directly treated by aerobic biochemical. When the high strength wastewater was directly discharged into WWTP, it would affect the normal operation and reduce the treatment efficiencies. Therefore, In order to reduce the pollution degree of wastewater, H_2O_2-$FeSO_4$ (Fenton reagent) was used to oxidize and decompose not easy biochemical degradation substances in wastewater after desulfurization and acidification so as to increase theirs biodegradability and reduce the pollution load for the downstream processes.

3.2 H_2O_2-$FeSO_4$ (Fenton Reagent) Effect

3.2.1 The Effect of H_2O_2 Dosage on Pollutants Removal Efficiencies

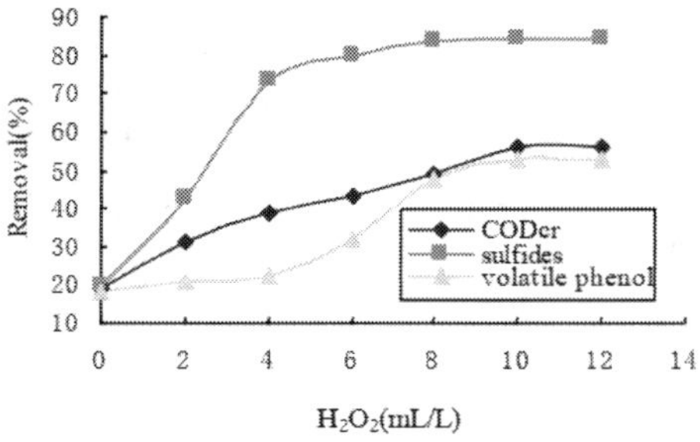

Figure1.The effect of H_2O_2 dosage on the Pollutant Removal Efficiencies

6 parts 200mL samples taken from wastewater after desulfurization and recovering phenol, which contain 5g/L $FeSO_4 \cdot 7H_2O$ and pH 6-7 determined after the preliminary testing and different dosages H_2O_2, were tested respectively in IR3-6 blender for coagulation at 200r/min for 60 min, then were tested again after adding 5mg / L PAM at 100r/min for 1 min, and determined supernatant liquid after precipitation 10 min. The influences of different dosage H_2O_2 on the removal of CODcr, sulfides and volatile phenol were investigated. Experiment Results were presented by Figure 1.

As could be seen from Figure 1, after Fenton oxidation-coagulation process, the wastewater had a good treatment performance, this was the reason that Fenton reagent could oxidize and polymerize water-soluble、colloidal、 difficult degraded and suspended pollutants in wastewater (Zhong Huawen et al. 2003; Li Ya-jun et al. 2009) to form polymerization products that would be easily removed by coagulation. Floccules occurred as a result from agglomeration and flocculation of these polymerization products (Chen Wensong et al. 2004). Experiment results showed that the optimal H_2O_2 dosage was 10mg/L.

3.2.2 The Effect of $FeSO_4 \cdot 7H_2O$ Dosage on Pulltants Removal Efficiencies

According to the above test results, the impacts of different dosages $FeSO_4 \cdot 7H_2O$ on the removal of CODcr, sulfides and volatile phenol were investigated when the concentrations of H_2O_2 was maintained at 10 mg/L (and other procedures and conditions as above). Experiment Results were presented by Figure 2.

From Figure 2, it showed that the removal efficiencies of CODcr, sulfide and phenol were higher when the concentration of $FeSO_4 \cdot 7H_2O$ was increased to 5g/L. However, when mass concentration was increased to 6g/L, wastewater would become turbidity and removal efficacies would decrease. This was the reason that more $Fe(OH)_3$ colloid would be generated by oxidation and hydrolysis with the increase dosage of $FeSO_4 \cdot 7H_2O$, and some of them and polymerization products formed were removed by co-precipitation, but $Fe(OH)_3$ colloid without causing co-precipitation led to turbidity of the solution. Therefore, the optimal $FeSO_4 \cdot 7H_2O$ dosage was 5g/L.

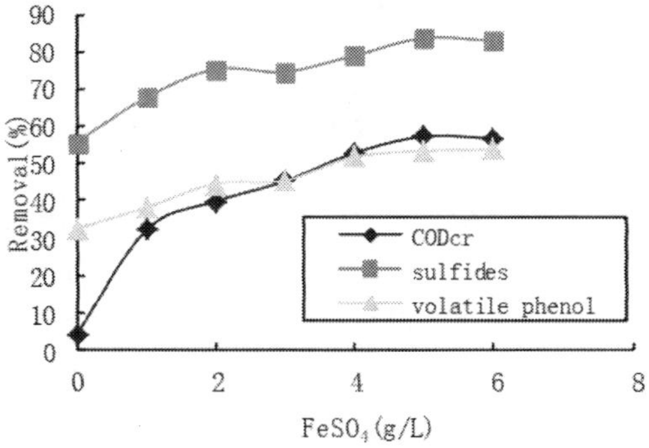

Figure2. The effect of $FeSO_4 \cdot 7H_20$ dosage on the Pollutant Removal Efficiencies

3.2.3 The Effect of pH Value on Pollutants Removal Efficiencies

Based on the experimental results above, when the concentrations of H_2O_2 and $FeSO_4 \cdot 7H_2O$ were maintained at 10 mg/L and 5 g/L, respectively (other procedures and conditions as above), the effect of different pH value on the removal efficiencies for various pollutants were investigated and shown in Figure 3.

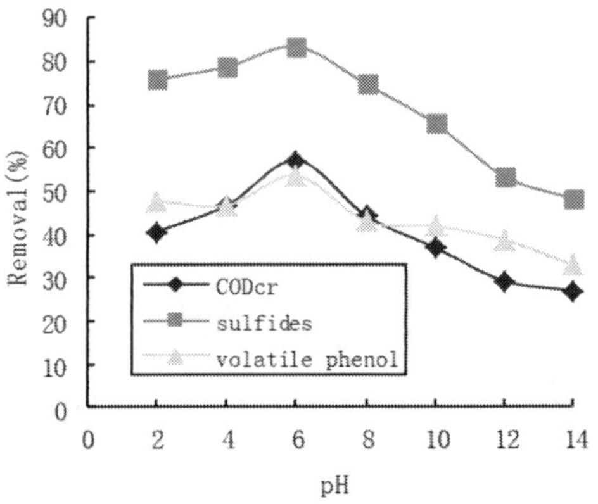

Figure3. The effect of pH values on Pollutant Removal Efficiencies

From Figure 3, it showed that pH values had strong influence on the removal efficacies for pollutants.

When PH≤6， pollutants removal efficiencies were relatively higher. When pH=6, the removal efficiencies reach the highest point even without occurring any floccules precipitations. When pH >6， the removal efficiencies for various pollutants were reduced with pH increasing. The reason was that varying pH values would determine either Fenton oxidation or Fenton coagulation to be dominated (Liu Yongdi et al. 1994). When pH value was lower, a kind of hydroxyl free radical (• OH) of strong oxidation capacity was produced by H_2O_2 under catalytic action of Fe^{2+} in Fenton reagent, which make carbon chain of organic pollutants in wastewater oxidized and cracked, and difficult biodegradable organic macromolecules cracked into readily biodegradable small molecules organic matters, or completely mineralized into CO_2 and H_2O. So, low pH value would cause Fenton oxidation and otherwise resist to Fenton coagulation that would cause floccules formation. The oxidation and coagulation would be occurred at pH=6-7, at this time， organic compounds in the wastewater could be oxidized and coagulated synchronously and form some polymerization products of relative small molecular mass so that the solubility and settledbility of these pollutants would be improved， associated with cohesion of Fe^{2+}, the co-action of Fenton oxidization and coagulation would result in an increasing of the removal efficiencies for CODcr, sulfides, phenol and chromaticity (Zhu Wanpeng et al. 1995). When pH>7, Fenton oxidization stopped and inhibited the oxidization and polymerization of soluble contaminations, so the solubility and settledbility remain the same, at this time, there were floccules, but the major floccules were $FeSO_4$ and PAM themselves. Therefore, the removal efficiencies were lower for all pollutants. Experiment results showed that the optimal pH=6-7.

3.2.4 The Effect of PAM Dosage on Pollutants Removal Efficiencies

when the concentrations of H_2O_2 and $FeSO_4 \cdot 7H_2O$ were maintained at 10 mg/L and 5 g/L, respectively (other procedures and conditions as above), the effect of different dosages PAM on the removal efficiencies for various pollutants were investigated. Experiment results showed that PAM of 5mg / L could result in higher removal rate of pollutants. But continue to increase PAM dosage would cause slightly decreasing of pollutant removal efficiencies. The reason was that the excessive PAM not only would result in toxicity and cause more CODcr, but also increased the cost of processing. Therefore, PAM concentration of 5mg / L was preferred.

3.2.5 The Biodegradability of Pre-treated Wastewater

Based on the study of constituent complexity and poor biodegradability of the caustic waste wastewater from catalytic gasoline process unit in refinery, the results obtained were satisfactory and were presented in Table 2.

Table 2 The performance results of oxidation-coagulation

items	pH	CODcr(mg/L)	Sulfide(mg/L)	Phenol(mg/L)	BOD_5(mg/L)	Chromaticity (multiples)	B / C
After desulfurization and acidification	6-7	65854	338	2172	15805	1050	0.24
Fenton treatment	6-7	30765	54.9	1011	15823	60	0.51
Removal (%)		53.28	83.73	53.45		94.28	
Total removal (%)		86.88	99.54	98.10		97.14	

From Table 2, it could be seen that, after the pretreatment of caustic waste wastewater, BOD/CODcr ratio increase from 0.24 (after desulfurization and acidification) to 0.51(after Fenton oxidization-coagulation).

This was main reason that most of complex large organic molecules were degraded into small molecular pollutants in wastewater after pretreatment by Fenton reagent, therefore, the biodegradability of wastewater was improved significantly. then wastewater that a small amount of the precipitate produced was separated was further processed in biochemical system, which would not impact the follow-up biochemical system and affect the normal operation of the WWTP. (Ma Chengyu et al. 2007).

4 Conclusions

Due to the complexity of constituents, heavy pollution and low biodegradability, the high- efficient desulfurization agent self-developed and the sulfuric acid were used to desulfurized and acidify caustic waste wastewater under the condition of micro-aeration. Results showed that the process was effective and the removal efficiencies of CODcr, sulfide, phenol and chromaticity reached 71.91%, 92.6%, 83.55% and 50%, respectively.

The wastewater after desulfurization and acidification (recycling phenol) were further oxidized and coagulated by Fenton reagent under the condition of pH = 6-7 and adding H_2O_2 of 10mL / L, $FeSO_4 \cdot 7H_2O$ of 5g / L and PAM of 5mg / L respectively, the removal efficiencies for various pollutants would be improved significantly. Not only the removal efficiencies of CODcr, sulfides, phenol and chromaticity in wastewater were again raised 53.28%, 83.73%, 53.45% and 94.28% respectively, but also the biodegradability was improved and the BOD/CODcr from 0.24(after desulfurization and acidification) to 0.51(after Fenton oxidization-coagulation) as well.

Acknowledgments This research was supported by the teachers of Engineering Research Center for Groundwater and Eco-Environment of Shaanxi Province, Here to express my sincere appreciation to them.

References

A Kolhatkar, K Sublette (1996). Biotreatment of refinery spent sulfidic caustic by specialized cultures and acclimated activated sludge. Applied Biochemistry and Biotechnology58(57): 945-946

Wujing(2007)Pretreatments technologies for high-concentrated caustic sludge wastewater of petroleum chemical industries. Modern Business Trade Industry 19(3): 153-154

YE Zong-jun, XU You-hao, WANG Xie-qing (2006)Study on the composition of products from catalytic cracking and thermal cracking for heavy fractions of FCC gasoline. ACTA PETROLEI SINICA (PETROLEUM PROCESSING SECTION). 22(3): 46-53.

Yu wei (2001) Refinery alkaline residue treatment and comprehensive utilization. ENVIRONMENT PROTECTION IN PETROCHEMICAL INDUSTRY 24(2)：43-47

Bai Yuehua, Bai Yunhong,Lei Xunmao(1999) Application of Additive Carbonized Method to Treating Reclaimable Alkali Residue ENVIRONMENT PROTECTION IN PETROCHEMICAL INDUSTRY 22(1)：55-57

Li Shuchun,Guo Xiwen(2004)Gasoline alkaline residue treatment and recycling production test ENVIRONMENTAL PROTECTION OF CHEMICAL INDUSTRY24(S1):316- 317

Song Yongxin.(2006) Refinery solid waste generation and disposal. SCIENCE & TECHNOLOGY INFORMATION4(20):125

Ni Binghua, Huang Fenglin, Liu Jurong(2006) KINETIC RESEARCH OF CAUSTIC REGENERATION IN FCC GASOLINE CAUSTIC REFINING PETROLEUM REFINERY ENGINEERING 36(4):55-59

ZHAO Hua，CAO Qing-qing，LIU Kan(2009) Status and development of treatment technology for alkaline waste from oil refinery. SCIENCE & TECHNOLOGY IN CHEMICAL INDUSTRY．17(2)71-74

Zhang Xiheng(1996)Wastewater anaerobic biological treatment engineering. China environmental science press .Beijing, China.331 pages

Liu Yogsong(1990)．Treatment of phenol—containing waste water by sequencing contact stabilization process[J]．Water Treatment 5(1)：77-86．

Ma Fengrong,Mu Tong,Bai Bing etal (2006)Application of high concentration wastewater treatment technology in diesel oil Caustic sludge treatment. ENVIRONMENTAL PROTECTION OF OIL & GAS FIELDS16(2):19-21

Ding Jiuliang(2001) The application of Biological treatment methods in the Boxi terminal oily wastewater COD treatment process. CHINA OFFSHORE OIL AND GAS(ENGINEERLNG)123(5):1-5

Li fei(2006) The innovation achievement of refining wastewater treated by Microbial method. China Environment News2(7)

Cai Qingling(2007) Status and Prospects of refining caustic sludge processing. LIAONING URBAN AND RURAL ENVIRONMENTAL SCIENCE & TECHNOLOGY27(2):19-24

CHAO Qun-fang,ZHOU Jun,FANG Xin-xiang,etal(2009) Studies on Isolation and Degrading Characteristics of Phenol-degrading Bacteria ZJ-1. BIOTECHNOLOGY19(2):57-59

P K Malik,S K Saha.(2003)Oxidation of direct dyes with hydrogenperoxide using ferrousion as catalyst. Separation and Purification Technology 31(3):241-250.

Deng Xiaohui,Zhang Haitao,Cao Guomin et al (2007)Progress of the research and application of Fenton Reagent for wastewater treatment. Shanghai Chemical Industry 32(8): 1-5

Kang N,Lee D S,Yoon J(2002)Kinetic modeling of Fenton oxidation of phenol and monochlorophenols.Chemo-sphere 47(9)：915-924.

Zhong Huawen, He Dongsheng, Liao Yan(2003)Pollution and pretreatment of refinery basic sewage.Petrochemical Technology & Application.21(4):288-290.

Li Ya-jun, ZHENG Jie(2009)Energy-use improvement of FCC process for crude gasoline upgrading. ACTA PETROLEI SINICA(PETROLEUM PROCESSING SECTION).25(2): 228-232.

Chen Wensong, Wei Chaohai(2004)Study on Fenton oxidation-coagulation process for the treatment of printing and dyeing wastewater.Industrial Water Treatment.24(4):39-41.

Liu Yongdi, Wang Huaxing, Zhu Yaxin(1994)Study on the mechanisms of phenol-bearing wastewater treatment using Fenton oxidization coupling coagulation".China Environmental Science.14(5): 341-345.

Zhu Wanpeng, Yang Zhihua, Wang Li(1995)Application of ferrous-hydrogen peroxide for treatmentof H-acid manufacturing process wastewater. China Environmental Science15(5): 368-372.

Ma Chengyu, Peng Yingli(2007)The treatment and control of high-concentrated refractory organicwastewater. Chemical Industry Press,Beijing,China.1

HYDROCRACKING REACTOR TEMPERATURE CONTROL FOR INCREASED SAFETY, RELIABILITY AND PERFORMANCE

Allan G. Kern, P.E., APCperformance.com, Red Lodge, MT, USA

Introduction

Most hydrocracking reactors in the oil refining industry today are under-automated. Conventional industry practice leaves many gaps in hydrocracking reactor temperature control, quench utilization, and depressure prevention.

Refiners pay a large cost for this automation shortfall at all levels – safety, reliability, and optimization. Hydrocracking reactor temperature excursions can occur frequently and incur costs of tens or hundreds of thousands of dollars per incident. A complete depressurization event brings acute penalties in refinery availability and operating cost, usually on the order of a million dollars per event, in addition to serious potential safety and environmental concerns.

The situation, which is typical across the oil refining industry, is not due to a technology limitation. The necessary automation capabilities and features are readily available in today's distributed control systems (DCS). The current situation is primarily the product of past assumptions regarding safety systems and advanced controls, that have turned out to be incorrect or incomplete.

Today, this presents industry with a safety and reliability challenge, and an automation opportunity. Large improvements in hydrocracking unit safety, reliability and optimization, remain available for relatively modest cost and effort, in most cases leveraging significant improvements and additional value from existing resources. This paper presents an updated hydrocracking reactor control paradigm for the refining industry going forward.

About Hydrocracking

Hydrocrackers are among oil refining's most demanding units, because they combine high temperature (ca. 400-800F) and high pressure (ca. 1400-2200 PSIG) with a strongly exothermic reaction. Achieving maximum cracking in the reactor, while avoiding an exothermic runaway reaction, is a particular challenge of hydrocracking unit operation and control. Exothermic runaway reactions, if not successfully brought under control, ultimately result in unit depressurization to flare, when reactor temperatures approach metallurgical design limits. Depressurization may be initiated manually by operators or automatically by the control system.

While depressurization is a necessary safety function, it must be avoided to the extent possible from an operation and economic standpoint. Depressurization brings the prospect of a several day unit restart, large and potentially damaging thermal and mechanical equipment stresses, the unsettling prospect of unexpected events during depressurization, environmental and community concerns related to the flaring, and an overall large economic cost penalty, typically in the million dollar range per event. Even minor temperature excursions that are brought under control can cost tens or hundreds of thousands of dollars in lost on-target production, due to loss of steady-state hydrocracking conditions and the time it takes to re-establish them.

The overall purpose of hydrocracking is to convert (or "crack") heavy oils, such as gas oil, cycle oil and coker oil, into lighter higher-value fuels, such as naphtha, kerosene and diesel. Figure 1 shows a typical hydrocracking reactor. Heated oil and hydrogen enter the top of a vertical down-flow reactor with multiple fixed catalyst beds. The catalyst promotes the desired hydrocracking and hydrogenation reactions. Temperature increases as flow passes through the bed, due to the exothermic nature of the reactions. Quench gas is introduced between each bed to cool the reaction mix, such that, ideally, each bed has similar bed inlet and bed outlet temperatures.

Due to the nonlinear heat release of exothermic reactions (reaction rates approximately double for each ten degree centigrade increase in temperature), the potential for a runaway reaction is always present. Common excursion initiators include full or partial loss of feed flow, loss of recycle compressor, heater disturbances, change in feed quality to a more readily cracked feed, change in catalyst activity, bed inlet or catalyst bed mal-distribution, overly aggressive start-up operations, and overly large feed rate or temperature adjustments during ongoing operation. Reference 1 describes additional potential initiators. Modern refineries with complex integration can experience these conditions on an essentially ongoing basis, rendering good control and intelligent automation response a necessity.

Despite its challenges, hydrocracking is economically very attractive in the modern hydrocarbon market environment, due to increasing global demand for jet fuel and diesel fuel over gasoline, and hydrocracking's ability to effectively convert today's low-cost natural gas into high value fuels. This is accomplished by converting natural gas into hydrogen and converting hydrogen into fuels in hydrocracking units. Hydrocracking unit new construction and upgrade projects are common around the world today.

Hydrocracking units are among the more complex refinery units, usually encompassing multiple reactors, recycle and make-up compressors, cascaded pressure let down and separation vessels, multiple feed heaters, high pressure sour gas scrubber, high pressure wash water system, and a complete (albeit low pressure) fractionation section. Figure 1 does not capture this process complexity, but does highlight the essential challenge of high pressure, high temperature and an exothermic reaction. Automation of the quench and temperature controls around this vessel, to control and avoid temperature excursions, is the focus of this paper.

Hydrocracking Reactors are Under-Automated

Most hydrocracking reactors in industry are under-automated from temperature control, excursion control, and depressure prevention perspectives. Figure 1 is a simplified hydrocracking reactor diagram with typical "as-purchased" temperature controls. Most hydrocracking units are initially designed, commissioned and operated with only these controls. This includes provision for manual, or increasingly automatic, depressurization on high temperature, and simple reactor bed *inlet* temperature controls. More advanced controls, such as multivariable control or the controls described in this paper, are usually added after the unit is in production, over the course of several years, and to varying extents.

Notably missing from Figure 1 are special temperature excursion controls, which are especially essential given the exothermic nature of hydrocracking reactions. Also missing are advanced auto-quench controls, designed to utilize quench as advantageously as possible to prevent reaching depressure conditions. And also missing are bed outlet controls, which are essential to stable operation and production, especially since equal bed outlet temperatures are one of the main criteria for hydrocracker operation from catalyst management and product quality standpoints, excursions aside.

Multivariable control (MPC) has often been applied to hydrocracking reactors to provide average bed temperature (ABT) and constraint control, but MPC does not provide effective excursion control, depressure prevention, and auto-quench. These features are readily accomplished in "smart" DCS function block control strategies, and are more appropriately applied at the base-layer DCS level for responsiveness, reliability, and the necessary smart functionality. MPC can still be appropriate for ABT, constraint, and conversion control.

Especially notable in the controls of Figure 1 is the complete lack of automatic control response to a temperature excursion. Excursions take hold as the hydrocarbon/hydrogen mix passes down through the catalyst bed, and the exothermic nature of the reaction results in a temperature "runaway". Quench flow to the bed must be increased quickly and nonlinearly or the excursion can proceed unchecked until it reaches depressure conditions, but the quench controls in Figure 1 lack such a mechanism. The controls will increase quench flow to the *subsequent* bed, based on high outlet temperature of the excursion bed, but there is no mechanism to increase quench flow where it is needed. This lack of base-layer bed outlet and excursion control, on an exothermic reactor operating at elevated temperature and pressure, is a common situation in industry today, but it no longer passes the sensibility test in modern refineries, or the explicit question, "can a runaway reaction occur", which is standard in most Hazop and management of change (MOC) questionnaires today.

Hydrocracking reactors are overall under-automated, in the sense that most of the units in industry lack some or all of these essential excursion control, auto-quench, and depressure prevention features, and because it is not due to any technology limitation. The missing control and automation features are readily accomplished in today's DCS control systems, presenting industry with a safety, reliability and performance challenge, as well as an automation opportunity. Achieving appropriate levels of automation is a matter of understanding the shortfalls in the past industry hydrocracker control paradigm, and making appropriate changes.

The Cost of Under-Automation is High

The penalty for hydrocracking reactor under-automation is substantial, and should be cause for concern at all levels – safety, reliability, and optimization.

A hydrocracking reactor over-temperature depressurization event carries a long list of serious safety concerns, including mechanical and thermal equipment stresses and potential equipment damage, environmental violations and community concerns due to the flaring, and the unsettling prospect of unexpected events or failures in the course of venting a high-temperature/high-pressure runaway reaction system to a flare. The best way to avoid a safety system failure on demand is to avoid demands in the first place.

From a refinery reliability standpoint, an uncontrolled excursion can send products off-specification for several hours, with incurred cost of tens or hundreds of thousands of dollars. A complete depressurization can require several days to restart and re-establish on-specification operating conditions, with total cost in the million dollar range for a typical refinery. Moreover, hydrocrackers are usually heavily integrated with other refinery units, thereby impacting overall refinery availability.

In addition to these "abnormal" (but not uncommon) situations, lack of bed outlet control takes an ongoing toll in the form of unnecessary and excessive product variance, the usual target of optimization controls.

Common excursion initiators include loss of feed, change in feed quality, heater operations, catalyst bed or quench internals maldistribution, and feed rate or bed temperature changes. Many additional excursion initiators are described in reference 1.

Not Due to a Technology Limitation

The current condition of under-automation and commensurate negative exposures in safety, reliability and optimization, are not due to a technology limitation. Today's distributed control systems (DCSs) are readily capable of meeting the requirements. The current industry-wide under-automation condition is more due to a past paradigm that has proven insufficient in both safety and advanced control aspects.

The dominant paradigm from a safety perspective for the past couple decades has been to install automatic over-temperature depressurization systems, with the perception that this addresses most or all critical safety needs from a reactor temperature runaway standpoint. This has been a major focus of hydrocracker automation, as reliance on manually initiated depressurization has been widely replaced with auto-depressure safety functions. This represents significant technological and safety progress, as more reliable thermocouple systems and more powerful safety system logic solvers have combined to provide sophisticated and reliable auto-depressure systems. However, this trend so far has failed to realize the need to include better excursion control and auto-quench features to avoid reaching depressure conditions in the first place. Today, many depressurization events unfold in the absence of any automatic control response.

The dominant paradigm from an advanced control perspective has been to install multivariable controllers, with the perception that this addresses all process automation and control requirements. But multivariable control is basically a single broad algorithm for overall gradual constraint management and optimization. It does include the type of smart critical control features needed for effective excursion control, auto-quench and depressure prevention.

These past paradigms regarding hydrocracking reactor safety and advanced control have brought important progress. But at the same time, over-reliance on them has resulted in persisting critical gaps. Because these gaps are primarily a matter of industry practice, rather than a technology limitation, closing the gaps is primarily a matter of adapting the industry paradigm to reflect the lessons learned.

Updated Hydrocracker Control Paradigm

Figure 2 is an updated hydrocracking reactor temperature control paradigm that takes into consideration the lessons learned to close the gaps in safety, reliability and performance. Multiple layers at each level bring robust and reliable performance. For example, excursions are addressed by bed outlet control, special excursion control, and auto-quench, before reaching depressure.

The calculation package stems from the large number of reactor bed temperature measurements and the need to reduce them to fewer more meaningful values, to apply alarm management principles to prevent alarm floods when excursions occur, to provide more timely "smart" alarms, to support higher levels of advanced control, and to assure overall data quality. There are important practical benefits to building the calculation package with the overall paradigm in mind.

Auto-depressure is now well established best practice, but basic decisions remain for each application, such as how temperature data will be shared between the safety system and control system, and whether to auto-depressure on high temperature rate-of-change, in addition to high absolute temperature.

Key considerations regarding auto-quench include heater and feed implications, achieving timely response without triggering unnecessarily, and striking a balance between robustly reliable logic and a graceful response that facilitates a rapid return to on-target operating conditions. Although auto-quench is a DCS control, conceptually it can be one of the most important functions in a refinery, in terms of economics and safety, because it is the final layer of depressure prevention.

The smart bed outlet control layer comprises a number of traditional advanced regulatory control (ARC) techniques applied to the bed inlet, bed outlet, and heater controllers, so that most excursions are handled as normal temperature disturbances and may go completely unnoticed, except perhaps to the console operator. (One purpose of auto-quench is to provide protection if the ARC controls are left in non-normal mode.) Bed outlet control also brings significant additional process and product stability during ongoing normal operations.

Average bed temperature (ABT) and constraint control have traditionally been provided by multivariable control (MPC), but in the new paradigm they can also be provided using ARC methods. Because bed outlet control is in the base-layer DCS controls in the updated paradigm, ABT and constraint control is simplified and becomes similar to a traditional ARC-based heater pass balancing algorithm.

Conversion control basically means coordinating the reaction and fractionation sections of a hydrocracking unit. This still lends itself to model-based control, due to the long response times involved (hours), although it is also important to include a steady-state algorithm to prevent fractionation section upsets from back-propagating to the reactors.

Hydrocracking reactor control and safety systems can be evaluated relative to the updated paradigm (Figure 2) to identify gaps in safety, reliability and optimization, and find solutions to close them. Figure 3 reflects a hydrocracking reactor with the updated paradigm applied, and Figure 4 shows excursion temperature improvement as these controls were deployed on an actual operating hydrocracker. Further discussion of this topic can be found in reference 2.

References

1. EPA Chemical Accident Investigation Report, Tosco Avon Refinery, Martinez California, November 1998
2. Hydrocracker Controls, By A. G. Kern, Hydrocarbon Processing Journal, October 2012 issue, or online at www.hydrocarbonprocessing.com.

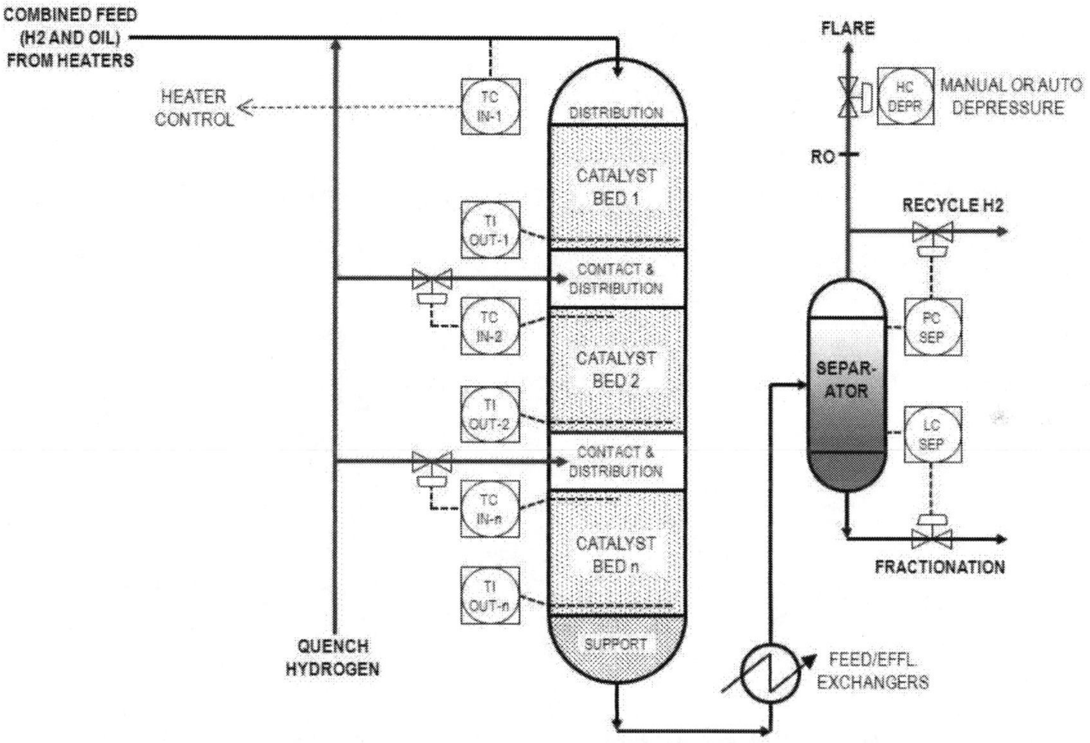

Figure 1: Simplified hydrocracking reactor diagram with "as purchased" controls.

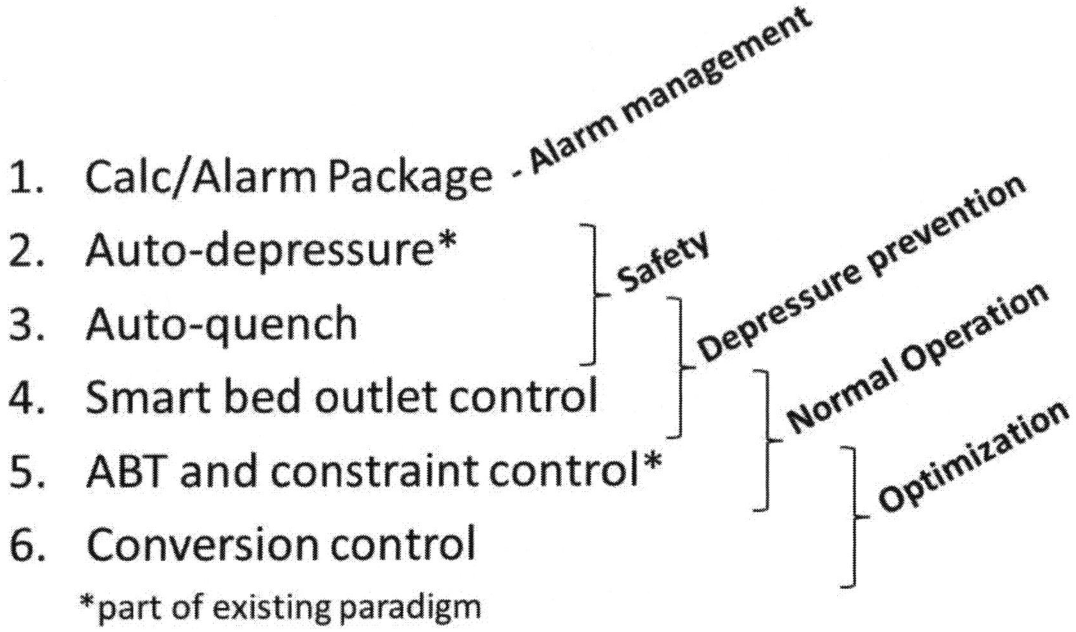

Figure 2: Updated hydrocracking reactor temperature control paradigm.

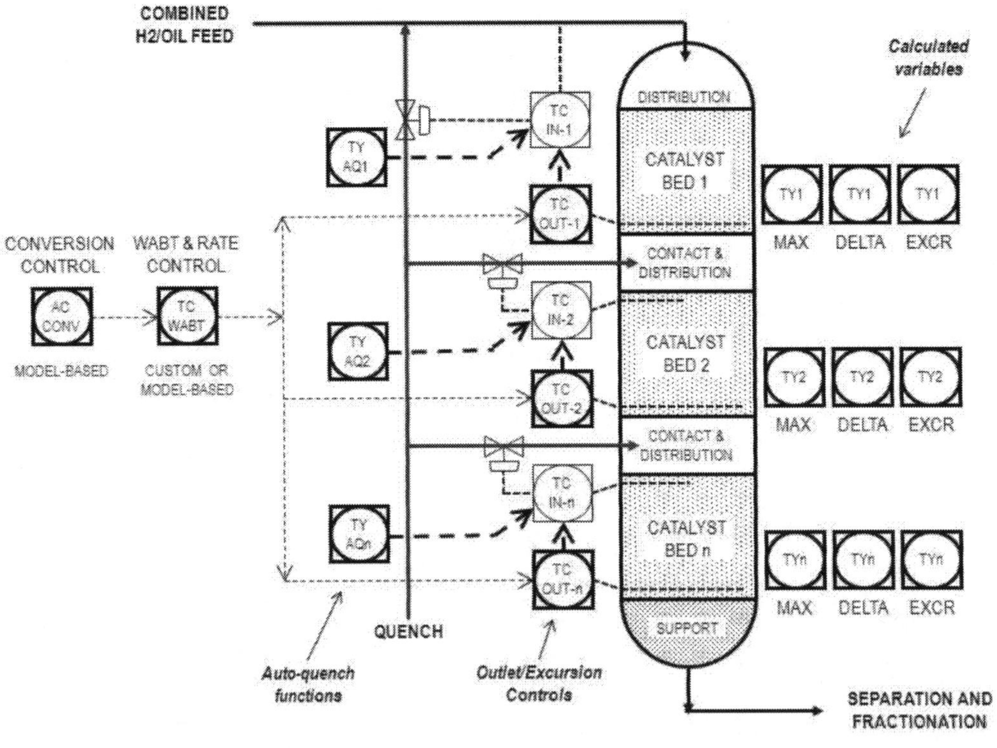

Figure 3: Hydrocracking reactor with updated control paradigm.

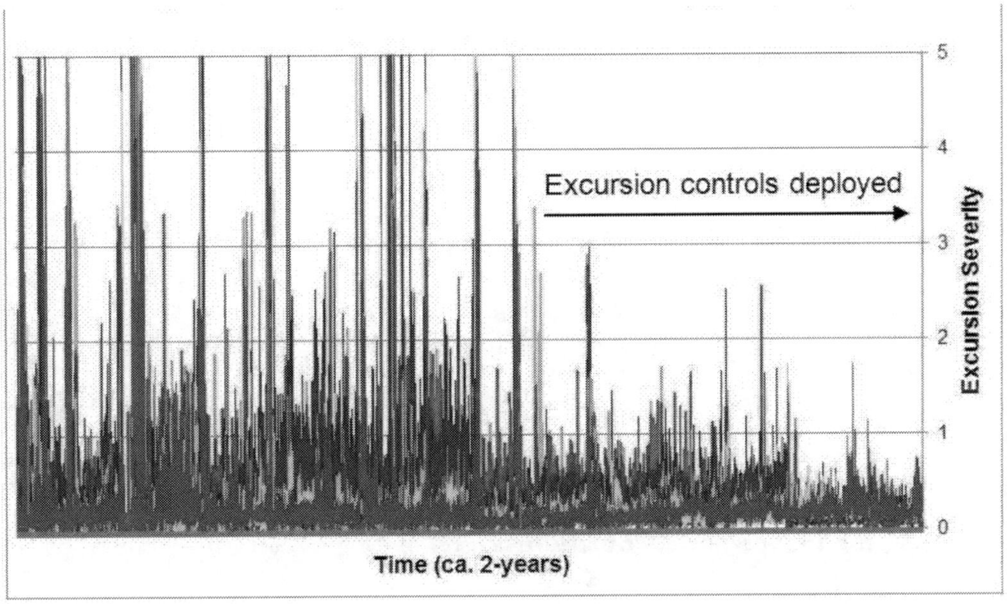

Figure 4: Temperature excursion improvement as new controls are implemented.

Abstract ID# 295208

Topical 7: 16th Topical on Refinery Processing
Heavy Oil and Bottom of the Barrel Processing
AIChE Spring National Meeting
April 28 – May 2, 2013
San Antonio, TX

Hydrogen Generation from Residuum

Girish Srinivas, Steven Gebhard, Robert Copeland and Jeff Martin*

TDA Research, Inc.
4663 Table Mountain Drive
Golden, Colorado 80403, USA
(303) 940-2321
gsrinivas@tda.com

* Author to whom correspondence should be addressed

Abstract

Refineries in the U.S. are processing increasingly heavy sour crudes that contain metals, sulfur, and high molecular weight aromatic hydrocarbons. Many sour crude oils originate in the Western Hemisphere, including heavy crudes from Venezuela, southern California, and the oil sands in Canada. Processing and upgrading these heavy feedstocks requires considerable hydrogen, but unfortunately revamping or installing new hydrogen capacity with conventional technologies such as steam methane reforming or petroleum coke gasification can be prohibitively expensive.

TDA Research Inc. is developing a new technology that will allow hydrogen to be produced in refineries at a cost that is considerably lower than hydrogen produced from conventional technologies or purchasing hydrogen from a third party. This technology converts "bottom of the barrel" residuum into hydrogen. In TDA's process, residuum is steam reformed over nickel based catalysts to produce hydrogen without catalyst deactivation and without the need for an oxygen plant; this greatly expands the range of feedstocks that can be used to generate hydrogen. We can steam reform residuum over nickel based catalysts without catalyst deactivation because the system uses a fluidized bed with periodic catalyst regeneration with air. The process is essentially chemical looping steam reforming (Adanez et al., 2012; Hoteit et al., 2011). Residuum and steam are fed into a fluidized bed reactor containing Ni steam reforming catalyst at 870°C to generate syngas ($CO + H_2$). Because the process uses contact times on the order of minutes, not enough carbon builds up on the catalyst to cause irreversible deactivation. The catalyst is then regenerated by burning the coke with air. In the laboratory, this is done using a single reactor with a nitrogen purge between reforming and regeneration steps; however, in an industrial setting, a circulating fluidized bed system would be used that would be somewhat similar to the early ESSO catalytic cracking reactors. Burning off the coke in the regenerator reheats the catalyst to about 900°C for the next reforming step (a small amount of residuum, heavy oil or other fuel can be added to the regenerator to increase the temperature if there is not enough coke on the catalyst). The hot nickel catalyst returning to the reforming reactor is present as NiO but is quickly reduced to active nickel metal by the hydrocarbons in the feed.

TDA has conducted an extensive series of tests with atmospheric tower bottoms (ATB) and two types of vacuum tower bottoms (VTB). Both VTB samples are solids at room temperature. The medium VTB was heated to be able to feed it into the laboratory reactor, and the very heavy VTB had to be cut with 20% xylene and heated. With ATB, it was possible to operate with steam to carbon (S/C) ratios as low as 3 without catalyst deactivation. Both VTB feeds required operating with a steam/carbon ratio of 5. Significant amounts of coke deposited around the feed injector in the reactor at lower S/C ratios. No catalyst deactivation has been observed with ATB or either VTB sample.

Keywords: catalysis, steam reforming, residuum, hydrogen

Introduction

More than half of the petroleum refined in the United States is imported, and approximately half of that can be classified as heavy crude that contains high concentrations of sulfur, metals and high molecular weight hydrocarbons. Sulfur, metals and other contaminants are removed by hydrotreating, and higher molecular weight hydrocarbons can be converted into lower molecular weight fractions by hydroprocessing and catalytic cracking. Very high molecular weight components such as asphaltenes that are found in residuum are usually processed by coking or used to make asphalt. The reduction in allowable aromatic hydrocarbons and sulfur in gasoline and diesel, along with the need to process heavier crude oils, has increased the demand for hydrogen in the refinery.

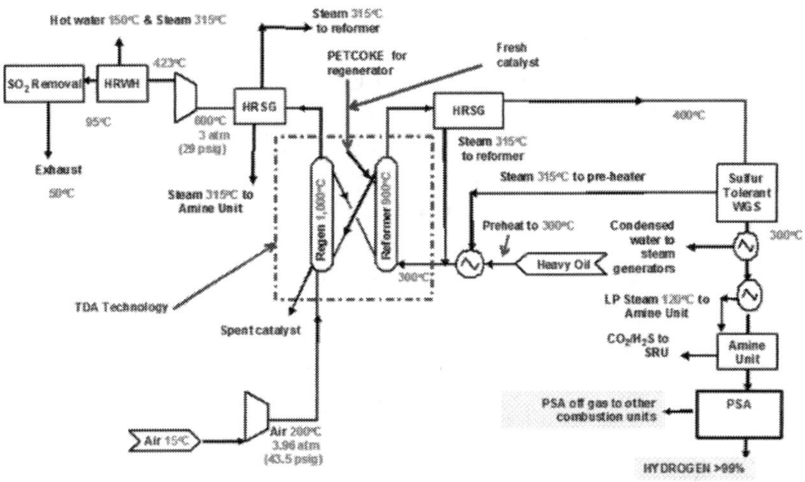

Figure 1. Schematic of TDA's heavy oil to hydrogen process.

TDA Research is developing a process to steam reform heavy hydrocarbon fractions such as atmospheric and vacuum residuum (also called atmospheric tower bottoms, ATB and vacuum tower bottoms, VTB) over nickel based catalysts without catalyst deactivation (Figure 1). The catalyst is not deactivated our process because it is regenerated in air at regular intervals. In an industrial system with circulating fluidized beds, the catalyst would be continuously regenerated. In our laboratory experiments, the catalyst is regenerated after about 90-120 minutes of reforming. In this sense our process functions like a chemical looping steam reformer (CLR) (Figure 2).

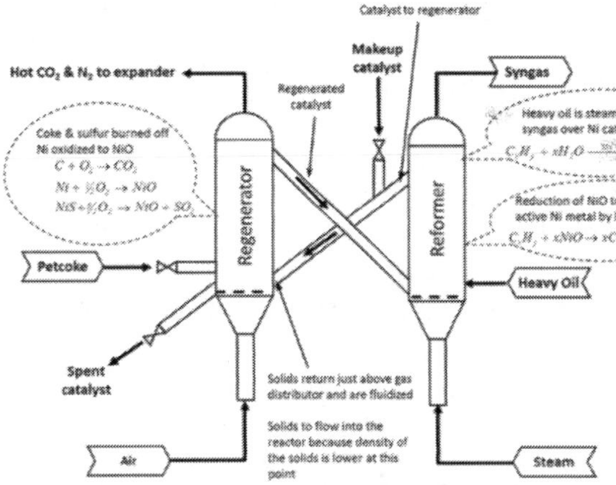

Figure 2. Reactions taking place in TDA residuum steam reforming process.

In our process, heavy oil and steam are fed into a fluidized bed reactor containing a Ni steam reforming catalyst at 870°C to generate syngas ($CO + H_2$). Because reforming is carried out for only about 90-120 min before the catalyst is regenerated, the amount of carbon (coke) that builds up on the catalyst surface is not sufficient to cause irreversible deactivation. The catalyst is then sent to a regenerator where the coke (and any sulfur) is burned off with air. The overall configuration in a full scale industrial plant would probably resemble the old ESSO Model III, FCC process (i.e. no riser reactor) where there would be two circulating fluidized bed reactors, one for steam reforming, and one for regeneration. Like the ESSO-FCC process, coke is burned off in air, which reheats the catalyst for the next reforming cycle (Gary and Handwerk 2001). Combustion of coke on the catalyst surface ($C + O_2 \rightarrow CO_2$, $\Delta H = -94$ kcal/mole) and the oxidation of nickel on the catalyst to nickel oxide ($Ni + \frac{1}{2} O_2 \rightarrow NiO$, $\Delta H = -56$ kcal/mole) supply most of the heat required to increase the catalyst temperature back up to ~900°C in the regenerator; however, additional heavy hydrocarbon feed can be burned in the regenerator if needed to increase the catalyst temperature. The hot catalyst returning to the reforming reactor is now in the form of nickel oxide, which is catalytically inactive for steam reforming; however, in the our process, the hydrocarbons in the feed quickly reduce the NiO back to catalytically active Ni for the next reforming cycle ($xNiO + C_xH_y \rightarrow xCO + y/2\ H_2 + xNi$). In an industrial sized unit, the sensible heat in the catalyst (and any inert solids added for additional thermal capacity) would provide most of the endothermic heat required by the steam-hydrocarbon reforming reactions. The catalyst flow, feed, steam and regenerator fuel rates would be adjusted to maintain the proper heat balance in the process. Figure 1 shows a concept of how TDA's heavy oil steam reforming would work at an industrial scale.

The economics of hydrogen generation in refineries strongly depends on feedstock cost. The price of natural gas for industrial consumers is currently low (Figure 3) making steam methane reforming economically attractive for generating hydrogen. The selling price of "bottom of the barrel" products depends on the price of coke (steel making demands in China have recently pushed up petcoke prices), whether there is a market for heavy fuel oil made by cutting residuum with a lighter distillate fraction, and asphalt prices. Nevertheless, increasing gas prices or lack of steam methane reforming infrastructure (e.g. smaller refineries), make TDA's hydrogen from residuum process more economically attractive.

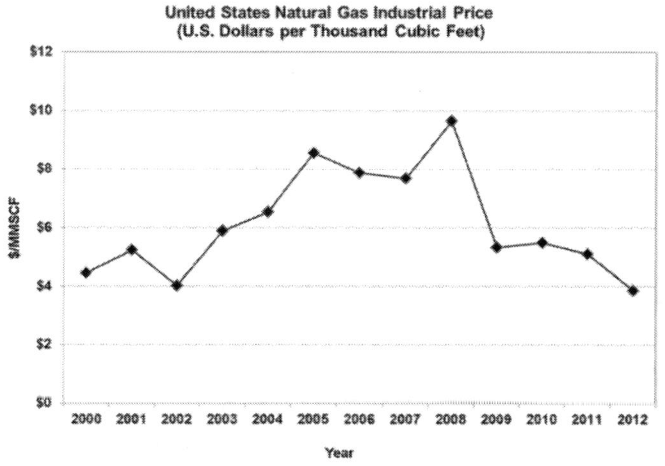

Figure 3. Historical NG prices from 2000 to present (Source: Energy Information Administration)

Another area where our residuum steam reforming process may be useful is for generating hydrogen for upgrading bitumen during syncrude production. Currently, syncrude is made by coking raw oil sand bitumen, which converts about 40-50 wt% of the carbon in the bitumen into coke. The liquid syncrude is then piped to refineries. Because coking the bitumen removes what would otherwise be the bottoms fraction of the crude, the syncrude is blended with conventional petroleum so that it can be processed in existing refineries. We are currently investigating whether our residuum steam reforming process can use raw oil sand bitumen to produce hydrogen. Since we can process extremely heavy vacuum residuum (see below) we anticipate that we will be able to process oil sand bitumen as well. By using bitumen to produce hydrogen, it may be possible to either reduce the amount of coking needed to produce syncrude by hydrotreating the bitumen directly, or upgrade syncrude produced by conventional methods before it enters the pipeline.

The main reason that our residuum steam reforming process is less expensive than partial oxidation (POX) or heavy oil gasification is that we get the advantages of oxygen gasification without the need for an expensive oxygen plant, and that we have a feedstock that can be easily handled and steam reformed at temperatures much lower than those required by gasification (our process operates at about 1500°F, whereas gasification operates at 2000+°F). Because we use a separate vessel for catalyst regeneration, no nitrogen is introduced into the syngas, which improves the efficiency of downstream water gas shift and hydrogen separation operations. All of the downstream processes shown in Figure 1 are proven technologies. Our preliminary economic analysis indicates that we can produce hydrogen for approximately $4-5/1000 SCF using our process compared with ~$5-9/1000 SCF from a small (0.5-10 MMSCFD) steam methane reforming plant, or purchasing delivered liquid hydrogen for ~$9-28/1000 SCF.

Experimental Methods

The laboratory experiments were done using the apparatus shown in Figure 4. The apparatus uses a single fluidized bed, which is cycled between reforming and regeneration. The reactor is briefly purged with nitrogen between steps to prevent mixing air with heavy oil/syngas. Electronic mass flow controllers are used to feed five gas streams to the reactor via ¼"- 316 stainless steel tubing. Each gas stream can be isolated by means of pneumatically actuated valves. There is also a high pressure liquid pump (ISCO stainless steel "syringe pump") that is used to feed water to a boiler (a tubing coil in a furnace) where it is vaporized into steam. A second ISCO high pressure liquid pump is used to feed heavy oil to a water-cooled injector where a nitrogen sweep aids in oil delivery. The heavy oil feedstocks tested were atmospheric tower bottoms (ATB) and two vacuum tower bottoms (VTB) obtained from three different refineries. The oil pump has a heated jacket filled with silicone oil at 70-90°C that is circulated by a heated oil bath to keep the viscosity of the oil low enough that it can be pumped.

Due to the high temperatures (800-900°C), the fluidized bed reactor is made from SCH40 Incoloy 800H pipe. Incoloy withstands high temperatures and has better sulfur resistance (and is less expensive) than Inconel. A section of 4" pipe is used as a "disengaging zone", which is welded to a 4"x 2" bell reducer that is attached to a section of 2" pipe that contains the catalyst bed. The catalyst bed contains about 500g of a mixture of Ni steam reforming catalyst and fluidizable α-Al_2O_3. Downstream from the reactor, a 10 μm filter collects any catalyst dust that is entrained in the gas flow, a heat exchanger/condensing coil to remove water, and a sodium citrate bubbler/scrubber to prevent H_2S (during reforming) and SO_2 (during regeneration) from entering the online H_2, CO, CO_2, CH_4 gas analyzer (Nova Analytical Systems). There is also a separate online paramagnetic O_2 analyzer. The system pressure is controlled using a computer-controlled, pressure control valve (PCV) made by Badger Meter Co. A PC running OPTO22 software controls the PCV as well as the heating tapes, tube furnace, mass flow controllers and other components of the system, as well as logs the data.

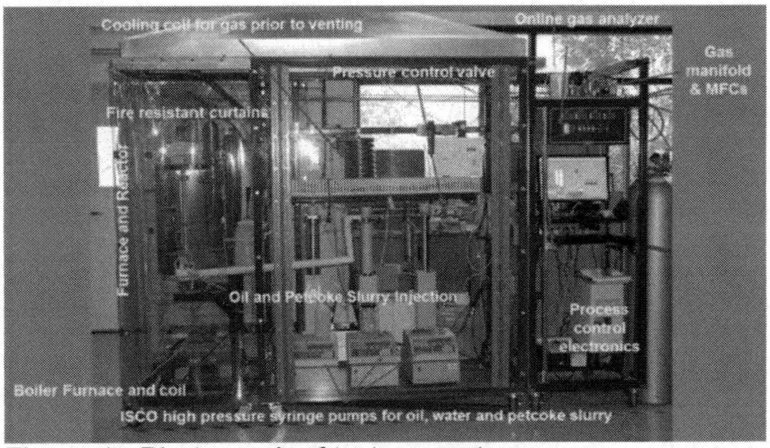

Figure 4. Photograph of test apparatus.

Results & Discussion

Tests with Atmospheric Tower Bottoms

The first heavy oil feedstock we tested was atmospheric tower bottoms (ATB) from a working refinery. A capillary column gas chromatogram of the oil dissolved at a concentration of 10 mg/mL in dichloromethane is shown in Figure 5. The baseline increases due to column bleed at high temperatures (the temperature at the highest point in Figure 5 occurs at T = 340°C (644°F). Even though the ATB stream is typically 650+°F material, small amounts of volatile hydrocarbons are still present that

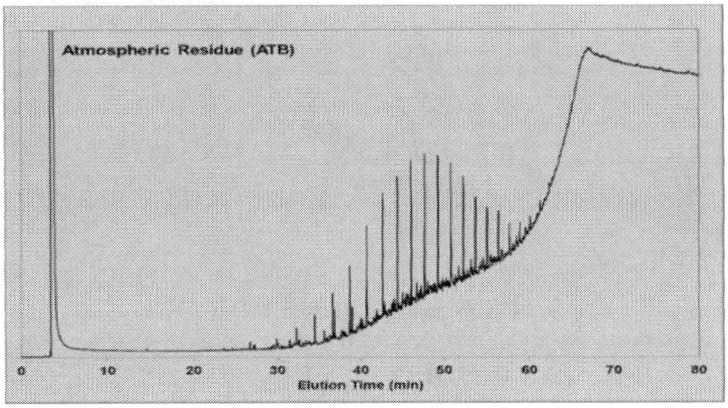

Figure 5. Gas chromatogram of ATB.

elute from the GC column between 30 and 60 min, which corresponds to about 250-600°F (roughly kerosine through gas oil; Gary and Handwerk 2001). The experiments were run with Ni catalyst concentrations of 25%, 50% and 75% (by weight) diluted with fluidizable α-Al_2O_3. The steam to carbon ratios were varied between 1.5 and 13.6. Reforming and regeneration were done at 865°C and 900°C respectively at a total system pressure of 50 psig. Depending on the test, reforming lasts about 90-120 minutes, and regeneration lasts about 40-60 min, so one cycle is about 3 hours long (e.g. ~25 cycles in ~80 hours).

Figure 6 shows the product gas composition measured using the online gas analyzers for a single steam reforming and regeneration cycle using ATB. In all cases the carbon balances were within experimental error 85-95%. The most important result is that steam reforming ATB does not require excessively high S/C ratios. In fact, after extensive testing, the optimum steam/carbon ratio for ATB was found to be about 3 and for VTB was about 5, which is encouraging because this is in the range of S/C ratios used industrially for steam naphtha reforming (Twigg 1996).

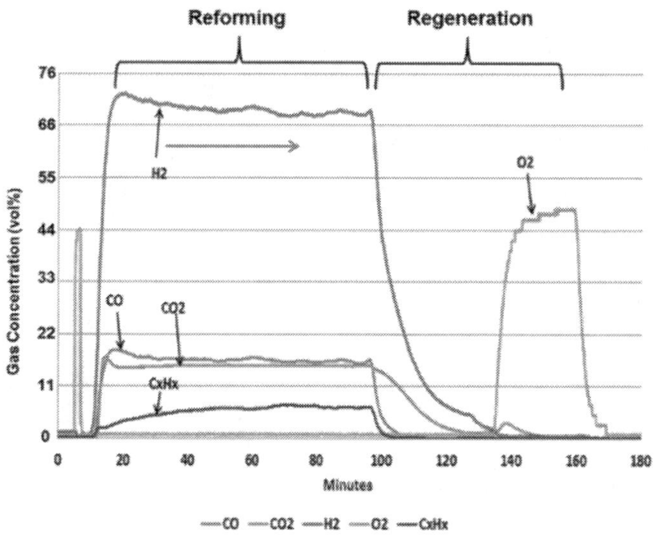

Figure 6. Single cycle of ATB reforming with S/C = 3 using 25 wt% Ni catalyst/75 wt% α-Al_2O_3 at 50 psig and 865°C.

Figure 7 shows the results for approximately 80 hours of cycling between steam reforming and regeneration using 75 wt% Ni catalyst/25 wt% α-Al_2O_3 with S/C = 5, at 865°C and 50 psig. Because the tops of the red line (H_2) return to the same level with each cycle, there was no catalyst deactivtion over the course of the test (26 cycles). If catalyst deactivation were occuring, there would be downward trend with time on stream as less and less hydrogen was produced during each cycle. The concentration of hydrogen in the gas remained relatively constant at 70-75 vol%.

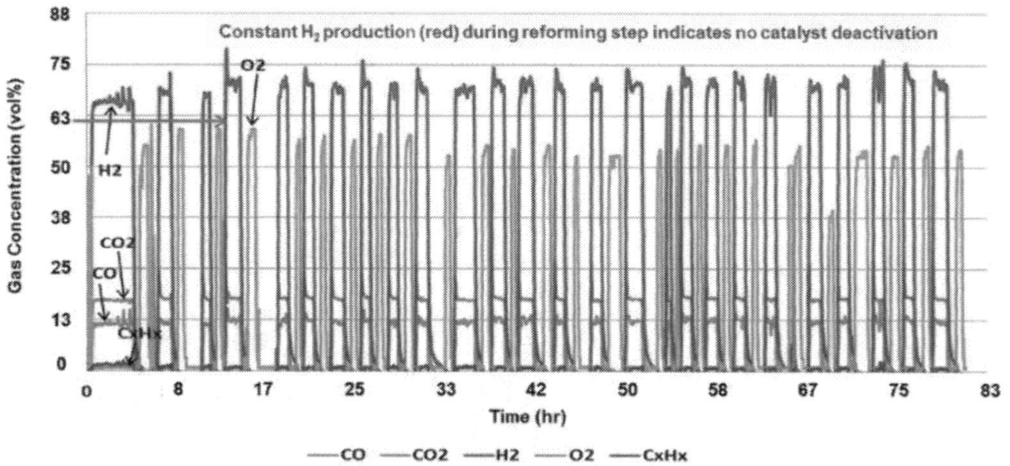

Figure 7. Gas composition for steam reforming ATB at 865°C, 50 psig, S/C = 5, 75 wt% Ni catalyst/25 wt% α-Al_2O_3.

Tests with "Medium" Vacuum Residuum

Our success with ATB encouraged us to try a more difficult and more economically attractive feed – vacuum residuum (a.k.a. vacuum tower bottoms – VTB). The VTB sample came from a working, but different refinery.

Figure 8 is a gas chromatogram for the "medium" VTB. While this material was a solid at room temperature (the ATB was a viscous liquid at room temp), its' GC analysis was surprisingly similar to the ATB in that kerosine through gas oil hydrocarbons were present (to about the same extent). Some heavier components appear as well. The main reason that the heavier component peaks are not larger (as might be expected) is that the "medium" VTB is less soluble in dichloromethane (CH_2Cl_2) than ATB. Nevertheless, comparing the gas chromatograms of the ATB and two VTB samples is qualitatively useful for characterizing their compositional differences.

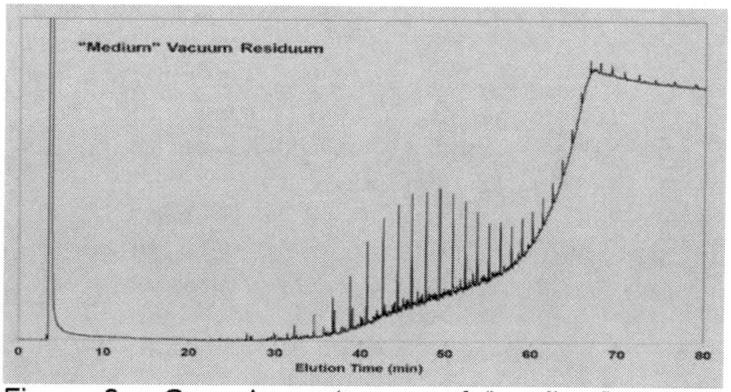

Figure 8. Gas chromatogram of "medium" vacuum residuum.

Figure 9 shows the results of 52 steam reforming-regeneration cycles (150 hours) with the "medium" VTB. As was the case with the ATB, no deactivation of the Ni catalyst was observed over the course of the 150 hours indicating that coke and sulfur deposited on the catalyst were burned off in the regenerator, and that than any metals (e.g. Ni and V) deposited had no measurable effect. As before, the hydrogen concentration was about 70 vol%.

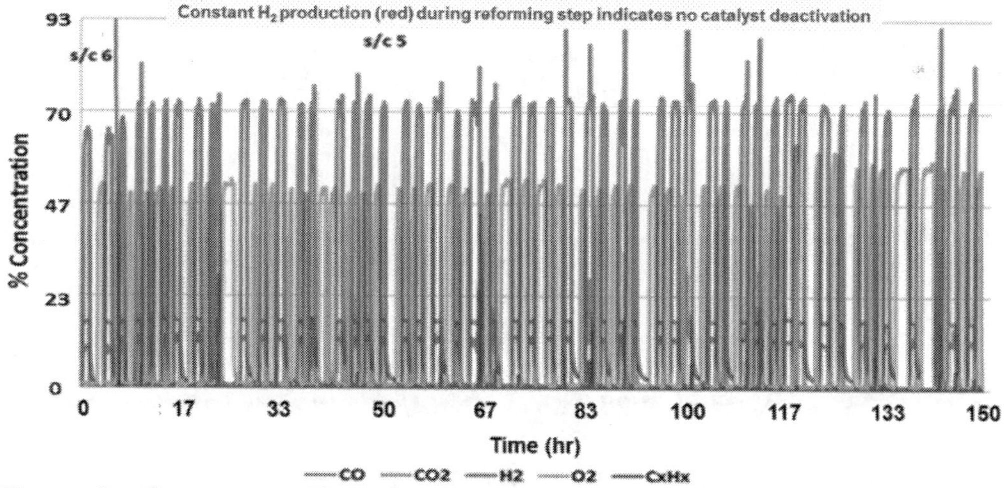

Figure 9. Gas composition when steam reforming "medium" VTB at 865°C, 50 psig, S/C =5, 75 wt% Ni catalyst/25 wt% α-Al_2O_3.

Tests with "Heavy" Vacuum Residuum

Figure 10 is a gas chromatogram for the "heavy" VTB. This material was a solid at room temperature and must be heated to almost 300°F to melt. Unfortunately, this temperature is too high to be used in our liquid feed system (an ISCO syringe pump) so we were forced to dilute it with 20 wt% xylene (and it still required heating). This material is much more tar-like than the "medium" VTB and has very little in the way of kerosine through gas oil hydrocarbons. Heavier components are present to a larger extent than in the "medium" VTB in spite of the fact that this material has even lower solubility in dichloromethane. The peaks have been enlarged by a factor of 5000 to make them more visible. The offset is where the background (column bleed) rose to very high levels. Cutting out the time between 60 and 70 min (where there were no peaks anyway) allows the high temperature data to be shown on the same graph. This "heavy" VTB is only slightly soluble in dichloromethane (CH_2Cl_2) so the peaks in the raw GC data are quite small. Nevertheless, the qualitative comparison between Figure 10 and Figure 9 shows that this VTB is considerably heavier than the medium VTB and therefore contains greater amounts of high molecular weight hydrocarbons. We will send the ATB and the two VTB samples to an outside laboratory for saturates, asphaltenes, resins and aromatics (SARA) analysis and API gravity determination to better correlate their compositions with their steam reforming performance.

Figure 11 shows the results of 22 steam reforming-regeneration cycles (66 hours) with the "heavy" VTB. Perhaps somewhat surprisingly, no deactivation of the Ni catalyst was observed over the course of the run, and as before, the hydrogen concentration was about 70 vol%. We are continuing our cycle testing with this VTB to see if any long term catalyst deactivation occurs. Nevertheless, this is an extremely encouraging result, and suggests that we may be able to generate hydrogen from Canadian oil sand bitumen and other heavy hydrocarbon feedstocks.

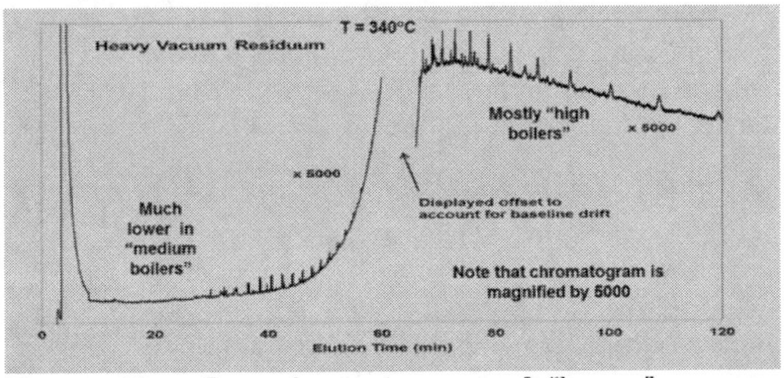

Figure 10. Gas chromatogram of "heavy" vacuum residuum.

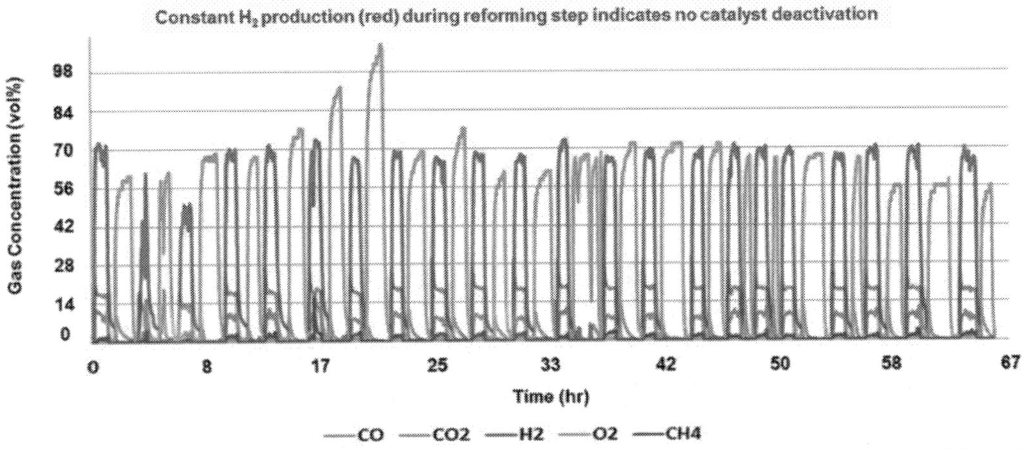

Figure 11. Gas composition when steam reforming "heavy" VTB at 865°C, 50 psig, S/C =5, 75 wt% Ni catalyst/25 wt% α-Al_2O_3.

Discussion and Conclusions

The reason that the Ni steam reforming catalyst does not deactivate in TDA's process is that cycling between reforming and regeneration removes coke and sulfur from the catalyst before it can build up to damaging levels. Figure 2 showed the reactions that take place in the reformer and regenerator. As mentioned earlier, hydrocarbons are steam reformed to syngas in the reformer vessel, and coke and sulfur that deposit on the catalyst are burned off in the regenerator, and Ni is oxidized to NiO. Catalyst from the regenerator (now NiO) is then reduced back to catalytically active Ni metal by the hydrocarbon feed when it returns to the reformer. In this sense, not only is the Ni a steam reforming catalyst, but it is also an oxygen transfer material. In a continuous system, balancing the circulation rate between the reformer and regenerator along with appropriate rates of heavy oil and steam, the system can produce syngas continuously using a Ni catalyst. This is in contrast to a fixed bed reformer, where unless oxygen is added (i.e. POX) the Ni catalyst will be rapidly coked.

Figure 12 is a comparison of the gas chromatograms for all three feeds that we have successfully used with our residuum steam reforming process. All three chromatograms are shown on the same scale. The ATB (black) has a considerable amount of dissolved hydrocarbons in the middle distillate boiling range and only a small amount of higher molecular weight components. In Figure 12, the peak height of each component is roughly proportional to its' concentration in a given sample. Thus, the "medium" VTB (pink) has almost as high a concentration of middle distillate hydrocarbons as the ATB, but the heavy VTB sample (blue) contains practically none. While these results are far from quantitative, they are useful for comparison purposes and show that the heavy VTB contains only very high molecular weight hydrocarbons, and therefore, the high conversion we observe when steam reforming heavy VTB is not simply due to reforming lighter components of the feed, because there are none. In addition, the conversions we measure are too high to be a result of just reforming the 20% xylene diluent added to make the VTB fluid.

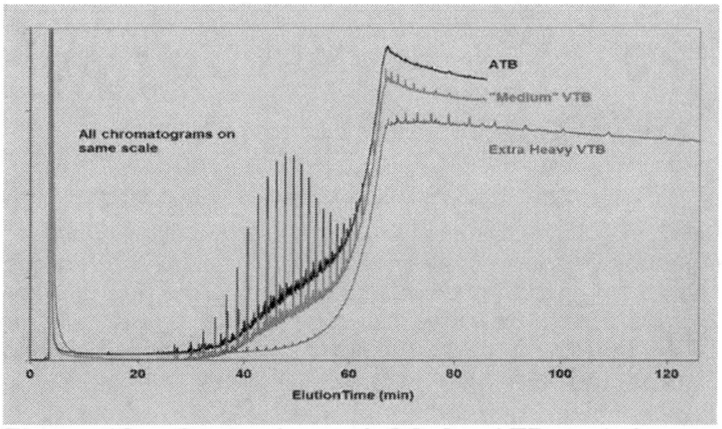

Figure 12. Comparison of GC for ATB and the two VTB samples.

In conclusion, we have successfully demonstrated that we can steam reform a wide variety of residuum feedstocks to generate hydrogen for refineries. Our economic calculations indicate that the process becomes competitive when the natural gas price approaches \$6/MM Btu (\$6/1000 SCF). These results show that unlike fixed bed steam reforming, TDA's process to produce hydrogen from heavy oil can be used with feedstocks as heavy as vacuum tower bottoms because of intermittent/continuous catalyst regeneration. This makes TDA's process very feedstock flexible because virtually any liquid hydrocarbon stream can be used. We are continuing to test the system with heavy VTB and are exploring the use of other feedstocks such as oil sand bitumen and biomass pyrolysis oil.

References

Adanez, J.; Abad, A.; Garcia-Labiano, F.; Gayan, P. and de Diego, L.F. (2012) Review: Progress in Chemical-Looping Combustion and Reforming Technologies," *Progress in Energy and Combustion Science* **38** (2012) 215-282

Gary, J.H. and Handwerk, G.E. (2001) Petroleum Refining, Technology and Economics, 4th ed., Marcel Dekker, Inc.

Hoteit, A.; Forret, A.: Pelletant, W.: Roesler, J. and Gauthier, T. (2011) "Chemical Looping Combustion with Different Types of Liquid Fuels," *Oil & Gas Science and Technology* - Rev. IPP Energies Nouvelles, **66** (2) pp. 193-199

Twigg, M.V. (1996) Catalyst Handbook, 2nd ed., Ch. 5, "Steam Reforming," Manson Publishing

(107c) Heavy Oil Upgrading Via Slurry Hydrocracking

Andy Towarnicky, Process Design & Development, UOP - A Honeywell Company, Des Plaines, IL

As crude oil supplies continue to become harder to process, fuel oil demand decreases, and demand for high-quality transportation fuel increases, the role of heavy-oil upgrading technologies will continue to expand. In this capacity, slurry hydrocracking offers a versatile solution to meet these needs and upgrade heavy bottom-of-the-barrel oils to transportation fuels and products more amenable to further processing. This talk will introduce slurry hydrocracking, review its process principles, and discuss recent developments in slurry hydrocracking as embodied in UOP's Uniflex ™ Technology.

(107d) Tools to Maximize Residuum Conversion

Stephen Warnock[1], Eloy Flores III[2] and Hsiang Yee Lai[1], (1)Chemistry and Chemical Engineering Division, Southwest Research Institute, San Antonio, TX, (2)Chemical Engineering Department, Division of Chemistry and Chemical Engineering, Southwest Research Institute, San Antonio, TX

As conventional crude oil reserves dwindle and global demand for oil rises, more companies are focusing on developing processes to refine unconventional oils such as bitumen, shale, and biofuels into conventional refinery feedstocks. Unconventional oils, such as heavy oil and shale, have higher amounts of residuum when compared to conventional crude oils. As conventional oils continue to be consumed at ever increasing rates, unconventional oils will take a higher market share in the future. We have evaluated many processes to maximize liquid yields from residuum. In order to further advance the conversion of residuum to useful liquid products, new analytical tools are being developed to evaluate the properties of the heaviest fraction of oil that is being upgraded or processed to form lighter more desirable liquid products. New analytical tools and how they can be applied to improve residuum conversion will be discussed with a brief status of the different residuum conversion technologies.

Optimizing Catalyst Performance of Continuous Catalyst Regeneration in Reforming Process

Abstract

The UOP CCR Platforming™ unit is the source of aromatics, hydrogen, and is an integral part of an aromatics complex production. Yields, on-stream reliability, and unit flexibility are key drivers that directly impact the aromatic complex profitability.

Fundamentally, UOP's Regenerator design differs from others. The UOP CCR Regenerator is uniquely designed with enhanced metals dispersion, ensuring optimal catalyst performance with reliable operations throughout a long catalyst life. A key feature of the UOP burn zone is recycling high concentrations of chlorine, minimizing Pt agglomeration. In addition, the UOP Chlorsorb system recovers chlorides by countercurrent flow of vent gas against the spent catalyst, further enhances the burn zone chlorine level. UOP has long recognized the benefits of maintaining a high chlorine concentration in the burn zone and the major impact on metals dispersion and overall catalyst life.

There are several reasons for replacing catalyst such as alumina phase change, metals poisoning or new catalyst products which can improve yields, activity and reduce coke make. It is important to highlight surface area change is not a criteria for replacing catalyst in a UOP CCR Platforming unit.

This talk will discuss the features and benefits of the UOP Regenerator burn zone and the impact on catalyst yields, cycle life, and unit reliability and operations.

(113b) A General Model for Estimating Air Emissions in Petroleum Refineries Using Emission Factors Technique

Mohammad S. Ba-Shammakh, Chemical Engineering, King Fahd University of Petroleum & Minerals, Dhahran, Saudi Arabia

Estimating the quantities of certain substances discharged into the atmosphere by several industrial activities is a necessary and fundamental stage in any environmental protection policy to tackle today's problems such as acid rain, the degradation of air quality, global warming and climate change and the depletion of the ozone layer. Any environmental protection policy requires comprehensive information on the pollution profiles of particular activities. In most countries and for petroleum refineries, particularly, such comprehensive information base does not exist. This paper proposed a general model for filling part of the information gap with respect to air emissions estimation from petroleum refineries. It presents a pollution-estimation model that draws from published sources of information and allows analysts to obtain order-of-magnitude estimates of the air pollution generated by refineries. As an illustration, the model is applied to a case study.

(113c) Benefits From Educated Selection of Advanced Materials for Radiant Coils

Dietlinde Jakobi, Reserach & Development Services, Schmidt + Clemens GmbH & Co. KG, D-51779 Lindlar, Germany and Jörg Weigandt, Reserach & Development Services, Schmidt + Clemens GmbH Co. KG, 51779 Lindlar, Germany

Radiant coils operate at severe conditions. They face high temperatures beyond 1100 °C, alternating reducing and oxidizing atmospheres, corrosion attacks, and have to provide excellent creep properties as well as minimizing the tendency of coke formation.

New insight on the interactions between protective scales and coke depositions allows S+C to draw conclusions on attack mechanisms. This puts us in a position to give recommendations on smart material selections to satisfy the demand for advanced material implementation. Each steam cracker is operated under different conditions and because of this there is always an individual solution required for optimum improvement. In order to achieve this, not only does the current situation of the cracker itself play an important role, but as well as or even more important, is to take into consideration what the operator actually wants to achieve.

Results of smart material selections on field applications will be exemplary presented.

(113d) Effect of Supercritical Hexanes Reaction Medium and H2/CO Molar Ratio On the Synthesis of Higher Alcohols From Syngas Over a K-Promoted Cu-Co-Zn Catalyst

Charlotte Stewart, Rui Xu, Sihe Zhang, David Roe and Christopher B. Roberts, Department of Chemical Engineering, Auburn University, Auburn University, AL

Optimizing the production of fuels and high-value chemicals from single-carbon compounds such as methane, carbon monoxide, and carbon dioxide has been an area of ongoing research for more than eighty years. Syngas, a mixture of carbon monoxide and hydrogen, can be generated by steam reforming natural gas or by gasifying coal or biomass and has been utilized industrially to produce a wide range of hydrocarbon products in a process known as Fischer-Tropsch Synthesis (FTS). FTS is typically carried out over either an iron or cobalt based catalyst and is a highly exothermic, surface-catalyzed polymerization reaction. By replacing the cobalt FT catalyst with a copper based catalyst, alcohols—as opposed to olefins and paraffins—can be primarily produced from syngas in a similarly unselective and exothermic alcohol synthesis reaction. Higher alcohols (e.g. ethanol and butanol) formed from the conversion of syngas are high-value feedstock chemicals that can also be used as both fuels and fuel additives due to their high octane ratings and low emissions compared to gasoline and diesel. However, traditional gas phase synthesis of higher alcohols from syngas has suffered from low productivity and selectivity.

In gas phase FTS, high rates of diffusion allow the reactants to readily access the active sites of the catalyst, but poor heat management in fixed-bed operation can result in deactivation of the catalyst and increased production of unwanted side products such as methane and carbon dioxide. On the other hand, in slurry phase FTS, the higher thermal conductivity of the liquid reaction medium allows for enhanced heat transfer; however, interphase mass transfer limitations reduce the accessibility of the reactants to the catalyst active sites. Supercritical fluids, which have physical properties intermediate between those of liquids and gases (e.g. vapor-like diffusivities and liquid-like thermal conductivities), have been shown to prevent catalyst deactivation, to facilitate heat removal, to reduce the formation of side products, and to promote carbon chain growth in FTS. Since supercritical phase operation has benefited FTS and since FTS and higher alcohol synthesis (HAS) are similar reactions, a supercritical reaction medium has the potential to improve the synthesis of higher alcohols compared to traditional gas phase operation. Additionally, incorporating cobalt into the copper based HAS catalyst has been shown in past studies to improve the selectivity towards higher alcohols.

In this study, the synthesis of higher alcohols from syngas was carried out in a fixed-bed reactor loaded with a Cu-Co based catalyst. The K-promoted Cu-Co-Zn catalyst was prepared using a co-precipitation and incipient wetness method, and hexanes were selected as the supercritical fluid reaction medium because the critical temperature of hexane falls just below the operating temperature for HAS. In the first part of this study, the effect of a supercritical environment on HAS was investigated by varying the hexane to syngas molar ratio while maintaining a constant H_2/CO ratio. Because a supercritical fluid medium will affect the relative rates of diffusion of the reactants to the catalyst active sites, the second part of this study involved varying the H_2/CO molar ratio in the feed in order to study the effect of syngas composition on catalytic performance under gas phase as well as supercritical phase conditions. Compared to gas phase operation at the same conditions, the use of supercritical hexanes resulted in a significant decrease in the selectivity towards methane and CO_2 and an increase in higher alcohol productivity.

Small and Mid-Scale LNG Pumps for an Evolving Market

Steve Rush
Director of Sales, Ebara International Corporation – Cryodynamics Division

Natural gas has become more and more prominent around the globe as a key fuel source. The need for a more clean burning fuel and the abundance of natural gas has increased the need for smaller systems to deliver LNG to customers for their use. As more countries develop an infrastructure to utilize natural gas as a fuel for trucks, buses, small power stations and local buffer storage for small communities, the need for small and mid-scale equipment is increasing.

In its infancy, the LNG industry began as a means of delivering large volumes of gas from an export country with excess supplies to another location where the gas was needed. These systems required large equipment, large ships and large plant sites to liquefy and transport the LNG. Equipment was developed for this purpose and one of the key components in these systems is the cryogenic pumps used to move the liquefied gases at very low temperatures.

ADVANTAGES OF SUBMERGED MOTOR PUMPS

Due to the hazardous nature of handling natural gas, the type of pump developed for the LNG industry and now widely used for this purpose is the submerged electric motor pump (Figure 1). Since LNG is basically non-conductive, this pump type utilizes a submerged electrical motor which is installed inside a vessel or storage tank. With no oxygen present during operation, this is the safest design available. The entire pump and motor is submerged in the LNG and the liquid is also used for cooling the motor and lubricating the bearings. The alternative to this design is an external motor type pump using shaft seals between the motor and pump sections. The motor is outside in the hazardous area and while an explosion proof housing can be used to maintain a safe environment, there is still risk in having the motor in this environment with the potential for flammable gases present. The external motor type pump also uses a mechanical seal on the shaft where the shaft penetrates the vessel into the liquid and this design adds risk as there is the potential for leakage of the LNG into the atmosphere.

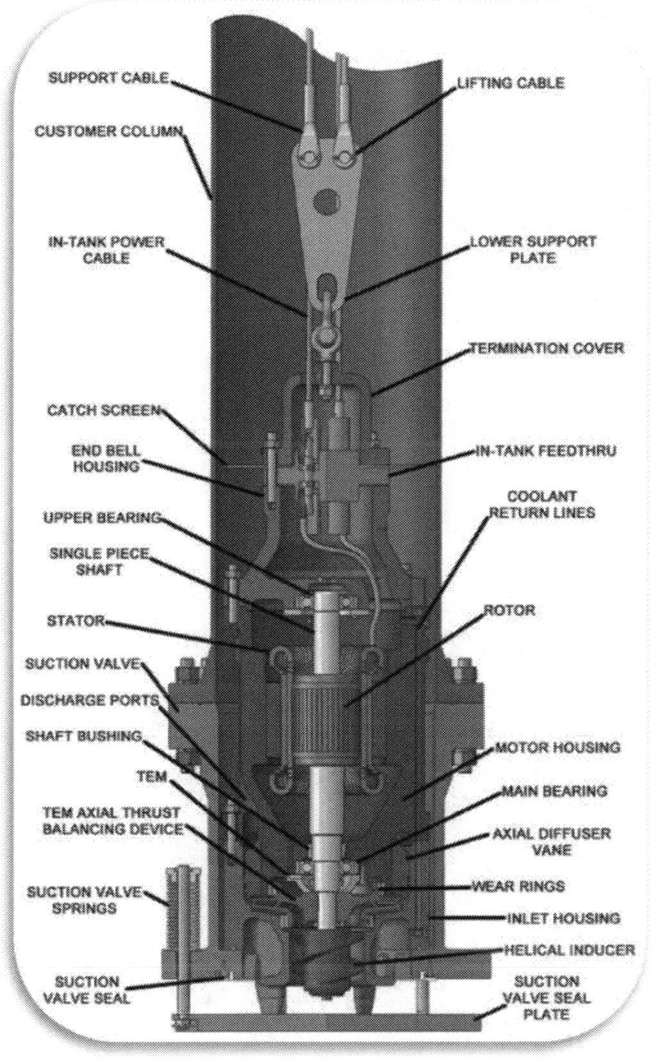

Figure 1: Typical In-Tank Retractable Type LNG Pump

Due to the safety advantages of the submerged motor pump design this pump type has become the most common type of pump used for pumping LNG as well as a variety of other liquefied gases which are non-conductive such as LPG, ethylene, ethane, propylene, LN2, etc.

THE NEED FOR MORE STANDARD SMALLER SCALE DESIGNS

The larger pumps used in many of the traditional LNG export, shipping and receiving terminal projects are typically designed to the project specifications and are manufactured when each project is built. This means that although many existing components in the manufacturer's inventory can be used, each pump is built for the project and is not a "standard" design. The delivery of these pumps is traditionally about one year from the time of order and can be very complex in their design. Many of these pumps are built in small quantities and do not lend themselves to standard manufacturing techniques. They are not built to stock and due to all of these issues, the costs are relatively high. However, the submerged electric motor

LNG pump designs have matured over more than 40 years of development and have become extremely reliable, often operating for as many as five years between overhauls.

Smaller scale LNG systems are being built with shorter delivery times and more standardized designs that require lower cost and more standardized equipment. These systems therefore require equipment that is less expensive and pre-engineered with parts that can be built to stock and assembled fairly easily and quickly.

Ebara International Corporation's Cryodynamics Division has recognized this need and has developed a line of submerged electric motor pumps which meet the increasing demand for the small to mid-scale market by adapting their experience from the larger scale pumps. This new pump model has no shaft seals like its larger cousin and uses Ebara's very reliable thrust balancing system (Thrust Equalizing Mechanism or TEM™).

EMPHASIS ON SAFETY

In the large scale LNG plants, the pumps and other equipment are monitored very closely by well-trained plant maintenance personnel. In most cases, the equipment has continuous monitoring systems which alert plant operators if any problems occur in the electrical systems or pump vibration. In smaller scale systems, which can sometimes be placed in more remote locations and may not be monitored continuously, the submerged motor design is an ideal choice as it lends itself to a very safe design with very low potential for leakage. You can imagine this type of system being similar to a gasoline station where the user is filling a vehicle with no personnel monitoring the system on a continuous basis. Although the liquid volumes are much smaller, using a pump with shaft seals and an external motor in small to mid-scale applications adds risk particularly when the equipment is not monitored in the same way as a large scale facility. This is particularly true in truck loading or other applications where a vehicle operator may be the one actually operating the system to fill a truck, bus, or other small tank.

The submerged LNG pumps developed by Ebara utilize pre-engineered and standardized designs to reduce cost and provide much shorter delivery times to customers building smaller scale systems while still providing the inherent safety and reliability of the larger scale pumps. These pumps can be assembled and shipped relatively quickly and while they can be shipped directly after assembly, if desired, they can also be performance tested in LNG in Ebara's test facility in Sparks, Nevada, in the western USA.

VARIABLE SPEED OPERATION

Another feature of the smaller scale pre-engineered pumps developed by Ebara is their ability to be connected to a variable frequency drive with no changes in the pump design or construction. In many smaller scale systems, wide ranges of pressure and flow are required due to variations in tank levels and temperatures particularly in applications such as truck trailers and truck and bus fuel tanks. When the tanks are close to empty, the temperature can be significantly warmer and a higher pressure is required during initial filling to collapse the pressure. Therefore, the pump needs to operate at a higher speed at first and after the temperature begins to drop in the tank, the pump speed can be reduced to continue filling the tank.

By introducing variable speed, fewer pump designs can be used to cover a wide range of flows and pressures using a relatively low number of common components which also helps reduce costs and parts inventory.

SMALLER SCALE PUMP APPLICATIONS AND TYPES

The LNG markets considered for this newly developed pump are as follows:

- Truck trailer loading
- Filling of truck and larger vehicle fuel tanks
- Peak-shaving
- Local municipal storage
- Small scale LNG ship or barge loading
- Small or mid-scale power plants

For many small scale systems, LNG storage tanks are built in the 5,000 to 20,000 m^3 size range. For this application, a removable type pump can be supplied in the same fashion as larger LNG storage tanks (Figure 2). This pump is installed in a pump column or tube with a suction valve mounted at the bottom. When the pump is lowered into the column, the bottom of the pump pushes the valve plate open allowing LNG to enter the pump inlet. The pump can then pump the liquid up to the top of the tank and into the discharge piping to its destination.

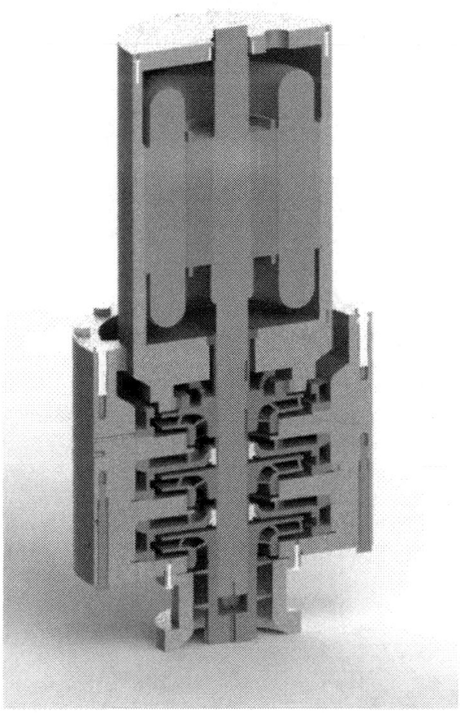

Figure 2: Small Scale In-Tank LNG Pump

If a pump is required for a pipeline type system the pump can be installed in its own stand-alone suction vessel which can be insulated using traditional insulation methods or can be supplied as a vacuum jacketed design. This type can be mounted anywhere in the piping system for LNG transfer.

In addition to the traditional land-based type pump systems, a variety of marine applications are also very suitable for the smaller scale pump designs. Ebara was the first manufacturer to introduce a design for the fuel gas pumping systems used on the Dual Fuel Diesel Electric (DFDE) LNG carriers being built today (Figure 3). This fuel gas pump can be easily adapted to any marine application where lower flows are required for off-loading LNG cargoes from barges or other smaller LNG marine vessels.

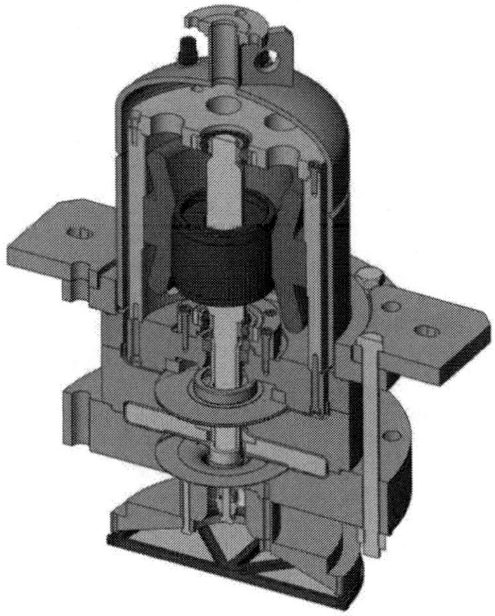

Figure 3: Small Scale Marine Pump

CONCLUSION

In conclusion, the newly developed submerged electric motor LNG pumps by Ebara International Corporation have been designed to provide safety, faster delivery and lower cost for the many customers building smaller scale systems. By providing the proven Thrust Equalizing Mechanism (TEM™) even in these lower cost pumps, customers can enjoy a significant level of reliability in addition to the inherent safety features of the submerged motor design with no rotating seals.

(116b) Relative Benefits Between Various Drivers for the Main Refrigeration Compressors for LNG Application

Patrice Bardon, Dresser-Rand, Houston, TX

The main refrigeration compressors in an LNG plant are consuming a lot of energy. The selection of the driver is a key decision and influences the future production pattern. Depending upon the driver selection the nominal plant size capacity can be affected. It is proposed to review the relative benefits of the various available drivers for the main refrigeration units: Electrical motors, Aero derivative gas turbines, industrial gas turbines and steam turbines. Two cases study will be reviewed in details: Steam turbine versus aero gas turbine for a 4 MTPA plant capacity and industrial gas turbine versus aero gas turbine for a 5 MTPA plant capacity.

(116c) Boil-off-Free LNG Transfer Between Floating and On-Shore LNG Storage Tanks

Hans E. Kimmel, Ebara International Corporation, Sparks, NV

The transfer of LNG from one tank to another tank is conducted by pumping at a discharge pressure of around 5 barg. The LNG in the tank is contained at a temperature around the boiling point stabilized by a continuous LNG boil-off to maintain vapor liquid equilibrium. Pumping the LNG at the discharge pressure increases the enthalpy of the LNG at the pump exit and during the isenthalpic expansion process inside the loading arm the LNG becomes a boiling two-phase fluid due to partial vaporization. The LNG vapor inside the pipe of the loading arm reduces the cross section of the pipe for liquid flow and substantially increases the flow velocity and the friction due to two-phase fluid flow. As a result and following the Bernoulli Equation, for a given differential pressure the mass flow will be reduced substantially.The reduction of the mass flow reduces the LNG transfer rate and increases the overall transfer time for a given quantity of LNG transfer. Also the vapor return line size will be reduced to a great extent. The paper presents a methodology and equipment to minimize or to eliminate this undesired LNG boil-off within the loading arm. It allows tight LNG flow inside inside loading arm which significantly shortens the LNG transfer time compared to current conventional methods.

() LNG Technology Development at Air Products

Yu-Nan Liu, Air Products & Chemicals, Allentown, PA

This presentation will trace the beginning of MCR (mixed components refrigerant) and use of Coil Wound Heat Exchanger, from helium recovery to the hybrid C3-MR process to AP-X® process and variation of it for natural gas liquefaction in baseload LNG plants and other applications, from Air Products' point of view.

(124a) Production of Middle Distillate Range Liquid Fuels From Syngas Using Fischer-Tropsch Synthesis and Associated Upgrading Technology Under Supercritical Phase Conditions and Multiple Reactor Configurations

David Roe, Sihe Zhang, Rui Xu, Charlotte Stewart and Christopher B. Roberts, Department of Chemical Engineering, Auburn University, Auburn University, AL

Fischer-Tropsch Synthesis (FTS) is a nearly century old method for producing low sulfur liquid fuels from a carbon monoxide and hydrogen gas feedstock. Dating back to Germany between WWI and WWII, FTS has been used industrially in South Africa for over 50 years. FTS is the highly exothermic surface-catalyzed hydrogenation and polymerization of carbon monoxide. Traditionally, the reaction is conducted on either an iron or cobalt based catalyst in either a gaseous or slurry phase. Under these reaction conditions, FTS has a few significant problems. As a polymerization reaction which has generally been shown to follow Anderson-Schulz-Flory (ASF) reaction kinetics, FTS inherently produces a broad range of products including gaseous light hydrocarbons, fuel range liquids, heavy waxes, CO_2 and water. Additionally, FTS products consist primarily of n-paraffins and 1-olefins with very few branched compounds, aromatics, or oxygenates. This leads to low octane numbers and poor cold flow properties, which are unfavorable for both gasoline and diesel engines. Traditional FTS also has a high selectivity towards CH_4 and CO_2 whichare unwanted side products whose production leads to lower carbon efficiency.

This study sought to enhance FTS productivity of fuel-range (C_5-C_{22}) hydrocarbons by modifying the traditional mode of operation in two ways. First, the reactor configuration was altered in order to shift the product distribution towards the target hydrocarbon range and also to produce more branched and aromatic compounds. Initially, two additional reaction beds were added sequentially below the FTS reaction bed. The three bed system consisted of a traditional FTS reactor loaded with Fe-based catalyst, an oligomerization bed loaded with amorphous silica alumina (ASA), and a hydrocracking/isomerization bed utilizing 1 wt. % Pd loaded onto ASA. The additional reactors served to oligomerize light olefins and crack heavy waxes down towards the target product range. The product distribution of this system was then compared to that of a single-bed FTS reactor as well as to that of a single bed reactor loaded with a mixture of Pd-ASA cracking catalyst and Fe based FT catalyst.

As a second modification to the traditional FTS process, the reactions were carried out in a supercritical fluid (hexanes) medium. Supercritical fluids (SCFs) have liquid-like densities and heat transfer properties, have high diffusivities, are miscible with gases, and can be excellent solvents for both liquids and waxy products. Gas phase FTS reactors suffer from poor heat management, wax buildup on catalyst active sites, and poor economies of scale. In contrast, slurry phase FTS reactors have better heat management, but suffer from interphase diffusion resistance, low conversion, catalyst abrasion, media maintenance complications, and difficult product separation. The unique properties of SCFs help to offset the problems of both gas and slurry phase systems by allowing gaseous reactants and liquid products to freely diffuse to and from surface reaction sites on the catalyst as well as by aiding in the removal of excess heat of reaction.

Relative to traditional gas phase FTS, the three bed system showed a significant shift of products towards middle distillates, an increase in aromatics and branched paraffin production, and a reduction in olefins. The further addition of SCF markedly reduced CH_4 and CO_2 selectivity, increased heavy product yields, and helped maintain the catalyst activity.

(124b) Quench Oil Viscosity Monitoring by Virtual Analyzer. Is It Possible and Useful?

Emerentino B. Quadro, Olefins Process Engineering, BRASKEM, Camaçari, Bahia, Brazil

Quench oil viscosity depends of a lot of variables and they change continuously. Monitoring it is an important concern of Olefins plant operation team. A too high quench oil viscosity could result in energy losses due its effect on heat exchangers efficiency, product off spec (some plants should specify the BMCI, viscosity and evaporation residue of the oil for the Carbon Black industry) and process safety risks in the scenery of emergency plant shutdown when the quench oil circulation is stopped.

Usually the quench oil monitoring is performed by laboratory analysis, every day and in some plants or situation, each eight hours. This procedure implies in operator and analyst time consumption, operator exposure every time he collects a sample and laboratory residues and equipments to clean.

At Braskem Olefins Plant 1 (Brazil – Camaçari site) it was developed a model to describe the quench oil viscosity, it´s called internally QUOVVA – Quench Oil Viscosity Virtual Analyzer. A Neural Network has been trained using thousands of operational data. Nowadays the QUOVVA average error is around 18%. Statistic calculation shows the maximum error expected, with 95% confidence, is 80 SSU (at 98.9ºC). The maximum recommended value to plant operation team is 280 SSU (at 98.9ºC).

Currently, laboratory analysis is only recommended case the QUOVVA indicates a quench oil viscosity higher than 200 SSU (at 98.9ºC). Once the usual value of quench oil viscosity varies between 100 and 160 SSU (at 98.9%), this model implementation resulted in 85% reduction of laboratory viscosity analysis.

(124d) A Novel Biosorption Resin for the Removal of Radioactive Cesium From Waste Water

E.B Naidoo, Chemistry, Vaal University of Technology, VanderbiljPark, South Africa

A NOVEL BIOSORBENT RESIN FOR THE REMOVAL OF RADIOACTIVE CESIUM FROM WASTEWATER

Naidoo E.B., Pholosi A

Biosorption and Water Treatment Research Laboratory, Department of Chemistry,

Faculty of Applied and Computer Science, Vaal University of Technology, P. Bag X021, Vanderbiljpark, 1900 South Africa.

*Corresponding author's address: E-mail bobby@vut.ac.za, Phone: +27 788932615

This study examines the design and application of a simple, cost effective biosorbent resin from pine cone powder for the removal of radioactive cesium from aqueous solution. The resin was designed to incorporate the cyanoferrate ligand which is known to have a high affinity for cesium ions in aqueous solution.

The biosorbent resin design involves the elimination of plant organic components, increasing the surface charge and iodine capacity by solvent extraction using toluene-ethanol mixture of ratio 2:1. Iron(III) ions were then immobilized on the extracted pine cone powder to produce the biosorbent-Fe^{3+} exchanger. Finally, the iron(III) loaded biosorbent was reacted with potassium ferricyanide so as to insert the cyanoferrate ligand on the biosorbent producing the biosorbent resin.

The results show that an optimum 1.90 mol/l of Fe^{3+} was immobilized on 5.0 g of pine cone powder surface at neutral solution pH. The optimum cyanoferrate ligand incorporated into the biosorbent was found to be 0.26 mol/l at neutral solution pH. The presence of Fe^{3+} and hexacyanoferrate ligand incorporated into the biosorbent was confirmed by FTIR and TGA analysis. Cesium removal experiments indicate that the new biosorbent resin was more efficient for cesium removal than the raw pine cone powder.

Keyword: Radioactive Cesium; Hexacyanoferrate: Pine cone: biosorbent resin

Vapor-Liquid Equilibrium of Trioxane-Formaldehyde-Water Mixtures --New Division of Components in UNIFAC Groups

Guizhen Li, Chunfeng Shi, Yufeng Hu and Yansheng Liu[1], *State Key Laboratory of Heavy Oil Processing Research, China University of Petroleum, Beijing, China*

Abstract

A thermodynamic model of trioxane-formaldehyde-water mixtures was revised to describe vapor -liquid and chemical equilibrium. In the model, division of components into UNIFAC groups was developed, according to different hygroscopicity of the CH_2O-groups with different positions in the poly (oxymethylene) glycol molecule. These new group-interaction parameters were calculated using experimental data in literatures. This leads to a distinctly improved agreement of model predictions with experimental results while largely keeping the simplicity of the functional group approach.

Introduction

Trioxane is a cyclic trimer of formaldehyde. It is an important intermediate in the production of polyacetal plastics from formaldehyde. Also, it is the basic raw materials in producing many fine chemicals such as pesticides, binders, disinfectants, antimicrobials [1]. In the production of such chemicals, trioxane with high purity is recovered by distillation from products of the formaldehyde conversion reaction in acidic aqueous solutions. The basic design of such processes requires, for example, a model for the vapor-liquid equilibrium of the chemical reactive system trioxane + formaldehyde + water.

A physicochemical model used in the description of vapor-liquid equilibrium in formaldehyde-containing mixtures was originally developed by Maurer (1986) [2]. This model has been continuously updated and its application range was extended as new experimental data for vapor-liquid equilibrium (Hasse et al., 1990 [3]; Hasse and Maurer, 1991 [4]; Albert et al., 1996 [5]; Albert et al., 1999 [6]; Albert et al., 2000 [7]; Albert et al., 2005 [8]). The model update is able to give accurate values for phase equilibrium properties. However, the agreement between the calculated results of the model and the experimental data for the vapor-phase concentrations of trioxane is considered to be unsatisfactory in the conditions of low temperature and low pressure. In comparison to the experimental data reported by Maurer (26 data points at around 369K and 100kPa [2]), the standard deviations and the maximum deviations are 7.32% and 21.73% for the vapor-phase concentrations of trioxane, respectively. A similar comparison for the experimental results reported by Albert (12 data points at 413K [8]) results in standard deviations and the maximum deviations of 3.553% and 9.2% for the vapor-phase concentrations of trioxane, respectively. Therefore, the present work was to revise the model and extend its application range. That extension was achieved by new division of components into UNIFAC groups. According to different hygroscopicity of the CH_2O-groups with different positions in the poly (oxymethylene) glycol molecule, the CH_2O-groups were divided into three types. These new group-interaction parameters were calculated using experimental data in literatures.

Vapor-Liquid Equilibrium Model

[1] Corresponding author. Tel.: +86-10-89733288.
E-mail address: wsuper@cup.edu.cn.

Formaldehyde is highly reactive and therefore forms different adducts with the solvents in aqueous solutions. The most important chemical-reaction products are methylene glycol ($HOCH_2OH$; here, also, MG) and poly(oxymethylene) glycols ($HO(CH_2O)_iH$, $i>1$;here, also, MG_i) [2]

$$\begin{aligned} &CH_2O + H_2O \rightleftharpoons HOCH_2OH \\ &HO(CH_2O)_{i-1}H + HOCH_2OH \rightleftharpoons HO(CH_2O)_iH + H_2O, i \geq 2 \end{aligned} \tag{1}$$

So, in the vapor-liquid equilibrium model for trioxane-formaldehyde-water mixtures, both chemical reactions and physical interactions between all species were taken into account [2].

Generally, assumed that the physical interactions in the gas phase and the effect of pressure on the fugacity and activity of any component in the liquid phase are disregarded at pressure less than 500kPa, in order to describe vapor-liquid equilibrium, activity coefficient model must be first established. Any activity coefficient models reported in literatures are obtained from excess Gibbs free energy, which is calculated from excess enthalpy (interaction between components in solutions) and excess entropy (differences between molecules). As shown in equation(2), the component content in solutions has an essential influence on excess Gibbs free energy. As for trioxane-formaldehyde-water mixture, the equilibrium distribution of formaldehyde to its reaction products with water is essential to calculate excess Gibbs free energy.

$$G^E = \sum x_i \bar{g}_i = \sum x_i (\bar{h}_i + T \cdot \bar{s}_i) \tag{2}$$

The Equilibrium Distribution Model

Basic assumptions of the equilibrium distribution model:

- The liquid phase is a mixture of trioxane, water, (very little molecular) formaldehyde, methylene glycol, and a large variety of poly(oxymethylene) glycols.
- The reactions in trioxane-formaldehyde-water mixtures are shown as the equations(1).
- The chemical equilibrium constants are defined by the equations:

$$K_1^L = \frac{x_{MG}}{x_{FA} \cdot x_W} \cdot \frac{\gamma_{MG}}{\gamma_{FA} \cdot \gamma_W}; K_2^* = \frac{x_{MG_2} \cdot x_W}{x_{MG} \cdot x_{MG}} \cdot \frac{\gamma_{MG_2} \cdot \gamma_W}{\gamma_{MG} \cdot \gamma_{MG}}; K_i^* = \frac{x_{MG_i} \cdot x_W}{x_{MG_{i-1}} \cdot x_{MG}} \cdot \frac{\gamma_{MG_i} \cdot \gamma_W}{\gamma_{MG_{i-1}} \cdot \gamma_{MG}}, i \geq 3 \tag{3}$$

And the constants have the relations:

$$K_3^* = K_4^* = \cdots = K_n^* = K^* \tag{4}$$

The equations of (3) are converted to the equations(5):

$$\frac{x_{MG}}{x_{FA} \cdot x_W} = K_1^L \Big/ \left(\frac{\gamma_{MG}}{\gamma_{FA} \cdot \gamma_W} \right) = K_1; \frac{x_{MG_2} \cdot x_W}{x_{MG} \cdot x_{MG}} = K_2^* \Big/ \left(\frac{\gamma_{MG_2} \cdot \gamma_W}{\gamma_{MG} \cdot \gamma_{MG}} \right) = K_2; \frac{x_{MG_i} \cdot x_W}{x_{MG_{i-1}} \cdot x_{MG}} = K_i^* \Big/ \left(\frac{\gamma_{MG_i} \cdot \gamma_W}{\gamma_{MG_{i-1}} \cdot \gamma_{MG}} \right) = K, i \geq 3 \tag{5}$$

Based on the basic assumptions, the true composition of formaldehyde and MG_i ($i>1$) is an expression of the true compositions of MG and water and chemical equilibrium constants:

$$x_{FA} = \frac{x_{MG}}{K_1 \cdot x_W}; x_{MG_i} = \frac{K_2 \cdot K^{(i-2)} \cdot x_{MG}^{\,i}}{x_W^{(i-1)}} = \frac{K_2 \cdot x_{MG}}{K} \cdot m^{i-1}, i \geq 2 \tag{6}$$

In the equations:

$$m = \frac{K \cdot x_{MG}}{x_W}.$$

Mass balance equations are:

$$\tilde{x}_{FA} = \frac{x_{FA} + x_{MG} + \sum_{i=2}^{n} i \cdot x_{MG_i}}{1 + x_{MG} + \sum_{i=2}^{n} i \cdot x_{MG_i}}, \tilde{x}_{TOX} = \frac{x_{TOX}}{1 + x_{MG} + \sum_{i=2}^{n} i \cdot x_{MG_i}}, x_{FA} + x_W + x_{TOX} + x_{MG} + \sum_{i=2}^{n} x_{MG_i} = 1 \quad (7)$$

Combine equation (6) with equation (7) and then obtain binary nonlinear functions of the true compositions of MG and water, in which $m^n \to 0$ when $n \to \infty$:

$$\tilde{x}_{FA} = \frac{\frac{m}{K_1 \cdot K} + x_{MG} + \frac{K_2 \cdot x_{MG}}{K}\left[\frac{m}{1-m} + \frac{m}{(1-m)^2}\right]}{1 + x_{MG} + \frac{K_2 \cdot x_{MG}}{K}\left[\frac{m}{1-m} + \frac{m}{(1-m)^2}\right]}; \tilde{x}_{TOX} = \frac{x_{TOX}}{1 + x_{MG} + \frac{K_2 \cdot x_{MG}}{K}\left[\frac{m}{1-m} + \frac{m}{(1-m)^2}\right]} \quad (8)$$

$$1 = \frac{m}{K_1 \cdot K} + \frac{K \cdot x_{MG}}{m} + x_{MG} + \frac{K_2 \cdot x_{MG}}{K} \cdot \frac{m}{1-m} + x_{TOX} \quad (9)$$

Through the substitution method, equations (8) ~ (9) are converted to a quartic equation of m:

$$c_4 \cdot m^4 + c_3 \cdot m^3 + c_2 \cdot m^2 + c_1 \cdot m + c_0 = 0 \quad (10)$$

In the equation (10),

$$c_0 = -K \cdot \tilde{x}_{FA}; c_1 = c + \tilde{x}_{FA} \cdot (2K-1) + \frac{K}{a}; c_2 = c \cdot (2b - 2 - 1/a) + \tilde{x}_{FA} \cdot (2 - K - b) + \left(\frac{1}{a} - \frac{2K}{a}\right); a = K_1 \cdot K, b = \frac{K_2}{K};$$

$$c_3 = c \cdot \left(1 - b - \frac{2b}{a} + \frac{2}{a}\right) + \tilde{x}_{FA} \cdot (b-1) + \left(\frac{K}{a} + \frac{b}{a} - \frac{2}{a}\right); c_4 = c \cdot \left(\frac{b}{a} - \frac{1}{a}\right) + \left(\frac{1}{a} - \frac{b}{a}\right); c = 1 - \tilde{x}_{FA} - \tilde{x}_{TOX}$$

Though the conversion, the composition distribution of trioxane-formaldehyde-water mixtures is easily calculated by use of MATLAB, which greatly simplifies the calculation process and therefore avoid using the complex algorithm of MATLAB [9].

Vapor-Liquid Equilibrium Model

In the physicochemical model of trioxane-formaldehyde-water system developed previously (Maurer)[2], it is assumed that the liquid phase is a mixture of trioxane, water, formaldehyde, methylene glycol, and a large variety of poly(oxymethylene) glycols and the vapor phase is an ideal mixture of trioxane, water, formaldehyde and methylene glycol in the industrial conditions. The vapor-liquid equilibrium of this system is calculated by extended Raoult's law:

$$P_i^S \gamma_i x_i = P y_i (i = FA、 Water、 MG、 Trioxane) \quad (11)$$

The activity coefficient in the liquid phase is calculated by UNIFAC. As the extension of previous model for the system, new division of components into UNIFAC groups is proposed in the present work. It is pointed out in the literature [10] that the position of the hydroxyl groups in the polyol molecule influences the interactions with water and the definition of structural groups within UNIFAC cannot account for such differences. Therefore, a new UNIFAC parameterization was developed in the literature, which discriminates between three types of alkyl groups based on their position in the molecule.

Similarly, in the trioxane-formaldehyde-water mixtures, the position of -CH_2O groups in the poly(oxymethylene) glycols molecule influences the interactions with water. The CH_2O groups are classified into the following three types:

- CH_2O in formaldehyde monomers. Formaldehyde is highly active and interacts greatly with water.

- CH_2O with a hydroxyl group in the poly(oxymethylene) glycols molecule. Compared to the first type, the interaction of this type of CH_2O group with water decreases. Trioxane is divided into three CH_2O groups of this type.
- CH_2O in the poly(oxymethylene) glycols molecule.

Division of components in trioxane-formaldehyde-water into UNIFAC groups is presented in Table 1. And the size and surface parameters of the groups are given in Table 2.

Table 1. Division of Components in UNIFAC Groups

UNIFAC Group / Component	H_2O	$HOCH_2OH$	CH_2O monomer	CH_2O with –OH group	CH_2O in the MG_i molecule	CH_2	OH
Water	1						
Formaldehyde			1				
Methylene glycol		1					
Poly(oxymethylene) glycols				1	i-2	1	2
Trioxane				3			

In the UNIFAC model, there are thirty unknown interaction parameters between group i and j, either of which represents one of the three new groups. To reduce the number of parameters in the model that require fitting, the UNIFAC parameters for interaction between either CH_2O monomer or CH_2O in the MGi molecule on one side and other groups on the other side were estimated by the parameters for interactions between

- CH_2O aldehyde group on one side and other groups on the other side and
- CH_2O in the ether molecular on one side and other groups on the other side, respectively.

These UNIFAC parameters are adopted from the literature [11].

Table 2. UNIFAC Size and Surface Parameters

Group	H_2O	$HOCH_2OH$	CH_2O monomer	CH_2O with –OH group	CH_2O in the MG_i molecule	CH_2	OH
No.	1	2	3	4	5	6	7
R	0.9200	2.6744	0.9183	0.9183	0.9183	0.6744	1.0000
Q	1.400	2.940	0.780	0.780	0.780	0.540	1.200

The interaction parameters between CH_2O with -OH group and other groups are fitted using experimental data for the vapor-liquid equilibrium for the ternary system (formaldehyde + trioxane +

water) at temperatures between 343K and 413K in literatures. The final set of UNIFAC interaction parameters is given in Table 3.

Table 3. UNIFAC Interaction Parameters α_{ij} /K

group *i* group *j*	1	2	3	4	5	6	7
1	0	373.49	9999.0	$\alpha_{1,4}$	2577.677	-89.71	-229.1
2	-271.28	0	241.883	$\alpha_{2,4}$	2094.934	-89.71	-229.1
3	9999.0	-90.999	0	$\alpha_{3,4}$	30.936	313.5	2553.0
4	$\alpha_{4,1}$	$\alpha_{4,2}$	$\alpha_{4,3}$	0	$\alpha_{4,5}$	-129.94	-260.61
5	-58.211	290.89	-23.836	$\alpha_{5,4}$	0	369.9	137.7
6	2314	2314	721.5	10460.002	230.5	0	972.8
7	353.5	353.5	-325.2	847.03	227.0	156.4	0

$\alpha_{1,4} = 14982.32/T - 133.88$, $\alpha_{2,4} = 21068.23/T + 306.032$, $\alpha_{3,4} = -50437.19/T + 6062.015$, $\alpha_{4,5} = 403.095/T + 940.805$

$\alpha_{4,1} = -22665.86/T + 392.49$, $\alpha_{4,2} = 25066.12/T + 322.16$, $\alpha_{4,3} = -227977.96/T - 145.4$, $\alpha_{5,4} = 433.587/T + 483$

As methylene glycol cannot be prepared in pure form, the vapor pressure of methylene glycol is determined from the vapor-liquid equilibrium data. Therefore, the vapor pressure of methylene glycol is an adjustable parameter. During the process of fitting the interaction parameters, the vapor pressure of methylene glycol is also refitted. The vapor pressures of components are given in Table 4.

Table 4. Pure-Component Vapor Pressure ($\ln P^S = A + B/(T/K + C)$)

Component	*A*	*B*	*C*	Component	*A*	*B*	*C*
Formaldehyde	14.4625	-2204.13	-30.15	Methylene glycol	19.5527	-6183.273	-4.123
Water	16.2886	-3816.44	-46.13	Trioxane	14.3796	-3099.47	-68.92

In the meantime, the chemical reaction equilibrium constants for the formation of methylene glycol from formaldehyde and water in the liquid phase and for the formation of poly(oxymethylene) glycols are slightly adjusted. The results are followed as:

$$K_1^L = 3.9158/100000 \times \exp\begin{pmatrix} 5233.2/T + 6183.273/(T - 4.123) - \\ 2204.13/(T - 30.15) - 3816.44/(T - 46.13) \end{pmatrix} \quad (12)$$

$$\ln K_2^* = 0.236335 + 756.02766/T ; \ln K_3^* = -0.025637 + 533.334/T \quad (13)$$

Comparison with Experimental Results

The comparison between the results of the correlation and the experimental data in literatures is given in Table 5. The standard deviations between the calculated and measured data reported by Albert

(12 data points at 413K) for the vapor-phase concentrations of formaldehyde and trioxane and the pressure amount to 5.08%, 5.29%, and 1.57%, respectively.

Table 5. Experimental and Correlation Results for the Vapor-Liquid Equilibrium of the Formaldehyde-Trioxane-Water System

T	$\tilde{x}_{FA}$	$\tilde{x}_{Tri}$	$\tilde{y}_{FA}$			$\tilde{y}_{Tri}$			P		
			Exp.	**Cal.**	**100D**	**Exp.**	**Cal.**	**100D**	**Exp.**	**Cal.**	**100D**
K	**mol%**		**mol%**		**%**	**mol%**		**%**	**kPa**		**%**
413.1	10.91	1.1	18.65	18.65	0.000	3.78	4.61	22.000	427.9	425.144	-0.644
413.2	10.56	4.78	17.36	16.58	-4.521	13.64	14.12	3.490	462.9	458.108	-1.035
413.2	13.68	5.87	19.94	19.39	-2.734	14.66	15.42	5.165	478.8	471.007	-1.628
413.1	4	9.25	7.89	8.04	1.930	19.53	20.90	6.991	461	458.301	-0.586
413.1	6.09	9.33	10.7	10.81	1.070	20.13	20.52	1.924	470.1	466.078	-0.856
413.1	8.54	9.53	13.85	13.63	-1.624	19.39	20.22	4.287	478.9	473.704	-1.085
413.1	4.68	20.84	8.56	9.16	7.042	25.09	25.76	2.688	478.8	476.199	-0.543
413.1	2.41	32.06	5.63	6.39	13.415	28.12	28.12	0.012	471.9	469.072	-0.599
413.2	3.57	46.91	9.43	9.43	0.032	31.96	30.10	-5.833	474	472.366	-0.345
413.2	3.08	52.8	9.01	9.10	0.995	30.3	31.54	4.097	465	466.034	0.222
413.2	3.06	65.41	9.36	10.50	12.139	37.66	35.66	-5.303	462.2	447.242	-3.236
413.3	2.24	89.75	16.73	14.14	-15.490	57.11	58.05	1.653	364.5	335.024	-8.087
σ					5.08			5.29			1.57

The partition coefficient of trioxane and formaldehyde as a function of the ratio of the sum of the liquid-phase formaldehyde and trioxane mole fraction to the liquid-phase water mole fraction at 100kPa is shown in a and b of Figure 1, respectively. Also, Figure 1 shows the comparison of the correlation between the revised model and previous model in literatures. In comparison to the experimental data reported by Maurer (26 data points at around 369 K and 100 kPa), the standard deviations are 3.07%, 5.56%, and 1.95% for the partition coefficient of formaldehyde and trioxane and the pressure, respectively. The agreement between the correlation results and the experimental data is considered to be satisfactory. Typically, the partition coefficient of trioxane and formaldehyde calculated by the revised model in the work is closer to the experimental data comparing with the model in literatures from Figure 1.

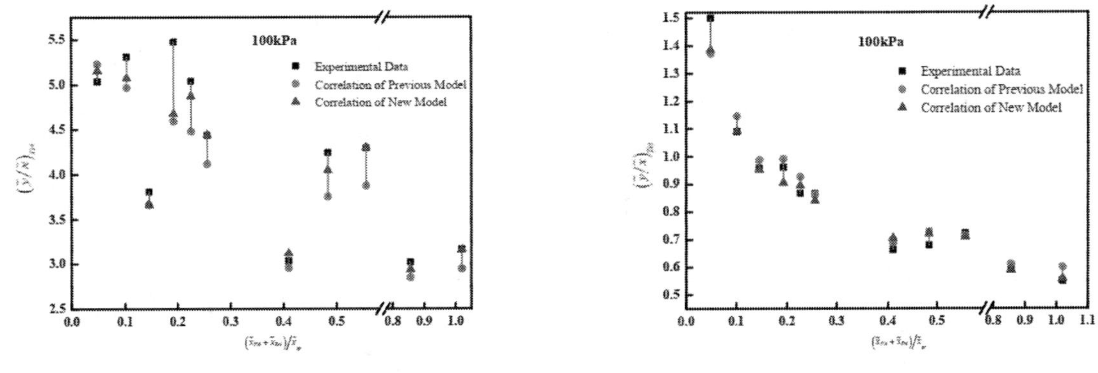

a) b)

Fig. 1 Partition coefficient of trioxane and formaldehyde in the trioxane-formaldehyde-water system at 100kPa

Conclusions

Comparing with the previous model for the vapor-liquid equilibrium of the trioxane-formaldehyde-water system, the new correlation gives better agreement with the literature data. Especially, the interaction parameters of UNIFAC groups in the new model represent the dependence upon temperature, which results in avoiding the non-balanced state that the prediction accuracy of model is bad at the lower temperature and on the contrary is well at the higher temperature. In other words, the application range of temperature of new model is extended, which provides reliable basis for the process design.

Acknowledgements

We gratefully acknowledge Doctor Wang Jun who guides and helps me in the research.

Nomenclature

A, B, C	Parameter of the vapor-pressure equation	P	pressure
a_{ij}	UNIFAC parameter for interaction between groups i and j	T	temperature
K_1^L,K_2^*,K_n^*	Equilibrium constants for reactions in the liquid phase	γ	activity coefficient
MG_n	poly(oxymethylene) glycol consisting of n formaldehyde groups	MG	Methylene glycol
$\tilde{x}_i$	stoichiometric mole fraction of component i in the liquid phase	FA	formaldehyde
x_i	mole fraction of species i in the liquid phase	W	water
$\tilde{y}_i$	stoichiometric mole fraction of component i in the vapor phase	Tri	trioxane

y_i	mole fraction of species i in the vapor phase	exp.	Experimental result
P_i^S	vapor pressure of component i	cal.	Calculated result
100D	Relative deviation between experimental result and a calculated number for property $z = 100(z_{cal.} - z_{exp.})/z_{exp}$	R	UNIFAC size parameter
σ	Standard deviation of property $z = \left(100 \times \sum_{i=1}^{N}\left(\left((z_{cal.} - z_{exp.})/z_{exp.}\right)_i^2\right)^{1/2}\right)\Big/N$	Q	UNIFAC surface parameter

References

1. Richard K. Haynes, and Simone C. Vonwiller. From Qinghao, Marvelous Herb of Antiquity, to the Antimalarial Trioxane Qinghaosu and Some Remarkable New Chemistry. *Acc. Chem. Res.* 1997, 30: 73-79
2. Maurer G. Vapor-Liquid Equilibrium of Formaldehyde and Water Containing Multicomponent Mixtures. *AIChE J.* 1986, 32: 932-948.
3. Hasse H., Hahnenstein, Maurer G. Revised Vapor-Liquid Equilibrium Model for Multicomponent Formaldehyde Mixtures. *AIChE J.* 1990, 36: 1807-1814.
4. Hasse H., Maurer G. Vapor-Liquid Equilibrium of Formaldehyde-Containing Mixtures at Temperatures below 320 K. *Fluid Phase Equilib.*1991, 64: 185-199.
5. Albert M., Hahnenstein, I., Hasse, H., Maurer, G. Vapor-Liquid Equilibrium of Formaldehyde Mixtures: New Experimental Data and Model Revision. *AIChE J.* 1996, 42: 1741-1752.
6. Albert M., Coto Garcia, B., Kreiter C., Maurer G. Vapor-liquid and chemical equilibria of formaldehyde-water mixtures. *AIChE J.* 1999, 45: 2024-2033.
7. Albert M., Coto Garcıa, B., Kuhnert Ch., Peschla R., Maurer G. Vapor- Liquid Equilibria for Aqueous Solutions of Formaldehyde and Methanol. *AIChE J.* 2000, 46: 1676-1687.
8. Albert M., Hasse H., Kuhnert Ch., Maurer G. New experimental results for the vapor-liquid equilibrium of the binary system (trioxane +water) and the ternary system (formaldehyde + trioxane + water). *J. Chem.Eng. Data* 2005, 50: 1218-1223.
9. Ryszard Pohorecki, Wladyslaw Moniuk, Piotr Machniewski. Calculations of the Vapor–Liquid Phase Equilibrium for the Formaldehyde-Water-Trioxane System. *Chemical and Process Engineering*. 2008, 29: 3-12.
10. C Marcolli and Th Peter. Water activity in polyols: new UNIFAC parameterization. *Atmos. Chem. Phys. Discuss.* 2005, 5: 1501-1527.
11. Bent L. Larsen, Peter Rasmussen and Aage Fredenslund. A Modified UNIFAC Group-Contribution Model for Prediction of Phase Equilibria and Heats of Mixing. *Ind. Eng. Chem. Res.* 1987, 26: 2274-2286.

FCC Riser Hydrodynamics Improvement Using Simulation Tools

Lev Davydov and Reza Mostofi, UOP/Honeywell, Des Plaines, IL

Past experiences of commercial FCC units indicate that hydrodynamics have a pronounced effect on the riser performance. For example, an increase of lift steam velocity has been shown to result in increased gasoline selectivity in many units. For a large FCC unit in Asia the installation of dual-radii feed distributors improved feed-catalyst contacting in the lower riser and also led to better gasoline yields.

A study of FCC riser hydrodynamics was undertaken to understand the phenomena observed in the field. It encompassed computational fluid dynamics (CFD), cold flow modeling, radioactive tracer studies, kinetic modeling and yield estimation. The ultimate goal of the study was to isolate the effects of riser diameter, lift steam velocity, and solids flux on the hydrodynamic behavior and yield performance of FCC risers.

Our CFD approach was successfully validated by cold flow results and applied to several commercial FCC units. The general trends of riser hydrodynamics were identified as follows. Lower vapor velocity, larger riser diameter and higher solids flux through the riser all contribute to hydrodynamic "pathologies", such as catalyst downflow at the wall and vapor channeling through the core of the riser. In practice, these "non-idealities" lead to "under-conversion" of feed in the core of the riser and overcracking of products in the annulus of the riser.

	Gas Superficial Velocity	Catalyst Flux	Air Density	Particle Size [1]	Particles Density
Condition	fps	lb/ft²/sec	lb/ft³	Microns	lb/ft³
A	22.9	149.9	0.12	80	90
B	50.7	250.4	0.10	80	90

(1) For dpdpm there is a particle size distribution

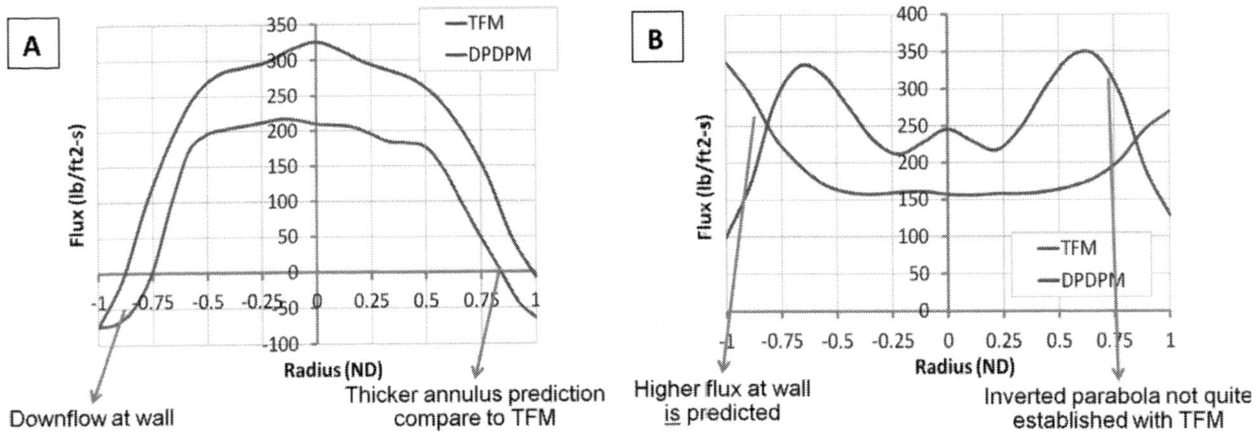

Figure 1: Comparison of solids flux profiles in the riser predicted by two different CFD models: *Red* – two-fluid model (TFM); *Green* – dense phase discrete particle model (DP-DPM).

(125b) Molecular Based Assays for the Practical Correlation and Prediction of Crude Oil and Petroleum Fraction Properties

Chau-Chyun Chen, R&D, Aspen Technology, Inc., Burlington, MA and Stephen Dziuk, Aspen Technology, Inc., Burlington, MA

Molecular Based Assays for the Practical Correlation and Prediction of Crude Oil and Petroleum Fraction Properties

Keywords: crude oil assay, molecular characterization, planning and scheduling, process simulation

Abstract:

For the first time, a comprehensive molecular based methodology has been developed for practical correlation and prediction of assays and properties for crude oils and petroleum fractions. Based on limited and commonly available crude oil assay characterization data such as distillation yield curve, API gravity, and viscosity, the method identifies chemical compositions of pre-determined surrogate model hydrocarbon molecules designed to mimic major chemical constituent components for the purpose of characterizing physical and chemical properties of crude oils. The resulting molecular representation of the crude oil then provides a self-consistent molecular basis to interpolate, extrapolate, and predict all physical and chemical properties of crude oils, petroleum fractions, and blends.

The determination of crude oil assay is a lengthy, tedious and costly process. The conventional approach to perform an assay consists of a limited set of measurements on the crude oil and its fractions. Most often only a few boiling points, densities, and other property measurements are available for selected distilled fractions or the whole crude. Therefore it is necessary for crude oil assay experts to predict or estimate missinq properties to meet various business needs, such as refinery planning and scheduling and refinery process simulation. Typically lower order polynomial expressions are used for interpolation and arithmetic probability scale is used for extrapolation of boiling point curves. Statistically derived predictive methods also have been extensively used in the industry for the prediction or estimation of crude oil assays. Analytical methods may predict crude oil properties by correlating the data obtained with rapid surrogate measurements (usually spectroscopic) to those of reference crude assays.

The key elements of this molecular based assay characterization approach include:

· identifying optimal classes of surrogate model hydrocarbon molecules, practical rules to construct these model molecules from molecular "repeat units" or "segments," and distribution functions of the model molecules required to account for properties of crude oils around the world,

· applying a thermodynamically-consistent framework to accurately estimate key physical properties of the individual surrogate model molecules and their mixtures, and

· developing an engineering methodology to robustly identify compositions of the surrogate model molecules from commonly available assay data.

Implemented in commercial software, molecular based assays have been successfully validated against hundreds of crude assays worldwide including many heavy crude oils. The methodology not only facilitates extraction of molecular insights for the crude oils, it also allows self-consistent interpolation, extrapolation, and prediction of a comprehensive set of physical and chemical properties such as distillation yield curve, API gravity, viscosity, Reid vapor pressure, Watson K, PNA contents, sulfur content, nitrogen content, acid content, asphaltene content, C/H ratio, etc. The methodology offers a superior alternative to conventional and empirical assay characterization technologies. Furthermore, it suggests a highly prized new frontier over the traditional "pseudocomponent," "hypothetical component," or "micro-cut" modeling approach for planning, scheduling, and process simulation of oil production and petroleum refining operations.

(125c) Petrobras in-House Process Simulator As a Strategic Advantage

Renato A. Barbosa[1], Jacques Niederberger[1], Letícia C. Santos[1], Carlos A. D. Moura[1], Joyce S. S. Aires[1], Lara S. S. Teixeira[1], Victor R. R. Ahon[2], Vinicius M. Silva[2], Carmen E. Vargas[2], José A. Silva[3], Anna E. B. Zobolli[2], Cláudio A. S. Aquino[1] and Ellen P. Silva[2], (1)Research and Development Center (CENPES), Petrobras - Petróleo Brasileiro S.A., Rio de Janeiro, Brazil, (2)GENPRO Engenharia S.A., (3)SPASSU Tecnologia e Serviços

PETROBRAS is an energy company, which operates in oil and gas production and refining, petrochemicals and biofuels production, but it is also a technology developer. In this sense, in-house process simulators are strategic to standardize and disseminate technological innovations amongst engineers of the entire company.

Due to the need to have a uniform environment for simulation and design, PETROBRAS has teams dedicated to provide support and to develop applications required by users, based on proprietary technologies and/or scientific literature.

PETROX is a steady state refining process simulator developed by PETROBRAS Research and Development Center (CENPES). It has been successfully applied in different business fields in PETROBRAS: revamp and design of new units such as: Atmospheric and Vacuum Distillation, Fluid Catalytic Cracking (FCC), Delayed Coking, Hydrotreating (HDT), Hydrocracking, Natural Gas Processing, Basic Petrochemicals Production, Offshore Production Facilities and Treating units like Sour Water Treatment and Sulfur Recovery. PETROX is the standard simulator for several groups in the areas of process design, operation simulation, and improvement of the company processes. Its usage allows significant improving on quality and productivity of process design, optimization, equipment sizing and operational conditions specification, as well as dealing with environmental issues.

PETROX incorporates the main characteristics of the modern commercial simulators. It has a graphic user-friendly Windows-based interface, with process flowsheet visualization. Unit operations and equipments are simulated by the sequential modular approach.

Owning an in-house process simulator allows customization and implementation of new features on a permanent basis, which could not be feasible with commercial software or, at least, would be considerably more difficult.

It has a pure component properties database, into which new components are frequently added as required by users. There is also a large, constantly updated database of national and international oils, including recent discovered crudes from the pre-salt area, containing complete assays measured by PETROBRAS or acquired from suppliers. There is a wide variety of methods, obtained from literature or developed by the PETROX team, to estimate physical properties of pure components, petroleum fractions (pseudocomponents) and mixtures. As many Brazilian crude oils are heavier than the average imported abroad, new correlations had to be developed in order to better simulate oil blend properties and process units.

Different thermodynamic models are available to simulate crude oil refining, gas processing, offshore production facilities, and petrochemical units. Those include equations of state, activity coefficient models and sour water model.

There are several unit operations models on PETROX, including rigorous distillation column, reactors, compressor and gas separation membrane. Among them, there are PETROBRAS proprietary conversion models, which are available as separated programs or as PETROX modules. They include Delayed Coking unit, FCC unit, Residue FCC unit, HDT reactor, and Shale Oil Retorting.

PETROX has also the usual features such as: sensitivity analysis, phase envelopes, heating curves calculation, and importing/exporting data.

It is possible to integrate PETROX to other process design and process analysis tools from PETROBRAS technology, partnership with Universities or commercial tools. Through this integration, engineers can simulate rigorous heat exchanger, design heat exchanger networks, export data to CAE systems, perform a column Pinch analysis or exchange data with a Furnace simulator, for example.

PETROX is accessible to the whole company. As it is based on a cloud network platform, our users always have access to the newest version of the programs, with updated data and features.

PETROBRAS process simulation technology has also been employed in real time application. For example: Fouling^{TR}, also developed by PETROBRAS teams, is a tool for monitoring heat exchangers and heat exchanger trains, which uses a PETROX application for on-line plant data acquisition and simulation.

As matter of fact, the effort of PETROX development during more than twenty years is a proof of PETROBRAS commitment with R&D investment.

RELIABILITY, AVAILABILITY & MAINTAINABILITY MODELING (RAM) AS A PROCESS DESIGN TOOL

Presented to AIChE Spring National Meeting
San Antonio, Texas May 1, 2013

WorleyParsons
resources & energy

USAC Western Operations

*** For Want of a Nail**

For want of a nail, the shoe was lost.
For want of a shoe, the horse was lost.
For want of a horse, the rider was lost.
For want of a rider, the message was lost.
For want of a message, the battle was lost.
For want of a battle win, the king was lost.
For want of a king, the kingdom was lost.
And all for the want of a horseshoe nail.

** ascribed to Richard III's fate at Bosworth Field on a foggy day. Author Unknown

*RAM Study Bases

Reliability is a difficult objective. Industrial equipment systems are specialized and complex. Equipment reliability must be built into eight critical functions: design, manufacture, selection, installation, operation, maintenance, monitoring and repair.

This effort also involves an understanding of how the system is interacting with the process and familiarity with hundreds of system components.

Equipment reliability requires understanding not only the equipment itself, but also the complexities of the system in which it is installed.

*RAM Study Bases

Building knowledge and understanding into front-line operations and maintenance personnel results in huge gains in process plant reliability.

A West Coast refinery, for example, recognized the critical role of people in equipment reliability. Building knowledge and problem-solving skill in their operators resulted in an increase in mean-time-between-repair (MTBR) from four years to over seven years for pumps.

Other rotating equipment - **typical life** of general-purpose **turbines is 15 years** between overhauls and **10 years between overhauls for compressors**. But only if the specified maintenance is routinely performed and the operating limits strictly adhered to.

* Setup procedures in a RAM study typically include:

- **Identify primary contributors** to downtime that restrict ability to provide product to customers.
- **Determine the effect of new capital** projects and changed/ augmented operating procedures that will raise the annual availability of the facility while minimizing flaring/venting /recycle/revenue/HES losses as much as possible
- **Assess the operations data** - logs, process historians, data repositories of the plant and build a reliability data model of the individual plant hierarchy to the best level that the maintenance records and operations staff testimony permit.
- **Consider all Hierarchies** involved and focus on those critical to your project to build a proper RAM application.

Plant > Unit level > Equipment groupings > Individual equipment items > Item-specific subcomponents

Reliability vs. Availability

- **Reliability** is the ability of the process facilities we design and operate to keep operating the plant safely and economically.
- **The real target is** unit or, better, plant wide **availability** - the on-stream operation of the facility in reasonable, if not always optimal, fashion.
- **Increased Availability is the goal which can only be attained by improved Reliabilities.** How you do this involves several strategies.

Availability improvement is a difficult objective.

1. Industrial equipment systems are specialized and complex. RAM started as military specialty but has broadened into a general equipment oriented protocol

2. Equipment reliability must be built into eight critical engineering functions:

design, manufacture,
selection, installation,
operation, maintenance,
monitoring repair.

Reliability improvement is a difficult objective.

3. This effort also involves an understanding of:

a. how the system is interacting with the process
and
b. familiarity with hundreds of components.

4. Equipment reliability requires understanding of:

c. the equipment and the process itself,
and
d. the complexities of the system in which it is installed.

*Why Is RAM a Process Design Function ?

- Traditionally left to EPC/Mechanical design groups to validate local optimum
- New or Improved Value Process design issues often ignored until long after FEED work is done.
- Life Cycle Analysis (LCA) is now a valid FEED activity. RAM studies make LCA spreadsheets more realistic in terms of STATISTICALLY VALID equipment replacements and costs.
- Gut feelings, simple consequence ascription *(... if this just happens once, the cost is worth it...),* and Rules of Thumb (ROT's) are still good tools, but should be proven out for the amazing amount of money being spent on your project.
- You can do this pretty quickly with good data and good software.

Some Interesting Math

- Typ. process design is for 7500 or 8000 hrs /yr uptime. Implies 7500/8760 =85.6% or 8000/8760=91.3 % availability
- Good process performance has been thought of as >90%.
- Say target is 400 hrs/yr max down or 95.4 %
- Average pump availability difference between quality PM and no PM is 0.997 vs 0.995. Big deal ? For a 20 pump sequential-hierarchy unit with normal driver redundancy

  $(0.995)^{20} = .905$ $(0.997)^{20} = .942$

  Delta= 324 hrs/yr

How easily the time loss can build

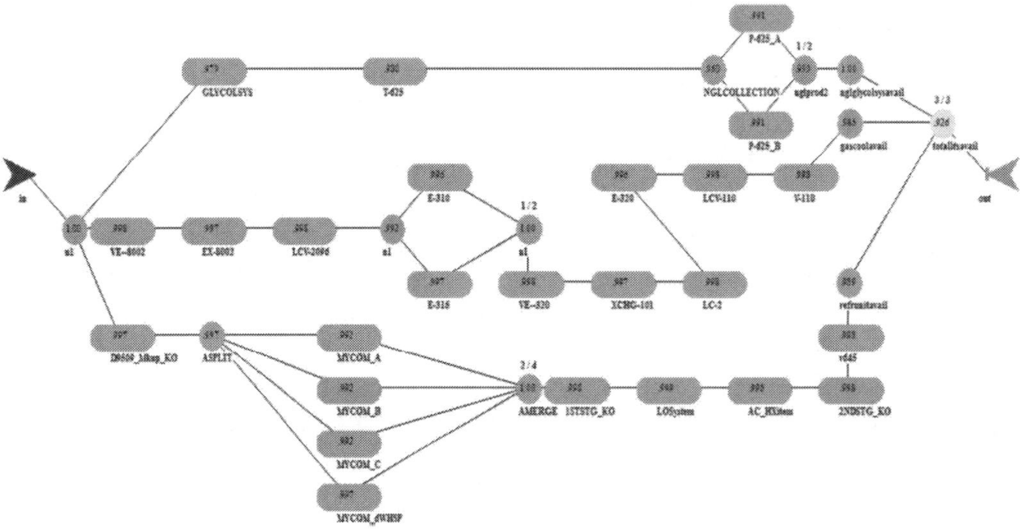

* RAM Study Data Needs – Existing Unit

- **Client/ Equipment group should provide what information they have** that can assembled from their MES/ESA applications. The following is the basic types of data needed to perform this level of study. Probably distributed in many different locations.
- **Event log showing time and duration of an "event" with some explanation of the cause or initiating event.**
- **Logical consequences of "event** – either noted in a log or likely based on PFD go-by sequence analysis.
- **Quality of Maintenance Culture** of the Plant

* RAM Study Data Needs – Existing Unit

- Use of standard industry values of MTBF/MTTR in event of inadequate plant history. Selection modified based on known plant history incidents and operating procedures.
- Calibrate to actual availability AMAP
- Basic understanding of process cause and effect.
- Cost of consequences / Rewards of non-failure
- Use of plant history for estimating MTBF/MTTR values of higher level (process grouping and major equipment level) availability.

* Highest Availability Time Loss Sources

- 1. **Rotating equipment = pumps –** verticals, recips, centrifugals in order of highest to lowest. No. 1 just because there are so many of them and the smaller ones often get ignored by PM group.
- 2. **Two phase reboilers/condensors/preheaters** – AKA "in-line corrosion coupons"
- 3. **Rotating equipment = compressors** – high pressure recips, high pressure centrifugal/axial
- 4 **Anything else that rotates and pumps fluid** with cavitation, boiling, or solids within.
- **5.** Anything where flow velocity is too high – erosion problems

* Highest Availability Time Loss Sources

- 6. **Anything where water** and Cl-, H2S/HS-, TAN, NH4+, or similar materials **can collect** stress corrosion causing solids in protected areas. Wash systems greatly affected. Un-protected oceanic exposure also bad
- 7. **Anything where** other c**racking, pitting or penetrating corrodents exist** in high materials-stress conditions
- 8. **Air coolers on hot days** may not cool adequately
- 9. **Communicattions or electronic data and control** equipment may fail during lightning storms. Brown outs, black outs or if there are significant stoppages in the supply chain routing.
- 10. Anything else can and **WILL** fail according to Murphy's law

Some Availability Rewards / Loss Evaluation Criteria

We'll use our recent go-by job for a recent RAM with a WPM client. Besides the equipment reliability history & details, we also considered the :

- Value of each MM BTU flared – two viable bases
  - loss of a cogen or other power plant production and/or
  - cost of alternate fuel purchased to maintain baseload rating
- Value of each MMSCF of sales gas not sold – normal market bid
- Value of each MMSCF of sales gas not sold – discount adhoc bid
- Value of each bbl of NGL or other products not sold

Some Availability Rewards / Loss Evaluation Criteria

- Value of power NOT sold from power plant
- Relationships between non-availability and flare utilization.
- **Some very strict emissions permitting and safety issues. Max limit exceeded may mean several $ MM in penalties, legal fees, and compliance capital/operating costs.**
- Cost of fixing the problem itself

Definition of Reliability – Basics

Mean Time Between Failures (MTBF) / Mean Time To Failure (MTTF)

Reliability is mostly quantified as MTBF (Mean Time Between Failures) for repairable product and MTTF (Mean Time To Failure) for non-repairable product. A correct understanding of MTBF is important.

An power supply unit with an MTBF of 40,000 hours does not mean that the power supply should last for an average of 40,000 hours. According to the theory behind the statistics of confidence intervals, the statistical average becomes the true average as the number of samples increase.

An MTBF of 40,000 hours for 1 year for 1 module, becomes 40,000/2 for two modules and 40,000/4 for four modules.

Sometimes failure rates are measured in percent failed per million hours of operation instead of MTBF.

The Failure Interval Time (FIT) is equivalent to one failure per billion device hours, which is equivalent to a MTBF of 1,000,000,000 hours. The formula for calculating the MTBF is

- MTBF = T/R.
- T = total time
- R = number of expected failures (everything fails after a while)

Definition of Reliability – Basics

Mean Time Between Failures (MTBF) / Mean Time To Failure (MTTF)

MTTF is stands for Mean Time To Failure. To distinguish between the two, the concept of suspensions must first be understood.

In reliability calculations, a suspension occurs when a destructive test or observation has been completed without observing a failure.

MTBF calculations do not consider suspensions whereas MTTF does. MTTF is the number of total hours of service of all devices divided by the number of devices.

It is only when all the parts fail with the same failure mode that MTBF converges to MTTF.

- MTTF = T/N
- T = total time
- N = Number of units under test.

Definition of Reliability (cont)

Example: Suppose 10 devices are tested for 500 hours. During the test 2 failures occur. Therefore:

MTBF = 10*500/2 = 2,500 hours / failure.

Whereas for MTTF

MTTF = 10*500/10 = 500 hours / failure.

If the MTBF is known, one can calculate the failure rate as the inverse of the MTBF.

MTTF = = 1/MTBF = r/T

where r is the number of failures

Definition of Reliability (cont)

Once MTBF is calculated, what is the probability that any one particular unit component will be operational at time equal to the MTBF?

We have the following equation:

R(t) = e-t/MTBF.

But when t = MTBF

R(t) = e-1 = 0.3677

This tells us that the probability that any one particular module will survive to its calculated MTBF is only 36.8%.

Modeling Tool - General Concepts

WPM WCO uses *Raptor*

- Raptor 7.0 is an industry-expansive simulation tool for analyzing a physical plant system's operational cost, reliability and capacity. An easy graphical interface and powerful simulation engine, Raptor allows one to quickly represent a system's operations and explore ways to optimize its performance.
- The system model can be developed at a high, intermediate or even low-level detail oriented mode. All or part of a plant can be divided into related blocks. The block is the basic modeling block that holds a system's reliability failure data such as MTBF and MTTR with appropriate failure mode probability distributions.
- The block itself can model a lumped system, such as say "NGL pumps & motor set with LO and all controls, or even an entire facilty, such as a powerplant, as long as one can describe the failure probabilities of that system adequately.
- Additionally, one can break the high level system into several lower level systems. For the NGL pumps, as an example it could be the pump, motor, seal, inlet/outlet valves and piping, outlet pressure transmitter, transducer, controller, control valve.
- This painstaking **lower level analysis can often be avoided if you have** GOOD plant records and a reasonably consistent maintenance policy. You can usually make a decent lumped system model in such situations.
- A detailed lower level analysis is usually necessary for at least some of a "new equipment" project unless all items have a reliability "track record" with the maintenance/reliability department.

* Modeling Analysis Methodology

- Block based Semi Markov modeling system using both analytic and stochastic calculation protocols over a multiple time periods of fixed time length.

- Uses basic analytical approaches to evaluating block reliability and availability.

- Behavior of each block can be independently fixed or related to other block (equipment ..) failure distributions

- One uses average or expected values of MTBF and MTTR in evaluating these items based on one of about a dozen failure PDF's

- A probability function, based on the ratio of mean to standard deviation, appropriate for the type of failure distribution, is used to "re-estimate" the MTBF and MTTR used for each case according to "random" response of independent failures.

Sources of Reliability Data

Simple block model still has MTBF, MTTR, failure distribution types & PM info, cost info, intraunit plant dependencies, sparing protocols

- CCPS process reliability data for process equipment components
- Mil-Std 217E and follow on standards for electrical equipment items
- OREDA consortia offshore plant data
- Client data of all sorts – high level to low level

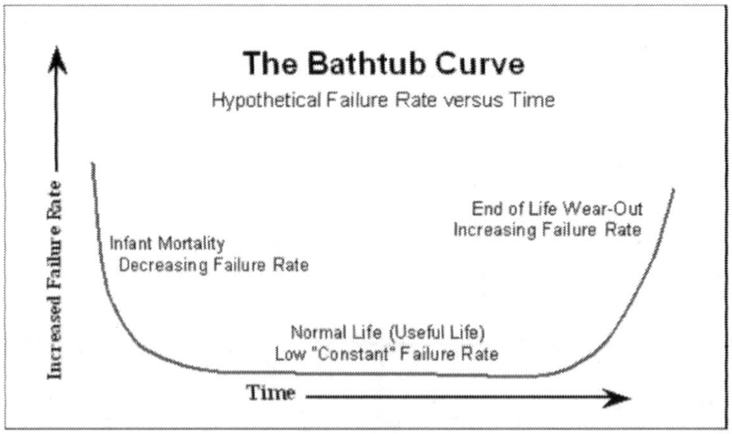

More Failure Modes

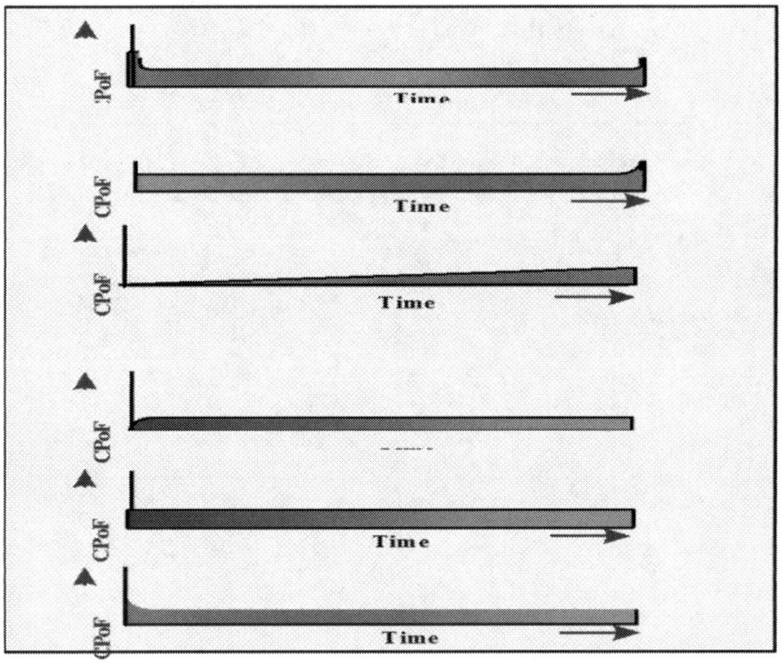

General Weibull Exponential Distribution

T is time **η** is the shape factor

$$f(\mathrm{T}) = \left(\frac{\beta}{\eta}\right)\left(\frac{\mathrm{T}}{\eta}\right)^{(\beta-1)} e^{\left(\frac{\mathrm{T}}{\eta}\right)^{\beta}}$$

The Weibull parameter β (beta) is the slope. It signifies the rate of failure. When $\beta < 1$, the Weibull distribution models early failures of parts. When $\beta = 1$, the Weibull distribution models the exponential distribution. The exponential distribution is the model for the useful life period, signifying that random failures are occurring. When $\beta = 3$, the Weibull distribution models the normal distribution. This is the early wearout time. When $\beta = 10$, rapid wearout is occurring.

Notice when all the curves are combined, the resultant graph is similar to the bathtub curve

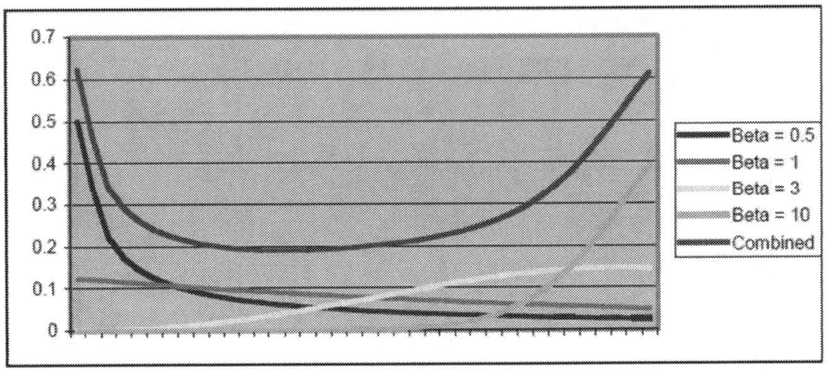

Mean Unit Time Between Repairs, Reciprocating Compressors
(Years)

Unit/Functional Category	Plant1	Plant2	Plant3	Plant4	Plant5	Plant6
Distilling		0.2		0.5		
Catalytic Reformers	4.0	0.6	1.5	no repairs	0.5	0.3
Hydrocrackers / Hydrogen	no repairs	0.3	1.5	1.0	0.1	0.4
Fuels Hydrotreaters	0.7	0.4	1.0	0.5	0.8	1.1
Lubricants	1.0		0.5		0.4	0.3
Logistics		0.3	2.5			
Flare Gas Recovery	1.0		0.7	0.2	0.3	0.3
Overall	1.0	0.4	1.1	0.6	0.3	0.6

Pure Refinery MTBR

Unit/Functional Category	Plant 1	Standalone	Plant4
Olefins Units	3		4.0
Solvents Units	0.5		
Detergent & Higher Olefins Units		0.5	
Flare Gas Recovery			1.0
Overall	0.9	0.5	4.0

Petrochem MTBR

Sample Pump Maintenance Cost – from Plant Records

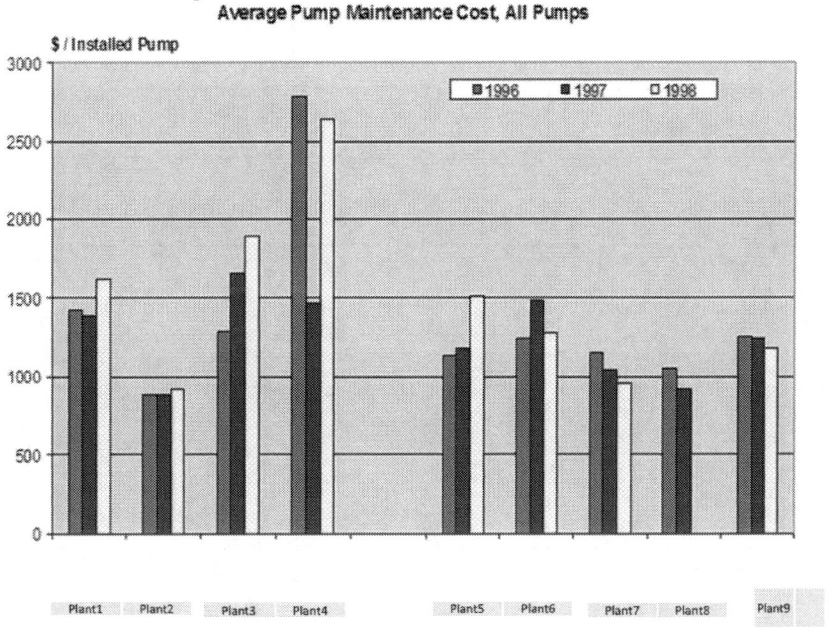

Equivalent Unit Downtime for Client 2 locations by unit category and equipment types..

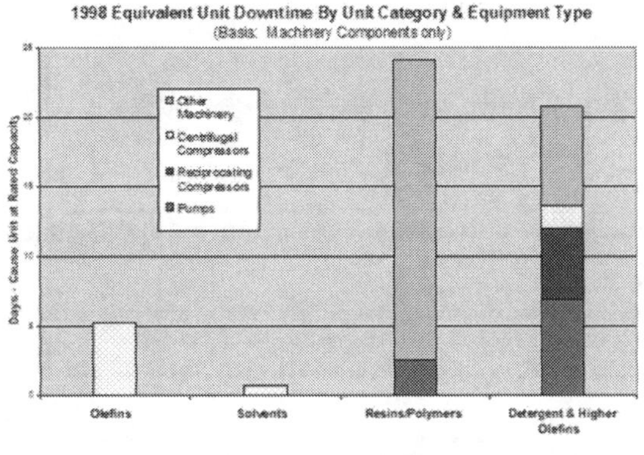

Table 1. Shell Chemical - 1998 Machinery-Related Equivalent Unit Downtime
(Days @ Rated Capacity -- Basis: Machinery Components only)

Unit Category	Plant1	Plant2	Plant3	Plant4
Olefins	0.0		5.2	5.2
Solvents	0.0		0.7	0.7
Resins/Polymers	20.0		0.0	20.0
Detergent & Higher Olefins		1.7		1.7
Overall	20.0	1.7	5.9	27.6

Equivalent Unit Downtime for Client 2 locations by unit category and equipment types..

Table 1. Chemical - 1998 Machinery-Related Equivalent Unit Downtime
(Days @ Rated Capacity -- Basis: Machinery Components only)

Unit Category	Plant1	Plant2	Plant3	Plant4
Olefins	0.0		5.2	5.2
Solvents	0.0		0.7	0.7
Resins/Polymers	20.0		0.0	20.0
Detergent & Higher Olefins		1.7		1.7
Overall	**20.0**	**1.7**	**5.9**	**27.6**

Equivalent Unit downtime (EUD) for Chemical unit categories by equipment type and criticality

Table 1. Chemical - 1998 Machinery-Related Equivalent Unit Downtime
(Days @ Rated Capacity -- Basis: Machinery Components only)

Unit Category	Plant1	Plant2	Plant 3	Total
Olefins	0.0		5.2	5.2
Solvents	0.0		0.7	0.7
Resins/Polymers	20.0		0.0	20.0
Detergent & Higher Olefins		1.7		1.7
Overall	**20.0**	**1.7**	**5.9**	**27.6**

Mean Unit Time Between Repairs, due to Reciprocating Compressors
(Years)

Unit/Functional Category	Plant1	Plant2	Plant3	Plant4	Plant5	Plant6
Distilling		0.3		0.5		
Catalytic Reformers	4.0	0.6	1.5	no repairs	0.5	0.3
Hydrocrackers/ Hydrogen	no repairs	0.3	1.5	1.0	0.1	0.4
Fuels Hydrotreaters	0.7	0.4	1.0	0.5	0.8	1.1
Lubricants	1.0		0.5		0.4	0.3
Logistics		0.3	2.5			
Flare Gas Recovery	1.0		0.7	0.2	0.2	0.3
Overall	1.0	0.4	1.1	0.6	0.3	0.6

Pure Refinery MTBR

Unit/Functional Category	Plant 1	Standalone	Plant4
Olefins Units	3		4.0
Solvents Units	0.5		
Detergent & Higher Olefins Units		0.5	
Flare Gas Recovery			1.0
Overall	0.9	0.5	4.0

Petrochem MTBR

Modeling Strategy

- **USE SEMI MARKOV MODELS WITH MONTE CARLO SIMULATION** TO EXPLORE THE FULL RANGE OF FAILURE Probability Distribution Functions.
- **Need 30-100 min cases** to get fully normalized stable statistical analysis. Prefer 1000+ if you have the time to wait for the answer.
- **Need loss penalties, cross block& hierarchy dependencies** to figure out total cost of a failure

- **Use plant data AMAP esp. for non-critical systems**
- **Use published data to augment plant history,**
- **Build new equipment models from lower level, and calibrate to system level availabilities** THERE IS A LOT OF INFO YOU NEED DO INPUT TO THE MODEL IF YOU REALLY WANT SOME GRANULARITY

Block Level RBD Submodel

- Example

Lower Level RBD – Pump System

- Demo

High level RBD – ex. Amine Plant

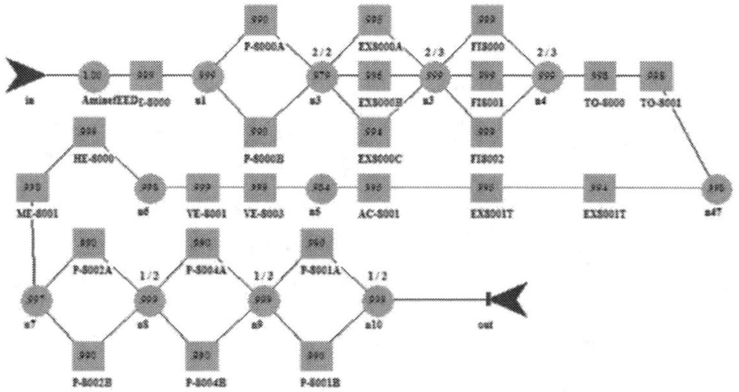

- Simple block model still has MTBF, MTTR, failure distribution types & PM info, Cost info, Intraunit plant dependencies, sparing protocols, & more

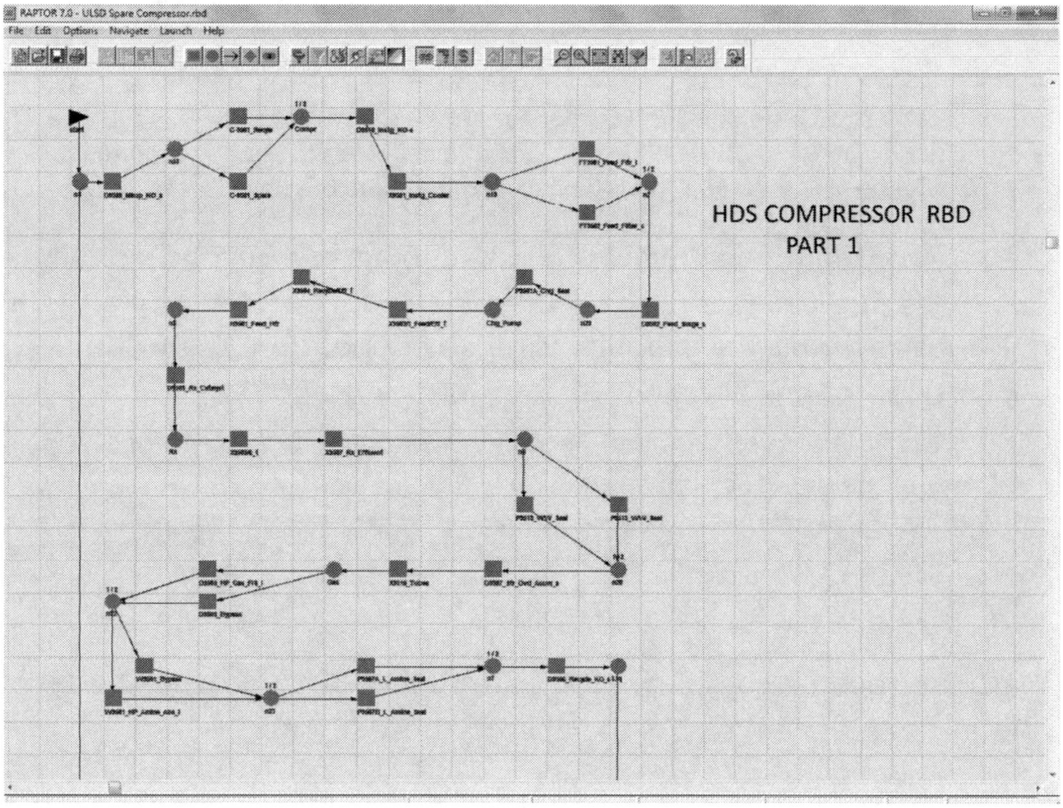

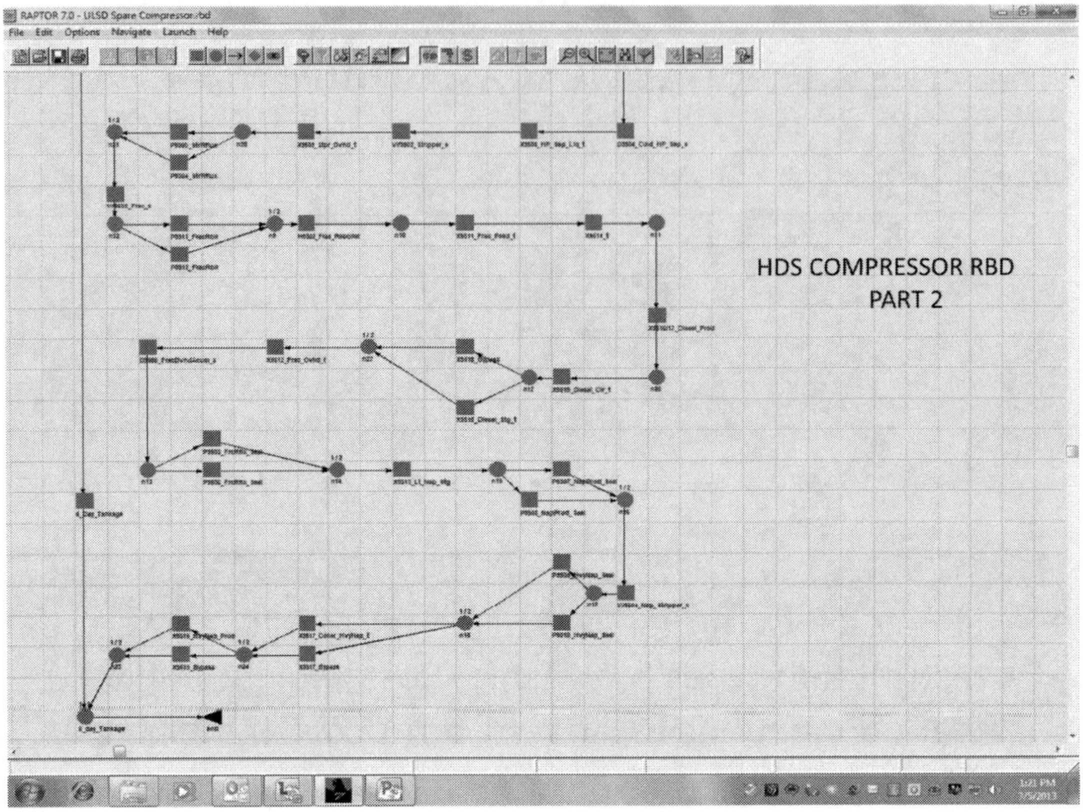

Every Operator wants 8400 hrs/yr online

8400 hrs online / 8760 calendar hrs = 95.9% availability

8000 hrs /yr => 91.3% availability – Typical of most process design guarantees

7500 hrs/yr +> 85.6% availability common for complex plants

- Every contractor that wants to get sued agrees unconditionally
- The basis for high availability is mostly dependent upon the plant operations practices –
  1. "Maintenance on demand" keeps new rotating equipment (most susceptible) MTBF between 1-2.5 years
  2. Modest PM (Preventative Maintenance) MTBF => 3- 4 years
  3. Rigorous (some say fanatical) PM gives you 5-8 years MTBF

Major petrochem company (CLIENT 2) tallied their varying global refineries and chem plants to validate these findings.

Alternative Availability Measures

- **Equivalent Unit Downtime (EUD) is a summation of unbudgeted loss incidents** for "Cause"
- Expressed in terms of equivalent days lost at rated capacity.
- **EUD is intended to measure** the **direct effect machinery** process function **outages have o**n their respective process units. Downstream affect on other units is purposely ignored.
- **EUD is a meaningful indicator when investigating machinery reliability issues** in individual process units **or when comparing machinery reliability in like units** across locations because it factors out differences in unit integration.

Alternative Availability Measures

- **Equivalent Distillation Capacity (EDC) Losses are as defined by Solomon & Associates and include total Cause Unit plus any Other Unit losses for each incident**.
- The **direct linkage between production** (revenue) losses resulting from machinery outages or impaired performance (slowdowns) and **net income** are clearly understood.
- **Improving reliability of a key machine can have an amplified affect** on EDC losses by avoiding impact on other units.
- **Original value linked to crude unit distillation capacity** (NOMINALLY IN BPD) since you can't run the crude unit unless the downstream units are available.

Alternative Availability Measures

- **Availability Definition 1**: *Ratio of total time a system is not operational due to planned or unplanned downing events or incidents to the Total plant time that system could have been viably been operated absent external constraints*

- **Availability Definition 2**: *Ratio of realized annual profit to total potential profit realized and not realized.* Includes effect of "non-optimal" revenue recovery due to reliability failures

- **Availability Definition 3:** *Number of non-flare incident hours / Number of hours gas plant* operates. This is not a true availability factor since the CLIENT facility can be on line delivering products, but have the flare operate if all the gas from the MAIN WELLS and other ancillary reservoirs cannot find a "home".

Example Problem Study Nodes/ Model Configuration

The gas gas plant was divided into three study nodes:

1. Feed Gas Compression (Ariel compressor)

2. Refrigeration and Gas Chilling System (Mycom unit, glycol, and chilling line)

3. Amine Plant

In addition, the following block sub-models are used for the model in order to assess dependency relations of downtime to the rest of the THUMS facility.

4. Gas Plant Feed Source (four deep wells and shallow gas area)

5. Product Distribution Systems (single block each)
   - Power Plant
   - Gas Sales
   - Oil Sales

Typical RAM Study Data Needs

The following is the basic types of data needed to perform this level of study

- **Event log** showing time and duration of an "event" with some explanation of the cause or initiating event.
- **Logical consequences of "event** – either noted in a log or likely based on PFD go-by sequence analysis.
- **Use of plant history for estimating MTBF/MTTR** values of higher level (process grouping and major equipment level) availability.
- **Use of standard industry values** of MTBF/MTTR in event of inadequate plant history. Values used are based on known plant history incidents and operating procedures
- **Basic understanding of process** cause and effect.
- **Cost of consequences / Rewards of non-failure**

Availability Rewards / Loss Evaluation Criteria included:

- Lost Value of each MM BTU flared – two viable bases
  - loss of power plant production ...and/or
  - cost of alternate fuel purchased to maintain baseload rating
- Value of each MMSCF of sales gas not sold – normal market bid
- Value of each MMSCF of sales gas not sold – discount adhoc bid
- Lost value of each bbl of NGL not sold
- Lost Value of each bbl of oil not sold
- Lost value of power NOT sold from powerplant – esp. in hot weather

Example RAM Study –
Modeling Analysis Methodology

- Not all failures are independent. Many are related, to some extent, to other plant failures.
  - Raptor has the ability to set up 30+ related failure protocols. Later versions are even more versatile.
- The vast majority of error distributions, however, are basically distributed by a random and independent probability distribution.
- Raptor uses Monte Carlo simulation is used to run, say, 30-10000 cases with failure mode parameters reset for each case run.
- Typically The Monte Carlo simulation will run between 300 and 1000 cases for a plant the size of this particular CLIENT.

Modeling Analysis Methodology

Electronic operator logs were the basis for evaluating gas plant component reliability performance.

WorleyParsons was able to derive enough data to model the availability of some major equipment items and plant subsystems (ex. Feed Compression Unit, Refrigeration).

Detailed reliability numbers for other units (Amine Unit & LTS) and minor equipment items such as piping, valving, drums, tanks, etc **were not adequately tallied and logged** by CLIENT, so the RAM analysis was only scoped for a "high level" approach, based on 9-10 MMSCFD production for over 2 years prior.

Records before this time were far less informative and thus ignored after initial examination.

Some potential situations that might occur during later gas production stages have also been noted, but quantitative analysis for this study is valid only for a basecase loading of 10 MMSCFD.

Example Problem Study Nodes/ Model Configuration

Since the scope of this study only involves the gas plant facility at existing basecase gas production rates, the complexity of the models in expnsion cases 4. and 5. above are somewhat reduced.

Data quality was an issue for these two nodes and another reason which kept them out of the detailed model. Lack of incident specifics and root cause analysis kept our analysis to simply matching a set of "lumped" overall island reliability functions that would describe the system availability in a reasonable manner. Quality data was mostly available only for the last three years, so the ability to formulate a " historical probability analysis" is limited.

While we put in a more detailed model for the amine unit, the usefulness of that extra detail is a bit lost for this project. The amine unit has had, to date, a distinct lack of critical on-line need unless sales gas is to be sold, so the MTTR values are skewed and the MTBF values are significantly dependent upon the fact the unit is kept off-line quite a bit. We believe there are some issues with the amine plant operations that THUMS may want to consider as a part of increasing its availability. Our recommendations are presented in section 6.4

Most of the equipment blocks we defaulted to log-normal failure distributions with and estimated MTBF and MTTR that let the model fit the average annual performance

RAM Modeling – Client Case Study

RAPTOR HIERARCHY MODELS

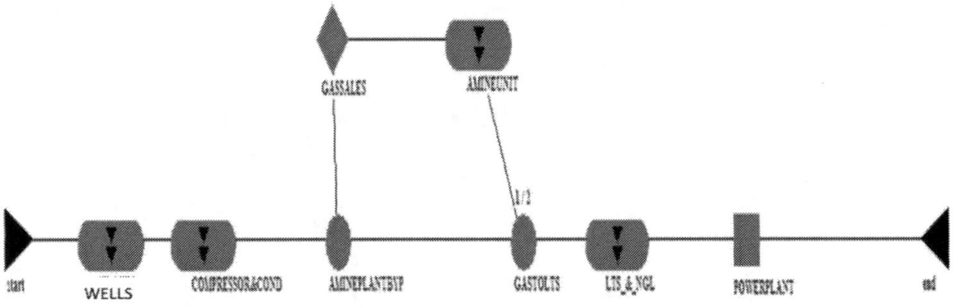

Upper level hierarchy for CLIENT availability model –OVERALL FACILITY

Reliability Block Diagram Submodel for CLIENT Reservoir and Production Model

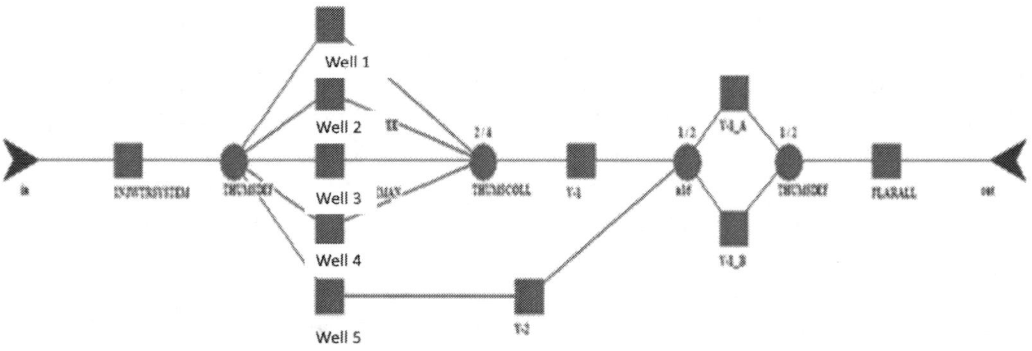

Reliability Block Diagram SubModel for CLIENT Gas Plant Feed Gas Compressor

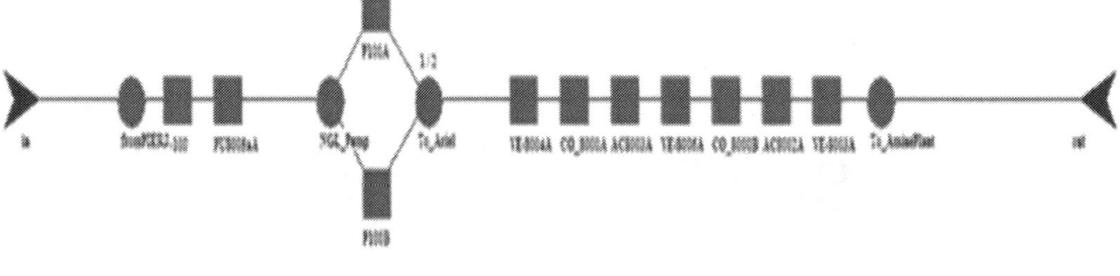

Reliability Block Diagram Submodel for CLIENT Refrigeration & Gas Chilling Plant

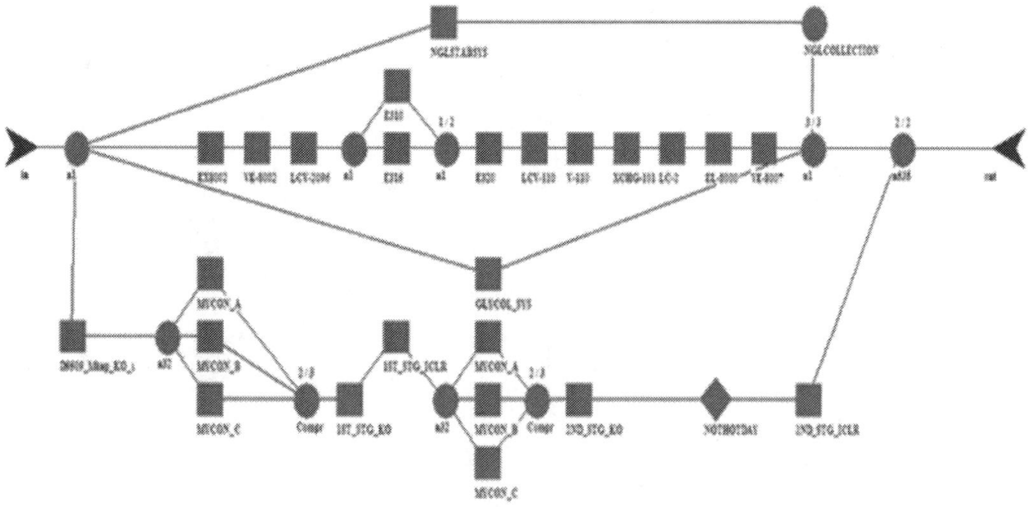

Reliability Block Diagram SubModel for CLIENT Amine Unit

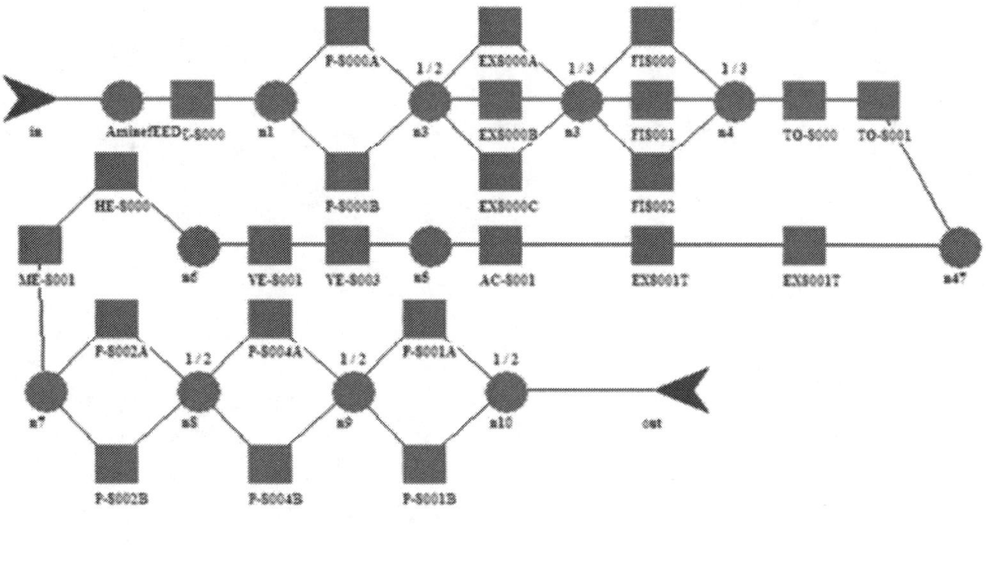

RAPTOR Demonstration

	Recommendations	Est. 3-yr Cost	Typ. Delta 3-yr Availability
1	Add motor amp, speed, vibration, winding temperature monitors to include new instrumentation and augmented EAM displays	$ 0.5 MM	30 hrs – 10% of added improvement
2	Add gas driver for field injection pumps	$?? MM Feasibility not yet full studied Out of Scope	20 hrs/flare avoidance and 40 + hrs more uptime for oil side
3	Add LO lab analysis to operations procedures	$ 0.1	30 hrs – 10% of added improvement
4	Add liquid level drainings, extended stripper ops, and inert blanketing during extended amine unit down time	$ 0.2	Unknown – not enough data to make a determination at this time
5	Install ISBL LACT meters (These needed to act as the facility "cash register" and not rely on client ticket for final value)	Out of Scope but ensures accuracy of custody transfer and could reduce down tome due to customer equipment failure	Might save 5-10 oil side downtime shipoing incidents Oil production down time loss if island pumping stops is $5500/hr

Ex. Sample Problem Solution

Ex. Sample Problem Solution

Disposition	Model	Model	Model	Model	Model	Model	Model	Comments
	Value	Rate	Value	Value	Value	Value	Value	
	BPD	MMSCFD	MW	$/BBL	$/MWH	$/MMSCF	$/MMBTU	
Gas to PP		5-10 MM	25-50		120			Avg Revenue or loss of revenue basis
Gas to Gas Sales		0 - +5				$4.50		intermittent
Gas to Flare	0 to -10					$4.50		intermittent
NGL to PP	0 to -125			70				intermittent
NGL to Gas Sales	0			70				controlled
NGL to Oil Sales	130			70				normal 5 BPD blend bonus putting NGL into crude
NGL to Flare	0 to -125			70				intermittent
Oil to Oil Sales	28,200			70				normal

Minimum Plant Design Situations and Standards that must be in place to maximize "real" availability

NON-PC TRANSLATION: Don't even bother with definitive RAM studies until you have a handle on all the significant failure modes that you CAN control.

- Equipment is designed to the required earthquake (seismic), wind, and storm requirements
- Fire personnel are adequately trained to respond to a fire condition
- A lock-out/tag-out procedure is in effect and strictly enforced by management.
- A work permit is required prior to starting any maintenance activity. The inadvertent opening of an isolation valve or blind removal during maintenance is not addressed because administrative controls are considered an adequate safeguard to prevent this situation.
- **NON-PC Translation – if you really think you need a PDF in the calculation chain to account for this situation, you have bigger problems than just figuring out reliability.**

Minimum Plant Design Situations and Standards that must be in place to maximize "real" availability

NON-PC TRANSLATION: Don't even bother with definitive RAM studies until you have a handle on all the significant failure modes that you CAN control.

- Basic training has been accomplished for operators/maintenance personnel and basic operating procedures have been developed and vetted AMAP prior to start-up.
- Procedures/training specific to equipment operation will be recommended during the review of each HAZOP scenario, as appropriate.
- It is assumed that at least minimal typical preventive maintenance will be provided. Safety equipment is routinely inspected.
- Unless otherwise specified, it is assumed that all process vessels are constructed to industry standards.

Minimum Plant Design Situations and Standards that must be in place to maximize "real" reliability

- It is assumed that the "as-built" plant will be thoroughly checked prior to startup and that it is built consistent with the design intent.
- It is assumed the Alarm set points are adequate to protect the equipment.
- A safeguard program covering "Operator Training and Procedures" is considered to both prevent an initiating event and prevent a "loss-heavy" (financial or HES) scenario from progressing after the initiating event has occurred.
- All Hazid/Hazop/PHA recommendations have either been implemented or acknowledged as non-critical and deferred.

THE END

REFERENCE SLIDES FOLLOW

REFERENCE SLIDES

Why WPM Selected Raptor?

Wide User Base

- Raptor is one of the world's best operations analysis software.
- Raptor has been in use for well over a decade and has thousands of customers worldwide
- Raptor is used across all major industries such as manufacturing, refining & petrochemical, telecommunications, aerospace, electronics Industry, and military (nicl. U.S. DOD's most advanced ballistic missile launch & defense systems)

Ease of Use and Advanced Analytics Features

- With Raptor, you can rapidly create analytical models and make operations decisions
  - based on rational analysis,
  - saving your company unnecessary operations costs,
  - unneeded logistics support,
  - and critical production capacity.

Two critical questions RAPTOR can help you define ..

- What is The Cost of Obtaining Capacity ?
- What is The Cost of Downtime ?

There is more than just availability analysis to the application

- Saving money is great, but if your tools don't have the ability to model your system, you'll never get to the savings. Raptor analyzes cost, capacity, sparing, and maintenance concepts, and the full Reliability, Availability, and Maintainability (RAM) suite, not just Reliability.
- Raptor's intuitive graphical interface avoids the difficulty of equation-based interfaces that limit system representation. For a fraction of the cost of other, less capable simulation tools, Raptor's multitude of impressive features give you incredible power.
- You can model from a From a Simple Electric Motor all the way up to a To a System of Any Magnitude

Modeling R&A - Basics

A block diagram model that consists of 3 subsystems (SS1, SS2, and SS3) follows:

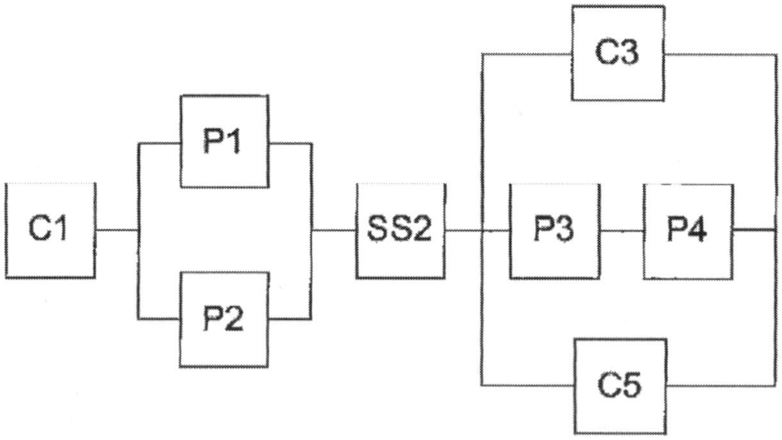

Reliability Block Diagram (RBD) Analysis Methodology

- Reliability Block Diagram (RBD) models are one of the tools that can be used to create a reliability model of equipment. One of the easiest ways to describe the basic ideas used in the creation of RBD models is to create a simple RBD which you will see in the demo.
- Construction of a reliability block diagram begins by defining what is meant by equipment failure; for example, equipment failure may be defined as any failure that causes the equipment to be down for 8 minutes or longer, caused flaring, or just results in economic loss.
- Once this is done, the next step is to determine the various ways that this failure can occur. This is initially done at a gross level; that is, 10 to 20 subsystems are defined that can lead to equipment failure.

A Reliability Block diagram model that consists of 3 subsystems (SS1, SS2, and SS3) follows:

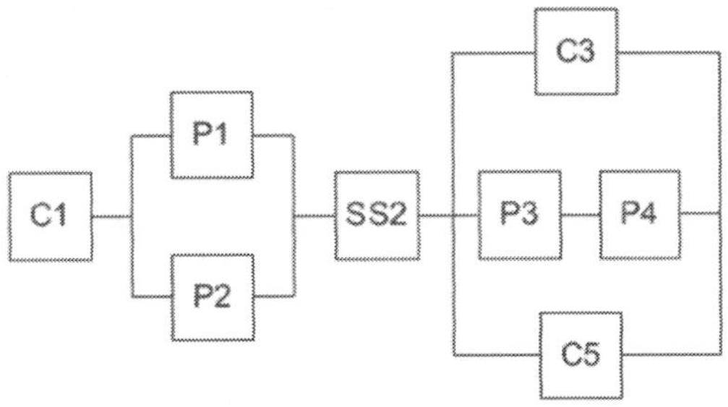

In this example SS2 is not a significant contributor to the unreliability of the equipment, so it will not be broken into any more detail. SS1 and SS3 however, are contributors to equipment unreliability. SS1 fails if component 1 or 2 (C1 or C2) fail. SS3 fails if components 3, 4, and 5 (C3, C4, and C5) all fail.

The RBD model now looks like:

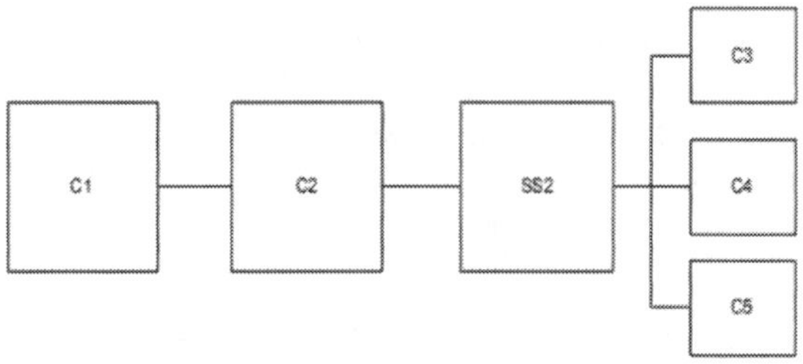

Further analysis reveals that C2 fails if parts 1 and 2 (P1 and P2) fail. C4 fails if parts 3 or 4 (P3 or P4) fail.

Series sequence denotes .AND. configuration
Parallel sequence denotes .OR. configuration

Simplify the block RBD as much as possible

Further analysis reveals that C2 fails if parts 1 and 2 (P1 and P2) fail. C4 fails if parts 3 or 4 (P3 or P4) fail. The block diagram model now looks like:

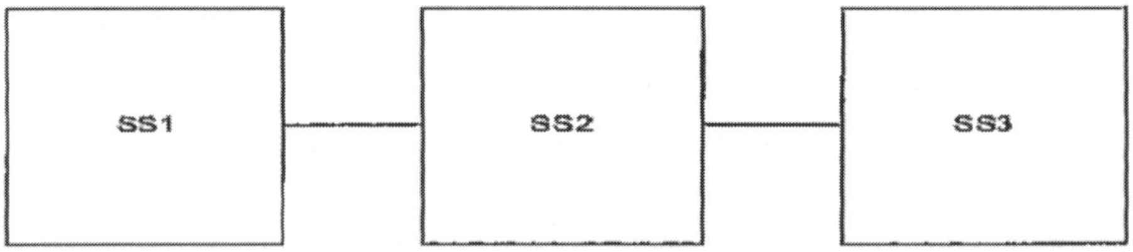

Equationalize the Model

Once construction of the model is conceptualized in block form, it is translated into a Boolean equation which is then used to quantify the equipment reliability. The references discuss Boolean algebra in detail, it will not be discussed here. The Boolean equation for the RBD is:

*Equipment Failure = C1 + P1 * P2 + SS2 + [C3 * (P3 + P4) * C5]*

Then expanding and using the associative and distributive laws,

*Equipment Failure = C1 + P1 * P2 + SS2 + C3 * P3 * C5 + C3 * P4 * C5.*

Each of the terms in this equation represent a way that the equipment can fail. For example, if part 1 and part 2 (P1 and P2) fail, the equipment fails.

Input component failure PDF's

- When the reliability block diagram has been translated into an equation-based model, it is now time to quantify the probability that the equipment fails as a function of its subsystems, components, and parts.
- Often the term *probability* is used when what is really meant is *frequency*, probabilities must lie between 0 and 1. A frequency can be any number greater than or equal to 0, depending on the number of failures and the time scale used. For example, if a component fails twice per year, its frequency is 2/yr, or 0.16/mo.
- Using the previous example, the probability of equipment failure can be written,

*P(Equipment Failure) = P(C1 + P1 * P2 + SS2 + C3 * P3 * C5 + C3 * P4 * C5).*

Simplify Again

But, how does one deal with the right-hand side of the equation? Considering the basic laws of probability and the small probability approximation, and assuming that the events are independent, the example equation becomes:

*P(Equipment Failure) = P(C1) + P(P1)*P(P2) + P(SS2) + P(C3)*P(P3)*P(C5) + P(C3)*P(P4)*P(C5).*

Terminology.

- **Availability Definition 1**: *Ratio of total time a system is not operational due to planned or unplanned downing events or incidents to the Total plant time that system could have been viably been operated absent external constraints*
- **Availability Definition 2**: *Ratio of realized annual profit to total potential profit realized and not realized.* Includes effect of "non-optimal" revenue recovery due to reliability failures
- **Availability Definition 3:** *Number of non-flare incident hours / Number of hours B&M plant* operates. This is not a true availability factor since the THUMS facility can be on line delivering products, but have the flare operate if all the gas from the Islands and other ancillary reservoirs cannot find a "home".
- **B&M** – Broadway & Mitchell factility, the main on-shore gas processing facility.
- **Causes:** These are the reasons would result in incidents occurring. Once a potential incident has been shown to have a creditable causes, it can be treated as a meaningful contributor to the reliability/availability function for the plant
- **Consequences:** These are the results and costs of the incidents should they occur.
- **Constraints:** Limits of the plant either set by mother nature (thermodynamics, heat and mass balance, pressure drops), by equipment capacity limits, or by instrumentation controls, including safeguards.
- **EAM** – Enterprise Asset Management system. This is the next step above MES that allows for full distribution of information beyond the control room and into the enterprise. Typically an EAM, such as Maximo, collects, presents, and organizes data of a specific kind and makes that data available by itself or in cooperation with other EAM systems that may be installed within the enterprise such as ERP and SIS/SIL systems or more general EAM apps such as Simatic XHQ or the Wonderware suite of applications.
- **FMEA** – Falure Mode Effects Analysis

Terminology

Incident/Downing Event An event that causes THUMS to incur actual economic value loss, lost opportunity cost, or a flaring incident.

- **IBWA:** Inspection by Walking Around. Constant operator attention to potential problems related to observable but abnormal conditions in the plant should always be included as a priority activity for each position.
- **LO** – Lube Oil
- **MES / HDA** - Manufacturing Execution System/ Historian & Data Acquisition package. PI/ Processbook is THUMS primary data retrieval and acquisition system along with a few other localized PLC displays attached to major equipment
- **MRO** –Maintenance, Reliability, and Operations
- **MTBDE:** Mean Time Between Downing Events – planned or unplanned.
- **MTD:** Mean time down (system) in a time range basis (typically. a year)
- **MTTF / MTBF:** Mean Time to (first incident)/ between(for recurring) Failure of a "down" equipment item
- **MTTR:** Mean time to Repair failure or Restore after incident cause
- **MW / MWH** – Megawatt/Megawatt-hour n reference to power generation capacity

Terminology

- **Outage - Unlanned** – A situation where a downing event, based on independent local equipment failure in a given unit or equipment item occurs without a planned shutdown. This failure data helps us determine MTTF and MTTR of a plant system and the equipment contained in that system.
- **Outage- Planned** – A situation where maintenance and operations plan to shut down a system or some of the equipment contained therein, resulting in an off-line (unavailable) situation for that system. The history of such outages is used to embed PM times inside the block model. A Planned outage does not typically permit valid MTTF determinations and may not provide the best data to determine MTTR data without careful analysis being given to the schedule of activities undertaken.
- **Outage - UnRelated** – A situation where a downing event is based on independent remote equipment failure in a different unit occurs without a planned shutdown. This failure data helps us determine MTBDE and MTTR of a plant system and the equipment contained in that system. It also allows us to establish and quantify dependencies of system failures so that the whole plant model is reasonably responsive to availability effects of equipment modifications.;
- **PM** – Preventative Maintenance
- **PP**- Powerplant
- **RAM** – Reliability and Availability Modeling
- **Risk:** The risk level is determined by the severity of the consequences and the likelihood it will occur. Risk can be reported as availability to operate or as an expected cost of consequence
- **Safeguards:** Any existing protective conditions and preventive devices which can prevent the cause, or protect against the adverse consequences.
- **Study Nodes**: The process is divided into manageable study nodes at which the process parameters are investigated for MTBF and MTTR contributions. Sub-nodes are implemented as needed

A Starting List of Failure Mode References

NORSOK Z-016 [1] (http://www.standard.no/PageFiles/960/Z-016.pdf)

OREDA 1st (1984), 2nd (1992), 3rd (1997), 4th (2002), 5th (2009) Edition [2](http://www.oreda.com/handbook.html) [3] (http://www.oreda.com/results.htm)

PARLOC [4] (http://www.hse.gov.uk/pipelines/parloc-2001-report.pdf)

PERD [5] (http://www.aiche.org/ccps/ActiveProjects/PERD/index.aspx)

NPRD[6](http://www.theriac.org/riacapps/search/?mode=displayresult&id=697

AIChE CCPS Database

Client Incident Record Logs

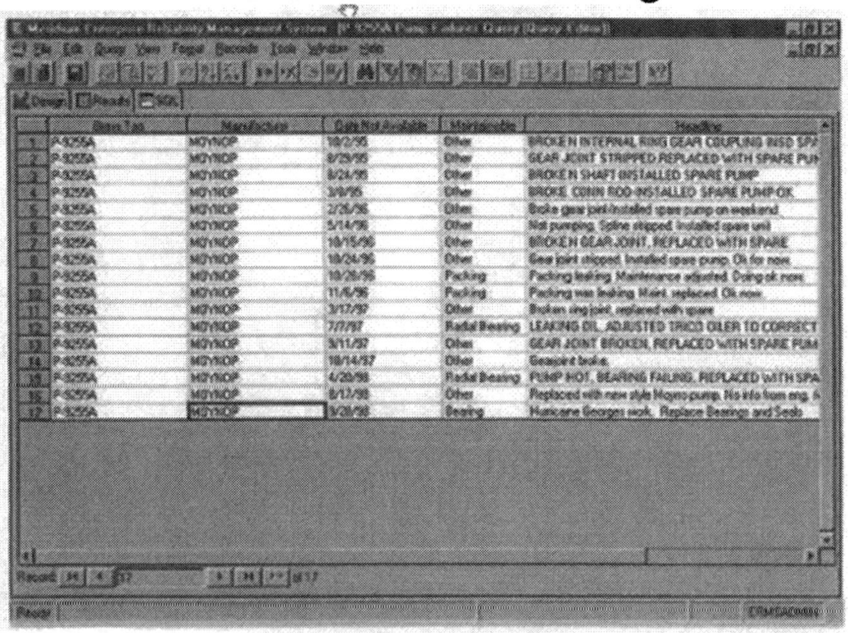

Figure 6-1 shows an example of a well documented maintenance log for a hypothetical piece of equipment.

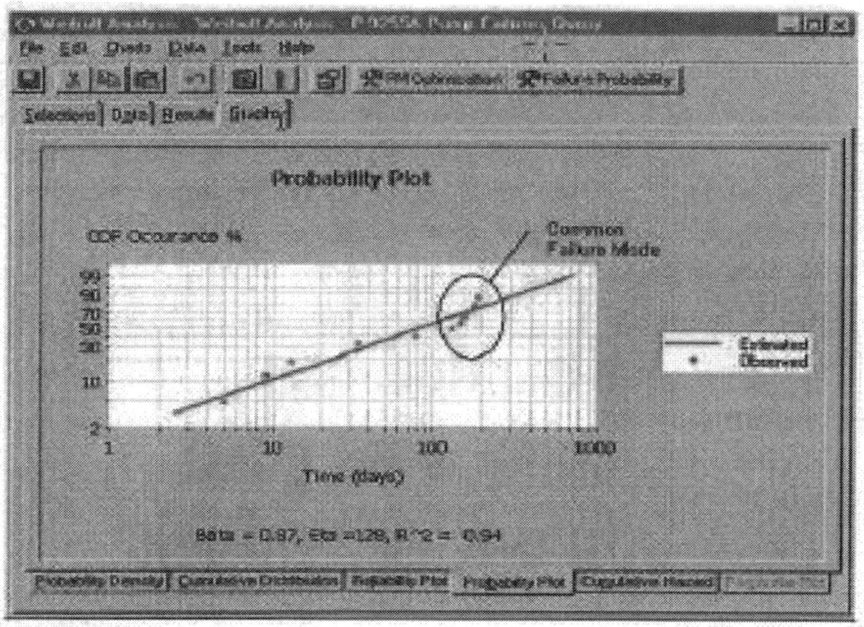

Figure 6-2 Weibull Analysis of Historical Failure Data

Be Sure the Problem is What You Think It Is !

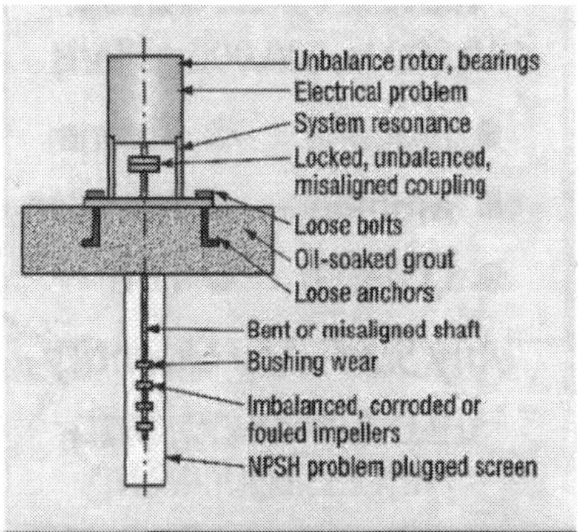

FIG. 1 Vertical pump vibration reported as a vibrating motor problem.

Average Pump Maintenance Cost, All Pumps

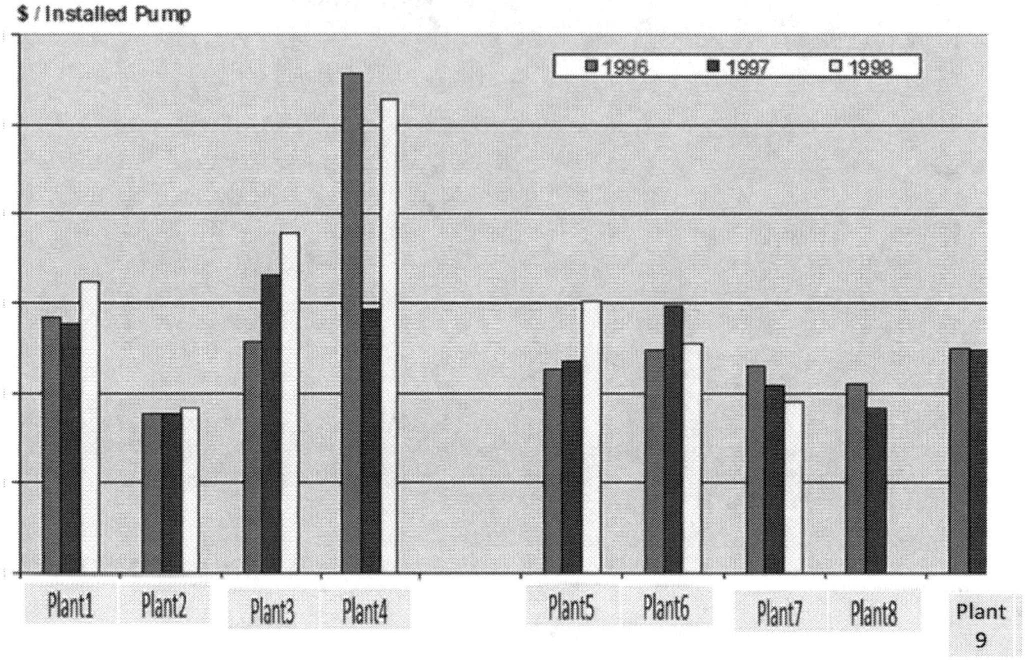

Solomon EDC as a Function of Plant Location

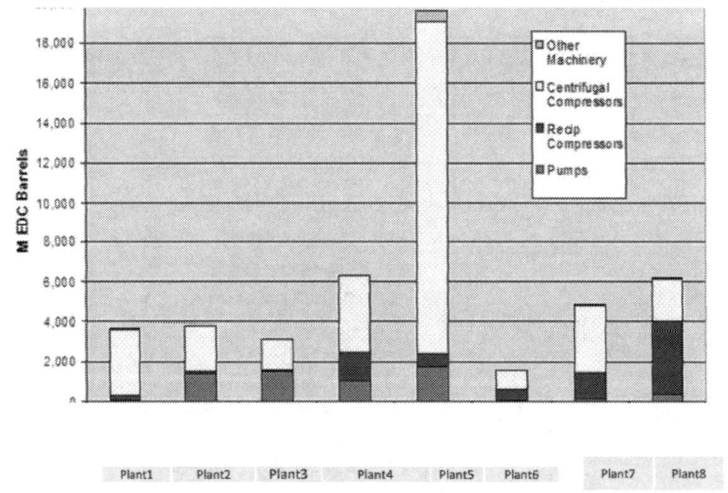

FAILURE CAUSE CATEGORIES

Machinery Components

Driven Machine:	Casing, internal parts & connections
	Bearings & rotor
	Seal(s) & mechanical seal system
	Coupling(s)
	Oil pump - shaft driven
Gearbox:	Gearcase, gear set, bearings, other internals, connections
	Coupling(s)
	Oil pump - shaft driven
Turbine Driver:	Casing, internal parts & connections
	Bearings & rotor
	Seal(s) & seal system
	Coupling(s)
	Oil pump - shaft driven
	Admission/fuel gas valves, actuators, linkages and hydraulic governors
	Overspeed trips, trip valves, trip/throttle valves, non-return valves (extraction)
Motor Driver:	Bearings, rotor, windings/stator
	Coupling
	Oil pump - shaft driven
Vibration:	Fouling/deposits or foreign objects
	Metal loss – corrosion, erosion, foreign object damage
	Component failure
	Resonance – train components, piping, structure
Lubrication:	Oil contamination, degradation, loss of level
Performance:	Damage/deterioration
	Must run lower capacity backup/spare machine
	Horsepower or Current (amp) limited

FAILURE CAUSE CATEGORIES

Support Systems

Machinery:	Lube Oil System – pumps, filters & non-instrument controls
	Seal Oil System – pumps, filters, seal pots & non-instrument controls
	Gas Sealing System – filters, & non-instrument controls
Electrical:	Starters, breakers, controls, control wiring, protective relays
	Power cables
	T-leads & junction box internals.
	Power distribution system, breakers, switchgear, etc. Lightning strikes.
Instrumentation:	Shutdown instrument loops & protective instrument systems
	Lube and seal oil level & pressure control loops
	Speed controls – electro-pneumatic or electronic
	Vibration monitoring systems
Heat Exchangers:	Intercoolers, aftercoolers, air coolers, radiators
Piping / Vessels:	Process piping and small bore piping not related to machinery
	Block valves, check valves, bleeder valves
	Suction/knockout drums, coalescers, separators, etc.

FAILURE CAUSE CATEGORIES

Process / Operational

Solids restricting flow (piping, valves, etc.
Liquid carryover/accumulator in suction drums

Low Flow

Coking up catalyst

Failure to act on SIS/SIL alarms

Ignoring basic level alarms

Failure of operators/supervisor to understand basic control principles

EFFECT OF PROCESS CONDITIONS ON FILM TEMPERATURE IN FIRED HEATERS

A. Morales-Fuentes[1], C. E. Gallegos-Buendía[1], M. Picón-Núñez[2], S. Méndez-Díaz[1], F. Sanchez-Cruz[1]
[1] *School of Mechanical and Electrical Engineering, Autonomous University of Nuevo León, México*
[2] *Department of Chemical Engineering, University of Guanajuato, México*

Abstract

Variation in process conditions in fired heaters arise due to multiple circumstances, of these, throughput, inlet temperature and percentage of excess air are the most recurrent. Film temperature is an important parameter to be observed during the operation of this type of equipment due to its influence in the development of coke on the heat transfer walls. The consequence of the building up of a fouling layer is the reduction of the thermo hydraulic performance which introduces additional expenses and eventually forces plant shut down for maintenance. In this work, an assessment of fired heaters performance is carried out at a fixed outlet temperature and pressure. Results show that plant throughput variation demand higher firing rates with the largest impact on film temperatures. Increased percentage of excess air also demands higher firing rates; however it does not show a strong impact on film temperatures. Finally the effect of inlet temperature appears not to shift the local film temperature profile from its original condition.

Keywords: Fired heater, Process conditions, Film temperature

Introduction

A fired heater is a key piece of equipment in a refining process since it provides the heat load at high temperature for crude separation in a distillation column. Lower temperature heat is supplied in a preheat train which consists of several units in parallel that converge in a fired heater. Operating conditions vary daily as a result of company policies and maintenance issues, disturbing the normal performance of the fired heater. Among these is the excess air used for combustion which for the case of gas fuels a 10-15% is recommended whereas 20-30% is recommended for liquid fuels. The amount of excess air is better determined in situ by measures of oxygen at the chimney flue gas. The lower the excess air used, the higher the heater efficiency as pointed out by Hamidi *et al.* (2006) who studied the heater efficiency for mixtures of different types of fuel and preheating conditions. Throughput is another process variable that is subject to changes and it is typically dictated by market demands, internal policies and operational issues. During startup throughput is only a fraction of the equipment capacity. Furnace inlet temperature variations arise when individual units in the preheat train are withdrawn for maintenance; the heat duty is shifted to the fired heater forcing the consumption of additional fuel. These variations on operating conditions affect the performance of fired heaters in several ways, for instance, they might increase fouling which is an important phenomenon studied in recent years due to its implication in energy consumption (ESDU 2000).

Fouling has a profound impact on plant production rate and furnace inlet temperature. Morales-Fuentes *et al.* (2012) studied the thermo-hydraulic performance of a fired heater running under fouling conditions. Cross (2009) analyzed the performance of fired heaters and the effects of fouling when increased capacity is demanded. Ishiyama *et al.* (2008) showed that fouling can lead to important flow variations in heat exchangers located in parallel. Furnace inlet temperatures can drastically vary during initial operation stages due to fouling as presented by Yeap *et al.* (2004) and Coletti & Macchietto (2009).

Panchal *et al.* (1999) presented a threshold fouling rate model composed of a deposition term which competes with a suppression rate where the temperature and fluid velocity are the two variables that control this phenomenon. Modifications to this model have been presented by Polley *et al.* (2002) and Yeap *et al.* (2004). A generalized fouling rate model was presented by Polley & Morales-Fuentes (2009) who described the fouling rate as function of film temperature T_f and shear stress on the tube wall τ_w as shown in Equation (1).

$$\frac{dR_d}{dt} = \frac{A}{\alpha}\exp\left(\frac{-E}{RT_f}\right) - B\tau_w \qquad (1)$$

Where A, B and E are parameters that depend on the deposition constant value, suppression constant and activation energy respectively; α stands for the heat transfer coefficient.

In this study the effect of operating conditions such as excess air, throughput and furnaces inlet temperature variations on local film temperature are studied assuming operation over a clean surface. The higher the film temperature the higher the deposition rate as shown by the fouling rate model presented in Equation (1). The local film temperature profile is reported for two geometries of fired heaters.

Geometries

Two vertical fired heaters used for heating of crude at atmospheric conditions (Geometry A) and vacuum residues (Geometry B) are presented. They consist of a cylindrical firebox with vertical tubes array next to the refractory wall. Above the radiant chamber there is a rectangular cabin containing bare and fined tubes in an offset array, striping the heat from combustion gases as shown in Figure 1. The main characteristic of these geometries are presented in Table 1. For some applications the tube diameter in the firebox is increased in order to reduce velocity and higher pressures due to the vapor fraction increment as observed in Geometry B.

The process stream enters at the top of the fired heater and descends through the convection section until it reaches the radiant chamber. The length is assumed 0 at the top of the convection section; whereas the maximum length is reached at the exit point of the radiant section. The convection section has a total length of 88 m for Geometry A and 126 m for Geometry B.

Table 1. Geometry description of fired heaters

		Geometry A	Geometry B	Units
Radiation section	Width	4.5	6	[m]
	Height	7.5	17	[m]
	No. tubes	42	96	
	Tube diameter	168	89, 114, 168, 219	[mm]
	Total heat transfer area	178	512	[m^2]
Convection section	Width	1.5	2.5	[m]
	length	4.5	6	[m]
	Height	3	3	[m]
	Bare tubes	8	40	
	Finned tubes	32	128	
	Tube diameter	0.168	0.089	[m]
	Total heat transfer area	291	282	[m^2]

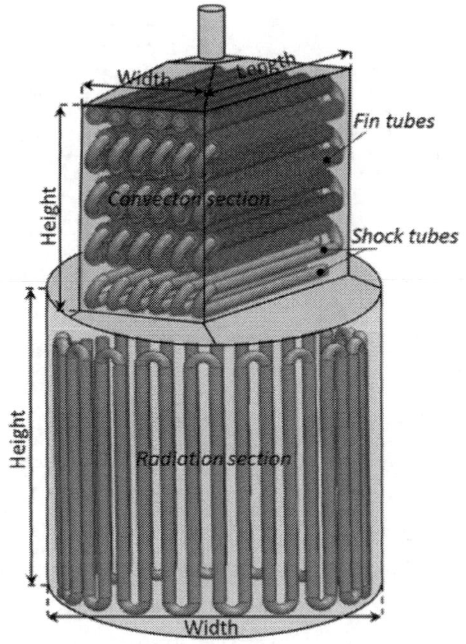

Figure 1. Vertical fired heater composed of a convection and radiant section

Process conditions

A firebox loss of 2 % is assumed. The operation of the fired heater is such that exit temperature and pressure must be kept at 330°C and 517 kPa for Geometry A and 432 °C and 41 kPa for Geometry B. Firing rate and outlet temperature are then adjusted according to the particular operating conditions. Clean operating conditions are assumed. The fuel used for both geometries is a heavy residual oil known as fuel oil No. 6; firing rate is adjusted according to the specific simulation.

Excess air variation

The effect of the local variation of film temperature as excess air increased is considered; heat duty, inlet and outlet temperatures are kept constant. A greater amount of air enhances the heat transfer in the convection section and therefore it could bring possible benefits/harms at specific applications. The crude mass flowrate for normal operation for Geometries A and B are 67 kg/s and 45 kg/s respectively. Excess air was studied at stoichiometric relation, 20% and 50 %. Profiles of local bulk and film temperatures are presented in Figure 2 for both geometries.

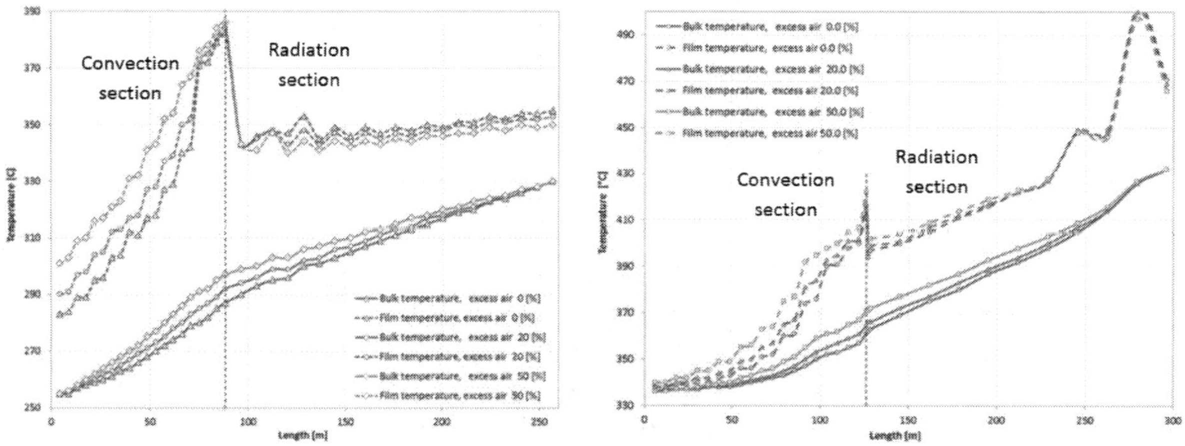

a) Geometry A b) Geometry B

Figure 2. Bulk and film temperature profiles along the heater for several excess air percent.

It is observed two different bulk temperature profiles for the heaters. Geometry A shows two different quasi-linear trends at convection and radiation section respectively. Bulk temperature at Geometry B differs only in convection section with a non-linear growing trend. Local film temperature also has two different behaviors. Geometry A shows a quick growth in convection section until a peak at tubes located on top of the firebox. At radiation section it does not show a big variation reducing the gap with the bulk temperature profile. On the other hand, the gap between film and bulk temperature in Geometry B is very narrow at the initial part of convection section and grows as it moves forward. Film temperature shows a growing trend all along the heater. As crude proceed from convection section to the firebox, there is a change on the heat transfer mechanism which is observed as a decreased film temperature in Figure 2.

The operation with higher amounts of excess air requires the combustion of additional fuel to reach the desired crude conditions. Higher amounts of excess air increases the heat transferred in the convective section leading to higher film temperatures in convection section as observed in Figure 2. The radiant section of Geometry A benefits a little reducing its film temperature; this situation is not observed in Geometry B.

Three adverse consequences of operation under excess are the reduction of furnace efficiency, a possible enhance of firing rates in convection section where film temperature overpass its initial condition and a slight increase of pressure drop since two phase flow might exist at previous zones. Table 2 shows the heaters efficiency, percent of heat absorbed in firebox and pressure drop as function of the excess air for both geometries.

Table 2. Operation conditions under different excess air conditions

	Geometry A (Throughput 67 kg/s)			Geometry B (Throughput 45 kg/s)		
Excess air [%]	Heater Efficiency	$\%Q_{Firebox}$ [%]	ΔP [kPa]	Heater Efficiency	$\%Q_{Firebox}$ [%]	ΔP [kPa]
0	76	62	263	85	77	33
20	70	57	269	83	73	39
50	61	50	280	79	69	45

Throughput variation

To assess the performance of fired heaters subject to throughput variations, inlet and outlet temperature are assumed constant. Exit pressure is set so inlet pressure is adjusted for each flow rate being analyzed. The excess air percent is kept to 20% in all cases. The range of flow rates studied resulted in a bulk velocity of 1.5, 2.5 and 3 m/s for Geometry A when single liquid phase is present and 1.6, 2.7 and 4.6 m/s for the Geometry B. Evolution of local bulk and film temperatures for several throughputs are presented in Figure 3.

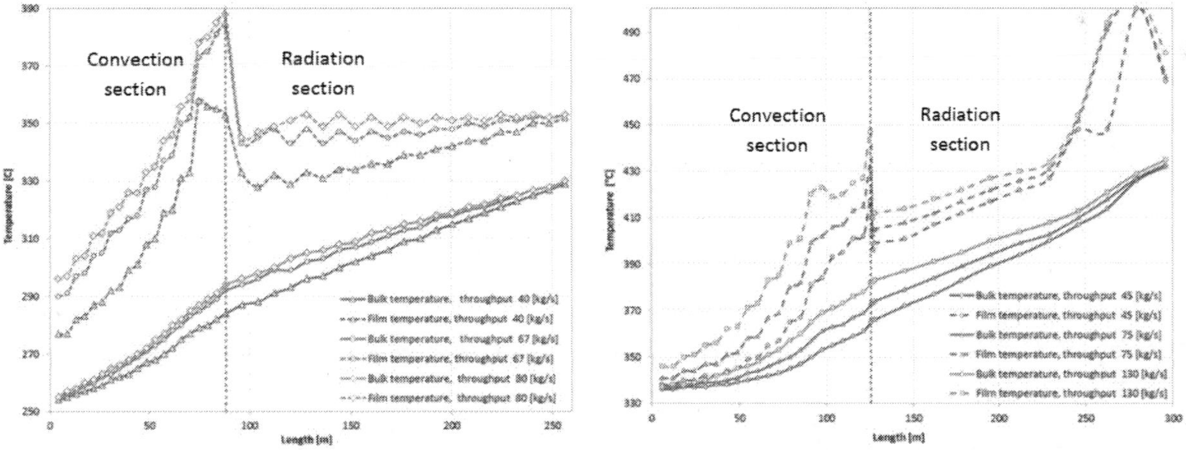

a) Geometry A a) Geometry B

Figure 3. Local bulk temperatures for different throughput specifications

A higher throughput demands additional fuel consumption leading to higher flue gas mass flux through the equipment. The film temperature effect is similar than the observed when additional excess air is evaluated. Heat transfer in firebox decreases and heater efficiency reduces. Since a greater amount of heat is transferred the heater efficiency does not decrease in magnitude as the excess air case (especially Geometry B). Heater efficiency of Geometry A is more sensible to changes in flue gas than Geometry B as observed in Table

3. This is in accordance with the fact that heat transfer area ratio firebox to convection is 0.61 for Geometry A and 1.81 for Geometry B.

Table 3. Operation conditions under different throughput conditions

Geometry A				Geometry B			
Throughput [kg/s]	Heater efficiency	%Q Firebox [%]	ΔP [kPa]	Throughput [kg/s]	Heater efficiency	%Q Firebox [%]	ΔP [kPa]
40	77	64	97	45	83	73	39
67	70	57	269	75	77	64	173
80	67	55	360	130	80	57	607

The zones with higher probability to foul derived from higher film temperatures are the final part of convection section for Geometry A, and the final part of radiant section for Geometry B. A possible consequence of increase throughput to obtain higher revenues might derive on premature fouling due to higher film temperatures leading to an unexpected maintenance shut down.

Furnace inlet temperature variation

Variations in the inlet temperature arise as a consequence of fouling in previous heaters or when individual heaters are taken off for maintenance. The heat load is shifted to the fired heater where the firing rate is increased to meet the required heat load. In this section, the inlet temperature for Geometry A is reduced 10 °C and 20 °C from its original condition. Figure 4 shows bulk and film temperature for Geometry A at a mass flow rate of 67 kg/s and excess air to 20%.

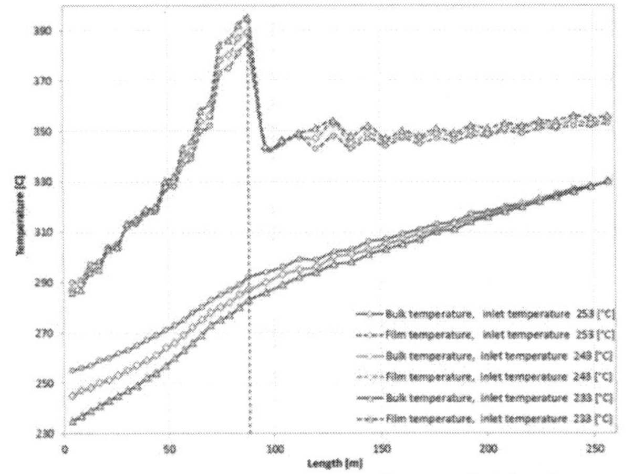

Figure 4. Local film and bulk temperatures under variable furnace inlet temperature

As inlet temperature decreases the firing rate is increased to meet the desired duty from its initial value of 24.2 MW to 27.5 MW and 30.9 MW for a reduction of 10 °C and 20 °C respectively. The additional duty is mainly absorbed in the convection section where heat transfer is enhanced as a result of increased firing conditions. Table 4 shows the heater efficiency, the heat percent absorbed in the firebox and the pressure drop. Although local bulk temperatures in convection section have significantly changed, the local film temperatures are very similar to those of the original case. This shows the dependence of film temperature on external conditions. The film temperatures at the shock tubes and radiation section are barely increased with firing rate.

Table 4. Operation conditions for reduced inlet temperature, Geometry A

Inlet temperature	Heater efficiency	%Q Firebox	□P
[°C]	[%]	[%]	[kPa]
253	70	57	269
243	69	56	262
233	67	55	257

Conclusions

In this work the effect of excess air, throughput and inlet temperature on film temperature are evaluated as a possible source of premature fouling. Simulations were carried out for two geometries assuming clean conditions and fixed inlet and outlet temperatures. Geometry A shows higher film temperatures at shock tubes meanwhile Geometry B does at the firebox exit point where higher bulk temperatures are present. It is observed that additional excess air will increase film temperature mainly in convection section where heat transfer is enhanced. Local film temperatures across the heater increase when higher production is demanded. Finally a reduction in inlet temperature will not modify the local film temperature. Those are aspects to consider during operation. The ratio radiation to convection heat transfer area dictates the amount of heat transferred in each section and how the film temperature changes with varying operation conditions. Therefore it is an important parameter to consider during design stage that helps to avoid premature fouling deposits due to hot spots.

References

Coletti, F. & Macchietto S., 2011, A dynamic, distributed model of shell and tube heat exchangers undergoing crude oil fouling, Ind. Eng. Chem. Res., 50, 4515–4533.

Cross A., Direct fired heaters evaluate thermal performance and the effect of fouling, Chemical Engineering 2009.

ESDU, 2000, Heat exchanger fouling in the pre-heat train of a crude oil distillation unit, data item 00016. IHS ESDU International plc., London.

Hamidi, A.A., Gargani, S., Dehghani, P., Optimizing crude oil fired heaters, 2006, Hydrocarbon Processing, 113-117.

Ishiyama, E.M., Paterson, W.R., Wilson, D.I., 2008, Thermo-hydraulic channeling in parallel heat exchangers subject to fouling, Chemical Engineering Scince, 63, 3400-3410

Morales-Fuentes, A., Polley, G.T., Picón-Nuñez, M., Martínez-Martínez, S., 2012, Modeling thermo-hydraulic performance of direct fired heaters for crude processing, Applied Thermal Engineering, 39, 157-162

Panchal, C.B., Kuru, W.C., Kiao, C.F., Ebert, W.A., Palen, J.W., 1999. Threshold conditions for crude oil fouling. In: Bott, T.R., Melo, L.F., Pancha, C.B., Somerscales, E.F.C. (Eds.), Understanding Heat Exchanger Fouling and its Mitigation. Begell House, New York, USA, pp. 273-279.

Polley, G.T., Morales-Fuentes, A., 2009, Observations Arising from the Analysis of Laboratory Data Relating to Fouling in Crude Oil Systems, Procedings of International Conference on Heat Exchanger Fouling and Cleaning VIII, www.heatexchanger-fouling.com

Polley, G.T., Wilson, D.I., Yeap, B.L., Pugh, S.J., 2002. Evaluation of laboratory crude oil threshold fouling data for application to refinery pre-heat trains. Applied Thermal Engineering 22, 777-788.

Yeap, B.L., Wilson, D.I., Polley, G.T., Pugh, S.J., 2004. Mitigation of crude oil refinery heat exchanger fouling through retrofits based on thermo-hydraulic fouling models. Chemical Engineering Research and Design 82 (A1), 53-71.

Trouble-shooting and Optimizing Ethylene Plant Cooling Water Networks by Using Advanced Process Simulation

Principal Author
Swapnil Pathak
Ingenero, Mumbai, India
spathak@ingenero.com

Other Author
James Brigman
Ingenero, Houston, TX, USA
jbrigman@ingenero.com

Abstract

Cooling Water Networks generally do not receive due attention from engineers and are usually overlooked in comparison to other critical equipment. Simulating the plant cooling water network can identify opportunities to increase operational efficiency, availability, and throughput. Flow networks are usually modeled using specialized hydraulic modeling tools. However, heat transfer effects on the process due to changes in cooling water network also need to be studied in detail. This paper describes how process simulation tools (Aspen Plus, Aspen Plate EDR, etc) were used for modeling the cooling water networks and further utilized for trouble-shooting and optimizing production at a major ethylene facility.

Case Study 1 – Optimization of the Cooling Water System for a petrochemical complex
An integrated simulation tool having a user friendly Graphics User Interface (GUI) in MS Excel was developed for seawater – cooling water system. Seawater and cooling water network hydraulic model was developed in ASPEN Plus to predict flow rates to the Plate heat exchangers. Seawater Plate heat exchangers were modeled in ASPEN Plate EDR, to evaluate their thermal operating performance and the seawater cooling tower model was developed in MS Excel to predict the basin temperature for given number of running fans and wet bulb temperature.

Case Study 2 – Trouble-shooting an Ethylene Plant Cooling Water Network
A hydraulic model for the ethylene plant cooling water network was developed using *Aspen Plus* and utilized to perform various what-ifs in order to ensure flow availability to critical cooling water exchangers. This model was further utilized to check the exchanger and network's adequacy at increased plant capacity and devise various modification options. The throughput through the entire plant could be increased due to the enhanced cooling.

1. Introduction

Cooling water system is one of the most important utilities in any process unit. However, cooling water networks generally do not receive due attention from engineers and are usually overlooked in comparison to other critical equipment. Lack of monitoring the cooling water system may hamper the plant operation in long run, which can even lead to loss in production form the process unit.

In the following sections, we will discuss why it is required to model plant-wide cooling water system, how to develop a typical cooling water network using a process simulation software, what are the pros & cons of using different simulation applications and finally two case studies will be presented to provide examples, as to how a cooling water network simulation was used effectively to facilitate the management's decision-making for capital intensive modifications in the cooling water network and how the simulation model was utilized to trouble-shoot & optimize an ethylene plant cooling water network.

2. Need For Modeling the Cooling Water Network

There are two different aspects that provide strong reasons to invest in an effort to develop a simulation model of the plant-wide cooling water network.

2.1 Design Aspect

During the engineering design phase, while calculating specifications of equipment, it is necessary to employ certain logical assumptions and ensure that the design is practical in terms of installation at its desired location and procuring the process equipment.

2.1.1 Design of Pipelines

Sizing of pipelines is usually based on total / maximum flow required. It also considers adequate design margins (such as 10%) on flows or values of various parameters specified in the design basis. Also, it is observed that the velocities considered while designing pipelines are conservative. Moreover, the design engineer has to specify a nearest available pipe size, which is commercially available. For example, for a calculated pipeline size of 7.4 inch, one has to select the pipe size to be 8 inch. This results in a cooling water piping network, which has different pressure drops & design margins in different sections of the plant.

2.1.2 Pump Specification

Pump sizing is usually performed considering the maximum head / pressure drop across the network that the pump may encounter during its course of operation. Cooling water supply pumps are designed based on the maximum pressure loss that may occur across the entire cooling water network from pump suction (from the cooling tower sump) to the top of cooling water distribution risers of the cooling tower.

2.1.3 Heat Exchanger Design

When designing an exchanger the design engineer has to consider the effect of process or utilities being handled by the equipment. This is usually addressed by providing allowances for fouling and over-design. As a result of these allowances, actual pressure drop across the exchanger is different than the specified design value. However, for performing pipeline and pump sizing, a fixed pressure drop (usually the design value) across the heat exchanger is considered.

As a result of these design aspects, sections of the (plant cooling water) network receive either less or more flow than that specified in the design. This uneven distribution of the cooling water flow is addressed by throttling the (globe) valves, bypassing the cooling water flow or by applying flow correction using restriction orifices.

2.2 Operational Aspect

Though cooling water is one of the most commonly used and critical utilities in any process unit, cooling water systems are either sparsely equipped with instrumentation (flow & pressure transmitters / gauges) or if available at certain locations, the instrumentation is found to be in non-working condition due to their limited use during normal plant operation.

Installing flow meters in various sections of the cooling water network is highly capital intensive due to large pipeline sizes. Another issue with having flow measurements in the cooling water network is increase in pressure drop, which in-turn will affect the pump discharge pressure requirement leading to higher capital expenditure for installing a bigger pump and also the operating cost.

Hence, having a tool, which replicates the plant cooling water network and allows for analysis of various operating scenarios, proves useful for the plant operation team for trouble-shooting and optimizing the plant operation.

3. Flow network-modeling options

Today, various simulation and modeling software / applications are commercially available, which can be used for simulating the flow networks. When one decides to develop a simulation model for the cooling water network, the next immediate question that arises is, which application should be used? Choice of the simulation application will depend on various factors such as, ready availability, cost of license, ease of use, available in-house expertise and so on.

For developing plant-wide cooling water network simulation, complete process simulation applications score over specialized flow network modeling software in terms of their ability to have heat transfer effects included, which help in observing the effect of changes in cooling water network parameters (flow, pressure & temperature) on the process, directly. Moreover, process simulation software are usually available with most of the manufacturers and will be readily available along with required expertise in-house, in comparison to procuring a specialized flow network modeling software and training the resources to use the new tool.

Let us now look at two case studies in detail, in which process simulation software were utilized as the basic simulation application for building cooling water flow network model.

4. Case Study #1 – Trouble-shooting the Cooling Water System of a major petrochemical complex

Client A (located in the Middle East) had a common cooling water system catering to various process units (Ethylene, Ethylene Glycol, Poly-Ethylene & Utility). Cooling water return from the process units was cooled using sea water in a number of plate heat exchangers (PHE) arranged in parallel. Hot sea water from plate heat exchangers was cooled with the help of a sea water cooling tower. During summer months, due to high ambient temperature, cooling water supply temperature delivered to the process units increased. Increase in the cooling water supply temperature had an adverse effect on the operation of process units & was limiting the plant throughput in ethylene manufacturing unit. It was decided to develop a simulation model of the common cooling water system in order to evaluate its performance and explore possible options to address the issue of high cooling water supply temperature to the process units.

4.1 System Description

Client A's petrochemical complex had a common cooling water system delivering cooling water to four process units (Ethylene, Ethylene Glycol, Poly-Ethylene & Utility). This system comprised of two main sections as follows:

- Seawater Network – consisting of the cooling tower, seawater pumps and 37 plate heat exchangers. 4 seawater pumps circulate the seawater across 34 plate heat exchanges (3 plate heat exchangers are on standby). In the plate heat exchangers, seawater exchanges heat with the cooling water and cools it. Hot seawater is then cooled in the cooling tower and pumped back to the Plate heat exchangers from the cooling tower basin. Seawater circulation rate for design condition was 104,667 m3/hr. Following figure shows the schematic of Seawater circulation loop:

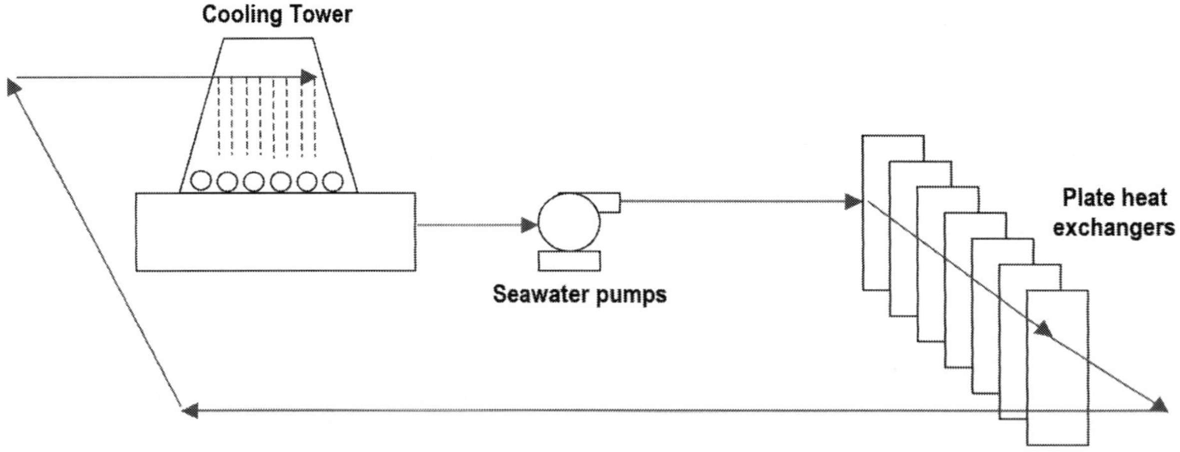

Figure 1: Seawater Circulation Loop

- Cooling Water Network – consisting of 8 cooling water pumps, cooling water users (four process units) and 37 plate heat exchangers. 6 cooling water pumps are online in normal operation and supply the cooling water to Ethylene unit, Ethylene glycol unit, Polyethylene unit and Utilities.

The cooling water return from these units then exchanges heat with seawater in the plate heat exchangers. Cooling water circulation rate for design condition was 94,122 m3/hr. Following figure shows the schematic of cooling water circulation loop:

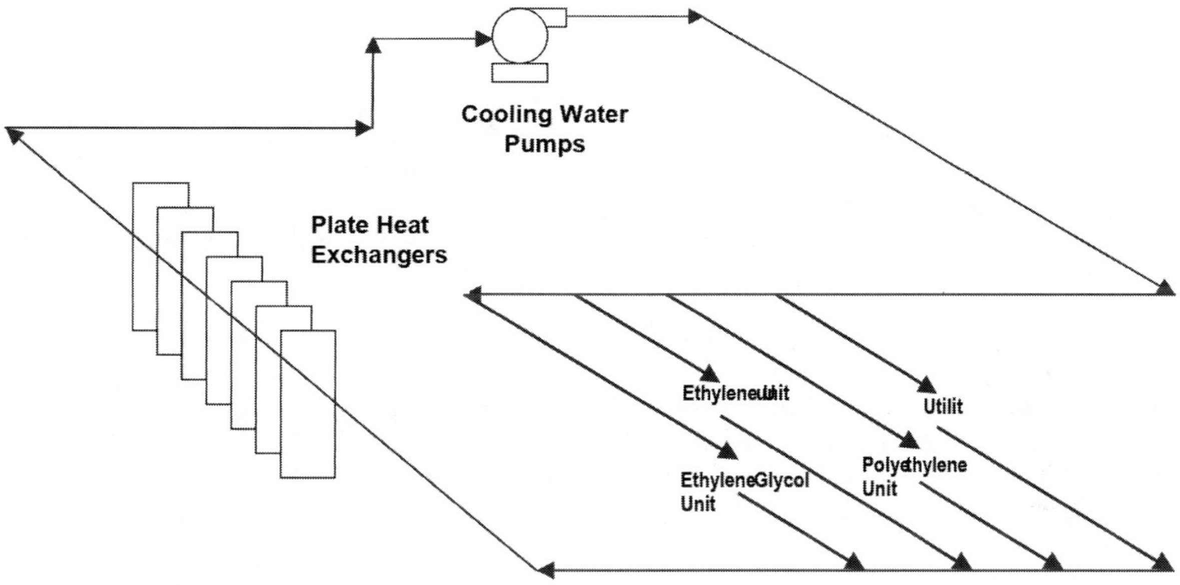

Figure 2: Cooling Water Circulation Loop

4.2 Developing an Integrated Simulation Model

To develop the simulation model for entire seawater – cooling water system and evaluate its performance, three different simulation models were developed:

4.2.1 <u>Cooling Tower Model (for seawater)</u>

A regression-based model of the cooling tower using the its design performance curves was developed in MS Excel. The Cooling Tower regression model required two inputs viz; the wet bulb temperature (WBT) and number of cooling tower fans in operation. The model was built to work for wet bulb temperatures between 0°C and 39°C and for 0 to 30 fans. For 0, 10, 20 and 30 fans the model used a linear function regressed from the cooling tower performance curves to predict the cooling tower basin temperature (for given WBT). If the number of fans in operation was other than 0, 10, 20 and 30, an interpolation between the appropriate regression expressions was used to predict the basin temperature.

4.2.2 <u>Seawater & Cooling Network Model</u>

A hydraulic network model for the seawater and cooling water network was developed in Aspen Plus for predicting seawater flow rates to the plate heat exchangers. This enabled the user to identify possible plate heat exchangers that would receive less than design / required flow rates. The user could then take corrective action by manipulating appropriate valves to balance the flows across the network. These two networks had the following components:

Sea Water Network Model Components:

- Plate heat exchangers (34nos Working + 3 Standby + 17nos for Future)
- Seawater pumps (4 Working + 1 Standby + 2 Future)
- Sea water Make-up pumps (2 Working + 1 Standby + 1 Future)
- All piping related to the network

Cooling Water Network Model Components:

- Plate heat exchangers (34nos Working+ 3 Standby+ 17nos for Future)
- Cooling water pumps - (6 Working + 2 Standby + 2 Future)
- CW system make-up water tank.
- Process Units (Ethylene, Ethylene Glycol, Polyethylene & Utilities)
- All piping related to the network.

4.2.3 Plate Heat Exchanger Model

Plate heat exchangers (37 numbers) were modeled in Aspen Plate EDR. Each Plate heat exchanger simulation was developed separately and was validated against the design data sheet. The plate heat exchanger model was used to perform thermal calculations (simulation) of individual exchanger for predicting the cooling water & sea water temperature at outlet of each exchanger.

4.2.4 User Interface

In order to enable the user to execute all three models (mentioned above) from a single interface, these models were integrated through an MS Excel interface. The user interface was developed in a way that all the simulation applications would be launched automatically and would run in the background. MS Excel VBA was used for this purpose.

4.3 Integrated Simulation Model Features

An integrated simulation model comprising three different systems as mentioned in the sections above was capable of:

- Predicting the seawater cooling tower basin temperature using the wet bulb temperature and number of fans in operation as inputs.
- Calculating *K* value *(pressure drop factor)* for all plate heat exchangers on seawater and cooling-water side. Flexibility to have different geometries for different PHEs was also included.
- Using this updated *K* in the Aspen Plus hydraulic models for seawater and cooling water and run them. These models estimated the flow distribution on seawater and cooling-water side based on the system hydraulics.
- Using individual flows of all plate heat exchangers obtained from the (Aspen Plus) Hydraulic model to evaluate thermal performance of all plate heat exchangers using Aspen Plate EDR.
- Predicting the fouling factor for all plate heat exchangers (using Aspen Plate EDR).
- Predicting the resultant cooling water supply temperature delivered to the process units.

4.4 Model validation with Operating Data

After developing the base (design) case of the simulation models they were validated against actual plant operating data. Model validation exercise was performed for each of the four simulation model building blocks.

- Cooling water Network – Cooling Water supply flow to Ethylene unit was not available for the actual plant operating case. Hence, total Cooling Water circulation rate was calculated so as to match the cooling water return header pressure with the cooling water supply pump suction pressure. Considering this calculated total cooling water circulation rate, flow distribution across plate heat exchangers (37 nos) was determined based on the Aspen Plus hydraulic simulation model.

- Seawater Network – Total seawater circulation rate was determined based on the heat balance of Cooling Water & Sea Water side. (Overall heat duty on cooling waterside was used to calculate seawater circulation rate). This calculated seawater circulation rate was input to the Aspen Plus model for plate heat exchangers and flow distribution across all plate heat exchangers (37 nos) was determined.

- Plate Heat Exchangers – PHE inlet flows of Cooling water & seawater predicted by the Aspen Plus simulation model were input to the individual PHE model in Aspen Plate (EDR). Thereafter all four temperatures for each PHE were matched with the actual plant operation data (tag values of respective temperature indicators) by adjusting fouling resistance of each PHE. Following graph presents a comparison of calculated fouling resistance versus its design value.

Figure 3 - PHE Fouling Resistance (Operating Case)

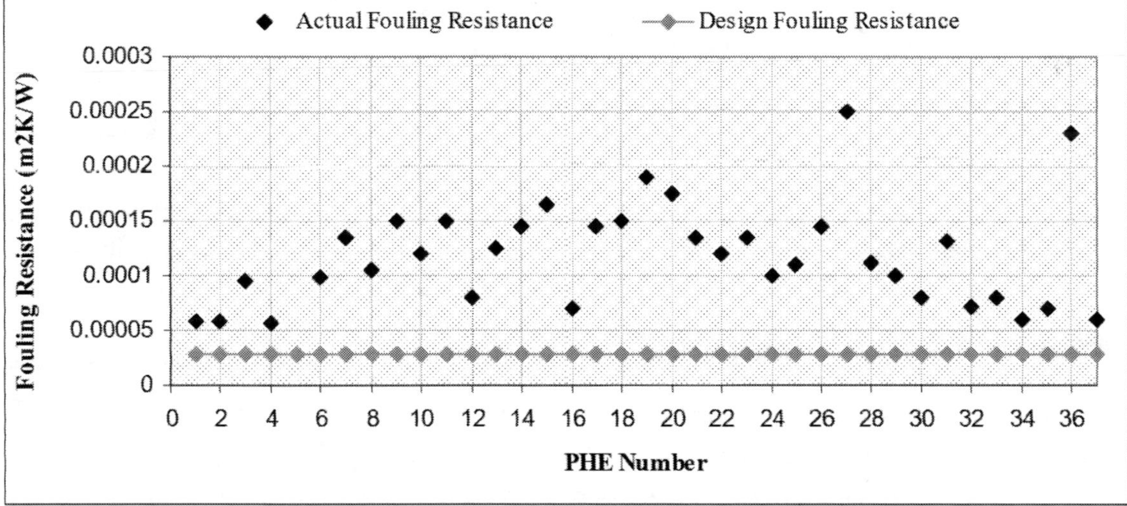

- Cooling Tower (seawater) – For operating case of the Cooling Tower simulation, maximum temperature recorded at the plant site was considered along with three different relative humidity conditions (Max-82.3%, Min-39.6% & Avg-70.5%). Using the data for dry bulb temperature & relative humidity respective wet bulb temperature was calculated based on the psychometric charts. For these wet bulb temperatures, cooling tower basin temperature (i.e. seawater supply temperature) was predicted using the MS Excel model of the Cooling Tower.

From sea water basin temperature predicted by the cooling tower model and calculated inlet flow of cooling water & sea water for each PHE, cooling water supply temperature to the process units was predicted by running plate heat exchanger (37 nos) simulation in Aspen Plate EDR. For this exercise, heat duty of the process plants as per individual plant operating data was considered. Sea Water basin temperature & Cooling Water supply temperature for each case predicted by the model are presented in the table below:

Table 1 – Operating Case – Simulation Results

Parameter	UOM	Maximum Temperature Reported at Plant Site		
Dry Bulb Temperature	°C	40	40	40
Relative Humidity	%	82.3	39.6	70.5
Wet Bulb Temperature	°C	37.0	27.6	34.7
Predicted basin Temperature	**°C**	**39.2**	**33.5**	**37.8**
Average Sea water $\Delta T_{(Plant\ Data)}$	°C	8.2		
Average cooling water $\Delta T_{(Plant\ Data)}$	°C	8.9		
Predicted CW supply temperature (PHE Outlet)	**°C**	**41.6**	**36.1**	**40.2**
Actual CW supply temperature	°C	40.6		
CW Return Temperature (PHE Inlet)	°C	50.7	45.2	49.3
PHE Outlet Sea Water Temperature	°C	47.3	41.7	46.0

4.4.1 Inference from the Operating Case Simulation

Following inferences were drawn from the results of model validation exercise for plant operating data:

- Cooling water and seawater hydraulic networks are adequately designed. Flow rate to the individual plate heat exchangers was also adequate.
- Possible fouling in the Plate Heat Exchangers or the seawater filters. Cleaning Seawater filters and plate heat exchangers periodically will help to improve the cooling water system performance.
- Cooling tower is designed for 41.6% relative humidity. However, on the day for which plant operating data was considered, average relative humidity reported was very high at 70%. This has resulted in higher seawater basin temperature, which in-turn resulted in higher cooling water supply temperature.

4.5 Analysis of Modification Options – Facilitating the Decision Making Process

After analyzing the actual plant operating data using the simulation model for the cooling water system, the simulation was further utilized to identify & evaluate possible solutions to mitigate the issue of high cooling water supply temperature.

During this phase of analysis the following modifications were employed in the simulation model:

- Sea water circulation rate was considered at 132,907 m3/hr (25% higher than the normal operation rate). This is the maximum allowable circulation rate with 3 quadrants operating in the Cooling Tower.
- All 37 plate heat exchangers were provided with 20% additional heat transfer area. 20% extra heat transfer area was corresponding to the maximum number of plates as per current frame-type of the exchangers.
- Total Cooling water circulation rate was calculated based on the hydraulic model so as to match the pump suction pressure. Calculated cooling water circulation rate was 99,502 m3/hr (around 6% higher than the normal operation).
- Both seawater & cooling water flow distribution (across all 37 PHE's) were determined based on the Aspen Plus hydraulic model. Subsequently, PHE thermal simulation runs were performed with these flow rates using PHE simulation model (in Aspen Plate EDR).
- As data (performance curves) for seawater-cooling tower was available only for one (design) circulation rate, simulation runs were carried out assuming three different approaches (viz.; 2°C, 3°C & 5°C) between the seawater cooling tower basin temperature & wet bulb temperature. For each of these approaches three different wet bulb temperatures corresponding to the maximum, minimum & average relative humidity were considered.
- Plate heat exchanger (37 nos) thermal simulation model runs were carried out for these nine different cases (three for each of the three cooling tower approach temperatures). Ambient conditions with maximum dry bulb temperature (40°C) were considered for this analysis.

Following table presents the simulation results. *Relative humidity, dry bulb temperature & total seawater circulation rate were inputs to the model. All other items in the table are outputs from the simulation model.*

Table 2 – Simulation Results: Analysis of Modification Options

Relative Humidity (%RH)	**Maximum**	**Minimum**	**Average**
	82.3	**39.6**	**70.5**
Wet Bulb Temperature	**37.0**	**27.6**	**34.7**
2°C Approach to Wet Bulb Temperature			
SW Basin Temperature	39.0	29.6	36.7
SW PHE Outlet Temperature	45.5	36.1	43.2
CWR Temperature (PHE Inlet)	49.6	40.4	47.4
CWS Temperature (PHE Outlet)	**40.8**	**31.6**	**38.6**
3°C Approach to Wet Bulb Temperature			
SW Basin Temperature	40.0	30.6	37.7
SW PHE Outlet Temperature	46.5	37.1	44.2
CWR Temperature (PHE Inlet)	50.6	41.4	48.4
CWS Temperature (PHE Outlet)	**41.8**	**32.6**	**39.6**
5°C Approach to Wet Bulb Temperature			
SW Basin Temperature	42.0	32.6	39.7
SW PHE Outlet Temperature	48.5	39.1	46.2
CWR Temperature (PHE Inlet)	52.6	43.4	50.4
CWS Temperature (PHE Outlet)	**43.8**	**34.6**	**41.5**

4.5.1 Inference from the Simulation & Possible Solutions

With the maximum seawater circulation rate (maximum possible with 3 quadrants operating in the cooling tower) and additional 20% heat transfer area in all 37 plate heat exchangers, cooling water supply temperature was still high at 38.6°C with cooling tower basin temperature approach (to the wet bulb temperature) of 2°C. Cooling water supply temperature was observed to further increase with increase in approach temperature (at 39.6°C for 3°C approach and 41.5°C for 5°C approach).

It was concluded from the operating case & network modification case simulation analysis that reduction in the cooling water supply temperature to the process units could be achieved by any of the following three approaches (or a combination of any of them):

- Providing for additional heat transfer area. (i.e. installing additional plate heat exchangers in parallel to the existing 37 exchangers).
- Modifications in the seawater cooling tower to reduce the basin temperature approach to wet bulb temperature.
- Addressing the issue inside individual plant battery-limits instead of modifying the cooling water system to reduce cooling water supply temperature.
- Considering the amount of capital expenditure required for modifying the common cooling water system (either plate heat exchangers or the cooling tower), it may be prudent to identify the limitations inside individual process / utility plant and address these issues inside individual battery-limits.
- A comparative evaluation of the possible economic benefits by addressing the issue of high cooling water supply temperature inside the battery-limits of process / utility units versus modifying the common cooling water system will be prudent and is expected to aid the decision making process further.

5. Case Study #2 – Trouble-shooting an Ethylene Plant Cooling Water Network

Client B was facing issue of high cooling water return temperature from the process gas compressor after-coolers. The problem became severe during summer months and started limiting the plant throughput due to high cooling water supply temperature. In order to address this issue, one of the options was to limit the cooling water return temperature in certain heat exchangers (process gas compressor after-coolers) to a maximum of 120°F. This could be achieved if the cooling water supply flow to the compressor after-coolers was increased. However, increasing the flow to process gas compressor after-coolers was going to reduce cooling water flow delivered to other sections of the plant cooling water network.

To evaluate & quantify how much increase in cooling water supply flow to the Process Gas Compressor after-coolers is possible, without negatively affecting the process due to reduced cooling water flow in other sections of the network, a simulation model was developed involving all ISBL & OSBL heat exchangers using cooling water as a cooling medium. Aspen Plus was used for developing the cooling water network simulation. Primary aim of developing the simulation was to perform various what-if scenarios so as to address the issue of high cooling water return temperature.

5.1 Developing the Simulation Model

Developing simulation model for the ethylene plant cooling water network, which could be used for performing various what-if scenarios, involved different stages as follows:

- First step was to extract all the required data from the cooling water piping isometric drawings; heat exchanger data-sheets and cooling water pump specification sheets.
- Based on these design data, a base case simulation was developed in Aspen Plus.
- Flow & pressure survey was performed across the plant cooling water network (using ultrasound flow measurement) and readings were taken at all practically possible locations.
- Based on the field survey results base case simulation model was reconciled by modifying resistance in various sections and/or exchangers in the cooling water network.
- Post reconciliation the simulation model predictions for flow & pressure matched (with reasonable accuracy) with the field survey results. The following table presents results from the reconciled simulation model:

Table 3 – Simulation Model Re-conciliation Results:

Exchanger Service	UOM	Flow Survey Reading	Simulation Result	Percentage Deviation
Propylene Refrigerant Condenser	GPM	29,900	29,833	0.2
PGC Surface Condenser		12,500	12,419	0.6
PRC/ERC Surface Condenser		17,400	17,414	0.1
Suspect Condensate Cooler		1,440	1,421	1.3
PGC 1st Stage Discharge Cooler		980	979	0.1
PGC 2nd Stage Discharge Cooler		815	816	0.1
PGC 3rd Stage Discharge Cooler		992	995	0.3
PGC 4th Stage Discharge Cooler		-	502	-
PGC 5th Stage Suction Cooler		-	636	-
PGC 5th Stage Discharge Cooler #1		1,592	1,569	1.4
PGC 5th Stage Discharge Cooler #2		1,258	1,266	0.6
Quench Water Cooler #1		6,020	6,092	1.2
Quench Water Cooler #2		2,870	2,881	0.4
Quench Water Cooler #3		3,150	3,211	1.9
ISBL Exchangers (Total Flow)		-	46,640	-
OSBL Exchangers (Total Flow)		2,000	2,060	3.0
Total CW Pump Discharge Flow		-	48,700	-
CW Pump Suction Pressure	PSIG	2.9	2.9	0.0
CW Pump Discharge Pressure		-	82.4	-
CW Return (Riser) Pressure		11.0	10.9	0.6

5.2 Utilizing the simulation model

Once the simulation model was reconciled using measurements taken during the flow survey, it was ready to be utilized as a tool.

In the reconciled simulation model, increased resistance (additional pipe length) was added to emulate throttling of cooling water inlet valves for various heat exchangers in the network. Similar approach was followed to indicate possible fouling in the piping segments. Following this approach, the cooling water network model was utilized to:

- Analyze various operating scenarios for the cooling water network
- Identify the sections in cooling water network which are deficient of cooling water flow
- Identify fouling in the heat exchangers and piping sections
- Observe effect of various possible modifications (pipeline, fittings, geometry change; etc) in the cooling water circuit on overall distribution of the cooling water flow
- Optimize cooling water distribution in order to maximize ethylene production

In addition to the above-mentioned analysis, the Cooling Water Network Simulation Model was also used to analyze a few cases of plant capacity expansion. The simulation was run considering cooling water pump discharge flow equivalent to the expansion flow rate (around 20% higher flow). With higher pump discharge flow, the cooling water return pressure (at the cooling tower riser) reduced considerably below the minimum required pressure of 13 psig. Remaining within the constraints of delivering the necessary cooling duty to the process & maintaining the cooling water return pressure at ~ 13 psig, following cases were evaluated using the simulation model:

- Throttling the cooling water inlet valves for those exchangers, which were having higher cooling water supply flow than required
- Reducing the number of tube-passes in process gas compressor after-coolers. In this case various permutations were analyzed by changing the geometry of either one or multiple after-coolers to satisfy the above-mentioned constraints
- Running an additional cooling water supply pump, which was used as a stand-by in normal plant operation
- Changing the pump impeller to get higher discharge head. Using the new impeller in either one or multiple pumps

This way the cooling water network simulation model could be utilized as a tool to optimize the plant operation during the normal run and also proved to be useful in evaluating adequacy of various options for plant capacity expansion.

6. Conclusion

Cooling water network is one of the vital utility networks in any process unit. With advances in process technology & design cooling water networks are growing in size and complexity. However, a humble cooling water network usually does not receive due attention in comparison to other critical equipment in the plant. Problems in the cooling water network are highlighted during start-up or plant expansion phase. Developing a simulation of the plant's cooling water network can provide a useful tool to process engineers in order to trouble-shoot and optimize the plant operation.

Utilizing a process simulation software can provide added advantage over the conventional flow modeling software by allowing for inclusion of heat transfer effects in the simulation model. It enables the user to observe the effect of changes in cooling water network parameters on the process and evaluate / quantify them.

In today's highly competitive market, one has to ensure that each unit of raw material going in the process unit is utilized fully and is most efficiently converted to high value products. In this context, trouble-shooting and optimizing the plant operation becomes utmost important. A simulation model of the plant's cooling water network, which can be used as a tool to perform various what-if scenarios, can certainly aid and strengthen the efforts of a process engineer to gain an edge in optimizing the overall plant performance.

7. References

- K. Mohanraj, "Need For Analyzing Cooling Water Networks", *Chemical Industry Digest,* March 2011

(126a) Proesa® Technology: The Platform to Enable Biomass Based Industry

Dario Giordano, Chief Technology Officer, Beta Renewables spa, Tortona (AL), Italy and **Delane Richardson**, Chemtex International, Wilmington, NC

Mankind is still strongly dependent on fossil fuels. As developing economies continue to grow, concerns are increasing about climate change and the potential economic and political impact of peak oil. To address these issues and reduce its dependency on fossil fuels, the EU has adopted measures to encourage the production and use of renewable resources for next generation bio-fuels.

The development of the bio-fuels industry is a strategic objective set by the Kyoto Protocol. The target imposed by the European Union requires that 10% of biofuels be of biological origin by 2020.

The development of the biofuels industry is a strategic goal and a real opportunity for the technological development in the Italian territory. Biofuels can provide a real solution to GHG reduction, and can reduce the Italian dependence from imported fossil fuels. Bioethanol, in particular, is the most common worldwide biofuel. The second-generation technology allows the production of bioethanol from lignocellulosic biomass avoiding the use of crops for food and reducing GHG emissions. It's is our Technology for the production of bioethanol from lignocellulosic biomass (second generation bioethanol).

In this scenario, M&G since 2007 has invested over 200M$ in research and development of a sustainable, scalable and simplified technology for the transformation of virtually any kind of lignocellulosic feedstock into low-cost, fermentable sugars for the production of bioethanol or biochemicals and it's now technology excellence to convert biomass into sugars.

Is a holistic solution that starts in the field and end with bioproducts and is a complex undertaking requiring integration of different sciences ranging from biochemistry and biotechnology, to agronomy and botany, from engineering to chemistry, from technology development to process modeling and from biomass supply chain logistic to product take off. This technology process is based on four main steps: pre-treatment, hydrolysis, fermentation and distillation.

The first step is an energy efficient pretreatment process carried out without chemical addition but using only water and vapor. Research and development activities are constantly working on the improvements of enzymatic hydrolysis as well its influence on fermentation step condition. At present some important achievements have been reached especially for what concerns reactor design and scale-up at pilot level and best enzyme package. Efficient enzyme cocktail have a key role for feasible bioethanol production from lignocellulosic material. Fermentation of sugar to ethanol is a well know process, but improvement of flexible process of fermentation were focused on adapting the process to the strain rather than developing a new strain in biological process. After final distillation step was obtained a "green bioethanol" that respect the standards for fuel. On the energy side the plant has been designed as an energy efficient process through optimization of steam consumption, energy–integrated process steps and above all the possibility to produce steam and electricity from by-product. This aim was achieved thanks to research activities on process development and optimization of adopted technology.

This process is flexible in feedstock supply, innovative in technological solutions and energy-integrated.

The Technology has been demonstrated on a continuous, 1 T/day pilot plant since June 2009, and is now being brought to full scale thanks to the construction of a 40'000 ton/yr industrial plant.

Composite Fuels Production for Combustion

ABSTRACT

Goal of the Project

The primary goal of the project is the creation of composite fuels for combustion processes containing biomass, coal, coal residues, crude oil processing residues, and other combustible material for effective firing in power plants. The project attempts to provide screening equations for the effective combustion of composite fuels and develop the combustion characteristics for effective burning. In addition a model for combustion of composite fuel emulsions will be proposed.

Modeling Composite Fuels

The literature is full of data for composite fuels but most include experimental values and not composite equations. Mixtures of small wood particles and heavy oil have been shown to improve combustion efficiency and lower some exhaust emission concentrations (Baojun). We have started with wood-oil and coal-oil composite fuels and will examine other potential fuels including mixtures containing wood, coal, biofuels mixed with an oil/water emulsion. Figure 1 depicts a visual model of a composite mixture of wood and oil. The oil film surrounds a wood particle.

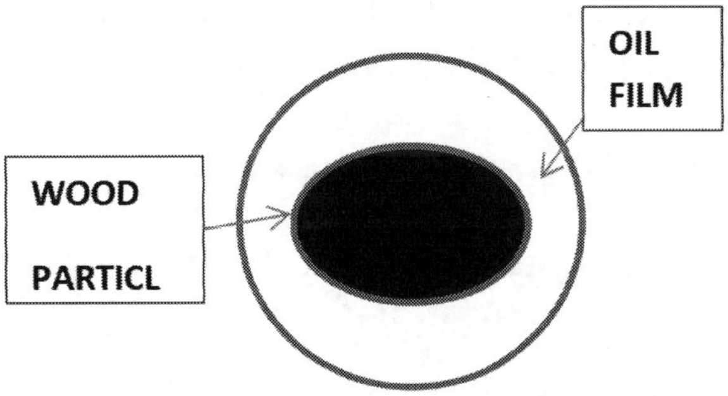

Figure 1. Visual Model: Oil surrounds Wood Particle

For our physical modeling we first define the steps of combustion for the solid-oil composite fuel mixture fed to a furnace. The solid particles will be in the range of diameters from 50-200 μm. The furnace operates in the steady state. The steps are:

1. Furnace is operating at 1000 F
2. Feed mixture is heated
3. Oil vaporizes and forms a combustible gas
4. Fuel gas undergoes combustion
5. Wood particle releases water
6. Wood particle releases volatile combustible gas
7. Wood gas undergoes combustion
8. Char undergoes combustion

Steps 2, 3 and 4 have a residence time of a small fraction of a second. Steps 5, 6, and 7 have a total residence time on the order of less than 2 seconds. The final step involving the combustion of char has a residence time depending on the amount of carbon in the wood. If a mixture is starting from room temperature and the combustion is time dependent then the total residence time can be on the order of 16 seconds for a torrefied wood water slurry or for a torrefied wood oil/water slurry (Vu and Bhaitacharya).

Coal and oil composite fuel mixtures follow a similar set of steps with a similar set of residence times. Ultimate analysis of coal results in around 60% carbon while wood has around 50% carbon. A portion of this carbon provides the char for the final step of combustion of the composite fuel

Experiments

Settling experiments were performed at two different temperatures about 24C and about 66C to understand the effect of temperature on obtaining relatively homogeneous mixtures. Anthracite coal-oil mixtures and pine wood-oil mixtures have been studied. The oil used was soybean oil. The density of these components at 25C is; anthracite coal 1.3 gm/cm^3, white pine wood 0.48 gm/cm^3, and soybean oil 0.92 gm/cm^3.

The normalized settling curves follow the equation

$$H(t) = 1.0 * e^{-\frac{t}{T}} \quad (1)$$

Where T is the time constant for exponential settling and the time where the solids have dropped to 37% of the original height. The settling data provided in Tables 1 and 2. The T is the time constant for settling.

Anthracite coal-oil mixtures and pine wood-oil mixtures have been studied. The oil used was soybean oil. The density of these components at 25C is; anthracite coal 1.3 gm/cm^3, white pine wood 0.48 gm/cm^3, and soybean oil 0.92 gm/cm^3.

Table 1. Experimental Parameters and Mass Fractions for Coal/Oil/Water Experiments and Time Constants for Exponential Settling

Coal Particle Size Range (mm)	Mass % Coal	Mass % Oil	Mass % Water	T (at Room temp), min	T (at elevated temp), min
0.5-1.0	0.319	0.681	0.000	2.717	2.717
0.5-1.0	0.255	0.543	0.203	9.091	12.658
0.297-0.5	0.329	0.671	0.000	2.717	2.717
0.297-0.5	0.263	0.537	0.201	3.876	5.848
0.178-0.25	0.318	0.682	0.000	1.730	3.165
0.178-0.25	0.252	0.542	0.206	7.874	14.925
0.075-0.125	0.313	0.687	0.000	55.556	40.000
0.075-0.125	0.253	0.553	0.194	142.857	111.111

Table 2. Experimental Parameters and Mass Fractions for Wood/Oil/Water Experiments and Time Constants for Exponential Settling

Average Wood Particle Diameter (mm)	Mass % Wood	Mass % Oil	Mass % Water	T (at Room temp), min	T (at elevated temp), min
0.001	0.098	0.902	0.000	8.696	6.536
0.001	0.048	0.736	0.217	7.407	7.407
0.00001	0.061	0.939	0.000	18.182	10.526
0.00001	0.077	0.708	0.215	111.111	13.889

The emulsions with approximately 20 % water in the mixture and markedly slowed down the settling time and kept the wood in suspension.

Modeling Equations

Viscosity, density, heating value, time for vaporization, exchange coefficient, temperature of combustion, sulfur content, burning rate constant, thermal diffusivity and flame temperature, and other parameters can be modeled for all these composite fuels.

The linear format of a three component mixture that can be used to calculate some properties is given by

$$MP = \alpha_1 * P_1 + \alpha_2 * P_2 + \alpha_3 * P_3 \quad (2)$$

where α is a weighting coefficient and P is the single component property of interest.

Figure 1 presents the viscosity of coal-oil mixtures (Timbalia). It has an exponential increasing form. An interesting thing to note is that the viscosity of small amounts of coal is less than the liquid oil viscosity of 0.65 Pa-s. The curve in Figure 1 can be fit to a relatively high accuracy with the equation

$$\mu = 0.01e^{15.27x} \quad (3)$$

This can be used as a model to calculate wood-oil properties.

Nonlinear equations of the form for the time of evaporation can be used as weighting factors for the appropriate model equation (Spaulding).

$$t_{vap} = \frac{D_o^2 \rho_{liq}}{8\Gamma_{vap} \ln\left(1 + \frac{m_{vap,o} - m_{vap,\infty}}{1 - m_{vap,o}}\right)} \quad (4)$$

Data of Timbalia have been fit to determine the relationship between viscosity and the mass fraction of coal. This fit is presented in figure 1 and can be used to determine residence times in steady state combustion in furnaces.

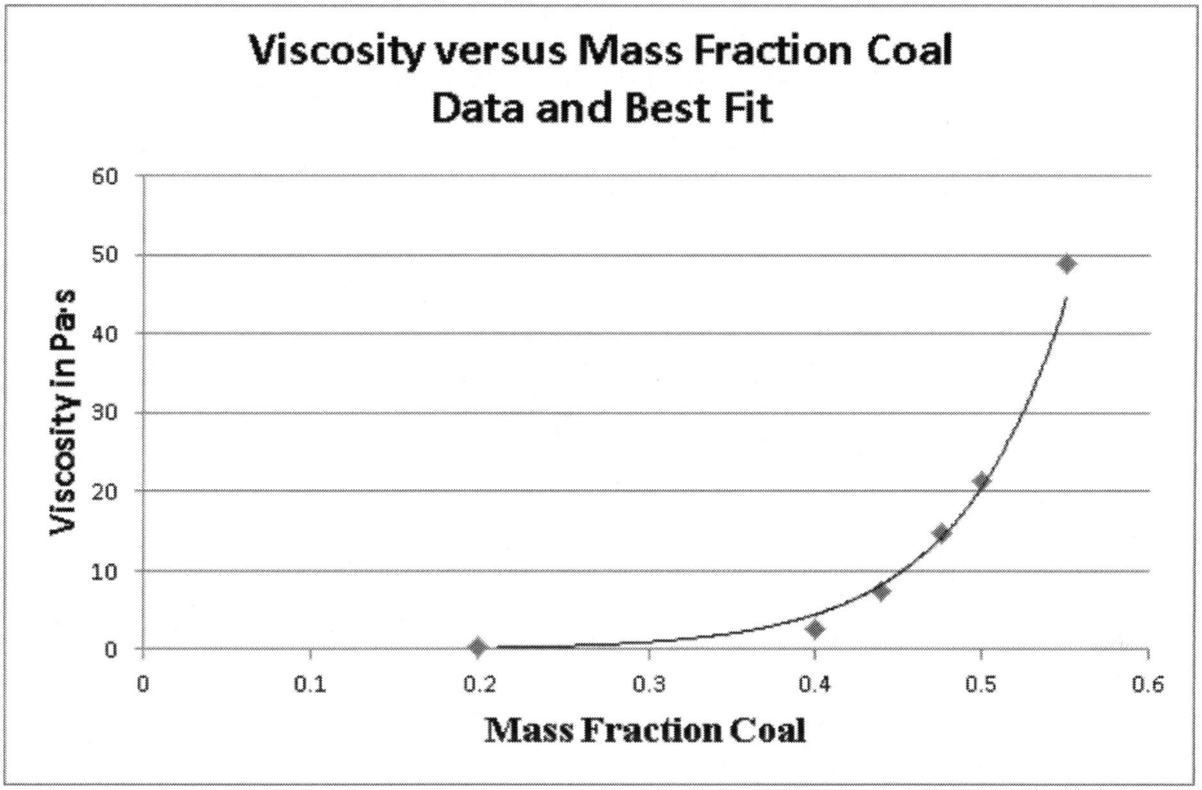

Figure 2. Viscosity of number 8 coal and number 6 oil-mixtures at 25C. (Timbalia)

Forming Emulsions

Stability of these emulsion mixtures will be dependent upon the energy used to create the emulsions. Hydrocavitators have been shown to be able to add the energy required to form relatively stable emulsions (Knickle). Special design of burner nozzles can be produced to take and effectively use these emulsions. Droplet size is important for efficient combustion. The optimization of the energy to create the emulsions will be able to maximize efficiency for these emulsions (Knickle). For composite fuel emulsions we can define an ideal droplet model as one containing an oil and water emulsion surrounding a solid such as a small particle of wood or coal. The solid particle diameter lies in the range of 50 to 200 micrometers. The optimum size of these model droplets will determine the efficiency of combustion of these composite fuels. Large droplets of higher viscosity generally burn less efficiently, while small drops of lower viscosity create a fine non-dispersive spray which also burns inefficiently because it does not mix well with the air flow and burns closer to the nozzle and not in the ideal burn zone.

A visual model for combustion of a wood oil water composite fuel is displayed in Figure 3. The oil/water emulsion surrounds the small wood particle.

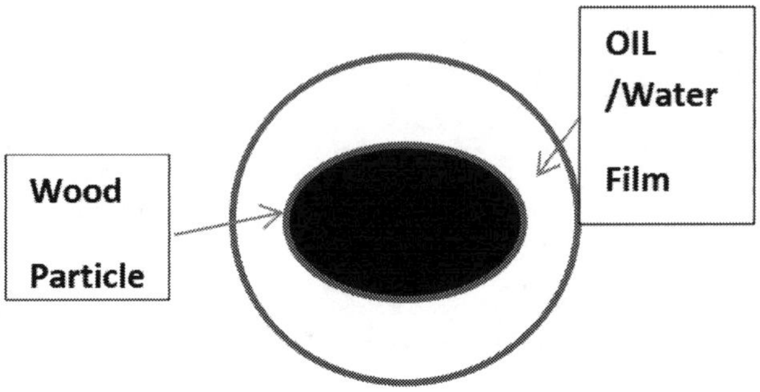

Figure 3 Visual Model: Oil/Water Emulsion Surrounds Wood Particle.

For our physical modeling we first define the steps of combustion for the wood oil/water emulsion composite fuel mixture fed to a furnace. The wood particles will be in the range of diameters from 10-200 μm. The furnace is assumed to operate in the steady state. The steps for the combustion of this composite mixture are:

1. Furnace is operating at 1000 F
2. Feed mixture is heated
3. Micro-explosions of the oil/water take place
4. Smaller droplets are created
5. Steam droplets are generated
6. Oil vaporizes and forms a combustible gas
7. Fuel gas undergoes combustion
8. Wood particle releases water
9. Wood particle releases volatile combustible gas
10. Wood gas undergoes combustion
11. Char undergoes combustion

The micro-explosions act to prevent the agglomeration of the solid in the composite fuel mixture.

Conclusions

Emulsions can last long enough to provide for combustion in a furnace providing heat and generating steam. Twenty percent water in the composite fuels is near an optimum value. Minimum energy input is required to form stable emulsions with coal and wood solids. Simplified modeling for composite fuels have been suggested including the properties of the composite fuels and the combustion of composite fuels.

Future Work

Work is continuing on the development of the modeling equation and their justification.

References

Knickle, H. et al "Innovative Hydrocavitation Technology Process to Create Composite Fuels" AIChE Spring Meeting, Session 141, Houston TX (2012)
Spaulding, D. B. "Combustion and Mass Transfer", (1979)
Timbalia, A. Master's Thesis-Ohio University June 1981 "Solvent Refined Coal and Coal-Oil Mixtures"

Vinh, Vu Quang and S. C. Bhaitacharya, "A Study on the Combustibility of Torrefied Wood Oil/ Water Mixture Droplets", Int. J. of Energy Research, Vol. 20,215-236 (1996)

Baojun, Wang et al "Combustion Characteristics of Wood Powder and A-heavy Oil Mixture Fuel", Proceedings of Thermal Engineering Conference, Pgs. 69-70 (2005)

Appendix

Typical parameters of conventional fuels are presented in Table A1.

Table A1 Combustion Parameters of Conventional Fuels[b]

	Temperature Combustion,	Moisture	Volatiles DB	Ash DB %	Excess Air	Form of	Energy (avg,)	Residence Time, s

	F[e]				Factor, wt.	Feed	Btu/lbm	
Methane	1475	small	100	small	1.05-1.1	Gas	24,000	small
Oil	1000	small	100	small	1.05-1.2	Liquid	16,500	small
Wood	1100-1500	To 50%	80%	1%	1.4-1.6	Particles	8,600	2-4
Coal	1000	To 20%	6-36%[c]	5-15	1.2-1.6	Particles[a]	8,000-14,000	2
Other[d]								

a) Typical values are 50-200 μm
b) Properties in this Table are representative but can have a large variation
c) Balance of heat from char
d) Wide variation in bio-matter
e) Depends greatly on oxygen quantity

(126c) Alcohol to Jet Fuel

Geoffrey Fichtl, Renewable Energy and Chemicals, UOP - Honeywell, Des Plaines, IL

In the interest of energy independence and reduction of greenhouse gas emissions, alternative fuels derived from renewable rather than fossil sources is a major topic of R&D within private industry, universities, and government. The Alcohol to Jet (AtJ) pathway is of major interest recently, especially as routes to higher-carbon bio-alcohols, such as bio-butanol, have become commercially viable. In the standard AtJ pathway, dehydration of bio-alcohol (usually bio-butanol) yields the versatile butene platform molecule, which can be oligomerized to form pure olefins that boil in the jet range. Upon further hydrogenation by well-established hydrotreating technology, the olefins are converted into stable paraffins with high yield of jet-boiling product.

(126d) Transportation Fuels From the Catalytic Hydrodeoxygenation of Biomass Pyrolysis Oil

Lance Baird[1], Tim Brandvold[2], Steve Lupton[2], Tom N. Kalnes[3] and Stanley Frey[4], (1)UOP LLC, A Honeywell Company, Des Plaines, IL, (2)UOP LLC - a Honeywell Company, Des Plaines, IL, (3)Research and Development, UOP LLC, Des Plaines, IL, (4)Renewable Energy & Chemicals, UOP LLC, Des Plaines, IL

Transportation Fuels from the Catalytic Hydrodeoxygenation of Biomass Pyrolysis Oil

L. Baird, T. Brandvold, S. Lupton, T. Kalnes, S. Frey

UOP LLC, A Honeywell Company

Des Plaines, IL USA

Thermal pyrolysis is an efficient process for converting solid biomass to a transportable, renewable liquid fuel. This fuel can be burned directly in boiler applications and for generating electricity. However, the pyrolysis oil has certain characteristics that limit its utility as a transportation fuel. In particular the high level of organic oxygenates, high water content and high acidity make the fuel incompatible with existing distribution infrastructure and conventional automobile or aviation engines.

In 2008 Honeywell's UOP LLC, the world's leading refining and petrochemicals process technology company, and Ensyn Corporation, the world's leader in commercial biomass pyrolysis technology, formed a joint venture, Envergent Technologies LLC, to commercialize the production of pyrolysis oil for energy applications using RTP™ Rapid Thermal Processing technology. In addition to delivering the technology to existing heat and electricity markets, the joint venture has been pursuing upgrading of biomass pyrolysis oil to transportation quality fuels.

One route for pyrolysis oil upgrading is hydroprocessing. Biomass is pyrolyzed via the RTP process to produce pyrolysis oil which is then catalytically hydrogenated to a hydrocarbon fuel having low oxygen, water and acid,content compatible with existing fuel infrastructure. Proof-of-principle validation for this approach to pyrolysis oil upgrading has been previously reported.

In this presentation we will discuss progress towards development and commerciallization of the integrated biomass to transporation fuels technology, including characterization of the products and Life Cycle Analysis.

DETAILED HYDROCARBON CHARACTERIZATION OF DIESEL FRACTIONS FROM CO-HYDROPROCESSING CANOLA OIL AND HVGO BLENDS

Jinwen Chen, Rafal Gieleciak, Hena Farooqi
CanmetENERGY, Natural Resources Canada
One Oil Patch Drive, Devon, AB, T9G 1A8, Canada

1. Introduction

Reducing the carbon footprint of industrial processes used to convert Canadian oil sands crudes into clean transportation fuels is a major challenge for upgraders, refiners and the R&D community. Co-processing biomass with petroleum/bitumen derived feedstocks may provide an alternative solution since the GHG emission from the production and processing of renewable biomass to produce biofuels is considered to be significantly lower than fuels derived solely from fossil fuel resources (Keesom et al., 2009; Sheehan et al., 1998). In addition, co-processing paraffinic biomass feeds and aromatic bitumen feeds may generate synergy to improve gasoline and diesel qualities.

The detailed hydrocarbon characterization of petrochemical samples exemplifies one of the most challenging tasks currently faced by separation science. Accurate analytical separation and speciation are necessary to determine what compounds and/or isomers are unaltered and at what concentration levels they are present. For decades, one dimensional gas chromatography (1D-GC) equipped with various detectors had been the preferred technique for the detailed analysis of hydrocarbons, sulfur and nitrogen compounds in petroleum products. However, detailed separation cannot be obtained by using this conventional GC technique because of the lack of chromatographic resolution. The number of hydrocarbon isomers increases exponentially with the increase in the number of carbon atoms, which limits the speciation capability that can be provided by 1D-GC for samples with more than 9 carbon atoms (Vendeuvre et al., 2007).

Comprehensive two-dimensional gas chromatography (GC×GC) offers a huge potential for examination of complex mixtures such as petroleum products due to its increased chromatographic resolving power. Moreover, this method has improved detection limits due to the focusing of analyte during the modulation process (Gieleciak et al., 2010; Adahchour et al., 2008; Adam et al., 2007; Hua et al., 2003; Blomberg et al., 1997).

The hydroprocessing of vegetable oil and petroleum feedstock blends has been studied by several research groups (Templis et al., 2011; Šimáček and Kubička, 2010; Tóth et al., 2010; Bezergianni et al., 2009; Walendziewski et al., 2009; Watkins et al., 2009; Huber et al., 2007; Viljava et al., 2001; Furimsky, 2000; Wong and Hong, 1999). Most of the published studies have focused on hydroprocessing performance rather than detailed analysis of the diesel products that result from co-processing petroleum and biomass feedstocks. In this study, twenty diesel samples were obtained by co-hydroprocessing low-grade canola oil and bitumen derived heavy vacuum gas oil (HVGO) with different canola oil /HVGO blending ratios, 0/100, 5/95, 10/90, and 20/80 under various reaction conditions. The two-dimensional GC×GC methodology was used to characterize these samples and provide an accurate and simultaneous determination of the hydrocarbons derived from both canola oil and HVGO.

2. Experimental

2.1 Feedstock description

The feeds used in this study were blends of bitumen derived HVGO and low-grade raw canola oil (Milligan Bio-Tech, SK, Canada) with 3 different mass blending ratios (HVGO/canola oil), 95/5, 90/10, and 80/20. For comparison, pure HVGO was also tested under the similar operating conditions.

2.2 The reactor system

A bench-scale reactor system was used for the experiments. The system consisted of a single tubular reactor, a high-pressure phase separator, and a product stabilizer. Hydrogen gas and liquid oil co-currently flowed upwards through the tubular reactor. The reactor tube, 100 cm in length and 9 mm in diameter (ID), was heated with a four-zone electric furnace. The liquid product from the stabilizer went to a collector while the gases leaving the separator and the stabilizer were combined and directed to the vent, from which a small slip-stream was analyzed by an online gas analyzer (HP5880 GC).

A commercial $NiMo/Al_2O_3$ hydrotreating catalyst (extrudes of 1.6 mm in diameter and 5 mm in averaged length) was used. A total of 20 ml of catalyst, with 20 ml of 0.2 mm diluting glass beads, was loaded in the reactor. Prior to the planned runs, the fresh catalyst was sulphided and then stabilized by processing the feed for about 170 hours to get relatively stable activity. Slight catalyst deactivation was observed and monitored by conducting check-back runs periodically at the same operating conditions.

The operating conditions (Table 1) used for the experiments were typical for hydrotreating/mild hydrocracking and covered a relatively wide range of temperature, pressure, and liquid hourly space velocity. For all experiments gas/oil ratio was fixed at 8000 NL/L.

Table 1. Hydroprocessing parameters

Case	Sample ID	Temperature (°C)	LHSV (h^{-1})	Pressure (psig)	Canola Oil (wt%)
1	1 - 3	360	1.5	1444	0/5/10
2	4 - 6	375	1.5	1444	0/5/10
3	7 - 10	385	1.5	1444	0/5/10/20
4	11 - 14	395	1.5	1444	0/5/10/20
5	15 - 17	395	1.0/2.0/2.5	1444	5
6	18 - 20	395	1.5	1155/1299/1589	5

2.3 Diesel fractions

The purpose of this study is to investigate the impact of bio-derived oils on the hydrocarbon composition, and indirectly, on the quality of diesel fractions obtained from co-processing of low-grade canola oil with HVGO. To obtain the desired diesel fraction (200 – 343°C), spinning band distillation was performed with the total liquid product (TLP).

The feeds and the products were analyzed for simulated distillation (SimDis), density, sulphur and nitrogen contents, and hydrocarbon type composition using ASTM methods and other in-house developed methods (Chen et al., 2003).

2.4 Comprehensive two-dimensional gas chromatography (GC×GC)

The diesel fractions were analyzed by two-dimensional gas chromatography (GC×GC). The GC×GC experiments were performed on two separate Agilent 6890 instruments (Agilent Technologies, Mississauga, ON, Canada) equipped with liquid nitrogen-cooled quad-jet cryogenic modulator (Leco Instruments, Mississauga, ON, Canada). Detection was performed by sulfur chemiluminescence detector (SCD) operated in tandem with flame ionization detector (FID). The GC×GC systems were equipped both with 'normal' and 'reversed' column set combination. In the former case the first column is non-polar and the second one is polar. In the 'reversed' mode the first column is polar and the second one is non-polar. The features and operating conditions for both 'normal' and 'reversed' GC×GC setup are listed in Table 2.

Table 2. Chromatographic conditions for 'normal' and 'reversed' GC×GC experiments

Sample Description	'normal'- GC×GC	'reversed'- GC×GC
1st column	DB-PONA, 50 m x 0.20 mm DF:0.5	VF17-MS, 30 m x 0.32 mm DF:0.15
Main oven program	50°C (5) to 270°C (15) at 2°C/min	40°C (5) to 350°C (1) at 2°C/min
2nd column	Solgelwax, 2.0 m x 0.1 DF:0.1	RTX-5, 1.5 m x 0.18 DF:0.2
Secondary oven program	20°C offset from main oven	20°C offset from main oven
Split ratio / Injection size	200:1 / 0.1 μL	300:1 / 0.1 μL
Carrier gas	He, constant flow, 1.2 mL/min	He, constant flow, 1.5 mL/min
Modulator temperature	60°C offset from main oven	70°C offset from main oven
Detector	FID, 300 °C with SCD adapter, 800 °C	FID, 350 °C
Modulation period	6 s	10 s

3. Results and Discussion

3.1 The GC×GC -FID hydrocarbon composition

Figure 1 shows 'normal' GC×GC -FID chromatograms of the diesel fraction obtained with pure HVGO (a) and a mixture of 80% HVGO and 20% canola oil (b) under condition specified in Table 1 (Case 4). Peaks are displayed as spots in the retention plane described as the first dimension retention time on the x-axis and the second-dimension retention time shown on the y-axis. Signal intensity is illustrated by the color intensity. Light gray color represents the baseline and dark gray/black color represents the most intense peaks in the chromatogram. Inserts placed in the upper left corner of each map show three-dimensional view of GC×GC separation. A summed signal on the second dimension, simulating a real 1D-GC chromatogram, is presented in the background of the 3D chromatograms. Comparing the 1D-GC and 2D-GC chromatograms it is easily to conclude that most likely two compounds co-eluting in the first separation can be separated in the second dimension.

Figure 1 shows one of the most important features of GC×GC separation - the structured nature of chromatograms, where compounds are arranged according to their chemical group and to their number of carbon atoms. The presence of ordered structures in GC×GC plots allows simple visual identification of the chromatograms. Co-hydroprocessing canola oil/HVGO mixture results in conversion of triglycerides to saturated hydrocarbons, mostly heptadecane and octadecane and to less extent pentadecane and hexadecane, as shown in Figure 1b.

To enhance visual aspects of GC×GC chromatograms, peak information (position and area) after preprocessing with ChromaTOF software was further processed in MATLAB® software

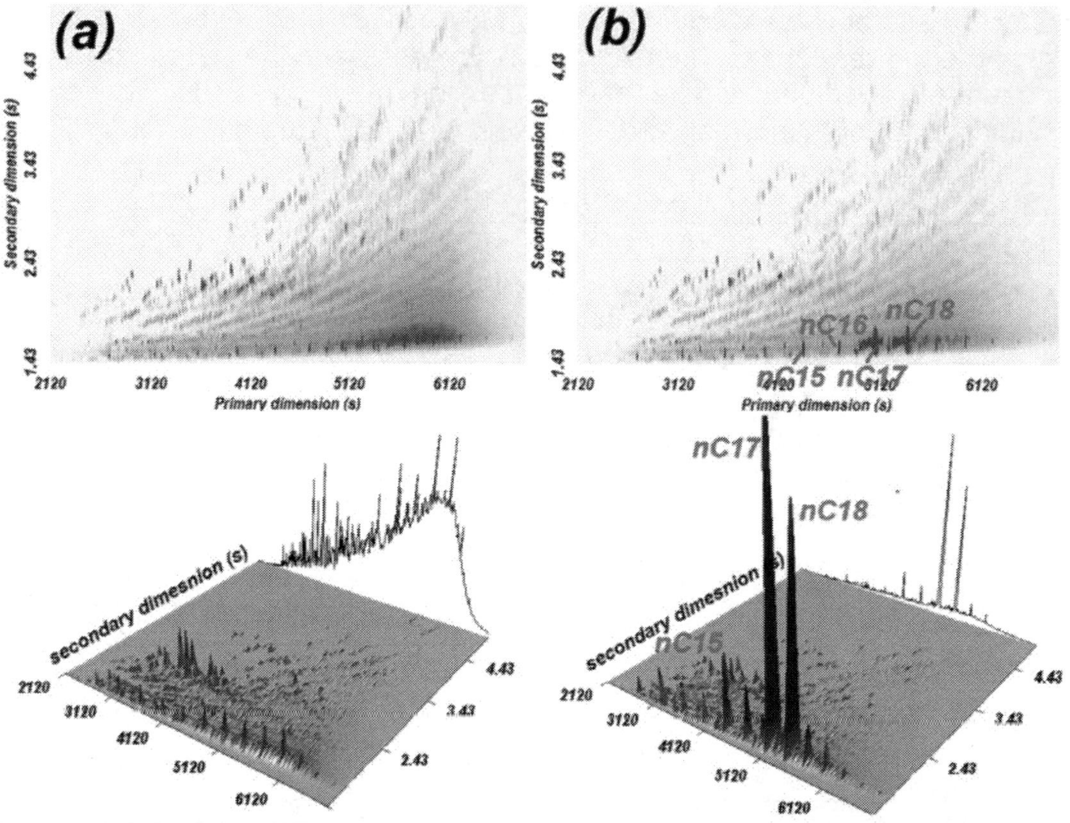

Figure 1. 'Normal' GC×GC – FID chromatograms of diesel fractions obtained by co-hydroprocessing (Table 1, Case 4) of HVGO (a), and 80% HVGO and 20% canola oil blend,(b)

(MATLAB, Release 2008b). The first dimension time was converted into a temperature scale using a correlation established between boiling point of n-paraffins and their retention time. This transformation allowed for presentation of 'normal'- GC×GC-FID chromatograms in the ASTM D2887 distillation temperature domain. The second separation is governed by polarity, which for these fuels means the number of aromatic rings. In addition, peaks found in chromatograms are presented in bubble plot form, in which the size of the bubble and its color are correlated to the compound concentration and relevant hydrocarbon type group, respectively. Figure 2 presents the upgraded version of GC×GC chromatogram of the diesel fraction from Figure 1b. The SimDis curve obtained from ASTM D2887 analysis was superimposed on the chromatogram as dashed magenta line. The black dashed vertical lines on the graph show the SimDis cuts at 10, 50, and 90%, respectively. Under the bubble plots, (right above the x-axis), there are indications of n-paraffinic carbon number labeled in green color.

The simulated distillation by gas chromatography (ASTM 2887) reported for petroleum or petroleum/vegetable oil blends is normally used for determining the volatility behavior and processability of refinery streams. The SimDis curves are usually characterized by several flat parts in their profile. For example, the SimDis profile (the magenta dotted line) presented in Figure 2 demonstrates that a significant amount of hydrocarbons have boiling points at around 305 °C and 320°C, which correspond to n-heptadecane and n-octadecane, respectively. This information can only obtained with GC×GC compositional data superimposed on SimDis results.

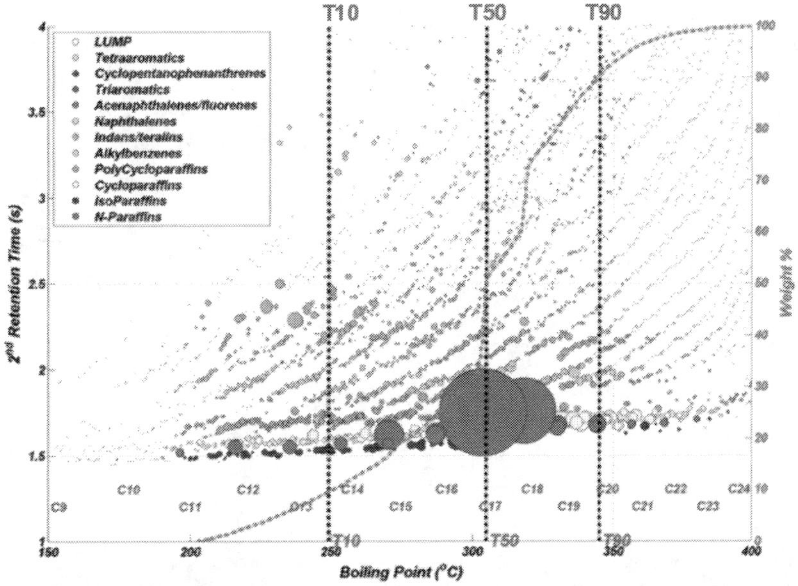

Figure 2. Bubble plot (after post-processing Figure 1b) as a function of polarity vs. boiling point for the diesel fraction obtained by co-hydroprocessing (Table 1, Case 4) 80% HVGO and 20% canola oil blend.

In addition to the 'normal' column setup, diesel fuel samples were subjected to GC×GC experiment with 'reversed' column setup (see Table 2). Figures 3a and 3b present 'reversed' - GC×GC-FID chromatograms of the diesel fractions obtained with pure HVGO, and the 80% HVGO and 20% canola oil blend, respectively, under the conditions specified in Table 1 (Case 4). The hydrocarbon groups are also observed in the 'reversed' GC×GC chromatograms, but in "reversed" order in comparison to that obtained with 'normal' column setup. For instance paraffins, which are less polar than aromatics, are located at the top part of the chromatograms. The 'reversed' column setup has been shown (Vendeuvre et al., 2005, Cannella et al., 2013) to improve the separation between saturates, olefins/cycloparaffins and aromatics; and potentially to increase capacity utilization. The use of the 'reversed' column setup leads to extended cross-border and thus, improved separation between saturates and aromatics as well as paraffinic and cycloparaffinc hydrocarbon types, which allows more accurate integration for these hydrocarbon types.

Despite unquestionable advantages of using 'reversed' system, some disadvantages are also present, such as the loss of correlation between retention time and boiling point, and relatively poor separation between aromatic hydrocarbons types in comparison to the 'normal' column setup.

As expected, the selected column combination (Table 1) provided good separation of individual compounds. Seven hydrocarbon classes (n-paraffins, isoparaffins, cycloparaffins, alkylbenzenes, indans/tetralins, naphthalenes, and acenaphthalenes/fluorenes) were identified and quantified using the FID detector. Averaged 'normal' and 'reversed' GC×GC hydrocarbon compositions for each case in Table 1 are presented in Figure 4, where the size of the bubbles and their color intensities are correlated to the total hydrocarbon class concentration. As seen in Figure 4, all the diesel fractions mainly consist of n-paraffins, cycloparaffins and indans/tetralins; the proportions of these fractions change with different HVGO/canola oil ratios. For instance,

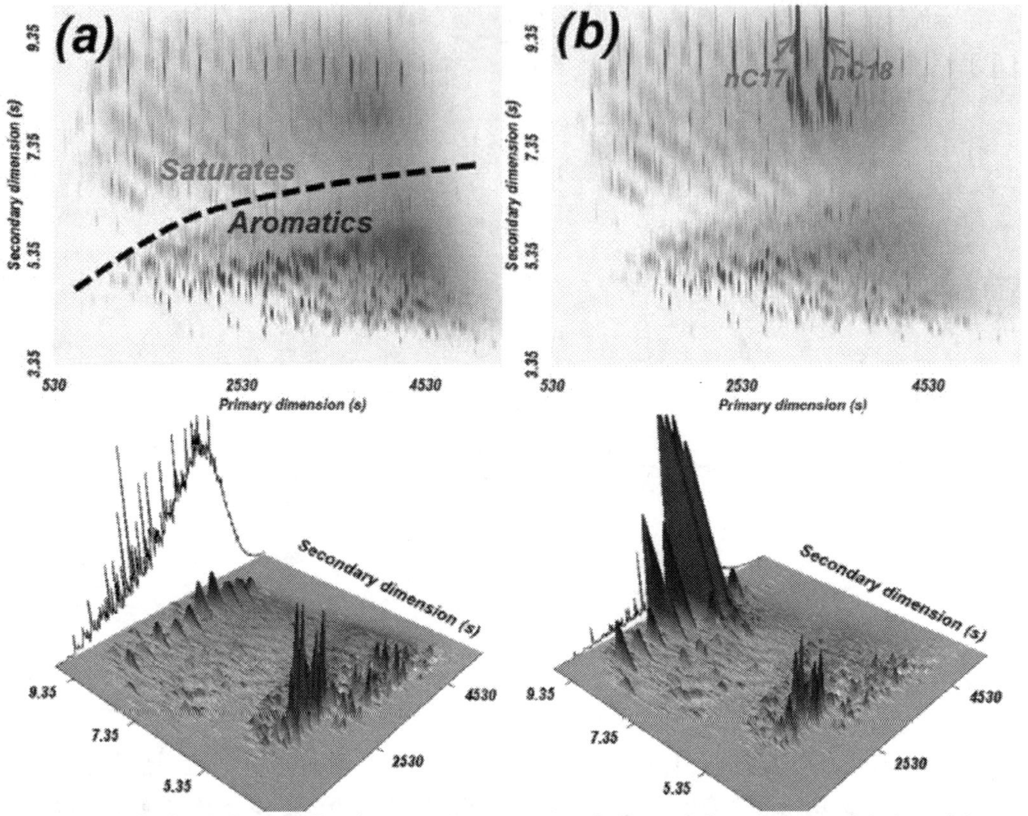

Figure 3. 'Reversed' GC×GC – FID chromatograms of diesel fractions obtained by hydroprocessing (Table 2, Case 4) of pure HVGO (a), and 80% HVGO and 20% canola oil blend (b)

Sample 10 (obtained from 80% HVGO and 20% canola oil blend) consists of 40.4 wt% n-paraffins (mostly n-C17 and n-C18), 27.6 wt% cycloparaffins, and 14.9 wt% indans/tetralins whereas Sample 7 (obtained from pure HVGO) consists of 5 wt% n-paraffins, 40.5 wt% cycloparaffins, and 25.5 wt% indans/tetralins. The numbers in horizontal axis of Figure 4 indicate the sample number.

Figure 4 reports only GC×GC bulk hydrocarbon class composition. However, for most of the hydrocarbon classes we are able to provide far more details, e.g. hydrocarbon class distribution by carbon number (Figure 5). Figure 5 shows the distribution of normal paraffins by carbon number (n-C14 – n-C19) for all the diesel fractions with the size of the squares and their color intensities representing the magnitude of concentration (wt%). In addition to presented compositional changes in n-paraffinic region, Figure 5 also depicts changes of decarboxylation selectivity for all cases in Table 1. The decarboxylation selectivity is defined as n-C17/(n-C17+n-C18)*100 (Huber et al., 2007) and reflects the relative hydrogen consumption during co-hydroprocessing canola oil and HVGO blends (Haldor Topsoe, 2010).

Most of the changes in hydrocarbon composition presented in Figure 4 seem to be caused by dilution effect of HVGO by canola oil. In addition to those of n-paraffins (especially n-C16 – n-C18) the yields of other hydrocarbon classes generally decrease or, at least are constant, with the increase in canola oil content in the feedstock. However, unlike the general trends discussed above, cycloparaffins containing 17 and 18 carbon atoms appear to have

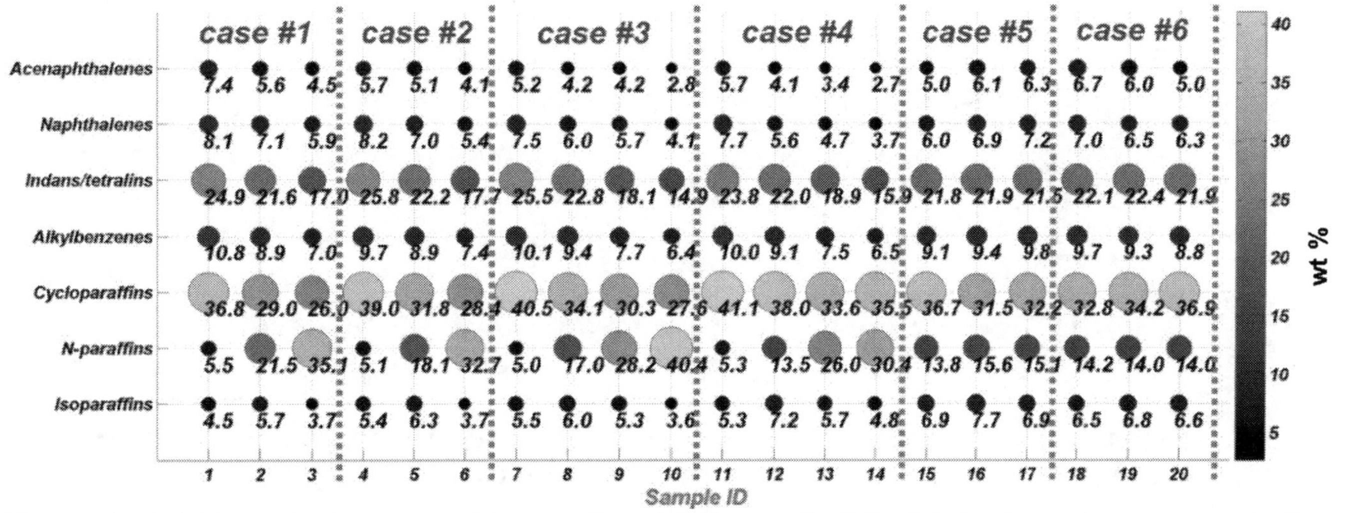

Figure 4. Bubble plot of averaged 'normal' and 'reversed'- GC×GC hydrocarbon concentrations in wt%. The details of each case are provided in Table 1.

increased content in parallel to n-paraffins (see Figures 6a and 6b). This in turn can be attributed to cycloisomerization of intermediate products of hydroprocessed fatty acids. Although this observation has been reported in some published papers, this paper is the first to show clear evidences of cycloisomerization of intermediate products produced through the hydroprocessing of fatty acids.

3.2 Sulfur speciation using GC×GC-SCD system

To investigate the impact of canola oil on hydrodesulfurization (HDS) GC×GC-SCD with normal column setup (cf. Table 2) was used in all quantitative analysis. Figure 7a shows an example of two-dimensional chromatogram of sulfur species in diesel range samples. It clearly Illustrates (Figure 7a) that thiophenes (Th), benzothiophenes (BTh) and dibenzothiophenes (DBTh) form three bands almost parallel to one another. After hydrodesulfurization (Figures 7b and 7c) only dibenzothiophenes are left in the sample. Based on GC×GC-SCD results performed on all samples it can be concluded that from the present study, canola oil does not affect the HDS performance. Hydrodesulfurization conversion increases with increasing temperature, but is not affected much by addition of canola oil (see Table 3).

3.3 Dependence of hydroprocessing performance on operating conditions

Table 3 summarizes all the analytical findings obtained in this study in the form of increasing/decreasing compositional trends in hydroprocessing reaction conditions. The first row shows the reaction conditions that are changing (eg. Increases for canola oil/HVGO ratios and temperature) whereas the second row presents major reaction conditions which are kept constant. Generally with the increase in canola oil content in the feedstock, diesel yield, n-paraffins content and cetane index increase whereas the nitrogen content, density, and other hydrocarbon classes content decrease. Interestingly, sulfur content in the diesel fractions are not affected by the content of canola oil in the feed. The changes in temperature, liquid hourly space velocity (LHSV), and pressure also affect the properties and compositions of the diesel samples as shown in Table 3, which are not discussed in detail in this paper.

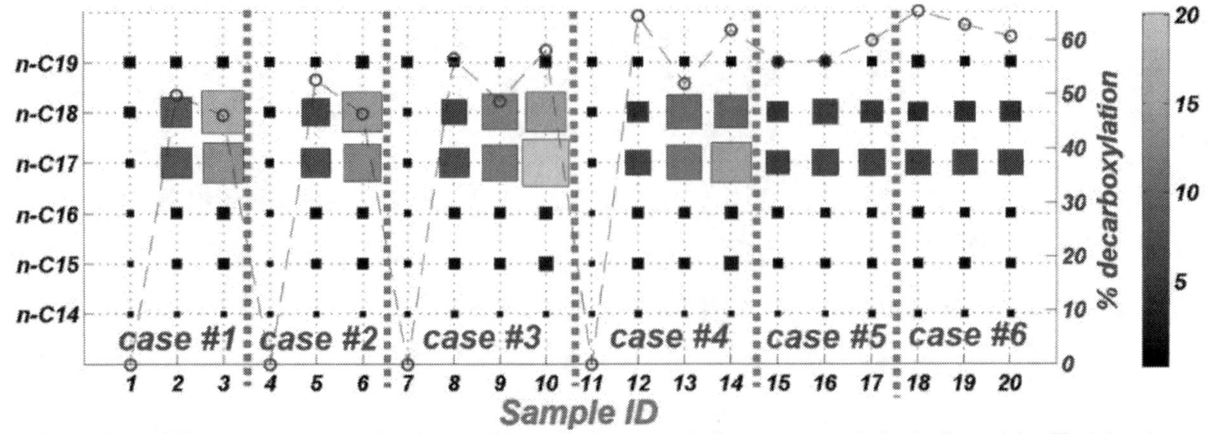

Figure 5. n-Paraffinic product yields (colored squares) for cases 1-6 defined in Table 1 and correlation between hydroprocessing conditions and decarboxylation selectivity (blue dotted line).

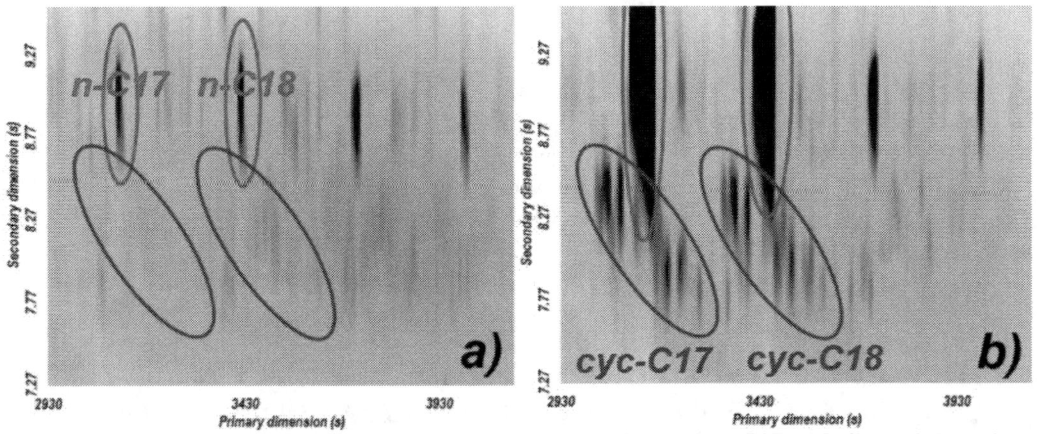

Figure 6. Zoom into saturates region of 'reversed' GC×GC – FID of diesel fractions obtained by hydroprocessing (Table 1, Case 4) of pure HVGO (a), and 80% HVGO and 20% canola oil blend (b)

4. Concluding Remarks

A commercial hydroprocessing catalyst and typical industrial operating conditions were used in this study to investigate the properties and hydrocarbon type compositions of diesel fractions obtained from co-hydroprocessing of HVGO and canola oil. These diesel samples were fully characterized by using standard ASTM methods and advanced two-dimensional gas chromatography (GC×GC).

The GC×GC technique, with one single analysis, can be used to visually distinguish the compositional fingerprints of diesel fractions obtained from hydroprocessed HVGO/canola oil mixtures. It provides complete quantitative characterization of these samples and, when used with advanced SimDis analysis, better modeling of reaction kinetics and thermodynamics in refinery processes can be conducted.

Characterization results showed that with the increase in canola oil content in the feed blends, the content of normal paraffins in the diesel fractions increased accordingly whereas the contents of aromatics and cycloparaffins decreased. The content of iso-paraffins remained relatively constant. The increase in the normal paraffins in the diesel samples was attributed to the hydrodeoxygenation and hydrodecarboxylation of triglycerides contained in the canola oil.

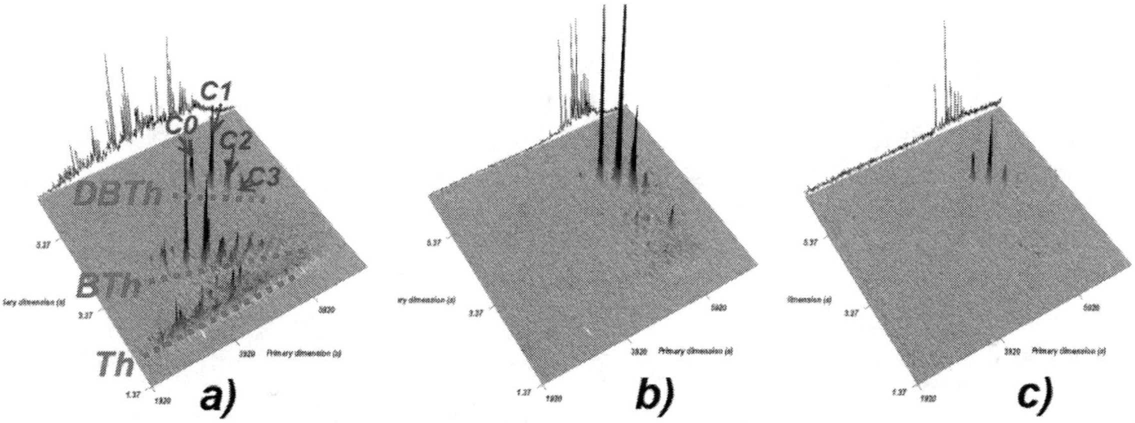

Figure 7. Examples of GC×GC-SCD chromatograms of sulfur class distribution in diesel range samples (3500 ppm sulfur) (a), diesel fractions obtained from pure HVGO hydrotreated at 365°C (620 ppm S) (b), and at 395°C (65 ppm S) (c)

Table 3. Dependence of hydroprocessing performance on operating conditions

Changed Conditions	Canola Oil/HVGO ratio ↗				Temperature ↗				LHSV ↗	Pressure ↗
remarks	360 °C	375 °C	385 °C	395 °C	100/0	95/5	90/10	80/20	Case 5	Case 6
Diesel Yield	↗	↗	↗	↗	↗	↗	↗	↗	↘	↔
Sulfur	↔	↔	↔	↔	↘	↘	↘	↘	↗	↘
Nitrogen	↘	↘	↘	↘	↘	↘	↘	↘	↗	↘
Cetane Index	↗	↗	↗	↗	↗	↘	↘	↘	↔	↗
Density	↘	↘	↘	↘	↘	↔	↔	↔	↔	↘
Saturates	↗	↗	↗	↗	↗	↗	↔	↔	↘	↗
Paraffins	↗	↗	↗	↗	↔	↘	↘	↘	↗	↔
IsoParaffins	↘	↘	↘	↘	↔	↗	↗	↗	↔	↔
N-paraffins	↗	↗	↗	↗	↔	↘	↘	↘	↗	↔
n-C17	↗	↗	↗	↗	↔	↘	↘	↘	↗	↔
n-C18	↗	↗	↗	↗	↔	↘	↘	↘	↔	↔
Cycloparaffins	↘	↘	↘	↘	↗	↗	↗	↗	↘	↗
Aromatics	↘	↘	↘	↘	↘	↘	↘	↘	↗	↘
Monoaromatics	↘	↘	↘	↘	↔	↔	↔	↗	↔	↔
Alkylbenzenes	↘	↘	↘	↘	↔	↔	↔	↔	↔	↔
Indans/tetralins	↘	↘	↘	↘	↔	↔	↔	↔	↔	↔
Diaromatics	↘	↘	↘	↘	↘	↘	↘	↔	↗	↘
Naphthalenes	↘	↘	↘	↘	↔	↘	↔	↔	↗	↘
Acenaphthalenes/fluorenes	↘	↘	↘	↘	↔	↘	↘	↔	↗	↔

The normal paraffins in the diesel samples were mostly normal heptadecane (product of hydrodecarboxylation) and normal octadecane (product of hydrodeoxygenation) with trace amounts of other lighter/heavier normal paraffins. The formation of normal heptadecane and normal octadecane resulted in changes in other physical and combustion properties of the diesel samples, such as density, boiling point distribution (or boiling curve), and cetane index/number.

This study uses advanced GCxGC techniques to provide detailed hydrocarbon characterization of diesel fractions from co-hydroprocessing canola oil and HVGO blends. These results are important for upgraders and refiners to understand how biomass derived feedstocks affect co-hydroprocessing performance, and composition and quality of the diesel products from co-hydroprocessing. Therefore technical guidelines can be generated for implementing co-processing biomass and bitumen derived feedstocks to reduce carbon footprint (or GHG emission) while clean transportation fuels are produced from oil sands.

Acknowledgments

Partial funding for this study was provided by the Canadian Interdepartmental Program of Energy Research and Development (PERD 1.1.3). The authors are grateful to the CanmetENERGY pilot plant staff for conducting the experiments and to the analytical lab staff for performing the analyses. Comments and suggestion on revising this manuscript from Drs. Edward Little and Huali Wang are greatly appreciated.

References

1. Adahchour, M., Beens, J., and Brinkman, U.A.Th., "Recent Developments in the Application of Comprehensive Two-dimensional Gas Chromatography", J. Chromatogr. A, 1186, 67-108, 2008.
2. Adam, F., Bertoncini, F., Brodusch, N., Durand, E., Thiébaut, D., Espinat, D., and Hennion, M.-C., "New Benchmark for Basic and Neutral Nitrogen Compounds Speciation in Middle Distillates Using Comprehensive Two-dimensional Gas Chromatography", J. Chromatogr. A., 1148, 55–64, 2007.
3. Bezergianni, S., Kalogianni, A., Vasalos, "Hydrocracking of Vacuum Gas Oil-Vegetable Oil Mixtures for Biofuels Production", Bioresource Technology, 100, 3036-3042, 2009.
4. Blomberg, J., Schoenmakers, P.J., Beens, J., and Tijssen, R., "Comprehensive Two-dimensional Gas Chromatography (GC×GC) and its Applicability to the Characterization of Complex (Petrochemical) Mixtures", J. High Resolut. Chromatogr, 20, 539–544, 1997.
5. Cannella, W., Fairbridge, C., Arboleda, P., Bays, T., Dettman, H., Foster, M., Gieleciak, R., Gunter, G., Hager, D., Lay, C., Lewis, S., Luecke, J., Sluder, S., Zigler, B. "Detailed Characterization of the Physical and Chemical Properties of the Reformulated FACE Diesel Fuels: FD2B, FD4B, and FD7B". CRC Report No. AVFL-19-1. Natural Resources Canada, CanmetENERGY-Devon-2013-302. http://www.crcao.com/reports/recentstudies2013/AVFL-19/AVFL-19-1%20Reformulated%20FACE%20Diesel%20Fuels%20Final%20Report%20-%20012213.pdf
6. Chen, J., Ring, Z., and Briker, Y., "Building a Hydrotreating Process Model for Modern Refineries", proceedings of 2003 AIChE Spring Conference, Paper 52e, March 30 –April 3, 2003, New Orleans, LA, USA.
7. Furimsky, E., "Catalytic Hydrodeoxygenation", Appl. Catal. A: Gen. 199, 147-190, 2000.
8. Gieleciak, R., Hager. D., Lay, C., and Fairbridge, C., "Hydrocarbon Characterization of Face Aromatic streams", Natural Resources Canada, CanmetENERGY-Devon 2010-082 (INT) http://www.crcao.com/publications/advancedVehiclesFuelsLubricants/FACE/index.html
9. Haldor Topsoe, "Turning over a new leaf in renewable diesel hydrotreating", 2010 http://www.topsoe.com/business_areas/refining/~/media/PDF%20files/Refining/topsoe_npra_2010_turning_over_a_new_leaf.ashx
10. Hua, R. X., Li, Y., Liu, W., Zheng, J., Wei, H., Wang, J., Lu, X., Kong, H., and Xu, G., "Determination of Sulfur-containing Compounds in Diesel Oils by Comprehensive Two-Dimensional Gas Chromatography with a Sulfur Chemiluminescence Detector", J. Chromatogr. A, 1019, 101–109, 2003.
11. Huber, G.W., O'Connor, P., and Corma, A., "Processing Biomass in Conventional Oil Refineries: Production of High Quality Diesel by Hydrotreating Vegetable Oils in Heavy Vacuum Oil Mixtures", Appl. Catal. A: Gen. 329, 120-129, 2007.

12. Keesom, W. H., Unnasch, S., and Moretta, J., "Life Cycle Assessment Comparison of North American and imported Crude Oils". Report by Jacobs Consultancy for Alberta Energy Research Institute (AERI), July 2009.
13. MATLAB, Release 2008b; The MathWorks, Inc.: Natick, MA, 2008.
14. Sheehan, J., Camobresco, V., Duffield, J., Graboski, M., Shapouri, H., "Life Cycle Inventory for Biodiesel and Petroleum Diesel for Use in an Urban Bus", Final Report by US Department of Agriculture and US Department of Energy, May 1998.
15. Šimáček, P., and Kubička, D., "Hydrocracking of Petroleum Vacuum Distillate Containing Rapeseed Oil: Evaluation of Diesel Fuel", Fuel, 89, 1508-1513, 2010.
16. Templis, Ch., Vonortas, A., Sebos, I., and Papayannakos, N., "Vegetable Oil Effect on Gasoil HDS in Their Catalytic Co-hydroprocessing", Applied Catalysis B: Environmental, 104 (3-4), 324-329, 2011.
17. Tóth, C., Baladincz, P., Kovács, S., and Hancsók, J., "Producing Diesel Fuel by Co-hydrogenation of Vegetable Oil with Gas Oil", Chem. Eng. Trans., 21, 1219-1224, 2010.
18. Vendeuvre, C., Ruiz-Guerrero, R., Bertoncini, F., Duval, L., Thiébaut, D., and Hennion, M.-C., "Characterisation of Middle-distillates by Comprehensive Two-dimensional Gas Chromatography (GC×GC): A Powerful Alternative for Performing Various Standard Analysis of Middle-Distillates", J. Chromatogr. A., 1086(1-2), 21–28, 2005.
19. Vendeuvre, C., Ruiz-Guerrero, R., Bertoncini, F., Duval, L., and Thiebaut, D., "Comprehensive Two-Dimensional Gas Chromatography for Detailed Characterization of Petroleum Products", Oil and Gas Science and Technology – Rev. IFP, 62 (1), 43-55, 2007.
20. Viljava, T, R., Saari, E. R. M., and Krause, A. O., I., "Simultaneous Hydrodesulfurization and Hydrodeoxygenation: Interactions between Mercapto and Methoxy Groups Present in the Same or in Separate Molecules", Appl. Catal. A: Gen. 209, 33-43, 2001.
21. Walendziewski, J., Stolarski, M., Łużny, R., and Klimek, B., "Hydroprocessing of Light Gas Oil – Rape Oil Mixtures", Fuel. Process. Tech., 90, 686-691, 2009.
22. Watkins, B., Olsen, C., Sutovich, K., Deady, J., Petti, N., and Wellach, S., "New Opportunities: for Coprocessing Renewable Feeds in the Refinery Process", Hydrocarbon Eng. 14(1), 49-58, 2009.
23. Wong, A., and Hong, E., "Biomass: A growth Opportunity in Green Energy and Value-added Products", the Fourth Biomass Conference of the Americas, Elsevier, Oakland, CA, 1999, p.837.

Key Findings of Experimental and Theoretical Studies on Forced Mitigation System for an LNG Spill Emergency

Byung Kyu Kim, Ray A. Mentzer, M. Sam Mannan[*]

Mary Kay O'Connor Process Safety Center, Artie McFerrin Department of Chemical Engineering, Texas A&M University System, College Station, Texas 77843-3122, USA
**Corresponding author. Tel.: +1 979 862 3985. E-mail address: mannan@tamu.edu (M.S. Mannan)*

Abstract

Mitigation measures have been categorized as one of the essential elements that require further investigation in determining the potential impact on public safety from an LNG spill. Several outdoor LNG spill experiments were conducted at the Brayton Fire Training Field to verify the effectiveness of the direct application of a water spray on LNG vapors as part of LNG Safety and Spill Response research at the Mary Kay O'Connor Process Safety Center (MKOPSC) at Texas A&M University (TAMU). Theoretical modeling was conducted of LNG forced mitigation using an upward-oriented conical water spray application. The computational fluid dynamics (CFD) model had been applied to investigate the dominant physical mechanisms involved in LNG forced mitigation and essential key design parameters were evaluated. Findings from the experimental and theoretical analysis on the LNG forced mitigation are summarized. Finally, recommendations on designing an effective forced mitigation system and potential hazards of applying the water curtain system are discussed. This work aims to provide a comprehensive understanding of the forced mitigation system with direct application to LNG facilities.

1. Introduction

The liquefaction capacity of natural gas is expected to double as the demand for natural gas expands substantially [1, 2]. The new emerging LNG supply from North America will also play a significant role in reorganizing the world gas supply chain in the near future [3]. With the transformation from an importer to exporter of LNG in North America, the new challenges rise from introduction of additional hazards from the liquefaction process and regulatory compliance to ensure safe operation [4]. NFPA 59A revised in 2013 requires all the new LNG facilities and any facilities that go through major modifications to conduct quantitative risk assessments to ensure that the facilities do not impose beyond tolerable risk to the nearby communities [5]. It is recommended to apply additional safety measures to meet the risk criteria. The effects of various mitigation techniques applicable to the LNG facilities still require further investigation to minimize the uncertainty in determining the impact of an LNG spill on the public safety and security [6]. This paper summarizes the research of water spray applications conducted at the Mary Kay O'Connor Process Safety Center (MKOPSC) at Texas A&M University (TAMU) as part of the LNG Safety Research Program.

2. Previous water spray research on LNG vapor clouds

The US Coast Guard conducted a small-scale LNG spill experiment to simulate a spill on an LNG transport ship in 1976 [7, 8]. It was concluded from the test results that the water spray curtain was effective in reducing the methane concentration and that the mechanical turbulence induced from the water droplets improved the mixing of the LNG vapors, enhancing the vapor dilution.

The Gas Research Institute conducted three-phase research, which consisted of theoretical analysis, small-scale spill tests, and wind tunnel experiments [9]. The downward spray application showed a reduction in concentration to 2–5% by volume in the downwind region. The upward sprays had diluted vapor concentrations approximately to 1 to 2% by volume; however, the effectiveness of vapor concentration reduction was not consistent with the theoretical estimation. It was concluded from the wind tunnel experiments that the upward sprays and downward-inclined sprays were the only two applications, which showed an effective dilution.

Meroney and Neff (1985) provided forced mitigation effects for both box and slab models assuming that the local entrainments increase with the application of the water curtain [10]. The enhanced air entrainment effects were incorporated into the model by applying a multiplicative or additive factor to the regular entrainment rates. The integral-type modeling provided a quick prediction compared to dense gas dispersion, and has shown reasonable predictions when compared to small-scale experiments. However, the integral-models present the fluid system in a way that limits the understanding of the complex interactions of water droplets with the gas flow. Also, there had been some concerns over the under-prediction from the integral-models in certain scenarios, mainly because of the simplified assumption [11].

3. Experimental work from MKOPSC

To investigate the complex physical mechanisms of LNG forced dispersion, the Mary Kay O'Connor Process Safety Center (MKOPSC) conducted outdoor LNG spill experiments at the Brayton Fire Training Field to verify the dominant mechanisms and evaluate the effectiveness in dilution the LNG vapor concentration [12, 13].

Fig. 1. LNG dispersion test: (a) without water curtain, (b) with water curtain (full cone) [14]*

Rana, Guo & Mannan (2010) discuss one of the experimental setups that had been conducted in 2007 [14]. Fig. 1 shows the experimental work of the water spray application on LNG vapor clouds. Two different types of water spray nozzles were mainly compared: 60° full-cone spray nozzle and 180° flat-fan spray nozzle [15]. The full-cone type produces finer droplets, while the flat-fan type creates a thin barrier in the vicinity of the nozzle and very course droplets as the bulk water travel and break into smaller droplets. The full-cone spray type provides more effective mixing with the air through high air entrainments (Fig. 2(a)). The full-cone type showed an effective dilution at all elevations. The flat-fan type creates a physical barrier in the pathway of LNG vapors and provides high momentum from the droplets created from the water pressure. The flat-fan type pushed the LNG vapors to higher elevations, resulting in an increase in concentration behind the water spray region (Fig. 2(b)).

* Reprinted figures with permission from Rana, M., Guo, Y., & Mannan, M. S. (2010). Use of water spray curtain to disperse LNG vapor clouds. Journal of Loss Prevention in the Process Industries, 23(1), 77-78. Copyright 2010 Elsevier.

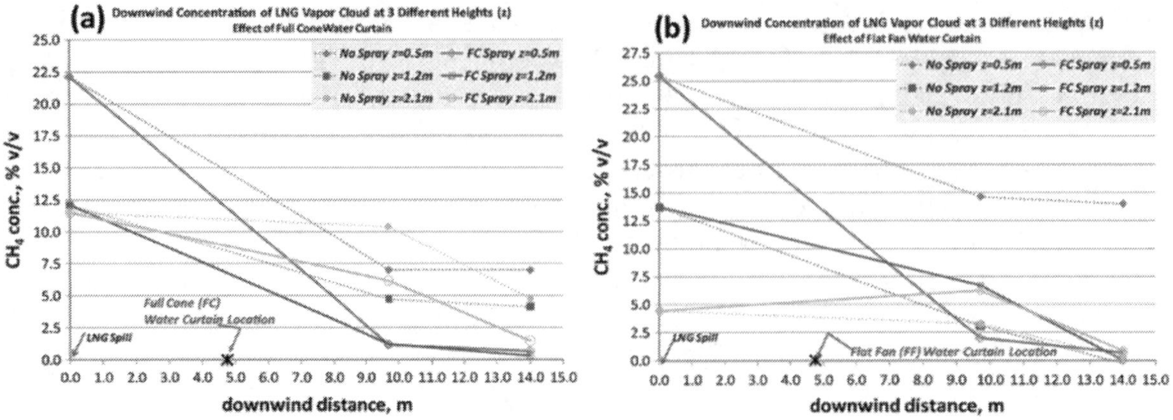

Fig. 2. Downwind concentration at three different heights with and without (a) full-cone water curtain and (b) flat fan type [14]*

The temperature changes collected from the experiments suggested that the water droplets provided heat transfer to the LNG vapors. The overall heat transfer evaluated from the experimental results showed that the droplets produced from the full-cone type nozzle provided more heat transfer to the LNG vapors than the flat-fan application (Fig. 3).

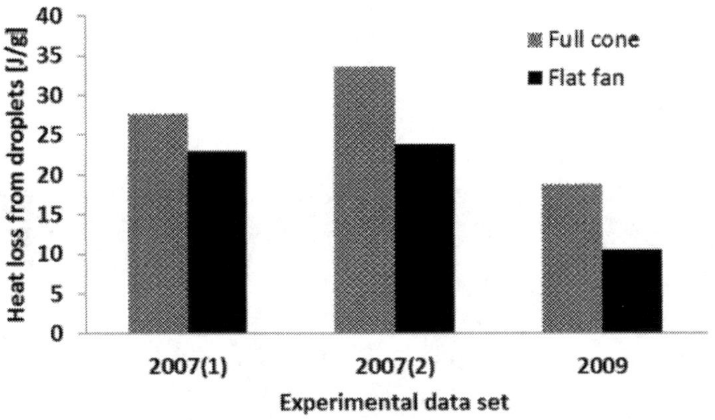

Fig. 3. Calculated heat loss by water curtain (Modified from [14])

The full-cone type nozzle produces smaller droplets, which increases the surface area. It was concluded that the mixing effects through the entrained air promotes the LNG vapors most effectively. The heat transfer from the droplets ensures that the vapors are sufficiently warmed to become positively buoyant.

4. Modeling approach from MKOPSC

The CFD tool was adopted to conduct a comprehensive assessment of the effects of the water curtain applications on LNG vapor clouds. In the present work, ANSYS Fluent was used to simulate the Eulerian–Lagrangian spray modeling coupled with the LNG vapor flows. The essential parameters for simulating the detailed description of LNG flow and interactions of the droplet-vapor system were calibrated against the experimental results obtained from the LNG spill work conducted at the Brayton

---

* Reprinted figures with permission from Rana, M., Guo, Y., & Mannan, M. S. (2010). Use of water spray curtain to disperse LNG vapor clouds. Journal of Loss Prevention in the Process Industries, 23(1), 77-78. Copyright 2010 Elsevier.

Fire Training Field. The physical parameters that are essential inputs in setting up the CFD simulation of the water-spray LNG system are discussed elsewhere [16].

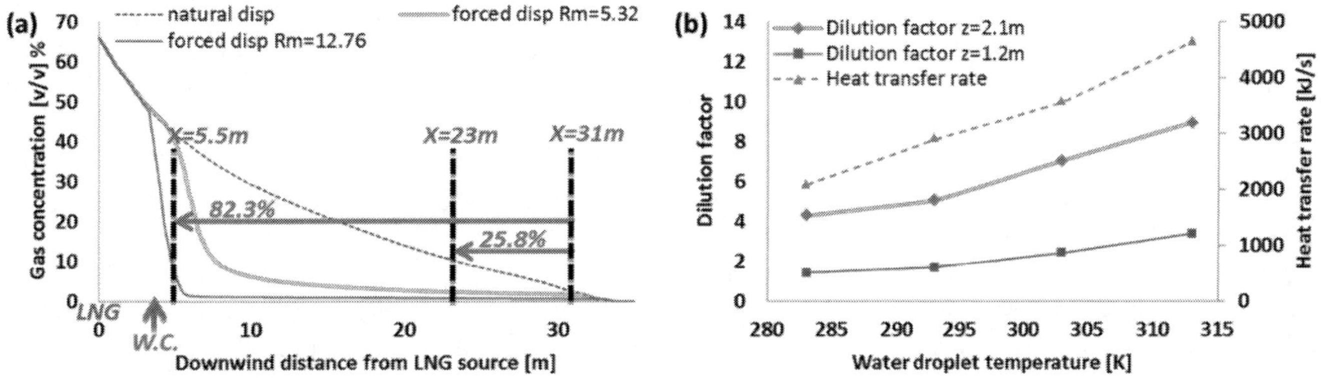

Fig. 4. (a) Volumetric concentration for natural/forced(RM=5.32)/forced(RM=12.76) dispersion and (b) heat transfer rate and dilution factor at different water droplet temperatures [16][**]

Fig. 4 shows the dilution effects of two different spray settings applied (R_M= 5.32, and 12.76) and heat transfer involved for various droplet temperatures simulated using the LNG forced dispersion model. R_M is the momentum ratio, which is the ratio of the momentum imparted from the droplets to the momentum of the gas clouds [17]. Different levels of safety distance reduction were observed for various momentum ratios applied from the water spray as shown in Fig. 4 (a). As the water nozzle discharged more momentum to the LNG vapors, the dilution effects increased. Also, the heat transfer rate increased as the droplet temperature increased as shown in Fig. 4 (b). The effectiveness of dilution improved as well, as more heat transfer was involved at different elevations (z = 1.2 and 2.1 m). These results indicate that the LNG forced dispersion modeling using the Eulerian-Lagrangian spray model is capable of simulating the dilution effects and influences on LNG vapor behavior by taking into account the detailed droplet characteristics.

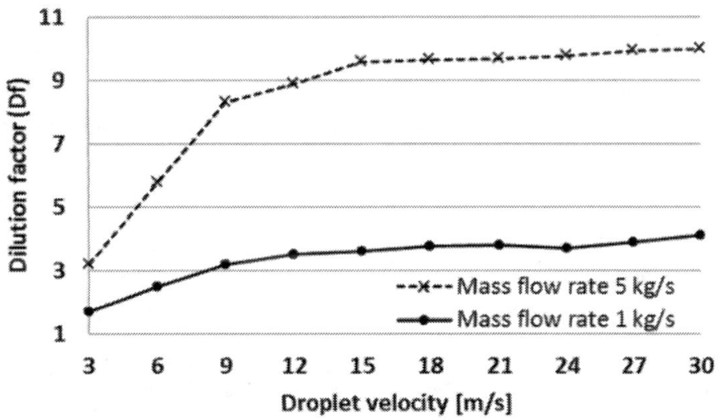

Fig. 5. Dilution effects from different mass flow rates and droplet velocities

Using LNG forced dispersion modeling, the physical mechanisms of the LNG forced dispersion were investigated. The effects of momentum imparting from the droplets to the air-vapor mixture,

[**] Reprinted figures with permission from Kim, B. K., Ng, D., Mentzer, R. A., & Mannan, M. S. (2012). Modeling of water spray application in the forced dispersion of LNG vapor cloud using a combined Eulerian-Lagrangian approach. Industrial & Engineering Chemistry Research, 51, 13803-13814. Copyright 2012 American Chemical Society.

thermal effects between two phases (vapor/droplet), and different level of air entrainments on the LNG vapor behavior were evaluated.

The dilution effects of LNG forced dispersion were evaluated with different ranges of droplet velocity (3–30 m/s) at two different mass flow rates (1 and 5 kg/s) as shown in Fig. 5. The dilution effects increased as the droplet velocity increased. However, the spray with mass flow rate of 1 kg/s showed only limited dilution effects even with a high droplet velocity applied. This is because even with the high droplet velocity to provide effective mixing of the air-vapor mixture, the insufficient water introduced from the nozzle limited the effective dilution of LNG vapors. This result indicates that both of the design elements, the droplet velocity and mass flow rate, must be set above a certain range to ensure an effective mitigation effect. From the water spray setting applied in this research, it was above 3 kg/s of mass flow rate with 12–15 m/s droplet velocity.

The thermal effects from the water droplets to the vapor clouds, particularly for cold gas releases, can enhance dissipation of the vapor cloud by warming to gain positive buoyancy [18]. The temperature of the water droplet was varied from 283 to 313 K to evaluate the turbulence effects induced from different levels of heat transfer effects of the droplets, as shown in Fig. 6.

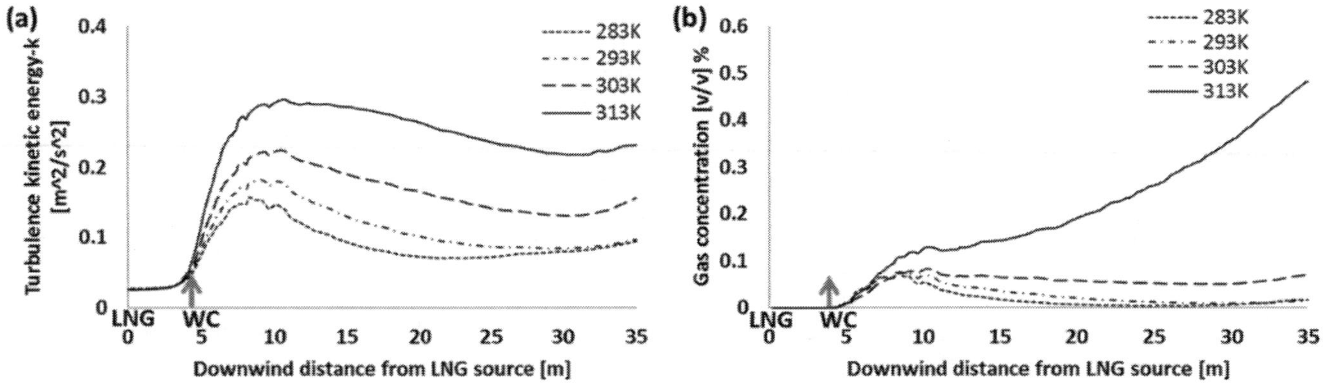

Fig. 6. (a) Turbulence kinetic energy and (b) vapor concentration from various droplet temperature applications at 8 m elevation

The turbulence kinetic energy showed the highest level being induced from the water droplet with highest temperature (313 K). The 8 m elevation was chosen to show the vapor behavior at high elevation. The concentration also increased significantly with the highest water droplet temperature, indicating that the vapors are sufficiently warmed to travel upward to the atmosphere. This result shows that the thermal effects from the droplet enhance the turbulent flow within the vapor clouds, resulting in an effective dilution of LNG vapors.

The air entrainment rate is determined by the spray characteristics: droplet sizes, droplet velocity, spray location, and configuration. Rana (2009) have verified that the entrained air rate into the spray increases as the nozzle angle size increases, while the operating pressure does not affect the rate of the air entrainment significantly [13]. A total of four different angle sizes (30°, 45°, 60°, and 75°) were simulated, as shown in Fig. 7.

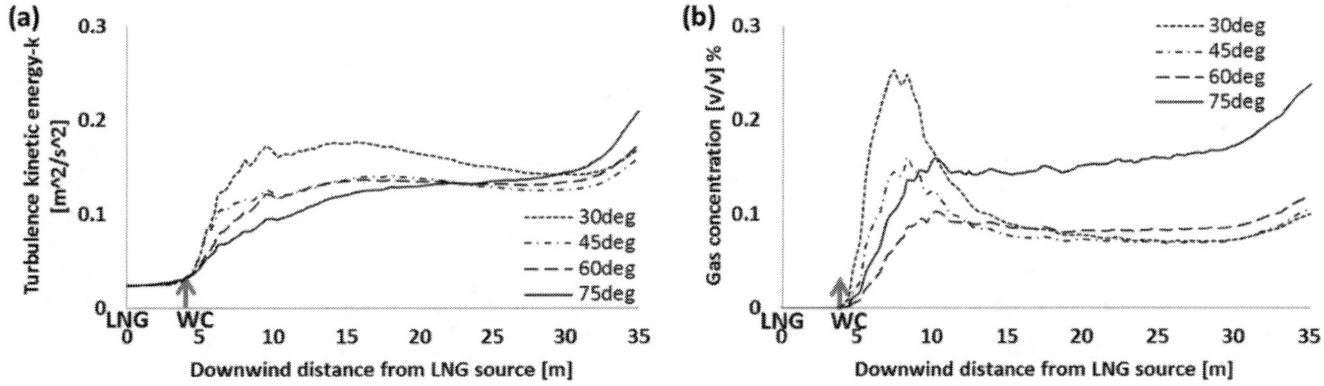

Fig. 7. (a) Turbulence kinetic energy and (b) vapor concentration from various nozzle angle size applications at 8 m elevation

The turbulence induced at 8 m elevation showed that the turbulence effects from various angle sizes were similar. On the other hand, the vapor concentration showed an increasing trend for the largest nozzle angle (75°) at 8 m elevation, while the concentration for the rest of other applications showed similar results once passed through the water spray. This indicates that the air entrainment setting that could induce a large amount of air is more effective in dispersing the LNG vapors through effective mixing, with turbulence effects playing a less significant role.

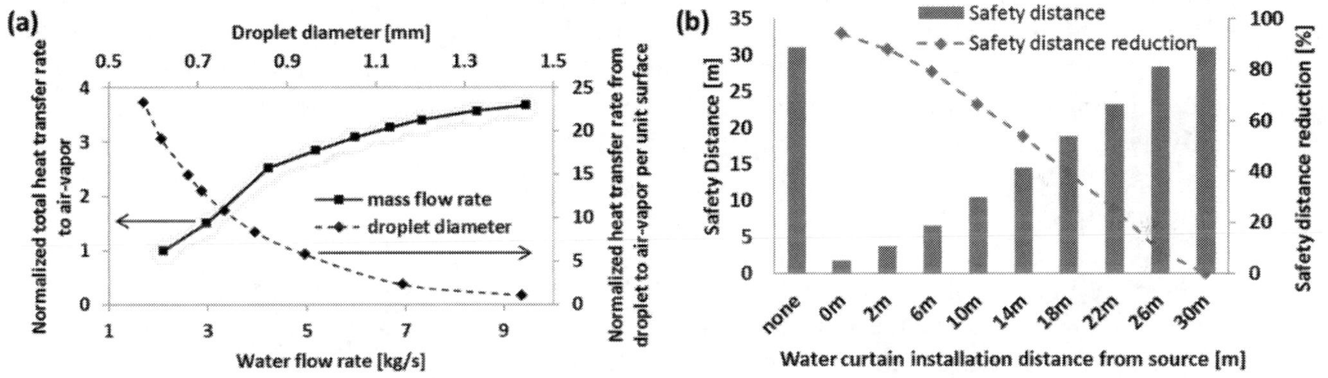

Fig. 8. (a) Heat transfer rate from droplets to air-vapor mixture and (b) safety distance evaluated from various installation distances [19]***

The size of droplets is determined by the water pressure and nozzle design, which is an intrinsic characteristic for a specific nozzle design. Different droplet sizes were simulated to evaluate the heat transfer rates (Fig. 8 (a)). The total heat transfer clearly indicated that the total amount of heat transferred from the droplet to LNG increased as the flow rate increased. However, the heat transfer from the surface decreased significantly as the droplet size increased, because of the larger droplets providing smaller surface area. The installation distance from the LNG source was investigated (Fig. 8(b)). The safe distance, defined as where the LFL is reduced by 50%, was reduced significantly as the water curtain was installed closer to the LNG source. These results can be used to evaluate the impact of the distance factor in calculating the dilution effects.

*** Reprinted figures with permission from Kim, Byung Kyu, Ng, Dedy, Mentzer, Ray A., & Mannan, M. Sam. (2013). Key parametric analysis on designing an effective forced mitigation system for LNG spill emergency. Journal of Loss Prevention in the Process Industries. doi: http://dx.doi.org/10.1016/j.jlp.2013.01.007. Copyright 2013 Elsevier.

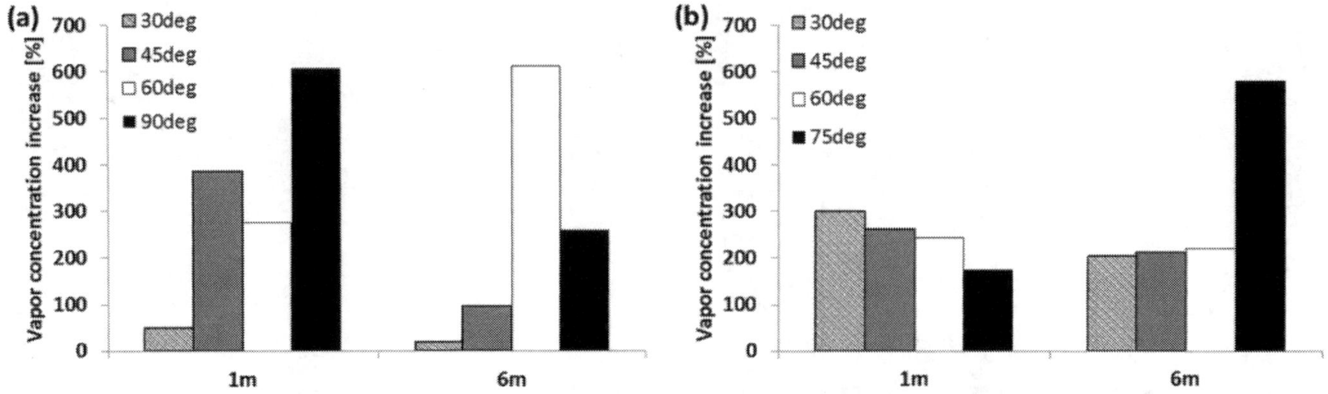

Fig. 9. Vapor concentration increase [%] at 1 m and 6 m with (a) modified nozzle installations (tilted at 30°, 45°, 60°, and 90°) and (b) different nozzle sizes (30°, 45°, 60°, and 75°) [19] ***

The tilted installation designs at various nozzle angles were evaluated (30°, 45°, 60°, and 90°). Per Fig. 9 (a), the vertical installation showed the most effective dilution at lower heights (1 m), while at higher elevation (6 m), the nozzles tilted at 60° showed the largest increase of vapor concentration. This indicates that the nozzles tilted at 60° discharge the droplets that could effectively envelop the LNG vapors and enhance dilution of the air-vapor mixture. Various levels of air entrainments were applied to evaluate the concentration increase in the atmosphere are plotted in Fig. 9 (b). The angle sizes of the nozzle were modified in 30°, 45°, 60°, and 75° to simulate different levels of entrainments. The smaller angle nozzles were more effective in pushing the vapors to lower heights. On the other hand, the nozzle with largest angle size (75°) showed the highest increase of vapor concentration at higher elevation (6 m). This indicates that the nozzle with the largest entrainment air setting provides enhanced mixing that could effectively dilute the LNG vapors to travel upward to the atmosphere.

5. Conclusion

The engineering guidelines in designing an effective water spray system for an LNG facility are currently lacking mainly due to the existing gaps in the experimental results and limited theoretical modeling developed using integral-type models. The CFD code had been applied in evaluating the forced dispersion of LNG vapors using a water curtain application. The proposed model uses a Eulerian–Lagrangian approach, which takes into account the detailed characteristics of the water droplets in evaluating the effects on the LNG vapor behavior. The model shows a promising solution in investigating the complex interaction of the droplet-vapor system. The CFD code also can serve as a rigorous tool in evaluating the design for a site-specified mitigation system, to predict any unforeseen hazards that might rise from under-sized mitigation. The modeling approach applied in this work can evaluate the risk reduction effects from applying the water spray application, which can directly be implemented for quantitative risk assessments, as indicated by NFPA 59A.

References

1. International Energy Outlook 2011 (US Energy Information Administration), 2011, US Energy Information Administration (EIA): Washington, DC.

*** Reprinted figures with permission from Kim, Byung Kyu, Ng, Dedy, Mentzer, Ray A., & Mannan, M. Sam. (2013). Key parametric analysis on designing an effective forced mitigation system for LNG spill emergency. Journal of Loss Prevention in the Process Industries. doi: http://dx.doi.org/10.1016/j.jlp.2013.01.007. Copyright 2013 Elsevier.

2. World Energy Outlook 2011: Are we entering a golden age of gas? 2011, International Energy Agency: Paris, France.
3. Henderson, J., The potential impact of North American LNG exports, 2012, The Oxford Institute for Energy Studies: Oxford, UK.
4. Chosnek, J. and V. Edwards. From LNG import to exports: Process safety and regulatory challenges. Presented in 15th Annual Symposium Mary Kay O'Connor Process Safety Center. 2012. College Station, TX.
5. NFPA 59A: Standard for the Production, Storage, and Handling of Liquefied Natural Gas (LNG), 2013, National Fire Protection Agency: Quincy, MA.
6. Public Safety Consequences of a Terrorist Attack on a Tanker Carrying Liquefied Natural Gas Need Clarification (Maritime Security GAO-07-316), United States Government Accountability Office, 2007: Washington, D.C.
7. Brown, L., et al., Small scale tests on control methods for some liquefied natural gas hazards, 1976, US Coast Guard: Washington D.C.
8. Martinsen, W.E., S.P. Muhlenkamp, and J. Olson, Disperse LNG vapors with water. Hydrocarbon Processing, 1977: p. 260-267.
9. Heskestad, G., et al. Effectiveness of water spray curtains in dispersing LNG vapor clouds. in In proceedings of the American Gas Association Transmission Conference Paper, Paper No 83-T-69, 169-183. 1983. Seattle, Washington.
10. Meroney, R. and D. Neff, Numerical modelling of water spray barriers for dispersing dense gases. Boundary - Layer Meteorology, 1985. 31(3): p. 233-247.
11. Gavelli, F., E. Bullister, and H. Kytomaa, Application of CFD (Fluent) to LNG spills into geometrically complex environments. Journal of Hazardous Materials, 2008. 159(1): p. 158-168.
12. MKOPSC, Data Report of MKOPSC LNG Spill Tests: 2005-2009, M.S. Mannan, Editor 2010, Mary Kay O'Connor Process Safety Center, Texas A&M Univeriesty: College Station, TX.
13. Rana, M., Forced dispersion of liquefied natural gas vapor clouds with water spray curtain application, Chemical Engineering, Texas A&M University, PhD Dissertation, 2009. p. 232.
14. Rana, M., Y. Guo, and M.S. Mannan, Use of water spray curtain to disperse LNG vapor clouds. Journal of Loss Prevention in the Process Industries, 2010. 23(1): p. 77-78.
15. Rana, M. and M.S. Mannan, Forced dispersion of LNG vapor with water curtain. Journal of Loss Prevention in the Process Industries, 2010. 23(6): p. 768-772.
16. Kim, B.K., et al., Modeling of water spray application in the forced dispersion of LNG vapor cloud using a combined Eulerian-Lagrangian approach. Industrial & Engineering Chemistry Research, 2012. 51: p. 13803-13814.
17. Hald, K., et al., Heavy gas dispersion by water spray curtains: A research methodology. Journal of Loss Prevention in the Process Industries, 2005. 18(4–6): p. 506-511.
18. CCPS, Guideline for postrelease mitigation technology in the chemical process industry. 1997, New York: Center for Chemical Process Safety, American Institute of Chemical Engineers.
19. Kim, B.K., et al., Key parametric analysis on designing an effective forced mitigation system for LNG spill emergency. Journal of Loss Prevention in the Process Industries, 2013(0).

9TH GLOBAL CONGRESS ON PROCESS SAFETY

A View of the Evolving LNG Regulations and Associated Exclusion Zones from an Industry Perspective

Alfonso F. Ibarreta
Exponent, Inc.
9 Strathmore Road
Natick, MA 01760
dust@exponent.com

Ryan Hart
Delmar "Trey" Morrison
Harri Kytomaa
Exponent, Inc.

Prepared for Presentation at
American Institute of Chemical Engineers
2013 Spring Meeting
9th Global Congress on Process Safety
San Antonio, Texas
April 28 – May 1, 2013

UNPUBLISHED

A View of the Evolving LNG Regulations and Associated Exclusion Zones from an Industry Perspective

Alfonso F. Ibarreta
Exponent, Inc.
9 Strathmore Road
Natick, MA 01760
dust@exponent.com

Ryan Hart
Delmar "Trey" Morrison
Harri Kytomaa
Exponent, Inc.

Keywords: LNG, FERC, vapor dispersion, exclusion zone

Abstract

Recent growth in availability of domestic natural gas has led to a shift in the U.S. gas industry from being a net LNG importer to the prospect of becoming a net LNG exporter. As a consequence, many industry players are actively pursuing natural gas liquefaction projects at existing LNG import terminals. In concert with this growth in proposed construction projects, the LNG facility permitting process driven by Federal Energy Regulatory Commission (FERC) has changed dramatically over the past few years. Although the requirements for LNG facility siting and design are codified in 49 CFR 193 and NFPA 59A, many of the exclusion zone analysis requirements are stated broadly and involve some level of expert interpretation. FERC has in the past clarified their interpretation of the federal requirements by means of formal letters and less formal precedent-setting memoranda. These interpretations continue to evolve over time and drive new tools and considerations.

This paper examines the ramifications of consequence-based accidental release scenarios that have been historically required, describes the latest requirements imposed by FERC and explains how these came into existence. The changing requirements have ostensibly been driven toward better definition of release sources; however, FERC's requirements have challenged the available and approved analytical tools. Also with this evolution, new types of scenarios coupled with new models have heavily influenced piping and equipment design, unit layout, unit siting, and mitigation strategies. This paper will discuss the evolution and the now dominant considerations for facility siting and unit design that drive the safety aspects of LNG facility permitting process from the industry's perspective.

Specific examples of scenarios will be provided as well as their consequences in terms of the resulting ½-LFL vapor clouds as well as the distance to the 1 psi boundary for the corresponding vapor cloud explosions. This paper will also provide a summary of the effectiveness of mitigation techniques, including the use of vapor fences based on numerous case studies.

1. Introduction

Recent growth in availability of domestic natural gas has led to a shift in the U.S. gas industry from being a net LNG importer to the prospect of becoming a net LNG exporter. As a consequence, many industry players are actively pursuing natural gas liquefaction projects at existing LNG import terminals. In concert with this growth in proposed construction projects, the LNG facility permitting process driven by Federal Energy Regulatory Commission (FERC) has changed dramatically over the past few years.

The codes and standards for which LNG facilities must be designed, constructed, and operated are found in 49 CFR Part 193 [1], and the authority having jurisdiction (AHJ) for these facilities is the Federal Energy Regulatory Commission (FERC). As part of the application process, the applicants (facility owners) must perform an environmental analysis in a series of thirteen (13) Resource Reports. Included in the assessment is a reliability and safety assessment (Resource Report 11) [2]. FERC then reviews the applications and must give approval prior to the commencement of any construction activities.

Although the requirements for LNG facility siting and design are codified in 49 CFR 193 and NFPA 59A [3], many of the exclusion zone analysis requirements are stated broadly and involve some level of expert interpretation. FERC has in the past clarified their interpretation of the federal requirements by means of formal letters and less formal precedent-setting memoranda. These interpretations continue to evolve over time, and drive new tools and considerations.

2. LNG Facility Permitting Exclusion Zone Requirements

The LNG facility permitting process requires that the designs meet the siting requirements set forth in:

- 40 CFR 68 - *Chemical Accident Prevention Provisions* [4].
- 49 CFR 193 - *Liquefied Natural Gas Facilities: Federal Safety Standards* [1].
- NFPA 59A (2001) *Standard for the Production, Storage, and Handling of Liquefied Natural Gas (LNG)* [3].

These standards require the presence of certain exclusion zones separating LNG and refrigerant storage, processing and transfer areas from potential ignition sources and locations not under the control of the LNG facility (off-site properties). The siting requirements set forth in Subpart B of 49 CFR 193 must be provided with hazard exclusion zones, ensuring that the following hazard areas do not extend beyond a property line that can be built upon:

- **Flammable vapor-gas dispersion hazard:** Regulations require that a ½-LFL flammable vapor cloud must not extend "beyond the property line that can be built upon" for worst-case 10-minute release scenarios [49 CFR § 193.2059 and NFPA 59A (2001) § 2.2.3.3]. Cases to be considered include:

  - A worst-case scenario LNG or refrigerant spill.
  - A worst-case scenario LNG or refrigerant jetting and flashing release.

- **Vapor explosion overpressure hazard:** Regulations require an offsite consequence analysis of the 1 psi overpressure boundary resulting from ignition of worst-case 10-minute releases [40 CFR 68]. Cases to be considered include:
  - A worst-case scenario refrigerant jetting and flashing release.

- **Thermal radiation (pool fire) hazard**: Regulations require that a 1,600 Btu/hr/ft^2 (5,000 W/m^2) level must not be exceeded "at a property line that can be built upon" for pool fire scenarios [49 CFR § 193.2057 and NFPA 59A (2001) § 2.2.3.2]. Cases to be considered include:
  - A worst-case scenario LNG or refrigerant liquid spill.

The hazard scenarios described above are based on a 10-minute LNG or refrigerant release at the maximum leak flow rate, based on pipe break scenarios persuaded by FERC.

Based on Exponent's previous experience, the flammable vapor dispersion hazard exclusion zones resulting from the worst-case jetting and flashing scenarios in process areas normally result in the largest stand-off distance requirements. Therefore, the following sections focus on the requirements imposed by FERC for the analysis of the flammable vapor-gas dispersion hazard described above.

3. Determination of Flammable Vapor-Gas Hazard Exclusion Zone and Effective Property Boundary

The federal regulation 49 CFR 193 requires that each LNG container and transfer system existing within an LNG facility must have an exclusion zone based upon vapor dispersion analyses conducted in accordance with the prescription of sections 2.2.3.3 and 2.2.3.4 of NFPA 59A (2001). The exclusion zone, according to 49 CFR 193, must be "an area surrounding an LNG facility in which an operator or government agency legally controls all activities in accordance with §193.2057 and §193.2059 for as long as the facility is in operation." Considering the regulatory definition for exclusion zone, the property line for which the vapor dispersion exclusion zone would apply is considered to be the boundary at which the LNG owner/operator ceases to control all activities at all times.

Additionally, according to the CFR requirements, the hazard exclusion zones cannot extend beyond "a property line that can be built upon." This is normally demarcated by the facility property boundary. This area can also be extended by either (a) proving the area of interest cannot be built upon (such as a waterway), or (b) contractually ensuring that the area is under the control of the LNG facility. In the case of a railroad or roadway traversing the property, FERC has previously set a precedent to allow classification as transient travel through an exclusion zone.

4. Determination of Design LNG and Refrigerant Spill Scenarios in Process and Transfer Areas

For process areas, NFPA 59A Table 2.2.3.5 requires the design spill be selected based on single accidental leakage source scenario. Unfortunately, NFPA 59A (2001) lists a few specific scenarios but does not generally define what a single accidental leakage source entails. FERC has therefore provided guidance for selecting single accidental liquid leakage sources to analyze in order to demonstrate compliance with 49 CFR Part 193.

4.1 Types of Leaks to Consider

Based on guidance provided by FERC, the releases of LNG and refrigerants are divided into two categories that are both to be considered in an exclusion zone analysis:

(1) Liquid spill scenarios consider full pipe ruptures where liquid spills onto the ground. The liquid conveyance in trenches and troughs, and collection in impoundments, resulting in subsequent vapor cloud formation and dispersion is then considered for evaluation of the exclusion zone.

(2) Conversely, pressurized jetting and flashing release scenarios are bounded by the constraint that little or no liquid from the fluid jet reaches the ground (i.e. no rainout). Depending upon the process conditions within the pipe and the release scenario, the resulting single accidental leakage source may produce some rainout. Only jetting and flashing release scenarios that produce little or no rainout are therefore analyzed, since they tend to create the largest vapor clouds.

4.2 Prescriptive Leak Size Determination

In order to officially establish what types of leak scenarios needed to be considered, FERC staff initially sent a letter to the DOT on April 19, 2005, requesting concurrence on the procedures for determining a single accidental leakage source [5]. FERC staff originally required an evaluation of all small diameter attachments to the transfer piping for instrumentation, pressure relief, recirculation, etc., and any flanges that may be used at valves or other equipment in order to determine the largest spill rate. The DOT affirmed this approach in a May 6, 2005, letter to FERC [5]. Based on these requirements, a wide variety of single accidental leakage sources, ranging from packing and flange leaks, to full guillotine ruptures of ship unloading lines, were submitted to FERC by applicants for several years. [5]

Subsequently, in the 2011 to 2012 timeframe, FERC identified more specific criteria for selecting leak sources for piping on the order of 100 m, and later for piping on the order of 1000 m in length. FERC initially provided the following criteria in 2011 for selecting the leak scenarios in "piping lengths on the order of 100 m:" [6]

- For piping less than or equal to 6 inches in diameter, a full pipe rupture at any point along the line; and
- For piping greater than 6 inches in diameter, the greater of:

- a full rupture of a connection less than or equal to 6 inches in diameter; or
- a hole of 2 inches in diameter.

FERC subsequently provided the following criteria in 2012 for selecting the leak scenarios in "piping lengths on the order of 1000 m:" [7]

- For piping less than or equal to 12 inches in diameter, a full pipe rupture at any point along the line; or
- For piping greater than 12 inches in diameter, a hole equivalent to 1/3 of the diameter.

4.3 Quantitative Leak Size Determination

The initial prescriptive approach for describing a single accidental leakage source did not provide any quantitative justification for the selection of the design spill [5]. In order to establish a quantitative criterion, FERC chose to set an acceptable failure rate limit for components based on a published risk analyses for the failure of an LNG storage tank outlet line. FERC staff determined that the rupture of a storage tank outlet line is on the order of one failure every 20,000 to 30,000 years ($5x10^{-5}$ to $3x10^{-5}$ failures per year) [5, 8]. Based on this, FERC required in May of 2012 that single accidental leakage sources be selected for analysis if the likelihood of failure is greater than $3x10^{-5}$ failures per year [9].

FERC provided a failure rate table to determine the likelihood of failure of all components in the LNG and refrigerant systems [9]. The table is meant to be used to determine whether a component failure rate is higher than the $3x10^{-5}$ failures per year criterion, meaning that it would have to be considered in the permitting analysis. Failure rates must currently be considered for:

- Transfer and process piping (failure per meter of piping)
  - Catastrophic rupture of the full pipe diameter
  - 1/3 diameter hole
  - 10% diameter hole, up to 2-inches
  - 1-inch hole
- Process vessels
- Valves and joints
- Transfer arms and hoses

Figure 1 illustrates how this process works. For a 10-inch diameter transfer line that is 45 meters in length, 3 separate scenarios must be first considered: a catastrophic rupture, a 3.33-inch diameter hole and a 1-inch diameter hole. The failure scenarios are first screened to determine if they meet the failure criterion set by FERC. The largest leak size that meets the FERC criterion needs to be considered in the vapor dispersion analysis.

Piping: 150mm (6-inch) <d< 299mm (12-inch)	Failures per year of operation
Catastrophic rupture	2E-7 per meter of piping
Release from hole with effective diameter of 1/3 diameter	4E-7 per meter of piping
Release from hole with effective diameter of 25mm (1-inch)	7E-7 per meter of piping

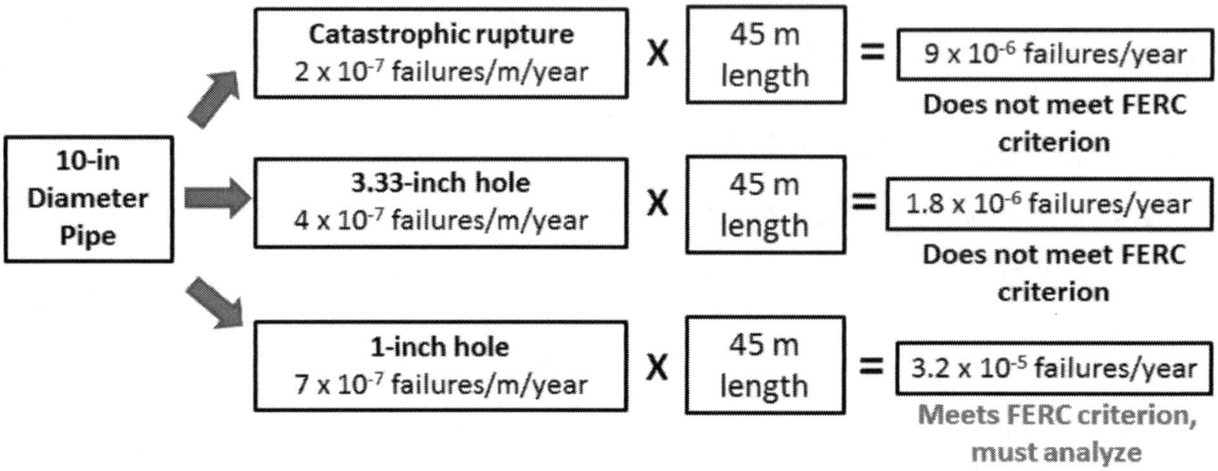

Figure 1 Example of quantitative leak size calculation logic tree.

5. Models Approved by FERC

Several computational models have been approved in the past for modeling the vapor dispersion resulting from LNG and refrigerant releases. In 2007, the Fire Protection Research Foundation issued a report which established a Model Evaluation Protocol (MEP) for LNG vapor dispersion models [10]. The computational models approved by FERC since then have been required to be first tested against the MEP. Three approved vapor dispersion models are discussed.

5.1 DEGADIS Integral Model

NFPA 59A (2001) *Standard for the Production, Storage, and Handling of Liquefied Natural Gas* §2.2.3.4 indicated that "flammable mixture dispersion distances shall be calculated in accordance with the model described in Gas Research Institute report GRI 0242, LNG Vapor Dispersion Prediction with the DEGADIS Dense Gas Dispersion Model." In the 2009 edition of NFPA 59A the prescription of DEGADIS was removed and it was required that the model used be acceptable to the AHJ based on an evaluation using the MEP [11]. FERC officially approved the model in 2011 [11]. It should be noted that 49 CFR 193 does not reference the 2009 edition of NFPA 59A.

The integral model DEGADIS was developed for the Gas Research Institute and the U.S. Coast Guard specifically to account for effects such as gravity spreading, negative or positive buoyancy effects on air entrainment, surface to cloud heat transfer, and phase change energy effects associated with air humidity in modeling dispersion of dense gases [11].

The DEGADIS model had previously been used in conjunction with the SOURCE5 model, developed originally by the Gas Research Institute. The SOURCE5 model allowed the determination of LNG evaporation rates for instantaneous or continuous releases on either land or water [12]. This model contained serious deficiencies, and DOT declared that it could no longer be used for LNG facility permitting in July 2010 [13].

5.2 PHAST Integral Model

The PHAST model was approved by PHMSA in 2011 for analyzing LNG vapor dispersion exclusion zones [14]. PHAST is commercial software developed and distributed by Det Norske Veritas (DNV). PHAST is widely used for the calculation of hazard distances from the release of several hazardous substances, including LNG. The PHAST code uses the Unified Dispersion Model (UDM) as an integral model to calculate the dispersion following a two-phase pressurized release or an unpressurised release. It consists of the following linked modules:

- Near-field jet dispersion;
- Non-equilibrium droplet evaporation and rainout, touchdown;
- Pool spread and vaporization;
- Heavy gas dispersion; and
- Far field passive dispersion.

The UDM allows for continuous, instantaneous, constant finite-duration and general time-varying releases. The UDM allows for possible plume lift-off when a grounded plume becomes buoyant. The UDM has been validated extensively with experimental data and is the subject of several peer-reviewed scientific papers.

5.3 FLACS CFD Model

The FLACS v9.1 model was validated against the experimental dataset in accordance with the LNG vapor dispersion Model Evaluation Protocol (MEP). A petition to have the FLACS v9.1 vapor gas dispersion model approved by the US DOT as an alternate model for LNG vapor dispersion under 49 CFR § 193.2059 was filed on October 27, 2010 [15]. The US DOT issued its Final Decision approving the use of FLACS v9.1 for vapor dispersion exclusion zone calculations on October 7, 2011 [15].

The FLACS CFD software is commercially-available and can model vapor dispersion scenarios and vapor cloud explosions. The CFD model provides a time-dependent solution to the Navier-Stokes equations. In the model, conservation equations for mass, momentum, enthalpy and gas concentrations are solved on a Cartesian grid using a finite volume method.

FLACS contains a liquid spill model which uses the shallow water equation to calculate the spread of LNG or refrigerants over the ground. FLACS also contains a pool evaporation model which calculates the heat transfer to the liquid from conductive heat transfer from the substrate, wind-induced convection and radiation. Currently FLACS is the only model approved by FERC that can be used to model the vapor clouds resulting from liquid spills onto trenches. FLACS is also the only approved model that can be used to determine the effect of structures and vapor fences on the flammable vapor cloud dispersion.

6. Significance of Changing Regulations

The changing LNG requirements have altered both, (a) the type and severity of scenarios considered, and (b) the tools used to analyze these scenarios. These changes have created a moving target for LNG facility permit applicants, where the latest FERC interpretations need to be reviewed and considered throughout the consequence modeling phase of a project.

Changes in the choice of leak scenarios have often resulted in having to consider larger and more severe LNG and refrigerant release scenarios. The jetting and flashing release scenarios, in particular, have transitioned to larger and larger leak sizes. This has resulted in the expansion of the size of the areas under facilities' control and the use of various types of mitigation strategies.

7. Mitigation Strategies

The growth in the severity of the scenarios to consider often results in a ½-LFL flammable vapor cloud that extends into an area that can be built upon. When this happens, some type of mitigation strategy must be employed to limit the extent of the analysis-based siting exclusion zones. Potential mitigation strategies include:

- Relocation of LNG and refrigerant storage and piping elements, in order to increase the distance to the property boundary.
- The use of insulated concrete to limit the amount of LNG and refrigerant evaporation during a spill.
- The use of vapor fences and other obstacles to contain the LNG and refrigerant vapor cloud during a release.
- Changes to the LNG and refrigerant flow design, by changing the size of piping, capacity and number of pumps, and process flow conditions in piping. This can reduce the worst case release flow-rate.
- Changes to the refrigerant storage capacity, which can limit the inventory and therefore the amount of refrigerant released.

6.1 Vapor Fences

Vapor fences can be used to significantly reduce the extent of a flammable vapor cloud. Typically, vapor fences are placed along a large portion of the facility boundary, to contain the vapor cloud within to the facility. In order to be effective, these vapor fences require large heights, up to 30 feet.

Figure 2 shows the effect of vapor fences at a sample LNG facility. The figure shows how the fences significantly shorten the downwind extent of the vapor cloud, but they increase the amount of fuel vapor present inside the facility.

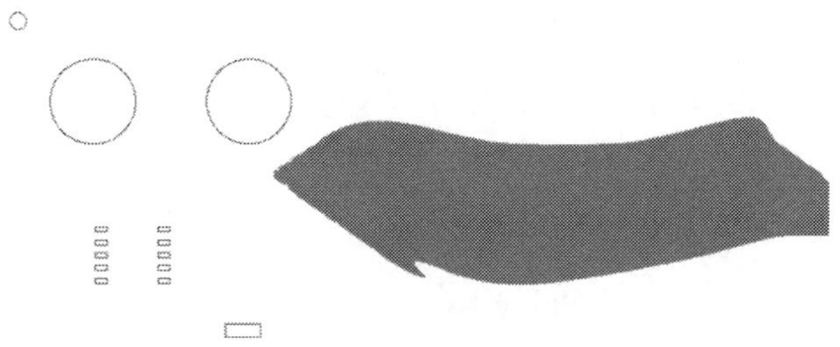

No Fences

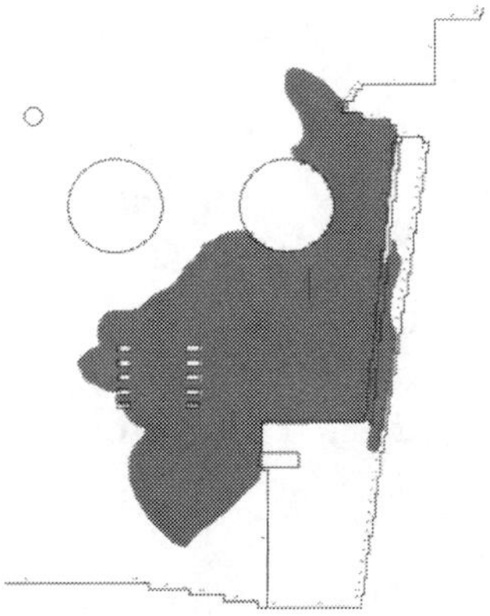

20-ft Fences

Figure 2 Effect of fences on the ½-LFL vapor cloud extent (in red) for a sample LNG jetting and flashing release scenario. Images were obtained using FLACS.

8. Conclusions

Although the requirements for LNG facility siting and design are codified in 49 CFR 193 and NFPA 59A, many of the exclusion zone analysis requirements are stated broadly and involve some level of expert interpretation. FERC has in the past clarified their interpretation of the federal requirements by means of formal letters and less formal precedent-setting memoranda. These interpretations continue to evolve over time, and drive new tools and considerations.

The changing LNG requirements have altered both, (a) the type and severity of scenarios considered, and (b) the tools used to analyze these scenarios. These changes have created a moving target for LNG facility permit applicants, where the latest FERC interpretations need to be reviewed and considered.

Changes in the choice of leak scenarios have often resulted in having to consider larger and more severe LNG and refrigerant release scenarios. The growth in the severity of the scenarios to consider has necessitated the use of mitigation strategies, including vapor fences, to limit the extent of the analysis-based siting exclusion zones.

9. References

[1] "Liquefied Natural Gas Facilities: Federal Safety Standards." Code of Federal Regulations Title 49, Pt. 193, 2012 edition.

[2] Federal Energy Regulatory Commission, "Guidance Manual for Environmental Report Preparation." FERC – Office of Energy Projects (August 2002).

[3] NFPA 59A, Standard for the Production, Storage, and Handling of Liquefied Natural Gas (LNG), 2001 Edition.

[4] "Chemical Accident Prevention Provisions." Code of Federal Regulations Title 40, Pt. 68, 2011 edition.

[5] Kohout A. U.S. "Regulatory Framework and Guidance for Siting Liquefied Natural Gas Facilities – A Lifecycle Approach". Proceedings of Mary Kay O'Connor Process Safety Center, 15th International Symposium, College Station, Texas, October 23-25, 2012.

[6] EcoEléctrica, L.P. Information Request on Exclusion Zone Modeling, Docket No. CP95-35-001, November 15, 2011.

[7] Freeport LNG Development, L.P. Environmental and Engineering Data Request, Docket No. CP12-29-000, March 20, 2012.

[8] McInerny EH, Hart RJ, Morrison DR, Kytomaa HK. "New Quantitative Risk Criteria for U.S. LNG Facilities." American Institute of Chemical Engineers, 2013 Spring Meeting, 9th Global Congress on Process Safety, San Antonio, Texas, April 28 – May 1, 2013.

[9] Corpus Christi Project LNG Engineering Conference Call, Docket No. PF12-3-000, May 7, 2012.

[10] Fire Protection Research Foundation "Evaluating Vapor Dispersion Models for Safety Analysis of LNG Facilities Research Project.",April 2007.

[11] Federal Energy Comission (FERC) "Evaluation of DEGADIS 2.1 Using Advisory Bulletin ADB-10-07." July, 2010.

[12] Fire Protection Research Foundation "LNG Source Term Models for Hazard Analysis: a Review of the State-of-the-Art and an Approach to Model Assessment." March 2009.

[13] Pipeline and Hazardous Materials Safety Administration (PHMSA) Interpretation Letter in Reference to the Proposed Waterfront Liquefied Natural Gas Plant in the City of Fall River, Massachusetts. Dated July 7, 2010.

[14] Federal Register Vol. 76, No. 137, July 18, 2011.

[15] Pipeline and Hazardous Materials Safety Administration (PHMSA) Final Decision, Docket No. 2011-0101. Dated October 7, 2011.

From LNG Imports to Exports: Process Safety and Regulatory Challenges

Jack Chosnek, Ph. D., P. E.
Knowledge*One*
President and Principal

Victor H. Edwards, Ph. D., P. E.
Director of Process Safety
IHI Engineering & Construction International Corporation

AIChE Spring Meeting
13th Topical Conference on Natural Gas Utilization
San Antonio, April 28-May 2, 2013

Introduction

As natural gas becomes more abundant in the U.S. the need for imports of gas, in the form of LNG, is disappearing and the need to find markets for the product is steadily increasing. LNG terminal operators are thus switching from LNG imports, which require re-gasification, to LNG exports, which require liquefaction resulting in dramatically changed processing of LNG. In LNG liquefaction for exporting, new flammable materials have been introduced in large quantities for cryogenic cooling: propane, ethane, ethylene, and perhaps propylene and butane. There are now present materials that can form vapor clouds that because of their low temperature and higher molecular weight will disperse more slowly reaching much further than LNG. The quantities of refrigerants that can be liberated from loss of containment can be as large as those of LNG.

In addition, as opposed to LNG vaporization when importing the material, there is significant processing which involves compression and distillation at very high pressures and low temperatures. Like LNG regasification, the natural gas and the LNG itself will be at very high pressures, on the order of 600-1,000 psig, and very high quantities will be flowing in large pipes or heat exchangers at those conditions. There is also processing of the received high-pressure pipeline gas that needs to be conditioned (e.g. removal of components such as mercury, H_2S and C2+ hydrocarbons) for the liquefaction process.

All of these represent additional hazards that typically are not known and addressed in LNG importing and re-gasification plants. There is a need to further develop data on dispersions of LNG and refrigerants at high pressures; to identify other hazards not typically found in LNG terminals; and to do facility siting studies to determine how to re-use the same facilities for two different purposes.

LNG is also heavily regulated posing challenges to producers. The Federal Energy Regulatory Commission (FERC) regulates LNG through the Department of Transportation's (DOT) Pipeline Hazardous Materials Safety Administration (PHMSA) (1-5). The main regulation is 49CFR 193 (3) which relies heavily on NFPA 59A (2) (and at that the 2001 edition and part of the 2006 edition are the only ones accepted). These regulations and standards are mainly consequence based instead of risk based. As long as there was straightforward vaporization of LNG and the only potential spills were with low pressure LNG (excluding the high pressure pumps feeding the vaporizers) there were no real problems that couldn't be solved with impoundment to comply with the regulations. This has changed significantly and guidance to and from government agencies is needed.

LNG Imports – Characteristics

LNG import facilities are comparatively simple as summarized below:

- Receive LNG from a ship
- Store LNG
- Pump to high pressure
- Heat to vaporize
- Put gas into pipeline

LNG is received from a ship and is pumped through unloading arms into pipes and then to large storage tanks that operate at low pressures. LNG is then pumped to high pressures (approximately 1,000 psi), regasified by passage through heat exchangers, and then put into pipelines.

LNG Exports – Characteristics

In contrast to LNG Imports, LNG exports are more complex. Here are typical steps in the liquefaction of natural gas:

- Receive natural gas (primarily methane) from pipeline at high pressure.

- Clean gas: Remove mercury, hydrogen sulfide, carbon dioxide, carbonyl sulfide
- Dry gas
- Remove heavies and fuel gas
- Liquefy (potentially one or more refrigeration cycles)
- Where natural gas supply contains significant nitrogen, strip nitrogen from LNG
- Send to storage tank
- Pump to ship

Natural gas, typically more than 90 percent methane is received from pipeline at high pressure (typically 1,000 psi). When processing dry pipeline natural gas, first mercury is removed by adsorption on a solid adsorbent. Then hydrogen sulfide, carbon dioxide and carbonyl sulfide are removed from the gas by absorption, typically using an aqueous amine solvent.

The wet gas is dried, usually by adsorption of water on molecular sieves. The dry gas is then cooled by heat exchange and expansion. Heavies are removed and sent to fractionation, where by-product condensate is sent out for sale and lights are used as fuel gas.

The lean gas is liquefied by refrigeration. When the lean gas contains significant amounts of nitrogen, the resulting LNG is stripped to remove nitrogen and the off-gas either used as fuel gas or flared.

LNG is then pumped to storage and then pumped to an LNG tanker for export.

Figure 1 is a block flow diagram of LNG production from natural gas.

Figure 1 - LNG Liquefaction - Linde's LIPROM Process

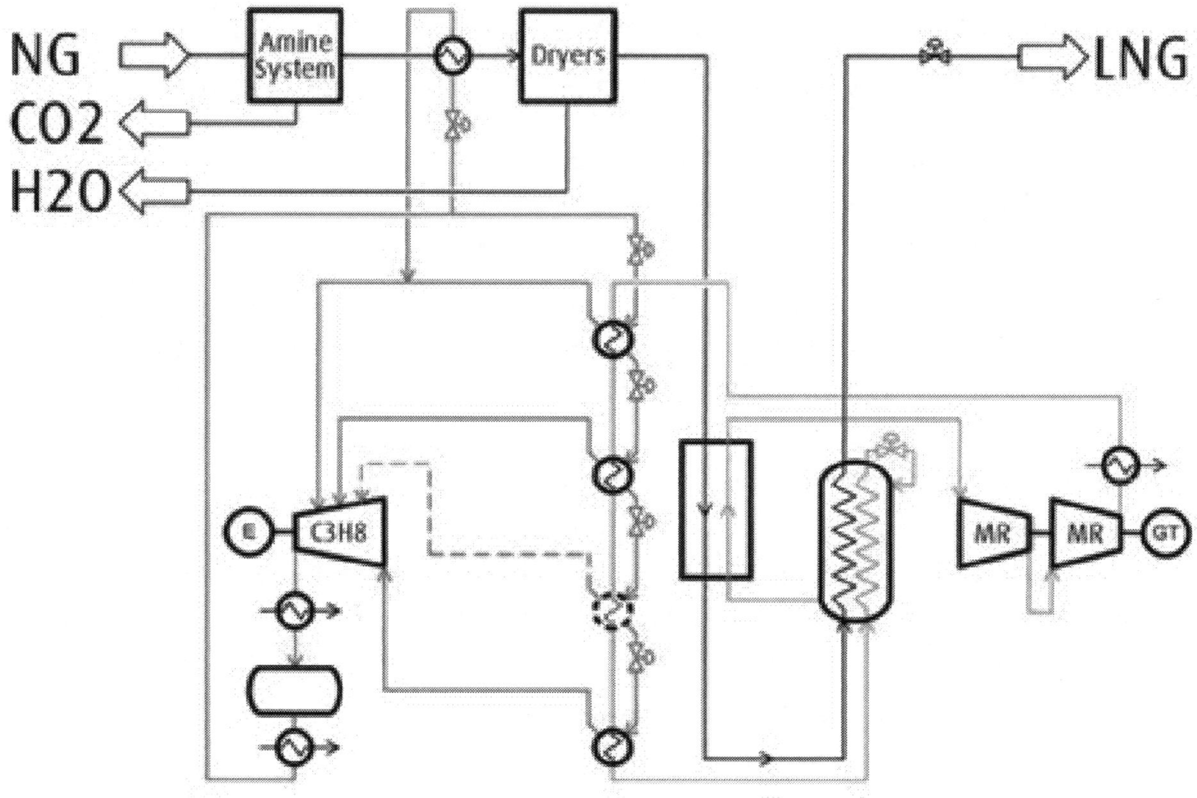

(Courtesy of Linde)

Process Chemicals – Contrasts

Table 1 contrasts the process chemicals in LNG regasification and natural gas liquefaction to LNG.

Table 1 – Contrasts in Process Chemicals in LNG Regasification and Natural Gas Liquefaction

Regasification	Liquefaction
Methane	Methane
Nitrogen	Amine
	Carbon dioxide
	Hydrogen sulfide
	Water
	Condensate (higher hydrocarbons)
	Propane, propylene
	Ethane, ethylene
	Butanes
	Mixed Refrigerant
	Nitrogen

Although LNG often contains small amounts of C2+ hydrocarbons, these are not found as a separate phase in regasification. In contrast, most natural gas liquefaction processes use and create a number of C2+ hydrocarbons streams.

Process Safety Challenges in LNG Importation

LNG import and regasification is not without its challenges. Listed below are the primary sources of process hazards:

- LNG handling, transfers and releases
- Cryogenic temperatures

- Fire
- Explosion (Low probability due to low congestion and low reactivity)
- Asphyxiation

Accidental releases of LNG pose all of the above hazards, but LNG facilities are designed to keep the risk of accidental releases very low. Explosion hazards are comparatively low because the simplicity of the process leads to low congestion and the high concentration of methane keeps reactivity low for most gas releases. (In one exception, a pool of LNG exhibits rapid differential evaporation of methane, leading to increasing concentrations of C2+ hydrocarbons in the residual pool. Explosion hazards are thus somewhat greater in weathered pools of LNG, but emergency response measures should be firmly established by the time significant weathering has occurred.)

Process Safety Challenges in LNG Exportation

Natural gas liquefaction has all of the challenges of LNG importation, plus a number of others:

- All of the process safety challenges of LNG Importation plus
- Handling of high pressure and low temperature refrigerants
- High processing temperatures (natural gas pre-treatment)
- Higher intensity fires
- Explosion (Higher probability; higher congestion and reactivity)
- Toxic exposure (H_2S) in those cases where there is significant amounts of hydrogen sulfide present in the incoming natural gas
- Training
- Facility Siting

Because natural gas liquefaction processes typically contain a number of different process streams rich in C2+ hydrocarbons, there is a higher risk of explosion caused by these more reactive compounds. In addition, the greater process complexity increases congestion due to addition of new and more complex facilities within an existing site, as well as more potential leak sites.

Where produced gas is being processed, hydrogen sulfide removed from the natural gas and concentrated during purification poses a toxic exposure hazard in the event of a release. Pipeline gas contains much less hydrogen sulfide, but in both cases common practice is to oxidize the hydrogen sulfide and the resulting sulfur oxides become an environmental emission.

The use of refrigerants or refrigerant mixtures adds the hazards of handling and storage of these materials. Propane or propylene may be used for initial refrigeration and used in mixtures with other materials such as ethane or ethylene in order to attain the low temperatures required for LNG liquefaction. These materials are typically used in closed loops where large quantities go through evaporation during heat exchange and subsequent recompression to high pressures. These materials represent also a much higher potential than methane both for the occurrence of fire and explosion and for a higher intensity it they were to occur.

The added complexity makes training of personnel more complex, with more opportunities for incidents.

The increased explosion hazards and the challenges of working within an existing site may create facility siting hazards with existing or new occupied buildings.

Regulatory Challenges – Regulatory Agencies

LNG facilities built within the United States must meet the requirements of a number of regulatory agencies (Table 2). The most specific requirements are those of the Federal Energy Regulatory Commission (FERC) and PHMSA (5). These agencies require that LNG facilities be designed to comply with NFPA (National Fire Protection Association) 59A – Standard for the Production, Storage, and Handling of Liquefied Natural Gas (LNG), as well as other applicable industry codes and standards.

Table 2 – Agencies Regulating LNG in the USA

U. S. FERC (Federal Energy Regulatory Commission)

U. S. DOE (export/import permits)

U. S. DOT (PHMSA-Pipeline & Hazardous Materials
Safety Administration)

U. S. Department of Homeland Security (Coast Guard)

U. S. EPA

U. S. Fish and Wildlife Service

NOAA (in some instances)

State and Local HSE and Utility Agencies

Advocate NGOs

Although NGOs (Non-Governmental Organizations) are not governmental regulatory agencies, they often actively promote the strict enforcement of existing regulations, the aggressive interpretation of existing law, and the creation of new law.

Regulatory Challenges - LNG Imports

Current US regulations focus on LNG import and regasification facilities (Table 3). FERC is the lead federal agency and with PHMSA regulates domestic LNG facilities through 49 CFR 193 (3). This regulation draws heavily on NFPA 59A (2).

Table 3 – Principal Regulations Addressing LNG in the USA

- **FERC (49 CFR 193 relies on NFPA 59A - 2001)**

  Federal - *Veto power*
- **U.S. DOT (PHMSA)**
- **U.S. Department of Homeland Security – U. S. Coast Guard Waterway Suitability Assessment (33 CFR 127 for safety and 33 CFR 105 for security)**
- **U.S. EPA**
- **State - *Veto power***
  - **Coastal Zone Management Act**
  - **Clean Water Act**
  - **Clean Air Act**

FERC requires a detailed life cycle approach to monitor and approve the siting, engineering design, construction, and operation of LNG facilities (5). Permits to operate LNG facilities are required by FERC, and FERC has veto power over these domestic facilities. FERC normally prepares the Environmental Impact Assessment for new LNG facilities.

In addition, the US Coast Guard has regulatory authority over the waterfront LNG import and export facilities (4).

States also have veto power over LNG facilities through delegated federal regulations such as the Coastal Zone Management Act, the Clean Water Act, and the Clean Air Act.

Regulatory Challenges – LNG Exports

The existing federal regulations (2,4-6)) rely heavily on the 2001 edition of NFPA 59A (3). They mention natural gas liquefaction only briefly, and were developed when liquefaction was used primarily at peak shaving facilities. The primary focus is for LNG imports.

FERC and NFPA 59A are consequence-based (Table 4).

NFPA 59A requires that a "design spill" doesn't surpass the limits shown in Table 4 at the property line with only passive mitigations (e.g. dikes or fixed barriers, gravity-flow impoundment, etc.). The LNG tanks in the facility are also included in the exclusion zone for radiation and overpressure. The "design spill" is created from a pipe break containing LNG that will last 10 minutes, and more if there is inventory left in the pipe. This approach made sense for regasification plants where the major hazard is an LNG spill which can be controlled by impoundment.

Table 4 – Consequence-based FERC Regulatory Criteria

Exclusion Zones (at the property line) from a release caused by a "design spill"

- ½ Lower Flammable Limit
- 1,600 Btu/hr-ft^2 thermal radiation
- 1 psi overpressure

Mainly consequence based

Only passive mitigations allowed (impoundment to basins, vapor fences, fans)

FERC recently clarified (6) the conditions for the piping ruptures, providing initiating frequencies for breaks and ruptures based on valve count, pipe lengths and diameters rather than full-bore breaks for all pipe sizes. FERC based this on published data of failures of piping to/from LNG tanks and deemed this failure rate as acceptable for all piping, modified in proportion to their failure rate based on diameter and length. This defined more reasonably a design spill than the traditional worst case scenario but let intact the consequence to be avoided and the restriction on using only passive mitigations.

A consequence-based approach is very difficult for complex processing operations where releases cannot be mitigated by passive means. For example, a high pressure release at elevation where there is no liquid pool formation (and therefore no possibility for impoundment of the spill) does not lend itself well to a consequence-based approach.

In addition, the restriction to allow only passive mitigations, such as impoundment and vapor fences, imposes a potential negative impact on design and facility siting when trying to mitigate releases from LNG export facilities.

Conclusions

Natural gas liquefaction and export is a safe and proven technology and poses less hazard than many other chemical manufacturing processes. However, conversion of LNG Import Terminals to natural gas liquefaction and purification facilities requires more complex processing and involves significant inventories of compounds that pose somewhat greater hazards than LNG Import Terminals. Therefore, the careful application of industry best practices in the conversion of LNG Import Terminals to LNG Export is essential.

Better regulations, based on dialog with regulatory agencies and a shift from consequence-based regulations to the risk-based approach to process safety of natural gas liquefaction plants advocated by Woodward and Pitblado (1) would be helpful.

In addition, improved modeling tools and a better understanding of potential effects on the community are needed.

Acknowledgements

Constructively critical reviews by Kamal Shah and Frank Del Nogal are appreciated.

References

1 Woodward, John L. and Robin M. Pitblado, "LNG Risk Based Safety-Modeling and Consequence Analysis", John Wiley & Sons, Inc., Hoboken, New Jersey, 2010.

2 NFPA 59A, "Standard for the Production, Storage, and Handling of Natural Gas (LNG)", National Fire Protection Association, Quincy, MA, 2001 and 2006.

3 Volume 49 Code of Federal Regulations, Part 193, "Liquefied Natural Gas Facilities: Federal Safety Standards", United States, Washington, DC, 2013.

4 Volume 33, Code of Federal Regulations, Part 127, "Facilities Handling Liquefied Natural Gas and Liquefied Hazardous Gases", United States, Washington, DC, 2013.

5 Kohout, Andrew, "U. S. Regulatory Framework and Guidance for Siting of Liquefied Natural Gas Facilities – A Lifecycle Approach", paper presented at 15th International Symposium, Mary Kay O'Connor Process Safety Center, College Station, TX, October 23-25, 2012.

6 FERC and Cheniere Meeting Summary, Corpus Christi Project, Docket PF12-3.000, May 7, 2012

Biography

Jack Chosnek, Ph. D., P. E., Knowledge*One*, www.knowledge1.net Jack is a consultant in Process Safety Management and Knowledge Management and has over thirty five years of experience in the petrochemical industry. He has consulted for companies in the chemical, oil and gas, offshore, waste management and mining industries, developing policies and implementing process safety management systems, facilitating PHAs, writing operating procedures and conducting incident investigations and process safety audits. He has developed commercial, full-featured software for PHA facilitation and Management of Change (MOC). Jack consulted as the principal risk engineer for two multi-billion dollar projects carried out by large engineering firms.

Before becoming a consultant Jack worked for Celanese Corp. for 25 years in various positions in process safety management, technology management, operations, process engineering, piloting, tolling and R&D. Jack was a member of a project team that developed one of the first commercial scale-ups in the biotechnology industry.

Jack is the Chairman of the Process Safety Workshops at the South Texas Section (STS) of AIChE. He chaired the Process Plant Safety Symposium (PPSS) at the 4th Global Congress on Process Safety, and has chaired and co-chaired other process safety sessions in other years. He is the chair of the Technical Advisory Committee at Texas A&M's Mary Kay O'Connor Process Safety Center.

Jack is the holder of three patents related to chemical production. He has a BS and MS from the Technion—Israel Institute of Technology, and a PhD from the University of Missouri at Rolla, all in Chemical Engineering, as well as an MBA from Texas A&M—Corpus Christi.

Victor H. Edwards, Ph. D., P. E. is Director of Process Safety for IHI Engineers and Constructors International, 1080 Eldridge Parkway, Houston, TX 77077 (Telephone: 1-832-379-7742; e-mail: vic.edwards@ihi-ec.com). Responsible for health, safety, and environment in design, Vic has worked for IHI and predecessor companies for 29 years. His experience includes process engineering, process safety management, and process, biochemical, environmental, and mineral processing technologies. His project assignments have included biochemical and chemical processes, biomass energy resources, coal gasification, coal liquefaction, liquefied natural gas, chemical weapons demilitarization, hazardous waste disposal, offshore hydrocarbons facilities, and extraction of metals.

Dr. Edwards has received three DuPont awards for safety and environmental engineering excellence and two DuPont awards for health, safety, and environmental excellence during 28 years as an alliance engineering contractor. In 1998, Kvaerner Engineers and Constructors named him employee of the year. In 2007, Vic's Houston HSE team was recognized by Aker Kvaerner with the Process and Construction Excellence Award for HSE in Design.

Vic's earlier experience included Assistant Professor of Chemical Engineering at Cornell University, Program Manager at the National Science Foundation (on leave from Cornell), Research Fellow at Merck, alternate energy research at United Energy Resources, Visiting Professor of Environmental Engineering at Rice University, and process engineering at Fluor.

Vic has contributed more than sixty publications to the engineering literature, including several book chapters. He chaired the 1992 Process Plant Safety Symposium, the 1997 International Plant Operations and Design Conference, and the 2013 Global Congress on Process Safety, as well as numerous technical sessions at other conferences. Vic received the 2003 Service Award from the Process Safety Center at Texas A & M University, where he chaired the Technical Advisory Committee from 2004 to 2011. In 2009, Vic was elected to the Centennial Council of the Department of Chemical and Biomolecular Engineering at Rice University. He also serves on the Editorial Advisory Board of CHEMICAL PROCESSING magazine. Edwards earned his B. A. from Rice University and his Ph. D. from the University of California at Berkeley; both degrees in chemical engineering. A registered professional engineer in Texas, he is an AIChE Fellow, and a member of ACS, AAAS, NFPA, NSPE, and the New York Academy of Sciences.

The mitigation conundrum: when reducing one hazard may increase another

Filippo Gavelli, Ph.D., P.E.
Principal Engineer, Head of Dispersion Consulting, GexCon US

Scott Davis, Ph.D., P.E.
Principal Engineer, President, GexCon US

AIChE Spring Meeting
13th Topical Conference on Natural Gas Utilization
San Antonio, April 28-May 2, 2013

Abstract

Siting regulations for onshore LNG facilities in the United States require the consequences of credible accidental scenarios to be confined within the facility's boundaries – that is, no harm should occur to the public or public property from the worst credible events. Compliance with regulations based on worst credible events is often challenging and may require mitigation methods. One example is that of vapor barriers being used to control the dispersion of a flammable cloud from a release within the facility, in order to keep the cloud from extending beyond the property boundaries before dissipating below ½-LFL.

However, using barriers to keep the flammable cloud from dispersing too far downwind is likely to result in a larger volume of gas remaining near the release location and possibly within congested areas inside the facility. This may therefore increase the hazard footprint from the worst credible overpressure scenarios. Similarly, the flow obstruction posed by the barriers may decrease the air flow through the area increasing the overpressure hazard from smaller releases. Since small releases tend to occur more frequently, increasing their potential severity is likely to increase the overall risk to which facility personnel are exposed. Fortunately, not all barrier installations will have an adverse impact on overpressure hazard and/or personnel risk.

This paper will examine the impact of vapor barriers on overpressure hazards for different barrier configurations. The case study will apply CFD modeling to realistic scenarios, inside a detailed 3D model of a typical LNG liquefaction facility.

Introduction

The siting of LNG facilities in the United States is regulated by prescriptive requirements for the hazard analysis. The tacit rationale behind prescriptive (or deterministic) regulations is that there may be no hazard to the public or public property in the event of an accident. Since it would be effectively impossible to protect the public against cataclysmic events to nearly any kind of facility (the example of ground-splitting earthquakes has been used in the past to argue against the safety of LNG terminals), one of the critical parts of prescriptive regulations is the definition of the maximum credible event (MCE - even though that term is not explicitly used in the regulatory language). Facility developers need to demonstrate that the hazard footprint, to specified threshold levels, for any accidental scenario of magnitude up to the MCE.

Over the last several years, and particularly since the recent market changes led to numerous LNG export facilities being proposed in the United States, the criteria for the selection of the MCE have evolved significantly: in addition to the "design spill" defined in NFPA 59A, which was the only scenario required for LNG import facilities until a few years ago, "flashing and jetting" release scenarios for both LNG and flammable refrigerants (if used) are now also required to be evaluated. Accidental leaks selection criteria have not yet been codified in the regulations, therefore, analysts currently need to rely upon practical guidance from regulatory staff.

Practical Scenarios

Even though every facility is different, flashing and jetting MCEs in excess of 100 kg/s are rather common. For these scenarios, the hazard distances for unimpeded vapor cloud dispersion is likely to extend beyond the property boundaries. In these cases, prescriptive regulations allow the use of passive mitigation measures in order to control the vapor cloud dispersion and thus ensure that the hazard footprint from the MCEs will not extend beyond areas under the facility owner's control.

The most common vapor dispersion mitigation measure adopted is that of vapor barriers. Therefore, most LNG facilities being proposed in the United States currently require one or more sets of vapor barriers to ensure public safety according to the regulations.

In certain cases, facilities have been able to demonstrate that installing a single row of vapor barriers along, or just inside, the property boundaries is sufficient to meet the regulatory requirements for vapor dispersion hazards. This is generally the case when there is sufficient distance between the accidental leakage source (e.g., high-pressure piping within the liquefaction area) and the boundary, so that the flashing jet release loses much of its momentum before reaching the barrier; the low-momentum, dense vapor cloud is unable to “climb” the barrier and instead spreads laterally along the length of the fence, remaining within the facility’s boundaries. Figure 1 shows such an example: the vertical cut plane through the vapor cloud shows a small increase in thickness as the cloud reaches the barrier, pushed primarily by the wind (as for a cloud produced by vaporization of a liquid spill).

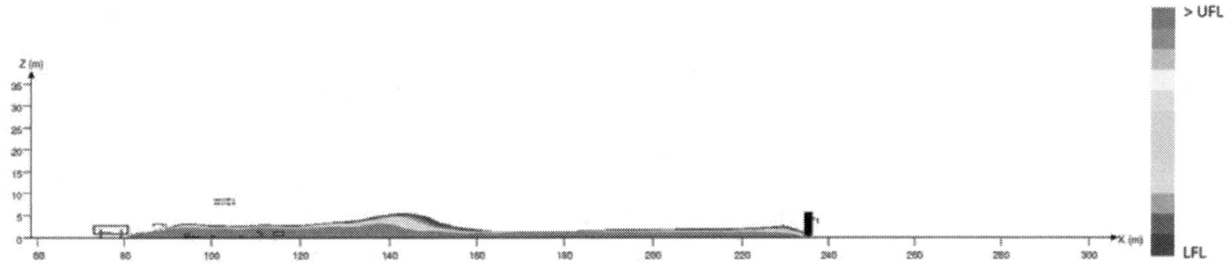

Figure 1. Low-momentum vapor cloud, contained by fenceline barrier

There are, however, situations in which the flashing jet release from an accidental leakage source reaches the vapor barrier with significant residual momentum. In these cases, the vapor cloud will manage to "jump" the barrier and spill over the opposite side, as shown in Figure 2.

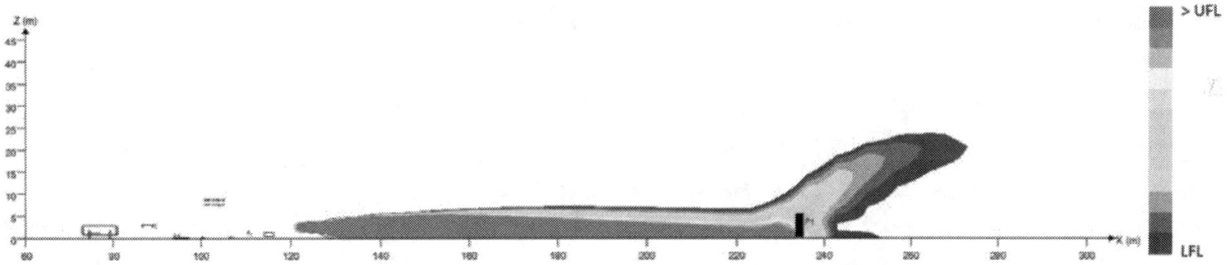

Figure 2. High-momentum vapor cloud, not contained by fenceline barrier

If the barrier is located along, or just inside, the property line, the barrier will be inadequate to control the vapor cloud dispersion in compliance with regulatory requirements. Potential solutions include: a) raising the vapor barrier height, or b) adding a second row of barriers between the leakage source and the property line. The barrier height necessary to control a momentum-driven dense vapor cloud can exceed 30 ft and become very expensive, due the need to sustain high wind loading, earthquake vibrations, etc. Therefore, adding a second row of barriers is often considered the most practical solution: the inner row intercepts the vapor cloud from a flashing jet release and reduces its momentum by forcing it to "jump" over the barrier; the inner row also reduces the mass flux of vapor that reaches the fenceline barrier. When the cloud reaches the fenceline barrier, it is smaller and with less momentum, making it possible for the barrier to contain it; in some cases, proper placement of the first barrier may render the fenceline barrier unnecessary, as shown in Figure 3.

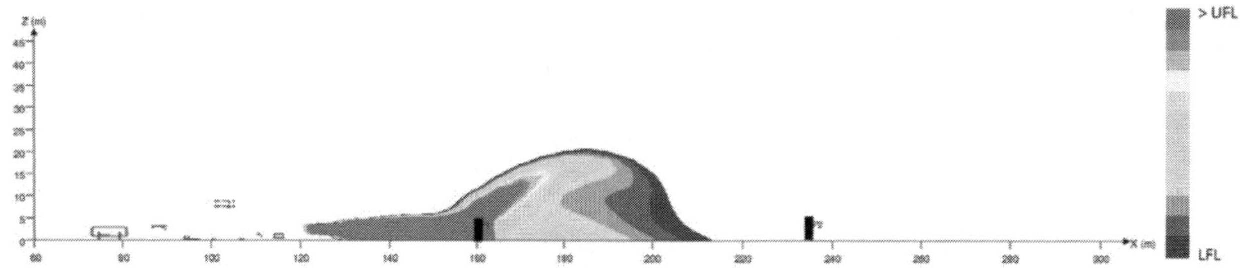

Figure 3. High-momentum vapor cloud, contained by two rows of barriers (the fenceline barrier is not strictly necessary in this case)

The conundrum

As shown in the examples above, by design vapor barriers tend to contain the vapor within a smaller area. This however may have an unintended adverse effect, particularly for the more reactive flammable refrigerants, as it may result in larger flammable volumes near the release location, which is often within or close to a congested area (e.g., the gas processing or liquefaction areas). Therefore, the use of vapor barriers to mitigate vapor dispersion hazards can – at least in principle – increase a facility's overpressure hazards. In practice, as will be discussed shortly, vapor barriers are not likely to result in increased overpressure hazards for the larger, less probable scenarios that must be evaluated for facility siting purposes; they are, however, likely to increase the hazard footprint of the smaller, more probable accidents.

Figure 4 shows an LNG liquefaction train from a small-scale LNG liquefaction facility. For the purpose of this example, the rest of the facility has been removed. Three different cases are considered, to evaluate the effect of vapor barrier location on potential overpressure hazards:

1. The reference case, without any vapor barriers (shown in Figure 4);
2. Vapor barriers on three sides, approximately 25 m away from the edges of the liquefaction train (shown in Figure 5);
3. Vapor barriers on three sides, approximately 10 m away from the edges of the liquefaction train (shown in Figure 6).

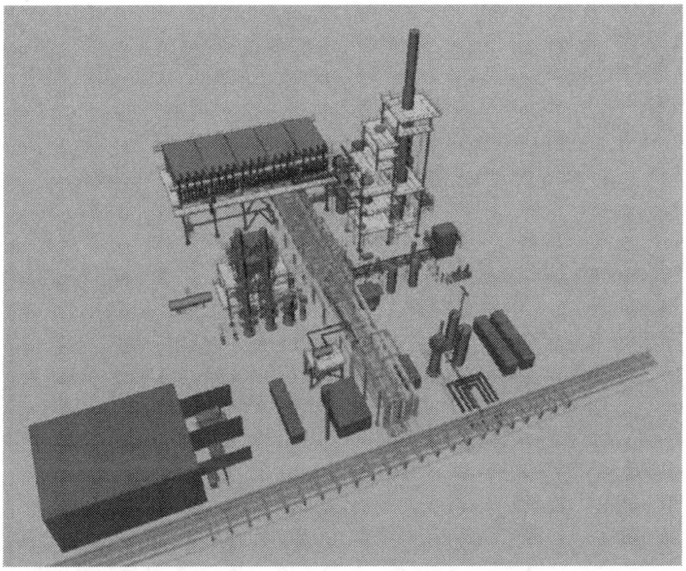

Figure 4. Reference case – no barriers

Figure 5. Wide barrier spacing (25 m outside process area)

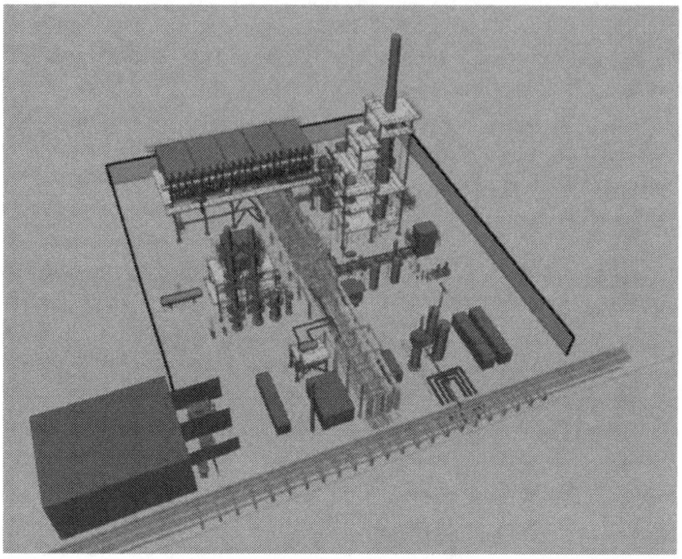

Figure 6. Tight barrier spacing (10 m outside process area)

A frequently used measure of the propensity for gas cloud accumulation within a space is the ventilation rate for a given wind speed and direction; this is often expressed in air changes per hour (ACH). Figure 7 shows the wind velocity vectors for the three barrier configurations, for an external wind at 2 m/s from the west. The vector plots show that the presence of the barriers affects the air flow through the liquefaction area, causing near-stagnant regions; these low-flow areas are more pronounced, as expected, for the tighter fence layout. The ACH rates for these scenarios are listed in Table 1; the values are normalized to the reference case (no barriers).

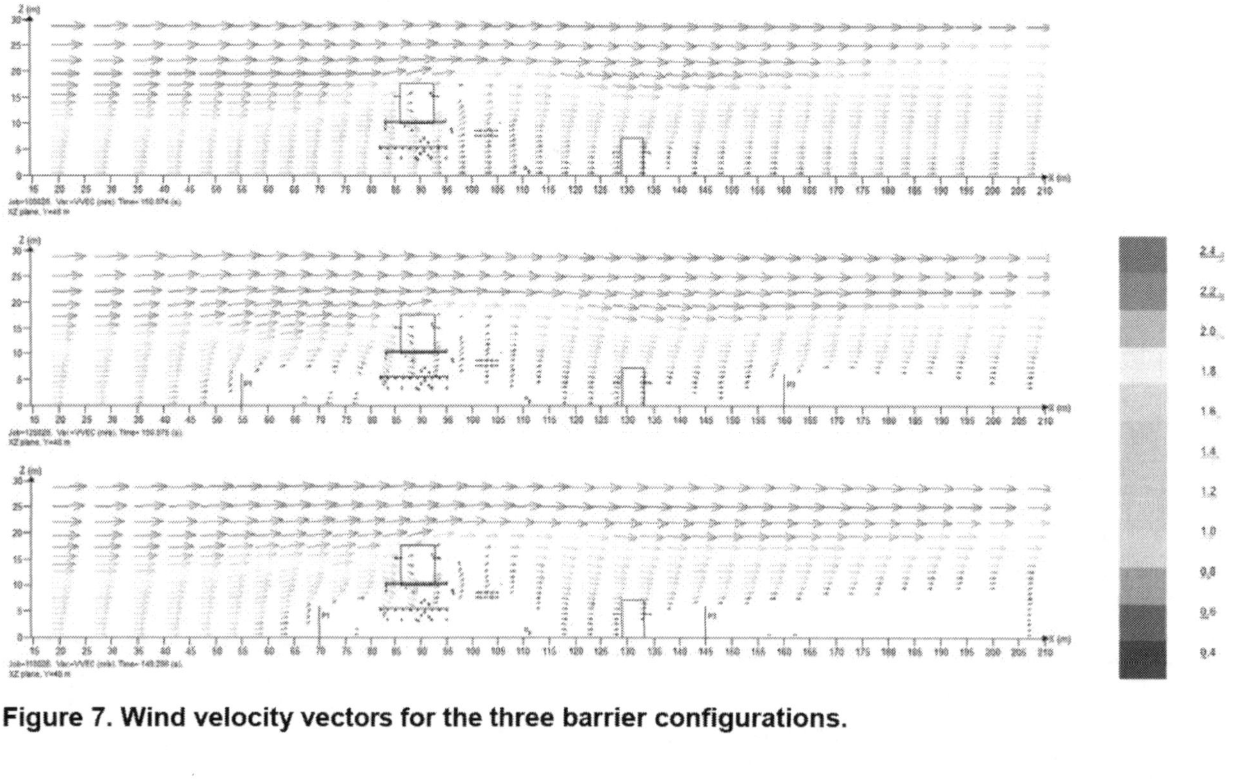

Figure 7. Wind velocity vectors for the three barrier configurations.

Table 1. Ventilation rates for different barrier spacings

Barrier gap	ACH (normalized)
No barriers	100%
25 m	91%
10 m	85%

The lower ACH due to the presence of vapor barriers will result, for any given release, in slower dilution of the vapor cloud. This, however, does not necessarily mean that overpressure hazards form the ignition of a vapor cloud will always be worse as ventilation is reduced. For example, Table 2 shows the max flammable volume obtained for a large release (over 100 kg/s) of a refrigerant within the liquefaction train. The results show that adding barriers increases the size of the flammable cloud in the congested area near the release, as the barriers limit the downwind dispersion of the cloud. However, the tighter barrier scenario (10 m out) results in a smaller flammable volume than the wider barrier gap (25 m); this occurs because, for a large release, the lower ventilation rate results in a richer cloud and therefore a smaller portion in the flammable range. In a prescriptive regulatory environment, the largest flammable volume determines the worst-case conditions for overpressure hazards and therefore the 10 m barrier distance would result in lesser consequences than the 25 m case.

Table 2. Max. flammable volumes for a large release

Barrier gap	Max. cloud volume (normalized)
No barriers	100%
25 m	119%
10 m	108%

Failure data for these facilities indicate that very large releases are very unlikely to occur; the more likely failure scenarios instead are small releases – e.g., from valve stem packing failure – on the order of 1 kg/s or less. While these releases do not pose a threat to the public, they may pose a threat to personnel safety. Table 3 shows the max flammable volume obtained for a 1 kg/s release of a refrigerant within the liquefaction train. The results show that as ventilation in reduced (tighter barrier spacing) the maximum flammable volume increases and therefore so does the overpressure hazard. This occurs because, for a small release, only a small portion of the cloud remains above the flammable range.

Table 3. Max. flammable volumes for a small release

Barrier gap	Max. cloud volume (normalized)
No barriers	100%
25 m	112%
10 m	127%

Conclusions

This paper examined the potential unintended effects of placing vapor barriers around congested areas of an LNG facility, to control large vapor cloud releases in order to comply with U.S. federal regulations. The analysis showed that, as expected, adding vapor barriers reduces the ventilation rates within the fenced-in area. The ventilation reduction generally results in larger flammable cloud sizes for small releases; the effect on large releases, however, is more difficult to predict and may actually result in smaller flammable volumes as much of the vapor cloud may remain above the flammable range.

Since U.S federal regulation require facility siting based on large releases and, for a given fuel and congested area, overpressures tend to increase with the size of the flammable cloud, the addition of vapor barriers to a facility does not necessarily result in a larger overpressure footprint for the LNG facility. However, the overpressure hazards for smaller releases are more likely to be increases, therefore, personnel hazards may be increased. Similarly, if a facility siting were subject to risk-based regulations, the increased consequences from the more frequent, smaller releases may indeed result in a larger risk footprint.

Given the large number of parameters that affect ventilation, gas dispersion and overpressures, the reader is cautioned to remember that the results presented in this paper are case-specific. As such, the numerical values and trend reported above should not be applied to any facility in lieu of a site-specific analysis.

Author Biographies

Dr. Filippo Gavelli is the Head of the Dispersion Consulting group at GexCon and is responsible for GexCon's LNG safety consulting business worldwide. He specializes in the analysis of heat transfer and fluid flow phenomena, including multiphase flows and cryogenic fluids. He has over 10 years of engineering consulting experience and 20 years of experience in computational fluid dynamics (CFD) modeling, using several research and commercial codes. He applies his expertise to modeling the atmospheric dispersion of hazardous gaseous releases, and has extensive experience modeling hazard scenarios for Liquefied Natural Gas (LNG) facilities, including vapor cloud dispersion, pool fires and vapor cloud deflagrations; his experience includes more than

20 LNG installations worldwide, including onshore, offshore and floating (FLNG) facilities for LNG import, export (liquefaction) and peak shaving.

Dr. Gavelli is a member of the technical committees for NFPA 59A (LNG facilities standard) and NFPA 2 (hydrogen technologies code) and was the lead in the model validation effort that led to FLACS' approval by the U.S. DOT under 49 CFR 193. Dr. Gavelli is the lead author of several gas dispersion and explosion safety-related papers and has been a regular contributor to LNG-related technical committees and expert panels for several years.

He is a member of the GexCon docents group, which develops and delivers safety seminars to facility owners and operators, safety engineers, and regulatory agencies, on the hazards associated with gas explosions, dust explosions and LNG. Dr. Gavelli is also responsible for software technical support and training for FLACS customers in North America and has provided training to over 100 users, including staff from regulatory and government agencies.

Dr. Scott Davis is the President at GexCon US and specializes in the engineering analysis and testing of combustion, thermal, and fluid processes. Dr. Davis is responsible for fire and explosion related activities, which include post-incident investigative work, worldwide training and experimentation, as well as performing risk assessments and safety studies for offshore and floating oil & gas installations, petrochemical facilities, and various other industries. These studies include explosion risk assessment, blast and venting analyses, assessment of combustible dust explosions, toxic/flammable gas releases and dispersion, hydrogen safety, ventilation, detector placement, and carbon monoxide dispersion with the assistance of the world-leading FLACS software developed by GexCon.

Dr. Davis serves on the committee responsible for NFPA 720 *Standard for the Installation of Carbon Monoxide (CO) Detection and Warning Equipment* and has served on that for NFPA 921 *Guide for Fire and Explosion Investigations*. He is also a member of GexCon's docents group, which develops and delivers worldwide industrial seminars to owners, operators, safety engineers, and regulatory agencies, on the hazards associated with gas explosions, dust explosions and LNG.

Operational Excellence —
Essential for Success in the 21st Century

John S. Mitchell
johnsmitchell2@gmail.com

13 AIChE
Spring Meeting, San Antonio, TX

Operational Excellence

- **Stated as a primary objective by many operating enterprises committed to achieving and sustaining best in field**
- **Great cachet at the highest executive levels**
- **What does it mean, how is it implemented, what are benefits?**

"Operational"

An operating enterprise; one that utilizes some physical means to produce and / or deliver a product; comply with a mission.

Applies to a broad range of industrial enterprises:

- Process, production, power generation and distribution, continuous and discrete manufacturing, mining, food processing, transportation, marine, etc.

Equally applicable to service industries, public transportation, municipalities (fresh and waste water treatment), military, financial institutions, not for profits, schools and other types of organizations where mission effectiveness and customer satisfaction, rather than profit, may be the primary objective

"Excellence"

Objective that must be achieved and sustained for continuing success

13 AIChE
Spring Meeting, San Antonio, TX

2

Defined by three global industry leaders

A critical driver for business success.
...the systematic management of
safety, health, environment, reliability and efficiency
to achieve world class performance.
Chevron Operational Excellence Management System

Ensuring the safety and reliability of our operations is fundamental to our business success and a critical challenge that ExxonMobil takes on every day.
ExxonMobil 2008 Annual Report
...we seek to maximize value and improve efficiency.
ExxonMobil 2011 Annual Report

"...an integrated management system that drives business productivity by applying proven practices and procedures in three "foundation blocks:"
Asset Productivity
Capital Effectiveness
Operations Risk Management
DuPont: Delivering Operational Excellence to the Global Market, 2005

13A
Spring Meeting, San Antonio, TX

3

A simpler definition

Safely Create Greatest Sustainable Value

- **Safety excellence above all**
  - *Includes health and environmentally sound operation (SHE)*
  - *Operational Excellence not possible without SHE excellence*
- **Value defined for the specific business / mission**
  - *Drives business / mission requirements into operating objectives*
    - Maximizes value delivered
      - *Profitability and shareholder value*
      - *Optimum mission effectiveness*
    - Results in effective operations / mission compliance
      - *Customer satisfaction*
- **Sustainable (permanent) strategic, operational, mission and business success**
  - *Physical plant not degraded for temporary gain (reducing costs below sustaining)*

13AIChE
Spring Meeting, San Antonio, TX

4

Characteristics

- **Strategic; enterprise wide imperative**
  - *Integrates business, technology, process and practice to attain greatest value*
- **Much more than a management system**
  - *Overall, holistic, continuing master program of excellence*
  - *High performance, cooperative, success oriented work culture*
  - *Elevates mindset, actions and activities to safely creating and sustaining greatest value*
- **Applies broadly to all processes, practices, activities and tasks**
  - *Includes, governs and coordinates all enterprises functional improvement programs to achieve a common set of business / mission value objectives*
  - *Drives cooperation, mutually reinforcing teamwork throughout all operational and support functions:*
    - Operations, maintenance, engineering, IT, human relations, finance, etc.
  - *Creates positive work culture, organization and human relationships*
    - Empowered, multi function, high performance teams,
    - Engaged employees: initiative, commitment to excellence, ownership, accountability
  - *Gains greatest results from proven procedures, practices and technologies*
    - Lean, Six Sigma, RCM, PM, CBM, etc.
- **Opportunity driven**
  - *Value prioritized improvements based on history (data, information) and risk*
- **Continually improving**
- **Sustaining**

13 AIChE Spring Meeting, San Antonio, TX

5

Governing principles

- **Vivid vision** — *a clear, compelling description of the necessity and benefits of the improvement initiative communicated and advocated by a committed, engaged senior executive actively supporting the Operational Excellence program*
- **Total understanding of requirements; commitment to excellence in all activities, continual improvement to performance and effectiveness** — *throughout the enterprise*
- **Identifying and eliminating deficiencies is the only way to permanently improve performance** — *reducing costs without eliminating the need for spending results in inefficiency and ultimately increased costs*
- **Ambitious, optimistic, and achievable improvement objectives** — *to SHE / EHS (Safety Health Environment), mission, business, operating performance and efficiency; reduced cost*
- **Strong management and operating leadership** — *enthusiastic, supportive, committed to success*
- **Ownership and empowerment** — *at all levels of the organization*
- **Responsibility at the working level** — *for identifying, valuing, developing, implementing improvement initiatives; accountability for success*
- **Linked, layered results metrics** — *demonstrating value contribution to business / mission objectives*
- **Complete data and information** — *accurate, secure, easily accessible and understood*

13 AIChE Spring Meeting, San Antonio, TX

The program

- **Demands — commitment to and ownership for excellence**
  - *Mutually supporting, cooperative, partnership relationships— minimize adversarial "silos"*
  - *Transformation of the work culture to— value directed, initiative, ownership, responsibility*
- **Emphasizes — risk and reliability**
  - *Minimize failures / incidents—increase safe operating lifetime*
- **Moves — from a task and activity oriented cost center to—** *results driven, profit center mentality*
  - *Work elevated from status quo / restoration to— eliminating defects; continuous improvement*
  - *Safely reduce requirements for work— permanently and sustainably reduce costs*
- **Extends — definition of reliability across all activities and the entire organization**

"Positive attitudes and behaviors are the key essentials to gaining success"
Dr. Peter G. Martin, Vice President; Invensys

13AIChE
Spring Meeting, San Antonio, TX

7

How implemented

- **Begins with a visionary, engaged senior executive totally committed to success**
  - *Establishes vision, business / mission objectives*
- **Directed strategically and facilitated by a facility / site Steering Team**
  - *Define program*
  - *Establish mission / business value prioritized improvement strategy*
  - *Appoint Program Leader / Champion*
- **Led by Operating Leadership Team(s)**
  - *Headed by Program Leader*
  - *Develop program objectives and strategy*
  - *Identify opportunities for improvement*
- **Implemented at the working level by multi-function, high performance Improvement Action Teams**
  - *Empowered members*
  - *Refine opportunities, develop and implement improvement action plans, identify, report and correct deficiencies, direct continuous improvement*

13AIChE
Spring Meeting, San Antonio, TX

8

Requirements for success

- **Total commitment to objectives for excellence; intolerant of deviations**
  - *Ambitious, clearly stated objectives*
- **Culture of excellence established and sustained**
  - *Importance, necessity and benefits fully understood by everyone in the organization*
    - Essential nature of organizational and individual conformance
    - Empowerment — initiative, ownership and accountability for success
- **Robust, reliable management and administrative systems**
  - *Clearly stated organizational objectives and individual responsibilities*
- **Full adherence to practices and procedures**
  - *Identify, report and correct deficiencies, continuously improve*
    - Learn from activities, successes, mistakes and near misses
- **Continuing learning, training and improvement**
  - *Concentrate on identifying opportunities for improvement; correcting deficiencies, strengthening weaknesses*

Do these characteristics and requirements remind you of something that exists today in your organization?

13 AIChE Spring Meeting, San Antonio, TX

9

Safety!

The Operational Excellence program

Program Highlights

- **Enterprise wide, value driven improvement initiative**
- **Equivalent to safety in terms of imperative and ownership for excellence; organizational and individual commitment**
  - *Demands initiative and accountability at all levels*
- **Sponsored and driven by a motivated, involved top executive**
  - *Responsible to establish, implement, nurture program—demand results*
- **Operational leadership preferably by production / operations**
- **Recommended implementation in a circular progression DIPICI**
  - *Similar to PDCA (Shewart Deming), DMAIC (Six Sigma), OODA (Boyd)*
  - *Iterative process*
- **Prime responsibility for improvement initiatives at working level**
  - *Detailed analysis, value prioritization, development and implementation of improvement plans*
  - *Essential for full ownership, commitment to success*
- **Includes assessment, feedback, improvement and follow up mechanisms to assure compliance with objectives**
  - *Formal assessment process*
- **Tangible rewards for successfully meeting objectives**
- **Produces real, sustainable value**

13 AIChE
Spring Meeting, San Antonio, TX

12

Descriptive elements — Operational Excellence Program

- **Typically begins with a directive from a top executive**
  - *States vision, necessity, business and financial objectives*
  - *Requires continuing visible engagement, drive and support*
- **Opportunity driven, master program—continuing journey**
  - *Identification and exploitation of opportunities to safely and sustainably increase business value, reduce risk and lost opportunity*
    - Improvement opportunities prioritized by financial value and return
    - Direct link from Operational Excellence program KPI's to business and financial objectives — demonstrate program contribution
- **Starts from where you are (not where you might think you are) — Gets you where you need to go**
  - *Comprehensive commencement assessment utilized to identify:*
    - Current conditions and strengths to build on
      - *Utilize institutional knowledge and ownership to greatest advantage*
    - Opportunities for improvement
    - Weaknesses to strengthen; barriers to overcome
    - Progress to success
- **Optimizes balance between reliability, risk and value**
  - *Increase reliability—Improve availability and effectiveness; reduce spending*
  - *Opportunity is the positivo aspect of risk*

13AIChE
Spring Meeting, San Antonio, TX

13

Descriptive elements — Operational Excellence Program

- **Working level improvement action teams responsible for identifying and valuing opportunities; planning and driving results**
  - *Strategy, analysis, discovery — all team developed*
    - Value and risk driven solutions
  - *Creates energy, ownership, commitment and responsibility*
- **Opportunistic, value oriented implementation — not sequential**
  - *Imaginative exploitation of opportunities*
- **Solid change (improvement) management — essential**
  - *Master the "soft" work culture and organizational issues: values, behavior, initiative, motivation, ownership*
- **Constructed on accurate data and information**
  - *Accurate, reliable, information system*
- **Gains greatest value from proven best practices: Lean, Six Sigma, RCA, PM, CBM, TPM, RCM, etc.**
  - *Prioritized implementation*
- **Demands "excellence at the basics"**
  - *Operations / Production efficiency*
  - *Work management and execution*
  - *Logistics management*

13AIChE
Spring Meeting, San Antonio, TX

14

Gains greatest value from proven processes and practices

- **Lean** — *optimize work flow, minimize waste*
- **Six Sigma** — *rigorous process; statistics applied to reduce performance deviations; control phase to assure sustainability*
- **Reliability Centered Maintenance (RCM)** — *establish maintenance requirements to address potential failure modes*
  - *Failure Modes and Effects (Criticality) Analysis — FME(C)A*
- **Total Productive Maintenance (TPM)** — *improve organization, teamwork*
  - *Maintenance / Production partnership, small group activities, orderliness*
- **Asset Integrity** — *expanded to require meeting all requirements for safe operation*
- **Root Cause Analysis (RCA)** — *identification of failure cause for elimination*
- **Condition Monitoring (CM)** — *define current performance and condition*
  - *vibration measurement and condition assessment*
  - *operating performance*
  - *fluid (lubricating oil) monitoring*
  - *Thermography*
  - *motor impedance, current analysis*
  - *ultrasonic's; active and passive*
- **Condition Based Maintenance (CBM)** — *maintain on need*
- **Time Based Preventive Maintenance, PM** — *maintain on time*
  - *Based on statistical and historical lifetime*
- **Proactive Maintenance** — *pre operation action to reduce failure*
- **Computerized Work Management Systems (CMMS)**
  - *Planning and Scheduling — optimize work management*
  - *Materials MRO — stores, repair parts management*

13AIChE
Spring Meeting, San Antonio, TX

Build from reliability

- **Definition:**
  Probability that a system, equipment or product will perform its required functions in a safe, satisfactory manner for a given period of time when used under specified operating conditions.
  Physical Asset Management Handbook
  Achievement of predictable results with as little variation as specific circumstances permit. *David L. Stringer, Brig Gen USAF; Keynote at RCM / EAM 2006*
- **Definition and requirements extend and apply to the entire organization and all activities:**
  - ***Performance*** — *safely meet requirements, predictable, minimum variation*
  - ***Organization*** — *completely defined (RASCI or equivalent), consistent decision process, empowered employees*
  - ***Work Culture*** — *commitment to and ownership for excellence, highest quality work and continuous improvement*
  - ***Skills*** — *requirements totally defined, training provided, qualifications up to date*
  - ***Processes and practices*** — *fully defined, totally repeatable, high quality results*
  - ***Systems*** — *fully capable of meeting all operational demands*
  - ***Equipment*** — *safe, cost effective conformance to operational requirements*
  - ***Data*** — *accurate, secure, up to date and accessible*

13AIChE
Spring Meeting, San Antonio, TX

16

Gain greatest effectiveness

- **Efficiency:**
  - *Performing a given task well — activity and task oriented.*
- **Effectiveness:**
  - *Performing the correct task efficiently — results oriented.*

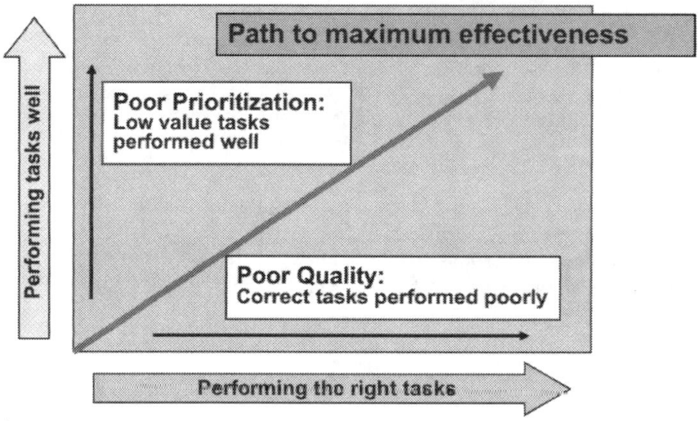

13AIChE
Spring Meeting, San Antonio, TX

17

Seek maximum value

- **Value defined for specific enterprise, business unit, facility, area / plant / unit**
  - *All unique*
- **Sources of value**
  - *Improve Safety, Health, Environmental performance*
  - *Achieve consistently greater business performance / mission compliance*
    - Increase throughput, availability, utilization
    - Improve quality—reduce rework, scrap (waste)
    - Reduce variation, interruptions, penalties, loss of good will
  - *Improve efficiency*
    - Reduce operating cost
  - *Reduce maintenance cost*
    - Increase reliability; reduce failures
  - *Improve capital effectiveness*
- **Adopt profit centered mentality**
  - *Invest for value rather than control costs to budget*
- **Minimize and control risk**

13AIChE
Spring Meeting, San Antonio, TX

18

Sources of business value

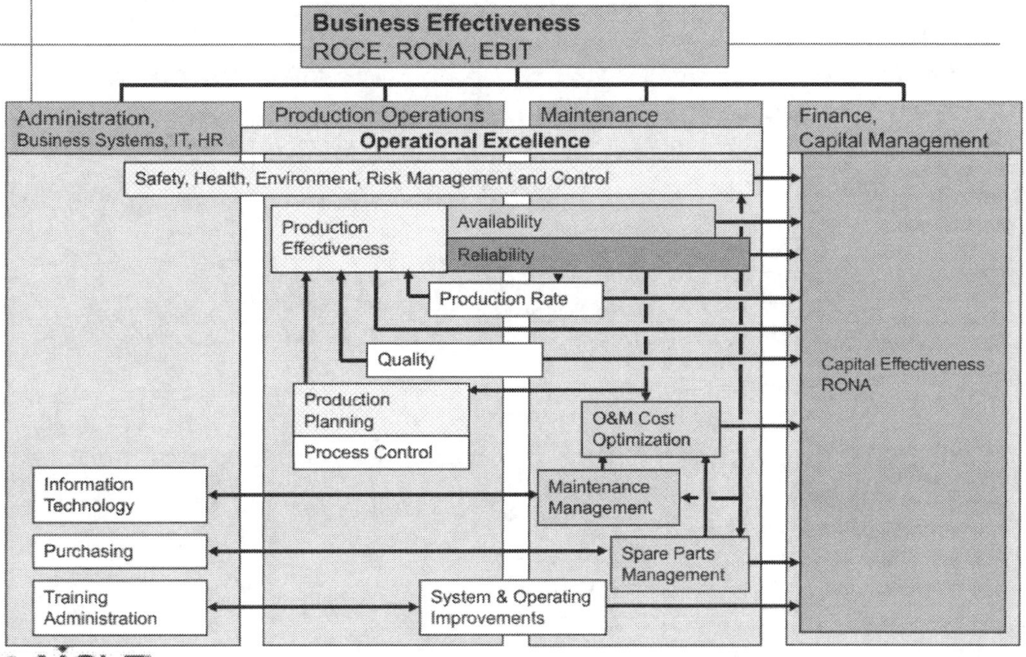

13 AIChE Spring Meeting, San Antonio, TX

19

Establishing value

- **Published benchmarks for overall guidance**
- **Operating history**
  - *Benefits of utilizing site / facility data*
    - Every business unit, facility, site different
    - Identifies real opportunities / deficiencies
      - *Creates enthusiasm, ownership*
  - *Base improvements by percentage increase in historical performance*
- **Analytical methodology**
  - *Production Weibull*
  - *Pareto*
- **Risk—the future**
  - *Failure Modes and Effects Analysis (FMEA)*
- **Business model**
  - *Accurate valuation of opportunities, results*
- **Dilemma—when excellence and value are mutually exclusive**

13 AIChE Spring Meeting, San Antonio, TX

20

Risk — make the future better than the past

- **History and Pareto look at the past; risk adds future threats**
- **Risk considers likelihood, vulnerability and impact—**
  probability times consequences
  - ***Probability*** *(how likely is an event?)*
    - Threat analysis to anticipate the future
      - *History—has it happened before*
      - *Current condition (mechanical and operating performance)*
    - Means and likelihood of detection — defect / fault identification
  - ***Consequences*** *(what happens when the event occurs?)*
    - Safety, environmental
    - Business interruption, lost production, quality
    - Operating and restoration cost, waste, penalties
- **Identify and assess future operational requirements**
  - *Regulations; safety, environmental*
  - *Operating intensity*
  - *Raw materials*
  - *External threats, market requirements, business and supply*
  - *Sustaining cost (investment) and decapitalization*

13AIChE Spring Meeting, San Antonio, TX

21

Risk Ranking Matrix

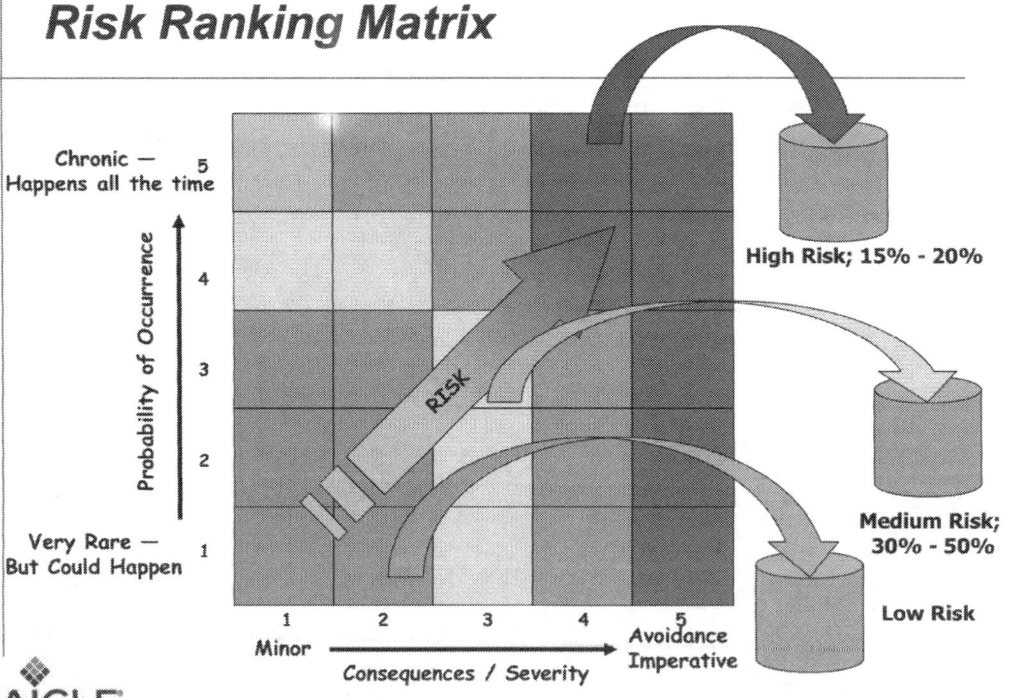

13AIChE Spring Meeting, San Antonio, TX

22

Failure Modes and Affects Analysis

- **Proactive analytical methodology from Reliability Centered Maintenance:**
  - *Identify and evaluate potential failure modes and effects of a system or component*
  - *In depth, risk prioritized, comprehensive understanding of the cause, probability, consequences and detectability of each potential failure*
  - *Means to develop effective countermeasures*
  - *Full, auditable documentation of the process including engineering logic utilized*
- **Definitions:**
  - ***Functional failure:*** *The inability to deliver a defined standard of performance:*
    - Full / partial
      - *Partial loss — reduced performance, efficiency*
    - Observable / hidden
      - *Hidden failure — not observable until a second failure occurs*
  - ***Mode:*** *Any potential occurrence (event) that could cause a functional failure*
  - ***Effect:*** *What happens when a failure occurs*
  - ***Risk:*** *How often a given failure occurs (probability);*
    how serious if it does occur (consequences)
  - ***Detection:*** *How are the symptoms of a failure observed; time available to manage a failure*
    - How recognized
    - Interval from recognition to action required
  - ***Controls:*** *Actions placed in effect to extend life, prevent / counter failure and / or identify a defect for mitigating corrective action prior to failure*

13 AIChE
Spring Meeting, San Antonio, TX

23

FMEA Template

Failure Modes and Effects Analysis Template

System: Cargo Transfer
Sub-System: Pumping
Equipment: Cargo Pump 1
Function: Transfer cargo from tanks off the vessel at rated flow and pressure without leaks
Date Created:
Date Revised:
Vessel:
FMEA Facilitator
Team Members:

Failure	Failure Mode		Occurrence / Probability	Detection	Failure Effect	Severity / Consequences	RPN Classification	Controls / Counter Actions		Responsibility	Due Date
	Component	Cause						Prevention	Detection		
Pump stopped	Pump High vibration	Failing bearing; caused by: 1. Unbalance 2. Shaft misalignment	3	2	Cargo transfer slowed	4	24	1. Assure rotor balanced at reassembly 2. Assure shafts aligned at reassembly	Implement vibration monitoring, quarterly interval	Dave Smith	
	Leak (controllable)	1. Gasket failed	3	2	Cargo transfer slowed	4	24	1. Assure proper make up at reassembly 2. Assure proper assembly	Visual inspection; hourly while in operation		
		2. Seal failing	3	2		4	24				
	Coupling fractured	1. Misalignment	2	1		5	10	1. Assure shaft alignment at assembly 2. inspect coupling during assembly	Vibration monitoring might identify crack precursors		
		2. Defective coupling	1	1		5	5				
	Shaft fractured	1. Defective shaft	1	1		5	5	None	Vibration monitoring		

13 AIChE
Spring Meeting, San Antonio, TX

24

FMEA Template

Failure Modes and Effects Analysis Template

System: Cargo Transfer — Date Created: ______ — Vessel: ______ — Team Members: ______
Sub-System: Pumping — Date Revised: ______ — FMEA Facilitator ______
Equipment: Cargo Pump 1
Function: Transfer cargo from tanks off the vessel at rated flow and pressure without leaks

Failure	Failure Mode: Component	Failure Mode: Cause	Occurrence / Probability	Detection	Failure Effect	Severity / Consequences	RPN Classification	Controls / Counter Actions: Prevention	Controls / Counter Actions: Detection	Responsibility	Due Date
Pump stopped	Pump High vibration	Failing bearing; caused by: 1. Unbalance 2. Shaft misalignment	3	2	Cargo transfer slowed	4	24	1. Assure rotor balanced at reassembly 2. Assure shafts aligned at reassembly	Implement vibration monitoring, quarterly interval	Dave Smith	
	Leak (controllable)	1. Gasket failed	3	2	Cargo transfer slowed	4	24	1. Assure proper make up at reassembly 2. Assure proper assembly	Visual inspection; hourly while in operation		
		2. Seal failing	3	2		4	24				
	Coupling fractured	1. Misalignment	2	1		5	10	1. Assure shaft alignment at assembly 2. Inspect coupling during assembly	Vibration monitoring might identify crack precursors		
		2. Defective coupling	1	1		5	5				
	Shaft fractured	1. Defective shaft	1	1		5	5	None	Vibration monitoring		

13 AIChE
Spring Meeting, San Antonio, TX

25

FMEA Template

Failure Modes and Effects Analysis Template

System: Cargo Transfer — Date Created: ______ — Vessel: ______ — Team Members: ______
Sub-System: Pumping — Date Revised: ______ — FMEA Facilitator ______
Equipment: Cargo Pump 1
Function: Transfer cargo [...]nks off the vessel at rated flow and pressure without leaks

FMEA Descriptive Information
- Vessel
- Date
- FMEA Team
- System
- Subsystem
- Equipment
- Function

Failure	Failure Mode: Component	Failure Mode: Cause	Occurrence / Probability	Detection	Failure Effect	Severity / Consequences	RPN Classification	Controls / Counter Actions: Prevention	Controls / Counter Actions: Detection	Responsibility	Due Date
Pump stopped	Pump				transfer	4	24	1. Assure rotor balanced at reassembly 2. Assure shafts aligned at reassembly	Implement vibration monitoring, quarterly interval	Dave Smith	
	Leak (				transfer	4	24	1. Assure proper make up at reassembly 2. Assure proper assembly	Visual inspection; hourly while in operation		
						4	24				
	Coupling fractured	1. Misalignment	2	1		5	10	1. Assure shaft alignment at assembly 2. Inspect coupling during assembly	Vibration monitoring might identify crack precursors		
		2. Defective coupling	1	1		5	5				
	Shaft fractured	1. Defective shaft	1	1		5	5	None	Vibration monitoring		

13 AIChE
Spring Meeting, San Antonio, TX

26

FMEA Template

Failure Modes and Effects Analysis Template

System: Cargo Transfer **Date Created:** ____ **Vessel:** ____ **Team Members:** ____
Sub-System: Pumping **Date Revised:** ____ **FMEA Facilitator** ____
Equipment: Cargo Pump 1
Function: Transfer cargo from tanks off the vessel at rated flow and pressure without leaks

Failure	Failure Mode		Occurrence / Probability	Detection	Failure Effect	Severity / Consequences	RPN Classification	Controls / Counter Actions		Responsibility	Due Date
	Component	Cause						Prevention	Detection		
Pump sto	High vibration	Failing bearing; caused by:	3	2	Cargo transfer slowed	4	24	1. Assure rotor balanced at reassembly 2. Assure shafts aligned at reassembly	Implement vibration monitoring, quarterly interval	Dave Smith	
					Cargo transfer slowed	4 4	24 24	1. Assure proper make up at reassembly 2. Assure proper assembly	Visual inspection; hourly while in operation		
	Coupling fractured	1. Misalignment 2. Defective coupling	2 1	1 1		5 5	10 5	1. Assure shaft alignment at assembly 2. Inspect coupling during assembly	Vibration monitoring might identify crack precursors		
	Shaft fractured	1. Defective shaft	1	1		5	5	None	Vibration monitoring		

Failure
Description of system or equipment inability to deliver required performance — functional failure

13 AIChE
Spring Meeting, San Antonio, TX

FMEA Template

Failure Modes and Effects Analysis Template

System: Cargo Transfer **Date Created:** ____ **Vessel:** ____ **Team Members:** ____
Sub-System: Pumping **Date Revised:** ____ **FMEA Facilitator** ____
Equipment: Cargo Pump 1
Function: Transfer cargo from tanks off the vessel at rated flow and pressure without leaks

Failure	Failure Mode		Occurrence / Probability	Detection	Failure Effect	Severity / Consequences	RPN Classification	Controls / Counter Actions		Responsibility	Due Date
	Component	Cause						Prevention	Detection		
Pump stopped	Pump High vibrati	ing bearing; d by:	3	2	Cargo transfer slowed	4	24	1. Assure rotor balanced at reassembly 2. Assure shafts aligned at reassembly	Implement vibration monitoring, quarterly interval	Dave Smith	
	Leak (c				ansfer	4 4	24 24	1. Assure proper make up at reassembly 2. Assure proper assembly	Visual inspection; hourly while in operation		
	Coupling fractured	1. Misalignment 2. Defective coupling	2 1	1 1		5 5	10 5	1. Assure shaft alignment at assembly 2. Inspect coupling during assembly	Vibration monitoring might identify crack precursors		
	Shaft fractured	1. Defective shaft	1	1		5	5	None	Vibration monitoring		

Failure Mode
Any potential occurrence (event) that could cause a functional failure.

13 AIChE
Spring Meeting, San Antonio, TX

FMEA Template

Failure Modes and Effects Analysis Template

System: Cargo Transfer **Date Created:** ____ **Vessel:** ____ **Team Members:** ____
Sub-System: Pumping **Date Revised:** ____ **FMEA Facilitator** ____
Equipment: Cargo Pump 1
Function: Transfer cargo from tanks off the vessel at rated flow and pressure without leaks

Failure	Failure Mode		Occurrence / Probability	Detection	Failure Effect	Severity / Consequences	RPN Classification	Controls / Counter Actions		Responsibility	Due Date
	Component	Cause						Prevention	Detection		
Pump stopped	Pump High vibration	Failing bearing; caused by: 1. Unbala… 2. Shaft… misalig…			Cargo transfer …owed	4	24	1. Assure rotor balanced at …	Implement vibration monitoring, quarterly interval	Dave Smith	
	Leak (controllable)	1. Gask…						…er	Visual inspection; hourly while in operation		
		2. Seal failing	3	2		4	24	…er			
	Coupling fractured	1. Misalignment	2	1		5	10	1. Assure shaft alignment at assembly	Vibration monitoring might identify crack precursors		
		2. Defective coupling	1	1		5	5	2. Inspect coupling during assembly			
	Shaft fractured	1. Defective shaft	1	1		5	5	None	Vibration monitoring		

Cause
Weakness or defect – the consequence of which is a Failure Mode.

13 AIChE Spring Meeting, San Antonio, TX

29

FMEA Template

Failure Modes and Effects Analysis Template

System: Cargo Transfer **Date Created:** ____ **Vessel:** ____ **Team Members:** ____
Sub-System: Pumping **Date Revised:** ____ **FMEA Facilitator** ____
Equipment: Cargo Pump 1
Function: Transfer cargo from tanks off the vessel at rated flow and pressure without leaks

Failure	Failure Mode		Occurrence / Probability	Detection	Failure Effect	Severity / Consequences	RPN Classification	Controls / Counter Actions		Responsibility	Due Date
	Component	Cause						Prevention	Detection		
Pump stopped	Pump High vibration	Failing bearing; caused by: 1. Unbalance 2. Shaft misalignment	3	2	…nsfer	4	24	1. Assure rotor balanced at reassembly	Implement vibration monitoring, quarterly interval	Dave Smith	
	Leak (controllable)	1. Gasket failed									
		2. Seal failing	3								
	Coupling fractured	1. Misalignment	2	1		5	10	1. Assure shaft alignment at assembly	Vibration monitoring might identify crack precursors		
		2. Defective coupling	1	1		5	5	2. Inspect coupling during assembly			
	Shaft fractured	1. Defective shaft	1	1		5	5	None	Vibration monitoring		

Likelihood / Probability
Probability that a specific Cause/Mechanism will occur; ranked 1 – 5

13 AIChE Spring Meeting, San Antonio, TX

30

FMEA Template

Failure Modes and Effects Analysis Template

System:	Cargo Transfer	**Date Created:**		**Vessel:**		**Team Members:**	
Sub-System:	Pumping	**Date Revised:**		**FMEA Facilitator**			
Equipment:	Cargo Pump 1						
Function:	Transfer cargo from tanks off the vessel at rated flow and pressure without leaks						

Failure	Failure Mode		Occurrence / Probability	Detection	Failure Effect	Severity / Consequences	RPN Classification	Controls / Counter Actions		Responsibility	Due Date
	Component	Cause						Prevention	Detection		
Pump stopped	Pump High vibration	Failing bearing; caused by: 1. Unbalance 2. Shaft misalignment	3	2	[illegible]	4	24	1. Assure rotor balanced at reassembly	Implement vibration monitoring, quarterly interval	Dave Smith	
	Leak (controllable)	1. Gasket failed	3	[illegible]	[illegible]	[illegible]	[illegible]	[illegible]	[illegible]		
		2. Seal failing	3	2	[illegible]	[illegible]	[illegible]	[illegible]	[illegible]		
	Coupling fractured	1. Misalignment	2	1		5	10	1. Assure shaft alignment at assembly	Vibration monitoring might identify crack precursors		
		2. Defective coupling	1	1		5	5	2. Inspect coupling during assembly			
	Shaft fractured	1. Defective shaft	1	1		5	5	None	Vibration monitoring		

Detection
How the specific failure
mode is detected;
ranked 1 - 5.

13 AIChE
Spring Meeting, San Antonio, TX

31

FMEA Template

Failure Modes and Effects Analysis Template

System:	Cargo Transfer	**Date Created:**		**Vessel:**		**Team Members:**	
Sub-System:	Pumping	**Date Revised:**		**FMEA Facilitator**			
Equipment:	Cargo Pump 1						
Function:	Transfer cargo from tanks off the vessel at rated flow and pressure without leaks						

Failure	Failure Mode		Occurrence / Probability	Detection	Failure Effect	Severity / Consequences	RPN Classification	Controls / Counter Actions		Responsibility	Due Date
	Component	Cause						Prevention	Detection		
Pump stopped	Pump High vibration	Failing bearing; caused by: 1. Unbalance 2. Shaft misalignment	3	2	Cargo transfer slowed	[illegible]	24	1. Assure rotor balanced at [illegible]	Implement vibration [illegible]	Dave Smith	
	Leak (controllable)	1. Gasket failed	3	2	Cargo [illegible] slowe[illegible]	[illegible]	[illegible]	[illegible]	[illegible]		
		2. Seal failing	3	2		4	24	[illegible]	[illegible]		
	Coupling fractured	1. Misalignment	2	1		5	10	1. Assure shaft alignment at assembly	Vibration monitoring might identify crack precursors		
		2. Defective coupling	1	1		5	5	2. Inspect coupling during assembly			
	Shaft fractured	1. Defective shaft	1	1		5	5	None	Vibration monitoring		

Failure Effect
Results of the Failure
Mode on the function,
observed by the user

13 AIChE
Spring Meeting, San Antonio, TX

32

FMEA Template

Failure Modes and Effects Analysis Template

System: Cargo Transfer — Date Created: ____ — Vessel: ____ — Team Members: ____
Sub-System: Pumping — Date Revised: ____ — FMEA Facilitator ____
Equipment: Cargo Pump 1
Function: Transfer cargo from tanks off the vessel at rated flow and pressure without leaks

Failure	Failure Mode: Component	Failure Mode: Cause	Occurrence / Probability	Detection	Failure Effect	Severity / Consequences	Classification	Controls / Counter Actions: Prevention	Controls / Counter Actions: Detection	Responsibility	Due Date
Pump stopped	Pump High vibration	Failing bearing; caused by: 1. Unbalance 2. Shaft misalignment	3	2	Cargo transfer slowed	4	24	[illegible]	Implement vibration [illegible]	Dave Smith	
	Leak (controllable)	1. Gasket failed	3	2	Cargo transfer slowed	[illegible]	[illegible]	[illegible]	[illegible]		
		2. Seal failing	3	2		[illegible]	[illegible]	[illegible]	[illegible]		
	Coupling fractured	1. Misalignment	2	1		5	10	1. Assure shaft alignment at assembly	Vibration monitoring might identify crack precursors		
		2. Defective coupling	1	1		5	5	2. Inspect coupling during assembly			
	Shaft fractured	1. Defective shaft	1	1		5	5	None	Vibration monitoring		

Severity / Consequences
The impact of the specific Failure Effect, ranked 1 – 5

FMEA Template

Failure Modes and Effects Analysis Template

System: Cargo Transfer — Date Created: ____ — Vessel: ____ — Team Members: ____
Sub-System: Pumping — Date Revised: ____ — FMEA Facilitator ____
Equipment: Cargo Pump 1
Function: Transfer cargo from tanks off the vessel at rated flow and pressure without leaks

Failure	Failure Mode: Component	Failure Mode: Cause	Occurrence / Probability	Detection	Failure Effect	Severity / Consequences	RPN Classification	Controls / Counter Actions: Prevention	Controls / Counter Actions: Detection	Responsibility	Due Date
Pump stopped	Pump High vibration	Failing bearing; caused by: 1. Unbalance 2. Shaft misalignment	3	2	Cargo transfer slowed	4	24	[illegible]	Implement vibration monitoring.	Dave Smith	
	Leak (controllable)	1. Gasket failed	3	2	Cargo transfer slowed	4	[illegible]	[illegible]	[illegible]		
		2. Seal failing	3	2		4	[illegible]	[illegible]	[illegible]		
	Coupling fractured	1. Misalignment	2	1		5	10	1. Assure shaft alignment at assembly	Vibration monitoring might identify crack precursors		
		2. Defective coupling	1	1		5	5	2. Inspect coupling during assembly			
	Shaft fractured	1. Defective shaft	1	1		5	5	None	Vibration monitoring		

RPN Classification
Likelihood (1 – 5) times
Detection (1 – 5) times
Severity (1 – 5)

FMEA Template

Failure Modes and Effects Analysis Template

System: Cargo Transfer **Date Created:** ____ **Vessel:** ____ **Team Members:** ____
Sub-System: Pumping **Date Revised:** ____ **FMEA Facilitator** ____
Equipment: Cargo Pump 1
Function: Transfer cargo from tanks off the vessel at rated flow and pressure without leaks

Failure	Failure Mode: Component	Failure Mode: Cause	Occurrence / Probability	Detection	Failure Effect	Severity / Consequences	Classification	Controls / Counter Actions: Prevention	Controls / Counter Actions: Detection	Responsibility	Due Date
Pump stopped	Pump High vibration	Failing bearing; caused by:	3	2	Cargo		24	1. Assure rotor balanced at reassembly 2. Assure shafts aligned at reassembly	Implement vibration monitoring, quarterly interval	Dave Smith	
	Leak (co							1. Assure proper make up at reassembly 2. Assure proper assembly	Visual inspection; hourly while in operation		
	Couplin	coupling						1. Assure shaft alignment at assembly 2. Inspect coupling during assembly	Vibration monitoring might identify crack precursors		
	Shaft fractured	1. Defective shaft	1	1		5	5	None	Vibration monitoring		

Controls — Prevention
Improvements to operating, preventive and/or repair procedures that prevent the Cause/Mechanism or Failure Mode/Effect from occurring, reduce the rate of occurrence and/or Severity

13 AIChE Spring Meeting, San Antonio, TX

35

FMEA Template

Failure Modes and Effects Analysis Template

System: Cargo Transfer **Date Created:** ____ **Vessel:** ____ **Team Members:** ____
Sub-System: Pumping **Date Revised:** ____ **FMEA Facilitator** ____
Equipment: Cargo Pump 1
Function: Transfer cargo from tanks off the vessel at rated flow and pressure without leaks

Failure	Failure Mode: Component	Failure Mode: Cause	Occurrence / Probability	Detection	Failure Effect	Severity / Consequences	RPN Classification	Controls / Counter Actions: Prevention	Controls / Counter Actions: Detection	Responsibility	Due Date
Pump stopped	Pump High vibration	Failing bearing; caused by: 1. Unb	3	2	Cargo transfer slowed	4		re rotor balanced at mbly	Implement vibration monitoring, quarterly interval	Dave Smith	
	Leak (controllable)								Visual inspection; hourly while in operation		
	Coupling fractured								Vibration monitoring might identify crack precursors		
	Shaft fractured	1. Defective shaft	1	1		5	5	None	Vibration monitoring		

Controls — Detection
Improved physical surveillance or analytical methods to detect the Cause/Mechanism or Failure Mode before a defect reaches a state where action must be taken (minimize consequences).

13 AIChE Spring Meeting, San Antonio, TX

36

FMEA Template

Failure Modes and Effects Analysis Template

System: Cargo Transfer | Date Created: | Vessel: | Team Members:
Sub-System: Pumping | Date Revised: | FMEA Facilitator
Equipment: Cargo Pump 1
Function: Transfer cargo from tanks off the vessel at rated flow and pressure without leaks

Failure	Failure Mode		Occurrence / Probability	Detection	Failure Effect	Severity / Consequences	RPN Classification	Controls / Counter Actions		Responsibility	Due Date
	Component	Cause						Prevention	Detection		
Pump stopped	Pump High vibration	Failing bearing; caused by: 1. Unbalance 2. Shaft misalignment	3	2	Cargo transfer slowed	4	24	1. Assure [illegible]	[illegible]	Dave Smith	
	Leak (controllable)	1. Gasket failed 2. Seal failing	[illegible]	[illegible]	[illegible]	[illegible]	[illegible]	[illegible]	[illegible]		
	Coupling fractured	1. Misalignment 2. Defective coupling	2 1	1 1		5 5	10 5	1. Assure shaft alignment at assembly 2. Inspect coupling during assembly	Vibration monitoring might identify crack precursors		
	Shaft fractured	1. Defective shaft	1	1		5	5	None	Vibration monitoring		

Responsibility
Individual, by name, with responsibility for the control / counter strategy for a specified Failure Mode

13 AIChE Spring Meeting, San Antonio, TX

37

Program Implementation

- Brief description
- Implementation—DIPICI
- Organization
- Exploiting opportunities
- Matrix implementation
- Assuring results
- Assessment process
- Benefits

Operational Excellence Program

- **An essential concept**
  Gain the effectiveness and sustainable performance necessary to succeed in an intense operating environment
- **Emphasizes the human, work culture and organizational issues that are essential for success**
  Organization wide commitment to excellence, initiative, ownership, support and participation mandatory to gain real success
- **Evolves from controlling costs to profit centered mentality**
  Invest for increases to mission / business value and success
- **Deviations from performance objectives identified, prioritized and corrected based on business value and risk**
  Performance assessments, benchmarking, statistical and risk analysis utilized to identify and prioritize opportunities for improvement:

Safe, sustainable improvement —
Not a command but rather the result of doing the right things

13AIChE
Spring Meeting, San Antonio, TX

39

Program—in brief

- **Define mission / business and operating objectives**
- **Appoint leadership**
- **Establish the organizational and administrative systems**
  - *Consider facilitation during program development and implementation*
    - Accelerate learning, gain results faster
  - *Identify requirements for training*
- **Identify and value prioritize improvement opportunities**
  - *Based on actual conditions — opportunities, strengths, weaknesses, barriers*
  - *In-depth analysis of causes of deviations from requirements and spending*
    - Relies on cause and correction data
    - Gathers institutional knowledge
  - *Creates excitement, energy, commitment and ownership*
- **Formulate and implement improvement action plans**
  - *Provide training as required*
  - *Introduce programs, practice and technology in a prioritized, opportunistic sequence to gain greatest value quickly*
- **Measure and monitor results; adjust plans as required**
- **Institutionalize and sustain results; continuously improve**

If you don't know where you are going — any road will do!

13AI
Spring Meeting, San Antonio, TX

40

Typical implementing methodology

- **Shewhart (Deming) — PDCA**
  - ***Plan*** — *what to accomplish*
  - ***Do*** — *initiate strategy and plan*
  - ***Check*** — *evaluate results*
  - ***Act*** — *on what has been accomplished*
- **Six Sigma — DMAIC**
  - ***Define*** — *improvement projects based on business objectives, customer needs and feedback, Critical To Quality (CTQ) characteristics*
  - ***Measure*** — *key processes, current performance and defects within the processes*
  - ***Analyze*** — *deficiencies (opportunities for improvement) defined by history, risk and lost opportunity — gaps between current and objective performance; establish why defects are generated*
  - ***Improve*** — *the process to stay within required ranges, implement optimized practices and technology, prioritized by value and risk criteria*
  - ***Control*** — *monitor, improve and sustain results to assure variables remain within acceptable ranges*
- **Boyd — OODA**
  - ***Observe*** — *opportunities*
  - ***Orient*** — *take advantage of opportunities*
  - ***Do*** — *take action to capitalize on opportunities*
  - ***Adjust*** — *corrective action if required*

13 AIChE
Spring Meeting, San Antonio, TX

41

Operational Excellence implementing methodology — DICIPI

- **Define**—the program
  - *Business and program objectives*
  - *Leadership, organizational and administrative requirements*
  - *Governing policy and strategy, business plan*
  - *Methodology for vital procedures such as risk assessment and business value model*
- **Identify**—opportunities for improvement
  - *Analyze—identify improvement opportunities by comparison to best performance, performance benchmarks*
  - *Prioritize—potential improvements based on value opportunity, cost, time to implement and probability of success*
- **Plan**—improvements to create maximum value
  - *Develop detailed action plans including resources and time required, responsibility and expected results*
  - *Gain approval*
- **Implement**—improvement action plans
  - *Deploy resources and technology*
  - *Execute improvement action plans*
- **Check**—measure and manage results
  - *Adjust actions as required to meet objectives*
- **Improve and institutionalize**—initiate continuous improvement
  - *Sustain gains and success*

13 AIChE
Spring Meeting, San Antonio, TX

42

Operational Excellence Program
Define
(need to do)
Establish
Business Requirements
Collect Facts
Identify
(opportunities
for improvement)
Based on
Value Potential
Plan
(going to do)
Formulate
Improvement Actions
Implement
(improvement action plans)
Increase
Business Value
Check
(what you did)
Measure Results
Improve
Initiate Continuous
Improvement
13AIChE
Spring Meeting, San Antonio
43

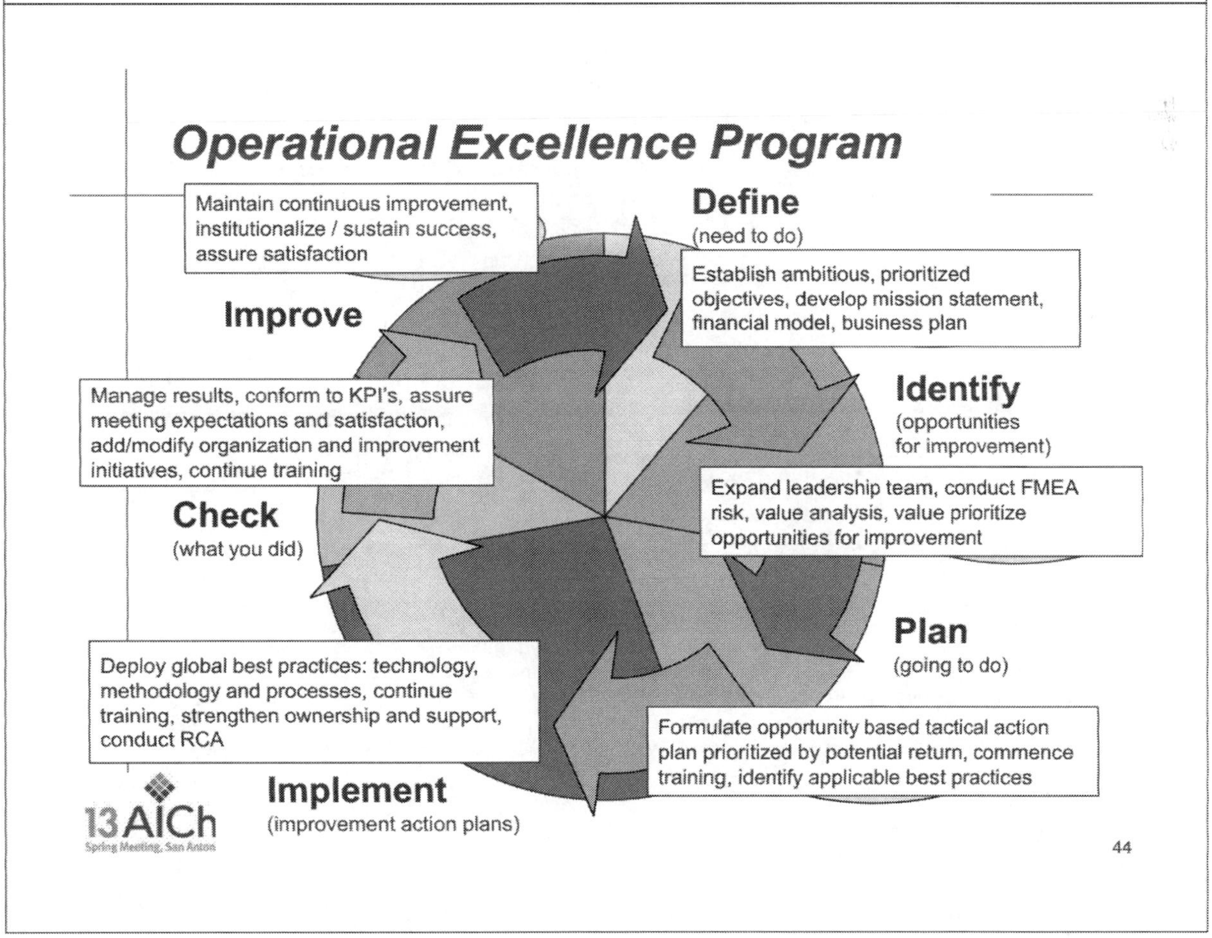
Operational Excellence Program
Define
(need to do)
Establish ambitious, prioritized objectives, develop mission statement, financial model, business plan
Identify
(opportunities
for improvement)
Expand leadership team, conduct FMEA risk, value analysis, value prioritize opportunities for improvement
Plan
(going to do)
Formulate opportunity based tactical action plan prioritized by potential return, commence training, identify applicable best practices
Implement
(improvement action plans)
Deploy global best practices: technology, methodology and processes, continue training, strengthen ownership and support, conduct RCA
Check
(what you did)
Manage results, conform to KPI's, assure meeting expectations and satisfaction, add/modify organization and improvement initiatives, continue training
Improve
Maintain continuous improvement, institutionalize / sustain success, assure satisfaction
13AIChE
Spring Meeting, San Antonio
44

Define the program

- **Executive champion**
  - *Provides compelling vision, necessity and benefits, business / mission objectives*
  - *Visibly engaged, continuing support and drive*
- **Appoint program leadership**
  - *Steering Team*
    - Strategic direction, guidance, support, facilitation and review
  - *Program Leader*
    - Operational leadership
  - *Communications team*
- **Establish program mission, objectives, principles, policy, strategy**
  - *Define value*
- **Formulate program charter**
  - *Business Plan*
- **Define organization, management and administrative systems**
- **Validate procedures for risk assessment and business value**
  - *Establish business and risk models*
  - *Value normalization*
- **Form Operating Leadership Team**

13 AIChE Spring Meeting, San Antonio, TX

45

Program organization

Site / Facility with multiple, dissimilar areas / units

Executive Champion	**Senior Corporate or Site Executive**
Steering Team	**Plant and Business Managers**, Site Production and Maintenance Superintendents Initiative Leader, Department Heads
Operating Leadership Team	**Initiative Leader**, Production and Maintenance Managers, Technical Professionals, First Line Supervisors, Multi Function Crafts
Improvement Action Teams	**Improvement Champion**, Technical Professionals, Multi-Function Working-Level Specialists as required

13 AIChE Spring Meeting, San Antonio, TX

46

Begins with—energized executive

- **States compelling vision, business objectives**
- **Identifies necessity for improvement and benefits—motivates participants**
- **Outlines overall program requirements**
- **Appoints respected operational leaders committed to success**
  - *Steering Team*
- **Provides visible engagement and continuing drive—fully committed to success**
- **Removes institutional and organizational barriers**

13 AIChE
Spring Meeting, San Antonio, TX

47

Strategic leadership—Steering Team

- **Composed of senior business and operating executives and functional managers with ownership for the program and committed to success**
  - *Operations / Production, Maintenance, Engineering, Finance, HR, IT*
  - *Preferably chaired by production executive*
  - *Appoints program leader—champion*
    - Program leader a member
- **Provides strategic leadership and direction**
  - *Champion, drive, actively lead / facilitate the program; assure results*
    - Energize implementation; serve as a facilitating resource
    - Decision authority to provide resources
  - *Involved in tactics only when reviewing results and acting to eliminate organizational barriers*
- **Defines program mission, overall strategy**
  - *Translate business objectives into program objectives; results required*
- **Approves plans to gain objective results**
  - *Validate the potential business value of proposed improvement initiatives*
  - *Review and approve; priorities and improvement initiatives*
  - *Review the function, progress and results of the various improvement action teams*
  - *Surmount challenges, remove barriers to success*
- **Monitor results, progress to objectives; oversee improvements**

13 AIChE
Spring Meeting, San Antonio, TX

48

Key individual— Program Leader / Champion

- **Appointed by and reports to the Steering Team**
  - *Individual with position and respect—at the superintendent level*
    - Credible within the organization
    - Possess knowledge, initiative, enthusiasm and drive necessary to achieve success
    - Excellent leadership, organizational and people skills
  - *Full time position, sole duty, not a collateral*
- **Working level responsibility, authority and accountability for program development, objectives, plans, implementation and results including:**
  - *Assure that program strategy and plan are consistent with and contribute value to organizational policy, strategy and objectives*
  - *Appoint members of Operating Leadership Team and subject matter experts; augment teams where necessary as improvement plans develop*
  - *Direct and facilitate the process of identifying, value prioritizing and developing improvement initiatives*
  - *Identify resource requirements*
  - *Assure that improvement plans are consistent and supportive with other organizational activities; resolve inconsistencies*
  - *Monitor progress and results, compliance with strategy and objectives; assure sustainability*

13 AIChE
Spring Meeting, San Antonio, TX

49

Operating Leadership Team

- **Established for each site / facility business unit, area, plant**
- **Appointed by Program Leader with agreement of Steering Team**
- **Multi level, function, discipline**
  - *Highly respected top people with direct knowledge and experience in areas of potential improvement*
    - Engineering, Operations, Maintenance augmented by IT, finance, safety as required
    - Technical professionals, working level supervisors, technicians, crafts and operators
- **Commence with introductory, kick off workshop**
- **Initial task: review and validate / modify program mission, principles, policy, strategy and business plan**
  - *Essential for full commitment and ownership*
- **Identify and develop potential opportunities for improvement**
- **Augmented as initiatives are identified and refined**
- **Spin off Improvement Action Teams**

Opportunity discovery process, developing and implementing improvement initiatives are essentials to gain the ownership, commitment and responsibility necessary for sustainable success

13 AIChE
Spring Meeting, San Antonio, TX

50

Conduct kick off workshop

- **Delivered to Operating Leadership Teams**
- **Detail program objectives, organization, function and process to participants**
- **Three to four days duration in two phases:**
  - *Program description and process*
    - Mission, strategy, principles, policy and business plan
    - Improvement discovery, prioritization and implementing process
  - *Working sessions to identify and develop preliminary improvement plans*
    - Identify opportunity for improvement
    - Describe future state; objectives
    - Define the value gained
    - Describe the step by step process to achieve the improvement; action plan
    - Identify strengths to build on; challenges and barriers to overcome
    - Define activity and results metrics to measure progress and success with objective time lines
    - List resources, people and funding required
    - Define potential risks that might impede the plan's success
  - *Present results of working sessions to all participants*
- **Highly beneficial for Executive Champion and Steering Team to participate in final presentations**

13AIChE
Spring Meeting, San Antonio, TX

51

Roll out the program

- **Communicate necessity and value of improvement**
  - *Meet changing business / mission / market conditions*
- **Identify major benefits for employees**
  - *Includes improved job security and satisfaction*
- **Summarize major principles including:**
  - *Excellence and value in all activities*
  - *Broad participation*
  - *Commitment equivalent to safety*
  - *Continuous sustainable improvement*
- **Describe inefficiency, cost and personal impact**
  - *Unnecessary / low value work and preventable disruptions to operations*
- **State program implementation will be fair, open and honest**
  - *Communications; feedback*
  - *Meet potential concerns head on; e.g. reduction in force*
    - Rumors always worse than reality
  - *Every effort to understand and respond quickly to questions and concerns*
- **Define transition plan**
  - *How to get from here to there!*

13AIChE
Spring Meeting, San Antonio, TX

52

Identify—opportunities for improvement

- **Validate systems**
- **Establish sources of mission / business value**
  - *Increase throughput / mission compliance*
  - *Reduce costs*
- **Identify opportunities for improvement**
  - *Industry benchmarks*
  - *Advantages of utilizing site / facility / unit specific data*
    - Site daily production—Production Weibull
    - Historical costs—Pareto
  - *Align with program objectives*
- **Conduct risk assessment**
  - *Failure Modes and Effects Analysis (FMEA)*
- **Normalize financial value**
  - *Throughput / cost value equivalency*
- **Estimate probability of success**
- **Prioritize by adjusted value**
  - *Considering probability of success*

13 AIChE Spring Meeting, San Antonio, TX

53

Identifying potential improvements begins with information

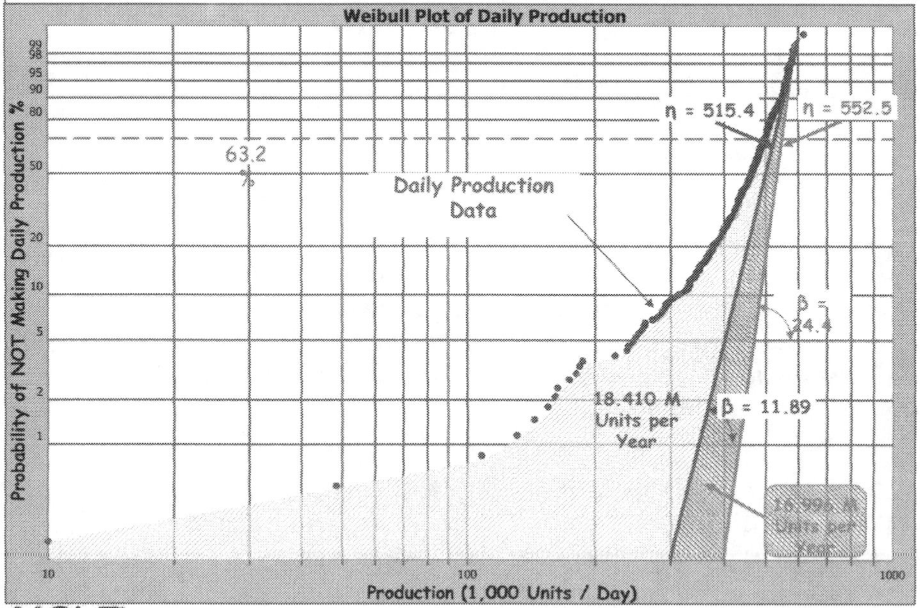

13 AIChE Spring Meeting, San Antonio, TX

54

Pareto — production losses by unit

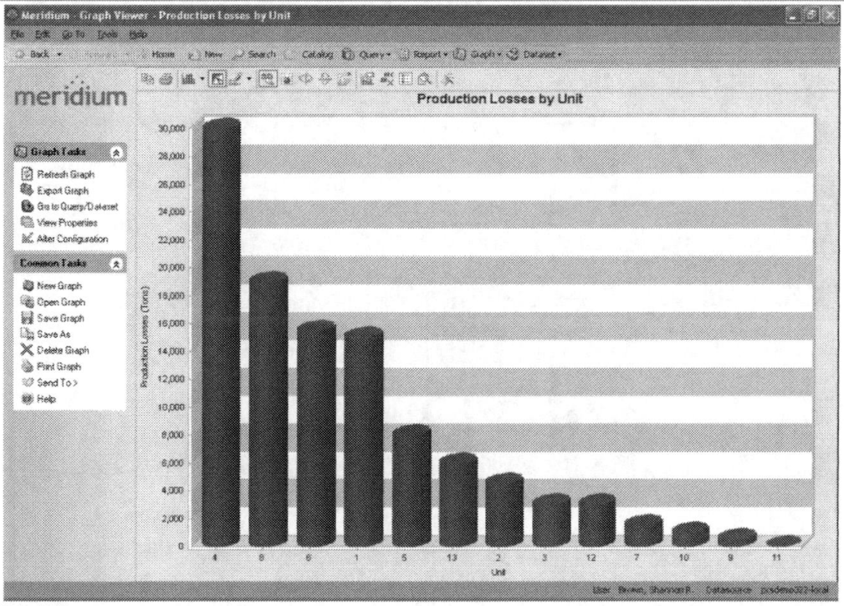

13 AIChE
Spring Meeting, San Antonio, TX

55

Consider all data to derive potential value gain

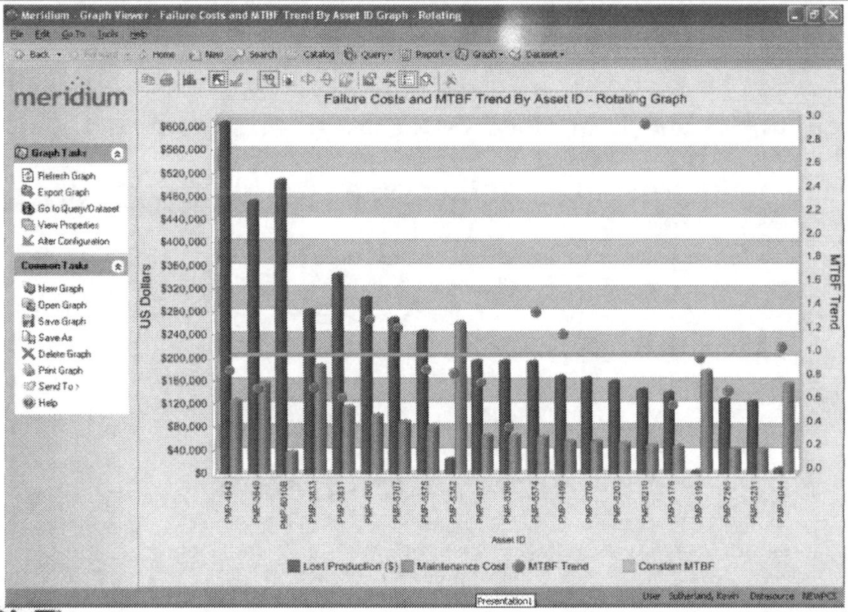

13 AIChE
Spring Meeting, San Antonio, TX

56

Improvement requires identifying cause

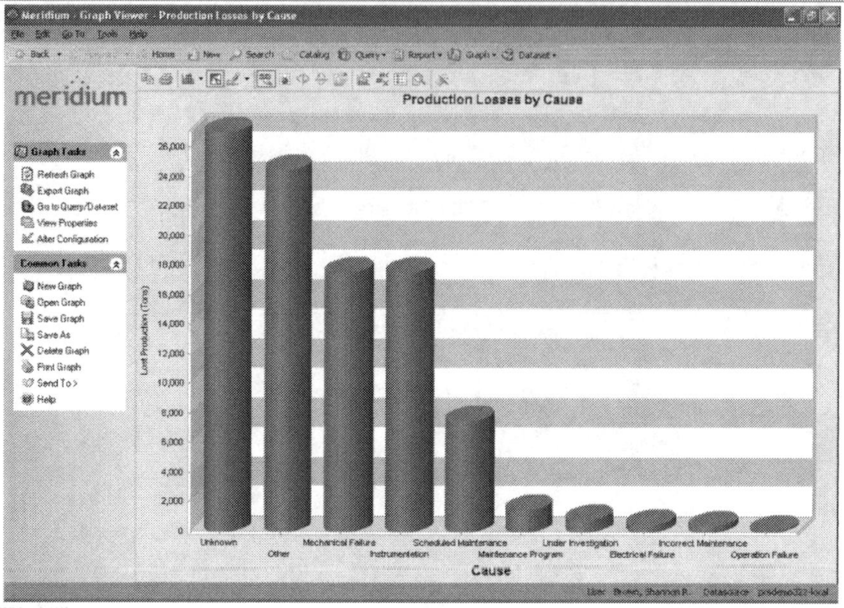

13AIChE
Spring Meeting, San Antonio, TX

57

Plan—Improvements

- **Form Improvement Action Teams**
  - *Prime responsibility for improvement initiatives*
  - *Champion(s) / team leader(s) from Operational Leadership Team*
- **Validate value prioritized potential improvements**
  - *Augment action teams with specific knowledge and experience*
- **Develop improvement action plans for highest value improvements**
  - *Refine improvement opportunity including value justification*
  - *Define specific objectives, future state, preliminary KPI's,*
  - *Validate value delivered; contribution to business / mission objectives*
  - *Develop detailed improvement action plan, specific activities and tasks to achieve objectives, including completion timelines*
  - *Identify strengths to build on, potential barriers, weaknesses; actions to overcome barriers and weaknesses*
  - *Identify resources (financial, technology, people, processes, systems, practices) and training necessary for implementation*
  - *Define final measures of performance; activity and result KPI's with timelines*
  - *Detail continuous improvement and sustaining plans*
  - *Estimate probability of success; detail actions and activities to maximize probability; minimize barriers and risk*
  - *Determine value delivered; adjusted by probability of success*
  - *Identify responsibility, by name for implementation and results*

13AIChE
Spring Meeting, San Antonio, TX

58

Improvement Action Teams — deliver success

- **Small collaborative, multi-function teams**
  - *5 to 7 members*
    - Too small can be intimidating; too large, difficult to gain consensus
  - *Led by champions from Operating Leadership Team*
  - *Multi-function members*
    - Knowledge, skills, institutional knowledge and experience necessary to develop and implement improvement initiatives
      - *Operations / production, maintenance and reliability*
      - *Professional, supervisory, hourly crafts*
      - *Multi-role and specialists (Subject Matter Experts)*
    - Computer expertise, literacy
      - *Data access, mining, analysis and reporting*
    - Credibility and respect within the organization
  - *Willing to accept ownership, responsibility, accountability*
  - *Ability to work together — teamwork*
- **Functional organizations provide assistance as necessary**
  - *Engineering, Safety, Finance, IT, HR, QA*
- **Knowledge, skills and resources made available as needed**
  - *Specialty tasks, expertise; subject matter experts*
- **Experienced facilitation and coaching to keep on track**

13AIChE
Spring Meeting, San Antonio, TX

59

Improvement Action Teams — deliver success

- **Identify, develop and implement detailed improvement action plans**
  - *Review and refine improvement opportunity — validate conclusions*
    - Value/risk
    - Objective end state, applicable strengths, weaknesses and barriers
    - Time to reach interim and final objectives
    - Resources required
  - *Formulate optimum solutions; specific action plans to capitalize on the opportunity*
    - Multiple initiatives with direct link to objective
  - *Directed toward quick results — create visibility, gain enthusiasm, support and momentum*
    - Short action horizon — 3 to 6 months from formation through implementation
    - Divide large populations into improved and unimproved
- **Establish activity and results metrics**
  - *Assure improvements are implemented on schedule (activity metrics)*
  - *Demonstrate contribution to mission / business objectives (results metrics)*
- **Track and publicize gains — create visibility, gain support and momentum**

13AIChE
Spring Meeting, San Antonio, TX

60

<u>Plan</u>—Improvements II

- **Assemble set of improvements to meet program objectives**
  - *Valuation adjusted by probability of success*
  - *Demonstrate contribution*
- **Consider pilot implementation for large scale improvements**
  - *Minimize risk*
  - *Work out any wrinkles on a small scale*
  - *Recognize any compromises that may be required due to infrastructure, old processes outside the pilot*
- **Develop transition plans where necessary**
  - *Transition between current and future conditions*
- **Select four to six initiatives totaling approximately 125% of overall objective for submission to Steering Team**
- **Finalize and submit action plans — obtain approval**
- **Publish improvement plan**

> **"Employee led leadership teams do unbelievably good strategic and tactical planning — if you give them the opportunity. Most important, they gain total buy-in for the plan, its implementation and results."** *Fortune 250 company CEO*

13 AIChE
Spring Meeting, San Antonio, TX

Matrix Implementation

Vivid Vision Communicated; High Level Agreement on the Need For Improvement; Commitment to Successful Implementation of OE Program					
Designate Executive Champion, Appoint Corporate and Site Leadership, Steering Team, Program Leader					**Define** Organize
Translate Corporate Business Objectives Into Overall OE Vision and Objectives, Business Drivers and Value Opportunities					
Develop OE Business and Mission Objectives; Compliance (SHE), Operating Effectiveness (Production Availability, Cost, LOE)					
Establish Business Aligned Overall Measures of Performance; Metrics, KPI's					
Develop the Site Overall Policy and Strategy					
Develop Business Model					
Form Communications Team; Develop Communications Strategy					
Initiate Consensus Development Within the Organization; Publicize Necessity for Improvement					
Design the Planning, Implementation and Sustainability Process					
Select OE Champions, Establish Improvement Action Team, Commence the OE Improvement Process					**Identify** Analyze Prioritize
Validate the OE Program Mission Statement					
Risk Rank Systems and Equipment	History	Probability	Consequences	Future	
Categorize Areas for Improvement	Culture, Values	Organization	Processes / Procedures	Material	
Establish Current Conditions	Strengths	Weaknesses	Barriers	Opportunities	
Benchmark to Refine OE Objectives	Results Expected	Value Potential	Probability of Success	Metrics	
Identify Gaps to OE Performance Objectives; Value Prioritize Opportunities for Improvement Based on Value, Probability of Success					
Highest Priority Value Opportunities	Results / Value	Time	Risk	Responsibility	**Plan**
Develop Value and Risk Prioritized OE Tactical Improvement Action Plans; Establish Measures of Performance; Control / Sustainability Plans					
Build and Train Organization; Expand Action Teams; Select OE Implementing Practices, Systems and Technology					
Develop Competencies	Practice	Systems	Methodology	Technology	
Deploy Risk / Value Prioritized OE Improvement Action Plans					**Implement**
Values, Culture, Relationships	Reliability Culture	Production / Maintenance Partnership		Change Management	
Basic Documentation	P&ID'S	Asset Register	Asset Hierarchy	Equipment History	
Foundation Processes	Risk / Value Prioritization	Work Management	Materials Management	Lubrication	
Organization and Administration	Champions	Leadership	Organizational Alignment	RASCI	
Information and Data Management	Integrity	Accuracy	Availability / Security	Revision Control	
Risk Management and Mitigation	Risk Determination Process	Reviewed and Validated	High Risk Mitigation Process	Risk Optimized OE Strategy	
Systems	IT Infrastructure	CMMS	Recommendation Management	Mobile Computing	
Reliability Improvement / Optimization	Performance Based	Defect Based -- RCA	Design Improvements	Lifecycle Analysis & Optimization	
Maintenance Optimization	RCM / FMEA	Time Based PM	Condition Based	Proactive	
Integrity Assurance	SIS Lifecycle Management	Electrical Distribution Integrity	Thickness Measurement	Corrosion Prevention	
Incident, Test and Inspection Management	Calibration Management	PRV Test Records	RBI / Inspection Management	Incident Management	
Quality Assurance	Skills Management -- Training	Work Quality Reviews	Configuration / Document Control	Data Security	
Reporting Programs	Automated Calculation	Posted	Published	Checked	
Skills Management -- Training	Safety	Team	Professional / Craft	Certification / Records	
Communicate and Promote Results	Wins	Lessons	People	Continuing Work	
Measure, Monitor and Review Results					**Check**
Measure and Manage Results	Actual Results	Metrics	Gap	Corrective Action?	
Adjust and Improve Action Plans; Improve and Maintain the OE Program Evergreen					**Improve** Sustain
Implement Continuous Improvement	Process	Practice	Technology	Organization	
Review Program Assumptions, Envelope	Operating	Raw Materials	Risk	Compliance	
Extend Ownership; Institutionalize and Sustain Success					
Emphasis on Continuous Improvement	Control Plan	Communicate	Publicize	Promote	
Safely Achieve Optimized Business and Market Value Return From Operations					

13 AIChE
Spring Meeting, San Antonio, TX

Implement—Improvement plans

- **Conduct training**
  - *Awareness; program necessity, principles, values, objectives*
  - *Details; technology, systems, processes and best practices*
- **Implement improvement action plans**
- **Apply resources**
  - *People*
  - *Technology, systems processes and best practices*
    - Deploy new
    - Improve existing
- **Deploy / optimize supporting processes**
  - *Information systems*
    - If adequate information structure is not in place at the beginning of implementation, it must be installed at commencement so improvements create history
    - Assure all changes are fully documented
- **Recognize that improvement initiatives will move at different paces**

13 AIChE
Spring Meeting, San Antonio, TX

63

Check—measure and manage results

- **Measure and monitor activity (task) KPI's to assure good start**
  - *Improve training, resource insertion as required to maintain expectations*
  - *Act promptly to eliminate deficiencies, remove barriers*
  - *Identify and resolve concerns*
- **Shift to results KPI's as soon as initial tasks / activities are completed**
  - *Confirm that improvements are succeeding*
    - Producing the expected increase in mission / business results; value gained
    - Divide large populations into improved and unimproved
- **Assure improvement initiatives are delivering objective results**
  - *Adjust, refine, modify, improve and extend improvement action plans and activities as necessary to gain objective, sustainable performance*
  - *Identify and eliminate deficiencies*
    - Improve sub-standard performance as soon as practicable
- **Identify additional high value areas for improvement**
  - *Value prioritize and address as soon as practicable*

13 AIChE
Spring Meeting, San Antonio, TX

64

Communicate—widely

- **Report progress and success to Steering Team; executive and business management**
  - *Financial results demonstrating contribution to business / mission objectives*
  - *Compliance with timeline*
- **Publicize progress, results and successes to entire enterprise**
  - *Create visibility, elevate enthusiasm, support and momentum*
    - Improvements accomplished / value created
    - Compliance with results timeline
    - Expectations
    - Identify and publicize team and individual successes
  - *Continue to build ownership, support and consensus*
- **Conduct regular two-way communications**
  - *Make certain vision, action and results required are understood*
    - Correct misunderstandings immediately
  - *Consider employee survey*
  - *Overcome resistance*
    - Emphasize the necessity to understand and live by new roles and responsibilities
    - Utilize rewards and peer pressure to assist achieving compliance with objectives
    - Eliminate adversarial conflicts
    - Provide counseling and additional training for underperformers

13 AIChE Spring Meeting, San Antonio, TX

65

Improve—continuously

- **Conduct formal performance assessments**
  - *Intervals determined by maturity; changes in factors that determine objectives and / or performance*
    - Utilize formal template
      - *Maximize objectivity, consistency*
  - *Strategy*
    - Validate all factors that influence the business / mission objectives, strategy and successful implementation of the program
  - *Program*
    - Producing continuing contribution to business / mission objectives
      - *Employ financially based measurements wherever possible*
    - Assure plans to reduce risk, gain optimum sustainable effectiveness and value remain effective and consistent with the strategy and objectives
    - Identify and value prioritize continuing improvement opportunities
- **Continually reexamine results, opportunities**
  - *Identify deficiencies in execution; improve action plans*
- **Improve active action plans; identify and develop additional improvements**

13 AIChE Spring Meeting, San Antonio, TX

66

Keep the Operational Excellence strategy current

- **Changing business / mission conditions elevate requirements for enterprise performance and effectiveness**
  - *Continuing pressure on performance (mission effectiveness, profit margins) results in corresponding pressure to reduce cost — Operational Excellence must conclusively demonstrate greater benefits from improved effectiveness*
  - *Provide continuing assurance that expenditures are being allocated correctly at the right time to gain optimum effectiveness*
    - Infuse business / mission objectives into day to day management decisions
  - *Make certain the 2nd and 3rd order effects (unintended consequences) associated with a change are fully identified and understood*
- **Best "benchmark" enterprises are continually improving their performance**
  - *What was excellent last year may be average next year!*
- **Periodically review and update Operational Excellence strategic objectives**

13 AIChE
Spring Meeting, San Antonio, TX

67

Sustain gains

- **Review effectiveness of the sustaining control plan**
  - *Actions and activities to maintain, build on and sustain improvements*
    - Continued monitoring of results; responsibility for action
- **Institutionalize gains and success**
  - *Assure sustaining plan followed*
  - *Extend enthusiasm, ownership and commitment*
    - Assure suggestions are considered; action reported
    - Fair, open and honest relations and communications
    - Create new institutional memory
  - *Everyone thoroughly understands the personal and organizational benefits of increased efficiency and objective performance*
- **Validate demobilization plan**
  - *Major pitfall — prematurely declaring victory; demobilizing efforts*
    - Must have continued effort until totally confident that improvements are sustainably institutionalized
- **The new way must be accepted as the only way!**
  - *Optimum way is the only way*
  - *No remaining memory of the "old way"*

13 AIChE
Spring Meeting, San Antonio, TX

68

Assessment—an essential process

- **Learning exercise**
- **Accomplished by a team**
  - *Composed of Subject Matter Experts in areas to be evaluated*
  - *Led by an experienced individual*
    - Should have participated in multiple assessments as a team member
  - *Team fully trained in the assessment process and grading*
- **Conducted in accordance with a template / scorecard**
  - *Covers all areas to be assessed*
  - *Lists best practices; elements necessary for compliance*
  - *Suggests questions to evaluate level of excellence*
  - *Guidance to assure assessment and scoring are as objective and consistent as possible*
    - Four quartiles—Reactive Unawareness to Strategic Excellence
- **Accomplished in a series of interviews**
  - *Begins with a kick-off meeting*
  - *Conducted in accordance with a detailed schedule*
    - Lists subject matter, knowledge, data and records to be evaluated
  - *Discussions recommended in interviewees office*
  - *Scored individually by assessment team*
- **Scores compiled by assessment team leader**
  - *Close out meeting conducted before departing site*
    - Summary, strengths, areas for improvement
  - *Final assessment report includes recommendations for improvements prepared by team members—more important than scores*

13AIChE Spring Meeting, San Antonio, TX

69

Typical assessment template— Organized in logical domains

Facility performance, 0 to 4; 4 being industry best, in the areas of:

1. **Performance, Results, Effectiveness**
2. **Leadership, Vision and Organization; Work Culture and Institutional Values**
3. **Business Strategy, Policy, Objectives and Plans**
4. **Program; Legal Requirements, Description and Implementation**
5. **Risk Identification, Assessment and Control**
6. **Reliability Improvement Processes: RCM / FMEA**
7. **Asset Integrity; Proof Tests and Inspections**
8. **Optimization Programs, Practices and Procedures: PM, CBM**
9. **Failure and Incident Analysis and Prevention: RCA**
10. **Work and Stores Effectiveness**
11. **Sustainability, Continuous Improvement**
12. **Information Technology; Reporting, Data and Information Management**
13. **Skills and Document Management**
14. **Organizational Improvement and Transformation; Change Management**
15. **Monitoring Results; Periodic Assessments: Audit**

13AIChE Spring Meeting, San Antonio, TX

70

Starts from where you are

The Assessment Template

Asset Operational Excellence — Program Evaluation Master Scorecard

Compiled by: John S Mitchell; johnsmitchell2@gmail.com
March 2010

Site or Facility Description	
Name:	Capacity:
Location:	Design:
Industry:	Average:
Year Commissioned:	Best achieved during: 24 Hours: 72 Hours:

Domains of Asset Operational Excellence	Best Practice: Levels of Excellence			
	First Quartile Strategic Excellence	Continuous Improvement Proactive Competence	Continuous Improvement Managed Improvement	Fourth Quartile [illegible]
1. Performance, Results, Effectiveness:	"World class" performance objectives, including HSE and asset operational effectiveness established, published and understood throughout the organization; site compliance demonstrated by KPI's			Few or no performance objectives established beyond HSE
2. Leadership, Vision and Organization; Work Culture and Institutional Values:	Strong, visible, committed leadership, effective organizational structure, decision process and work culture; initiative, commitment and ownership throughout the organization			Leadership and commitment not apparent, cooperative, ownership lacking
3. Business Strategy, Policy, Objectives and Plans:	Strong vision, policy and mission statement for Operational Excellence, published, understood and shared; solid, consensus strategy and detailed action plans linked to gaining organizational objectives			Little or no stated vision, ineffective organization, inconsistent work culture
4. Program; Legal Requirements, Description and Implementation:	Operational Excellence program well defined and documented; implementing processes defined and documented; all reviewed periodically, updated as necessary			No defined improvement program
5. Risk Identification, Assessment and Control:	Risk identification, assessment and control process documented and in place; governs application and implementation of asset integrity, reliability optimization, equipment management and surveillance programs; improvement and repair priorities			Little or no knowledge or use of risk analyses or optimization
6. Reliability Improvement Processes: RCM / FMEA	Strategic, proactive reliability plan documented, fully utilized, periodically reviewed and updated: unexpected events "surprises", deviations from strategic plan, expectations rare			Little or no strategy; activities largely ad hoc response to events
7. Asset Integrity; Proof Tests and Inspections:	All statutory, regulatory and site requirements for tests and inspections fully identified and documented; compliance assured by automated reporting variations from normal, out of limit measurements and test intervals			Statutory and regulatory requirements not fully defined: no reliable reporting system
8. Optimization Programs, Practices and Procedures: PM, CBM	Predictive (condition based: operating performance, mechanical condition) preventive and proactive programs fully documented including application, implementation, reporting and follow up requirements to assure effectiveness			Support programs uncoordinated, little or no assurance of consistency or effectiveness
9. Failure and Incident Analysis and Prevention: RCA	Robust Failure Analysis (RCA), tracking and prevention program fully defined, in place and effective; failures treated as a learning experience, opportunity to improve, eliminate recurrence			Failure analysis accomplished inconsistently or not at all; ineffective follow up
10. Work and Stores Effectiveness:	Work and stores managed and conducted effectively in accordance with written, understood and utilized procedures; activities primarily determined by strategic plan; minimal ad hoc response to unexpected events required			Work and stores management system non existent or ineffective
11. Sustainability, Continuous Improvement:	Continuous improvement program documented, in place and functioning effectively; capital replacement process documented and followed; Change Management process established, in-place and functioning			No formal program for improvement; change management
12. Information Technology; Reporting, Data and Information Management:	Robust data and information infrastructure and management system; fast, accessible network, solid graduated security procedures; ready access: simplified, effective reporting			Poor IT infrastructure, data and information inconsistent, ineffective controls
13. Skills and Document Management:	Document and skills management system; written procedure in place, fully functional, easy to use and access including configuration control of all documentation, drawings, manuals, plans, etc; skills matrix in place with training, certification and succession plans			Inconsistent training; little or no document and skills management system in place
14. Organizational Improvement and Transformation; Change Management:	Formalized program for organizational improvement and transformation documented. Individual appointed to lead and develop support for change throughout the organization, identifying opinion leaders, and establishing ambassador network.			
15. Monitoring Results; Periodic Assessments: Audit	Periodic assessments conducted to evaluate performance, effectiveness and results to objectives. Assessment reports include recommendations for improvements			No formal evaluations
Scoring	3.5 - 4.0	2.1 - 3.4	0.5 - 2.0	0 - 0.5

1. Performance, Results, Effectiveness: HSE Compliance, Asset Effectiveness				
Compliance Safety Health Environmental (HSE) objectives and requirements Full compliance with all legislative and statutory requirements for a safe, environmentally benign operation meeting all local expectations for good corporate citizenship. Clear objectives established, understood and reinforced.	# Focus on eliminating / minimizing potential event sources. # Minimizing probability and / or consequences of events by continuing, comprehensive risk analyses and mitigation.	# Continuing emphasis on learning and avoidance action from "near misses." # Cultural commitment to HSE compliance.	# Emphasis on HSE compliance: safety message and lessons learned discussed at every employee meeting. # Performance to HSE metrics publicized and prominently displayed.	# Low cultural commitment to HSE compliance. # Compliance by directive; inconsistent adherence. # Environmental or safety incidents with significant consequences have occurred.
# Safety: OSHA Recordables or equivalent – Sites OSHA recordable rate over the last year; Five years. Objectives posted and stressed.	TRR <.8	TRR >.8 <1.8	TRR >=1.8 <2.5	TRR >2.5
# Environmental incidents and/or local limits exceeded over the last year. Five years .	No significant incidents.	Minor incidents have occurred, without significant release of any kind. Action taken to prevent recurrence	Moderate-scale releases, events have occurred . Action taken to prevent recurrence	Large-scale release event or fire have occurred resulting to property damage and / or injury.

13 AIChE
Spring Meeting, San Antonio, TX

71

Overall results

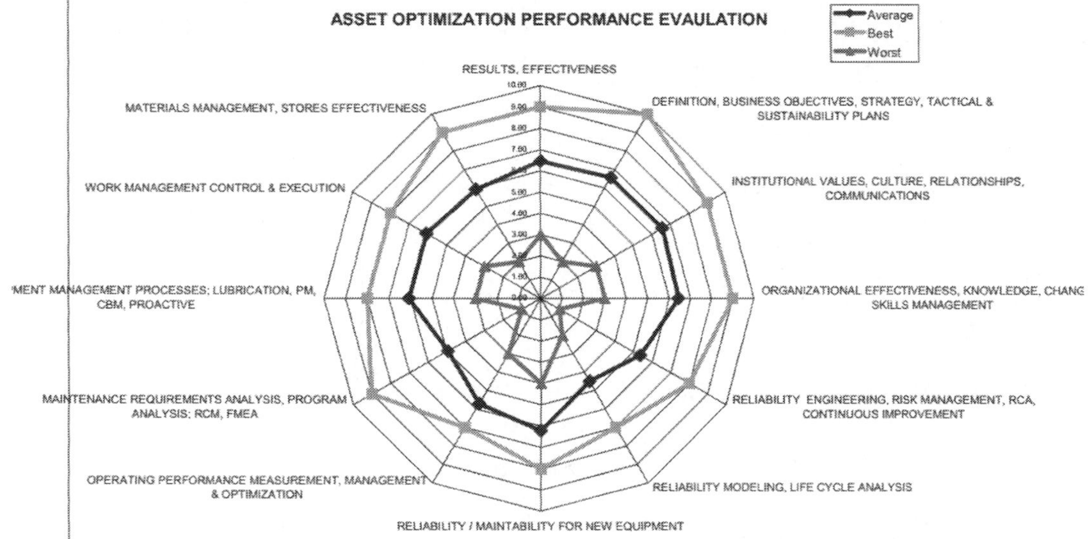

13 AIChE
Spring Meeting, San Antonio, TX

72

In Closing: Operational Excellence:

- Requires essential characteristics
- Produces major benefits
  - *Creates maximum value*
  - *Promotes effectiveness in all activities*
  - *Extends reliability*
  - *Minimizes risk*

Essentials for success

- **Clear vision stated, understood and accepted**
- **Solid values established, communicated and adopted by all in the organization**
  - *Initiative, ownership, commitment to excellence, responsibility and proactive continuous improvement throughout the organization*
- **Effective operational leadership and organization; constructive work culture**
- **Ambitious, optimistic, and achievable operational objectives**
  - *Full compliance with business, mission and organizational objectives*
  - *Established and owned by implementation action teams*
- **Optimally applied; value driven risk and reliability optimizing strategy and programs**
- **Energized, empowered and motivated champions**
  - *Real ownership, initiative, responsibility and accountability by all program participants*
- **Value and opportunity driven improvement initiatives**
  - *Formulated at the working level for maximum commitment and ownership*
  - *Contribution to business objectives / mission compliance compellingly demonstrated*
    - Increased profitability / mission compliance, availability, production output, quality, and reduced cost
- **Detailed improvement plans**
  - *Developed and owned by the action teams*
  - *Prioritized and implemented in value order*
- **Array of layered results oriented metrics and KPI's**
  - *Linked to overall business and organizational objectives*
  - *Interim and final program operational objectives*
  - *Measure to assure compliance, demonstrate value produced*

13 AIChE
Spring Meeting, San Antonio, TX

74

Essentials for success

- **Financial system in place to accurately assess real costs and value produced**
- **Effective training**
  - *Initiated at launch, and continued throughout*
  - *Skills matrix established and maintained current*
- **Robust, accurate and accessible support programs, practices and procedures, technology**
  - *Deployed to address specific requirements*
- **Complete, accurate data and information structure**
  - *Identify opportunities for improvement*
  - *Monitor and display performance; communicate results*
  - *Full document, configuration and data control*
- **Effective management of change, personnel skills and certification**
- **Full commitment to continuous review, improvement and sustainability**
  - *Identification of additional opportunities for improvement*
  - *Results institutionalized*
- **Periodic assessments to measure performance; identify additional opportunities for improvement**
- **Tangible rewards for results and value created**

13AIChE
Spring Meeting, San Antonio, TX

75

Major benefits

- **Achieves highest safety and environmental (SHE) performance**
  - *Concentrates efforts on highest priority opportunities for improvement*
- **Delivers maximum overall business / mission value and operating effectiveness**
  - *Maximizes business / mission value delivered*
    - Permanent improvements in availability, efficiency and reduced spending
    - Success regardless of economic conditions!
      - *Demand driven (availability) and excess capacity (costs)*
  - *Improves operational / mission effectiveness*
    - Gaps to objective performance identified, defined and closed
    - Greater predictability — reduced variation
      - *Minimal "surprises" Lost Opportunity (unplanned) Events (LOE)*
  - *Identifies and eliminates defects — improves operation, maintenance and design*
    - Safely reduce requirements for work — only way to safely, permanently and sustainably reduce costs
      - *Increase reliability, extend operating life*
      - *Minimize low value and unnecessary work*
  - *Demands optimum operating surveillance — leads to effective preventive, predictive and proactive tasks*
    - Deficiencies identified and corrected prior to failure; minimize unexpected interruptions
  - *Results demonstrated in credible business and financial terms*
    - Gains greatest credibility and organizational support
- **Gains optimum balance between risk, reliability and cost effectiveness**
  - *Identification, management and improvement prioritized by potential value / return*
  - *Only path to safe, sustainable improvements*

13AIChE
Spring Meeting, San Antonio, TX

76

Major benefits

- **Gains effective institutional culture and resource utilization**
  - *Totally defined organization; roles and responsibilities*
  - *Optimal resource effectiveness*
    - People, material and financial
  - *Maximum contribution from engaged, energetic work culture committed to excellence*
    - Gained from inclusion, motivation, enthusiasm, commitment, ownership and responsibility
- **Assures best processes, practices and technology**
  - *Established and sustained in a most effective, risk / value prioritized sequence*
- **Delivers maximum value and return from improvements**
  - *Identified, prioritized and implemented based on business / mission objectives; value delivered:*
    - Attainment of objective / expected / required performance
    - Reduced risk
    - Resources required; cost and time
    - Potential value gain
      - *Includes the probability that improvement initiatives will produce expected results*
- **Increases capital effectiveness**
  - *Extend operating lifetime*
    - Operate safely for longer intervals at lower costs
    - Minimizes requirement for additional capacity
  - *Improve logistics*
    - Reduce work in process (WIP)
    - Reduce stocked spare parts
- **Sustaining**

13AIChE
Spring Meeting, San Antonio, TX

77

A few quotes

From the Physical Asset Management Handbook

"Employee knowledge of the business environment, its ramifications for company success and job preservation through participation in the improvement process has promoted ownership and a desire to maximize opportunities."

"Ownership and commitment is essential to successful change and takes time and effort to gain. Employees resist change. They will typically 'ride out' management directed changes because experience indicates the change will be forgotten in six months or the change itself will be changed. Employee involvement and ownership must be part of the overall strategy. Employees must know and believe why they must change and improve. A clear strategy paves the way to the necessity for changes."

"When a manager inquired why no one had ever mentioned the huge waste and inefficiencies within the old system, workers replied: You never asked!"

13AIChE
Spring Meeting, San Antonio, TX

78

Thank You!

Questions, comments?

13 AIChE
Spring Meeting, San Antonio, TX

79

(134b) APM/Asset Performance Management: More Than the Sum of Its Parts

Paula Hollywood, ARC Advisory Group, Dedham, MA

APM/Asset Performance Management: More than the Sum of Its Parts
Unplanned downtime is the nemesis of all manufacturers. In North America, unscheduled shutdowns account for two to five percent of planned process plant production. Improving asset availability, efficiency, flexibility, and responsiveness, without sacrificing safety and consistent product quality are among the challenges facing manufacturers. How can manufacturers better manage existing assets to be more efficient, available, safe, and reliable? The rich information found in asset performance management (APM) solutions provides actionable information to improve plant performance.

Going Beyond Asset Monitoring for Improved Reliability: Remote Asset Monitoring and Alerts

Nikki Bishop, Emerson Process Management, Atlanta, GA
Aaron Crews, Emerson Process Management, Austin, TX

Introduction

In recent years, the influx of communication technologies has enabled nearly instantaneous communication with virtually anyone located anywhere in the world. These technologies can also be applied to the plant floor to allow plant assets to communicate their status with plant personnel. Assets can now "talk" to the control room. Even further, assets can now alert the right person, right away when attention is necessary. However, before jumping into a discussion of remote monitoring, it is necessary to discuss how to select the most effective asset monitoring strategy. A good automated monitoring strategy lays the foundation for a good remote monitoring infrastructure.

Asset Management Strategies

It's no surprise that a good predictive maintenance strategy improves overall reliability and helps to meet production availability targets. However, not all predictive maintenance strategies are created equal. Predictive maintenance based on periodic – possibly infrequent – data acquisition fails to give real-time insight into asset health. Periodic data may come in the form of "clipboard rounds", where personnel are sent into the field to manually collect data at some scheduled interval. This may occur once a shift, once a day or possibly less frequently. This method provides only a snapshot of asset data and early warnings of impending problems may be missed. Even further, sending personnel out to the field for manual data collection might also mean putting them in a hazardous environment. With little or no insight to know which assets actually need attention, resources may be wasted servicing assets that don't need it while missing assets that do. In fact, studies have shown that more than 60% of a typical instrument technician's trips to the field result in either no action or a minor configuration adjustment that could be done from their office. So what's the answer for a good predictive maintenance strategy? Automated monitoring provides online indication of an asset's health and can even detect process conditions that may be unintentionally and unknowingly inducing a fault on equipment, allowing operators to make adjustments so that process related equipment faults can be avoided altogether. With advanced warning, maintenance staff can spend their time servicing assets that actually need it rather than wasting time searching for problems through manual inspection.

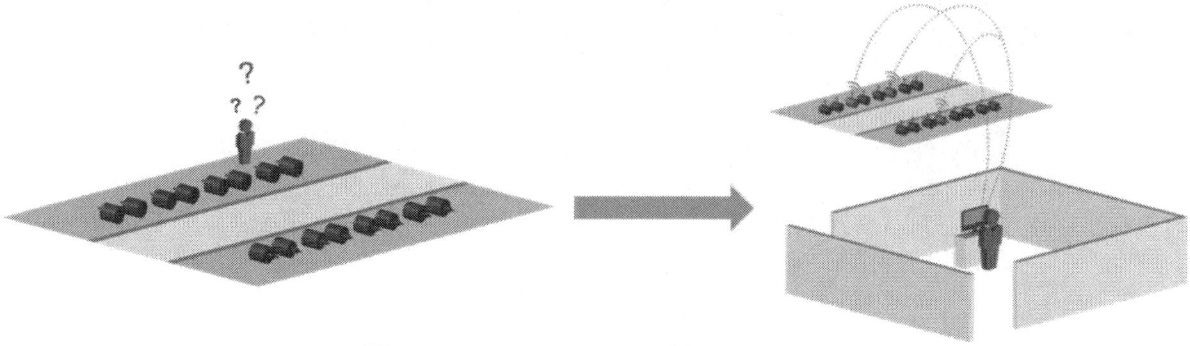

Figure 1. Automated Monitoring

Automated monitoring allows you to accurately determine when service is necessary and brings asset monitoring into the control room.

The assessed criticality of an asset often determines how an asset is managed. While real-time monitoring (and protection) of critical process equipment, such as large compressors or turbines, is standard practice at most facilities, on-line monitoring of second-tier equipment, such as pumps, heat exchangers, blowers, small compressors, cooling towers and air cooled heat exchangers ("fin-fans") has traditionally been deemed cost-prohibitive or too difficult. Even though these unmonitored or manually monitored assets may not have been originally classified as "critical", an outage or failure can cause a serious process disturbance or shutdown, resulting in process downtime and increased load on site personnel for repairs. These assets can be referred to as "essential assets". Essential assets present an opportunity to add online monitoring solutions and improve overall reliability while lowering maintenance cost.

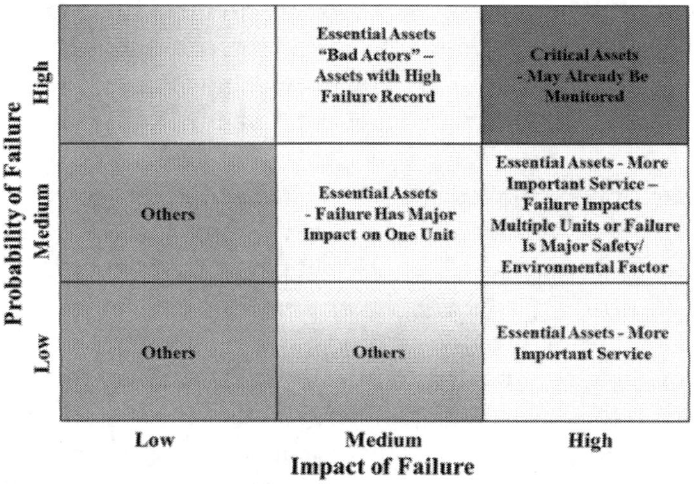

Figure 2. Asset Criticality Assessment

Essential assets are a class of assets that typically do not have monitoring systems in place already, but consequences can be high if a failure occurs.

Components of Successful Strategies

Asset monitoring is not only about collecting data. Data collection simply lays the foundation for an asset monitoring strategy. Existing measurements can be used or new wireless measurements can be added easily. Once the measurement framework is in place, pre-engineered monitoring solutions are plug-and-play and take raw measurement data and transform it into meaningful alerts through analysis. Process and asset data can be aggregated to detect process conditions that may be inducing a fault on equipment. Adjustments to process conditions can be made so that process-related equipment faults can be avoided altogether.

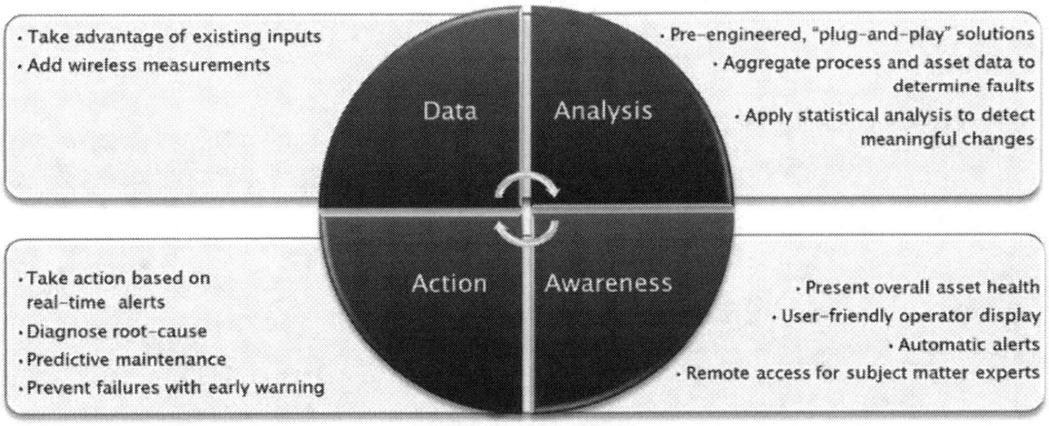

Figure 3. Asset Monitoring Strategy

Data collection alone is not sufficient for a monitoring strategy. A combination of data collection, analysis, awareness and action is necessary for a successful program.

Meaningful alerts generated by data analysis and aggregation are only useful if they are communicated to the right personnel at the right time. Providing awareness of the alerts is a critical component of an automated monitoring system. While there are a number of ways to provide awareness of alerts, the most effective method is through automatic notification. Alerts via text message or email ensure that the alert gets to the right person, right away. Once the alert has been received, action must be taken by the person to resolve any issues. Remote access via tablet or smartphone allows for almost instantaneous diagnosis and action. If necessary, remote subject matter experts can be alerted and can log in remotely to help diagnose the problem. With an automated alert notification system in place, periodic reports can also be generated and distributed. These reports can include trends of asset health where asset degradation can be seen and action taken to prevent subsequent failure. Automated monitoring combined with automatically generated alerts and remote access capability provides a powerful means of monitoring asset performance.

Essential Asset and Remote Monitoring in Action

One site taking advantage of remote asset monitoring is the Separations Research Program at the University of Texas-Austin's J.J Pickle Research Campus. The Separations Research Program at UT is a cooperative industry and university program that performs fundamental research of interest to chemical, biotechnological, petroleum refining, gas processing, pharmaceutical, and food companies. One current focus project is around carbon dioxide removal from stack gas. This process includes absorber and stripper columns and the associated assets such as pumps, blowers and heat exchangers. The process does not have redundancy in the assets so it is important to keep the assets well-maintained and in operation. Losing one asset means the entire process is down until the repair is complete. In order to mitigate the risk of unplanned downtime, they have successfully implemented essential asset monitoring strategies for pumps, heat exchangers and blowers. They now have online insight into the health of their assets. They can see when process conditions may be contributing to asset health degradation and make adjustments to prevent further damage or failure. For example, alerts for increasing vibration warn of impending failures and allow time for servicing before failure occurs.

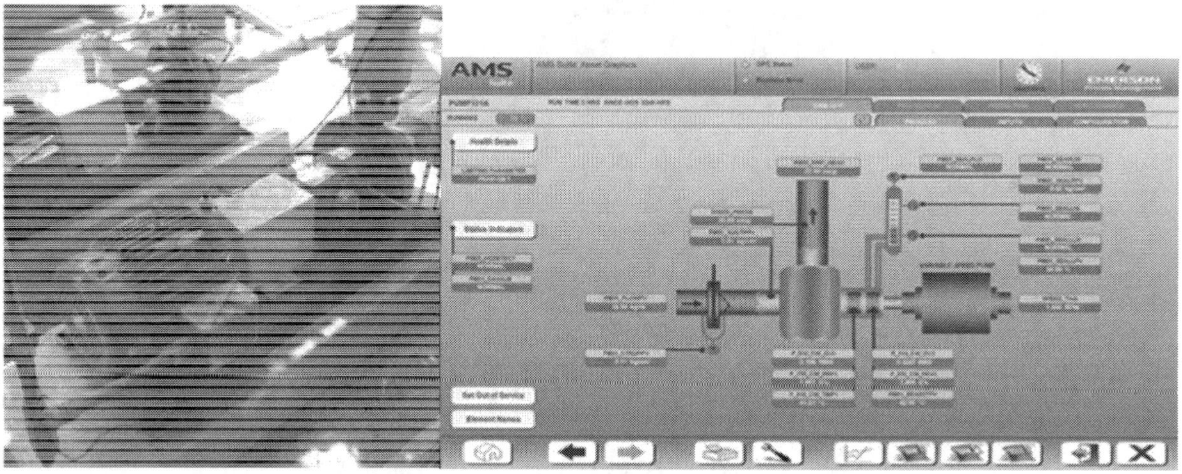

Figure 4. Pump Health Monitoring

A wireless vibration transmitter installed on the pump provides valuable data for an automated pump monitoring solution, where pump health and status are clearly indicated.

Recognizing a need to communicate alerts to the proper personnel at the right time, UT has taken one step further by setting up a remote monitoring infrastructure. Alerts for conditions such as heat exchanger fouling, resonance speed detection, hydrocarbon leaks and pump cavitation can be automatically routed to site personnel as well as trained subject matter experts at the onset of the fault condition. Besides monitoring process assets, the remote monitoring system, also known as iOps, monitors for control system health alerts such as an overloaded PC or failed backup controller. These alerts can be automatically routed via text message or email. The remote connection also allows experienced subject matter experts to log in using secure VPN access to help diagnose problems with assets and assist with the appropriate corrective action. With log in access via tablet or smart phone, diagnosis support is available instantaneously. Having a remote monitoring infrastructure in place, customizable, periodic reports can be generated and automatically distributed. These reports contain trends of asset and system health and clearly identify which assets or systems require attention. The automated, remote system monitoring means that UT has trained subject matter experts armed and ready to take action when adverse conditions arise, whether it be a cavitating pump or an overloaded PC.

The diagram below shows the remote monitoring process at UT. The plant and the control room with the operators in the control room are shown in the center. Essential asset monitoring strategies are implemented for pumps, heat exchangers and blowers. These monitoring solutions use field data to generate alerts and communicate those alerts to the control room. But what happens if the operator is not in the control room or not focused on the screen? Even if the control room operator is not present, the iOps center is monitoring for any alerts 24/7 via the remote monitoring appliance installed. If there is an issue with the pump, such as cavitation, the essential asset monitoring solution will collect, aggregate and analyze the asset and process data. A meaningful alert and percent health value will be communicated to the remote monitoring appliance. The appliance transmits the alert to the iOps center. The iOps center can contact the local site representative as well as contact a remote subject matter expert, if necessary. The remote expert can log in to the site to help diagnose the problem and suggest corrective action. The local site representative and remote expert collaborate on required corrective action. The UT operator can then carry out the corrective action and fix the fault before it becomes a failure. This method ensures that the fault is not missed and that issues are resolved quickly and efficiently.

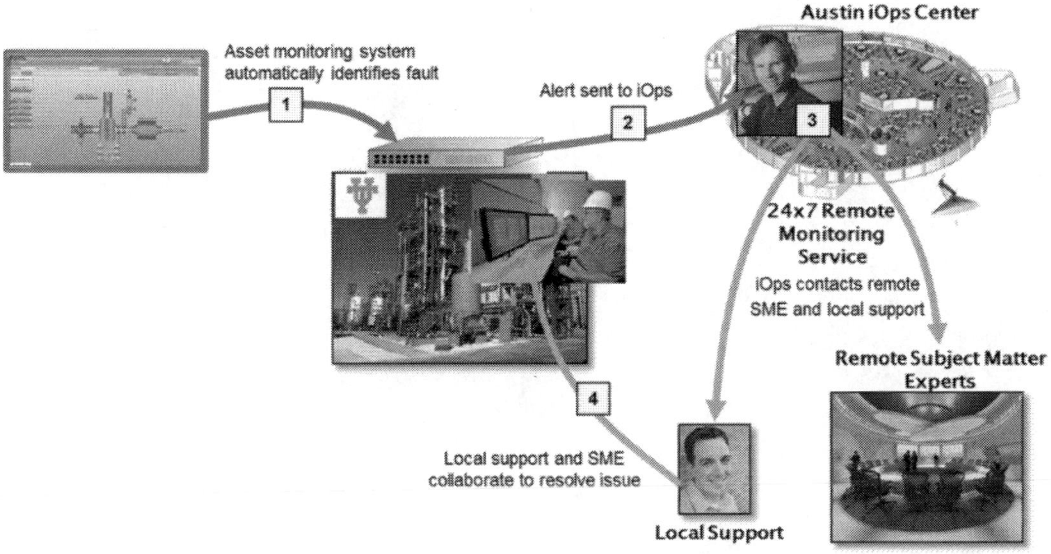

Figure 5. Remote Monitoring Process

Automated, remote system monitoring means that UT has trained subject matter experts armed and ready to take action when adverse conditions arise, whether it be an asset or control system component

Conclusions

Using the latest developments in wireless and communications technologies, the era of on-line remote monitoring of plant assets is now a reality. Wireless technology makes it easy and cost-effective to add missing measurements for essential assets. Pre-engineered monitoring solutions are plug-and-play and provide an easy means of data aggregation and analysis. Remote monitoring and automated alerts ensure that alerts generated by monitoring solutions are not missed and that corrective action can be taken to prevent unplanned downtime due to a failed asset.

About the Authors

Nikki Bishop, PE, is a Senior Application Consultant at Emerson Process Management. With over 10 years of experience in the process control industry, her experience includes automation projects in industrial energy, pharmaceuticals, power generation, pulp and paper and refining. In her current role, she designs, develops and manages integrated application solution packages. She holds a BSChE degree from Georgia Tech and is a registered professional engineer in the state of Georgia. She can be contacted via email at Nikki.bishop@emerson.com.

Aaron Crews is a Lead Process Systems Engineer with Emerson Process Management. Aaron's over 8 years of experience includes control system modernization projects for refining and chemical industries, with specialties in DCS architecture, operator interface design and knowledge management. In his current role, he leads a team of engineers focused on the integration of instrumentation, control, asset management, and enterprise applications. He holds a chemical engineering degree from Texas A&M University and it currently pursuing his MBA from the University of Texas at Austin. He can be contacted at aaron.crews@emerson.com.

(134d) Effective and Optimized Asset Strategies Through Combining Condition Data and Diagnostics

John Renick, Meridium Inc, Roanoke, VA

There are many systems, processes and people in place to support gathering asset condition data and predictive diagnostics all in the context of reliability. Usually these systems are not leveraged to deliver their full potential and comprehensive value. This presentation will show the benefits of compiling data from multiple systems and work processes to optimize strategies and drive smart decisions. Sources of information that will be discussed will include: device alerts, vibration, lubrication, operator rounds, calibration and process data.

RELIABILITY MONITORING FOR ADVANCED PROCESS CONTROL IMPLEMENTATIONS

J. Ben Rodin
Honeywell Process Solutions
Duluth, Georgia

Introduction

Advanced Process Control / Model Predictive Control (APC/MPC) has gained wide acceptance in many industries. APC/MPC is used to tackle supervisory control problems ranging from very long deadtimes with accumulating disturbances to infrequent measurements such as lab analyses being incorporated into control strategies.

There are numerous presentations on MPC and the benefits of its implementation. This presentation is instead focused on one approach to helping keep MPC in use and operating, remote usage and performance monitoring. Usage monitoring indicates what percentage of operating time the APC controls are working, both at the master level and at the individual handle level. Performance monitoring goes further, by looking at constraints and determining which constraints may be tying down enhanced economic performance. We focus on what information to glean from the APC and how to handle this information to help improve operation.

APC/MPC Fundamentals

Model Predictive Control, as far as this presentation is concerned, is a methodology to address supervisory level control problems. MPC requires a foundation of properly tuned low level regulatory controls such as flow controls to achieve its objectives. MPC gives us a toolbox to consider dead times, accumulating disturbance feed forwards, complex process dynamics, and other process circumstances in our control solutions and take perhaps multiple variable actions to achieve optimal control. MPC generally works by determining an optimal steady state control solution along with an appropriate controls path to achieve this solution.

MPC is discussed in numerous books, articles, and presentations. This section is presented only as a quick review on its important concepts.

The three primary process variable types addressed in MPC problems are Controlled Variables (CV's), Manipulated Variables (MV's), and Disturbance Variables (DV's). Controlled Variables are those process quantities for which we wish to maintain a value. The value may be a range or a set value. For example, in a distillation column, overhead product concentration might be one of the key variables required to be in a range.

Manipulated Variables are those process quantities that we can use as handles to try to maintain the CV's. An example of an MV could be a reflux flow. An Advanced (Multivariable)

Process Controller would manipulate the reflux flow setpoint, while the underlying regulatory controls would manipulate valve outputs to maintain the setpoint. Lastly, Disturbance Variables are those process quantities that affect the values of CV's, but, unlike MV's, we cannot manipulate these variables to achieve our desired CV values. Ambient temperature may be a DV.

When setting up an MPC solution, one approach is to set up different types of operating ranges. If we aim to maintain CV's in a range, we might have two different ranges, one being "hard limits" which are values that we absolutely do not want the CV's to exceed both high and low. These hard limits are values that the controller can do the best it can with available MV's to maintain regardless of any optimization goals. Within the hard limits, a controller can have "soft limits." Soft limits set a smaller range with the high and low range established by the hard limits that provide a range where we wish to keep CV's for optimization purposes. If there is a configured cost function, this soft range establishes our range of optimal control. Having dual range controls allow us many advantages, one key one is to optimize a CV while leaving room for process swings that might take a CV out of its soft range for a time. We use soft limits to direct our optimizer to essentially maintain a buffer away from hard limits.

Soft limits can also apply to Manipulated Variables. If a cost function for optimization is configured, only optimal control within the soft range may be performed. Hard limits establish the total range a controller can adjust a Manipulated Variable.

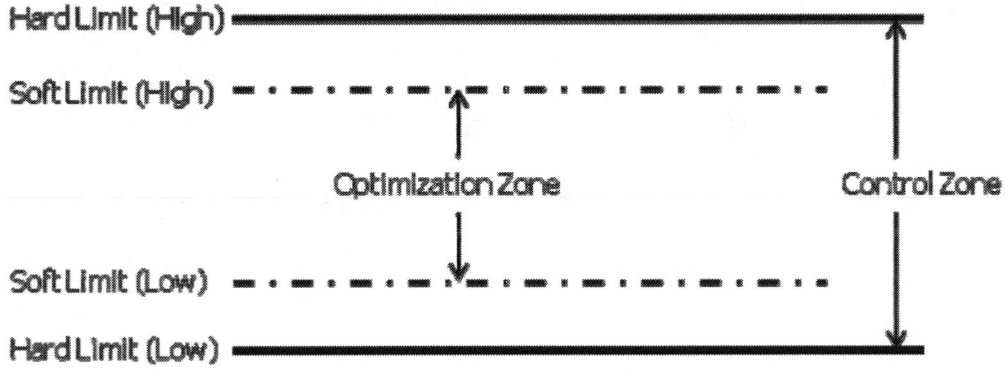

Figure 1. Limits Description for Model Predictive Control. Hard Limits represent absolute limit of Controlled or Manipulated Variables. Soft Limits represent optimization limits.

Reliability Monitoring

Generally, these applications are installed at industrial facilities along with all the other controls applications at industrial applications. Some of the models relating CV's to their underlying MV's and DV's can be quite complex and they might need to be changed with time (e.g. fouling). A CV might have several underlying MV's and DV's. MPC implementations may only be designed to run under one set of conditions (e.g. a distillation process operating for a certain grade) but the operating conditions typically seen may be changed.

These and other possible challenges may arise which may induce operators to simply "switch off" the MPC because it stopped performing like it originally did at installation. Another possibility is that the MPC is no longer doing its job as well as it could. Operators might look and see that a CV is in control range and think all is well. However, optimization may not be appropriate. Opportunities might be missed to modify constraint values to achieve greater economic benefit.

Remote reliability monitoring allows engineers to watch and analyze an MPC implementation from offsite. This remote monitoring enables process experts to simultaneously monitor many sites. The engineer could be getting a lot of data but must focus on that information which will help come up with improvements for implementation. Thanks to the internet, automated process data transmittal and report generation enables engineers to work potentially thousands of miles away and find process improvements. Even if the engineer is local to the implementation, automated reporting is still very useful.

Automated reporting also enables the process examiner to do month to month comparisons and see if degradation of MPC performance is occurring.

Overall Monitoring

The primary goal of any reliability reporting is overall system monitoring. The first measure, percentage of time running (percentage uptime), is a critical indicator. If the APC/MPC application is switched off, then it is obviously not doing its job. Other measures could include the following:

- Percentage of time the system is able to optimize. In other words, this time would be when all the MV's and CV's can have a solution within soft limits and an economic objective function is able to be minimized.
- Percentage of time the system is unable to find a control solution to keep all CV's within hard limits.
- Percentage of time the system is unable to find a control solution to keep all CV's within soft limits.

CV monitoring

The first set of variables to be examined by a controls engineer is that of the status and values of CV's. A typical report over a given time frame (such as monthly) might include tables such as those shown in Tables 1 and 2.

Table 1. Controlled Variable Sample Report Part 1. Only two CV's are shown for illustration purposes.

Tagname	Percentage Time SS value at Low Soft Limit	Percentage Time SS value at High Soft Limit	Average Value When Controller On	Average Steady State Predicted Value when Controller On
CV1: 101AI101.PV	2.2	4.01	3.56	3.44
CV2: 101DI102.PV	2.56	65.33	0.78	0.607

Table 2. Controlled Variable Sample Report Part 2. Only two CV's are shown for illustration purposes.

Tagname	Percentage SS Infeasible in Hard Limits	Percentage Violate (Percentage actual time outside hard limits)	Percent Drop (Percentage time Variable Dropped)	Average Low Limit	Average High Limit
CV1: 101AI101.PV	2.2	4.01	3.56	3.44	3.44
CV2: 101DI102.PV	2.56	65.33	0.78	0.607	0.607

In the above tables, percentage time SS value at low soft limit and percentage time SS value at high soft limit represent the amount of time that a steady state predicted solution will end up at the respective soft limit constraints on each variable. The average value ON is what the CV is reading while the controller is on. The steady state predicted value would be that obtained if the MPC is allowed to ideally take the process to an optimal steady state. The percentage SS infeasible in hard limits is the time that no predicted steady state can be found within hard constraint limits. The percentage violate is when the actual value is outside hard limits. Percentage drop represents the fraction of time when the variable is dropped from control by operators from control by the MPC package. Lastly, the average limits are the averaged hard limits entered by engineers or operators.

A gleaning of data by a process expert would first focus on the percentage drop. If it is high, the process expert must find out why the operators are disabling that specific point. One example may be that some CV's are only applicable during specific grade operation but another example is that the operator encountered a process problem.

If both the controller has a high uptime and the CV has a low percentage dropped, then a process expert can look at improving situations. First, one would examine percentage violating. This is of strong concern as high percentages indicate the MPC is unable to maintain an important limit. A high percentage violating could indicate unreliability of the model or it could indicate insufficient range on an MV. If it seems the MV most closely associated with this CV is seen to have large and frequent movements, the model at the base of the controller may not be of high quality.

A large percentage of time for a steady state at a soft limit, be it at a low or high soft limit, is usually positive. This situation indicates that a CV is being pushed to some operational optimal envelope and is able to be improved. However, in a process with multiple CV's, some of the CV's may not be able to be pushed to a soft limit as it might be infeasible for all of the CV's to be at a soft limit.

In Table 2, a large percentage of time is seen violating CV2's set limits. One must examine further why this occurred. Three possibilities are inappropriateness of limits, insufficient range of MV's associated with that CV, or model degradation.

Generally, a quick examination of a system will look for uptime and percentage drop concerns. If associated MV's are at limits most the time, CV optimization is usually occurring in an appropriate manner. Again, this depends on the process and the CV/MV interactions.

MV monitoring

The second set of variables to be examined by a controls engineer is that of the status and values of MV's. A typical report over a given time frame might include tables such as those shown in Tables 3 and 4.

Table 3. Manipulated Variable Sample Report Part 1. Only two MV's are shown for illustration purposes.

Tagname	Percentage Time SS value at Low Soft Limit	Percentage Time SS value at High Soft Limit	Average Value When Controller On	Standard Deviation of Value When Controller On	Average Steady State Predicted Value When Controller On
MV1: 101TIC103.SP	0.732	1.9	702.63	286.81	704.37
MV2: 101FIC104.SP	21.29	0	5.39	0.642	5.39

Table 4. Manipulated Variable Sample Report Part 2. Only two MV's are shown for illustration purposes.

Tagname	Percentage of Time at Low Hard Limit	Percentage of Time at High Hard Limit	Percentage of Time in Operator Mode	Average Low Limit	Average High Limit
MV1: 101TIC103.SP	0.219	0.512	13.97	100	1569.1
MV2: 101FIC104.SP	15.07	0	3.07	4.93	10

In the above tables, percentage low soft and percentage high soft represent the amount of time that a steady state predicted solution will end up at the respective soft limit constraints on each variable. The average values are what the MV is set while the controller is on or off. Similarly, the standard deviation indicates when the variation of the respective MV is high or low. The steady state predicted value would be that obtained if the MPC is allowed to ideally take the process to a steady state. The percentages hard at limits, low and high, represent the time percentages that the respective MV is taken to the respective low and high constraint hard limits. The percentage operator mode is when the MV is not manipulated by the MPC package. Lastly, the average limits are the hard limits entered by engineers or operators.

A gleaning of data by a process expert would first focus on the operator mode percentage. If it is high, the process expert must find out why the operators are disabling that specific point.

If both the controller has a high uptime and the MV has a low time percentage in operator mode, then a process expert can look at improving situations. First, one would examine percentages at hard limits. In conjunction with CV results, expansion of MV ranges can be considered. Another variable to examine is the standard deviation. If an MV has a high standard deviation, this indicates significantly high manipulation. Typically, it is best to have relatively non-moving MV's. An optimal solution should be typical and thus MV's should be relatively stable. If the standard deviations are high for several MV's, some process problem may be indicated.

In Table 3, it is seen that MV1 has a relatively high standard deviation. If it is associated with a problematic CV such as CV2 seen in Tables 1 and 2 (high percentage violation), one explanation may be that the underlying models may have degraded with time since implementation. In other words, the controller may no longer be predicting the correct effect changing an MV has on controlling a CV. A controller may be causing overshoots on CV's by underestimating effects from an MV on a DV. The converse problem, where the gain is set too high for an MV's effect on a CV, can be seen by significant CV violations with poor, relatively small, MV responses. Building new controller models may be a possible remedy if model degradation is suspected. It should be noted that other explanations are possible including significant DV movement affecting an associated CV.

DV Monitoring

Lastly, the Disturbance Variables should be considered. If a lot of movement is seen in the MV's, it may be caused by large DV swings. A typical report over a given time frame might include tables such as that shown in Table 5.

Table 5. Disturbance Variable Sample Report. Only two DV's are shown for illustration purposes.

Tagname	Average Value When Controller is On	Standard Deviation of Value When Controller is On	Percentage ot Time Bad Value	Percentage Time Operator Drop From Consideration
DV1: 101AI105	30.07	6.23	0	0
DV2: 101TI106	39.44	8.06	0	0

Typically, information on average and standard deviation can be used to determine a cause of large MV moves. Bad Value percentage can easily indicate measurement problems.

The percentage time when the variable is dropped by the operator from being included in the models should also be considered.

Overall APC/MPC Reliability Examination

The first thing a controls engineer should look at in a monthly or some other regular report is the percentage uptime. If this is showing declines on an MPC controller, it is obvious there are problems in the MPC, in the process, or in operator training or awareness. If uptime is indicating near-continuous operation, then percentage times in drop or operator mode (for CV's and MV's respectively) should be examined.

Indications of near limit operation on different MV's and CV's will indicate to a process expert what is constraining a process. If the constraint can be relaxed, improvement may be possible.

Generally, if an MPC implementation is successful, but month to month reporting indicates increased percentages of violation, infeasibility, or operator disabling, this should raise red flags. If large swings in an MV (as seen by large values of standard deviation) are more frequent this could indicate that model reliability is coming into question.

Enhancement Possibilities

The above described tables only give basics of what data can be garnered by a controls engineer from automated reporting of an MPC installation. With automated reporting, historian manipulation, and internet transfer of data used in proper coordination more is possible and has been implemented. Some examples are given below.

Grade Specific Statistics

Grade specific statistics can also be generated. In other words, all the above values described for MV's, CV's, DV's, uptimes, and overall controller mode can usually be segregated by grades. The same tables can be multiplied several times, once for each grade. For this to be possible, some variable must be sent to the report generation to indicate what grade is in current operation. This depends on the reporting package, platform data formats, and other factors.

An example of the usefulness of this would be if low uptime on one grade is seen while the other grades have high uptime percentages. This situation may indicate problems with a specific MV that is more active for that grade. If an MV's underlying regulatory controls are poorly tuned, that problem may translate up to poor system response which might lead operators to turn off the controls for a specific grade.

Mathematical Model Quality Analysis

Assuming the MPC package has standard input output models relating the effects of changing an MV or a DV to a change in a CV, model quality analysis can be performed. Simply put, the mathematical models already used to describe the process can be used for pure prediction. Comparisons to real responses over a given timeframe can be made in an automated manner and one can determine how well the models are predicting the system behavior.

Conclusions

An examination of the basics of reliability monitoring has been presented. Automated history analysis and reporting enables technical workers to examine how well an MPC implementation is performing and if any apparent reliability issues arise. Month-to-month report comparison enables technical workers to see if more and more manipulated variable movement is seen, if violations of constraints are increasing, or other problems are arising.

Remaining Life Assessments of Refinery Coker Furnace Tubes

Gerald W. Wilks
CITGO[(1)] Petroleum Inc.
135th Street and New Avenue
Lemont, Illinois 60430
USA

ABSTRACT

Process simulations are used in refineries to determine the optimum operating parameters for a wide range of processes. The development of Omega[1] technology by the oil industry in the 1990's led a new type of simulation that made it possible to assess both the effects of corrosion and past service on the remaining life of high temperature refinery equipment. Furnace tubes are a type of high temperature equipment that is well suited to Omega simulation analyses. However, in coker furnaces the accumulation of coke inside the tubes leads to gradual changes in temperature and pressure that make accurate simulations of the coker furnace difficult. This paper describes how corrosion data and process data can be used in Omega simulations to obtain accurate assessments of coker furnace tube remaining life. Included in this paper is an explanation of how anticipated changes in process conditions can be evaluated with Omega simulations to assess the impact of those changes on coker furnace tube life.

Keywords: Creep, Omega, refinery, cokers, coking, furnace, tubes, corrosion, sulfidation, pigging, K90941, S32100 stainless steel, S34700 stainless steel

INTRODUCTION

The refinery coking process is essentially a way to upgrade heavy oil (called "resid") into lighter products that have higher value. Resid typically consists of a mixture of crude unit vacuum tower bottoms, FCU fractionator bottoms, and coker fractionator bottoms. In a coker these low value products are converted into light gasses that can be used as refinery fuel gas, naphtha, a range of distillate oils (kerosene, diesel, heating oils, etc.), and gas oils that are used as feed to catalytic cracking units. The coking process uses heat to thermally crack the resid, and coke is produced as a byproduct. To cause the desired cracking to occur the resid must be heated to above 480°C, and this is typically done in a refinery furnace that is typically called a "coker furnace". A diagram that shows the coking process and major equipment on a coking unit is shown in Figure 1. The feed to the unit is pumped into the bottom of the fractionator, and from there it goes into the coker furnace where it will be heated to high enough temperature to cause some of the large molecules to crack. This produces the more valuable hydrocarbon molecules that are the goal of the coking process. The furnace tubes are

---

(1) Company name, CITGO Petroleum Corporation, 1293 Eldridge Parkway, Houston, Texas 77077

exposed to the highest temperature in this process, and there is also significant risk of high temperature sulfidic corrosion. Therefore there are two degradation mechanisms working together to reduce furnace tube life: creep and corrosion. Of course other damage mechanisms could be active, and an example of such another damage mechanism will be discussed later in this report.

From the furnace the hydrocarbons travel to the coke drums where coke forms on the wall of the drums. The lighter hydrocarbons that were formed when cracking occurred inside the furnace tubes leave the top of the coke drums and are carried back to the fractionator where they are separated into specific oil products. Figure 1 shows several of the types of products that are produced in the fractionator: fuel gas, coker naphtha, light coker gas oil, and heavy coker gas oil. However, other products can also be produced in the fractionator. Another product that originates in the coker is coke. Refinery coke is typically used for power generation, but it can also be used in steel production. Coke can also be further processed to extract more useful hydrocarbons. Additional processing of coker unit products is a topic that is beyond the scope of this paper. This paper concentrates on what is occurring in the coker furnace and the life of coker furnace tubes.

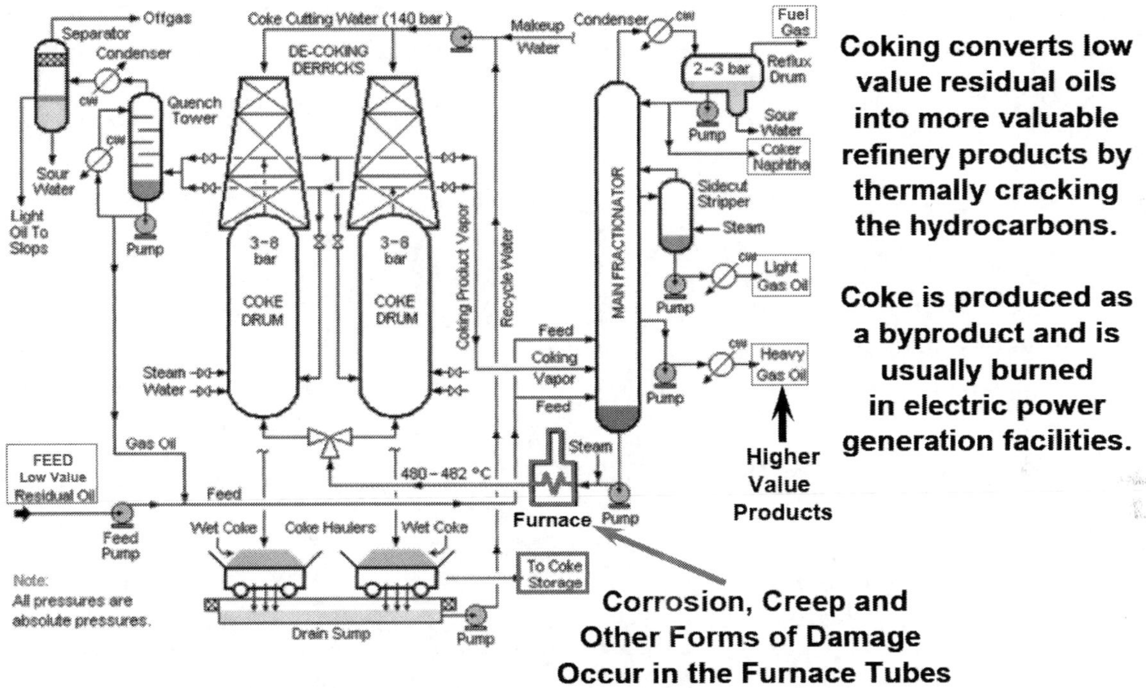

Figure 1: Diagram of major components of a coking unit.

The tubes in the coker furnace are typically made of high alloys because they must be able to tolerate high temperatures to heat the hydrogen to the coking temperatures: 480-482°C. Heating the hydrocarbon to these temperatures requires that the furnace tube be significantly hotter to obtain the desired heat transfer. Unfortunately, the hot tube surface promotes the formation of coke on the tube inside surface. The coke layer that forms interferes with heat transfer resulting in the tube wall temperature increasing as the coke layer builds. The coke layer formation also results in increased resistance to flow through the furnace tubes. As a result the furnace inlet pressure must be increased as the coke builds up inside the tubes.

As a result of these changes to the temperature and pressure that occur as coke accumulates inside the tubes, analyzing the tubes to determine remaining life is relatively difficult. Long term averages of temperatures and pressures cannot be used to determine the remaining life because important short term increases in temperature and pressure that occur during only part of a coker furnace cycle might be missed. Also, the combined effect of temperature and stress is very important for the analysis of creep life since creep damage is dependent upon both the stress in the tube wall and the tube wall temperature. The increasing temperature also affects the corrosion rate that the tube wall experiences. The coke accumulation is increasing the effects of both the creep and corrosion degradation mechanisms in the furnace tubes. Also neither of these degradation mechanisms has a linear effect on tube life so their combined effect cannot be ascertained by measuring creep and corrosion independently. A solution to this analysis problem was developed by the oil industry during the 1990's when Omega technology was invented during a joint industry project. Omega technology makes it possible analyze the impact of several degradation mechanisms on remaining life using short term process data. The examples provided in this paper show how the process history can be used along corrosion damage and other data on other forms of degradation in Omega simulations to make accurate predictions of coker furnace tube remaining life.

COKER FURNACE DESCRIPTION

The furnaces that were analyzed are part of a coker unit in a refinery in Lemont, Illinois. These were the box type furnaces with the tubes oriented in the horizontal direction. Figure 2 is a sketch showing the furnace components and how they are arranged inside the furnace. In this furnace the tubes are numbered with the inlet tube as "Tube 1" and the outlet tube as "Tube 40". Tube 15 is located at the top of the firebox. Coker furnaces of this type consist of a refractory line steel box with high alloy steel tubes inside that carry the hydrocarbon through the furnace.

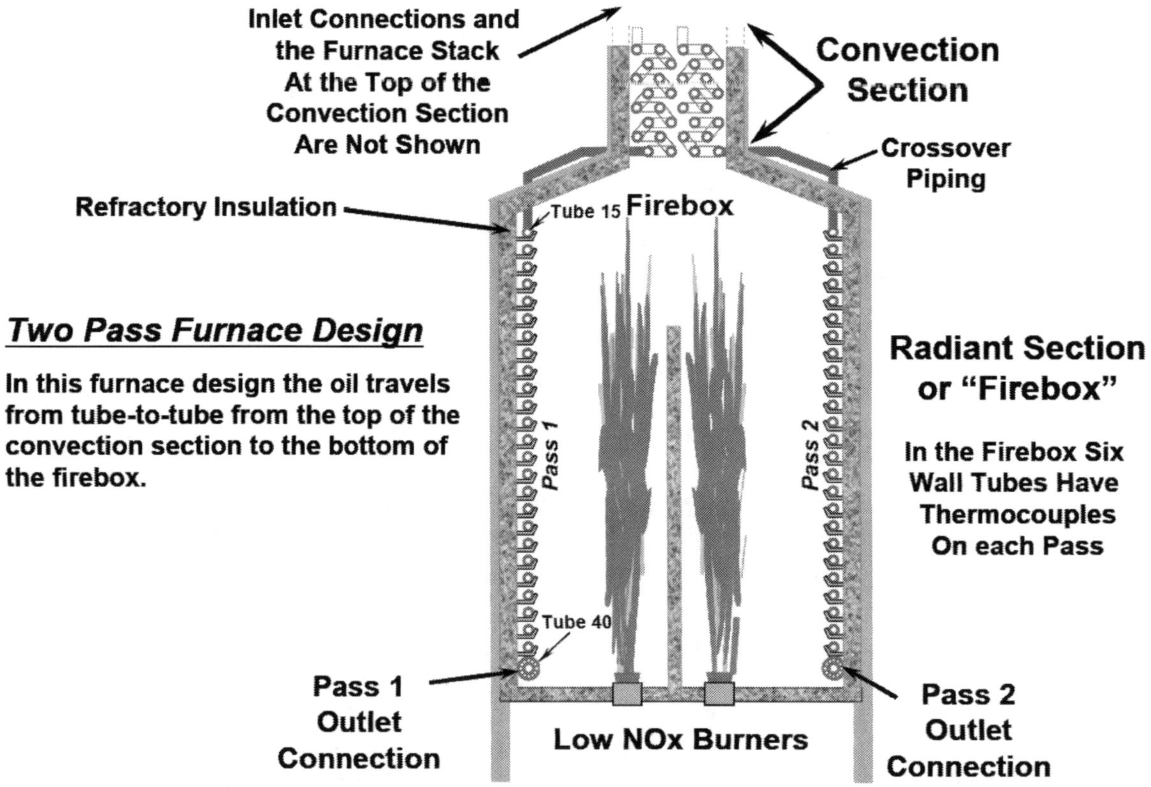

Figure 2: Sketch showing the cross section of a coking furnace.

The typical alloys used for coker furnace tubes include: K90941, S32100 stainless steel, or S34700 stainless steel. Heat transfer to the outside of the tubes is by radiation and convection from the burner flames and from the hot refractory walls. Heat is transferred through the tubes walls to the resid inside the tubes primarily by conduction. In coker furnaces the tube walls must be heated 65°C to 125°C above the desired temperature of the resid to achieve sufficient heat transfer. As was mentioned earlier, the hot metal surface causes coke to form on the tube ID surface and that coke formation leads to gradual increases in tube metal temperature and furnace pressure.

Figure 3 is a graph showing the daily average tube wall temperature and the daily average pass inlet pressure for one of the coker furnace tubes. The temperature data represent the daily average of the maximum temperatures measured using thermocouples on 12 of the tubes in the furnace firebox. The pressure graph represents the calculated pressure in the tubes at the top of the furnace firebox based upon the daily averages of the furnace inlet pressures along with information about the pressure drop occurring in the furnace. This tube pressure is determined based upon a pressure simulation of the furnace.

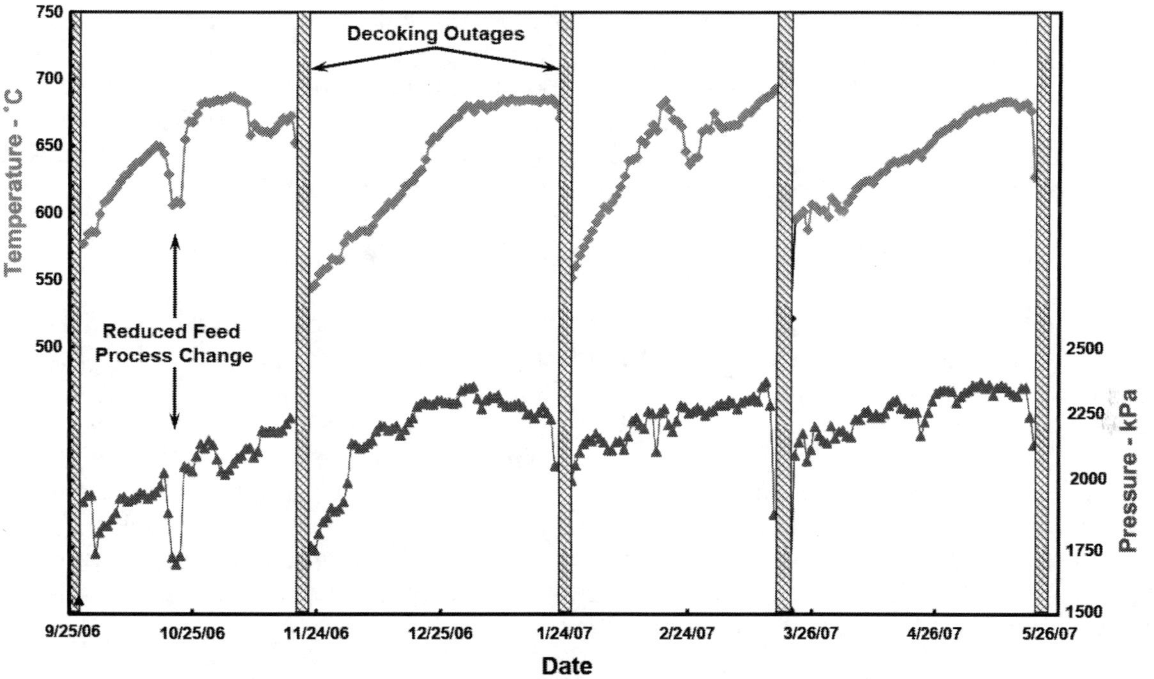

Figure 3: Graph showing daily average temperatures and pressures for the tubes at the top of the firebox. Outages for decoking are also indicated on this graph.

This furnace has a convection section located above the firebox so the tubes at the top of the firebox are approximately half way through the furnace. However, the temperature of the hydrocarbon in the convection section isn't high enough to cause significant coke to form inside the convection section tubes. The coke tends to form toward the furnace outlets where the tube wall temperature is higher. As a result of this the tubes toward the top of the firebox experience the highest combination of temperature and pressure (stress), and they are the tubes most likely to fail by creep. The flame pattern in the firebox also tends to cause the tubes in the top of the firebox to have the highest heat flux.

It should be noted that when analyzing elevated temperature equipment for creep damage, it is important to use either the complete stress, temperature, and exposure interval history or use stress and temperature values that are averaged in time over short periods of time during the service interval[2]. If long term service periods are used, significant short term events that impact creep could be missed. Many refineries maintain databases that go back ten years or more, but the data in these databases are typically in the form of daily averages. Consequently, it is usually convenient to use daily averages of process parameters when conducting creep analysis of refinery equipment.

The data plotted on the graph in Figure 1 illustrate how the temperature and pressure tend to increase together as coke builds up in the furnace tubes. Of course, there are process changes on the coker unit and on other units that can result in reductions of the flow through the unit, and when a flow reduction occurs the pressure is reduced and the tube wall temperatures decrease. With reduced flow the burners in the firebox are cut back to maintain the same furnace outlet temperature which usually results in lower tube metal temperatures.

In general there is a correlation between the increase in tube wall temperatures and an increase in pressure that the furnace tube experiences. However, this isn't always the case because there are other factors that can cause local heating. For example, flame impingement and the coking that sometimes results can cause a particular tube or a particular spot on a tube to become hotter. Coker furnaces are usually surveyed with infrared cameras toward the end of each cycle to detect localized hot spots. Also, as the tubes approach their temperature limits, sometimes the run length is extended by cutting flow through that furnace. In many refineries the cokers have several furnaces, so it is possible to adjust the flow distribution between the furnaces within certain limits. When the flow to a particular furnace is reduced, the maximum tube temperature remains nearly constant, but the pressure drops off so less creep damage occurs. The second cycle and fourth cycles in Figure 1 are examples of this. This figure also shows where decoking outages occurred. These furnaces are decoked by pigging which typically requires that the furnace be out of service for two or three days.

OMEGA TECHNOLOGY AND OMEGA SIMULATIONS

Omega is the name of an approach for dealing with creep damage and remaining life assessments based upon work done through an API[(2)] joint industry project with the MPC[(3)]. Work began on the development of Omega technology in 1986, and Martin Prager published his first paper on Omega technology in 1994. Prior to the development of Omega technology the Oil Industry used the traditional Larson-Miller parameter approach for analysis of high temperature equipment. API 530[3] is the guideline for furnace tube design in refineries and the Larson-Miller approach is described in this document. While API 530 it works fairly well for design, it is difficult to use for assessing the remaining life of existing equipment. API 530 also is not designed to deal with degradation mechanisms other than creep. Both API 530 and Omega use measured material properties as the basis for predicting how materials will respond to high temperatures and stress.

Omega is defined as a creep damage coefficient, and it is related to the strain and the strain rate by Equation 1 shown below. Omega is also related mathematically to the rate at which strain rate accelerates as a result of creep strain. Equation 2 provides the basis of Omega technology. For a more complete explanation of Omega equations refer to API 579, Appendix F[4] or Martin Prager's 1994 paper.

$$\Omega = \partial \ln \varepsilon / \partial \varepsilon \qquad (1)$$

$$\varepsilon = \ln(1-\varepsilon_0 \Omega t)/\Omega \qquad (2)$$

ε - true strain (today) t - time elapsed ε_0 - initial creep strain rate

If the strain that occurs at rupture, the time it takes to rupture and the initial strain rate for an alloy are known, this equation can be used to calculate the Omega parameter for that alloy. Then that Omega parameter can be applied to other applications of that alloy to predict remaining life based upon the history of process temperatures and stresses. As part of the development of Omega technology, MPC obtained data on many alloys and compiled a matrix database of the Omega parameters for a range of materials. MPC compiled Omega data into tables containing Omega data for steel grades that represented average, high side and low side properties. An example of an Omega data table is provided in the articles by Martin Prager that was mentioned earlier. Of course there is some variation in properties of a steel grade from heat-to-heat so the actual Omega database will contain a range of Omega values for each type and grade of steel. Typically the values representing the average

(2) American Petroleum Institute (API), 1220 L St. NW, Washington, DC 20005-4070
(3) Materials Properties Council (MPC), 20600 Chagrin Blvd., Suite 1200, Shaker Heights, Ohio 44122

properties are used for analysis of equipment, but minimum properties or high side properties can also be used. OmegaLife[4] simulation software makes it possible to select the properties you want to use for an analysis. The software used for the simulations in this paper was OmegaLife Version 1.1.026.

OmegaLife software was developed by MPC to facilitate rapid simulations of equipment so that the equipment's response to creep conditions could be applied to a wide range of alloys. The OmegaLife software also makes it possible to download past operating conditions and predict how the equipment will respond to similar combinations of stress and temperature in the future. For example, furnace tubes are modeled as cylinders in Omega, and the tube diameter and wall thickness are included in the simulation. The stress is then calculated based upon the internal pressure and other loads can be included as needed. Also, tube wall damage rate information such as the corrosion rate can be included in the simulation. Past temperature and pressure histories can be included in spreadsheet format that makes it easy to check for data entry errors or other forms of incorrect data.

After all the appropriate aspects of the equipment and the process have been loaded into the program, the software predicts the remaining life of the furnace tubes. The Omega software also produces a graph that shows how small changes in temperature affect remaining life. Examples of Omega life curves for the tubes in one of the coker furnaces at the refinery in Lemont are provided in Figure 4. Each tube in the furnace has its own curve since the pressure is different in every tube. These curves can be used to predict future performance as long as the only variable that changes is temperature. If the wall damage rate or the pressure changes as a result of adjustments to the tube wall temperature, the life that this curve predicts will be inaccurate. In coker furnaces and in many other refinery furnaces, small changes in tube wall temperature result in significant changes in wall loss rate since corrosion is frequently temperature dependent. However, the Omega software has capabilities for varying corrosion rate with temperature so corrosion can be included in the Omega analysis. Omega cannot include pressure variations. Consequently, using the Omega curves to predict the impact of process changes can lead to inaccurate assessments if the process changes that are being considered cause too much change in other process variables. A method for dealing with this so that pressure changes can also be included in the Omega analysis will be discussed later in this paper.

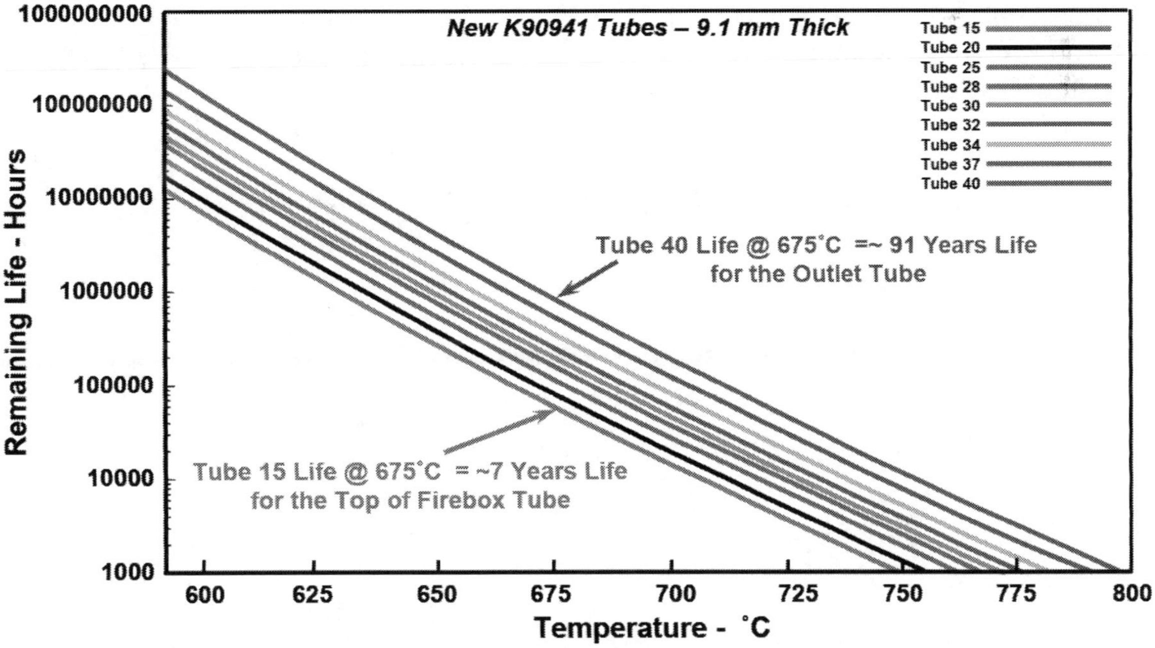

Figure 4: Examples of Omega Creep Curves for a refinery furnace. Each curve represents a tube at a different location in the furnace.

---

[4] Trademark

The Omega software uses a spreadsheet for data entry, and as a result, data representing very short time periods can be used in Omega simulations even if the equipment history covers a long time span. For example, in the analyses described in this paper average temperatures and pressures for each day were used. Since the short term process data available at the refinery spans 15 years, in this instance 5500 pairs of temperature and pressure entries were used. These process data were first downloaded directly from the refinery computer system into a spreadsheet. Then the data was checked to identify false data or corrupt data that are in the database. False data are usually created when the instrumentation fails or when a process upset occurs. For example, when thermocouples fail or become detached from the tube walls, they read either very high or very low. The incorrect data would tend to make the Omega analysis inaccurate so it needs to be removed. Also, if data are used when temperatures are so low that no creep damage occurs, that data causes the Omega software to divide by zero and lock the program. Due to this software characteristic, data from portions of start-ups and shut-downs need to be removed before conducting the analysis.

PREPARING PROCESS DATA FOR OMEGA SIMULATIONS

As was mentioned earlier, there are only about 15 years of data available for the coker furnaces in the refinery where these analyses were conducted even though the tubes in these furnaces had been in service since 1968. One way to deal with this lack of process data is to interview people who were on the unit during the time periods in question to find out if the unit was operated differently in the past. Also, sometimes there are written records in storage that could be reviewed to determine if there have been major process changes in the past. After all the past time periods have been reviewed, estimates of what the process temperatures and pressure were in the past can be used to fill in the missing data. It is important to try to obtain information on past process changes or unusual past operations, such as periods of time when the unit was being pushed, or when it was being used to compensate for problems in other units.

Examples of significant process changes that played a major role in the analysis of this coker furnace are provided in Figure 5. Fortunately, these process changes were made after the computer system was active so the process details of the changes were recorded. From interviews it was learned that processing from 1991 till 1995 was typical of the way this furnace was operated during the 1968-1991 time period. Therefore, that earlier furnace history was filled in by simply duplicating the 1991-1995 data several times to provide data for the missing years.

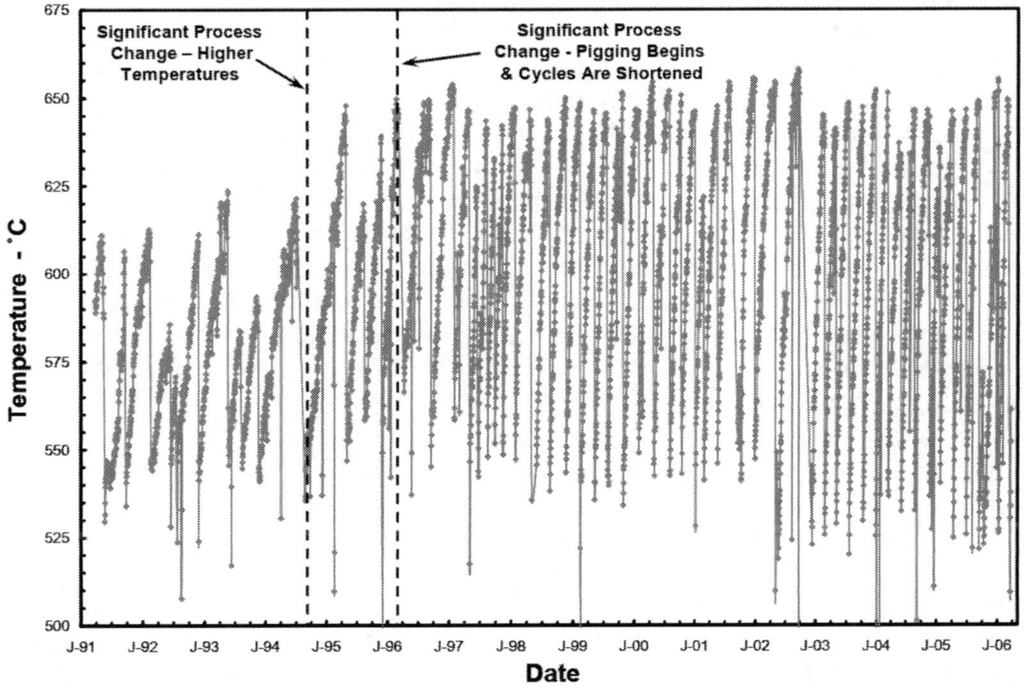

Figure 5: Graph showing the maximum daily average tube temperatures from 1991 to 2006. The vertical dashed lines indicate where major process changes occurred.

Figure 5 shows when two major process changes occurred: 9/94 and 3/96. In September of 1994 the decision was made to experiment with higher tube metal temperatures, and a test run to a maximum near 650°C was conducted. During the next coker furnace cycle temperatures went to the previous maximum, and from then on higher maximums were used. Then in March 1996 the decoking method was changed from steam air decoking to pigging. This turned out to be a significant process change because it resulted in increased tube wall loss. The effects of pigging on furnace tube wall loss are described in a presentation[5] that can be downloaded from the AIChE[(5)] Chicago website. Basically, there is tooling on the pigs that mechanically removes steel from the ID surface of the furnace tubes, and return bends tend to be more severely damaged due to the forces generated as the pigs are deformed passing through the return bends.

Another important factor in creep analysis of refinery furnace tubes is when tube inspections are conducted. In general coker furnace tubes are inspected during every major unit shut down, however there are exceptions. Coker units typically run 3 to 5 years between major shut downs, so the tube conditions are typically assessed every 3 or 5 years. It is believed that prior to 1996 the coker furnaces were not experiencing significant wall loss, but no inspections were conducted in 1996. The inspections conducted in 1994 showed no wall loss had occurred since the tubes were installed in 1968 while the inspection conducted in 1997 indicated that since 1994 the tubes had lost 0.58 mm for a wall loss rate of 0.193 mm/year. The next inspection which took place in 2002 indicated additional wall loss of 1.73 mm for a wall loss rate of 0.346 mm/year for the 1997-2002 time period. Inspections conducted after 2002 all indicated that the tube wall loss rate was ~0.34 mm/year. The lower wall loss rate during the 1994-1997 time period occurred because pigging wasn't being used during that interval between inspections. Pigging began about half way through that time period. Due to the variations in rates of wall loss that can occur in refinery process equipment, it is frequently necessary to divide the furnace life into segments between inspections. However, since in this case the change in wall loss rate was due to a change in the decoking technique, the date of the decoking procedure change could be used as a process change point in the furnace history.

[(5)] American Institute of Chemical Engineers (AIChE), 3 Park Avenue, New York, NY 10016-5991

An additional factor of importance in refinery furnace analyses is changes that impact the corrosion rate. For example, changes in the sulfur content would tend to influence the wall loss rate and would therefore be important to remaining life calculations. Those changes would serve as another dividing point in the furnace history around which furnace process data could be arranged. Another more subtle change that can affect remaining life would be a change in corrosives in the process. For example, many types of crude contain naphthenic acids. Some types of naphthenic acids are corrosive while others are not. A change in the crude blend could result in increases in the amount of corrosive naphthenic acids present while the total naphthenic acid content of the crude blend didn't change. The parameter used to assess the risk from naphthenic acid corrosion measures the total acids present in the crude by titration. This hydrocarbon characteristic is called the total acid number or TAN. However, that acid content measurement technique cannot assess the how much corrosion will occur due to the acids. These aspects of the process can all be dealt with by using the measured wall loss rate instead of estimates based upon hydrocarbon composition and temperature. However, it is important that the process data be divided into time segments to represent the process before and after the process change so that the Omega assessment of the equipment is accurate.

RESULTS OF OMEGA ANALYSIS OF A COKER FURANCE

After the process data has been downloaded, reviewed and divided into logical time periods representing process changes, it can then be loaded into the Omega software for analysis. It is usually better to load each time period separately, and then add the creep damage from the earlier time period to the following Omega analysis - the Omega software allows you to do this. This is usually beneficial because the Omega software doesn't easily facilitate changes in wall loss damage rates. It is possible to adjust corrosion rates in the Omega software based upon temperature, but frequently there are other factors that effect corrosion rates – for example: changes in decoking operations, changes in hydrocarbon composition, the introduction of erosive agents in the hydrocarbon, etc. Figure 5 shows the results of analysis of one of the coker furnaces which was conducted in 2006.

For this analysis the data could be divided into three time periods based upon differences in wall loss rate for the three time periods. Actual daily averages of maximum tube wall temperature and internal pressure were used for the 1994-1997 and 1997-2006 time periods. For the 1968-1994 time period the existing data available from 1991-1994 was duplicated several times to fill the entire 1968-1994 time period. Consequently, daily averages for actual process data were used for all time periods, but for the 1968-1990 time period the process data used for the analysis may not represent what actually occurred on the unit.

Figure 6 shows three Omega curves each corresponding to a different time period, and along with each curve is the amount of the life consumed previously. The 1968-1994 time period had the lowest loss of life because there was no wall loss occurring during this time period. All of the 8.7% of life lost was due to creep damage. The other two curves pertain to time periods when significant wall loss was occurring, and as a result of the wall loss the rate of life consumption was greater even though the time periods are much shorter.

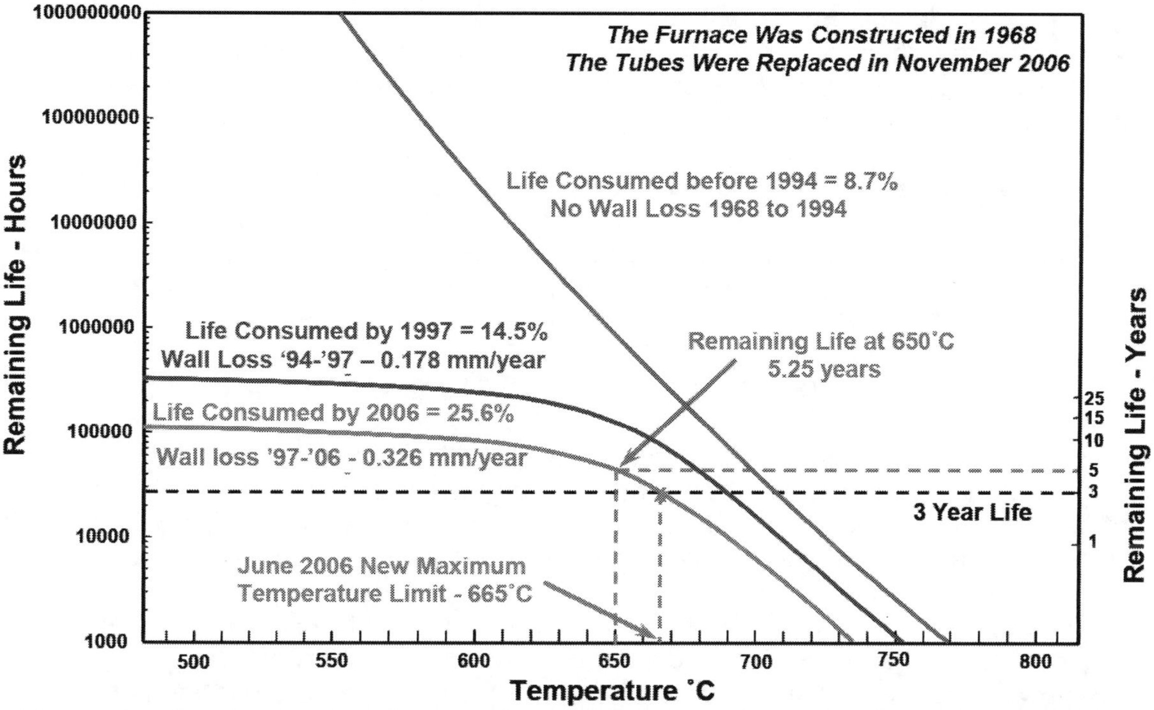

Figure 6: Graph showing the results of Omega analyses for three time periods. The dashed lines pertain to previous operating conditions (red) and future operating conditions (black) if all that is required is a three year life.

Figure 6 shows that the remaining life using the previous maximum tube metal temperature of 650°C was about 5.25 years – refer to the vertical dashed line above 650°C. It intersects the 2006 Omega curve at about 46000 hours or 5.25 years. This Omega curve can also be used to predict the effects of increasing the tube temperatures. For example, suppose you knew that the furnace was going to be shut down in three years to replace the tubes. Then you could plan to use most of the tube life during those three years. The Omega curve predicts that consuming the tube life in three years would allow you to raise the maximum tube wall temperature to 665°C. However, this process change should only be made if safety factors were included in the analysis to account for uncertainties pertaining to process control and uncertainties pertaining to the properties of the tube steel.

An example of a safety factor applied during furnace analyses is to subtract about 15°C from the calculated maximum temperature to account for temperature measurement errors. Another way to add safety factors is to base the analysis on daily maximum tube temperatures instead of daily average temperatures. As was mentioned earlier, the Omega curves do not include the effects of changes in pressure, and Omega software cannot automatically deal with changes in tube wall damage rates other than corrosion damage that can be directly related to temperature changes. Consequently, using Omega curves to predict how the tubes will respond to process changes can be sometimes be inaccurate – the accuracy depends upon what damage mechanisms are active in the equipment. Small changes in temperature that would not result in significant changes in pressure or wall loss rate could be used to assess the risk of process changes, but to accurately determine the impact of process changes on remaining life it would be necessary to use something other than the automatically generated Omega output curves. This will be discussed in the following section of this paper.

One important aspect of the Omega curves that are shown in Figure 6 is the change in the shape of the curves that occurred once significant wall loss started. Compare the shape of the 1968-1994 curve to that of the curves representing the complete furnace history prior to 2006 – the red Omega curve. This change in shape is caused by the change in the tube wall loss rate. The 1968-1994 wall loss rate was near zero while the wall loss rate from 1997 till 2006 was approximately 0.326 mm/year. The Omega curves flatten at lower temperatures because remaining life at lower temperatures depends mostly upon the wall loss rate. This causes the Omega curve to turn toward the y-axis and flatten. In these furnaces the wall loss rate was determined primarily by the frequency of decokes by pigging so tube life wasn't varying much in the lower temperature ranges where creep was not a significant factor.

This coker has three furnaces and the curves shown pertain to the 13B-1 furnace. As a result of these Omega analyses, the tubes in two of the furnaces were replaced late in 2006. The unit wanted to plan for a five year run, and the tube life in those two furnaces was not sufficient to guarantee a five year run so tubes were replaced. The third furnace had thicker tubes, and Omega analysis indicated it had significantly longer remaining, life and it was scheduled for tube replacements later.

IMPROVED OMEGA ANALYSIS FOR ASSESSING THE IMPACT OF PROCESS CHANGES

In the previous sections it was mentioned that there is some inherent error involved with using Omega analyses to predict what will happen when process changes are made in the future. Omega does a good job of analyzing the impact of past process changes particularly if you download all the important process data associated with those changes. However, what many operators of refinery equipment want to know is what happens in the future if we make significant process changes? Typical questions include: how hot can we go, how high a pressure can we use, will the proposed changes cause tube failures, will the changes cause us to need more shut downs to inspect the equipment, etc. To adequately answer these questions, an accurate assessment of the tube life needs to be made that includes the past service history while taking into account the anticipated future service. The following example provides a method for using Omega simulations to predict the impact of process changes on the life of refinery furnaces.

The first step in determining the impact of process changes on tube life is to build a process model showing what the changes would look like in terms of the tube temperatures and the pressures. This can sometimes be done with process simulations or by running step tests on the furnace and recording the changes in process conditions that occur. For such a step test in a coker furnace the test would require the full coking cycle time period (two or three months). You can't use a short term increase in temperature to assess what happens because part of the process is coke formation in the tubes, and a short step test wouldn't allow for enough coke accumulation. Also, it is sometimes difficult to take risks with the equipment simply to learn how the equipment responds to increases in temperature and pressure. Consequently, other approaches typically are employed. One approach that sometimes is successful is to review the available equipment history to find past events that can be used to predict what would happen if the process changes are made.

Figure 7 shows tube temperature data from several coker cycles plotted over the same time period. This data consists of maximum daily averages for the hottest tubes in the furnace – tubes at or near the top of the firebox. Note that the increase in temperature is nearly linear early in each coker cycle, and the tube temperature heating rate tapers down later in the cycle. This occurs because the operators adjust the furnace to extend the cycle as the temperature limits are approached. The operators adjust the flow through the passes and through the three furnaces on this unit to slow the heat up rate of the hottest tubes. During the cycle the inlet pressure is raised to maintain almost constant flow through the furnace, but the pressure drop across the furnace is increasing due to coke formation inside the tubes. If operations were changed to allow heating the tubes to higher temperatures, the linear portion of these curves would be extended to higher temperatures, but the curves would bend in a similar way as the new limit is approached.

A review of past coker furnace cycles identified one of the cycles where the tubes were allowed to heat to 668°C, and that cycle was also the longest cycle with duration of 95 days. This cycle is marked on the graph. Since the decoking process takes a furnace out of service for a few days and reduces unit throughput, extending the length of the coker cycles is desirable. Extending the cycle duration can be accomplished by raising the temperature limit on the furnace tubes, so raising the tube temperature limit has become a frequent request from unit operations in many refineries.

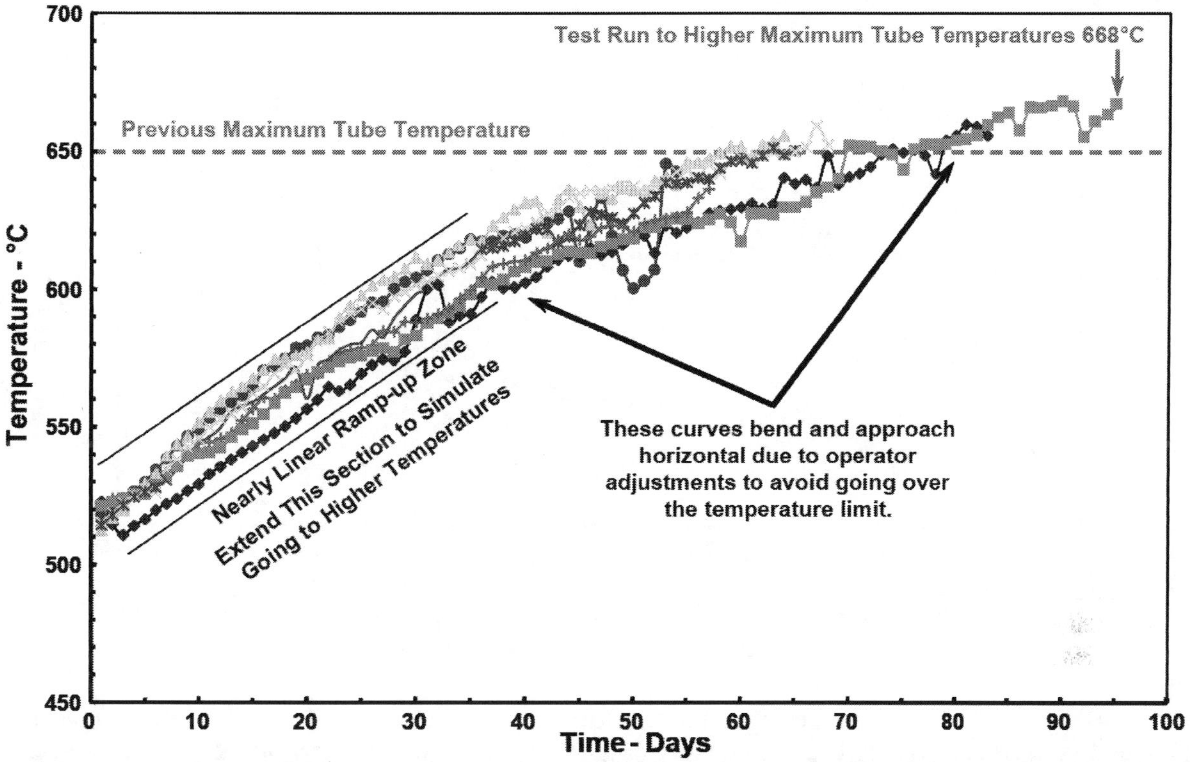

Figure 7: Graph of maximum tube temperatures from one of the coker furnaces. Two of these cycles went to temperatures above 650°C providing data pertaining to operations at higher tube temperatures.

By averaging the data shown in Figure 7 a temperature curve can be developed that represents how the furnace is typically operated. This representative curve can then be extended to higher temperatures by simply making the linear region of the curve longer. The same can be done with the data for the furnace pressures to develop a representative curve for pressures in the tubes. Figure 8 shows the results of averaging the data from individual coker furnace runs along with extension of the linear portions of both the temperature and pressure curves. These curves represent the conditions that the top tube of the firebox is expected to experience.

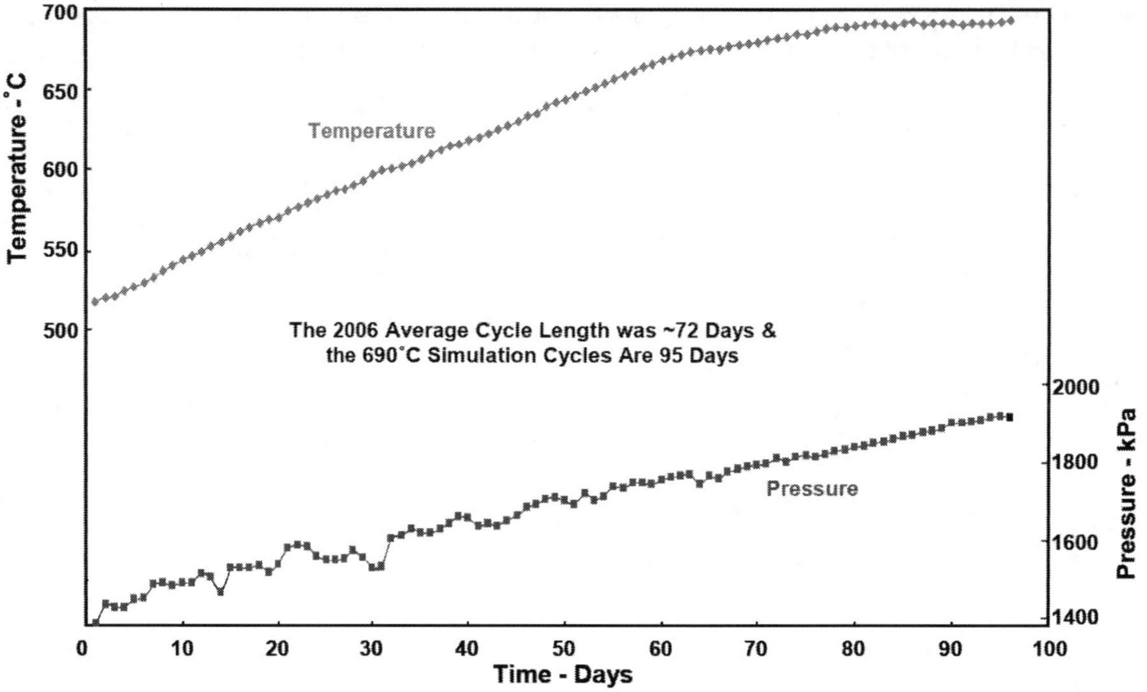

Figure 8: Graph showing the temperature and pressure curves used for the Omega simulations of future operations.

After these temperature and pressure curves were developed, the next step was to determine how long the tubes in the furnace needed to be in service. This question was posed to operations, and their answer was based upon the plans for future cycle ending shut downs. Based upon the shut down schedule it was determined that it would be best if the tubes in two of the furnaces were replaced in the fall of 2017. That would mean that the tube life needed to be about 11 years. The tubes in these two furnaces had been in service since 2006 so the process data was downloaded to cover the 2006 to mid-2008 time period. Then the cycles shown in Figure 8 were repeated to fill the 11 year time period. This resulted in a set of data in Omega that could be used to assess the effect of raising the maximum tube temperature on the life of the vessel. Also, similar extended cycles to higher temperatures and longer furnace cycles were created to make it possible to evaluate the impact of higher temperatures. Running simulations for series of process conditions made it possible to determine what maximum temperature was appropriate for future operations.

For all of these analyses the tube wall loss rate was used was that which the tubes were experiencing. This was conservative because the wall loss was primarily determined by the frequency of pigging. If the coker furnace cycle time was increased there would be less pigging during the 11 years, and the tube wall loss rate should decrease. Also, shut downs and decokes weren't included in the simulation, and this is another factor that makes the simulation conservative. Also, the 690°C simulation resulted in 82% life consumed so there were about a 2.4 years of life remaining after October 2017.

Figure 9 is a graph showing one set of temperature data that was used in the Omega simulations. This represents the furnace cycles that were ultimately selected – maximum tube temperatures of 690°C. This maximum would be for tubes at the furnace outlet where the pressure is relatively low. Tubes at the top of the firebox would need to operate at lower temperatures to have the desired life. It was determined that operating the tubes at the top of the firebox at 675°C would be appropriate. Figure 10 shows the pressure drop that was occurring for the process data since 2006 and the way the future pressure history was adjusted to deal with the long term pressure drop issue.

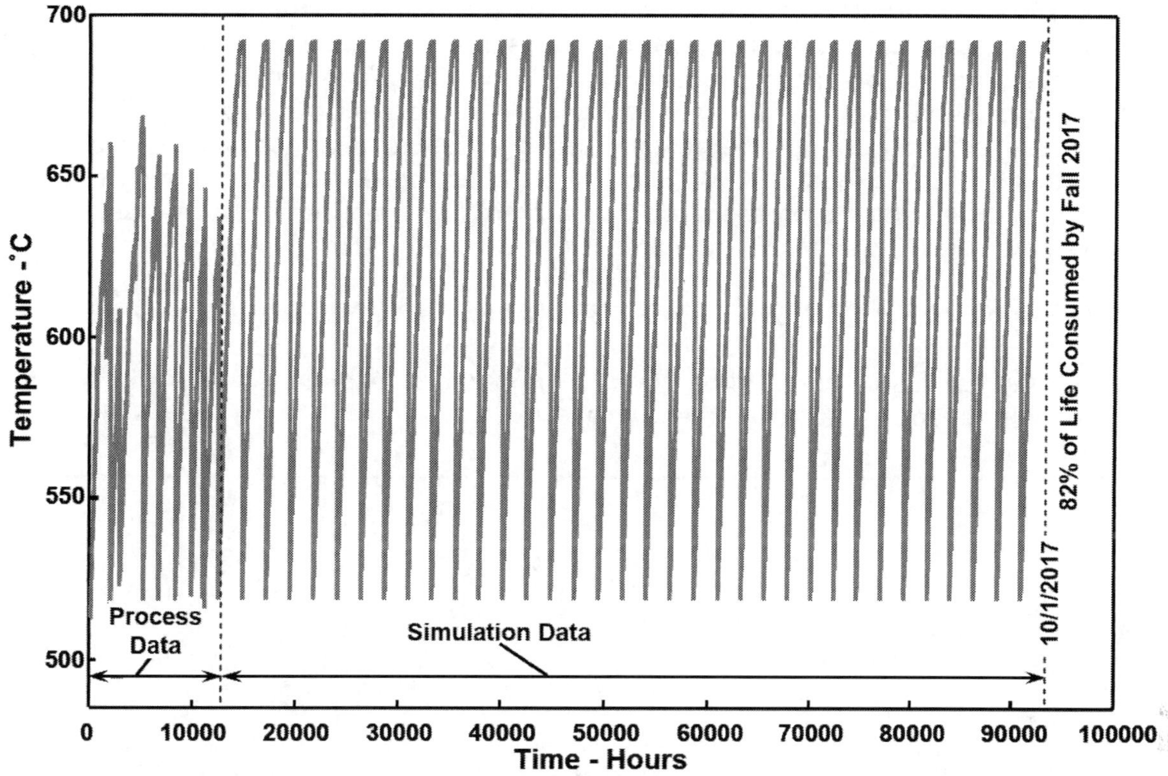

Figure 9: Graph of the past and possible future temperature data used for an Omega simulation to determine the impact of process changes.

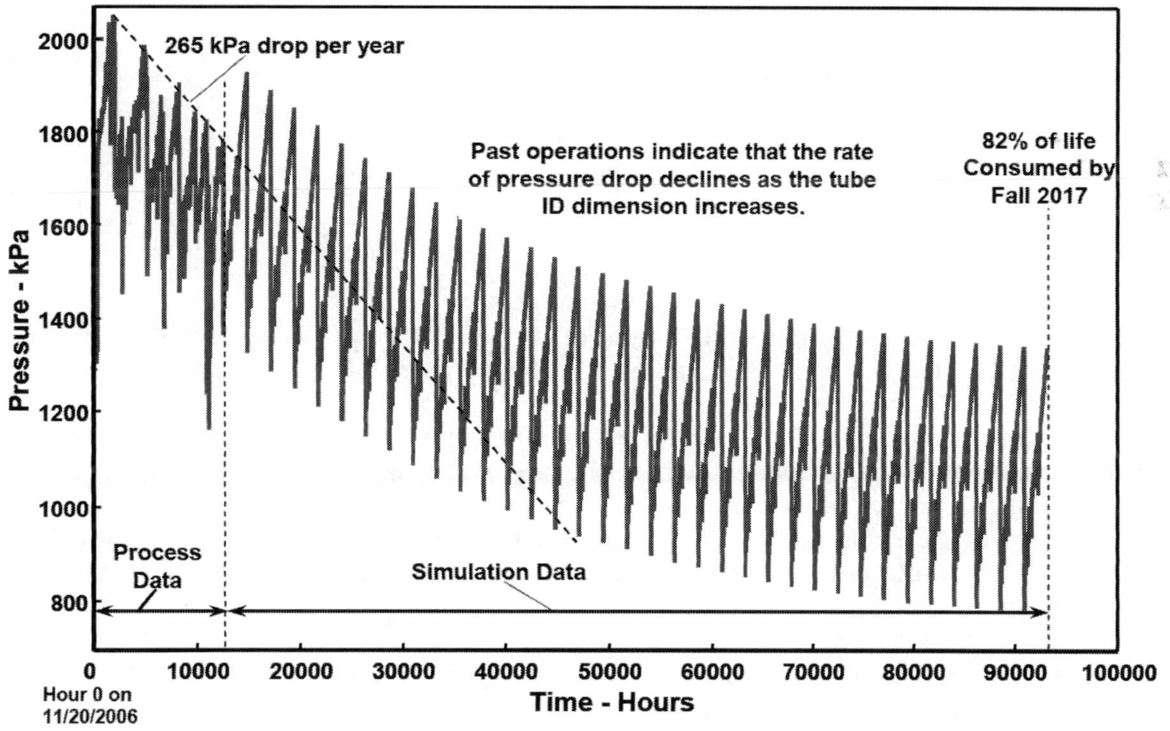

Figure 10: Graph of past and possible future pressure data used for an Omega simulation to determine the impact of future process changes.

The decrease in pressure shown in Figure 10 is caused by the gradual change in the area of the inside of the tube due to the wall thickness being reduced by pigging. The dashed line that is on an angle to the graph axis shows the trend that was established by the process data since the new tubes were installed in 2006. The predicted future pressure cycles begin at the vertical dashed line on the left side of the graph. These are the predicted cycles when the tubes have a maximum temperature of 690°C. Analysis of past cycles indicated that the rate of decrease in pressure decreased as time passes. To accurately represent what occurred in the past the rate of pressure decrease was gradually reduced to produce the cycles shown in Figure 10. However, subsequent observations pertaining to the changes in pressure drop revealed that the pressure drop wasn't exactly what was predicted due to accelerated damage done by the pigging operations – the pressure dropped more than expected.

THE RESULTS OF INCREASING THE MAXIMUM TUBE TEMPERATURE

The obvious question at this point would be how did the process respond to the changes in maximum tube metal temperature, and the results are shown in Figure 11.

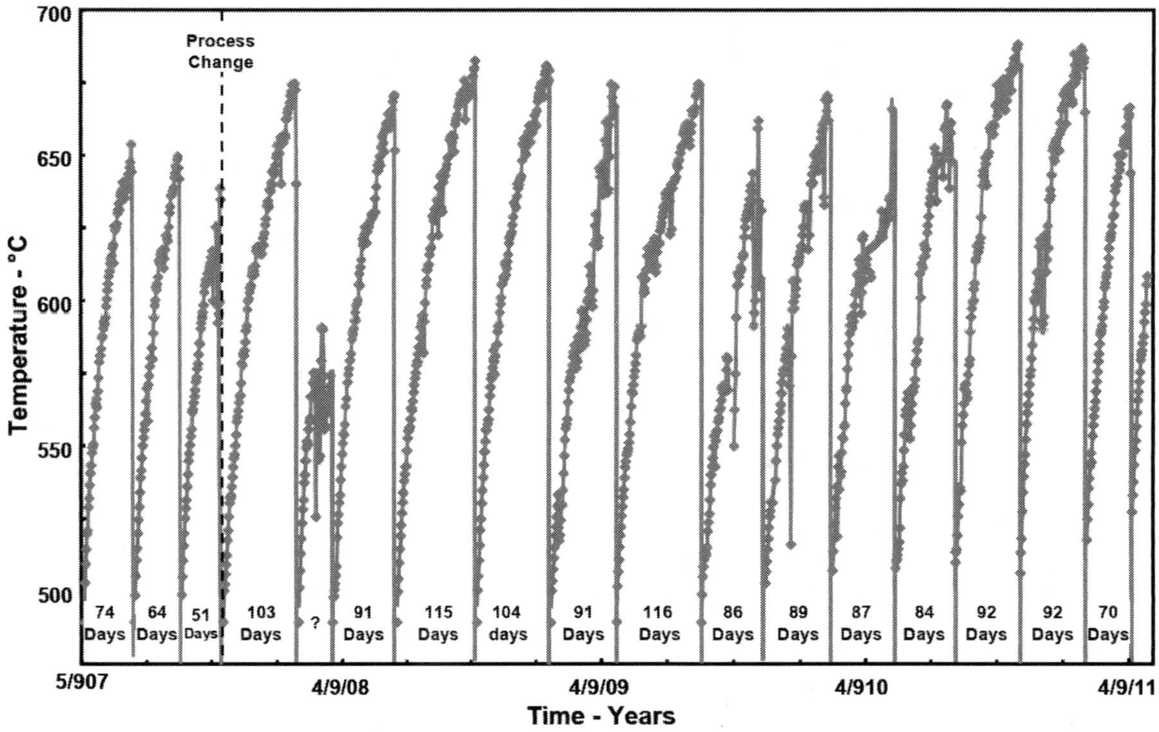

Figure 11: This graph shows the temperature cycles for the hottest tube in the coker furnace after the maximum tube temperature was increased.

The three cycles on the left side of this graph represent operations with the previous maximum tube temperature of 650°C. The remaining cycles on the graph show what has happened since the maximum tube temperatures were increased. Since the hottest tube is at the top of the firebox, the maximum tube temperature was set at 675°C. The higher pressure in the tube at that location caused the maximum temperature to be less than 690°C. Cycle duration is shown for most of the cycles at the bottom of the graph. The cycle with the "?" indicates that that the duration of that cycle was not measured since the process had been interrupted. The data show that the time between decoking operations could be significantly increased by going to higher temperatures. There was a possibility that going to higher temperatures would not result in a major improvement because the higher temperatures could make coking occur faster. However, in most cases the improvement was significant. The average cycle duration since the process change occurred was 93.85 days which is not too far from the estimated cycle duration that was discussed earlier (95 days).

Another goal of this analysis was to have the furnace tubes remain in service until 2017, and that goal was not achieved due to changes in the rate of damage that is occurring due to pigging. The return bends in this furnace are losing wall at a rate of 0.772 mm/year - more than twice what the tubes experience. Also some of the return bends are being severely damaged while others are hardly attacked. Since the pigging process is not conducted by the refinery, the cause of the wall loss rate increase was not determined. While this approach to analyzing the furnace tubes did achieve some of the goals, not including the return bends was a mistake. Return bends in this furnace are outside of the firebox, and therefore they aren't heated to high enough temperature to experience creep damage. Analysis of the return bends wall loss based upon smart pig inspection results would have revealed the return bend problem earlier, but smart pig inspections weren't conducted until 2010. Also, the pigging damage in the return bends consisted of deep grooving on the inside sweep of the bends and this deep grooving may not have been accurately measured with the smart pig inspections.

CONCLUSIONS

This paper demonstrates that it is possible to use Omega simulations to analyze coker furnace tube degradation that occurs due to temperature, pressure, and corrosion aspects of the process. Below are aspects of coker furnace tube analysis that should be included in Omega simulations of coker furnaces:

1. To obtain an accurate analysis it is best to use as much process data as possible and avoid averaging process data over long periods of time. Daily averages of temperature and pressure were used in this analysis.

2. The Omega simulation software developed by MPC is able to take variations in corrosion rate and pressure changes into account automatically.

3. Past significant process changes need to be taken into account and the available process data divided into time periods representing spans of time between major changes. Then each time span should be run separately in the Omega simulation to reveal the specific effects of the process changes.

4. The impact of future process changes can be analyzed using Omega simulations by creating process data that represents the anticipated process conditions.

5. Return bends in this furnace were not included in these Omega simulations because the return bends are outside of firebox, and they are not hot enough to experience creep damage. It would have been better if a separate analysis of return bends had been conducted based upon data from smart pig inspections.

REFERENCES

1. Prager, M. Development of the MPC Omega Method for Life Assessment in the Creep Range. ASME, PVP-Vol. 288, Service Experience and Reliability Improvement: Nuclear, Fossil and Petrochemical Plants, 1994.

2. ECCC RECOMMENDATIONS - VOLUME 5 part III [Issue 2], Guidance for the Assessment of Creep Rupture Data, Recommendations for the Assessment of Post Exposure (Ex-Service) Creep Data, ECCC, 31/8/2005.

3. API Standard 530, 5th Edition, January 2003, ISO 13704:2001(E), Calculation of Heater-Tube Thickness in Petroleum Refineries, January 2003.

4. API Standard 579-1/ASME FFS-1, Fitness-For-Service, Appendix F, section F.7, June 5, 2007.

5. Wilks, G. W. Remaining Life Assessments of Refinery Furnace Tubes Using Omega Simulations. AIChE Chicago 2006 Symposium, Refining Presentations, October 9-10, 2006, http://www.aiche-chicago.org/symposium06/wilks.pdf.

(137a) Catalytic Conversion Wood Syngas to 'Wide Cut' Diesel Over A Multifunctional Catalyst

Qiangu Yan, Department of Agricultural & Biological Engineering, Mississippi State University, Mississippi State University, MS and Fei Yu, Department of Agricultural & Biological Engineering, Mississippi State University, Mississippi State, MS

A continuous process involving gasification, syngas cleaning, and Fischer–Tropsch (FT) synthesis was developed to efficiently produce 'wide cut' diesel. Oak-tree wood chips were first gasified to syngas over a commercial pilot plant downdraft gasifier. The raw wood syngas contains about 47% N_2, 21% CO, 18% H_2, 12% CO_2, 2% CH_4 and trace amounts of impurities. A purification reaction system was designed to remove the impurities in the syngas such as moisture, oxygen, sulfur, ammonia, and tar. The purified syngas meets the requirements for catalytic conversion to liquid fuels. A multi-functional catalyst was developed and tested for the catalytic conversion of wood syngas to diesel. It was demonstrated that liquid fuels similar to commercial diesel was successfully synthesized from bio-syngas.

MODELO DE CÁLCULO DE LA PERMITIVIDAD DE CARBONES

Peña Benavides, Manuel G. y Vidal Teherán, Dairo A.
Universidad del Atlántico, Barranquilla, Colombia.

Resumen
El carbón posee muchas propiedades que, según su estado, lo hace aplicable para un uso específico. Una de esas propiedades es la permitividad eléctrica, que es la capacidad que posee un material de polarizarse. Esta propiedad es la que fija el comportamiento que van a tener los carbones cuando sean expuestos a una señal electromagnética, marca como se propaga la energía en el material, como se transforma la energía de microondas en calor al interior del mismo y, finalmente, nos indica si un material es susceptible de ser tratado por la vía de microondas. El método de medición de la potencia absorbida, es el que se propone en el presente estudio; este consiste en aplicar al material de análisis una onda electromagnética de una cierta potencia conocida como potencia suministrada para medir la cantidad de esa potencia que absorbe y posteriormente, a través de un modelo matemático, calcular el valor de la permitividad eléctrica.

Palabras claves: *Propiedad dieléctrica, permitividad, microondas, guía de ondas, carbón.*

1. Introducción

Permitividad

La capacidad que posee un material para calentarse en presencia de la acción de las microondas está en función de la permitividad. La unidad de medida en el Sistema Internacional SI, es el faradio por metro (F/m).

Tomada en función de la frecuencia, puede tomar valores tanto reales como complejos, ecuación (1). Por lo general no es constante, debido a que varía con la posición del medio, la frecuencia del campo que se aplique, la humedad del material y la temperatura.

$$\varepsilon = \varepsilon' - i\varepsilon'' \quad (1)$$

De donde ε' es la constante dieléctrica o permitividad real y, ε'' es el factor de pérdidas dieléctricas o permitividad imaginaria.

Un factor de pérdidas alto indica que el material es susceptible de ser tratado por la vía de microondas.

En un medio que no sea lineal, la permitividad depende de la magnitud del campo eléctrico. En electromagnetismo al campo de desplazamiento eléctrico se le define como D, que representa como un campo eléctrico E influye en la organización de las cargas eléctricas del medio. La relación de estos campos con la permitividad se define con la siguiente expresión:

$$D = \varepsilon \cdot E \quad (2)$$

En donde el campo de desplazamiento D se mide en culombios por metro cuadrado (C/m^2), mientras que el campo eléctrico E se mide en voltios por metro (V/m) y la permitividad eléctrica ε en faradio por metro (F/m).
En el caso en que ese campo eléctrico se dé en el vacío, la ecuación queda de la siguiente forma [7]:

$$D = \varepsilon_0 \cdot E \quad (3)$$

De donde, ε_0 es la permitividad del vacío. Dicha permitividad está relacionada con la permeabilidad en el vacío μ_o y la velocidad de la luz C, por la siguiente expresión:

$$\varepsilon_o C^2 \mu_o = 1 \quad (4)$$

Despejando la permitividad en el vacío,

$$\varepsilon_o = \frac{1}{C^2 \mu_o} \quad (5)$$

De donde C es la velocidad de la luz ($3*10^8$ m/s) y μ_o es la permeabilidad magnética del vacío ($4\pi*10^{-7}$ Tm/A), definidas en unidades del SI. Se obtiene un valor de la permitividad del vacío de aproximadamente

$$\varepsilon_o = 8{,}8541878176 * 10^{-12}\, F/m$$

Efecto de las microondas sobre la materia

La interacción de las partículas cargadas que poseen ciertos materiales con el campo eléctrico que es generado a causa de la radiación electromagnética que provocan las microondas, hace que estos materiales se calienten.

La irradiación de microondas interactúa con la materia como una fuente interna de energía, haciendo que se caliente directamente desde la profundidad. La eficacia de esta técnica de calentamiento, depende, en gran medida, de las propiedades dieléctricas de la materia.

En el caso de moléculas polares, el campo eléctrico de las microondas hace que los dipolos permanentes presentes en dichas moléculas y los que son inducidos roten al intentar alinearse con el campo oscilante [9]. Para el caso en estudio en el presente trabajo, como se trabaja a una frecuencia de 2,45 GHz, la molécula rota 2450 millones de veces por segundo, y se encontró que a esta frecuencia es ideal para trabajar, debido a que la molécula de agua resuena aproximadamente en ese valor [28].

El carbón y las microondas

La mayoría de los materiales de carbono son excelentes absorbentes de microondas. Las microondas inducen una corriente en fase con el campo electromagnético del material. Los electrones encuentran resistencia al libre desplazamiento por el material, haciendo que cierta cantidad de la energía se disipe en forma de calor.

En los carbones puede ocurrir un interesante fenómeno adicional, la energía cinética de algunos electrones puede llegar a ser suficiente para que salten fuera del material provocando una ionización de la atmósfera que lo rodea [9]. Este fenómeno se puede apreciar con la aparición de chispas o arcos eléctricos.

Los carbones poseen pérdidas eléctricas elevadas y por ese motivo pueden ser sometidos a calentamiento por microondas. Además se pueden utilizar para recibir las microondas y calentar indirectamente otros materiales que no son buenos receptores de microondas, utilizándose en algunos procesos como la pirolisis de biomasa y residuos orgánicos o en reacciones catalíticas heterogéneas.

Guía de onda rectangular

Para la realización del calentamiento de los carbones se hizo uso de una guía de onda rectangular, diseñada y construida especialmente para este estudio.

Tienen varias aplicaciones como en sistemas de radiofrecuencia, microondas terrestres y satelitales. Son fáciles de fabricar y presentan varias ventajas puesto que poseen un gran ancho de banda y presentan pocas pérdidas.

Las corrientes inducidas en sus paredes generan un aumento en las pérdidas de la potencia suministrada, y para contrarrestar esto, la resistencia de las paredes de la guía se debe hacer lo más pequeño posible.

Las paredes deben ser especialmente laminadas, debido a que las corrientes tienden a concentrarse cerca del interior de la superficie de dichas paredes. Estas deben construirse con un material que sea un conductor perfecto.

2. Modelo Matemático para el Cálculo de la Permitividad en Carbones

En el modelo que se plantea con este estudio se realiza un calentamiento por vía de microondas al carbón y se compara con el agua calentada de la misma manera. Se registran los datos de temperatura y tiempo que se grafican y se aplican en el modelo, para el cálculo de la permitividad de los carbones.

Toma como punto de partida el balance de energía del sistema que, en general, está dado por la siguiente ecuación:

$$\{Flujo\ de\ calor\ por\ el\ ingreso\ de\ masa\} - \{Flujo\ de\ calor\ por\ salida\ de\ masa\} + \left\{\begin{matrix} Generación\ de\ energía\ térmica\ por \\ fricción\ molecular \end{matrix}\right\} - \{Calor\ por\ conducción\} = \{Acumulación\ de\ energía\}$$

Reemplazando los términos respectivo

$$F_0 C_P T_0 - F C_P T + 2\pi f V \varepsilon_0 E^2 \varepsilon_c'' + 2\pi f V_c \varepsilon_0 E^2 \varepsilon_v'' - UA(T - T_a) = M C_P \frac{dT}{dt} \qquad (6)$$

Pero como no hay flujo de masa que ingrese ni que salga del sistema, en el caso en estudio, porque se da un calentamiento tipo batch:

$$F_0 = F = 0 \qquad (7)$$

El calor generado por la fricción molecular del vidrio, provocado por la acción de las microondas, se puede despreciar debido a que es insignificante con respecto al generado por el carbón. La permitividad eléctrica relativa del vidrio de borosilicato 3.3 que es el tipo del vidrio Duran, es de 0,01702 [38].

El término de acumulación de energía térmica, en función del carbón y del vidrio queda expresado de la siguiente manera:

$$(M_v C_{Pv} + M_c C_{Pc}) \frac{dT}{dt} \qquad (8)$$

Por todo lo anterior la ecuación del modelo matemático resulta de la siguiente forma:

$$(M_v C_{Pv} + M_c C_{Pc}) \frac{dT}{dt} = 2\pi f V_c \varepsilon_0 E^2 \varepsilon_c'' - UA(T_C - T_a) \qquad (9)$$

De donde,
M_v= Masa del vidrio del tubo de ensayo que contiene la muestra del carbón y que se encuentra expuesto a la acción de las microondas, [Kg]
C_{Pv}= Capacidad calorífica del vidrio, [J/Kg °C]
M_c= Masa del carbón, [Kg]
C_{Pc}= Capacidad calorífica del carbón, [J/Kg °C]
$\frac{dT}{dt}$= La variación de la temperatura con respecto al tiempo, [°C/s]
f= Frecuencia de las microondas, [Hz]
V_c= Volumen de carbón, [m^3]
ε_0= Permitividad en el vacío, [F/m]
E^2= Campo eléctrico, [V/m]
ε_c''= Permitividad relativa del carbón
UA= Coeficiente externo de transferencia de calor, [J/s °C]
T_C= Temperatura del carbón, [°C]
T_a= Temperatura ambiente, [°C]

Para el agua se realiza el mismo análisis que se hizo para el carbón obteniéndose la siguiente expresión:

$$(M_v C_{Pv} + M_{H_2O} C_{PH_2O}) \frac{dT}{dt} = 2\pi f V_{H_2O} \varepsilon_0 E^2 \varepsilon_{H_2O}'' - UA(T_{H_2O} - T_a) \qquad (10)$$

De donde,
M_{H2O}= Masa del agua, [Kg]
C_{PH2O}= Capacidad calorífica del agua, [J/Kg °C]
V_{H2O}= Volumen de agua, [m^3]
ε_{H2O}''= Permitividad relativa del agua
T_{H_2O}= Temperatura del agua, [°C]

La anterior ecuación posee dos condiciones en sus extremos:

1) Cuando t = 0, entonces, $T_{H_2O} = T_a$

Por consiguiente el segundo término de la derecha de la ecuación (10) se hace cero (0) y se obtiene la siguiente expresión para el t = 0:

$$(M_v C_{Pv} + M_{H_2O} C_{PH_2O})\frac{dT}{dt}\Big|_{t=0} = 2\pi f V_{H_2O}\varepsilon_0 E^2 \varepsilon''_{H_2O} \qquad (11)$$

2) Cuando t = $t_{equilibrio}$, entonces, $\frac{dT}{dt} = 0$

Por consiguiente el término de la izquierda en la ecuación (10) se hace cero (0) y se obtiene la siguiente expresión para el t = $t_{equilibrio}$:

$$2\pi f V_{H_2O}\varepsilon_0 E^2 \varepsilon''_{H_2O} = UA(T_{H_2O} - T_a) \qquad (12)$$

De la ecuación (11) se despeja el campo eléctrico y se calcula, ya que los demás términos de la ecuación se pueden determinar o ya se conocen.

$$E^2 = \frac{(M_v C_{Pv} + M_{H_2O} C_{PH_2O})}{2\pi f V_{H_2O}\varepsilon_0 \varepsilon''_{H_2O}}\frac{dT}{dt} \qquad (13)$$

De la ecuación (12) se despeja la temperatura de equilibrio del agua y reemplazando el valor del campo eléctrico calculado con la ecuación (13) y haciendo las respectivas operaciones se obtiene el valor de dicha temperatura.

$$T_{H_2O} = \frac{2\pi f V_{H_2O}\varepsilon_0 E^2 \varepsilon''_{H_2O}}{UA} + T_a \qquad (14)$$

Esta temperatura del agua debe ser más alta que la temperatura de equilibrio del carbón, debido a que las propiedades dieléctricas del agua hacen que la acción de las microondas sea más fuerte en ella que en el carbón, por esta misma razón se espera que el valor de la permitividad del carbón sea menor que la del agua.

Cálculo de la permitividad del carbón

Hasta el momento se han determinado las ecuaciones que explican el sistema, pero aún falta determinar el valor de la permitividad del carbón.

Para el carbón se dan las mismas condiciones que para el agua, ecuaciones (11) y (12), que para el carbón quedan expresadas de la siguiente manera:

En el arranque $(T_c - T_a) = 0$

$$(M_v C_{Pv} + M_c C_{Pc})\frac{dT}{dt}\Big|_{t=0} = 2\pi f V_c \varepsilon_0 E^2 \varepsilon''_c \qquad (15)$$

En el equilibrio $\left(\frac{dT}{dt} = 0\right)$

$$2\pi f V_c \varepsilon_0 E^2 \varepsilon''_c = UA(T_c - T_a) \qquad (16)$$

Despejando de la ecuación (15) la *permitividad en el arranque* del carbón, resulta la siguiente expresión:

$$\varepsilon''_c\Big|_{t=0} = \frac{(M_v C_{Pv} + M_c C_{Pc})\frac{dT}{dt}\Big|_{t=0}}{2\pi f V_c \varepsilon_0 E^2} \qquad (17)$$

De donde,

$\varepsilon''_c\Big|_{t=0} = permitividad\ relativa$
$en\ el\ arranque$

Despejando de la ecuación (16) la *permitividad en la temperatura de equilibrio* del carbón, resulta la siguiente expresión:

$$\varepsilon''_c\Big|_{e.e} = \frac{UA(T_c - T_a)}{2\pi f V_c \varepsilon_0 E^2} \qquad (18)$$

De donde,

$\varepsilon''_c\Big|_{e.e} = permitividad\ relativa\ en\ la$
$temperatura\ de\ equilibrio$

Para poder resolver la ecuación anterior es necesario hacer una condición, que el campo eléctrico se mantiene constante para el carbón y para el agua, ya que la potencia suministrada es mínima, tan solo el 10% de la potencia total del equipo, así mismo, la potencia absorbida por el carbón es pequeña.

3. Gráficas Obtenidas

Se realizaron tres pruebas al agua y a cada uno de los carbones estudiados, bituminoso y antracítico.

Para el Agua

Gráfica 1. Agua Desionizada.

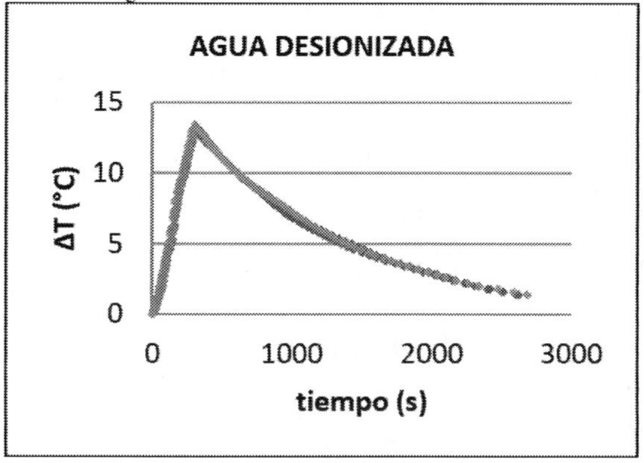

En cada una de las pruebas el calentamiento se realizó durante 5 minutos y se dejó enfriar hasta que el agua alcanzara su temperatura inicial. Se puede observar como el calentamiento se da en forma uniforme para las tres pruebas y que el enfriamiento se da de manera exponencial guardando el mismo perfil.

Para el Carbón Bituminoso

Gráfica 2. Carbón Bituminoso.

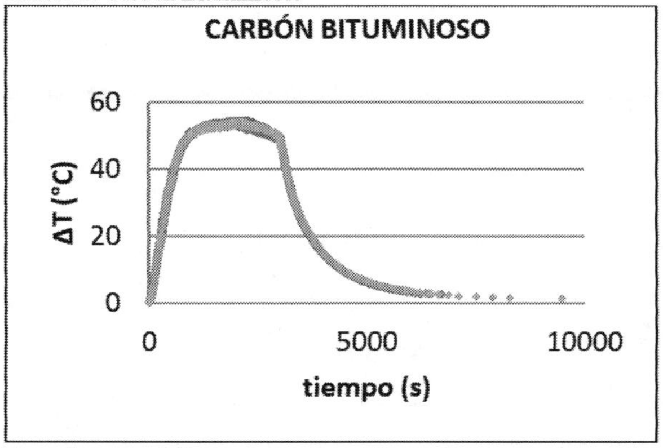

En la gráfica se puede observar como el calentamiento se mantiene constante y después de un tiempo se mantiene a la temperatura de equilibrio, a una diferencia de temperatura entre el carbón y el ambiente se encuentra alrededor de 53°C, y el carbón a 81°C, aproximadamente.

Para el Carbón Antracítico

Gráfica 3. Antracita.

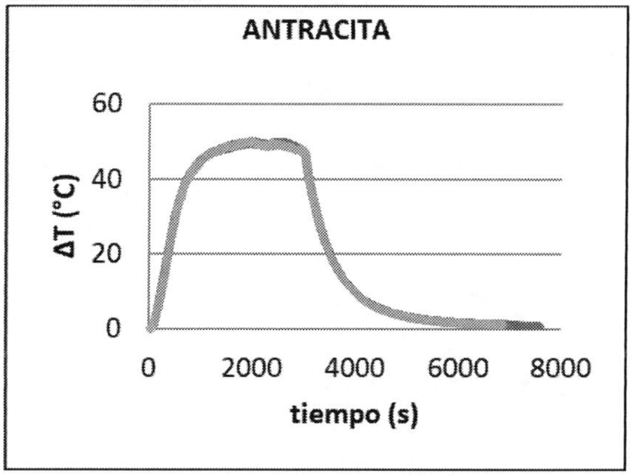

Se puede observar que el calentamiento es de pendiente constante hasta que se comienza a llegar a la temperatura de equilibrio, en donde se mantiene por unos minutos. Pasado un tiempo, el carbón empieza a sufrir daños, haciendo que la temperatura descienda aún con el calentamiento de las microondas.

El enfriamiento se da de manera casi exacta para las tres pruebas, conservando el mismo perfil con buena precisión.

4. Resultados

Luego de que se realizaron las pruebas, se elaboraron las gráficas, se analizaron, se extrajo información de ellas y se aplicó el modelo matemático se obtuvieron los siguientes resultados:

Carbón Bituminoso. Prueba 1.

Campo eléctrico, ecuación (13)

$$E = 370{,}5\,\frac{V}{m}$$

Temperatura de equilibrio del agua, ecuación (14)

$$T_{H_2O} = 90{,}21740144\ °C$$

Permitividad del carbón en el arranque, ecuación (17)

$$\varepsilon_c''|_{t=0} = 7{,}09196$$

Permitividad del carbón en el estado estacionario, ecuación (18)

$$\varepsilon_c''|_{e.e} = 2{,}19229$$

Carbón Bituminoso. Prueba 2.

Campo eléctrico, ecuación (13)

$$E = 379{,}8\ \frac{V}{m}$$

Temperatura de equilibrio del agua, ecuación (14)

$$T_{H_2O} = 94{,}74207057\ °C$$

Permitividad del carbón en el arranque, ecuación (17)

$$\varepsilon_c''|_{t=0} = 7{,}02674$$

Permitividad del carbón en el estado estacionario, ecuación (18)

$$\varepsilon_c''|_{e.e} = 2{,}10455$$

Carbón Bituminoso. Prueba 3.

Campo eléctrico, ecuación (13)

$$E = 376{,}6\ \frac{V}{m}$$

Temperatura de equilibrio del agua, ecuación (14)

$$T_{H_2O} = 92{,}2738915\ °C$$

Permitividad del carbón en el arranque, ecuación (17)

$$\varepsilon_c''|_{t=0} = 6{,}91878$$

Permitividad del carbón en el estado estacionario, ecuación (18)

$$\varepsilon_c''|_{e.e} = 2{,}02441$$

Antracita. Prueba 1.

Campo eléctrico, ecuación (13)

$$E = 370{,}5\ \frac{V}{m}$$

Temperatura de equilibrio del agua, ecuación (14)

$$T_{H_2O} = 90{,}21740144\ °C$$

Permitividad del carbón en el arranque, ecuación (17)

$$\varepsilon_c''|_{t=0} = 4{,}83344$$

Permitividad del carbón en el estado estacionario, ecuación (18)

$$\varepsilon_c''|_{e.e} = 2{,}54616$$

Antracita. Prueba 2.

Campo eléctrico, ecuación (13)

$$E = 379{,}8\ \frac{V}{m}$$

Temperatura de equilibrio del agua, ecuación (14)

$$T_{H_2O} = 94{,}74207057\ °C$$

Permitividad del carbón en el arranque, ecuación (17)

$$\varepsilon_c''|_{t=0} = 4{,}69888$$

Permitividad del carbón en el estado estacionario, ecuación (18)

$$\varepsilon_c''|_{e.e} = 2{,}41124$$

Antracita. Prueba 3.

Campo eléctrico, ecuación (13)

$$E = 376{,}6\,\frac{V}{m}$$

Temperatura de equilibrio del agua, ecuación (14)

$$T_{H_2O} = 92{,}2738915\,°C$$

Permitividad del carbón en el arranque, ecuación (17)

$$\varepsilon_c''|_{t=0} = 4{,}74887$$

Permitividad del carbón en el estado estacionario, ecuación (18)

$$\varepsilon_c''|_{e.e} = 2{,}51280$$

5. Conclusiones

Fue necesario plantear un modelo matemático, que permitiera plasmar la influencia de la energía transmitida por las microondas a la muestra en la temperatura de esta en el tiempo para poder observar el comportamiento de las mismas, y realizar una deducción de las permitividades de los carbones.

Al comparar los resultados obtenidos con el equipo y modelo matemático planteado, con los obtenidos con anteriormente para el agua, se puede afirmar la validez de los resultados obtenidos en el presente trabajo, ya que se puede apreciar la consistencia de la relación de las permitividades en el estado estacionario e inicial y se mantiene la tendencia a disminuir la permitividad, con un aumento de temperatura a la frecuencia trabajada.

Aunque los resultados obtenidos por medio del equipo construido y el modelamiento matemático planteado, pueden ser analizados y comparados con una tendencia parecida a la del agua, el margen de error que pueda existir está delimitado por la ausencia de un cálculo detallado –con modelos no ideales en el comportamiento del agua en el aumento de temperatura pero esto implica mayor complejidad y tiempo de cálculo.

6. Referencias

[1] ABDELGHANI-IDRISSI, M. A. Experimental investigations of occupied volume effect on the microwave heating and drying kinetics of cement powder in a mono-mode cavity. 2000.

[2] AGUILAR, J. A., GARCIA, R., VALDEZ, Z. Simulación del calentamiento de manganitas mediante microondas. 2007.

[3] AJIACO, F. Evaluación del comportamiento térmico de carbones del Cerrejón, carbones coquizantes y sus mezclas en la producción de coque metalúrgico. Universidad Nacional de Colombia. Bogotá, 2011.

[4] AKHTAR, M. J., FEHER, L., THUMM, M. A Multi-Layered Waveguide Technique for Determining Permittivity and Conductivity of Composite Materials. 2005.

[5] ALMANSA MORALES, Alejandro. Estudio experimental para la optimización energética y termográfica en aplicaciones de calentamiento por microondas. Universidad Politécnica de Cartagena. 2007, 19.

[6] ARIZTIMUÑOZ, A. GONZALEZ, F. RISUEÑO, A. El carbón como materia prima. Escuela Universitaria de Ingeniería Técnica Industrial.

[7] ASTIGARRAGA, Julio. Hornos de alta

frecuencia y microondas. Mc Graw Hill. Bogotá, 1999.
[8] BAKER-JARVIS, J., JANEZIC, M., GROSVENOR, J., GEYER, R. Transmission/Reflection and Short-Circuit Line Methods for Measuring Permittivity and Permeability. National Institute of Standards and Technology. Boulder, Colorado. 1993.
[9] BEVERLY MICROWAVE DIVISION. Magnetron theory of operation. Massachusetts, Estados Unidos.
[10] CARMONA MARTINEZ, F. Análisis y optimización de circuitos de microondas en guía de onda. Universidad Autónoma de Madrid. Madrid. 2009.
[11] CONSTANTINI, S. Guías de ondas. Universidad Metropolitana de Telecomunicaciones.
[12] DILDINE, Bob. Measuring complex permittivity and permeability at RF and microwave frequencies. Hewlett Packard. Yokogawa, Japón. 1989.
[13] FERNÁNDEZ, Y. *et al.* Procesos Térmicos Asistidos por Microondas cobre Materiales Carbonosos. Grupo de Materiales Carbonosos para Aplicaciones Tecnológicas. Oviedo, España. 2010.
[14] GRUPO ESPAÑOL DEL CARBÓN. Boletín del Grupo Español del Carbón. N°16, Mayo 2010.
[15] GUZMAN RABANILLO, D. Sistema de de deposición química en fase vapor para crecimiento homoepitaxial de diamante. Universidad de Cádiz. 2004.
[16]http://translate.google.com.co/translate?hl=es&langpair=en%7Ces&u=http://www.radartutorial.eu/08.transmitters/tx08.en.html
[17] http://www.catamutun.com/produc/carbon/
[18]http://www.lacomunidadpetrolera.com/showthread.php/2046-Carbonificaci%C3%B3n.
[19] http://www.metalloyd.co.uk/spanish/products_raw_materials.html
[20]http://www.oviedocorreo.es/personales/carbon/carbon_mineral/carbon%20mineral.htm
[21] http://www.pdv.com/lexico/museo/minerales/carbon.htm
[22] JOHNSON, M., HANSEN, W., RATL, W. Wave guide construction. United States Patent Ofice. Patent dec. 30, 1947.
[23] KEYSON, D. *et al.* Domestic Microwave Oven Adapted for Fast Heat Treatment of $Ba_{0.5}Sr_{0.5}(Ti_{0.8}Sn_{0.2})O_3$ Powders. Journal of Materials Processing Technology 189 (2007). Pág. 316 - 319.
[24] KLINBUN, W. RATTANADECHO, P. PAKDEE, W. Microwave heating of saturated packed bed using a rectangular waveguide (TE10 mode): Influence or particle size, sample dimension, frequency, and placement inside the guide. International Journal of Heat and Mass Transfer. Vol. 54. 2011. Pág. 1763 – 1774.
[25] LEON, Elard F. La importancia del carbón mineral en el desarrollo. Revista del Instituto de Investigaciones FIGMMG. Vol. 9, N° 18, 91 – 97. 2006. San Marcos, Perú.
[26] NAKAMURA, T., NIKAWA, Y. Precise measurement of complex permittivity of materials for telecommunications devices. Journal of Telecommunications and Information Technology. 2001.
[27] NEBOT, Eduardo. Caracterización de dieléctricos a frecuencia de microondas. Universidad de Zaragoza. 2004, 3.
[28] PICERNO, Alberto. Todo sobre hornos de microondas. Saber electrónica. Edición internacional N° 207. México, 2007. Páginas 5 – 22.
[29] PIRA, C. Densificazione di scariche a bagliore in dispositivi magnetron sputtering tramite accoppiamiento con sorgenti Hollow Cathode. Universitá degli Studi de Padova. 2007.
[30] RATTANADECHO, P. The simulation of microwave heating of wood using a rectangular wave guide: Influence of frequency and sample size. Chemical Engineering Science. Vol. 61. 2006. Pág. 4798 – 4811.
[31] RODRIGUEZ, J.M. et al. Medidas de propiedades dieléctricas de materiales de construcción utilizando una guía rectangular. Universidad Politécnica de Madrid. 2008.
[32] ROGOV, I.A. GORBATOV, A.V. Métodos físicos de procesamiento de productos

alimenticios. Industria Alimenticia. Moscú. 1974.
[33] STANLEY, R. Wave coupling arrangement. United States Patent Office. 1959
[34] STURM, G.S.J. , STEFANIDIS, G.D. , VERMEIJ, M.D. , VAN GERVEN, T.D.T. , STANKIEWICZ, A.I. Design principles of microwave applicators for small-scale process equipment. Chemical Engineering and Processing: Process Intensification. 2010. Pág 912 - 922.
[35] UNIDAD DE PLANEACION MINERO ENERGETICA. La cadena del carbón. Ministerio de Minas y Energías. República de Colombia. 2005. Bogotá, Colombia.
[36] USLU, T. ATALAY, U. AROL, A.I. Effect of Microwave Heating on Magnetic Separation of Pyrite. Colloids and Surfaces A: Physicochem. Eng. Aspects 225 (2003). Pág. 161 - 167.
[37] ZHENFENG, L. RAGHAVAN, G.S.V. ORSAT, V. Temperature and power control in microwave drying. Journal of Food Engineering 97. 2010. Pág. 478 - 483.
[38] http://www.vidrasa.com/esp/productos/duran/duran_pf.html
[39] J. B. Hasted, Liquid water: Dielectric properties, in *Water A comprehensive treatise*, Vol 1, Ed. F. Franks (Plenum Press, New York, 1972) Pág. 255-309.
[40] http://www.lsbu.ac.uk/water/microwave.html
[41] I.Takei, Dielectric relaxation of ice samples grown from vapor-phase or liquid-phase water, in *Physics and Chemistry of Ice*, ed. W. Kuhs (Royal Society of Chemistry, Cambridge, 2007) Pág. 577-584

(137c) Novel Warm Gas Multi-Contaminant Removal System

Gokhan Alptekin, Ambalavanan Jayaraman, Margarita Dubovik and Mike Cesario, TDA Research Inc., Wheat Ridge, CO

Gasification technologies convert coal and biomass feedstocks into synthesis gas feed streams that can either be used as a fuel for highly efficient power generation cycles or converted into value-added chemicals and transportation fuels. However, the coal/biomass-derived synthesis gas contains a myriad of trace contaminants (e.g. mercury, arsenic, selenium) and ammonia (NH_3) that must be removed to eliminate power plant emissions; NH_3 will convert into NO_xin the gas turbine combustor (an acid rain precursor), while Hg is designated as an hazardous air pollutant and being regulated by the U.S. EPA. These contaminants also poison the catalysts used in the manufacturing processes that converts synthesis gas into fuels or chemicals.

Although conventional clean-up technologies can effectively remove these contaminants, they typically require aqueous quenching and cooling of the syngas to around 100°F, followed by scrubbing with chemical or physical solvents, and finally absorption/adsorption of trace contaminants on solid sorbents. However, cooling the syngas and condensation of steam significantly penalizes the cycle efficiency. The power plant efficiency can be improved 3 to 4% if the synthesis gas is not cooled below its dew point because the energy in high pressure steam can be recovered by expanding it steam in the gas turbine and eliminating the energy penalty associated with the re-heating of the synthesis gas. Warm gas clean-up also eliminates the need for the expensive heat exchangers (to cool the synthesis gas to the operating temperature of the clean-up system and then to re-heat it back to before the combined cycle) and the gray water treatment problems associated with processing the large amounts of condensate.

In a DOE funded SBIR project, TDA Research, Inc. (TDA) is developing a chemical absorbent-based gas clean-up technology to remove the NH_3 and hydrogen cyanide (HCN) as well as all trace metal contaminants (e.g., Hg, As, Se and Cd) from the synthesis gas at high temperatures (500°F) in a single process step. The sorbent can be operated in a regenerable manner to remove NH_3 and Hg, while irreversibly absorbing all other contaminants. The regenerable operation for the Hg (which has very low affinity to any surface at high temperature) and NH_3 (which is present in very high concentration in the gas) ensures high utilization of the sorbent, while its high capacity to other contaminants and low cost ensures its economical use as an expendable sorbent. We will present the results from this sorbent development effort at the meeting.

(137d) "Clean" and "Dirty" Tar Reforming of Biomass Gasification Gas

Klas J. Andersson, Reforming Department, Haldor Topsøe A/S, Lyngby, Denmark and Poul Erik Højlund Nielsen, Chemicals Department, Haldor Topsøe A/S, Lyngby, Denmark

A prerequisite for the establishment of an efficient biomass-to-chemicals production or CHP applications utilizing engines is a good thermodynamic utilization of the hot synthesis gas exiting a biomass gasifier. A second and important criteria is the extraction or conversion of tars (polyaromatic hydrocarbons), which, if left untreated, leads to fouling of downstream equipment and issues around waste water treatment. Both above criteria are met by steam reforming the tar components at high temperatures in the sulfur-laden exit gas from the biomass gasifier. We discuss the benefits of different approaches depending on pressurized or near atmospheric gasification, leading to the concepts of "clean" and "dirty" tar reforming.

The basic difference between the two approaches is how dust is handled. In the dirty tar reforming processes the residual dust in the gas - after having passed cyclones - is removed downstream the tar reformer. In the clean tar reformer the dust removal takes place upstream the tar reformer and preferentially the dust removal takes place at temperatures in the vicinity of the gasifier exit temperature to minimize loss of exergy. The catalyst and processes are widely different. In the dusty tar reformer monolithic catalysts are a necessity whereas the clean reformer can use pelletized catalysts with a much higher content of active material.

We will discuss the virtues of the different approaches and look into the requirements for producing CHP, SNG or fuels like gasoline or diesel, supplemented by our achievements in clean and dirty tar reforming in industrial and pilot settings.

CFD-Based Probabilistic Explosion Hazard Analysis As an Early Tool to Improve FLNG Design

Scott Davis, Remi Martini and Filippo Gavelli

GexCon US

Abstract

Process areas on floating LNG facilities are typically designed to be very compact, due to the high costs of "floating real estate". Given the high congestion levels of FLNGs, explosion design accidental loads (DALs) can become very large and are very sensitive to the layout of piping and equipment, the size of safety gaps, specifics of the liquefaction process, etc. For these reasons, the analysis of explosion hazards on FLNGs is a critical step that should accompany the facility development from the very earliest design phase. This paper will discuss how a probabilistic explosion hazard analysis methodology based on computational fluid dynamics (CFD) modeling can be used to optimize the layout of an FLNG facility with respect to DALs. The methodology uses a detailed 3D model of the FLNG geometry (down to 1-2 inch pipe diameters) within which gas dispersion and vapor cloud explosion scenarios are simulated using a CFD tool; leak rate probability data and the TDIIM ignition model are then used to calculate probability of exceedance curves and to determine the DALs. The paper will present results from a study, which compared the effectiveness of a number of design alternatives, such as:

- Grated vs. solid decks;
- Weak barriers vs. blast walls;
- Safety gap size;
- Different liquefaction processes.

The paper will demonstrate how this methodology can provide FLNG designers with critical safety-related design implications previously not available.

Introduction

The advanced methodology implemented in the present work, which incorporates advanced techniques to supplement an "incomplete" model of the FLNG geometry in Concept or FEED phase to levels anticipated in the "as-built" condition, provides a powerful tool to assess FLNG layouts/technologies in the earliest design phase when evaluating *Inherently Safer Designs*. This paper will provide some case studies on how valuable insights were gained by applying a detailed 3D model in early Concept phase of the design.

Plated vs. Grated process deck

This section demonstrates the hazard of introducing grating on the process deck of on FLNG when using heavier-than-air gases/refrigerants in the liquefaction process. When some typical heavier-than-air refrigerants (e.g., mixed refrigerants) are released in the liquefaction area on a FLNG vessel with a completely plated process deck (solid deck above the hull deck), the released vapour tends to either remain on the process deck and its modules or falls overboard. However, when using grated sections between the liquefaction modules on the process deck, this heavy vapor tends to disperse and fall into the hull deck, which is a confined and is not well ventilated.

Even if the hull deck height up to the process deck is only between 5 and 8m, the flammable vapour can extend all along the length and width of the vessel, allowing very large flammable vapour clouds to form from such releases. Figure 1 and Figure 2 show that by changing the process deck from partly grated to fully plated, the size of the flammable vapour clouds are significantly reduced due to very limited quantities of flammable gas migrating to the hull deck area when the process deck is plated. When these results are combined with leak frequency data and statistical wind data, the exceedance frequency of large clouds is drastically reduced, as shown in Figure 3.

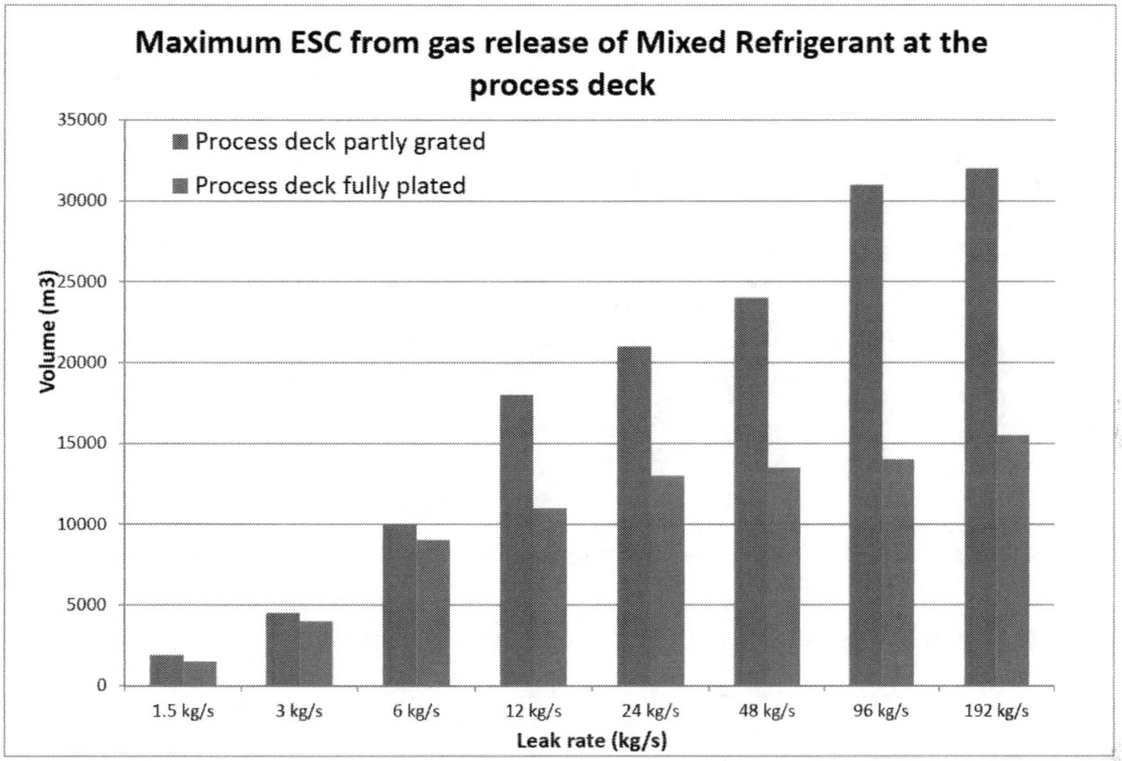

Figure 1: Maximum Equivalent Stoichiometric Cloud (ESC) from a release of mixed refrigerant at the process deck

Regarding explosion consequences, while it is intuitive to think that with increased grating on the process deck the explosion consequences in the hull deck will be reduced due to the increased venting. This however is not the case for the present study as the cloud size resulting from releases on the process deck result in clouds that occupy volumes both on the process deck and hull deck. The grating therefore does not the effect of venting for two reasons: when the flammable vapour cloud is below and above the grating, this allows the flame to accelerate through upper congested liquefaction modules, and into the confined hull deck, resulting in minimal benefits of venting through the grating (see); and in the case of having the flammable cloud only upon the process deck, the hull deck is typically too confined to adequately vent the explosion due to its height (5 to 8 meters).

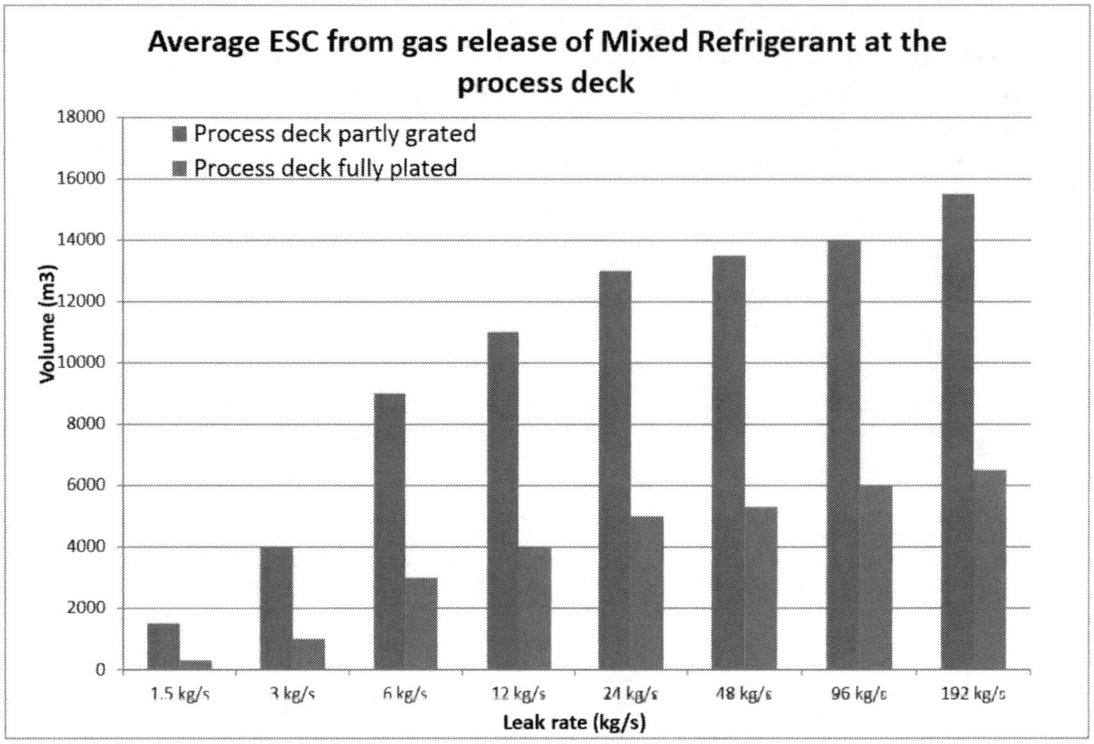

Figure 2: Average Equivalent Stoichiometric Cloud (ESC) from a release of mixed refrigerant at the process deck

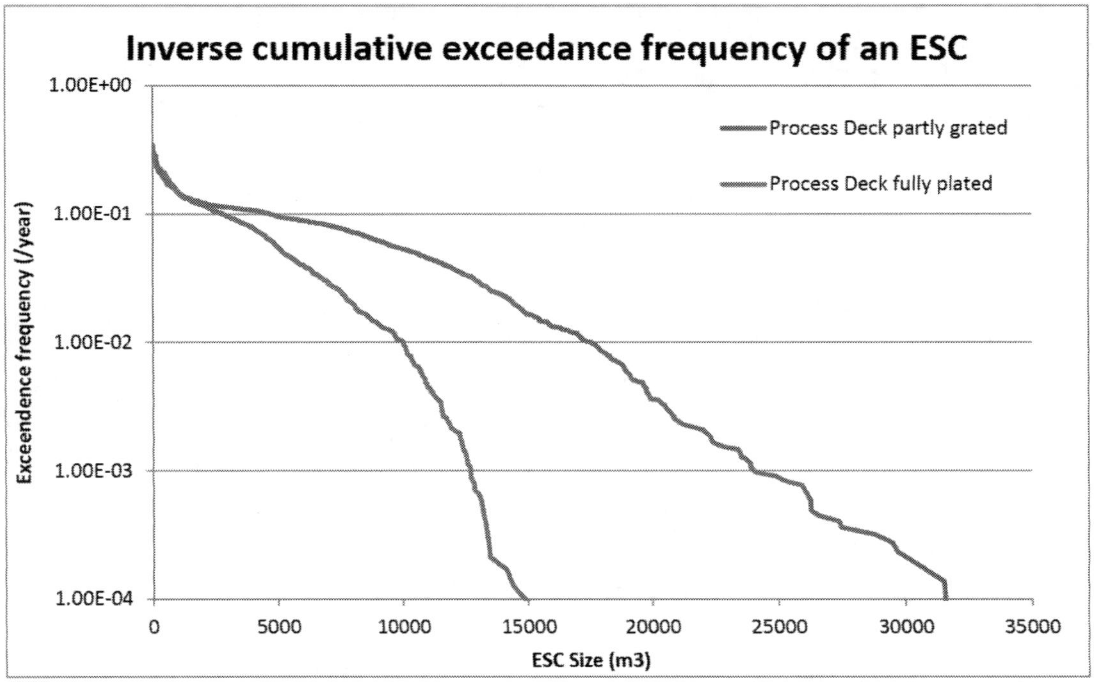

Figure 3: Inverse cumulative exceedance frequency of an explosion gas cloud

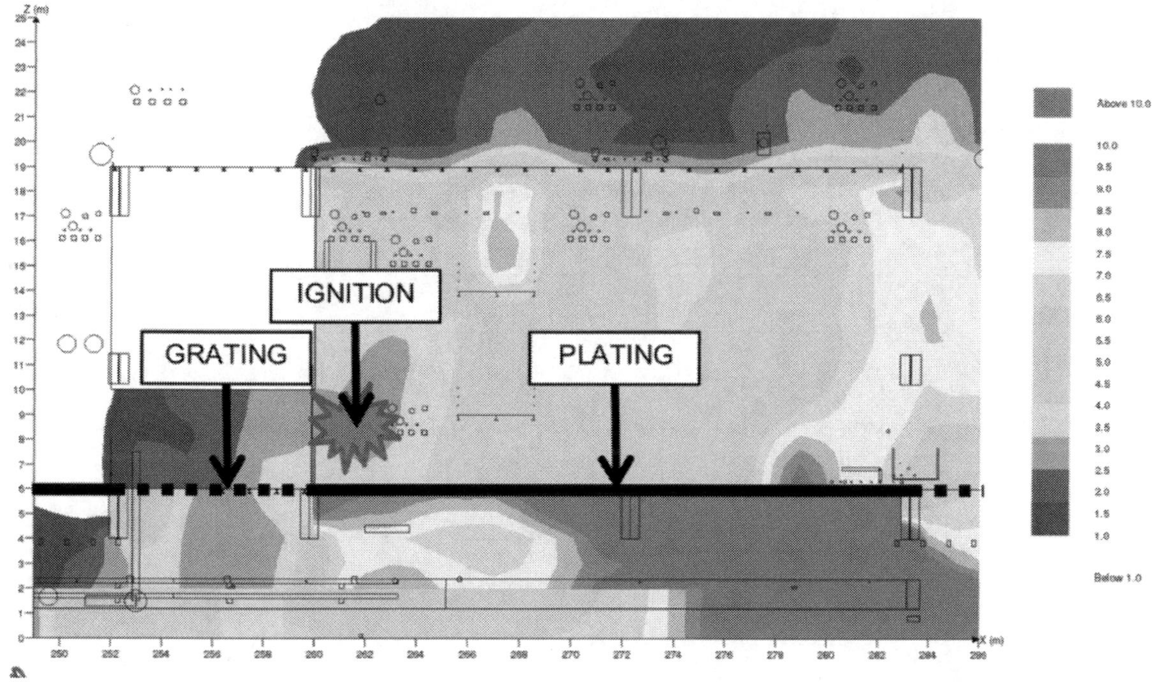

Figure 4: Example and an explosion simulation given more than 10 barg overpressure on the process deck

When assessing explosion risk, an ignition probability model is used to calculate the explosion frequency of a gas mixture. When using a time dependant ignition model, part of the ignition probability is calculated based on area (m2) in contact with the flammable mixture. Due to the presence of a flammable mixture in the hull deck, the whole deck area has to be considered as potential ignition source for the risk calculation. Even if the ignition sources within the hull deck are limited, they will slightly increase the ignition probability.

While a fully plated process deck does not have a significant impact on reducing the likelihood of an explosion (mainly due to the limited number of ignition sources at the hull deck), it significantly reduces the magnitude of the overpressures occurring with a frequency of 1×10^{-4}/[year], due to the limited size and location of the gas cloud as shown in Figure 5.

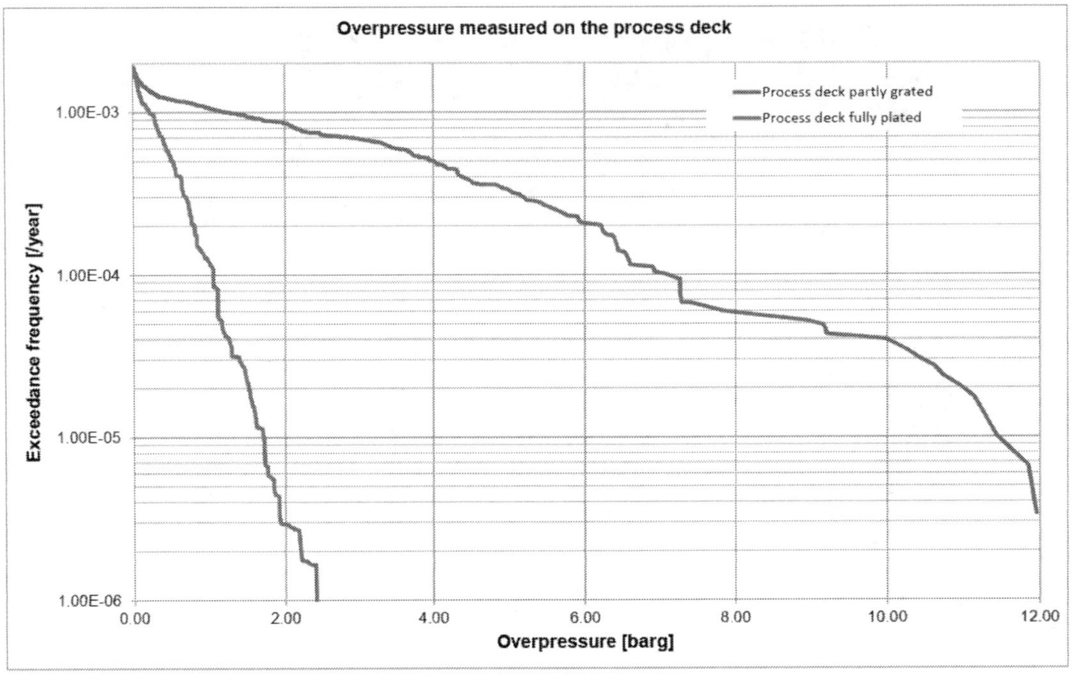

Figure 5: Inverse cumulative exceedance frequency of an explosion gas cloud

Gas Detection

Gas detection is a critical safety device that can initiate certain mitigation measures. The time to detect a leak will largely affect two parameters: the size of the gas cloud due to initiation of ESD and blowdown; and the time of exposure to ignition sources (ignition source isolation after confirmed gas detection). Studies have shown that using an efficient gas detector layout, an accidental release can be detected within few seconds. This present case study only compares generic detection times for all leak rates and assumes that all accidental releases are detected. The difference in the two detection philosophies are that the confirmed gas alarm is reach in 30 seconds for the base case and 15 seconds in the optimized case.

Two detection philosophies are compared:

	TIME (s)				
	Leak starts	Low alarm	High Alarm*	ESDV closed	Blowdown
Base Case	0	15	30	60	150
Optimized	0	10	15	45	135

*Start of Emergency ShutDown (ESD)

Despite its generic nature, this example enlightens the importance of detection time in terms of explosion risk reduction. For the present case study, the explosion loads occurring once every 10,000 years (10^{-4}) were reduced from 2.2 barg to 1.7 barg.

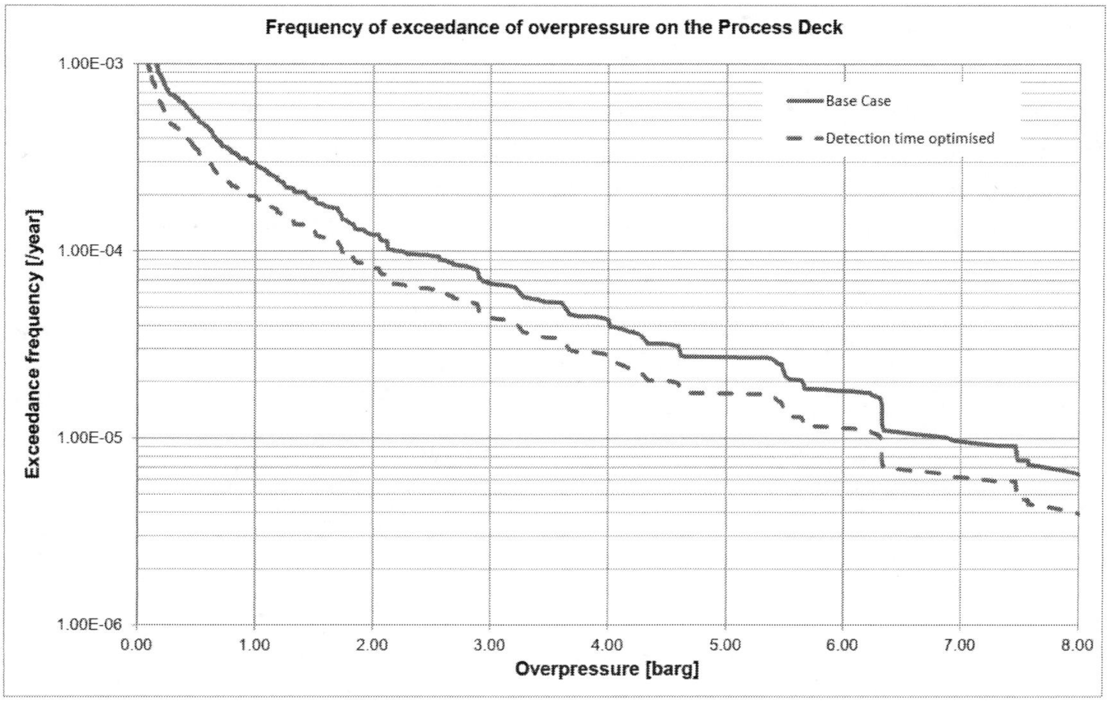

Figure 6: Inverse cumulative exceedance frequency of an explosion gas cloud

Safety Gaps

This case study evaluated the effectiveness of two different safety gaps (12.5 meters and 20 meters) between the liquefaction modules (see).

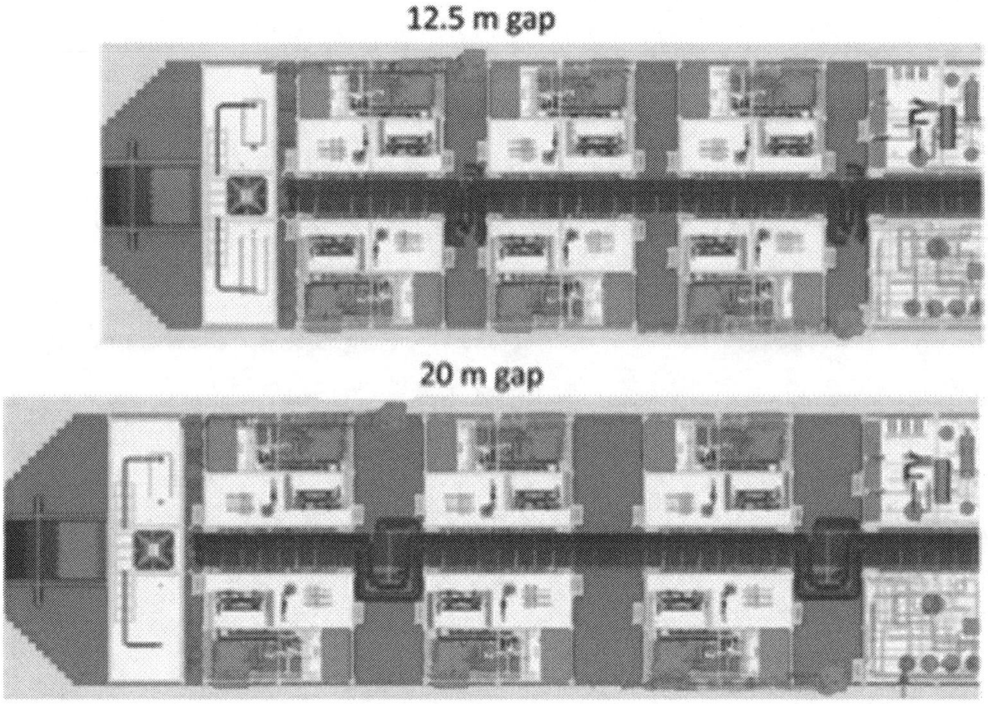

Figure 7: Safety gap configurations

The results showed that the ventilation rate within the module (measured by ACH) remained essentially unchanged as the safety gap was increased from 12.5 to 20 meters. Therefore, increasing the gap size from 12.5 to 20 meters did not significantly reduce the size of a gas cloud that remained inside the module where the leak originated. In contrast, significant reductions in cloud sizes were observed for clouds that migrated out of the origin module and into neighboring modules (see Figure 8). This was significant with respect to overall gas cloud size and ignition frequency as discuss in a previous section.

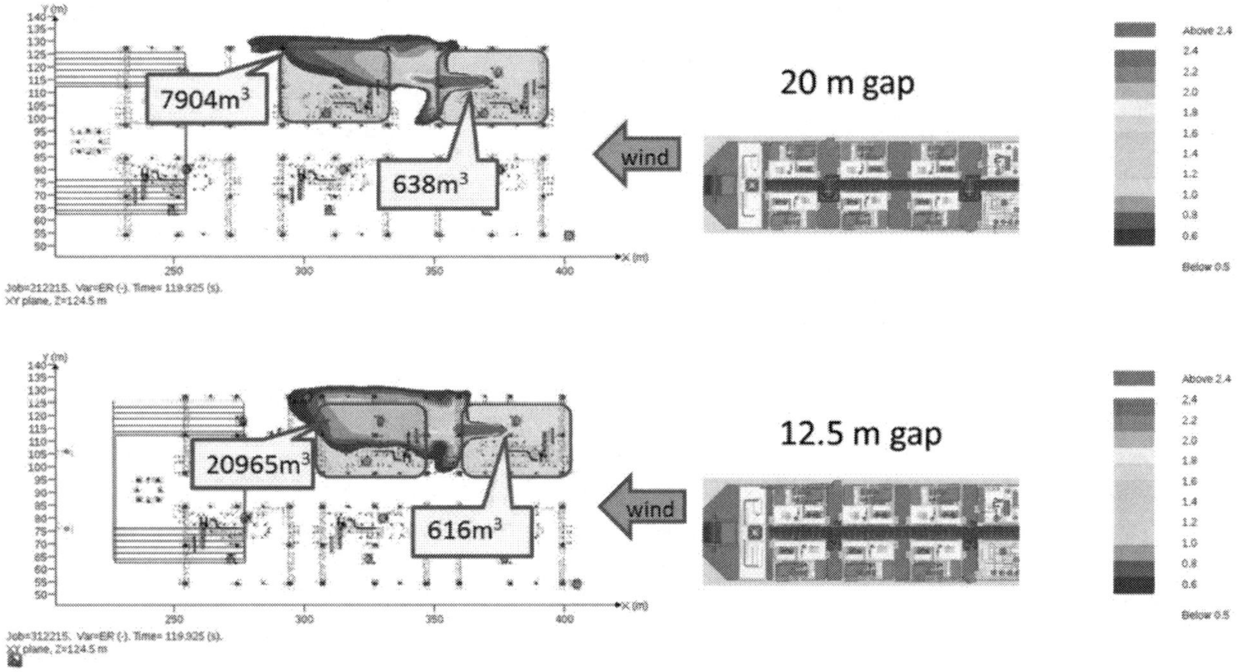

Figure 8: Dispersioin example for the 20 meter gap and 12.5 meter gap

Conclusions

This paper examined the benefits of using advance CFD simulations for explosion risk in early design phase of FLNG vessels. Many insights were obtained early enough as to positively influence the design towards an inherently safer design.

Forced Dispersion Analysis of LNG Vapor Mitigation Using Fire Dynamic Simulator

*Wilson Y. Molina-Torres, Byung Kyu Kim, Ray A. Mentzer and M. Sam Mannan**
Mary Kay O'Connor Process Safety Center, Artie McFerrin Department of Chemical Engineering,
Texas A&M University System, College Station, TX
*Corresponding author. Tel.: +1 979 862 3985. E-mail address: mannan@tamu.edu (M.S. Mannan)

Abstract

With the expected demand of liquefied natural gas (LNG) to double by year 2020 world-wide, the mitigation techniques for an LNG spill emergency require further investigation to provide accurate information in evaluating the consequences of an LNG spill and the potential impact on the public safety and security. This work applies the Fire dynamic simulator (FDS) to set up and simulate LNG forced dispersion using a water spray application. A comparison study with a previous methodology has been conducted and potential application of FDS in the LNG safety research is discussed.

1. Introduction

The demand of the natural gas in the US had exceeded the production level for last couple of the decades [1]. However, with the increases of natural gas production in US from the development of the shale gas industry, supplies of natural gas is expected to surpass the domestic demands. Once a liquefied natural gas (LNG) importer, the US natural gas industry is in the phase of transiting to LNG exporter, and expected to play a significant role in redefining the global natural gas market [2].

With the expectation of the LNG production in US, it becomes critical to understand all the hazards related to the liquefaction and transportation of the LNG to ensure public safety. It was concluded that the safety mitigation measures require further study to minimize the uncertainty in determining the influence on reducing the hazards imposed to the society [3]. The Fire dynamic simulator (FDS) is an open source computational fluid dynamics (CFD) model of fire-driven fluid flow developed by the National Institute of Standards and Technology (NIST). This work applies the FDS code to evaluate the LNG forced mitigation phenomena. A preliminary comparison analysis had been conducted using the Fluent model. The potentials of applying FDS and further work to minimize the uncertainties of the vapor prediction are discussed.

2. Numerical modeling with FDS code

The FDS code is a publically available CFD model developed by the NIST. FDS was developed as software to model fire-driven fluid flow with emphasis on smoke and heat transport [4]. The FDS code is commonly used in the fire protection engineering industry and the hydrodynamic model embedded in the code can be used to model the low-speed fluid flow, such as vapor dispersion. FDS provides solutions for the Navier Stokes equations of motion, energy and species concentration in space and time. The vertical profiles in plume properties (velocity, concentration, species concentration, etc.) are not simplified or averaged, and the turbulence model components are provided by adjusting the damping effects of thermal or density stratification.

The FDS dispersion code had been evaluated for LNG vapor dispersion [5]. FDS provided fairly reasonable dispersion prediction. A safety factor of 2 was recommended to predict the unobstructed flow fields, and factor of 3 for the obstructed flow field from the results.

2.1 LES turbulence modeling

The FDS codes apply the Large Eddy Simulation (LES) to evaluate the turbulent flow induced within the vapor fields [4]. The LES model provides a qualitative description of the flow physics and applies the quantitative model to produce large-scale motions directly [6]. LES requires a flow field where only the large-scale components are present, and this process can be achieve through filtering process. With the available sub-grid scale models, large-scale fields are solved, while the smaller scale motions are approximated through localized filter functions of either a Gaussian or top-hat filter.

While the Reynolds Averaged Navier-Stokes (RANS) approach approximates and averages out the fluid fluctuations for all turbulent length scales, LES provides only the turbulent fluctuations that are resolvable by its mesh. This enables better fidelity of fluid flow behavior from turbulent fluctuations. Theoretically, this approach may assist in reducing the potential hazards that might rise from approximation or averaging the concentration. Although, LES is more computationally expansive, the model can provide an accurate turbulent vapor behavior [5].

2.2 Lagrangian spray model

The Lagrangian particles are utilized in FDS to simulate the fine droplets for the water sprinkler or spray [4]. A prescribed number of droplets are introduced from the nozzle every second, and the droplet characteristics, such as the density, specific heat, diameter, temperature, and mass flow rate can be defined. The force balance of an individual droplet is evaluated by the governing equation (1) and overall momentum imparted from the droplets to the gas flow in a grid cell is calculated using equation (2) [7].

$$\frac{du_d}{dt} = g - \frac{1}{2}\frac{\rho C_D \pi r_d^2}{m_d}(u_d - u)|u_d - u| \qquad (1)$$

$$f_b = \frac{1}{V}\sum_{d\in V}\left[\left[\frac{1}{2}\rho C_D \pi r_d^2 (u_d - u)\right]|u_d - u| - \frac{dm_d}{dt}(u_d - u)\right] \qquad (2)$$

u_d is the velocity of the droplet, $C_d$the drag coefficient, $r_d$is the droplet radius, u is the velocity of the gas, ρ is the density of the gas, V is the volume of the grid cell, m_d is the mass of the droplet, and g is the gravity constant. The initial droplet trajectory is evaluated with the predefined droplet characteristics and the overall momentum effects of overall droplets in a computational grid cell are calculated to determine the level of interaction of droplet/vapor system.

In additional to the momentum effects, the FDS solver also considers the heating and evaporation of the droplet particles.

$$m_L C_L \frac{dT_L}{dt} = Ah(T_g - T_L) + Ah_s(T_s - T_L) + \dot{q}_r + \frac{dm_L}{dt}h_V \qquad (3)$$

A is the surface area of the liquid droplet, h_m is the mass transfer coefficient, ρ is the density of the gas, $C_L$is the liquid specific heat, h is the heat transfer coefficient between the liquid and wall, $m_L$is the mass of the liquid droplet, h_S is the heat transfer coefficient between the liquid and solid surface, $h_V$is the latent heat of vaporization of the liquid, and $\dot{q}_r$is the radiative heating of the droplet. As the droplets travel in the gas flow, the droplet temperatures are constantly evaluated by considering the convective thermal transfer with the gas phase. The heat transfer with the ground is not considered in this work as the droplet interactions get terminated once the droplets travel back to ground. Also, the radiative heat exchange is neglected, as the heat transfer in this work does not involve high thermal source, such as fire or flame. The Lagrangian spray model also considers the evaporation effects of the droplets to the surroundings.

$$\frac{dm_i}{dt} = -Ahm\rho\left(Y_L - Y_g\right) \quad (4)$$

where the liquid equilibrium vapor mass fraction Y_L, the local gas phase vapor mass fraction Y_g, the assumed uniform liquid temperature T, and the local temperature T_g. However, considering the short duration of the droplets traveling in the atmosphere, the effects of evaporation will be relatively less significant compared to the heat transfer effects described as equation 3.

3. Simulation specifications

3.1 Computational domain and water spray modeling

The water spray settings and the computational domain are described in Fig. 1.

Parameter	Setting
Nozzle size (m)	1.27×10^{-3}
Cone angle (deg)	60
Droplet material	Water
Droplet diameter (mm)	0.935
Temperature (k)	300
Mass flow rate (kg/s)	4.6

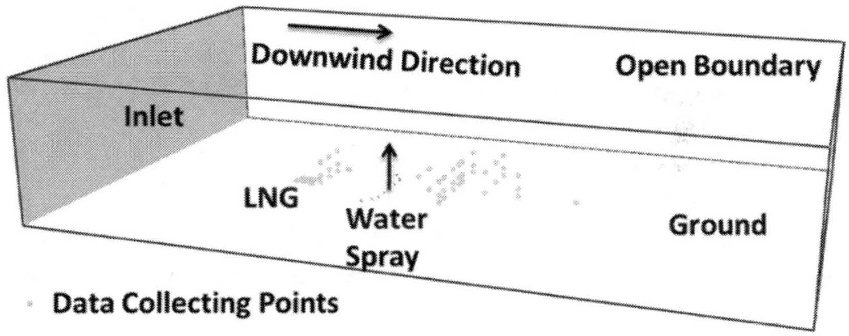

Fig. 1. Water spray applied for FDS modeling and computational domain setup

The computational domain was created in FDS platform with the downwind direction in the x direction. The size of the domain was 50 m x 30 m x 10 m. The For the simulation grid size the cell size was refined to acquire grid-independent setting to a 0.2-meter in a cube-like cell. A steady state result was achieved for the concentration profiles at different distances and approximately 1,875,000 grids were created. The simulation starts running without the LNG release for 300 s in order to obtain a constant (steady) wind profile on the domain. The natural dispersion was set for 600 s, and forced dispersion of LNG vapors with water sprays was applied for 500 s. These scenarios are based on the experimental work from March 2009 test [9].

For the LNG inlet boundary condition, the LNG source term was set using a circular inlet condition at the ground level. The methane was injected at its boiling temperature of 111 K. The size of the inlet was set at 0.857 m with an average mass flux rate set at 0.2 kg/(m^2-s). The atmospheric conditions used for the simulation were adopted from the experimental data [8].

The data collecting points and spray location, LNG pool size was adopted from the March 2009 MKOPSC LNG spill experiments [9]. The spray settings were also adopted from the experimental setup. A total of 100 droplets are injected each time step, which was set as 1 s, per nozzle. The droplet type is set as water droplets, which takes into account of the evaporation, and radiative absorption. However, these functions are negligible, as the droplet travel time is relatively short and temperature applied in the simulation is not high enough for the radiative heat transfer to be dominant. The coalescence and collision of the droplets are not considered in this modeling and the droplets are discharged in the shape of upward-oriented full cone type nozzle.

3.2 Wind Profile

The FDS is capable of simulating the atmospheric condition using the steady or unsteady wind profiles [4]. The magnitude of velocity profile and direction can be defined by specifying the velocity along the x, y, and z directions within the domain boundaries [5]. The FDS has two different functions available to describe the atmospheric condition. The ramp function can be used to make changes in the velocity component in the normal direction. As the ramp function can only simulate the wind field in the normal direction, it is difficult to simulate the transient wind. Another option is to apply the vent condition in the inlet and create corresponding ramp function. The wind profile can be described using a parabolic wind profile or a typical atmospheric wind profile in normal direction. For this study, a parabolic wind profile combined with the vent function was applied to provide a closer approximation of the wind behavior presented in the field experimentation. The parabolic wind profile used in this work is as follow:

$$u = u_O \left(\frac{z}{z_O}\right)^p \qquad (5)$$

where u is the wind velocity at height z , u_O is the wind velocity known at an specific height z_O and p is an empirical constant that varies depending on the stability of the atmosphere.

It is recommended to use low wind speed with stable atmospheric condition, as it will result in longer distances to downwind concentration. The requirement outlined in 49 CFR Part 193.2059 is the lowest wind speed that occurs 90% of time for the area or 2 m/s, to replicate the worst-case scenario. It is expected that 2 m/s wind speed will demonstrate the worst-case approach, with the farthest downwind concentration.

4. Model comparison

To evaluate the LNG force dispersion modeling of FDS, the simulation predictions were compared with the Fluent results. The physical parameters for Fluent were adopted from the simulation work done by Kim et al. (2012) and the experimental data from MKOPSC LNG spill experiments were also compared [8, 10].

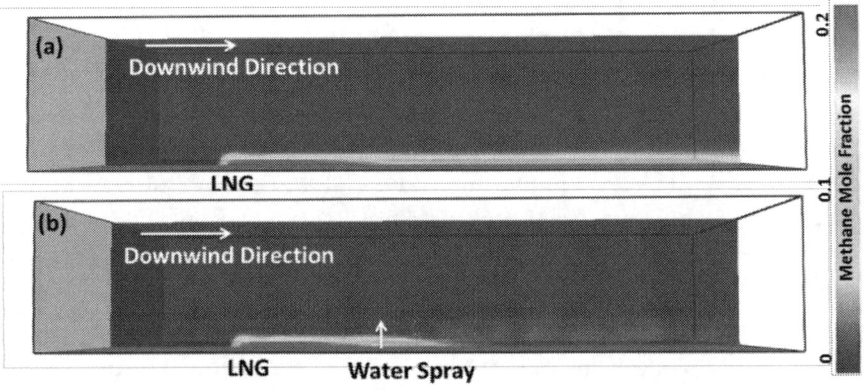

Fig. 2. Vapor contour simulated for (a) LNG natural dispersion and (b) forced dispersion with water spray

After acquiring the steady state solution of the wind profile, LNG vapors were injected to the domain as shown in Fig. 2 (a). The LNG vapors dispersed in the prevailing downwind direction, and diluted as results of heat transfer and air entrainments into the vapor clouds. The water spray application on the LNG vapors is illustrated in Fig. 2 (b). As the water spray starts to interact with LNG vapors, the droplets start lifting the LNG vapors to a higher elevation, increasing the buoyancy of the cloud. The momentum and heat transfer effects from the water spray resulted in decrease of the vapor concentration in the post-spray region.

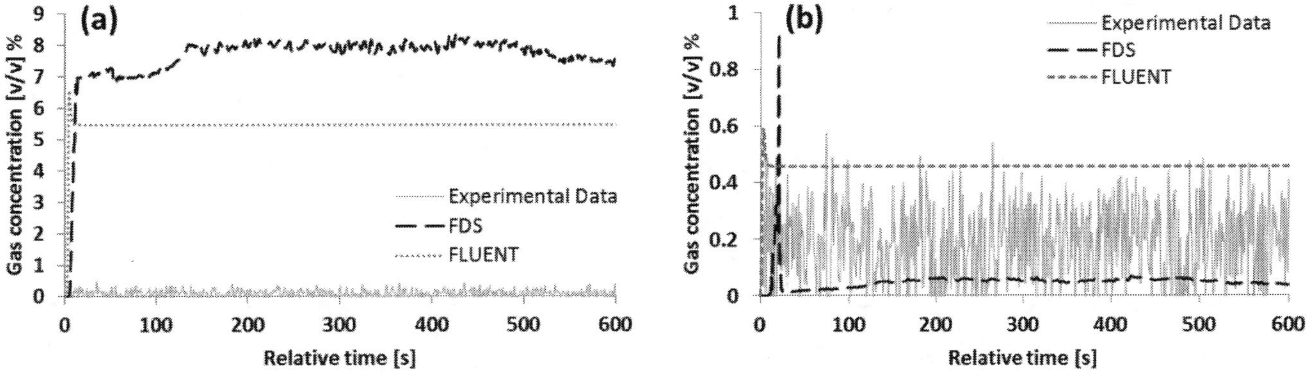

Fig. 3. Comparison of LNG natural dispersion prediction (FDS/Fluent) with experimental data at downwind distance 0 m at (a) 0.5 m elevation and (b) 1.2 m elevation

Fig. 3 shows the comparison of the LNG natural dispersion with the experimental data to the simulation prediction (FDS/Fluent) at 0 m distance. At the lower elevation, the results obtained by both FDS and Fluent over-predicted the LNG vapor concentration as shown in Fig. 3 (a). For the 1.2 m elevation case, Fluent showed agreeable prediction to the experimental result, while FDS provided an under-predicted vapor concentration as shown in Fig. 3 (b). This is mainly due to the limited vapor movement in the FDS. The FDS solver was capable to disperse the LNG vapors only up to approximately 1 m elevation. Further investigation is required to implement better heat transfer model and grid setting to improve the vapor movement prediction using FDS. The high peak observed during the initial dispersion of the LNG vapors shown in Fig. 3 (b) was caused by the LNG vapor behavior. When LNG was injected into the domain, the prevailing wind dominated the LNG vapor movement, as there was only a small amount of LNG vapors initially. Although heavier than air, the LNG introduced into the domain during the initial stage might had been carried by the wind as bulk of vapor, which might had caused the concentration to increase for short amount of time. As more vapors are introduced to the domain, the LNG vapor concentration reached steady state after approximately 200 s.

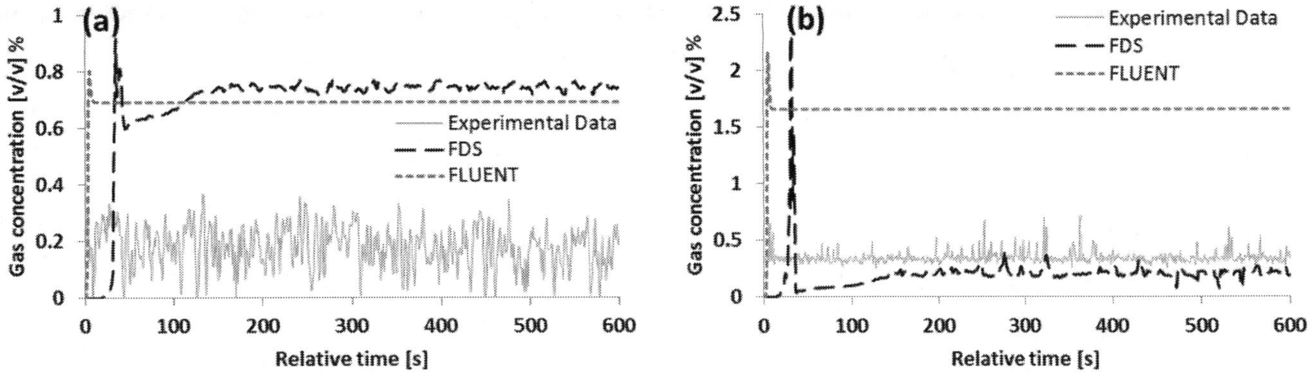

Fig. 4. Comparison of LNG natural dispersion prediction (FDS/Fluent) with experimental data at downwind distance 9.7 m at (a) 0.5 m elevation and (b) 1.2 m elevation

Fig. 4 shows the vapor concentration plotted for 9.7 m away from the LNG source at two elevations. These data will show the LNG vapor behavior after certain amount of time, as the vapors had traveled 9.7 m from the LNG source. Similar to Fig. 3, both FDS and Fluent over-predicted the experimental data at 0.5 m elevation. However, the FDS under-predicted the experimental data at 1.2 m elevation, as shown in Fig. 4 (b). Further investigation will address the issue of limited heat transfer and other essential physical parameters to improve the LNG natural dispersion using FDS.

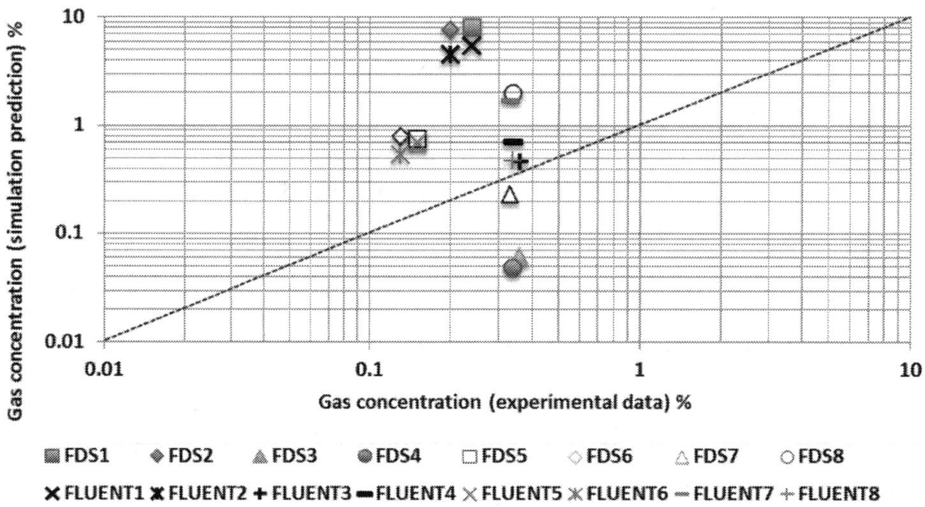

Fig. 5. Comparison of concentration prediction (FDS/Fluent)

Fig. 5 compares the FDS and Fluent prediction. It can be seen from these predictions that FDS provided under-prediction, which may require further safety precaution in real applications. Kohout (2010) recommended to implement the safety factor of 2 for an unobstructed flow fields [5]. Further study will address the degree of under-prediction quantitatively to provide detail guidelines on developing and implementing the safety factors for the application of FDS.

Fig. 6 shows the overall concentration prediction of LNG natural dispersion and forced mitigation using the water spray application. Both FDS and Fluent achieved a steady state solution after 600 s, and when water spray was applied, the water droplets influenced the vapor behavior. The FDS prediction showed less significant changes of vapor concentration at 0.5 m elevation; however, the vapor concentration at 1.2 m elevation showed approximately 5 times more vapor increased at this level. This indicates that the FDS predicted the LNG vapors to travel towards higher elevation (1.2 m elevation) after the water spray application, pushing the vapor cloud beyond the 0.5 m elevation. On the other hand, Fluent predicted the vapor behavior in the opposite trend, where the vapor concentration in both elevations decreased.

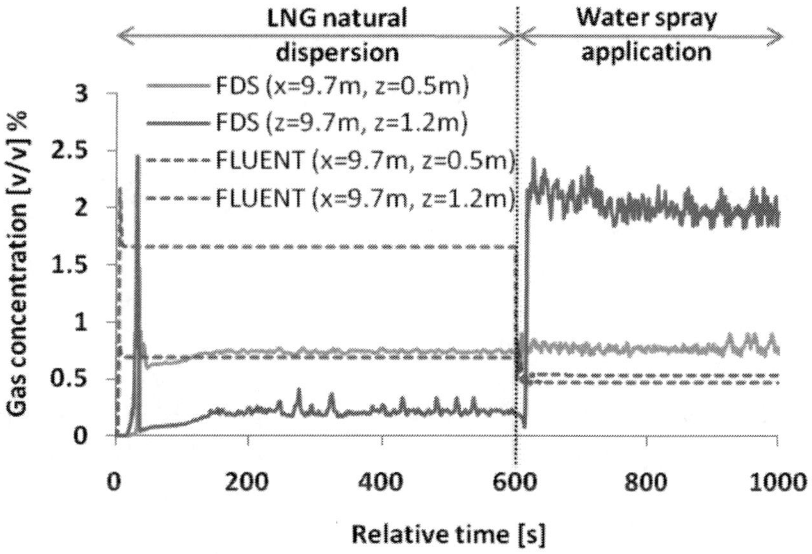

Fig. 6. Vapor concentration of LNG natural dispersion and forced mitigation using water spray application

The opposite trend can be explained by the two different turbulence models applied for FDS and Fluent. The LES predicts the vapor behavior different from the k- ε turbulence model, which was used for Fluent application. It can be seen from the Fluent results that the solver predicted the vapors to travel beyond the 1.2 m elevation, while the FDS solver predicted that the vapors would travel up to 1.2 m elevation. Currently, further investigation is planned to compare with the experimental results to determine the accuracy of both turbulence model.

5. Conclusion

This work presented preliminary results of applying FDS in setting up and simulating the LNG forced dispersion phenomena using the water spray application. The results of LNG vapor dispersion using FDS showed under-prediction of the vapor concentration at higher elevation. Further analysis will determine the exact level of the under-prediction and implementation of the safety factors. The forced dispersion simulated using the FDS produced opposite trend compared to the Fluent prediction. This is mainly because of the discrepancy in describing the turbulent flow induced from the interaction of water droplets. Detail study is required to determine the accuracy of the turbulence model, which could provide better description of droplet/vapor interaction system.

FDS can provide a rigorous prediction of the complex behavior of LNG vapor dispersion. Currently, FDS requires further analysis on the following to improve the vapor dispersion simulation: grid sensitivity, implementation of more specific models for heat transfer model, and the description of a specific fully developed atmospheric profile (wind speed and temperature).

References

1. EIA, *Annual Energy Outlook 2012 with Projections to 2035*, 2012, US Department of Energy.
2. IEA, *World Energy Outlook 2011: Are we entering a golden age of gas?* , 2011, IEA: Paris, France.
3. USGAO, *Public Safety Consequences of a Terrorist Attack on a Tanker Carrying Liquefied Natural Gas Need Clarification (Maritime Security GAO-07-316), United States Government Accountability Office*, 2007: Washington, D.C.
4. McGrattan, K.F., et al., *Fire Dynamics Simulator (Version 5) Use's Guide (Special Publication 1019-5)*, 2007, National Institute of Standards and Technology: Gaithersburg, MD.
5. Kohout, A.J., *Evaluation of Fire Dynamics Simulator for liquefied natural gas vapor dispersion hazards*, in *Fire Protection Engineering*2011, University of Maryland: College Park, MD.
6. Tu, J., G. Yeoh, and C. Liu, *Computational fluid dynamics: a practical approach* 2008, Boston, MA: Butterworth-Heinemann.
7. NIST, *Fire dynamics simulator technical reference guide (Vol 1: Mathematical model-NIST Special Publication 1018)*, 2012, National Institute of Standards and Technology: Gaithersburg, MD.
8. MKOPSC, *Data Report of MKOPSC LNG Spill Tests: 2005-2009*, M.S. Mannan, Editor 2010, Mary Kay O'Connor Process Safety Center, Texas A&M Univeristy: College Station, TX.
9. Rana, M., *Forced dispersion of liquefied natural gas vapor clouds with water spray curtain application, Chemical Engineering, Texas A&M University, **PhD Dissertation***, 2009. p. 232.
10. Kim, B.K., et al., *Modeling of water spray application in the forced dispersion of LNG vapor cloud using a combined Eulerian-Lagrangian approach.* Industrial & Engineering Chemistry Research, 2012. **51**: p. 13803-13814.

HSE Evolution in Conversion of LNG Import Terminals to Export

Jack Chosnek, Ph. D., P. E.
President and Principal
Knowledge*One*

Filippo Gavelli, Ph. D., P.E.
Head of Dispersion Consulting
GexCon

Victor H. Edwards, Ph. D., P. E.(TX)
Director of Process Safety
IHI E&C International Corporation

AIChE Spring Meeting
13th Topical Conference on Natural Gas Utilization
San Antonio, April 28-May 2, 2013

Introduction

With the abundance of natural gas (NG) in the U.S. in recent years, companies that had built marine terminals for the import of Liquefied Natural Gas (LNG) and subsequent re-gasification for pipeline transportation are now considering converting those facilities for liquefaction of pipeline gas in order to export LNG. This has brought challenges in the Health, Safety & Environment (HSE) area as well as in the regulatory arena regarding plot plan development and the handling of high pressure liquefied hydrocarbons. Besides the need to address the increased hazards of NG liquefaction versus LNG re-gasification, government agencies are faced with application of regulations and codes to the liquefaction of NG that were intended mainly to address LNG re-gasification terminals earlier. As the regulations try to apply the consequence-based codes to the variety of materials used in the liquefaction process, such as refrigerants that may include propane, ethane, ethylene, butanes, and other heavier hydrocarbons that may also be present in the NG, the need for tools and techniques to evaluate safe alternatives becomes evident. In addition, the environmental perspective also changes from regasification to the liquefaction. This paper will describe the methodology used in an NG liquefaction project to address these needs, which among other aspects include a combination of traditional vapor dispersion, vapor cloud explosion, and thermal radiation modelling tools, with computational fluid dynamics (CFD) modelling. The environmental factors, as well as the use of barriers to screen equipment noise and to limit the extent of vapor clouds will also be discussed.

Many US LNG Import Facilities are Being Converted to LNG Export

Domestic supplies of natural gas are growing rapidly due to availability of extensive shale gas resources. As a result, US natural gas prices have dropped to a fraction of their peak several years ago and US LNG imports have decreased dramatically.

At the same time, LNG prices in Japan, in other parts of the Asia Pacific region, and in Europe have remained high. Contributing factors are the Japanese decision to shut down nuclear power generation after the Fukushima tragedy, and the growing global demand for natural gas. Yet another factor is the environmental advantages of natural gas as a fuel when compared with coal.

Consequently, many US LNG import facilities are being converted to either LNG export facilities or to bidirectional LNG facilities.

Conversion from LNG Import to LNG Export poses Increased Hazards

Chosnek and Edwards (1) discussed the hazards of conversion of LNG Import Facilities to LNG Export. These hazards include the following:

- Natural gas liquefaction facilities required for LNG export are more complex and congested due to higher number of processing equipment involving different processing steps with multiple products, and require more space than corresponding LNG regasification facilities.
- Plot space at existing LNG import sites for new LNG export facilities may be limited.
- The risks of fire and explosion may be increased by the presence of more reactive chemicals (e.g., propane, ethylene) and higher congestion typical of liquefaction and other gas processing facilities.
- These risks are also increased by the use of more reactive heavy hydrocarbon streams as by-products from natural gas purification.
- Liquefaction often requires large inventories of refrigerants such as ethane, ethylene, propane, propylene, and butanes or their mixtures, at comparatively high pressures.
- Hydrogen sulfide streams may result from purification of natural gas prior to liquefaction, although the primary impact is on potential emissions of sulfur oxides resulting from oxidation of the hydrogen sulfide by-product.

Current US LNG Regulations are Consequence-based

Chosnek and Edwards (1) highlighted that current US LNG regulations are consequence-based instead of a preferred risk-based approach (2). Summarized below are the current criteria (3-6):

- Exclusion Zones (at property line) from a release caused by a "design spill"

  - ½ Lower Flammable Limit for flammable vapor dispersion

  - 1,600 BTU/hr-ft^2 Thermal Radiation Limit for pool fires

  - 1 psi Overpressure Limit for Vapor Cloud Explosions (although not in the regulations and never an issue for LNG import facilities, this is now an important criterion for liquefaction facilities in the US)

These existing US regulations were adequate for LNG import, although the constraint that only passive mitigations (impoundments, vapor fences) could be used to reduce the hazard footprint was a disadvantage.

Consequence modeling of existing US import terminals was done in most cases using the following two software products:

- LNGFIRE3 for Thermal Radiation

- DEGADIS for Flammable Vapor Dispersion

These tools model the design spill as an LNG pool on an otherwise empty horizontal plane, which is a reasonable approximation when the major hazard is an LNG spill that can be controlled by impoundment.

Advanced Consequence Modeling is needed to Safely Design LNG Export Facilities

As mentioned above, natural gas liquefaction facilities introduce new hazards, such as overpressures from vapor cloud explosions, because of C2+ hydrocarbons and increased congestion.

Comparatively recently (October 2011), FERC approved two additional software packages for LNG vapor dispersion hazard analyses: PHAST™ from DNV, an analytical tool described by Witlox, *et al.* (7), and FLACS from GexCon, a CFD tool described by Hansen, *et al.* (8).

The two models now available for LNG hazard analysis offer a number of advantages over the previously allowed DEGADIS:

- Model verification and extensive validation with field data

- Wide variety of source terms

- Mixtures of gases can be considered

- Both vapor dispersion and vapor cloud explosion can be modeled

-Tabular and graphic outputs can be generated (2D in the case of PHAST, 2D or 3D in the case of FLACS)

The main differences between the two models are as follows:

- PHAST is based on analytical data and correlations (Unified Dispersion Model), which allows for rapid calculations; FLACS is a 3D CFD model that solves the Navier-Stokes and turbulence closure equations in each of several hundred thousand computational grid cells, therefore requires longer times to complete a calculation;

- The correlations implemented in PHAST do not account for the presence of obstacles or terrain features, therefore its results are less accurate and it does not allow all mitigation measures to be evaluated; FLACS can include any obstacle or obstruction (as long as they are built into the geometry model used for the simulations) and therefore provides more realistic results.

While both tools have pros and cons, an efficient and effective approach can be achieved by combining the use of PHAST and FLACS in order to take advantage of each model's strengths. This approach can make thorough and detailed consequence modeling possible within the typical time frame of LNG facility design development, leading to a safer and regulatory-compliant plot plan:

- Use PHAST to rapidly screen a large number of scenarios, involving releases of hazardous fluids from different process streams and locations (approximately 100 scenarios can be examined for a single LNG project).
- Rearrange equipment on the plot plan to minimize adverse consequences of releases of hazardous fluids.
- Validate the safer plot plan using FLACS.
- Establish and evaluate passive protections, such as vapor fences, using FLACS.
- Quantify location specific overpressures from vapor cloud explosions using FLACS.

Atmospheric Emissions will Increase and Amended Permits may Face Stricter Regulations

Natural gas liquefaction requires energy, so greenhouse gas (CO_2 and CH_4) emissions may increase if power is generated onsite.

- Emissions of ozone precursors (NOx and VOCs) may increase.
- Greater fugitive emissions may result due to increased process size and complexity.
- To a smaller extent in the US pipeline gas, emissions of SOx may increase.
- Amendment of existing permits may require meeting new and more demanding regulations.
- Best Available Control Technology (BACT) emission controls may be required.

Modest Increases in Aqueous Discharges

- LNG import terminals have few aqueous streams to treat
  - Storm water with potential lube oil contamination
  - Demineralization by-product water
  - Fire water system test water
  - Hydro-test water
  - Sanitary waste
- LNG export terminals have a few more:
  - All of the above aqueous streams
  - Higher potential for lube oil contamination of storm water (larger plot plan and significantly more rotating equipment)
  - Storm water with potential amine contamination
  - Amended permit may need to achieve newer and stricter regulations

It should be possible to minimize storm water contamination through segregation, containment and rigorous maintenance and housekeeping practices.

Noise Sources and Mitigation

New export facilities add new piping and equipment to the existing site that are new noise sources. Examples include air-cooled heat exchangers, compressors, and turbines. FERC has very strict limitations on noise from LNG facilities.

In addition to buildings, enclosures, silencers, and sound insulation, a noise barrier wall may be required. Computer modeling of noise from equipment and piping will be required during both FEED and Detailed Design. Noise modeling during FEED can give guidance on plot plan modification to minimize noise exposure to any nearby noise sensitive areas.

Design should also incorporate features to permit easy provision of additional noise controls if as-built facilities are noisier than predicted during design phase. Examples include building foundations for noise barrier walls strong enough for an increased wall height and providing spool pieces in compressor piping to allow easy provision of silencers if needed.

Conclusions

LNG import and export facilities are among the safest and most environmentally friendly fossil fuel processes. Conversion of a number of US LNG import facilities to either export or bi-directional facilities is being proposed. Fitting major new natural gas liquefaction facilities into an existing site can be challenging.

Challenges include the following:

- Limited plot space
- Increased fire and explosion hazards
- Increased emissions and stricter regulations
- Increased noise levels

This paper discusses how these challenges can be overcome.

Acknowledgments

We gratefully acknowledge constructively critical reviews by Kamal Shah and Frank Del Nogal.

References

1. Chosnek, Jack and Victor H. Edwards, “LNG Imports to Exports: Process Safety and Regulatory Challenges”, paper presented at the AIChE Spring Meeting,13th Topical Conference on Natural Gas Utilization San Antonio, April 28-May 2, 2013

2. Woodward, John L. and Robin M. Pitblado, "LNG Risk Based Safety-Modeling and Consequence Analysis", John Wiley & Sons, Inc., Hoboken, New Jersey, 2010.
3. NFPA 59A, "Standard for the Production, Storage, and Handling of Natural Gas (LNG)", National Fire Protection Association, Quincy, MA, 2001 and 2006.
4. Volume 49 Code of Federal Regulations, Part 193, "Liquefied Natural Gas Facilities: Federal Safety Standards", United States, Washington, DC, 2013.
5. Volume 33, Code of Federal Regulations, Part 127, "Facilities Handling Liquefied Natural Gas and Liquefied Hazardous Gases", United States, Washington, DC, 2013.
6. Kohout, Andrew, "U. S. Regulatory Framework and Guidance for Siting of Liquefied Natural Gas Facilities – A Lifecycle Approach", paper presented at 15th International Symposium, Mary Kay O'Connor Process Safety Center, College Station, TX, October 23-25, 2012.
7. H.W.M. Witlox, Mike Harper, and Robin Pitblado, "Validation of PHAST Dispersion Model As Required for USA LNG Siting Applications", LNG Plant Safety and Applications, AIChE Spring Meeting, April 2012.
8. Hansen, O. R., Gavelli, F., Ichard, M., Davis, S. G., Validation of FLACS against experimental data sets from the model evaluation database for LNG vapor dispersion, J. Loss Prev. Proc. Ind., 23:867-877, 2010.

Biography

Jack Chosnek, Ph. D., P. E., Knowledge*One*, www.knowledge1.net Jack is a consultant in Process Safety Management and Knowledge Management and has over thirty five years of experience in the petrochemical industry. He has consulted for companies in the chemical, oil and gas, offshore, waste management and mining industries, developing policies and implementing process safety management systems, facilitating PHAs, writing operating procedures and conducting incident investigations and process safety audits. He has developed commercial, full-featured software for PHA facilitation and Management of Change (MOC). Jack consulted as the principal risk engineer for two multi-billion dollar projects carried out by large engineering firms.

Before becoming a consultant Jack worked for Celanese Corp. for 25 years in various positions in process safety management, technology management, operations, process engineering, piloting, tolling and R&D. Jack was a member of a project team that developed one of the first commercial scale-ups in the biotechnology industry.

Jack is the Chairman of the Process Safety Workshops at the South Texas Section (STS) of AIChE. He chaired the Process Plant Safety Symposium (PPSS) at the 4th Global Congress on Process Safety, and has chaired and co-chaired other process safety sessions in other years. He is the chair of the Technical Advisory Committee at Texas A&M's Mary Kay O'Connor Process Safety Center.

Jack is the holder of three patents related to chemical production. He has a BS and MS from the Technion—Israel Institute of Technology, and a PhD from the University of Missouri at Rolla, all in Chemical Engineering, as well as an MBA from Texas A&M—Corpus Christi

Filippo Gavelli, Ph. D., P. E.

Dr. Filippo Gavelli is the Head of the Dispersion Consulting group at GexCon and is responsible for GexCon's LNG safety consulting business worldwide. He specializes in the analysis of heat transfer and fluid flow phenomena, including multiphase flows and cryogenic fluids. He has over 10 years of engineering consulting experience and 20 years of experience in computational fluid dynamics (CFD) modeling, using several research and commercial codes. He applies his expertise to modeling the atmospheric dispersion of hazardous gaseous releases, and has extensive experience modeling hazard scenarios for Liquefied Natural Gas (LNG) facilities, including vapor cloud dispersion, pool fires and vapor cloud deflagrations; his experience includes more than 20 LNG installations worldwide, including onshore, offshore and floating (FLNG) facilities for LNG import, export (liquefaction) and peak shaving. Dr. Gavelli is a member of the technical committees for NFPA 59A (LNG facilities standard) and NFPA 2 (hydrogen technologies code) and was the lead in the model validation effort that led to FLACS' approval by the U.S. DOT under 49 CFR 193. Dr. Gavelli is the lead author of several gas dispersion and explosion safety-related papers and has been a regular contributor to LNG-related technical committees and expert panels for several years. He is a member of the GexCon docents group, which develops and delivers safety seminars to facility owners and operators, safety engineers, and regulatory agencies, on the hazards associated with gas explosions, dust explosions and LNG. Dr. Gavelli is also responsible for software technical support and training for FLACS customers in North America and has provided training to over 100 users, including staff from regulatory and government agencies.

Victor H. Edwards, Ph. D., P. E. is Director of Process Safety for IHI E&C International Corporation, Houston, TX 77077 (832-379-7742; e-mail: vic.edwards@ihi-ec.com). Responsible for health, safety, and environment in design, Vic has worked for IHI and predecessor companies for 29 years. His experience includes process engineering, process safety management, and process, biochemical, environmental, and mineral processing technologies. His project assignments have included biochemical and chemical processes, biomass energy resources, coal gasification, coal liquefaction, liquefied natural gas, chemical weapons demilitarization, hazardous waste disposal, offshore hydrocarbons facilities, and extraction of metals.

Dr. Edwards has received three DuPont awards for safety and environmental engineering excellence and two DuPont awards for health, safety, and environmental excellence during 28 years as an alliance engineering contractor. In 1998, Kvaerner Engineers and Constructors named him employee of the year. In 2007, Vic's Houston HSE team was recognized by Aker Kvaerner with the Process and Construction Excellence Award for HSE in Design.

Vic's earlier experience included Assistant Professor of Chemical Engineering at Cornell University, Program Manager at the National Science Foundation (on leave from Cornell), Research Fellow at Merck, alternate energy research at United Energy Resources, Visiting Professor of Environmental Engineering at Rice University, and process engineering at Fluor.

Vic has contributed more than sixty publications to the engineering literature, including several book chapters. He chaired the 1992 Process Plant

Safety Symposium, the 1997 International Plant Operations and Design Conference, and the 2013 Global Congress on Process Safety, as well as numerous technical sessions at other conferences. Vic received the 2003 Service Award from the Process Safety Center at Texas A & M University, where he chaired the Technical Advisory Committee from 2004 to 2011. In 2009, Vic was elected to the Centennial Council of the Department of Chemical and Biomolecular Engineering at Rice University. He also serves on the Editorial Advisory Board of CHEMICAL PROCESSING magazine. Edwards earned his B. A. from Rice University and his Ph. D. from the University of California at Berkeley; both degrees in chemical engineering. A registered professional engineer in Texas, he is an AIChE Fellow, and a member of ACS, AAAS, NFPA, NSPE, and the New York Academy of Sciences.

9TH GLOBAL CONGRESS ON PROCESS SAFETY

Guidelines for Relative Hazard Ranking of Refrigerants and Siting Considerations for LNG Liquefaction Units

Ryan J. Hart, Ph.D., P.E.
Exponent, Inc.
4580 Weaver Parkway, Suite 100, Warrenville, Illinois 60555
rhart@exponent.com

Delmar "Trey" Morrison III, Ph.D., P.E.
Exponent, Inc., Warrenville, Illinois 60555

Alfonso F. Ibarreta, Ph.D., P.E. and Harri K. Kytömaa, Ph.D., P.E.
Exponent, Inc., Natick, Massachusetts

Prepared for Presentation at
American Institute of Chemical Engineers
2013 Spring Meeting
9th Global Congress on Process Safety
San Antonio, Texas
April 28 – May 1, 2013

UNPUBLISHED

Guidelines for Relative Hazard Ranking of Refrigerants and Siting Considerations for LNG Liquefaction Units

Ryan J. Hart, PhD
Exponent, Inc.
4580 Weaver Parkway, Suite 100
Warrenville, IL 60555
rhart@exponent.com

Delmar "Trey" Morrison, PhD
Alfonso F. Ibarreta, PhD
Harri K. Kytomaa, PhD
Exponent, Inc.

Keywords: LNG, refrigerant, exclusion zone, vapor cloud

Abstract

Recent growth in the availability of domestic natural gas has led the industry to push for the development of liquefaction facilities, at both existing import/regasification facilities and new locations. As a consequence of this evolving need for North American liquefaction capacity, the Federal Regulators (FERC) have added interpretations and expectations for new accidental release case studies that are specific to liquefaction facilities. The requirements for LNG facility siting and design are codified in the 49 CFR 193 and NFPA 59A, but many of the exclusion zone analysis requirements are stated broadly and involve some level of expert interpretation.

Accidental release case studies required for the Resource Report 11 filings must now include LNG and flammable refrigerants. The hazards posed by refrigerants had not been previously relevant for receiving terminals. One goal of the required analysis is to calculate the extent of flammable vapor clouds, to demonstrate and confirm that these do not cross the proposed facility's property boundaries. Given economic and geographic limitations, there is often a trade-off between liquefaction unit capacity, siting, and even technology. Flammable refrigerants also introduce the new risk of vapor cloud explosions and associated overpressures. As a result, refrigerant vapor dispersion analysis is also now required, as is the quantification of the associated overpressure from potential vapor cloud explosions.

However, the analysis for liquefaction units must include both the LNG and the flammable liquid refrigerants. Two primary liquefaction refrigeration technologies are typically considered: propane pre-cooled mixed refrigerant and mixed refrigerant (PRICO). We have found that different considerations may drive the flammable vapor exclusion zones due to liquid spills and jetting and flashing of pressurized releases. The combination of physical properties, mixture composition, inventory, and operating parameters can lead to widely different results. The tools that are available and approved for performing these analyses impose limitations upon the mixtures and the number of constituents that can be modeled.

This paper will provide case studies and discussion illustrating the dominant parameters for liquefaction unit accidental release modeling. Guidance for identifying and selecting refrigerant release scenarios will be provided. Of most interest may be the interaction of the relative hazard distances for LNG and refrigerant vapor dispersion with refrigerant overpressure analysis.

1 Introduction

The liquefaction of natural gas for storage and transportation is, historically, a unique process for facilities located in the United States. Outside of peakshaving facilities, which are primarily used to liquefy natural gas during low periods of demand and re-gasify the stored liquid during high demand periods, large scale liquefaction facilities are a rarity in the US natural gas infrastructure. This has been driven by the fact in recent US history the amount of natural gas produced domestically has trailed domestic demand considerably. The graph provided in Figure 1 illustrates this point by providing the import of natural gas (MMcf, million cubic feet) into the US dating back into the 1970's [1].

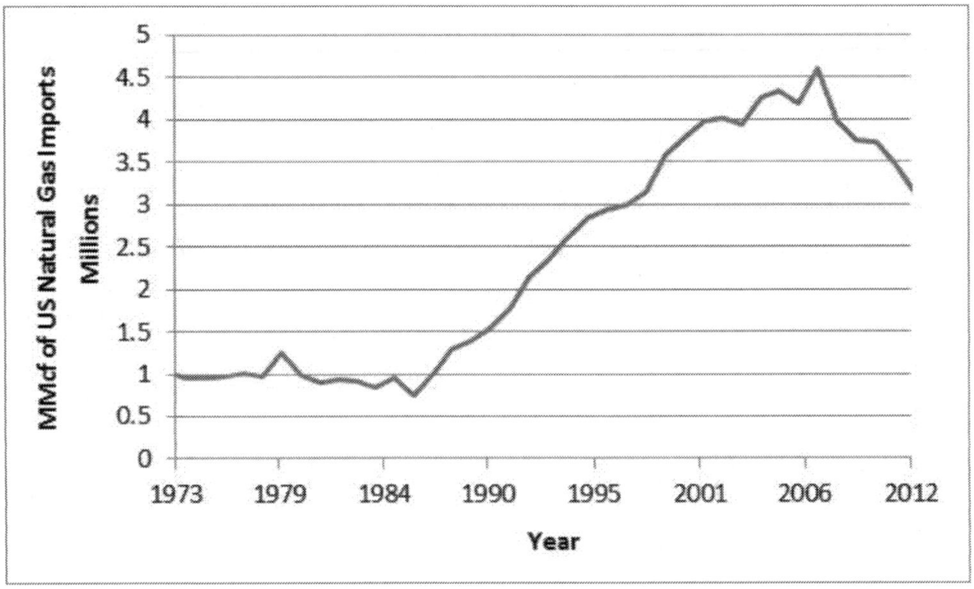

Figure 1. Historical US imports of natural gas (total, in millions of MMcf).

US imports of natural gas peaked in 2007 at just over 4,600,000 MMcf of natural gas. Since that time, the rate of natural gas imports has decreased by an average of 7.3% each year. This trend correlates with the rate of domestic natural gas production, which has been increasing steadily since 2006 (Figure 2, blue line, left axis) [2]. The increase in domestic production of natural gas is largely attributed to the increase in production from unconventional resources – mainly shale gas – as evidenced in Figure 2 (red line, right axis) [3]. In fact, the rate of shale gas production has been increasing at a rate of 44% per year since production began in 2007.

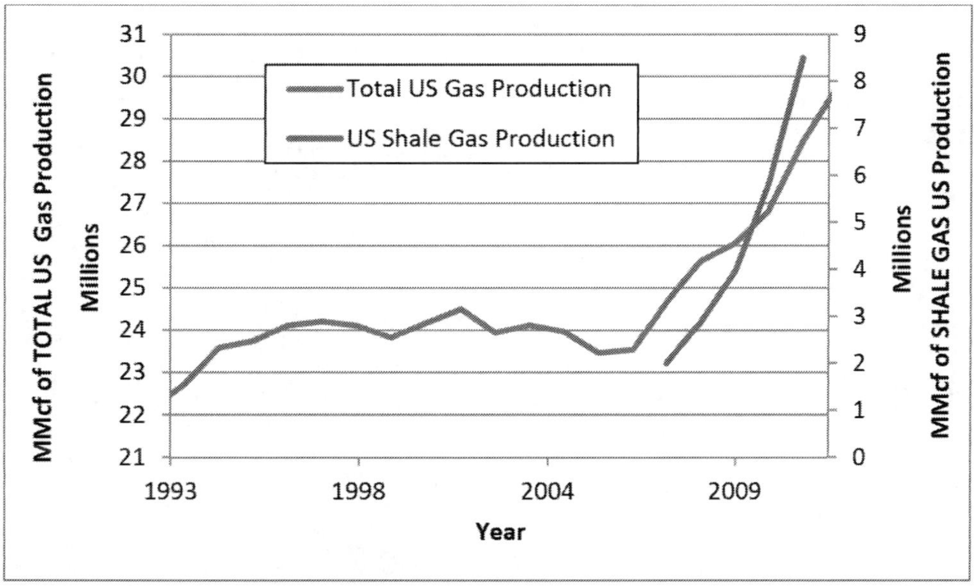

Figure 2. **Historical US production of total natural gas (blue line, in millions of MMcf) and shale gas (red line, in millions of MMcf).**

Because the domestic underground storage capacity of natural gas has remained relatively constant over the past two decades (between 8 and 9 million MMcf) [4], increases in domestic production can only be absorbed by the US infrastructure through decreasing imports, increasing exports, or converting our gas storage to liquid storage. If the rate of domestic production continues at its current pace, it is expected that production capacity will outpace domestic consumption at some point in the near future. This leaves only two options to handle increasing production rates: domestic storage of liquid natural gas (LNG) or the export and sale of gas. Exporting natural gas to countries not adjacent to the US (i.e. Canada and Mexico) is typically accomplished via trans-oceanic shipments of LNG. In fact, there have already been a number of proposed or potential LNG export facilities in the US, as shown in Figure 3 [5].

The codes and standards for which the facilities must be designed, constructed, and operated are found in 49 CFR Part 193 [6], and the authority having jurisdiction (AHJ) for these facilities is the Federal Energy Regulatory Commission (FERC). As part of the application process, the applicants (facility owners) must perform an environmental analysis in a series of thirteen (13) Resource Reports. Included in the assessment is a reliability and safety assessment (Resource Report 11) [7]. FERC then reviews the applications and must give approval prior to the commencement of any construction activities.

Resource Report 11 concerns reliability and safety of the facility, which includes a facility siting study. Although seemingly innocuous, the facility siting typically requires thorough analyses of potential loss of containment (LOC) events and the consequences of said events. 49 CFR Part 193, Subpart B addresses the siting requirements (including thermal radiation analysis and flammable vapor cloud analysis); however, the federal code references NFPA 59A (2001) throughout much of the subpart for specific guidance.

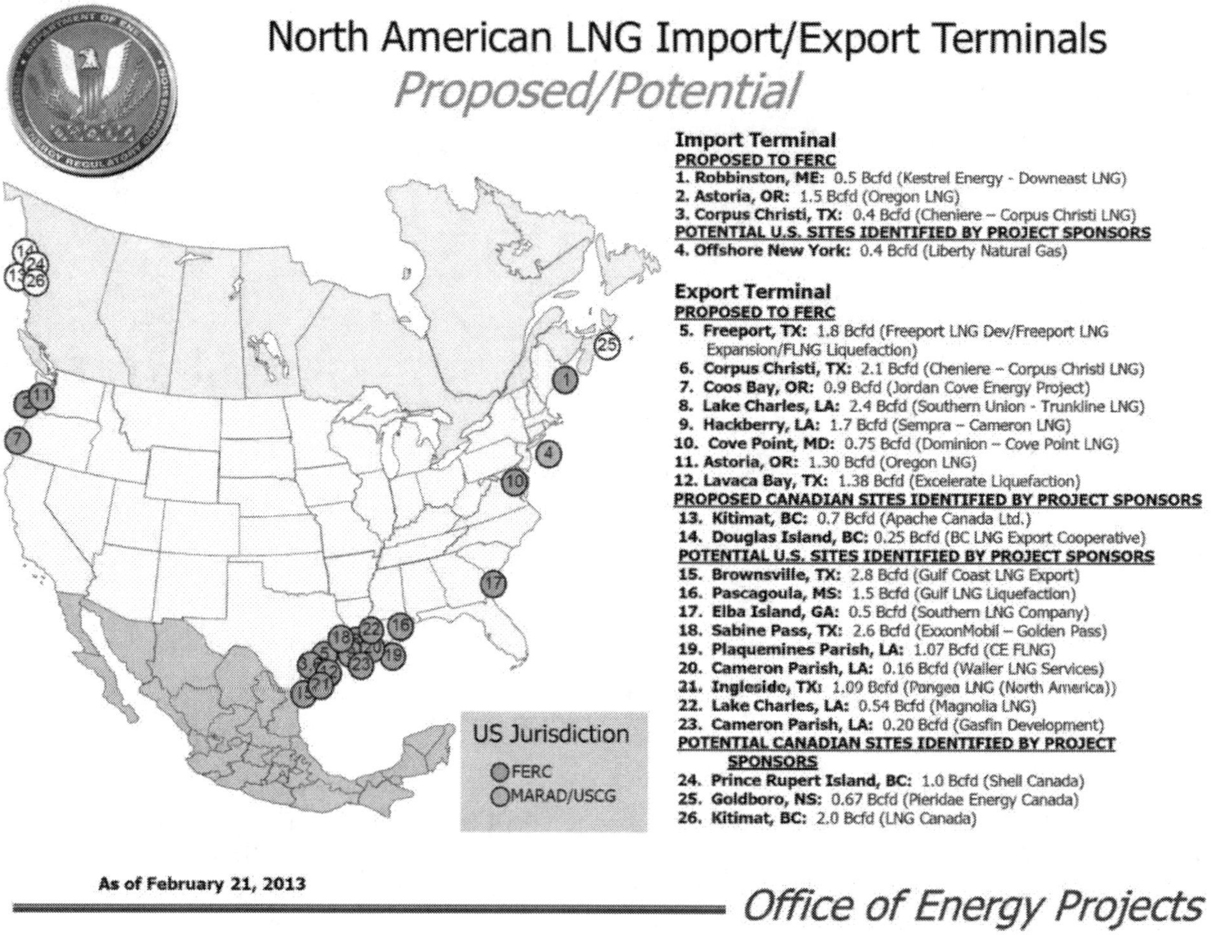

Figure 3. Proposed and potential liquefied natural gas import/export terminals, as identified by the US Department of Energy Federal Energy Regulatory Commission (FERC).

NFPA 59A is the "Standard for the Production, Storage, and Handling of Liquefied Natural Gas (LNG)" [8]. The standard provides general guidance for facility siting considering thermal and flammable vapor dispersion exclusion zones. The flammable vapor dispersion exclusion zones are found via calculation of the extent of flammable cloud vapor dispersion resulting from a LOC event. Although not required for import terminals, the exclusion zones at export terminals must consider the thermal radiation and flammable vapor dispersion exclusion zones for flammable refrigerants. Additionally, the AHJ (in this case FERC) has required that explosion overpressures be considered for the ignition of flammable refrigerant vapor clouds, based on 40 CFR 68 "Chemical Accident Prevention Provisions" requirements.

The exclusion zones related to refrigerant release scenarios may sometimes be larger than those associated with the traditional LNG-release exclusion zones. In the present analysis, we will use a FERC-approved commercial modeling tool, PHAST v6.7 [9], to evaluate the thermal radiation and flammable vapor dispersion exclusion zones associated with typical LNG liquefaction facility refrigerants. Additionally, the overpressures resulting from ignition of the flammable clouds will also be discussed. The results for the refrigerant vapor dispersion and thermal radiation analyses will be compared against LNG (methane).

2 LNG Liquefaction Technology

The transport of LNG is favorable due to the high density of liquids relative to gases and is a well understood process as a large portion of historical US imports has been received as LNG. However, unlike propane where the fluid can be liquefied by simply increasing the pressure, natural gas (primarily methane) must be refrigerated as part of its liquefaction process. The critical temperature of methane is -82.7°C (-117°F) and the critical temperature of propane is 96.6°C (206°F) [10]. Thus, as part of the liquefaction process for natural gas, the fluid must be refrigerated. The typical storage temperature of LNG is around -162°C (-260°F) [11], which is right at the normal boiling point for methane (-161.6°C, [10]).

The liquefaction of natural gas is accomplished using modified versions of the refrigeration cycles most commonly applied in residential appliances. All cycles use a working fluid that is expanded via Joules-Thomson (J-T) or isentropic means. The working fluid can vary greatly depending on the type of technology employed. The two primary technologies for the liquefaction of natural gas utilize a single-stage mixed refrigerant (such as the PRICO® process) or the two-stage propane/mixed refrigerant process [12]. More information on commercial LNG liquefaction technology is available elsewhere [13-15].

Either the two-stage or single-stage liquefaction process utilizes a mixed refrigerant as the working fluid at some point in the process. The mixed refrigerant composition can vary depending on the developer of the technology but is typically a mixture of the following components: methane, ethylene, propane, butane or higher molecular weight hydrocarbons, and nitrogen. As nitrogen is an inherently inert material, the contribution to thermal radiation and flammable vapor dispersion exclusion zones is non-existent. Selected properties of typical refrigerant materials can be found in Table 1 [16]. For comparison, the density of air is 0.075 lb/ft^3 (1.20 kg/m^3) at 68°F (20°C).

Table 1. Selected properties for the refrigerant components examined in this study. All properties were taken from the PHAST v6.7 materials database.

Material	Methane	Ethylene	Propane	n-Butane
Molecular Weight	16.0	28.1	44.1	58.1
Normal Boiling Point (n.b.p.)	-260°F (-162°C)	-155°F (-104°C)	-43.7°F (-42.0°C)	31.1°F (-0.50°C)
Liquid Density at n.b.p.	26.4 lb/ft^3 (423 kg/m^3)	35.5 lb/ft^3 (569 kg/m^3)	36.4 lb/ft^3 (583 kg/m^3)	37.6 lb/ft^3 (602 kg/m^3)
Vapor Density at 20°C	0.042 lb/ft^3 (0.668 kg/m^3)	0.073 lb/ft3 (1.17 kg/m3)	0.116 lb/ft^3 (1.86 kg/m^3)	0.155 lb/ft^3 (2.49 kg/m^3)
Sat. Pressure at 20°C	3630 psi (25.0 MPa)	910 psi (6.27 MPa)	121 psi (0.834 MPa)	30.2 psi (0.208 MPa)

3 Hazards of Refrigerants

The composition, temperature, and pressure of the mixed refrigerant may vary depending on the type of facility, liquefaction technology employed, and point along the refrigeration cycle. Therefore, a broad range of fluid temperatures and pressures must be considered in the exclusion zone analysis. While specific mixed refrigerant operating conditions and compositions are not made publicly available, all of these materials may be stored on-site as pure species (for refrigerant make-up) and may be present as a near pure species in the liquefaction cycle. Thus, for the exclusion zone analysis provided here we will examine pure component releases covering a range of temperatures and pressures. The specific species examined include: methane, ethylene, propane, and butane.

3.1 Thermal Radiation

The thermal radiation exclusion zone considers the ignition of a quantity of a spilled liquid material in an impoundment area. The size of the thermal radiation exclusion zone will therefore depend on the type of material and size of impoundment. Although the size of the impoundment area is unique to each and every facility, the resulting thermal radiation exclusion zone is expected to scale with the size of the impoundment.

The "Pool Fire" model in PHAST v6.7 was used to calculate the thermal radiation distances to the 1,600 BTU/ft^2.hr (5.0 kW/m^2) heat flux boundary for 50 ft (15 m) diameter circular pools of methane, ethylene, propane, and butane. The results are provided in Table 2. For all scenarios the wind speed was 4.5 mph (2 m/s), atmospheric stability class F, and the ambient temperature and surface temperature were 68°F (20°C).

Table 2. Pool fire thermal radiation distances to the 1,600 BTU/ft^2.hr (5.0 kW/m^2) heat flux boundary.

Material	Distance to 1,600 BTU/ft^2.hr
Methane	254 ft (77.3 m)
Ethylene	206 ft (62.8 m)
Propane	230 ft (70.0 m)
Butane	236 ft (71.8 m)

The results indicate that for equivalent pool diameters the methane liquid spills result in the farthest distance to the 1,600 BTU/ft^2.hr heat flux boundary. Propane and butane pool fires result in similar exclusion zones and ethylene resulted in the smallest exclusion zone.

3.2 Flammable Vapor Dispersion

The flammable refrigerants involved in the liquefaction cycle for LNG production pose a significant hazard in the event of a LOC. High pressure and possibly cold refrigerants may be discharged from piping or appurtenances in the event of a release. If the release involves the fluid in its liquid state, these liquids may flash or vaporize completely upon release, they may spill into collection areas (impoundments), or both may occur. Even if the fluid released is in the vapor state, a high momentum cloud will be formed (although at lower discharge rate for equivalent leak size) which can disperse consider distances if aligned with the prevailing wind direction.

For the flammable vapor dispersion exclusion zone analysis, two separate sets of scenarios are examined: (1) refrigerant releases under typical storage conditions and (2) refrigerant releases under "product LNG conditions" (i.e. conditions at which LNG is produced). For the typical storage conditions, it was assumed that the refrigerants were stored at their vapor pressure at 68°F (20°C); however, the pressure for methane was capped at 1000 psi. For the typical process conditions, it was assumed that the refrigerants were processed at -260°F (-162°C) and 100 psi (0.689 MPa); however, the process temperature for butane was selected just above its melting point at -216°F (-138°C). A summary of the release conditions examined are provided in Table 3. The LFL values reported are the PHAST v6.7 default values for the pure species.

Table 3. Release scenarios and fluid conditions considered for the flammable vapor dispersion exclusion zone analysis.

Material	Temperature	Pressure	LFL
	Refrigerant Storage Releases		
Methane	68°F (20°C)	1,000 psi (6.89 MPa)	4.4%
Ethylene	68°F (20°C)	910 psi (6.27 MPa)	2.3%
Propane	68°F (20°C)	121 psi (0.834 MPa)	2.0%
Butane	68°F (20°C)	30.2 psi (0.208 MPa)	1.5%
	Refrigerant Process Releases		
Methane	-260°F (-162°C)	100 psi (0.689 MPa)	4.4%
Ethylene	-260°F (-162°C)	100 psi (0.689 MPa)	2.3%
Propane	-260°F (-162°C)	100 psi (0.689 MPa)	2.0%
Butane	-216°F (-138°C)	100 psi (0.689 MPa)	1.5%

The model input parameters examined in the PHAST v6.7 vapor dispersion calculations are provided in Table 4. The surface roughness factor selected corresponds to dispersion along open, flat terrain. The leak elevation that was chosen corresponds to the height of a typical pipe rack. The model uses the orifice equation to calculate discharge rate, and a discharge coefficient of 0.6 corresponds to a typical square-edged orifice (what one might expect in the event of clean break of piping connection). The leak size was chosen to be 1 in, which is an expected piping connection for an instrument tap. All releases were allowed to continue until a full steady-state cloud was formed. The releases were in the horizontal orientation for all cases.

Table 4. PHAST v6.7 model input parameters for vapor dispersion exclusion zone analysis of the releases described in Table 3.

Parameter	Value
Ambient Temperature	68°F (20°C)
Surface Temperature	68°F (20°C)
Relative Humidity	70%
Wind Speed	4.5 mph (2 m/s)
Pasquill-Gifford Atmospheric Stability	Category F
Surface Roughness Factor (m)	12 in (0.03 m)
Leak Elevation	12 ft (3.7 m)
Discharge Coefficient	0.6
Leak Size	1 in (2.5 cm)

The results of the flammable vapor dispersion for the releases described above, reported as distances from the source to the edge of the ½-LFL vapor cloud, are provided in Table 5. A portion of the propane process and butane process releases rained out (liquid droplets reached the ground), yielding rainout values of 7.4% and 20%, respectively. For all other releases, there was no liquid that rained out.

Certain trends are observed in the calculated vapor dispersion results. Firstly, the flammable cloud dispersion distances correlate with the discharge rate (as one might expect) for all components. Secondly, it appears that the clouds with higher density vapors (which disperse closer to the ground where the wind speeds are lower) travel greater distances. Lastly, it appears as though within a given set of releases (i.e. storage vs. process) the materials with the higher molecular weights (and coincidentally higher densities and lower LFL values) travel farther distances. The only exception to this is the ethylene process release, which was found to travel the greatest distance of any of the materials.

Table 5. PHAST v6.7 calculated flammable vapor cloud release rates and dispersion distances to the ½-LFL for the high pressure storage and process releases described in Table 3.

Material	Discharge Rate	Distance to ½-LFL	Height at ½-LFL
	Refrigerant Storage Releases		
Methane	8.73 lb/s (3.96 kg/s)	108 ft (33.0 m)	12.2 ft (3.73 m)
Ethylene	16.6 lb/s (7.53 kg/s)	234 ft (71.4 m)	6.93 ft (2.11 m)
Propane	20.2 lb/s (9.16 kg/s)	244 ft (74.3 m)	0.545 ft (0.166 m)
Butane	10.8 lb/s (4.90 kg/s)	356 ft (109 m)	0 ft (0 m)
	Refrigerant Process Releases		
Methane	16.2 lb/s (7.36 kg/s)	699 ft (214 m)	0 ft (0 m)
Ethylene	20.5 lb/s (9.28 kg/s)	850 ft (259 m)	0 ft (0 m)
Propane	21.8 lb/s (9.89 kg/s)	755 ft (230 m)	0 ft (0 m)
Butane	22.3 lb/s (10.1 kg/s)	770 ft (235 m)	0 ft (0 m)

Although only two of the pressurized refrigerant releases resulted in appreciable rainout of liquid fractions, the collection of liquids in impoundments from low pressure releases should also be considered in flammable vapor dispersion exclusion zone analyses. Next we will consider the resulting vaporization, cloud formation, and flammable vapor dispersion from liquid spills of the four refrigerant materials into a circular impoundment area that is 50 ft (15 m) in diameter. The impoundment material chosen for this analysis was regular concrete, and the PHAST default thermal conductivity ($\kappa = 1.21$ W/m.K) and thermal diffusivity ($\alpha = 5.72 \times 10^{-7}$ m^2/s) were used. We have found that the choice of concrete thermal properties can significantly affect the rate of vaporization for the spilled cryogenic liquids, but we only provide the default values here. Each of the four materials was assumed to spill at a rate of 11.0 lb/s (5.00 kg/s) at their normal point boiling temperature (provided in Table 1). The weather conditions described in Table 4 were used and the spills occurred for 10 minutes.

Table 6. PHAST v6.7 flammable vapor dispersion distances to the ½-LFL boundary for liquid spills into a 50 ft (15 m) diameter regular concrete impoundment.

Material	Discharge Rate	Distance to ½-LFL	Height at ½-LFL
Methane	11.0 lb/s (5.0 kg/s)	1175 ft (358 m)	0 ft (0 m)
Ethylene	11.0 lb/s (5.0 kg/s)	674 ft (205 m)	0 ft (0 m)
Propane	11.0 lb/s (5.0 kg/s)	344 ft (105 m)	0 ft (0 m)
Butane	11.0 lb/s (5.0 kg/s)	328 ft (100 m)	0 ft (0 m)

Unlike the high pressure process and storage refrigerant release scenarios, the liquid spills into impoundment indicate that the materials with the lowest normal boiling points produce the longest dispersing clouds. In fact, an 11 lb/s (5 kg/s) liquid methane spill resulted in a cloud that travelled nearly the length of four football fields. With methane being the lone exception, the liquid refrigerant spills resulted in clouds that travelled less distance than the high pressure refrigerant process releases.

3.3 Overpressure Analysis

Several explosion overpressure calculation models exist and are available for determination of the flammable clouds described above. In PHAST v6.7, the available models include TNT, TNO Multi-energy (TNO), and Baker-Strehlow-Tang (BST). For the purposes of evaluating the resulting overpressure profiles of the releases described previously, we will be focusing on only the BST model results. Although all of the models have their advantages and disadvantages, the BST model is the only one of the three that allows for consideration of material reactivity. Since we are examining overpressures as the result of ethylene vapor cloud explosions, and ethylene is widely considered to be a highly reactive fuel (the others are generally accepted to have moderate to low reactivity), the BST model was chosen.

The BST model takes confinement, congestion, and material reactivity into consideration. However, unlike the TNO model the blast strength is not defined by the user. The following inputs need to be prescribed [17]:

> ***Material reactivity***: The material reactivity can be defined as Low, Medium or High. Low reactivity corresponds to components such as methane, carbon monoxide or ammonia, while High reactivity corresponds to very reactive materials such as hydrogen, ethylene or acetylene. The reactivity values for the refrigerant species are provided in Table 7, below.

Table 7. The material reactivities used for the Baker-Strehlow-Tang explosion model.

Refrigerant	Material Reactivity
Methane	Low
Ethylene	High
Propane	Medium
Butane	Medium

Flame expansion: The flame expansion describes the number of dimensions in which the blast can propagate. For example, if a blast occurs within a pipe, the overpressures can only propagate in a single direction. For a blast that is confined between two floors, the overpressures can only propagate in 2 dimensions. In the case of most leaks at a liquefaction facility, the overpressures are able to propagate in all three dimensions. A flame expansion parameter of 3 was therefore selected.

Obstacle density: The obstacle density is selected based on the blockage ratio in congested areas.

- A "low" obstacle density corresponds to a blockage ratio of less than 10%,
- A "medium" obstacle density corresponds to a blockage ratio between 10% and 40%,
- A "high" obstacle density corresponds to a blockage ratio above 40%.

At the typical liquefaction facility, most confined areas correspond to pipe racks which are not highly congested; pipes are typically aligned parallel to one another along pipe supports, rather than staggered. A "medium" obstacle density was selected for all scenarios.

Ground reflection factor: The ground reflection factor is a number between 1 (no reflection) and 2 (perfect ground reflection). In order to provide for the most conservative overpressure calculations, a ground reflection of 2 was selected. This assumption corresponds to perfect ground reflection and highest possible overpressures.

Confined volume: The confined volume corresponds to the total flammable cloud volume that exists in congested regions. For the analyses represented here, it was conservatively assumed that the entire flammable cloud was congested for each scenario. The flammable cloud volume is defined as the portion of the vapor cloud found between the LFL and UFL boundaries.

The resulting flammable vapor clouds described in Section 3.2 were examined using the methods discussed above. The ignition location was placed at the farthest LFL boundary, and the individual distances from the release source to the 1 psi overpressure boundary are provided in Table 8.

Table 8. PHAST v6.7 calculated distances to the LFL and BST model calculated distances to 1 psi overpressure boundary for the refrigerant storage, process, and liquid spill release scenarios described in Section 3.2.

Material	Discharge Rate	Distance to LFL	Distance to 1 psi Overpressure
	Refrigerant Storage Releases		
Methane	8.73 lb/s (3.96 kg/s)	56.3 ft (17.2 m)	60.9 ft (18.6 m)
Ethylene	16.6 lb/s (7.53 kg/s)	108 ft (33.0 m)	346 ft (106 m)
Propane	20.2 lb/s (9.16 kg/s)	113 ft (34.4 m)	289 ft (88.0 m)
Butane	10.8 lb/s (4.90 kg/s)	154 ft (46.8 m)	371 ft (113 m)
	Refrigerant Process Releases		
Methane	16.2 lb/s (7.36 kg/s)	312 ft (95.0 m)	443 ft (135 m)
Ethylene	20.5 lb/s (9.28 kg/s)	497 ft (152 m)	1381 ft (421 m)
Propane	21.8 lb/s (9.89 kg/s)	327 ft (99.8 m)	776 ft (237 m)
Butane	22.3 lb/s (10.1 kg/s)	336 ft (102 m)	832 ft (254 m)
	Refrigerant Liquid Spill Releases		
Methane	11.0 lb/s (5.0 kg/s)	613 ft (187 m)	852 ft (260 m)
Ethylene	11.0 lb/s (5.0 kg/s)	334 ft (102 m)	1307 ft (398 m)
Propane	11.0 lb/s (5.0 kg/s)	236 ft (71.8 m)	786 ft (240 m)
Butane	11.0 lb/s (5.0 kg/s)	232 ft (70.6 m)	806 ft (246 m)

Ignition of the flammable cloud for the low reactivity fuel, methane, resulted in BST calculated overpressure distances to the 1 psi boundary that didn't exceed the vapor cloud dispersion distance to the ½-LFL value. For the medium reactivity fuels propane and butane, the BST calculated distances to the 1 psi overpressure boundary were similar to their ½-LFL vapor dispersion distances (butane was slightly higher and propane was slightly lower for most cases). However, for the high reactivity fuel ethylene the BST calculated distance to the 1 psi boundary greatly exceed the vapor dispersion to the ½-LFL boundary.

4 Conclusions

The hazard consequence of LOC events related to refrigerant storage and processing and LNG liquefaction facilities can be successfully examined using modern software tools. The model tool PHAST v6.7 was used here to examine the thermal radiation, flammable vapor dispersion, and overpressure exclusion zones. Such information is not only required by the FERC for regulatory permit application and approval, but can provide useful information for facility siting.

The results indicate that for typical refrigerant materials the ½-LFL vapor cloud dispersion exclusion zone represents the largest expected exclusion zone. The only exception is for highly reactive fuels, such as ethylene, which may produce overpressures to the 1 psi overpressure boundary that greatly exceed the predicted ½-LFL vapor cloud extents. Thus, fuel characteristics and release conditions both can greatly influence the exclusion zone analyses.

Proposed liquefaction facilities would be benefited by screening the potential exclusion zones from anticipated LOC events at early stages in the design of the facility. Consider that some of the calculated exclusion zones for typical LOC events at a liquefaction facility may result in exclusion zones that extend well beyond 1300 ft. Such information will play a pivotal role in facility siting, and should be conducted prior to the full arrangement of the facility piping and equipment systems (at which point it may be impractical to move large structures or capital equipment).

5 References

[1] U.S. Natural Gas Imports (MMcf), released on February 28, 2013. Accessed via <www.eia.gov>.

[2] U.S. Natural Gas Gross Withdrawals (MMcf), released on February 28, 2013. Accessed via <www.eia.gov>.

[3] U.S. Natural Gas Gross Withdrawals from Shale Gas (Million Cubic Feet), released on February 28, 2013. Accessed via <www.eia.gov>.

[4] U.S. Total Natural Gas Underground Storage Capacity (MMcf), released on February 28, 2013. Accessed via <www.eia.gov>.

[5] North American LNG Import/Export Terminals – Proposed/Potential. US Department of Energy Federal Energy Regulatory Commission, February 21, 2013. Accessed via <ferc.gov/industries/gas/indus-act/lng.asp>.

[6] "Liquefied Natural Gas Facilities: Federal Safety Standards." Code of Federal Regulations Title 49, Pt. 193, 2011 edition.

[7] Federal Energy Regulatory Commission, "Guidance Manual for Environmental Report Preparation." FERC – Office of Energy Projects (August 2002).

[8] NFPA 59A, Standard for the Production, Storage, and Handling of Liquefied Natural Gas (LNG), 2001 Edition.

[9] PHMSA Docket No. 2011-0075, Final Decision on petition for approval for PHAST UDM Version 6.6 and 6.7, October 7, 2011.

[10] Air Liquide Gas Encyclopedia, Air Liquid (2009). Accessed via <encyclopedia.airliquide.com>.

[11] U.S. Department of Energy, "Liquefied Natural Gas: Understanding the Basic Facts." U.S. Department of Energy – Office of Fossil Energy (August 2005).

[12] Kidnay, AJ, WR Parrish, DG McCartney, "Fundamentals of Natural Gas Processing, 2nd ed." CRC Press (2011).

[13] Ransbarger, W, "A fresh look at LNG process efficiency." LNG Industry (Spring 2007).

[14] Shukri, T, "LNG technology selection," Hydrocarbon Engineering (February 2004).

[15] Shukri, T and M Barclay, "Single mixed refrigerant process has appeal for growing offshore market," LNG Journal (June 2007).

[16] PHAST v6.7 Material Properties, DNV Software.

[17] Guidelines for Vapor Cloud Explosions, Pressure Vessel Burst, BLEVE and Flash Fire Hazards. Second Edition. Wiley (2010).

AUTHOR INDEX

AUTHOR INDEX

AUTHOR INDEX

AUTHOR INDEX